CONTENTS

TECHNICAL DRAWING

BOOKS BY THE AUTHORS

BASIC TECHNICAL DRAWING by H. C. Spencer (Macmillan, 1956)

THE BLUEPRINT LANGUAGE by H. C. Spencer and H. E. Grant (Macmillan, 1947)

ENGINEERING PREVIEW by L. E. Grinter, H. C. Spencer, et al. (Macmillan, 1945)

HOT WATER HEATING, RADIANT HEATING, AND RADIANT COOLING by F. E. Giesecke (Technical Book Co., 1947)

LETTERING EXERCISES FROM TECHNICAL DRAWING PROBLEMS by F. E. Giesecke, A. Mitchell, and H. C. Spencer (Macmillan, 1948)

TECHNICAL DRAWING by F. E. Giesecke, A. Mitchell, and H. C. Spencer (Macmillan, 1958)

TECHNICAL DRAWING PROBLEMS by F. E. Giesecke, A. Mitchell, and H. C. Spencer (Macmillan, 1947)

TECHNICAL DRAWING PROBLEMS, SERIES 2 by H. C. Spencer and H. E. Grant (Macmillan, 1948)

TECHNICAL LETTERING PRACTICE by H. C. Spencer and H. E. Grant (Macmillan, 1950)

Courtesy Ford Motor Co.

FOURTH EDITION

REVISED BY

HENRY CECIL SPENCER

TECHNICAL DRAWING

FREDERICK E. GIESECKE

M.E., B.S. in ARCH., C.E., PH.D.

Late Professor of Drawing, Agricultural and Mechanical College of Texas.

ALVA MITCHELL

B.C.E.

Late Professor Emeritus of Engineering Drawing, Agricultural and Mechanical College of Texas.

HENRY CECIL SPENCER

A.B., B.S. in ARCH., M.S.

Professor of Technical Drawing. Director of Department, Illinois Institute of Technology.

THE MACMILLAN COMPANY

NEW YORK

FIRST PRINTING

PRINTED IN THE UNITED STATES OF AMERICA

LIBRARY OF CONGRESS CATALOG CARD NUMBER 58-5164

PREFACE

This volume is intended as a class text and reference book in technical drawing. It contains a very great number of problems covering every phase of the subject, and it constitutes a complete teaching unit in itself. In addition to the problems in the text, two complete workbooks have been prepared especially for use with this text: *Technical Drawing Problems*, by Giesecke, Mitchell, and Spencer, and *Technical Drawing Problems, Series 2*, by Spencer and Grant. Thus, there are available three alternate sources of problems, and problem assignments can be varied easily from year to year. In general, the teacher who uses this text and one of the workbooks will supplement the workbook sheets by assignments from the text to be drawn upon blank paper. Many of the new text problems are designed for $8\frac{1}{2}'' \times 11''$ sheets.

The extensive use of this text during the last twenty-five years in college classes and in industrial drafting rooms has encouraged the authors in their original aim, which was and still is to prepare a book which *teaches the language of the engineer*, and to keep it in step with developments in industry. The idea has been to illustrate and explain each basic principle from the standpoint of the student, to make it so clear that the student is certain to understand, and to make it interesting enough so that he will read and study on his own initiative. Thus, it was hoped to free the teacher of a great deal of repetitive labor in teaching every student individually those things which the textbook can teach, and to permit him to give his attention to students having real difficulties.

This fourth edition constitutes the most sweeping revision that the text has undergone. Very few of the original illustrations and very little of the original text remain. A large number of new illustrations and problems have been added, and several hundred of the old ones have been redrawn. The entire book has been revised to reflect the latest American Standards, especially the various sections of ASA Y14 *American Drafting Standards Manual*.

An important basic improvement is the adoption of a larger format which permits the reproduction of illustrations and problems large enough for easy use. Many drawings have been redrawn in order to take full advantage of the more generous page size. An important objective has been to maintain and, where possible, to improve the quality of the drafting in the illustrations. It is logical that in a drawing book the drawings are more important than anything else.

The two former chapters on "Instruments and Materials" and "Instrumental Drawing" have been combined into one chapter on "Mechanical Drawing," which makes it possible to bring together all related topics for more effective use in class.

In the chapter on "Geometrical Constructions," practically every figure has been redrawn, with many changed to the step-by-step variety.

An outstanding feature is the emphasis on technical sketching throughout the text, and the expanded chapter given early in the book. This chapter is unique in integrating the basic concepts of views with freehand rendering, so that the subject of multiview drawing can be introduced through the medium of sketches. Practically all sketching and multiview problems are new.

The chapter on Dimensioning has been completely rewritten and expanded into two chapters, "Dimensioning" and "Tolerancing." Both are in accord with the new ASA Y14.5-1957 *Dimensioning and Notes*. Particularly important is the addition of complete information on geometric and positional tolerances.

The chapter on "Threads, Fasteners, and Springs" has been revised to conform with the latest American Standards. A simple system for drawing schematic thread symbols has been introduced. In addition, cap screws and machine screws may be drawn by a simplified method of proportions based on diameter. Methods of drawing springs are shown in detail.

The chapter on "Shop Processes" has been completely revised by Professors J. G. H. Thompson and J. G. McGuire, of Texas A & M College, so as to relate more closely shop processes to drafting. The chapter on "Gearing and Cams" has been revised by Professor B. L. Wellman of Worcester Polytechnic Institute. The chapter on "Graphs" has been completely rewritten, and a completely new chapter on "Engineering Graphics" has been prepared—both by Mr. E. J. Mysiak, design engineer. These two chapters are in line with current trends toward more emphasis upon graphics as a tool. Special appreciation is also due to Mr. Mysiak for his skilled drafting of many figures in this book.

The chapters on "Structural Drawing" and "Topographic Drawing" have been completely revised by Dr. E. I. Fiesenheiser, of Illinois Institute of Technology. The chapter on "Piping Drafting" has been rewritten by Mr. D. G. Reid and Mr. L. A. Anderson, of Sargent & Lundy Co. The chapter on "Aeronautical Drafting" has been revised once more by Mr. William N. Wright, of Boeing Airplane Company. Mr. C. G. Hebruck, of The Lincoln Electric Co., gave valuable assistance in the revision of the chapter on "Welding Representation."

Thus, every effort has been made to bring the book completely abreast of the many technological developments that have occurred in the past few years. In specialized fields, writers who have special qualifications have been called upon to contribute authoritative material.

Through the cooperation of leading engineers and manufacturers, this volume includes many commercial drawings of value in developing the subject. The authors wish to express their thanks to these persons and others too numerous to mention here who have contributed to the production of this book.

The authors wish to express special appreciation to Professor H. E. Grant for many valuable suggestions and for illustrations and problem material, to Professors I. L. Hill, R. O. Loving, A. P. McDonald, W. F. Brubaker, and Albert Jorgensen for many valuable suggestions, and to Professors Walter Downard and D. W. Fleming for assistance in connection with the chapter on "Shop Processes."

Special appreciation is expressed to my wife whose encouragement and untiring labors have contributed so much. Finally, it is fitting to acknowledge a continued indebtedness to the late Professors F. E. Giesecke and A. Mitchell for their contributions to previous editions and for the benefit of their wisdom through the years.

HENRY CECIL SPENCER

Chicago, Illinois

TECHNICAL DRAWING

CHAPTER 1

THE GRAPHIC LANGUAGE

1. A World Language. Many of the troubles of the world today are caused by the fact that the various peoples do not understand one another. The infinite number of languages and dialects which contributed to this condition resulted from a lack of intercommunication of peoples widely separated in various parts of the world. Even today when communication is so greatly improved, the progress toward a world language is painfully slow—so slow, indeed, that we cannot foresee the time when it will be a fact. All efforts to create a world language, such as Esperanto, have failed, and it is necessary for people who come in contact with other countries to learn a number of foreign languages. In Europe today it is not uncommon for educated persons to be able to speak a half-dozen languages or more.

2. The Universal Language. Although men have not been able to get together on a world language of words and sentences, there has actually been a universal

Fig. 1 Egyptian Hieroglyphics.

language in use since the earliest times: the *graphic language.* The idea of communicating thoughts from one person to another by means of *pictures* occurred to even the earliest cave-dwellers, and we have examples still in existence to prove it. These earliest men communicated orally, undoubtedly by grunts and other elementary sounds, and when they wished to *record* an idea, they made *pictures* upon whatever materials they could find, such as skins or stone. The earliest forms of writing were through picture-forms, such as the Egyptian hieroglyphics, Fig. 1. Later these forms were simplified and became abstract symbols used in our languages today. Thus, even the letter-characters in present word-languages have their basis in drawings.

A drawing is a *graphic representation* of a real thing. Drawings may take many forms, but since they are always based on real things, the method of representation is a basic natural form of communication of ideas, which is universal and timeless in character.

3. Two Types of Drawings. Man has developed drawing along two distinct lines, according to his purpose: (1) Artistic and (2) Technical.

From the beginning of time, artists have used drawings to express aesthetic, philosophic, or other abstract ideas. In ancient times nearly everybody was illiterate. There was no printing; hence no newspapers or books as we know them today. The books were hand-lettered on papyrus or on parchment, and were not available to the general public. People learned by listening to their superiors, and by looking at sculptures, pictures, or drawings in public places. Everybody could understand pictures, and they were a principal source of information. In our museums and in ruins of antiquity are thousands of examples of story-telling or teaching by means of drawings. If someone wished to preserve his own image or a friend's, he had to have the job done in stone, bronze, in oil on canvas, or in some other art-medium—there were no photographs. The artist was not just an artist in the aesthetic sense—he was a teacher or philosopher, a means of expression.

The other line along which drawing has developed has been the technical. From the beginning of recorded history, man has used drawings to *represent* objects to be built or constructed. Of these earliest drawings no trace remains today, but we

Bettmann Archive.

Fig. 2 The Circus Maximus in Rome.

definitely know that drawings were used, for man could not have built as he did without using fairly accurate drawings. In the Bible the statement is made that Solomon's Temple was "built of stone made ready before it was brought thither."* Each stone and timber was carved or hewn into shape, brought to the site, and noiselessly fitted together. It is evident that accurate drawings must have been used, showing the exact shapes and sizes of the component parts.

Moreover, we can see today the ruins of fine old buildings, aqueducts, bridges, and other structures of antiquity which could not possibly have been erected without carefully made drawings to guide the builders. Many of these structures are still regarded as "Wonders of the World," such as the Temple of Ammon at Karnak in ancient Egypt, completed in about 980 B.C., which took seven centuries to construct. In sheer mass of stone, this building exceeded any roofed structure ever built, so far as we know, being 1200 feet long and 350 feet wide at its greatest width. Likewise, The Circus Maximus in Rome was a large structure, Fig. 2: according to the historian Pliny, it seated a total of 250,000 spectators. Compare this figure with perhaps our greatest stadium, Soldiers Field in Chicago, which seats about 110,000 people. History is full of examples of amazing structures which could not possibly have been built without accurate drawings of some kind.

4. Earliest Technical Drawings. Perhaps the earliest-known technical drawing in existence is the plan view of a fortress drawn by the Chaldean engineer Gudea,

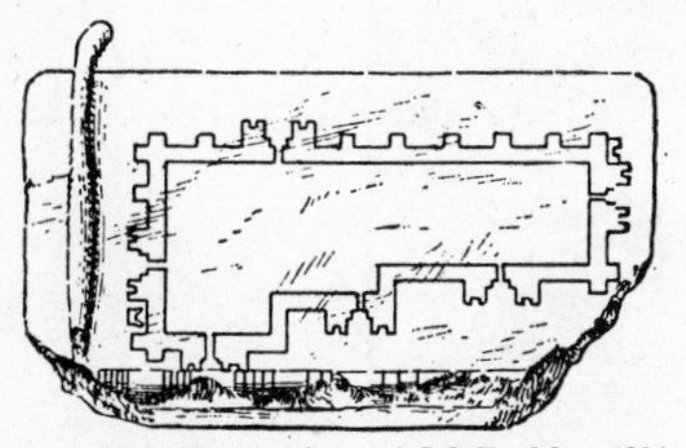

From Transactions, *A.S.C.E., May, 1891.*

Fig. 3 Plan of a Fortress. This stone tablet is part of a statue now in the Louvre, in Paris, and is classified in the earliest period of Chaldean art, about 4000 B.C.

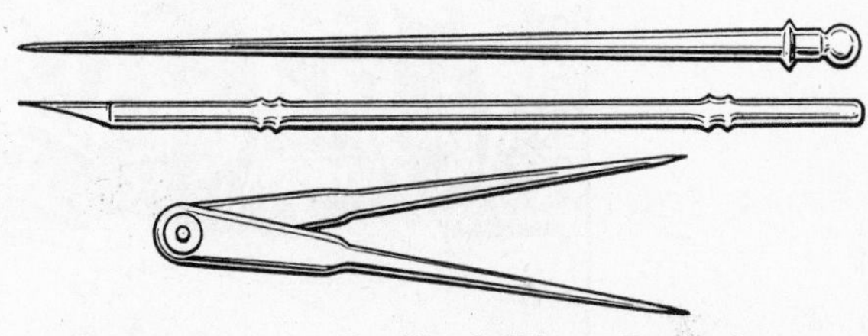

From Historical Note on Drawing Instruments, *published by V & E Manufacturing Co.*

Fig. 4 Roman Stylus, Pen, and Compass.

and engraved upon a stone tablet, Fig. 3. It is remarkable how similar this plan is to those made by architects today, although "drawn" thousands of years before paper was invented.

The first written evidence of the use of technical drawings was in 30 B.C. when the Roman architect Vitruvius wrote a treatise on architecture in which he said, "The architect must be skilful with the pencil and have a knowledge of drawing so that he readily can make the drawings required to show the appearance of the work he proposes to construct." He went on to discuss the use of the rule and compasses in geometric constructions, in drawing the plan and elevation views of a building, and in drawing perspectives.

In the museums we can see actual specimens of these early drawing instruments.

*I. Kings, 6, 6.

Compasses were made of bronze and were about the same size as those used today. As shown in Fig. 4, the old compass resembled the dividers of today. Pens were cut from reeds.

The theory of projections of objects upon imaginary planes of projection (to obtain *views*) apparently was not developed until the early part of the fifteenth century—by the Italian Architects Alberti, Brunelleschi, and others. It is well known that Leonardo da Vinci used drawings to record and transmit to others his ideas on

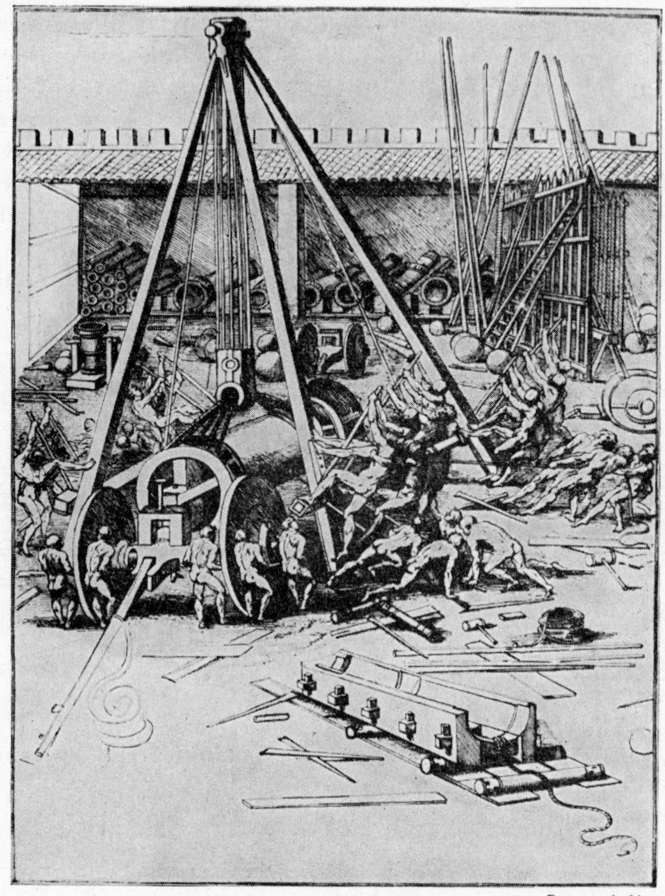

Bettman Archive.

Fig. 5 An Arsenal, by Leonardo da Vinci.

mechanical constructions, and many of these drawings are in existence today, Fig. 5. It is not clear whether Leonardo ever made mechanical drawings showing orthographic views as we know them today, but it is probable that he did.

Leonardo's treatise on painting, published in 1651, is regarded as the first book ever printed on the theory of projection drawing; however, its subject was perspective and not orthographic projection.

The compass of the Romans remained very much the same during Leonardo's time, Fig. 6. Circles were still scratched with metal points, since graphite pencils were not invented until the eighteenth century, when the firm of Faber was established in Nuremburg, Germany. By the seventh century reed pens had been replaced by quills made from bird feathers, usually those of geese (hence: goose-quill pens).

The scriber-type compass gave way to the compass with a graphite lead shortly after graphite pencils were developed. At Mount Vernon we can see the drawing instruments used by the great civil engineer, George Washington, and bearing the date 1749. This set, Fig. 7, is very similar to the conventional drawing instruments used today, having a pencil attachment and a pen attachment for the compass, and a ruling pen with parallel blades similar to the modern pens.

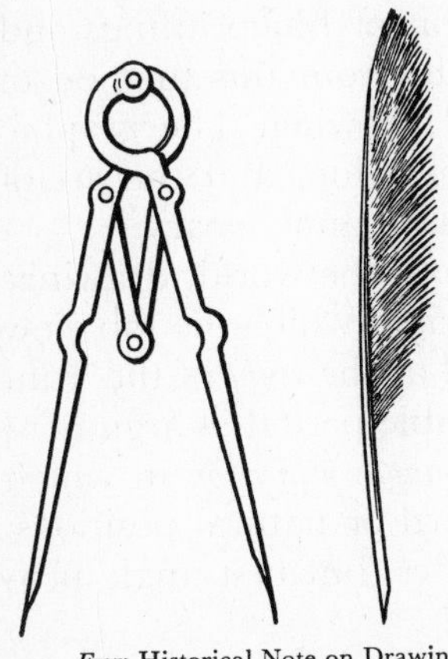

From Historical Note on Drawing Instruments, *published by V & E Manufacturing Co.*

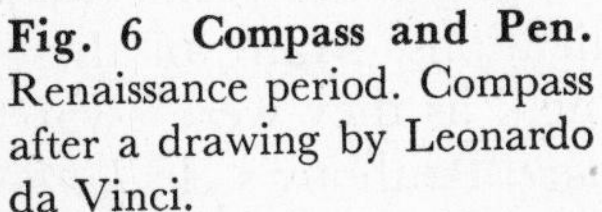

Fig. 6 Compass and Pen. Renaissance period. Compass after a drawing by Leonardo da Vinci.

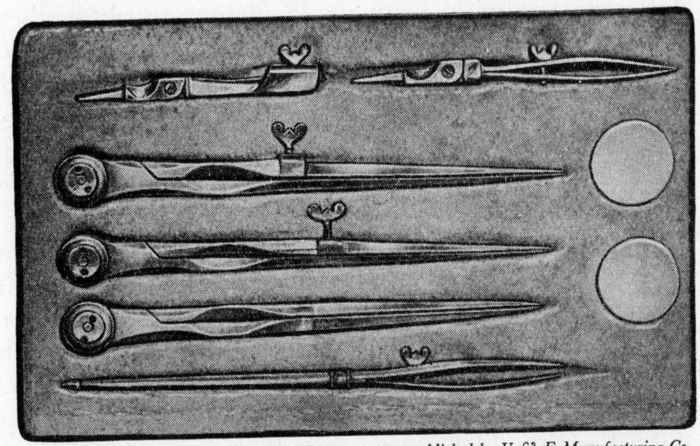

From Historical Note on Drawing Instruments, *published by V & E Manufacturing Co.*

Fig. 7 George Washington's Drawing Instruments.

5. Early Descriptive Geometry. The beginnings of descriptive geometry are associated with the problems encountered in building construction and in fortifications. Gaspard Monge (1746–1818) is considered the "inventor" of descriptive geometry, although his efforts were preceded by publications on stereotomy, architecture, and perspective, in which many of the principles were used. It was while he was a professor at the Polytechnic School in France near the close of the eighteenth century that Monge developed the principles of projection which are today the basis of our technical drawing. These principles of descriptive geometry were soon recognized to be of such military importance that Monge was compelled to keep his principles a secret until 1795, following which they became an important part of technical education in France and Germany, and later in the United States. His book *la Geometrie Descriptive* is still regarded as the first text to expound the basic principles of projection drawing.

Monge's principles were brought to the United States from France by Claude Crozet, who became a professor at the United States Military Academy at West Point in 1816. He published the first text on the subject of descriptive geometry in the English language in 1821. In the years immediately following, these principles became a regular part of early engineering curricula at Rensselaer Polytechnic Institute, Harvard University, Yale University, and others. During the same period, the idea of interchangeable manufacturing in the early arms industries was being developed, and the principles of projection drawing were applied to these problems.

6. Modern Technical Drawing. Perhaps the first text on technical drawing in this country was *Geometrical Drawing*, published in 1849 by William Minifie, a high school teacher in Baltimore. In 1850 the Alteneder family organized the first drawing instrument manufacturing company in this country (Theo. Alteneder & Sons, Philadelphia). In 1876 the blueprint process was introduced in this country at the Philadelphia Centennial Exposition. Up to this time the graphic language was more or less an "art," characterized by fine line drawings made to resemble copper-plate engraving, by the use of shade lines, and by the use of water color "washes." These techniques became unnecessary after the introduction of blueprinting, and drawings gradually were made plainer to obtain best results from this method of reproduction. This was the beginning of modern technical drawing. The graphic language now became a relatively exact method of representation, and the art of model-making as a regular preliminary to construction became unnecessary.

Up to about the turn of the nineteenth century throughout the world, drawings were generally made in what is called *first-angle projection*, §245, in which the top view was placed *under* the front view, the left side view was placed at the *right* of the front view, etc. At this time in the United States, after a considerable period of argument pro and con, practice gradually veered to the present *third-angle projection* in which the views are situated in what we regard as their more logical or natural positions. Today, third-angle projection is standard in the United States, but first-angle projection is still used throughout most of the rest of the world.

During the early part of the twentieth century many books on the subject were published, in which the graphic language was analyzed and explained in connection with its rapidly changing engineering and industrial applications. Many of these writers were not satisfied with the term "mechanical drawing," as they were aware of the fact that technical drawing was really a *graphic language*. Anthony's *An Introduction to the Graphic Language*, French's *Engineering Drawing*, and our text *Technical Drawing* were all written with this point of view.

7. Drafting Standards. In all of the above books there has been a definite tendency to standardize the characters of the graphic language, to eliminate its provincialisms and dialects, and to give industry and engineering a uniform, effective graphic language. Of prime importance in this movement has been the work of the American Standards Association, which with the American Society for Engineering Education and the American Society of Mechanical Engineers as sponsors, has prepared and published the *American Drafting Standards Manual*, ASA Y-14, published in seventeen separate sections. These publications are now in process of development and will be published as approved standards as they are completed.

These booklets outline the most important idioms and usages in a form that is acceptable to the majority, and are considered the most authoritative guide to uniform drafting practices in this country today. The Standard gives the *characters* of the graphic language, and it remains for the textbooks to explain the *grammar* and the *penmanship*.

8. Definitions. After surveying briefly the historical development of the graphic language and before starting a serious study of its theory and applications, it is well to consider the definitions of a few terms.

Descriptive Geometry is the grammar of the graphic language; it is the three-dimensional geometry forming the background for the practical applications of the language, and through which many of its problems may be solved graphically.

Mechanical Drawing is the term that should be applied only to a drawing made with drawing instruments. The term has been used to denote all industrial drawings, which is unfortunate because such drawings are not always mechanically drawn, and because it tends to belittle the broad scope of the language by naming it superficially by its principal mode of execution.

Engineering Drawing and *Engineering Drafting* are broad terms widely used to denote the graphic language. However, since the language is not used by engineers only, but also by a much larger group of people in diverse fields who are concerned with technical work or with industrial production, the term is still not broad enough.

Technical Drawing is a broad term that adequately suggests the scope of the graphic language. It is rightly applied to any drawing used to express technical ideas. The term has been used by various writers since Monge's time at least, and is still widely used, mostly in Europe.

Engineering Graphics is a term sometimes applied generally to drawings for technological use, but in recent years it has come to mean more specifically that part of drawing which is concerned with graphical computations and with charts and graphs. See Chapters 20 and 21.

Technical Sketching is the freehand expression of the graphic language, while *mechanical drawing* is the instrumental expression of it. Technical sketching is a most valuable tool for the engineer and others engaged in technical work, because through it most technical ideas can be expressed quickly and effectively without the use of special equipment.

*Blueprint Reading** is the term applied to the "reading" of the language from drawings made by others. Actually, the blueprint process is only one of many forms by which drawings are reproduced today (see §§730–743), but the term "blueprint reading" has been accepted through usage to mean the interpretation of all ideas expressed on technical drawings, whether the drawings are blueprints or not.

9. What the Engineering Student Should Know. The development of technical knowledge from the dawn of history has been accompanied, and to a large extent made possible, by a corresponding graphic language. Today the intimate connection between engineering and its universal graphic language is more vital than ever before, and the engineer who is ignorant of, or deficient in, the principal mode of expression in his technical field is *professionally illiterate*. That this is true is shown by the fact that descriptive geometry and technical drawing are required subjects in virtually every engineering school in the world.

The old days of fine-line drawings and of shading and "washes" is gone forever; no artistic talent is necessary for the modern technical student to learn his graphic language. For its mastery he needs precisely the aptitudes and abilities he will need to learn the science and engineering courses which he studies concurrently and later. The student who is poor in technical drawing and descriptive geometry is likely to be poor in his other technical courses.

The well-trained engineer must be able to make a correct graphical representation of engineering structures, which means that he must understand the fundamental principles, or the *grammar* of the language, and must be able to execute the work with

*See *The Blueprint Language*, by H. C. Spencer and H. E. Grant (New York: The Macmillan Company, 1947).

reasonable skill, which is *penmanship*. In developing this skill, the left-handed student need not feel that he is handicapped in any way. Some of the finest draftsmen are left-handed.

Drawing students often try to excuse themselves for inferior results (usually caused by lack of application) by arguing that after graduation they do not expect to do any drafting at all; they expect to have others make any needed drawings under their direction. Such a student presumptuously pictures himself, immediately after graduation, as the accomplished engineer concerned with bigger things, and forgets that his first job may well be "on the board" and that he will be the one who will make the drawings under the direction of a really experienced engineer. Though he may not realize it, entering the engineering profession via the drawing board is fortunate for him, since it affords an unexcelled opportunity "to learn the ropes" in the industry in which he has started. Even if the young engineer has not been too successful in developing a skilful penmanship in the graphic language, he still will have great use for its grammar, since the ability to *read* a drawing is of utmost importance, and he will need this ability throughout his professional life.

Further, the young engineer is apt to overlook the fact that in practically all the subsequent engineering courses he will take in college, he will encounter technical drawings in most of his textbooks, and he will be called upon by his instructors to supplement his calculations with mechanical drawings or sketches. Thus, a mastery of his course in technical drawing will aid him materially not only in engineering practice after graduation, but more immediately in his other technical courses, and will have a definite bearing on his scholastic progress.

Besides the direct values to be obtained from a serious study of the graphic language, there are a number of very important training values which, though they may be considered by-products, are fully as essential as the language itself. Many a student learns for the first time in his drawing course the meaning of *neatness*, *speed*, and *accuracy*—basic habits that every successful engineer must have or acquire.

All authorities agree that the ability to *think in three dimensions* is one of the most important requisites of the successful engineer. This training to visualize objects in space, to use the "constructive imagination," is one of the principal values to be obtained from a study of the graphic language. The ability to *visualize* is possessed in an outstanding degree by persons of extraordinary creative ability. It is difficult to think of Edison, Steinmetz, or De Forest as being deficient in constructive imagination.

With the increase in technological development and the consequent crowding of drawing courses by the other engineering and science courses in our colleges, it is doubly necessary for the engineering student to make the most of the limited time

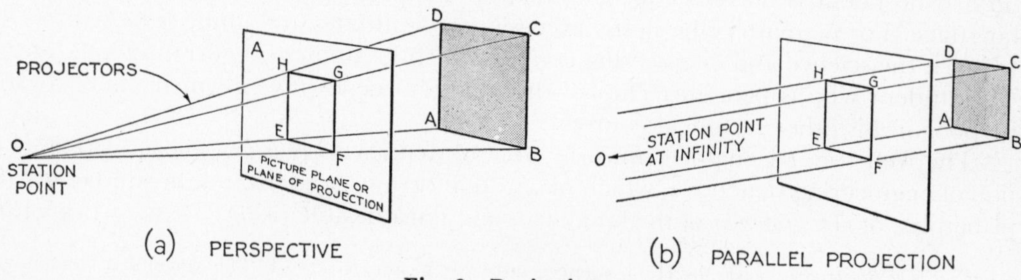

Fig. 8 Projections.

devoted to the language of his profession, to the end that he will not be professionally illiterate, but will possess an ability to express himself quickly and accurately through the correct use of the graphic language.

10. Projections. Behind every drawing of an object is a space relationship involving four imaginary things: the *observer's eye* or *station point*, the *object*, the *plane* or *planes of projection*, and the *projectors**. For example, in Fig. 8 (a) the drawing EFGH is the projection on the plane of projection A of the square ABCD as viewed by an observer whose eye is at the point O. The projection or drawing upon the plane is produced by the piercing points of the projectors in the plane of projection. In this case, where the observer is relatively close to the object and the projectors form a "cone" of projectors, the resulting projection is known as a *perspective.*

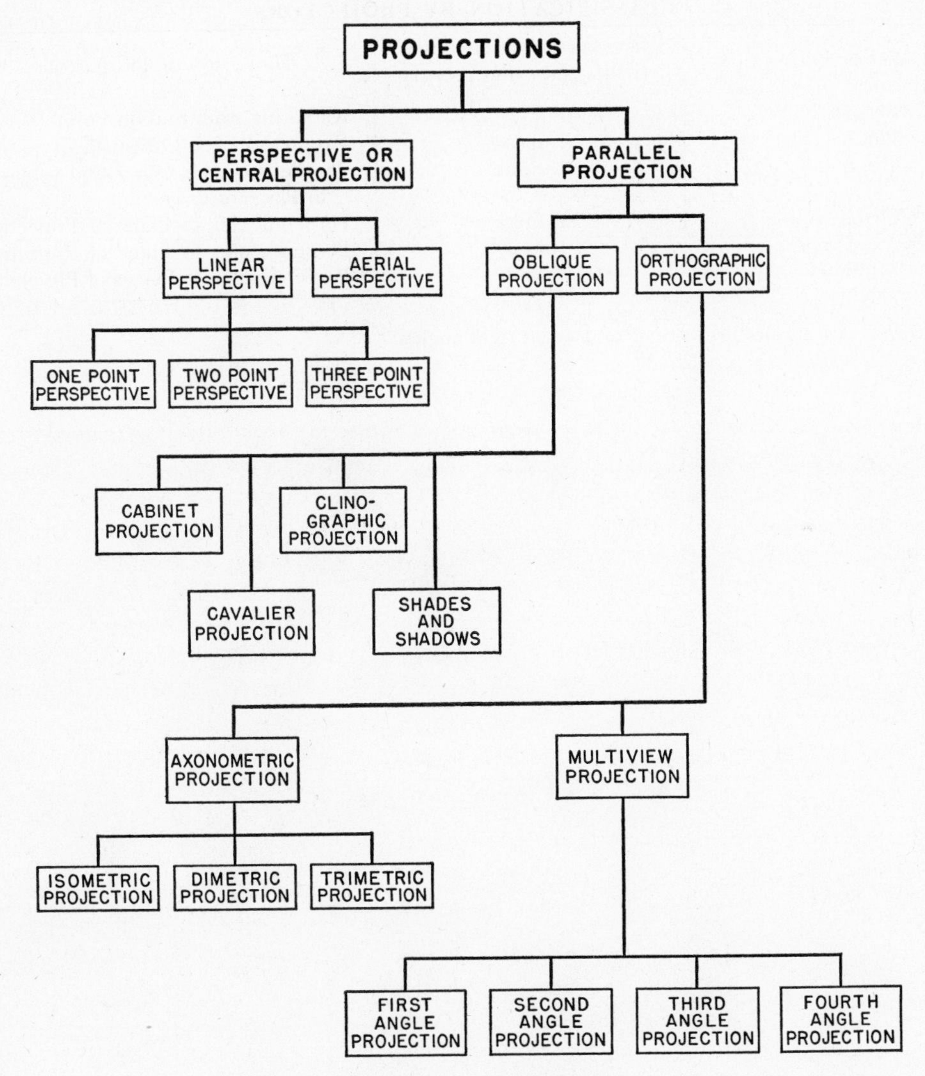

Fig. 9 Classification of Projections.

*Also called *visual rays* and *lines of sight.*

If the observer's eye is imagined as infinitely distant from the object and the plane of projection, the projectors will be parallel, as shown in Fig. 8 (b); hence, this type of projection is known as a *parallel projection.* If the projectors, in addition to being parallel to each other, are perpendicular to the plane of projection, the result is an *orthographic** *projection.* If they are parallel to each other but oblique to the plane of projection, the result is an *oblique projection.*

These two main types of projection—perspective and parallel projection—are further broken down into many sub-types as shown in Fig. 9, and will be treated at length in the various chapters that follow.

A classification of the main types of projection according to their projectors is shown in the table below.

CLASSIFICATION BY PROJECTORS

Classes of Projection	Distance from Observer to Plane of Projection	Direction of Projectors
Perspective	Finite	Radiating from Station Point
Parallel	Infinite	Parallel to Each Other
Oblique	Infinite	Parallel to Each Other and Oblique to Plane of Projection
Orthographic	Infinite	Perpendicular to Plane of Projection
Axonometric	Infinite	Perpendicular to Plane of Projection
Multiview	Infinite	Perpendicular to Planes of Projection

**Orthographic* means "to write or to draw at right angles."

CHAPTER 2

MECHANICAL DRAWING

Details are trifles, but trifles make perfection, and perfection is no trifle.
BEN FRANKLIN

11. Typical Equipment. The principal items of equipment needed by students in technical schools and by draftsmen in professional practice are shown in Fig. 10. To secure the most satisfactory results, the drawing equipment should be of high grade. When drawing instruments (item 3) are to be purchased, the advice of an

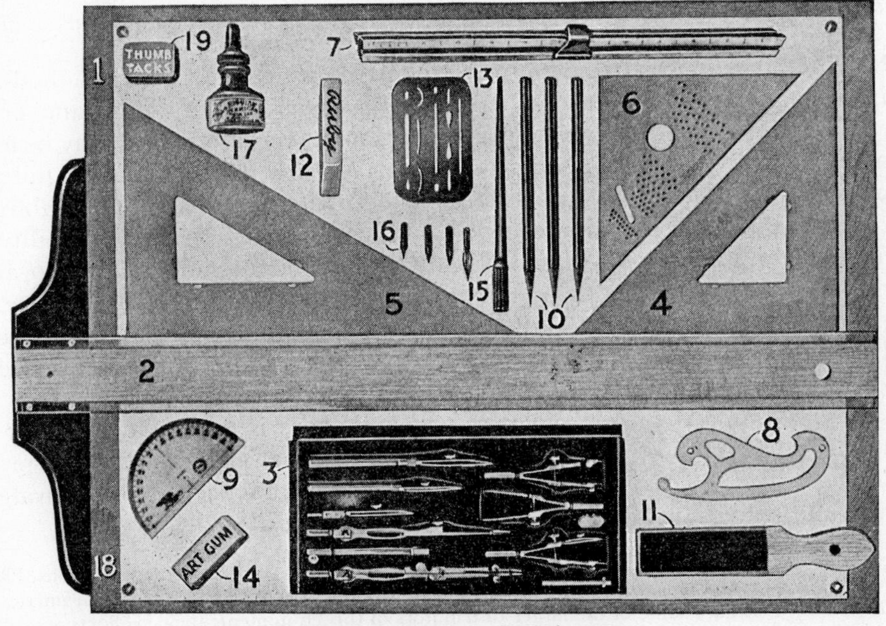

Fig. 10 Principal Items of Equipment.

experienced draftsman or a reliable dealer* should be sought because it is impossible for beginners to distinguish high-grade instruments from those which are inferior. Each student should scratch his initials or some mark of identification on each instrument.

A complete list of equipment, which should provide a satisfactory selection for students of technical drawing, follows. The numbers refer to Fig. 10:

1. Drawing board (approx. 20″×24″).
2. T-square (24″, transparent edge).
3. Set of Instruments, §§42 and 43.
4. 45° triangle (8″ side).
5. 30° × 60° triangle (10″ long side).
6. Lettering triangle or Ames Lettering Instrument.
7. Triangular Architects scale (or flat Mechanical Draftsmans scale), Fig. 43.
8. Irregular curve.
9. Protractor.
10. Drawing pencils: HB, F, 2H, and 4H to 6H, or mechanical pencil and leads, Fig. 18 (b).
11. Pencil pointer (sandpaper pad, file, or Tru-Point sharpener).
12. Pencil eraser.
13. Erasing shield.
14. Art gum or other cleaning eraser.
15. Pen staff.
16. Pen points (Gillott's 303, 404; Hunt's 512; Leonardt's ball-pointed 516F).
17. Drawing ink (black waterproof) and pen wiper.
18. Drawing paper, tracing paper, or tracing cloth.
19. Backing sheet (drawing paper—preferably white—to be used under tracing paper, vellum, or cloth).
20. Thumbtacks, drafting tape, or draftsman's stapler.

In addition (*not illustrated above*), *these items may be included if necessary*:

Drop pen, detail pen, proportional dividers, beam compass, contour pen.
Arkansas oil stone (for sharpening ruling pens).
Pocket knife (not a razor blade!).
Slide rule.
Dusting brush.
Dust cloth (not your handkerchief).

12. Objectives in Drafting. On the following pages the correct methods to be used in mechanical drawing are explained. The student should learn and practice correct manipulation of the drawing instruments so that correct habits may be formed and maintained. Eventually he should draw correctly by habit so that his full attention may be given to the problems at hand. The instructor will insist upon absolutely correct form at all times, making exceptions only in cases of physical disability.

The following are the important objectives the student should strive to attain:

1. *Accuracy.* No drawing is of maximum usefulness if it is not accurate. The young draftsman or engineer must learn from the beginning that he cannot be successful in his college career or later in his professional employment if he does not acqure the habit of accuracy in his work.

2. *Speed.* "Time is money" in industry, and there is no demand for the slow draftsman. However, speed is not attained by hurrying; it is an unsought by-product of *intelligent and continuous work.* It comes with study and practice. The slow draftsman is usually dull, while the fast worker is usually mentally alert.

*Keuffel & Esser Co., New York; Eugene Dietzgen Co., New York; Theo. Alteneder & Sons, Philadelphia; Frederick Post Co., Chicago; V & E Manufacturing Co., Pasadena, Calif.; and the Gramercy Guild Group, Inc., New York, are some of the larger distributors of this equipment; their products are available through local dealers.

3. *Legibility.* The draftsman should remember that his drawing is a means of communication to others, and that it must be clear and legible in order to serve its purpose well. Care should be given to details, especially to lettering (Chapter 3).

4. *Neatness.* If a drawing is to be accurate and legible, it must also be clean; therefore the student should constantly strive to acquire the habit of neatness. Untidy drawings are the result of sloppy and careless methods, §23, and will not be accepted by the instructor.

13. Drafting at Home or School. If a draftsman is to turn out drawings rapidly and accurately, he must work in quiet surroundings without distractions. Technical drawing requires "headwork," and the draftsman who is whistling, singing, talking, eating, or smoking, is probably "pushing the pencil" without much mental activity behind it.

In the school drafting room, as in the industrial drafting room, the student is expected to give thoughtful and continuous attention to the problems at hand. If he does this, he will not have time to annoy others. The efficient draftsman sees to it that he has the correct equipment and refrains from borrowing—which is a nuisance to everyone. While the student is drawing, the textbook, his chief source of information, should be available in a convenient position, Fig. 11.

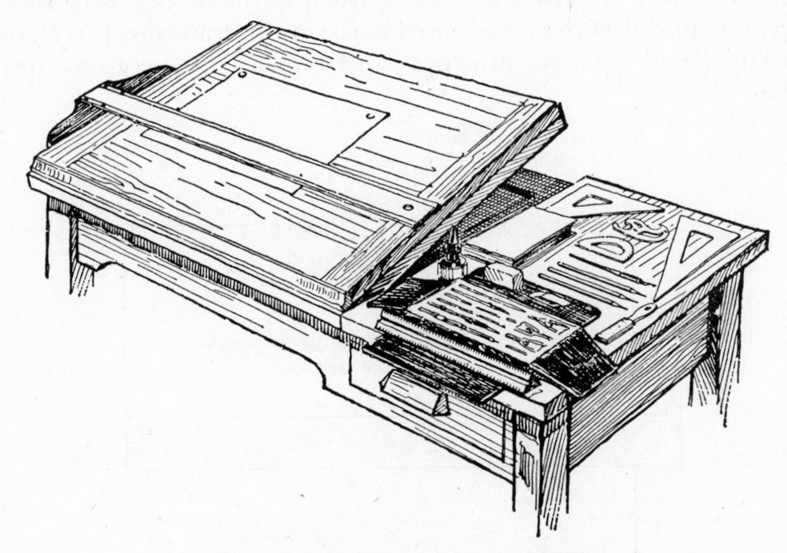

Fig. 11 Orderliness Promotes Efficiency.

When questions arise, first use the index of your text and endeavor to find the answer for yourself. Try to develop self-reliance and initiative. On the other hand, if you really need help, ask your instructor for assistance. The student who goes about his work intelligently, with a minimum waste of time, *first* studies the assignment carefully to be sure that he understands the principles involved, *second*, makes sure he has the correct equipment in proper condition (such as sharp pencils), and *third*, makes an effort to dig out answers for himself (the only true education).

One of the principal means of promoting efficiency in drafting is *orderliness*. All needed equipment and materials should be placed in an orderly manner so that everything is in a convenient place and can readily be found when needed, Fig. 11.

The drawing area should be kept clear of books, compasses, slide rules, and other equipment not in direct use. Form the habit of placing each item in a regular place outside the drawing area when it is not being used.

When drawing at home, it is best to work in a room by yourself, if possible. A book can be placed under the upper portion of the drawing board to give the board a convenient inclination, or the study-table drawer may be pulled out and used to support the drawing board at a slant.

It is best to work in natural north light coming from the left and slightly from the front. Never work on a drawing in direct sunlight or in dim light, as either will be injurious to the eyes. If artificial light is needed, the light source should be such that shadows are not cast where lines are being drawn, and such that there will be as little reflected glare from the paper as possible. Special draftsman's fluorescent lamps are available with adjustable arms so that the light source may be moved to any desired position.* Drafting will not hurt eyes that are in normal condition, but the exacting work will often disclose deficiencies not previously suspected.

Left-Handers. Place the head of the T-square on the right, and arrange the light source from the right and slightly from the front.

14. Drawing Boards. If the left edge of the drafting table top has a true straight edge and if the surface is hard and smooth (such as masonite), a drawing board is unnecessary, provided drafting tape is used to fasten the drawings. It is recommended that a "backing sheet" of heavy drawing paper be placed between the drawing and the table top.

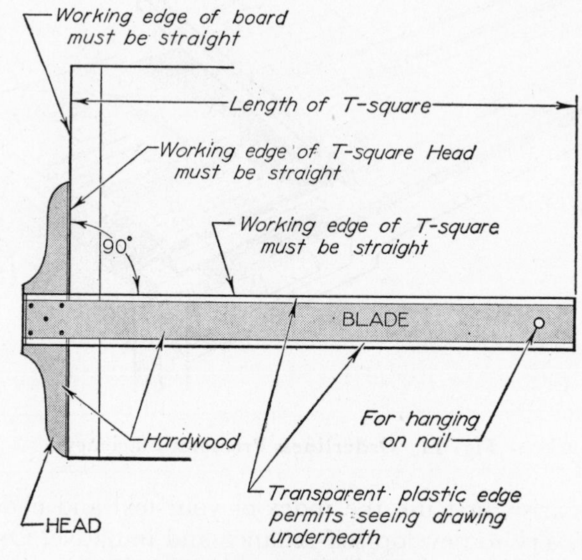

Fig. 12 The T-Square.

However, in most cases a drawing board will be needed. These vary from 9″ × 12″ (for sketching and field work) up to 48″ × 72″ or larger. The recommended size for students is 20″ × 24″, Fig. 10, which will accommodate the largest sheet likely to be used, which is 17″ × 22″.

Drawing boards traditionally have been made of soft woods, such as white pine,

*Dazor Mfg. Corp., St. Louis 10, Mo.

so that thumbtacks can be easily pushed down. However, after considerable use, the board is likely to be full of thumbtack holes, which are objectionable. Many draftsmen now prefer to use Scotch drafting tape, which in turn permits hard surfaces such as hardwood, masonite, linoleum, or other materials to be used.

The left-hand edge of the board is called the *working edge*, because the T-square head slides against it, Fig. 12. This edge must be straight, and you should test the edge with a framing square, or with a T-square blade which has been tested and found straight, Fig. 13. If the edge of the board is not true, it must be run through a jointer or planed with a jack plane.

15. T-Square. The T-square, Fig. 12, is composed of a long strip called the *blade*, fastened rigidly at right angles to a shorter piece called the *head*. The upper edge of the blade and the inner edge of the head are *working edges*, and must be straight. The working edge of the head must not be convex, so that the T-square

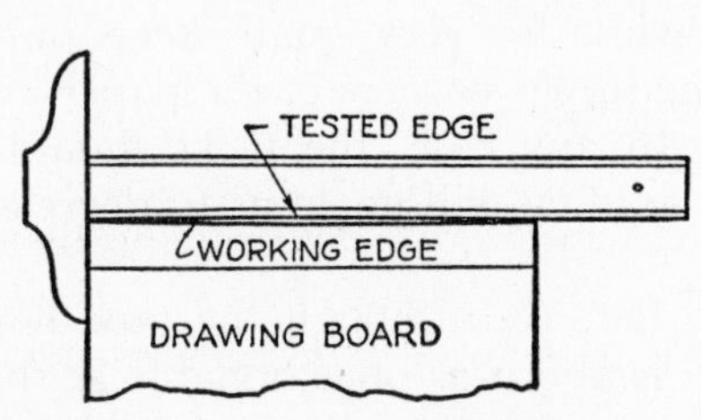

Fig. 13 Testing the Working Edge of the Drawing Board.

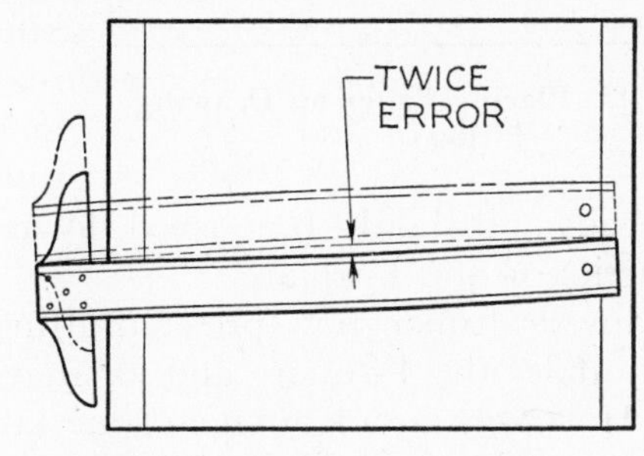

Fig. 14 Testing the T-Square.

"rocks" when the head is placed against the board. The blade should have transparent plastic edges and should be free of nicks along the working edge. Transparent edges are desirable, since they permit the draftsman to see the drawing in the vicinity of the lines being drawn.

Do not use the T-square to drive tacks into the board or for any rough purpose. Never cut paper along its working edge, as the plastic is easily cut and even a slight nick will ruin the T-square.

16. Testing and Correcting the T-Square. To test the working edge of the head, see if the T-square "rocks" when the head is placed against a straight edge, such as a framing square or a drawing board working edge that has already been tested and found true. If the working edge of the head is convex, remove the head and run it through a jointer, or plane it by hand until it tests straight. In replacing the blade on the head, use furniture glue in addition to the screws.

To test the working edge of the blade, Fig. 14, draw a sharp line very carefully with a hard pencil along the entire length of the working edge; then turn the T-square over and draw the line again along the same edge. If the edge is straight, the two lines will coincide; otherwise the space between the lines will be twice the error of the blade.

It is difficult to correct a crooked T-square blade, and if the error is considerable, it may be necessary to discard the T-square and obtain another. However, if care is taken the blade can be made true by scraping the edge with a scraper or a sharp knife, as in truing a triangle, Fig. 33, and then sanding with #0 or #00 sandpaper wrapped around a block.

17. Fastening Paper to the Board. The drawing paper should be placed close enough to the working edge of the board to reduce to a minimum any error resulting from a slight "give," or bending of the blade of the T-square, and close enough to the upper edge of the board to permit space at the bottom of the sheet for using the T-square and supporting the arm while drawing, Fig. 15.

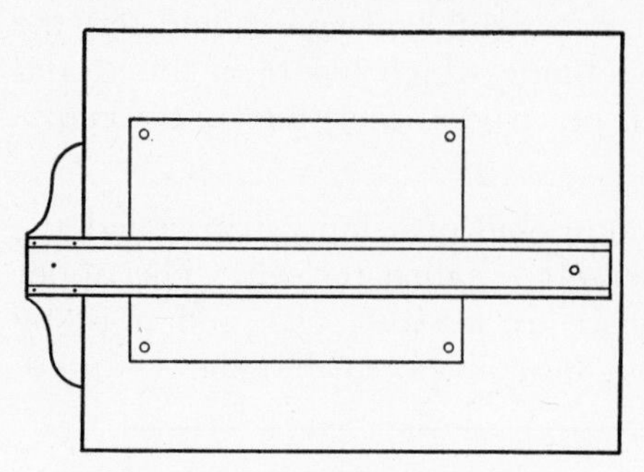

Fig. 15 Placing Paper on Drawing Board.

To thumbtack the paper in place, press the T-square head firmly against the working edge of the drawing board with the left hand, while paper is adjusted with the right hand until the top edge "lines up" with the upper edge of the T-square. Place the first tack in the upper left corner, and the second in the upper or lower right corner, depending upon the condition of the paper. Generally four tacks are required, but for very large sheets more than four may be required, while for very small sheets only two tacks in the upper corners of the paper may be sufficient. In any case, the tacks should have thin heads and should be pushed down as far as they will go so as to decrease the obstruction to the T-square.

Many draftsmen now prefer drafting tape, Fig. 16, although the tape tends to roll up under the T-square and for that reason is about as objectionable as thumbtacks. However, an advantage of the tape is that it does not damage the board. In removing a taped drawing, *pull the tape back slowly*. If pulled rapidly, the tape will take with it the top surface of the paper.

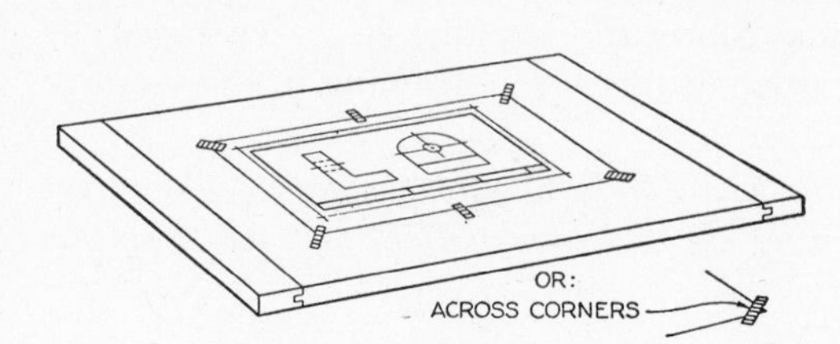

Fig. 16 Drafting Tape.

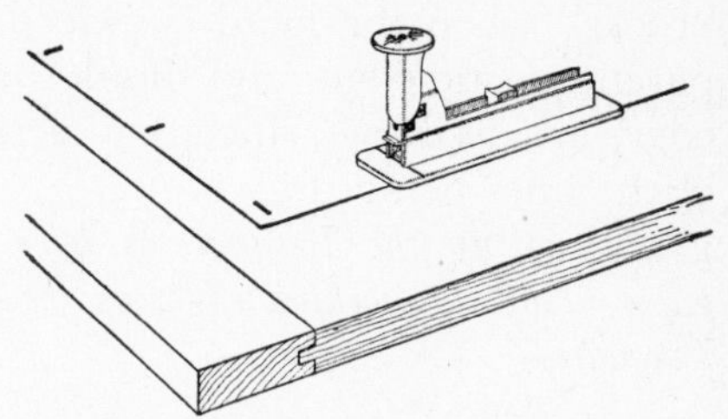

Fig. 17 Draftsman's Stapler.

Some draftsmen prefer to use cellophane tape, which is perfect for the purpose except that it is a little more tedious to remove and it tends to take off even more of the top surface of the paper if not handled carefully. Therefore, such tape should be placed outside the trim line of the drawing, if possible.

Another method for fastening paper to the board is to use *wire staples*. A special draftsman's stapler is shown in Fig. 17.

Tracing paper should not be fastened directly upon the board because small imperfections in the surface of the board will interfere with the line work. Always fasten a larger backing sheet of heavy white drawing paper on the board first; then fasten the drawing over this sheet.

18. Drawing Pencils. High-quality *drawing pencils*, Fig. 18 (a), should be used in technical drawing—never ordinary writing pencils.

Drawing pencil leads are made of graphite with kaolin (clay) added in varying amounts to make eighteen grades from 9H (the hardest) down to 7B (the softest), Fig. 19. The uses of these different grades are shown in the figure. Note that the harder pencils have small-diameter leads, while the softer grades have large-diameter leads to give more strength; hence, the degree of hardness can be roughly judged by a comparison of the diameters. Drawing pencils are hexagonal in shape to fit between the thumb, forefinger, and second finger, and to prevent them from rolling off the the table easily.

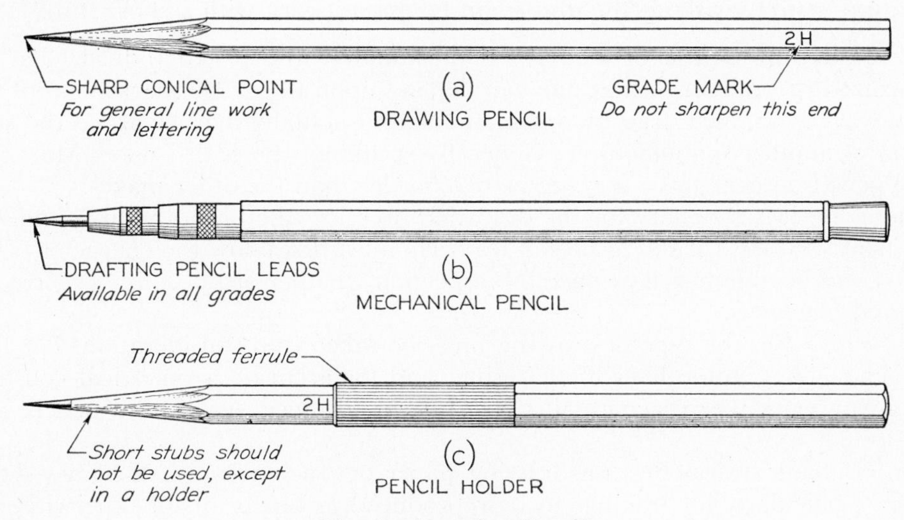

Fig. 18 Drawing Pencils.

Many makes of mechanical pencils are available, Fig. 18 (b), together with refill drafting leads in all grades. Choose those which "feel" well in the hand and which grip the lead firmly without slipping. Mechanical pencils have the advantage of maintaining a constant length of lead (no more trying to draw with a stub!), of permitting use of a lead practically to the end, of being easily refilled with new leads, of affording a ready source for compass leads, and of having no wood to be

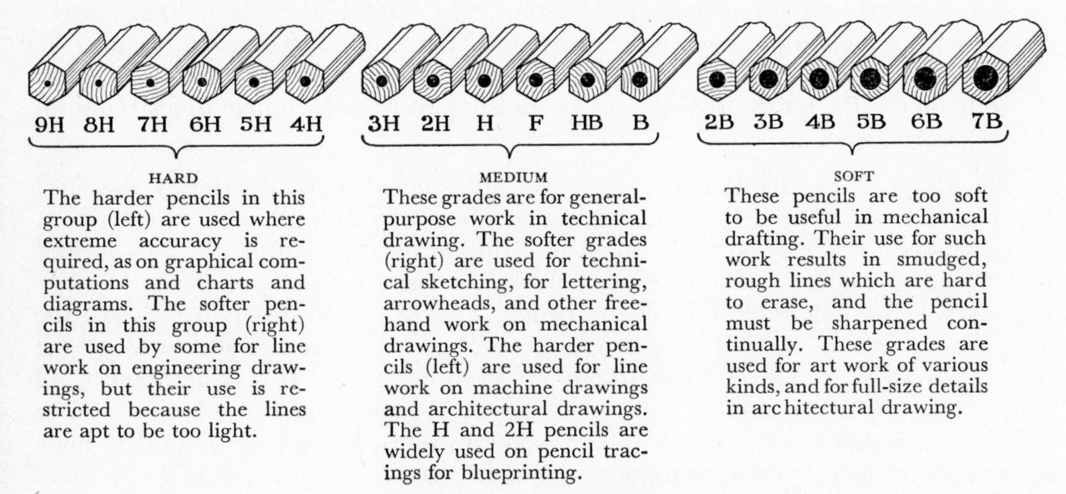

9H 8H 7H 6H 5H 4H
3H 2H H F HB B
2B 3B 4B 5B 6B 7B
HARD
The harder pencils in this group (left) are used where extreme accuracy is required, as on graphical computations and charts and diagrams. The softer pencils in this group (right) are used by some for line work on engineering drawings, but their use is restricted because the lines are apt to be too light.
MEDIUM
These grades are for general-purpose work in technical drawing. The softer grades (right) are used for technical sketching, for lettering, arrowheads, and other freehand work on mechanical drawings. The harder pencils (left) are used for line work on machine drawings and architectural drawings. The H and 2H pencils are widely used on pencil tracings for blueprinting.
SOFT
These pencils are too soft to be useful in mechanical drafting. Their use for such work results in smudged, rough lines which are hard to erase, and the pencil must be sharpened continually. These grades are used for art work of various kinds, and for full-size details in architectural drawing.

Fig. 19 Pencil Grade Chart.

sharpened. In the long run, the total expense for mechanical pencils is probably less than for wooden pencils.

A new type of mechanical pencil has been introduced which uses flat leads. Lines of uniform width—suitable for visible or hidden lines—may be drawn without the necessity of ever sharpening the lead. In drawing a line, the flat face of the lead is held against the ruling edge.

Pencil holders are also available, Fig. 18 (c), by means of which a pencil may be used up entirely without the draftsman having to work with a short stub.

19. Choice of Grade of Pencil. Unfortunately the pencil industry has not standardized its product so that one can depend upon the grade marks, except in a general way. Thus, an F lead by one manufacturer actually may be about the same as a 2H of another manufacturer. Generally speaking, the Koh-i-noor, Mars, and Castell pencils appear to be a grade or two harder than the other makes.

Therefore, it is necessary for the draftsman to select the brand he likes and then experiment with the various grades of lead. He must first know the character of line required and be able to tell at once by inspection whether or not a line is correct in width and blackness.

To begin with, the type of drawing must be taken into consideration. For light construction lines, guide lines for lettering, and for accurate geometrical constructions or any work where accuracy is of prime importance, use a hard pencil, such as 4H to 6H.

For mechanical drawings on drawing paper or tracing paper, the lines should be *black*, particularly for tracings to be reproduced as blueprints or otherwise. The pencil chosen must be soft enough to produce jet-black lines, but hard enough not to smudge too easily or permit the point to crumble under normal pressure. This pencil will vary from F to 2H, roughly, depending upon the paper and weather conditions. The same comparatively soft pencil is preferred for lettering and arrowheads.

Another factor to consider is the texture of the paper. If the paper is hard and has a decided "tooth," it will be necessary generally to use harder leads. For smoother surfaces, softer leads can be used. Hence, to obtain dense black lines, the paper should not have too much "tooth."

A final factor to consider is the weather. On humid days the paper absorbs moisture from the atmosphere and becomes soft. This can be recognized because the paper expands and becomes wrinkled. It is necessary to select softer leads to offset the softening of the paper. If you have been using an F lead, change to an HB until the weather clears up.

20. Sharpening the Pencil. *Keep your pencil sharp!* This is certainly the most frequent instruction needed by the beginning student. A dull pencil produces fuzzy, sloppy, indefinite lines and is the mark of a dull and careless student. Only a *sharp pencil* is capable of producing clean-cut black lines that sparkle with clarity.

Sharpen the unlettered end of the pencil, Fig. 18 (a), in order to preserve the grade mark and to make it easy to identify the pencil in use. First, cut away the wood with a sharp knife, starting about $1\frac{1}{2}''$ from the end, and leaving about $\frac{3}{8}''$ of lead uncut, Fig. 20 (a) and 21 (a). Or use a simple pencil sharpener, Fig. 22 (a), or an automatic sharpener, (b), to cut the wood away as shown in Fig. 20 (b). Then shape the lead to a sharp conical point on a pencil pointer (sandpaper pad or a small file), Figs. 20 (c) and 21 (b), and wipe the lead clean to remove loose particles of graphite. *Never sharpen your pencil over the drawing or any of your equipment.* Many draftsmen then "burnish" the point on a piece of hard paper to obtain a smoother, sharper point. However, for drawing visible lines the point should not be needle-sharp, but very slightly rounded. First, sharpen the lead to a needle point, then stand the pencil vertically and with a few rotary motions on the paper, wear the point down slightly

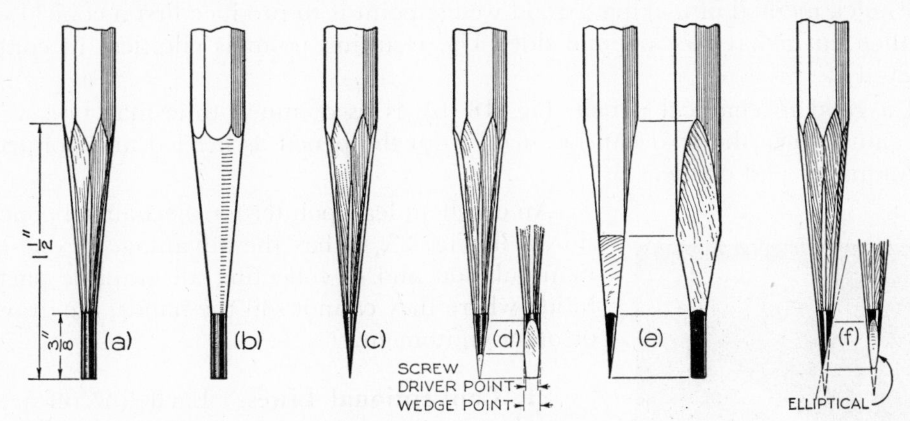

Fig. 20 Pencil Points.

to the desired point. Keep the sandpaper pad, when not in use, in a container, such as an envelope with one end cut open, Fig. 22 (c), to prevent the particles of graphite from falling upon the drawing or drawing equipment.

Keep the pencil pointer close by, as frequent pointing of the pencil will be necessary.

For straight-line drawing, some draftsmen prefer the wedge point, Fig. 20 (d). This point is produced by cutting the wood away as at (a) or (b), and then sharpening on opposite sides, as shown at (d). In drawing a line with this point, the flat face

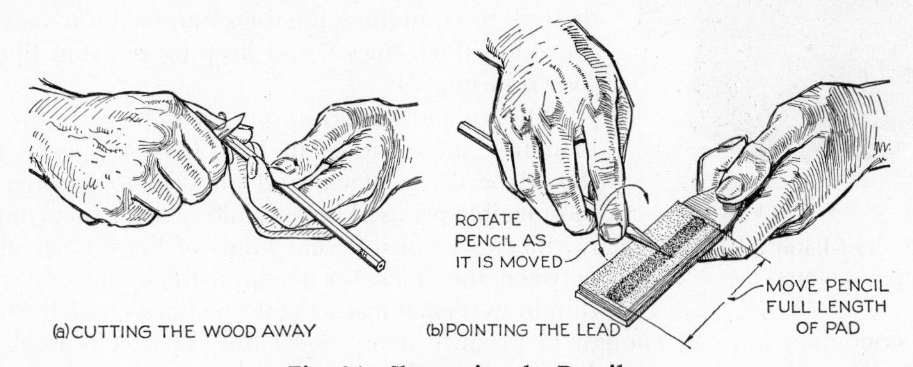

Fig. 21 Sharpening the Pencil.

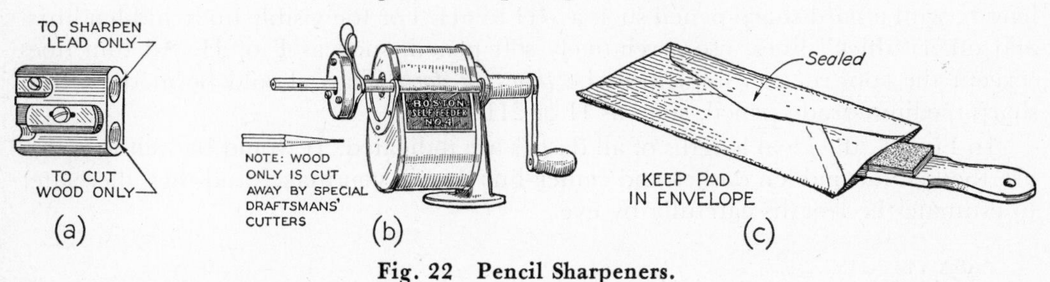

Fig. 22 Pencil Sharpeners.

of the lead is held against the straightedge. Some prefer to sharpen slightly on the edges of the wedge so as to produce a "screwdriver" point, as shown. If pronounced flat cuts are made in the wood on the same sides as the flat cuts in the lead as shown at (e), the draftsman will be able easily to "feel" when a flat face of the lead is against the straightedge.

A quick method of making a good wedge point is to produce first a conical point and then cut a flat face on one side. The resulting point is elliptical in contour, Fig. 20 (f).

If a good mechanical pencil, Fig. 18 (b), is used, much time may be saved in sharpening, since the lead can be "fed" from the pencil as needed and pointed on the sandpaper pad or file.

Courtesy Elward Manufacturing Co.

Fig. 23 Tru-Point Pencil Lead Pointer.

An excellent lead pointer for mechanical pencils is shown in Fig. 23. It has the advantage of one-hand manipulation and of collecting all graphite particles inside where they cannot soil the hands, the drawing, or other equipment.

21. Conventional Lines. Each line on a technical drawing has a definite meaning and is drawn in a certain way. The conventional lines adopted by the American Standards Association are shown in Fig. 24, together with illustrations showing the various applications. According to the American Standards,* "Three widths of lines, *thick*, *medium*, and *thin* are recommended . . . exact thicknesses may vary according to the size and type of drawing. . . ." The standard further provides that pencil lines may be made in two widths, if desired, by combining the thick and medium lines into "medium-thick lines," and keeping the thin lines as shown in Fig. 24.

The student should simply try to produce his lines to match those shown in Fig. 24. "All pencil lines should be dense black, and uniform to produce the most legible prints. There should be a distinct contrast in thicknesses of different kinds of lines, particularly between the thick (or medium-thick) lines and the thin lines. Pencil leads should be hard enough to prevent smudging, but soft enough to produce dense-black lines of best reproducing quality."* Make construction lines so light that they can barely be seen at arm's length, with a hard sharp pencil such as 4H to 6H. For the visible lines, hidden lines, and other "thick" lines, use a relatively soft pencil, such as F or H. All thin lines (except the construction line) must be *thin, but dark.* They should be made with a sharp medium-grade pencil, such as H or 2H.

In Fig. 24, the ideal lengths of all dashes are indicated. It would be well to measure the first few hidden dashes and center-line dashes you make, and then thereafter to estimate the lengths carefully by eye.

*ASA Y14.2-1957.

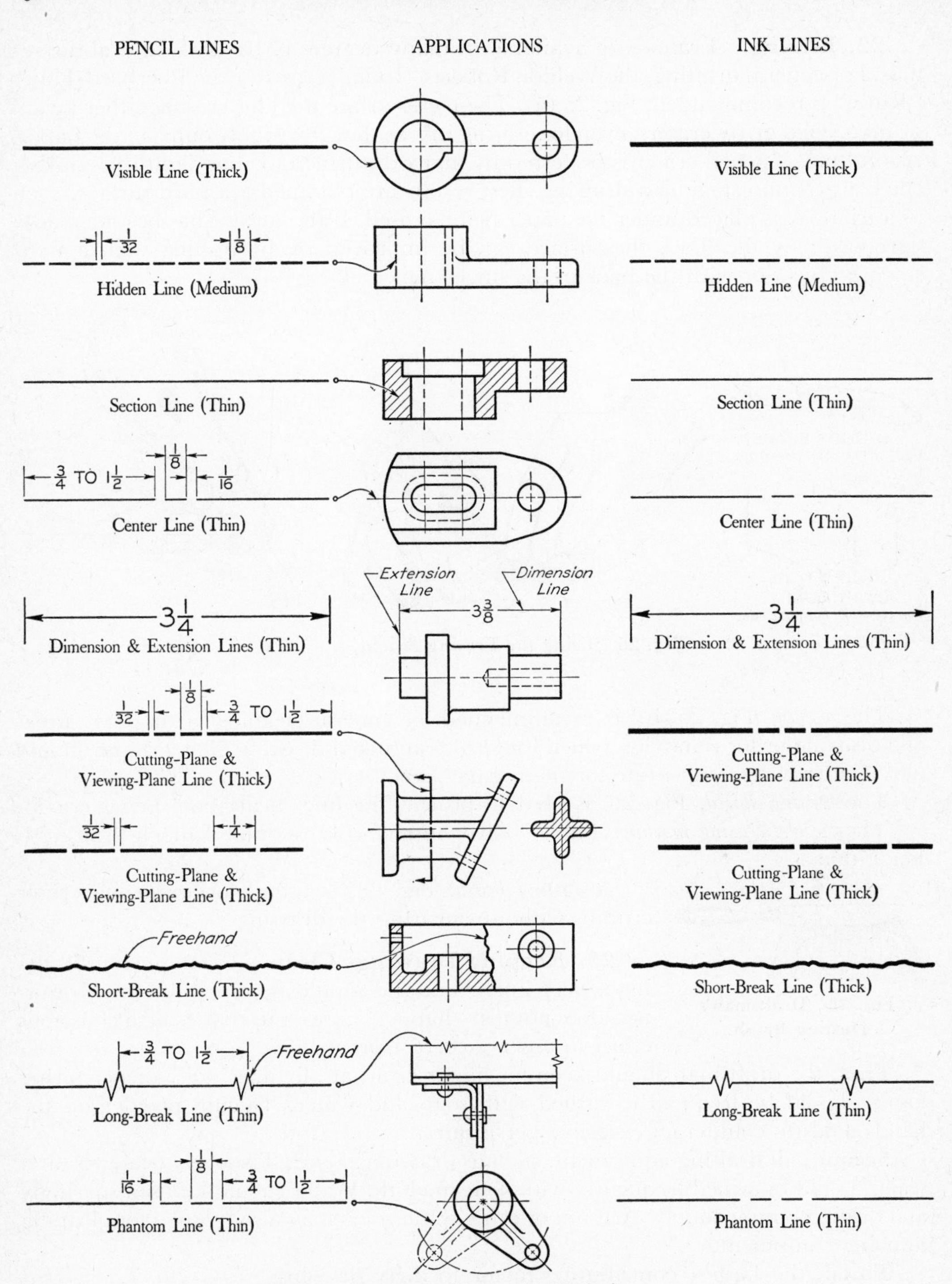

Fig. 24 Alphabet of Lines (Full Size).

Note that the thick and medium ink lines are considerably wider than the corresponding pencil lines, while the thin lines are about the same width in pencil or in ink. For inking procedures, see §§57–64.

22. Erasing. Erasers are available in many degrees of hardness and abrasiveness. For general drafting, the Weldon Roberts "India" eraser or the Eberhard-Faber "Ruby" is recommended, Fig. 25 (a). These erasers are used for erasing either pencil or ink. Avoid gritty erasers, even for erasing ink, as they invariably damage the paper. A soft "pink" pencil eraser is preferred by many draftsmen to erase light lines during the construction stage of a drawing. Best results are obtained if a hard surface, such as a triangle, is placed under the paper being erased. If the surface has become badly "grooved" by the lines, the surface can be improved by burnishing with a hard smooth object or with the back of the finger nail.

Fig. 25 Erasers. **Fig. 26 Using the Erasing Shield.** **Fig. 27 Electric Erasing Machine.**

The *artgum*, Fig. 25 (b), is recommended for general cleaning of the large areas of a drawing or for removing pencil lines from an inked drawing. The artgum should never be used as a substitute for the regular pencil eraser.

The *erasing shield*, Fig. 26, is used to protect the lines near those being erased.

The *electric erasing machine*, Fig. 27, saves time and is essential if much drafting is being done.

A *dusting brush*, Fig. 28, is useful for removing eraser crumbs without smearing the drawing.

Fig. 28 Draftsman's Dusting Brush.

23. Keeping Drawings Clean. Cleanliness in drafting is very important and should become a habit. Cleanliness does not just "happen"; it results only from a conscious effort to observe correct procedures.

First, the draftsman should keep his hands clean at all times. Oily or perspiring hands should be frequently washed with soap and water. Talcum powder on the hands tends to counteract excessive perspiration.

Second, all drafting equipment, such as drawing board, T-square, triangles, and scale, should be wiped frequently with a clean cloth. Water should be used sparingly and dried off immediately. Artgum or other soft erasers may also be used for cleaning drawing equipment.

Third, the largest contributing factor to dirty drawings is *not dirt, but graphite* from the pencil; hence the draftsman should observe the following precautions:

1. Never sharpen a pencil over the drawing or any equipment.
2. Always wipe the pencil point with a clean cloth, after sharpening or pointing, to remove small particles of loose graphite.

3. Never place the sandpaper pad in contact with any other drawing equipment unless it is completely enclosed in an envelope or similar cover, Fig. 22 (c).

4. Never work with the sleeves or hands resting upon a penciled area. Keep such parts covered with clean paper (not a cloth). In lettering a drawing, always place a piece of paper under the hand.

5. Avoid sliding anything across the drawing. A certain amount of sliding of T-square and triangles is necessary, but this can be minimized if triangles are picked up by their tips and the T-square blade tilted upward slightly before moving.

6. Never rub across the drawing with the palm of the hand to remove eraser particles; use a dust brush, Fig. 28, or flick—don't rub—the particles off with a clean cloth.

When the drawing is completed, it is not necessary to "clean it" if the above rules have been observed. The practice of making a pencil drawing, scrubbing it with artgum, and then retracing the lines, is poor technique and a waste of time, and this habit should not be acquired.

At the end of the period or of the day's work, the drawing should be covered to protect it from dust.

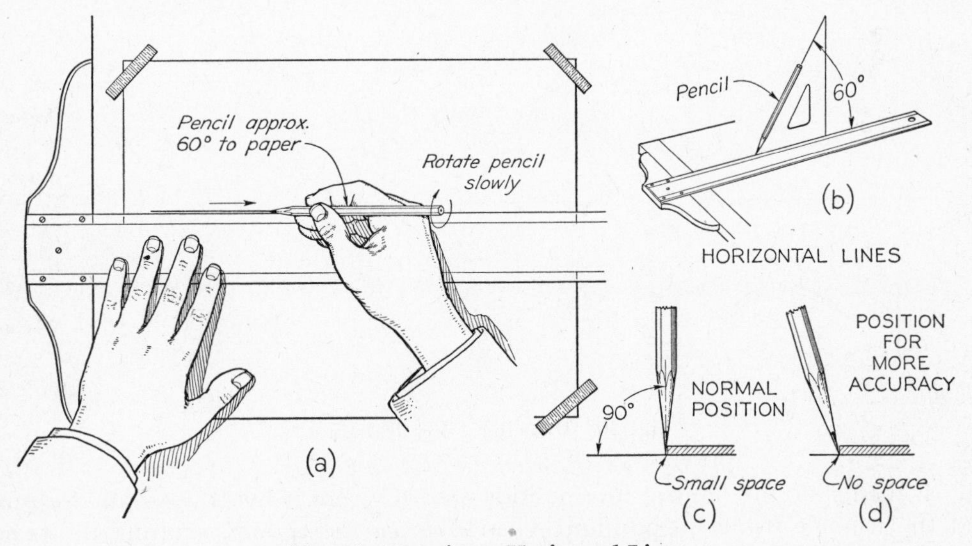

Fig. 29 Drawing a Horizontal Line.

24. Horizontal Lines. To draw a horizontal line, Fig. 29 (a), press the head of the T-square firmly against the working edge of the board with the left hand; then slide the left hand to the position shown, so as to press the blade tightly against the paper. Lean the pencil in the direction of the line at an angle of approximately 60° with the paper, (b), and draw the line from left to right. Keep the pencil in a vertical plane, (b) and (c); otherwise, the line may not be straight. While drawing the line, let the little finger of the hand holding the pencil glide lightly on the blade of the T-square, and rotate the pencil slowly between the thumb and forefinger so as to distribute the wear uniformly on the lead and maintain a symmetrical point.

When great accuracy is required, the pencil may be "toed in" as shown at (d) to produce a perfectly straight line.

Left-Handers. In general, reverse the above procedure. Place the T-square head against the right edge of the board, and with the pencil in the left hand, draw the line from right to left.

Triangles and T-squares, especially when new, often have very sharp edges which tend to cut into the pencil lead and cause a trail of graphite to be distributed along the line. To prevent smearing of these particles, blow them off at intervals. If the edges of the triangles or T-square are too sharp, they can be sanded *very lightly* with #00 sandpaper—just enough to remove the sharp edges.

25. Vertical Lines. Use either the 45° triangle or the 30° × 60° triangle to draw vertical lines. Place the triangle on the T-square with the *vertical edge on the left* as shown in Fig. 30. With the left hand, press the head of the T-square against the

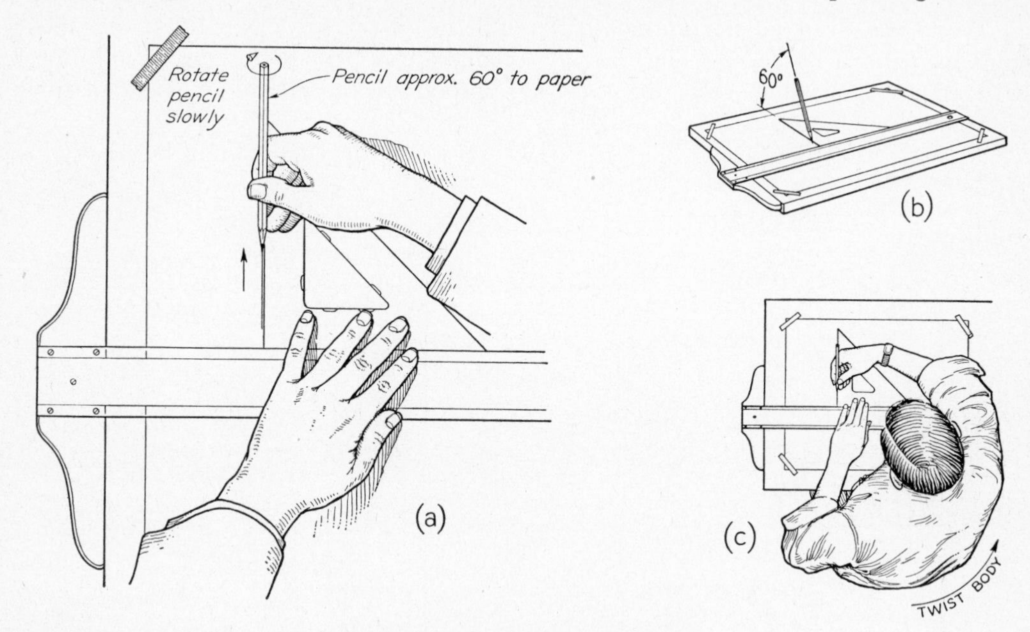

Fig. 30 Drawing a Vertical Line.

board, then slide the hand to the position shown where it holds both the T-square and the triangle firmly in position. Then *draw the line upward*, rotating the pencil slowly between the thumb and forefinger.

Lean the pencil in the direction of the line at an angle of approximately 60° with the paper and in a vertical plane, (b). Meanwhile, the upper part of the body should be twisted to the right, as shown at (c).

Left-Handers. In general, reverse the above procedure. Place the T-square head on the right and the vertical edge of the triangle on the right; then, with the right hand, hold the T-square and triangle firmly together, and with the left hand draw the line upward.

The only time it is permissible for right-handers to turn the triangle so that the vertical edge is on the right is when drawing a vertical line near the right end of the T-square. In this case, the line would be drawn downward.

26. The Triangles. Most inclined lines in mechanical drawing are drawn at standard angles with the *45° triangle* and the *30° × 60° triangle*, Fig. 31. The triangles are made of transparent plastic so that lines of the drawing can be seen through them. A good combination of triangles is the 30° × 60° triangle with a long side of 10″, and a 45° triangle with each side 8″ long.

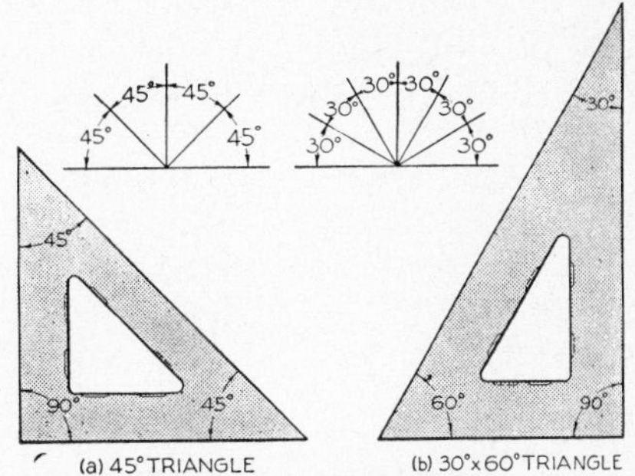

Fig. 31 Triangles.

27. Testing and Correcting the Triangles. Triangles are subject to warping, sometimes even before they are sold by the dealer. Therefore, the purchaser should test his new triangles immediately after purchase to determine if they are "true." If they are not found to be correct, they should be returned at once to the dealer.

Test the sides of the triangles for straightness in the same manner as for the T-square blade, §16. To test the right angle of either of the triangles, Fig. 32 (a), place the triangle on the T-square and draw a vertical line; then turn the triangle over (like turning a page in a book) and draw the line again along the same edge. If the two lines thus drawn do not coincide, the right angle is not 90° and the error is half the angle between the two lines.

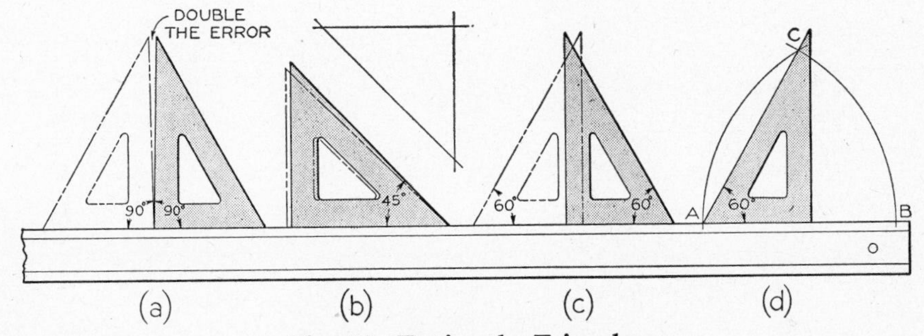

Fig. 32 Testing the Triangles.

To test the 45° angle, place the triangle on the T-square, as at (b), and draw a line along the hypotenuse; then turn the triangle over, and using the other 45° angle of the triangle, draw a line along the hypotenuse. If the two lines do not coincide, there is an error in one or both 45° angles. A direct test of the 45° angle can be made by drawing a right triangle. The sides adjacent to the 90° angle will be equal if the two 45° angles are correct (assuming the 90° angle to be correct).

To test the 60° angle of the 30° × 60° triangle, draw an equilateral triangle, as shown at (c). If the sides are not exactly equal in length, the 60° angle is incorrect. Another method of testing the 60° angle, (d), is to draw a horizontal line AB slightly shorter than the hypotenuse of the triangle, and to draw arcs with A and B as centers and AB as radius, intersecting at C. When the triangle is placed as shown, its hypotenuse should pass through C.

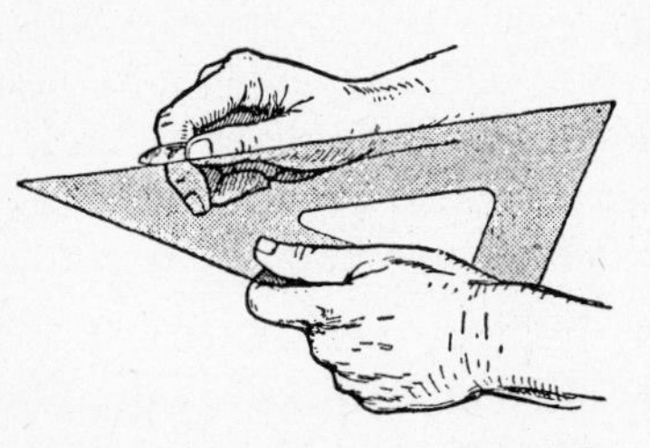

Fig. 33 Scraping the Triangle.

To true up the edge of a triangle, make a "rough cut" by scraping the edge with a knife or with a scraper, Fig. 33. Or place the triangle in a vise and plane with a sharp block plane set for a very shallow cut. Then hold the triangle flat against the edge of a table top, with the edge of the triangle level with it, and sand the edge with #0 or #00 sandpaper wrapped around a block.

28. Inclined Lines. The positions of the triangles for drawing lines at all of the possible angles are shown in Fig. 34. In the figure it is understood that the triangles in each case are resting upon the blade of the T-square. Thus, it is possible

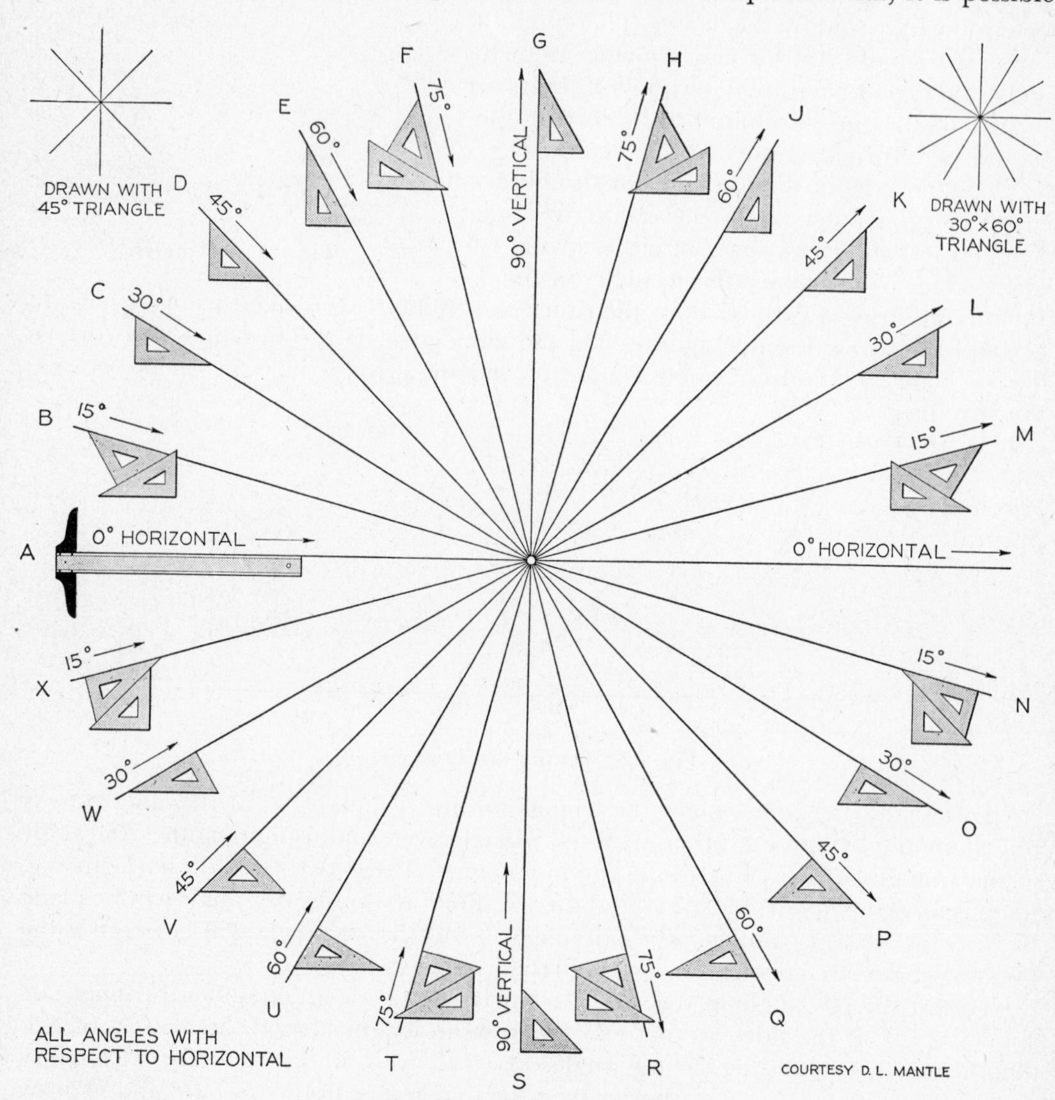

Fig. 34 The Triangle Wheel.

to divide 360° into twenty-four 15° sectors with the triangles used singly or in combination. Note carefully the directions for drawing the lines, as indicated by the arrows. Note that all arrows in the left half point *toward the center*, while those in the right half point *away from the center*.

29. Protractors. For measuring or setting off angles other than those obtainable with the triangles, the *protractor* is used. The best protractors are made of nickel silver and are capable of most accurate work, Fig. 35 (a). For ordinary work, the plastic or the sheet-metal protractor is satisfactory and is much cheaper, (b). To set off angles with greater accuracy, use one of the methods presented in §127.

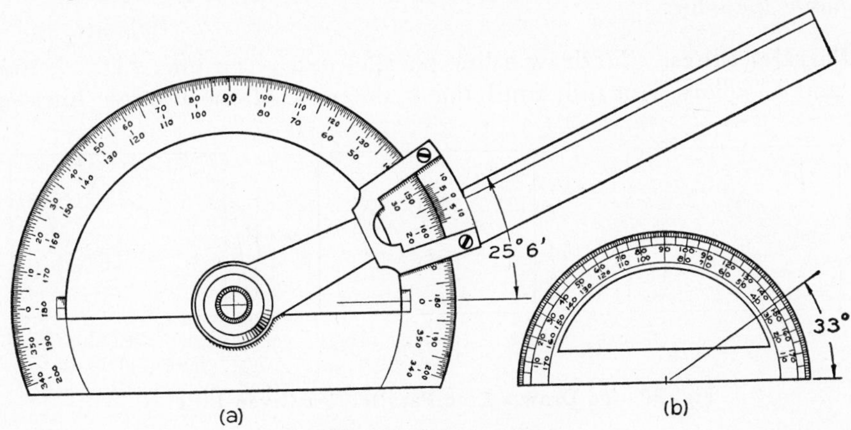

Fig. 35 Protractors.

30. Drafting Angles. A variety of devices combining the protractor with triangles or scales to produce great versatility of use are available, two types of which are shown in Fig. 36. The *Draft-Scale-Angle*, (a), is a practical drafting aid whose applications are apparent in the figure. An adjustable triangle is shown at (b).

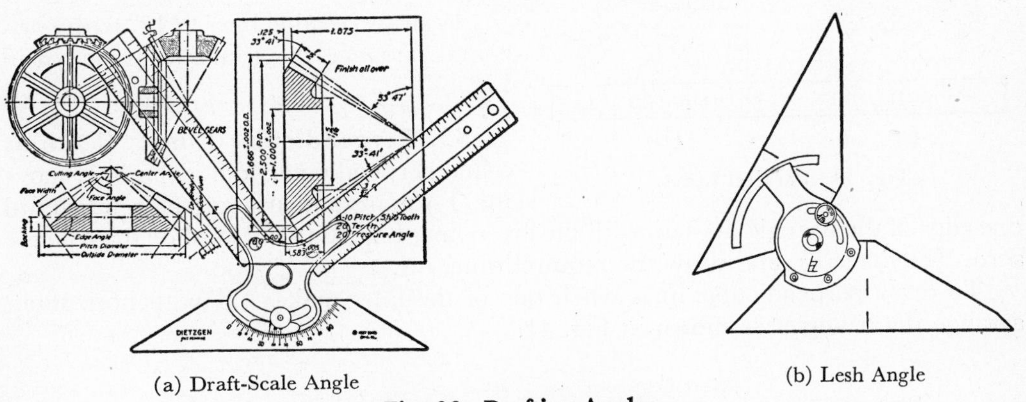

(a) Draft-Scale Angle (b) Lesh Angle

Fig. 36 Drafting Angles.

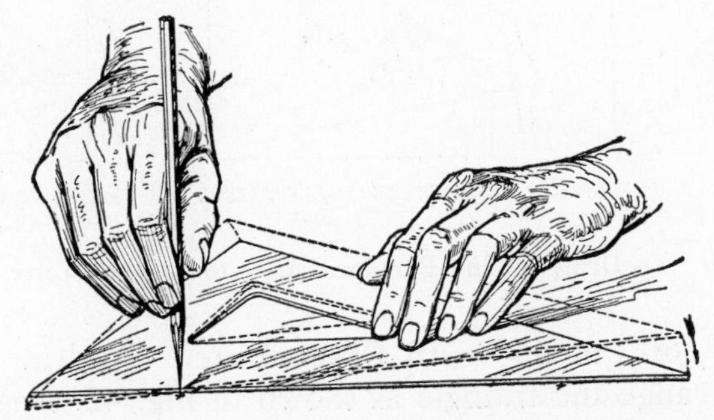

Fig. 37 To Draw a Pencil Line through Two Given Points.

31. To Draw a Line Through Two Points. To draw a line through two points, Fig. 37, place the pencil vertically at one of the points, and move the straight-edge about the pencil point as a pivot until it lines up with the other point; then draw the line along the edge.

32. Parallel Lines. To draw a line parallel to a given line, Fig. 38, move the triangle and T-square as a unit until the hypotenuse of the triangle lines up with

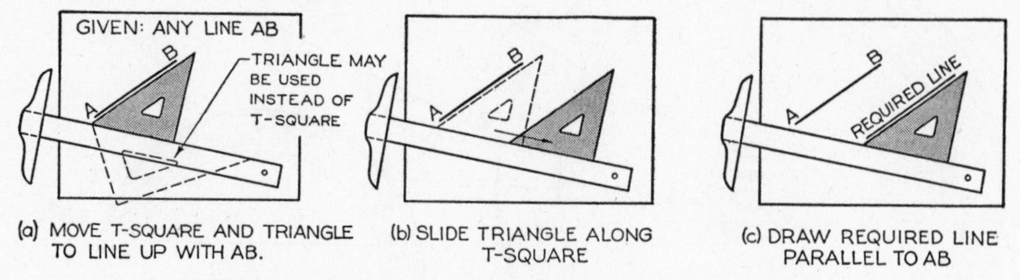
GIVEN: ANY LINE AB
TRIANGLE MAY BE USED INSTEAD OF T-SQUARE
A
B
REQUIRED LINE
(a) MOVE T-SQUARE AND TRIANGLE TO LINE UP WITH AB.
(b) SLIDE TRIANGLE ALONG T-SQUARE
(c) DRAW REQUIRED LINE PARALLEL TO AB

Fig. 38 To Draw a Line Parallel to a Given Line.

the given line, (a); then, holding the T-square firmly in position, slide the triangle away from the line, (b), and draw the required line along the hypotenuse, (c).

Obviously any straightedge, such as one of the triangles, may be substituted for the T-square in this operation, as shown at (a).

To draw parallel lines at 15° with horizontal, arrange the triangles as shown in Fig. 39.

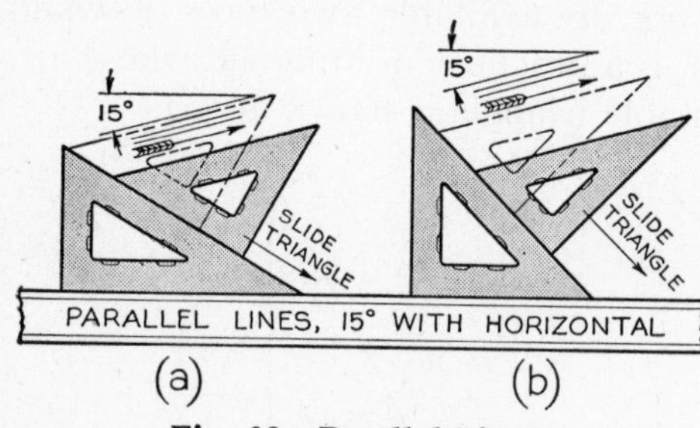
15°
SLIDE TRIANGLE
PARALLEL LINES, 15° WITH HORIZONTAL
(a)
(b)

Fig. 39 Parallel Lines.

33. Perpendicular Lines. To draw a line perpendicular to a given line, move the T-square and triangle as a unit until one edge of the triangle lines up with the given line, Fig. 40 (a); then slide the triangle across the line, (b), and draw the required line, (c).

To draw perpendicular lines when one of the lines makes 15° with horizontal, arrange the triangles as shown in Fig. 41.

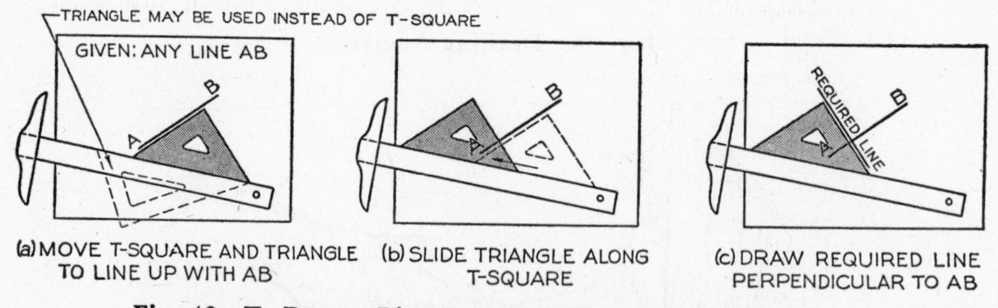
TRIANGLE MAY BE USED INSTEAD OF T-SQUARE
GIVEN: ANY LINE AB
A
B
REQUIRED LINE
(a) MOVE T-SQUARE AND TRIANGLE TO LINE UP WITH AB
(b) SLIDE TRIANGLE ALONG T-SQUARE
(c) DRAW REQUIRED LINE PERPENDICULAR TO AB

Fig. 40 To Draw a Line Perpendicular to a Given Line.

34. Lines at 30°, 60°, or 45° with Given Line. To draw a line making 30° with a given line, arrange the triangle as shown in Fig. 42. Angles of 60° and 45° may be drawn in a similar manner.

35. Scales. A drawing of an object may be the same size as the object (full size), or it may be larger or smaller than the object; in most cases, if not drawn full size, the drawing is made smaller than the object represented. The ratio of reduction depends upon the relative sizes of the object and of the sheet of paper upon which

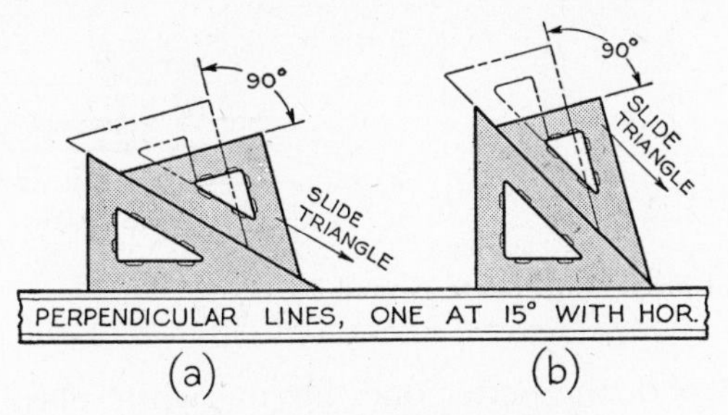

Fig. 41 Perpendicular Lines.

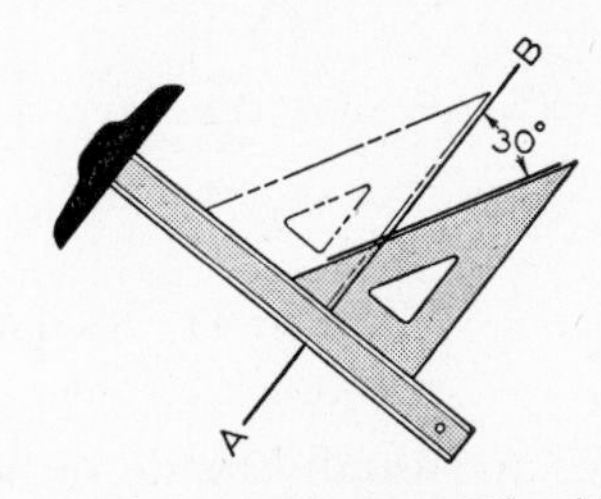

Fig. 42 Line at 30° with Given Line.

the drawing is to be made. For example, a machine part may be half size ($\frac{1}{2}'' = 1''$); a building may be drawn $\frac{1}{48}$ size ($\frac{1}{4}'' = 1'\text{-}0''$); a map may be drawn $\frac{1}{1200}$ size ($1'' = 100'\text{-}0''$); or a gear in a wrist watch may be ten-times size ($10'' = 1''$ or $\frac{10}{1}$).

Scales are classified as *architects scale*, Fig. 43 (a), *engineers scale*, (b), *mechanical draftsmans scale*, (c), and the *decimal scale*, (d).

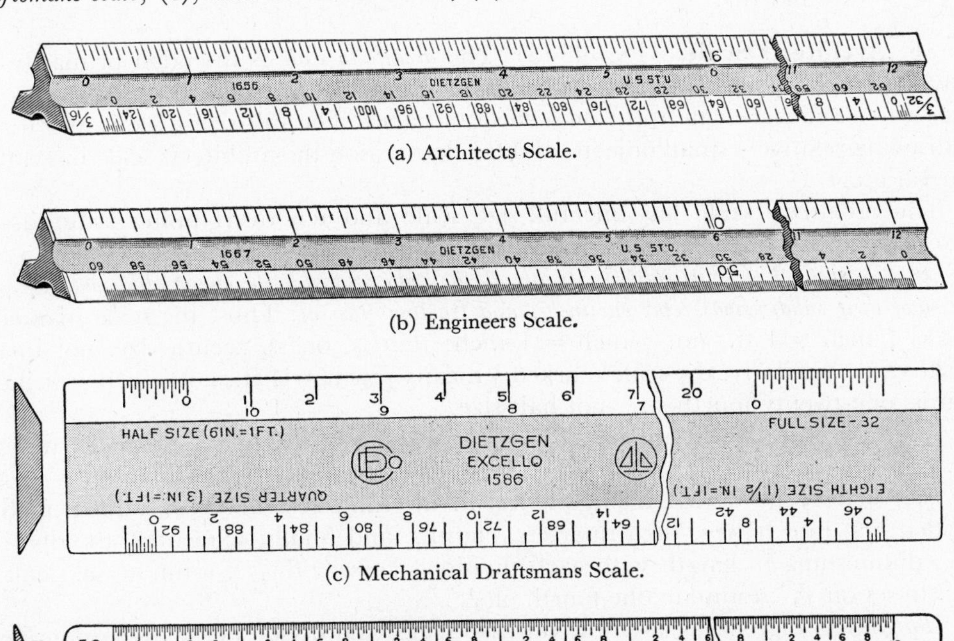

(a) Architects Scale.

(b) Engineers Scale.

(c) Mechanical Draftsmans Scale.

(d) Decimal Scale.

Courtesy Eugene Dietzgen Co.

Fig. 43 Types of Scales.

A *full-divided scale* is one in which the basic units are subdivided throughout the length of the scale, Fig. 43 (a), (b), and (d). An *open-divided scale* is one in which only the end unit is subdivided, as at (c).

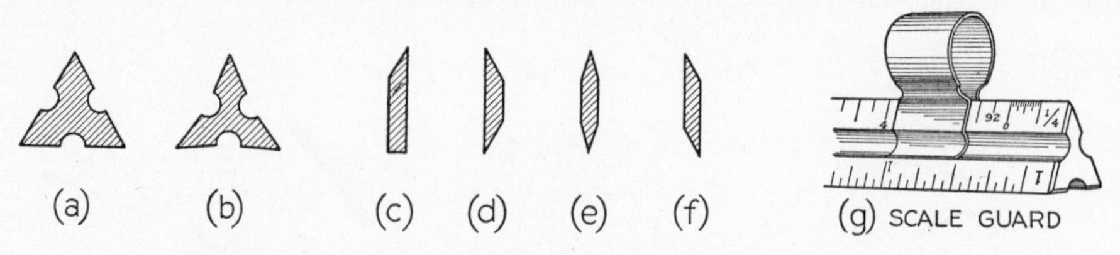

Fig. 44 Sections of Scales and Scale Guard.

Scales are usually made of boxwood, the better ones having white plastic edges. Scales are either triangular, Fig. 44 (a) and (b), or flat, (c) to (f). The triangular scales have the advantage of combining many scales on one stick, but the user will waste much time looking for the required scale if a *scale guard*, (g), is not used. The flat scale is almost universally used by professional draftsmen because of its convenience, but several flat scales are necessary to replace one triangular scale, and the total cost is greater. Since machine drawings are made full, half, quarter, and eighth size, these scales may be obtained on one or two flat scales of the type shown in Fig. 43 (c) and (d).

36. Architects Scale. Fig. 43 (a). The *architects scale* is intended primarily for drawings of buildings, piping systems, and other large structures which must be drawn to a reduced scale to fit on a sheet of paper. The full-size scale is also useful in drawing relatively small objects, and for that reason the architects scale has rather general usage.

The architects scale has one full-size scale and ten overlapping reduced-size scales. By means of these scales a drawing may be made to various sizes from full size to $\frac{1}{128}$ size. *Note particularly: In all of the reduced scales the major divisions represent feet, and their subdivisions represent inches and fractions thereof.* Thus, the scale marked $\frac{3}{4}$ means $\frac{3}{4}$ inch = 1 ft., not $\frac{3}{4}$ inch = 1 inch; that is, one-sixteenth size, not three-fourths size. Similarly, the scale marked $\frac{1}{2}$ means $\frac{1}{2}$ inch = 1 ft., not $\frac{1}{2}$ inch = 1 inch; that is, one-twenty-fourth size, not half-size.

All of the scales, from full-size to $\frac{1}{128}$ size, are shown in Fig. 45. Some are upside-down, just as they may occur in use. These scales are described as follows:

Full Size. Fig. 45 (a). Each division in the full-size scale is $\frac{1}{16}''$. Each inch is divided first into halves, then quarters, eighths, and finally sixteenths, the division lines diminishing in length with each division. To set off $\frac{1}{32}''$, estimate one half of $\frac{1}{16}''$; to set off $\frac{1}{64}''$, estimate one fourth of $\frac{1}{16}''$.

Half Size. Fig. 45 (a). Use the full-size scale, and divide every dimension mentally by two (do not use the $\frac{1}{2}''$ scale, which is intended for drawing to a scale of $\frac{1}{2}''$ equals 1′, or one-twenty-fourth size). To set off 1″, measure $\frac{1}{2}''$; to set off 2″, measure 1″; to set off $3\frac{1}{4}''$, measure $1\frac{1}{2}''$ (half of 3), then $\frac{1}{8}''$ (half of $\frac{1}{4}''$); to set off $2\frac{13}{16}''$ (see figure), measure 1″, then $\frac{13}{32}''$ ($\frac{6\frac{1}{2}}{16}''$ or half of $\frac{13}{16}''$).

Quarter Size. Fig. 45 (b). Use the 3-inch scale in which 3″ equals 1′. The subdivided portion to the left of zero represents one foot, and is divided into inches, then half inches, quarter inches, and finally eighth inches. The entire portion representing one foot would actually measure three inches; therefore, 3″ equals 1′. To set off anything less than twelve inches, start at zero and measure to the left.

To set off $10\frac{1}{8}$″, read off 9″ from zero to the left, then add $1\frac{1}{8}$″ and set off the total $10\frac{1}{8}$″, as shown. To set off more than 12″, for example, 1′-$9\frac{3}{8}$″ (see your scale), find the 1′ mark to the right of zero, and the $9\frac{3}{8}$″ mark to the left of zero; the required distance is the distance between these marks, and represents 1′-$9\frac{3}{8}$″, being actually one-fourth of 1′-$9\frac{3}{8}$″.

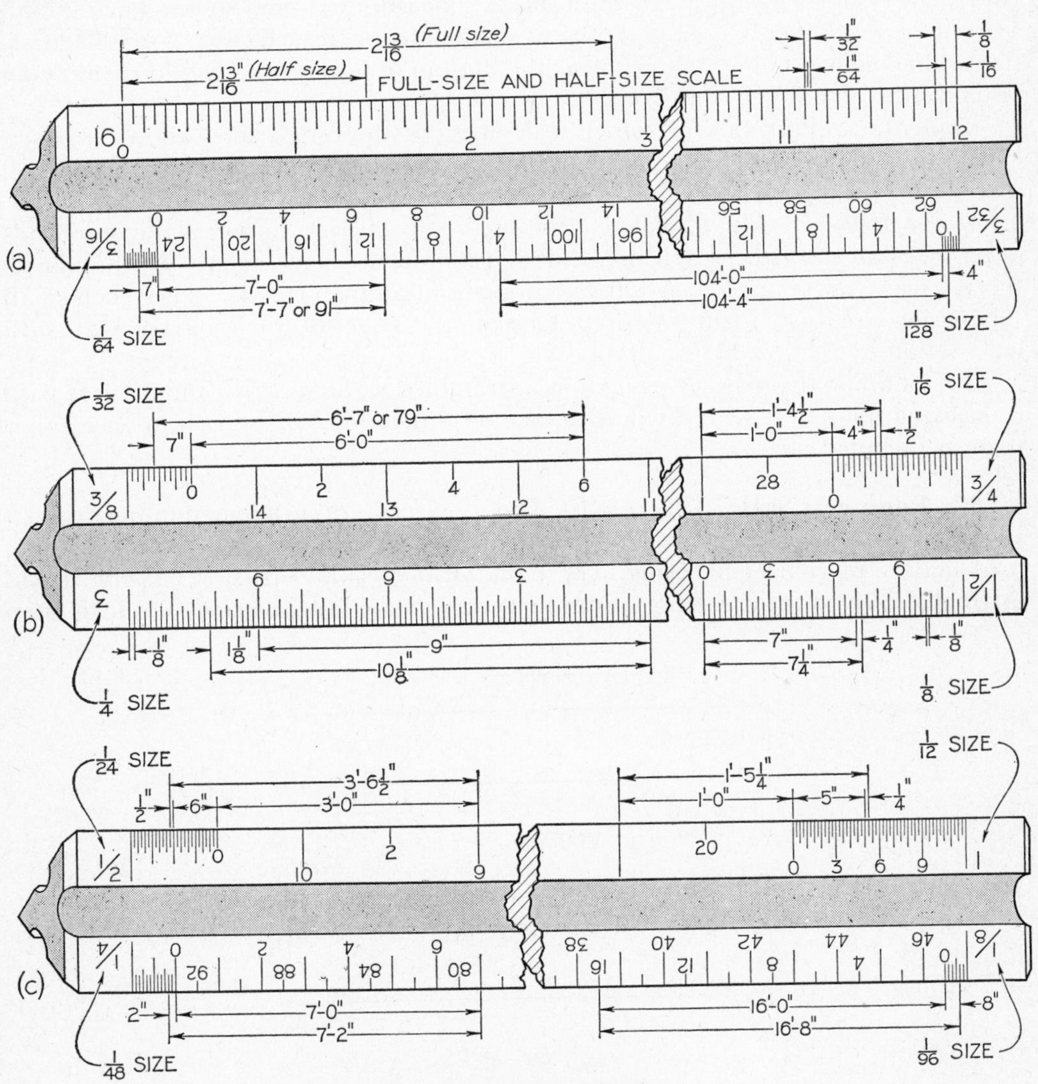

Fig. 45 Architects Scales.

Eighth Size. Fig. 45 (b). Use the $1\frac{1}{2}$″ scale in which $1\frac{1}{2}$″ equals 1′. The subdivided portion to the right of zero represents 1′, and is divided into inches, then half inches, and finally quarter inches. The entire portion, representing 1′, actually is $1\frac{1}{2}$″; therefore: $1\frac{1}{2}$″ equals 1′. To set off anything less than twelve inches, start at zero and measure to the right.

To set off $7\frac{1}{4}''$, (see figure), read off $7''$ from zero to the right, then add $\frac{1}{4}''$ and set off the total $7\frac{1}{4}''$, as shown.

To set off more than twelve inches, for example $3'$-$10\frac{3}{4}''$ (see your scale), find the $3'$ mark to the left of zero, and the $10\frac{3}{4}''$ mark to the right of zero; the required distance is the distance between these marks, and represents $3'$-$10\frac{3}{4}''$, being one-eighth of $3'$-$10\frac{3}{4}''$.

Double Size. Use the full-size scale, and multiply every dimension mentally by two. To set off $1''$, measure $2''$; to set off $3\frac{1}{4}''$, measure $6\frac{1}{2}''$, and so on. The double-size scale is occasionally used to represent small objects. In such cases, a small actual-size outline view should be shown near the bottom of the sheet to help the shop man visualize the actual size of the object.

Other Sizes. Fig. 45. The other scales besides those described above are used chiefly by architects. Machine drawings are customarily made only double size, full size, $\frac{1}{2}$ size, $\frac{1}{4}$ size, or $\frac{1}{8}$ size.

Special Methods. The $\frac{3}{8}''$ scale can be used conveniently to set off thirty-seconds of an inch full size, since each small subdivision on this scale equals $\frac{1}{32}''$. Similarly, the $\frac{3}{16}''$ scale can be used to set off sixty-fourths of an inch full size, since each small subdivision equals $\frac{1}{64}''$. If it is desired, for example, to set off the radius of a $\frac{13}{64}''$ drill hole, set off $6\frac{1}{2}$ (half of 13) small divisions on the $\frac{3}{16}''$ scale.

Do not abuse the scale by using it as a straightedge, hammering thumbtacks with it, pricking holes in it with dividers to take off dimensions, or using it in ways other than its intended use.

37. Engineers Scale. Fig. 43 (b). The *engineers scale* is graduated in the decimal system. It is also frequently called the *civil engineers scale* because it was originally used mainly in civil engineering. The name *chain scale* also persists because it was originally derived from the use of the surveyors' chain composed of 100 links, used

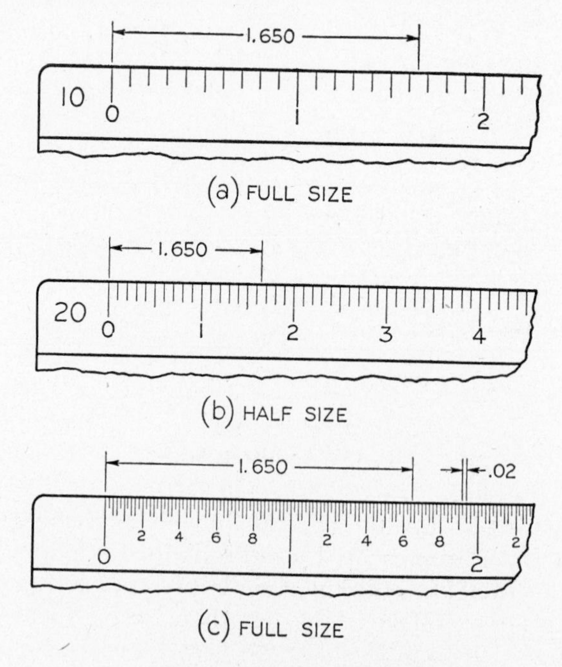

Fig. 46 Decimal Dimensions.

for land measurements. The name *engineers scale* is perhaps best, because the scale is used generally by engineers of all kinds.

The engineers scale is especially suitable for map drawing, because when a tract of land is surveyed, distances are measured with a flexible steel tape graduated in feet and tenths of a foot. The engineers scale is graduated in units of one inch divided into 10, 20, 30, 40, 50, and 60 parts. These scales are used in drawing maps to scales of 1″ = 50′, 1″ = 500′, 1″ = 5 miles, etc., and in drawing stress diagrams or other graphical constructions to such scales as 1″ = 20 lbs., 1″ = 4000 lbs., etc.

The engineers scale is also convenient in machine drawing to set off dimensions expressed in decimals. For example, to set off 1.650″ full size, Fig. 46 (a), use the "10-scale" and simply set off one main division plus $6\frac{1}{2}$ subdivisions. To set off the same dimension half size, use the "20-scale," (b), since the "20-scale" is exactly half the size of the "10-scale." Similarly, to set off a dimension quarter size, use the "40-scale."

38. Mechanical Draftsmans Scale. Fig. 43 (c). The objects represented in machine drawing vary in size from small parts, an inch or smaller in size, to machines of large dimensions. By drawing these objects full size, $\frac{1}{2}$ size, $\frac{1}{4}$ size, or $\frac{1}{8}$ size, the drawings will readily come within the limits of the standard-size sheets. For this reason the mechanical draftsmans scales are divided into units representing inches to full size, $\frac{1}{2}$ size, $\frac{1}{4}$ size, or $\frac{1}{8}$ size. To make a drawing of an object to a scale of $\frac{1}{4}$ size, for example, use the mechanical draftsmans scale marked $\frac{1}{4}$ size, which is graduated so that every $\frac{1}{4}$ inch represents 1 inch. Thus, the $\frac{1}{4}$-size scale is simply a full-size scale "shrunk" to $\frac{1}{4}$ size.

These scales are also very useful in dividing dimensions. For example, to draw a $3\frac{11}{16}$″ dia. circle full size, we need half of $3\frac{11}{16}$″ to use as radius. Instead of using arithmetic to find half of $3\frac{11}{16}$″, it is easier to set off $3\frac{11}{16}$″ on the half-size scale.

Triangular combination scales are available which include the full and half-size mechanical draftsmans scales, several architects scales, and an engineers scale.*

39. Decimal Scale. Fig. 43 (d). The increasing use of decimal dimensions has brought about the development of a scale specifically for that use, approved by the ASA.† On the full-size scale, each inch is divided into fiftieths of an inch, or .02″, as shown in Fig. 46 (c); and on the half- and quarter-size scales the inches are "shrunk" to half size or quarter size, and then are divided into ten parts, so that each subdivision stands for .1″.

The complete decimal system of dimensioning, in which this scale is used, is described in §368.

40. To Indicate the Scale on a Drawing. For machine drawings, the most common practice is to letter FULL SIZE, $\frac{1}{2}$ SIZE, $\frac{1}{4}$ SIZE, or $\frac{1}{8}$ SIZE, or to abbreviate with the words FULL, HALF, QUARTER, or EIGHTH, or to equate inches to inches: 1″ = 1″, $\frac{1}{2}$″ = 1″, $\frac{1}{4}$″ = 1″, $\frac{1}{8}$″ = 1″, or 2″ = 1″ (double size). For examples of how scales are shown on machine drawings, see Figs. 718-721.

For drawings of buildings, piping, and other structures in which the dimensions are in feet and inches, the architects scales should be given in terms of inches to feet, as 3″ = 1′-0″ (quarter size), $\frac{1}{2}$″ = 1′-0″ (twenty-fourth size), etc.

Map scales are indicated in terms of fractions, as Scale $\frac{1}{62500}$, or graphically as 400 0 400 800 Ft. See also §656.

*Frederick Post's No. 1306 and Keuffel & Esser's No. 8885.
†ASA Z75.1-1955.

41. Accurate Measurements. Accurate drafting depends considerably upon the correct use of the scale in setting off distances. Do not take measurements directly off the scale with the dividers or compass, as damage will result to the scale. Place the scale on the drawing with the edge parallel to the line on which the measurement is to be made and, with a sharp pencil having a conical point, make a short dash at right angles to the scale and opposite the correct graduation mark, as shown in

Fig. 47 Accurate Measurements.

Fig. 47 (a). If extreme accuracy is required, a *tiny* prick mark may be made at the required point with the needle point or stylus, as shown at (b), or with one leg of the dividers.

Avoid cumulative errors in the use of the scale. If a number of distances are to be set off end-to-end, all should be set off at one setting of the scale by adding each successive measurement to the preceding one, if possible. Avoid setting off the distances individually by moving the scale to a new position each time, since slight errors in the measurements may accumulate and give rise to a large error.

42. Drawing Instruments. *Drawing instruments* are generally sold in "sets," in velvet-lined cases, but they may be purchased separately. The principal parts of high-grade instruments are made of nickel silver, which has a silvery luster, is corrosion-resistant, and can be readily machined into desired shapes. Tool steel is used for the blades of ruling pens, for spring parts, for divider points, and for the various screws.

In technical drawing, accuracy, neatness, and speed are essential, §12. These ob-

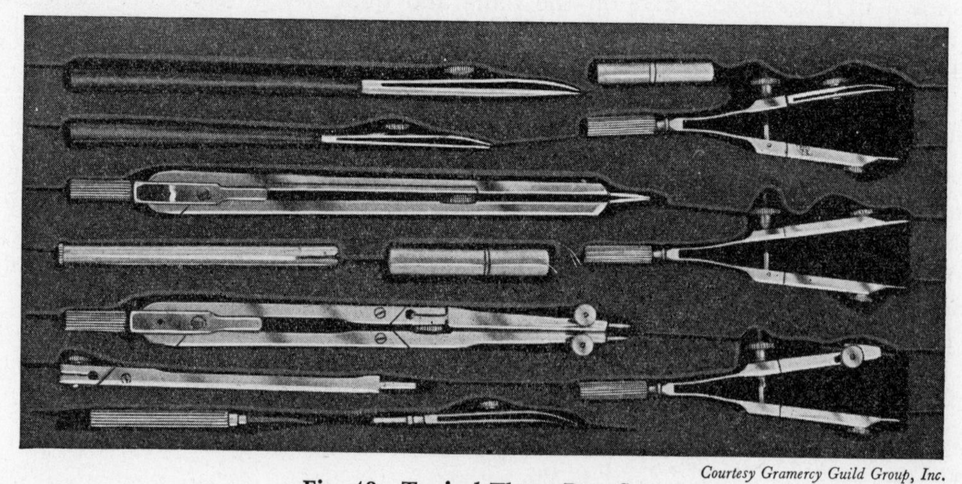

Courtesy Gramercy Guild Group, Inc.

Fig. 48 Typical Three-Bow Set.

jectives are not likely to be obtained with cheap or inferior drawing instruments. For the student or the professional draftsman it is advisable, and in the end more economical, to purchase the best instruments that can be afforded. Good instruments will satisfy the most rigid requirements, and the satisfaction, saving in time, and improved quality of work which good instruments can produce will more than justify the higher price.

Unfortunately, the qualities of high-grade instruments are not likely to be recognized by the beginner, who is not familiar with the performance characteristics required and who is apt to be attracted by elaborate sets containing a large number of shiny low-quality instruments. Therefore, the student should obtain the advice of his drafting instructor, of an experienced draftsman, or of a reliable dealer.*

A typical set of traditional-style drawing instruments is shown in Fig. 48. This set contains a compass, dividers, bow pencil, bow pen, bow dividers, two ruling pens, and various auxiliary parts.

43. Giant Bow Set. Formerly it was general practice to make pencil drawings on detail paper and then to make an inked tracing from it on tracing cloth. As reproduction methods and transparent tracing papers were improved, it was found that a great deal of time could be saved by making drawings directly in pencil with dense black lines on the tracing paper and making prints therefrom, thus doing away with the preliminary pencil drawing on detail paper. Today, though inked tracings are made when a fine appearance is necessary and where the greater cost is justified, the overwhelming proportion of drawings are made directly in pencil on tracing paper, vellum, or pencil tracing cloth.

These developments have brought about the present "giant-bow" sets which are offered now by all the major manufacturers, Fig. 49.* The sets contain various combinations of instruments, but are all featured by a large bow compass in place of the traditional large compass. The large bow instrument is much sturdier and is capable of taking the heavy pressure necessary to produce dense black lines.

Most of the large bows are of the center-wheel type, Fig. 50 (a). Several manu-

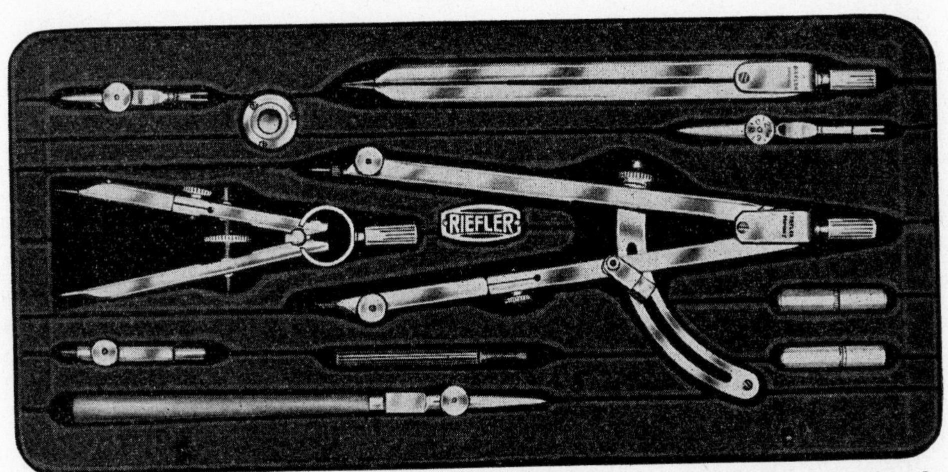

Courtesy Gramercy Guild Group, Inc.

Fig. 49 Giant Bow Set.

*See footnote, p. 12

facturers now offer different varieties of quick-acting bows. The large bow shown at (b) can be adjusted by simply opening and closing the legs in the same manner as for the old-style compass. The friction in the pivot joint is sufficient to hold a setting except when the pressure must be very heavy and the radius large. In that case the compass can be rigidly locked in position.

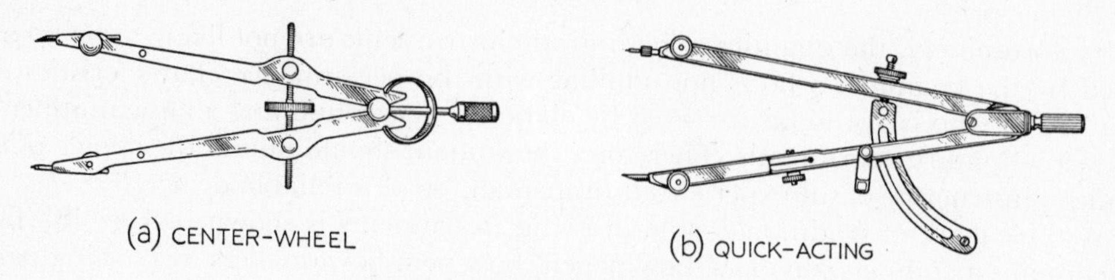

Fig. 50 Giant Bow Compass.

44. The Compass. A typical old-style compass is shown in Fig. 51. All such instruments have a socket joint in one leg which permits the insertion of either pencil or pen attachments. A *lengthening bar*, (c), is used to increase the radius. This type of compass is excellent for drafting where it is not necessary to exert great pressure in order to produce heavy dark lines. For production drafting, in which it is necessary to make dense black lines to produce clear legible prints, the giant bow, Fig. 50, is preferred.

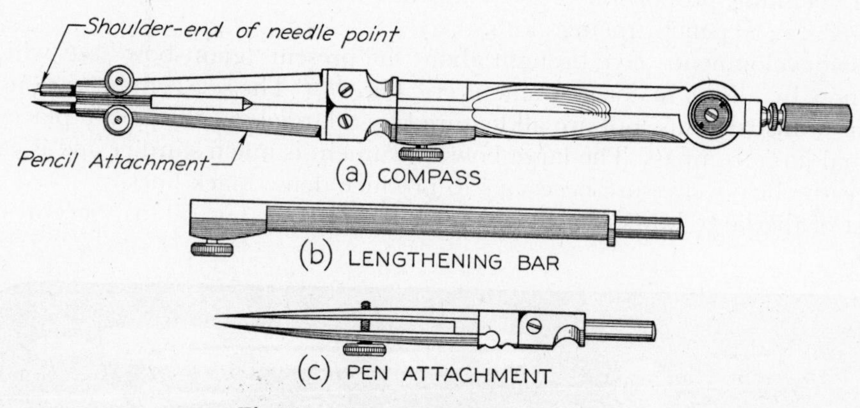

Fig. 51 Compass and Attachments.

45. Using the Compass. These instructions apply generally both to the old-style and the giant bow compasses. The compass, with pencil and inking attachments, is used for drawing circles of approximately 1″ radius or larger, Fig. 52. Most compass needle-points have a plain end for use when the compass is converted into dividers, and a "shoulder-end" for use as a compass. Adjust the needle point with the shoulder-end out and so that the small point extends *slightly* farther than the pencil lead or pen nibs, Fig. 55 (d).

To draw a penciled circle, Fig. 52: (1) set off the required radius on one of the center lines, (2) place the needle point at the exact intersection of the center lines, (3) adjust the compass to the required radius (1″ or more), (4) lean the compass forward and draw the circle clockwise while rotating the handle between the thumb and forefinger. To obtain sufficient weight of line, it may be necessary to repeat the movement several times.

Any error in radius will result in a doubled error in diameter; hence, it is best to draw a trial circle first on scrap paper or on the backing sheet and then check the diameter with the scale.

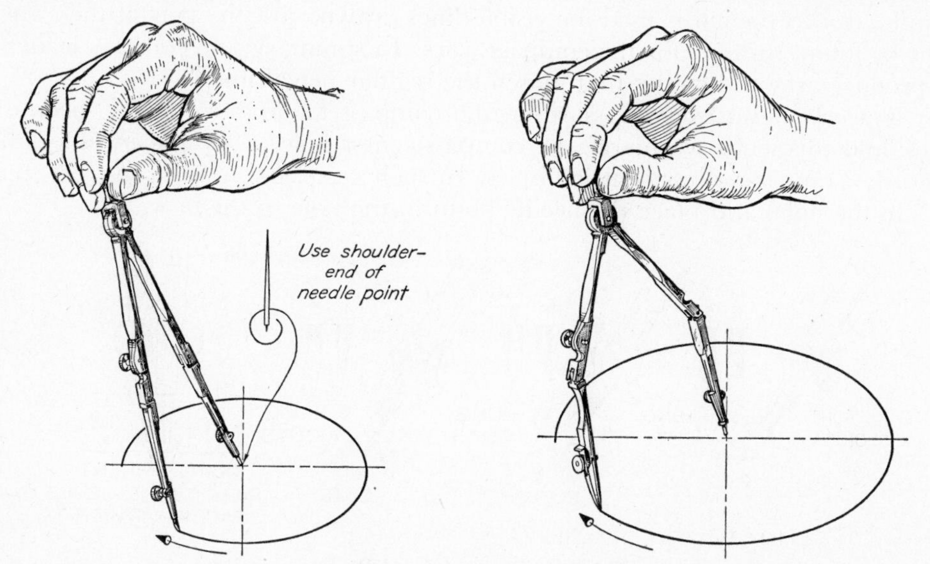

Fig. 52 Using the Compass. **Fig. 53 "Breaking" the Legs of the Compass.**

When drawing inked circles, "break" the legs of the compass, Fig. 53, so that they will stand approximately perpendicular to the paper; otherwise both nibs of the pen attachment will not touch the paper equally.

Manipulate the compass with one hand, as shown for handling the dividers in Fig. 58. On drawings having circular arcs and tangent straight lines, draw the arcs first, whether in pencil or in ink, as it is much easier to connect a straight line to an arc than the reverse.

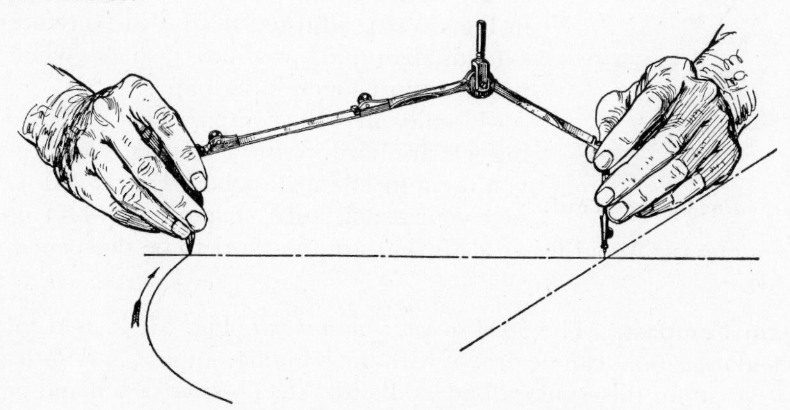

Fig. 54 Drawing Circles of Large Radii.

For very large circles, use the lengthening bar to increase the compass radius. Use both hands, as shown in Fig. 54, but be careful not to jar the compass and thus change the adjustment.

When using the compass to draw construction lines, use a 4H to 6H lead so that the lines will be very dim. For required lines, the arcs and circles must be black, and softer leads must be used. However, since heavy pressure cannot be exerted on the compass as it can on a pencil, it is usually necessary to use a compass lead that is about one grade softer than the pencil used for the corresponding line work. For example, if an F pencil is used for visible lines drawn with the pencil, then an HB might be found suitable for the compass work. In summary, use compass leads that will produce arcs and circles that match the regular pencil lines.

It is necessary to exert pressure on the compass to produce heavy "printable" circles, and this tends to enlarge the compass center hole in the paper, especially if there are a number of concentric circles. In such cases, use a *horn center* or "center tack" in the hole, and place the needle point in the hole in the tack.

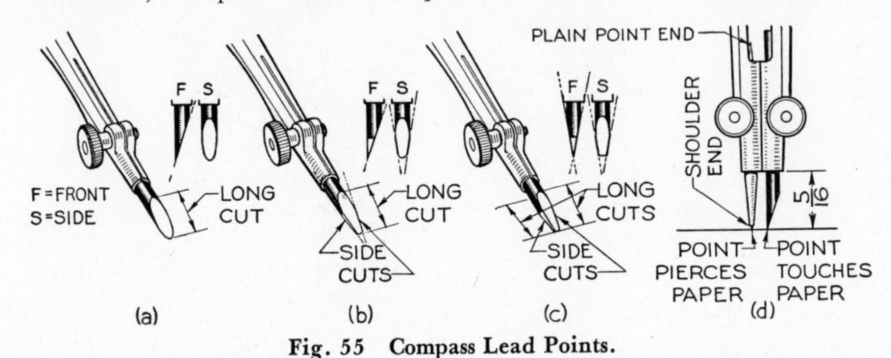

Fig. 55 Compass Lead Points.

46. Sharpening the Compass Lead. Various forms of compass lead points are illustrated in Fig. 55. At (a) a single elliptical face has been formed by rubbing on the sandpaper pad, as shown in Fig. 56. At (b) the point is narrowed by small side cuts. At (c) two long cuts and two small side cuts have been made so as to produce a point similar to that on a screwdriver.

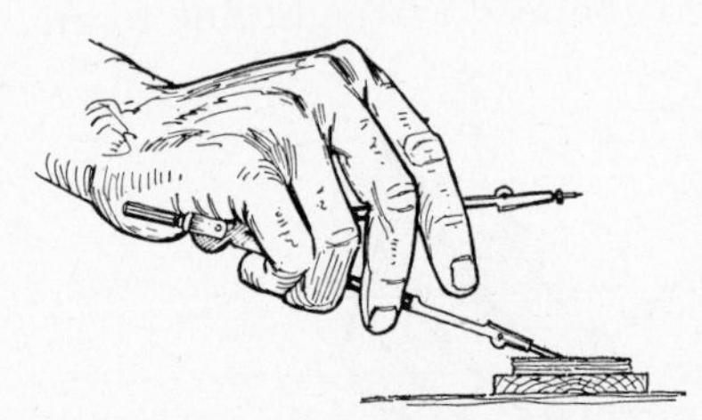

Fig. 56 Sharpening Compass Lead.

In using the compass, *never use the plain end or the needle point*. Instead, use the shoulder-end as shown in Fig. 55 (d), adjusted so that the tiny needle point extends about half-way into the paper when the compass lead just touches the paper.

Usually all of the compass leads provided in drawing sets are hard. Softer leads are readily available if you use a mechanical pencil, Fig. 18 (b). Otherwise, save your pencil stubs, strip off the wood, and use the leads in the compass. Avoid using leads that are too short to be sharpened as shown in Fig. 55 (d).

47. Beam Compass. The *beam compass* or *trammel*, Fig. 57, is used for drawing arcs or circles larger than can be drawn with the regular compass, and for transferring distances too great for the regular dividers. Besides steel points, pencil and pen attachments are provided. The "beams" may be made of nickel silver, steel, or wood, and are procurable in various lengths.

48. Dividers. The dividers are similar to the compass in construction, and are made in square, flat, and round forms. In general, the tension in the pivot joint should be slightly less than in the compass, for very little pressure is ever exerted on the dividers. The pivot joint should be loose enough to permit easy manipulation with one hand, as shown in Fig. 58. If the pivot joint is too tight, the legs of the

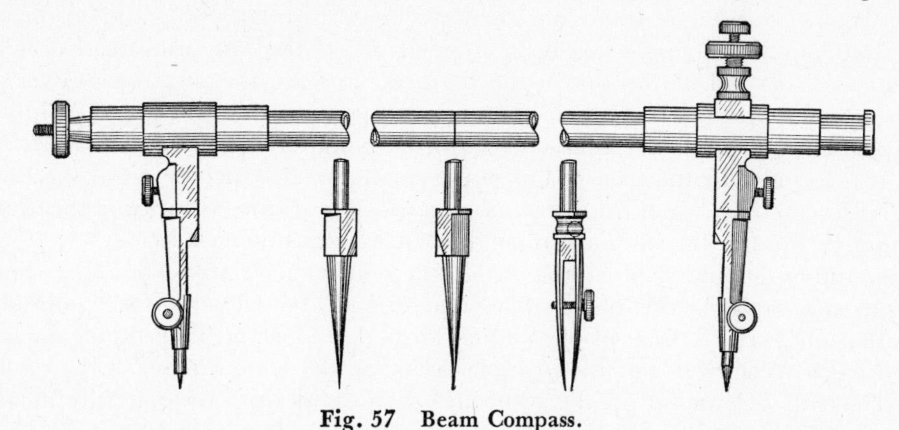

Fig. 57 Beam Compass.

compass tend to "spring back" instead of stopping at the desired point when the pressure of the fingers is released. To adjust the tension, use the small screw driver in the same manner as for the compass.

Most dividers are provided with a "hair spring" so that minute adjustments can be made by turning the small thumbscrew.

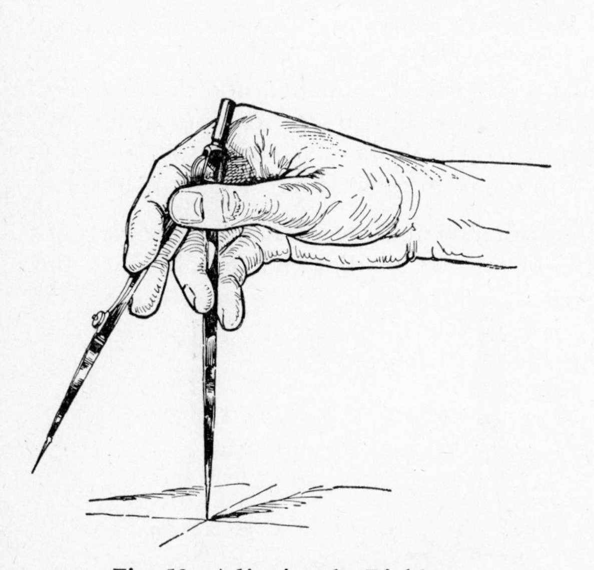

Fig. 58 Adjusting the Dividers.

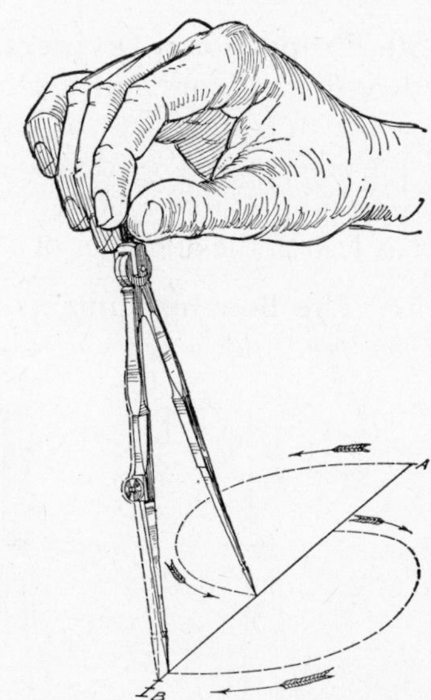

Fig. 59 Using the Dividers.

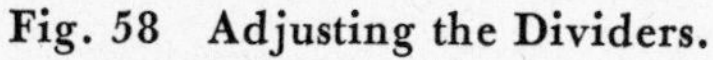

49. Using the Dividers. The dividers, as the name implies, are used for *dividing* distances into a number of equal parts. They are used also for *transferring distances* or for *setting off* a series of equal distances. The dividers are used for spaces of approximately 1″ or less. For less than 1″ spaces, use the bow dividers, Fig. 61 (a). *Never use the large dividers for small spaces when the bow dividers can be used, as the latter are more accurate.*

To divide a given distance into a number of equal parts, Fig. 59, the method is one of trial and error. Adjust the dividers with the fingers of the hand that holds them, to the approximate unit of division, estimated by eye. Rotate the dividers counterclockwise through 180°, and then clockwise through 180°, and so on, until the desired number of units has been stepped off. If the last prick of the dividers falls short of the end of the line to be divided, increase the distance between the divider points proportionately. For example, to divide the line AB, Fig. 59, into three equal parts, the dividers are set by eye to approximately one-third the length AB. When it is found that the trial radius is too small, the distance between the divider points is increased by one-third the remaining distance. If the last prick of the dividers is beyond the end of the line, a similar decreasing adjustment is made.

The student should avoid *cumulative errors* which may result when the dividers are used to set off a series of distances end-to-end. To set off a large number of equal divisions, say 15, first set off 3 equal large divisions and then divide each into 5 equal parts. Wherever possible in such cases, use the scale instead of the dividers, as described in §41, or set off the total and then divide into the parts by means of the parallel-line method, §§121 and 122.

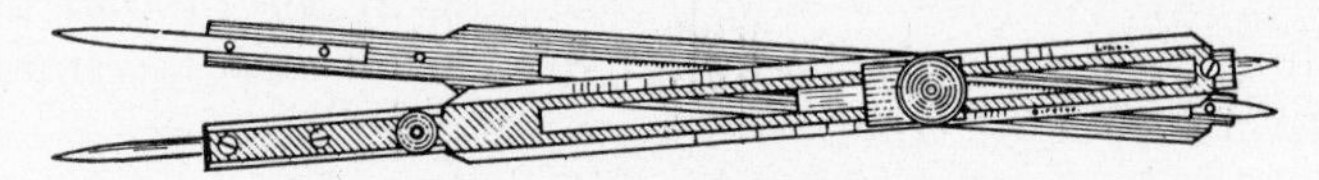

Fig. 60 Proportional Dividers.

50. Proportional Dividers. For enlarging or reducing a drawing, proportional dividers, Fig. 60, are convenient. They may be used also for dividing distances into a number of equal parts, or for obtaining a percentage reduction of a distance. For this purpose, points of division are marked on the instrument so as to secure the required subdivisions readily. Some instruments are calibrated to obtain special ratios, such as $1:\sqrt{2}$, the diameter of a circle to the side of an equal square, feet to meters, etc.

51. The Bow Instruments. The bow instruments are classified as the *bow dividers*, *bow pen*, and *bow pencil*, Fig. 61. Except for the handles and thumbscrews, the

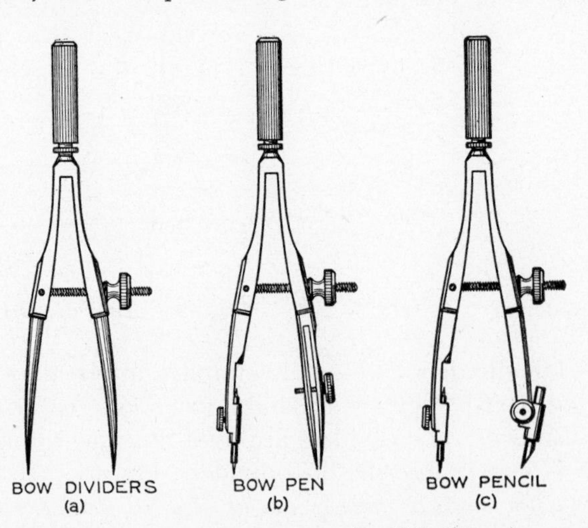

Fig. 61 Bow Instruments with Side Wheel.

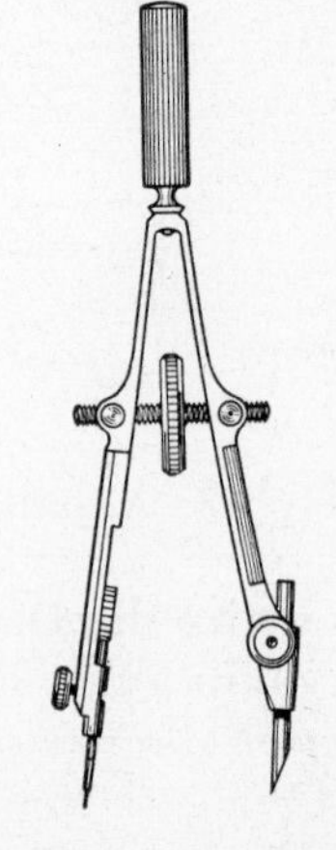

Fig. 62 Bow Pencil with Center Wheel.

bow instruments are made of tool steel. There are two general types, the side-wheel bows, Fig. 61, and the center-wheel bows, Fig. 62. Fine instruments are available in both styles, and hence the choice is purely a matter of personal preference.

52. Using the Bow Instruments. The bow pencil is used for drawing penciled circles having a radius of approximately 1″ or smaller. Whether the center-wheel instrument or the side-wheel instrument is used, the adjustment should be made with the fingers of the hand that holds the instrument, Fig. 63 (a).

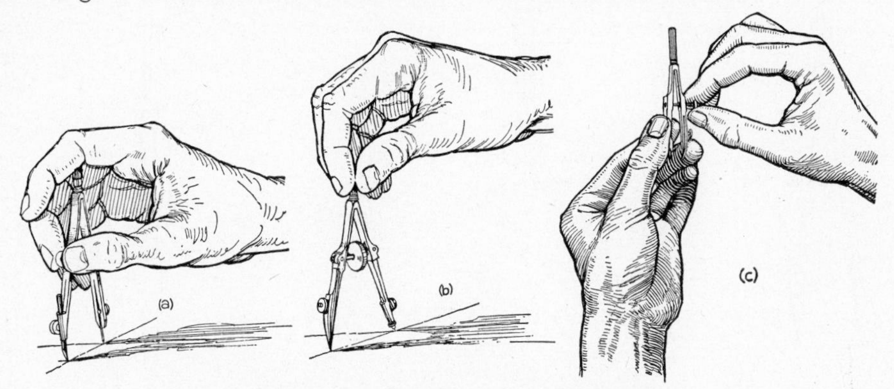

Fig. 63 Using the Bow Instruments.

The bow pen, (b), is used for drawing inked circles of approximately 1″ radius or less.

When adjustment of any of the side-screw bow instruments must be made directly from a large to a small radius, or vice-versa, it is best to press the two legs together with the fingers of the left hand, (c), thus relieving the pressure upon the thumb-screw, so that it may be spun to position quickly and without unnecessary wear upon the threads.

The lead is sharpened in the same manner as for the large compass, §46, except that for small radii, the inclined cut may be turned *inside* if preferred, Fig. 64 (a). For general use, the lead should be turned to the outside, as shown at (b). In either case, always keep the compass lead sharpened. *Avoid stubby compass leads*, which cannot be properly sharpened. At least $\frac{1}{4}$″ of lead should extend from the compass at all times.

In adjusting the needle point of the bow pencil or bow pen, be sure to have the needle extending slightly longer than the pen or lead, Fig. 55 (d), the same as for the large compass.

In drawing small circles, greater care is necessary in sharpening and adjusting the lead and the needle point, and especially in accurately setting the desired radius. If a $\frac{1}{4}$″ diameter circle is to be drawn, and if the radius is "off" only $\frac{1}{32}$″, the total error on diameter is 25 per cent, which is far too much error.

The bow dividers are used for the same purposes as the large dividers, but for smaller (approximately 1″ or less) spaces and more accurate work. Always use the bow dividers in preference to the large dividers, if the spacings are within the range of the bow dividers, since they are more accurate.

53. Drop Spring Bow Pencil and Pen. These compasses, Fig. 65, are designed for drawing multiple identical small circles, such as drill holes or rivet heads. A central pin is made to move easily up and down through a tube to which the pen or pencil unit is attached. To use the instrument, hold the knurled head of the tube between the thumb and second finger, placing the first finger on top of the knurled head of the pin. Place the point of the pin at the desired center, lower the pen or pencil until it touches the paper, and twirl the instrument clockwise with the thumb and second finger. Then lift the tube independently of the pin, and finally lift the entire instrument.

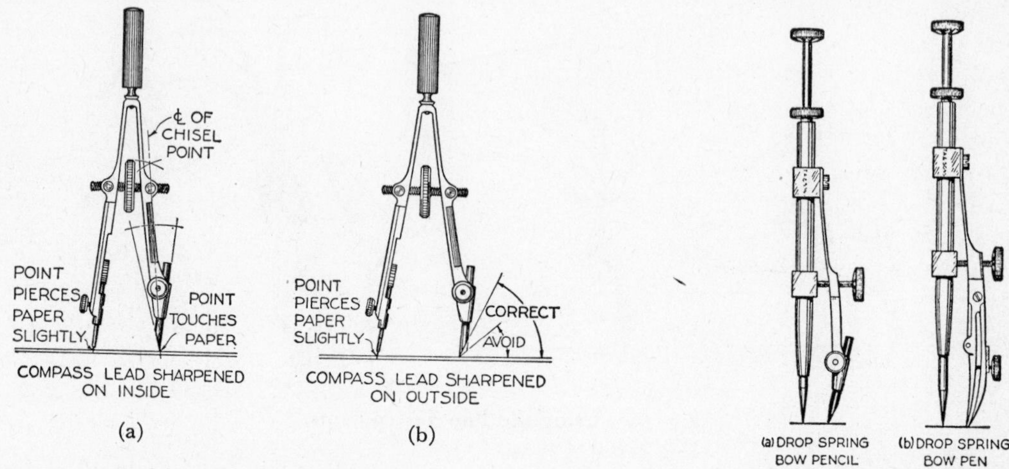

Fig. 64 Compass Lead Points.

Fig. 65 Drop Spring Bow Instruments.

54. To Lay Out a Sheet. Fig. 66. After the sheet has been attached to the board, as explained in §17, proceed as follows:

I. Using the T-square, draw a horizontal *trim line* near the lower edge of the paper; and then using the triangle, draw a vertical trim line near the left edge of the paper. Both should be *light construction lines.*

II. Place the scale along the lower trim line with the full-size scale up. Draw short light dashes *perpendicular* to the scale at the required distances. See Fig. 47 (a).

III. Place the scale along the left trim line with the full-size scale to the left, and mark the required distances with short light dashes perpendicular to the scale.

IV. Draw horizontal construction lines with the aid of the T-square through the marks at the left of the sheet.

V. Draw vertical construction lines, *from the bottom upward*, along the edge of the triangle through the marks at the bottom of the sheet.

VI. Retrace the border and the title strip to make them heavier. Notice that the layout is made independently of the edges of the paper.*

55. Technique of Pencil Drawing. By far the greater part of commercial drafting is executed in pencil. Most prints are made from pencil tracings, and all ink tracings must be preceded by pencil drawings. It should therefore be evident that skill in drafting chiefly implies skill in pencil drawing.

"Technique" is a style or quality of drawing imparted by the individual drafts-

*In industrial drafting rooms the sheets are available, cut to standard sizes, with border and title strips already printed. Drafting supply houses can supply such papers, printed to order, to schools for little or no extra cost.

man to his work. It is characterized by bright sparkling line work and lettering. Technique in lettering is discussed in §86.

1. *Dark Accented Lines.* The pencil lines of a finished pencil drawing or tracing should be very dark, Fig. 67. Dark "bright" lines are necessary to give "punch" or "snap" to the drawing. Ends of lines should be accented by a little extra pressure on the pencil, (a). Curves should be as dark as other lines, (b). Hidden-line dashes and center-line dashes should be carefully estimated as to length and spacing, and should be of uniform width throughout their length, (c) and (d).

Dimension lines, extension lines, section lines, and center lines also should be dark. The difference between these lines and visible lines is mostly in width—there is very little difference, if any, in blackness.

A simple way to determine whether your lines on tracing paper or cloth are dense black is to hold the tracing up to the light, Fig. 68. Lines that are not opaque will not print clearly by blueprinting or otherwise.

Construction lines should be made with a sharp, hard pencil and *should be so light that they need not be erased* when the drawing is completed.

2. *Contrast in Lines.* Contrast in pencil lines should be similar to that of ink lines; that is, the difference between the various "weights" should be mostly in the *width* of the line, with little if any difference in the degree of darkness, Fig. 69. The visible lines should contrast strongly with the thin lines of the drawing. If necessary, draw over a visible line several times to get the desired thickness and darkness. A short retracing stroke backwards (to the left), producing a "jabbing" action, results in a darker line.

56. Pencil Tracing. While some pencil tracings are made of a drawing placed underneath the tracing paper (usually when a great deal of erasing and changing is necessary on the original drawing), most drawings today are made directly in pencil on tracing paper, pencil tracing cloth, or vellum. These are not "tracings" but pencil drawings, and the methods and technique are the same as previously described for pencil drawing.

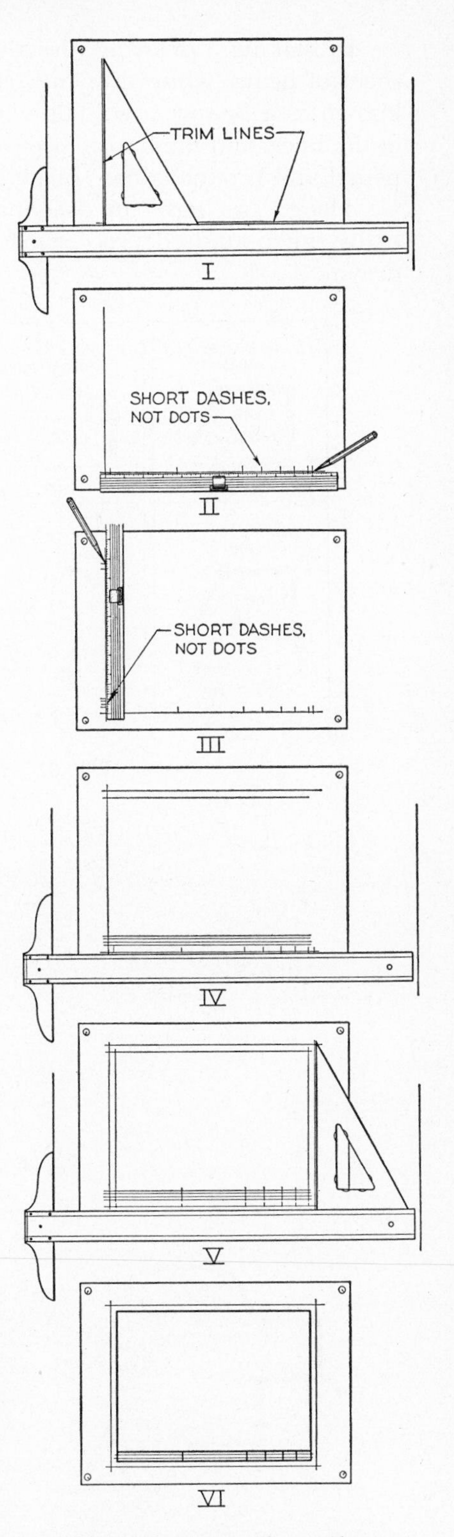

Fig. 66 To Lay Out a Sheet.

In making a drawing directly on tracing paper, vellum, or cloth, a stiff smooth sheet of heavy white drawing paper should be placed underneath. Such a sheet is known as a *backing sheet.* The whiteness of the backing sheet improves the visibility of the lines, and the hardness of the surface makes it possible to exert pressure on the pencil and produce dense black lines without excessive grooving of the paper.

These "tracings," or drawings, are intended to be reproduced by blueprinting or by other kindred processes, Chapter 29, and all lines must be dark and cleanly drawn.

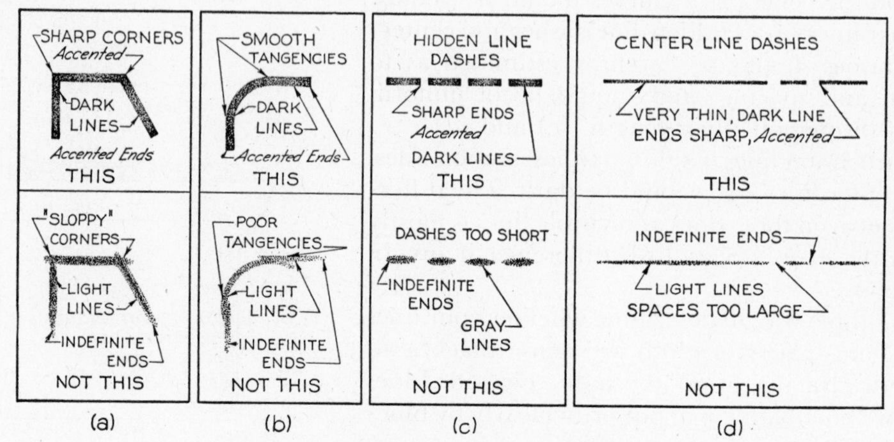

Fig. 67 Technique of Lines (Enlarged).

Fig. 68 Testing Density of Lines.

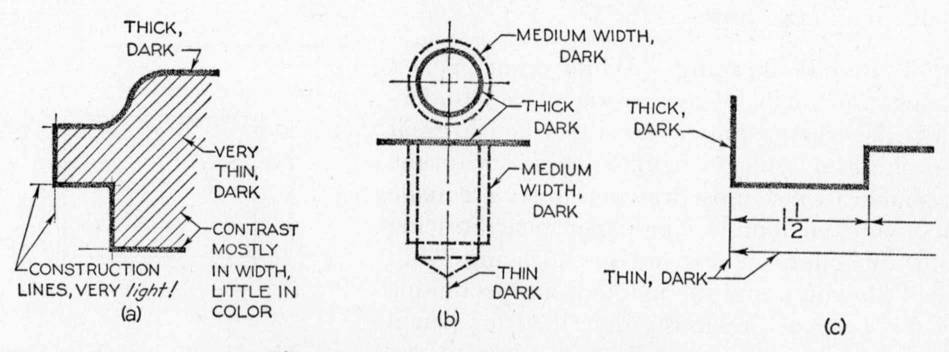

Fig. 69 Contrast of Lines (Enlarged).

57. Ruling Pens. The *ruling pen*, Fig. 70, should be of the highest quality, with blades of high-grade tempered steel sharpened properly at the factory. The nibs should be sharp, but not sharp enough to cut the paper. See §63 for sharpening the ruling pen. Various devices have been devised to permit the blades to open for easy cleaning, as shown in Fig. 70 (a) to (e).

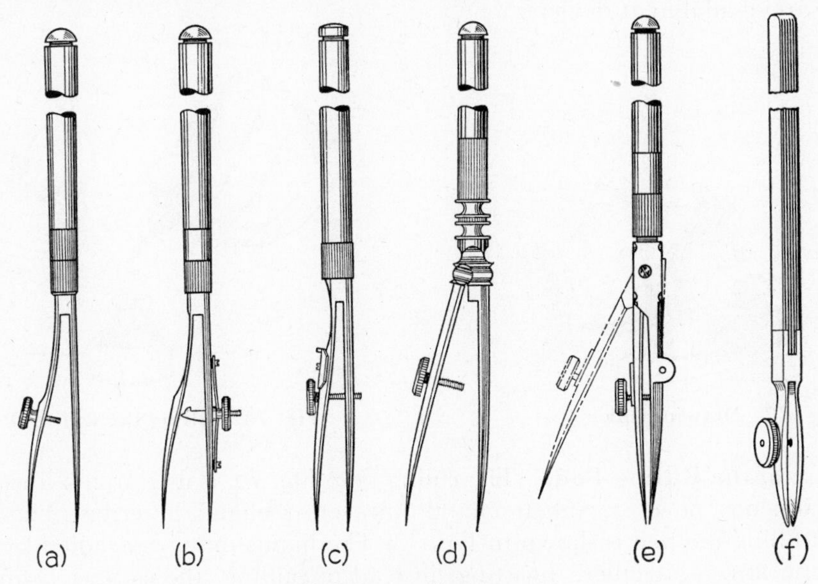

Fig. 70 Ruling Pens.

The *detail pen*, capable of holding a considerable quantity of ink, is extremely useful for drawing long heavy lines, (f), and is preferred by some for general use.
For methods of using the ruling pen, see §60.

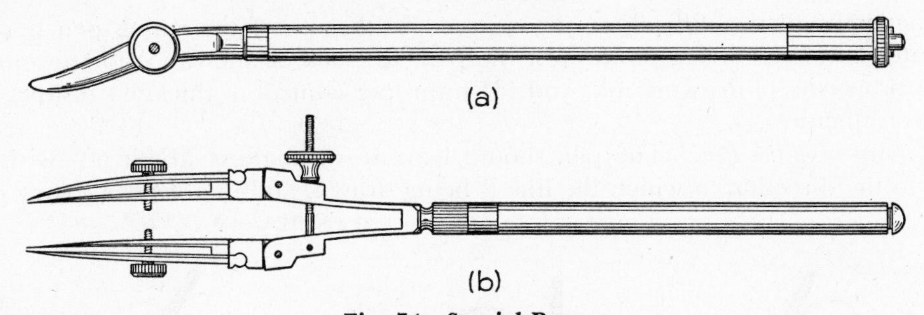

Fig. 71 Special Pens.

58. Special Pens. The *contour pen*, Fig. 71 (a), is used for tracing freehand curves, such as contour lines on maps. The *railroad pen*, (b), is used for drawing two parallel lines straight or moderately curved, as for roads and railroads.

59. Drawing Ink. Drawing ink, Fig. 72, is composed chiefly of carbon in colloidal suspension, and gum. The fine particles of carbon give the deep, black luster to the ink, and the gum makes it waterproof and quick to dry. The ink bottle should not be left uncovered, as evaporation will cause the ink to thicken. Thickened ink may be thinned by adding a few drops of a solution of four parts of aqua ammonia to one part of distilled water. Ink left over in the ruling pen can be saved by holding the pen in a horizontal position with the point over the open bottle and tapping the edges of the nibs against the bottle. A convenient pen-filling ink stand, which requires the use of only one hand, is shown at (b).

For removing dried waterproof drawing ink from pens or instruments, pen cleaning fluids are available at dealers.*

Fig. 72 Drawing Ink.

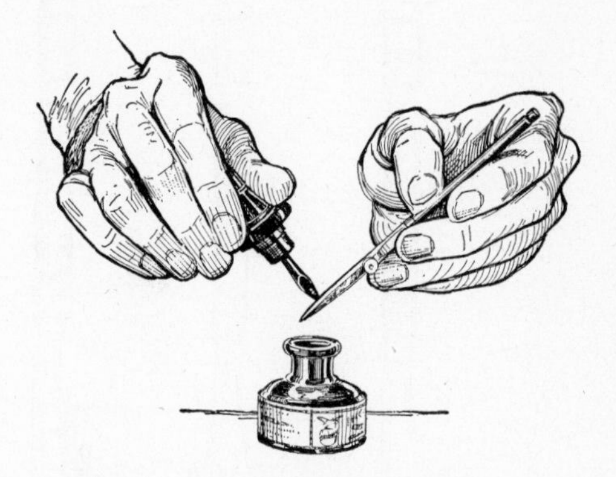

Fig. 73 Filling the Ruling Pen.

60. Use of the Ruling Pen. The ruling pen, Fig. 70, is used to ink lines drawn with instruments, never to ink freehand lines or freehand lettering. The proper method of filling the pen is shown in Fig. 73. The hands may be steadied by touching the little fingers together. Twisting, instead of pulling, the stopper from a new bottle of ink, or one that has not been used for some time, will often save the stopper from being broken. After the pen has been filled, the ink should stand about $\frac{1}{4}''$ deep in the pen.

Horizontal lines and vertical lines are drawn in the same manner as for the corresponding pencil lines, Figs. 29 (a) and 30 (a).

Practically all the difficulties encountered in the use of the ruling pen may be attributed to (1) incorrect position of the pen, (2) lack of allowance for the quick-drying properties of drawing ink, and (3) improper control of thickness of lines and incorrect junctures.

1. *Position of the Pen.* The pen should lean at an angle of about 60° with the paper in the direction in which the line is being drawn, and in a vertical plane con-

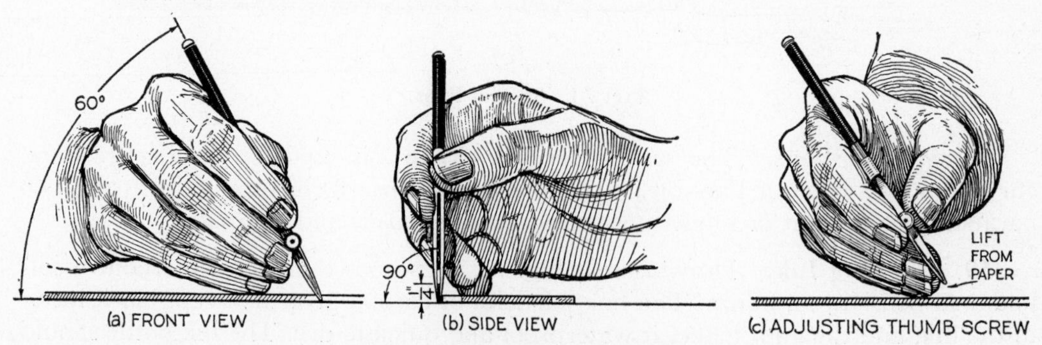

Fig. 74 Using the Ruling Pen.

*Higgins Pen Cleaner or Leroy Pen Cleaning Fluid.

taining the line, Fig. 74 (a) and (b). In general, the more the pen is leaned toward the paper, the thicker the line will be; and the more nearly vertical the pen is held, the thinner the line will be. The thumbscrew is faced away from the straight edge, and is adjusted, Fig. 74 (c), with the thumb and forefinger of the same hand that holds the instrument. The correct position of the pen and the resulting correct line are shown in Fig. 75 (a).

If the nibs are pressed tightly against the T-square or triangle, the effect is to close the nibs and thus reduce the thickness of the lines, (b). If the pen is held as shown at (c), the ink will come in contact with the T-square and paper at the same time and will run under the T-square and cause a blot on the drawing. The same result may occur if, in filling the pen, ink is deposited on the outside of the nib that touches the T-square. If the pen is held as shown at (d), the outside nib of the ruling pen may not touch the paper and the line is apt to be ragged.

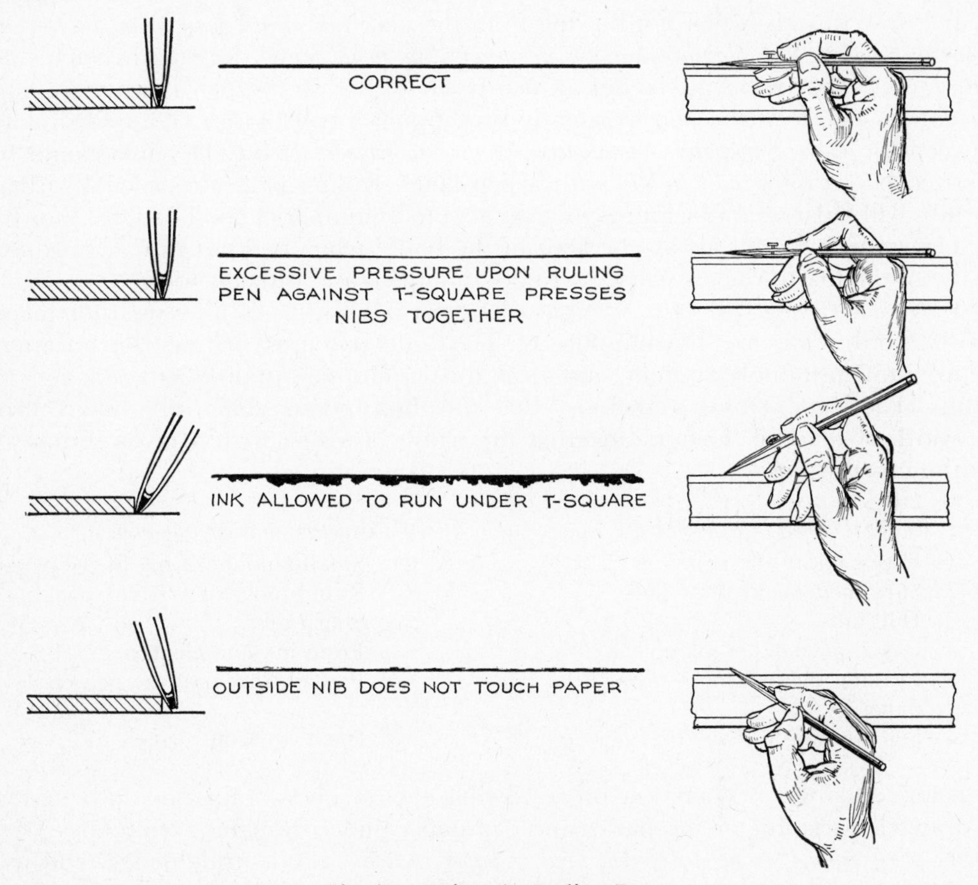
CORRECT
EXCESSIVE PRESSURE UPON RULING PEN AGAINST T-SQUARE PRESSES NIBS TOGETHER
INK ALLOWED TO RUN UNDER T-SQUARE
OUTSIDE NIB DOES NOT TOUCH PAPER

Fig. 75 Using the Ruling Pen.

When the line has been correctly drawn, care must be exercised not to touch the wet ink when removing the T-square or triangle. The triangle or T-square should be carefully drawn away from the line before being picked up. If more than $\frac{1}{4}''$ of ink is placed in the pen, the ink will flow too readily, thus increasing the danger of a blot.

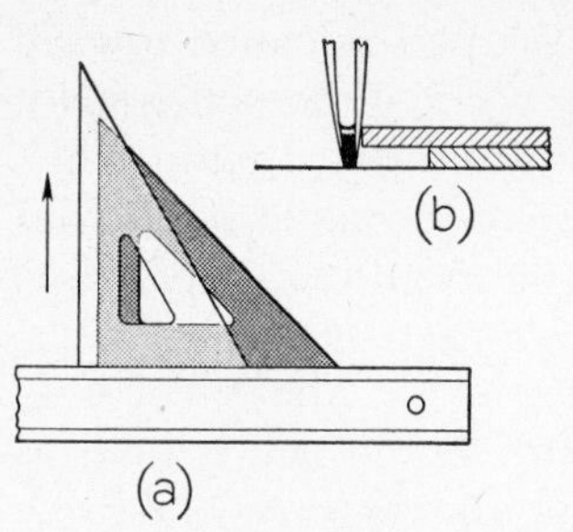

Fig. 76 Inking.

Some draftsmen prefer to place another triangle under the one being used, as shown in Fig. 76, to raise the first triangle above the paper and thus prevent ink from running under the edge. This is especially useful if the lines are heavy and blots may easily occur.

2. *Correct Use of Drawing Ink.* One of the most common difficulties is that the pen will not "feed." A clogged pen often may be started by marking on drafting tape or on the back of a finger. If the pen will not make a fine line, the nibs have been screwed too close together, the ink has been allowed to partially dry and thicken in the pen, the ink in the bottle is too thick from age and exposure to air, or the pen is dull and needs sharpening.

The pen should never be filled until the draftsman is ready to use it, because the ink dries quickly when not flowing from the pen. *Ink should never be allowed to dry in any instrument. Never lay a ruling pen down with ink in it.* Some drawing inks have an acid content that will "pit" a ruling pen if left to dry in the pen repeatedly. The student should clean the pen frequently by slipping a stiff blotter or a folded cloth between the nibs. *Sandpaper should never be used to remove dry ink.* Dry ink should be removed by scraping *very lightly* with a pen knife. Ruling pens constructed so that the nibs will separate for cleaning are available in a number of good designs, Fig. 70.

The stopper should always be kept in the bottle when it is not in use, since exposure of the ink to the air causes it to become thick and difficult to use.

3. *How to Control Thickness of Lines.* The various widths of lines used for inked drawings or tracings are shown in Fig. 24. The draftsman must first develop a trained eye to distinguish fine variations, and must also acquire skill in producing the desired widths. The student must remember that the thumbscrew alone does not control the width of the line. Factors affecting the width of a line with a given setting of the thumbscrew are:

Factors that tend to make line heavier:

(1) Excess ink in the pen.
(2) Slow movement of the pen.
(3) Dull nibs.
(4) Caked particles of ink on the nibs.
(5) Leaning the pen more toward the paper.
(6) Soft working surface.

Factors that tend to make line finer:

(1) Small amount of ink in the pen.
(2) Rapid movement of the pen.
(3) Sharp nibs.
(4) Fresh ink and clean pen.
(5) Pen approaching the perpendicular.
(6) Hard working surface.

Before making a new ink line on a drawing, the thickness of line should be tested by drawing a test line on a separate piece of paper under the same conditions. *Never test the pen freehand, or on a different kind of paper.* Always use a straightedge, and use identical paper.

If excess ink is in the pen or if wet lines are allowed to intersect previously drawn lines that are still wet, teardrop ends and rounded corners will result, Fig. 77 (a) and (b).

The ruling pen is used in inking irregular curves, as well as straight lines, as shown in Figs. 87 and 88. The pen should be held more nearly perpendicular when used with an irregular curve than when used with the T-square or a triangle. The ruling pen should lean only slightly in the direction in which the line is being drawn. It should be in a vertical plane containing a tangent to the curve at the position of the pen.

Some draftsmen insert a triangle under the irregular curve, back from the line, in order to raise the curve from the paper and prevent ink running under the edge. Another effective method is to glue several thin pieces of plastic to the faces of the curve or to the triangles.

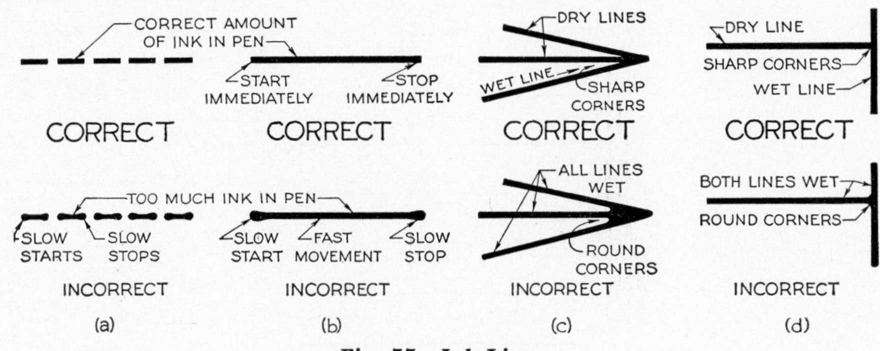

Fig. 77 Ink Lines.

61. Tracing in Ink. To make a tracing in ink, tracing paper, §71, or tracing cloth, §72, is fastened over the drawing, and the copy is made by tracing the lines. When a drawing is important enough to warrant the use of ink, it is generally made on tracing cloth. Although the glazed side of the cloth formerly was intended as the working surface, most draftsmen prefer the dull side, because it takes ink better and can be marked with a pencil. It is common practice to make the pencil drawing directly upon the tracing cloth and then trace it with ink, thus eliminating the traditional pencil drawing on detail paper.

Before the ink is applied, the cloth should be dusted with a small quantity of *pounce*, which should be rubbed in lightly with a soft fabric and then thoroughly removed with a clean cloth. Instead of the special drafting pounce, any slightly abrasive powder, such as talcum or chalk dust (calcium carbonate), applied with an ordinary blackboard eraser, may be used.

A greater difference in the widths of lines is necessary on tracings than on pencil drawings, because the contrast between blue and white on blueprints is not so great as that between black and white on drawings. Visible lines should be very bold. Extension lines, dimension lines, section lines, and center lines should be very *fine*, but strong enough to insure positive printing. The *draftsman's line gage*, Fig. 78, prepared by Prof. C. V. Mann, is convenient when referring to lines of various widths.

In inking or tracing a pencil line, the ink line should be centered over the pencil line, as shown in Fig. 79 (a), and not along one side as at (b). If this is done correctly in the case of tangent arcs, the line thicknesses will overlap at the points of tangency, as at (c), resulting in smooth tangencies. Incorrect practice is shown in exaggerated form at (d). Tangent points should be constructed in pencil, §§140-148, to assist in making smooth connections.

Make visible lines full thickness if the lines are spaced well apart, as shown at (e) and (g). When they are close together, the lines should be made thinner, (f) and (h).

Pencil guide lines for lettering should be ruled directly upon the tracing paper or cloth, since guide lines on the drawing underneath cannot be seen distinctly enough to furnish an accurate guide for letter-heights. For conventional ink lines, see Fig. 24.

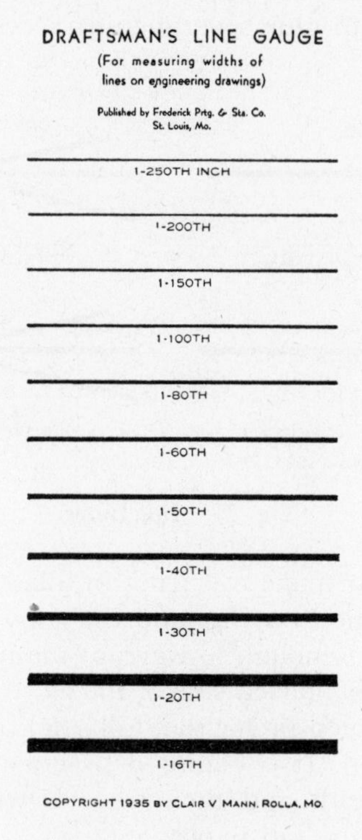

Fig. 78 Mann's Line Gauge.

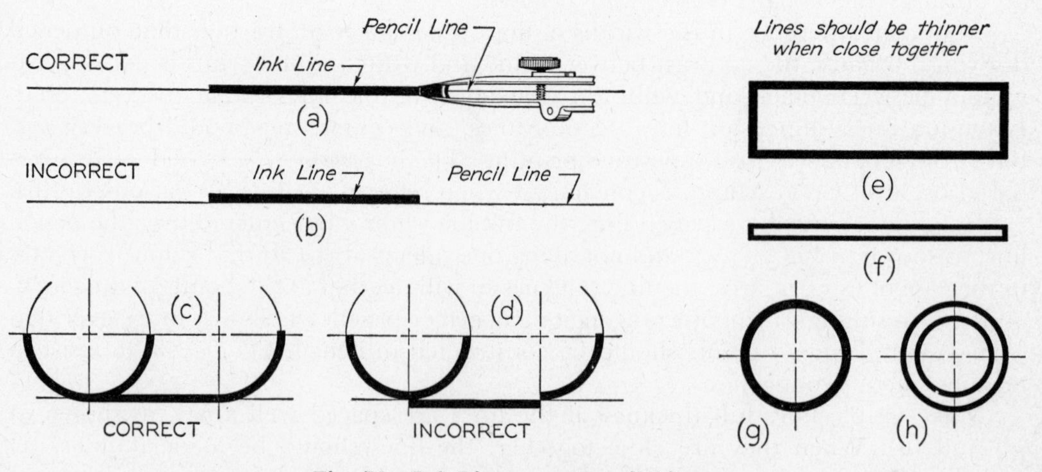

Fig. 79 Ink Lines over Pencil Lines.

62. Order of Inking. Fig. 80. A definite order should be followed in inking a drawing or tracing, as follows:

1

(a) Mark all tangent points in pencil directly on tracing.
(b) Indent all compass centers (with pricker or divider point).
(c) Ink visible circles and arcs.
(d) Ink hidden circles and arcs.
(e) Ink irregular curves, if any.

2

(a) Ink visible straight lines.
(b) Ink hidden straight lines.

3

(a) Ink center lines, extension lines, dimension lines, leader lines, and section lines (if any).

(Steps 2 and 3: 1st: Horizontal; 2nd: Vertical; 3rd: Inclined)

4

(a) Ink arrowheads and dimension figures.
(b) Ink notes, titles, etc. (pencil guide lines directly on tracing).

Some draftsmen prefer to ink center lines before indenting the compass centers because of the possibility of ink going through the holes and causing blots on the back of the sheet.

63. To Sharpen the Ruling Pen. If a ruling pen is subjected to frequent or extended use, its nibs will become so worn that good lines cannot be drawn with it. The correct point is shown in Fig. 81 (a), and a characteristic worn point is shown at (b). The nibs at (c) are too pointed, and as a result the ink tends to hang suspended in the pen and not touch the paper. The nibs at (d) are too rounded, or blunt, in which case the ink tends to run out of the pen onto the paper too readily.

An examination of the tips of the nibs of a worn ruling pen reveals a bright point on each nib, which reflects light like a small mirror. Such a condition first manifests itself to the student in his inability to draw fine lines. Another indication of a defective ruling pen is a scratchy contact with the paper, or the necessity of pressing the points firmly into the paper to get the ink to flow. The reason for this condition is illustrated in Fig. 82 (b), which shows one nib much longer

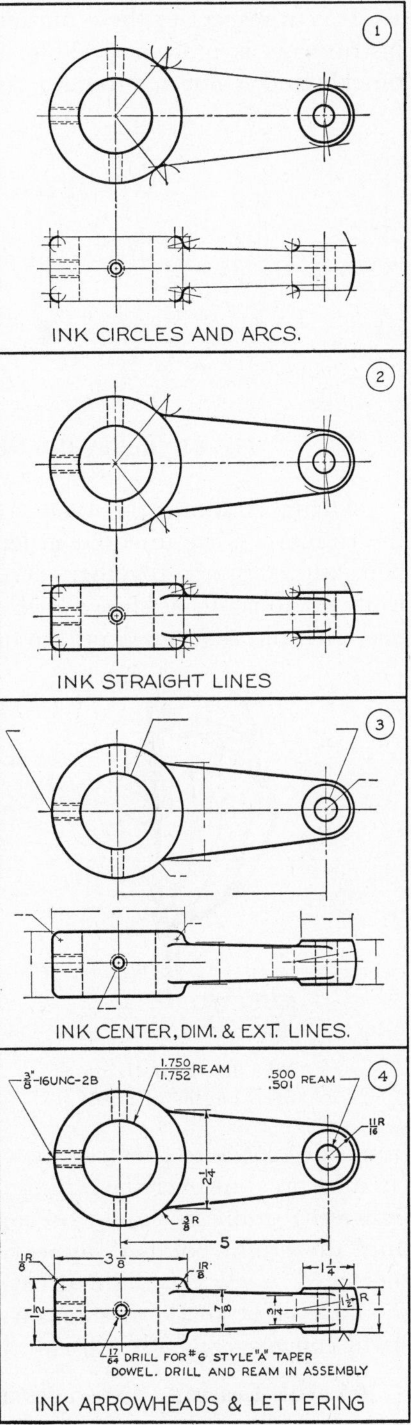

Fig. 80 Order of Inking.

than the other. This may result, to some extent, from wear or from dropping the pen and chipping away small segments of the nibs.

It is in respect to these matters that the difference between inferior and superior instruments is most noticeable. The material in good ruling pens does not wear quickly and is not apt to chip, and the pen is more easily sharpened.

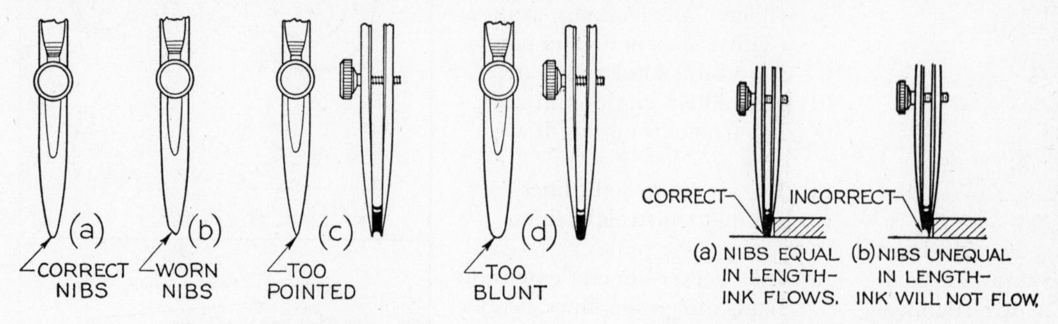

Fig. 81 Ruling Pen Nibs.

Fig. 82 Correct and Faulty Nibs.

A hard Arkansas oil stone is excellent for sharpening ruling pens. If the nibs of the ruling pen are unequal in length, they should first be equalized by moving the pen, with the nibs together, across the stone lightly with an oscillating movement from left to right, as shown in Fig. 83. To sharpen the nibs, they should be opened and each nib sharpened on the outside, as shown in Fig. 84, rolling the pen slightly

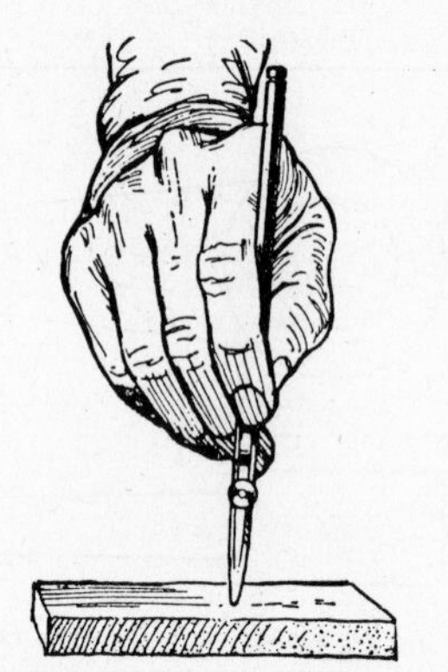

Fig. 83 Equalizing the Lengths of Nibs.

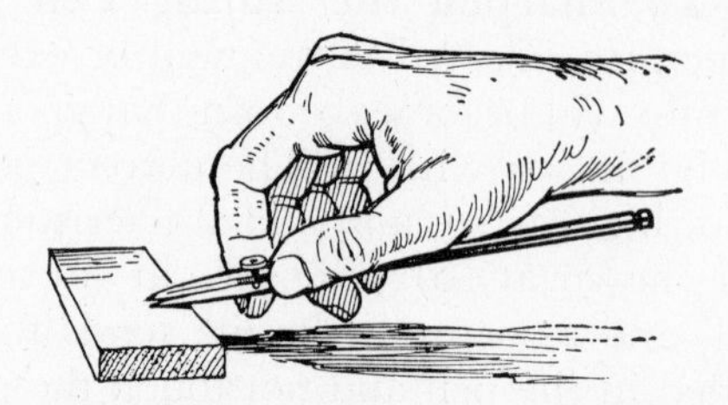

Fig. 84 Sharpening the Ruling Pen.

from side to side to preserve the convex surface of the nib. Great care must be exercised to prevent oversharpening one nib and thus shortening it. The bright points, indicating dullness, should be carefully watched and the nibs should be sharpened until the bright points disappear. Finally, to make sure that one nib has not been shortened, a few very light strokes should again be taken, as in Fig. 83. No attempt should ever be made to sharpen the inside of the nibs, for this always results in a slight convexity, which will ruin the pen.

64. Ink Erasing. Mistakes are certain to occur in inking, and correct methods of erasing should be considered a part of the technique. For general ink or pencil erasing, the Weldon Roberts "India" or the Eberhard-Faber "Ruby" eraser is recommended. Ink erasers are usually gritty and too abrasive, and their use tends to destroy the surface of the paper. If this occurs, it may be impossible to ink over the erased area. Best results are obtained if a smooth hard surface, such as a triangle, is placed under the area being erased.

An application of pounce or chalk dust will improve the surface and prevent running of the ink. The erasing shield, Fig. 26, should be used to protect lines adjacent to the area to be erased.

When an ink blot is made, the excess ink should be taken up with a blotter, or smeared with the finger if a blotter is not available, and not allowed to soak into the paper. When the spot is thoroughly dry, the remaining ink can be erased easily.

For cleaning untidy drawings or for removing the original pencil lines from an inked drawing, a sponge rubber, kneaded rubber, or artgum is useful. The artgum is recommended for general use. Pencil lines or dirt can be removed from tracing cloth by rubbing lightly with a cloth moistened with carbon tetrachloride (Carbona) or benzine (Energine).

When erasure on cloth damages the surface, it may be restored by rubbing the spot with soapstone and then applying pounce or chalk dust. If the damage is not too great, an application of the powder will be sufficient.

When a gap in a thick ink line is made by erasing, the gap should be filled in with a series of fine lines that are allowed to run together. A single heavy line is likely to "run" and cause a blot.

In commercial drafting rooms, the electric erasing machine, Fig. 27, is usually available to save the time of the draftsman.

65. Irregular Curves. When it is required to draw mechanical curves other than circles or circular arcs, an *irregular* or *French curve* is generally employed. Many different forms and sizes of curves are manufactured, as suggested by the more common forms illustrated in Fig. 85.

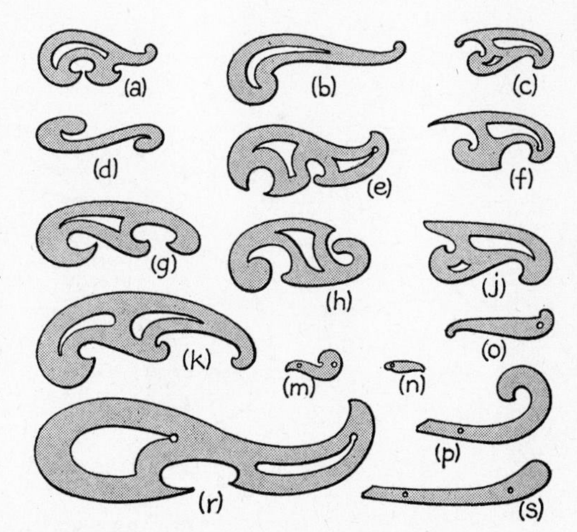

Fig. 85 Irregular or French Curves.

The curves are composed largely of successive segments of the geometric curves, such as the ellipse, parabola, hyperbola, involute, etc. The best curves are made of highly transparent plastic. Many special types of curves are available, including hyperbolas, parabolas, ellipses, logarithmic spirals, ship curves, railroad curves, etc.

Adjustable curves, Fig. 86, are also available. The curve shown at (a) consists of a core of lead, enclosed by a coil spring attached to a flexible strip. The one at (b) consists of a *spline*, to which *ducks* (weights) are attached. The spline can be bent to form any desired curve, limited only by the elasticity of the material (see §713). An ordinary piece of solder wire can be used very successfully by bending the wire to the desired curve.

66. Using the Irregular Curve. The irregular curve is a device for the *mechanical drawing of curved lines and should not be applied directly to the points*, or used for purposes of producing an initial curve. The proper use of the irregular curve requires skill, especially when the lines are to be drawn in ink. After points have been plotted through which the curve is to pass, a light pencil line should be sketched freehand smoothly through the points.

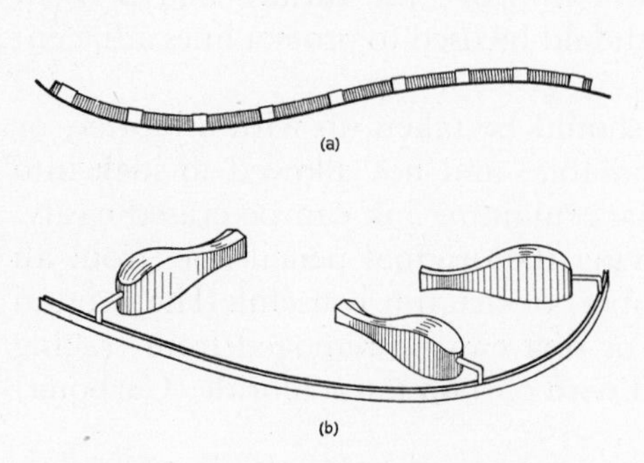

Fig. 86 Adjustable Curve Rulers.

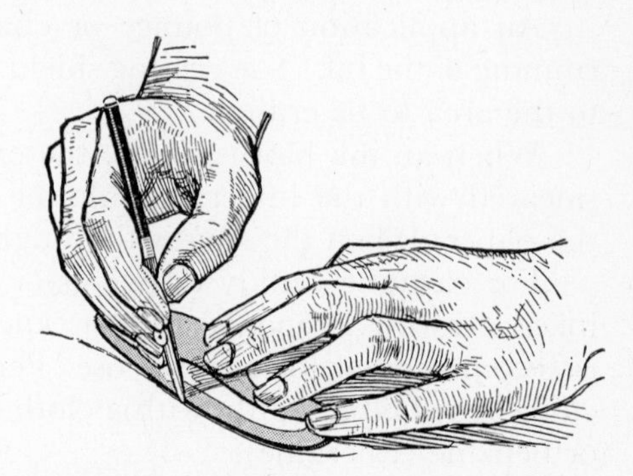

Fig. 87 Using the Irregular Curve.

To draw a mechanical line over the freehand line with the aid of the irregular curve, it is only necessary to match the various segments of the irregular curve with successive portions of the freehand curve and to draw the line with pencil or ruling pen along the edge of the curve, Fig. 87. It is very important that the irregular curve match the curve to be drawn for some distance at each end beyond the segment to be drawn for any one setting of the curve, as shown in Fig. 88. When this rule is

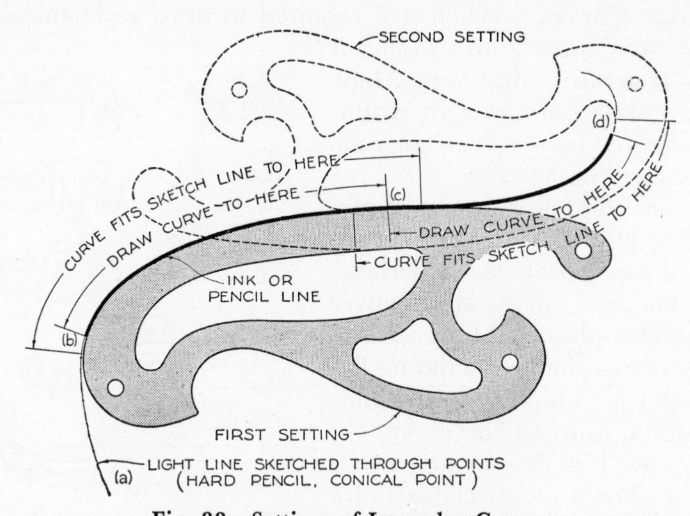

Fig. 88 Settings of Irregular Curve.

observed, the successive sections of the curve will be tangent to each other, without any abrupt change in the curvature of the line. In placing the irregular curve, the short-radius end of the curve should be turned toward the short-radius part of the curve to be drawn; that is, the portion of the irregular curve used should have the same curvilinear tendency as the portion of the curve to be drawn. This will prevent abrupt changes in direction.

The draftsman should change his position with respect to the drawing when necessary, so that he always works on the side of the curve away from him; that is, he should avoid working on the "under" side of the curve.

When plotting points to establish the path of a curve, it is desirable to plot more points, and closer together, where sharp turns in the curve occur.

Free curves may also be drawn with the compass, as shown in Fig. 194.

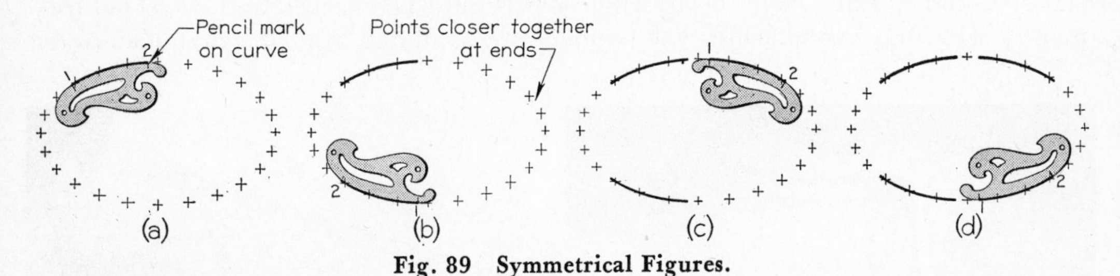

Fig. 89 Symmetrical Figures.

For symmetrical curves, such as an ellipse, Fig. 89, use the same segment of the irregular curve in two or more opposite places. For example, at (a) the irregular curve is matched to the curve and the line drawn from 1 to 2. Light pencil dashes are then drawn directly on the irregular curve at these points (the curve will take pencil marks well if it is lightly "frosted" by rubbing with a hard pencil eraser). At (b) the irregular curve is turned over and matched so that the line may be drawn from 2 to 1. In similar manner, the same segment is used again at (c) and (d). The ellipse is completed by filling in the gaps at the ends by using the irregular curve or, if desired, the compass.

67. Templates. A great variety of *templates* is available for specialized needs. A template may be found for drawing almost any ordinary drafting symbols or repetitive features. The *Engineers' Triangle*, Fig. 90 (a), is useful for drawing hexagons or

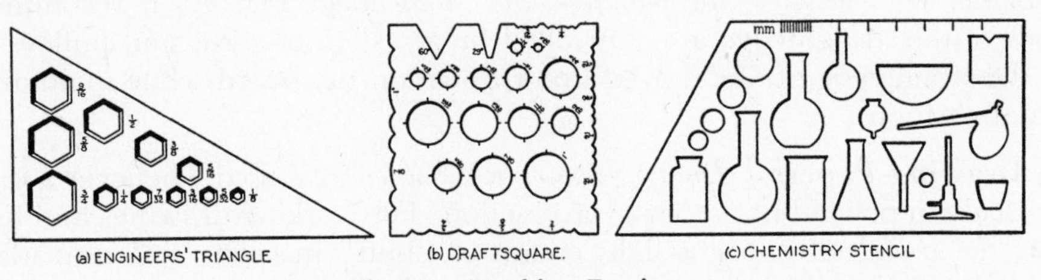

Fig. 90 Drafting Devices.

for bolt heads and nuts; the *Draftsquare*, (b), is convenient for drawing the curves on bolt heads and nuts, for drawing circles, thread forms, etc.; and the Chemistry Stencil, (c), is useful for drawing chemical apparatus in schematic form.

Ellipse templates, §163, are perhaps more widely used than any other type. Circle templates are useful in drawing small circles quickly, and in drawing fillets and rounds, and are used extensively in tool and die drawings.

68. Drafting Machines. The *drafting machine*, Fig. 91, is an ingenious device that replaces the T-square, triangles, scales, and protractor. The links, or bands, are arranged so that the controlling head is always in any desired fixed position regardless of where it is placed on the board; thus the horizontal straightedge will remain horizontal if so set. The controlling head is graduated in degrees (including a vernier in certain machines), which allows the straightedges, or scales, to be set and locked at any angle. There are automatic stops at the most frequently used angles, such as 15°, 30°, 45°, 60°, 75°, and 90°.

Drafting machines* have been greatly improved in recent years. The chief advantage of the drafting machine is that it speeds up drafting. Since its parts are made of metal, their accurate relationships are not subject to change, whereas T-squares, triangles, and working edges of drawing boards must be checked and corrected frequently. Drafting machines for left-handers are available from the manufacturers.

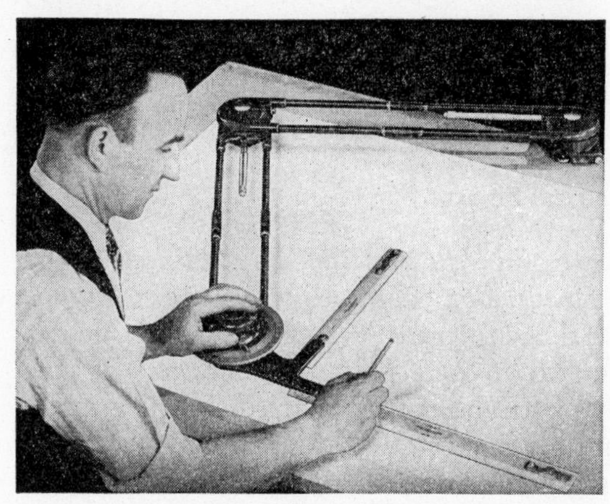

Courtesy Universal Drafting Machine Co.

Fig. 91 Drafting Machine.

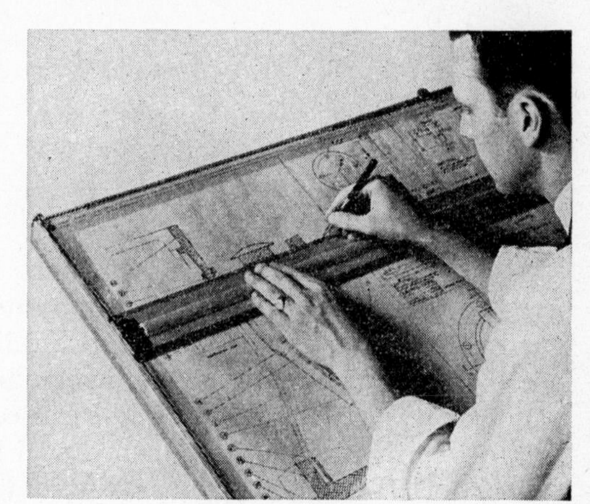

Courtesy Eugene Dietzgen Co.

Fig. 92 Parallel Ruling Straightedge.

69. Parallel-Ruling Straightedge. For large drawings, the long T-square becomes unwieldy, and considerable inaccuracy may result from the "give" or swing of the blade. In such case the *parallel-ruling straightedge*, Fig. 92, is recommended. The ends of the straightedge are controlled by a system of cords and pulleys which permit the straightedge to be moved up or down on the board while maintaining a horizontal position.

70. Drawing Papers. *Drawing paper*, or *detail paper*, is used whenever a drawing is to be made in pencil but not for reproduction. For working drawings and for general use, the preferred paper is light cream or "buff" in color, and is available in rolls of widths 24″, 36″, etc., and in cut sheets of standard sizes 9″×12″, 12″×18″, 18″×24″, etc., or $8\frac{1}{2}$″×11″, 11″×17″, 17″×22″, etc. Most industrial drafting rooms use standard sheets with printed borders and title strips, and since the cost for printing adds so little to the price per sheet, many schools have also adopted printed sheets.

The best drawing papers have upwards of 100 per cent pure rag stock, have stronger fibres that afford superior erasing qualities, folding strength, and toughness, and will not discolor or grow brittle with age. The paper should have a fine grain or "tooth" which will pick up the graphite and produce clean, dense black lines. However, if the paper is too rough it will wear down the pencil excessively, and will produce ragged, grainy lines. The paper should have a hard surface so that it will not "groove" too easily when pressure is applied to the pencil.

*Universal Drafting Machine Co., Charles Bruning Co., Keuffel & Esser Co., Eugene Dietzgen Co., Frederick Post Co., and V & E Manufacturing Co. are some of the manufacturers of drafting machines.

For ink work, as for catalog and book illustrations, white papers are used. The better papers, such as Bristol Board and Strathmore, come in several thicknesses, as 2-ply, 3-ply, 4-ply, etc.

71. Tracing Papers. *Tracing paper* is a thin transparent paper upon which drawings are made for the purpose of reproducing by blueprinting or by other similar processes. Tracings are usually made in pencil, but may be made in ink. Most tracing papers will "take" pencil or ink, but some are especially suited to one or to the other.

Tracing papers are of two kinds: (1) those treated with oils, waxes, or similar substances to render them more transparent, called *vellums*; (2) those not so treated, but which may be quite transparent, owing to the high quality of the raw materials and the methods of manufacture. Some treated papers deteriorate rapidly with age, becoming brittle in many cases within a few months, but some excellent vellums are available. Untreated papers made entirely of good rag stock will last indefinitely and will remain tough. For a discussion of tracing methods, see §§61 and 62.

72. Tracing Cloth. *Tracing cloth* is a thin transparent muslin fabric, (cotton, not "linen" as commonly supposed) sized with a starch compound or plastic to provide a good working surface for pencil or ink. It is much more expensive than tracing paper. Tracing cloth is available in rolls of standard widths, as 30″, 36″, and 42″, and also in sheets of standard sizes, with or without printed borders and title forms.

For pencil tracings, special pencil tracing cloths are available. Many concerns make their drawings in pencil directly on this cloth, dispensing entirely with the preliminary pencil drawing on detail paper, thus saving a great deal of time. These cloths generally have a surface that will produce dense black lines when hard pencils are used. Hence, these drawings do not easily smudge and will stand up well with handling.

A disadvantage of tracing cloth has been the ease with which it could be damaged by water or perspiration, but excellent waterproof cloths have been developed and are now widely used.

73. Standard Sheets. Two systems of sheet sizes are approved by the ASA* as follows:

A	8½″ × 11″	A	9″ × 12″
B	11″ × 17″	B	12″ × 18″
C	17″ × 22″	C	18″ × 24″
D	22″ × 34″	D	24″ × 36″
E	34″ × 44″	E	36″ × 48″
F	28″ × 40″	F	28″ × 40″

The use of the basic sheet size 8½″×11″, and multiples thereof, permits filing of small tracings and of folded prints in standard files with or without correspondence. These sizes can be cut without waste from the standard 36″ rolls of paper or cloth.

The alternate system based on size 9″×12″ is widely used in the automotive industry and has the advantage of slightly larger areas.

For American Standard sheet layouts, see Fig. 93. Note that in both systems for sheets size C and larger, zoning letters may be used. The sizes of the zone spacings may be varied as desired.

For title blocks, revision blocks, and list of materials blocks, see the back endpaper of this book. See also §433.

*ASA Y14.1-1957.

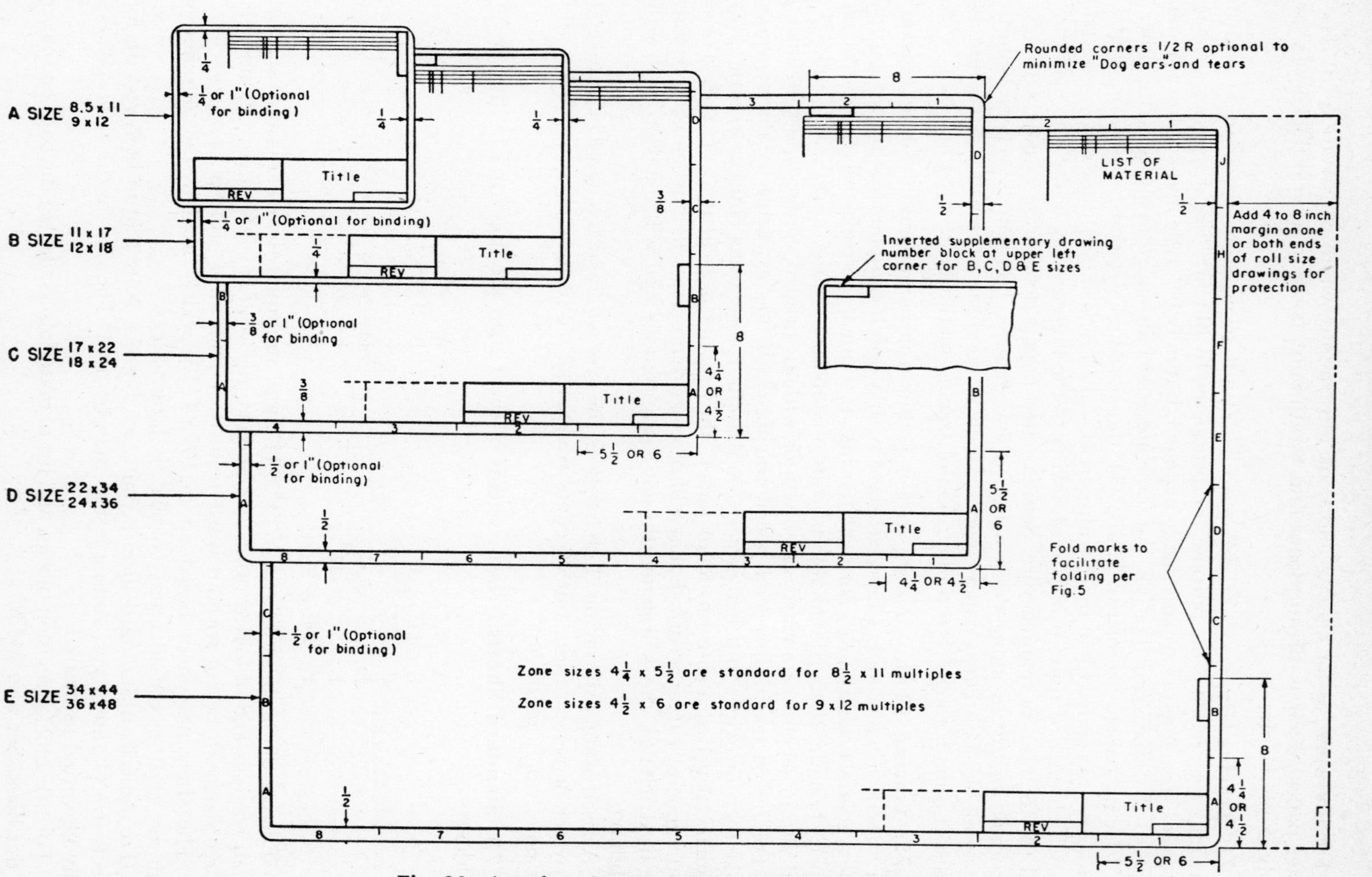

Fig. 93 American Standard Sheets (ASA Y14.1-1957).

74. Mechanical Drawing Problems. All of the following constructions are to be drawn in pencil on Layout A-2 (see the back endpaper of this book). The steps in drawing this layout are shown in Fig. 66. Draw all construction lines *lightly*, using a hard pencil (4H to 6H), and all required lines dense black with a softer pencil (F to H). If construction lines are drawn properly—that is, *lightly*—they need not be erased.

The drawings in Figs. 99-104 are to be drawn in pencil, preferably on tracing paper or vellum; then prints should be made to show the effectiveness of the students' technique. If ink tracings are required, the originals may be drawn on detail paper and then traced on vellum or tracing cloth. For any assigned problem, the instructor may require that all dimensions and notes be lettered in order to afford further lettering practice.

The problems of Chapter 4, "Geometrical Constructions," provide excellent additional practice to develop skill in the use of drawing instruments.

Problems in convenient form for solution may be found in *Technical Drawing Problems*, by Giesecke, Mitchell, and Spencer, and in *Technical Drawing Problems, Series 2*, by Spencer and Grant, both designed to accompany this text, and published by The Macmillan Company.

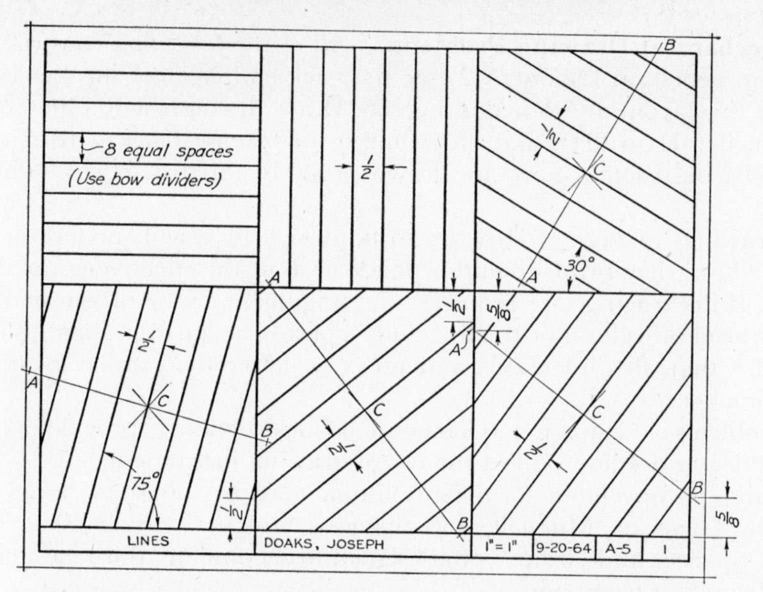

Fig. 94 Using Layout A-2, divide working space into six equal rectangles, and draw visible lines, as shown. Draw construction lines AB through centers C at right angles to required lines; then along each construction line, set off $\frac{1}{2}''$ spaces and draw required visible lines. Omit dimensions and instructional notes.

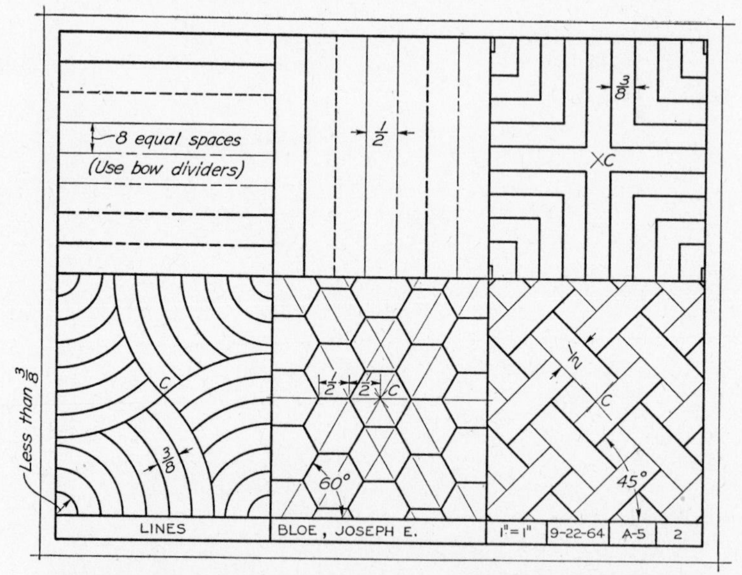

Fig. 95 Using Layout A-2, divide working space into six equal rectangles, and draw lines shown. In first two spaces, draw conventional lines to match those in Fig. 24 (pencil lines). In remaining spaces, locate centers C by diagonals, and then work constructions out from them. Omit dimensions and instructional notes.

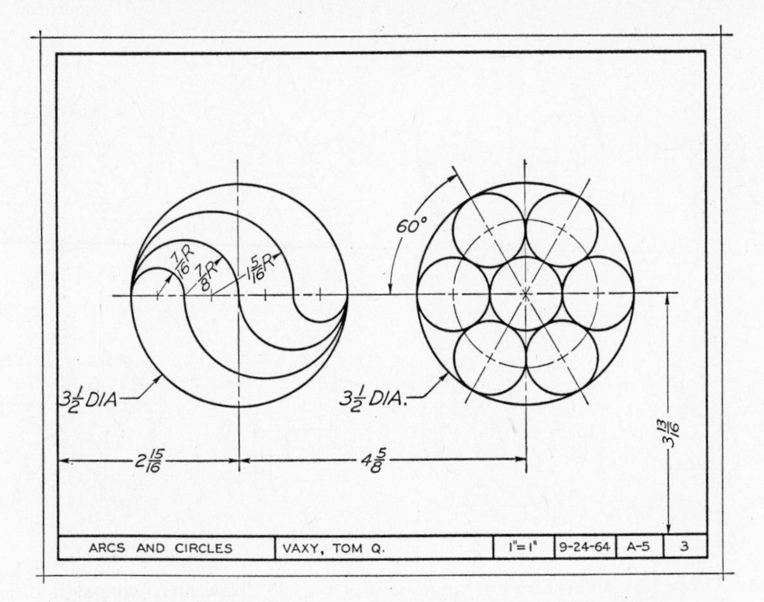

Fig. 96 Using Layout A-2, draw figures in pencil, as shown. Use bow pencil for all arcs and circles within its radius-range. Omit all dimensions.

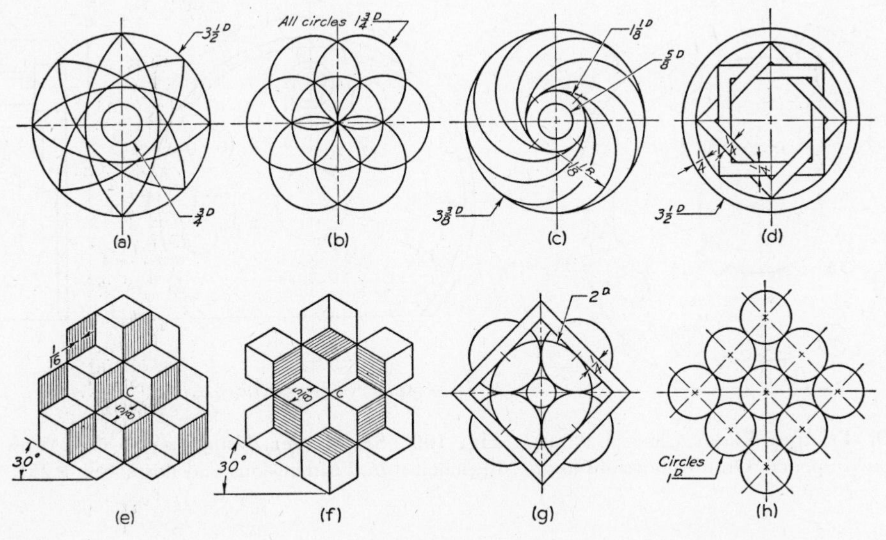

Fig. 97 Using Layout A-2 and arrangement of Fig. 96, draw any two assigned figures in pencil, as shown. Omit all dimensions and instructional notes.

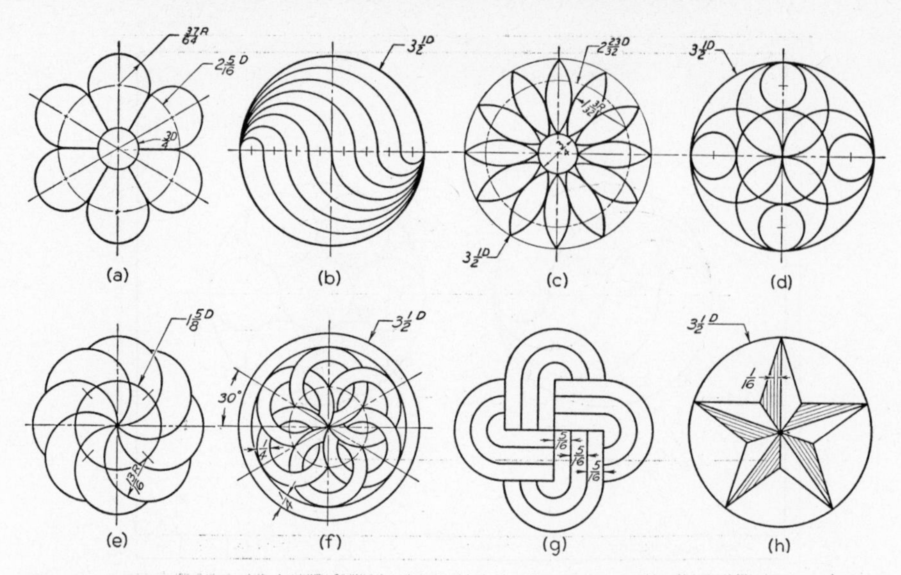

Fig. 98 Using Layout A-2 and arrangement of Fig. 96, draw any two assigned figures in pencil, as shown. Omit all dimensions and instructional notes.

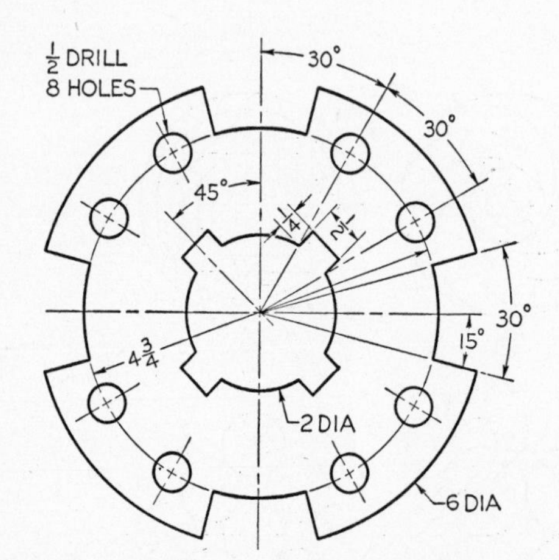

Fig. 99 Friction Plate. Using Layout A-2, draw in pencil. Omit dimensions and notes.

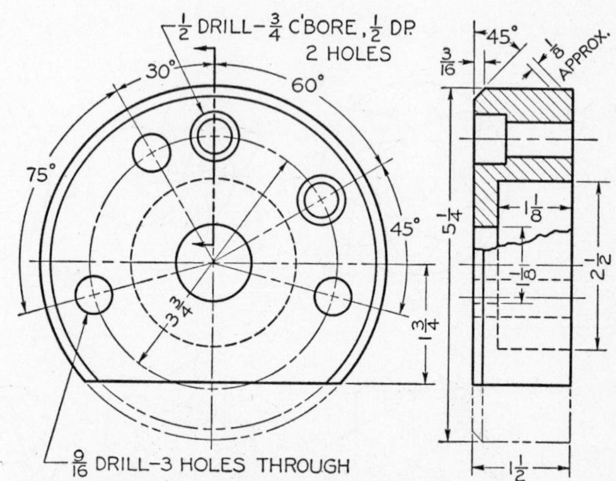

Fig. 100 Seal Cover. Using Layout A-2, draw views in pencil. Omit dimensions and notes. See § 251.

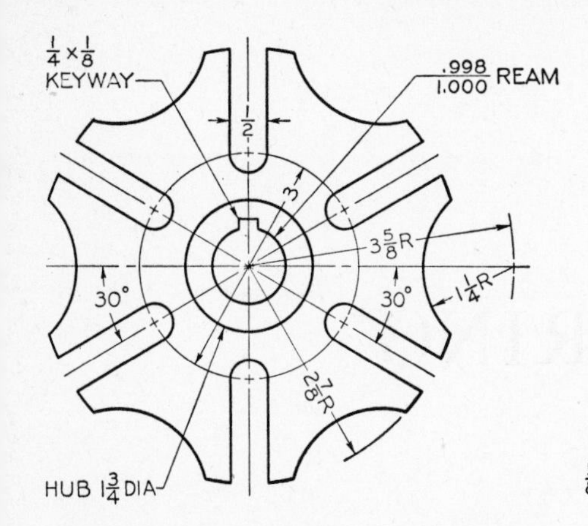

Fig. 101 Geneva Cam. Using Layout A-2, draw in pencil. Omit dimensions and notes.

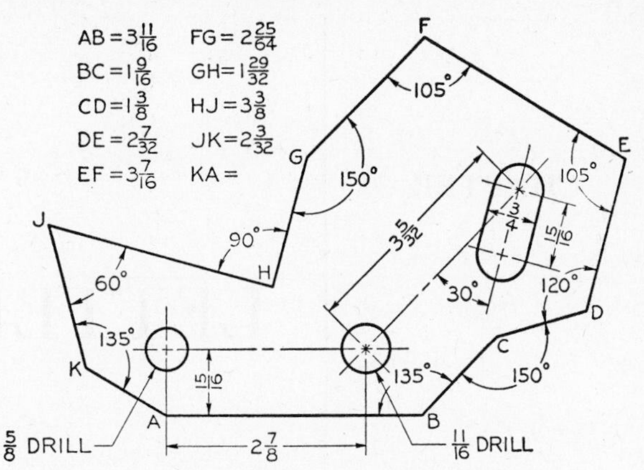

Fig. 102 Shear Plate. Using Layout A-2, draw accurately in pencil. Give length of KA. Omit other dimensions and notes.

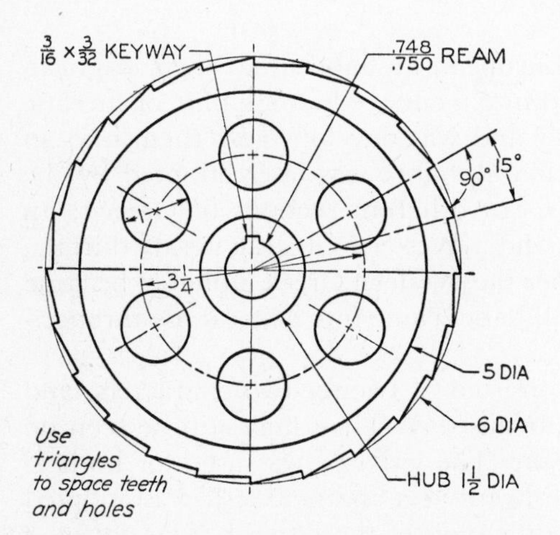

Fig. 103 Ratchet Wheel. Using Layout A-2, draw in pencil. Omit dimensions and notes.

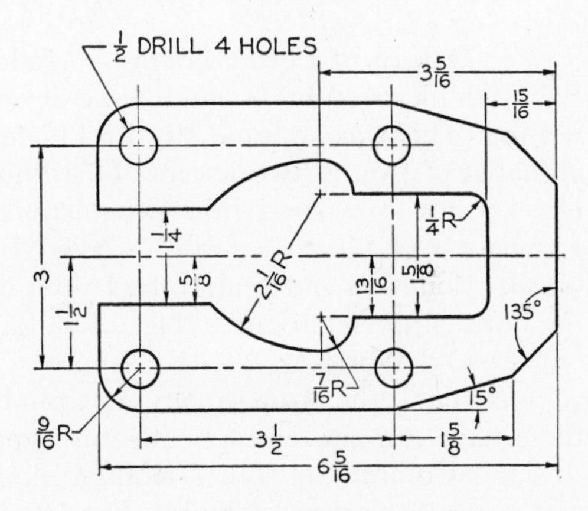

Fig. 104 Latch Plate. Using Layout A-2, draw in pencil. Omit dimensions and notes.

CHAPTER 3

LETTERING*

75. Origin of Letter Forms. Modern European alphabets had their origin in Egyptian hieroglyphics, which were developed into a cursive hieroglyphic or hieratic writing. This was adopted by the Phoenicians and was developed by them into an alphabet of twenty-two letters. This Phoenician alphabet was later adopted by the Greeks, but it evolved into two distinct types in different sections of Greece: an Eastern Greek type, used also in Asia Minor, and a Western Greek type, used in the Greek colonies in and near Italy. In this manner the Western Greek alphabet became the Latin alphabet about 700 B.C. The Latin alphabet came into general use throughout the Old World.

Originally the Roman capital alphabet consisted of twenty-two characters, and these have remained practically unchanged to this day. They may still be seen on Trajan's Column and other Roman monuments. The letter V was used for both U and V until the tenth century. The last of the twenty-six characters, J, was adopted at the end of the fourteenth century as a modification of the letter I. The dot over the lower-case j still indicates its kinship to the i; in Old English the two letters are very similar. The numerous modern styles of letters were derived from the original Roman capitals.

Before the invention of printing by Gutenberg in the fifteenth century, all letters were made by hand and were modified and decorated according to the taste of the individual writer. These letters were introduced into England, where they became known as Old English. The early German printers adopted these letters, and they are still in use in Germany. The early Italian printers used Roman letters, which were later introduced into England, where they gradually replaced the block letters

***Lettering*, not "printing," is the correct term for making letters by hand. *Printing* means the production of printed material on a printing press.

of German origin. Thus the Roman capital has come down to us virtually in its original form.

76. Letter Styles. A general classification of letter styles is shown in Fig. 105. They were all made with Speedball pens, as indicated, and are therefore largely single-stroke letters.

Classification of letter styles by division of groups

ABCDEFGH
abcdefgh
GOTHIC All letters having the elementary strokes of even width are classified as Gothic —
Made with Style A or B Speedball Pen

ABCDEFGH
abcdefghij
Roman All letters having elementary strokes "accented" or consisting of heavy and light lines, are classified as Roman.
Made with Style C or D Speedball Pen

ABCDEFGHI
abcdefghijklm
Italic- All slanting letters are classified as Italics- These may be further designated as Roman-Italics, Gothic Italics or Text Italic.
Made with Style C or D Speedball Pen

ABCDEFG
abcdefghijkl
Text—This term includes all styles of Old English, German text, Bradley text or others of various trade names — Text styles are too illegible for commercial purposes.
Made with Style C or D Speedball Pen

Fig. 105 Classification of Letter Styles.

If the letters are drawn in outline and filled in, they are referred to as "filled-in" letters, Figs. 143 and 146. The plainest and most legible style is the GOTHIC from which our single-stroke engineering letters are derived. The term ROMAN refers to any letter having wide downward strokes and thin connecting strokes, as would result from the use of a wide pen, while the ends of the strokes are terminated with spurs called *serifs*. Roman letters include Old Roman and Modern Roman, and may be vertical or inclined. Inclined letters are also referred to as *Italic*, regardless of the letter style; those shown in Fig. 105 are inclined Modern Roman. *Text* letters are often loosely referred to as **Old English**, though these letters as well as the other similar letters such as German Text are actually Gothic. The Commercial Gothic shown at the top of Fig. 105 is a relatively modern development, which originates from the earlier Gothic forms. German Text is the only commercially used form of medieval Gothic in use today.

For more extensive and detailed information regarding the styles of letters, see §§101-103.

77. Extended and Condensed Letters. To meet design or space requirements, letters may be narrower and spaced closer together, in which case they are called "compressed" or "condensed" letters. If the letters are wider than normal, they are referred to as "extended" letters, Fig. 106.

CONDENSED LETTERS
EXTENDED LETTERS
Condensed Letters
Extended Letters

Fig. 106 Extended and Condensed Letters.

78. Lightface and Boldface Letters. Letters also vary as to the thickness of the stems or strokes. Letters having very thin stems are called LIGHTFACE, while those having heavy stems are called **BOLDFACE,** Fig. 107.

LIGHTFACE
BOLDFACE

Fig. 107 Lightface and Boldface Letters.

79. Single-Stroke Gothic Letters. During the latter part of the nineteenth century the development of industry and of technical drawing in the United States made evident a need for a simple legible letter that could be executed with single strokes of an ordinary pen. To meet this need C. W. Reinhardt, formerly Chief Draftsman for the *Engineering News*, developed alphabets of capital and lower-case inclined and "upright" letters,* based upon the old Gothic letters, Fig. 142. For each letter he worked out a systematic series of strokes. The single-stroke Gothic letters used on technical drawings today are based upon Reinhardt's work.

80. Standardization of Lettering. The first step toward standardization of technical lettering was made by Reinhardt when he developed single-stroke letters with a systematic series of strokes, §79. However, since that time there has been an unnecessary and confusing diversity of lettering styles and forms, and the American Standards Association in 1935 suggested letter forms that are now generally considered as standard. The present Standard (ASA Y14.2-1957) is practically the same as that given in 1935, except that vertical lower-case letters have since been added.

The letters in this chapter and throughout this text conform to the American Standard. Vertical letters are perhaps slightly more legible than inclined letters, but are more difficult to execute. Both vertical and inclined letters are standard, and the engineer or draftsman may be called upon to use either. Students should, therefore, learn to execute both forms well, though they may give more attention to the style they like and in which they can do better.

According to the American Standard, "The most important requirement for lettering as used on working drawings is legibility. The second is ease and rapidity of execution. . . . It is not desirable to vary the size of the lettering according to the size of the drawing except when a reduced photographic reproduction of the drawing is to be made. Letters should be compactly spaced so that the background areas

*Published in the *Engineering News* about 1893, and in book form in 1895.

are approximately equal, and words should be clearly separated. . . . Lettering should not be underlined except for particular emphasis."*

81. Uniformity. In any style of lettering, *uniformity* is essential. Uniformity in height, proportion, inclination, strength of lines, spacing of letters, and spacing of words insures a pleasing appearance, Fig. 108.

Uniformity in height and inclination is promoted by the use of light guide lines, §88. Uniformity in strength of lines can be obtained only by the skillful use of properly selected pencils and pens, §§84 and 85.

82. Optical Illusions. Good lettering involves artistic design, in which the white and black areas are carefully balanced to produce a pleasing effect. Letters are designed to *look* well, and some allowances must be made for incorrect perception. Some of the more striking optical illusions are shown in Fig. 109. Note that in Fig. 122 the standard H is narrower, and the W is wider than those shown in Fig. 109; and that the numeral 8 is narrower at the top than is that shown in Fig. 109. Note that the very acute angles in the W give the letter a compressed appearance; such acute angles should be avoided in letter design. Other optical illusions are shown in Fig. 97 (e) and (f).

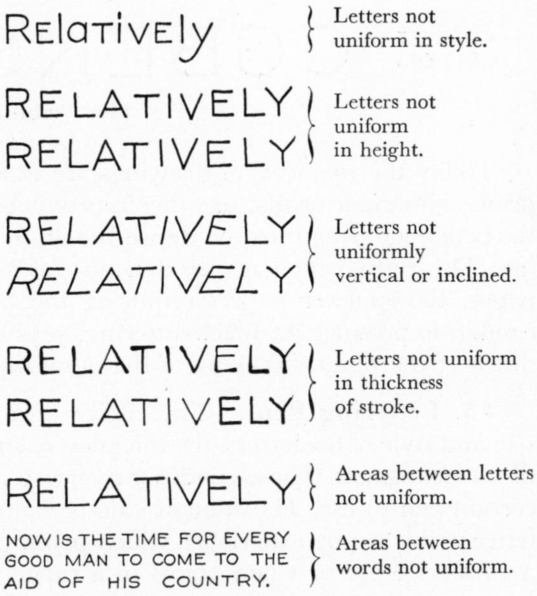

RElaTivEly — Letters not uniform in style.

RELATIVELY
RELATIVELY — Letters not uniform in height.

RELATIVELY
RELATIVELY — Letters not uniformly vertical or inclined.

RELATIVELY
RELATIVELY — Letters not uniform in thickness of stroke.

RELATIVELY — Areas between letters not uniform.

NOW IS THE TIME FOR EVERY
GOOD MAN TO COME TO THE
AID OF HIS COUNTRY. — Areas between words not uniform.

Fig. 108 Uniformity in Lettering.

83. Stability. If the upper portions of certain letters and numerals are equal in width to the lower portions, the characters appear top-heavy. To correct this, the upper portions are reduced in size where possible, thereby producing the effect of *stability* and a more pleasing appearance, Fig. 110.

If the central horizontal strokes of the letters B, E, F, and H are placed at mid-height, they will appear to be below center. To overcome this optical illusion, these strokes should be drawn slightly above the center.

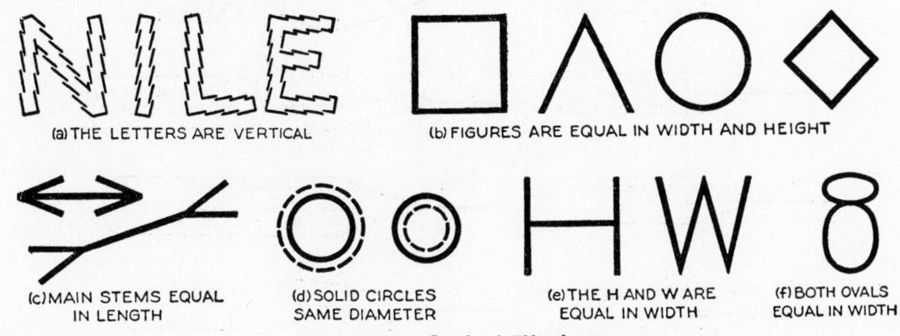
NILE
(a) THE LETTERS ARE VERTICAL
(b) FIGURES ARE EQUAL IN WIDTH AND HEIGHT
(c) MAIN STEMS EQUAL IN LENGTH
(d) SOLID CIRCLES SAME DIAMETER
HW 8
(e) THE H AND W ARE EQUAL IN WIDTH
(f) BOTH OVALS EQUAL IN WIDTH

Fig. 109 Optical Illusions.

*ASA Y14.2-1957.

84. Lettering Pencils. Pencil letters can be best made with a medium soft pencil with a conical point, Fig. 20 (c). First, sharpen the pencil to a needle point; then dull the point *very slightly* by marking on paper while holding the pencil vertically and rotating the pencil to round off the point. An F or H pencil is suitable for use on ordinary drawing paper of smooth surface. Between letters, turn the pencil occasionally to new positions in order to keep the point symmetrical.

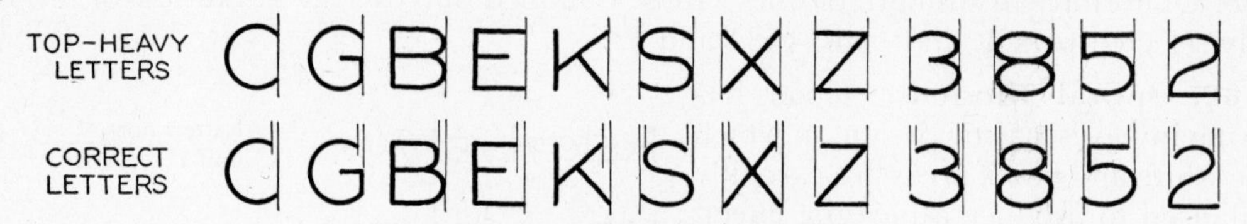

Fig. 110 Stability of Letters.

Today the majority of drawings are finished in pencil and reproduced as blueprints, ammonia prints, or other reproductions. To reproduce well by any process, the pencil lettering must be **dense black,** as should all other final lines on the drawing. The right pencil to use depends largely upon the amount of "tooth" in the paper, the rougher papers requiring the harder pencils. The lead should be soft enough to produce jet-black lettering, yet hard enough to prevent excessive wearing down of the point, breaking of the point, and smearing of the graphite.

85. Lettering Pens. The choice of a pen for lettering is determined by the size and style of the letters, the thickness of stroke desired, and the personal preference of the draftsman. These conditions vary so much that it is impossible to specify any certain pen to use. The student who is zealous in his efforts to develop his ability to letter will learn by experience which pen is best suited to his purpose. Fig. 111 shows a variety of the best pen points in a range from the *tit-quill*, the finest, to the *ball-*

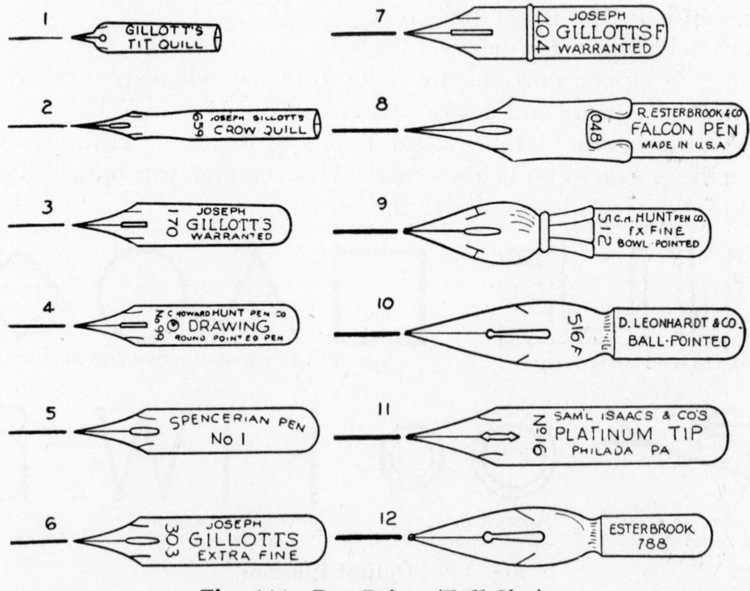

Fig. 111 Pen Points (Full Size).

pointed, the coarsest. The widths of the lines made by the several pens are shown full-size. The medium widths, represented by the *Gillott's* 303 and 404 (or equivalent) are most widely used for lettering notes and dimensions on drawings, in which case the letters are usually $\frac{1}{8}''$ high. For lettering $\frac{3}{16}''$ to $\frac{1}{4}''$ high, as for titles, the ball-pointed pens are commonly used.

A very flexible pen should not be used for lettering, because the downward strokes are apt to be shaded. A good lettering pen is one with which it is easy to make a stroke of uniform width. New pen points have a thin film of oil, which should be removed with a cloth—not burned off with a match flame. The best results are secured from a pen which has been used for some time, that is, "broken in" with use. Hence, when a pen point has proved satisfactory, it should be carefully wiped after using and taken care of as a valuable instrument.

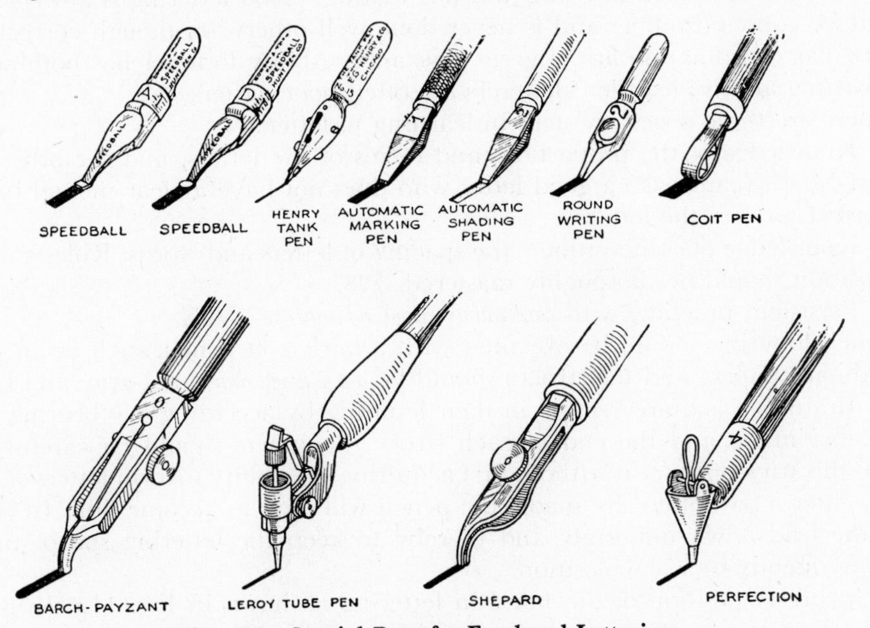

Fig. 112 Special Pens for Freehand Lettering.

Letters more than $\frac{1}{2}''$ in height generally require a special pen, Fig. 112. The *Speedball* pens are excellent for Gothic letters, Fig. 105, and are often used for titles and for the large drawing numbers in the corner of the title block, Figs. 718-722. Other styles of Speedball pens are suitable for Roman or text letters. These pens have the additional advantage of being low in cost. The *Barch-Payzant Lettering Pen* is available in eleven sizes ranging from 000 (very coarse) to 8 (very fine). The size 8 pen produces a stroke fine enough to be used for the usual lettering on technical drawings, being satisfactory for letters from $\frac{1}{8}''$ to $\frac{3}{16}''$ high.

The *Henry Tank Pen* is available in both plain and ball points, and has a simple device under the pen to hold ink and prevent the nibs from spreading.

The *Leroy Pen* is also available in a wide range of sizes, and is highly recommended. It can be used in a regular pen staff for freehand lettering, or in a "scriber" for mechanical lettering, Fig. 135.

Several other types of pens are shown in Fig. 112. *Any lettering pen must be kept clean.* Drawing ink corrodes the point of the pen if allowed to dry, and "builds up" the width of the point so that it has to be cleaned anyway. To remove dried drawing ink from any instrument, scrape carefully with a knife or use a special pen-cleaning fluid.*

Esterbrook and Venus fountain pens, with removable points for cleaning, are now available for use with drawing ink. The Koh-I-Noor "Rapidograph" is a different type in which the point is a tiny tube. A small automatic plunger rod keeps the ink flowing. The pen is suitable for letters $\frac{1}{8}''$ high or slightly higher. All of these pens should be frequently cleaned with cleaning fluid to keep them in service.

86. Technique of Lettering. *Any normal person can learn to letter if he is persistent and intelligent in his efforts.* While it is true that "practice makes perfect," it must be understood that practice alone is not enough; it must be accompanied by *continuous effort to improve.*

Lettering is freehand drawing and not writing. Good lettering is always accomplished by conscious effort and is never done well otherwise, though correct habits of muscular coordination are of great assistance. Ability to letter has nothing to do with writing ability; excellent letterers are often poor writers.

There are three necessary steps in learning to letter:

1. Knowledge of the proportions and forms of the letters, and the order of the strokes. No one can make a good letter who does not have a clear mental image of the correct form of the letter.

2. Knowledge of composition—the spacing of letters and words. Rules governing composition should be thoroughly mastered, §98.

3. Persistent practice, with *continuous effort to improve.*

Pencil lettering should be executed with a fairly soft pencil, such as an F or H for ordinary paper; and the strokes should be *dark* and *sharp*, not gray and blurred. Many draftsmen acquire "snap" in their lettering by accenting or "bearing down" at the beginning and the end of each stroke. Beginners should be careful not to overdo this trick or to try it without first acquiring the ability to form letters of correct shape. After a few letters are made, the pencil will tend to become dull. In order to wear the lead down uniformly and thereby to keep the lettering sharp, turn the pencil frequently to a new position.

The correct position of the hand in lettering is shown in Fig. 113. In general, draw vertical strokes downward or toward you with a finger movement, and draw horizontal strokes from left to right with a wrist movement without turning the paper.

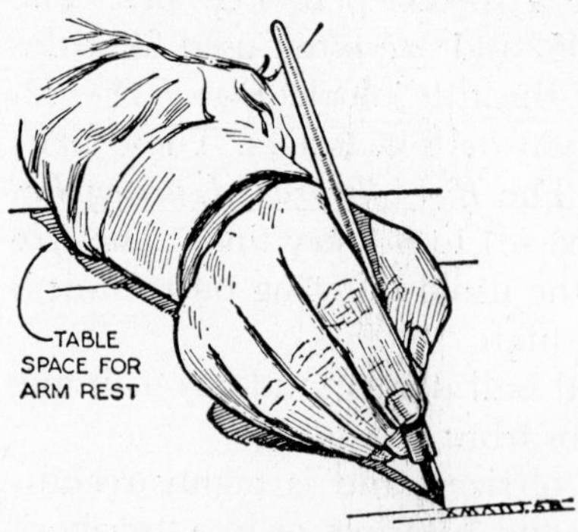

Fig. 113 Position of Hand in Lettering.

The forearm should be approximately at right angles to the line of lettering, and *resting on the board*—never suspended in mid-air. If the board is small, revolve it counterclockwise about 45°, until the line of lettering is approximately perpendicular to the forearm. If the board is larger and cannot be moved, shift the left side of your body around toward the board to approximate this position as nearly as possible.

Since practically all pencil lettering will be reproduced by blueprinting or otherwise, the letters should be **dense black.** Avoid hard pencils which, even with considerable pressure, produce gray lines.

*Higgins Ink Co. or Keuffel & Esser Co.

Use a fairly soft pencil and keep it sharp by frequent dressing of the point on the sandpaper pad or file. An example (full size) of pencil lettering exhibiting correct technique is shown in Fig. 114.

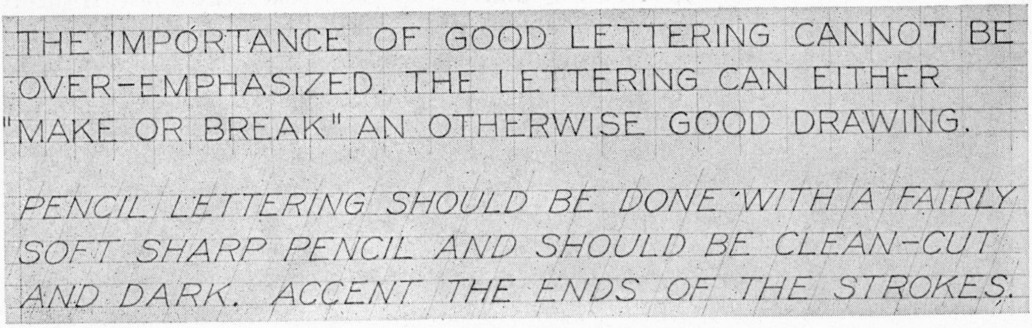

THE IMPORTANCE OF GOOD LETTERING CANNOT BE
OVER-EMPHASIZED. THE LETTERING CAN EITHER
"MAKE OR BREAK" AN OTHERWISE GOOD DRAWING.

PENCIL LETTERING SHOULD BE DONE WITH A FAIRLY
SOFT SHARP PENCIL AND SHOULD BE CLEAN-CUT
AND DARK. ACCENT THE ENDS OF THE STROKES.

Fig. 114 Pencil Lettering (Full Size).

In ink lettering, most beginners have a tendency toward excessive pressure on the pen point, thus producing strokes of varying widths. Select a pen point that will make strokes of correct thickness without spreading the nibs. The correct position of the pen is shown in Fig. 113. Move the pen with light uniform pressure, and allow the ink to *flow off* the point instead of being forced off by pressure upon the point.

Some draftsmen "ink" the pen with a quill from the ink bottle, but this is generally unnecessary unless some ink-holding device (such as that on the Henry Tank Pen, Fig. 112) is used. Excess ink, however, should be removed from the point by lightly touching it against the opening of the bottle.

Before an inked tracing is lettered, all guide lines should be drawn in pencil directly upon the tracing paper or cloth if the guide lines underneath are too indistinct to serve their purpose well as a guide for inking.

87. Left-Handers. All evidence indicates that the left-handed draftsman is just as skillful as the right-hander, and this includes skill in lettering. The most important step in learning to letter is to learn the correct shapes and proportions of letters, and these can be learned as well by the left-hander as by anyone else. The left-hander does have a problem of developing a system of strokes that seems most suitable for himself. The strokes shown in Figs. 122 and 123 are for right-handers. The left-hander should experiment with each letter to find out which strokes are best for him. The habits of left-handers vary so much that it is futile to suggest a standard system of strokes for all left-handers.

The left-hander, in developing his own system of strokes, should decide upon strokes he can make best with the pen, and he should then use the same strokes for pencil lettering. Pen strokes can be drawn in the direction the pen is leaned, or at right angles to this, or in curved paths between the two. The pen should never be "pushed" in the direction contrary to the way the pen is leaned, as the point has a tendency to dig into the paper. The strokes should, therefore, be those which are in harmony with the natural and intended use of the pen point.

The regular left-hander assumes a natural position exactly opposite to that of the right-hander, but he will be able to use many of the same strokes as shown for right-handers, with perhaps some minor differences. As prescribed for right-handers, he will draw all vertical strokes downward, and may also draw all horizontal strokes from left to right. He may, however, prefer to draw horizontal strokes from right to left, and he should do this if it seems more natural to him. Also he may wish to change the order of drawing horizontal strokes, so that in the case of the E, for example, the top stroke would be drawn first and the bottom stroke drawn last. If this is done, the pen or pencil will not tend to hide strokes already drawn. Curved strokes will be essentially the same as for right-handers, with perhaps some adjustments of the starting and ending points of the curves.

The hooked-wrist left-hander has a more serious problem, and each such person will have to adopt a system that seems best for his own particular habits. Vertical strokes may be drawn downward as for right-handers, but many hooked-wrist left-handers will find it easier to draw vertical strokes upward. Horizontal strokes will most certainly be drawn from right to left with a finger movement, for the pencil or pen will dig into the paper if pushed in the other direction. Furthermore, the order of horizontal strokes will be to do those at the bottom first, and those at the top last, as described above for the letter E. Since a sheet is lettered from the top downward, the hook-wrist left-hander must move his hand over lines of lettering previously made. Therefore, a piece of paper should be placed over the lettered areas so that smearing of the graphite cannot occur.

If you are left-handed, advise your instructor at once. On examinations in which lettering is tested, use strokes that you have found most suitable for your own use, and letter a statement in the margin to the effect that you are left-handed.

88. Guide Lines. Extremely light horizontal guide lines are necessary to regulate the height of letters. In addition, light vertical or inclined guide lines are needed to keep the letters uniformly vertical or inclined. Guide lines are absolutely essential for good lettering, and should be regarded as a welcome aid, not as an unnecessary requirement. Paradoxically, the better draftsman always uses guide lines, while the unskilled letterer who needs them most is inclined to slight this important step. See Fig. 115.

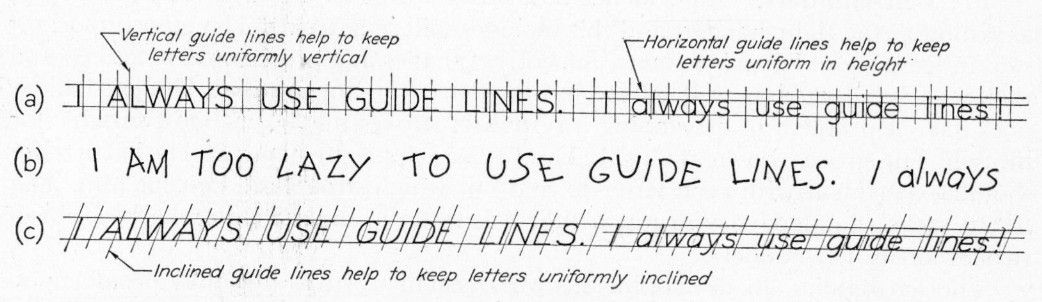

Fig. 115 Guide Lines.

Make guide lines for finished pencil lettering *so lightly, that they need not be erased*, as indeed they cannot be after the lettering has been completed. Use a relatively hard pencil, such as a 4H to 6H, with a long, sharp, conical point, Fig. 20 (c). If the letters are inked, the guide lines may be removed with the artgum after the ink is dry, Fig. 25 (b).

In preparation for ink lettering, complete guide lines should be drawn, and the letters first drawn lightly in pencil. Experienced letterers often draw the complete guide lines, and then letter directly in ink, without first penciling the letters.

89. Guide Lines for Capital Letters. Guide lines for vertical capital letters are shown in Fig. 116. On working drawings, capital letters are commonly made $\frac{1}{8}''$ high, with the space between lines of lettering from three-fifths to the full height of the letters. The vertical guide lines are not used to space the letters (as this should always be done by eye while lettering), but only to keep the letters uniformly vertical, and they should accordingly be drawn at *random*. Where several lines of letters are to be made, these vertical guide lines should be continuous from top to bottom of the lettered area, as shown.

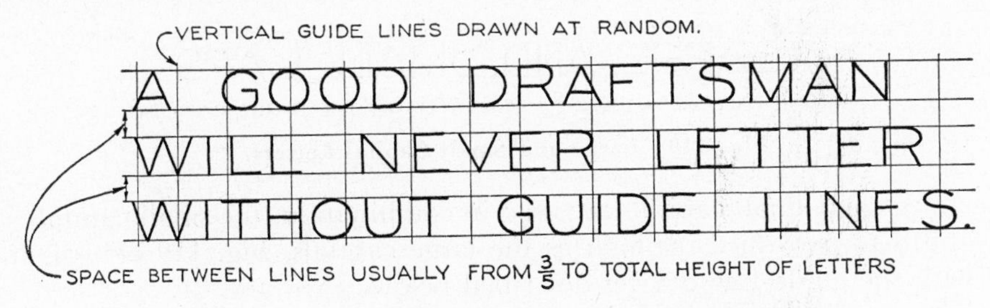

Fig. 116 Guide Lines for Vertical Capital Letters.

Guide lines for inclined capital letters are shown in Fig. 117. The spacing of horizontal guide lines is the same as for vertical capital lettering. The American Standard slope of 2 in 5 (or $67\frac{1}{2}°$ with horizontal) may be established by drawing a "slope triangle," as shown at (a), and drawing the guide lines at random with the T-square and triangle, as shown at (b). Special triangles for the purpose may be used, as shown at (c), or the lines may be drawn with the Braddock-Rowe Lettering Triangle, Fig. 120, or the Ames Lettering Instrument, Fig. 121.

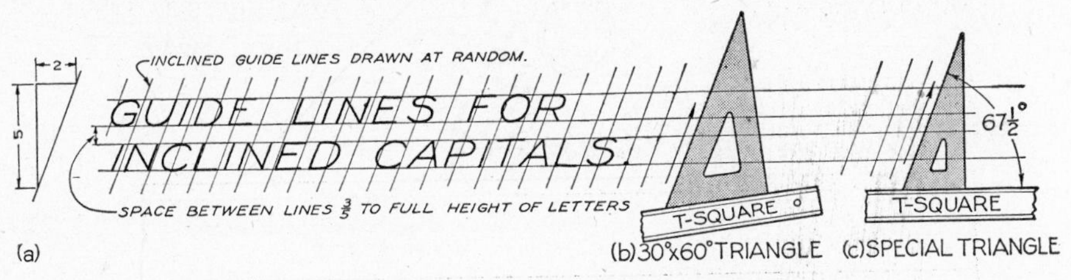

Fig. 117 Guide Lines for Inclined Capital Letters.

A simple method of spacing horizontal guide lines is to use the scale, as shown in Fig. 118 (a), and merely set off a series of $\frac{1}{8}''$ spaces, making both the letters and the spaces between lines of letters $\frac{1}{8}''$ high. Another method of setting off equal spaces, $\frac{1}{8}''$ or otherwise, is to use the bow dividers, as shown at (b).

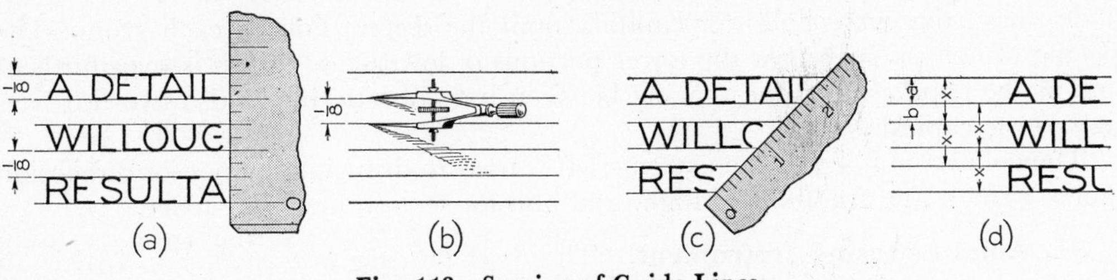

Fig. 118 Spacing of Guide Lines.

If it is desired to make the spaces between lines of letters less than the height of the letters, the methods shown at (c) and (d) will be convenient. At (c) the scale is placed diagonally, the letters in this case being four units high and the spaces between lines of lettering being three units. If the scale is rotated clockwise about the zero mark as a pivot, the height of the letters and the spaces between lines of letters diminish but remain proportional. If the scale is moved counterclockwise, the spaces are increased. The same unequal spacing may be accomplished with the bow dividers, as shown at (d). Let distance x = a + b, and set off x-distances, as shown.

Fig. 119 Large and Small Capital Letters.

When large and small capitals are used in combination, the small capitals should be three-fifths to two-thirds as high as the large capitals, Fig. 119. This is in conformity with the guide-line devices described below, §§90 and 91.

90. Lettering Triangles. *Lettering triangles*, which are available in a variety of shapes and sizes, are provided with sets of holes in which the pencil is inserted and the guide lines produced by moving the triangle with the pencil point along the T-square. The *Braddock-Rowe Lettering Triangle*, Fig. 120, is convenient for drawing

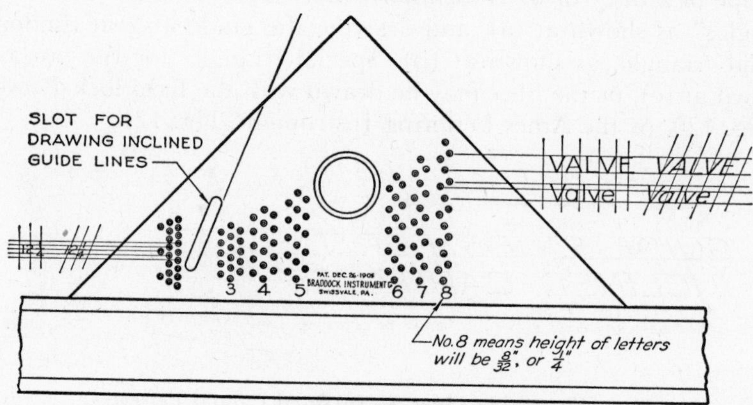

Fig. 120 Braddock-Rowe Lettering Triangle.

guide lines for lettering and dimension figures, and also for drawing section lines. In addition, the triangle is used as a utility 45° triangle. The numbers at the bottom of the triangle indicate heights of letters in thirty-seconds of an inch. Thus, to draw guide lines for $\frac{1}{8}''$ capitals, use the No. 4 set of holes. For lower-case letters, draw guide lines from every hole; for capitals, omit the second hole in each group. The spacing of holes is such that the lower portions of lower-case letters is two-thirds as high as the capitals, and the spacing between lines of lettering is also two-thirds as high as the capitals.

The column of holes at the extreme left is used to draw guide lines for dimension figures $\frac{1}{8}''$ high and fractions $\frac{1}{4}''$ high, and also for section lines $\frac{1}{16}''$ apart.

91. Ames Lettering Instrument. The *Ames Lettering Instrument*, Fig. 121, is an ingenious transparent plastic device composed of a frame holding a disc containing three columns of holes. The vertical distances between the holes may be adjusted quickly to the desired spacing for guide lines or section lines by simply turning the disc to one of the settings indicated at the bottom of the disc. These numbers indicate heights of letters in thirty-seconds of an inch. Thus, for $\frac{1}{8}''$ high letters, the No. 4 setting would be used. The center column of holes is used primarily to draw guide lines for numerals and fractions, the height of the whole number being two units and the height of the fraction four units. The No. 4 setting of the disc will provide guide lines for $\frac{1}{8}''$ whole numbers, with fractions twice as high, or $\frac{1}{4}''$, as shown at (a). Since the spaces are equal, these holes can also be used to draw equally-spaced guide lines for lettering, or to draw section lines.

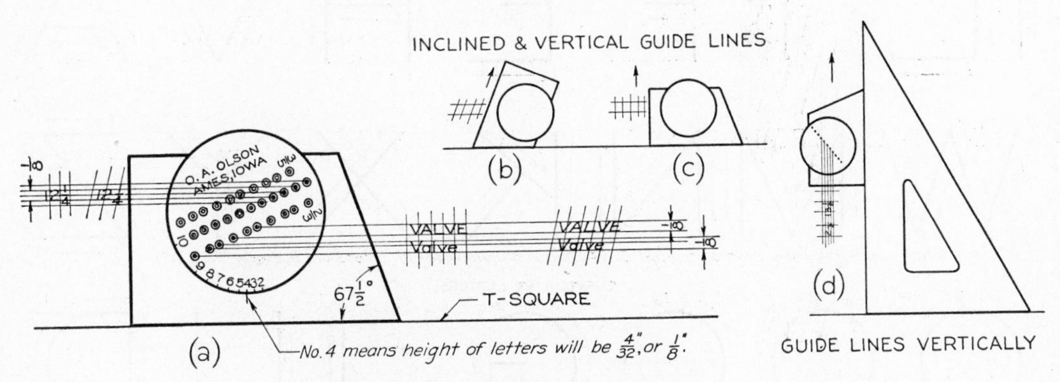

Fig. 121 Ames Lettering Instrument.

The two outer columns of holes are used to draw guide lines for capitals or lower-case letters, the column marked three-fifths being used where it is desired to make the lower portions of lower-case letters three-fifths the total height of the letters, and the column marked two-thirds being used where the lower portion is to be two-thirds the total height of the letters. In each case, for capitals, the middle hole of each set is not used. The two-thirds and three-fifths also indicate the spaces between lines of letters.

The sides of the instrument are used to draw vertical or inclined guide lines, as shown at (b) and (c). The left side of the regularly produced instrument has an angle of 75°, but this angle is not standard and has little use. The instrument is now available (on special order) with the vertical left side, which is convenient for drawing vertical guide lines. This special instrument is the one shown in the figure.

92. Vertical Capital Letters and Numerals. Fig. 122. For convenience in learning the proportions of the letters and numerals, each character is shown in a grid 6 units high. Numbered arrows indicate the order and direction of strokes. The widths of the letters can be easily remembered. The letter I, or the numeral 1, has no width. The W is 8 units wide ($1\frac{1}{3}$ times the height), and is the widest letter in the alphabet. All the other letters or numerals are either 5 or 6 units wide, and it is easy to remember the 6-unit letters because when assembled they spell TOM Q. VAXY. All numerals, except the 1, are 5 units wide.

All horizontal strokes are drawn to the right, and all vertical strokes are drawn downward. Most of the strokes are natural and easy to remember. All the strokes and proportions should be thoroughly learned in the beginning, and it is recommended that this be done by practice-sketching the vertical capital letters on cross-section paper, making the letters 6 squares high.

As shown in Fig. 122, the letters are classified as *straight-line letters* or *curved-line letters.* On the third row, the letters O, Q, C, and G are all based on the circle. The lower portions of the J and U are semi-ellipses, and the right sides of the D, P, R, and B are semicircular. The 8, 3, S, and 2 are all based on the figure 8, which is composed of a small ellipse over a larger ellipse. The 6 and 9 are based on the elliptical zero. The lower part of the 5 is also elliptical in shape.

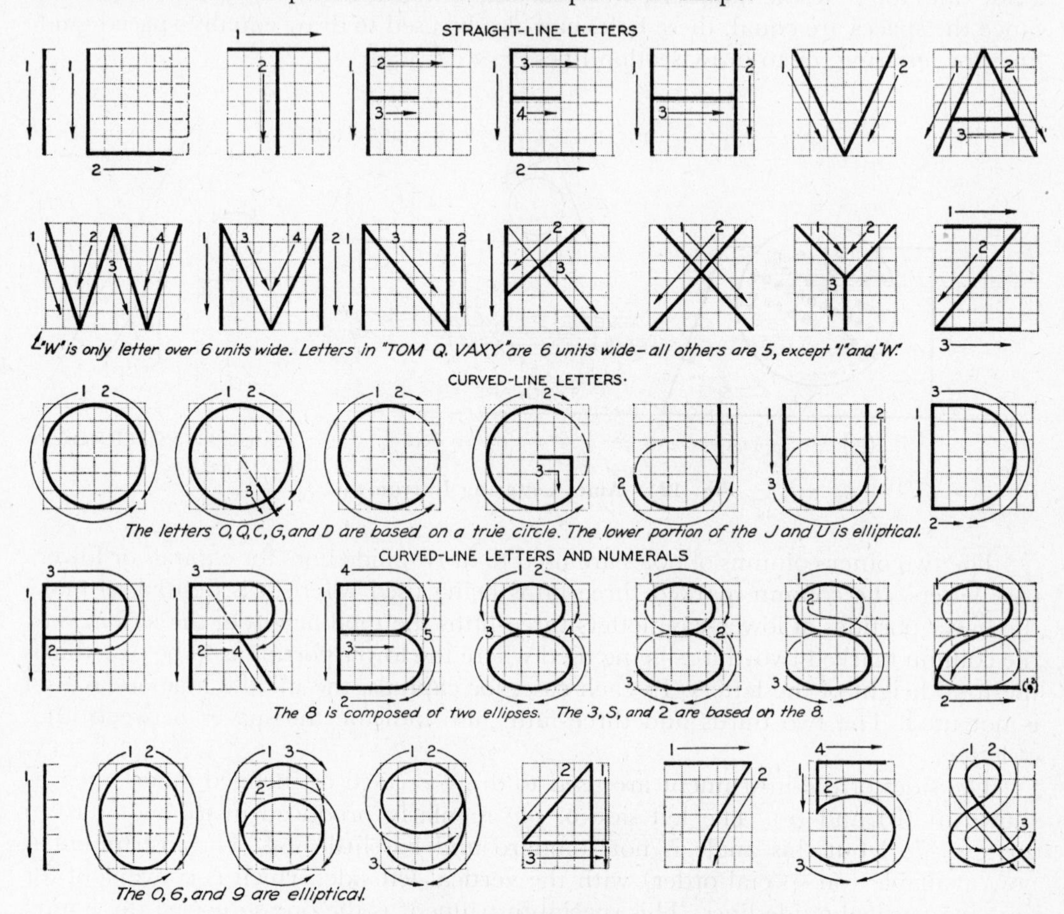

Fig. 122 Vertical Capital Letters and Numerals.

93. Inclined Capital Letters and Numerals. Fig. 123. The order and direction of the strokes and the proportions of the inclined capital letters and numerals are the same as those for the vertical characters. The methods of drawing guide lines for inclined capital letters are given in §89, and for numerals in §94.

Inclined capitals may be regarded as oblique projections, §§481 and 482, of vertical capitals. In the inclined letters, the circular parts become elliptical, the major axes of the ellipses based on the *O* making an angle of 45° with horizontal. The letters are classified as *straight-line letters* or *curved-line letters*, most of the curves being elliptical in shape. Therefore, skill in inclined lettering depends somewhat upon the ability to form smooth ellipses that appear to "lean" properly to the right.

The letters having sloping sides, such as the *V*, *A*, *W*, *X*, and *Y* are also a source of difficulty for most beginners. The letters should be made symmetrically about an imaginary inclined center line. If this is done, the left side of the *V* and the right side of the *A* will be practically vertical, while the opposite sides will slope at less than 60° with horizontal.

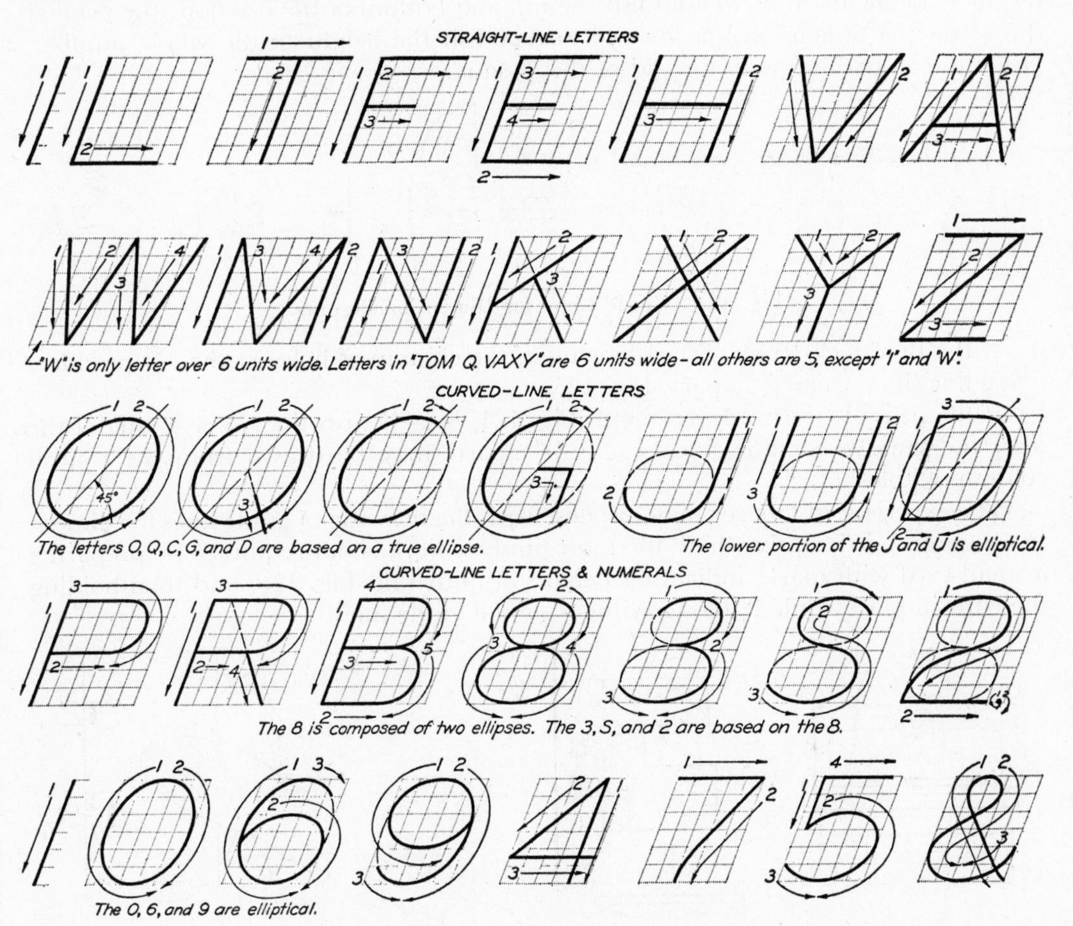

Fig. 123 Inclined Capital Letters and Numerals.

94. Guide Lines for Whole Numbers and Fractions. Complete guide lines should be drawn for whole numbers and fractions, especially by beginners. This means that both horizontal and vertical guide lines, or horizontal and inclined guide lines, should be drawn. Even the expert letterer will be able to do better lettering if he uses guide lines, and for this reason he is more likely to use them than the beginner who considers them "too much trouble." The guide lines, of course, should be drawn extremely lightly, with a hard pencil, 4H to 6H.

Draw five equally-spaced guide lines for whole numbers and fractions, Fig. 124. Thus, fractions are twice the height of the corresponding whole numbers. Make the numerator and the denominator each about three-fourths as high as the whole number, to allow ample clear space between them and the fraction bar, as shown.

For dimensioning, the most commonly-used height for whole numbers is $\frac{1}{8}''$, and for fractions $\frac{1}{4}''$, as shown at (a) and (c). These spaces may be easily set off directly with the scale, as shown. After the horizontal guide lines have been drawn, add vertical or inclined guide lines spaced at random, (b) and (c).

Another simple method is to set the bow dividers at $\frac{1}{8}''$ as shown at (d); then, with one point on the dimension line, set off $\frac{1}{8}''$ above the line and swing down and set off $\frac{1}{8}''$ below the line, to establish the top and bottom of the fraction; then center the $\frac{1}{8}''$ on the dimension line by eye to establish the height of the whole number. Draw the guide lines with the aid of the T-square.

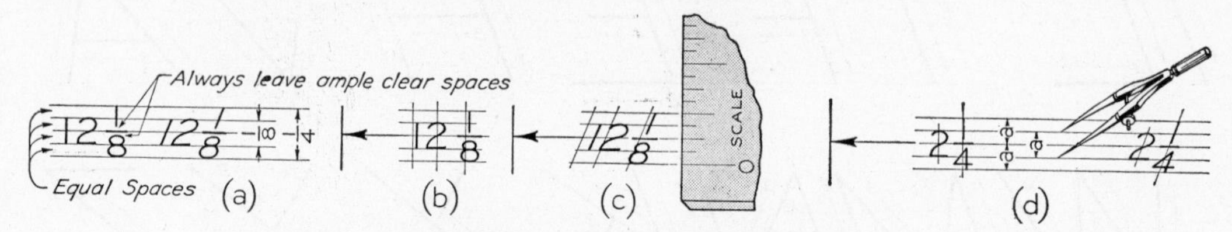

Fig. 124 Guide Lines for Dimension Figures.

If the Braddock-Rowe Triangle is used, the column of holes at the left produces five guide lines, each $\frac{1}{16}''$ apart, Fig. 125.

If the Ames Lettering Instrument, Fig. 121, is used with the No. 4 setting of the disc, the same five guide lines, each $\frac{1}{16}''$ apart, may be drawn from the central column of holes.

The experienced letterer may dispense with the drawing of guide lines for dimension figures, particularly where the most finished work is not required, by preparing a small card with marks indicating heights of numerals, Fig. 126, and then holding the card in place while lettering without actual guide lines.

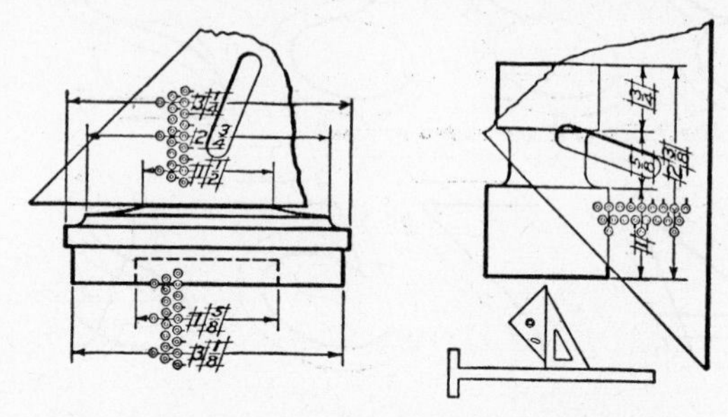

Fig. 125 Use of Braddock-Rowe Triangle.

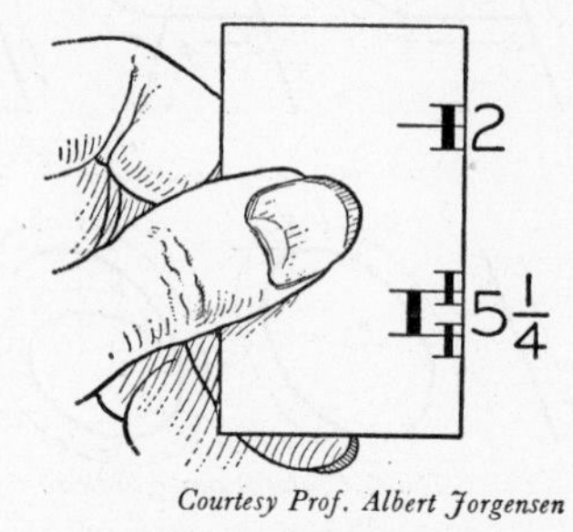

Courtesy Prof. Albert Jorgensen

Fig. 126 Letter-Height Indicator.

Some of the most common errors in lettering fractions are illustrated in Fig. 127. Never let numerals touch the fraction bar, (a). Center the denominator under the numerator, (b). Never use an inclined fraction bar, (c), except when lettering in a narrow space, as in a parts list. Make the fraction bar slightly longer than the widest part of the fraction, (d).

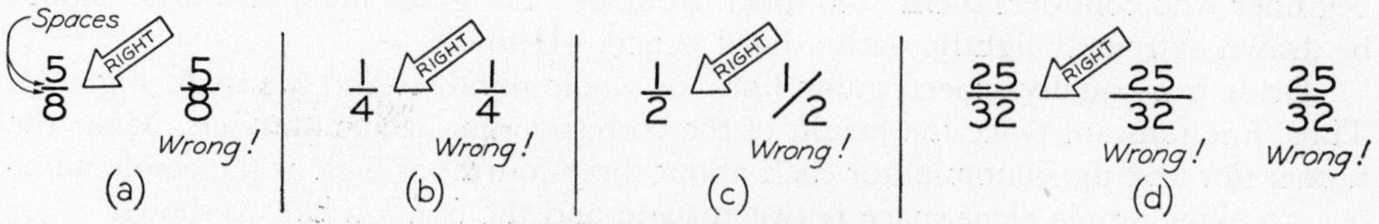

Spaces — RIGHT — Wrong! (a) | RIGHT — Wrong! (b) | RIGHT — Wrong! (c) | RIGHT — Wrong! Wrong! (d)

Fig. 127 Common Errors.

95. Guide Lines for Lower-Case Letters. Lower-case letters have four horizontal guide lines, called the *cap line*, *waist line*, *base line*, and *drop line*, Fig. 128 (a). Strokes of letters that extend up to the cap line are called *ascenders*, and those that extend down to the drop line, *descenders*. Since there are only five letters that have descenders, the drop line is little needed and is usually omitted. In spacing horizontal guide lines, space a may vary from three-fifths to two-thirds of space b. Spaces c are equal, as shown.

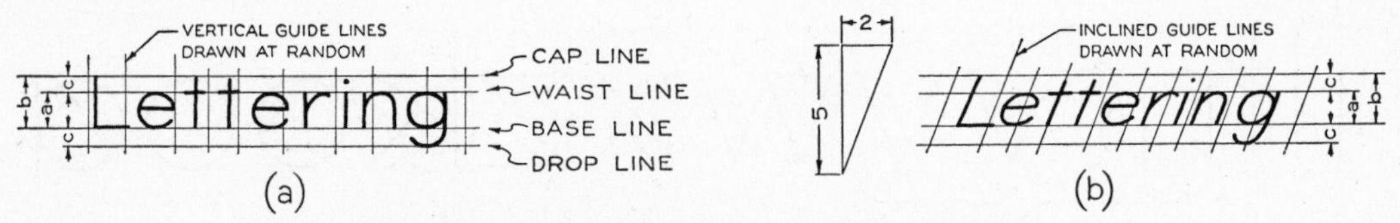

Fig. 128 Guide Lines for Lower-Case Letters.

If it is desired to set off guide lines for letters $\frac{3}{16}''$ high with the scale (using the two-thirds ratio), it is only necessary to set off equal spaces each $\frac{1}{16}''$, Fig. 129 (a). The lower portion of the letter thus would be $\frac{1}{8}''$, and the space between lines of letters would also be $\frac{1}{8}''$. If the scale is placed at an angle, the spaces will diminish but remain equal, (b). Thus, this method may be easily used for various heights of lettering.

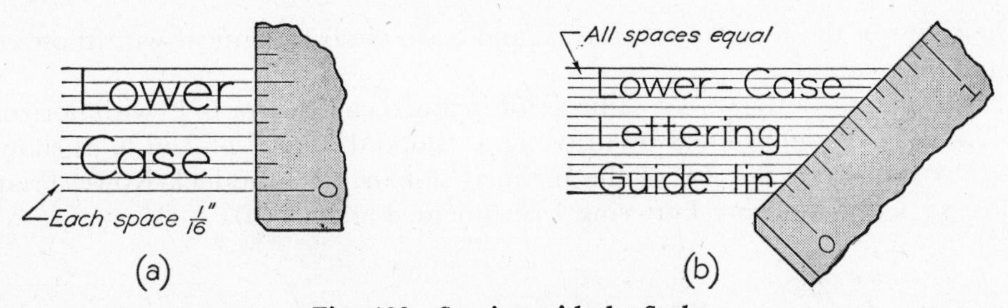

Fig. 129 Spacing with the Scale.

The Braddock-Rowe Triangle, Fig. 120, and the Ames Lettering Instrument, Fig. 121, produce guide lines for lower-case letters as described here, and are highly recommended.

In addition to horizontal guide lines, vertical or inclined guide lines, drawn at random, should always be used to keep the letters uniformly vertical or inclined, Fig. 128.

96. Vertical Lower-Case Letters. Fig. 130. Vertical lower-case letters are used largely on map drawings, and very seldom on machine drawings. The shapes are based upon a repetition of the circle or circular arc and the straight line, with some variations. The lower part of the letter is usually two-thirds the height of the capital letter.

Stroke 3 of the e is slightly above mid-height. The crosses on the f and t are on the waist line and are symmetrical with respect to strokes 1. The curved strokes of h, m, n, and r intersect strokes 1 approximately two-thirds of the distance from the base line to the waist line.

The descenders of the g, j, and y terminate in curves that are tangent to the drop line, while those of p and q terminate in the drop line without curves.

97. Inclined Lower-Case Letters. Fig. 131. The order and direction of the strokes and the proportions of inclined lower-case letters are the same as those of vertical lower-case letters. The inclined lower-case letters may be regarded, like the inclined capital letters, as oblique projections of vertical letters, in which all circles in the vertical alphabet become ellipses in the inclined alphabet. As in the inclined capital letters, all ellipses have their major axes inclined at an angle of 45° with horizontal.

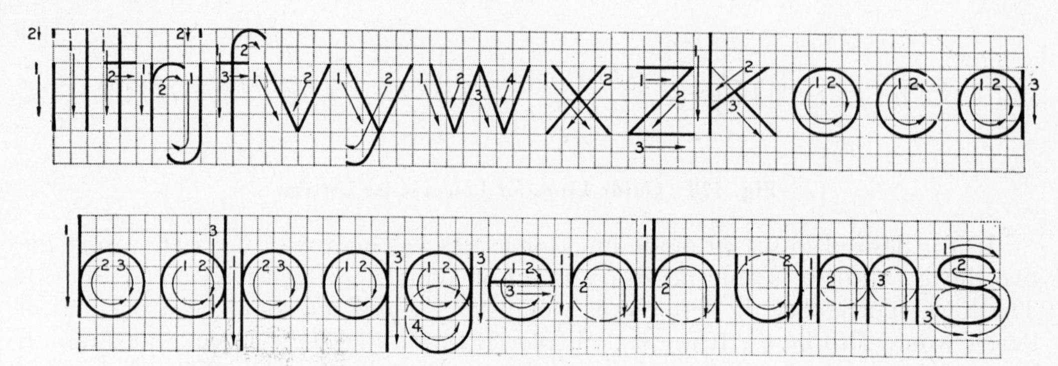

Fig. 130 Vertical Lower-Case Letters.

The forms of the letters *c*, *o*, *s*, *v*, *w*, *x*, and *z* are almost identical with those of the corresponding capitals.

The slope of the letters is the same as for inclined capitals, or $67\frac{1}{2}°$ with horizontal. The slope may be determined by drawing a "slope triangle" of 2 in 5, as shown in Fig. 128 (b), or with the aid of the inclined slot in the Braddock-Rowe Triangle, Fig. 120, or with the Ames Lettering Instrument, Fig. 121 (b).

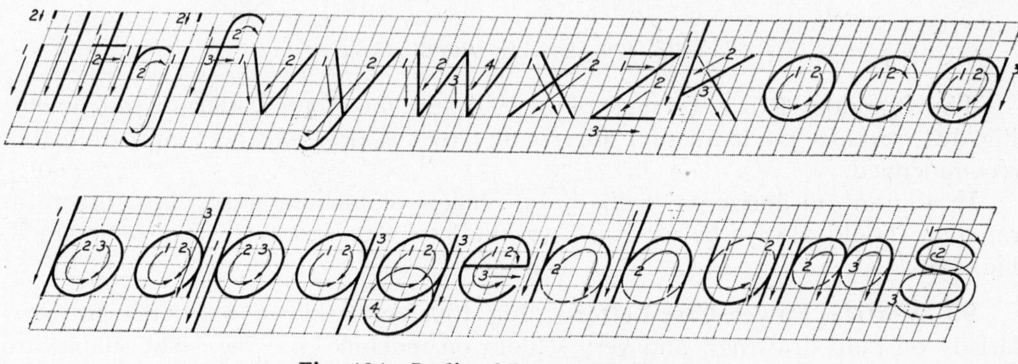

Fig. 131 Inclined Lower-Case Letters.

98. Spacing of Letters and Words. Uniformity in spacing of letters is a matter of equalizing spaces by eye. *The background areas between letters, not the distance between them, should be approximately equal.* In Fig. 132 (a) the actual distances are equal, but the letters do not appear equally spaced. At (b) the distances are intentionally unequal, but the background areas between letters are approximately equal, and the result is an even and pleasing spacing.

Some combinations, such as LT and VA, may even have to be slightly overlapped to secure good spacing. In some cases the width of a letter may be decreased. For example, the lower stroke of the L may be shortened when followed by A.

Space words well apart, but space letters closely within words. Make each word a compact unit well-separated from adjacent words. For either upper-case or lower-case lettering, make the spaces between words approximately equal to a capital O, Fig. 133. Avoid spacing letters too far apart and words too close together, as shown at (b). Samples of good spacing are shown in Fig. 114.

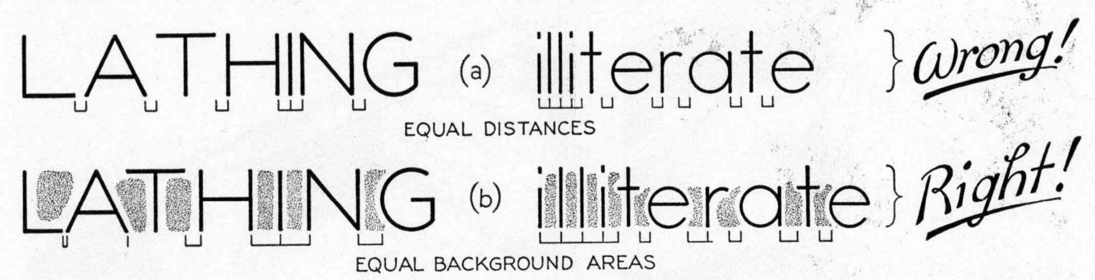

Fig. 132 Spacing Between Letters.

When it is necessary to letter to a stop-line as in Fig. 134 (a), space each letter from *right* to *left*, as shown in step II, estimating the widths of the letters by eye. Then letter from *left to right*, as shown at III, and finally erase the spacing marks.

When it is necessary to space letters symmetrically about a center line, (b), which is frequently the case in titles, Figs. 139-141, number the letters as shown, with the space between words considered as one letter. Then place the middle letter on center, making allowance for narrow letters (I's) or wide letters (W's) on either side. The X at (b) is placed slightly to the left of center to compensate for the letter I, which has no width. Check with the dividers to make sure that distances a are exactly equal.

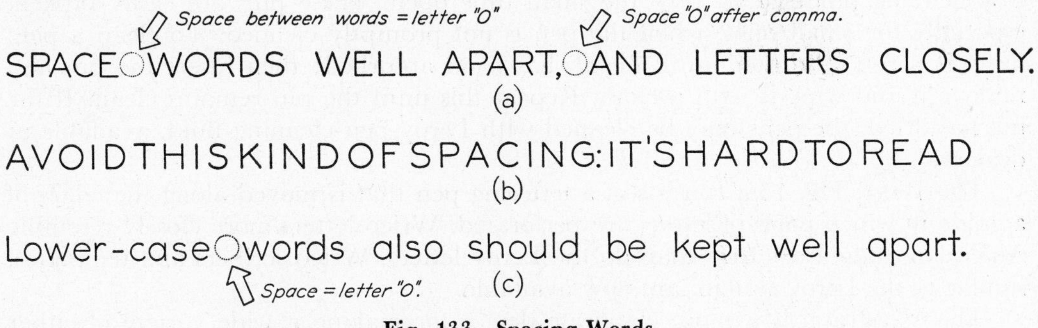

Space between words = letter "O" *Space "O" after comma.*

SPACE○WORDS WELL APART,○AND LETTERS CLOSELY.

(a)

AVOIDTHISKINDOFSPACING:IT'SHARDTOREAD

(b)

Lower-case○words also should be kept well apart.

Space = letter "O". (c)

Fig. 133 Spacing Words.

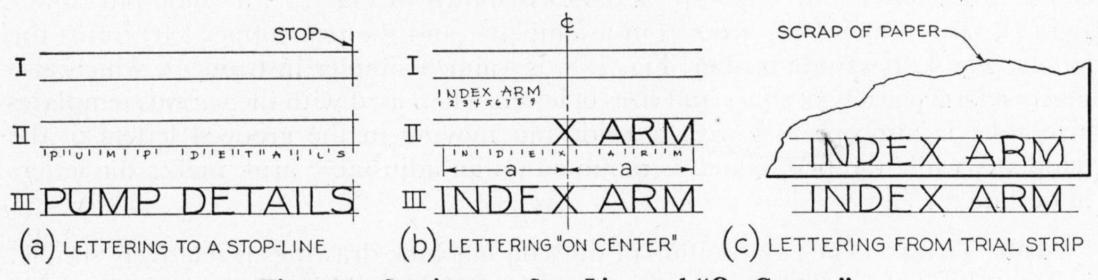

Fig. 134 Spacing to a Stop-Line and "On Center."

Another method is to letter roughly a trial line of lettering along the bottom edge of a scrap of paper, place it in position immediately above, as shown at (c), and then letter the line in place. Be sure to use guide lines for the trial lettering.

If the lettering is being done on tracing paper or cloth, the trial letters can be placed underneath, arranged for lettering to a stop line or "on center," and then lettered directly over or with slight improvement as may be desired, Fig. 139.

Courtesy Keuffel & Esser Co.

Fig. 135 Leroy Lettering Instrument.

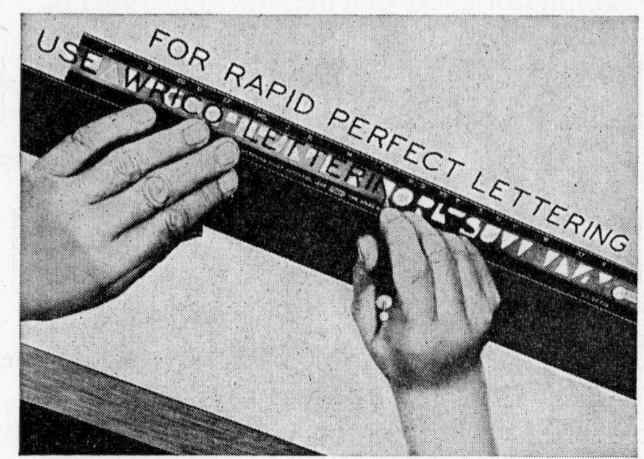

Courtesy Wood-Regan Instrument Co., Inc.

Fig. 136 Use of Wrico Pen and Guide.

99. Lettering Devices. The *Leroy Lettering Instrument*, Fig. 135, is perhaps the most widely used lettering device. A guide pin follows grooved letters in a template, and the inking point moves on the paper. By adjusting the arm on the instrument, the letters may be made vertical or inclined. A number of templates and sizes of pens is available, including templates for a wide variety of "built-up" letters similar to those made by the Varigraph and Letterguide, described below. Inside each pen is a cleaning pin used to keep the small tube open. These pins are easily broken, especially the small ones, when the pen is not promptly cleaned. To clean a pen, draw it across a blotter until all ink has been absorbed; then insert the pin and remove it and wipe it with a cloth. Repeat this until the pin remains clean. If the ink has dried, the pens may be cleaned with Leroy pen-cleaning fluid, available at dealers.

The *Wrico*, Fig. 136, consists of a lettering pen that is moved along the edges of a guide in which parts of letters are perforated. Wrico letters more closely resemble American Standard letters than do the Leroy letters. Wrico scribers and templates, similar to the Leroy system, are now available.

The Varigraph is a more elaborate device for making a wide variety of either single-stroke letters or "built-up" letters. As shown in Fig. 137, a guide pin (lower right) is moved along the grooves in a template, and the pen (upper left) forms the letters. The Letterguide scriber, Fig. 138, is a much simpler instrument, which also makes a large variety of styles and sizes of letters when used with the various templates available. It also operates with a guide pin moving in the grooved letters of the template, while the pen, which is mounted on an adjustable arm, makes the letters in outline.

100. Titles. The composition of titles on machine drawings is relatively simple. In most cases, the title and related information are lettered in "title boxes" or "title strips," which are printed directly on the drawing paper, tracing paper, or cloth, Figs. 719-722. The main drawing title is usually centered in a rectangular space. This may be done by the method shown in Fig. 134 (b); or if the lettering is being done on tracing paper or cloth, the title may be lettered first on scrap paper and then placed underneath the tracing, as shown in Fig. 139, and then lettered directly over.

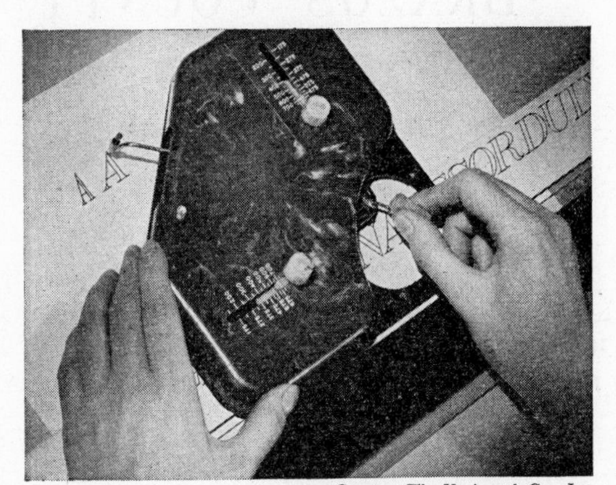

Courtesy The Varigraph Co., Inc.

Fig. 137 The Varigraph.

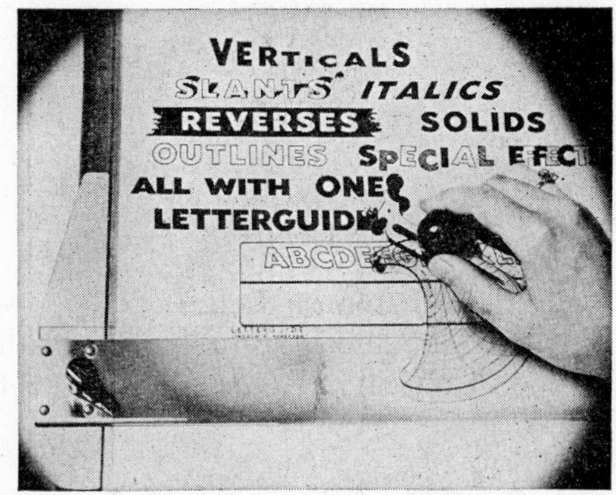

Courtesy Letterguide

Fig. 138 Letterguide.

If a title box is not used, the title of a machine drawing may be lettered in the lower-right corner of the sheet as a "balanced title," Fig. 140 (a). A balanced title is simply one that is arranged symmetrically about an imaginary center line. These titles take such forms as the rectangle, the oval, the inverted pyramid, or any other simple symmetrical form.

On display drawings, or on highly finished maps or architectural drawings, titles may be composed of filled-in letters, usually Gothic or Roman, Fig. 141.

In any kind of title, the most important words are given the most prominence by making the lettering larger, heavier, or both. Other data, such as scale, date, etc., are not so important, and should not be prominently displayed.

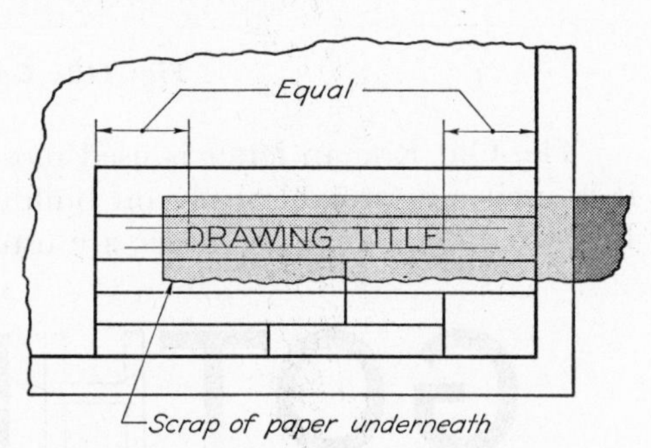

Fig. 139 Centering Title in Title Box.

101. Gothic Letters. Fig. 142. Among the many forms of Gothic styles, including Old English, German Gothic, etc., the so-called *sans-serif* Gothic letter is the only one of interest to draftsmen. It is from this style that the modern single-stroke engineering letters, discussed in the early part of this chapter, are derived. While they are admittedly not as beautiful as many other styles, they are very legible and comparatively easy to make.

Sans-serif Gothic letters should be used, therefore, on drawings where legibility and not beauty is the determining factor. They should be drawn in outline and then filled in, Fig. 143, the thickness of the stems being from one-fifth to about one-tenth the height of the letter. An attractive letter may be produced by making heavy outlines, and not filling in, as for the letter H shown at (a). A slight spur may be added to the ends of the stem, as for the letter T at (a). An example of condensed Gothic is shown at (b).

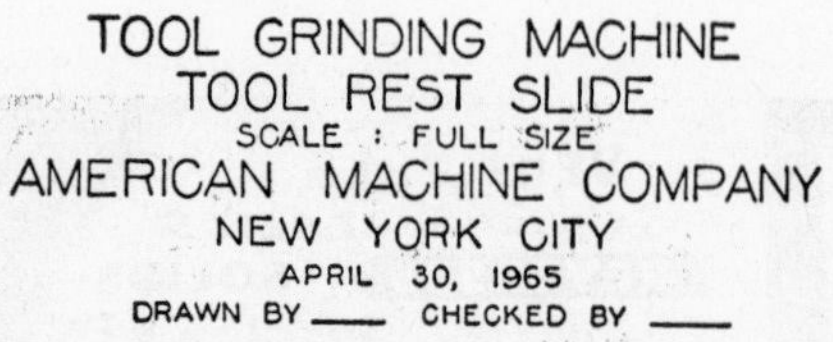

TOOL GRINDING MACHINE
TOOL REST SLIDE
SCALE : FULL SIZE
AMERICAN MACHINE COMPANY
NEW YORK CITY
APRIL 30, 1965
DRAWN BY ___ CHECKED BY ___

Fig. 140 Balanced Machine-Drawing Title.

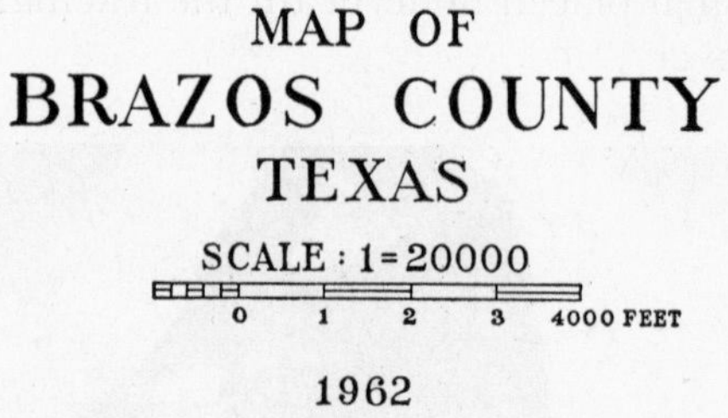

MAP OF
BRAZOS COUNTY
TEXAS
SCALE : 1=20000
0 1 2 3 4000 FEET
1962

Fig. 141 Balanced Map Title.

102. Old Roman Letters. Fig. 144. The Old Roman letter is the basis of all of our letters, and is still regarded as the most beautiful. The letters on the base of Trajan's Column in Rome are regarded by many as the finest example of Old Roman letters.

ABCDEFGHIJKLM
NOPQRSTUVWXYZ
&1234567890MAR

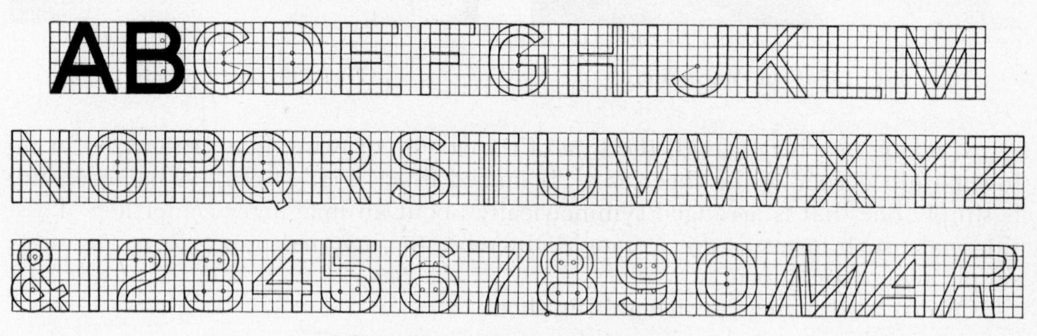

Fig. 142 Gothic Capital Letters.

The Old Roman letter is used mostly by architects. Because of its great beauty, it is used almost exclusively on buildings and for inscriptions on bronze or stone. Full-size "details" of the letter are usually drawn for such inscriptions.

GOTHIC GOTH
(a)
(b)

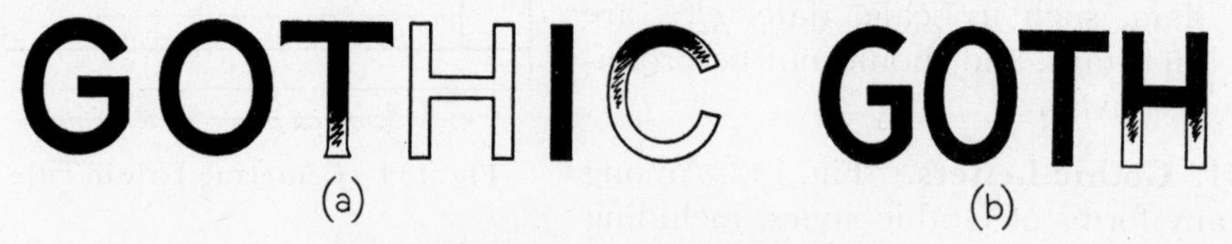

Fig. 143 Gothic Letter Construction.

Originally this letter was made on manuscript with a broad-point reed pen, Fig. 6; the wide stems were produced by downward strokes, and the narrow portions by horizontal strokes. A brief examination of any Roman letter will show why certain strokes are wide, while others are narrow.

Several styles of steel broad-nib pens are available and are suitable for making Roman, Gothic, or Text letters, Figs. 105 and 145. If necessary, an ordinary pen may be used to "touch up" after using the broad-nib pen, or to add fine-line flourishes, as in Text letters.

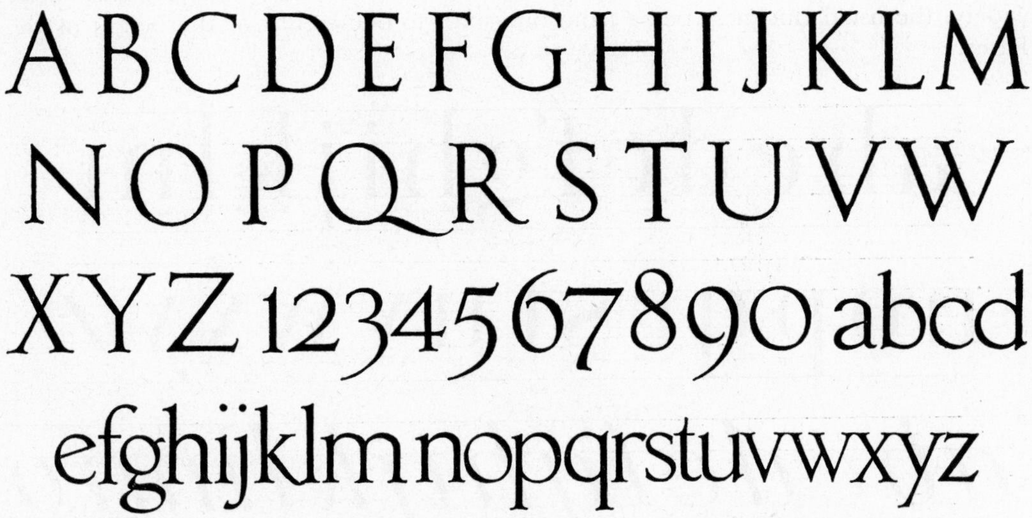

ABCDEFGHIJKLM
NOPQRSTUVW
XYZ 1234567890 abcd
efghijklmnopqrstuvwxyz

Fig. 144 Old Roman Capitals, with Numerals and Lower-Case of Similar Design.

As in the case of Gothic, Old Roman letters may be drawn in outline and filled in, or may be left in outline.

103. Modern Roman Letters. Figs. 146 and 147. The Modern Roman, or simply "Roman," letters were evolved during the eighteenth century by the type founders; the letters used in most modern newspapers, magazines, and books are of this style. The text of this book is set in Modern Roman Letters. These letters are often used on maps, especially for titles. They may be drawn in outline and then filled in, as shown in Fig. 146, or they may be produced with one of the broad-nib pens shown in Fig. 112.

Roman

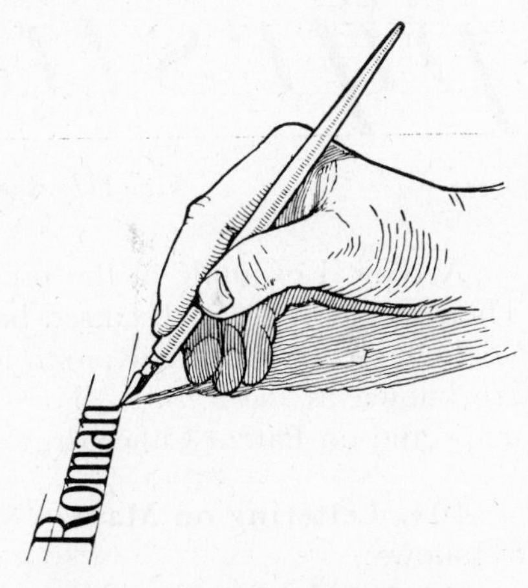

Fig. 145 Use of Broad-Nib Pen.

ABCDEFGHIJKLM
NOPQRSTUVWXYZ
&1234567890ABLK

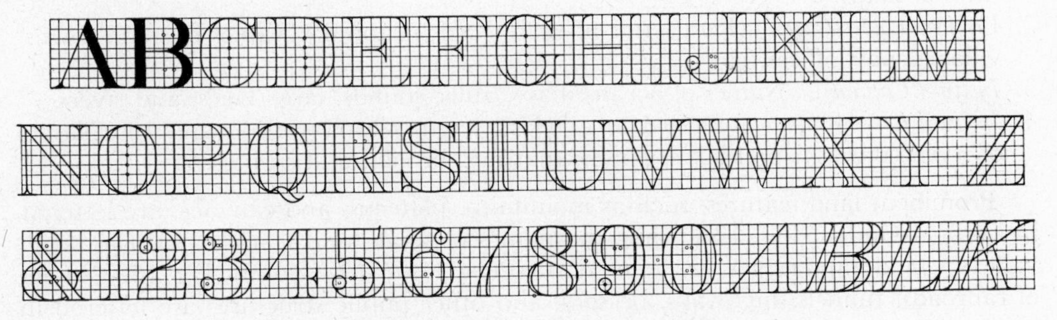

Fig. 146 Modern Roman Capitals and Numerals.

If drawn in outline and filled in, the straight lines may be drawn with the ruling pen, and the circular curves drawn with the bow pen. The fillets and other non-circular curves are drawn freehand. The thickness of the stem, or broad stroke, varies widely, the usual thickness being from one-sixth to one-eighth of the height of the letter.

Fig. 147 Lower-Case Modern Roman Letters.

A typical example of the use of Modern Roman in titles is shown in Fig. 141. Their use on maps is discussed below.

Lower-case Modern Roman letters are shown in Fig. 147. The lower-case italics are known as *stump letters.* They are easily made freehand, and are often used on maps and on Patent Office drawings, §724.

104. Lettering on Maps. Modern Roman letters are generally used on maps, as follows:

1. *Vertical Capitals.* Names of states, countries, townships, capitols, large cities, and titles of maps. See §100.
2. *Vertical Lower-Case.* (First letter of each word a capital.) Names of small towns, villages, post offices, etc.
3. *Inclined Capitals.* Names of oceans, bays, gulfs, sounds, large lakes, and rivers.
4. *Inclined Lower-Case, or "Stump Letters."* (First letter of each word a capital.) Names of rivers, creeks, small lakes, ponds, marshes, brooks, and springs.

Prominent land features, such as mountains, plateaus, and canyons, are lettered in vertical Gothic, Fig. 122, while the names of small land features, such as small valleys, islands, and ridges, are lettered in vertical lower-case Gothic, Fig. 130. Names of railroads, tunnels, highways, bridges, and other public structures are lettered in inclined Gothic capitals, Fig. 123.

105. Greek Alphabet. Greek letters are often used as symbols in both mathematics and technical drawing by the engineer. A Greek alphabet, showing both upper-case and lower-case letters, is given for reference purposes in Fig. 148.

A α alpha	Z ζ zeta	Λ λ lambda	Π π pi	
B β beta	H η eta	M μ mu	P ρ rho	Φ φ phi
Γ γ gamma	Θ θ theta	N ν nu	Σ ς, σ sigma	X χ chi
Δ δ delta	I ι iota	Ξ ξ xi	T τ tau	Ψ ψ psi
E ε epsilon	K κ kappa	O ο omicron	Υ υ upsilon	Ω ω omega

Fig. 148 Greek Alphabet.

106. Lettering Exercises. Layouts for lettering practice are given in Figs. 149-152. Draw complete horizontal and vertical or inclined guide lines *very lightly.* Draw the vertical or inclined guide lines through the full height of the lettered area of the sheet. For practice in ink lettering, the last two lines and the title strip on each sheet may be lettered in ink, if assigned by the instructor. Omit all dimensions.

Lettering sheets in convenient form for lettering practice may be found in *Technical Drawing Problems*, by Giesecke, Mitchell, and Spencer, and in *Technical Drawing Problems, Series 2*, by Spencer and Grant, both designed to accompany this text, and published by The Macmillan Company. The lettering sections of these two workbooks are also published separately as *Lettering Exercises* and *Technical Lettering Practice*, respectively.

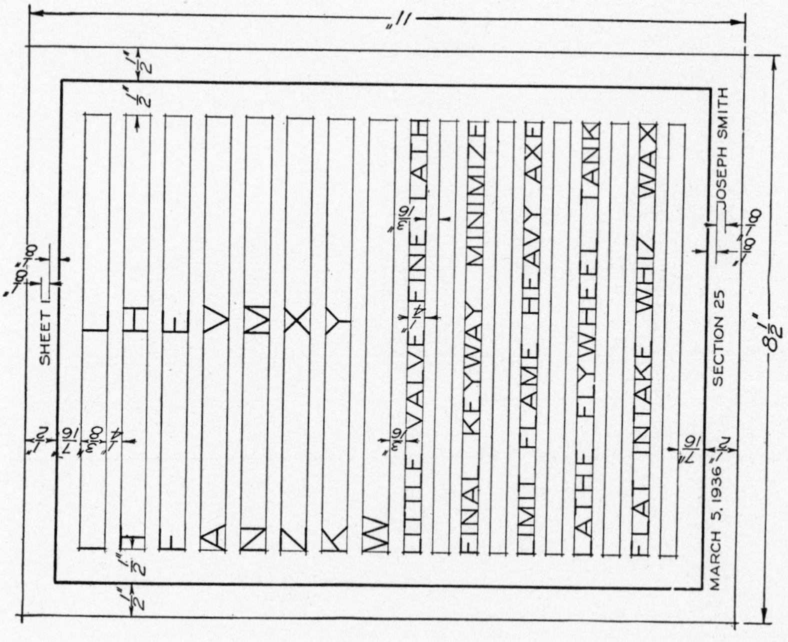

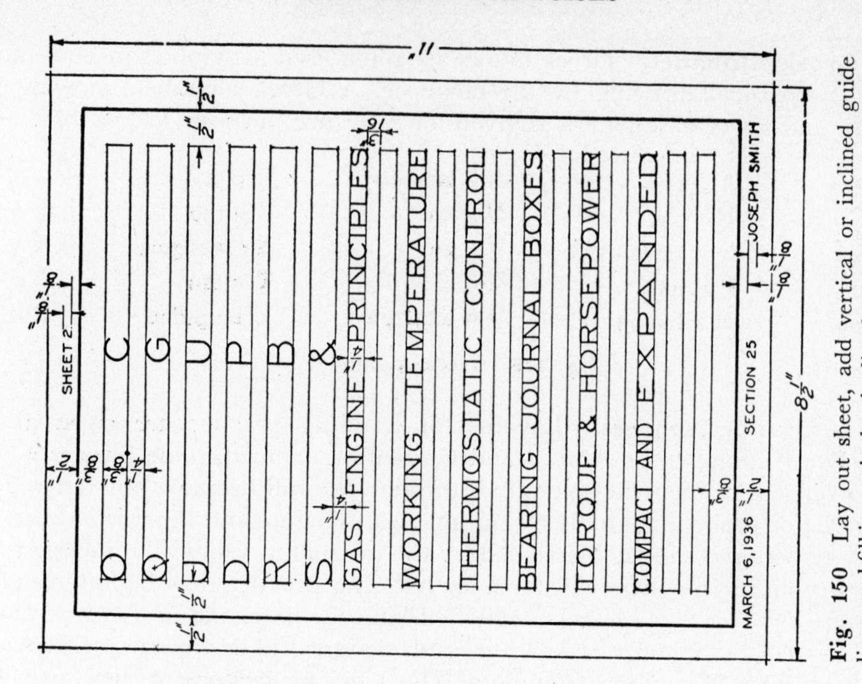

Fig. 150 Lay out sheet, add vertical or inclined guide lines, and fill in vertical or inclined capital letters as assigned.

Fig. 149 Lay out sheet, add vertical or inclined guide lines, and fill in vertical or inclined capital letters as assigned.

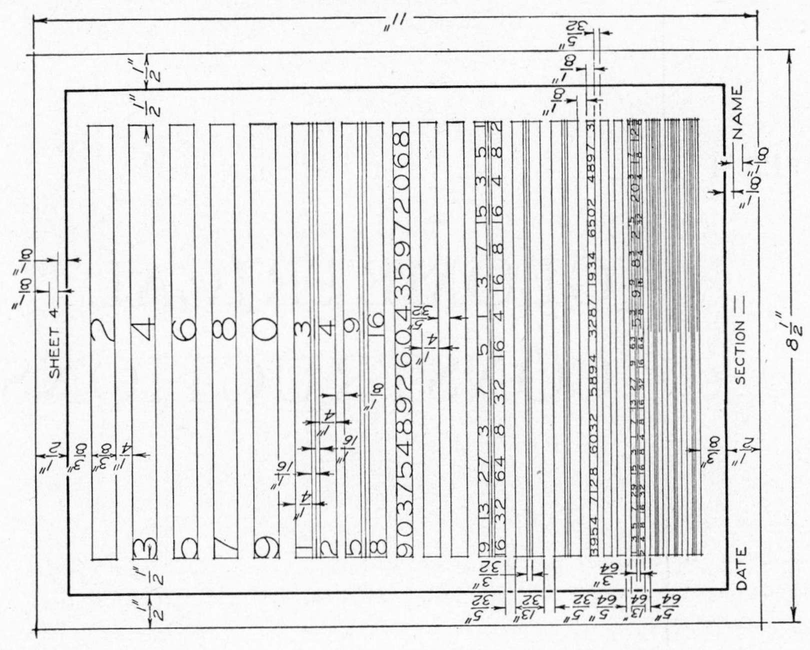

Fig. 152 Lay out sheet, add vertical or inclined guide lines, and fill in vertical or inclined numerals as assigned.

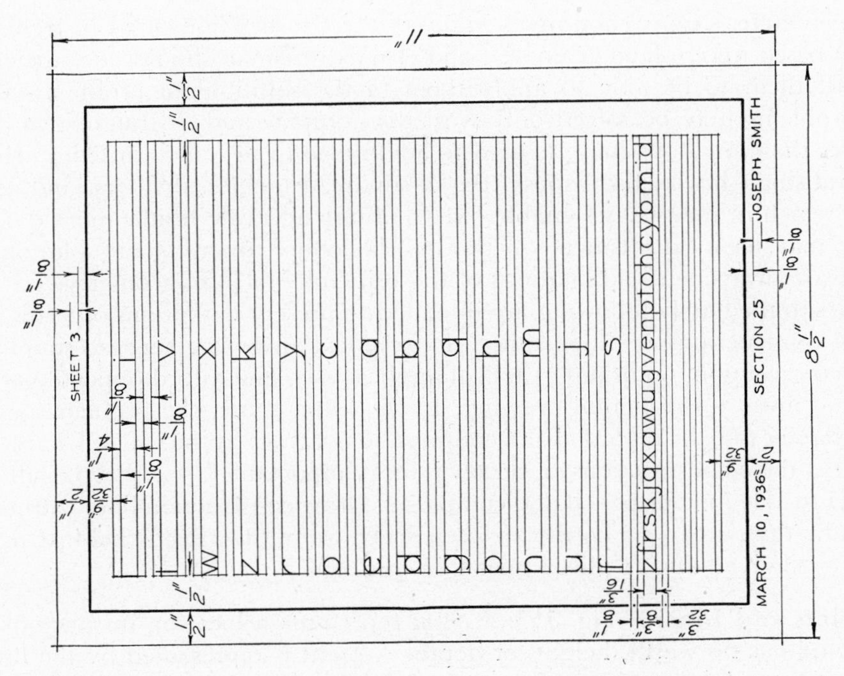

Fig. 151 Lay out sheet, add vertical or inclined guide lines, and fill in vertical or inclined lower-case letters as assigned.

CHAPTER 4

GEOMETRICAL CONSTRUCTIONS

107. Geometrical Constructions. Many of the constructions used in technical drawing are based upon plane geometry, and every draftsman should be sufficiently familiar with them to be able to apply them to the solutions of problems. Pure geometry problems may be solved only with the compass and a straightedge, and in some cases these methods may be used to advantage in technical drawing. However, the draftsman has at his disposal the T-square, triangles, dividers, and other equipment, which in many cases enable him to obtain accurate results more quickly by what we may term "draftsmen's methods." Therefore, many of the solutions in this chapter are draftsmen's adaptations of the principles of pure geometry.

The pages immediately following are intended to present definitions of terms and geometrical constructions of importance in technical drawing, suggest simplified methods of construction, point out practical applications, and afford opportunity for practice in accurate instrumental drawing. In the latter sense, the problems at the end of this chapter may be regarded as a continuation of those at the end of Chapter 2.

In drawing these constructions, accuracy is most important. Use a sharp medium-hard lead (H to 3H) in your pencil and compasses. Draw construction lines extremely light—so light that they can hardly be seen when your drawing is held at arm's length. Draw all given and required lines thin but dark.

108. Points and Lines. Fig. 153. A *point* represents a location in space or on a drawing, and has no width, height, or depth. A point is represented by the intersection of two lines, (a), by a short cross-bar on a line, (b), or by a small cross, (c). Never represent a point by a simple dot on the paper.

A line is defined by Euclid as "that which has length without breadth." A *straight line* is the shortest distance between two points, and is commonly referred to simply as a "line." If the line is indefinite in extent, the length is a matter of convenience, and the end points are not fixed, (d). If the end points of the line are significant, they must be marked by means of small mechanically drawn cross-bars, (e). Other common terms are illustrated from (f) to (h). Either straight lines or curved lines

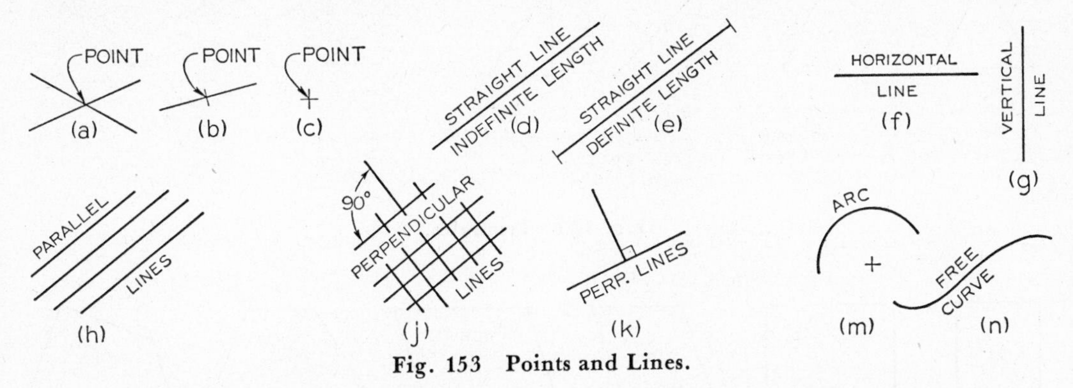

Fig. 153 Points and Lines.

are parallel if the shortest distance between them remains constant. A common symbol for parallel lines is ||, and for perpendicular lines is ⊥ (singular) or ⊥s (plural). Two perpendicular lines may be marked with a "box" to indicate perpendicularity, as shown at (k). Such symbols may be used on sketches, but not on production drawings.

109. Angles. Fig. 154. An *angle* is formed by two intersecting lines. A common symbol for angle is ∠ (singular) or ∠s (plural). There are 360 *degrees* (360°) in a full circle, as shown at (a). A degree is divided into 60 *minutes* (60′), and a minute is divided into 60 *seconds* (60″). Thus, 37° 26′ 10″ is read: 37 degrees, 26 minutes, and 10 seconds. When minutes alone are indicated, the number of minutes should be preceded by 0°, as 0° 20′.

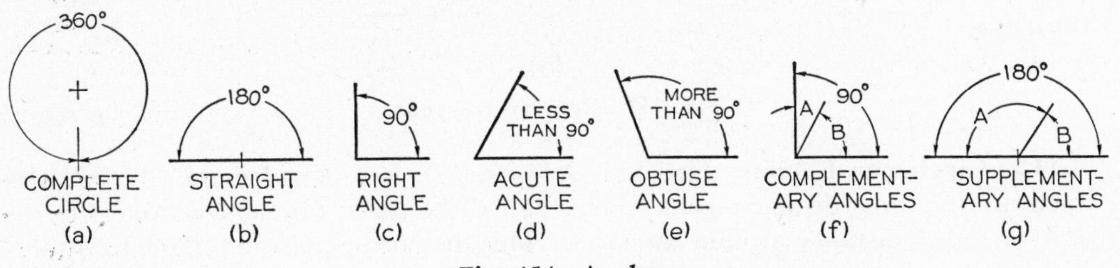

Fig. 154 Angles.

The different kinds of angles are illustrated from (b) to (e). Two angles are *complementary*, (f), if they total 90°, and are *supplementary*, (g), if they total 180°. Most angles used in technical drawing can be drawn easily with the T-square and triangles, Fig. 34. To draw odd angles, use the protractor, Fig. 35. For considerable accuracy, use a *vernier protractor*, or the *tangent*, *sine*, or *chord* methods, §127.

110. Triangles. Fig. 155. A *triangle* is a plane figure bounded by three straight sides, and the sum of the interior angles is always 180°. A right triangle, (d), has one 90° angle, and the square of the hypotenuse is equal to the sum of the squares of the two sides, (e). As shown at (f), any triangle inscribed in a semicircle is a *right triangle* if the hypotenuse coincides with the diameter.

111. Quadrilaterals. Fig. 156. A *quadrilateral* is a plane figure bounded by four straight sides. If the opposite sides are parallel, the quadrilateral is also a *parallelogram*.

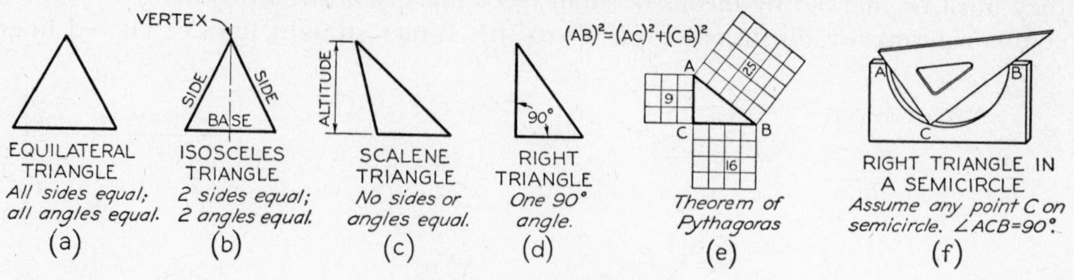

Fig. 155 Triangles.

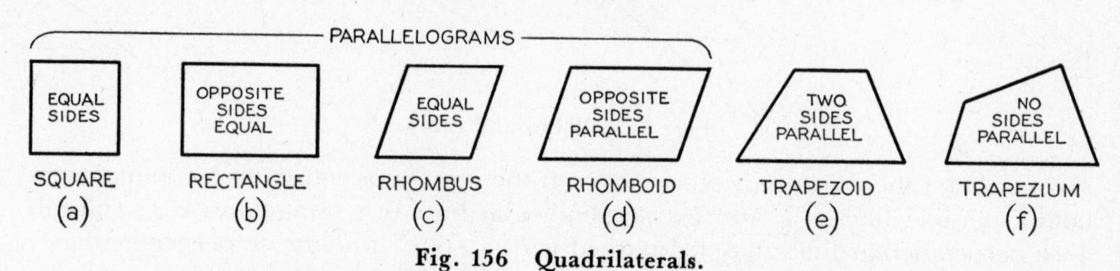

Fig. 156 Quadrilaterals.

112. Polygons. Fig. 157. A *polygon* is any plane figure bounded by straight lines. If the polygon has equal angles and equal sides, it can be *inscribed* in or *circumscribed* around a circle, and is called a *regular polygon*.

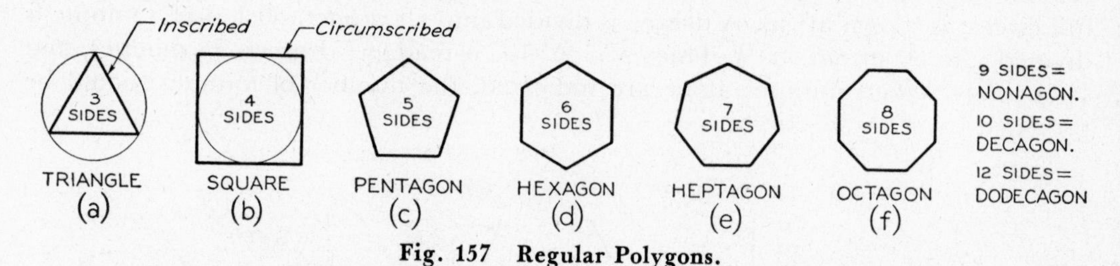

Fig. 157 Regular Polygons.

113. Circles and Arcs. Fig. 158. A *circle*, (a), is a closed curve all points of which are the same distance from a point called the *center*. *Circumference* refers to the circle or to the distance around the circle. This distance equals the diameter multiplied by π (called *pi*, pronounced *pie*) or 3.1416. Other definitions are illustrated in the figure.

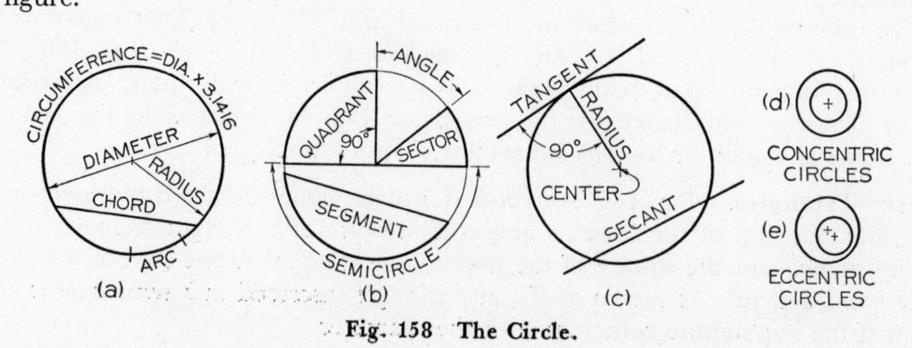

Fig. 158 The Circle.

114. Solids. Fig. 159. Solids bounded by plane surfaces are *polyhedra*. The surfaces are called *faces*, and if these are equal regular polygons, the solids are *regular polyhedra*.

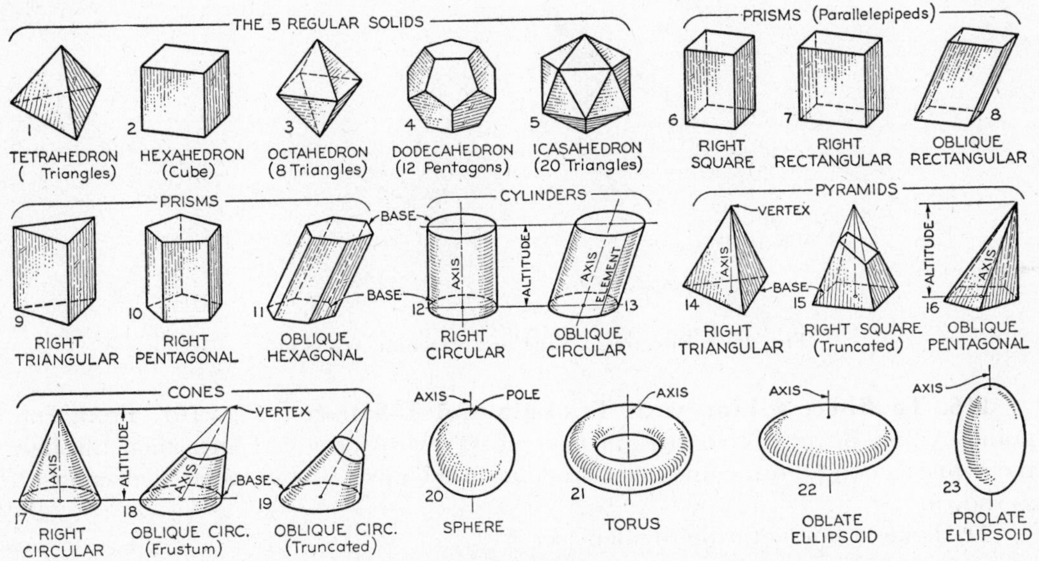

Fig. 159 Solids.

A *prism* has two *bases*, which are parallel equal polygons, and three or more lateral faces, which are parallelograms. A *triangular prism* has a triangular base; a *rectangular prism* has rectangular bases, etc. If the bases are parallelograms, the prism is a *parallelepiped*. A *right prism* has faces and lateral edges perpendicular to the bases; an *oblique prism* has faces and lateral edges oblique to the bases. If one end is cut off to form an end not parallel to the bases, the prism is said to be *truncated*.

A *pyramid* has a polygon for a base, and triangular lateral faces intersecting at a common point called the *vertex*. The center line from the center of the base to the vertex is the *axis*. If the axis is perpendicular to the base, the pyramid is a *right pyramid;* otherwise it is an *oblique pyramid*. A *triangular pyramid* has a triangular base; a *square pyramid* has a square base, etc. If a portion near the vertex has been cut off, the pyramid is *truncated*, or referred to as a *frustum*.

A *cylinder* is generated by a straight line, called the *generatrix*, moving in contact with a curved line and always remaining parallel to its previous position or to the axis. Each position of the generatrix is called an *element* of the cylinder.

A *cone* is generated by a straight line moving in contact with a curved line, and passing through a fixed point, the vertex of the cone. Each position of the generatrix is an *element* of the cone.

A *sphere* is generated by a circle revolving about one of its diameters. This diameter becomes the *axis* of the sphere, and the ends of the axis are *poles* of the sphere.

A *torus* is generated by a circle (or other curve) revolving about an axis which is eccentric to the curve.

For a classification of solids, see §514.

115. To Bisect a Line or a Circular Arc. Fig. 160 (a). Given line or arc AB, to be bisected.

I. From A and B draw equal arcs with radius greater than half AB.
II and III. Join intersections D and E with a straight line to locate center C.

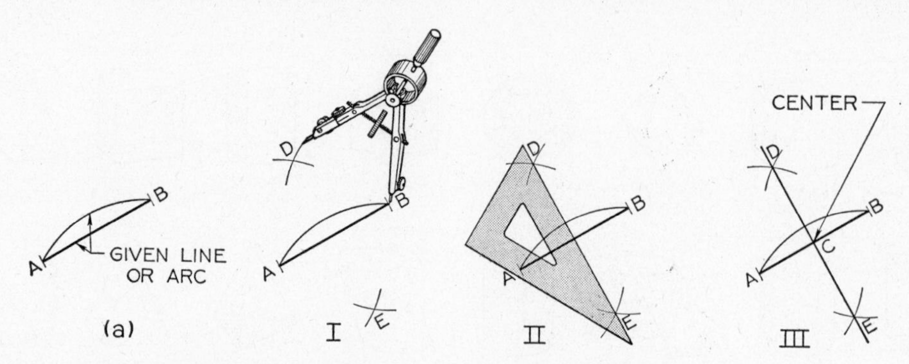

Fig. 160 Bisecting a Line or a Circular Arc (§115).

116. To Bisect a Line with Triangle and T-Square. Fig. 161. From end points A and B, draw construction lines at 45° with the given line; then through their intersection D, draw line perpendicular to the given line to locate the center C, as shown.

To divide a line with the dividers, see §49.

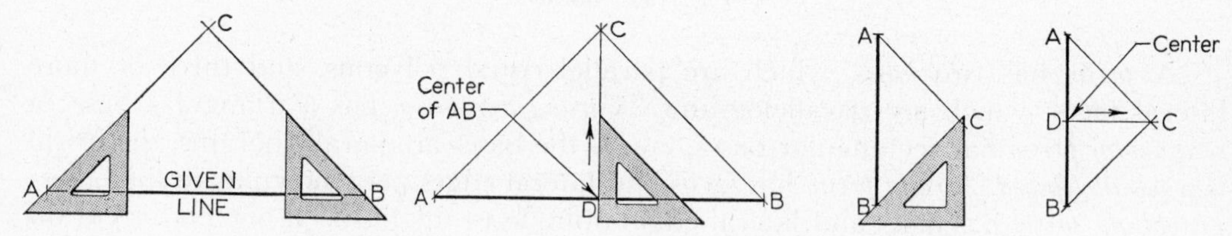

Fig. 161 Bisecting a Line with Triangle and T-square (§116).

117. To Bisect an Angle. Fig. 162 (a). Given angle BAC, to be bisected.
I. Strike large arc R.
II. Strike equal arcs r with radius slightly larger than half BC, to intersect at D.
III. Draw line AD which bisects angle.

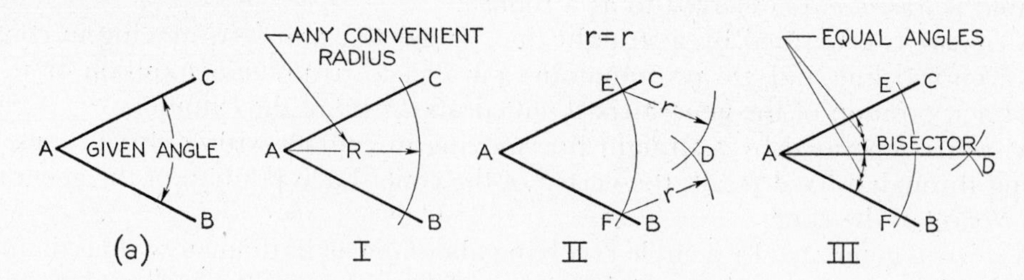

Fig. 162 Bisecting an Angle (§117).

118. To Transfer an Angle. Fig. 163 (a). Given angle BAC, to be transferred to the new position at A′B′.

I. Use any convenient radius R, and strike arcs from centers A and A′.
II. Strike equal arcs r, and draw side A′C′.

119. To Draw a Line through a Point Parallel to a Line. Fig. 164 (a). With given point P as center, and any convenient radius R, strike arc CD to intersect the

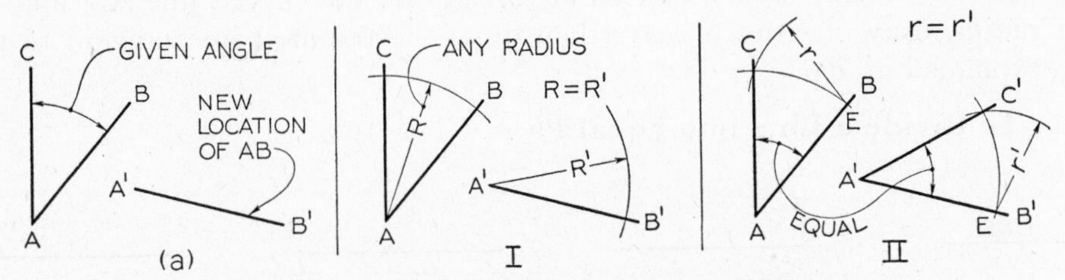

Fig. 163 Transferring an Angle (§118).

given line AB at E. With E as center and the same radius, strike arc R′ to intersect the given line at G. With PG as radius, and E as center, strike arc r to locate point H. The line PH is the required line.

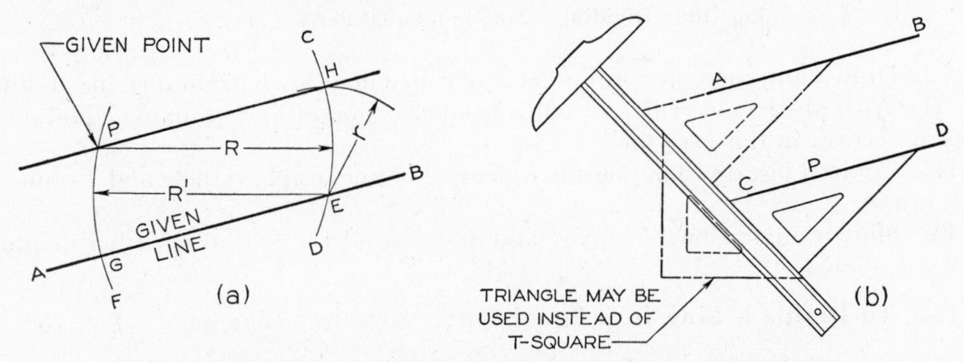

Fig. 164 Drawing a Line through a Point Parallel to a Line (§119).

Fig. 164 (b). Preferred by Draftsmen. Move the triangle and T-square as a unit until the triangle lines up with given line AB; then slide the triangle until its edge passes through the given point P. Draw CD, the required parallel line. See also §32.

120. To Draw a Line Parallel to a Line and at a Given Distance. Let AB be the line and CD the given distance.

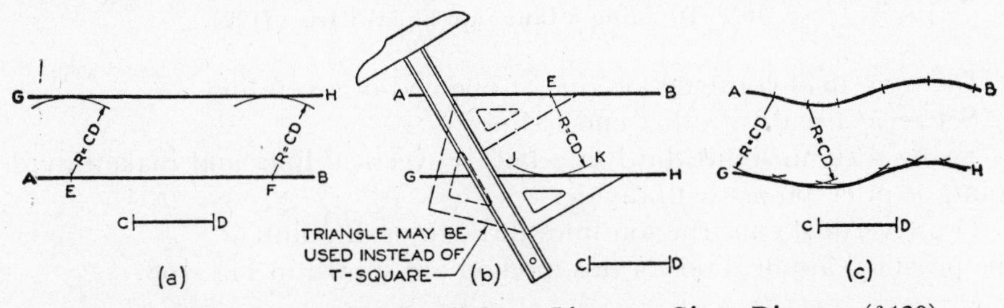

Fig. 165 Drawing a Line Parallel to a Line at a Given Distance (§120).

Fig. 165 (a). With points E and F near A and B respectively as centers, and CD as radius, draw two arcs. The line GH, tangent to the arcs, is the required line.

Fig. 165 (b). Preferred by Draftsmen. With any point E of the line as center and CD as radius, strike an arc JK. Move the triangle and T-square as a unit until the triangle lines up with the given line AB; then slide the triangle until its edge is tangent to the arc JK, and draw the required line GH.

Fig. 165 (c). With centers selected at random on the curved line AB, and with CD as radius, draw a series of arcs; then draw the required line tangent to these arcs as explained in §66.

121. To Divide a Line into Equal Parts. Fig. 166.

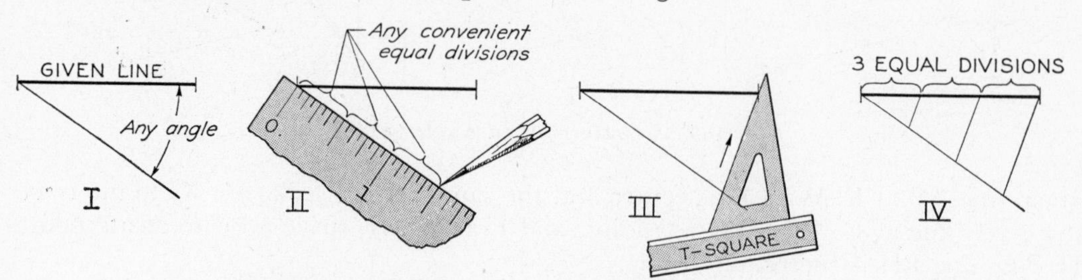

Fig. 166 Dividing a Line into Equal Parts (§121).

I. Draw light construction line at any convenient angle from one end of line.

II. With dividers or scale, set off from intersection of lines as many equal divisions as needed, in this case three.

III. Connect last division point to other end of line, using triangle and T-square, as shown.

IV. Slide triangle along T-square and draw parallel lines through other division points, as shown.

122. To Divide a Line into Equal Parts. *Preferred by Draftsmen.* Fig. 167.

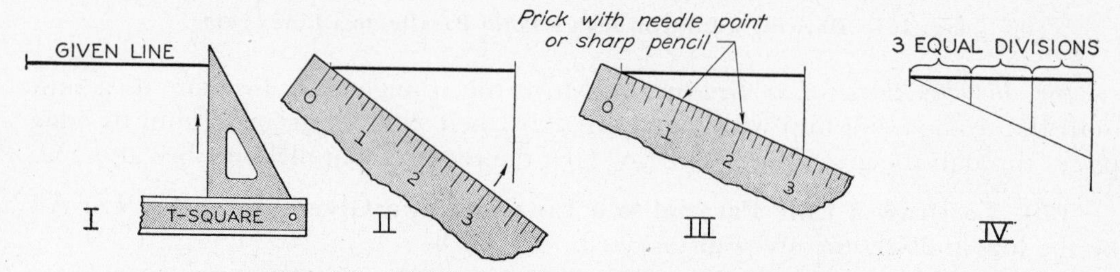

Fig. 167 Dividing a Line into Equal Parts (§122).

I. Draw vertical construction line at one end of given line.

II. Set zero of scale at other end of line.

III. Swing scale up until third unit falls on vertical line, and make tiny dots at each point, or prick points with dividers.

IV. Draw vertical construction lines through each point.

Some practical applications of this method are shown in Fig. 168.

123. To Divide a Line into Proportional Parts. Fig. 169 (a) and (b). Let it be required to divide the line AB into three parts proportional to 2, 3, and 4.

Fig. 169 (a). Preferred by Draftsmen. Through a point A, draw line CD at any convenient angle. On this line, set off 2, 3, and 4 units of convenient size by means of the scale as shown. Draw the line 9B, and then parallel to it draw the lines 5-5 and 2-2, as explained for Fig. 166 (a).

Fig. 169 (b). Draw a line CD parallel to AB and at any convenient distance. On this line, set off 2, 3, and 4 units, as shown. Draw lines through the ends of the two lines to intersect at the point O. Draw lines through O and the points 2 and 5 to divide AB into the required proportional parts.

Constructions of this type are useful in the preparation of graphs, Chapter 20.

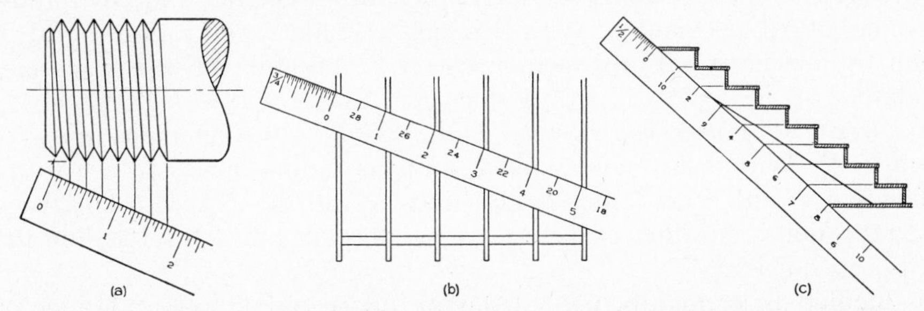

Fig. 168 Practical Applications of Dividing a Line into Equal Parts (§122).

Fig. 169 (c). Given AB, to divide into proportional parts, in this case proportional to the square of X, where X = 1, 2, 3, etc. Set zero of scale at end of line and set off divisions 4, 9, 16, etc. Join the last division to the other end of the line, and draw parallel lines as shown. This method may be used for any power of X. The construction is used in drawing nomographic charts, §587.

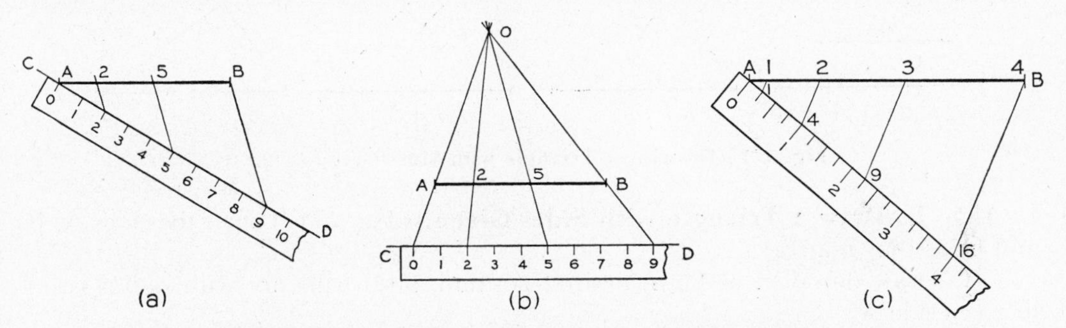

Fig. 169 Dividing a Line into Proportional Parts (§123).

124. To Draw a Line Through a Point Perpendicular to a Line. Given the line AB and a point P.

Fig. 170 (a). When the Point Is Not on the Line. From P draw any convenient in-

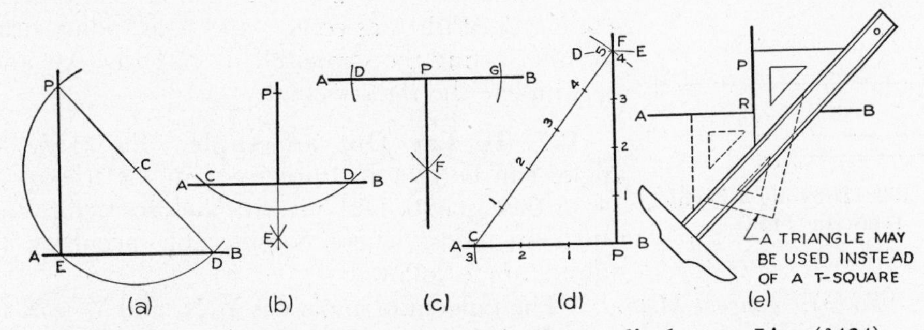

Fig. 170 Drawing a Line Through a Point Perpendicular to a Line (§124).

clined line, as PD. Find center C of line PD, and draw arc with radius CP. The line EP is the required perpendicular.

Fig. 170 (b). When the Point Is Not on the Line. With P as center, strike an arc to intersect AB at C and D. With C and D as centers, and radius slightly greater than half CD, strike arcs to intersect at E. The line PE is the required perpendicular.

Fig. 170 (c). When the Point Is on the Line. With P as center and any radius, strike arcs to intersect AB at D and G. With D and G as centers, and radius slightly greater than half DG, strike equal arcs to intersect at F. The line PF is the required perpendicular.

Fig. 170 (d). When the Point Is on the Line. Select any convenient unit of length, for example $\frac{1}{4}''$. With P as center, and 3 units as radius, strike an arc to intersect given line at C. With P as center, and 4 units as radius, strike arc DE. With C as center, and 5 units as radius, strike an arc to intersect DE at F. The line PF is the required perpendicular.

This method is frequently used in laying off rectangular foundations of large machines, buildings, or other structures. For this purpose a steel tape may be used and distances of 30, 40, and 50 feet measured as the three sides of the right triangle.

Fig. 170 (e). Preferred by Draftsmen. Move the triangle and T-square as a unit until the triangle lines up with AB; then slide the triangle until its edge passes through the point P (whether P is on or off the line), and draw the required perpendicular.

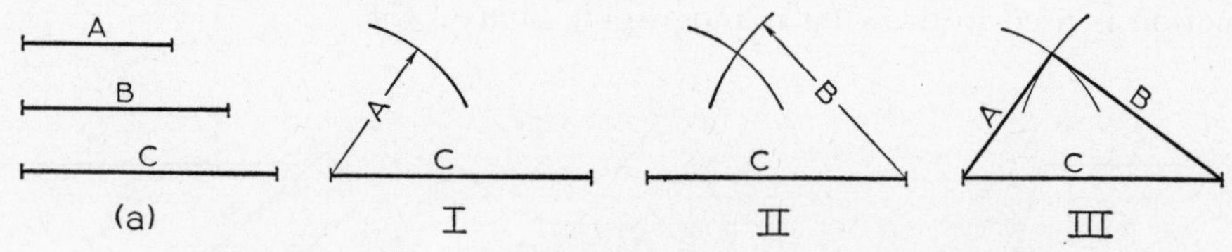

Fig. 171 Drawing a Triangle with Sides Given (§125).

125. To Draw a Triangle with Sides Given. Fig. 171. Given the sides A, B, and C, as shown at (a):

I. Draw one side, as C, in desired position, and strike arc with radius equal to given side A.

II. Strike arc with radius equal to given side B.

III. Draw sides A and B from intersection of arcs, as shown.

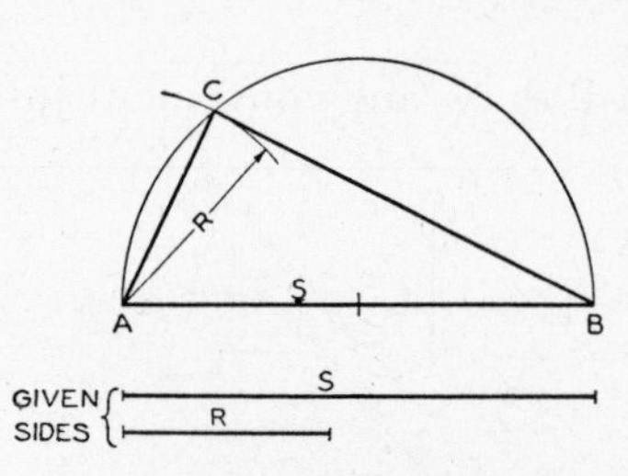

Fig. 172 Drawing a Right Triangle (§126).

126. To Draw a Right Triangle with Hypotenuse and One Side Given. Fig. 172. Given sides S and R. Draw AB equal to S and with AB as diameter, draw semicircle. With A as center, and R as radius, draw an arc intersecting the semicircle at C. Draw AC and CB to complete the right triangle.

127. To Lay Out an Angle. Fig. 173. Many angles can be laid out directly with the triangle, Fig. 34; or they may be laid out with the protractor, Fig. 35. Other methods, where considerable accuracy is required, are as follows:

Fig. 173 (a). Tangent Method: The tangent of angle θ is Y/X, and $Y = X \tan \theta$. To construct the angle, assume a convenient value for X, preferably 10″, as shown.

Find the tangent of angle θ in a table of natural tangents, multiply by 10, and set off Y = 10 tan θ. *Example:* To set off $31\frac{1}{2}°$, find the natural tangent of $31\frac{1}{2}°$, which is .6128. Then Y = 10″ (.6128) = 6.128″.

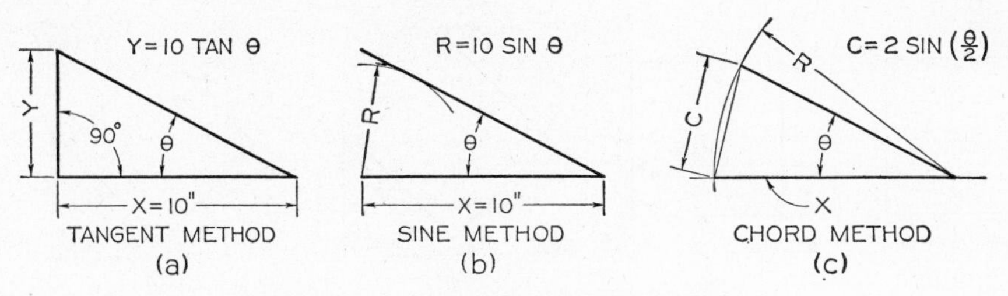

Fig. 173 Laying Out Angles (§127).

Fig. 173 (b). Sine Method. Draw line X to any convenient length, preferably 10″ as shown. Find the sine of angle θ in a table of natural sines, multiply by 10″, and strike arc R = 10″ sine θ. Draw the other side of the angle tangent to the arc, as shown. *Example:* To set off $25\frac{1}{2}°$, find the natural sine of $25\frac{1}{2}°$, which is .4305. Then R = 10″ (.4305) = 4.305″.

Fig. 173 (c). Chord Method. Draw line X to any convenient length, draw arc with any convenient radius R, say 10″. Find the chordal length C in a table of chords (see a machinists' handbook), and multiply the value by 10″, since the table is made for a radius of 1″. *Example:* To set off 43° 20′, the chordal length C for 1″ radius, as given in a table of chords = .7384, and if R = 10″, then C = 7.384″. If a table is not available, the chord C may be calculated by the formula C = 2 sine $(\theta/2)$.

Example: Half of 43° 20′ = 21° 40′. The sine of 21° 40′ = .3692. C = 2 (.3692) = .7384 for a 1″ radius. For a 10″ radius, C = 7.384″.

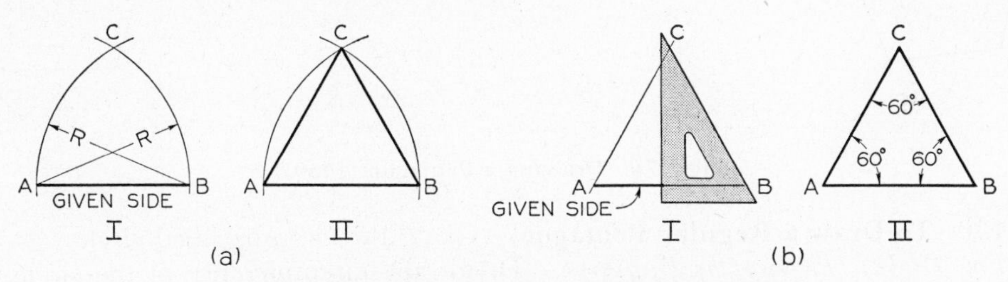

Fig. 174 Drawing an Equilateral Triangle (§128).

128. To Draw an Equilateral Triangle. Given side AB:

Fig. 174 (a). With A and B as centers and AB as radius, strike arcs to intersect at C. Draw lines AC and BC to complete the triangle.

Fig. 174 (b). Preferred by Draftsmen. Through points A and B draw lines making angles of 60° with the given line and intersecting at C, as shown.

129. To Draw a Square. Fig. 175.

Fig. 175 (a). Given one side AB. Through point A, draw a perpendicular, Fig. 170 (c). With A as center, and AB as radius, draw the arc to intersect the perpendicular at C. With B and C as centers, and AB as radius, strike arcs to intersect at D. Draw lines CD and BD.

Fig. 175 (b). Preferred by Draftsmen. Given one side AB. Using the T-square and

45° triangle, draw lines AC and BD perpendicular to AB, and the lines AD and BC at 45° with AB. Draw line CD with the T-square.

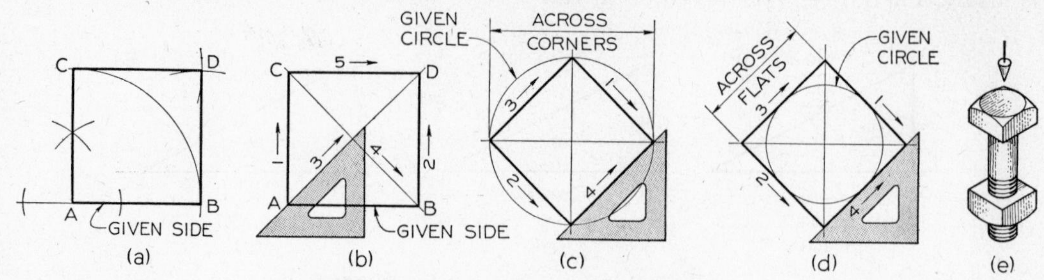

Fig. 175 Drawing a Square (§129).

Fig. 175 (c). Preferred by Draftsmen. Given the circumscribed circle (distance "across corners"). Draw two diameters at right angles to each other. The intersections of these diameters with the circle are vertexes of an inscribed square.

Fig. 175 (d). Preferred by Draftsmen. Given the inscribed circle (distance "across flats," as in drawing bolt heads). Using the T-square and 45° triangle, draw the four sides tangent to the circle. See Fig. 686.

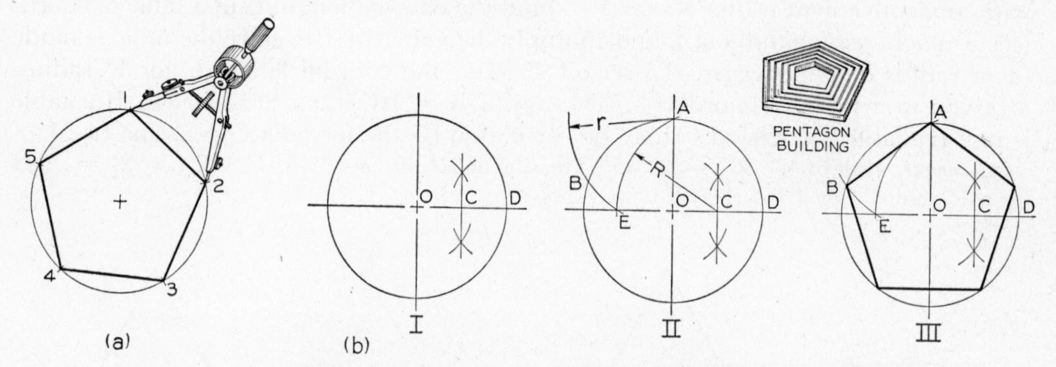

Fig. 176 Drawing a Pentagon (§130).

130. To Draw a Regular Pentagon. Given the circumscribed circle.

Fig. 176 (a). Preferred by Draftsmen. Divide the circumference of the circle into five equal parts with the dividers, and join the points with straight lines.

Fig. 176 (b). Geometrical Method.

I. Bisect radius OD at C.

II. With C as center, and CA as radius, strike arc AE. With A as center, and AE as radius, strike arc EB.

III. Draw line AB, then set off distances AB around the circumference of the circle, and draw the sides through these points.

131. To Draw a Hexagon. Given the circumscribed circle.

Fig. 177 (a). Each side of a hexagon is equal to the radius of the circumscribed circle. Therefore, using the compass or dividers and the radius of the circle, set off the six sides of the hexagon around the circle, and connect the points with straight lines. As a check on the accuracy of the construction, make sure that opposite sides of the hexagon are parallel.

Fig. 177 (b). Preferred by Draftsmen. This construction is a variation of the one shown at (a). Draw vertical and horizontal center lines. With A and B as centers and radius equal to that of the circle, draw arcs to intersect the circle at C, D, E, and F, and complete the hexagon as shown.

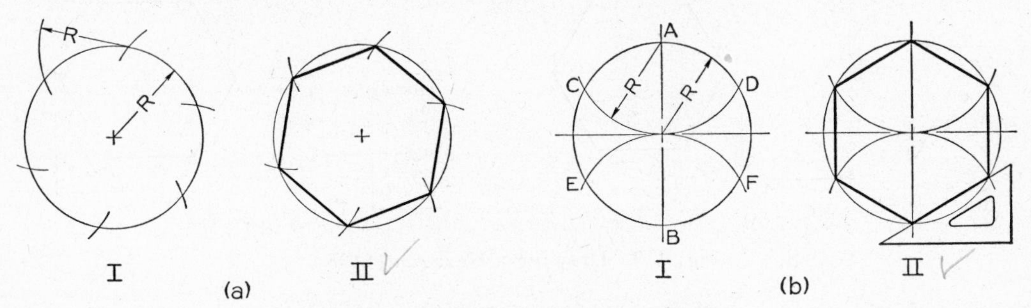

Fig. 177 Drawing a Hexagon (§131).

132. To Draw a Hexagon. Given the circumscribed or inscribed circle. *Both Preferred by Draftsmen.*

Fig. 178 (a) and (b). Given the circumscribed circle (distance "across corners"). Draw vertical and horizontal center lines, and then diagonals AB and CD at 30° or 60° with horizontal; then with the 30° × 60° triangle and the T-square, draw the six sides as shown.

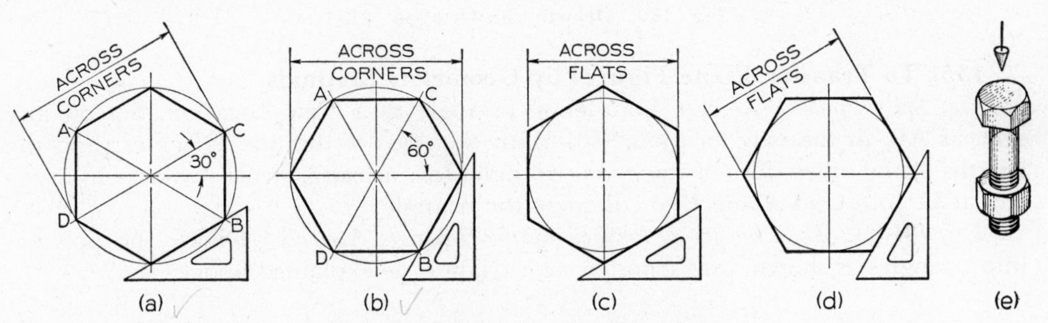

Fig. 178 Drawing a Hexagon (§132).

Fig. 178 (c) and (d). Given the inscribed circle (distance "across flats"). Draw vertical and horizontal center lines; then with the 30° × 60° triangle and the T-square, draw the six sides tangent to the circle. This method is used in drawing bolt heads and nuts, §415. For maximum accuracy, diagonals may be added as at (a) and (b).

133. To Draw a Hexagon. Fig. 179. Using the 30° × 60° triangle and the T-square, draw lines in the order shown at (a) where the distance AB "across corners" is given, or as shown at (b) where a side CD is given.

134. To Draw an Octagon.

Fig. 180 (a). Preferred by Draftsmen. Given inscribed circle, or distance "across flats." Using the T-square and 45° triangle, draw the eight sides tangent to the circle, as shown.

Fig. 180 (b). Given circumscribed square, or distance "across flats." Draw diagonals of square; then with the corners of the given square as centers, and with half the diagonal as radius, draw arcs cutting the sides as shown at I. Using the T-square and 45° triangle, draw the eight sides as shown at II.

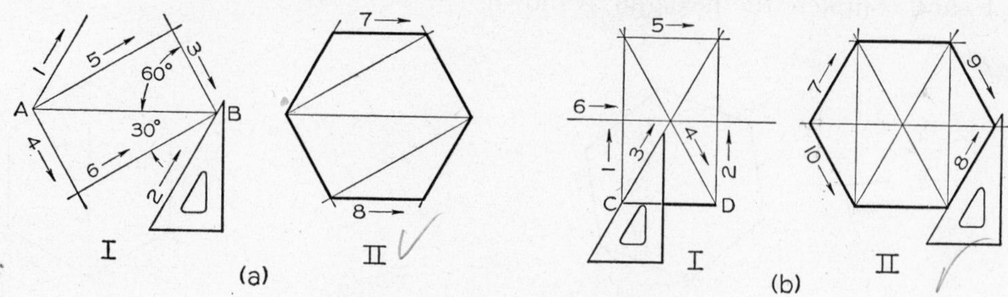

Fig. 179 Drawing a Hexagon (§133).

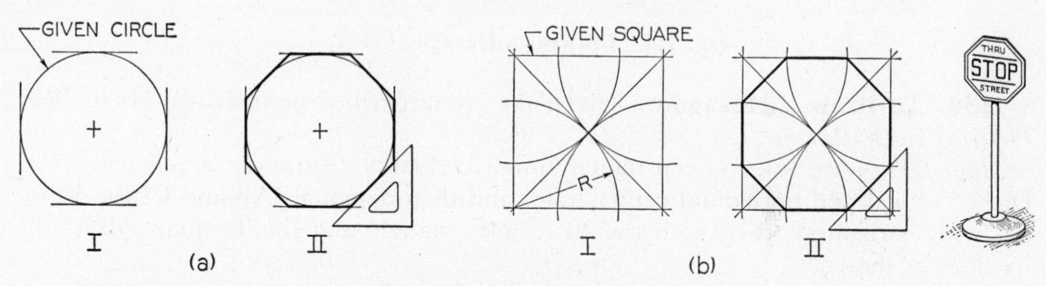

Fig. 180 Drawing an Octagon (§134).

135. To Transfer Plane Figures by Geometric Methods.

Fig. 181 (a) and (b). To Transfer a Triangle to a New Location. Set off any side, as AB, in the new location, (b). With the ends of the line as centers and the lengths of the other sides of the given triangle, (a), as radii, strike two arcs to intersect at C. Join C to A and B to complete the triangle.

Fig. 181 (c). To Transfer a Polygon by the Triangle Method. Divide the polygon into triangles as shown, and transfer each triangle as explained above.

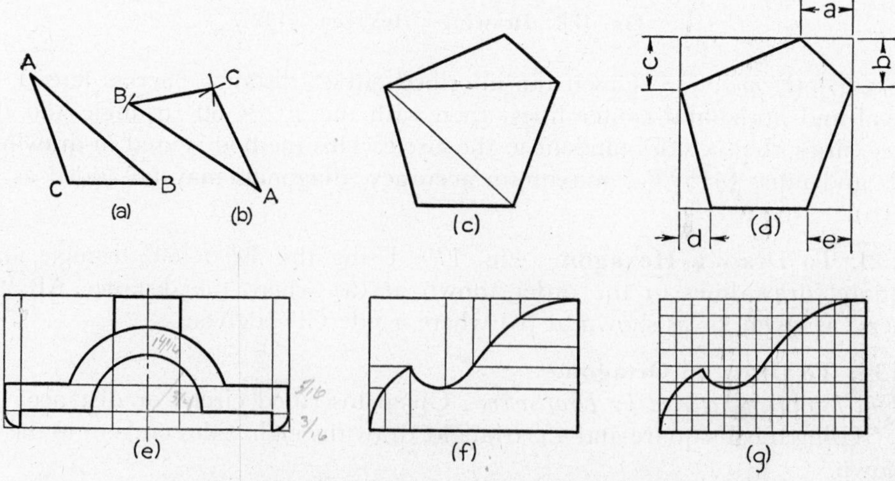

Fig. 181 Transferring a Plane Figure (§135).

Fig. 181 (d). To Transfer a Polygon by the Rectangle Method. Circumscribe a rectangle about the given polygon. Draw a congruent rectangle in the new location and locate the vertexes of the polygon by transferring location measurements a, b, c, etc., along the sides of the rectangle to the new rectangle. Join the points thus found to complete the figure.

Fig. 181 (e). To Transfer Irregular Figures. Figures composed of rectangular and circular forms are readily transferred by enclosing the elementary features in rectangles and determining centers of arcs and circles. These may then be transferred to the new location.

Fig. 181 (f). To Transfer Figures by Offset Measurements. Offset location measurements are frequently useful in transferring figures composed of free curves. When the figure has been enclosed by a rectangle, the sides of the rectangle are used as reference lines for the location of points along the curve.

Fig. 181 (g). To Transfer Figures by a System of Squares. Figures involving free curves are easily copied, enlarged, or reduced by the use of a system of squares. For example, to enlarge a figure to double size, draw the containing rectangle and all small squares double their original size. Then draw the lines through the corresponding points in the new set of squares.

136. To Transfer by Tracing-Paper Methods. To transfer a drawing to an opaque sheet, the following procedure may be used:

Prick-Point Method. Lay tracing paper over the drawing to be transferred. With a sharp pencil, make a small dot directly over each important point on the drawing. Encircle each dot so as not to lose it. Remove the tracing paper, place it over the

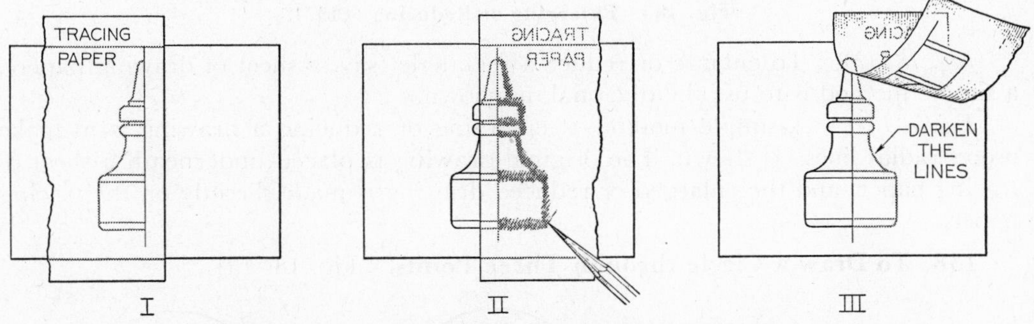

Fig. 182 Transferring a Symmetrical Half (§136).

paper to receive the transferred drawing, and maneuver the tracing paper into the desired position. With a needle-point (such as a point of the dividers), prick through each dot. Remove the tracing paper and connect the prick-points to produce the lines as on the original drawing.

To transfer arcs or circles, it is only necessary to transfer the center and one point on the circumference. To transfer a free curve, transfer as many prick-points on the curve as desired.

Tracing Method. Lay tracing paper over the drawing to be transferred, and make a pencil tracing of it. Turn the tracing paper over and mark over the lines with short strokes of a soft pencil so as to provide a coating of graphite over every line. Turn tracing face up and fasten in position where drawing is to be transferred. Trace over all lines of the tracing, using a hard pencil. The graphite on the back acts as a carbon paper and will produce dim but definite lines. Heavy-in the dim lines to complete the transfer.

Fig. 182. If one-half of a symmetrical object has been drawn, as for the ink bottle at I, the other half may be easily drawn with the aid of tracing paper.

I. Trace the half already drawn.

II. Turn tracing paper over and maneuver to the position for the right half. Then trace over the lines freehand or mark over the lines with short strokes as shown.

III. Remove the tracing paper, revealing the dim imprinted lines for the right half. Heavy-in these lines to complete the drawing.

137. To Enlarge or Reduce a Drawing.

Fig. 183 (a). The construction shown is an adaptation of the parallel-line method, Figs. 166 and 167, and may be used whenever it is desired to enlarge or reduce any group of dimensions to the same ratio. Thus if full-size dimensions are laid off along the vertical line, the enlarged dimensions would appear along the horizontal line, as shown.

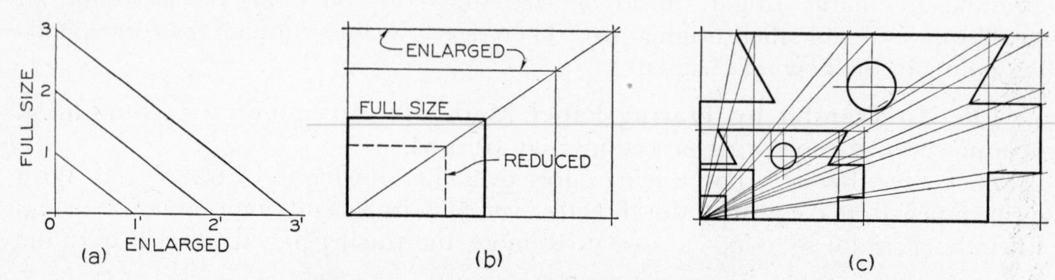

Fig. 183 Enlarging or Reducing (§137).

Fig. 183 (b). To enlarge or reduce a rectangle (say, a sheet of drawing paper), a simple method is to use the diagonal, as shown.

Fig. 183 (c). A simple method of enlarging or reducing a drawing is to make use of radial lines, as shown. The original drawing is placed underneath a sheet of tracing paper, and the enlarged or reduced drawing is made directly on the tracing paper.

138. To Draw a Circle through Three Points. Fig. 184 (a).

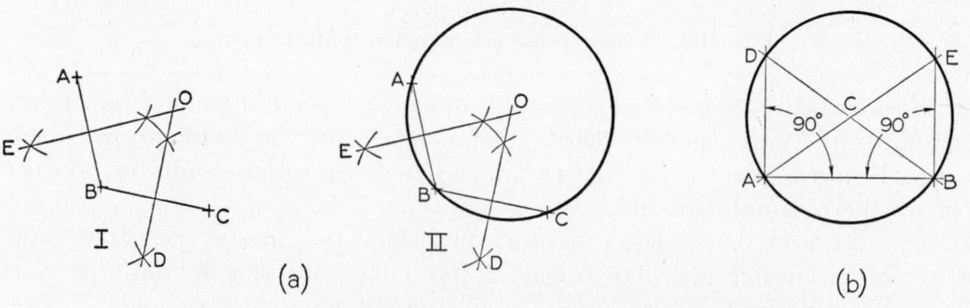

Fig. 184 Finding Center of Circle (§§138 and 139).

I. Let A, B, and C be the three given points not in a straight line. Draw lines AB and BC, which will be chords of the circle. Draw perpendicular bisectors EO and DO, Fig. 160, intersecting at O.

II. Through center O, draw required circle through the points.

139. To Find the Center of a Circle. Fig. 184 (b). Draw any chord AB, preferably horizontal, as shown. Draw perpendiculars from A and B, cutting circle at D and E. Draw diagonals DB and EA whose intersection C will be the center of the circle.

Another method, slightly longer, is to reverse the procedure of Fig. 184 (a). Draw any two non-parallel chords and draw perpendicular bisectors. The intersection of the bisectors will be the center of the circle.

140. To Draw a Circle Tangent to a Line at a Given Point. Fig. 185. Given a line AB and a point P on the line, as shown at (a).

I. At P erect a perpendicular to the line.
II. Set off the radius of the required circle on the perpendicular.
III. Draw circle with radius CP.

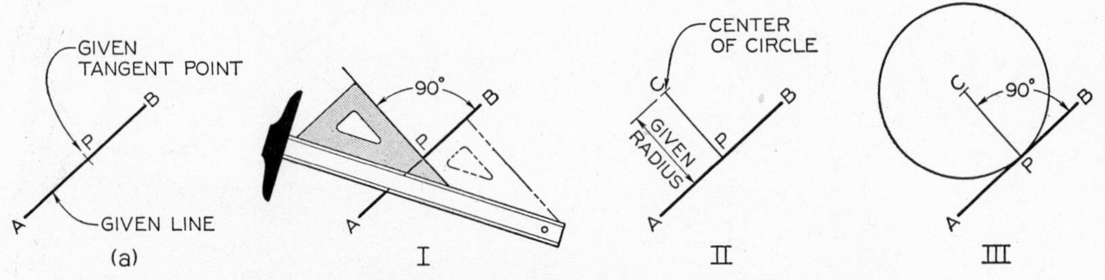

Fig. 185 Drawing a Circle Tangent to a Line (§140).

141. To Draw a Tangent to a Circle through a Point. *Preferred by Draftsmen.*

Fig. 186 (a). Given Point P on the Circle. Move the T-square and triangle as a unit until one side of the triangle passes through the point P and the center of the circle; then slide the triangle until the other side passes through point P, and draw the required tangent.

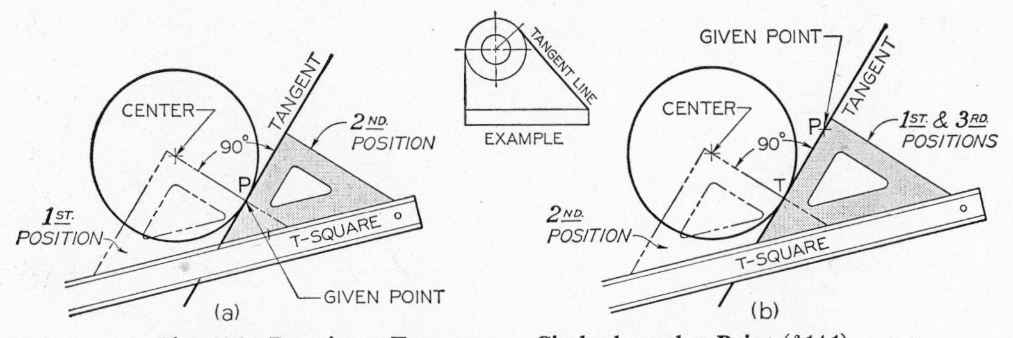

Fig. 186 Drawing a Tangent to a Circle through a Point (§141).

Fig. 186 (b). Given Point P Outside the Circle. Move the T-square and triangle as a unit until one side of the triangle passes through point P and, by inspection, is tangent to the circle; then slide the triangle until the other side passes through the center of the circle, and lightly mark the point of tangency T. Finally move the triangle back to its starting position, and draw the required tangent.

In both constructions either triangle may be used. Also, a second triangle may be used in place of the T-square.

142. To Draw Tangents to Two Circles. Fig. 187 (a) and (b). Move the triangle and T-square as a unit until one side of the triangle is tangent, by inspection, to the two circles; then slide the triangle until the other side passes through the center of one circle, and lightly mark the point of tangency. Then slide the triangle until the side passes through the center of the other circle, and mark the point of tangency. Finally slide the triangle back to the tangent position, and draw the tangent lines between the two points of tangency. Draw the second tangent line in a similar manner.

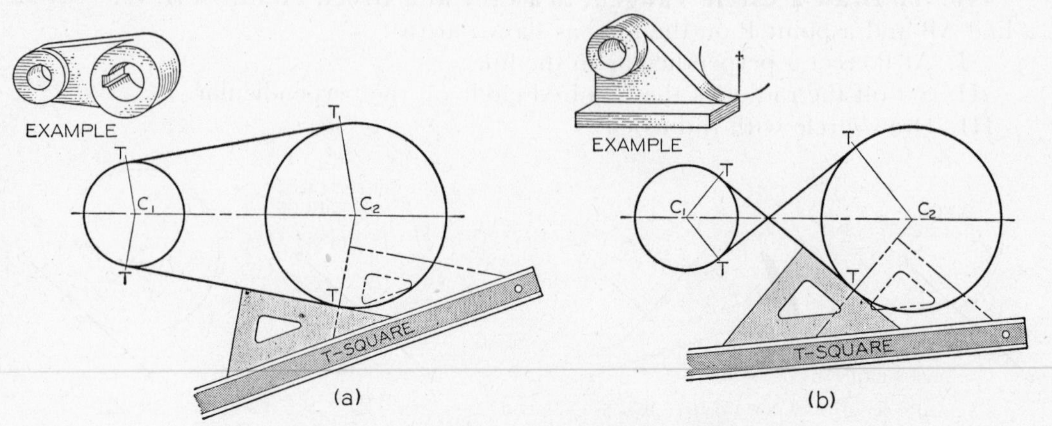

Fig. 187 Tangents to Two Circles (§142).

143. To Draw an Arc Tangent to a Line or Arc and Through a Point. *Fig. 188 (a). Given Line AB, Point P, and Radius R.* Draw line DE parallel to given line at distance R from it. From P draw arc with radius R, cutting line DE at C, the center of the required tangent arc.

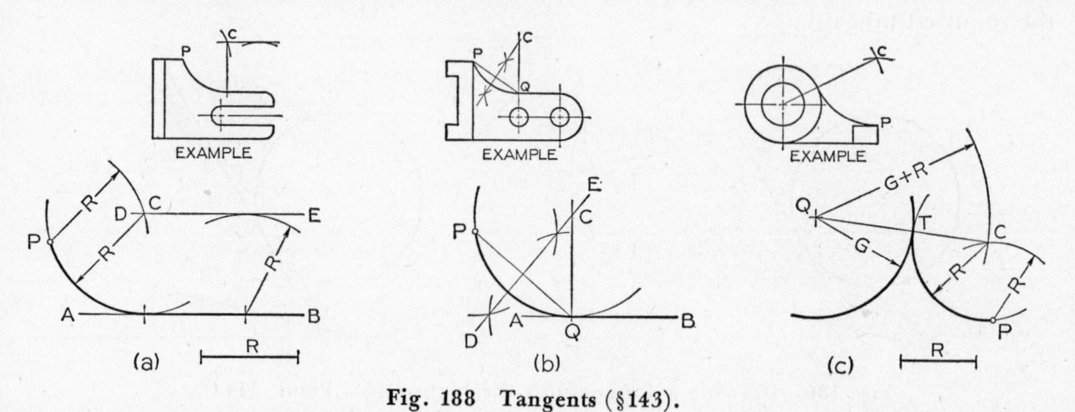

Fig. 188 Tangents (§143).

Fig. 188 (b). Given Line AB, with Tangent Point Q on the Line, and Point P. Draw PQ, which will be a chord of the required arc. Draw perpendicular bisector DE, and at Q erect a perpendicular to the line to intersect DE at C, the center of the required tangent arc.

Fig. 188 (c). Given Arc with Center Q, Point P, and Radius R. From P strike arc with radius R. From Q strike arc with radius equal to that of the given arc plus R. The intersection C of the arcs is the center of the required tangent arc.

144. To Draw a Tangent Arc to Two Lines at Right Angles. Fig. 189 (a).

I. Given two lines at right angles to each other.

II. With given radius R, strike arc intersecting given lines at tangent points T.

III. With given radius R again and with points T as centers, strike arcs intersecting at C.

IV. With C as center and given radius R, draw required tangent arc.

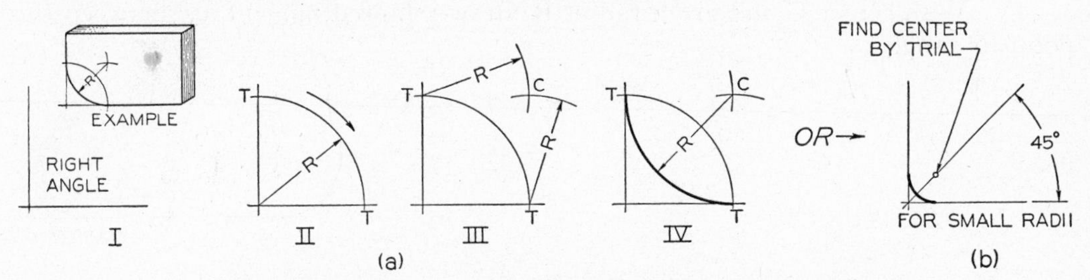

Fig. 189 Drawing a Tangent Arc in a Right Angle (§144).

Fig. 189 (b). For Small Radii. For small radii, such as $\frac{1}{8}$R for fillets and rounds, it is not practicable to draw complete tangency constructions. Instead, draw a 45° bisector of the angle and locate the center of the arc by trial along this line, as shown.

145. To Draw a Tangent Arc to Two Lines at Acute or Obtuse Angles. Fig. 190 (a) or (b).

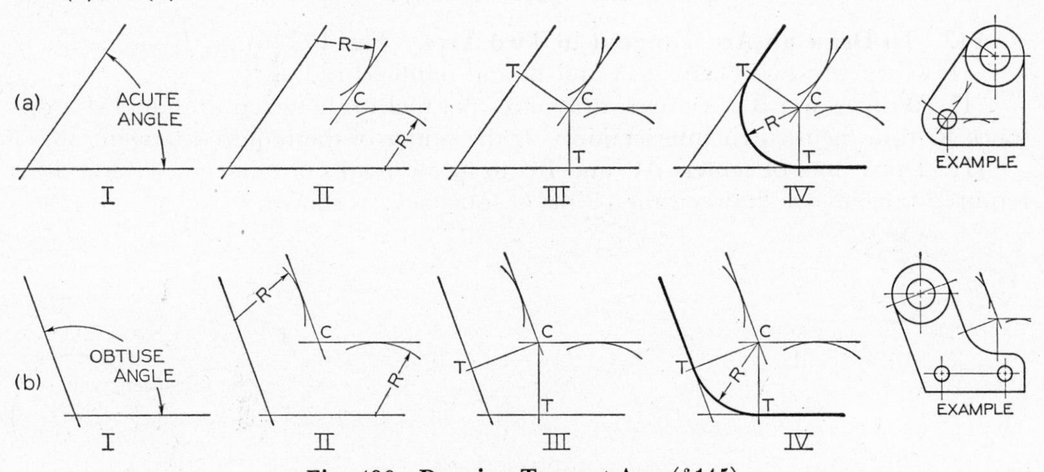

Fig. 190 Drawing Tangent Arcs (§145).

I. Given two lines not making 90° with each other.

II. Draw lines parallel to given lines at distance R from them, to intersect at C, the required center.

III. From C drop perpendiculars to the given lines respectively to locate points of tangency T.

IV. With C as center and with given radius R, draw required tangent arc between the points of tangency.

146. To Draw Tangent Arc to an Arc and a Straight Line. Fig. 191 (a) or (b).

I. Given arc with radius G and straight line AB.

II. Draw straight line and an arc parallel respectively to the given straight line and arc at the required radius distance R from them, to intersect at C, the required center.

III. From C drop perpendicular to given straight line to obtain one point of tangency T. Join centers C and O with a straight line to locate the other point of tangency T.

IV. With center C and given radius R, draw required tangent arc between the points of tangency.

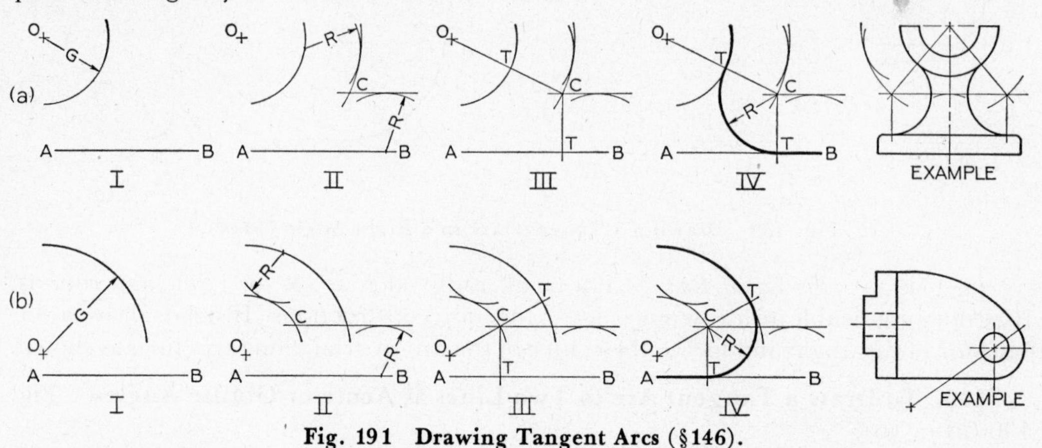

Fig. 191 Drawing Tangent Arcs (§146).

147. To Draw an Arc Tangent to Two Arcs.

Fig. 192 (a) or (b).

I. Given arcs with centers A and B, and required radius R.

II. With A and B as centers, draw arcs parallel to the given arcs and at a distence R from them; their intersection C is the center of the required tangent arc.

III. Draw lines of centers AC and BC to locate points of tangency T, and draw required tangent arc between the points of tangency, as shown.

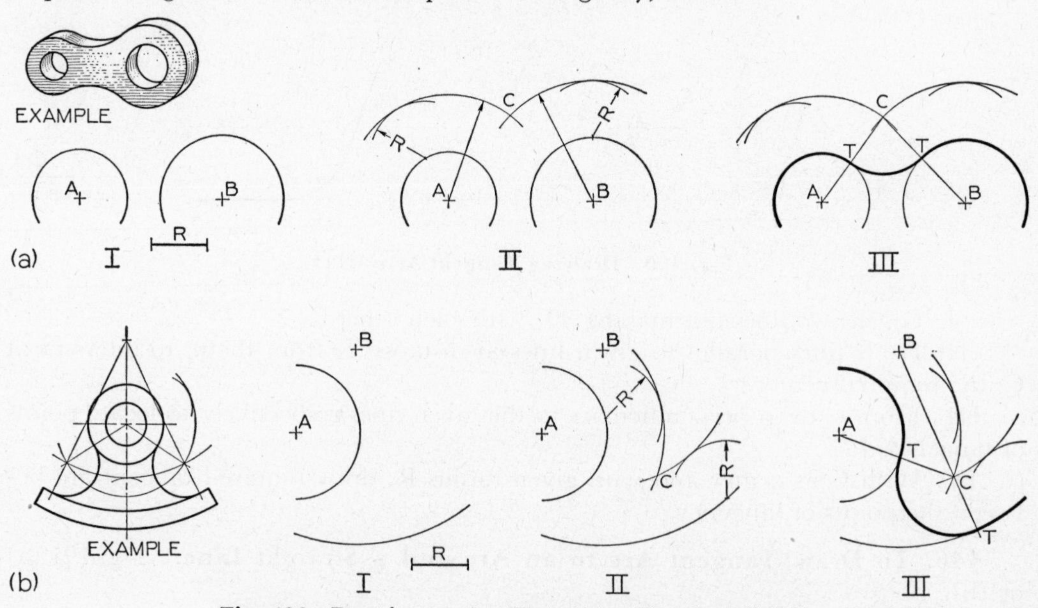

Fig. 192 Drawing an Arc Tangent to Two Arcs (§147).

148. To Draw an Arc Tangent to Two Arcs and Enclosing One or Both. *Fig. 193 (a). The Required Arc Encloses Both Given Arcs.* With A and B as centers, strike arcs HK − r (given radius minus radius of small circle) and HK − R (given radius minus radius of large circle) intersecting at G, the center of the required tangent arc. Lines of centers GA and GB (extended) determine points of tangency T.

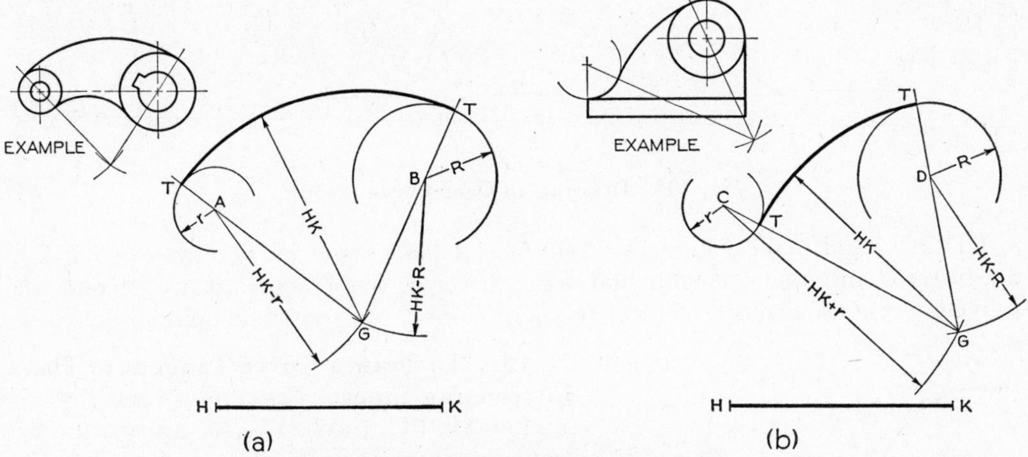

Fig. 193 Drawing Tangent Arcs (§148).

Fig. 193 (b). The Required Arc Encloses One Given Arc. With C and D as centers, strike arcs HK + r (given radius plus radius of small circle) and HK − R (given radius minus radius of large circle) intersecting at G, the center of the required tangent arc. Lines of centers GC and GD (extended) determine points of tangency T.

149. To Draw a Series of Tangent Arcs Conforming to a Curve. Fig. 194. First sketch lightly a smooth curve as desired. By trial, find a radius R and a center C, producing an arc AB which closely follows that portion of the curve. The successive centers D, E, etc., will be on lines joining the centers with the points of tangency, as shown.

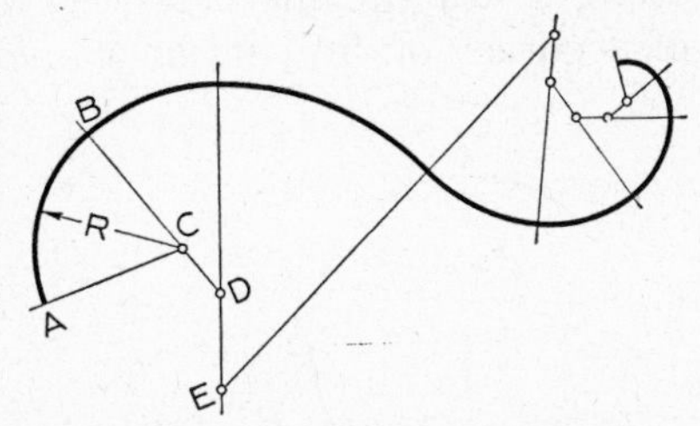

Fig. 194 A Series of Tangent Arcs (§149).

150. To Draw an Ogee Curve.

Fig. 195 (a). Connecting Two Parallel Lines. Let NA and BM be the two parallel lines. Draw AB, and assume inflection point T (at midpoint if two equal arcs are desired). At A and B erect perpendiculars AF and BC. Draw perpendicular bisectors of AT and BT. The intersections F and C of these bisectors and the perpendiculars, respectively, are the centers of the required tangent arcs.

Fig. 195 (b). Connecting Two Parallel Lines. Let AB and CD be the two parallel lines, with point B as one end of the curve, and R the given radii. At B erect perpendicular to AB, make BG = R, and draw arc as shown. Draw line SP parallel to CD at distance R from CD. With center G, draw arc of radius 2R, intersecting line SP at O. Draw perpendicular OJ to locate tangent point J, and join centers G and O to locate point of tangency T. Using centers G and O and radius R, draw the two tangent arcs as shown.

Fig. 195 (c). Connecting Two Non-Parallel Lines. Let AB and CD be the two non-parallel lines. Erect perpendicular to AB at B. Select point G on the perpendicular so that BG equals any desired radius, and draw arc as shown. Erect perpendicular

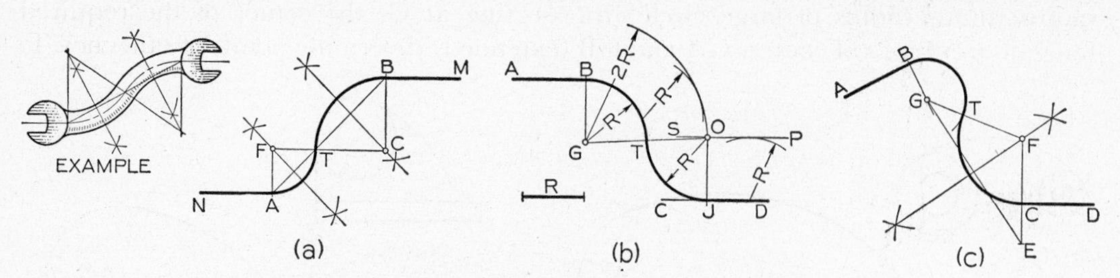

Fig. 195 Drawing an Ogee Curve (§150).

to CD at C and make CE = BG. Join G to E and bisect it. The intersection F of the bisector and the perpendicular CE, extended, is the center of the second arc. Join centers of the two arcs to locate tangent point T, the inflection point of the curve.

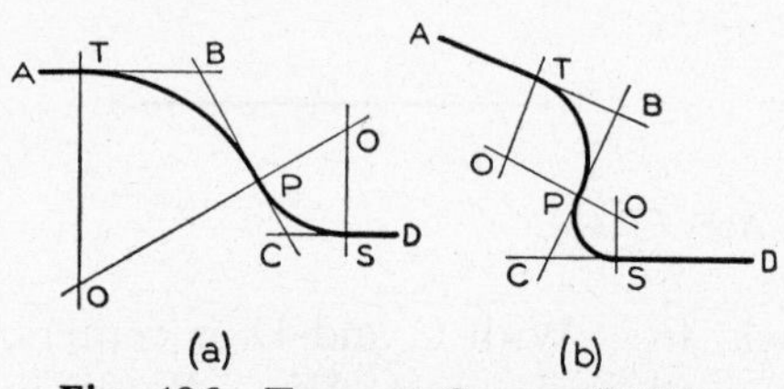

Fig. 196 Tangent Curves (§151).

151. To Draw a Curve Tangent to Three Intersecting Lines. Fig. 196 (a) and (b).

Let AB, BC, and CD be the given lines. Select point of tangency P at any point on line BC. Make BT equal to BP, and CS equal to CP, and erect perpendiculars at the points P, T, and S. Their intersections O are the centers of the required tangent arcs.

152. To Rectify a Circular Arc. To *rectify* an arc is to lay out its true length along a straight line. The constructions are approximate, but well within the range of accuracy of drawing instruments.

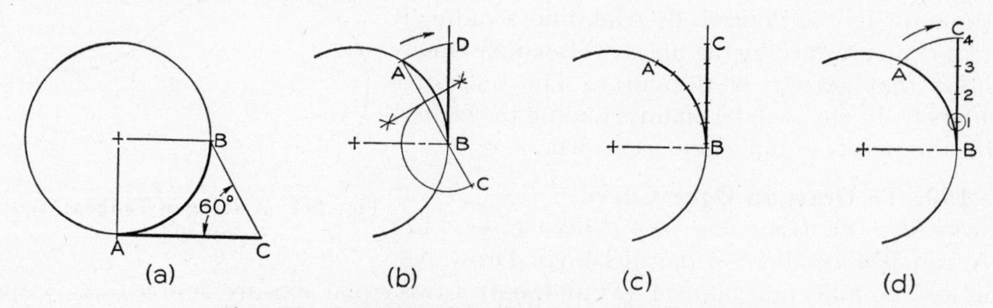

Fig. 197 Rectifying Circular Arcs (§§152 and 153).

Fig. 197 (a). To Rectify a Quadrant of a Circle, AB. Draw AC tangent to the circle and BC at 60° to AC, as shown. The line AC is almost equal to the arc AB, the difference in length being about 1 in 240.

Fig. 197 (b). To Rectify Arc AB. Draw tangent at B. Draw chord AB and extend it to C, making BC equal to half AB. With C as center and radius CA, strike the arc AD. The tangent BD is slightly shorter than the given arc AB. For an angle of 45° the difference in length is about 1 in 2866 ±.

Fig. 197 (c). To Rectify Arc AB. Use the bow dividers, and beginning at A, set off equal distances until the division point nearest to B is reached. At this point, reverse the direction and set off an equal number of distances along the tangent to determine point C. The tangent BC is slightly shorter than the given arc AB. If the angle subtended by each division is 10°, the error is 1 in 830 ±.

Note: If the angle θ subtending an arc of radius R is known, the length of the arc $= 2\pi R(\frac{\theta}{360°}) = 0.01745R\theta$. See also §712.

153. To set off a given length along a given arc.

Fig. 197 (c). Reverse the method described above so as to transfer distances from the tangent line to the arc.

Fig. 197 (d). To set off the length BC along the arc BA, draw BC tangent to the arc at B. Divide BC into four equal parts. With center at 1, the first division point, and radius 1C, draw the arc CA. The arc BA is practically equal to BC for angles less than 30°. For 45° the difference is 1 in 3232 ±, and for 60° it is 1 in 835 ±.

154. The Conic Sections. Fig. 198. The *conic sections* are curves produced by planes intersecting a right circular cone. Four types of curves are produced: the

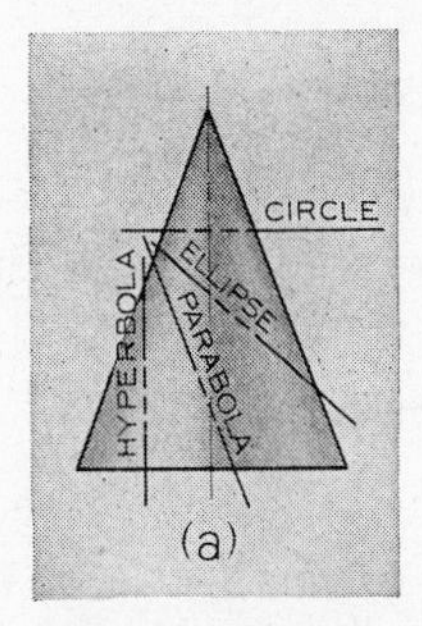

(b)

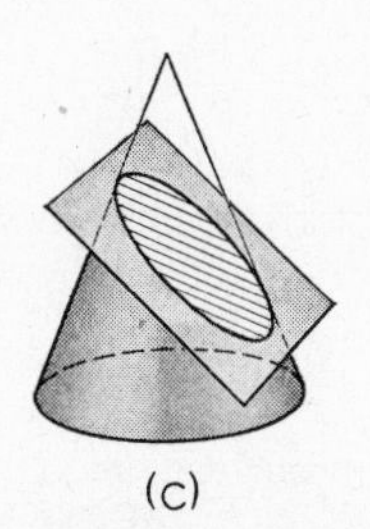

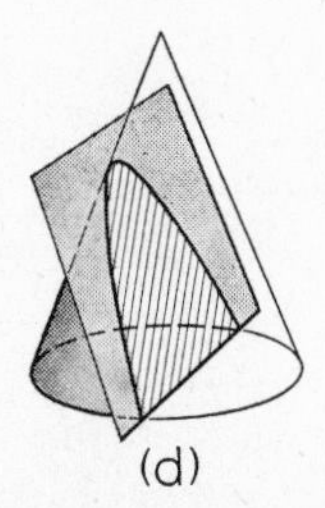

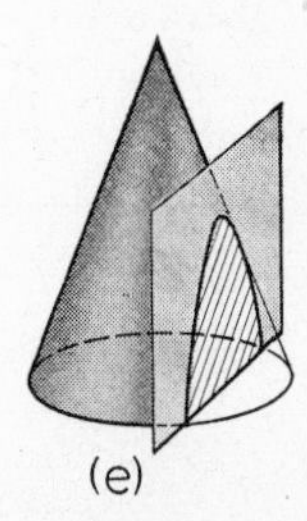

View of cone showing cutting planes.

Circle. Plane perpendicular to axis.

Ellipse. Plane oblique to axis, but making greater angle with axis than elements do.

Parabola. Plane oblique to axis, and making same angle with axis as elements do.

Hyperbola. Plane making smaller angle with axis than elements do.

Fig. 198 Conic Sections (§154).

circle, *ellipse*, *parabola*, and *hyperbola*, according to the position of the planes, as shown. These curves were studied in detail by the ancient Greeks, and are of great interest in mathematics, as well as in technical drawing. For equations, see any text on analytic geometry.

155. Ellipse Construction. The long axis of an ellipse is the *major axis*, and the short axis is the *minor axis*, Fig. 199 (a). The *foci* E and F are found by striking arcs with radius equal to half the major axis and with center at the end of the minor axis. Another method is to draw a semicircle with the major axis as diameter, then to draw GH parallel to the major axis, and GE and HF parallel to the minor axis, as shown.

An ellipse may be generated by a point moving so that the sum of its distances from two points (the foci) is constant and equal to the major axis. For example, Fig. 199 (b), an ellipse may be constructed by placing a looped string around the foci E and F, and around C, one end of the minor axis, and moving the pencil point P along its maximum orbit while the string is kept taut.

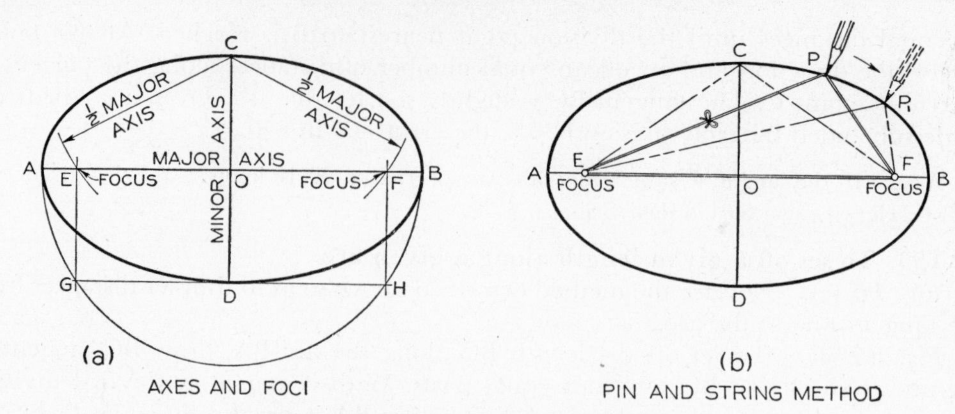

Fig. 199 Ellipse Constructions (§155).

156. To Draw Foci Ellipse. Fig. 200. Let AB be the major axis, and CD the minor axis. This method is the geometrical counterpart of the pin-and-string method.

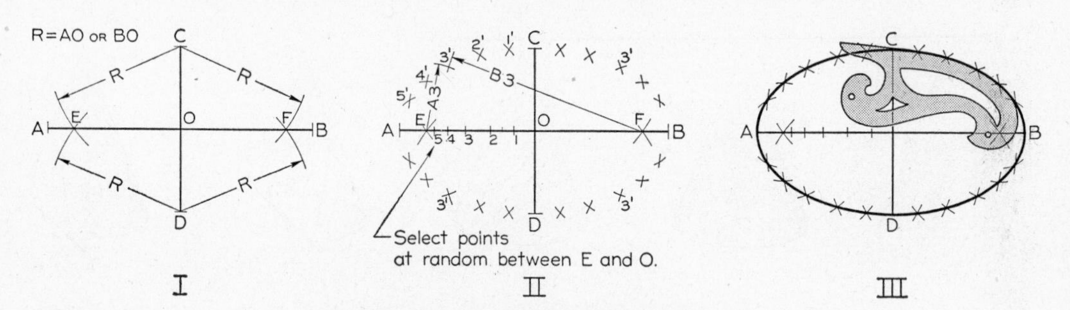

Fig. 200 Drawing Foci Ellipse (§156).

Keep the construction very light, as follows:

I. To find foci E and F, strike arcs R with radius equal to half the major axis and with centers at the ends of the minor axis.

II. Between E and O on the major axis, mark at random a number of points (spacing those on the left more closely), equal to the number of points desired in each quadrant of the ellipse. In this figure, five points were deemed sufficient. For large ellipses, more points should be used—enough to insure a smooth, accurate curve. Begin construction with any one of these points, such as 3. With E and F as centers and radii A3 and B3, respectively (from the ends of the major axis to point 3), strike arcs to intersect at four points 3′, as shown. Using the remaining points 1, 2, 4, and 5, for each find four additional points on the ellipse in the same manner.

III. Sketch the ellipse lightly through the points; then heavy-in the final ellipse with the aid of the irregular curve, §66.

157. To Draw Trammel Ellipse. Fig. 201. A "long trammel" or a "short trammel" may be prepared from a small strip of stiff paper or thin cardboard, as shown. In both cases, set off on the edge of the trammel distances equal to the semi-major and semi-minor axes. In one case these distances overlap; in the other they are end-to-end. To use either method, place the trammel so that two of the points are on the respective axes, as shown; the third point will then be on the curve and can be marked with a small dot. Find additional points by moving the trammel to other positions, always keeping the two points exactly on the respective axes. Extend the axes to use the long trammel. Find enough points to insure a smooth and symmetrical ellipse. Sketch the ellipse lightly through the points; then heavy-in the ellipse with the aid of the irregular curve, §66.

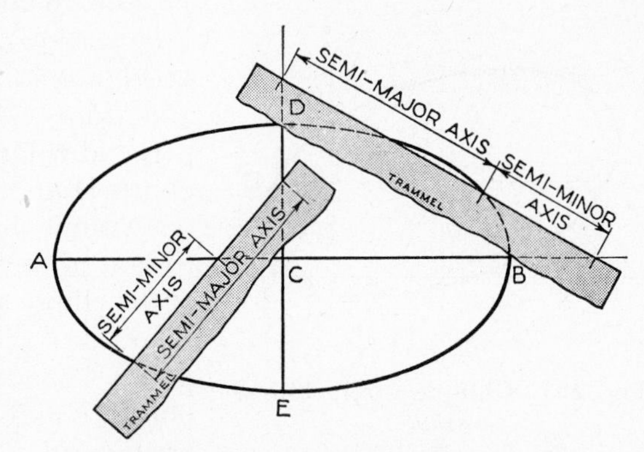

Fig. 201 Drawing Trammel Ellipse (§ 157).

158. To Draw Concentric-Circle Ellipse. Fig. 202. If a circle is viewed "head-on"—that is, so that the line of sight is perpendicular to the plane of the circle, as shown for the silver dollar at (a), the circle will appear as a circle, in true size and shape. If the circle is viewed at an angle, as shown at (b), it will appear as an ellipse. If the circle is viewed edgewise, it appears as a straight line, as shown at (c). The case shown at (b) is the basis for the construction of an ellipse by the *concentric-circle method*, as follows (keep the construction very light):

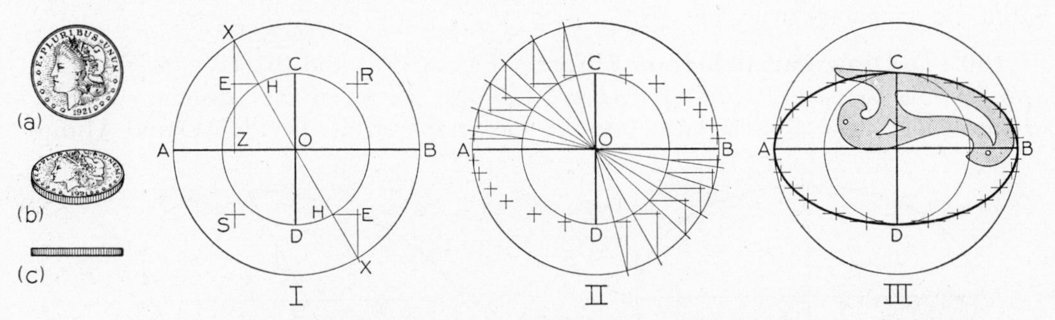

Fig. 202 Drawing Concentric-Circle Ellipse (§158).

I. Draw circles on the two axes as diameters, and draw any diagonal XX through center O. From the points X, in which the diagonal intersects the large circle, draw lines XE parallel to the minor axis; and from points H, in which it intersects the small circle, draw lines HE parallel to the major axis. The intersections E are points on the ellipse. Two additional points, S and R, can be found by extending lines XE and HE, giving a total of four points from the one diagonal XX.

II. Draw as many additional diagonals as needed to provide a sufficient number of points for a smooth and symmetrical ellipse, each diagonal accounting for four points on the ellipse. Notice that where the curve is sharpest (near the ends of the ellipse), the points are constructed closer together to better determine the curve.

III. Sketch the ellipse lightly through the points; then heavy-in the final ellipse with the aid of the irregular curve, Fig. 89.

Note: It is evident at I that the ordinate EZ of the ellipse is to the corresponding ordinate XZ of the circle as b is to a, where b represents the semi-minor axis and a the semi-major axis. Thus, the area of the ellipse is equal to the area of the circumscribed circle multiplied by b/a; hence it is equal to πab.

159. To Draw an Ellipse on Conjugate Diameters. Oblique Circle Method. Fig. 203.

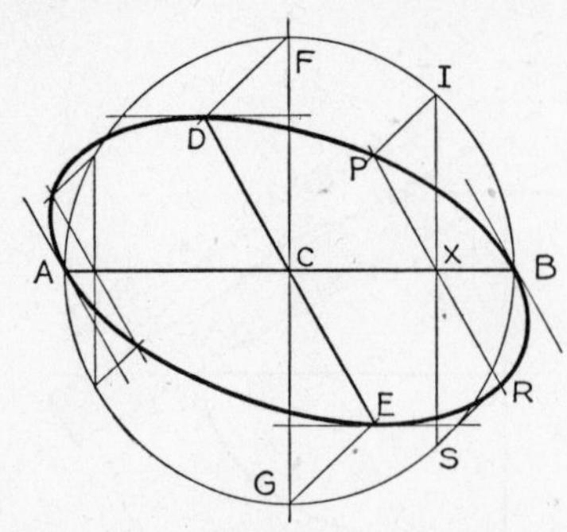

Fig. 203 Oblique Circle Ellipse (§159).

Let AB and DE be the given conjugate diameters. *Two diameters are conjugate when each is parallel to the tangents at the extermities of the other.* With center at C and radius CA, draw a circle; draw the diameter GF perpendicular to AB, and draw lines joining the points D and F, and G and E.

Assume that the required ellipse is an oblique projection of the circle just drawn; the points D and E of the ellipse are the oblique projections of the points F and G of the circle, respectively; and similarly, the points P and R are the oblique projections of the points I and S, respectively. The points P and R are determined by assuming the point X at any point on AB and drawing the lines IS and PR, and IP and SR, parallel respectively to GF and DE, and FD and GE.

Determine at least five points in each quadrant (more for larger ellipses) by assuming additional points on the major axis and proceeding as explained for point X. Sketch the ellipse lightly through the points; then heavy-in the final ellipse with the aid of the irregular curve, Fig. 89.

160. To Draw Parallelogram Ellipse. Fig. 204 (a) and (b). *Given the major and minor axes, or the conjugate diameters AB and CD.* On the given axes, draw a rectangle or parallelogram with sides parallel to the axes, respectively. Divide AO and AJ into

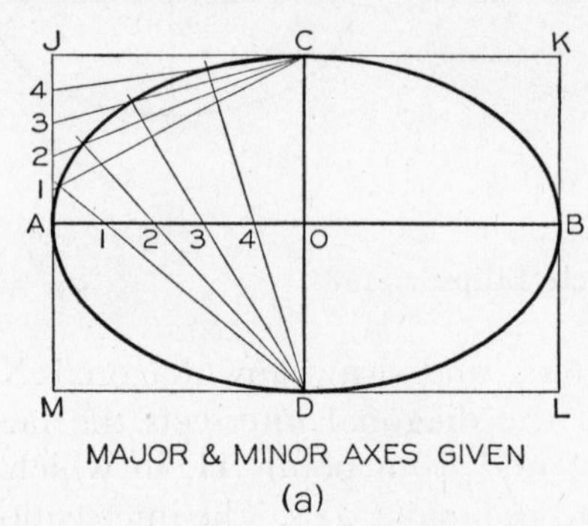

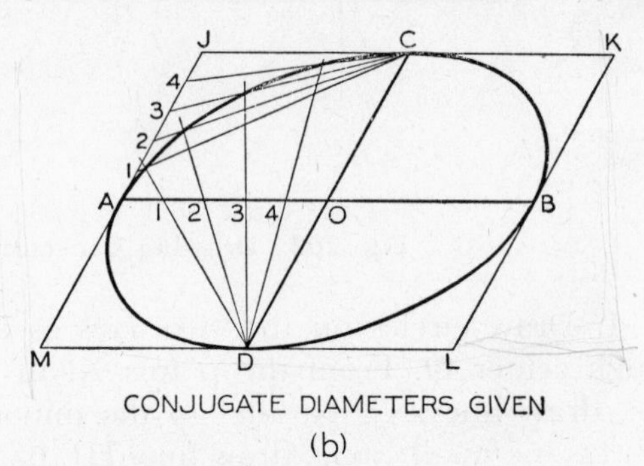

Fig. 204 Parallelogram Ellipse (§160).

the same number of equal parts, and draw *light* lines through these points, as shown. The intersection of like-numbered lines will be points on the ellipse. Locate points in the remaining three quadrants in a similar manner.

Sketch the ellipse lightly through the points; then heavy-in the final ellipse with the aid of the irregular curve, Fig. 89.

161. To Find the Axes of an Ellipse, with Conjugate Diameters Given.

Fig. 205 (a). Conjugate Diameters AB and CD and the Ellipse Are Given. With intersection O of the conjugate diameters (center of ellipse) as center, and any convenient radius, draw a circle to intersect the ellipse in four points. Join these points with straight lines, as shown; the resulting quadrilateral will be a rectangle whose sides are parallel, respectively, to the required major and minor axes. Draw the axes EF and GH parallel to the sides of the rectangle.

Fig. 205 (b). Ellipse Only Is Given. To find the center of the ellipse, draw a circumscribing rectangle or parallelogram about the ellipse; then draw diagonals to intersect at center O, as shown. The axes are then found as shown at (a).

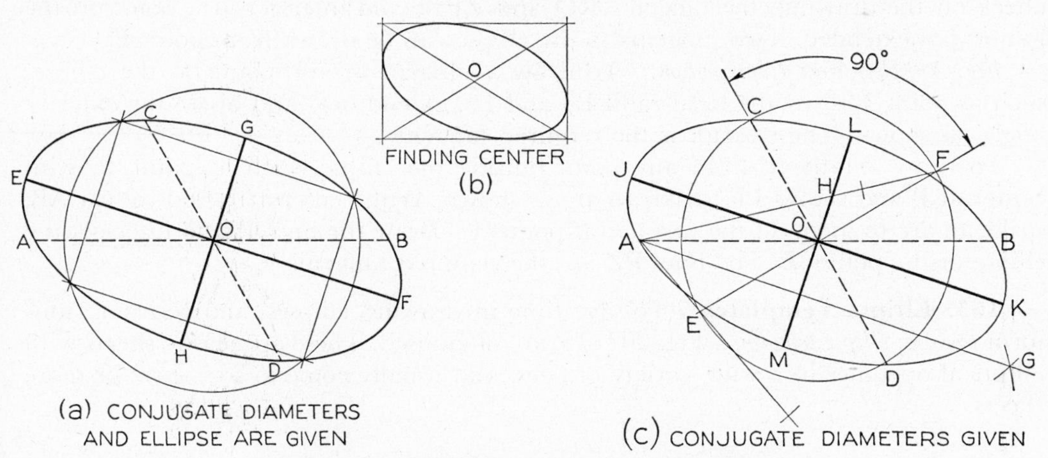

Fig. 205 Finding the Axes of an Ellipse (§161).

Fig. 205 (c). Conjugate Diameters AB and CD Only Are Given. With O as center and CD as diameter, draw a circle. Through center O and perpendicular to CD, draw line EF. From points E and F, where this perpendicular intersects the circle, draw lines FA and EA to form angle FAE. Draw the bisector AG of this angle. The major axis JK will be parallel to this bisector, and the minor axis LM will be perpendicular to it. The length AH will be one-half the major axis, and HF one-half the minor axis. The resulting major and minor axes are JK and LM, respectively.

162. To Draw a Tangent to an Ellipse.

Fig. 206 (a). Concentric Ellipse Construction. To draw a tangent at any point on

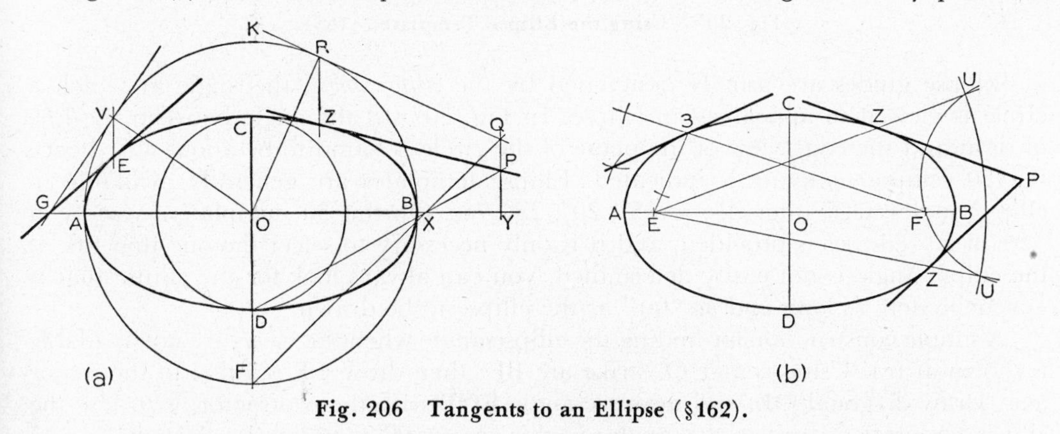

Fig. 206 Tangents to an Ellipse (§162).

the ellipse, as E, draw the ordinate at E to intersect the circle at V. Draw a tangent to the circle at V, §141, and produce it to intersect the major axis produced at G. The line GE is the required tangent.

To draw a tangent from a point outside the ellipse, as P, draw the ordinate PY and extend it. Draw DP, intersecting the major axis at X. Draw FX and extend it to intersect the ordinate through P at Q. Then, from similar triangles, QY:PY = OF:OD. Draw tangent to the circle from Q, §141, find the point of tangency R, and draw the ordinate at R to intersect the ellipse at Z. The line ZP is the required tangent. As a check on the drawing, the tangents RQ and ZP should intersect at a point on the major axis extended. Two tangents to the ellipse can be drawn from point P.

Fig. 206 (b). Foci Construction. To draw a tangent at any point on the ellipse, such as point 3, draw the focal radii E3 and F3, extend one, and bisect the exterior angle, as shown. The bisector is the required tangent.

To draw a tangent from any point outside the ellipse, such as point P, with center at P and radius PF, strike an arc as shown. With center at E and radius AB, strike an arc to intersect the first arc at points U. Draw the lines EU to intersect the ellipse at the points Z. The lines PZ are the required tangents.

163. Ellipse Templates. To save time in drawing ellipses, and to insure uniform results, *ellipse templates*, Fig. 207 (a), are often used. These are plastic sheets with elliptical openings in a wide variety of sizes, and usually come in sets of six or more sheets.

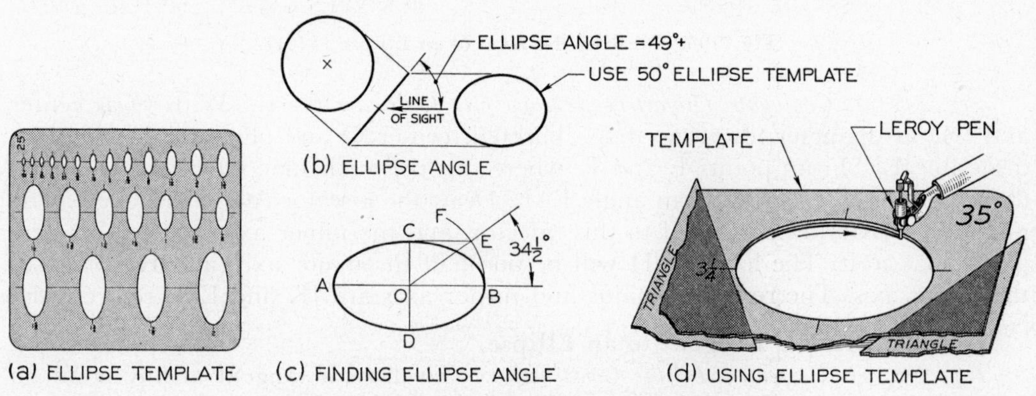

Fig. 207 Using the Ellipse Template (§163).

Ellipse guides are usually designated by the *ellipse angle*, the angle at which a circle is viewed to appear as an ellipse. In Fig. 207 (b) the angle between the line of sight and the edge view of the plane of the circle is found to be about 49°; hence the 50° ellipse template is indicated. Ellipse templates are generally available in ellipse angles at 5° intervals, as 15°, 20°, 25°, etc. On this 50° template a variety of sizes of 50° ellipses is provided, and it is only necessary to select the one that fits. If the ellipse angle is not easily determined, you can always look for the ellipse that is approximately as long and as "fat" as the ellipse to be drawn.

A simple construction for finding the ellipse angle when the views are not available is shown at (c). Using center O, strike arc BF; then draw CE parallel to the major axis. Draw diagonal OE, and measure angle EOB with the protractor, §29. Use the ellipse template nearest to this angle; in this case a 35° template is selected.

Since it is not practicable to have ellipse openings for every exact size that may be required, it is often necessary to use the template somewhat in the manner of an irregular curve. For example, if the opening is too long and too "fat" for the required ellipse, one end may be drawn and then the template shifted slightly to draw the other end. Similarly, one long side may be drawn and then the template shifted slightly to draw the opposite side. In such cases, leave gaps between the four segments, to be filled in freehand or with the aid of an irregular curve. When the differences between the ellipse openings and the required ellipse are small, it is only necessary to lean the pencil or pen slightly outward or inward from the guiding edge to offset the differences.

For inking the ellipses, the Leroy or Wrico pens are recommended. The Leroy pen is shown in Fig. 207 (d). Place triangles under the ellipse template, as shown, so as to lift the template from the paper and prevent ink from spreading under the template; or better still, place a larger opening of another ellipse guide underneath.

164. To Draw an Approximate Ellipse. Fig. 208. For many purposes, particularly where a small ellipse is required, the approximate circular-arc method is perfectly satisfactory. Such an ellipse is sure to be symmetrical and may be quickly drawn. *Given axes AB and CD.*

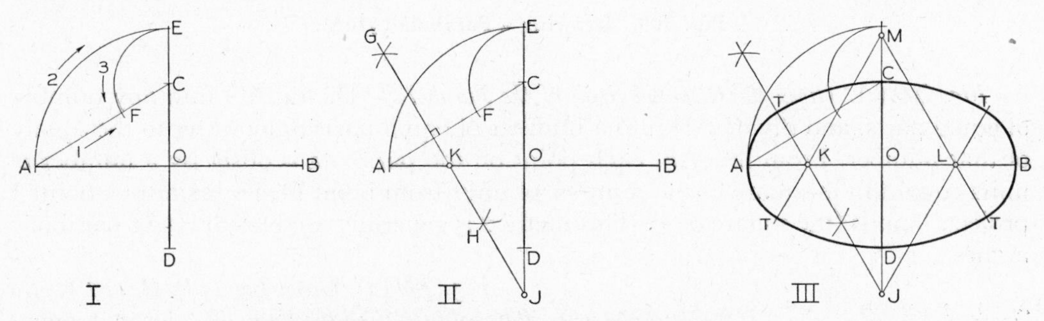

Fig. 208 Drawing Approximate Ellipse (§164).

I. Draw line AC. With O as center and OA as radius, strike the arc AE. With C as center and CE as radius, strike the arc EF.

II. Draw perpendicular bisector GH of the line AF; the points K and J, where it intersects the axes, are centers of the required arcs.

III. Find centers M and L by setting off OL = OK and OM = OJ. Using centers K, L, M, and J, draw circular arcs as shown. The points of tangency T are at the junctures of the arcs on the lines joining the centers.

165. To Draw a Parabola. The curve of intersection between a right circular cone and a plane parallel to one of its elements, Fig. 198 (d), is a *parabola. A parabola may be generated by a point moving so that its distances from a fixed point, the focus, and from a fixed line, the directrix, remain equal.* For example:

Fig. 209 (a). Given Focus and Directrix. A parabola may be "generated" by a pencil guided by a string, as shown. Fasten the string at F and C; its length is GC. The point C is selected at random, its distance from G depending on the desired extent of the curve. Keep the string taut and the pencil against the T-square, as shown.

Fig. 209 (b). Given Focus and Directrix. Draw a line DE parallel to the directrix and at any distance CZ from it. With center at F and radius CZ, strike arcs to intersect the line DE in the points Q and R, which are points on the parabola. Determine as many additional points as are necessary to draw the parabola accurately, by drawing additional lines parallel to line DE and proceeding in the same manner.

A tangent to the parabola at any point G bisects the angle formed by the focal line FG and the line SG perpendicular to the directrix.

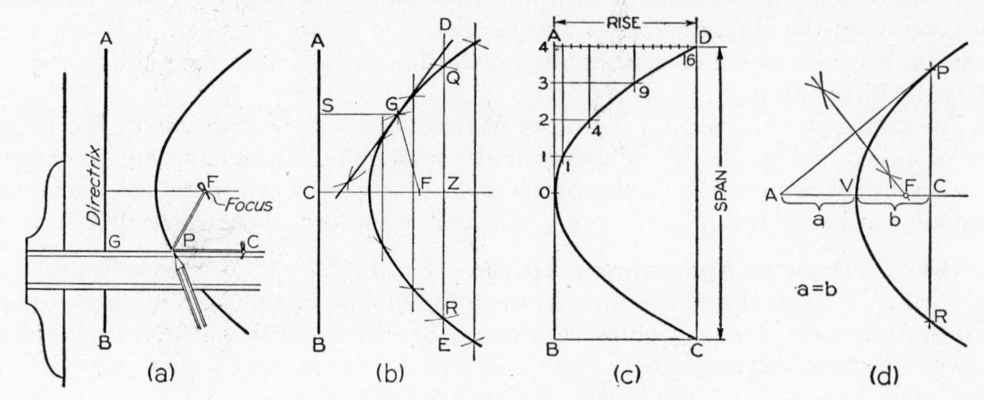

Fig. 209 Drawing a Parabola (§165).

Fig. 209 (c). Given the Rise and Span of the Parabola. Divide AO into any number of equal parts, and divide AD into a number of equal parts amounting to the square of that number. From line AB, each point on the parabola is offset by a number of units equal to the square of the number of units from point O. For example, point 3 projects 9 units (the square of 3). This method is generally used for drawing parabolic arches.

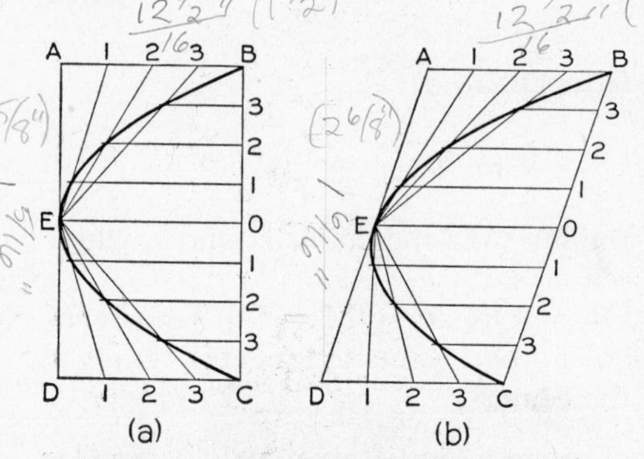

Fig. 210 Drawing a Parabola (§165).

Fig. 209 (d). Given Points P, R, and V of a Parabola, to Find the Focus F. Draw tangent at P, making a = b. Draw perpendicular bisector of AP, which intersects the axis at F, the focus of the parabola.

Fig. 210 (a) or (b). Given Rectangle or Parallelogram ABCD. Divide BC into any even number of equal parts, and divide the sides AB and DC each into half as many parts, and draw lines as shown. The intersections of like-numbered lines are points on the parabola.

Practical Applications. The parabola is used for reflecting surfaces for light and sound, for vertical curves in highways, for forms of arches, and approximately for forms of the curves of cables for suspension bridges. It is also used to show the bending moment at any point on a uniformly loaded beam or girder.

166. To Join Two Points by a Parabolic Curve. Fig. 211. *Let X and Y be the given points.* Assume any point O, and draw tangents XO and YO. Divide XO and YO into the same number of equal parts, number the division points as shown, and connect corresponding points. These lines are tangents of the required parabola and form its envelope. Sketch a light smooth curve, and then heavy-in the curve with the aid of the irregular curve, §66.

These parabolic curves are more pleasing in appearance than circular arcs and are useful in machine design. If the tangents OX and OY are equal, the axis of the parabola will bisect the angle between them.

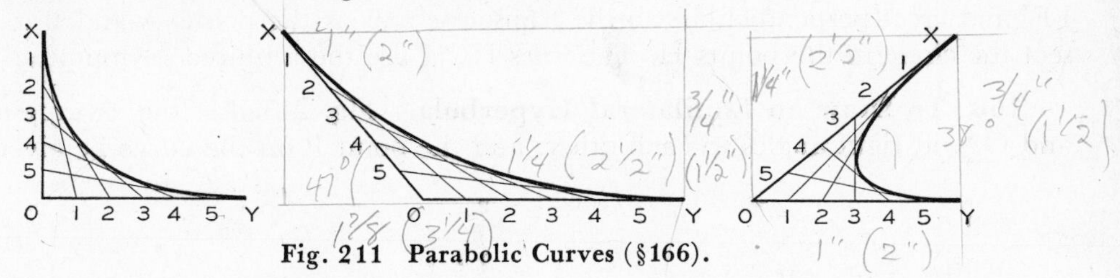

Fig. 211 Parabolic Curves (§166).

167. To Draw a Hyperbola. The curve of intersection between a right circular cone and a plane making an angle with the axis smaller than that made by the elements, Fig. 198 (e), is a *hyperbola. A hyperbola is generated by a point moving so that the difference of its distances from two fixed points, the foci, is constant and equal to the transverse axis of the hyperbola.*

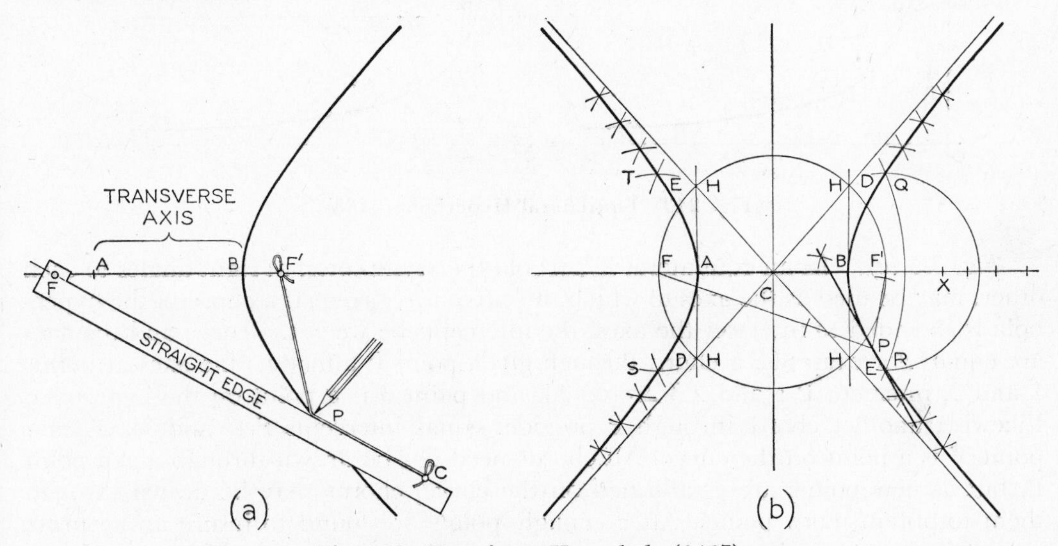

Fig. 212 Drawing a Hyperbola (§167).

Fig. 212 (a). Let F and F′ be the foci and AB the transverse axis. The curve may be generated by a pencil guided by a string, as shown. Fasten a string at F′ and C; its length is FC minus AB. The point C is chosen at pleasure; its distance from F depends on the desired extend of the curve.

Fasten the straightedge at F. It it is revolved about F, with the pencil point moving against it and with the string taut, the hyperbola may be drawn as shown.

Fig. 212 (b). To construct the curve geometrically, select any point X in the transverse axis produced. With centers at F and F′ and BX as radius, strike the arcs DE. With the same centers and AX as radius, strike arcs to intersect the arcs first drawn in the points Q, R, S, and T, which are points of the required hyperbola. Find as many additional points as necessary to draw the curves accurately by selecting other points similar to point X along the transverse axis, and proceeding as described for point X.

To draw the tangent to a hyperbola at a given point P, bisect the angle between the focal radii FP and F′P. The bisector is the required tangent.

To draw the asymptotes HCH of the hyperbola, draw a circle with the diameter FF′ and erect perpendiculars to the transverse axis at the points A and B to intersect the circle in the points H. The lines HCH are the required asymptotes.

168. To Draw an Equilateral Hyperbola. Fig. 213. Let the asymptotes OB and OA, at right angles to each other, and the point P on the curve be given.

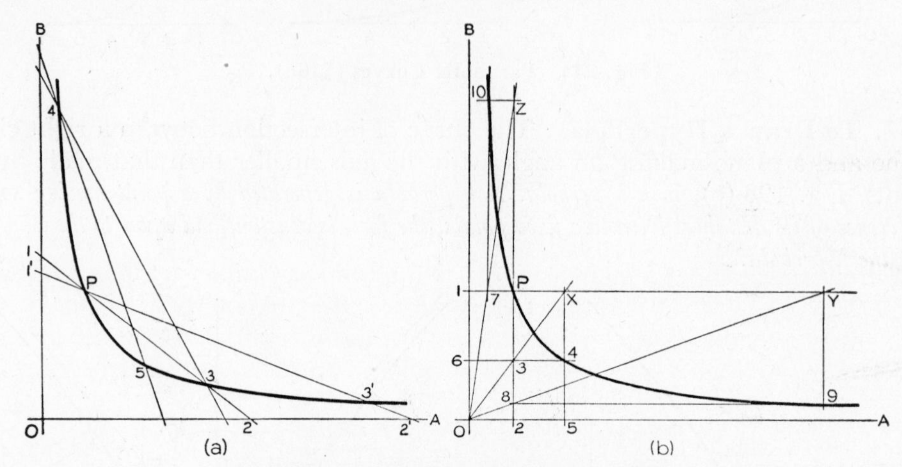

Fig. 213 Equilateral Hyperbola (§168).

Fig. 213 (a). In an equilateral hyperbola the asymptotes, at right angles to each other, may be used as the axes to which the curve is referred. If a chord of the hyperbola is extended to intersect the axes, the intercepts between the curve and the axes are equal. For example, a chord through given point P intersects the axes at points 1 and 2, intercept P,1 and 2,3 are equal, and point 3 is a point on the hyperbola. Likewise, another chord through P provides equal intercepts P,1′ and 3′, 2′, and point 3′ is a point on the curve. All chords need not be drawn through given point P, but as new points are established on the curve, chords may be drawn through them to obtain more points. After enough points are found to insure an accurate curve, the hyperbola is drawn with the aid of the irregular curve, §66.

Fig. 213 (b). In an equilateral hyperbola, the coordinates are related so that their products remain constant. Through given point P, draw lines 1,P,Y and 2,P,Z parallel, respectively, to the axes. From the origin of coordinates O, draw any diagonal intersecting these two lines at points 3 and X. At these points draw lines parallel to the axes, intersecting at point 4, a point on the curve. Likewise, another diagonal from O intersects the two lines through P at points 8 and Y, and lines through these points parallel to the axes intersect at point 9, another point on the curve. A third diagonal similarly produces point 10 on the curve, and so on. Find as many points as necessary for a smooth curve, and draw the parabola with the aid of the irregular curve, §66. It is evident from the similar triangles O,X,5 and O,3,2 that P,1 × P,2 = 4,5 × 4,6.

The equilateral hyperbola can be used to represent varying pressure of a gas as the volume varies, because the pressure varies inversely as the volume; i.e., pressure × volume is constant.

169. To Draw a Spiral of Archimedes. Fig. 214. To find points on the curve, draw lines through the pole C, making equal angles with each other, such as 30° angles, and beginning with any one line, set off any distance, such as $\frac{1}{16}''$; set off twice that distance on the next line, three times on the third, and so on. Through the points thus determined, draw a smooth curve, using the irregular curve, §66.

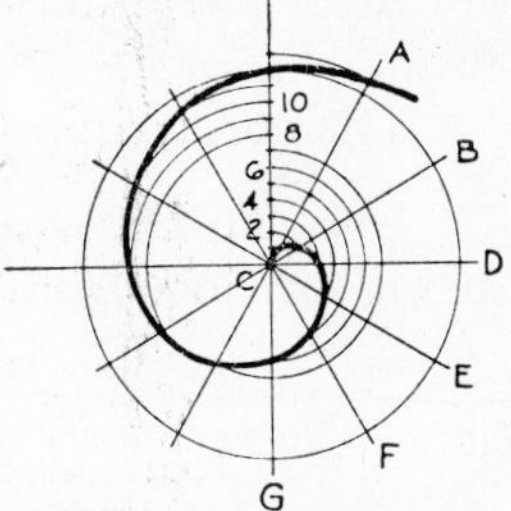

Fig. 214 Spiral of Archimedes (§169).

170. To Draw a Helix. Fig. 215. *A helix is generated by a point moving around and along the surface of a cylinder or cone with a uniform angular velocity about the axis, and with a uniform linear velocity in the direction of the axis.* A cylindrical helix is generally known simply as a "helix." The distance measured parallel to the axis traversed by the point in one revolution is called the *lead.*

If the cylindrical surface upon which a helix is generated is rolled out onto a plane, the helix becomes a straight line as shown in Fig. 215 (a), and the portion below the helix becomes a right triangle, the altitude of which is equal to the lead of the helix and the length of the base equal to the circumference of the cylinder. Such a helix can, therefore, be defined as the shortest line which can be drawn on the surface of a cylinder connecting two points not on the same element.

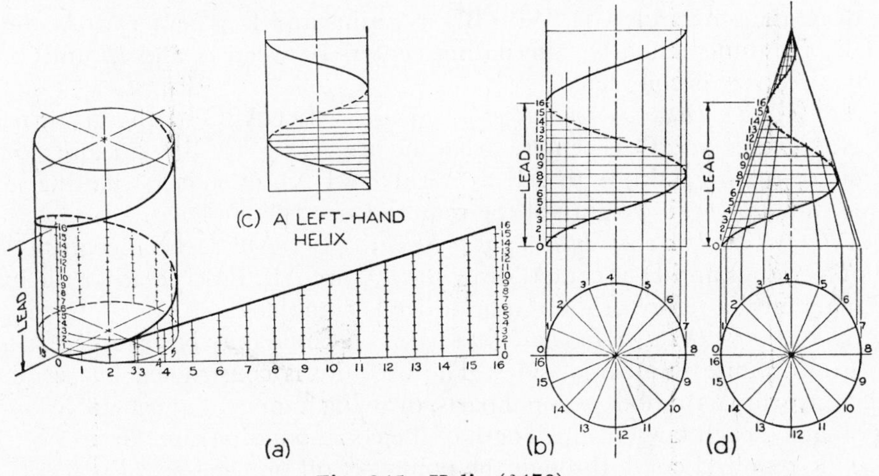

Fig. 215 Helix (§170).

To draw the helix, draw two views of the cylinder upon which the helix is generated, (b), and divide the circle of the base into any number of equal parts. On the rectangular view of the cylinder, set off the lead and divide it into the same number of equal parts as the base. Number the divisions as shown, in this case sixteen. When the generating point has moved one-sixteenth of the distance around the cylinder, it will have risen one-sixteenth of the lead; when it has moved half-way around the cylinder, it will have risen half the lead, and so on. Points on the helix are found by projecting up from point 1 in the circular view to line 1 in the rectangular view, from point 2 in the circular view to line 2 in the rectangular view, and so on.

The helix shown is a *right-hand helix.* In a *left-hand helix,* the visible portions of the curve are inclined in the opposite direction, i.e., downward to the right, as shown at (c). The helix shown at (b) can be converted into a left-hand helix by interchanging the visible and hidden lines.

The helix finds many applications in industry, as in screw threads, worm gears, conveyors, "spiral" stairways, and so on. The stripes of a barber pole are helical in form.

The construction for a right-hand conical helix is shown at (d).

171. To Draw an Involute. Fig. 216. The path of a point on a string, as the string unwinds from a line, a polygon, or a circle, is an *involute*.

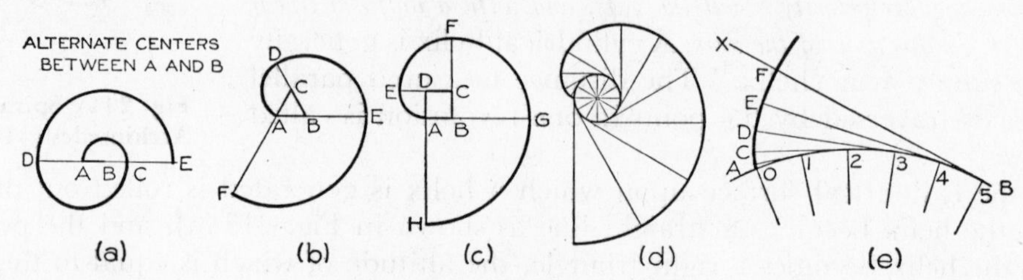

Fig. 216 Involutes (§171).

Fig. 216 (a). To Draw an Involute of a Line. Let AB be the given line. With AB as radius and B as center, draw the semicircle AC. With AC as radius and A as center, draw the semicircle CD. With BD as radius and B as center, draw the semicircle DE. Continue similarly, alternating centers between A and B, until a figure of the required size is completed.

Fig. 216 (b). To Draw an Involute of a Triangle. Let ABC be the given triangle. With CA as radius and C as center, strike the arc AD. With BD as radius and B as center, strike the arc DE. With AE as radius and A as center, strike the arc EF. Continue similarly until a figure of the required size is completed.

Fig. 216 (c). To Draw an Involute of a Square. Let ABCD be the given square. With DA as radius and D as center, draw the 90° arc AE. Proceed as for the involute of a triangle until a figure of the required size is completed.

Fig. 216 (d). To Draw an Involute of a Circle. A circle may be regarded as a polygon with an infinite number of sides. The involute is constructed by dividing the circumference into a number of equal parts, drawing a tangent at each division point, setting off along each tangent the length of the corresponding circular arc, §152, and drawing the required curve through the points set off on the several tangents.

Fig. 216 (e). The involute may be generated by a point on a straight line which is rolled on a fixed circle. Points on the required curve may be determined by setting off equal distances 0-1, 1-2, 2-3, etc., along the circumference, drawing a tangent at each division point, and proceeding as explained for (d).

The involute of a circle is used in the construction of involute gear teeth, §548. In this system, the involute forms the face and a part of the flank of the teeth of gear wheels; the outlines of the teeth of racks are straight lines.

172. To Draw a Cycloid. Fig. 217. *A cycloid may be generated by a point P in the circumference of a circle which rolls along a straight line.*

Given the generating circle and the straight line AB tangent to it, make the distances CA and CB each equal to the semi-circumference of the circle, Fig. 197 (a). Divide these distances and the semi-circumference into the same number of equal parts, six for instance, and number them consecutively as shown. Suppose the circle to roll to the left; when point 1 of the circle reaches point 1′ of the line, the center of the circle will be at D, point 7 will be the highest point of the circle, and the generating point 6 will be at the same distance from the line AB as point 5 is when the circle is in its central position. Hence, to find the point P′, draw a line

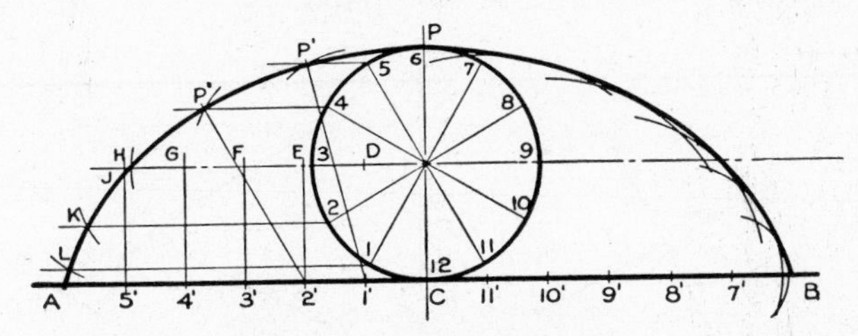

Fig. 217 Cycloid (§172).

through point 5 parallel to AB and intersect it with an arc drawn from the center D with a radius equal to that of the circle. To find point P″, draw a line through point 4 parallel to AB, and intersect it with an arc drawn from the center E, with a radius equal to that of the circle. Points J, K, and L are found in a similar manner.

Another method which may be employed is shown in the right half of the figure. With center at 11′ and the chord 11-6 as radius, strike an arc. With 10′ as center and the chord 10-6 as radius, strike an arc. Continue similarly with centers 9′, 8′, and 7′. Draw the required cycloid tangent to these arcs.

The student may use either method; the second is the shorter one, and is preferred. It is evident, from the tangent arcs drawn in the manner just described, that the line joining the generating point and the point of contact for the generating circle is a normal of the cycloid; the lines 1′P′ and 2′P″, for instance, are normals; this property makes the cycloid suitable for the outlines of gear teeth.

173. To Draw an Epicycloid or a Hypocycloid. Fig. 218. If the generating point P is on the circumference of a circle which rolls along the convex side of a

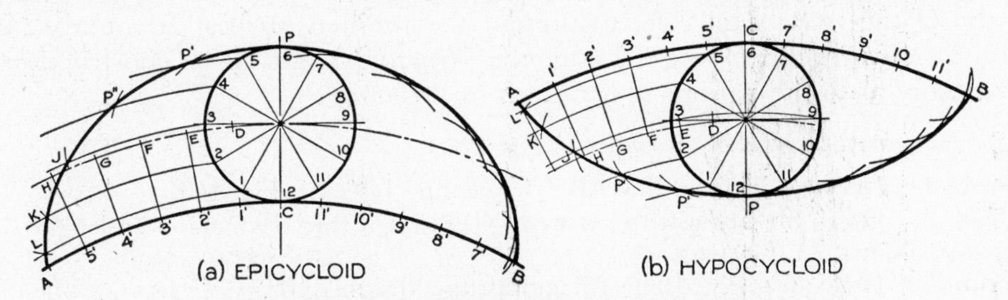

Fig. 218 Epicycloid and Hypocycloid (§173).

larger circle, (a), the curve generated is an *epicycloid.* If the circle rolls along the concave side of a larger circle, (b), the curve generated is a *hypocycloid.* These curves, like the cycloid, are used to form the outlines of certain gear teeth, and are therefore of practical importance in machine design.

174. Geometric Construction Problems. Fig. 219. The following geometric constructions should be made very accurately, with a hard pencil (2H to 4H) having a long sharp conical point. Draw given and required lines dark, and medium in thickness, and draw construction lines *very light*. Do not erase construction lines. Indicate points and lines as described in §108.

Geometric construction problems in convenient form for solution may be found in *Technical Drawing Problems*, by Giesecke, Mitchell, and Spencer, and in *Technical Drawing Problems, Series 2*, by Spencer and Grant, both designed to accompany this text, and published by The Macmillan Company.

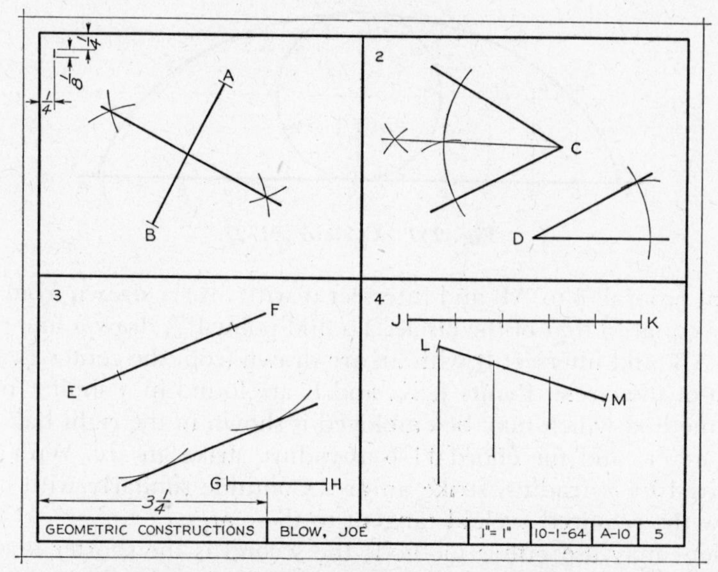

Fig. 219 Geometric Constructions—Layout A-2.

Listed below are a large number of problems from which the instructor will make assignments. Use Layout A-2 (see back endpaper) divided into four parts, as shown in Fig. 219. Additional sheets, with other problems selected and drawn on the same sheet layout, may be assigned by the instructor.

The student should exercise care in setting up each problem so as to make the best use of the space available, to present the problem to best advantage, and to produce a pleasing appearance. Letter the principal points of all constructions in a manner similar to the various illustrations in this chapter.

The first four problems below are shown in Fig. 219.

Prob. 1. Draw an inclined line AB $2\frac{1}{2}''$ long and bisect it, Fig. 160.

Prob. 2. Draw any angle with vertex at C. Bisect it, Fig. 162, and transfer one-half in reversed position at D, Fig. 163.

Prob. 3. Draw an inclined line EF and assume distance GH = $1\frac{5}{8}''$. Draw a line parallel to EF and at the distance GH from it, Fig. 165 (a).

Prob. 4. Draw the line JK $3\frac{3}{4}''$ long and divide it into five equal parts with the dividers, §49. Draw a line LM $2\frac{13}{16}''$ long and divide it into three equal parts by the parallel-line method, Fig. 167.

Prob. 5. Draw a line OP $3\frac{5}{8}''$ long and divide it into three proportional parts to 3, 5, and 9, Fig. 169 (a).

Prob. 6. Draw a line $3\frac{7}{16}''$ long and divide it into parts proportional to the square of X where X = 1, 2, 3, and 4, Fig. 169 (c).

Prob. 7. Draw a triangle having sides 3″, $3\frac{1}{4}''$, and $2\frac{1}{2}''$, Fig. 171. Bisect the three interior angles, Fig. 162. The bisectors should meet at a point. Draw the inscribed circle.

Prob. 8. Draw a right triangle having legs $2\frac{1}{2}''$ and $1\frac{9}{16}''$, Figs. 155 and 172, and draw a circle through the three vertexes, Fig. 160.

Prob. 9. Draw an inclined line QR $3\frac{5}{16}''$ long. Select a point P on the line $1\frac{1}{4}''$ from Q, and erect a perpendicular, Fig. 170 (c). Assume a point S about $1\frac{3}{4}''$ from the line, and erect a perpendicular from S to the line, Fig. 170 (b).

Prob. 10. Draw two lines making an angle of $35\frac{1}{2}°$ with each other, using the tangent method, Fig. 173 (a). Check with protractor, §29.

Prob. 11. Draw two lines making an angle of 33°16′ with each other, using the sine method, Fig. 173 (b). Check with protractor, §29.

Prob. 12. Draw an equilateral triangle, Fig. 155 (a), having $2\frac{1}{2}''$ sides, Fig. 174 (a). Bisect the interior angles, Fig. 162. Draw the inscribed circle, using the intersection of the bisectors as center.

Prob. 13. Draw inclined line TU $2\frac{3}{16}''$ long, and draw a square on TU as a given side, Fig. 175 (a).

Prob. 14. Draw a $2\frac{1}{8}''$ dia. circle (lightly); then inscribe a square in the circle and circumscribe a square on the circle, Fig. 175 (c) and (d).

Prob. 15. Draw a $2\frac{1}{2}''$ dia. circle (lightly), find the vertexes of a regular inscribed pentagon, Fig. 176 (a), and join the vertexes to form a five-pointed star, Fig. 98 (h).

Prob. 16. Draw a $2\frac{1}{2}''$ dia. circle (lightly), inscribe a hexagon, Fig. 177 (b), and circumscribe a hexagon, Fig. 178 (d).

Prob. 17. Draw a square (lightly) with $2\frac{1}{2}''$ sides, Fig. 175 (b), and inscribe an octagon, Fig. 180 (b).

Prob. 18. Draw a triangle similar to that in Fig. 181 (a), having sides 2″, $1\frac{1}{2}''$, and $2\frac{7}{8}''$ long, and transfer the triangle to a new location and turned 180° similar to that in Fig. 181 (b). Check by "prick-point" method, §136.

Prob. 19. In center of space, draw a rectangle $3\frac{1}{2}''$ wide and $2\frac{3}{8}''$ high. Show construction for reducing this rectangle until it is $2\frac{5}{8}''$ wide and again when $2\frac{3}{4}''$ wide, Fig. 183 (b).

Prob. 20. Draw three points arranged approximately as in Fig. 184 (a), and draw a circle through the three points.

Prob. 21. Draw a $2\frac{1}{4}''$ dia. circle. Assume a point S on the left side of the circle and draw a tangent at that point, Fig. 186 (a). Assume a point T to the right of the circle 2″ from its center, and draw two tangents to the circle through the point, Fig. 186 (b).

Prob. 22. Through center of space, draw horizontal center line; then draw two circles 2″ dia. and $1\frac{1}{4}''$ dia., respectively, with centers $2\frac{1}{8}''$ apart. Locate the circles so that the construction will be centered in the space. Draw "open belt" tangents to the circles, Fig. 187 (a).

Prob. 23. Same as Prob. 22, except draw "crossed belt" tangents to the circle, Fig. 187 (b).

Prob. 24. Draw vertical line VW $1\frac{3}{8}''$ from left side of space. Assume point P $1\frac{3}{4}''$ farther to the right and 1″ down from top of space. Draw a $2\frac{1}{4}''$ dia. circle through P, tangent to VW, Fig. 188 (a).

Prob. 25. Draw vertical line XY $1\frac{3}{8}''$ from left side of space. Assume point P $1\frac{3}{4}''$ farther to the right and 1″ down from top of space. Assume point Q on line XY and 2″ from P. Draw circle through P and tangent to XY at Q, Fig. 188 (b).

Prob. 26. Draw $2\frac{1}{2}''$ dia. circle with center C $\frac{5}{8}''$ directly to left of center of space. Assume point P at the lower right and $2\frac{3}{8}''$ from C. Draw an arc with 1″ radius through P and tangent to the circle, Fig. 188 (c).

Prob. 27. Draw a vertical line and a horizontal line, each $2\frac{1}{2}''$ long, Fig. 189 (a). Draw arc with $1\frac{1}{2}''$ radius, tangent to the lines.

Prob. 28. Draw horizontal line $\frac{3}{4}''$ up from bottom of space. Select a point on the line 2″ from the left side of space, and through it draw a line upward to the right at 60° to horizontal. Draw arcs within obtuse angle and acute angle, respectively, tangent to the two lines, Fig. 190.

Prob. 29. Draw two intersecting lines making an angle of 60° with each other as in Fig. 190 (a). Assume a point P on one line at a distance of $1\frac{3}{4}''$ from the intersection. Draw an arc tangent to both lines with one point of tangency at P, Fig. 185.

Prob. 30. Draw vertical line AB $1\frac{1}{4}''$ from left side of space. Draw arc of $1\frac{5}{8}''$ radius with center 3″ to right of line and in lower right portion of space. Draw arc of 1″ radius tangent to AB and to the arc, Fig. 191.

Prob. 31. With centers $\frac{3}{4}''$ up from bottom of space, and $3\frac{3}{8}''$ apart, draw arcs of radius $1\frac{3}{4}''$ and $\frac{15}{16}''$, respectively. Draw arc of $1\frac{1}{4}''$ radius tangent to the two arcs, Fig. 192.

Prob. 32. Draw two circles as in Prob. 22. Draw arc of $2\frac{3}{4}''$ radius tangent to upper sides of circles and enclosing them, Fig. 193 (a). Draw arc of 2″ radius tangent to the circles but enclosing only the smaller circle, Fig. 193 (b).

Prob. 33. Draw two parallel inclined lines $1\frac{3}{4}''$ apart. Choose a point on each line and connect them with an ogee curve tangent to the two parallel lines, Fig. 195 (a).

Prob. 34. Draw an arc of $2\frac{1}{8}''$ radius that subtends an angle of 90°. Find the length of the arc by two methods, Fig. 197 (a) and (c). Calculate the length of the arc and compare with the lengths determined graphically. See note at end of §152.

Prob. 35. Draw major axis 4″ long (horizontally) and minor axis $2\frac{1}{2}''$ long, with their intersection at the center of the space. Draw ellipse by foci method with at least five points in each quadrant, Fig. 200.

Prob. 36. Draw axes as in Prob. 35, but draw ellipse by trammel method, Fig. 201.

Prob. 37. Draw axes as in Prob. 35, but draw ellipse by concentric-circle method, Fig. 202.

Prob. 38. Draw axes as in Prob. 35, but draw ellipse by parallelogram method, Fig. 204 (a).

Prob. 39. Draw conjugate diameters intersecting at center of space. Draw $3\frac{1}{2}''$ diameter horizontally, and $2\frac{3}{4}''$ diameter at 60° with horizontal. Draw oblique-circle ellipse, Fig. 203. Find at least 5 points in each quadrant.

Prob. 40. Draw conjugate diameters as in Prob. 39, but draw ellipse by parallelogram method, Fig. 204 (b).

Prob. 41. Draw axes as in Prob. 35, but draw approximate ellipse, Fig. 208.

Prob. 42. Draw a parabola with a vertical axis, and the focus $\frac{1}{2}''$ from the directrix, Fig. 209 (b). Find at least 9 points on the curve.

Prob. 43. Draw a hyperbola with a horizontal transverse axis 1″ long and the foci $1\frac{1}{2}''$ apart, Fig. 212 (b). Draw the asymptotes.

Prob. 44. Draw horizontal line near bottom of space, and vertical line near left side of space. Assume point P $\frac{5}{8}''$ to right of vertical line and $1\frac{1}{4}''$ above horizontal line. Draw equilateral hyperbola through P and with reference to the two lines as asymptotes. Use either method of Fig. 213.

Prob. 45. Using the center of the space as the pole, draw a spiral of Archimedes with the generating point moving in a counterclockwise direction and away from the pole at the rate of 1″ in each convolution, Fig. 214.

Prob. 46. Through center of space, draw horizontal center line, and on it construct a right-hand helix 2″ dia., $2\frac{1}{2}''$ long, and with a lead of 1″, Fig. 215. Draw only a half-circular end view.

Prob. 47. Draw the involute of an equilateral triangle with $\frac{1}{2}''$ sides, Fig. 216 (b).

Prob. 48. Draw the involute of a $\frac{3}{4}''$ dia. circle, Fig. 216 (d).

Prob. 49. Draw a cycloid generated by a $1\frac{1}{4}''$ dia. circle rolling along a horizontal straight line, Fig. 217.

Prob. 50. Draw an epicycloid generated by a $1\frac{1}{4}''$ dia. circle rolling along a circular arc having a radius of $2\frac{1}{2}''$, Fig. 218 (a).

Prob. 51. Draw a hypocycloid generated by a $1\frac{1}{4}''$ dia. circle rolling along a circular arc having a radius of $2\frac{1}{2}''$, Fig. 218 (b).

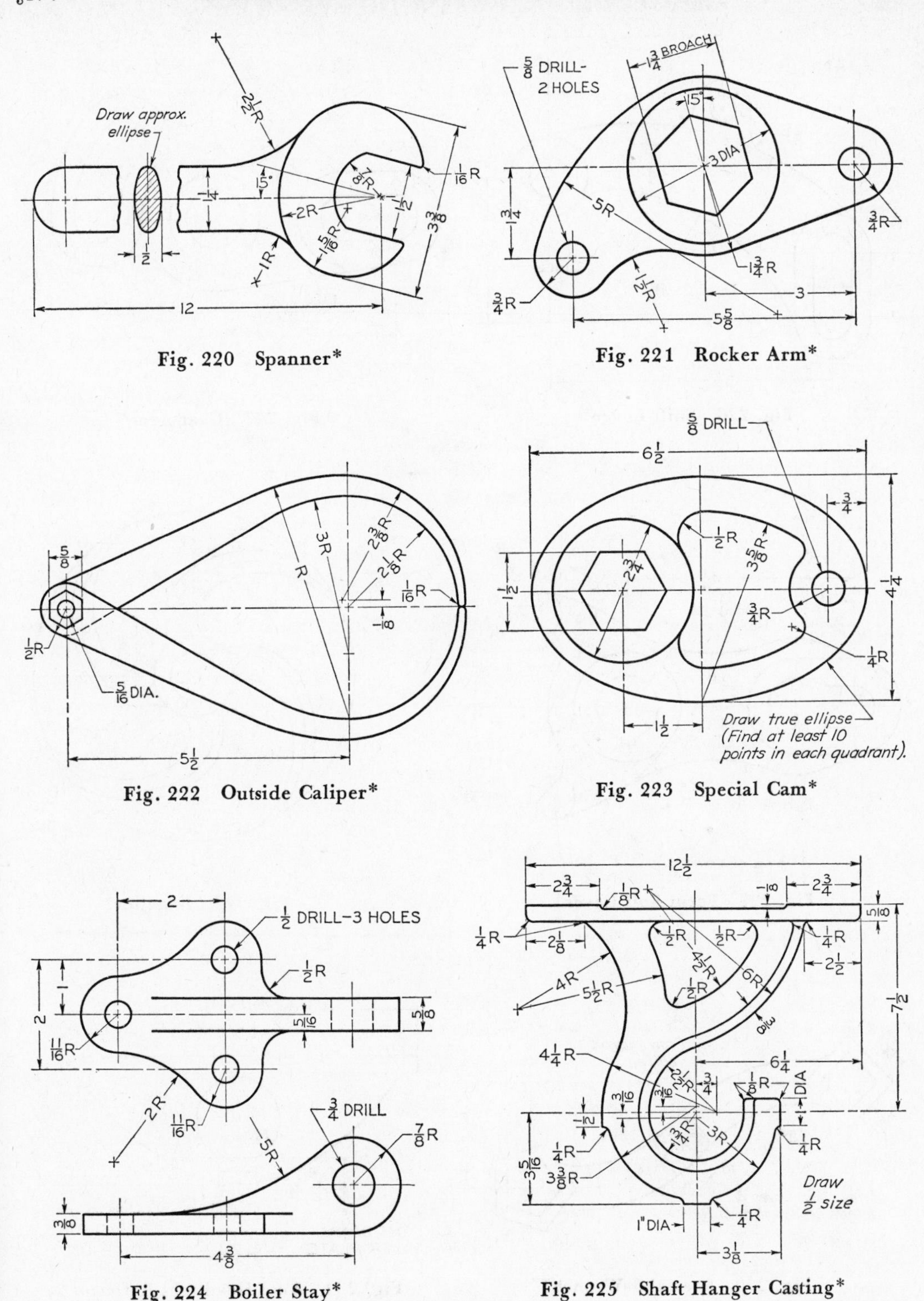

Fig. 220 Spanner*

Fig. 221 Rocker Arm*

Fig. 222 Outside Caliper*

Fig. 223 Special Cam*

Fig. 224 Boiler Stay*

Fig. 225 Shaft Hanger Casting*

*Using Layout A-2, draw assigned problem with instruments. Omit dimensions and notes unless assigned by instructor.

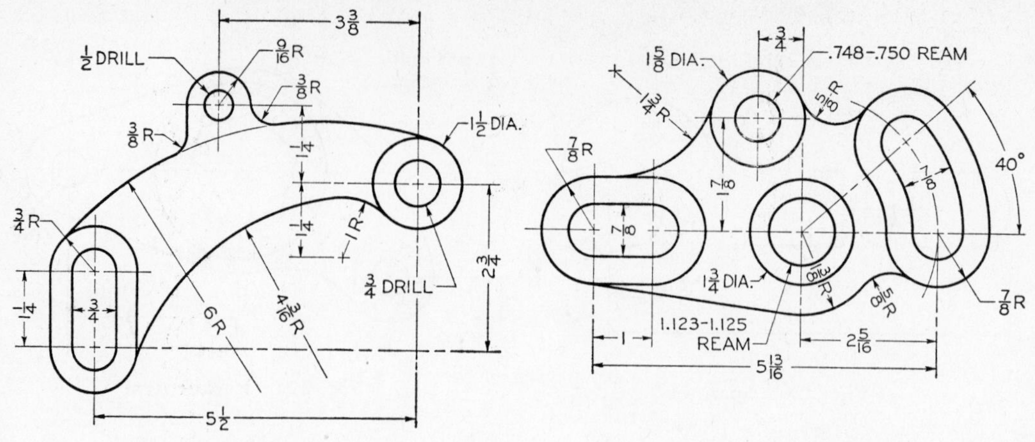

Fig. 226 Shift Lever*

Fig. 227 Gear Arm*

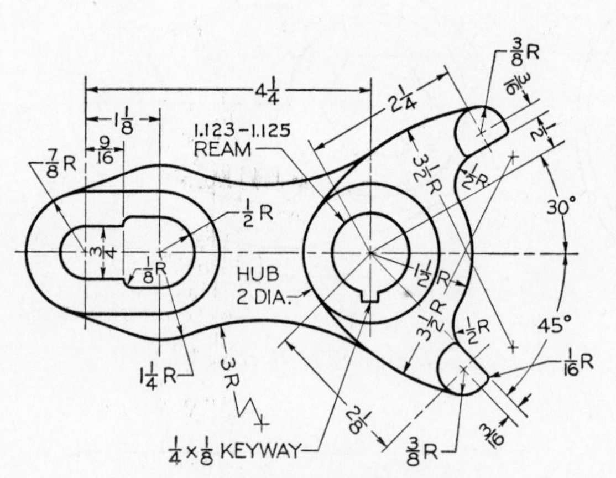

Fig. 228 Form Roll Lever*

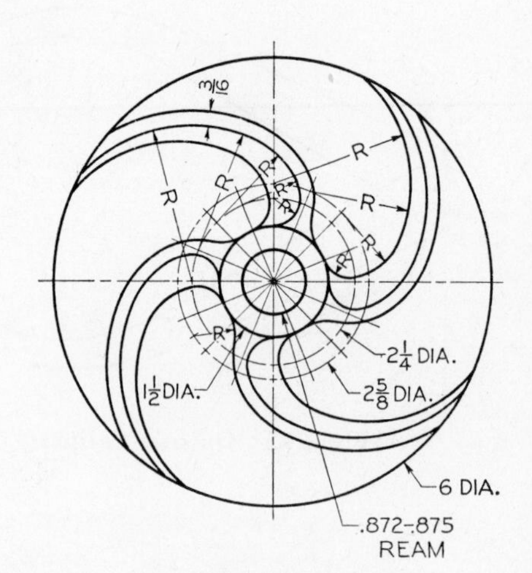

Fig. 229 Impeller*

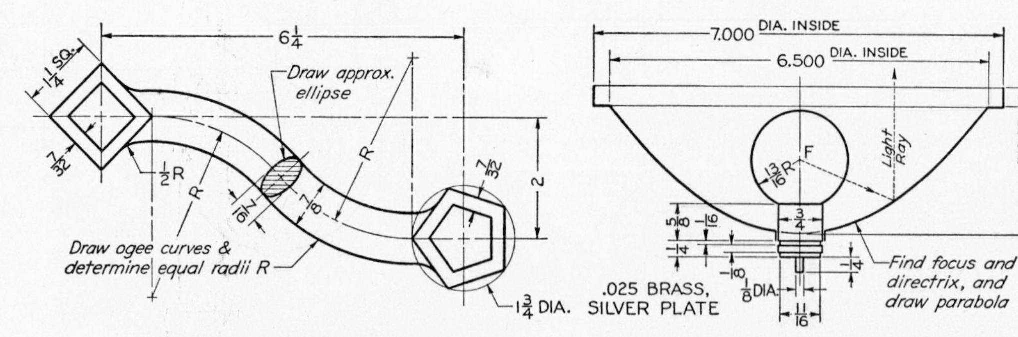

Fig. 230 Special S-Wrench*

Fig. 231 Auto Headlight Reflector*

*Using Layout A-2, draw assigned problem with instruments. Omit dimensions and notes unless assigned by instructor.

CHAPTER 5

TECHNICAL SKETCHING AND SHAPE DESCRIPTION

175. Importance of Technical Sketching. The importance of freehand sketching to the engineer cannot be overestimated. To the person who possesses a complete knowledge of drawing as a language, the ability to execute quick, accurate, and clear sketches constitutes a valuable means of expression. There is an old Chinese saying that "one picture is worth a thousand words."

Most original mechanical ideas find their first expression through the medium of a freehand sketch, §430. It is a valuable means of amplifying and clarifying, as well as recording, verbal explanations. Executives resort to it daily to explain their

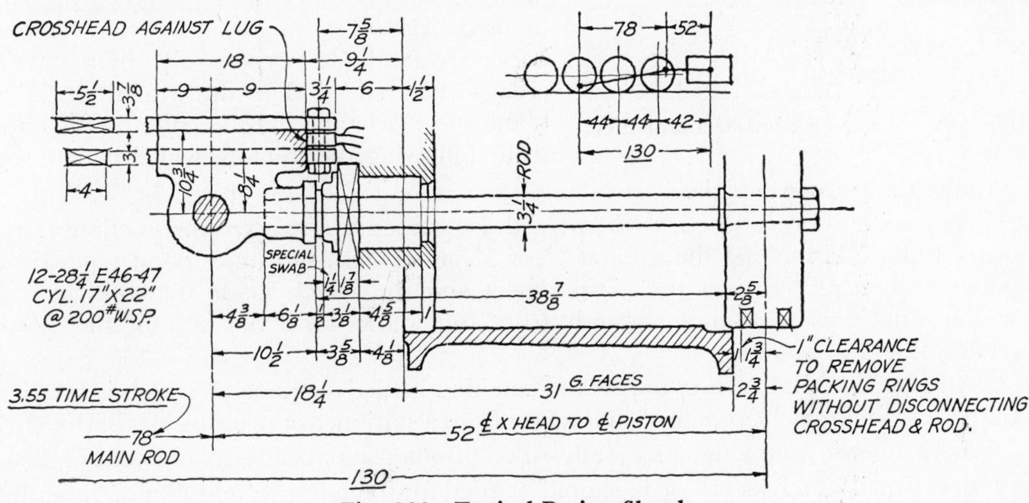

Fig. 232 Typical Design Sketch.

ideas to subordinates. Engineers often prepare their designs and turn them over to their detailers or draftsmen in this convenient form, Fig. 232. Freehand sketches are of great assistance to the designer in organizing his thoughts and recording his ideas. They are an effective and economical means of formulating various solutions to a given problem so that a choice can be made between them at the outset. Often much time can be lost if the designer starts his scaled layout before adequate preliminary study with the aid of sketches. Information concerning changes in design or covering replacement of broken parts or lost drawings is usually conveyed through sketches. Many engineers consider the ability to render serviceable sketches of even greater value to them than skill in mechanical drawing. The draftsman will find daily use for this valuable means of formulating, expressing, and recording ideas in his work.

The degree of perfection required in a given sketch depends upon its use. Sketches hurriedly made to supplement oral description may be rough and incomplete. On the other hand, if a sketch is the medium of conveying important and precise information to engineers or to workers, it should be executed as carefully as possible under the circumstances.

176. Sketching Materials. One of the advantages of freehand sketching is that it requires only pencil, paper, and eraser—items that anyone has for ready use.

When sketches are made in the field, where an accurate record is required, a small notebook or sketching pad is frequently used. Often clip boards holding manila or bond paper are employed.

Cross-section paper is helpful to the sketcher, especially to the beginner or to the person who cannot sketch reasonably well without guide lines. It is available in rolls, sheets, and in pad form. Paper with $\frac{1}{8}''$ or $\frac{1}{4}''$ squares is recommended. Such paper is convenient for sketching to scale, since values can be assigned to the squares, and the squares counted to secure proportional distances, as shown in Fig. 233.

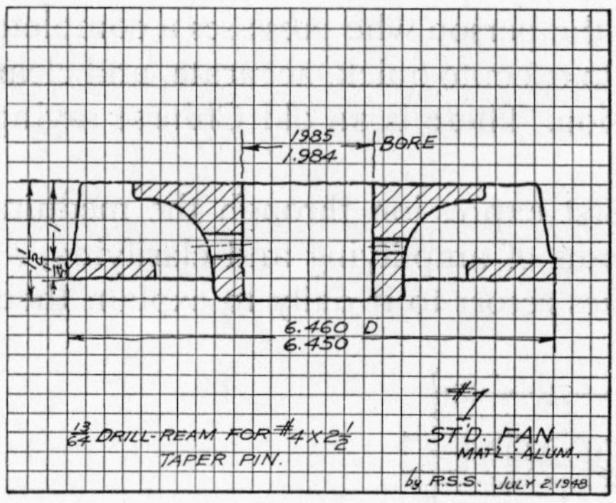

Fig. 233 Sketch on Cross-Section Paper.

Where multiple copies of a sketch are needed, a special tracing paper with cross-section lines that do not reproduce on a blueprint can be used. Also, sketching pads of plain tracing paper are available, accompanied by a master cross-section sheet. The draftsman places a blank sheet over the master grid and sketches while following the squares, which he can see through the transparent sheet.

An excellent procedure is to draw, with instruments, a master cross-section sheet, using India ink, making the squares $\frac{1}{4}''$ or $\frac{1}{8}''$, as desired. Ordinary bond typewriter paper is then placed over the master sheet and the sketch made thereon. Such a sketch is not only more uniform and "true," but shows up better because the cross-section lines are absent.

An excellent plain white paper for sketching is "ledger paper," which is heavier than bond and may be procured at low cost from a stationer or printing establishment.

For isometric sketching, a specially-ruled "isometric paper" is available, Fig. 276.

Soft pencils, such as HB or F, should be used for freehand sketching. For carefully made sketches, two erasers are recommended, an artgum and an ordinary soft pencil eraser, §22.

177. Types of Sketches. Since technical sketches are made of three-dimensional objects, the form of the sketch conforms approximately to one of the four standard types of projection, as shown in Fig. 234. In *multiview* projection, (a), the object is

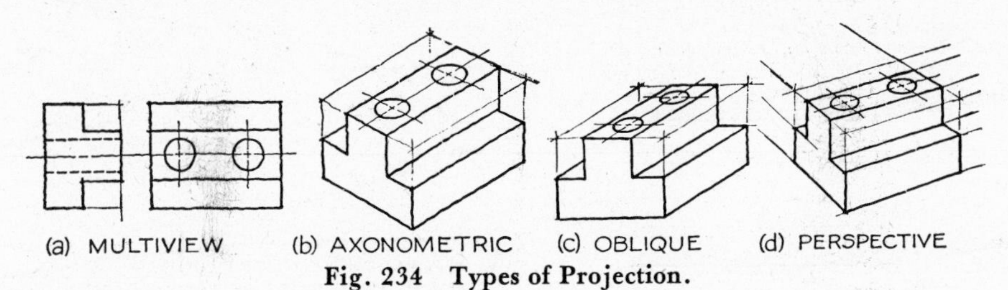

Fig. 234 Types of Projection.

described by its necessary views, as discussed in §§185-191. Or the object may be shown pictorially in a single view, by *axonometric*, *oblique*, or *perspective*, (b) to (d), as discussed in §§199-205.

178. Scale. *Sketches usually are not made to any scale.* Objects should be sketched in their correct proportions as accurately as possible, by eye. However, cross-section paper provides a ready scale (by counting squares) which may be used if only to assist in sketching to correct proportions. The size of the sketch is purely optional, depending upon the complexity of the object and the size of paper available. Small objects are often sketched oversize so as to show the necessary details clearly.

179. Technique of Lines. The chief difference between a mechanical drawing and a freehand sketch lies in the character or *technique* of the lines. A good freehand line should not be rigidly straight or exactly uniform, as a mechanical line. While the effectiveness of a mechanical line lies in exacting uniformity, the quality of a freehand line lies in its *freedom* and *variety*, Figs. 235 and 238.

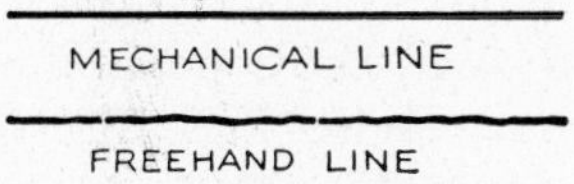

MECHANICAL LINE

FREEHAND LINE

Fig. 235 Comparison of Lines.

Conventional lines, drawn mechanically, are shown in Fig. 24, and the corresponding freehand renderings are shown in Fig. 236. The freehand construction line is a very light rough line in which some strokes may overlap. All other lines should be dark and clean-cut. Accent the ends of all dashes, and maintain a sharp contrast between the three line-thicknesses. Especially, make visible lines **heavy** so the outline will stand out clearly, and make center lines, dimension lines, and extension lines *very thin*.

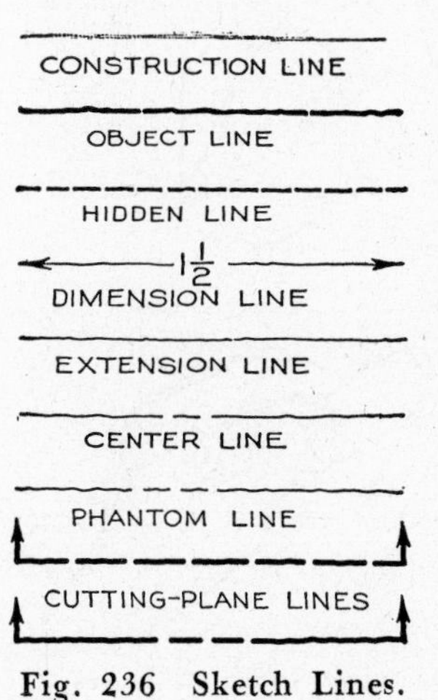

Fig. 236 Sketch Lines.

The term "freehand sketch" is too often understood to mean a crude or sloppy freehand drawing in which no particular effort has been made. On the contrary, a freehand sketch should be made with care and with attention to good line weights.

180. Sharpening Sketching Pencils. Use a soft pencil, such as HB or F, and sharpen it to a conical point, as shown in Fig. 20 (c). Use this needle-sharp point for center lines, dimension lines, and extension lines. For visible lines, hidden lines, and cutting-plane lines, round off the point slightly to produce the desired thickness of line, Fig. 237. Make all lines dark, with the exception of construction lines, which should be very light.

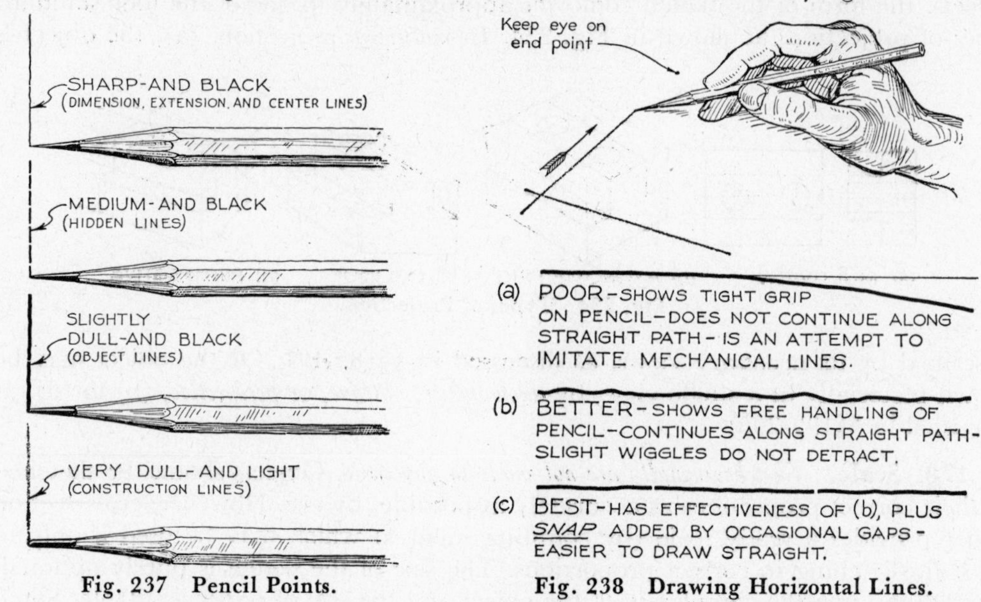

Fig. 237 Pencil Points. **Fig. 238 Drawing Horizontal Lines.**

181. Straight Lines. Since the majority of lines on the average sketch are straight lines, it is necessary to learn to make them well. Hold the pencil naturally about $1\frac{1}{2}''$ back from the point, and approximately at right angles to the line to be drawn. Draw horizontal lines from left to right with a free and easy wrist-and-arm movement, Fig. 238. Draw vertical lines downward with finger-and-wrist movements, Fig. 239.

Inclined lines may be made to conform in direction to horizontal or vertical lines by shifting position with respect to the paper, or by turning the paper slightly; hence, they may be drawn with the same general movements, Fig. 240.

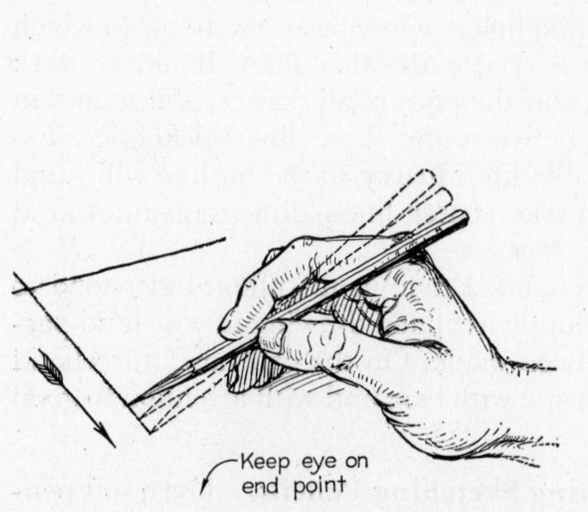

Fig. 239 Drawing Vertical Lines.

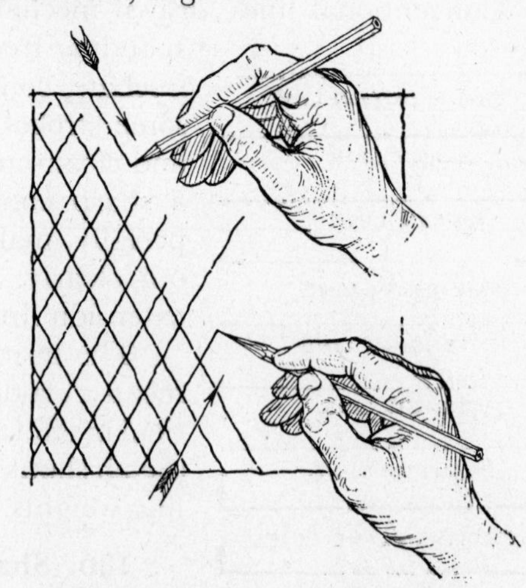

Fig. 240 Drawing Inclined Lines.

In sketching long lines, mark the ends of the line with light dots, then move the pencil back and forth between the dots in long sweeps, keeping the eye always on the dot toward which the pencil is moving, the point of the pencil touching the paper lightly, and each successive stroke correcting the defects of the preceding strokes. When the path of the line has been established sufficiently, apply a little more pressure, replacing the trial series with a distinct line. Then, dim the line with the artgum and draw the final line clean-cut and dark, keeping the eye now on the point of the pencil.

An easy method of blocking in horizontal or vertical lines, Fig. 241 (a), is to hold the hand and pencil rigidly and glide the finger tips along the edge of the pad or board, as shown.

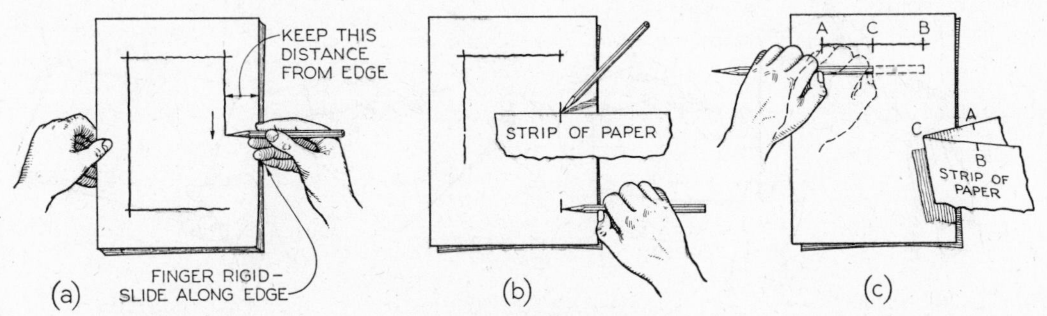

Fig. 241 Blocking-In.

Another method, (b), is to mark the distance on the edge of a card or a strip of paper, and to transfer this distance at intervals, as shown; then to draw the final line through these points. Or the pencil may be held as shown at the lower part of (b), and distance-marks made on the paper at intervals by tilting the lead down to the paper. It will be seen that both methods of transferring distances are substitutes for the dividers and will have many uses in sketching.

To find the mid-point of a line AB at (c), hold the pencil in the left hand with the thumb gaging the estimated half-distance. Try this distance on the left and then on the right until the center is located by trial, and mark the center C, as shown. Another method is to mark the total distance AB on the edge of a strip of paper and then to fold the paper to bring points A and B together, thus locating center C at the crease. To find quarter points, the folded strip can be folded once more.

182. Circles and Arcs. Small circles and arcs can be easily sketched in one or two strokes, as for the circular portions of letters, without any preliminary "blocking in."

One method of sketching a larger circle, Fig. 242, is first to sketch lightly the

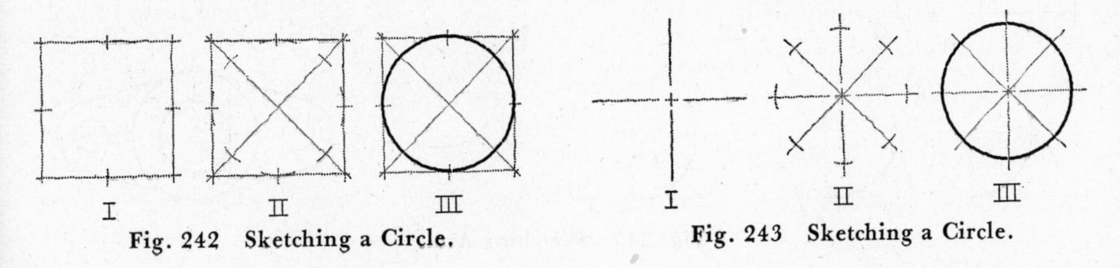

Fig. 242 Sketching a Circle. **Fig. 243 Sketching a Circle.**

enclosing square, mark the mid-points of the sides, draw light arcs tangent to the sides of the square, and then to heavy-in the final circle.

Another method, Fig. 243, is to sketch the two center lines, add light 45° radial lines, sketch light arcs across the lines at the estimated radius-distance from the center, and finally to sketch the required circle heavily.

In both methods, dim all construction lines with the artgum before heavying-in the final circle.

An excellent method, particularly for large circles, Fig. 244 (a), is to mark the estimated radius on the edge of a card or scrap of paper, to set off from the center as many points as desired, and to sketch the final heavy circle through these points.

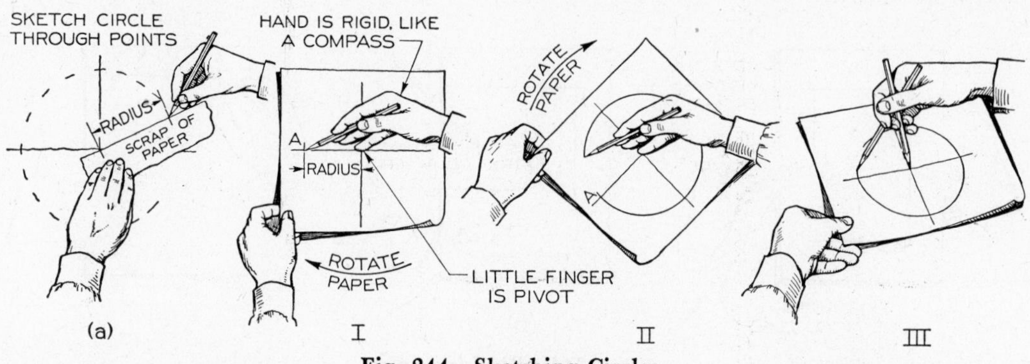

Fig. 244 Sketching Circles.

The clever draftsman will prefer the method at I and II, in which the hand is used as a compass. Place the tip of the little finger, or the knuckle-joint of the little finger, at the center; "feed" the pencil out to the desired radius, hold this position rigidly, and carefully revolve the paper with the other hand, as shown. If you are using a sketching pad, place the pad on your knee and revolve the entire pad on the knee as a pivot.

At III two pencils are held rigidly like a compass, and the paper is slowly revolved.

Methods of sketching arcs, Fig. 245, are adaptations of those used for sketching

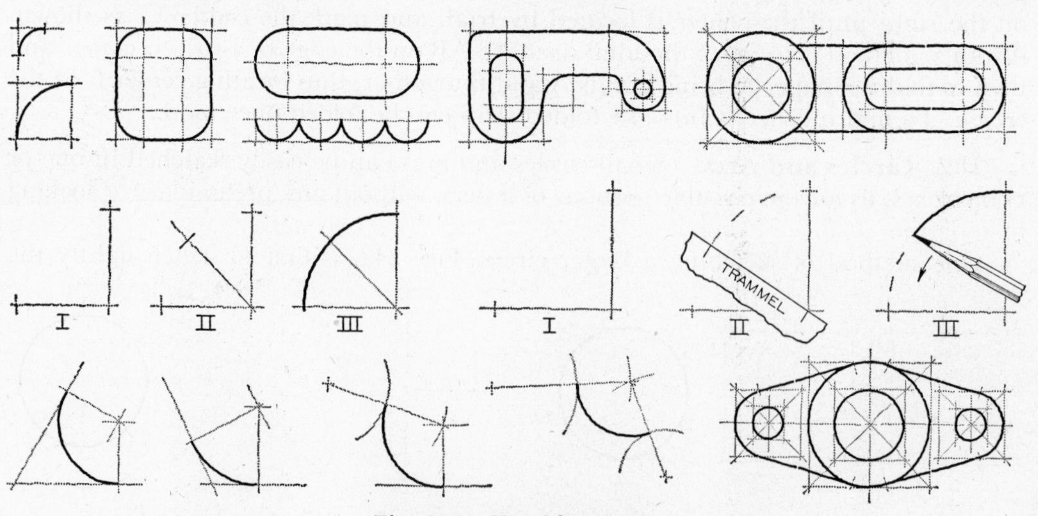

Fig. 245 Sketching Arcs.

circles. In general, it is easier to sketch arcs with the hand and pencil on the concave side of the curve. In sketching tangent arcs, always keep in mind the actual geometric constructions, carefully approximating all points of tangency.

183. Ellipses. If a circle is viewed obliquely, Fig. 202 (b), it appears as an ellipse. With a little practice, you can learn to sketch small ellipses with a free arm movement, Fig. 246 (a). Hold the pencil naturally, rest the weight on the upper part

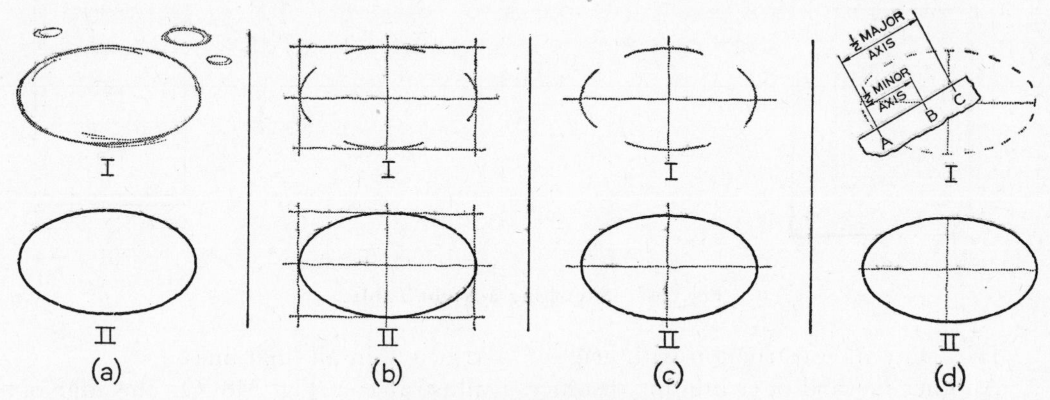

Fig. 246 Sketching Ellipses.

of the forearm, and move the pencil rapidly above the paper in the elliptical path desired; then lower the pencil so as to describe several light overlapping ellipses, as shown at I. Dim all lines with the artgum and heavy-in the final ellipse, II.

Another method, (b), is to sketch lightly the enclosing rectangle, I, mark the mid-points of the sides, and sketch light tangent arcs, as shown. Then, II, complete the ellipse lightly, dim all lines with the artgum, and heavy-in the final ellipse.

The same general procedure shown at (b) may be used in sketching the ellipse upon the given axes, as shown at (c).

The *trammel method*, (d), is excellent for sketching large ellipses. Prepare a "trammel" on the edge of a card or strip of paper, move it to different positions, and mark points on the ellipse at A. The trammel method is explained in §157. Sketch the final ellipse through the points, as shown.

For sketching isometric ellipses, see §201.

184. Proportions. *The most important rule in freehand sketching is to keep the sketch in proportion.* No matter how brilliant the technique or how well the small details are drawn, if the proportions—especially the large over-all proportions—are bad, the sketch will be bad. First, the relative proportions of the height to the width must be carefully established; then as you proceed to the medium-size areas and the small details, constantly compare each new estimated distance with established distances already set down.

If you are working from a given picture, such as the night table in Fig. 247 (a), it is first necessary to establish the relative width compared to the height. One way is to use the pencil as a measuring stick, as shown. In this case, the height is about 1$\frac{3}{4}$ times the width. Then:

I. Sketch the enclosing rectangle in the correct proportion. In this case, the sketch is to be slightly larger than the given picture.

II. Divide the available drawer space into three parts with the pencil by trial, as shown. Sketch light diagonals to locate centers of drawers, and block-in drawer handles. Sketch all remaining details.

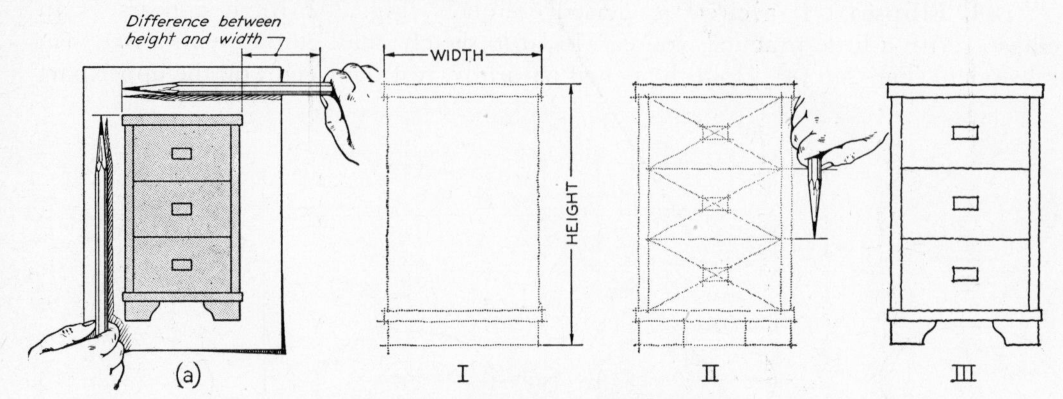

Fig. 247 Sketching a Night Table.

III. Dim all construction with artgum, and heavy-in all final lines.

Another method of estimating distances is illustrated in Fig. 248. On the edge of a card or strip of paper, mark an arbitrary unit. Then see how many units wide and how many units high the desk is. If you are working from the actual object, you could use a foot rule, a piece of paper, or the pencil itself as a unit to determine the proportions.

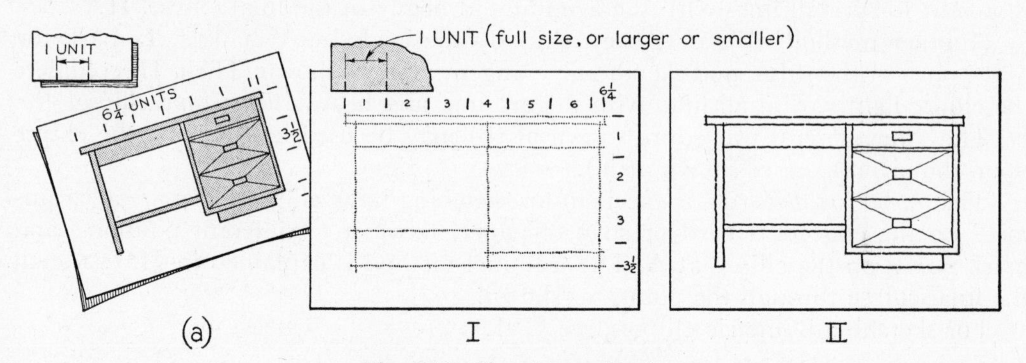

Fig. 248 Sketching a Desk.

To sketch an object composed of many curves to the same scale or to a larger or smaller scale, the method of "squares" is recommended, Fig. 249. On the given picture, rule grid lines to form squares of any convenient size. It is best to use a scale and some convenient spacing, such as $\frac{1}{2}''$. On the new sheet, rule a similar grid, making the spacing of the lines proportional to the original, but reduced or enlarged as desired. Make the final sketch by drawing the lines in and across the grid lines as in the original, as near as you can estimate by eye.

In sketching from an actual object, you can easily compare various distances on the object by using the pencil to compare measurements as shown in Fig. 250. While doing this, do not change your position, and always hold your pencil at arm's length. The length sighted can then be compared in similar manner with any other dimension of the object. If the object is small, such as a machine part, you can compare distances in the manner of Fig. 247, by actually placing the pencil against the object itself.

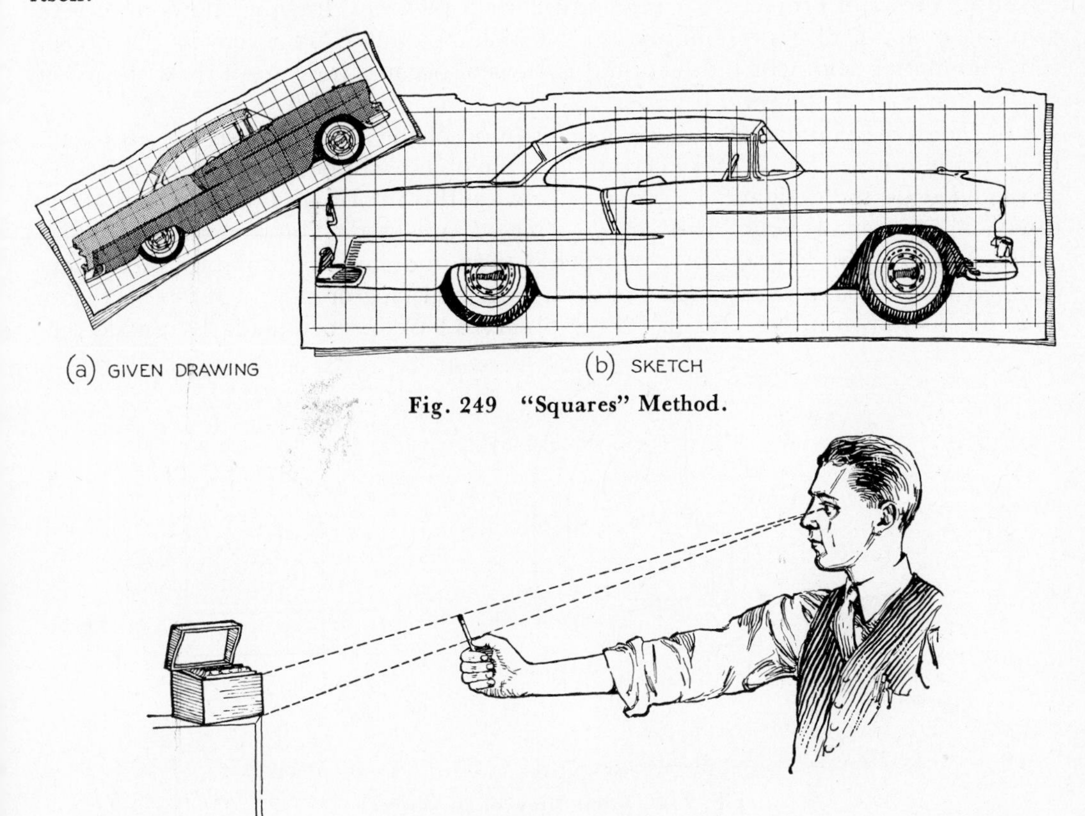

Fig. 249 "Squares" Method.

Fig. 250 Estimating Distances.

In establishing proportions, the "blocking-in" method is recommended, especially for irregular shapes. The steps for blocking in and completing the sketch of a Shaft Hanger are shown in Fig. 251. As always, first give attention to the main proportions, next to the general sizes and direction of flow of curved shapes, and finally to the "snappy" lines of the completed sketch.

In making sketches from actual machine parts, it is necessary to use the measuring

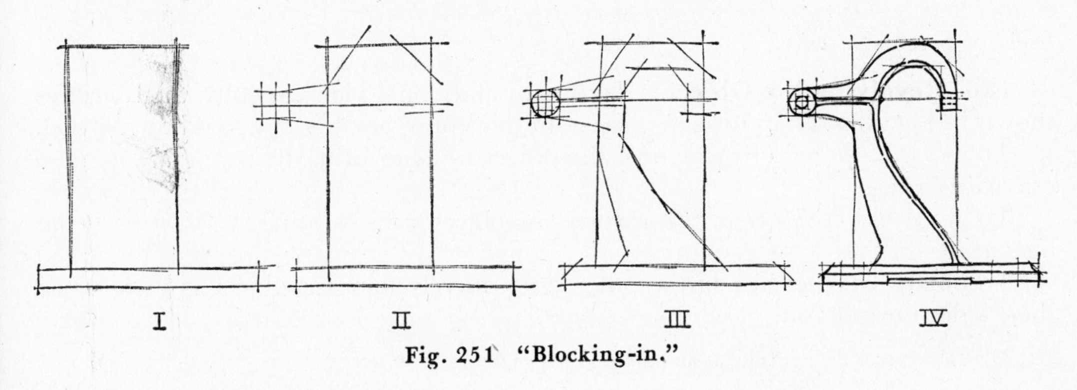

Fig. 251 "Blocking-in."

tools used in the shop, especially those needed to determine dimensions which must be relatively accurate. For a discussion of these methods, see §317.

185. Views of Objects. A photograph or a pictorial drawing shows an object as it *appears* to the observer, but not as it *is*. Such a picture cannot describe the object fully, no matter from which direction it is viewed, because it does not show the exact shapes and sizes of the several parts.

In industry, a complete and clear description of the shape and size of an object to be made is necessary to make certain that the object will be manufactured exactly as intended by the designer. In order to provide this information clearly and accurately, a number of *views, systematically arranged*, are used. This system of views is called *multiview projection*. Each view provides certain definite information if the view is taken in a direction perpendicular to a principal face or side of the object. For example, as shown in Fig. 252 (a), if the observer looks perpendicularly toward one

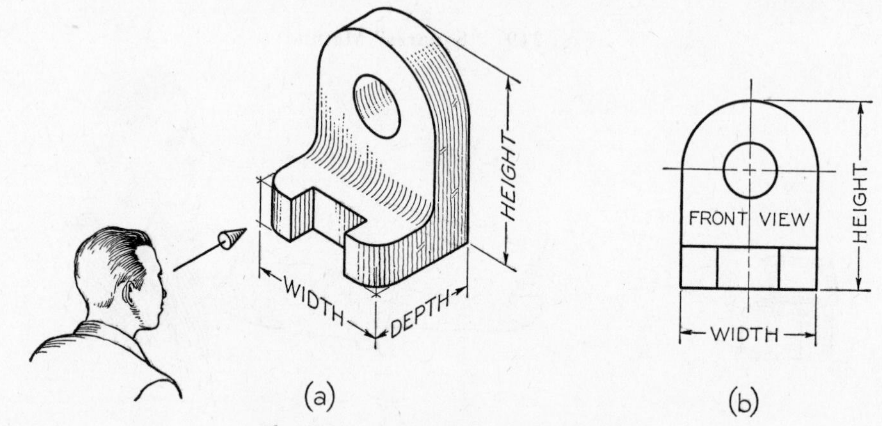

Fig. 252 Front View of an Object.

face of the object, he obtains a true view of the shape and size of that side. This view as seen by the observer is shown at (b).*

An object has three principal dimensions: *width*, *height*, and *depth*, as shown at (a). In technical drawing, these fixed terms are used for dimensions taken in these directions, regardless of the shape of the object. The terms "length" and "thickness" are not used because they cannot be applied in all cases. Note at (b) that the front view shows only the height and width of the object, and not the depth. In fact, *any one view of a three-dimensional object can show only two dimensions; the third dimension will be found in an adjacent view.*

186. Revolving the Object. To obtain additional views, revolve the object as shown in Fig. 253. First, hold the object in the front-view position, as shown at (a).

To get the *top view*, (b), revolve the object so as to bring the *top of the object up and toward you.*

To get the *right-side view*, (c), revolve the object so as to bring the *right side to the right and toward you.*

To obtain views of any of the other sides, merely turn the object so as to bring those sides toward you.

*The observer is theoretically at an infinite distance from the object.

The top, front, and right-side views, arranged closer together, are shown at (d). These are called the *three regular views* because they are the views most frequently used.

At this stage we can consider spacing between views as purely a matter of appearance. The views should be spaced well apart and yet close enough to "hang together." The space between the front and top views may or may not be equal to the space between the front and side views. If dimensions, Chapter 11, are to be added to the sketch, sufficient space for them between views will have to be allowed.

An important advantage that a view has over a photograph of an object is that hidden features can be clearly shown by means of *hidden lines*, Fig. 24. In Fig. 253 (d), surface 7-8-9-10 in the front view appears as a visible line 5-6 in the top view, and as a hidden line 15-16 in the side view. Also, hole A, which appears as a circle in the

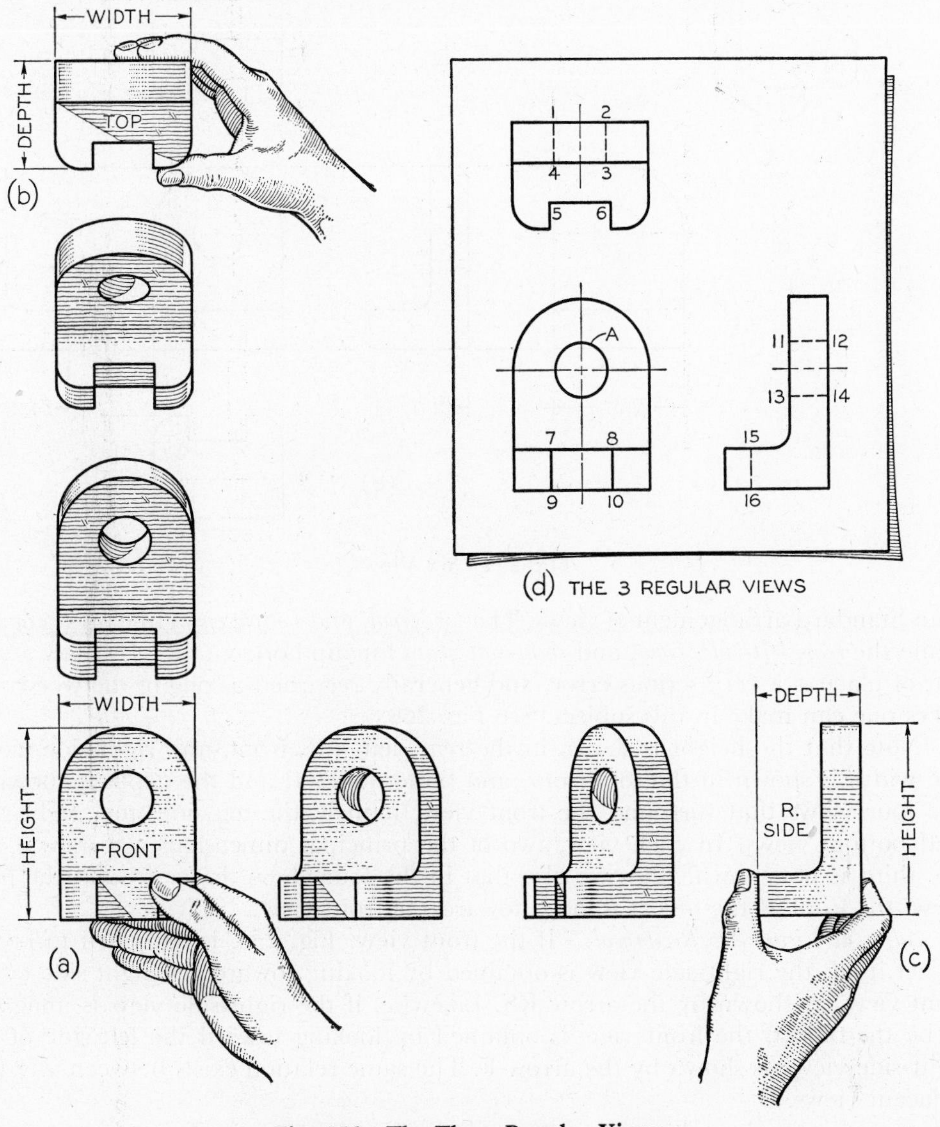

Fig. 253 The Three Regular Views.

front view, shows as hidden lines 1-4 and 2-3 in the top view, and 11-12 and 13-14 in the side view. For a complete discussion of hidden lines, see §192.

Also, at (d) note the use of center lines for the hole. See §193.

187. The Six Views. Any object can be viewed from six mutually perpendicular directions, as shown in Fig. 254 (a). Thus, six views may be drawn if necessary, as shown at (b). These six views are always arranged as shown,* which is the American Standard arrangement of views. The *top*, *front*, and *bottom views* line up vertically, while the *rear*, *left-side*, *front*, and *right-side views* line up horizontally. To draw a view out of place is a very serious error, and generally regarded as one of the worst mistakes one can make in this subject. See Fig. 268.

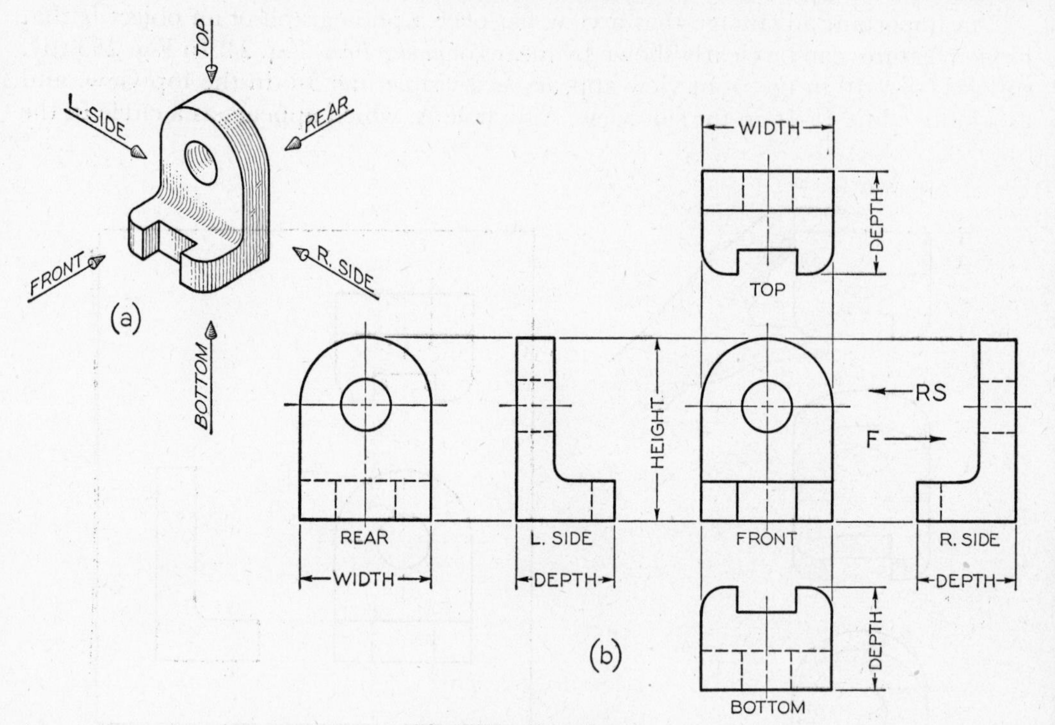

Fig. 254 Six Views.

Note that the height is shown in the rear, left-side, front, and right-side views; the width is shown in the top, front, and bottom views; and the depth is shown in the four views that surround the front view, namely the left-side, top, right-side, and bottom views. In each view, two of the principal dimensions are shown, and the third is not shown. Observe also that in the four views that surround the front view, the front of the object is faced toward the front view.

Adjacent Views Are Reciprocal. If the front view, Fig. 254, is imagined to be the object itself, the right-side view is obtained by looking toward the right side of the front view, as shown by the arrow RS. Likewise, if the right-side view is imagined to be the object, the front view is obtained by looking toward the left side of the right-side view, as shown by the arrow F. The same relation exists between any two adjacent views.

*Except as explained in § 214.

Obviously the six views may be obtained either by shifting the object with respect to the observer, as we have seen, Fig. 253, or by shifting the observer with respect to the object, Fig. 254. Another illustration of the second method is given in Fig. 255, showing six views of a house. The observer can walk around the house and

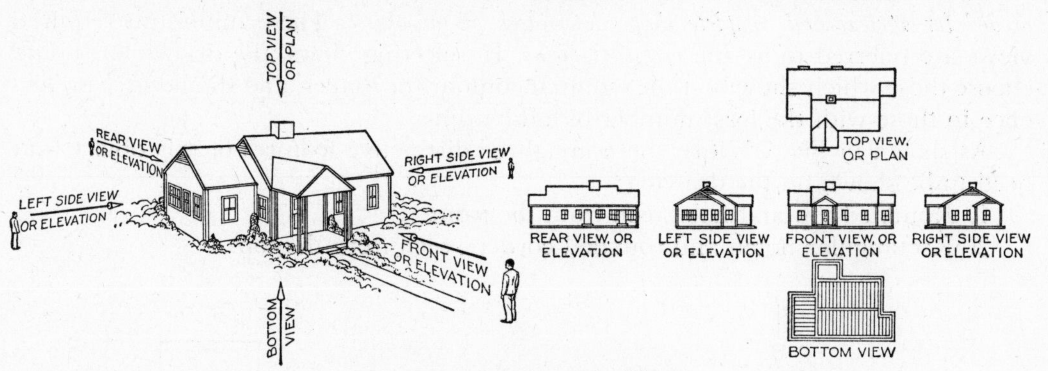

Fig. 255 Six Views of a House.

view its front, sides, and rear, and he can imagine the top view as seen from an airplane and the bottom or "worm's-eye view" as seen from underneath.* Notice the use of the term *plan* for the top view, and *elevation* for all views showing the height of the building. These terms are regularly used in architectural drawing, and occasionally they are used with reference to drawings in other fields.

188. Orientation of Front View. Six views of an automobile are shown in Fig. 256. The view chosen for the front view in this case is the side, not the front of

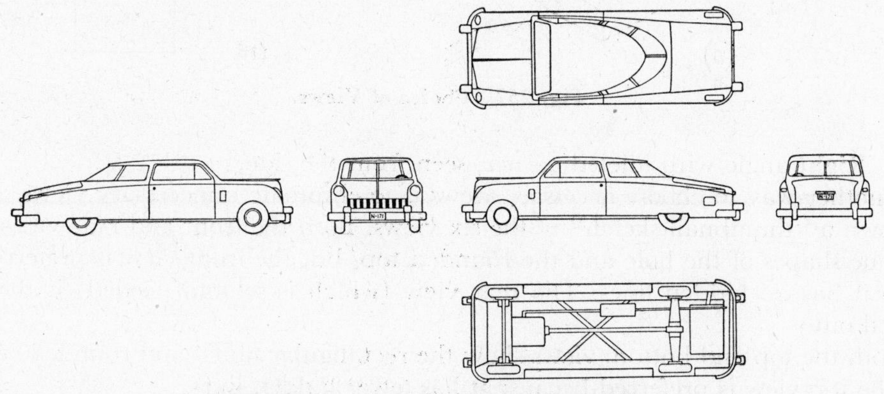

Fig. 256 Six Views of an Automobile.

the automobile. In general, the front view should show the object in its operating position, particularly of familiar objects such as the house and automobile shown above. A machine part is often drawn in the position it occupies in the assembly. However, in most cases this is not important, and the draftsman may assume the object in any convenient position. For example, an automobile connecting rod is usually drawn horizontally on the sheet, Fig. 727. Also, it is customary to draw

*Architects usually draw the views of a building on separate sheets because of the large sizes of the drawings. When two or more views are drawn together, they are usually drawn in first-angle projection, § 243.

screws, bolts, shafts, tubes, and other elongated parts in a horizontal position, not only because they are usually manufactured in this position, but because they can be presented more satisfactorily on paper in this position.

189. Choice of Views. *A drawing for use in the shop should contain only those views needed for a clear and complete shape description of the object.* These minimum required views are referred to as the *necessary views.* In selecting views, the draftsman should choose those which show best the essential contours or shapes, and should give preference to those with the least number of hidden lines.

As shown in Fig. 257 (a), there are three distinctive features of this object that need to be shown on the drawing:

1. Rounded top and hole, seen from the front.
2. Rectangular notch and rounded corners, seen from the top.

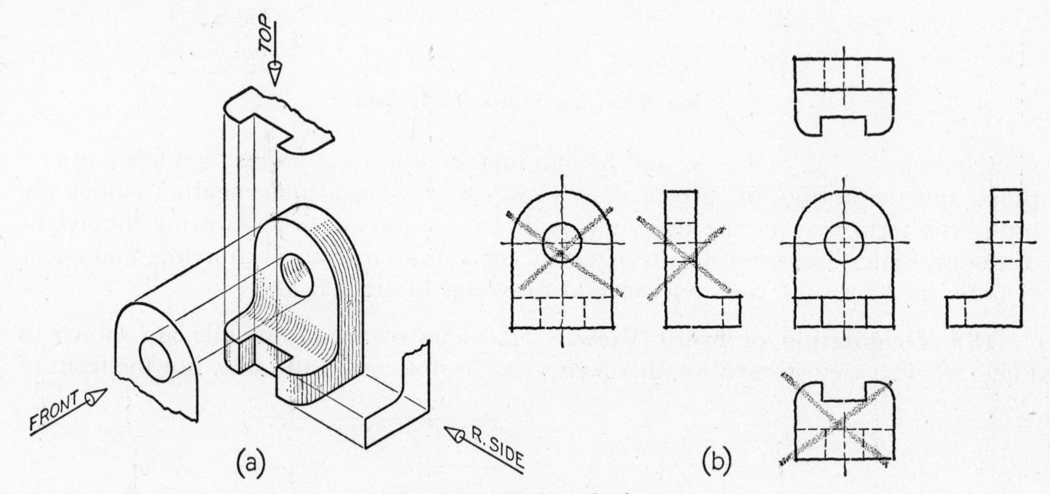

Fig. 257 Choice of Views.

3. Right angle with filleted corner, seen from the side.

Another way to choose necessary views is to eliminate unnecessary views. At (b) is shown a "thumbnail sketch" of the six views. Both the front and rear views show the true shapes of the hole and the rounded top, but the front view is preferred because it has no hidden lines. The rear view (which is seldom needed) is therefore crossed out.

Both the top and bottom views show the rectangular notch and rounded corners, but the top view is preferred because it has fewer hidden lines.

Both the right-side and left-side views show the right angle with the filleted corner. In fact, in this case the side views are identical, except reversed. In such instances, it is customary to choose the right-side view.

The necessary views then are the three remaining views: the top, front, and right-side views. These are the "three regular views" referred to in connection with Fig. 253.

More complicated objects may require more than three views, or in many cases special views such as partial views, §215, sectional views, Chapter 7, and auxiliary views, Chapter 8.

190. Two-View Drawings. Often only two views are needed to describe clearly the shape of an object. In Fig. 258 (a), the right-side view shows no significant contours of the object, and is crossed out. At (b) the top and front views are identical,

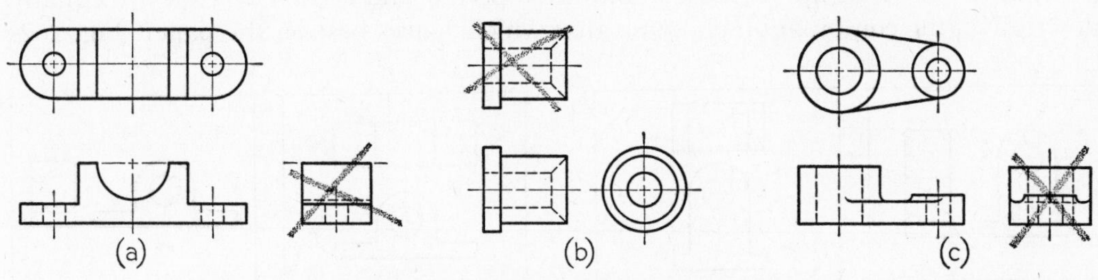

Fig. 258 Two Necessary Views.

so the top view is eliminated. At (c), no additional information not already given in the front and top views is shown in the side view, so the side view is unnecessary.

The question often arises: What are the absolute minimum views required? For example, in Fig. 259, the top view might be omitted, leaving only the front and right-side views. However, it is more difficult to "read" the two views or visualize the object, because the characteristic "Z" shape of the top view is omitted. In addition, one must assume that corners A and B (top view) are square and not filleted. In this example, all three views are necessary.

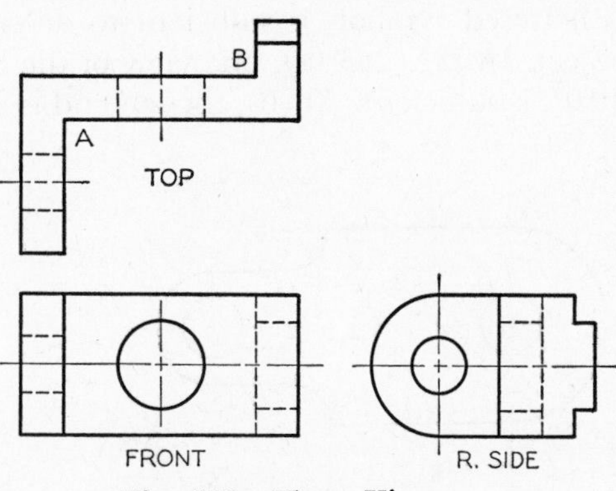

Fig. 259 Three Views.

If the object requires only two views, and the left-side and right-side views are equally descriptive, the right-side view is customarily chosen, Fig. 260. If contour A were omitted, then the presence of slot B would make it necessary to choose the left-side view in preference to the right-side view.

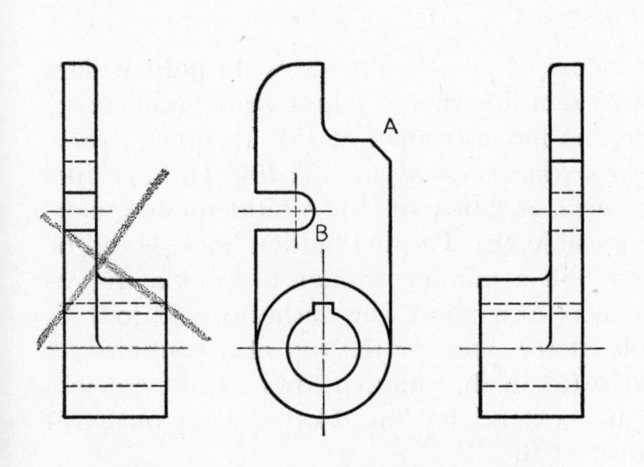

Fig. 260 Choice of Right-Side View.

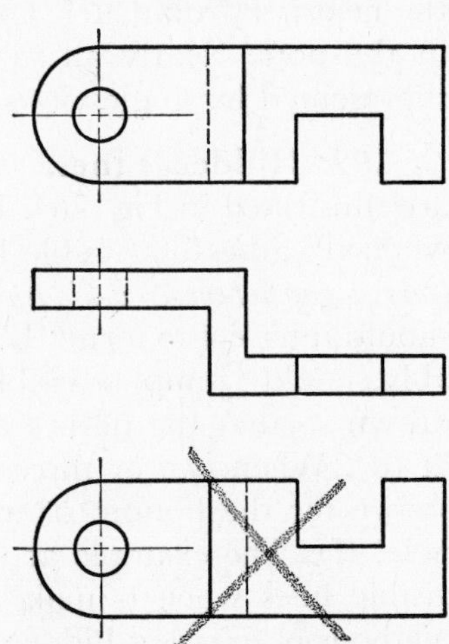

Fig. 261 Choice of Top View.

If the object requires only two views, and the top and bottom views are equally descriptive, the top view is customarily chosen, Fig. 261.

If only two views are necessary, and the top view and right-side view are equally descriptive, the combination chosen is that which spaces best on the paper, Fig. 262.

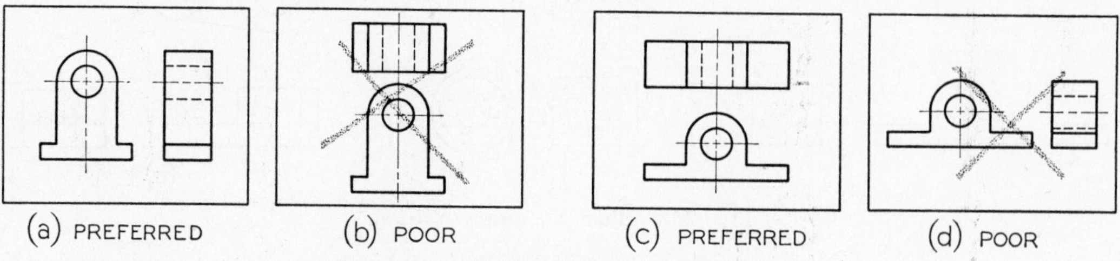

Fig. 262 Choice of Views to Fit Paper.

191. One-View Drawings. Frequently a single view supplemented by a note or lettered symbols is sufficient to describe clearly the shape of a relatively simple object. In Fig. 263 (a), one view of the Shim, plus a note indicating the thickness as .010″, is sufficient. At (b), the left end is $2\frac{1}{2}''$ square, the next portion is $1\frac{15}{16}''$ diameter,

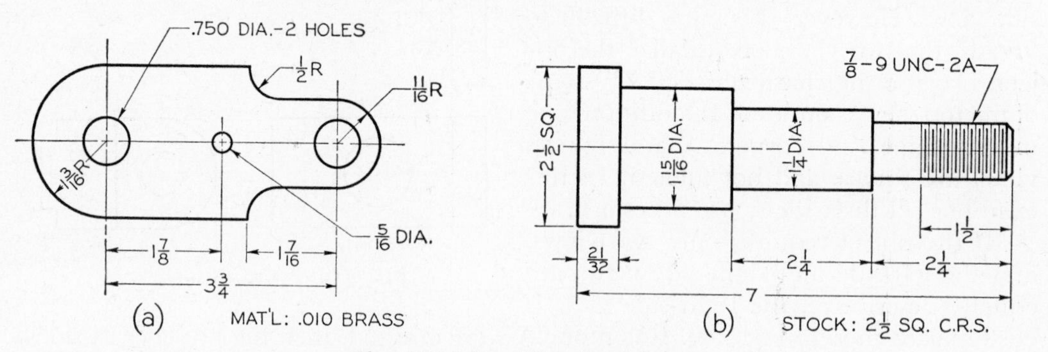

Fig. 263 One-View Drawings.

the next is $1\frac{1}{4}''$ diameter, and the portion with the thread is $\frac{7}{8}''$ diameter, as indicated in the note. Nearly all shafts, bolts, screws, and similar parts can and should be represented by single views in the above manner.

192. Hidden Lines. Correct and incorrect practices in drawing hidden lines are illustrated in Fig. 264. In general, a hidden line should join a visible line except when it causes the visible line to extend too far, as shown at (a). In other words, *leave a gap whenever a hidden line dash forms a continuation of a visible line.* Hidden lines should intersect to form "L" and "T" corners, as shown at (b). A hidden line preferably should "jump" a visible line when possible, (c). Parallel hidden lines should be drawn so that the dashes are "staggered," in a manner similar to bricklaying, as at (d). When two or three hidden lines meet at a point, the dashes should join, as shown for the bottom of the drilled hole at (e), and for the top of a countersunk hole, (f). The example at (g) is similar to (a) in that hidden lines should not join visible lines when it makes the visible line extend too far. Correct and incorrect methods of drawing hidden arcs are shown at (h).

Poorly drawn hidden lines can easily spoil a drawing. Each dash should be carefully drawn about $\frac{1}{8}''$ long and spaced only about $\frac{1}{32}''$ apart, by eye. Accent the beginning and end of each dash by pressing down on the pencil, whether drawn freehand or mechanically.

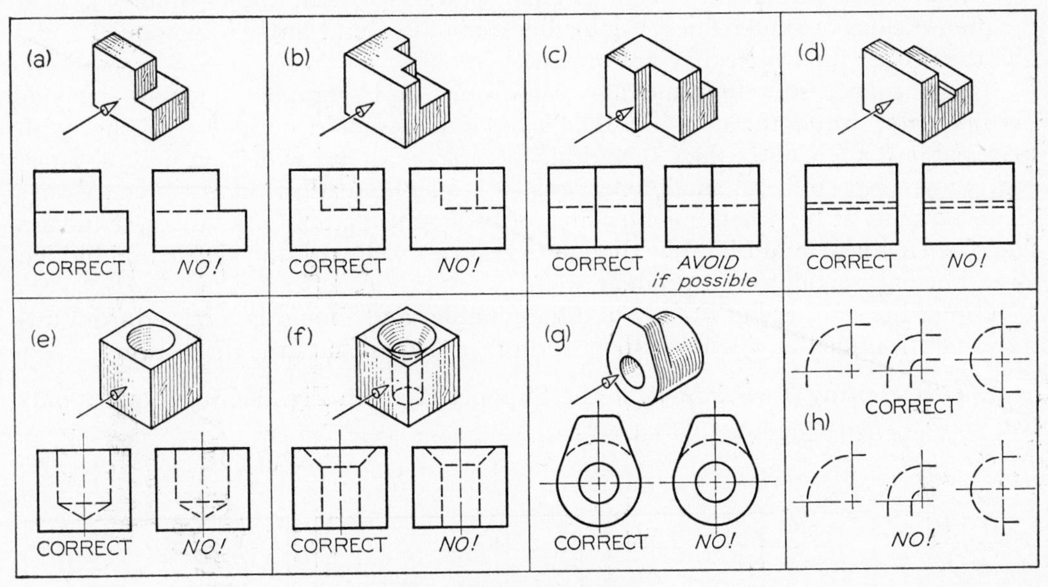

Fig. 264 Hidden Lines.

In general, views should be chosen that show features with visible lines, so far as possible. After this has been done, hidden lines should be used wherever necessary to make the drawing clear. Where they are not needed for clearness, hidden lines should be omitted, so as not to clutter the drawing any more than necessary and in order to save time. The beginner, however, would do well to be cautious about leaving out hidden lines until experience shows him when they can be safely omitted.

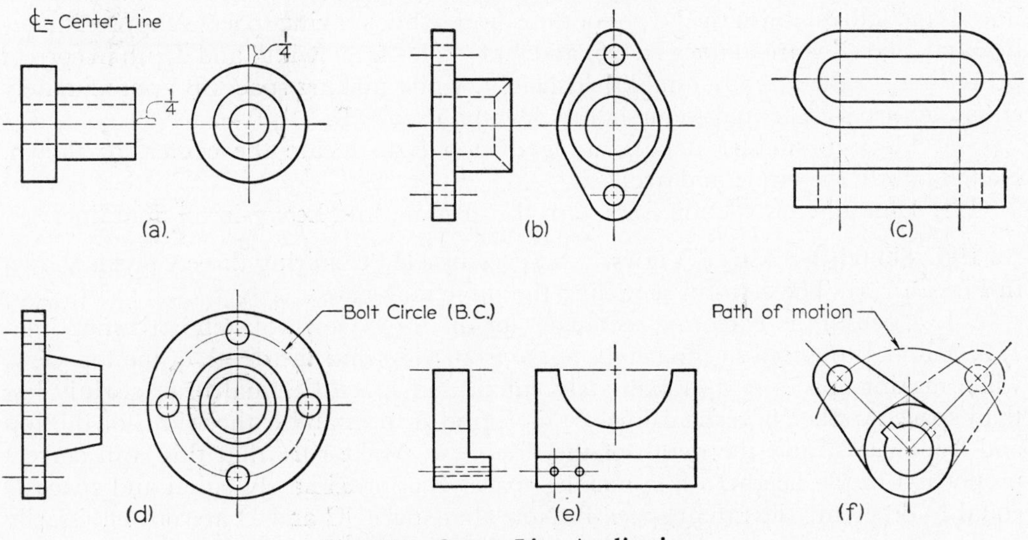

Fig. 265 Center Line Applications.

193. Center Lines. *Center lines* (symbol: ℄) are used to indicate axes of symmetrical objects or features, bolt circles, and paths of motion. Typical applications are shown in Fig. 265. As shown at (a), a single center line is drawn in the longitudinal view and crossed center lines in the circular view. The small dashes should cross at the intersections of center lines. Center lines should extend uniformly about $\frac{1}{4}''$ outside the feature for which they are drawn.

The long dashes of center lines may vary from $\frac{3}{4}''$ to $1\frac{1}{2}''$ or more in length, depending upon the size of the drawing. The short dashes should be about $\frac{1}{8}''$ long, with spaces about $\frac{1}{16}''$. Center lines should always start and end with long dashes. Short center lines, especially for small holes, as at (e), may be made solid as shown. Always leave a gap as at (e) when a center line forms a continuation of a visible or hidden line. Center lines should be thin enough to contrast well with the visible and hidden lines, but dark enough to reproduce well.

Center lines are useful mainly in dimensioning, and should be omitted from unimportant rounded or filleted corners and other shapes that are self-locating.

194. Sketching Two Views. The Support Block in Fig. 266 (a) requires only two views. The steps in sketching are:

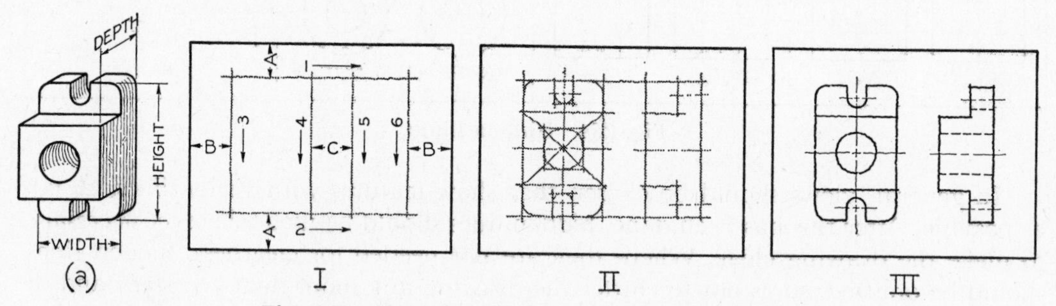

Fig. 266 Sketching Two Views of Support Block.

I. Block-in lightly the enclosing rectangles for the two views. Sketch horizontal lines 1 and 2 to establish the height of the object, while making spaces A approximately equal. Sketch vertical lines 3, 4, 5, and 6 to establish the width and depth in correct proportion to the already established height, while making spaces B approximately equal, and space C equal to or slightly less than space B.

II. Block-in smaller details, using diagonals to locate the center, as shown. Sketch lightly the circle and arcs.

III. Dim all construction lines with the artgum, and heavy-in all final lines.

195. Sketching Three Views. A Lever Bracket requiring three views is shown in Fig. 267 (a). The steps in sketching the three views are:

I. Block-in the enclosing rectangles for the three views. Sketch horizontal lines 1, 2, 3, and 4 to establish the height of the front view and the depth of the top view, while making spaces A approximately equal, and space C equal to or slightly less than space A. Sketch vertical lines 5, 6, 7, and 8 to establish the width of the top and front views, and the depth of the side view. Make sure that this is in correct proportion to the height, while making spaces B approximately equal and space D equal to or slightly less than space B. Note that spaces C and D are not necessarily equal, but are independent of each other. Similarly, spaces A and B are not necessarily equal. To transfer the depth dimension from the top view to the side view, use the edge of a card or strip of paper, as shown; or transfer the distance by using the pencil as a measuring stick, as shown in Fig. 241 (b) and (c). *Note that the depth in the top and side views must always be equal.*

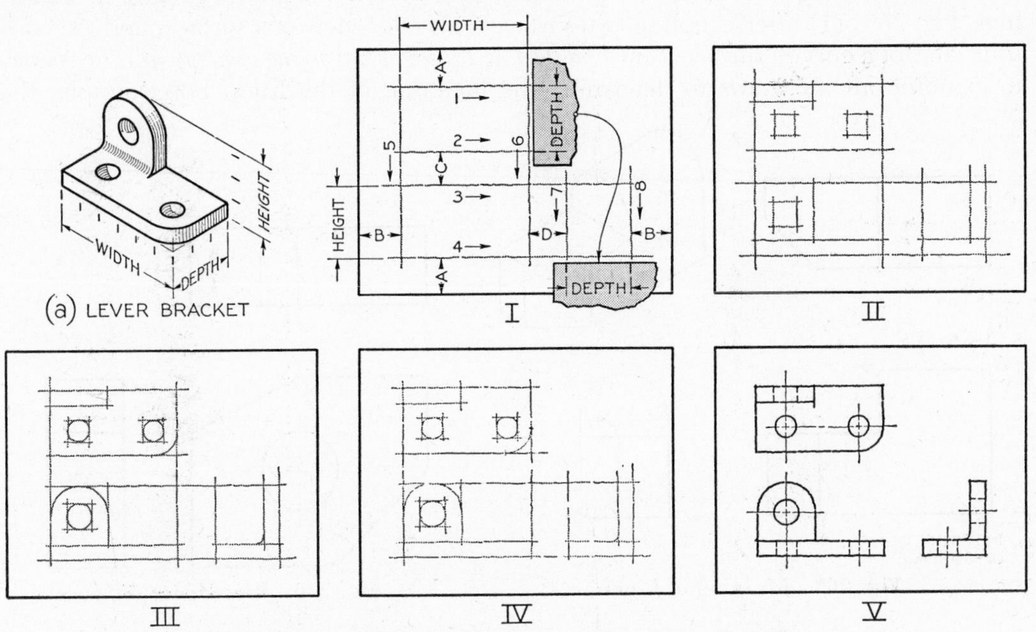

Fig. 267 Sketching Three Views of a Lever Bracket.

II. Block-in all details lightly.
III. Sketch all arcs and circles lightly.
IV. Dim all construction lines with artgum.
V. Heavy-in all final lines so that the views will stand out clearly.

196. Alignment of Views. Errors by beginners in arranging the views are so common that it is necessary to repeat that the views must be drawn in accordance with the American Standard arrangement, Fig. 254. In Fig. 268 (a) is shown an

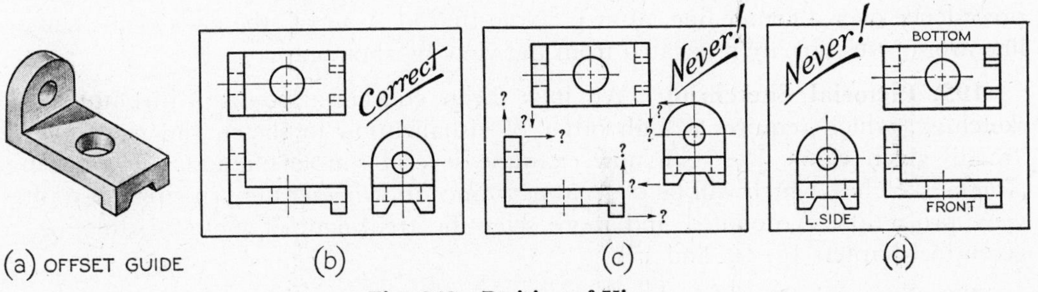

Fig. 268 Position of Views.

Offset Guide, which requires three views. These three views, correctly arranged, are shown at (b). The top view must be directly above the front view, and the right-side view directly to the right of the front view—not out of alignment, as at (c). Also, never draw the views in reversed positions, with the bottom over the front view, or the right-side to the left of the front view, as shown at (d), even though the views do line up with the front view.

197. Meaning of Lines. A visible line or a hidden line has three possible meanings, Fig. 269: (1) intersection of two surfaces, (2) edge view of a surface, and (3) contour view of a curved surface. Since *no shading is used on a working drawing*, it is necessary to examine all the views to determine the meaning of the lines. For example, the

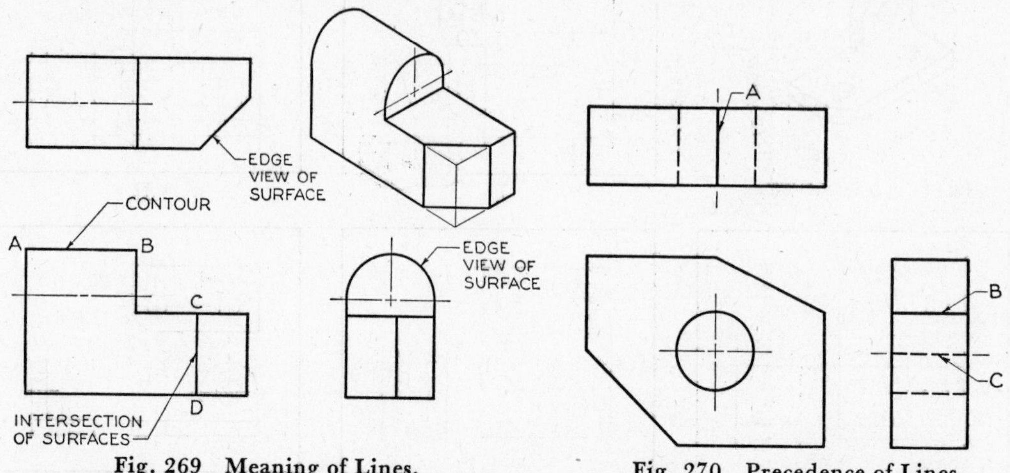

Fig. 269 Meaning of Lines.

Fig. 270 Precedence of Lines.

line AB at the top of the front view might be regarded as the edge view of a flat surface if we look at only the front and top views and do not observe the curved surface on top of the object as shown in the right-side view. Similarly, the vertical line CD in the front view might be regarded as the edge view of a plane surface if we look at only the front and side views. However, the top view shows that the line represents the intersection of an inclined surface.

198. Precedence of Lines. Visible lines, hidden lines, and center lines often coincide on a drawing, and it is necessary for the draftsman to know which line to show. As shown in Fig. 270, a visible line always "takes precedence over" (covers up) a center line or a hidden line, as shown at A and B. A hidden line always takes precedence over a center line, as at C. Note that at A and C the ends of the center line are shown, but are separated from the view by short gaps.

199. Pictorial Sketching. We have been concerned so far with multiview sketching, which requires a high order of visualization to sketch the views or to "read" given views. We shall now examine several simple methods of preparing *pictorial* sketches, which will be of great assistance in learning the principles of multiview projection. A detailed and more scientific treatment of pictorial drawing is given in Chapters 15, 16, and 17.

200. Isometric Sketching. To make an *isometric* sketch from an actual object, hold the object in your hand and tilt it toward you, as shown in Fig. 271 (a). In this position, the front corner will appear vertical, and the two receding bottom edges and those parallel to them, respectively, will appear at about 30° with horizontal, as shown. The steps in sketching are:

I. Sketch the enclosing box lightly, making AB vertical, and AC and AD approximately 30° with horizontal. These three lines are the *isometric axes*. Make AB,

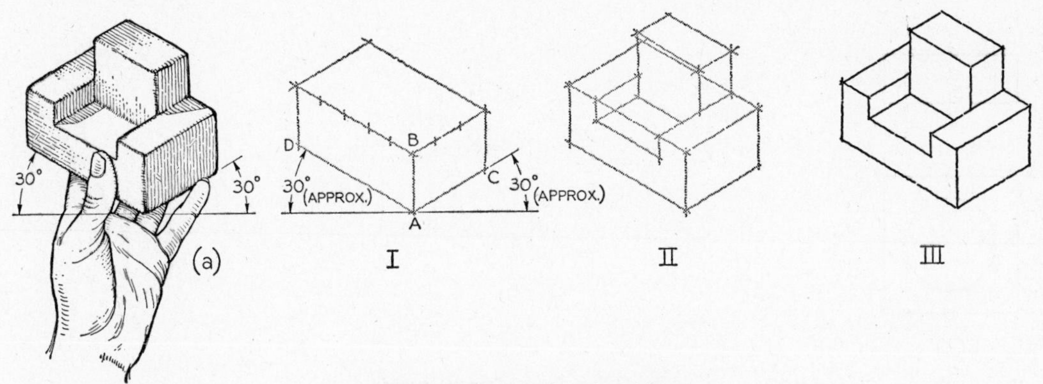

Fig. 271 Isometric Sketching.

AC, and AD approximately equal in length to the actual corresponding edges on the object. Sketch the remaining lines parallel, respectively, to these three lines.

II. Block-in the recess and the projecting block.

III. Dim all construction lines with the artgum, and heavy-in all final lines.

Note: The angle of the receding lines may be less than 30°, say 20° or 15°. Although the result will not be an isometric sketch, the sketch may be more pleasing and effective in many cases.

201. Isometric Ellipses. As shown in Fig. 202 (b), a circle viewed at an angle appears as an ellipse. When objects having cylindrical or conical shapes are placed in the isometric or other oblique positions, the circles will be viewed at an angle and will appear as ellipses, Fig. 272.

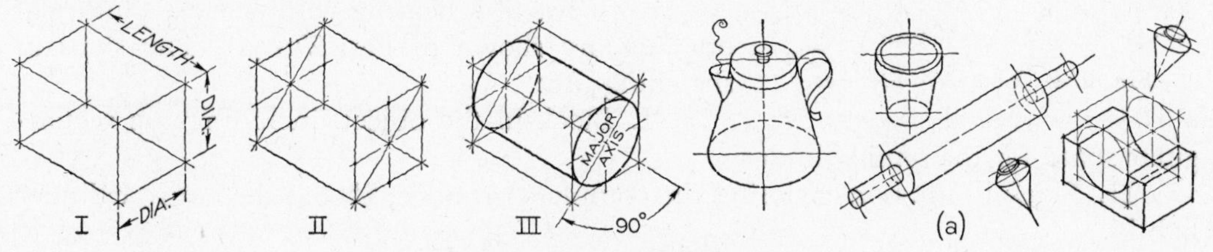

Fig. 272 Isometric Ellipses.

The most important rule in sketching isometric ellipses is: *The major axis of the ellipse is always at right angles to the center line of the cylinder, and the minor axis is at right angles to the major axis and coincides with the center line.*

Two views of a block with a large cylindrical hole are shown in Fig. 273 (a). The steps in sketching the object are:

I. Sketch the block and the enclosing parallelogram for the ellipse, making the sides of the parallelogram parallel to the edges of the block and equal in length to the diameter of the hole. Draw diagonals to locate the center of the hole, and then draw center lines AB and CD. Points A, B, C, and D will be midpoints of the sides of the parallelogram, and the ellipse will be tangent to the sides at those points.

The major axis will be on the diagonal EF, which is at right angles to the center line of the hole, and the minor axis will fall along the short diagonal. Sketch long "flat" elliptical sides CA and BD, as shown.

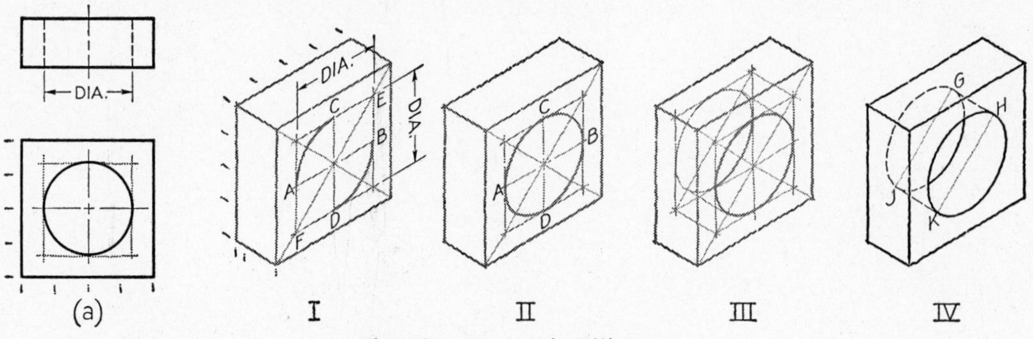

Fig. 273 Isometric Ellipses.

II. Sketch short small-radius arcs CB and AD to complete the ellipse. Avoid making the ends of the ellipse "squared off," or pointed like a football.

III. Sketch lightly the parallelogram for the back ellipse, and sketch the ellipse in the same manner as the front ellipse.

IV. Draw lines GH and JK tangent to the two ellipses. Dim all construction with the artgum, and heavy-in all final lines.

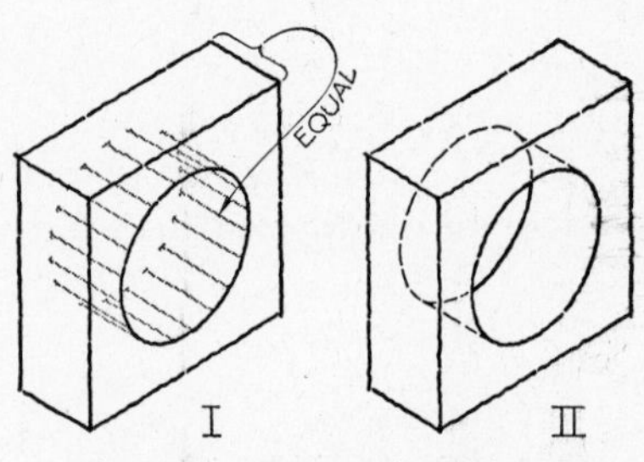

Fig. 274 Isometric Ellipses.

Another method for determining the back ellipse is shown in Fig. 274.

I. Select points at random on the front ellipse and sketch "depth lines" equal in length to the depth of the block.

II. Sketch the ellipse through the ends of the lines, as shown.

Two views of a Bearing with a semi-cylindrical opening are shown in Fig. 275 (a). The steps in sketching are:

I. Block-in the object, including the rectangular space for the semi-cylinder.

II. Block-in the box enclosing the complete cylinder. Sketch the entire cylinder lightly.

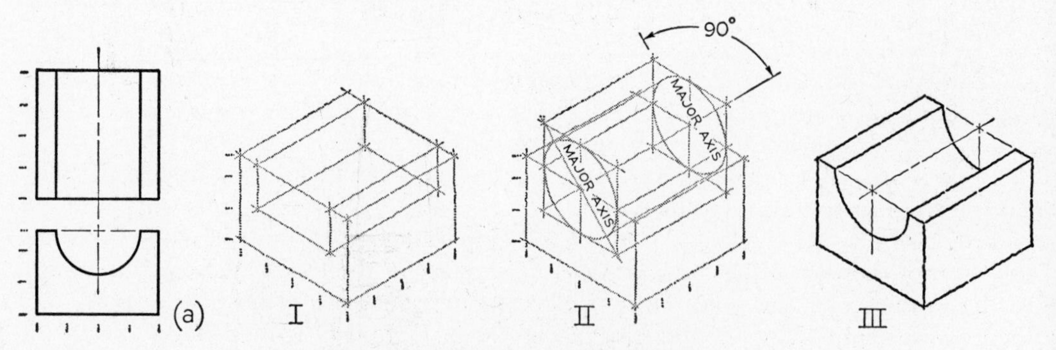

Fig. 275 Sketching Semi-Ellipses.

III. Dim all construction lines, and heavy-in all final lines, showing only the lower half of the cylinder.

202. Sketching on Isometric Paper. Two views of a Guide Block are shown in Fig. 276 (a). The steps in sketching illustrate not only the use of isometric paper,

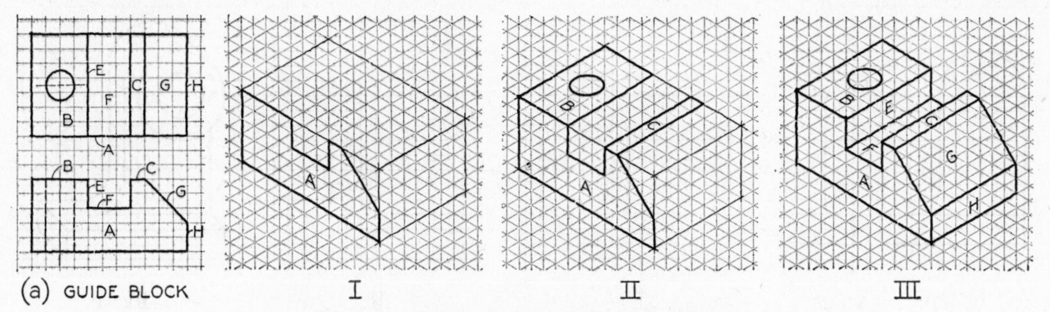

Fig. 276 Sketching on Isometric Paper.

but the sketching of individual planes or faces of the object in order to build up pictorially a visualization of the given views.

As shown at I, sketch isometric of enclosing box, counting off the isometric grid spaces to equal the corresponding squares on the given views. Sketch surface A, as shown.

Then, as shown at II and III, sketch additional surfaces B, C, E, etc., and the small ellipse, to complete the sketch.

203. Oblique Sketching. Another simple method for sketching pictorially is *oblique* sketching, Fig. 277. Hold the object in your hand, as shown at (a).

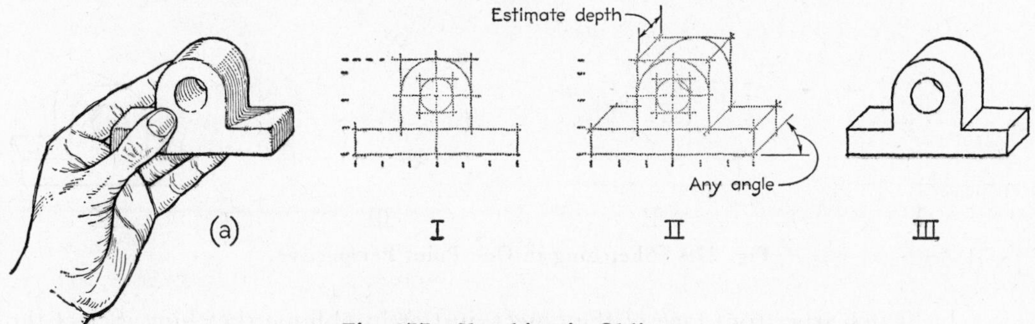

Fig. 277 Sketching in Oblique.

I. Block-in the front face of the object, as if you were sketching a front view.

II. Sketch receding lines parallel to each other and at any convenient angle, say 30° or 45° with horizontal, approximately. Cut off receding lines so that the depth appears correct. These lines may be full length, but a more natural appearance results if they are cut to three-quarters or one-half size, approximately. If they are full length, the sketch is a *cavalier* sketch. If half size, the sketch is a *cabinet* sketch.

III. Dim all construction and heavy-in the final lines.

Note: "Oblique" is not a suitable method for any object having circles in, or parallel to, more than one plane of the object, because distorted ellipses result when circles are viewed obliquely. Therefore, in oblique sketching always turn the circular shapes toward you so that they will appear in true size and shape, as in Fig. 277.

204. Oblique Sketching on Cross-Section Paper. Ordinary cross-section paper is suitable for oblique sketching. Two views of a Bearing Bracket are shown in Fig. 278 (a). The dimensions are determined simply by counting the squares.

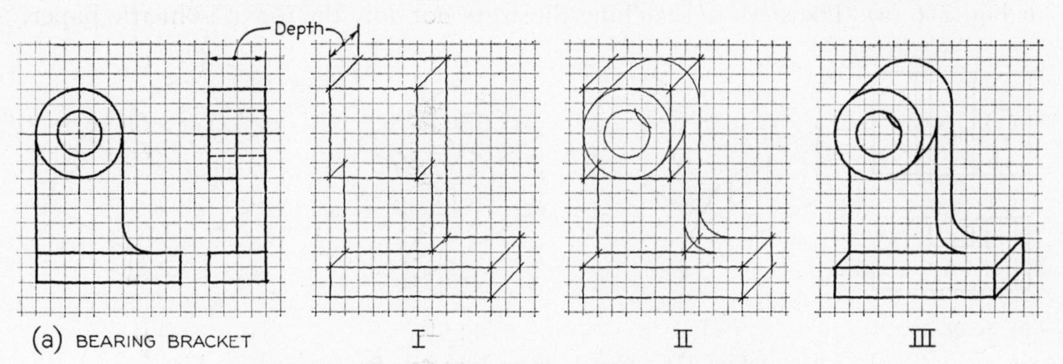

Fig. 278 Oblique Sketching on Cross-Section Paper.

I. Sketch lightly the enclosing box construction. Sketch the receding lines at 45° diagonally through the squares. To establish the depth, sketch the receding lines diagonally through half as many squares as the given number shown at (a).

II and III. Sketch all arcs and circles, and heavy-in all final lines.

205. Perspective Sketching. The Bearing sketched in oblique in Fig. 277 can easily be sketched in *one-point perspective* (one vanishing point), as shown in Fig. 279:

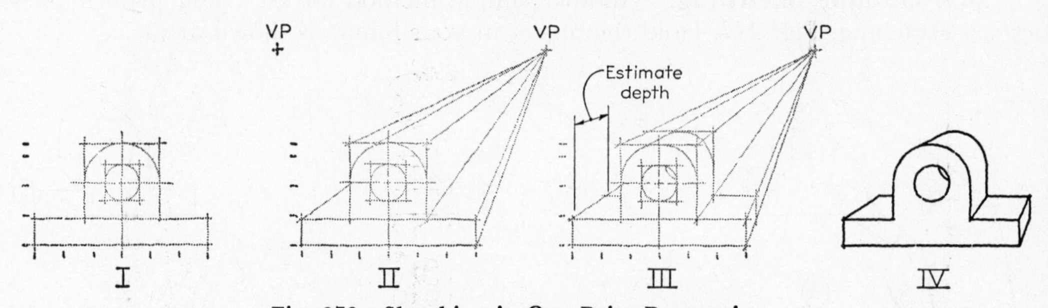

Fig. 279 Sketching in One-Point Perspective.

I. Sketch true front face of the object, just as in oblique sketching. Select the vanishing point (VP) for the receding lines. In most cases, it is desirable to place VP above and to the right of the picture, as shown, but it can be placed anywhere in the vicinity of the picture. If placed too close to the center, the lines will converge too sharply, and the picture will be distorted.

II. Sketch receding lines toward VP.

III. Estimate the depth to look well, and sketch in the back portion of the object. Note that the back circle and arc will be slightly smaller than the front circle and arc.

IV. Dim all construction with the artgum, and heavy-in all final lines. Note the similarity between the perspective sketch and the oblique sketch in Fig. 277.

Two-point perspective (two vanishing points) is the most "true to life" of all pictorial methods, but requires some natural sketching ability or considerable practice for best results. A simple method is shown in Fig. 280 that can be used successfully by the non-artistic student:

I. Sketch front corner of desk in true height, and locate two vanishing points on a *horizon* line (eye level). The distance CA may vary—the greater it is, the higher the eye level will be and the more we will be looking down on top of the object. A good rule-of-thumb is to make C-VPL one-third to one-fourth of C-VPR.

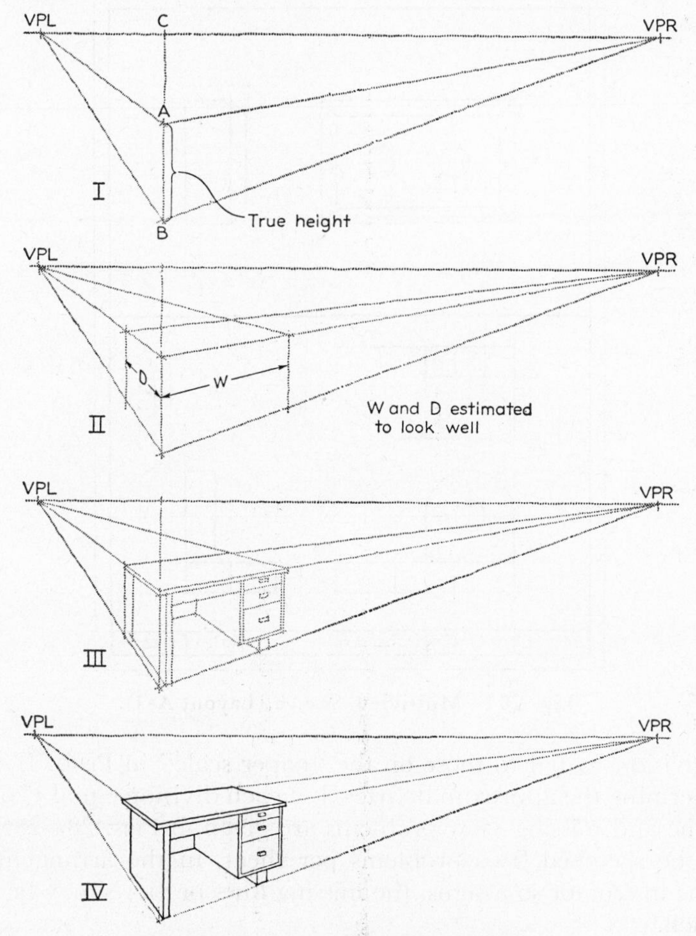

Fig. 280 Two-Point Perspective.

II. Estimate depth and width, and sketch enclosing box.

III. Block-in all details. Note that all parallel lines converge toward the same vanishing point.

IV. Dim the construction lines with the artgum as necessary, and heavy-in all final lines. Make the outlines thicker and the inside lines thinner, especially where they are close together.

206. Sketching Problems. In Figs. 282 and 283 are given a variety of objects from which the student is to sketch the necessary views. Using $8\frac{1}{2}'' \times 11''$ cross-section paper, sketch a border and title strip, and divide the sheet into two parts as shown in Fig. 281. Sketch two assigned problems per sheet, as shown. On the problems in

Fig. 282, ticks are given that indicate $\frac{1}{2}''$ or $\frac{1}{4}''$ spaces. Thus, measurements may be easily spaced off on cross-section paper having $\frac{1}{8}''$ or $\frac{1}{4}''$ grid spacings.

On the problems in Fig. 283, no indications of size are given. The student is to sketch the necessary views of assigned problems to fit the spaces comfortably, about as shown in Fig. 281. It is suggested that the student prepare a small paper "scale,"

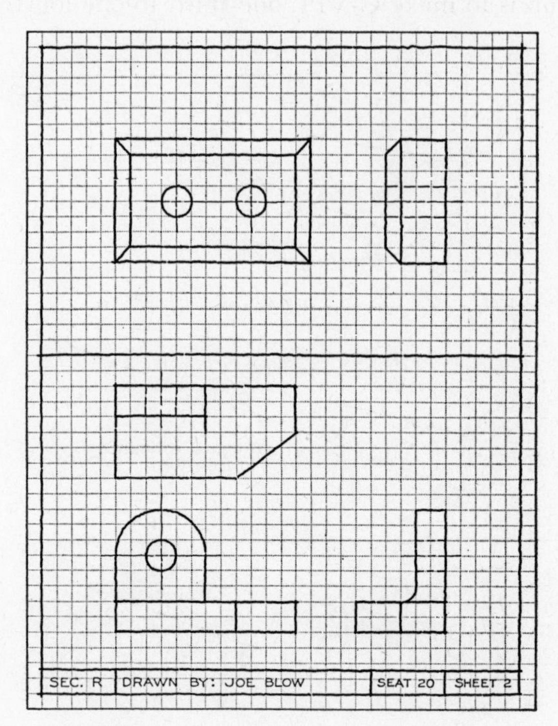

Fig. 281 Multiview Sketch (Layout A-1).

making the divisions equal to those on the "paper scale" in Prob. 1. This scale can be used to determine the approximate sizes. Let each division equal $\frac{1}{2}''$ on your sketch.

Missing-line and missing-view problems are given in Figs. 284-285, respectively. These are to be sketched, two problems per sheet, in the arrangement shown in Fig. 281. If the instructor so assigns, the missing lines or views may be sketched with a colored pencil.

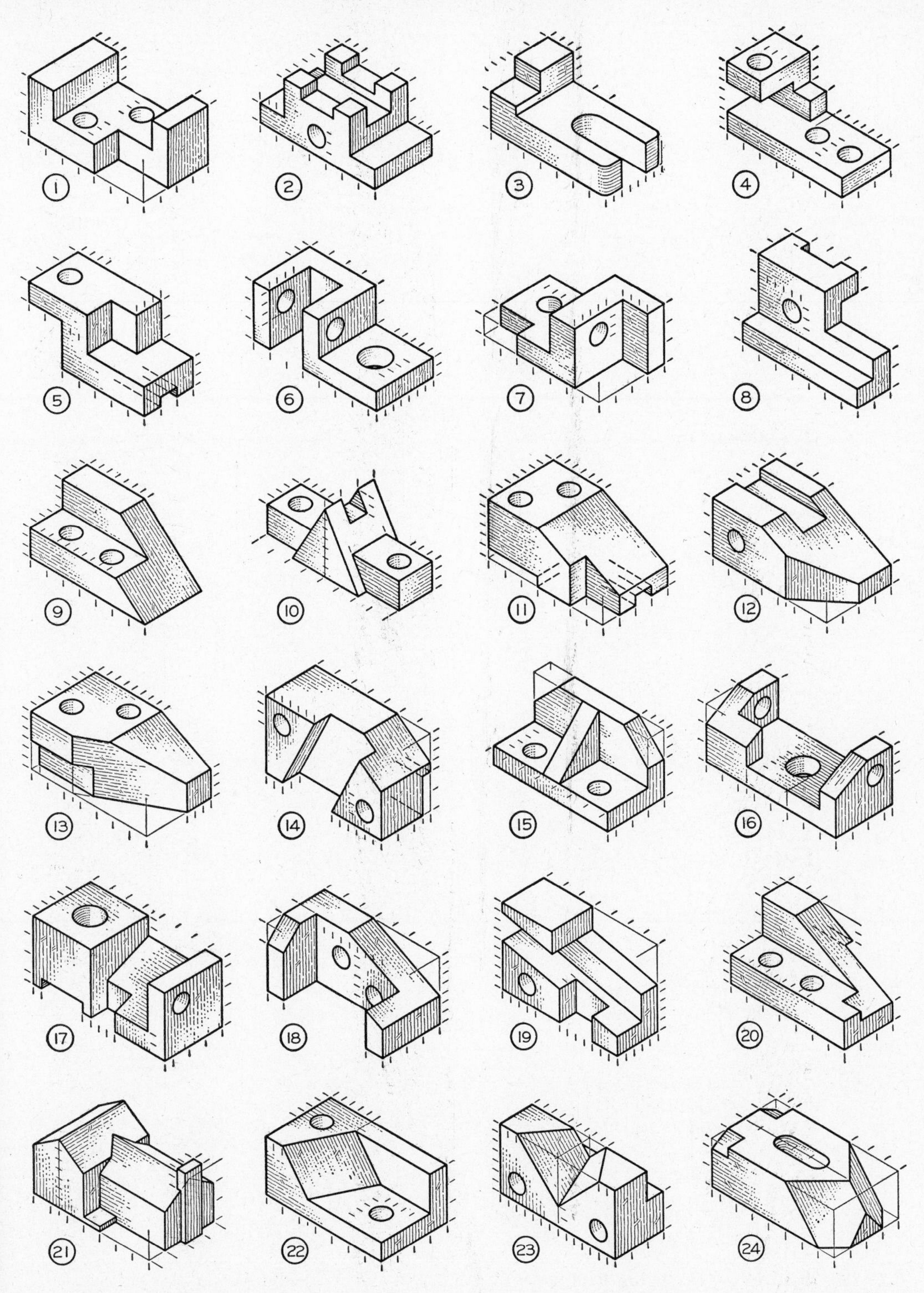

Fig. 282 Multiview Sketching Problems. Sketch necessary views, using Layout A-1 (freehand), on cross-section paper or plain paper, two problems per sheet as in Fig. 281. The units shown are $\frac{1}{2}''$ or $\frac{1}{4}''$. All holes are through holes. See §206.

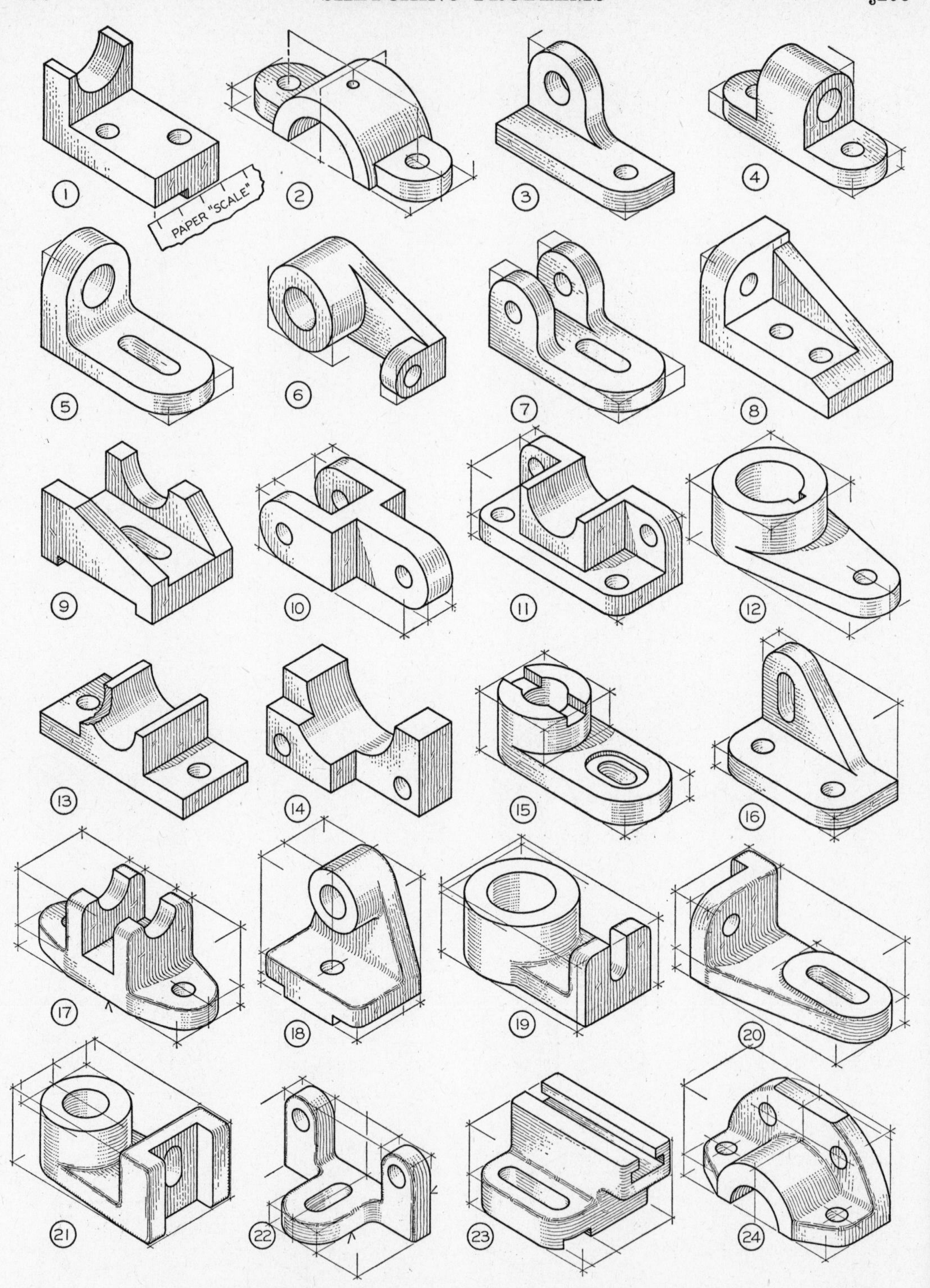

Fig. 283 Multiview Sketching Problems. Sketch necessary views, using Layout A-1 (freehand), on cross-section paper or plain paper, two problems per sheet as in Fig. 281. Prepare paper "scale" with divisions equal to those in Prob. 1, and apply to problems to obtain approx. sizes. Let each division = $\frac{1}{2}''$ on your sketch. Study Fig. 320 and for Probs. 17-24, study §§239 and 240.

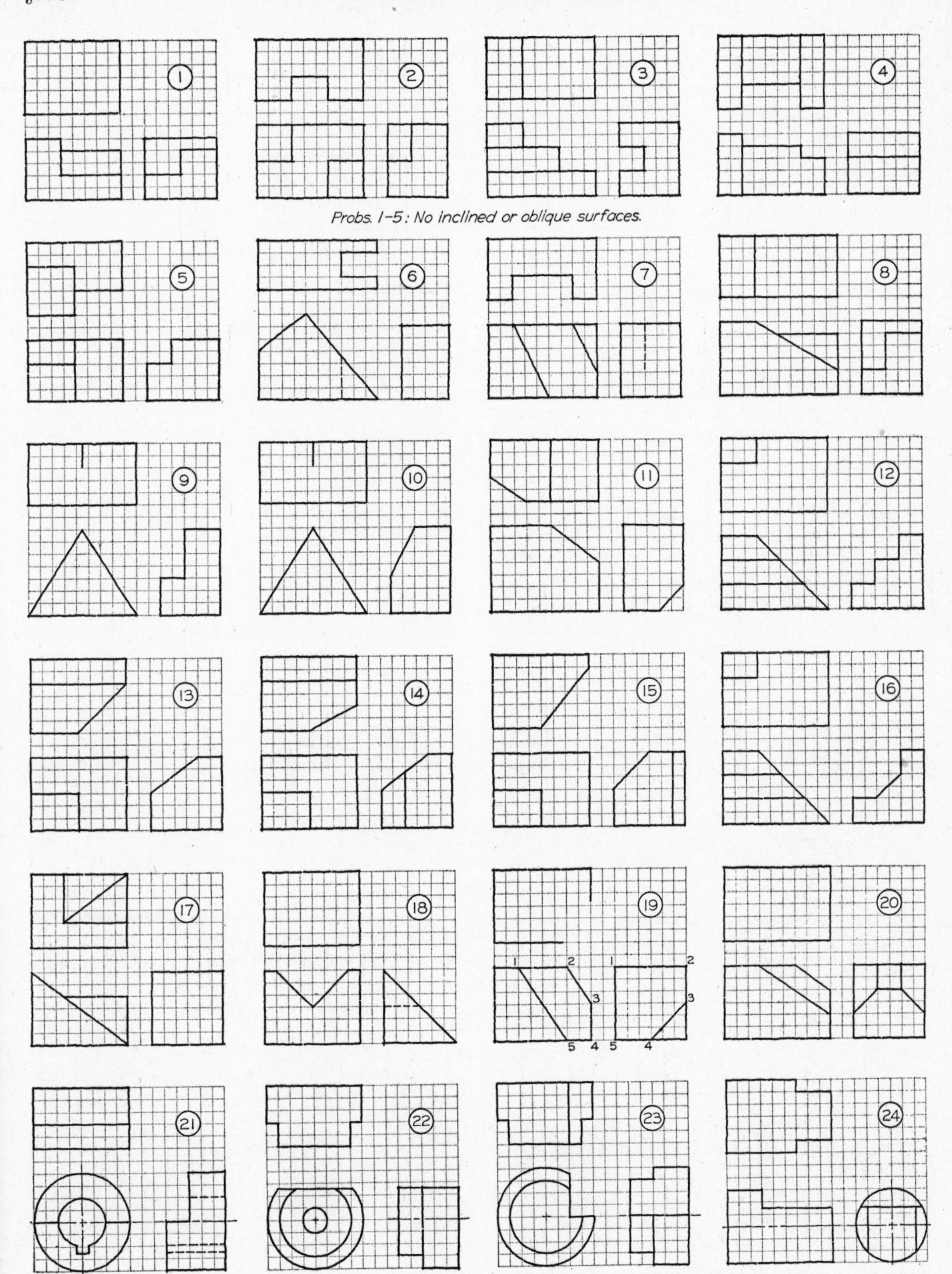

Fig. 284 Missing-Line Sketching Problems. (1) Sketch given views, using Layout A-1 (freehand), on cross-section paper or plain paper, two problems per sheet as in Fig. 281. The squares are each $\frac{1}{4}''$. (2) Sketch in isometric on isometric paper, or in oblique on cross-section paper.

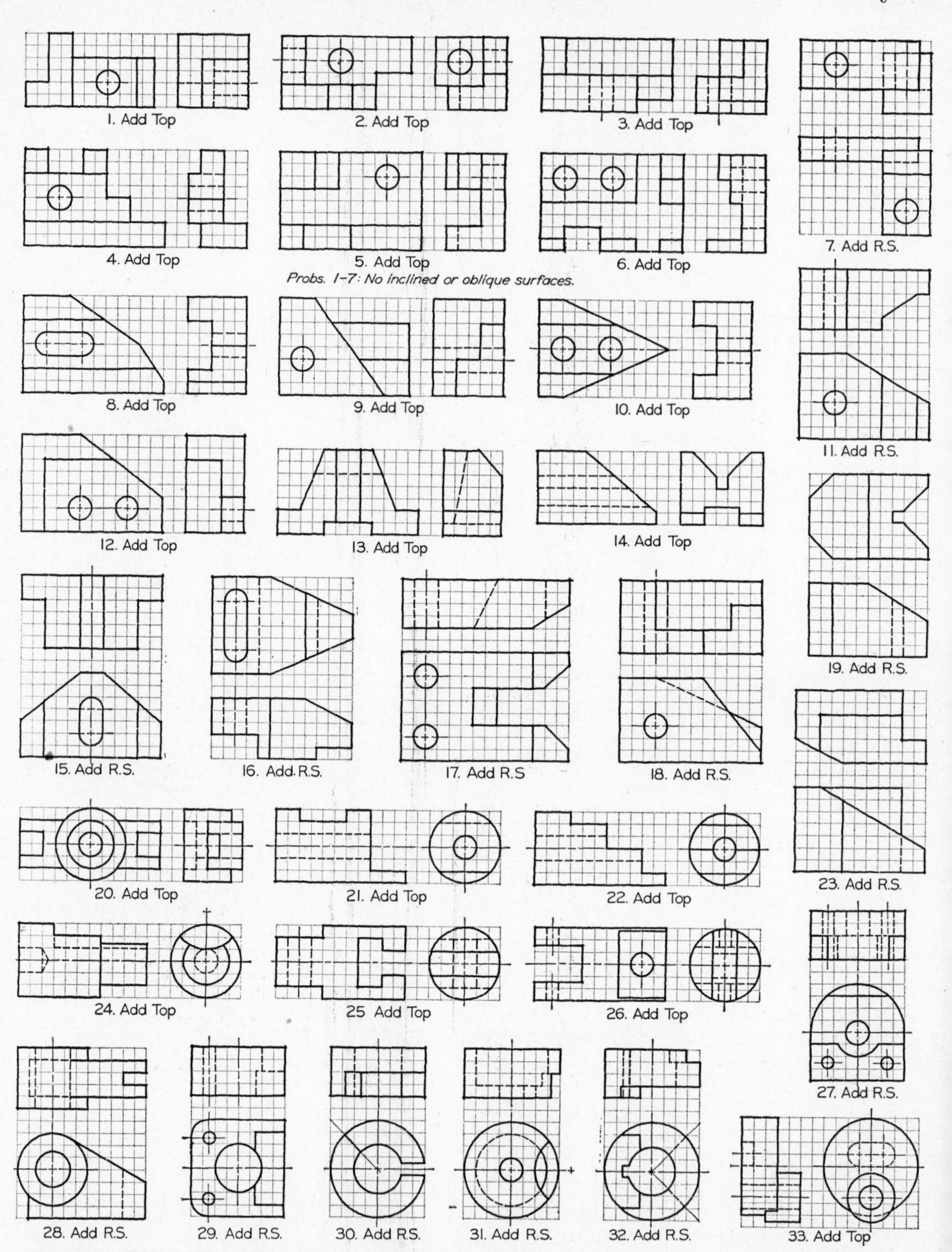

Fig. 285 Third-View Sketching Problems. (1) Using Layout A-1 (freehand), on cross-section paper or plain paper, two problems per sheet as in Fig. 281, sketch given two views and add the missing views, as indicated. The given views are either front and right-side views, or front and top views. Hidden holes with center lines are drilled holes. (2) Sketch in isometric on isometric paper, or in oblique on cross-section paper.

CHAPTER 6

MULTIVIEW PROJECTION*

207. Multiview Projection. A view of an object is known technically as a *projection.* A projection is a view conceived to be drawn or "projected" onto a plane known as the *plane of projection.*

The method of viewing an object to obtain a *multiview projection,* in this case a front view, is shown in Fig. 286 (a). Between the observer and the object a trans-

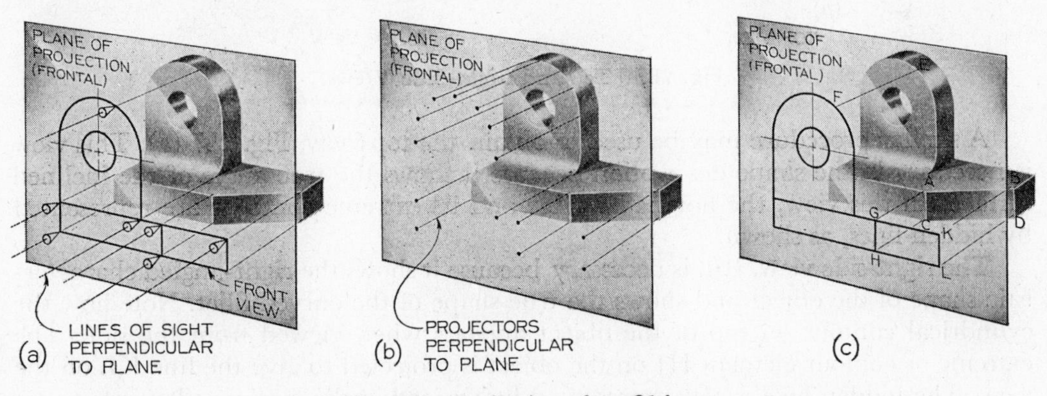

Fig. 286 Projection of an Object.

parent plane, or pane of glass, representing a plane of projection, is placed parallel to the front surfaces of the object. On the pane of glass is shown in outline how the object appears to the observer from that direction. Theoretically the observer is at an infinite distance from the object, so that the *lines of sight* are parallel.

*See ASA Y14.3-1957.

In more precise terms, this view is obtained by drawing perpendiculars, called *projectors*, from all points on the edges or contours of the object to the plane of projection, (b). The collective piercing points of these projectors, being infinite in number, form lines on the pane of glass, as shown at (c).

Thus, as shown at (c), a projector from point A on the object pierces the plane of projection at point G, which is a view or projection of the point. The same procedure applies to point B, whose projection is point K. Since A and B are end points of a straight line on the object, the projections G and K are joined to give the projection of the line GK. Similarly, if the projections of the four corners A, B, C, and D are found, the projections G, K, L, and H may be joined by straight lines to form the projection of the rectangular surface.

The same procedure can be applied to curved lines, as for example, the top curved contour of the object. A point E on the curve is projected to the plane at F. The projection of an infinite number of such points on the plane of projection results in the projection of the curve, a few of which are shown at (b). If this procedure of projecting points is applied to all edges and contours of the object, a complete view or projection of the object results. This view is necessary in the shape description because it shows the true curvature of the top and the true shape of the hole.

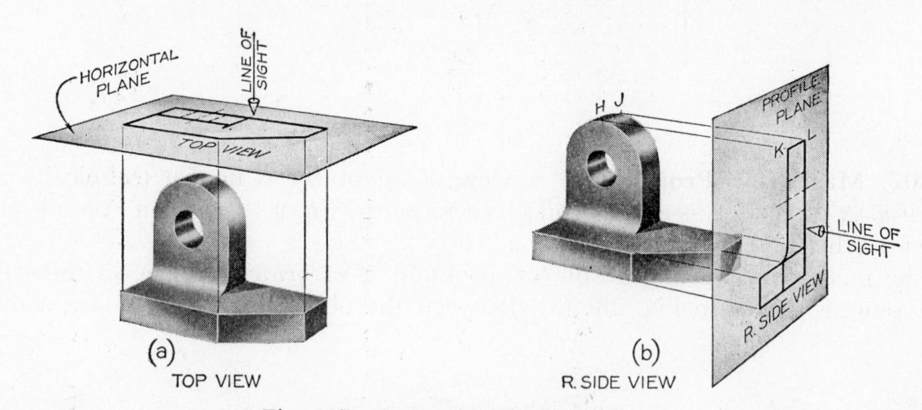

Fig. 287 Top and Right-Side Views.

A similar procedure may be used to obtain the top view, Fig. 287 (a). This view is necessary in the shape description because it shows the true angle of the inclined surface. In this view, the hole is invisible and its extreme contours are represented by hidden lines, as shown.

The right-side view, (b), is necessary because it shows the right-angled characteristic shape of the object and shows the true shape of the curved fillet. Note how the cylindrical contour on top of the object appears when viewed from the side. The extreme or contour element HJ on the object is projected to give the line KL on the view. The hidden hole is also represented by projecting the extreme elements.

The plane of projection upon which the front view is projected is called the *frontal plane*, that upon which the top view is projected, the *horizontal plane*, and that upon which the side view is projected, the *profile plane*.

208. The Glass Box. If planes of projection are placed parallel to the principal faces of the object, they form a "glass box," as shown in Fig. 288 (a). Notice that the observer is always *on the outside looking in*, so that he sees the object through the planes of projection. Since the glass box has six sides, six views of the object are obtained.

Note that the object has three principal dimensions: *width*, *height*, and *depth*. These are fixed terms used for dimensions in these directions, regardless of the shape of the object. See §185.

Since it is required to show the views of a solid or three-dimensional object on a flat sheet of paper, it is necessary to unfold the planes so that they will all lie in the same plane, Fig. 288 (b). All planes except the rear plane are hinged upon the frontal plane, the rear plane being hinged to the left-side plane.* Each plane revolves outwardly from the original box position until it lies in the frontal plane, which remains stationary. The "hinge lines" of the glass box are known as *folding lines*.

The positions of these six planes, after they have been revolved, are shown in Fig. 289. Carefully identify each of these planes and corresponding views with its original position in the glass box, and repeat this mental procedure, if necessary, until the revolutions are thoroughly understood.

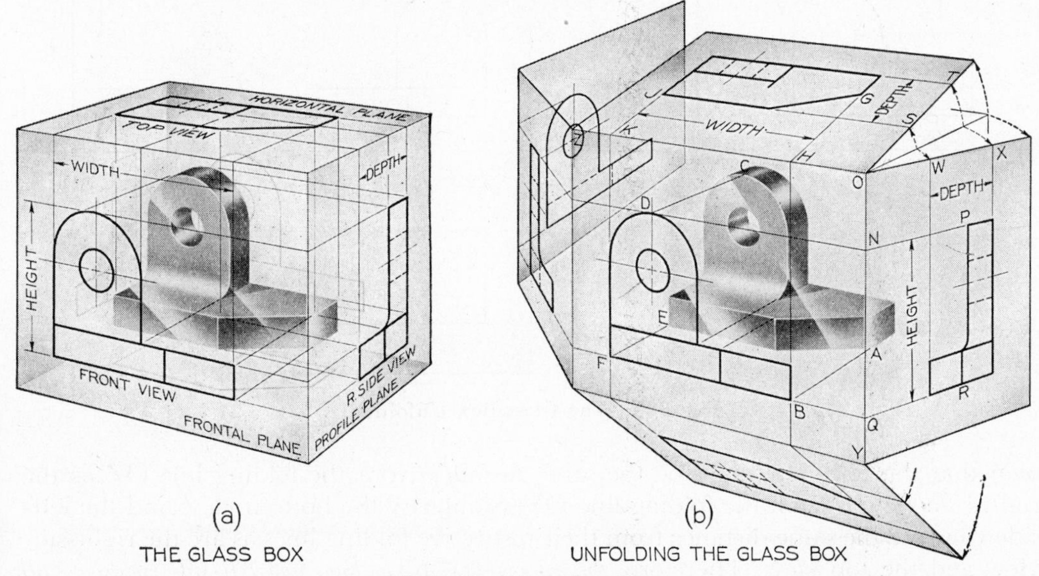

Fig. 288 The Glass Box.

In Fig. 288 (b), observe that lines extend around the glass box from one view to another upon the planes of projection. These are the *projections of the projectors* from points on the object to the views. For example, the projector BA is projected on the horizontal plane at HG and on the profile plane at QR. When the top plane is folded up, lines JK and GH will become vertical and line up with KF and HB, respectively. Thus, JK and KF form a single straight line JF, and GH and HB form a single straight line GB, as shown in Fig. 289. This explains why the top view is the same width as the front view and why it is placed directly above the front view. The same relation exists between the front and bottom views. Therefore, *the front, top, and bottom views all line up vertically and are the same width.*

In Fig. 288 (b), when the profile plane is folded out, lines DN and NP become a single straight line DP, and lines BQ and QR become a single straight line BR,

*Except as explained in § 214.

as shown in Fig. 289. The same relation exists between the front, left-side, and rear views. Therefore, *the rear, left-side, front, and right-side views all line up horizontally, and are the same height.*

In Fig. 288 (b), note that lines OS and OW and lines ST and WX are respectively equal. These equal lines are shown in the unfolded position in Fig. 289. Thus, it is

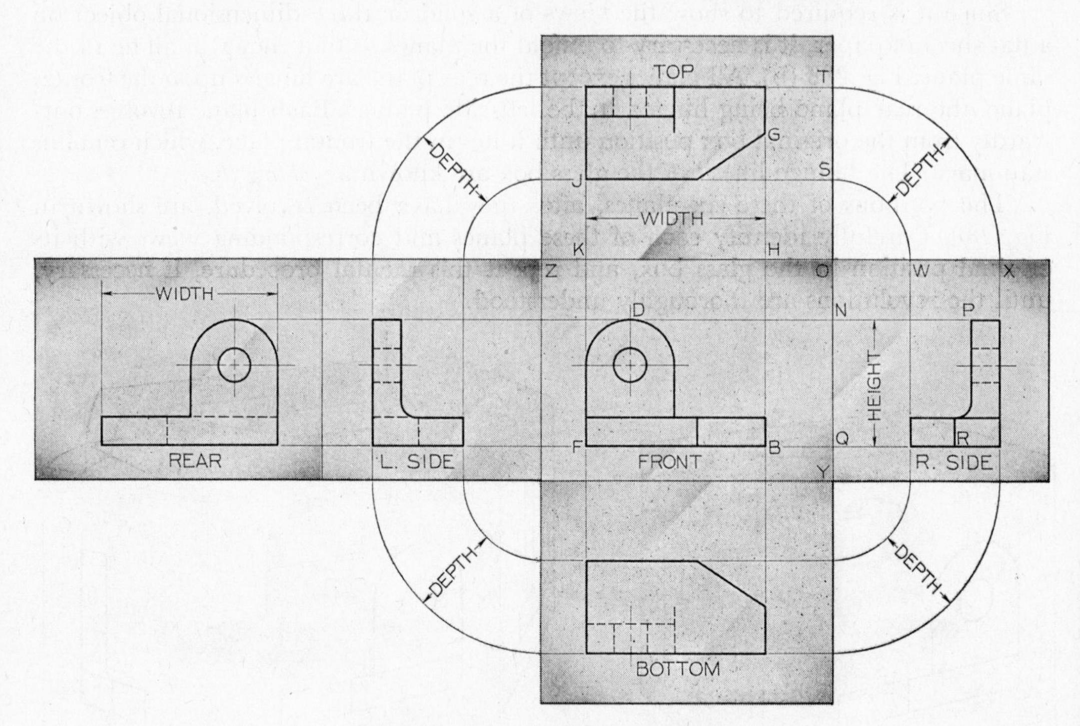

Fig. 289 The Glass Box Unfolded.

seen that the top view must be the same distance from the folding line OZ as the right-side view is from the folding line OY. Similarly, the bottom view and the left-side view are the same distance from their respective folding lines as are the right-side view and the top view. Therefore, *the top, right-side, bottom, and left-side views are all equidistant from the respective folding lines, and are the same depth.* Note that in these four views that surround the front view, the front surfaces of the object are faced inward, or toward the front view. Observe also that the left-side and right-side views and the top and bottom views are the reverse of each other in outline shape. Similarly, the rear and front views are the reverse of each other.

209. Folding Lines. Three views of the object discussed above are shown in Fig. 290 (a), with folding lines between the views. These folding lines correspond to the "hinge lines" of the glass box, as we have seen. The H/F folding line, between the top and front views, is the intersection of the horizontal and frontal planes. The F/P folding line, between the front and side views, is the intersection of the frontal and profile planes.

The distances X and Y, from the front view to the respective folding lines, are not necessarily equal, since they depend upon the relative distances of the object from the horizontal and profile planes. However, as shown in §208, distances D_1, from the top and side views to the respective folding lines, must always be equal. Therefore, the views may be any desired distance apart, and the folding lines may be drawn anywhere between them, so long as distances D_1 are kept equal and the folding lines are at right angles to the projection lines between the views.

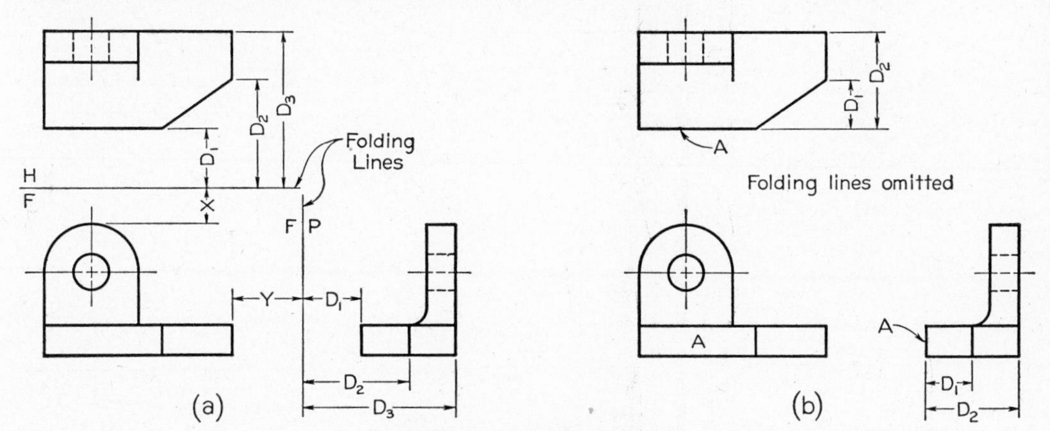

Fig. 290 Folding Lines.

It will be seen that distances D_2 and D_3, respectively, are also equal, and the folding lines H/F and F/P are in reality reference lines for making equal *depth* measurements in the front and side views. Thus, any point in the top view is the same distance from HF as the corresponding point in the side view is from F/P.

While it is necessary to understand the folding lines, particularly because they are useful in solving graphical problems in descriptive geometry, they are as a rule omitted in industrial drafting. The three views, with the folding lines omitted, are shown in Fig. 290 (b). Again, the distances between the top and front views and between the side and front views are not necessarily equal. Instead of using the folding lines as reference lines for setting off depth measurements in the top and side views, we use the front surface A of the object as a reference line. In this way, D_1, D_2, and all other depth measurements are made to correspond in the two views in the same manner as if folding lines were used.

210. Two-View Mechanical Drawing. Let it be required to draw, full size with instruments, the necessary views of the Operating Arm shown in Fig. 291 (a). In this case, as shown by the arrows, only the front and top views are needed.

I. Determine the spacing of the views. The width of the front and top views is 6″, and the width of the working space is $10\frac{1}{2}$″. As shown at (b), subtract 6″ from $10\frac{1}{2}$″ and divide the result by 2 to get the value of space A. To set off the spaces, place the scale horizontally along the bottom of the sheet and make short vertical marks.

The depth of the top view is $2\frac{1}{2}$″ and the height of the front view is $1\frac{3}{4}$″, while the height of the working space is $7\frac{5}{8}$″. Assume a space C, say 1″, between views that will look well and that will provide sufficient space for dimensions, if any.

As shown at (b), add $2\frac{1}{2}$″, 1″, and $1\frac{3}{4}$″, subtract the total from $7\frac{5}{8}$″, and divide the result by 2 to get the value of space B. To set off the spaces, place the scale vertically along the left side of the sheet with the full-size scale on the left, and make short marks perpendicular to the scale.

II. Locate center lines from spacing marks. Construct arcs and circles lightly.

III. Draw horizontal and then vertical construction lines in the order shown. Allow construction lines to cross at corners.

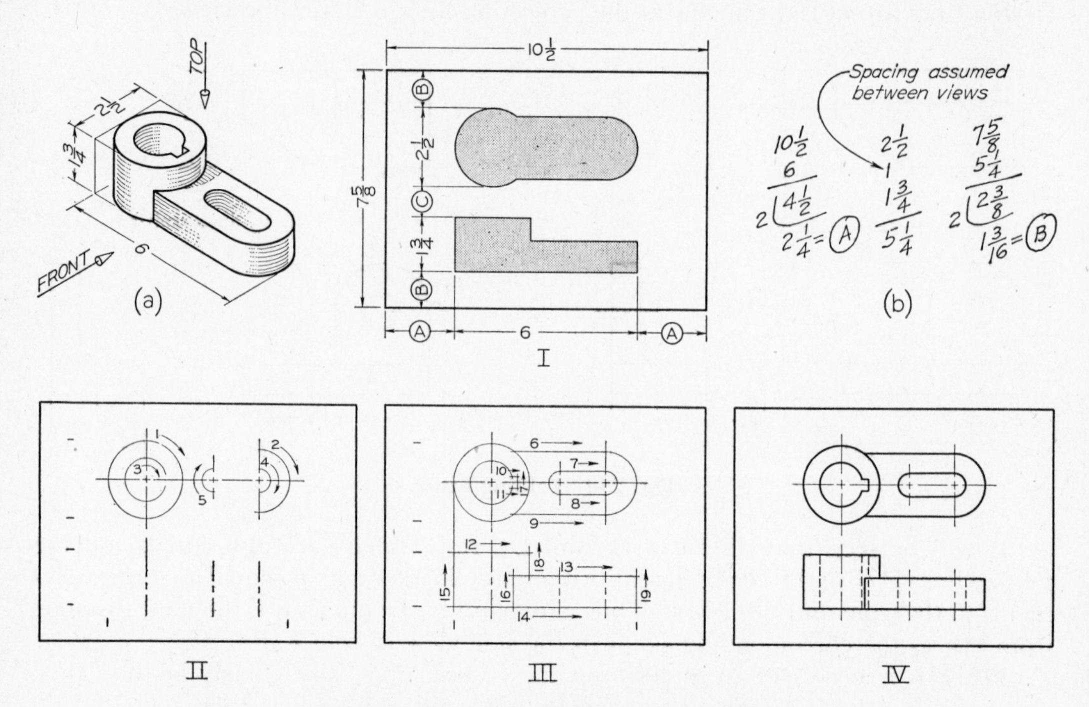

Fig. 291 Two-View Mechanical Drawing.

IV. Add hidden lines and heavy-in all final lines, clean-cut and dark. The visible lines should be heavy enough to make the views stand out. The hidden lines and center lines should be sharp in contrast to the visible lines, but dark enough to reproduce well. See §55 for technique of pencil drawing. Construction lines need not be erased if drawn lightly. If you are working on tracing paper, hold the sheet up to the light to see if the density of your lines is sufficient to reproduce well. See §55 and Fig. 68.

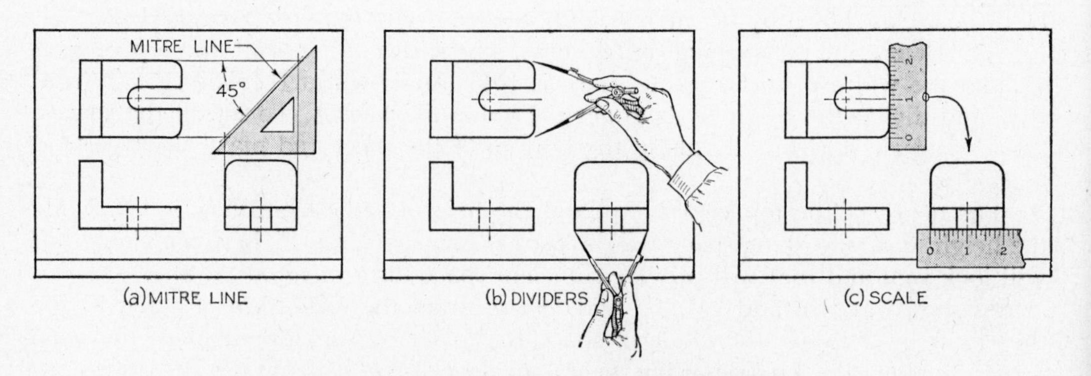

Fig. 292 Transferring Depth Dimensions.

211. Transferring Depth Dimensions. Since all depth dimensions in the top and side views must correspond point-for-point, accurate methods of transferring these distances, such as d_1, d_2, etc., Fig. 290 (b), must be used.

The 45° *mitre line* method, Fig. 292 (a), is a convenient method, especially when transferring a large number of points, as when plotting a curve, Fig. 318. Note that the right-side view may be moved to the right or left, or the top view may be moved upward or downward, by shifting the 45° line accordingly. It is not necessary to draw continuous lines between the top and side views via the mitre line. Instead, make short dashes across the mitre line and project from these.

In practice it is generally recommended, for the sake of accuracy, that the depth dimensions be transferred with the aid of the dividers, (b), or scale, (c). These methods are best when only a small number of very accurate measurements are to be transferred, as is usually the case. The scale method is especially convenient when the drafting machine, Fig. 91, is used, because both vertical and horizontal scales are readily available.

212. Projecting a Third View. In Fig. 293 (top) is a pictorial drawing of a given object, three views of which are required. Each corner of the object is given a number, as shown. At I, the top and front views are shown, with each corner properly numbered in both views. Each number appears twice, once in the top view and once again in the front view.

If a point is *visible* in a given view, the number is placed outside the corner, but if the point is *invisible*, the numeral is placed *inside* the corner. For example, at I point 1 is visible in both views, and is therefore placed outside the corners in both views. However, point 2 is visible in the top view and the number is placed outside, while in the front view it is invisible and is placed inside.

This numbering system, in which points are identified by the same numbers in all views, is useful in projecting known points in two views to unknown positions in a third view. Note that in this numbering system a given point has the same number in all views, and should not be confused with the numbering system used in Fig. 306 and others, in which a point has different numbers in each view.

In Fig. 293, before starting to project the right-side view, try to visualize the view as seen in the direction of the arrow (see pictorial drawing).

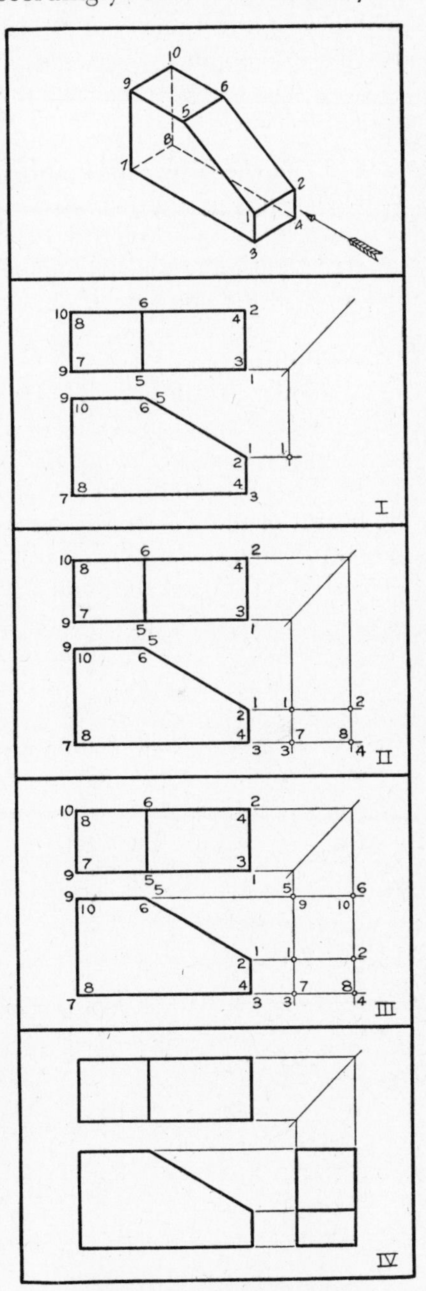

Fig. 293 Use of Numbers.

Then construct the right-side view point-by-point, using a hard pencil and very light lines.

As shown at I, locate point 1 in the side view by projecting from point 1 in the top view and point 1 in the front view. In Space II, project points 2, 3, and 4 in a similar manner to complete the vertical end surface of the object. In Space III, project points 5 and 6 to complete the side view of the inclined surface 5-6-2-1. This completes the right-side view, since invisible points 9, 10, 8, and 7 are directly behind visible corners 5, 6, 4, and 3, respectively. Note that in the side view also, the invisible points are lettered *inside*, and the visible points *outside*.

As shown in Space IV, the drawing is completed by heavying-in the lines in the right-side view.

213. Three-View Mechanical Drawing. Let it be required to draw, full size with instruments, the necessary views of the V-Block in Fig. 294 (a). In this case, as shown by the arrows, three views are needed.

I. Determine the spacing of the views. The width of the front view is $4\frac{1}{4}''$, and the depth of the side view is $2\frac{1}{4}''$, while the width of the working space is $10\frac{1}{2}''$. Assume a space C between views, say $1\frac{1}{4}''$, that will look well, and that will allow sufficient space for dimensions, if any.

As shown at (b), add $4\frac{1}{4}''$, $1\frac{1}{4}''$, and $2\frac{1}{4}''$, subtract the total from $10\frac{1}{2}''$, and divide the result by 2 to get the value of space A. To set off these horizontal spacing measurements, place the scale along the bottom of the sheet and make short vertical marks.

The depth of the top view is $2\frac{1}{4}''$, and the height of the front view is $1\frac{3}{4}''$, while the height of the working space is $7\frac{5}{8}''$. Assume a space D between views, say $1''$. As shown in §209, space D need not be the same as space C. As shown at (b), add $2\frac{1}{4}''$, $1''$, and $1\frac{3}{4}''$, subtract the total from $7\frac{5}{8}''$, and divide the result by 2 to get the value

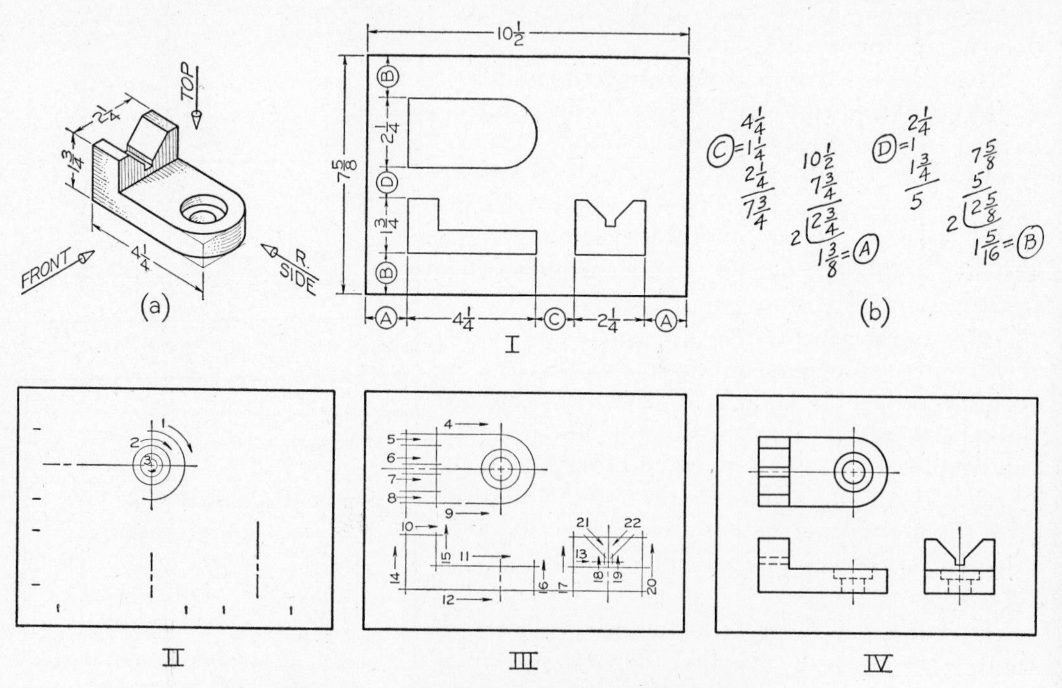

Fig. 294 Three-View Mechanical Drawing.

of space B. To set off these vertical spacing measurements, place the scale along the left side of the sheet with the scale used on the left, and make short marks perpendicular to the scale. Allow for dimensions, if any.

II. Locate the center lines from the spacing marks. Construct lightly the arcs and circles.

III. Draw horizontal, then vertical, then inclined construction lines, in the order shown. Allow construction lines to cross at the corners. Do not complete one view at a time; construct the views simultaneously.

IV. Add hidden lines and heavy-in all final lines, clean-cut and dark. A convenient method of transferring a hole diameter from the top view to the side view is to use the compass with the same setting used for drawing the hole. The visible lines should be heavy enough to make the views stand out. The hidden lines and center lines should be sharp in contrast to the visible lines, but dark enough to reproduce well. Construction lines need not be erased if they are drawn lightly. If you are working on tracing paper, hold the sheet up to the light to see if the density of your lines is sufficient to reproduce well. See §55 and Fig. 68.

214. Alternate Positions of Views. If three views of a wide flat object are drawn, using the conventional arrangement of views, Fig. 295 (a), a large wasted

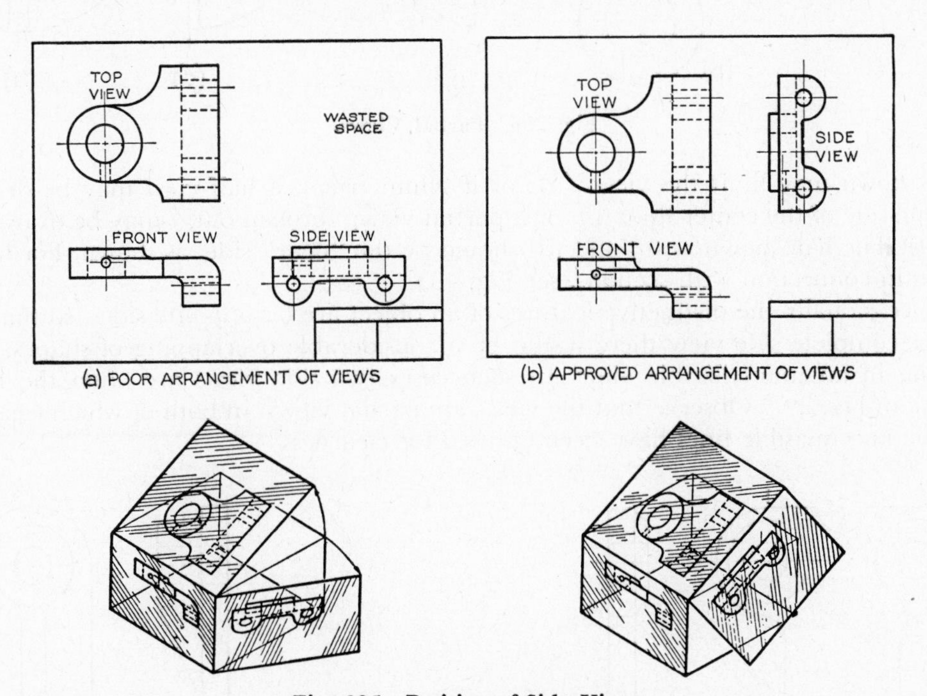

Fig. 295 Position of Side View.

space is left on the paper, as shown. In such cases, the profile plane may be considered hinged to the horizontal plane instead of the frontal plane, as shown at (b). This places the side view beside the top view, which results in better spacing and in some cases makes the use of a reduced scale unnecessary.

It is also permissible in extreme cases to place the side view across horizontally from the bottom view, in which case the profile plane is considered hinged to the bottom plane of projection. Similarly, the rear view may be placed directly above the top view or under the bottom view, if necessary, in which case the rear plane is considered hinged to the horizontal or bottom plane, as the case may be, and then rotated into coincidence with the frontal plane.

215. Partial Views. A view may not need to be complete but may show only what is necessary in the clear description of the object. Such a view is a *partial view*, Fig. 296. A break line, (a), may be used to limit the partial view; the contour of the

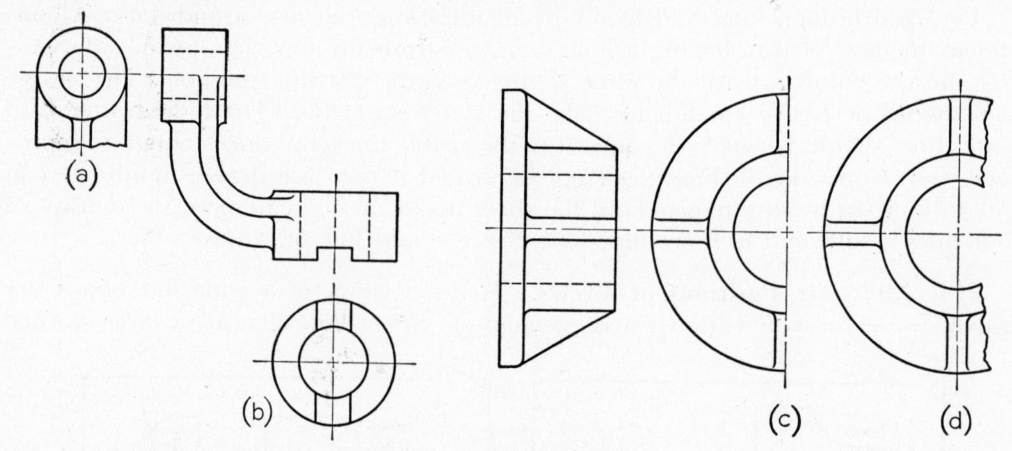

Fig. 296 Partial Views.

part shown may limit the view, (b); or if symmetrical, a half-view may be drawn on one side of the center line, (c), or a partial view, "broken out," may be drawn as at (d). The half shown at (c) and (d) should be the "near" side, as shown. For half-views in connection with sections, see Fig. 433.

Occasionally the distinctive features of an object are on opposite sides, so that in either complete side view there would be a considerable overlapping of shapes, resulting in an unintelligible view. In such cases two side views are often the best solution, Fig. 297. Observe that the views are partial views, in both of which certain visible and invisible lines have been omitted for clearness.

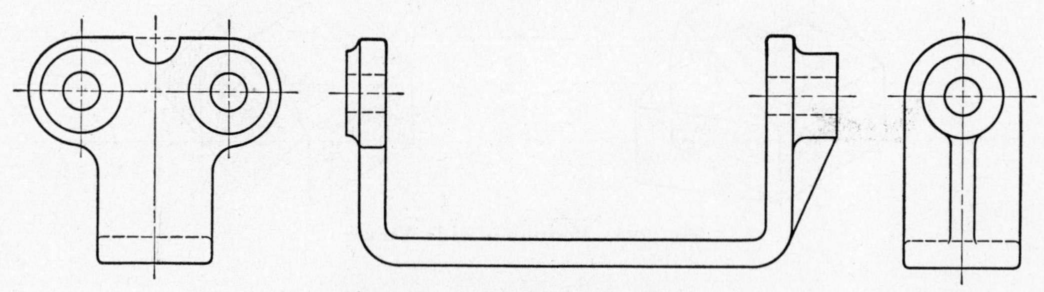

Fig. 297 Incomplete Side Views.

216. Removed Views. A *removed view*, Fig. 298, is a complete or a partial view removed to another place on the sheet so that it no longer is in direct projection with any other view. Such a view may be used to show some feature of the object more clearly, possibly to a larger scale, or to save drawing a complete regular view.

A viewing-plane line is used to indicate the part being viewed, the arrows at the ends showing the direction of sight. See §248. The removed view should be labeled VIEW A-A or VIEW B-B, etc., the letters referring to those placed at the ends of the viewing-plane line.

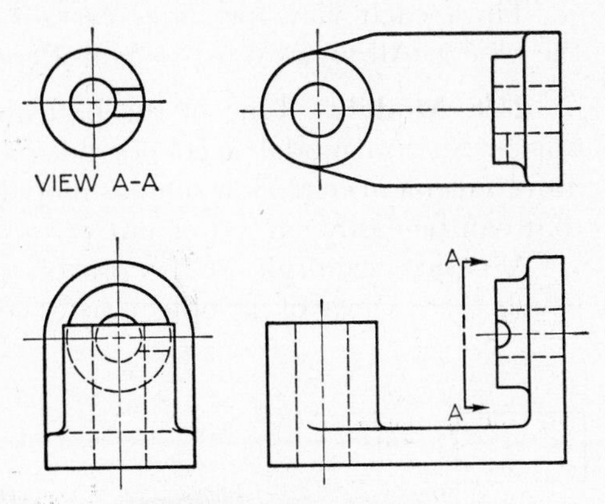

Fg. 298 Removed View.

217. Visualization. As stated in §9, the ability to *visualize* or *think in three dimensions* is one of the most important requisites of the successful engineer. In practice, this means the ability to study the views of an object and to form a mental picture of it—to *visualize* its three-dimensional shape. To the designer it means the ability to *synthesize* or form a mental picture before the object even exists, and the ability to express this image in terms of views. The engineer is the master planner in the construction of new machines, structures, or processes. The ability to visualize, and to use the language of drawing as a means of communication or recording of mental images, is indispensable.

Even the experienced engineer cannot look at a multiview drawing and instantly visualize the object represented (except for the simplest shapes), any more than he can grasp the ideas on a book page merely at a glance. It is necessary to *study* the drawing, to "read" the lines in a logical way, to piece together the little things until a clear idea of the whole emerges.

How this is done is described in the following paragraphs.

218. Visualizing the Views. A method of reading drawings that is essentially the reverse mental process to that of obtaining the views by projection is illustrated in Fig. 299. The given views of an Angle Bracket are shown at (a). The front view

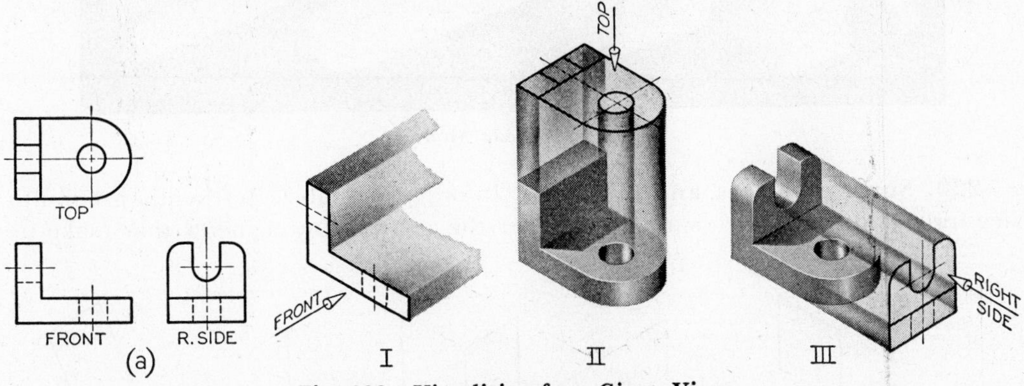

Fig. 299 Visualizing from Given Views.

shows, Step I, that the object is L-shaped, how wide and how high the object is, and the thickness of the members. The meaning of the hidden lines and center lines is not yet clear, nor do we yet know the depth of the object.

The top view tells us, Step II, that the horizontal member is rounded on the end and has a round hole. Some kind of slot is indicated at the left end.

The right-side view tells us, Step III, that the left end of the object has rounded corners at the top and has an open-end slot in a vertical position.

Thus, each view provides certain definite information regarding the shape of the object. All views must be considered in order to visualize the object completely.

219. Models. One of the best aids to visualization is an actual model of the object. Such a model need not be made accurately to scale, and may be made of any convenient material, such as modeling clay, laundry soap, wood, or any material that can be easily carved or cut.

A typical example of the use of soap or clay models is shown in Fig. 300, in which three views of an object are given and the student is to supply a missing line.

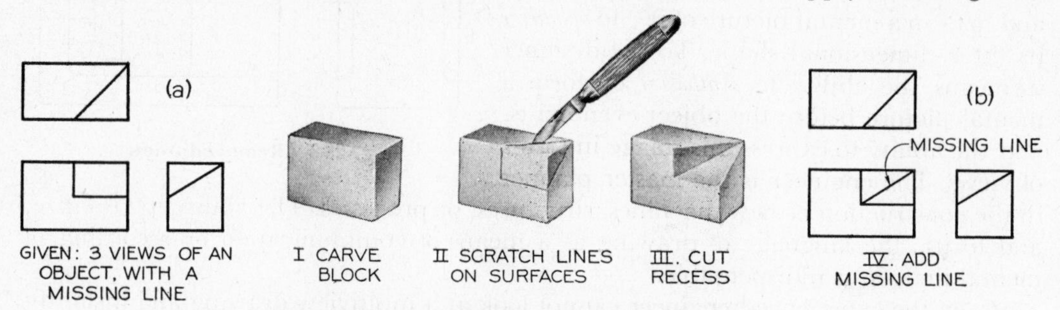

Fig. 300 Use of Model to Aid Visualization.

The model is carved as shown in Steps I, II, and III, and the "missing" line, discovered in the process, is added to the drawing as shown in Step IV.

Some typical examples of soap models are shown in Fig. 301.

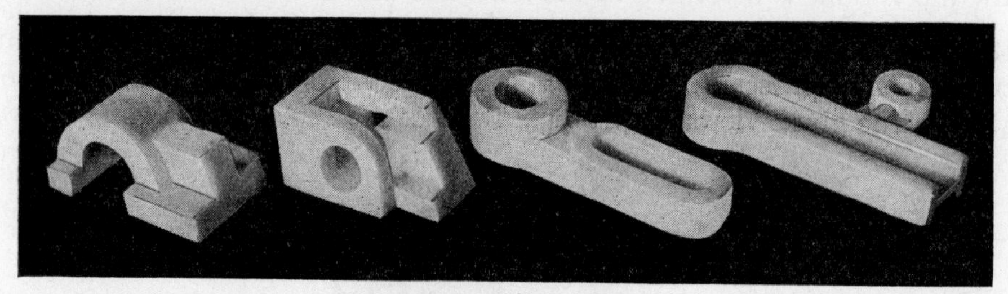

Fig. 301 Soap Models.

220. Surfaces, Edges, and Corners. In order to analyze and synthesize multiview projections, it is necessary to consider the component elements that make up

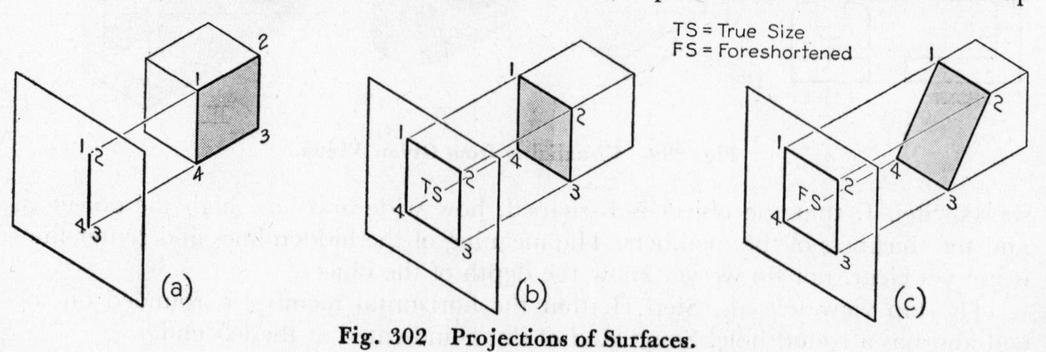

Fig. 302 Projections of Surfaces.

most solids. A *surface* (plane) may be bounded by straight lines or curves, or a combination of them. A surface may be *frontal*, *horizontal*, or *profile*, according to the plane of projection to which it is parallel.

If a plane surface is perpendicular to a plane of projection, it appears as a line, Fig. 302 (a). If it is parallel, it appears as a surface, *true size*, (b). If it is situated at an angle, it appears as a surface, *foreshortened*, (c). Thus, *a plane surface always projects as a line or as a surface.*

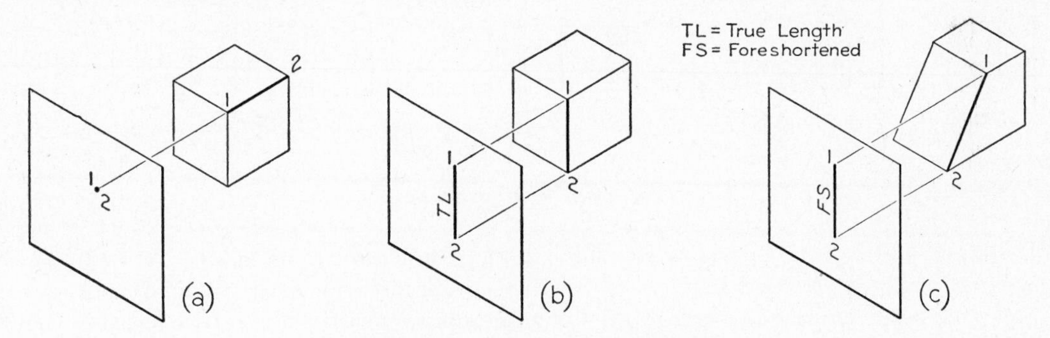

Fig. 303 Projections of Lines.

The intersection of two plane surfaces produces an *edge*, or a straight line. Such a line is common to both surfaces and forms a boundary line for each. If an edge is perpendicular to a plane of projection, it appears as a point, Fig. 303 (a); otherwise it appears as a line, (b) and (c). If it is parallel to the plane of projection, it shows true length, (b); if not parallel, it shows foreshortened, (c). Thus, *a straight line always projects as a straight line or as a point.* A line may be *frontal*, *horizontal*, or *profile*, according to the plane of projection to which it is parallel.

A *corner*, or point, is the common intersection of three or more surfaces or edges. A corner always appears as a point in every view.

221. Adjacent Areas. Consider a given top view, as shown at Fig. 304 (a). Lines divide the view into three areas. Each of these must represent a surface *at a different level*. Surface A may be high and B and C lower, as shown at (b). Or B may

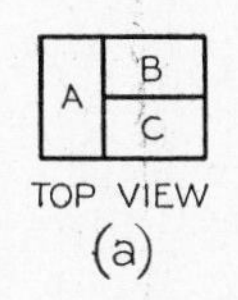

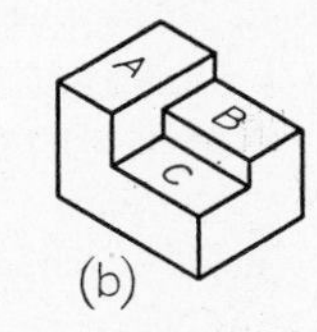

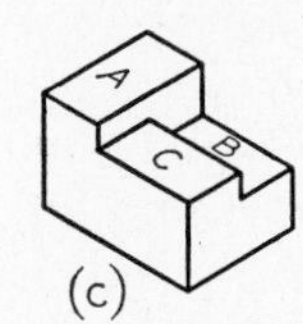

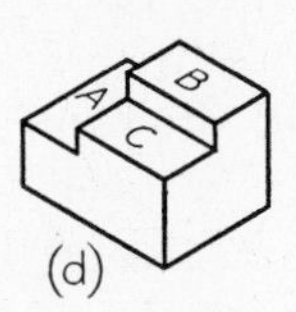

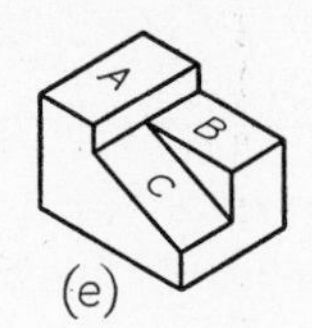

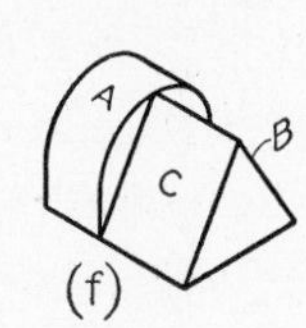

Fig. 304 Adjacent Areas.

be lower than C, as shown at (c). Or B may be highest, with C and A each lower, (d). Or one or more surfaces may be inclined, as at (e). Or one or more surfaces may be cylindrical, as at (f), etc. Hence the rule: *No two adjacent areas can lie in the same plane.*

The same reasoning can apply, of course, to the adjacent areas in any given view. Since an area (surface) on a view can be interpreted in several different ways, it is necessary to observe other views also in order to determine which interpretation is correct.

222. Similar Shapes of Surfaces. If a surface is viewed from several different positions, it will in each case be seen to have a certain number of sides and to have a certain characteristic shape. An L-shaped surface, Fig. 305 (a), will appear as an

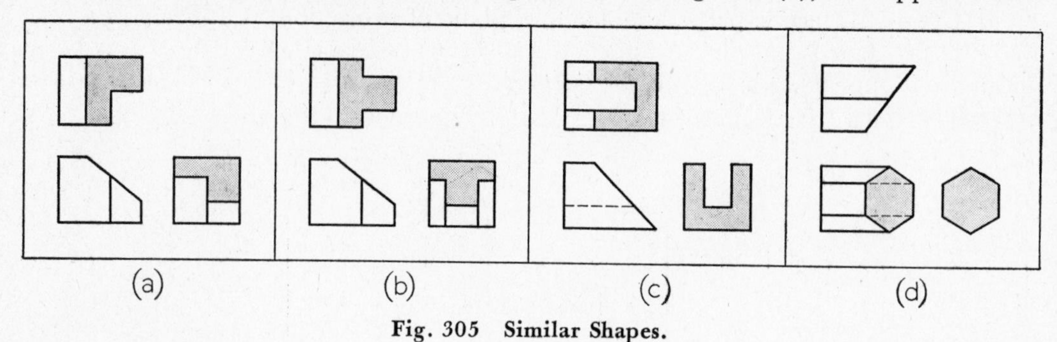

Fig. 305 Similar Shapes.

L-shaped figure in every view in which it does not appear as a line. A T-shaped surface, (b), a U-shaped surface, (c), or a hexagonal surface, (d), will in each case have the same number of sides and the same characteristic shape in every view in which it appears as a surface.

This repetition of shapes is one of our best means for analyzing the views.

223. Reading a Drawing. Let it be required to "read" or visualize the object shown by three views in Fig. 306. Since no lines are curved, the object is made up of plane surfaces.

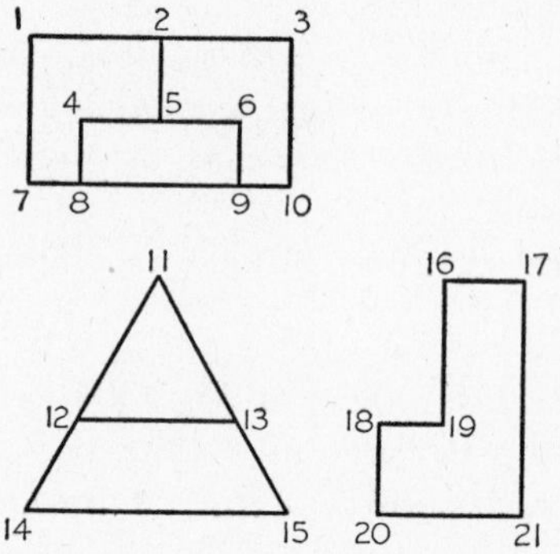

Fig. 306 Reading a Drawing.

Surface 2-3-10-9-6-5 in the top view is an L-shaped surface of six sides. It appears in the side view at 16-17-21-20-18-19, and is L-shaped and six-sided. No such shape appears in the front view, but we note that points 2 and 5 line up with 11 in the front view, points 6 and 9 line up with 13, and points 3 and 10 line up with 15. Evidently line 11-15 in the front view is the edge view of the L-shaped surface.

Surface 11-13-12 in the front view is triangular in shape, but no corresponding triangles appear in either the top or the side view. We note that point 12 lines up with 8 and 4 and that point 13 lines up with 6 and 9. However, surface 11-13-12 of the front view cannot be the same as surface 4-6-9-8 in the top view because the former has three sides and the latter has four. Obviously, the triangular surface appears as a line 4-6 in the top view and as a line 16-19 in the side view.

Surface 12-13-15-14 in the front view is trapezoidal in shape. But there are no trapezoids in the top and side views, so the surface evidently appears in the top view as line 7-10, and in the side view as line 18-20.

The remaining surfaces can be identified in the same manner, whence it will be seen that the object is bounded by seven plane surfaces, two of which are rectangular, two triangular, two L-shaped, and one trapezoidal.

Note that the numbering system used in Fig. 306 is different from that in Fig. 293 in that different numbers are used for all points and there is no significance in a point being "inside" or "outside" a corner.

224. Normal Surfaces. *A normal surface is a plane surface that is parallel to a plane of projection.* It appears in true size and shape on the plane to which it is parallel, and as a vertical or a horizontal line on adjacent planes of projection.

In Fig. 307 are shown four stages in machining a block of cold-rolled steel to produce the final Tool Block in Space IV. All the surfaces are normal surfaces. In Space I, normal surface A is parallel to the horizontal plane and appears true size

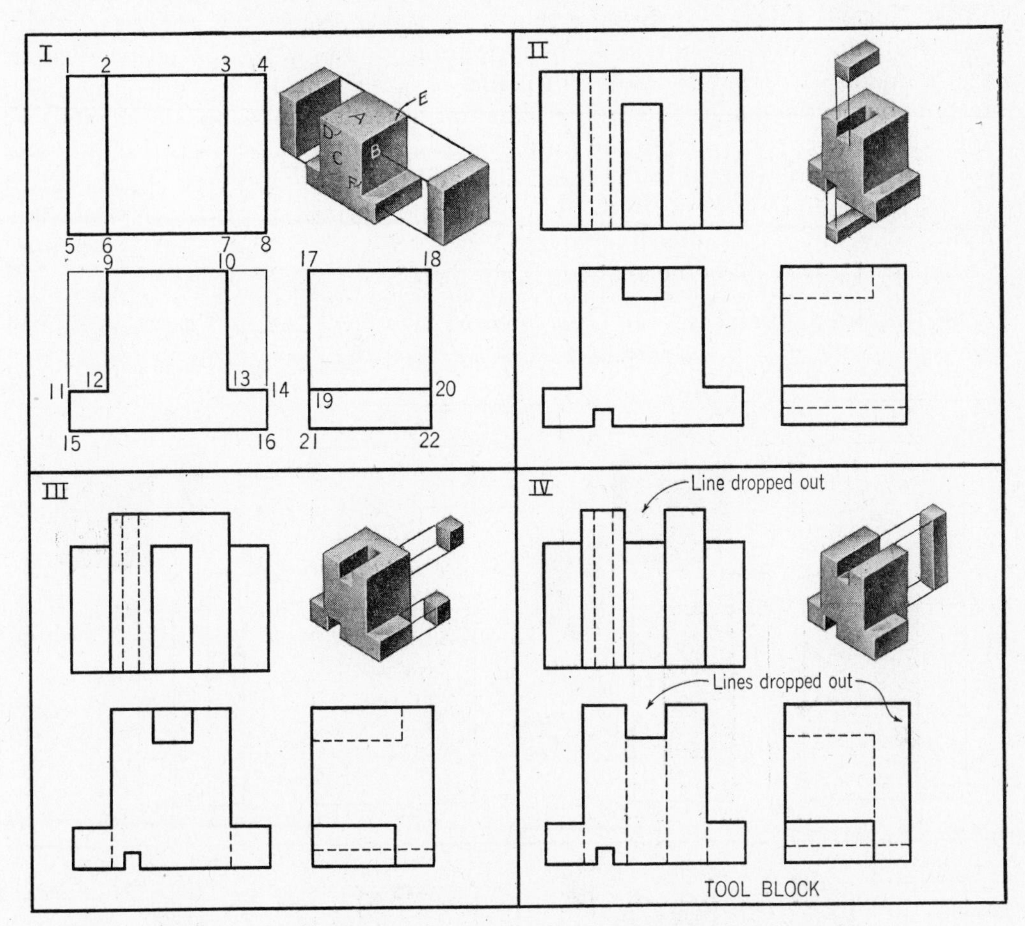

Fig. 307 Machining a Tool Block—Normal Surfaces and Edges.

in the top view at 2-3-7-6, as line 9-10 in the front view, and as line 17-18 in the side view. Normal surface B is parallel to the profile plane and appears true size in the side view at 17-18-20-19, as line 3-7 in the top view, and as line 10-13 in the front view. Normal surface C, an inverted T-shaped surface, is parallel to the frontal plane and appears true size in the front view at 9-10-13-14-16-15-11-12, as line 5-8 in the top view, and as line 17-21 in the side view.

All other surfaces of the object may be visualized in a similar manner. In the four stages of Fig. 307, observe carefully the changes in the views produced by the machining operations, including the introduction of new surfaces, new visible edges and hidden edges, and the "dropping out" of certain lines as the result of a new cut.

The top view in Space I is cut by lines 2-6 and 3-7, which means that there are three surfaces, 1-2-6-5, 2-3-7-6, and 3-4-8-7. In the front view, surface 9-10 is seen to be the highest, and surfaces 11-12 and 13-14 are at the same lower level. In the side view both of these latter surfaces appear as one line 19-20. Surface 11-12 might appear as a hidden line in the side view, but surface 13-14 appears as a visible line 19-20, which covers up the hidden line and takes precedence over it.

In Space IV, how many normal surfaces are there altogether?

225. Normal Edges. *A normal edge is a line that is perpendicular to a plane of projection.* It will appear as a point on the plane of projection to which it is perpendicular, and as a line in true length on adjacent planes of projection. In Space I of Fig. 307, edge D is perpendicular to the profile plane of projection and appears as point 17 in the side view. It is parallel to the frontal and horizontal planes of projection, and is shown true length at 9-10 in the front view and 6-7 in the top view. Edges E and F are perpendicular respectively to the frontal and horizontal planes of projection, and their views may be similarly analyzed.

In Space II, how many normal edges are there?

226. Inclined Surfaces. *An inclined surface is a plane surface that is perpendicular to one plane of projection but inclined to adjacent planes.* It will project as a straight line

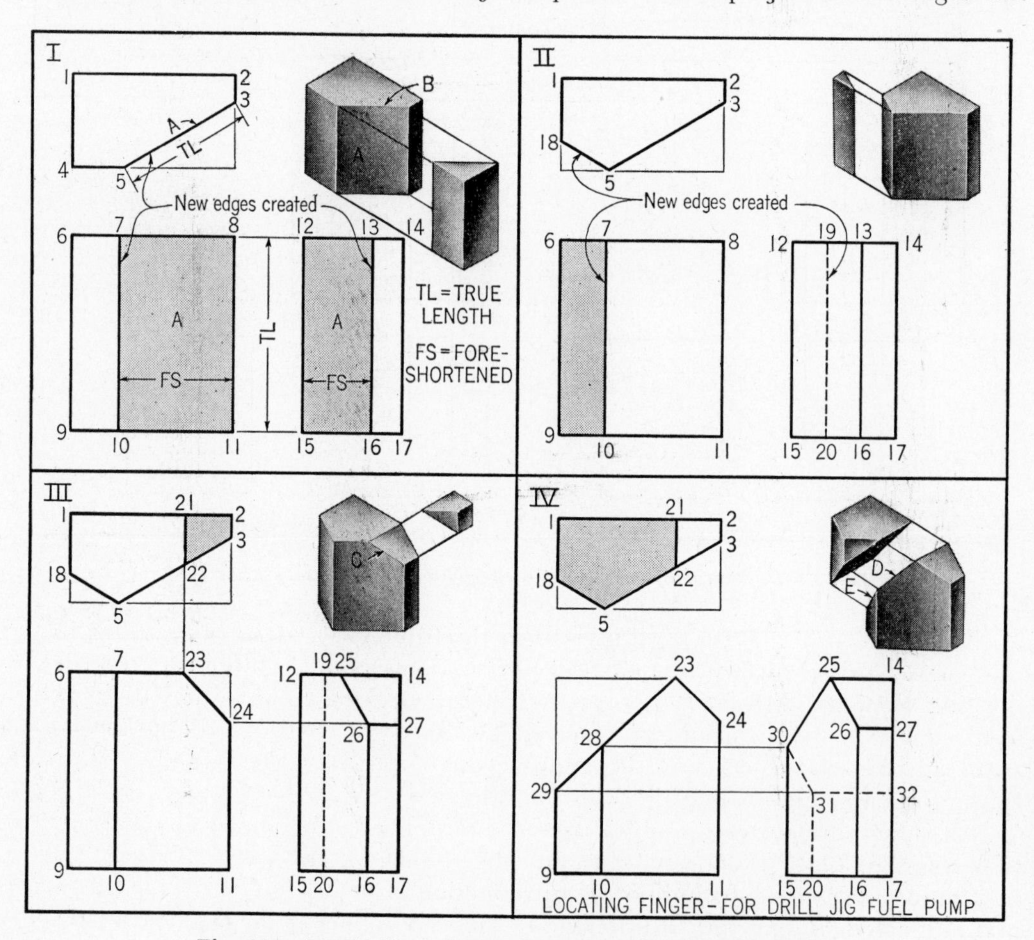

Fig. 308 Machining a Locating Finger—Inclined Surfaces.

on the plane to which it is perpendicular, and will appear foreshortened (FS) on planes to which it is inclined, the degree of foreshortening being proportional to the angle of inclination.

In Fig. 308 are shown four stages in machining a Locating Finger, producing several inclined surfaces. In Space I, inclined surface A is perpendicular to the horizontal plane of projection and appears as line 5-3 in the top view. It is shown as a foreshortened surface in the front view at 7-8-11-10, and in the side view at 12-13-16-15. Note that the surface is more foreshortened in the side view than in the front view because the plane makes a greater angle with the profile plane of projection than with the frontal plane of projection.

In Space III, edge 23-24 in the front view is the edge view of an inclined surface which appears in the top view as 21-2-3-22, and in the side view as 25-14-27-26. Note that 25-14 is equal in length to 21-22, and that the surface has the same number of sides (four) in both views in which it appears as a surface.

In Space IV, edge 29-23 in the front view is the edge view of an inclined surface which appears in the top view as visible surface 1-21-22-5-18, and in the side view as invisible surface 25-14-32-31-30. While the surface does not appear true size in any view, it does have the same characteristic shape and the same number of sides (five) in the views in which it appears as a surface.

In Space IV, how many normal surfaces are there? Inclined surfaces?

In order to obtain the true size of an inclined surface, it is necessary to construct an auxiliary view, §262, or to revolve the surface until it is parallel to a plane of projection, §286.

227. Inclined Edges. *An inclined edge is a line that is parallel to a plane of projection but inclined to adjacent planes.* It will appear true length on the plane to which it is parallel, and foreshortened on adjacent planes, the degree of foreshortening being proportional to the angle of inclination. The true-length view of an inclined line is always inclined, while the foreshortened views are either vertical or horizontal lines.

In Space I of Fig. 308, inclined edge B is parallel to the horizontal plane of projection, and appears true length in the top view at 5-3. It is foreshortened in the front view at 7-8, and in the side view at 12-13. Note that plane A produces two normal edges and two inclined edges.

In Spaces III and IV some of the sloping lines are not inclined lines. In Space III, the edge that appears in the top view at 21-2, in the front view at 23-24, and in the side view at 14-27, is an inclined line. However, the edge that appears in the top view at 22-3, in the front view at 23-24, and in the side view at 25-26, is not an inclined line by the definition above. Actually, it is an *oblique line*, §229.

In Space IV, how many normal edges are there? Inclined edges?

228. Oblique Surfaces. *An oblique surface is a plane that is oblique to all planes of projection.* Since it is not perpendicular to any plane, it cannot appear as a line in any view. Since it is not parallel to any plane, it cannot appear true size in any view. Thus, an oblique surface always appears as a foreshortened surface in all three views.

In Space II of Fig. 309, oblique surface C appears in the top view at 25-3-6-26, and in the front view at 29-8-31-30. What are its numbers in the side view? Note that any surface appearing as a line in any view cannot be an oblique surface. How many inclined surfaces are there? How many normal surfaces?

To obtain the true size of an oblique surface, it is necessary to construct a secondary auxiliary view, §281, or to revolve the surface until it is parallel to a plane of projection, §292.

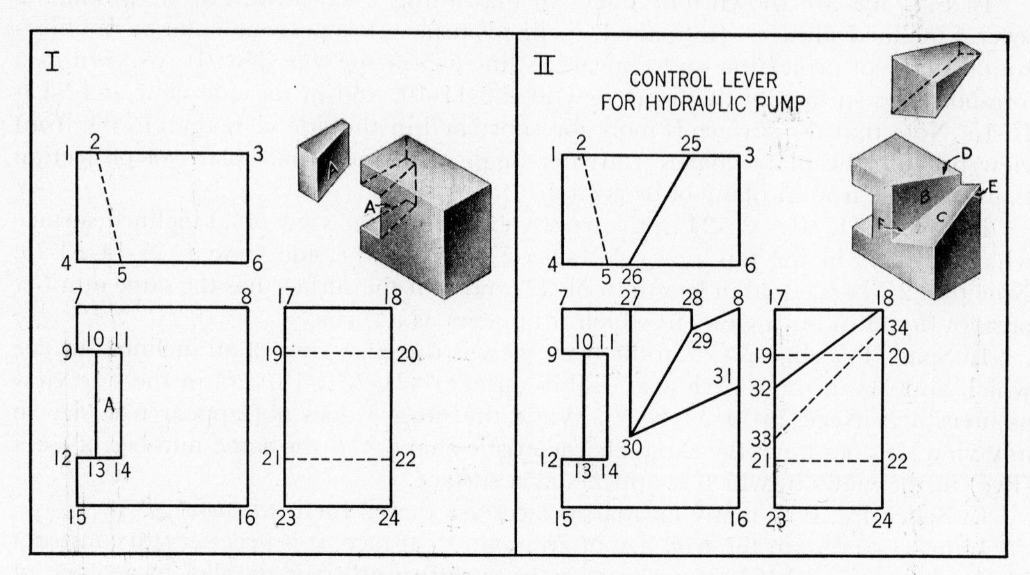

Fig. 309 Machining a Control Lever—Inclined and Oblique Surfaces.

229. Oblique Edges. *An oblique edge is a line that is oblique to all planes of projection.* Since it is not perpendicular to any plane, it cannot appear as a point in any view. Since it is not parallel to any plane, it cannot appear true length in any view. An oblique edge appears foreshortened, and in an inclined position, in every view.

In Space II of Fig. 309, oblique edge F appears in the top view at 26-25, in the front view at 30-29, and in the side view at 33-34. Are there any other oblique lines in this figure? What are the oblique lines in Fig. 308 (IV)?

230. Parallel Edges. If a series of parallel planes is intersected by another plane, the resulting lines of intersection will be parallel, Fig. 310 (a). At (b) the top plane of the object intersects the front and rear planes, producing the parallel edges 1-2 and 3-4. If two lines are parallel in space, their projections in any view are parallel. The example at (b) is a special case in which the two lines appear as points

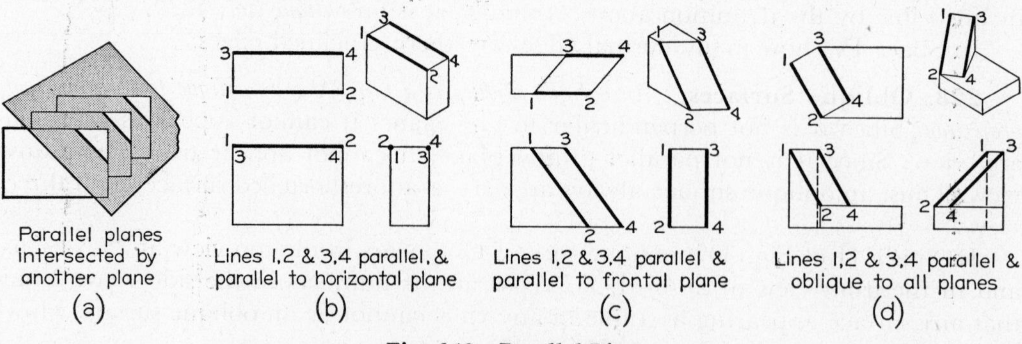

Fig. 310 Parallel Lines.

in one view and coincide as a single line in another, and should not be regarded as an exception to the rule. Note that even in the pictorial drawings the lines are shown parallel.

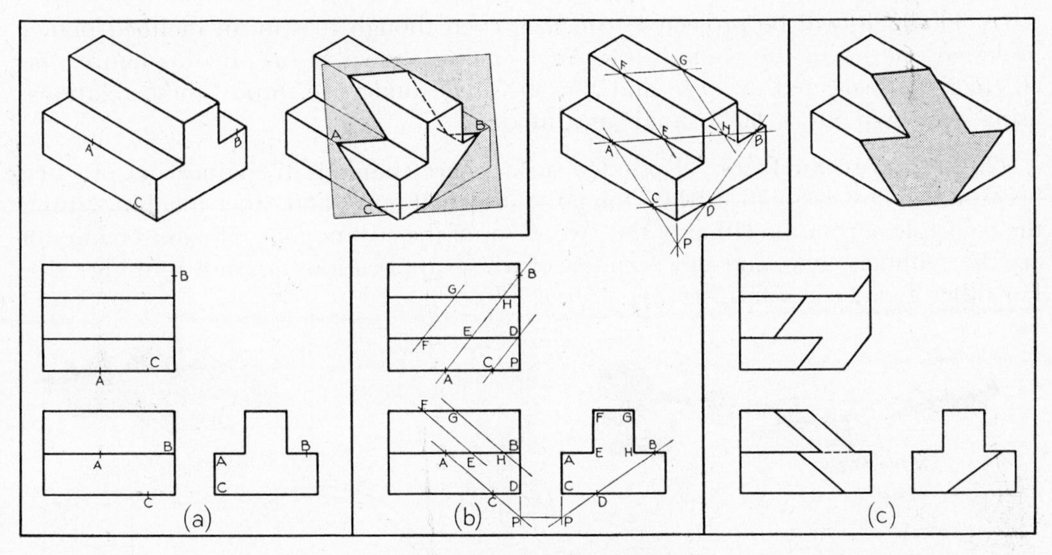

Fig. 311 Oblique Surface.

Parallel inclined lines are shown at (c), and parallel oblique lines at (d).

In Fig. 311 it is required to draw three views of the object after a plane has been passed through the points A, B, and C. As shown at (b), only points that lie in the same plane are joined. In the front view, join points A and C, which are in the same plane, extending the line to P on the vertical front edge of the block extended. In the side view, join P to B, and in the top view, join B to A. Complete the drawing by applying the rule: *parallel lines in space will be projected as parallel lines in any view.* The remaining lines are thus drawn parallel to lines AP, PB, and BA.

231. Angles. If an angle is in a normal plane—that is, parallel to a plane of projection—the angle will be shown true size on the plane of projection to which it is parallel, Fig. 312 (a).

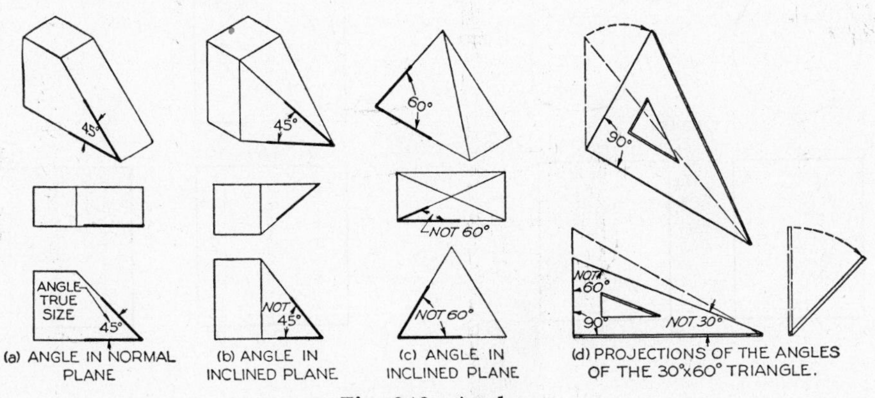

Fig. 312 Angles.

If the angle is in an inclined plane, (b) and (c), the angle may be projected either larger or smaller than the true angle, depending upon its position. At (b) the 45° angle is shown *oversize* in the front view, and at (c) the 60° angle is shown undersize in both views.

A 90° angle will be projected true size, even though it is in an inclined plane, provided one leg of the angle is a normal line, as shown at (d). In this figure, the 60° angle is projected oversize and the 30° angle undersize. Study these relations, using your own 30° × 60° triangle as a model.

232. Curved Surfaces. Rounded surfaces are common in engineering practice because they are easily formed on the lathe, the drill press, and other machines using the principle of rotation either of the "work" or of the cutting tool. The most common are the cylinder, cone, and sphere, a few of whose applications are shown in Fig. 313. For other geometric solids, see Fig. 159.

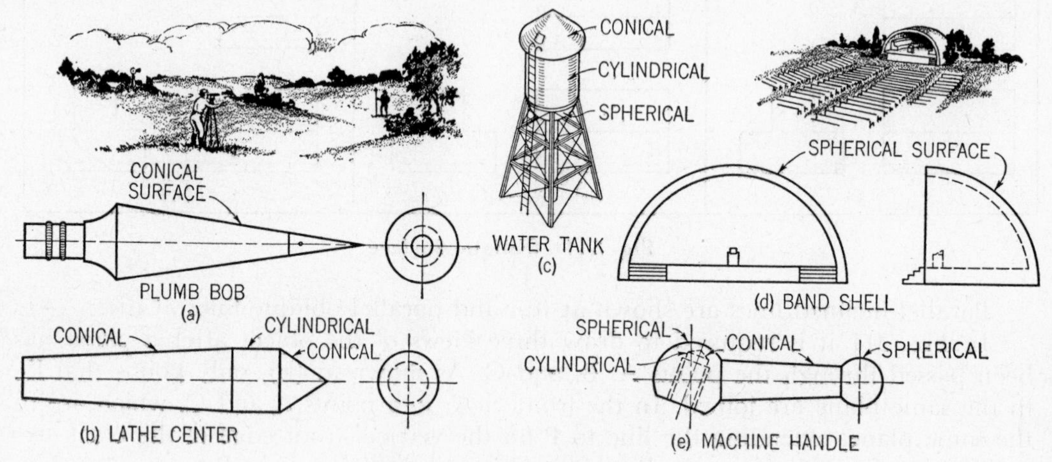

Fig. 313 Curved Surfaces.

233. Cylindrical Surfaces. Three views of a *right-circular cylinder*, the most common type, are shown in Fig. 314 (a). The single cylindrical surface is intersected by two plane (normal) surfaces, forming two curved lines of intersection or *circular edges*

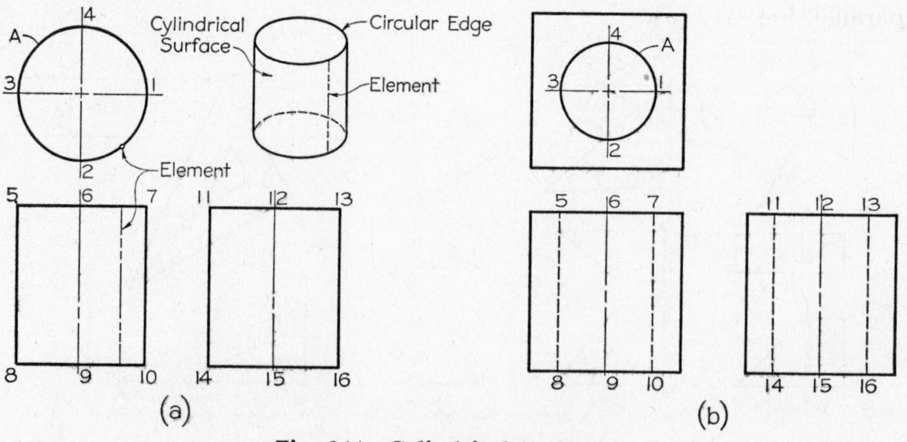

Fig. 314 Cylindrical Surfaces.

(the bases of the cylinder). These circular edges are the only actual edges on the cylinder.

The cylinder is represented on a drawing by its circular edges and the contour elements. An *element* is a straight line on the cylindrical surface, parallel to the axis, as shown in the pictorial view of the cylinder at (a). In this figure, at both (a) and (b), the circular edges appear in the top views as circles A, in the front views as horizontal lines 5-7 and 8-10, and in the side views as horizontal lines 11-13 and 14-16.

The contour elements 5-8 and 7-10 in the front views appear as points 3 and 1 in the top views. The contour elements 11-14 and 13-16 in the side views appear as points 2 and 4 in the top views.

In Fig. 315 are shown four possible stages in machining a Cap, producing several cylindrical surfaces. In Space I the removal of the two upper corners forms cylindrical surface A which appears in the top view as surface 1-2-4-3, in the front view as arc 5, and in the side view as surface 8-9-Y-X.

In Space II a large reamed hole shows in the front view as circle 16, in the top view as cylindrical surface 12-13-15-14, and in the side view as cylindrical surface 17-18-20-19.

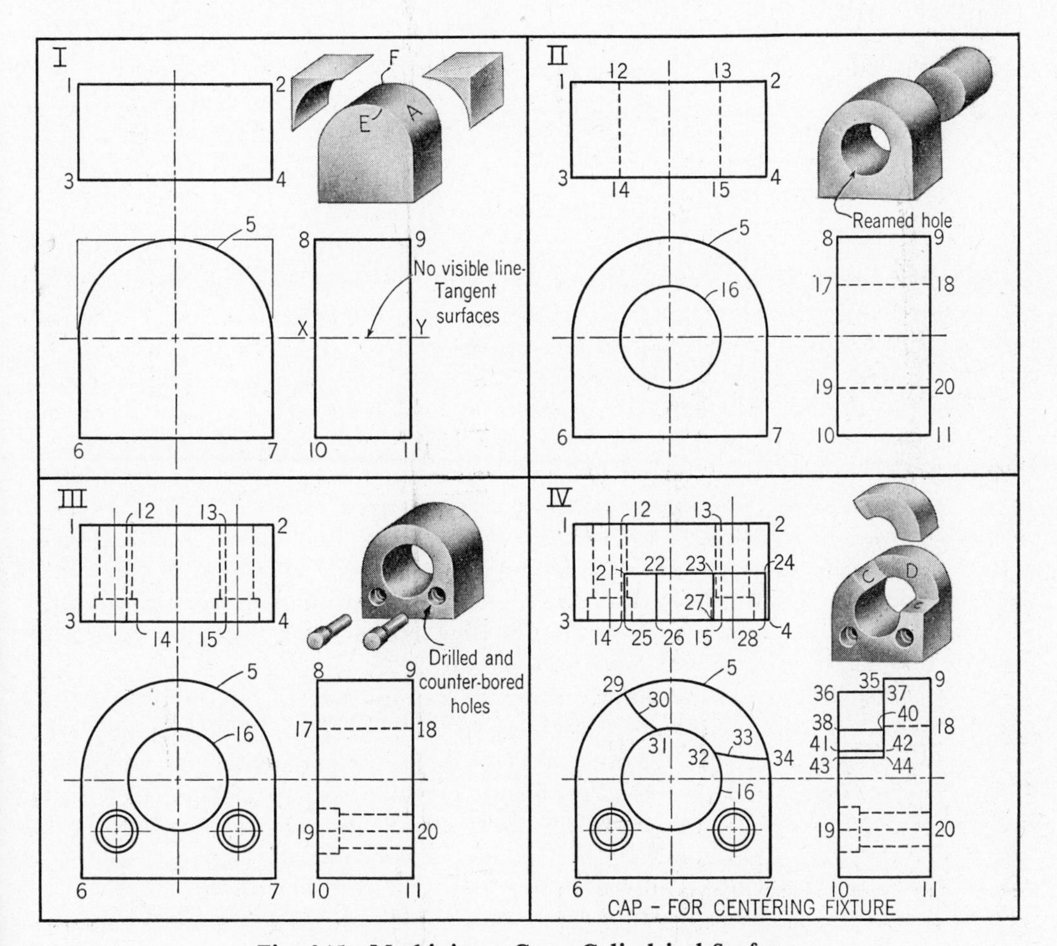

Fig. 315 Machining a Cap—Cylindrical Surfaces.

In Space III two drilled and counterbored holes are added, producing four more cylindrical surfaces and two normal surfaces. The two normal surfaces are those at the bottoms of the counterbores.

In Space IV a cylindrical cut is added, producing two cylindrical surfaces that appear edgewise in the front view as arcs 30 and 33, in the top view as surfaces 21-22-26-25 and 23-24-28-27, and in the side view as surfaces 36-37-40-38 and 41-42-44-43.

234. Deformities of Cylinders. In shop practice, cylinders are usually machined or formed so as to introduce other surfaces, usually plane surfaces. In Fig. 316 (a) a cut is shown that introduces two normal surfaces. One surface appears as

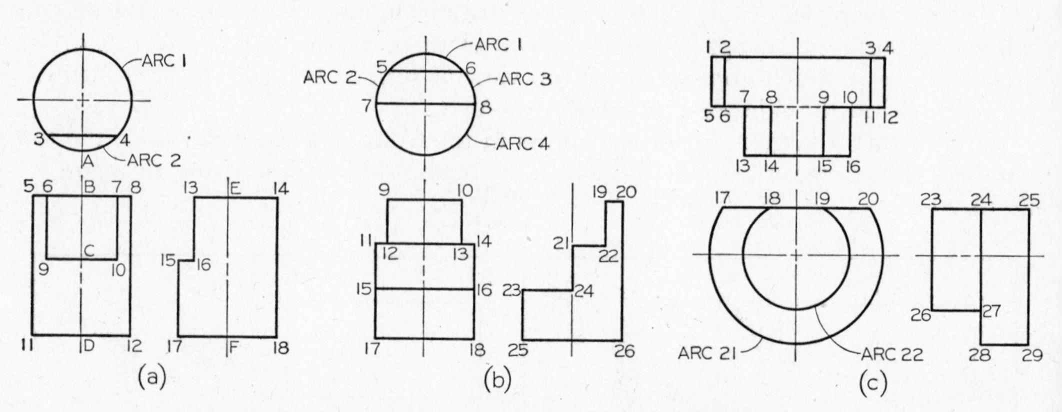

Fig. 316 Deformities of Cylinders.

a line 3-4 in the top view, as surface 6-7-10-9 in the front view, and as line 13-16 in the side view. The other appears as line 15-16 in the side view, line 9-10 in the front view, and surface 3-4, arc 2 in the top view.

All elements touching arc 2, between 3 and 4 in the top view, become shorter as a result of the cut. For example, element A, which shows as a point in the top view, now becomes CD in the front view, and 15-17 in the side view. As a result of the cut, the front half of the cylindrical surface changed from 5-8-12-11 (front view) to 5-6-9-10-7-8-12-11. The back half remains unchanged.

At (b) two cuts introduce four normal surfaces. Note that surface 7-8 (top view) is through the center of the cylinder, producing in the side view line 21-24, and in the front view surface 11-14-16-15 equal in width to the diameter of the cylinder. Surface 15-16 (front view) is read in the top view as 7-8-arc 4. Surface 11-14 (front view) is read in the top view as 5-6-arc 3-8-7-arc 2.

At (c) two cylinders on the same axis are shown, intersected by a normal surface parallel to the axis. Surface 17-20 (front view) is 23-25 in the side view, and 2-3-11-9-15-14-8-6 in the top view. A common error is to draw a visible line in the top view between 8 and 9. However, this would produce two surfaces 2-3-11-6 and 8-9-15-14 not in the same plane. In the front view, the larger surface appears as line 17-20 and the smaller as line 18-19. These lines coincide; hence they are all one surface, and there can be no visible line joining 8 and 9 in the top view.

The surface that appears in the front view at 17-18-arc 22-19-20-arc 21 appears in the top view at 5-12, which explains the hidden line 8-9 in the top view.

235. Cylinders and Ellipses. If a cylinder is cut by an inclined plane, as in Fig. 317 (a), the inclined surface is bounded by an ellipse. The ellipse appears as circle 1 in the top view, as straight line 2-3 in the front view, and as ellipse ADBC in the side view. Note that circle 1 in the top view would remain a circle regardless of the angle of the cut. If the cut is 45° with horizontal, the ellipse will appear as a circle in the side view (see phantom lines) since the major and minor axes in that view would be equal. To find the true size and shape of the ellipse, an auxiliary view will be required, with the line of sight perpendicular to surface 2-3 in the front view, §273.

The ellipse in the side view can be drawn by selecting points in the circular view and projecting them to the front and side views as shown. To locate points in the top view, draw a series of vertical lines at random, each intersecting the circle at two points. In projecting from the top to the side view, distances d will be equal.

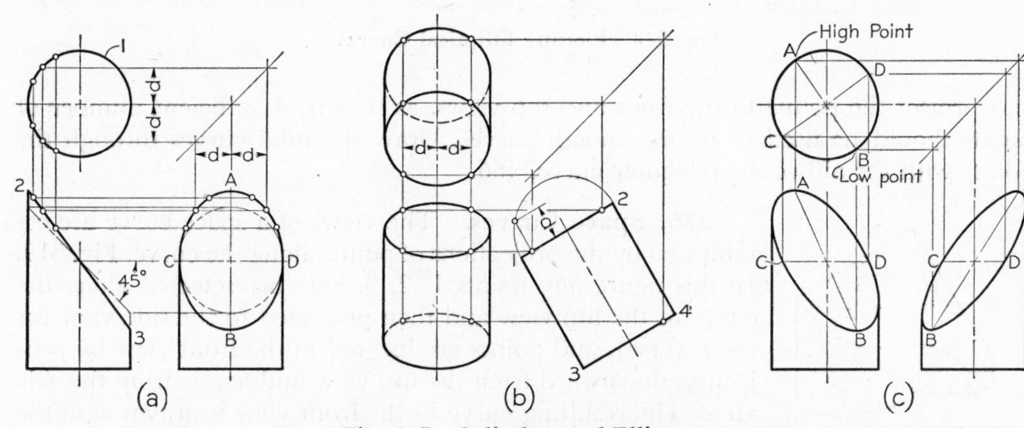

Fig. 317 Cylinders and Ellipses.

Since the major and minor axes AB and CD are known, the ellipse can be drawn by any of the methods in Figs. 200-202 and 204 (a) (true ellipses) or by the method of Fig. 208 (approximate ellipse). Or the ellipse may be drawn with the aid of an ellipse guide, §163.

If the cylinder is tilted forward, (b), the bases or circular edges 1-2 and 3-4 (side view) become ellipses in the front and top views. Points on the ellipses can be plotted, in which use is made of a semi-circular "end view" of the cylinder, as shown, distances d being equal. Since the major and minor axes for each ellipse are known, the ellipses can be drawn with the aid of an ellipse guide, or by any of the true ellipse methods, or by the approximate method.

If the cylinder is cut by an oblique plane, (c), the elliptical surface appears as an ellipse in all three views. In the top view, points A and B are selected, diametrically opposite, as the high and low points in the ellipse, and CD is drawn perpendicular to AB. These are the projections of the major and minor axes, respectively, of the actual ellipse in space. In the front and side views, points A and B are assumed at the desired altitudes. Since CD appears true length in the top view, it will appear horizontal in the front and side views, as shown. These axes in the front and side views are the *conjugate axes* of the ellipses. The ellipses may be drawn upon these axes by the methods of Figs. 203 or 204 (b), or by trial with the aid of an ellipse template, §163.

In Fig. 318, the intersection of a plane and a "quarter-round" molding is shown at (a), and with a "cove" molding at (b). In both figures, assume points 1, 2, 3, etc., at random in the side views in which the cylindrical surfaces appear as curved lines,

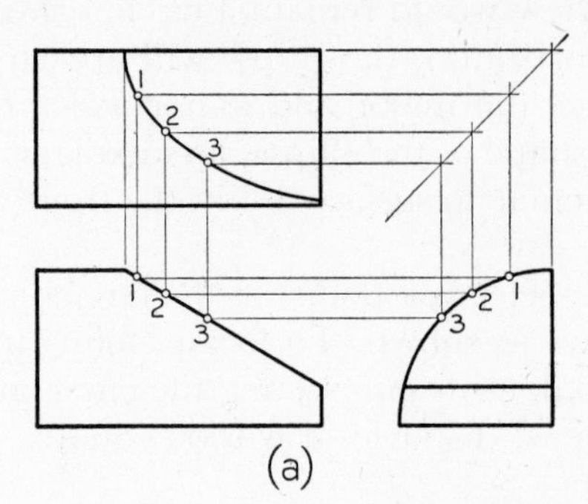

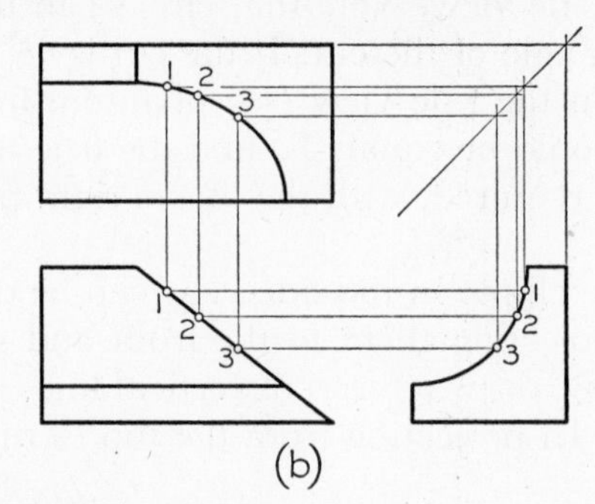

Fig. 318 Plotting Elliptical Curves.

and project the points to the front and top views, as shown. A sufficient number of points should be used to insure smooth curves. Draw the final curves through the points with the aid of the irregular curve, §66.

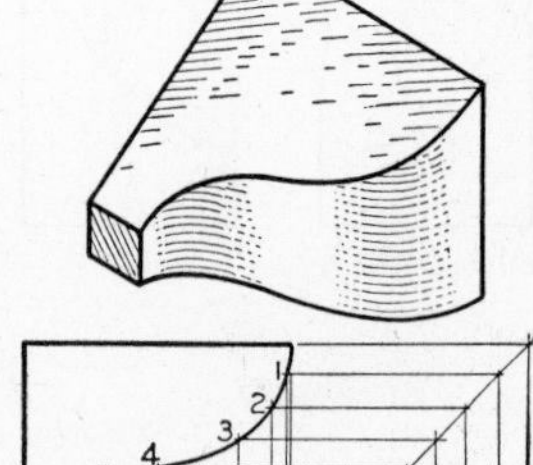

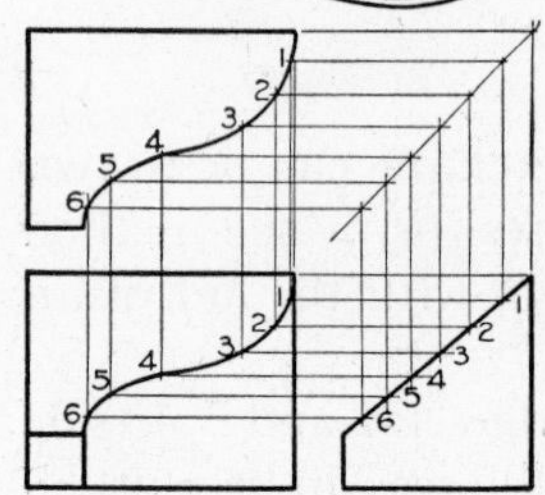

Fig. 319 Space Curve.

236. Space Curves. The views of a space curve are established by the projections of points along the curve, Fig. 319. In this figure any points 1, 2, 3, etc., are selected along the curve in the top view and then projected to the side view (or the reverse), and points are located in the front view by projecting downward from the top view and across from the side view. The resulting curve in the front view is drawn with the aid of the irregular curve, §66.

237. Intersections and Tangencies. No line should be drawn where a curved surface is tangent to a plane surface, Fig. 320 (a), but when a curved surface *intersects* a plane surface, a definite edge is formed, (b). If curved surfaces are arranged as at (c), no lines appear in the top view, as shown. If the surfaces are arranged as at (d), a vertical surface in the front view produces a line in the top view. Other typical intersections and tangencies of surfaces are shown from (e) to (h). To locate the point of tangency A in (g), refer to Fig. 186 (b).

The intersection of a small cylinder with a large cylinder is shown in Fig. 321 (a). The intersection is so small that it is not plotted, a straight line being used instead. At (b) the intersection is larger, but still not large enough to justify plotting the curve. The curve is approximated by drawing an arc whose radius r is the same as radius R of the large cylinder.

The intersection at (c) is significant enough to justify constructing the true curve. Points are selected at random in the circle in the side or top view, and these are then projected to the other two views to locate points on the curve in the front view, as shown. A sufficient number of points should be used, depending upon the size of the intersection, to insure a smooth and accurate curve. Draw the final curve with the aid of the irregular curve, §66.

At (d), the cylinders are the same diameter. The figure of intersection consists of two semi-ellipses that appear as straight lines in the front view.

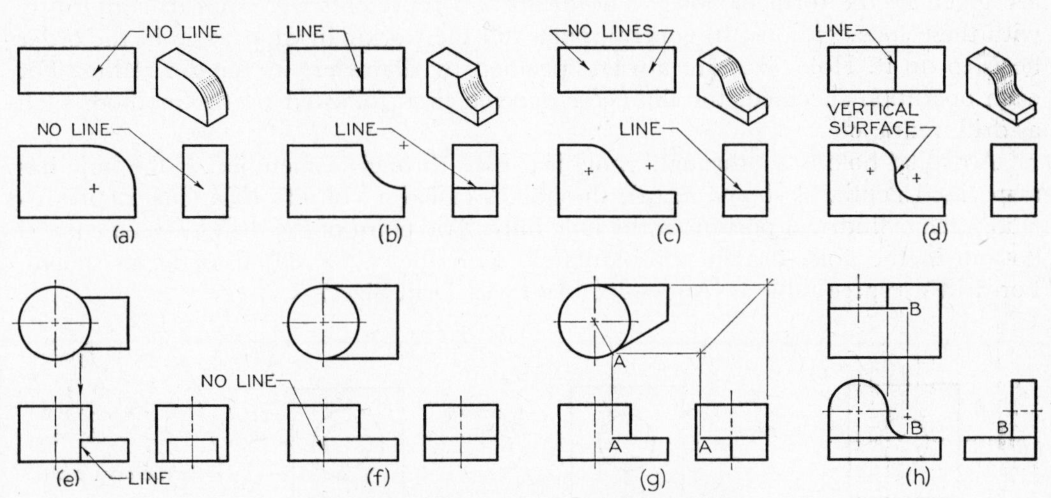

Fig. 320 Intersections and Tangencies.

If the intersecting cylinders are holes, the intersections would be similar to those for the external cylinders in Fig. 321.

For other intersections of solids, see §§533-543.

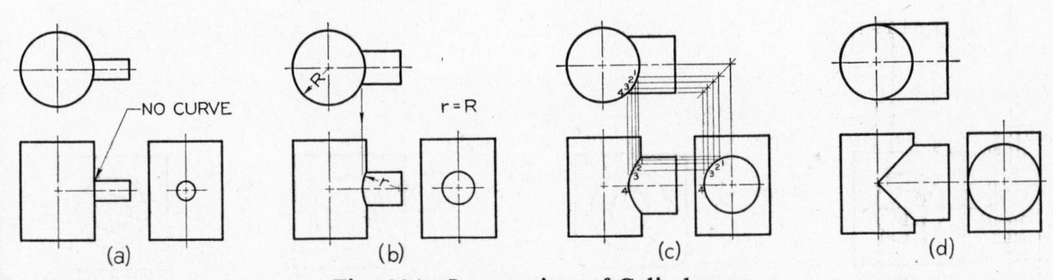

Fig. 321 Intersections of Cylinders.

In Fig. 322 (a), a narrow prism intersects a cylinder, but the intersection is insignificant and is ignored. At (b) the prism is larger and the intersection is noticeable enough to warrant construction, as shown. At (c) and (d) are shown a keyseat and a small drilled hole, respectively; in both cases the intersection is not important enough to construct.

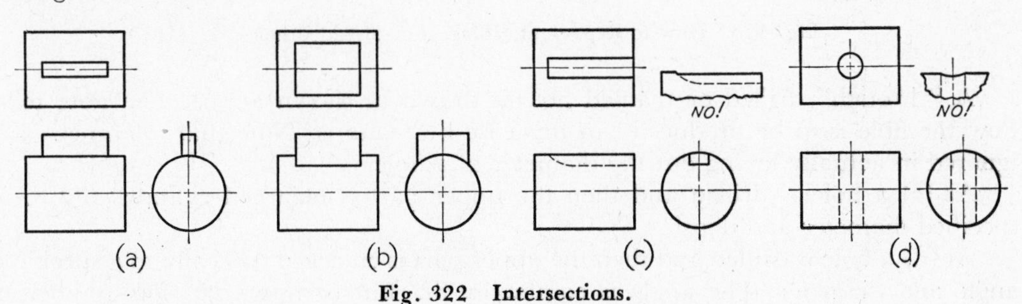

Fig. 322 Intersections.

238. How to Draw Holes. The correct methods of representing the most common types of machined holes are shown in Fig. 323. Instructions to the shop worker are given in the form of notes. The draftsman represents the holes in conformity with these specifications. In general, notes tell the workman what to do in the order he is to do it. Hole sizes are always specified by diameter—never the radius. For each operation specified, the diameter is given first, followed by the method, such as drill, ream, etc.

A drilled hole is a "through" hole if it goes through a member. If the hole has a specified depth, as shown at (a), the hole is called a "blind" hole. The depth includes the cylindrical portion of the hole only. The point of the drill leaves a conical bottom in the hole, drawn approximately with the 30° × 60° triangle, as shown. For drill sizes, see table in Appendix 14 (Twist Drill Sizes).

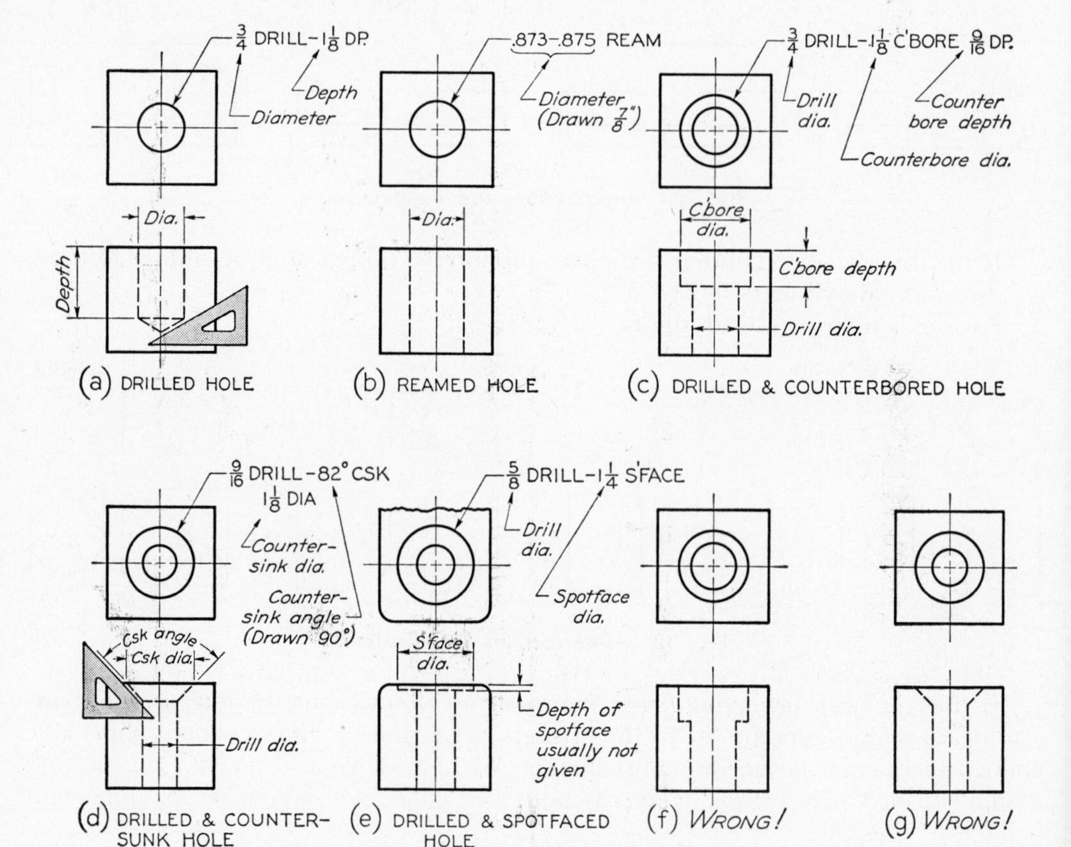

Fig. 323 How to Represent Holes. (For threaded holes, see §413.)

A "through" drilled or reamed hole is drawn as shown at (b). The note tells how the hole is to be produced—in this case by reaming. Note that tolerances are ignored in actually laying out the diameter of a hole.

At (c) a hole is drilled and then the upper part is enlarged cylindrically to a specified diameter and depth.

At (d) a hole is drilled and then the upper part is enlarged conically to a specified angle and diameter. The angle is commonly 82° but is drawn 90° for simplicity.

At (e) a hole is drilled and then the upper part is enlarged cylindrically to a specified diameter. The depth usually is not specified, but is left to the shop to determine. For average cases, the depth is drawn $\frac{1}{16}''$.

For complete information about how holes are made in the shop, see §316. For further information on notes, see §349.

239. Fillets and Rounds. A rounded interior corner is called a *fillet*, and a rounded exterior corner, a *round*, Fig. 324 (a). Sharp corners should be avoided in designing parts to be cast or forged, not only because they are difficult to produce, but in the case of interior corners are a source of weakness and failure. See §302 for shop processes involved.

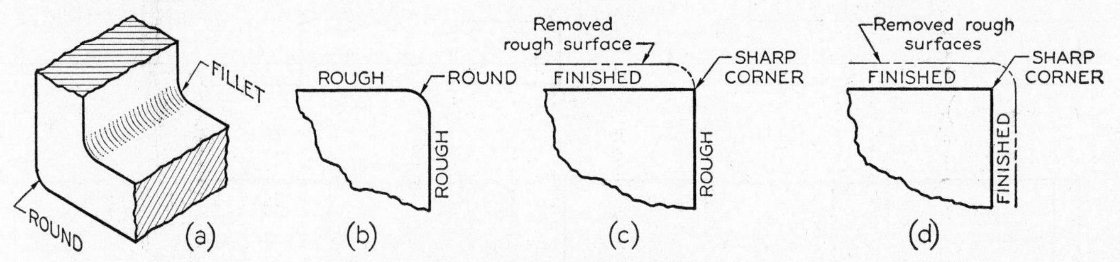

Fig. 324 Rough and Finished Surfaces.

Two intersecting rough surfaces produce a rounded corner, (b). If one of these surfaces is machined, (c), or if both surfaces are machined, (d), the corner becomes sharp. Therefore, on a drawing a rounded corner means that both intersecting surfaces are rough, and a sharp corner means that one or both surfaces are machined. On working drawings, fillets and rounds are never shaded. The presence of the curved surfaces is indicated only where they appear as arcs, except as shown in Fig. 328.

Fillets and rounds should be drawn with the bow pencil or bow pen if they are $\frac{1}{8}''$ radius or larger. Those smaller than $\frac{1}{8}''$ R should be made carefully freehand. As an aid in drawing these smaller arcs, some prefer to use the ends of the slots in the erasing shield, the filleted corners of the triangle, a special fillets-and-rounds template, or a circle template.

240. Runouts. The correct method of representing fillets in connection with plane surfaces tangent to cylinders is shown in Fig. 325. These small curves are called *runouts*. Note that the runouts F have a radius equal to that of the fillet and a curvature of about one-eighth of a circle, (d).

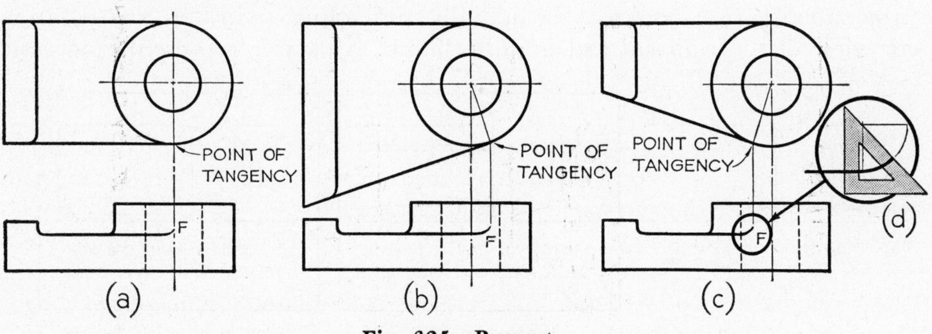

Fig. 325 Runouts.

Typical filleted intersections are shown in Fig. 326. The runouts from (a) to (d) differ because of the different shapes of the horizontal intersecting members. At (e) and (f) the runouts differ because the top surface of the web at (e) is flat, with only

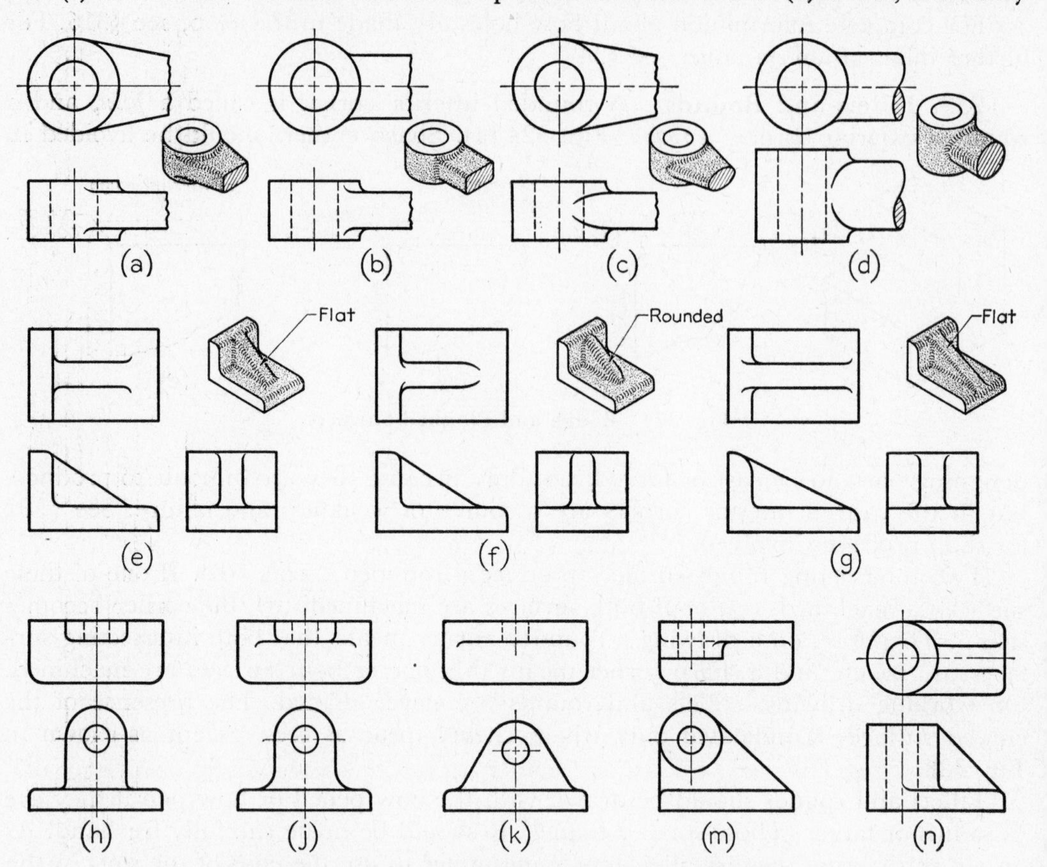

Fig. 326 Conventional Fillets, Rounds, and Runouts.

slight rounds along the edge, while the top surface of the web at (f) is considerably rounded. When two different sizes of fillets intersect, as at (g) and (j), the direction of the runout is dictated by the larger fillet, as shown.

241. Conventional Edges. Rounded and filleted intersections eliminate sharp edges and sometimes make it difficult to present a clear shape description. In fact, true projection in some cases may be actually misleading, as in Fig. 327 (a), in which the side view of the railroad rail is quite blank. A clearer representation results if

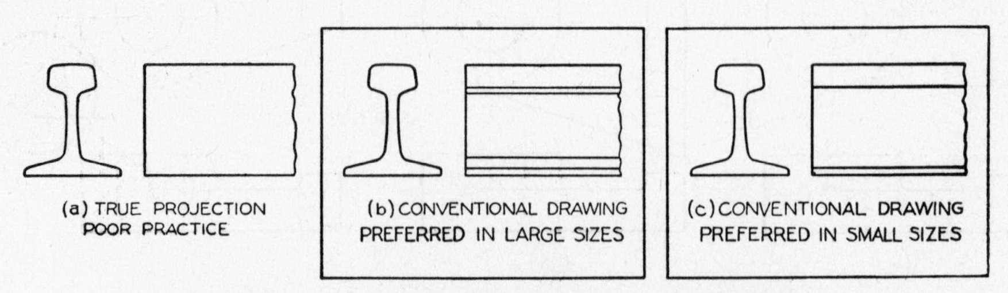

Fig. 327 Conventional Representation of a Rail.

lines are added for rounded and filleted edges, as shown at (b) and (c). The added lines are projected from the actual intersections of the surfaces as if the fillets and rounds were not present.

In Fig. 328, two top views are shown for each given front view. The upper top views are nearly devoid of lines that contribute to the shape descriptions, while the

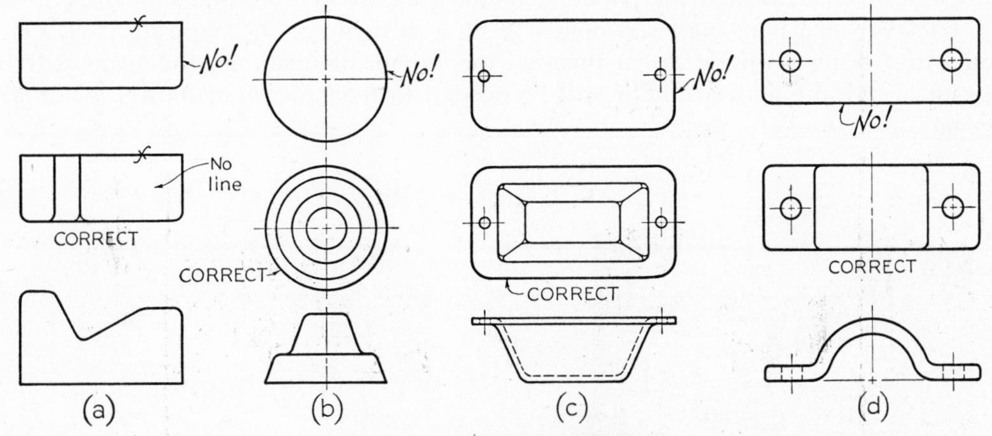

Fig. 328 Conventional Edges.

lower top views, in which lines are used to represent the rounded and filleted edges, are quite clear. Note, in the lower top views at (a) and (c), the use of small freehand Y's where rounded or filleted edges meet a rough surface. If such an edge intersects a finished surface, no Y is shown.

242. Right-Hand and Left-Hand Parts. In industry many individual parts are located symmetrically so as to function in pairs. These opposite parts are often exactly alike, as for example, the hub caps used on the left and right sides of the automobile. In fact, whenever possible, for economy's sake the designer will design identical parts for use on both the right and left. But opposite parts often cannot be exactly alike, as for example, a pair of gloves or a pair of shoes. Similarly, the right front fender of an automobile cannot be the same shape as the left front fender. Therefore, a left-hand part is not simply a right-hand part turned around; the two parts will be opposite and different in shape.

A left-hand part is referred to as an L.H. part, and a right-hand part as an R.H. part. In Fig. 329 (a), the image in the mirror is the "other-hand" of the part shown.

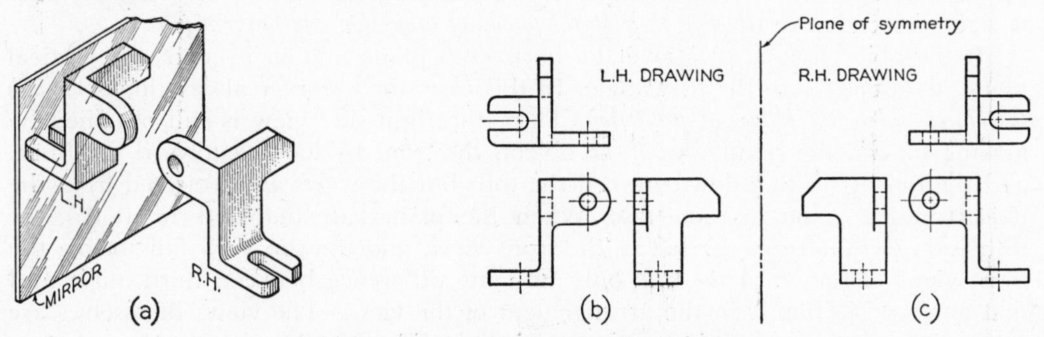

Fig. 329 Right-Hand and Left-Hand Parts.

If the part in front of the mirror is an R.H. part, the image shows the L.H. part. No matter how the object is turned, the image will show the L.H. part. At (b) and (c) are shown L.H. and R.H. drawings of the same object, and it will be seen that the drawings are also symmetrical with respect to a reference-plane line between them.

If you hold a drawing faced against a window pane or a light table so that the lines can be seen through the paper, you can trace the reverse image of the part on the back or on tracing paper, which will be a drawing of the opposite part. Or if you run a tracing upside down through the blueprint machine, the print will be reversed, and although a mirror will be needed to read the lettering, the print will be that of the opposite part.

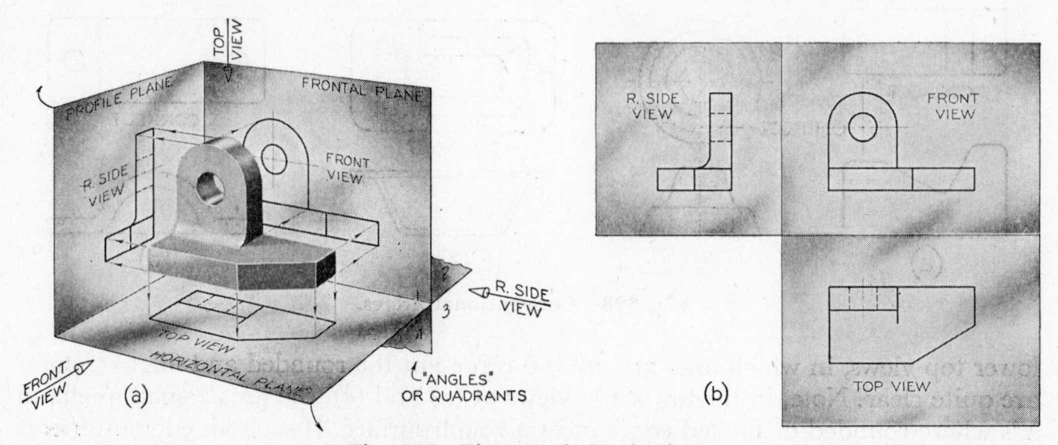

Fig. 330 First-Angle Projection.

It is customary in most cases to draw only one of two opposite parts, and to label the one that is drawn with a note, such as: L.H. PART SHOWN, R.H. OPPOSITE. If the opposite-hand shape is not clear, a separate drawing must be made for it and properly identified.

243. First-Angle Projection. If the vertical and horizontal planes of projection are considered indefinite in extent and intersecting at 90° with each other, the four dihedral angles produced are the *first*, *second*, *third*, and *fourth* angles, Fig. 330 (a). The profile plane intersects these two planes and may extend into all angles. If the object is placed below the horizontal plane and behind the vertical plane as in the "glass box," Fig. 288, the object is said to be in the third angle. In this case, as we have seen, the observer is always "outside, looking in," so that for all views the lines of sight proceed from the eye *through the planes of projection and to the object.*

If the object is placed above the horizontal plane and in front of the vertical plane, the object is in the first angle. In this case, the observer always looks *through the object and to the planes of projection.* Thus, the right-side view is still obtained by looking toward the right side of the object, the front by looking toward the front, and the top by looking down toward the top; but the views are projected from the object onto a plane in each case. When the planes are unfolded, as at (b), the right-side view falls on the left of the front view, and the top view falls below the front view, as shown. Thus, the only ultimate difference between third-angle and first-angle projection is in the arrangement of the views. The views themselves are the same in both systems. Compare the views in Figs. 330 (b) and 290 (b).

In the United States and Canada and to some extent in England, third-angle projection is standard, while in most of the rest of the world, first-angle projection is used. First-angle projection was originally used all over the world, including the United States, but in this country it was abandoned about seventy-five years ago.

244. Multiview Projection Problems. The following problems are intended primarily to afford practice in instrumental drawing, but any of them may be sketched freehand on cross-section paper or plain paper. Sheet layouts (see back endpaper) are suggested, but the instructor may desire to use a different sheet size or arrangement.

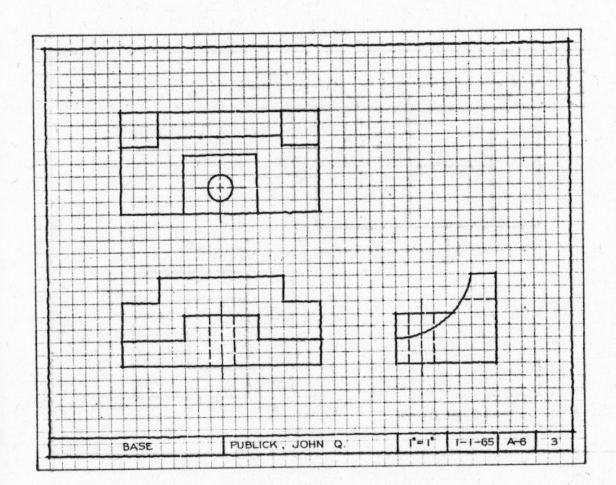

Fig. 331 Freehand Sketch (Layout A-2).

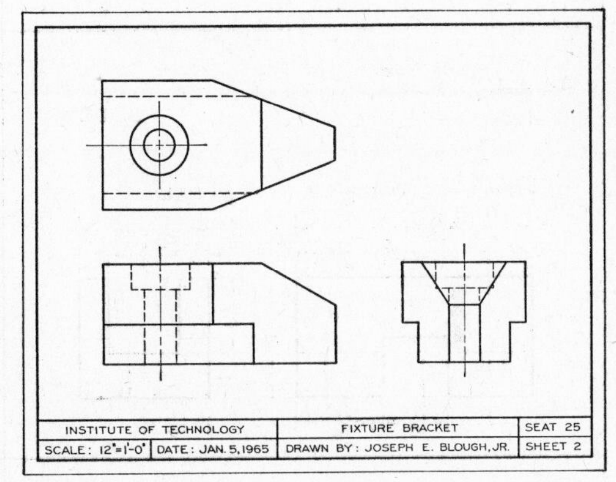

Fig. 332 Mechanical Drawing (Layout A-3).

Dimensions may or may not be required by the instructor. If they are assigned, the student should study §§327-350. In the given problems, whether in multiview or in pictorial form, it is often not possible to give dimensions in the preferred places or, occasionally, in the standard manner. The student is expected to move dimensions to the preferred locations and otherwise to conform to the dimensioning practices recommended in Chapter 11.

For the problems in Fig. 336-395, it is suggested that the student make a thumbnail sketch of the necessary views in each case, and obtain his instructor's approval before starting the mechanical drawing.

Problems in convenient form for solution are available in *Technical Drawing Problems*, by Giesecke, Mitchell, and Spencer, and in *Technical Drawing Problems, Series 2*, by Spencer and Grant, both designed to accompany this text, and published by The Macmillan Company.

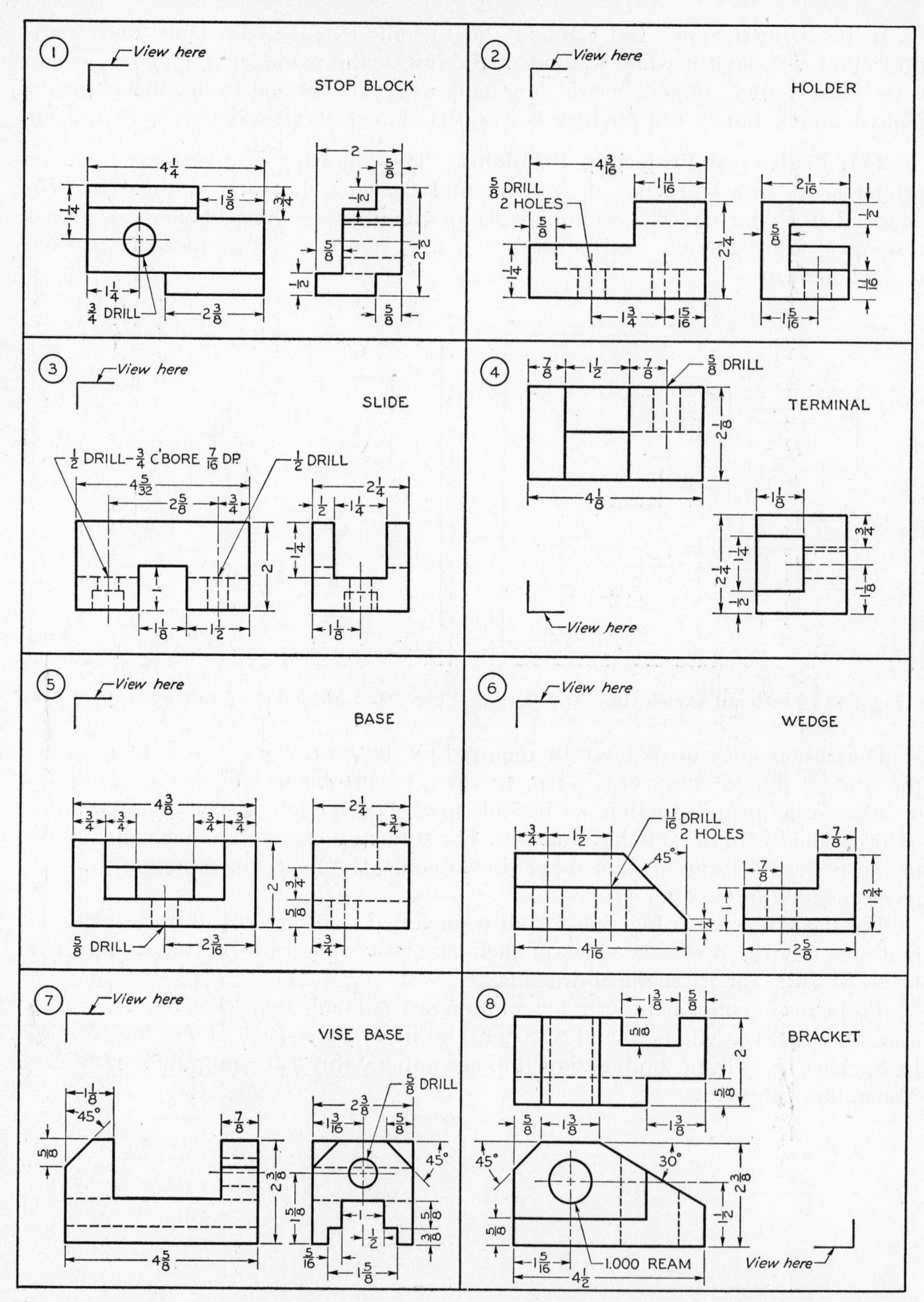

Fig. 333 Missing-View Problems. Using Layout A-2 or A-3, sketch or draw mechanically the given views, and add the missing view, as shown in Figs. 331 and 332. If dimensions are required, study §§327-350. Move dimensions to better locations where possible. In Probs. 1-5, all surfaces are normal surfaces.

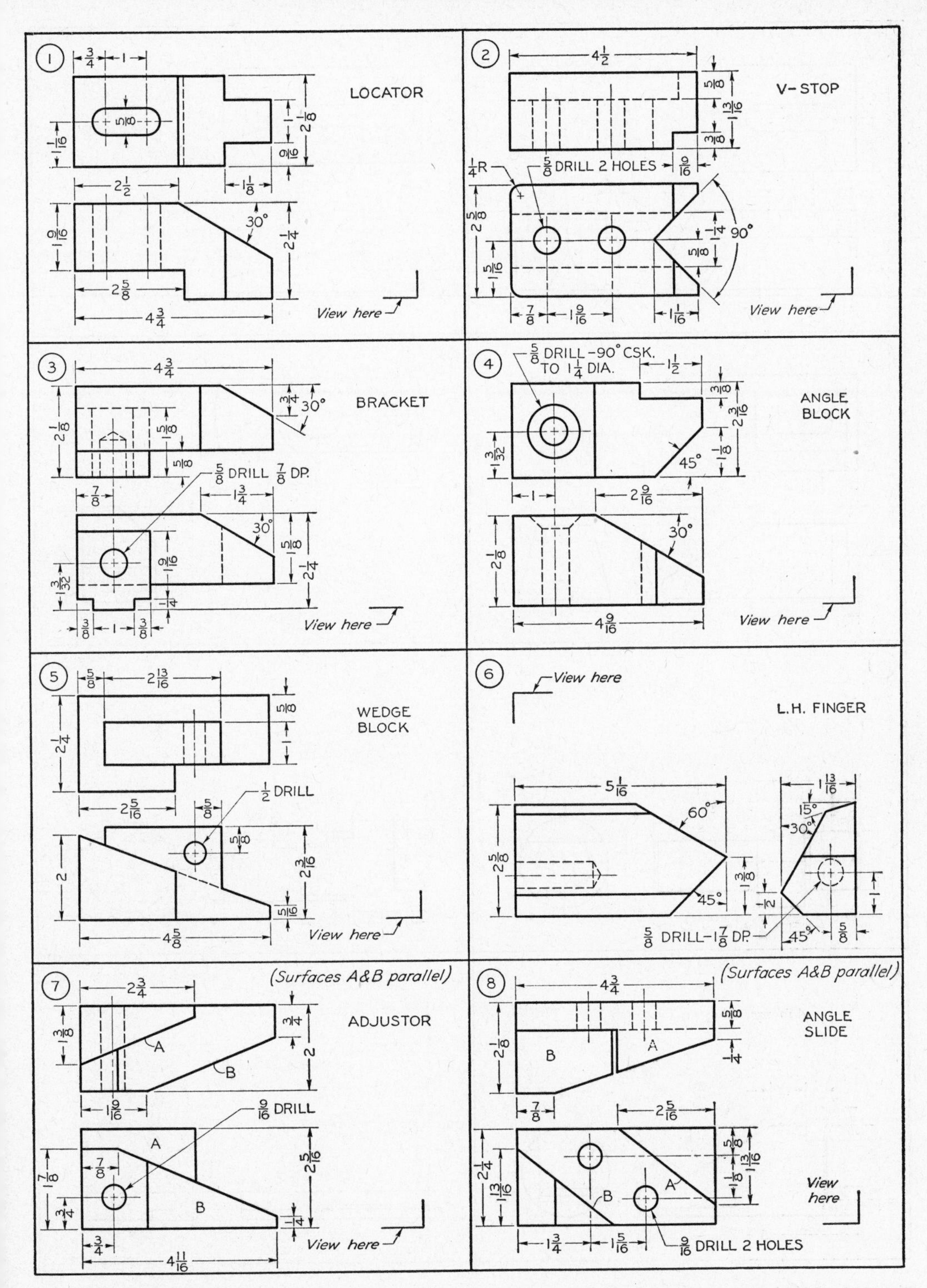

Fig. 334 Missing-View Problems. Using Layout A-2 or A-3, sketch or draw mechanically the given views, and add the missing view, as shown in Figs. 331 and 332. If dimensions are required, study §§327-350. Move dimensions to better locations where possible.

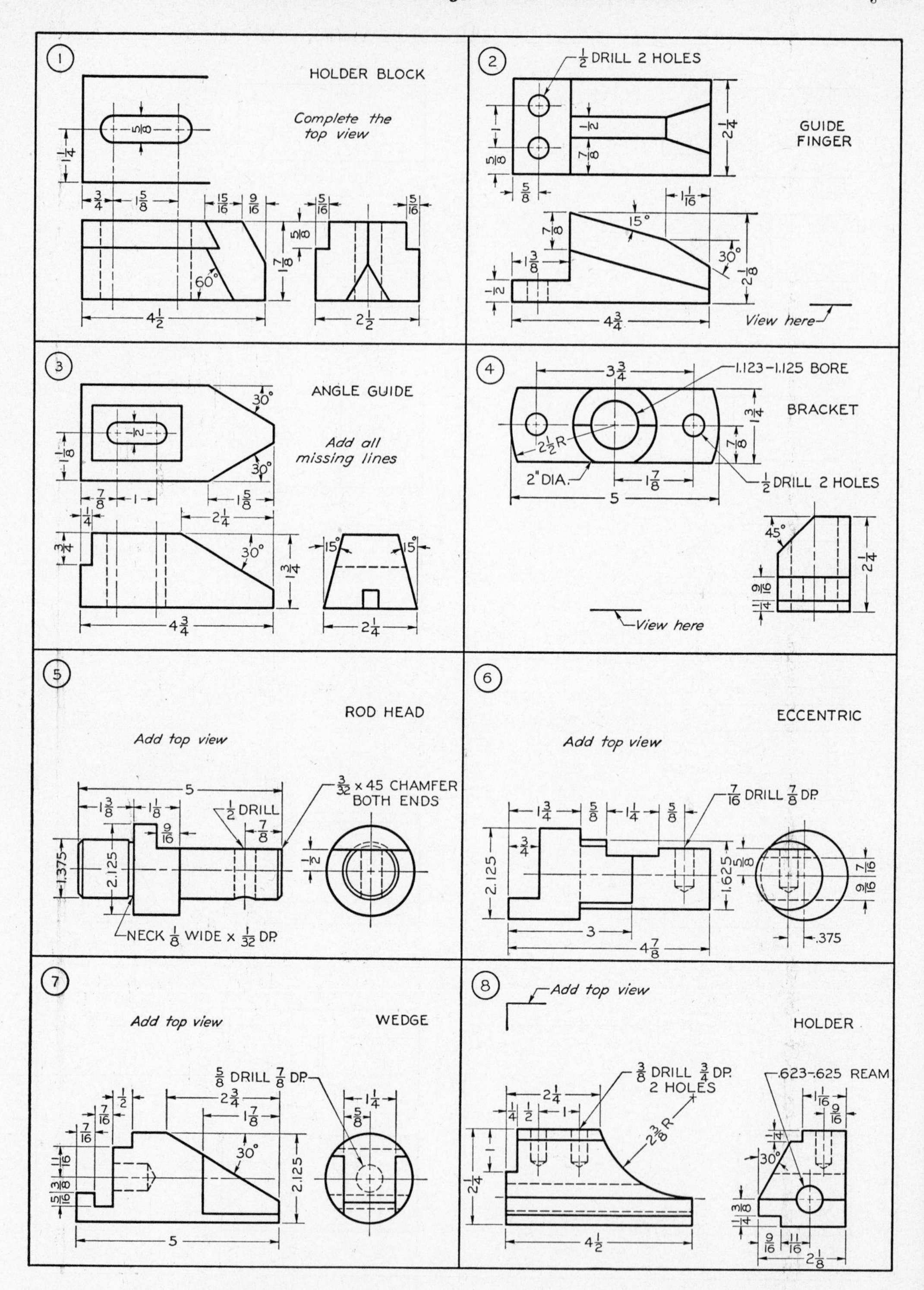

Fig. 335 Missing-View Problems. Using Layout A-2 or A-3, sketch or draw mechanically the given views, and add the missing view, as shown in Figs. 331 and 332. If dimensions are required, study §§327-350. Move dimensions to better locations where possible.

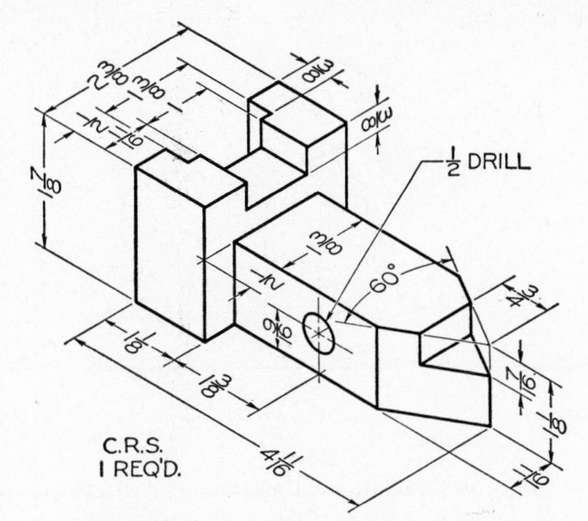

Fig. 336 Safety Key (Layout A-3).*

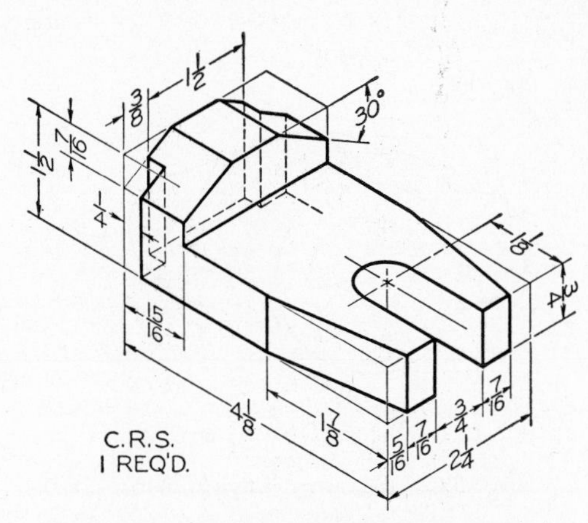

Fig. 337 Finger Guide (Layout A-3).*

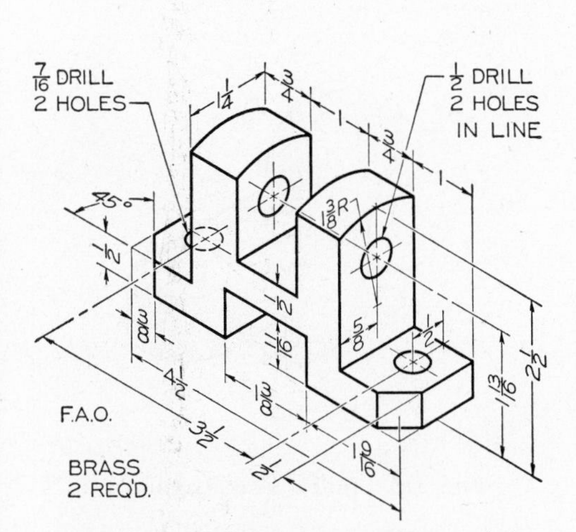

Fig. 338 Rod Support (Layout A-3).*

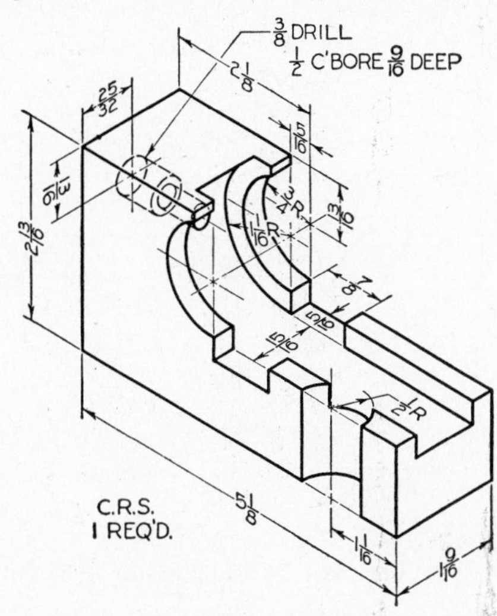

Fig. 339 Terminal Block (Layout A-3).*

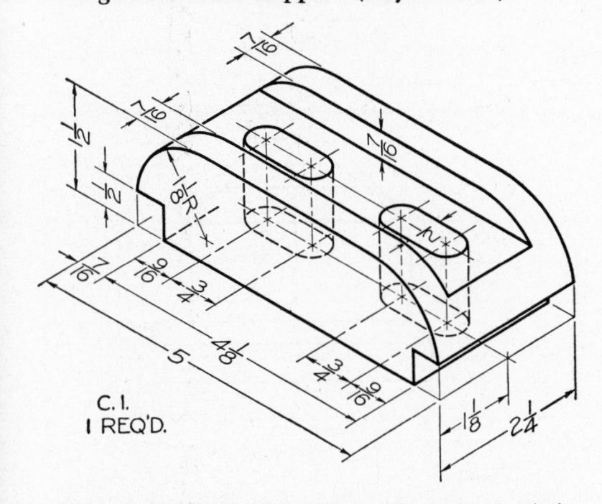

Fig. 340 Tailstock Clamp (Layout A-3).*

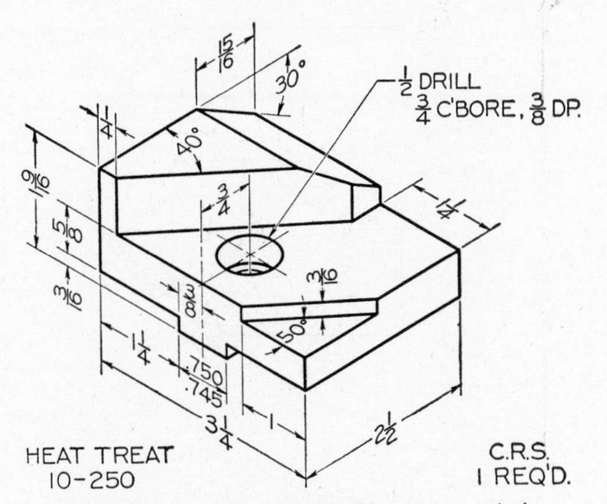

Fig. 341 Index Feed (Layout A-3).*

*Draw or sketch necessary views. If dimensions are required, study §§327-350.

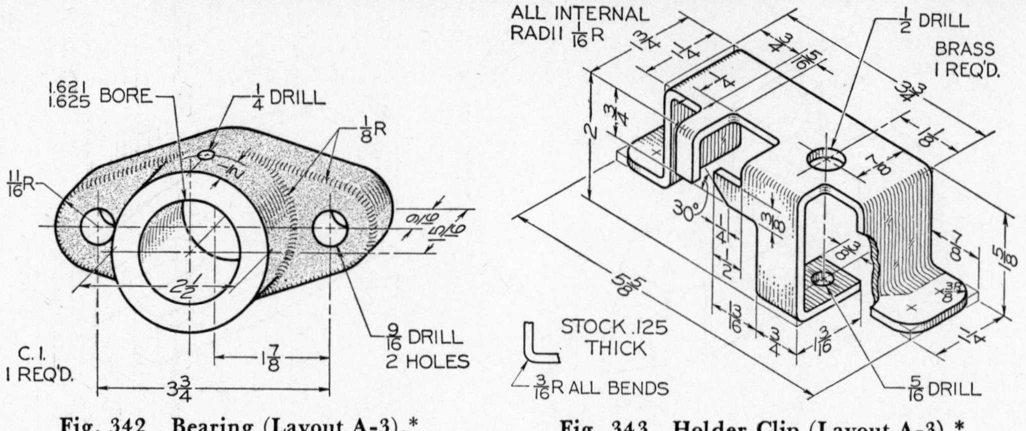

Fig. 342 Bearing (Layout A-3).* **Fig. 343** Holder Clip (Layout A-3).*

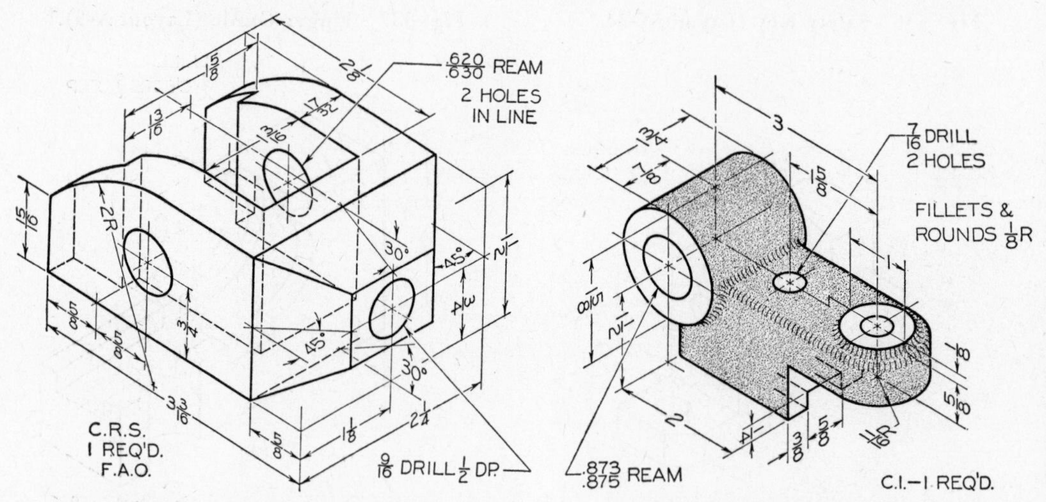

Fig. 344 Cam (Layout A-3).* **Fig. 345** Index Arm (Layout A-3).*

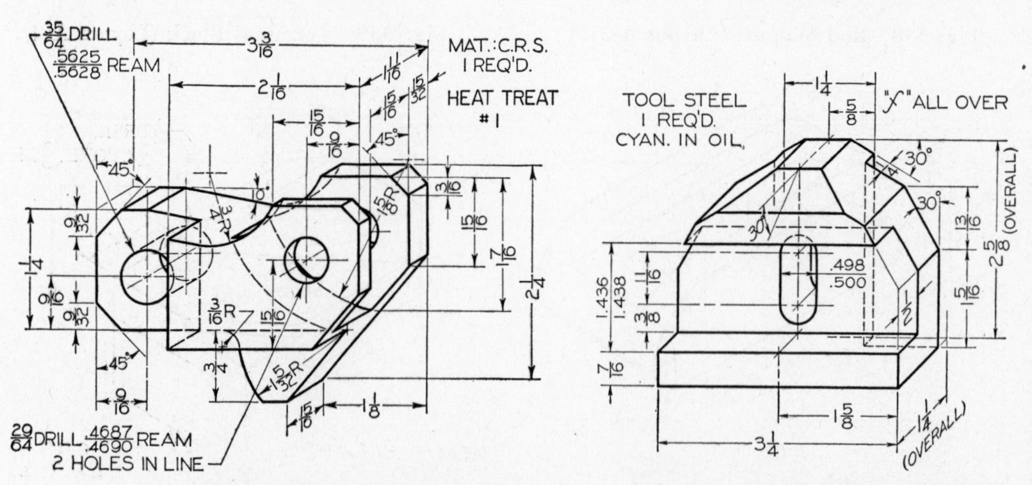

Fig. 346 Roller Lever (Layout A-3).* **Fig. 347** Support (Layout A-3).*

*Draw or sketch necessary views. If dimensions are required, study §§327-350.

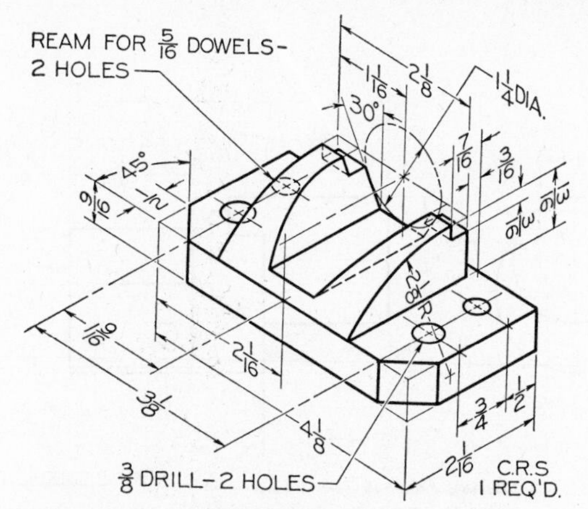

Fig. 348 Locating Finger (Layout A-3).*

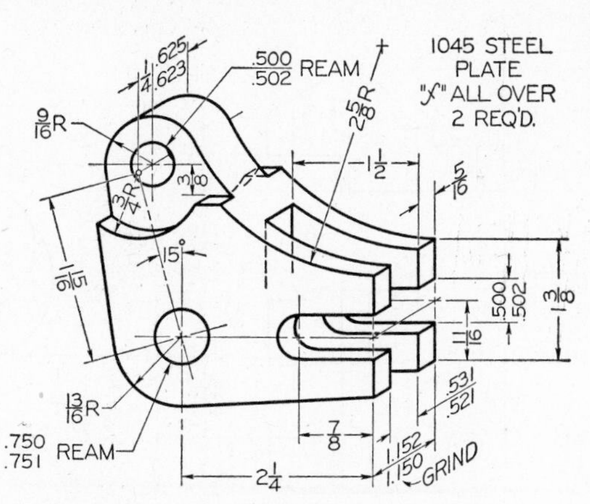

Fig. 349 Toggle Lever (Layout A-3).*

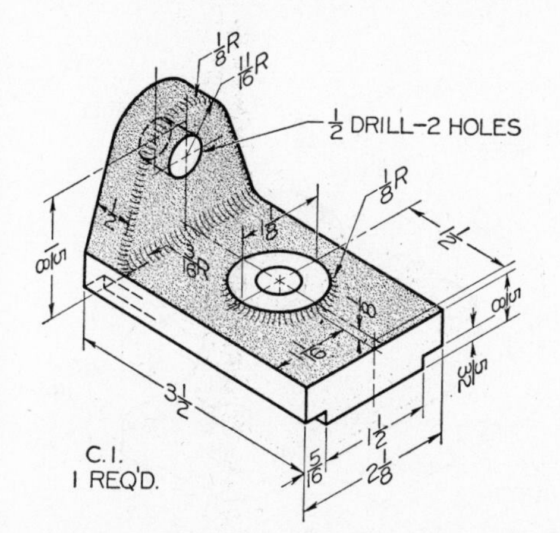

Fig. 350 Cut-Off Holder (Layout A-3).*

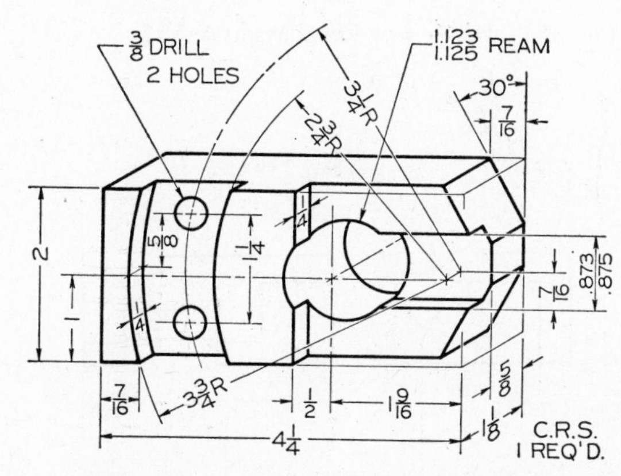

Fig. 351 Index Slide (Layout A-3).*

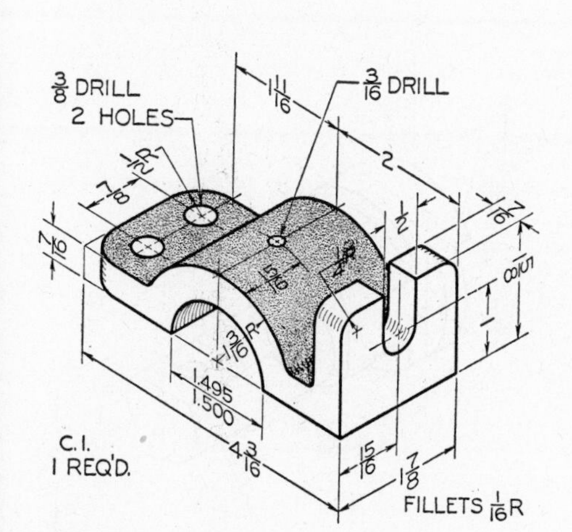

Fig. 352 Frame Guide (Layout A-3).*

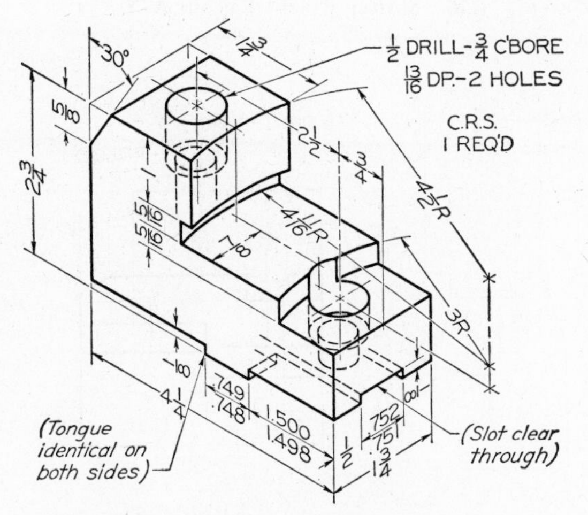

Fig. 353 Chuck Jaw (Layout A-3).*

*Draw or sketch necessary views. If dimensions are required, study §§327-350.

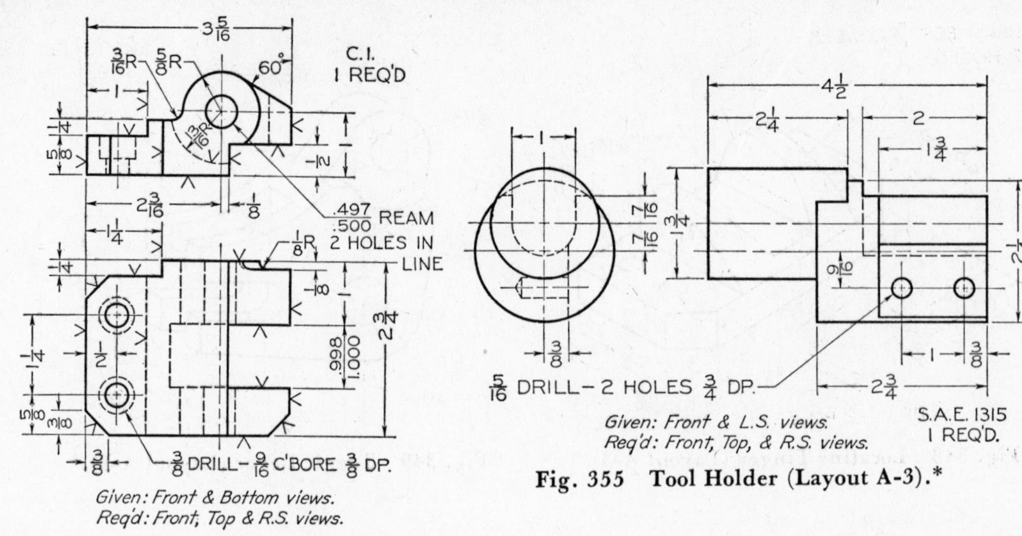

Fig. 354 Hinge Bracket (Layout A-3).*

Fig. 355 Tool Holder (Layout A-3).*

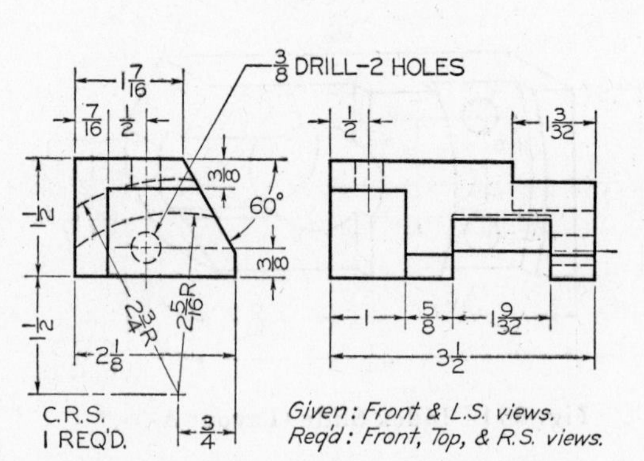

Fig. 356 Shifter Block (Layout A-3).*

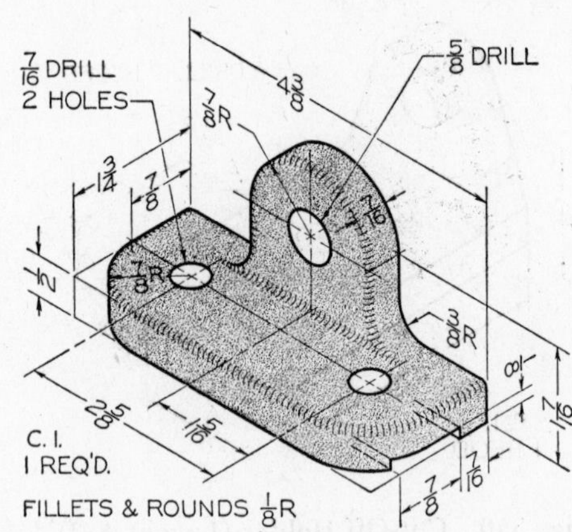

Fig. 357 Cross-Feed Stop (Layout A-3).*

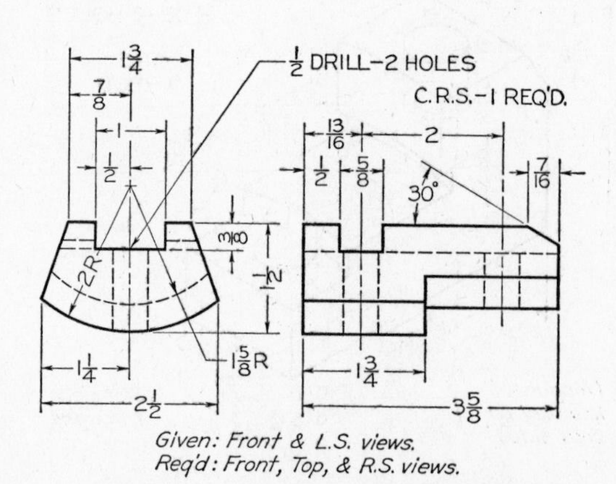

Fig. 358 Cross Cam (Layout A-3).*

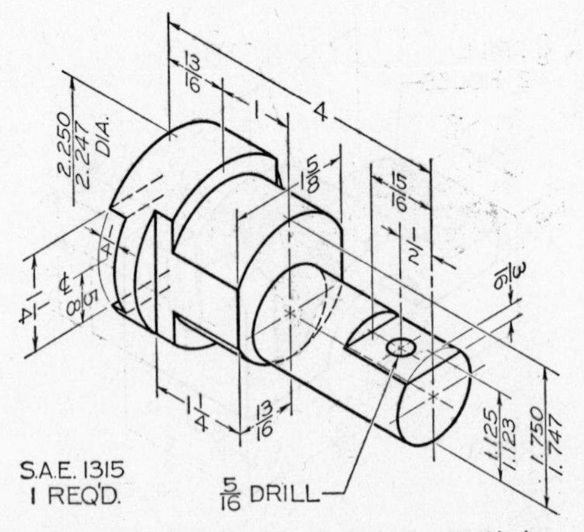

Fig. 359 Roller Stud (Layout A-3).*

*Draw or sketch necessary views. If dimensions are required, study §§327-350.

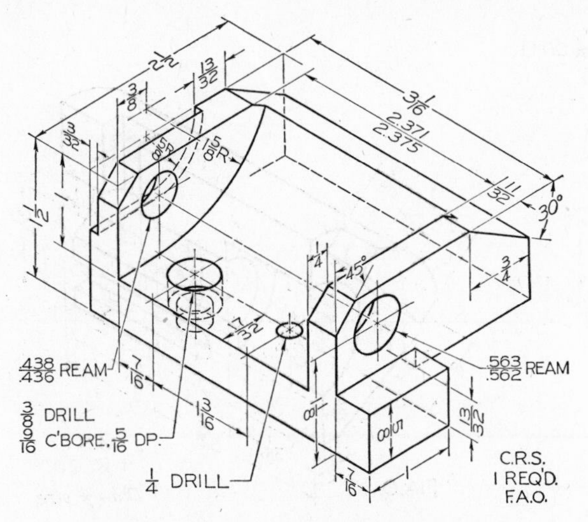

Fig. 360 Hinge Block (Layout A-3).*

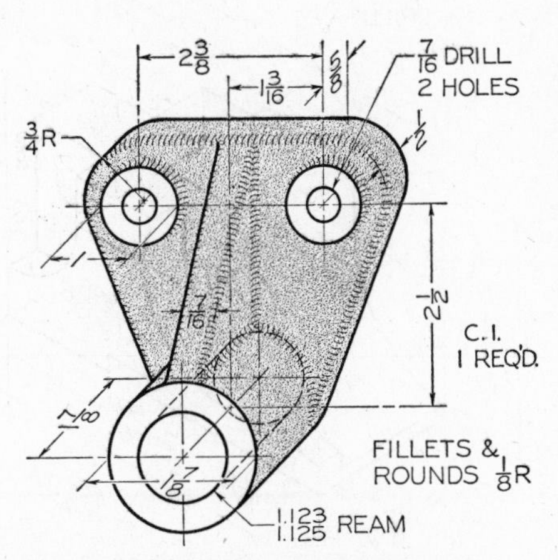

Fig. 361 Feed Rod Bearing (Layout A-3).*

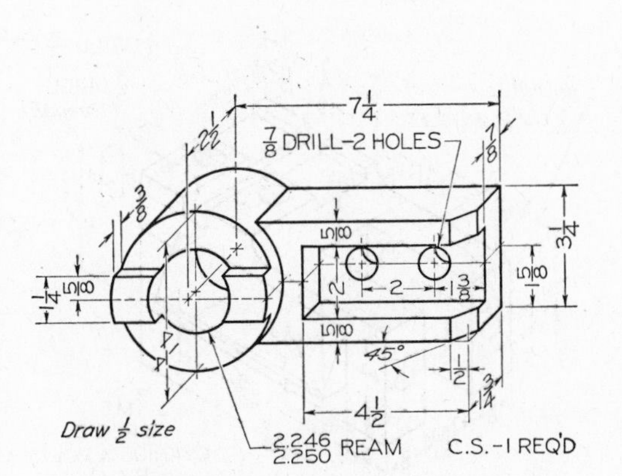

Fig. 362 Lever Hub (Layout A-3).*

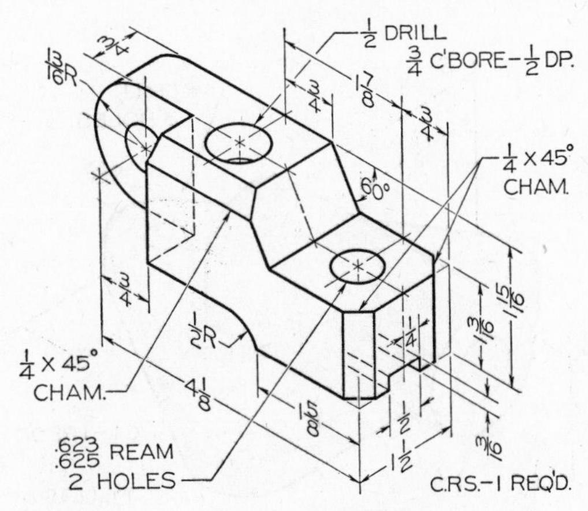

Fig. 363 Vibrator Arm (Layout A-3).*

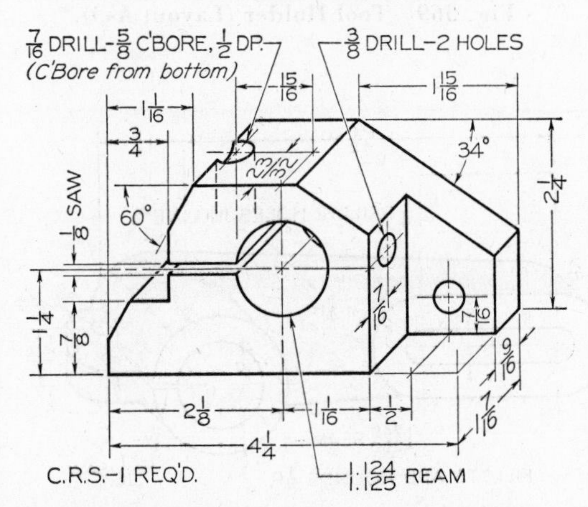

Fig. 364 Clutch Lever (Layout A-3).*

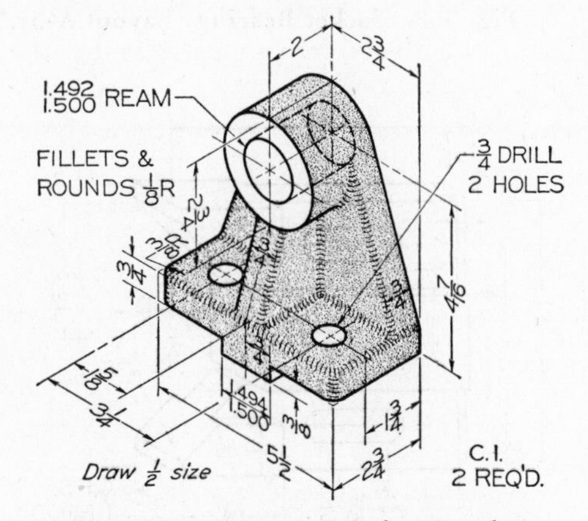

Fig. 365 Counter Bearing Bracket (Layout A-3).*

*Draw or sketch necessary views. If dimensions are required, study §§327-350.

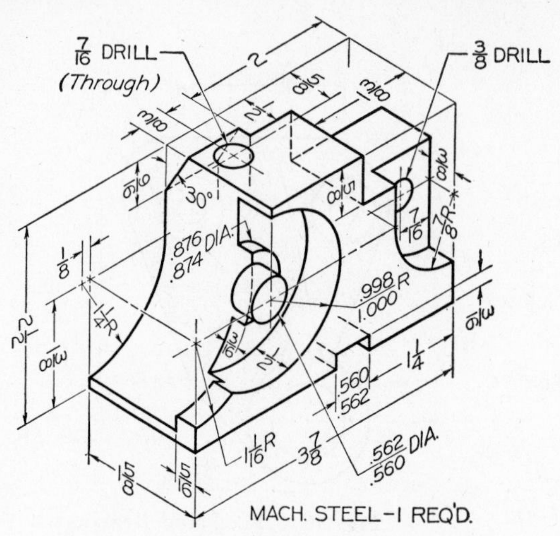

Fig. 366 Tool Holder (Layout A-3).*

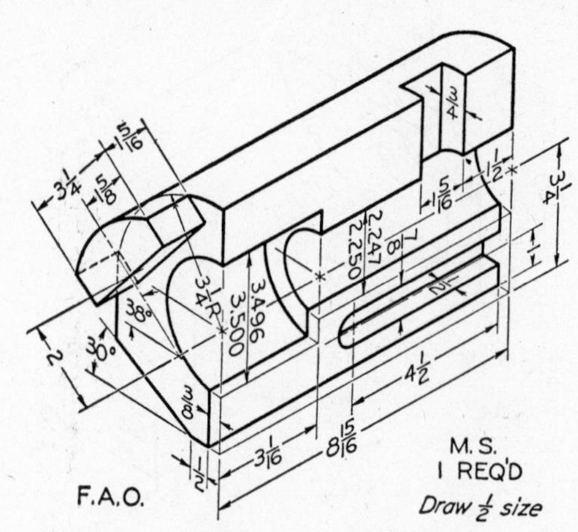

Fig. 367 Control Block (Layout A-3).*

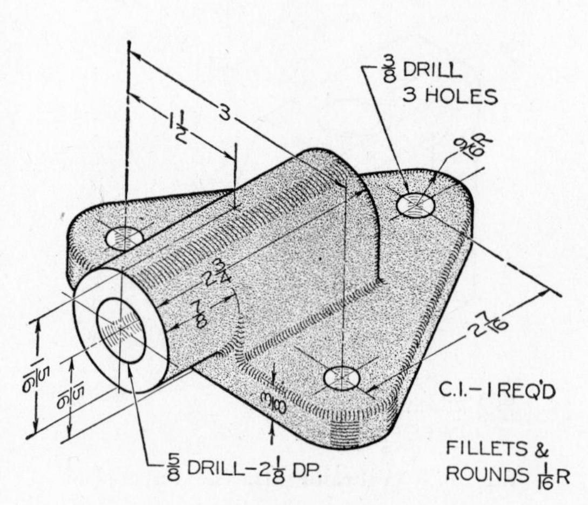

Fig. 368 Socket Bearing (Layout A-3).*

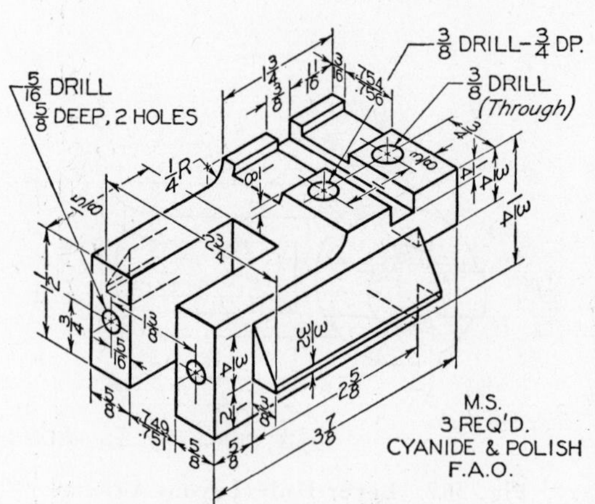

Fig. 369 Tool Holder (Layout A-3).*

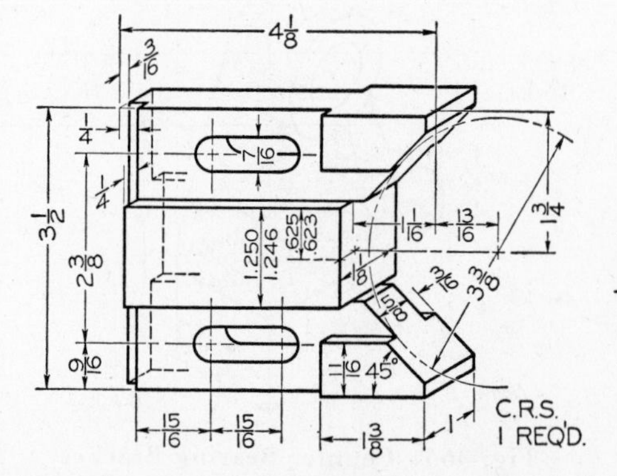

Fig. 370 Locating V-Block (Layout A-3).*

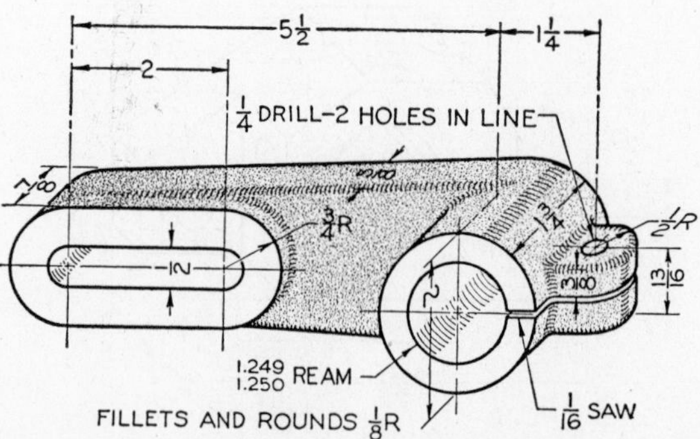

Fig. 371 Anchor Bracket (Layout A-3).*

*Draw or sketch necessary views. If dimensions are required, study §§327-350.

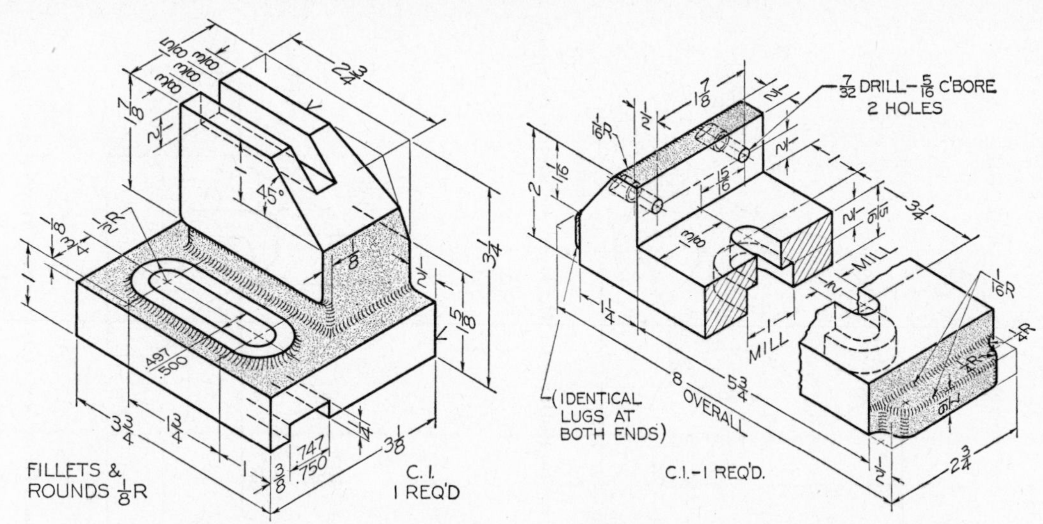

Fig. 372 Door Bearing (Layout B-3).*

Fig. 373 Vise Base (Layout B-3).*

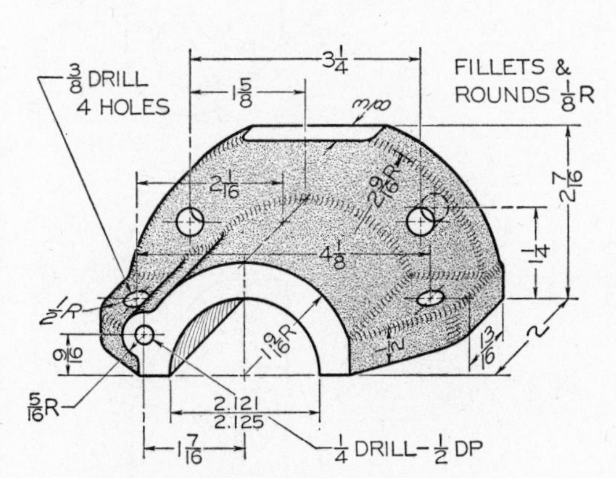

Fig. 374 Dust Cap (Layout B-3).*

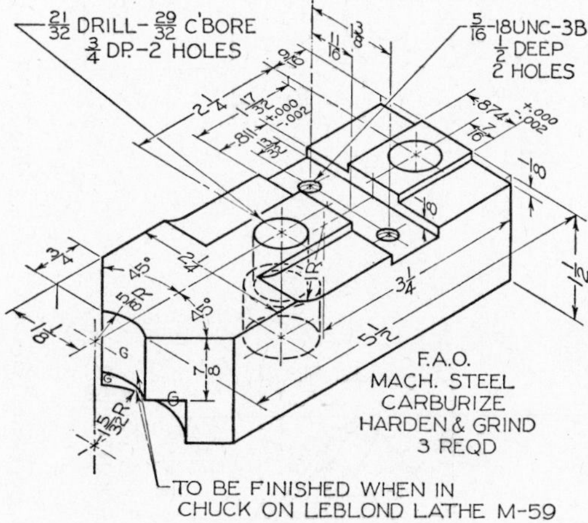

Fig. 375 Chuck Jaw (Layout B-3).*For threads, see §402.

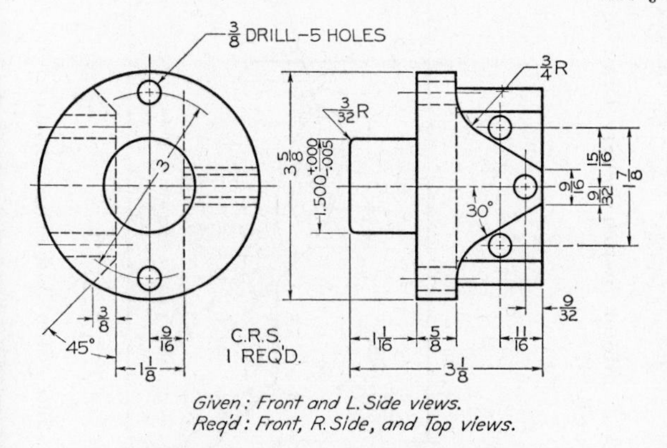

Fig. 376 Holder (Layout B-3).*

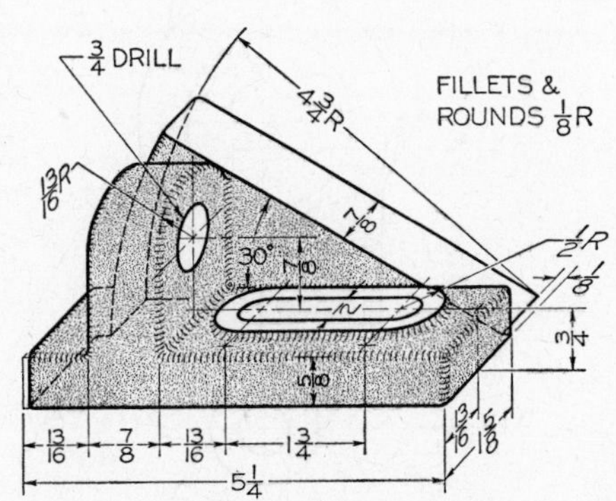

Fig. 377 Centering Wedge (Layout B-3).*

*Draw or sketch necessary views. If dimensions are required, study §§327-350.

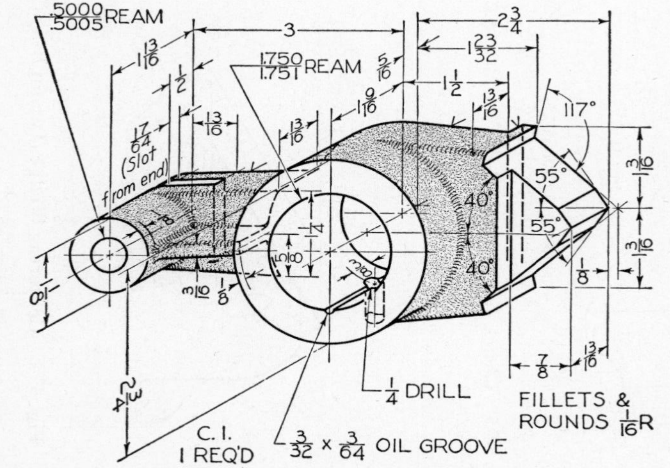

Fig. 378 Motor Switch Lever. Draw or sketch necessary views (Layout B-3).

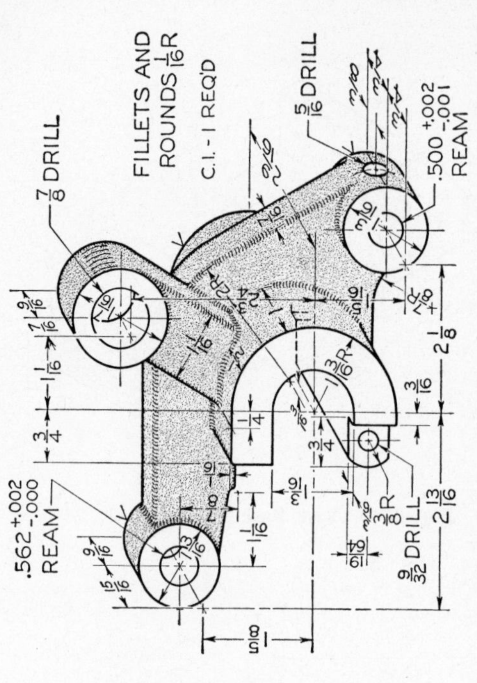

Fig. 379 Socket Form Roller—L. H. Draw or sketch necessary views (Layout B-4).

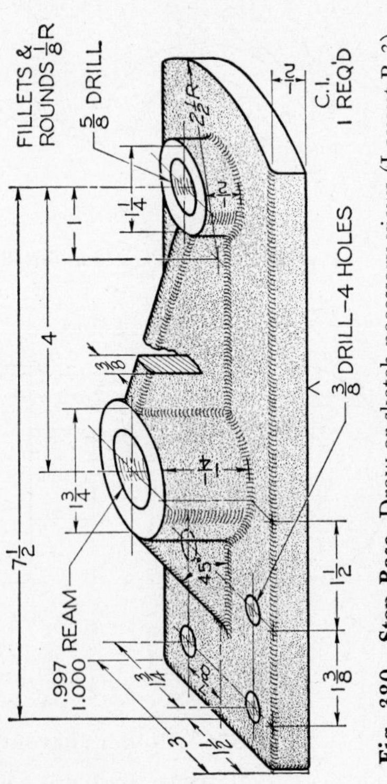

Fig. 380 Stop Base. Draw or sketch necessary views (Layout B-3).

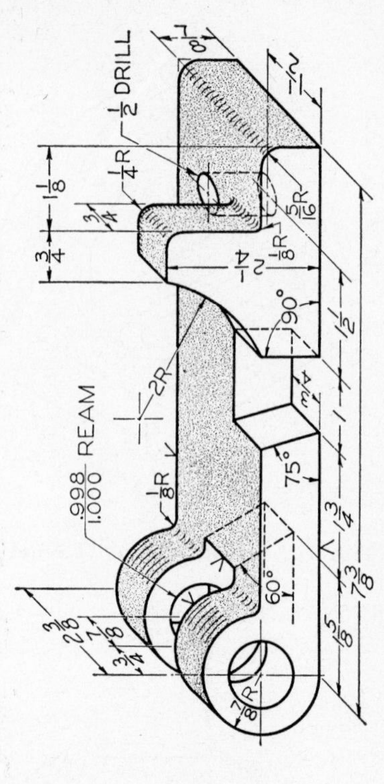

Fig. 381 Hinge Base. Draw or sketch necessary views (Layout B-3).

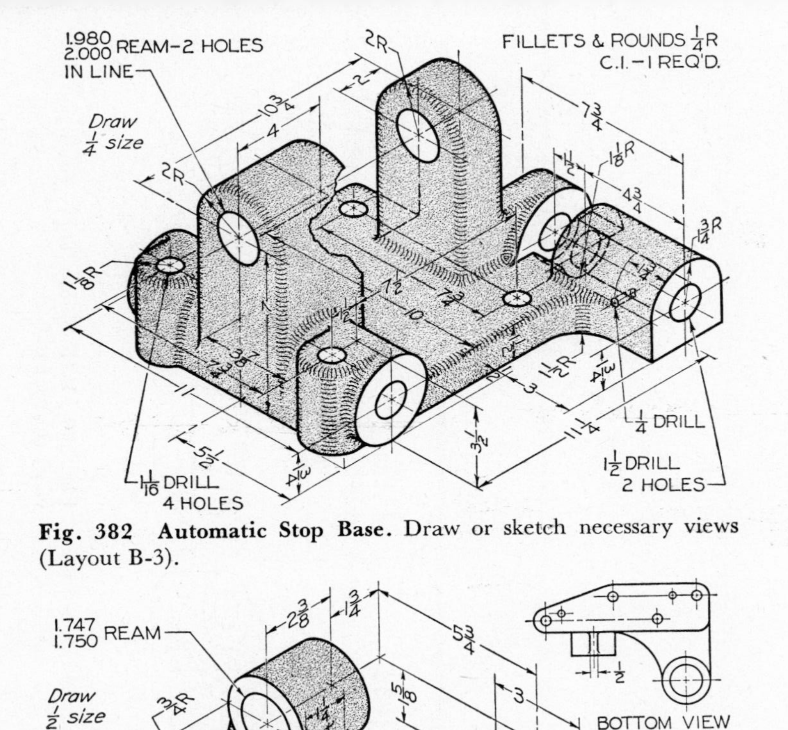

Fig. 382 Automatic Stop Base. Draw or sketch necessary views (Layout B-3).

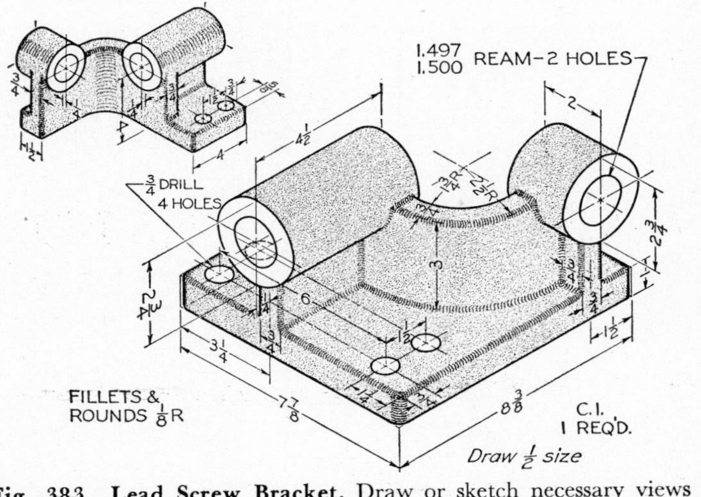

Fig. 383 Lead Screw Bracket. Draw or sketch necessary views (Layout B-3).

Fig. 384 Lever Bracket. Draw or sketch necessary views (Layout B-3).

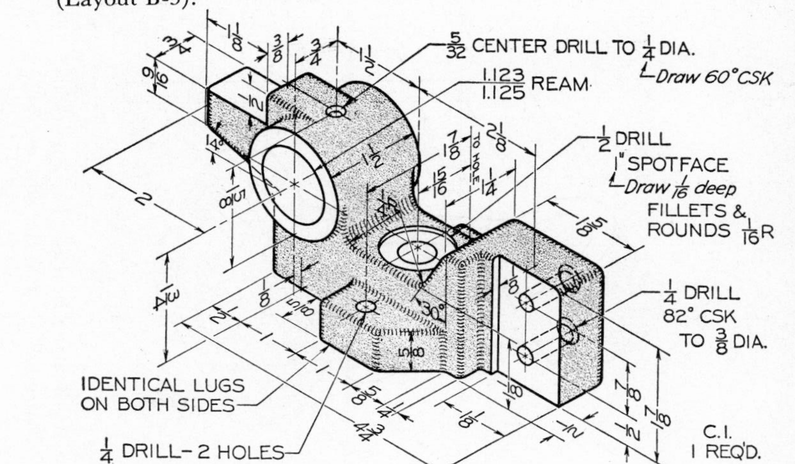

Fig. 385 Gripper Rod Center. Draw or sketch necessary views (Layout B-3).

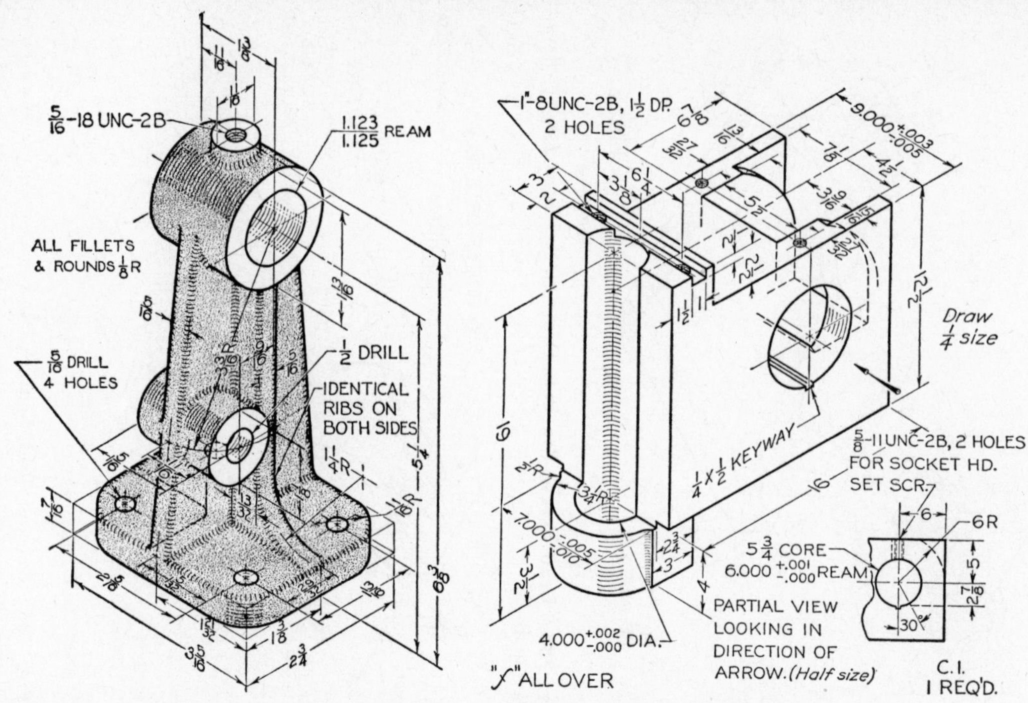

Fig. 386 Bearing Bracket. Draw necessary views (Layout B-3). For threads, see §402.

Fig. 387 Link Arm Connector. Draw necessary views (Layout B-3). For threads, see §402.

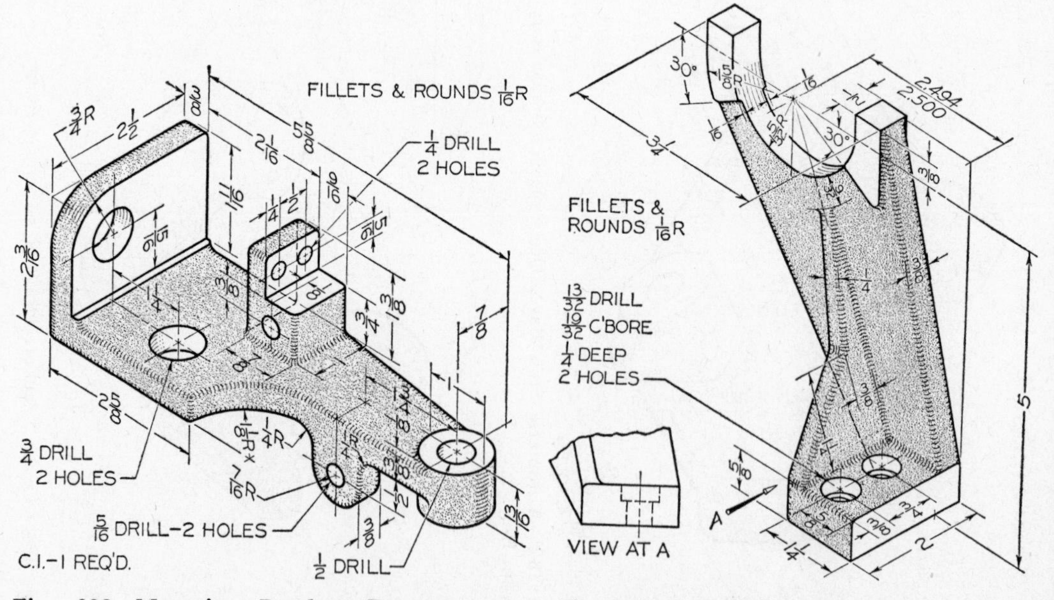

Fig. 388 Mounting Bracket. Draw necessary views (Layout B-3).

Fig. 389 L.H. Shifter Fork. Draw necessary views (Layout B-3).

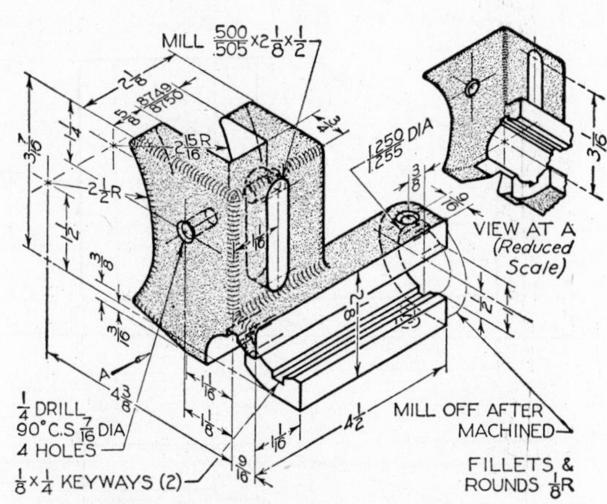

Fig. 390 Gear Shift Bracket (Layout C-4).*

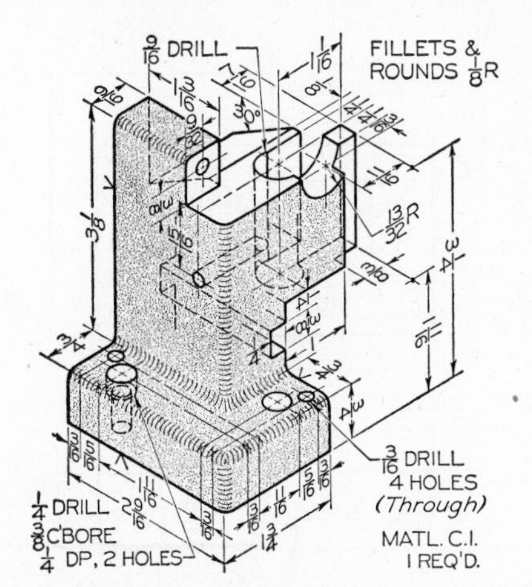

Fig. 391 Fixture Base (Layout C-4).*

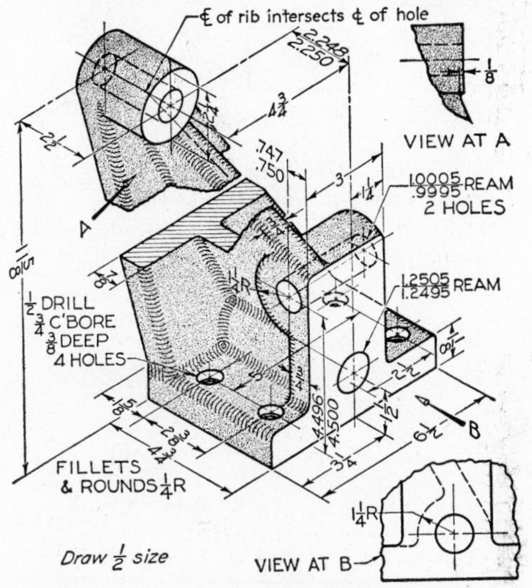

Fig. 392 Ejector Base (Layout C-4).*

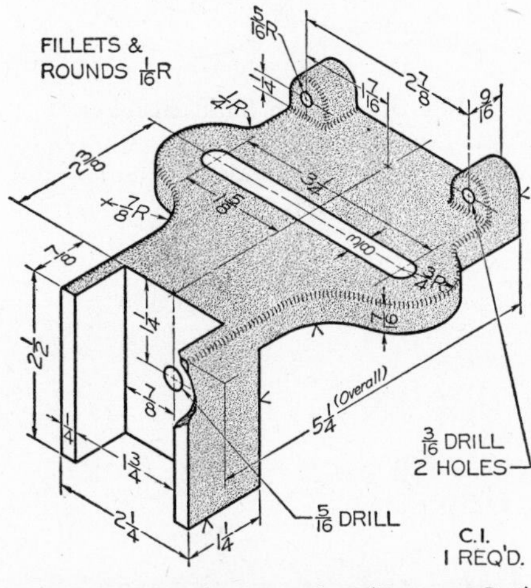

Fig. 393 Tension Bracket (Layout C-4).*

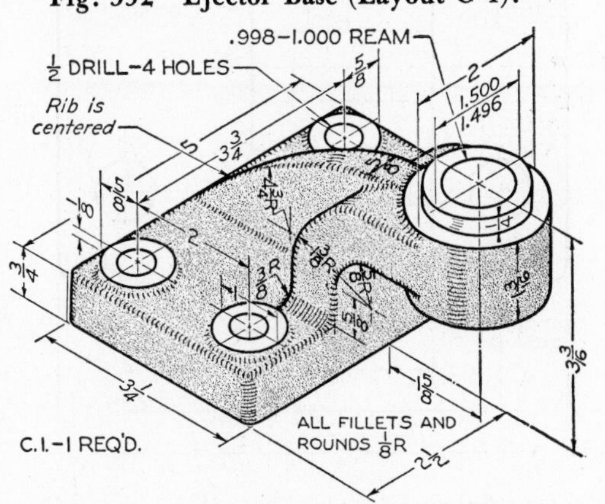

Fig. 394 Offset Bearing (Layout C-4).*

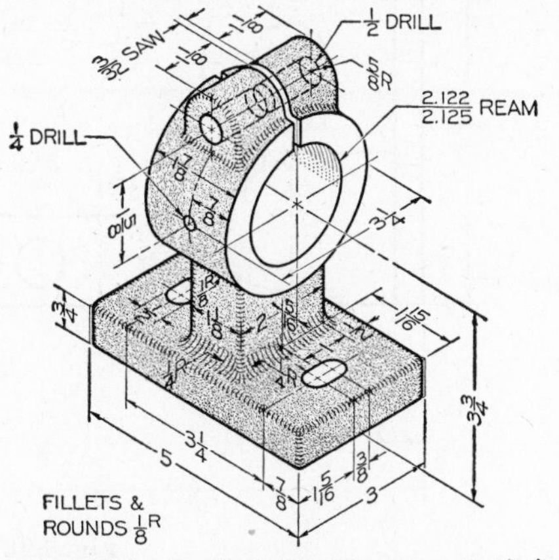

Fig. 395 Feed Guide (Layout C-4).*

*Draw necessary views. If dimensions are required, study §§327-350.

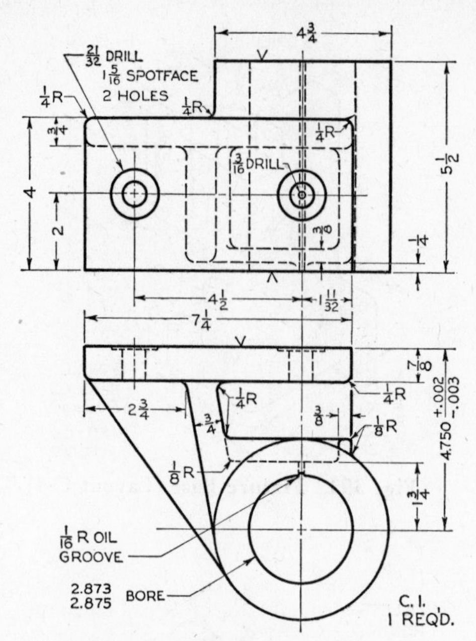

Fig. 396 Feed Shaft Bracket.
Given: Front and Top views.
Required: Front, Top, and Right-Side views, half size (Layout B-3).

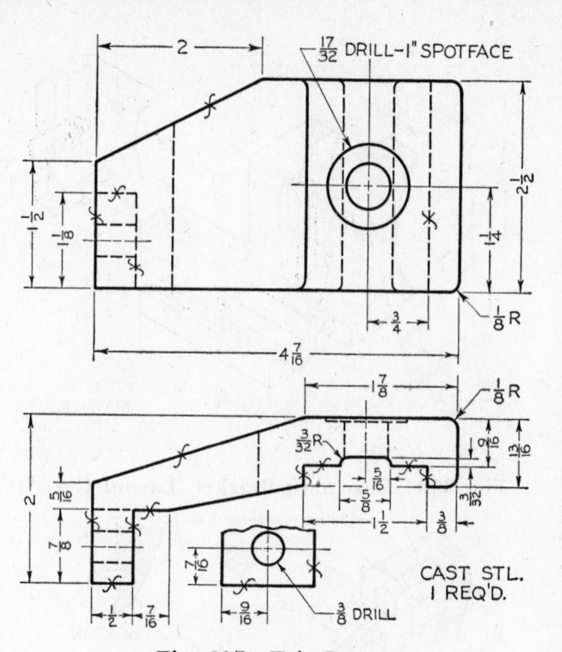

Fig. 397 Trip Lever.
Given: Front, Top, and Partial Side view.
Required: Front, Bottom, and Left-Side views, drawn completely (Layout B-3).

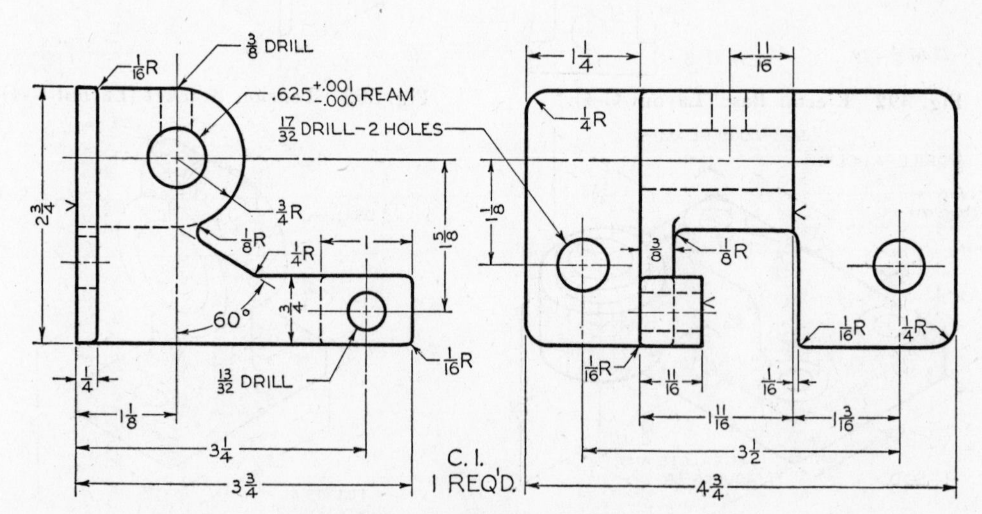

Fig. 398 Knurl Bracket Bearing.
Given: Front and Left Side views.
Required: Take Front as Top view on new drawing, and add Front and Right-Side views. (Layout B-3).

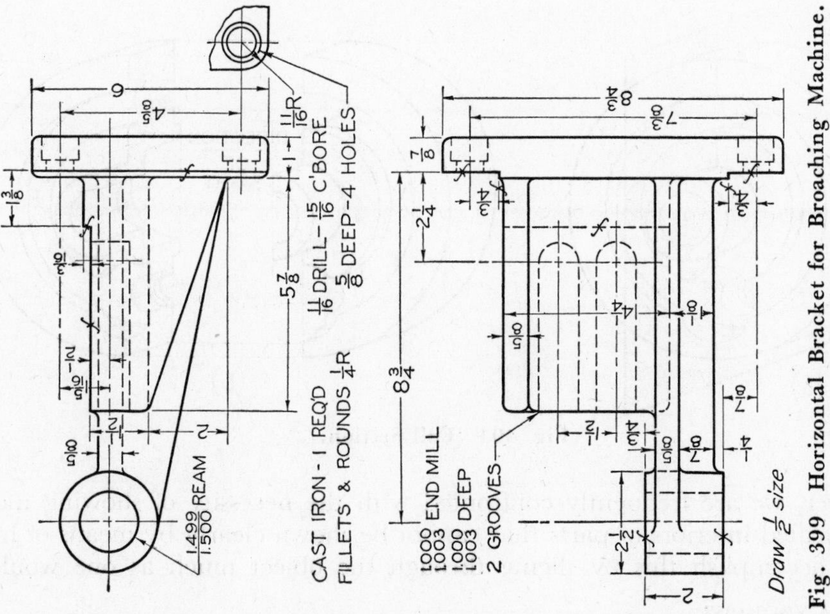

Fig. 399 Horizontal Bracket for Broaching Machine.
Given: Front and Top views.
Required: Take Top as Front view in new drawing; then add Top and Left-Side views (Layout C-4).

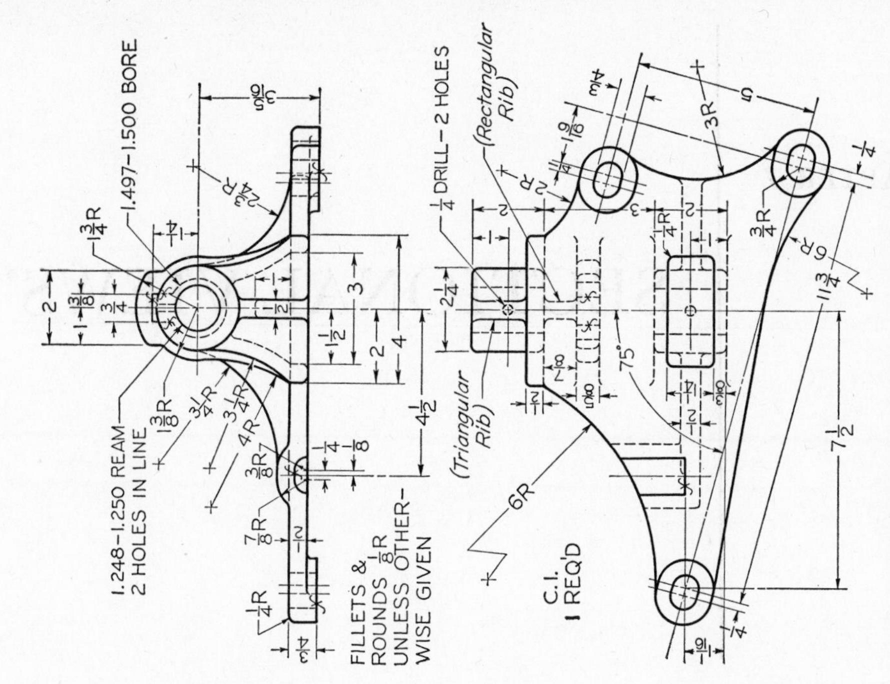

Fig. 400 Boom Swing Bearing for a Power Crane.
Given: Front and Bottom views.
Required: Front, Top, and Left-Side views (Layout C-4).

CHAPTER 7

SECTIONAL VIEWS*

245. Full Sections. The basic method of representing objects by views, or projections, has been explained in previous chapters. By means of a limited number of carefully selected views, the external features of the most complicated objects can be thus fully described.

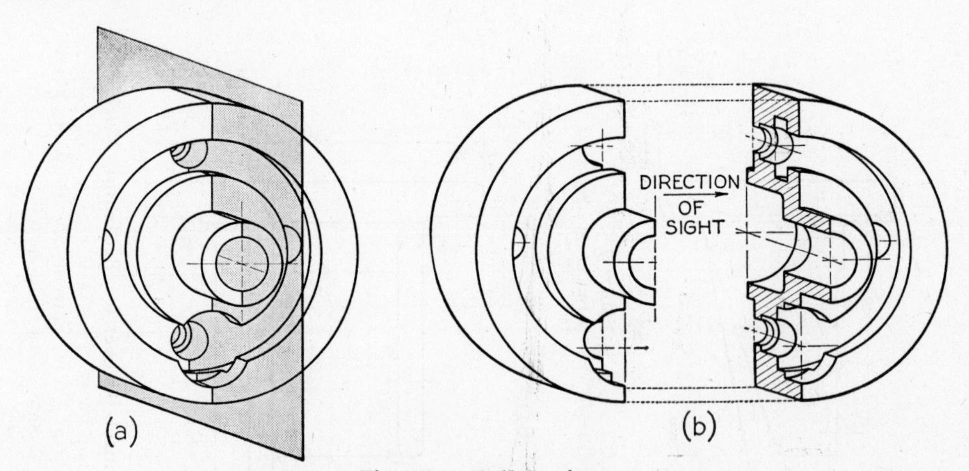

Fig. 401 Full Section.

However, we are frequently confronted with the necessity of showing more or less complicated interiors of parts that cannot be shown clearly by means of hidden lines. We accomplish this by slicing through the object much as one would cut

*See ASA Y14.2-1957.

through an apple or a watermelon. A cutaway view of the object is then drawn; it is called a *sectional view*, a *cross-section*, or simply a *section*.

More exactly, a *cutting plane*, §248, is assumed to be passed through the object, as shown in Fig. 401 (a). Then, at (b) the cutting plane is removed and the two halves drawn apart, exposing the interior construction. In this case, the direction of sight is toward the right half, as shown, and for purposes of the section, the left half is mentally discarded. The sectional view will be in the position of a left-side view.

The sectional view thus obtained is shown in Fig. 402 (a). A comparison of this sectional view with the right-side view, (c), emphasizes the advantage in clearness

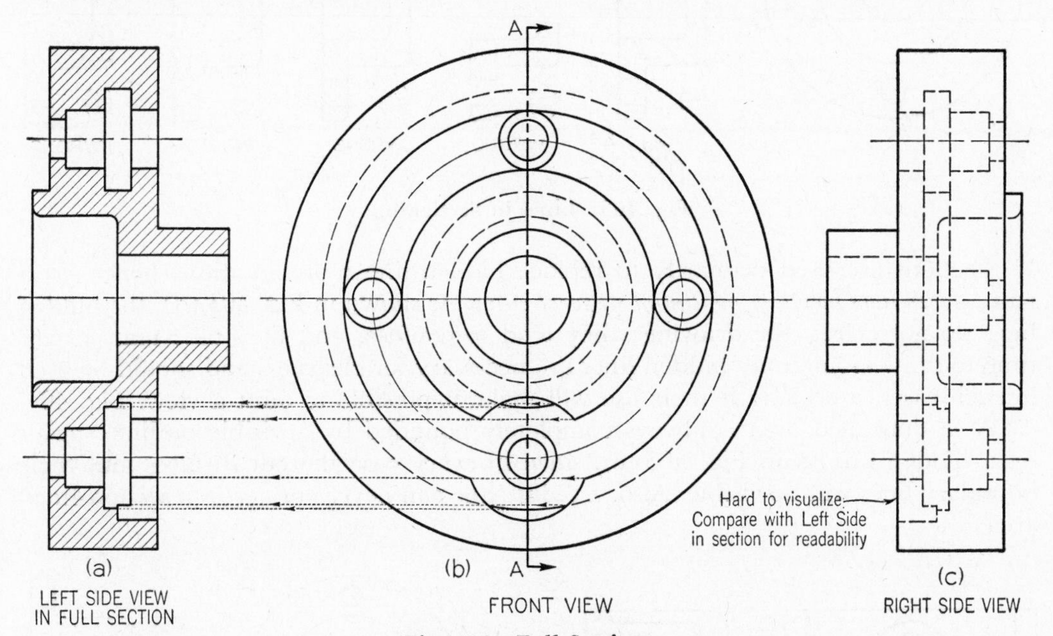

Fig. 402 Full Section.

of the former. The right-side view would naturally be omitted. In the front view, the cutting plane appears as a line, called a *cutting-plane line*, §248. The arrows at the ends of the cutting-plane line indicate the direction of sight for the sectional view. This type of section is called a *full section*, because the cutting plane passes "fully" through the object.

Note that in order to obtain the sectional view, the left half is only *imagined* to be removed, and not actually shown removed anywhere except in the sectional view itself. In the sectional view, the section-lined areas are those portions that have been in actual contact with the cutting plane. Those areas are *crosshatched* with very sharp parallel section lines spaced carefully by eye. In addition, the visible parts behind the cutting plane are shown, but not crosshatched.

As a rule, the location of the cutting plane is obvious from the section itself, and the cutting-plane line is therefore omitted. It is shown in Fig. 402 for illustration only. Cutting-plane lines should, of course, be used wherever necessary for clearness, as in Figs. 420, 421, 424, and 425.

246. Lines in Sectioning. A correct front view and sectional view are shown in Fig. 403 (a) and (b). In general, *all visible edges and contours behind the cutting plane* *should be shown;* otherwise a section will appear to be made up of disconnected and unrelated parts, as shown at (c). Occasionally, however, visible lines behind the cutting plane are not necessary for clearness and should be omitted, Fig. 432.

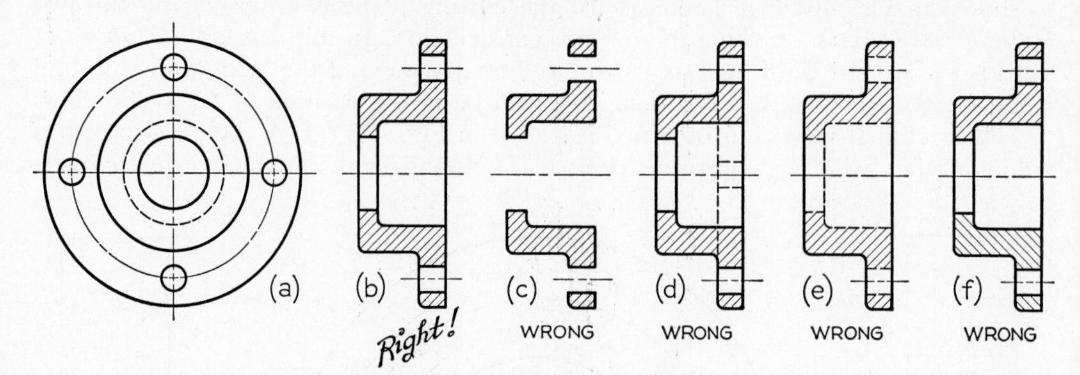

Fig. 403 Lines in Sectioning.

Sections are used primarily to replace hidden-line representation; hence, as a rule, *hidden lines should be omitted in sectional views*. As shown in Fig. 403 (d), the hidden lines do not clarify the drawing; they tend to confuse, and they take unnecessary time to draw. Sometimes hidden lines are necessary for clearness and should be used in such cases, especially if their use will make it possible to omit a view, Fig. 404.

A section-lined area is always completely bounded by a visible outline—never by a hidden line as in Fig. 403 (e), since in every case the cut surfaces and their boundary lines will be visible. Also, a visible line can never cut across a section-lined area.

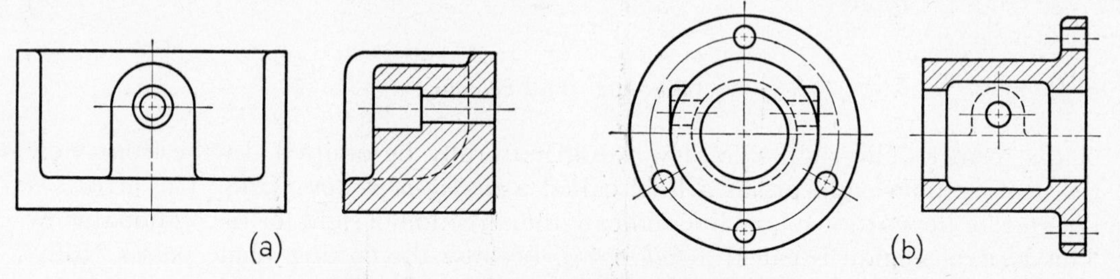

Fig. 404 Hidden Lines in Sections.

In a sectional view of a part, alone or in assembly, the section lines in all sectioned areas must be parallel, not as shown in Fig. 403 (f). The use of section lining in opposite directions is an indication of different parts, as when two or more parts are adjacent in an assembly drawing, Fig. 731.

247. Section Lining. In the past, different section lining symbols, Fig. 733, have been used to indicate the material to be used in producing the object. These symbols represented the general types only, such as cast-iron, brass, steel, etc. But now there are so many different materials and each general type has so many sub-types, that a general name or symbol is not enough. For example, there are hundreds of different kinds of steel alone. Since detailed specifications of material must be lettered in the form of a note or in the title strip, the ASA recommends that on detail drawings (single parts) the general-purpose (cast-iron) section lining be used for all materials.

Symbolic section lining may be used in assembly drawings in cases where it is desirable to distinguish the different materials; otherwise, the general-purpose symbol is used for all parts. For assembly sections, see §441.

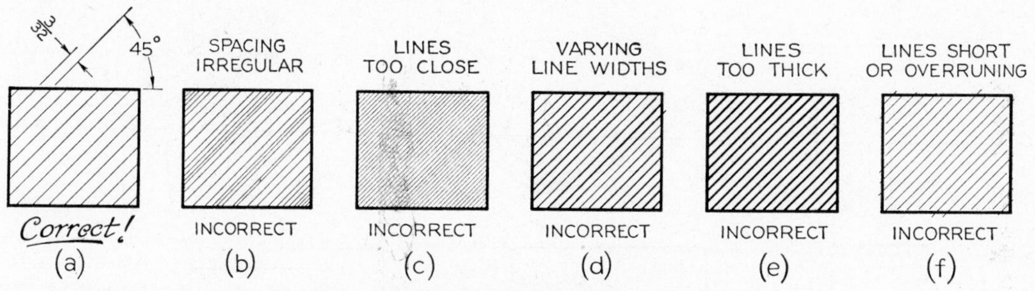

Fig. 405 Section-Lining Technique.

The correct method of drawing section lines is shown in Fig. 405 (a). Draw the section lines with a very sharp, medium-grade pencil (H or 2H) with a conical point as shown in Fig. 20 (c). Always draw the lines at 45° with horizontal as shown, unless there is some advantage in using a different angle. Space the section lines as uniformly as possible by eye from about $\frac{1}{32}''$ to $\frac{1}{8}''$ or more apart, depending on the size of the drawing or of the sectioned area. For average drawings, space the lines about $\frac{3}{32}''$ apart. As a rule, space the lines as generously as possible and yet close enough to distinguish clearly the sectioned areas.

After the first few lines have been drawn, look back repeatedly at the original spacing to avoid gradually increasing or decreasing the intervals, Fig. 405 (b). Beginners almost invariably draw section lines too close together, (c). This is very tedious business, because with small spacing the least inaccuracy in spacing is conspicuous.

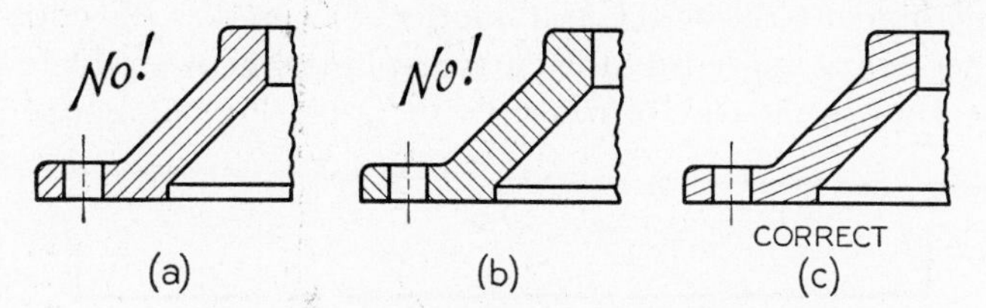

Fig. 406 Direction of Section Lines.

Section lines should be uniformly thin, never irregular in thickness, as at (d). There should be a marked contrast in thickness of the visible outlines and the section lines. Section lines should not be too thick, as at (e). Also avoid running section lines beyond the visible outlines or stopping the lines too short, as at (f).

If section lines drawn at 45° with horizontal would be parallel or perpendicular (or nearly so) to a prominent visible outline, the angle should be changed to 30°, 60°, or some odd angle, Fig. 406.

Dimensions should be kept off sectioned areas, but when this is unavoidable the section lines should be omitted where the dimension figure is placed.

248. The Cutting Plane. The *cutting plane* is indicated in a view adjacent to the sectional view, Fig. 407. In this view, the cutting plane appears edgewise or as a line, called the *cutting-plane line.* Alternate styles of cutting-plane lines are approved by the ASA, as shown in Fig. 408. The first form, composed of alternate long dashes and pairs of short dashes, has been in general use for a long time. The second form, composed of equal dashes each about $\frac{1}{4}''$ long, is standard in the automotive industry,

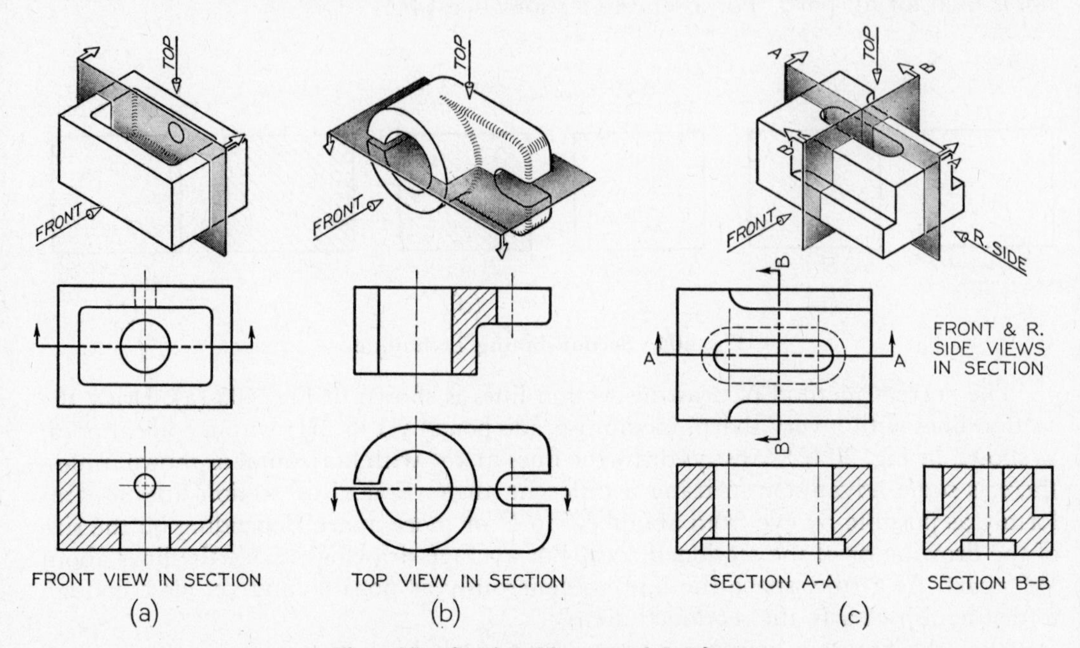

Fig. 407 Cutting Planes and Sections.

and has the advantage of showing up clearly on complicated drawings. Both lines are drawn the same thickness as visible lines. Arrowheads indicate the direction in which the cutaway object is viewed. Capital letters are used at the ends of the cutting-plane line when necessary to identify the cutting-plane line with the indicated section. This most often occurs in the case of multiple sections, Fig. 425, or removed sections, Fig. 420.

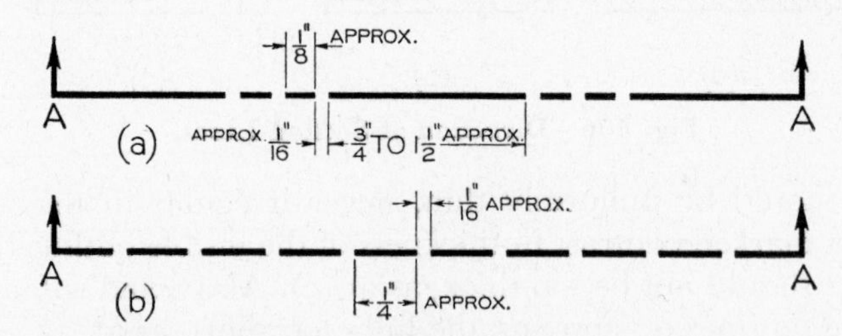

Fig. 408 Cutting-Plane Lines (Full Size).

As shown in Fig. 407, sectional views occupy normal projected positions in the standard arrangement of views. At (a) the cutting plane is a frontal plane, §220, and appears as a line in the top view. The front half of the object (lower half in the top view) is imagined removed. The arrows at the ends of the cutting-plane line point in the direction of sight for a front view; that is, away from the front view or section. Note that the arrows do not point in the direction of withdrawal of the removed portion. The resulting full section may be referred to as the "front view in section," since it occupies the front view position.

In Fig. 407 (b), the cutting plane is a horizontal plane, §220, and appears as a line in the front view. The upper half of the object is imagined removed. The arrows point toward the lower half in the same direction of sight as for a top view, and the resulting full section is a "top view in section."

In Fig. 407 (c), two cutting planes are shown, one a frontal plane and the other a profile plane, both of which appear edgewise in the top view. Each section is completely independent of the other and drawn as if the other were not present. For section A-A, the front half of the object is imagined removed. The back half then is viewed in the direction of the arrows for a front view, and the resulting section is a "front view in section." For section B-B, the right half of the object is imagined removed. The left half then is viewed in the direction of the arrows for a right-side view, and the resulting section is a "right-side view in section." The cutting-plane lines are preferably drawn through an exterior view, in this case the top view, instead of a sectional view, as shown.

The cutting-plane lines in Fig. 407 are shown for purposes of illustration only. They are generally omitted in cases such as these, in which the location of the cutting plane is obvious. When a cutting-plane line coincides with a center line, the cutting-plane line takes precedence.

Correct and incorrect relations between cutting-plane lines and corresponding sectional views are shown in Fig. 409.

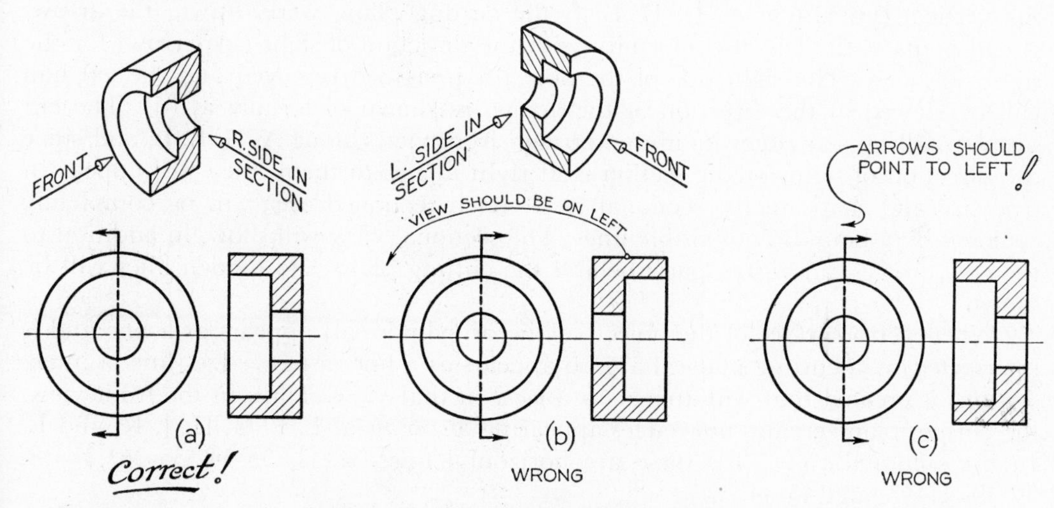

Fig. 409 Cutting Planes and Sections.

249. Visualizing a Section. Two views of an object to be sectioned, having a drilled and counterbored hole, are shown in Fig. 410 (a). The cutting plane is assumed along the horizontal center line in the top view, and the front half of the object (lower half of the top view) is imagined removed. A pictorial drawing of the remaining back half is shown at (b). The two cut surfaces produced by the cutting plane are 1-2-5-6-10-9 and 3-4-12-11-7-8. However, the corresponding section at I is incomplete because certain visible lines are missing.

If the section is viewed in the direction of sight, as shown at (b), arcs A, B, C, and D will be visible. As shown at II, these arcs will appear as straight lines 2-3, 6-7, 5-8, and 10-11. These lines may also be accounted for in other ways. The top and bottom surfaces of the object appear in the section as lines 1-4 and 9-12. The bottom surface of the counterbore appears in the section as line 5-8. Also, the semi-cylindrical surfaces for the back half of the counterbore and of the drilled hole will appear as rectangles in the section at 2-3-8-5 and 6-7-11-10.

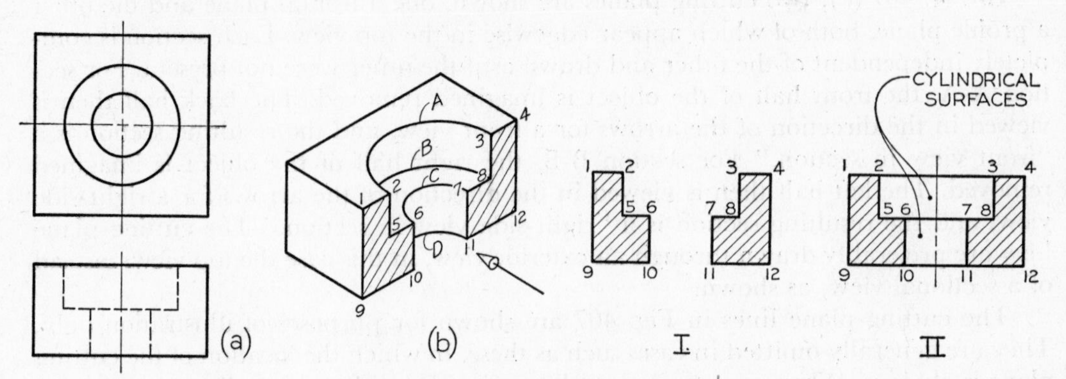

Fig. 410 Visualizing a Section.

The front and top views of a "Collar" are shown in Fig. 411 (a), and a right-side view in full section is required. The cutting plane is understood to pass along the vertical center line A, D, E, L. If the cutting plane were drawn, the arrows would point to the left in conformity with the direction of sight (see arrow) for the right-side view. The right side of the object is imagined removed and the left half will be viewed in the direction of the arrow, as shown pictorially at (d). The cut surfaces will appear edgewise in the top and front views along A, D, E, L, and since the direction of sight for the section is at right angles to them, they will appear in true size and shape in the sectional view. Each sectioned area will be completely enclosed by a boundary of visible lines. The sectional view will show, in addition to the cut surfaces, all visible parts behind the cutting plane. No hidden lines will be shown.

Whenever a surface of the object (plane or cylindrical) appears as a line and is intersected by a cutting plane that also appears as a line, a new edge (line of intersection) is created that will appear as a *point* in that view. Thus, in the front view, the cutting plane creates new edges appearing as points at E, F, G, H, J, K, and L. In the sectional view, (b), these are horizontal lines 31-32, 33-34, 35-36, 37-38, 39-40, 41-42, and 43-44.

Whenever a surface of the object appears as a surface (i.e., not as a line) and is cut by a cutting plane that appears as a line, a new edge is created that will appear as a line in the view, coinciding with the cutting-plane line, and as a line in the section.

In the top view, D is the *point view* of a vertical line KL in the front view and 41-43 in the section at (b). Point C is the point view of a vertical line HJ in the front view and 37-39 in the section. Point B is the point view of two vertical lines EF and GH in the front view, and 31-33 and 35-38 in the section. Point A is the point view of three vertical lines EF, GJ, and KL in the front view, and 32-34, 36-40, and 42-44 in the section. This completes the boundaries of three sectioned areas 31-32-

34-33, 35-36-40-39-37-38, and 41-42-44-43. It is only necessary now to add the visible lines beyond the cutting plane.

The semi-cylindrical left half F-21-G of the small hole (front view) will be visible as a rectangle in the sections at 33-34-36-35, as shown at (c). The two semi-circular arcs will appear as straight lines in the section at 33-35 and 34-36.

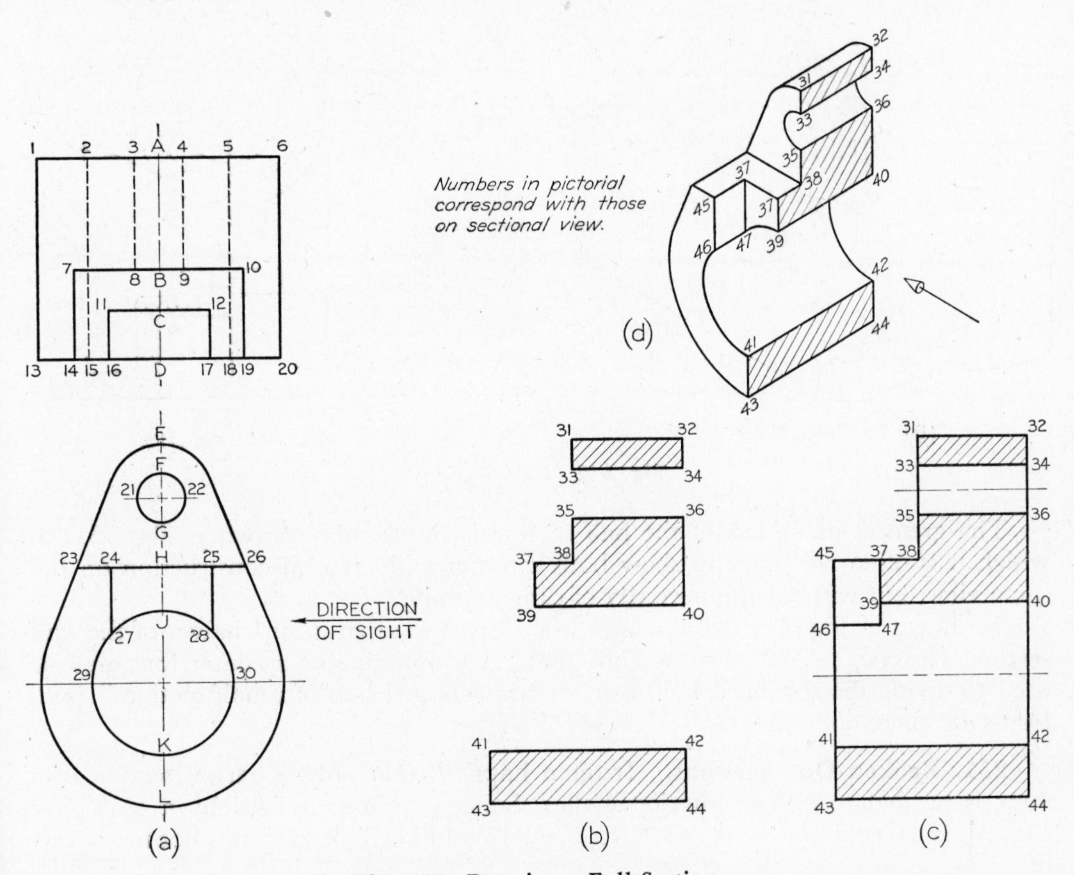

Fig. 411 Drawing a Full Section.

Surface 24-27, appearing as a line in the front view, appears as a line 11-16 in the top view, and as surface 45-37-47-46, true size, in the section at (c).

Cylindrical surface J-29-K, appearing as an arc in the front view, appears in the top view as 2-A-C-11-16-15, and in the section as 46-47-39-40-42-41. Thus, arc 27-29-K (front view) appears in the section, (c), as straight lines 46-41; and arc J-29-K appears as straight line 40-42.

All cut surfaces here are part of the same object; hence the section lines must all run in the same direction, as shown.

250. Half Sections. If the cutting plane passes halfway through the object, the result is a half section, Fig. 412. A half section has the advantage of exposing the interior of one half of the object and retaining the exterior of the other half. Its usefulness is, therefore, largely limited to symmetrical objects. It is not widely used in detail drawings (single parts) because of this limitation of symmetry and also because of difficulties in dimensioning internal shapes, part of which are shown in the sectioned half and part in the unsectioned half.

In general, hidden lines should be omitted from both halves of a half section. However, they may be used in the unsectioned half if necessary for dimensioning.

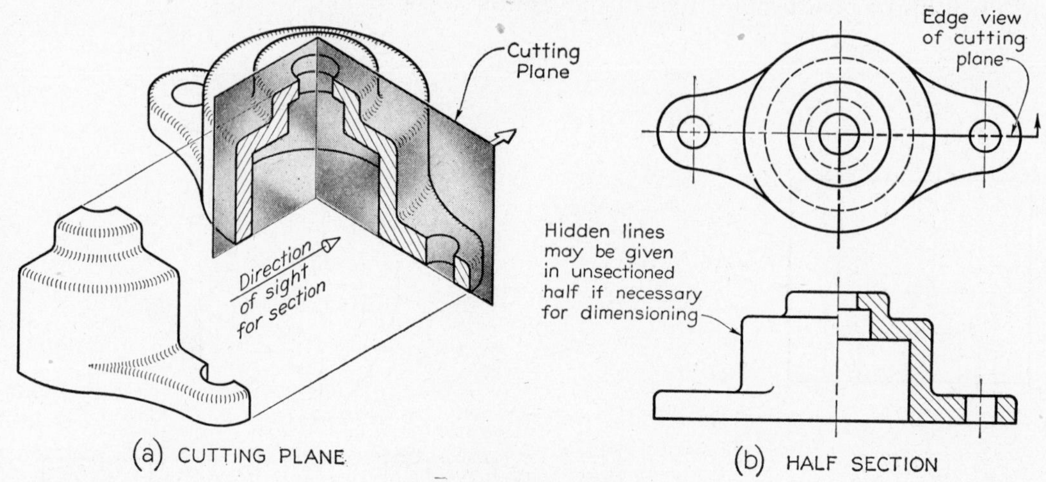

Fig. 412 Half Section.

The greatest usefulness of the half section is in assembly drawing, Fig. 731, in which it is often necessary to show both internal and external construction on the same view, but without the necessity of dimensioning.

As shown in Fig. 412 (b), a center line is used to separate the halves of the half section. However, ASA* recommends that "A visible line or a center line may be used to divide the sectioned half from the unsectioned half of a half section." Both forms are therefore correct.

251. Broken-Out Sections. It often happens that only a partial section of a view is needed to expose interior shapes. Such a section, limited by a *break line*, Fig. 24, is called a *broken-out section*. In Fig. 413 a full or half section is not necessary,

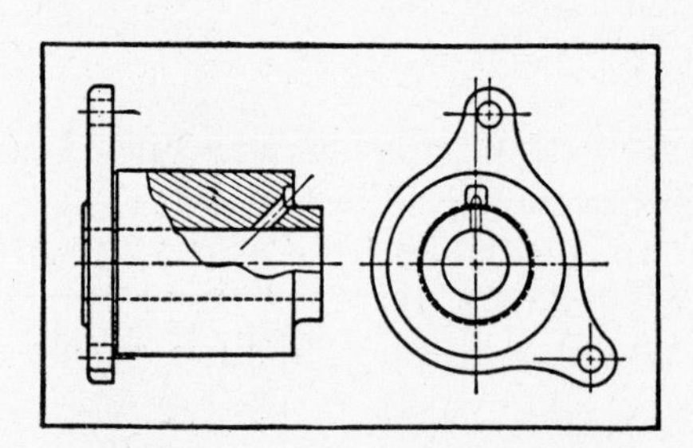

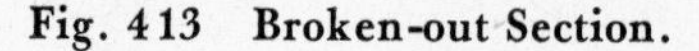

Fig. 413 Broken-out Section.

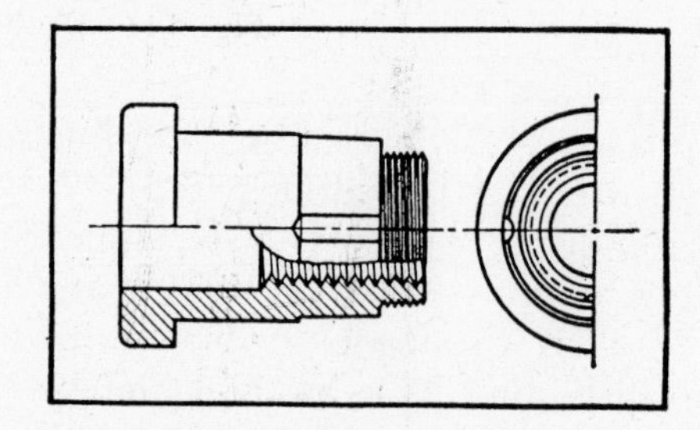

Fig. 414 Break Around Keyway.

a small broken-out section being sufficient to explain the construction. In Fig. 414, a half section would have caused the removal of half the keyway. The keyway is preserved by "breaking-out" around it. Note that in this case the section is limited partly by a break line and partly by a center line.

*ASA Y14.2-1957.

252. Revolved Sections. The shape of the cross section of a bar, arm, spoke, or other elongated object may be shown in the longitudinal view by means of a *revolved section*, Fig. 415. Such sections are made by assuming a plane perpendicular

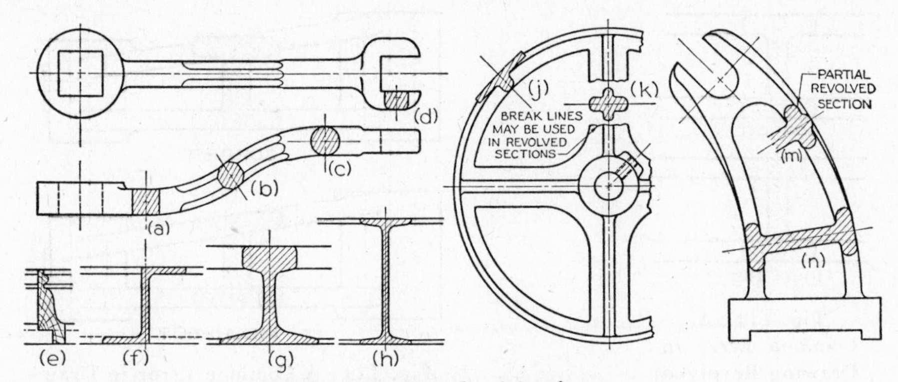

Fig. 415 Revolved Sections.

to the center line or axis of the bar or other object, as shown in Fig. 416 (a), then revolving the plane through 90° about a center line at right angles to the axis, as at (b) and (c).

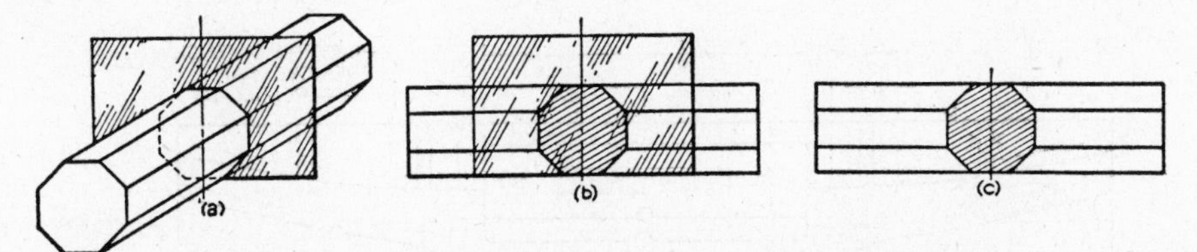

Fig. 416 Use of the Cutting Plane in Revolved Sections.

The visible lines adjacent to a revolved section may be "broken-out" if desired, as shown in Figs. 415 (k) and 417.

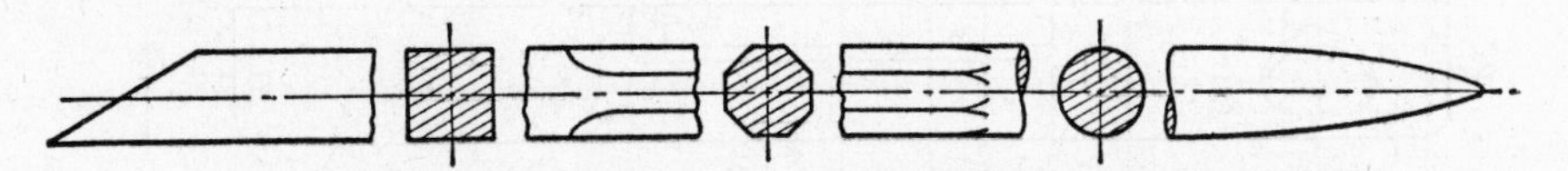

Fig. 417 Conventional Breaks Used with Revolved Sections.

The superimposition of the revolved section requires the removal of all original lines covered by it, Fig. 418. The true shape of a revolved section should be retained after the revolution of the cutting plane, regardless of the direction of the lines in the view, Fig. 419.

253. Removed Sections. "A section may be removed, if necessary, from its normal projected position in the standard arrangement of views, in which case it becomes a *removed section*,"* Fig. 420.

Removed sections should be labeled, such as SECTION A-A, SECTION B-B, etc., corresponding to the letters at the ends of the cutting-plane line. They should be arranged in alphabetical order from left to right on the sheet. Section letters

*ASA Y14.2-1957.

should be used in alphabetical order, but letters I, O, and Q should not be used because they are easily confused with the numeral 1 or the zero.

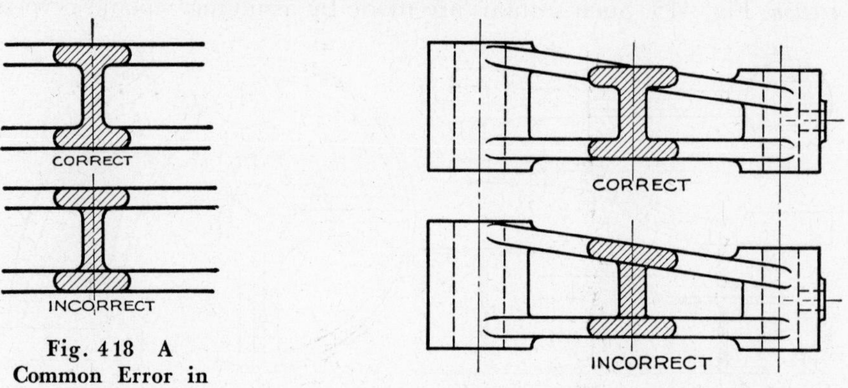

Fig. 418 A Common Error in Drawing Revolved Sections.

Fig. 419 A Common Error in Drawing Revolved Sections.

A removed section is often a partial section. Such a removed section, Fig. 421, is frequently drawn to an enlarged scale, as shown. This is often desirable in order

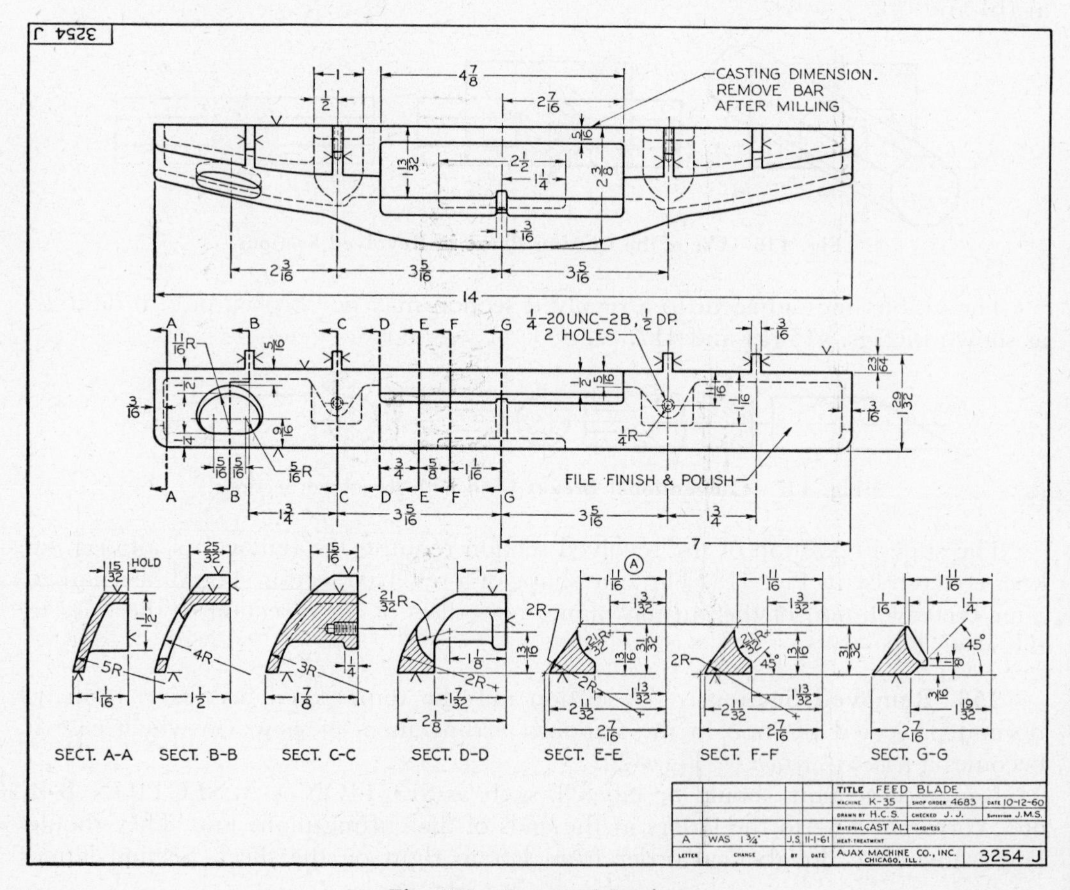

Fig. 420 Removed Sections.

to show clear delineation of some small detail and to provide sufficient space for dimensioning. In such case the enlarged scale should be indicated beneath the section title.

A removed section should be placed so that it no longer lines up in projection with any other view. It should be separated clearly from the standard arrangement of views.

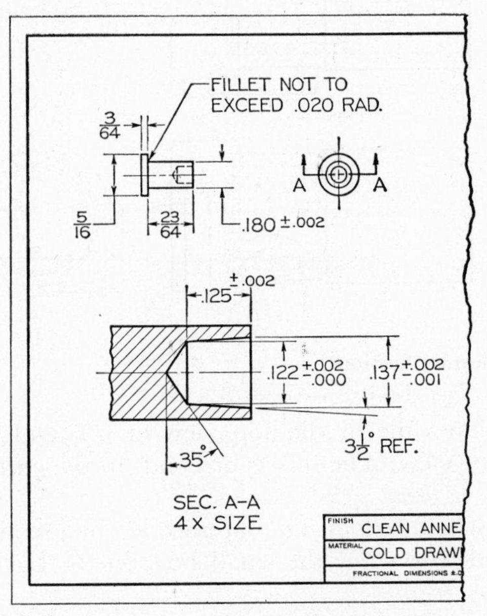

Fig. 421 Removed Section.

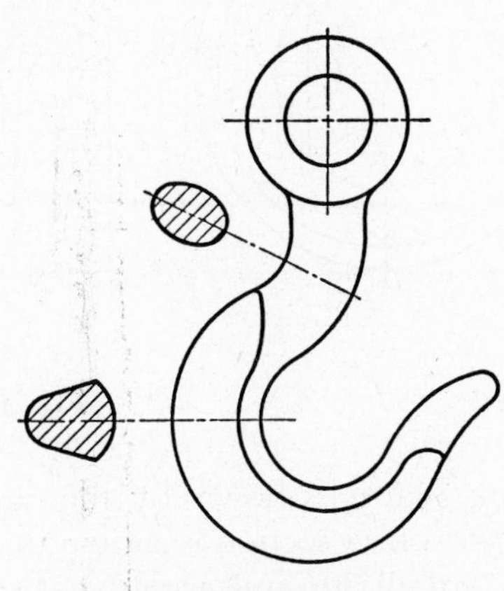

Fig. 422 Removed Sections (ASA Y14.2-1957).

Whenever possible, removed sections should be on the same sheet with the regular views. If a section must be placed on a different sheet, cross-references should be given on the related sheets. A note should be given below the section title, such as:

SECTION B-B ON SHEET 4, ZONE A3

A similar note should be placed on the sheet on which the cutting-plane line is shown, with a leader pointing to the cutting-plane line and referring to the sheet on which the section will be found.

Sometimes it is convenient to place removed sections on center lines extended from the section cuts, Fig. 422.

254. Phantom Sections. A *phantom section* is a section superimposed on an external view so as to show the interior shapes without removing the front portion of the object, Fig. 423. The section lining is similar to that for cast iron except that the lines are broken into short thin dashes so as to appear hidden. An example is shown at (a) in which an additional view is saved by "phantom hatching" the side view. At (b) phantom sectioning is used to show the shape and position of a mating part, in this case a bushing.

255. Offset Sections. In sectioning through irregular objects, it is often desirable to show several features that do not lie in a straight line, by "offsetting" or bending the cutting plane. Such a section is called an *offset section*. In Fig. 424 (a) the cutting plane is offset in several places in order to include the hole at the left end, one of the parallel slots, the rectangular recess, and one of the holes at the right end. The front portion of the object is then imagined to be removed, (b). The path of the

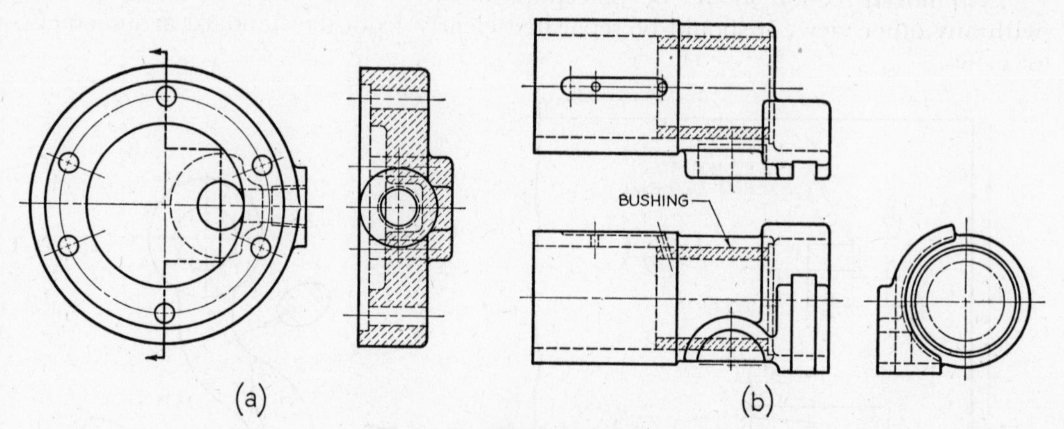

Fig. 423 Phantom Sections.

cutting plane is shown by the cutting-plane line in the top view at (c), and the resulting offset section is shown in the front view. The offsets or bends in the cutting plane are all 90° and are *never shown in the sectional view.*

Fig. 424 also illustrates an example in which hidden lines are needed in a section. In this case, an extra view would be needed to show the small boss on the back if hidden lines were not shown.

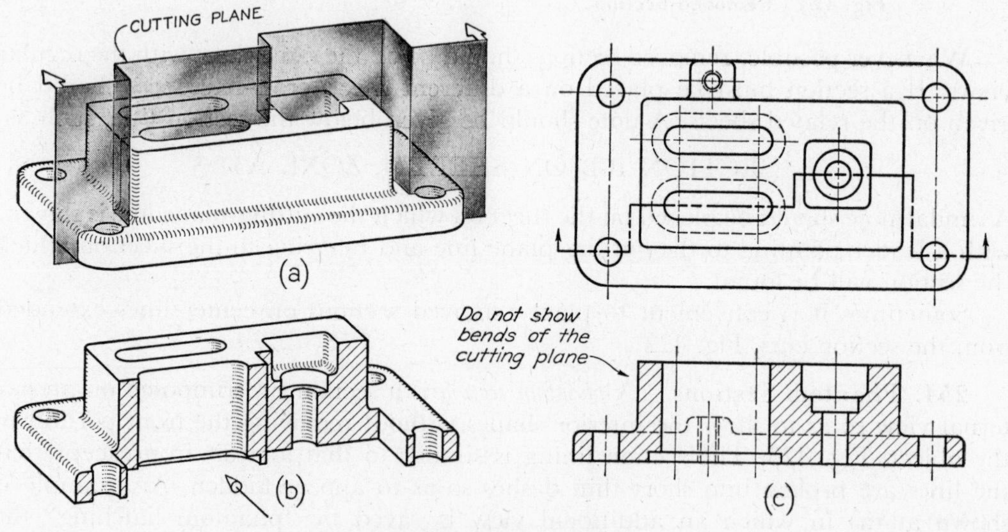

Fig. 424 Offset Section.

An example of multiple offset sections is shown in Fig. 425.

256. Ribs in Section. "When the cutting plane passes flatwise through a web, rib, gear tooth, or similar flat element, in order to avoid a false impression of thickness or solidity the element should not be sectioned."* For example, in Fig. 426, the cutting plane A-A passes flatwise through the vertical web, or rib, and the web is

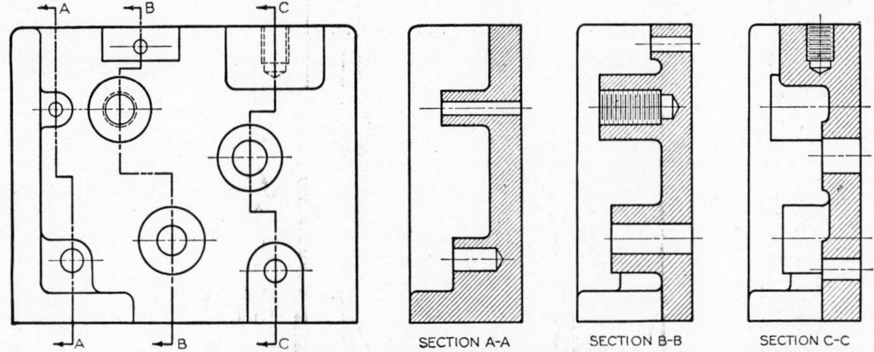

Fig. 425 Three Offset Sections.

not section-lined, (a). *Such thin features should not be section-lined, even though the cutting plane passes through them.* The incorrect section is shown at (b). Note the false impression of thickness or solidity resulting from section-lining the rib.

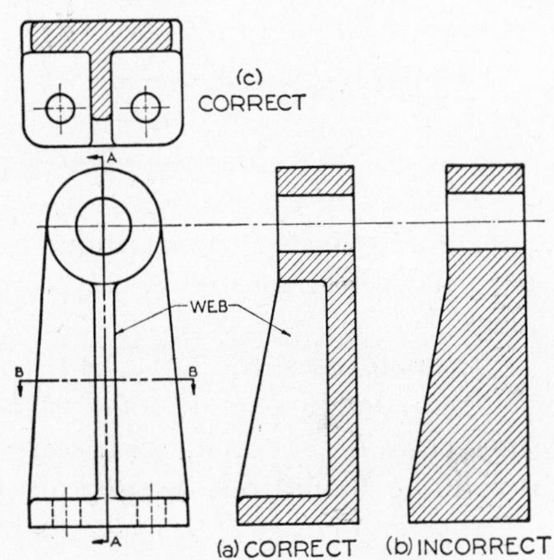

Fig. 426 Webs in Section.

If the cutting plane passes *crosswise* through a rib or any thin member, as does the plane B-B in Fig. 426, the member should be section-lined in the usual manner, as shown in the top view at (c).

In some cases, if a rib is not section-lined when the cutting plane passes through it flatwise, it is difficult to tell whether the rib is actually present, as for example, ribs A in Fig. 427 (a) and (b). It is difficult to distinguish spaces B as open spaces and spaces A as ribs. In such cases, *alternate section lining* of the ribs should be used, (c). This consists simply in continuing alternate section lines through the ribbed areas, as shown.

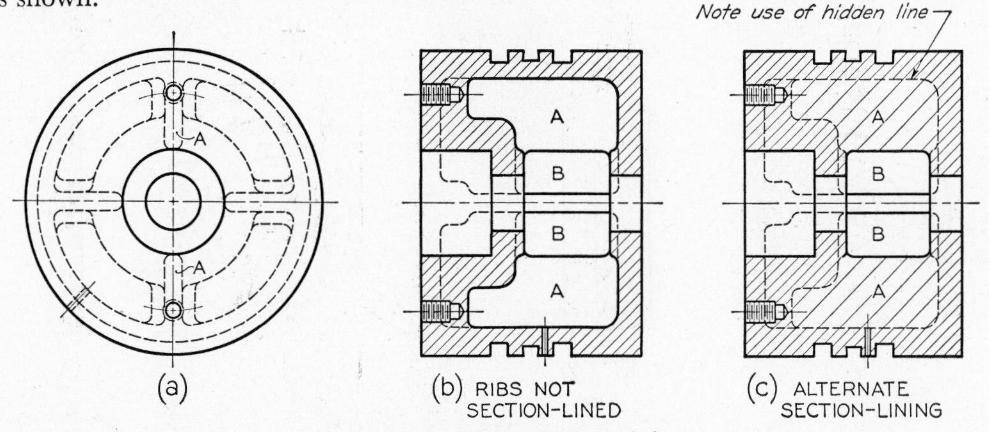

Fig. 427 Alternate Section Lining.

*ASA Y14.2-1957.

257. Aligned Sections. In order to include in a section certain angled elements, the cutting plane may be bent so as to pass through those features. The plane and feature are then imagined to be revolved into the original plane. For example, in Fig. 428, the cutting plane was bent to pass through the angled arm, and then revolved to a vertical position (aligned), from where it was projected across to the sectional view.

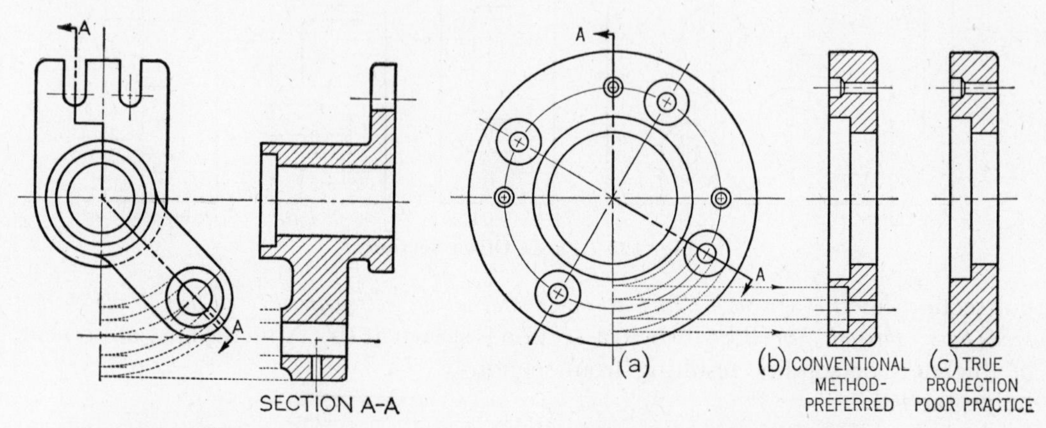

Fig. 428 Aligned Section. **Fig. 429 Aligned Section.**

In Fig. 429 the cutting plane is bent so as to include one of the drilled and counterbored holes in the sectional view. The correct sectional view at (b) gives a clearer and more complete description than does the section at (c), which was taken along the vertical center line of the front view—that is, without any bend in the cutting plane.

In such cases, the angle of revolution should always be less than 90°.

The student is cautioned *not to revolve* features when clearness is not gained. In some cases the revolving of features will result in a loss of clarity. Examples in which revolution should not be used are Fig. 439, Probs. 17 and 18.

In Fig. 430 (a) is an example in which the projecting lugs were not sectioned on the same basis that ribs are not sectioned. At (b) the projecting lugs are located so that the cutting plane would pass through them crosswise; hence they are sectioned.

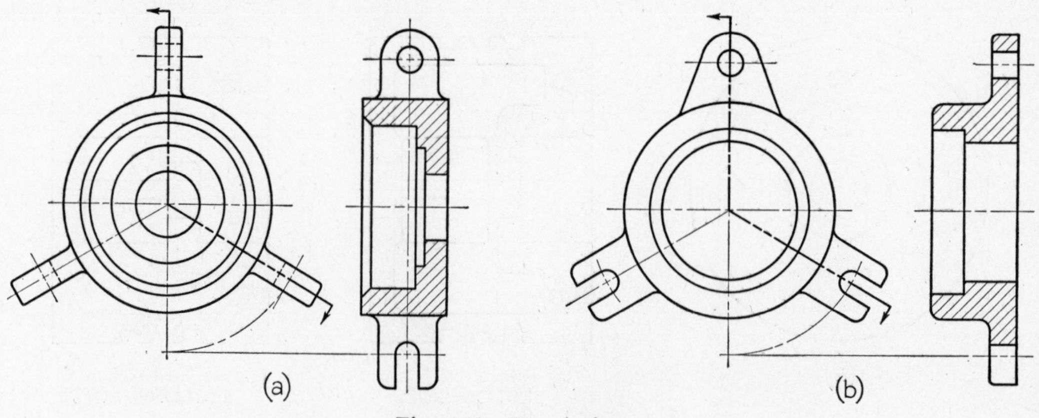

Fig. 430 Revolutions.

Another example involving rib-sectioning and also revolution is shown in Fig. 431. In the circular view, the cutting plane is offset in circular-arc bends to include the upper hole, an upper rib, the keyway and center hole, the lower rib, and one of the lower holes. These features are then imagined to be revolved until they line up

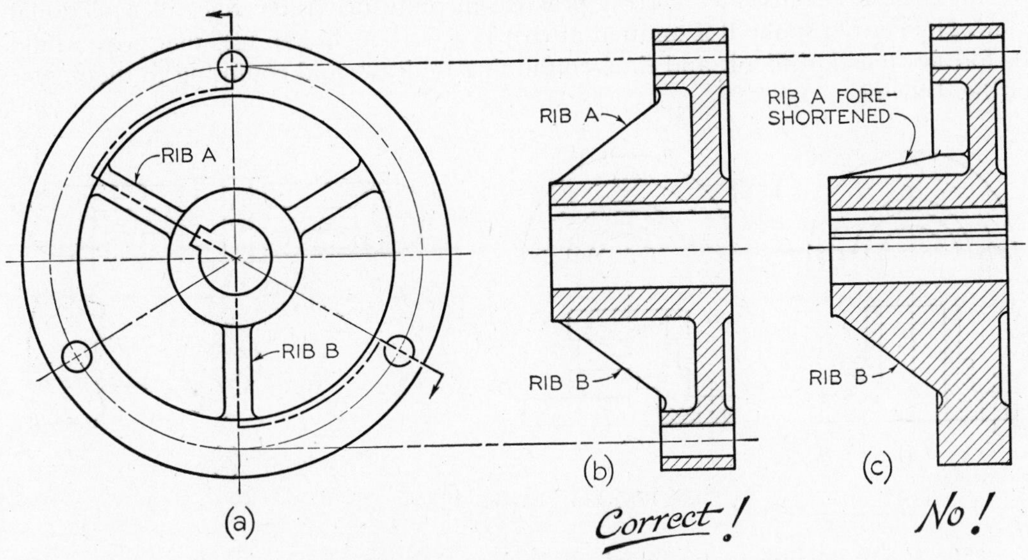

Fig. 431 Symmetry of Ribs.

vertically, and then projected from that position to obtain the section at (b). Note that the ribs are not sectioned. If a regular full section of the object were drawn, without the use of conventions discussed here, the resulting section, (c), would be both incomplete and confusing, and in addition would take more time to draw.

In sectioning a pulley or any spoked wheel, it is standard practice to revolve the spokes if necessary (if there is an odd number), and not to section-line the spokes,

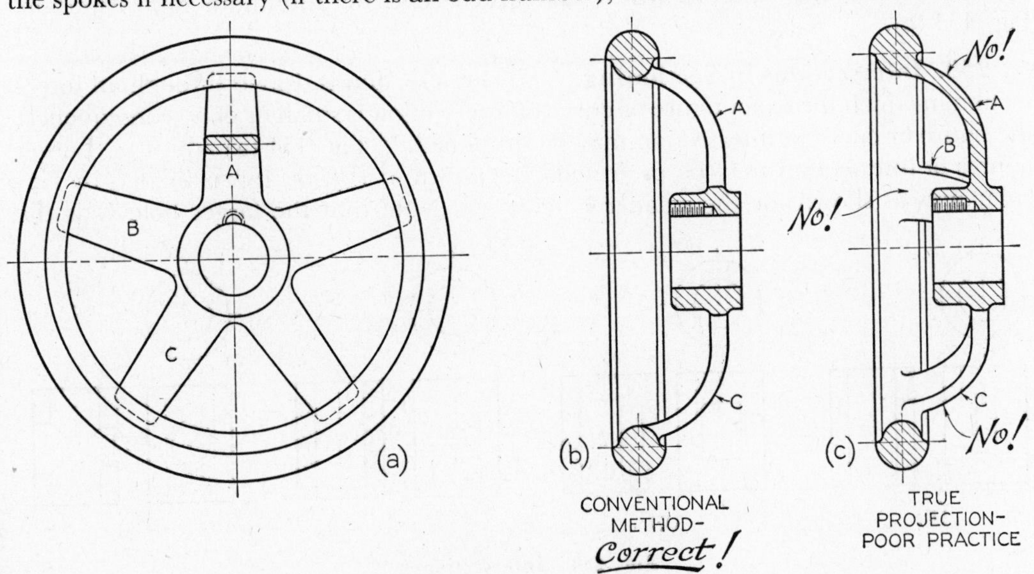

Fig. 432 Spokes in Section.

Fig. 432 (b). If the spoke is sectioned, as shown at (c), the section gives a false impression of continuous metal. If the lower spoke is not revolved, it will be foreshortened in the sectional view in which it presents an "amputated" and wholly misleading appearance.

Fig. 432 also illustrates correct practice in omitting visible lines in a sectional view. Notice that spoke B is omitted at (b). If it were included, (c), the spoke would be foreshortened, difficult and time-consuming to draw, and confusing to the reader of the drawing.

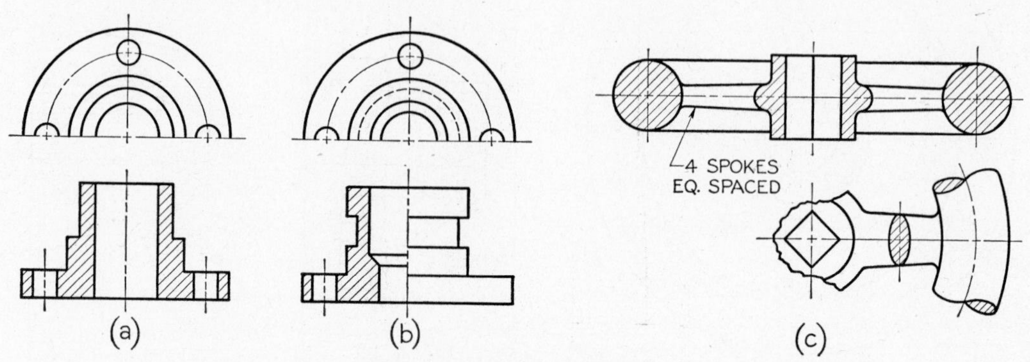

Fig. 433 Partial Views.

258. Partial Views. If space is limited on the paper or if it is necessary to save drafting time, *partial views* may be used in connection with sectioning, Fig. 433. *Half-views* are shown at (a) and (b) in connection with a full section and a half section, respectively. Note that in each case the back half of the object in the circular view is shown, in conformity with the idea of removing the front portion of the object in order to expose the back portion for viewing in section. See also §215.

Another method of drawing a partial view is to break out much of the circular view, retaining only those features that are needed for minimum representation, Fig. 433 (c).

259. Intersections in Sectioning. "When a section is drawn through an intersection in which the exact figure or curve of intersection is small or of no consequence, the figure or curve of intersection may be simplified,"* Fig. 434 (a) and (c). "Larger figures of intersection may be projected," as shown at (b), or approximated by circular arcs as shown for the smaller hole at (d). Note that the larger hole K is the

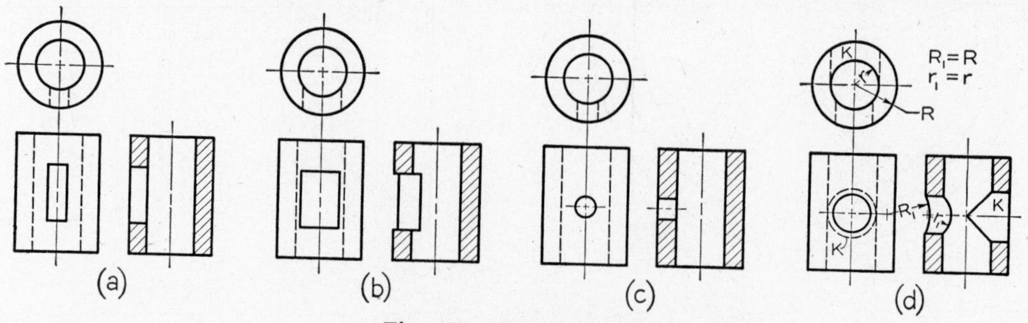

Fig. 434 Intersections.

*ASA Y14.2-1957.

same diameter as the vertical hole. In such cases the curves of intersection (ellipses) appear as straight lines, as shown. See also Figs. 321 and 322.

260. Conventional Breaks. Two views of a garden rake are shown in Fig. 435 (a), drawn to a small scale in order to get it on the paper. At (b) the handle was "broken," a long central portion removed, and the rake then drawn to a larger scale, producing a much clearer delineation.

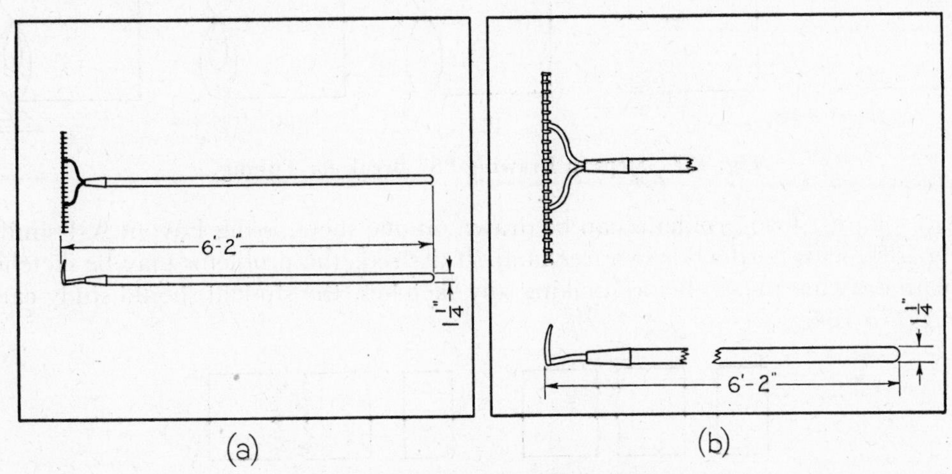

Fig. 435 Use of Conventional Breaks.

Parts thus broken must have the same section throughout, or if tapered, they must have a uniform taper. Note at (b) the full-length dimension is given, just as if the entire rake were shown.

The breaks used on cylindrical metal shafts or tubes are often referred to as "S-breaks" and in the industrial drafting room are usually drawn entirely freehand or partly freehand and partly with the irregular curve or the compass. By these methods, the result is often very crude, especially when attempted by beginners. Simple methods of construction for use by the student or the industrial draftsman are shown in Figs. 436 and 437, and will always produce a professional result.

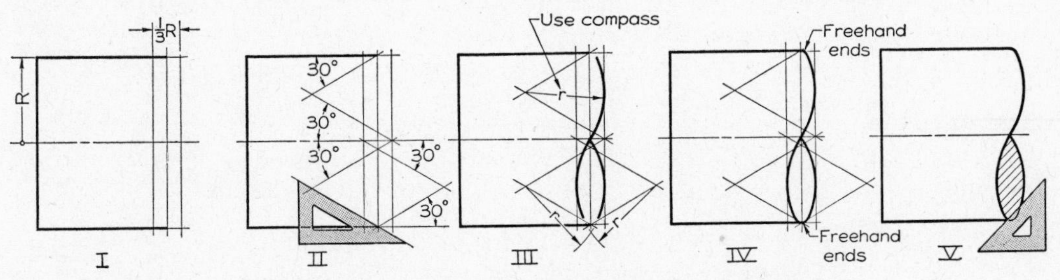

Fig. 436 Steps in Drawing "S" Break for Solid Shaft.

Breaks for rectangular metal and wood sections are always drawn freehand, as shown in Fig. 438. See also Fig. 417, which illustrates the use of breaks in connection with revolved sections.

261. Sectioning Problems. Any of the following problems may be drawn freehand or with instruments, as assigned by the instructor. However, the problems in Fig. 439 are especially suitable for sketching on $8\frac{1}{2}'' \times 11''$ cross-section paper with

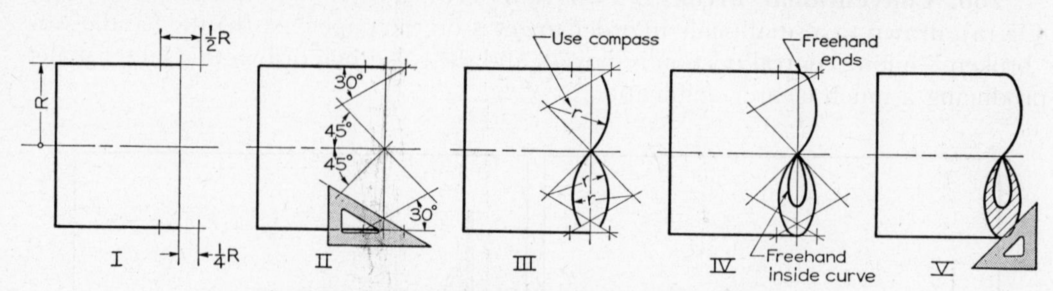

Fig. 437 Steps in Drawing "S" Break for Tubing.

$\frac{1}{4}''$ grid squares. Two problems can be drawn on one sheet, using Layout A-1 similar to Fig. 281, with borders drawn freehand. If desired, the problems may be sketched on plain drawing paper. Before making any sketches, the student should study carefully §§175-184.

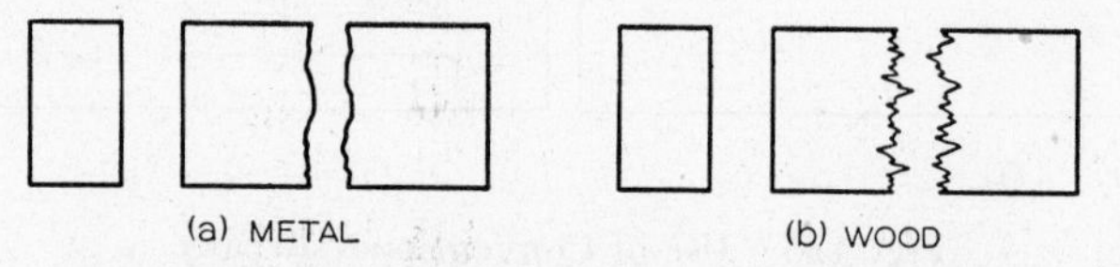

Fig. 438 Rectangular Metal and Wood Breaks.

The problems in Figs. 440-459 are intended to be drawn mechanically, but may be drawn freehand, if desired. If dimensions are required, the student should first study §§327-350. If an ink tracing is required, the student is referred to §§59-62 and 64.

Sectioning problems in convenient form for solution are available in *Technical Drawing Problems*, by Giesecke, Mitchell, and Spencer, and in *Technical Drawing Problems, Series 2*, by Spencer and Grant, both designed to accompany this text, and published by The Macmillan Company.

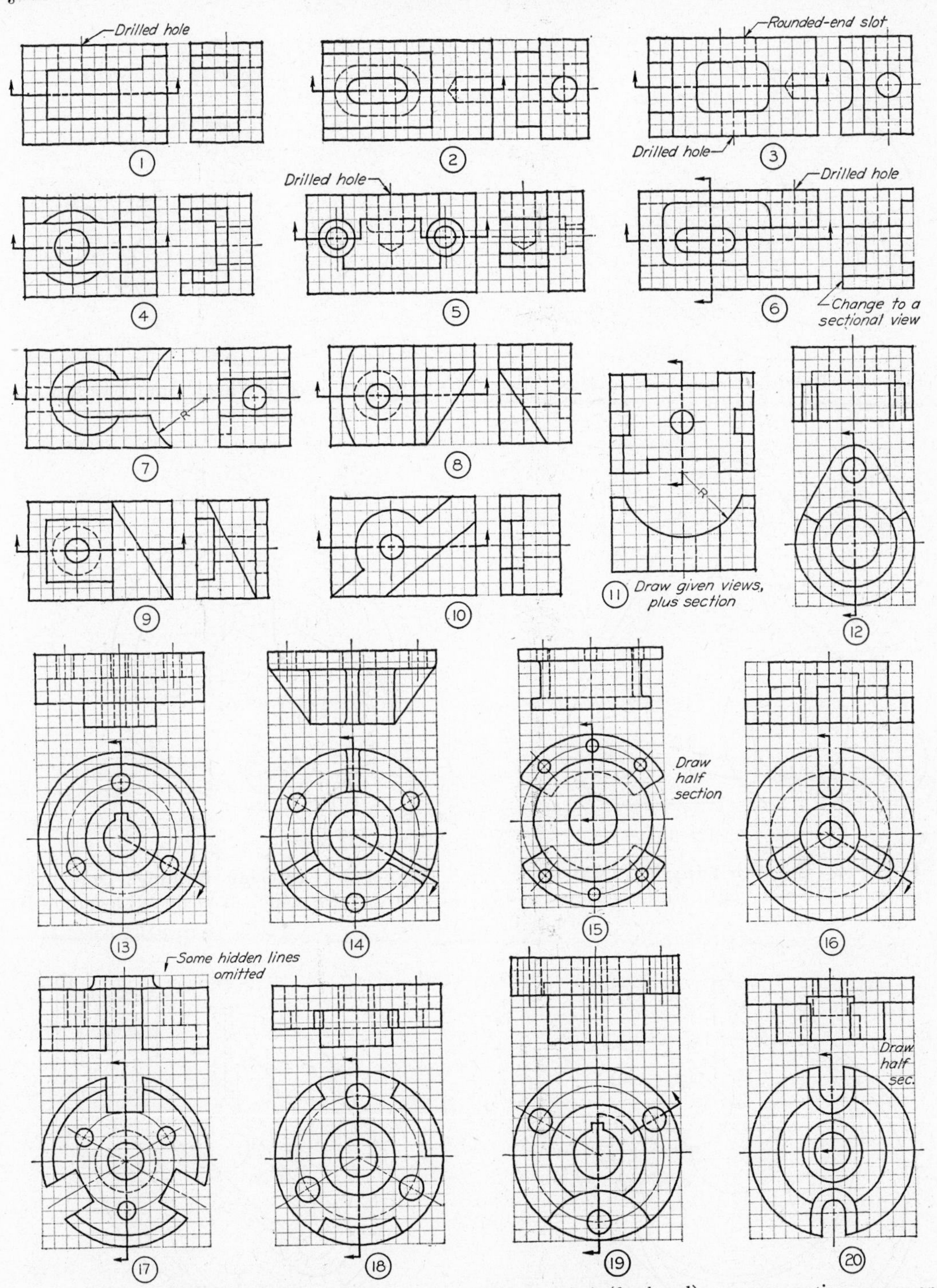

Fig. 439 Freehand Sketching Problems. Using Layout A-1 (freehand) on cross-section paper or plain paper, two problems per sheet as in Fig. 281, sketch views with sections as indicated. Each grid square = $\frac{1}{4}''$. In Probs. 1-10, move R. Side views to line up horizontally with Front Sectional views. In Probs. 12-20, draw given Front views plus Sectional views, omitting given Top views. Omit cutting planes except in Probs. 5 and 6.

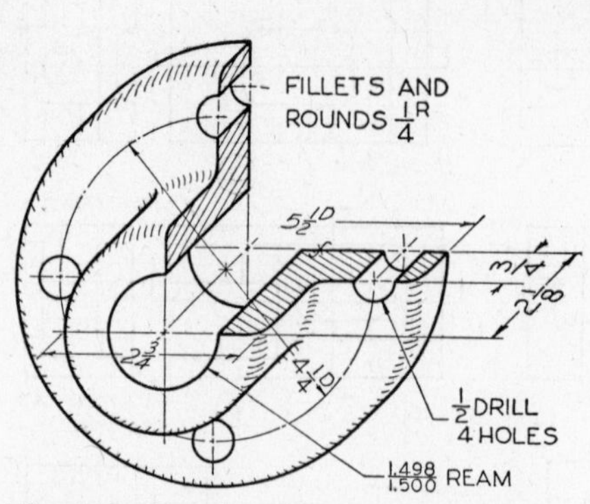

Fig. 440 Bearing. Draw necessary views, with full section (Layout A-3).

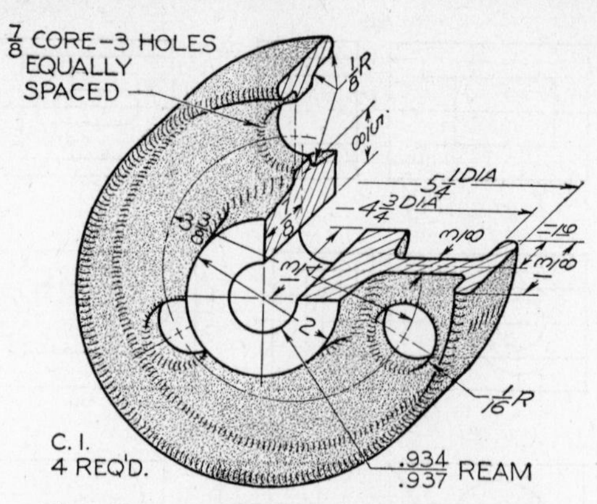

Fig. 441 Truck Wheel. Draw necessary views, with half section (Layout A-3).

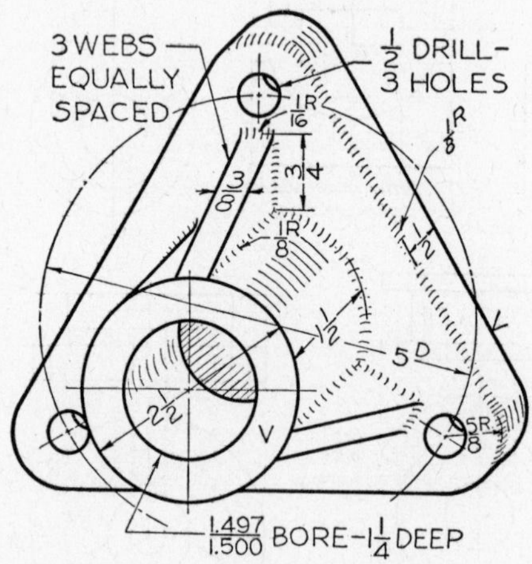

Fig. 442 Column Support. Draw necessary views, with full section (Layout A-3).

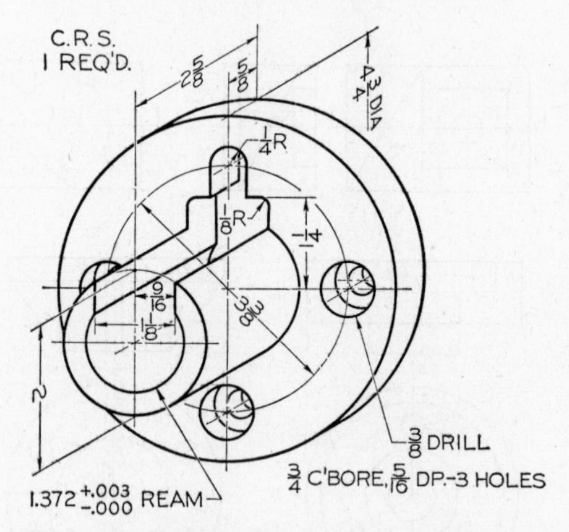

Fig. 443 Centering Bushing. Draw necessary views, with full section (Layout A-3).

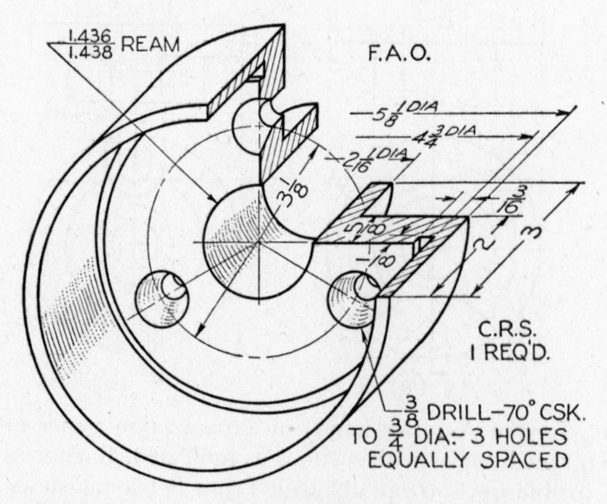

Fig. 444 Special Bearing. Draw necessary views, with full section (Layout A-3).

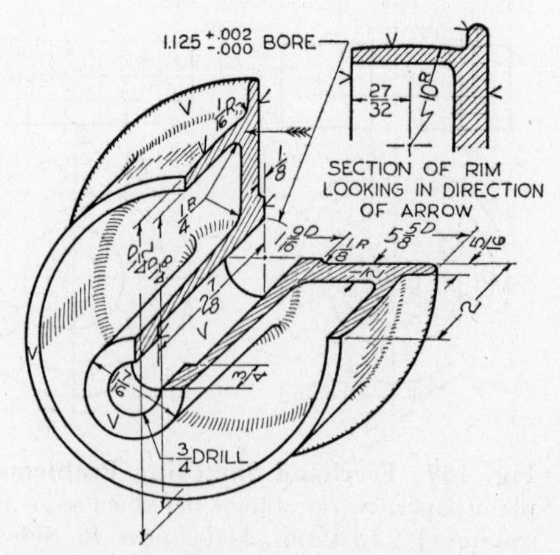

Fig. 445 Idler Pulley. Draw necessary views, with full section (Layout A-3).

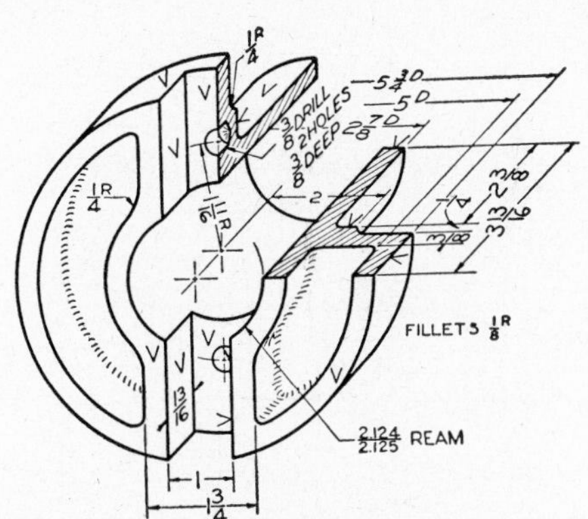

Fig. 446 Stock Guide. Draw necessary views, with half section (Layout B-4).

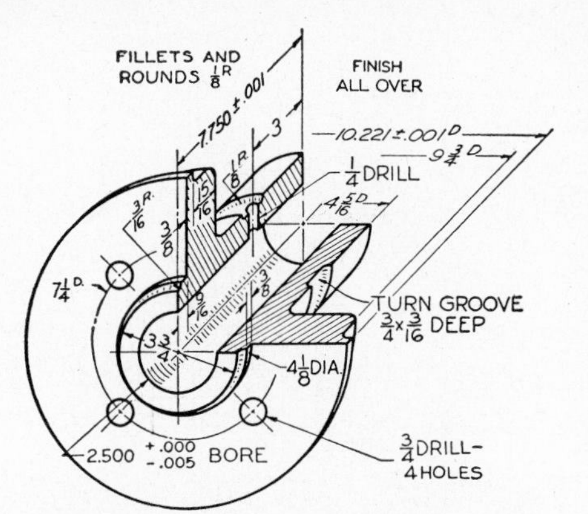

Fig. 447 Bearing. Draw necessary views, with half section. Scale: half size (Layout B-4).

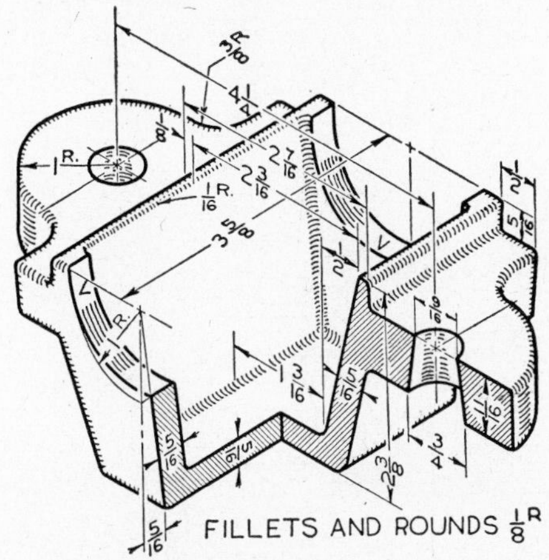

Fig. 448 Bearing Housing. Draw necessary views, with half sections in front and Side views (Layout B-4).

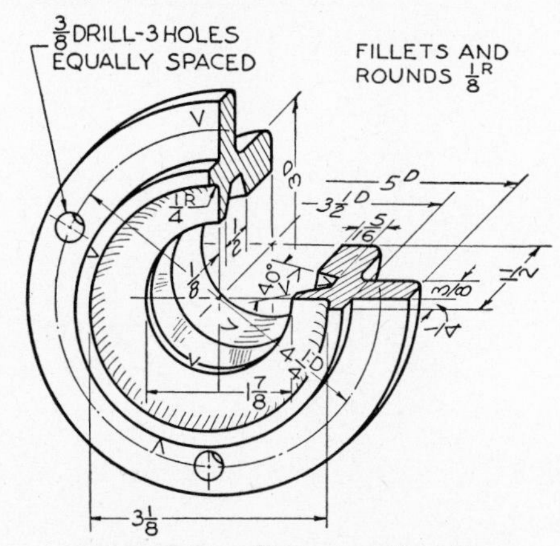

Fig. 449 Fixed Bearing Cup. Draw necessary views, with full section (Layout B-4).

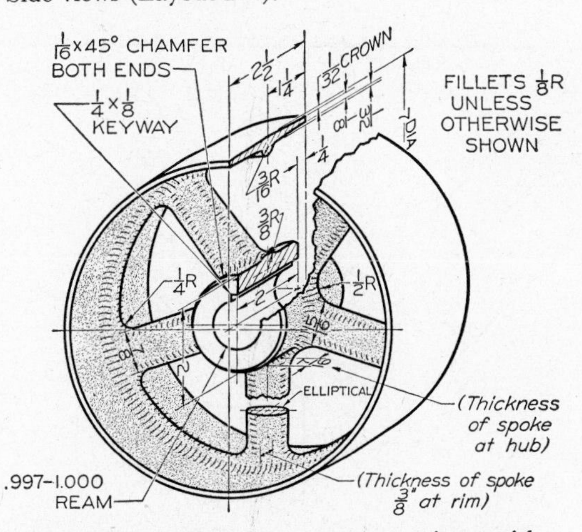

Fig. 450 Pulley. Draw necessary views, with full section, and revolved section of spoke (Layout B-4).

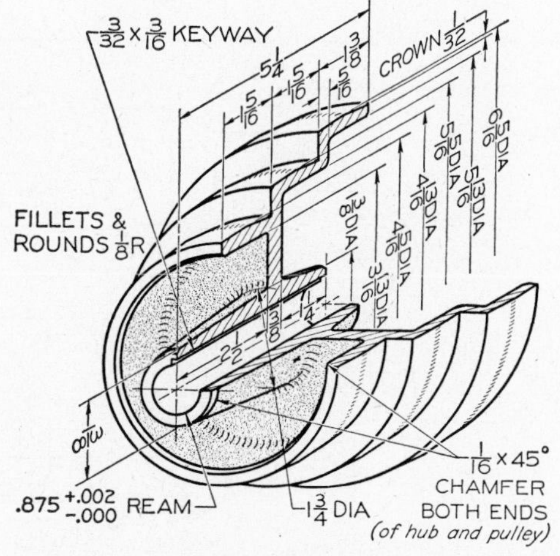

Fig. 451 Step-Cone Pulley. Draw necessary views, with full section (Layout B-4).

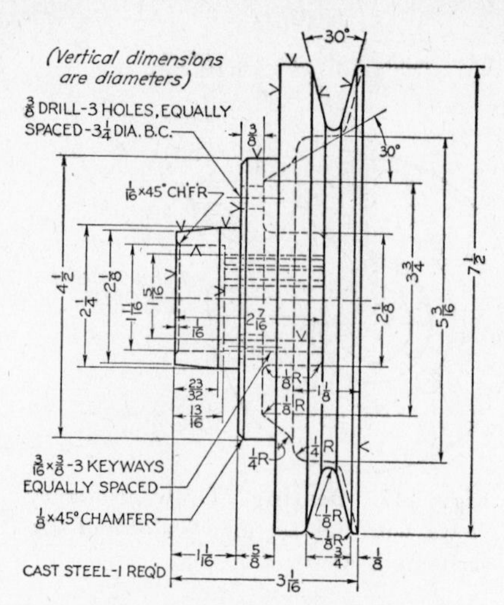

Fig. 452 Sheave. Draw two views, including half section (Layout B-4).

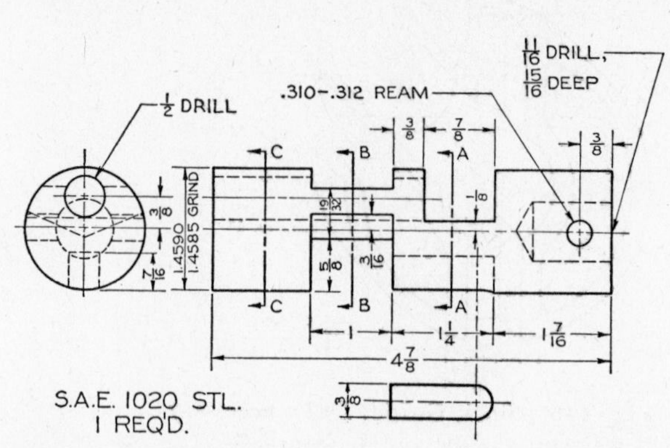

Fig. 453 Operating Valve. Draw present Front view and add R. Side view, Bottom view, and indicated removed sections. Omit present L. Side view and Partial Bottom view (Layout B-4).

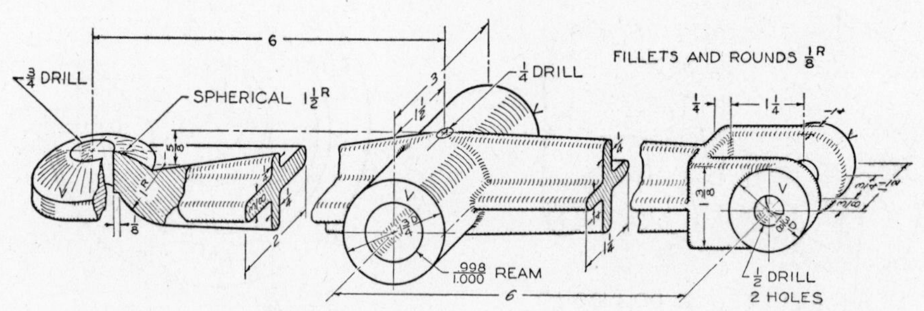

Fig. 454 Rocker Arm. Draw necessary views, with revolved sections (Layout B-4).

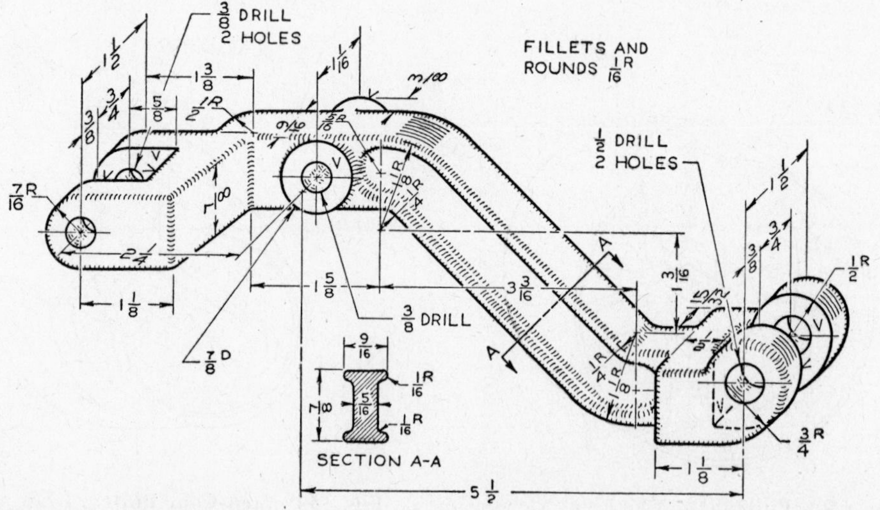

Fig. 455 Dash Pot Lifter. Draw necessary views, using revolved section instead of removed section (Layout B-4).

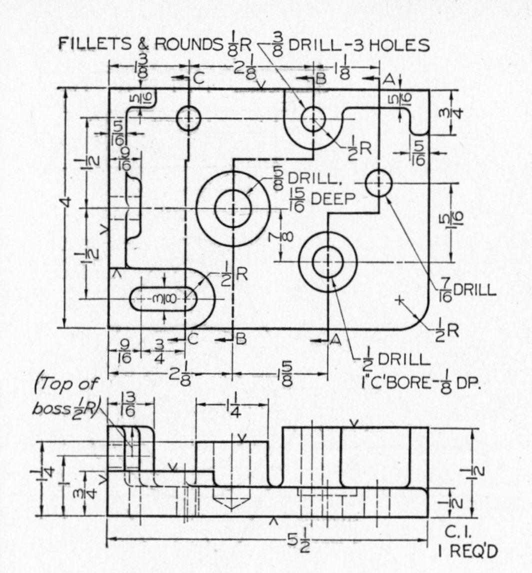

Fig. 456 Adjuster Base.
Given: Front and Top views.
Req'd: Front, Top, and Sec's A-A, B-B, and C-C. Show all visible lines (Layout B-4).

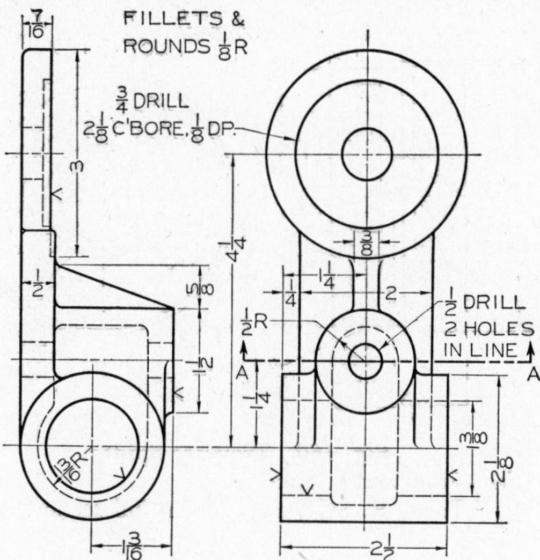

Fig. 457 Mobile Housing.
Given: Front and L. Side views.
Req'd: Front, R. Side in full sec., and removed Sec. A-A. (Layout B-4).

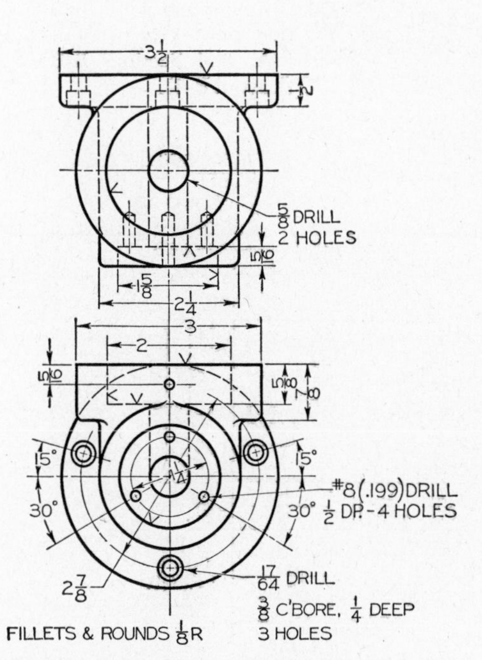

Fig. 458 Hydraulic Fitting. Draw given Front and Top views, plus R. Side view in full section (Layout B-4).

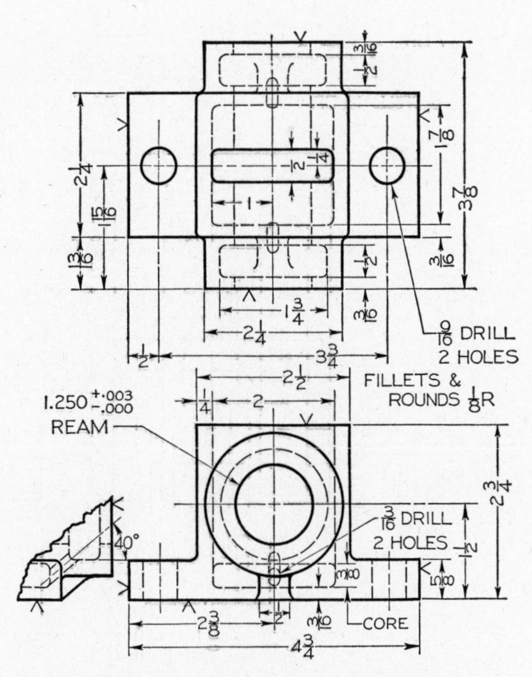

Fig. 459 Aux. Shaft Bearing. Draw given Front and Top views, plus R. Side view in full section (Layout B-4).

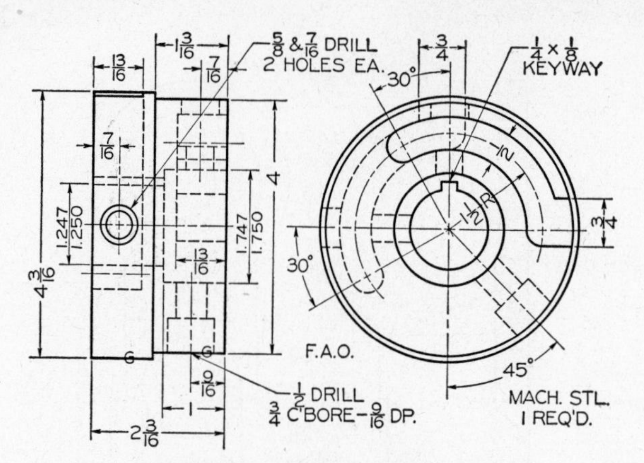

Fig. 460 Traverse Spider.
Given: Front and L. Side views.
Req'd: Front, R. Side, and Top in full section (Layout B-4).

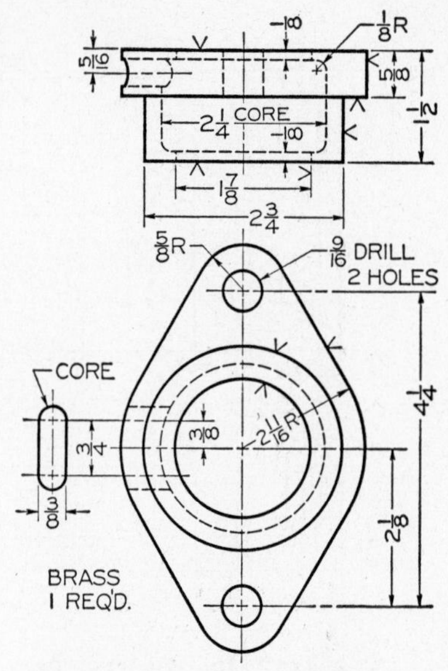

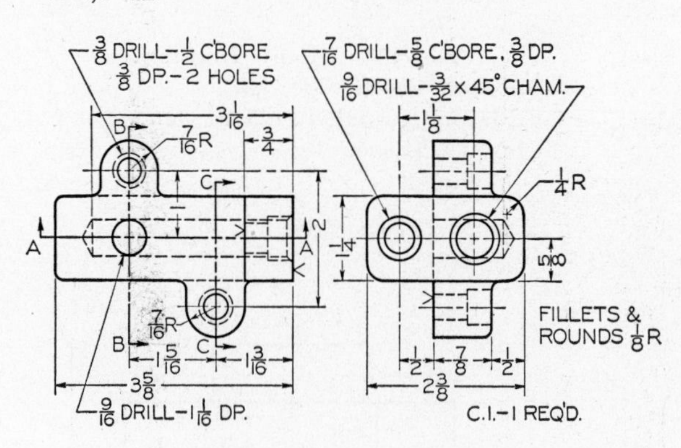

(ABOVE) **Fig. 461 Gland.**
Given: Front, Top, and Partial L. Side views.
Req'd: Front and R. Side in full section (Layout A-3).

(LEFT) **Fig. 462 Bracket.**
Given: Front and R. Side views.
Req'd: Take Front as new Top; then add R. Side view, Front in full Section A-A, and Sections B-B and C-C (Layout B-4).

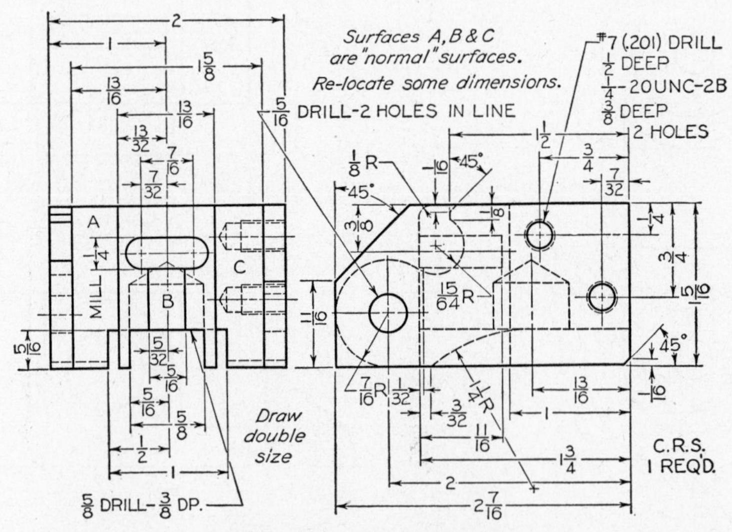

Fig. 463 Cocking Block.
Given: Front and R. Side views. *Req'd:* Take Front as new Top view; then add new Front view, L. Side view, and R. Side view in full section. Draw double size on Layout C-4.

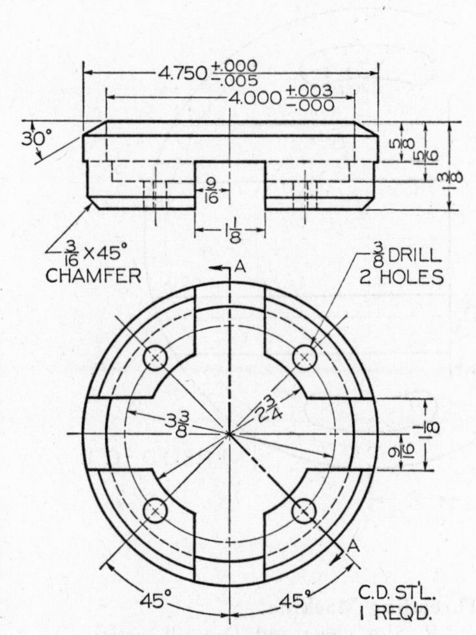

Fig. 464 Packing Ring.

Given: Front and Top views.
Req'd: Front view and Sec. A-A (Layout A-3).

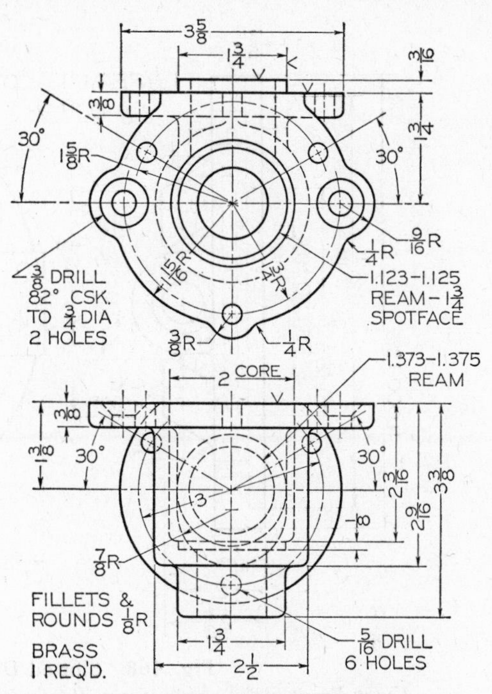

Fig. 465 Strainer Body.

Given: Front and Bottom views.
Req'd: Front, Top, and R. Side in full section (Layout C-4).

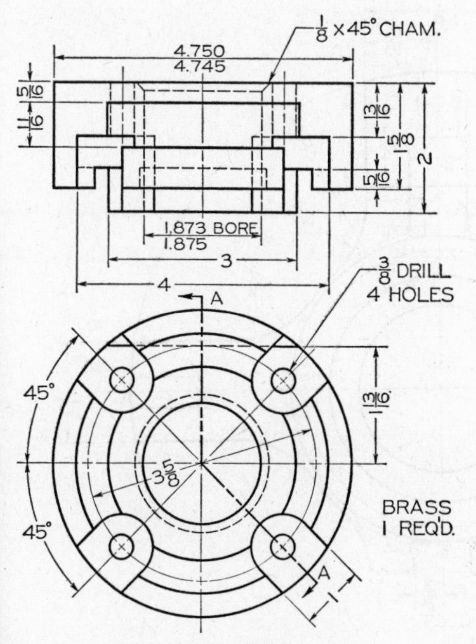

Fig. 466 Oil Retainer.

Given: Front and Top views.
Req'd: Front view and Sec. A-A (Layout B-4).

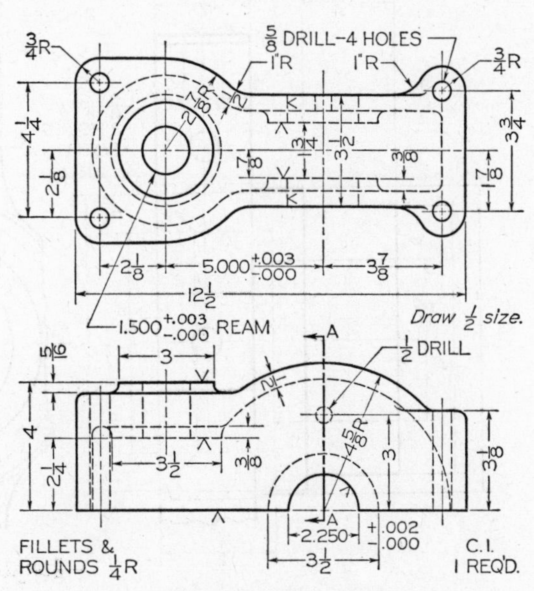

Fig. 467 Gear Box.

Given: Front and Top views.
Req'd: Front in full section, Bottom view, and R. Side Sec. A-A. Half size (Layout B-4).

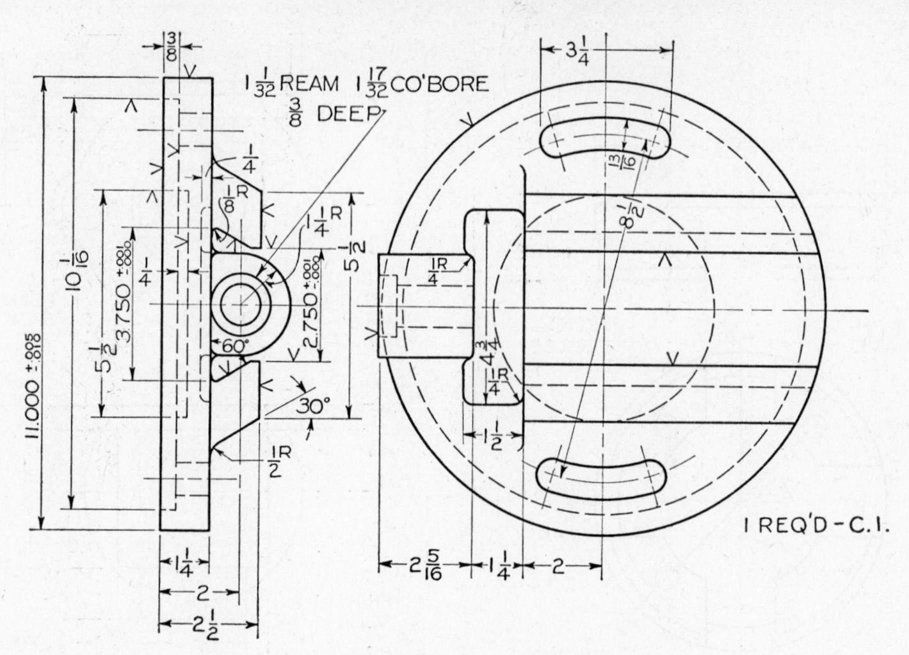

Fig. 468 Slotted Disc for Threading Machine.
Given: Front and L. Side views. *Required:* Front and R. Side views and Top full-section view. Half size (Layout B-4).

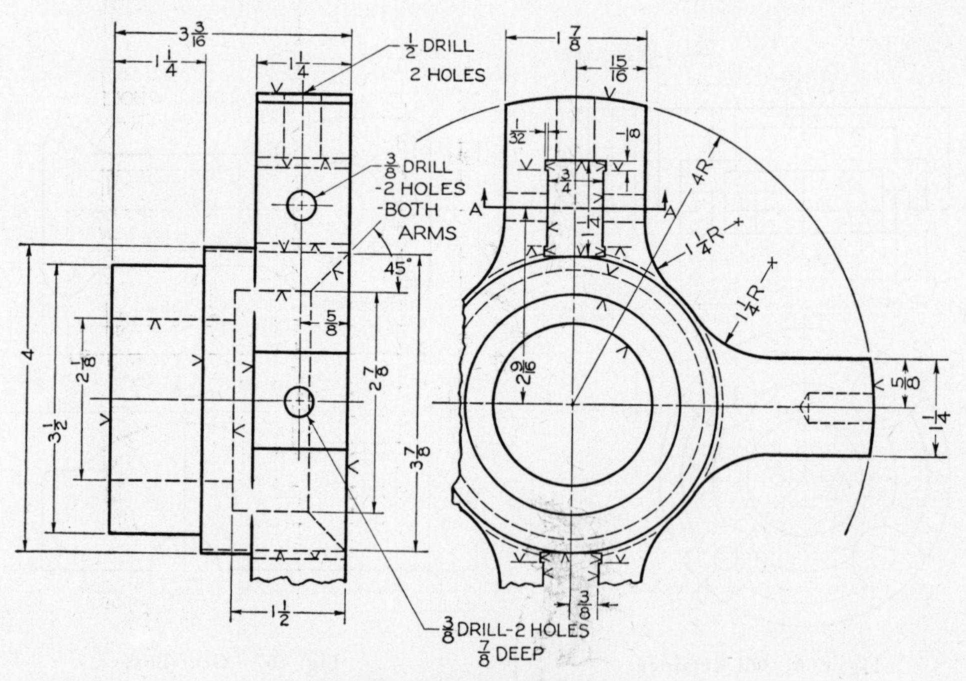

Fig. 469 Web for Lathe Clutch.
Given: Partial Front and L. Side views. *Required:* Complete Front view, R. Side view in full section, and removed Section A-A (Layout C-4).

CHAPTER 8

AUXILIARY VIEWS

262. Auxiliary Views. Many objects are of such shape that their principal faces cannot always be assumed parallel to the regular planes of projection. For example, in Fig. 470 (a), the base of the Bearing is shown in its true size and shape, but the rounded upper portion is situated at an angle with the planes of projection and does not appear in its true size and shape in any of the three regular views.

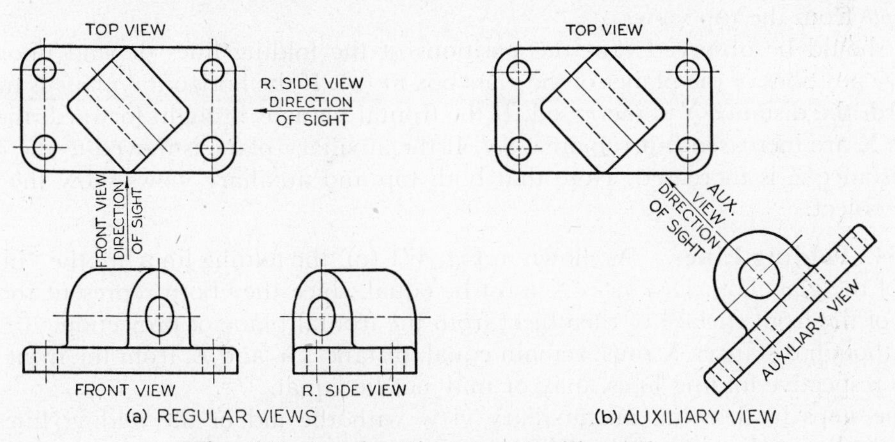

Fig. 470 Regular Views and Auxiliary Views.

In order to show the true circular shapes, it is necessary to assume a direction of sight that is perpendicular to the planes of those curves, as shown at (b). The resulting view is called an *auxiliary view*. This view, together with the top view, completely describes the object, and the front and right-side views are not necessary.

263. The Auxiliary Plane. A view projected on any plane other than one of the six regular planes of projection, §208, is an auxiliary view. For example, in Fig. 471 (a), the object has an inclined surface that does not appear in its true size and shape in any regular view. The auxiliary plane is assumed parallel to the inclined surface P, that is, perpendicular to the line of sight, which is at right angles to that surface. The auxiliary plane is then perpendicular to the frontal plane of projection, and hinged to it.

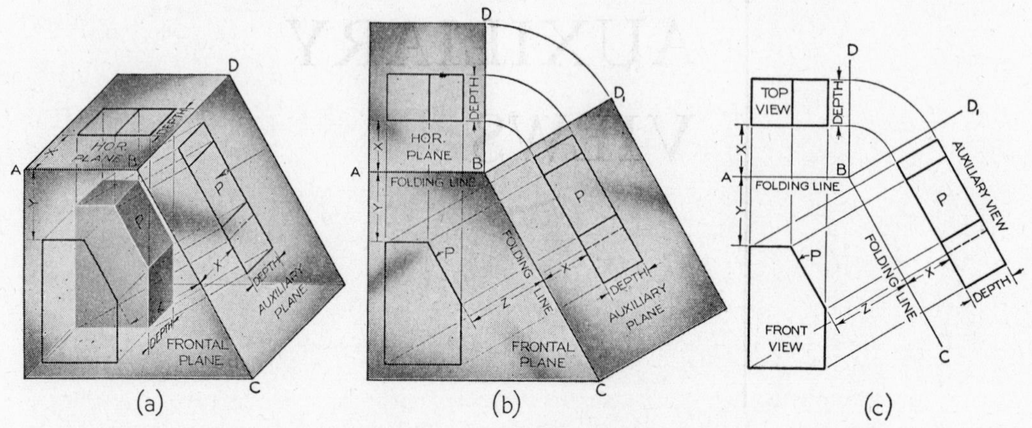

Fig. 471 An Auxiliary View.

When the horizontal and auxiliary planes are unfolded to lie in the plane of the front view, as shown at (b), the *folding lines* represent the "hinge" lines joining the planes. The drawing is simplified by omitting the planes, as shown at (c), and as will be shown later, the folding lines may themselves be omitted in the actual drawing. The inclined surface P is shown in its true size and shape in the auxiliary view, the long dimension of the surface being projected directly from the front view, and the *depth* from the top view.

It should be observed that the positions of the folding lines depend upon the relative positions of the planes of the glass box at (a). If the horizontal plane is moved upward, the distance Y is increased. If the frontal plane is brought forward, the distances X are increased but remain *equal*. If the auxiliary plane is moved to the right, the distance Z is increased. Note that both top and auxiliary views show the *depth* of the object.

264. Folding Lines. As shown in Fig. 471 (c), the folding lines are the "hinge" lines of the glass box. Distances X must be equal, since they both represent the distance of the front surface of the object from the frontal plane of projection.

Although distances X must remain equal, distances Y and Z, from the front view to the respective folding lines, may or may not be equal.

The steps in drawing an auxiliary view with the aid of the folding lines are shown in Fig. 472, and are described as follows:

I. The front and top views are given. It is required to draw an auxiliary view showing the true size and shape of inclined surface P. Draw the folding line H/F between the views at right angles to the projection lines. Distances X and Y may or may not be equal, as desired.

II. Draw arrow, indicating direction of sight, perpendicular to surface P. Draw light projection lines from the front view parallel to the arrow, or perpendicular to surface P.

III. Draw folding line F/1 for the auxiliary view at right angles to the projection lines and at any convenient distance from the front view.

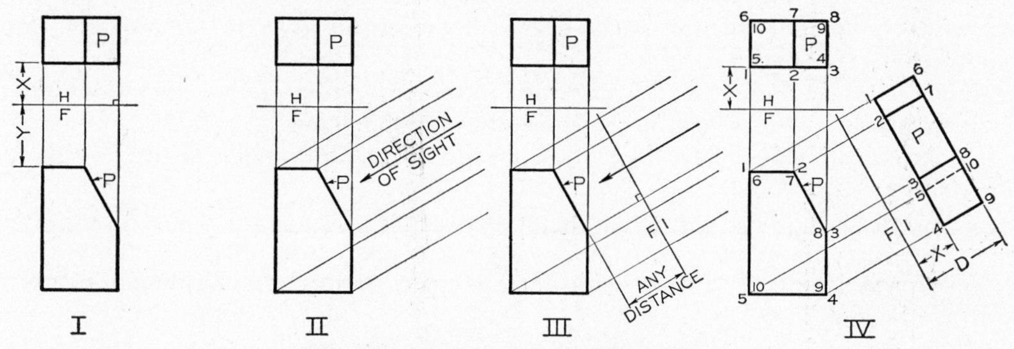

Fig. 472 To Draw an Auxiliary View.

IV. Draw the auxiliary view, using the numbering system explained in §212. Locate all points the same distances from folding line F/1 as they are from folding line H/F in the top view. For example, points 1 to 5 are distance X from the folding lines in both the top and auxiliary views, and points 6 to 10 are distance D from the corresponding folding lines. Since the object is viewed in the direction of the arrow, it will be seen that edge 5-10 will be hidden in the auxiliary view.

265. Reference Planes. In Fig. 471 (c) and 472 the folding lines are "edge views" of the frontal plane of projection. In effect, the frontal plane is used as a *reference plane*, or *datum plane*, for transferring distances (*depth measurements*) from the top view to the auxiliary view.

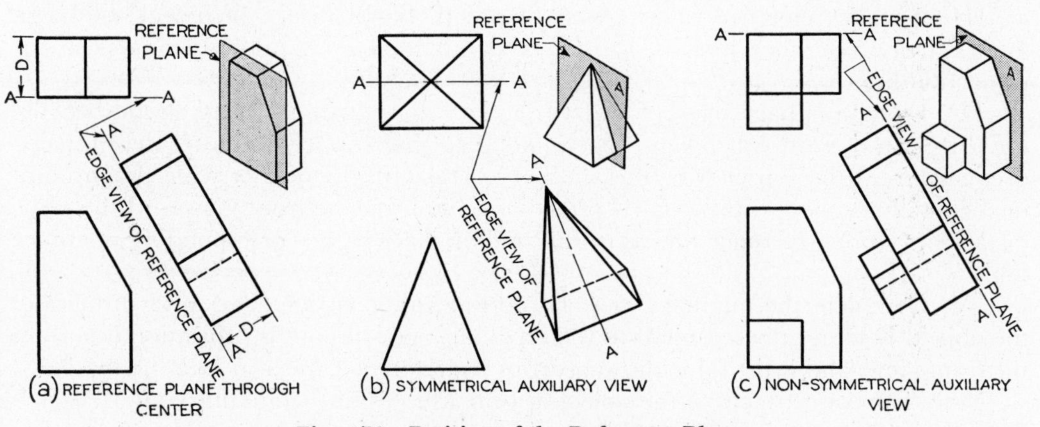

Fig. 473 Position of the Reference Plane.

Instead of using one of the planes of projection as a reference plane, it is often more convenient to assume a reference plane inside the glass box parallel to the plane of projection and touching or cutting through the object. For example, Fig. 473 (a), a reference plane is assumed to coincide with the front surface of the object. This plane appears edgewise in the top and auxiliary views, and the two reference lines are then used in the same manner as folding lines. Dimensions D, to the reference lines, are equal. The advantage of the reference-plane method is that fewer measurements are required, since some points of the object lie in the reference plane.

The reference plane may coincide with the front surface of the object as at (a), or it may cut through the object as at (b) if the object is symmetrical, or the reference plane may coincide with the back surface of the object as at (c), or through any intermediate point of the object.

The reference plane should be assumed in the position most convenient for transferring distances with respect to it. Remember the following:

1. Reference lines, like folding lines, are always at right angles to the projection lines between the views.

2. A reference plane appears as a line in two alternate views, never in adjacent views.

3. Measurements are always made at right angles to the reference lines, or parallel to the projection lines.

4. In the auxiliary view, all points are at the same distances from the reference line as the corresponding points are from the reference line in the *second previous view*, or alternate view.

266. To Draw an Auxiliary View, Using Reference Plane. The object shown in Fig. 474 (a) is numbered as explained in §212. To draw the auxiliary view, proceed as follows:

I. Draw two views of the object, and of an arrow indicating the direction of sight for the auxiliary view.

Note: In the following steps, manipulate the triangle (either triangle) as shown in Fig. 475 to draw lines parallel or perpendicular to the inclined face.

II. Draw projection lines parallel to the arrow.

III. Assume reference plane coinciding with back surface of object as shown at (a). Draw reference lines in the top and auxiliary views at right angles to the projection lines; *these are the edge views of the reference plane.*

IV. Draw auxiliary view of surface A. It will be true size and shape because the direction of sight was taken perpendicular to that surface. Transfer depth measurements from the top view to the auxiliary view with dividers or scale. Each point in the auxiliary view will be on its projection line from the front view and the same distance from the reference line as it is in the top view to the corresponding reference line.

V. Complete the auxiliary view by adding other visible edges and surfaces of the object. Each numbered point in the auxiliary view lies on its projection line from the front view, and is the same distance from the reference line as it is in the top view.

Note that two surfaces of the object appear as lines in the auxiliary view. Which surfaces are these (give numbers)? Does the bottom surface of the object appear true size and shape in the auxiliary view? Why? Before you draw the bottom surface in the auxiliary view, how do you know its general configuration and exact number of sides? Which edges of surface 2-5-6-3 are foreshortened and which are true length in the auxiliary view? Does surface 5-8-9-6 appear true size in the auxiliary view? Which edges of surface 5-8-9-6 appear true size in the front view? In the auxiliary view?

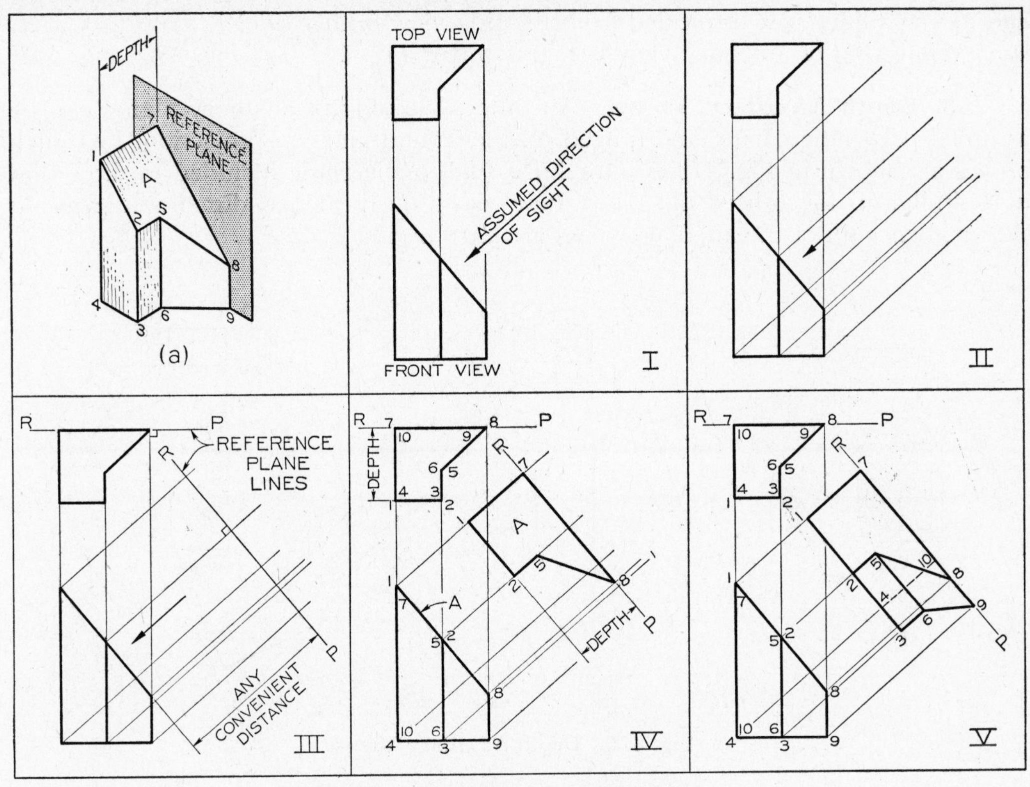

Fig. 474 To Draw an Auxiliary View.

267. Classification of Auxiliary Views. Auxiliary views are classified and named according to the principal dimensions of the object shown in the auxiliary view. For example, the auxiliary view in Fig. 474 is a *depth auxiliary view* because it shows the principal dimension of the object, *depth*. Any auxiliary view projected from the front view will show the depth of the object and is a depth auxiliary view.

Similarly, any auxiliary view projected from the top view is a *height auxiliary view*,

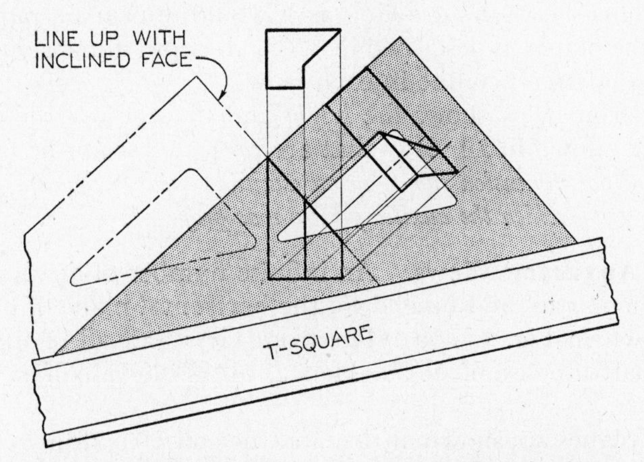

Fig. 475 Drawing Parallel or Perpendicular Lines.

and any auxiliary view projected from the side view (either side) is a *width auxiliary view*. What kind of auxiliary view is Fig. 470 (b)? Fig. 482 (b)? Fig. 496? Fig. 502?

268. Depth Auxiliary Views. An infinite number of auxiliary planes can be assumed perpendicular to, and hinged to, the frontal plane of projection. Five such planes are shown in Fig. 476 (a), the horizontal plane being included to show that it is similar to the others. In all of these views the principal dimension, *depth*, is shown; hence all of the auxiliary views are *depth auxiliary views*.

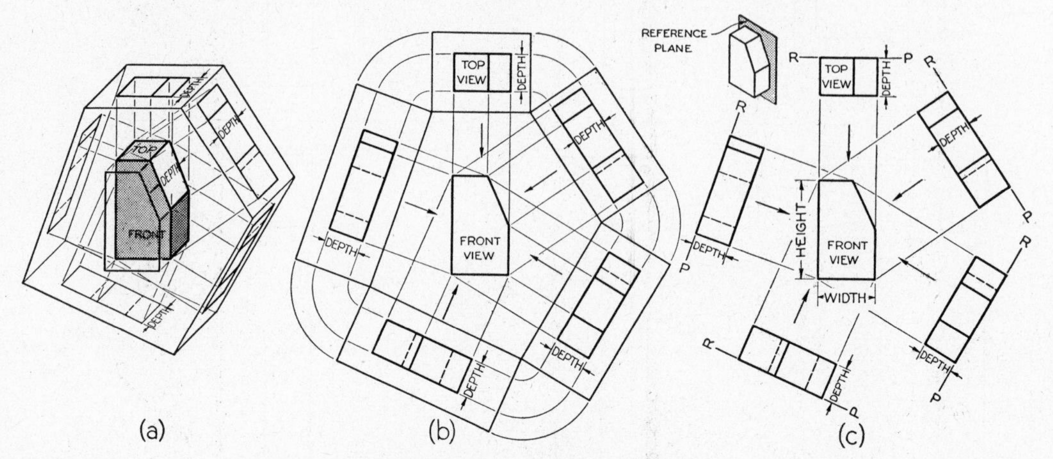

Fig. 476 Depth Auxiliary Views.

The unfolded auxiliary planes are shown at (b), where also is shown how the depth dimension may be projected from the top view to all auxiliary views. The arrows indicate the directions of sight for the several views, and the projection lines are respectively parallel to these arrows. The arrows may be assumed but need not be actually drawn, since the projection lines determine the direction of sight. The folding lines are perpendicular to the arrows and to the corresponding projection lines. Since the auxiliary planes can be assumed at any distance from the object, it follows that the folding lines may be any distance from the front view.

The complete drawing, with the planes of projection and folding lines omitted, is shown at (c). This shows the drawing as it would appear on paper, in which use is made of reference planes as described in §265, all depth dimensions being measured perpendicular to the reference line in each view.

Note that the front view shows the *height* and the *width* of the object, *but not the depth*. The depth is shown in all views that are projected from the front view; hence, this rule: *The principal dimension shown in an auxiliary view is that one which is not shown in the adjacent view from which the auxiliary view was projected.*

269. Height Auxiliary Views. An infinite number of auxiliary planes can be assumed perpendicular to, and hinged to, the horizontal plane of projection, several of which are shown in Fig. 477 (a). The front view and all of the auxiliary views show the principal dimension, *height*. Hence, all of the auxiliary views are *height auxiliary views*.

The unfolded planes are shown at (b), and the complete drawing, with the planes of projection and the folding lines omitted, is shown at (c). All reference lines are perpendicular to the corresponding projection lines, and all height dimensions are measured parallel to the projection lines, or perpendicular to the reference lines, in each view. Note that in the view projected from, the top view, the only dimension *not shown* is height.

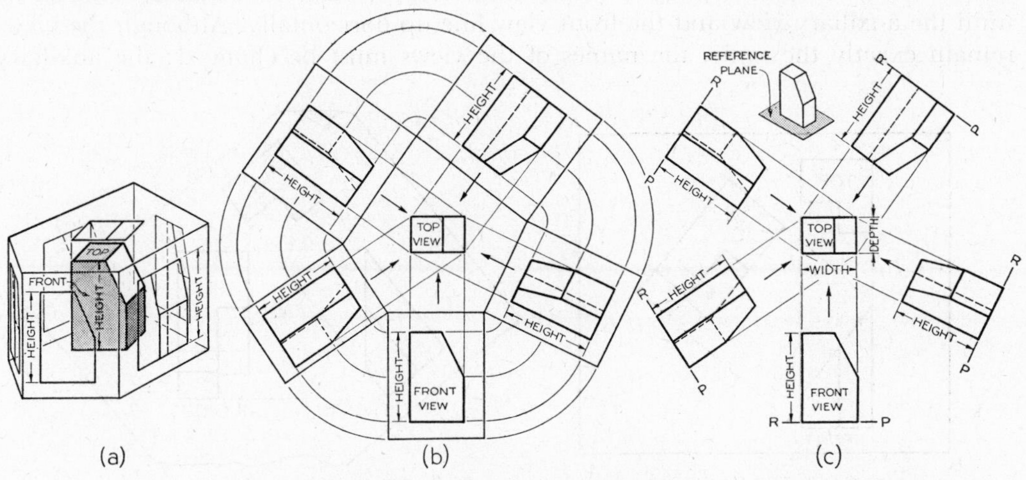

Fig. 477 Height Auxiliary Views.

270. Width Auxiliary Views. An infinite number of auxiliary planes can be assumed perpendicular to, and hinged to, the profile plane of projection, several of which are shown in Fig. 478 (a). The front view and all of the auxiliary views show the principal dimension, *width.* Hence, all of the auxiliary views are *width auxiliary views.*

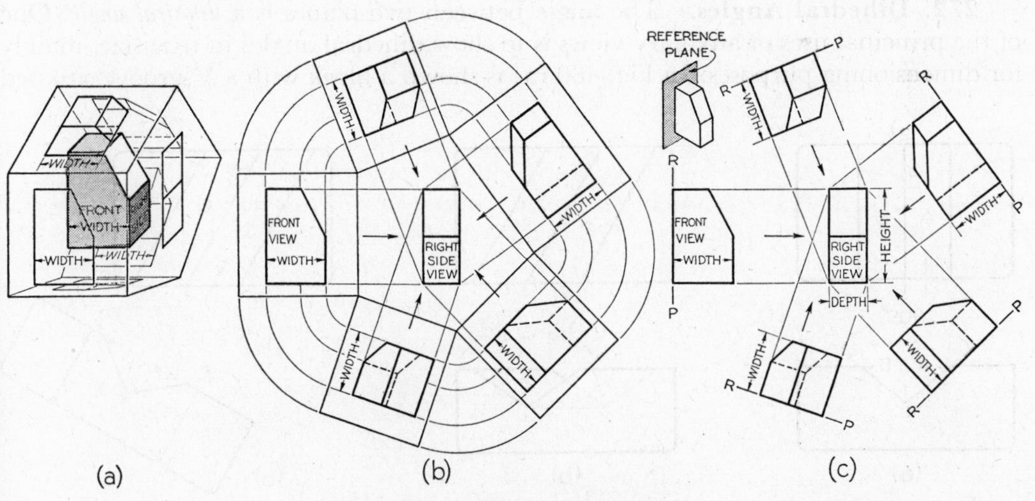

Fig. 478 Width Auxiliary Views.

The unfolded planes are shown at (b), and the complete drawing with the planes of projection and the folding lines omitted is shown at (c). All reference lines are perpendicular to the corresponding projection lines, and all width dimensions are measured parallel to the projection lines, or perpendicular to the reference lines, in each view. Note that in the view projected from, the side view, the only dimension *not shown* is width.

271. Revolving a Drawing. In Fig. 479 (a) is a drawing showing top, front, and auxiliary views. At (b) the drawing is shown revolved, as indicated by the arrows, until the auxiliary view and the front view line up horizontally. Although the views remain exactly the same, the names of the views must be changed; the auxiliary

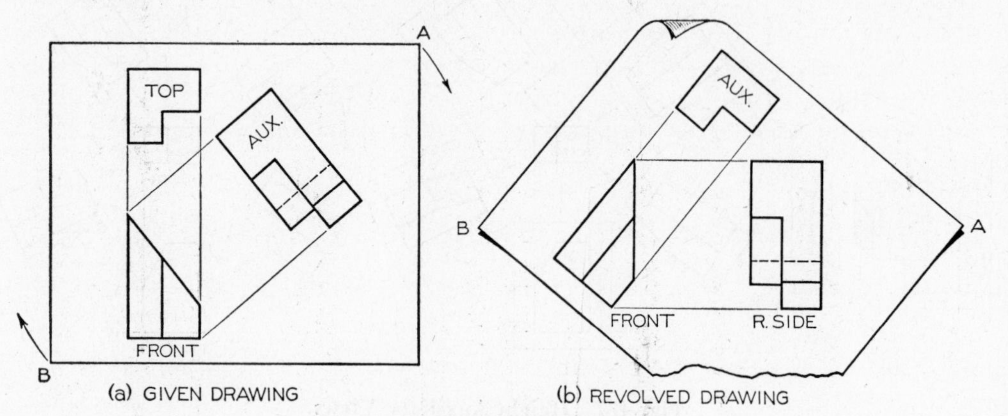

Fig. 479 Revolving a Drawing.

view now becomes a right-side view, and the top view becomes an auxiliary view. Some students find it easier to visualize and draw an auxiliary view when revolved to the position of a regular view in this manner. In any case, it is clear that an auxiliary view basically is like any other view.

272. Dihedral Angles. The angle between two planes is a *dihedral angle*. One of the principal uses of auxiliary views is to show dihedral angles in true size, mainly for dimensioning purposes. In Fig. 480 (a) is shown a block with a V-groove situated

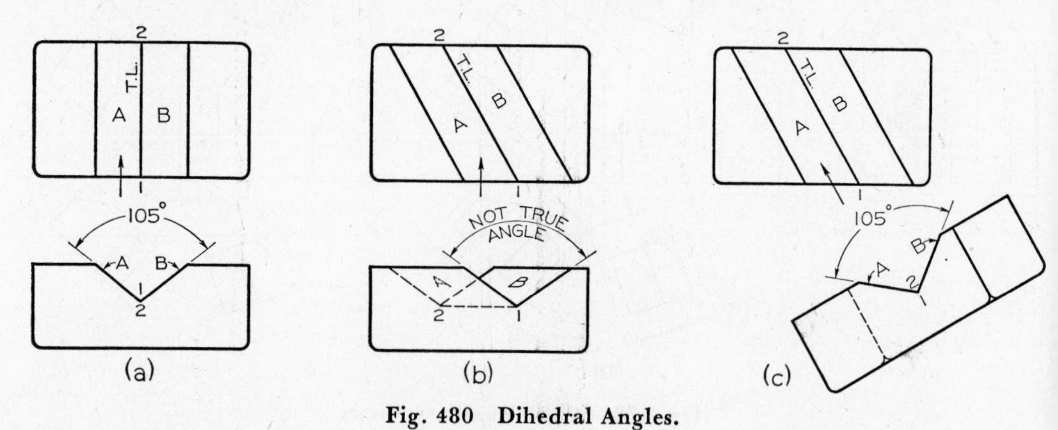

Fig. 480 Dihedral Angles.

so that the true dihedral angle between inclined surfaces A and B is shown in the front view. Why does this view show the true angle?

Assume a line in a plane. For example, draw a straight line on a sheet of paper; then hold the paper so as to view the line as a point. You will observe that when the line appears as a point, the plane containing the line appears as a line. Hence, this rule: *To get the edge view of a plane, get the point view of any line in that plane.*

In Fig. 480 (a), line 1-2 is the line of intersection of planes A and B. Now, line 1-2 lies in both planes at the same time; therefore, a point view of this line will show both planes as lines, and the angle between them is the dihedral angle between the planes. Hence, this rule: *To get the true angle between two planes, get the point view of the line of intersection of the planes.*

At (b), the line of intersection 1-2 does not appear as a point in the front view; hence, planes A and B do not appear as lines, and the true dihedral angle is not shown. Assuming that the actual angle is the same as at (a), does the angle show larger or smaller than at (a)? The drawing at (b) is unsatisfactory. The true angle does not appear because the direction of sight (see arrow) is not parallel to the line of intersection 1-2.

At (c) the direction of sight arrow is taken parallel to line 1-2, producing an auxiliary view in which line 1-2 appears as a point, planes A and B appear as lines, and the true dihedral angle is shown. *To draw a view showing a true dihedral angle, assume the direction of sight parallel to the line of intersection between the planes of the angle.*

273. Plotted Curves. As shown in §235, if a cylinder is cut by an inclined plane, the inclined surface is elliptical in shape. In Fig. 317 (a), such a surface is produced, but the ellipse does not show true size and shape because the plane of the ellipse is not seen at right angles in any view.

In Fig. 481 (a), the line of sight is taken perpendicular to the edge view of the inclined surface, and the resulting ellipse is shown in true size and shape in the

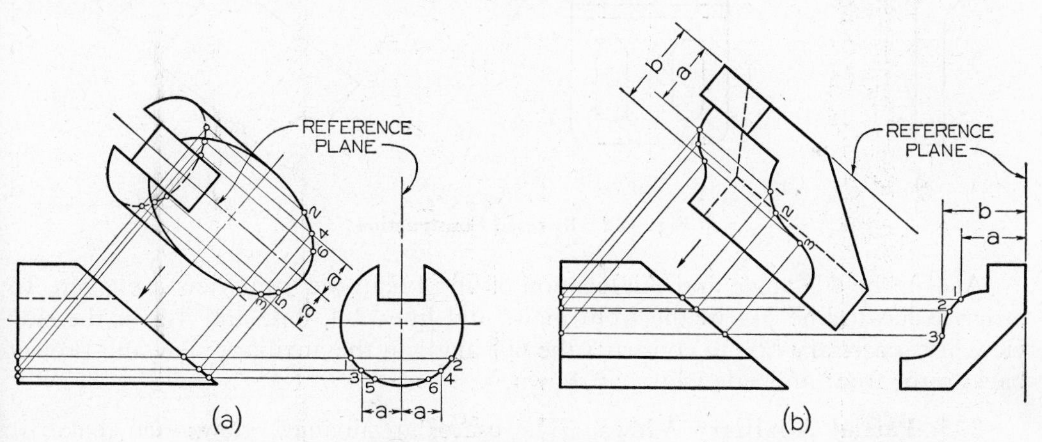

Fig. 481 Plotted Curves.

auxiliary view. The major axis is found by direct projection from the front view, and the minor axis is equal to the diameter of the cylinder. The left end of the cylinder (a circle) will appear as an ellipse in the auxiliary view, the major axis of which is equal to the diameter of the cylinder.

Since this is a symmetrical object, the reference plane is assumed through the center, as shown. To plot points on the ellipses, select points on the circle of the side view, and project them across to the inclined surface or to the left-end surface, and then upward to the auxiliary view. In this manner, two points can be projected each time, as shown for points 1-2, 3-4, and 5-6. Distances *a* are equal and are transferred from the side view to the auxiliary view with the aid of dividers. A sufficient number of points must be projected to establish the curves accurately. Use the irregular curve as described in §66.

Since the major and minor axes are known, any of the "true" ellipse methods of Figs. 200-202 or 204 (a) may be used. Or if an approximate ellipse is adequate for the job in hand, the method of Fig. 208 can be used. But the quickest and easiest method is to use an ellipse template, as explained in §163.

In Fig. 481 (b), the auxiliary view shows the true size and shape of the inclined cut through a piece of molding. The method of plotting points is similar to that above.

274. Reverse Construction. In order to complete the regular views, it is often necessary to construct an auxiliary view first. For example, in Fig. 482 (a) the upper portion of the right-side view cannot be constructed until the auxiliary view is drawn and points established on the curves and then projected back to the front view, as shown.

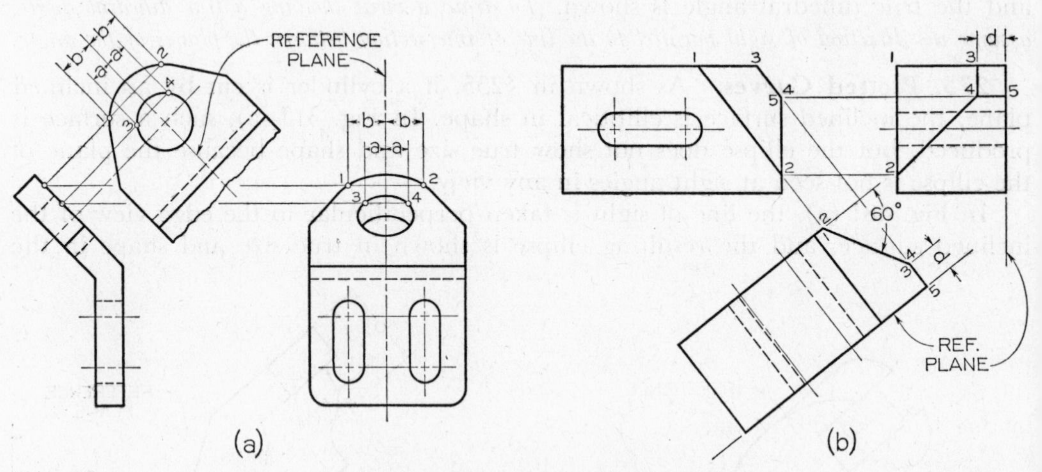

Fig. 482 Reverse Construction.

At (b), the 60° angle and the location of line 1-2 in the front view are given. In order to locate line 3-4 in the front view, and lines 2-4, 3-4, and 4-5 in the side view, it is necessary first to construct the 60° angle in the auxiliary view and project back to the front and side views, as shown.

275. Partial Auxiliary Views. The use of an auxiliary view often makes it possible to eliminate one or more regular views and thus to simplify the shape description, as shown in Fig. 470 (b).

In Fig. 483 are shown three complete auxiliary-view drawings. Such drawings take a great deal of time to draw, particularly when ellipses are involved as is so often the case, and the completeness of detail may add nothing in clearness or may even detract from it because of the clutter of lines. However, in these cases some portion of every view is needed—no view can be completely eliminated, as was done in Fig. 470 (b).

As described in §215, *partial views* are often sufficient, and the resulting drawings are considerably simplified and easier to read. Similarly, as shown in Fig. 484, partial regular views and partial auxiliary views are used with the same result. Usually a break line is used to indicate the imaginary "break" in the views. *Do not draw a break line coinciding with a visible line or a hidden line.*

In order to clarify the relation of views, the auxiliary views should be "tied to" the views from which they are projected, either with a center line, or with one or two projection lines. This is particularly important with regard to partial views that often are small and can easily appear to be "lost" and not related to any view.

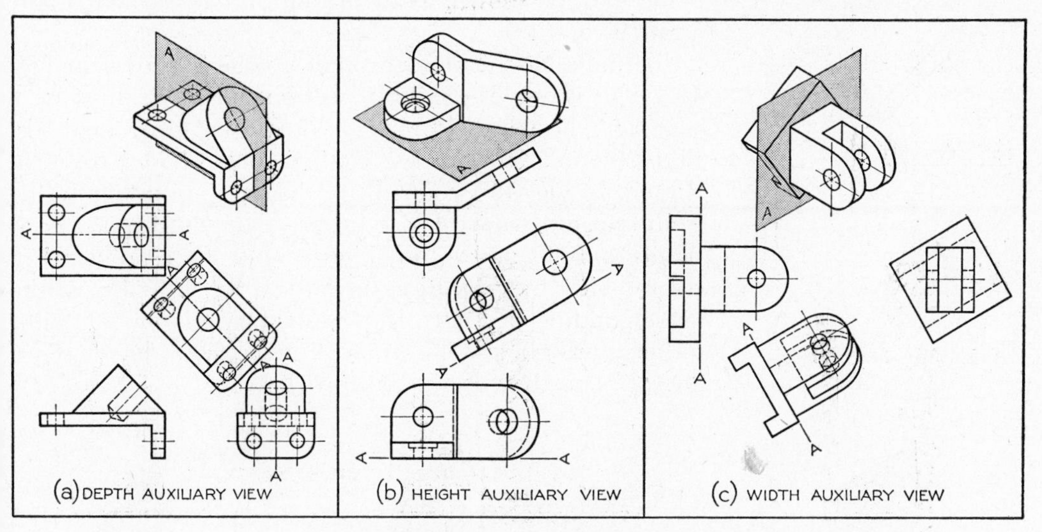

Fig. 483 Primary Auxiliary Views.

276. Half Auxiliary Views. If an auxiliary view is symmetrical, and if it is necessary to save space on the drawing or to save time in drafting, only a half of the auxiliary view may be drawn, as shown in Fig. 485. In this case, a half of a regular view is also shown, since the bottom flange is also symmetrical. See §215. Note that in each case the *near half* is shown.

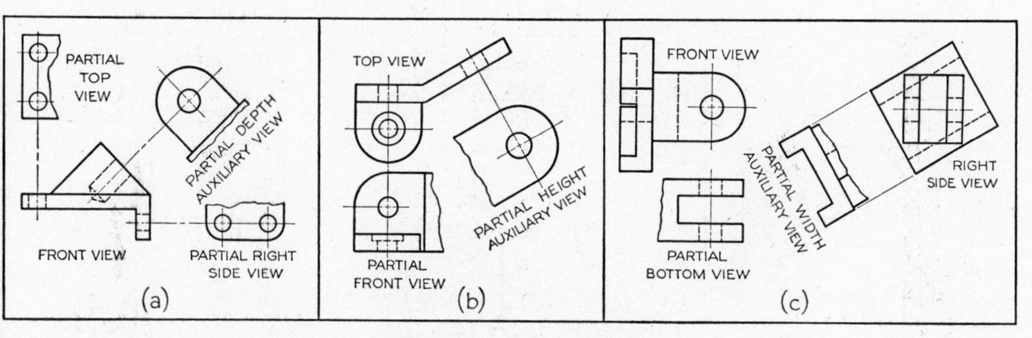

Fig. 484 Partial Views.

277. Hidden Lines in Auxiliary Views. In practice, hidden lines should be omitted in auxiliary views, as in ordinary views, §192, unless they are needed for clearness. The beginner, however, should show all hidden lines, especially if the auxiliary view of the entire object is shown. Later, in advanced work, it will become clearer as to when hidden lines can be omitted.

278. Auxiliary Sections. An *auxiliary section* is simply an auxiliary view in section. In Fig. 486 (a), note the cutting-plane line and the terminating arrows that indicate the direction of sight for the auxiliary section. Observe that the section lines are drawn at approximately 45° with the visible outlines. In drawing an auxiliary section, the entire portion of the object behind the cutting plane may be shown, as at (a), or the cut surface alone, as at (b) and (c).

An auxiliary section through a cone is shown in Fig. 487. This is one of the conic sections, §154, in this case a *parabola*. The parabola may be drawn by other methods, Figs. 209 and 210, but the method shown here is by projection. In Fig. 487, elements of the cone are drawn in the front and top views. These intersect the cutting plane at points A, B, C, and so on. These points are established in the top view by projecting upward to the top views of the corresponding elements. In the auxiliary section, all points on the parabola are the same distance from the reference line X-X as they are in the top view.

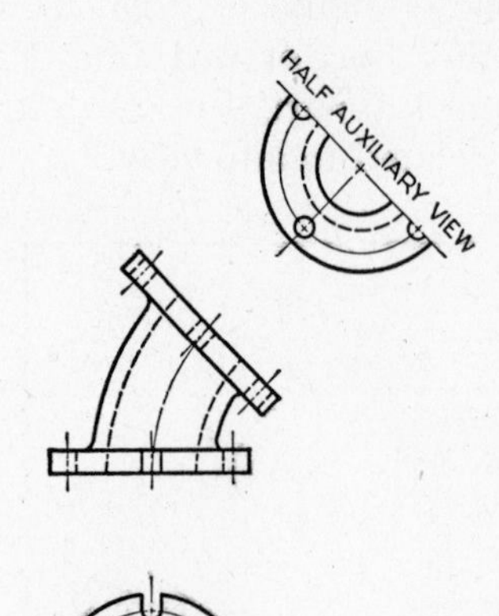

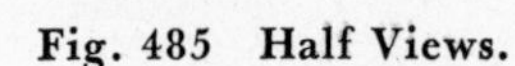

Fig. 485 Half Views.

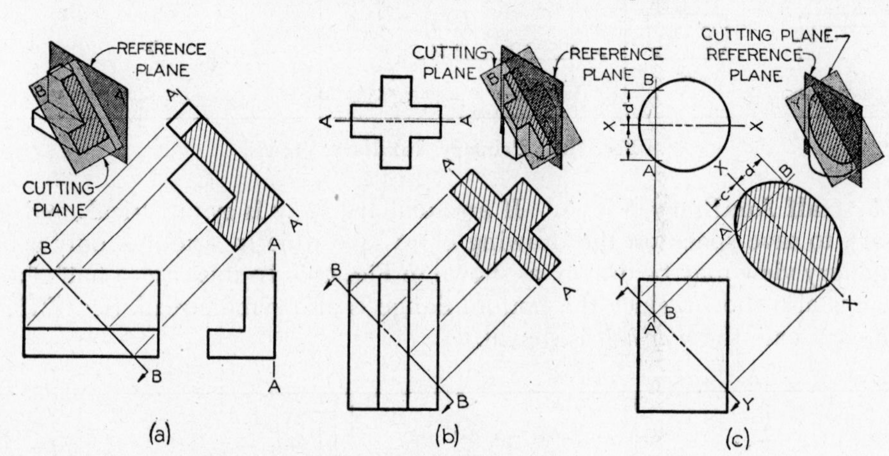

Fig. 486 Auxiliary Sections.

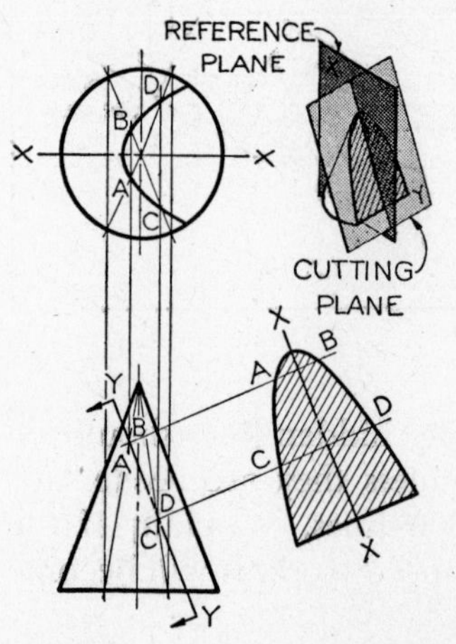

Fig. 487 Auxiliary Section.

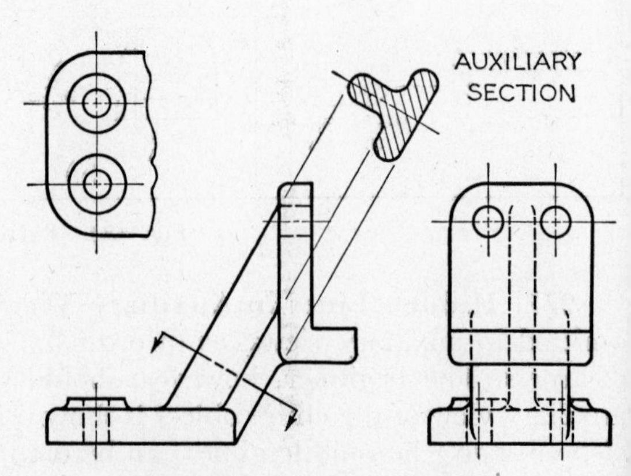

Fig. 488 Auxiliary Section.

A typical example of an auxiliary section in machine drawing is shown in Fig. 488. Here, there is not sufficient space for a *revolved section*, §252, although a *removed section*, §253, could have been used instead of an auxiliary section.

279. True Length of Line. In Fig. 489, let it be required to find the true length of the hip rafter AB by means of a depth auxiliary view.

I. Draw an arrow perpendicular to AB (front view) indicating the direction of sight, and place the reference plane through the center of the roof, as shown.

II. Draw the reference line X-X perpendicular to the arrow and at any convenient distance from AB (front view), and project the points A and B toward it.

III. Set off the points A and B in the auxiliary view at the same distance from the reference line as they are in the top view. The triangle ABC in the auxiliary view shows the true size and shape of the roof section ABC, and the distance AB in the auxiliary view is the true length of the hip rafter AB.

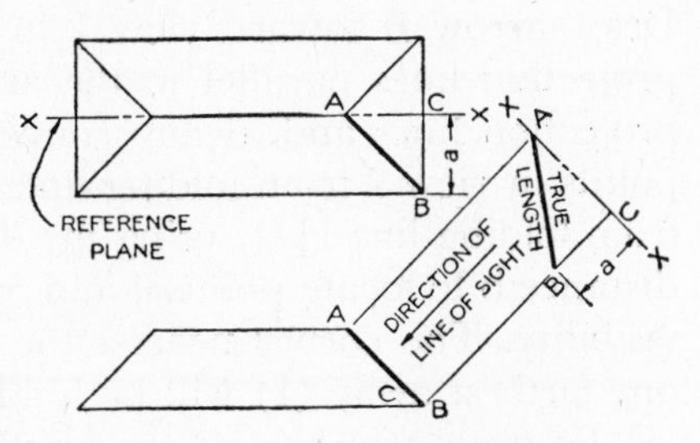

Fig. 489 True Length of a Line by Means of an Auxiliary View.

To find the true length of a line by revolution, see §291.

280. Successive Auxiliary Views. Up to this point we have dealt with *primary auxiliary views*—that is, single auxiliary views projected from one of the regular views. In Fig. 490, auxiliary view 1 is a primary auxiliary view, projected from the top view.

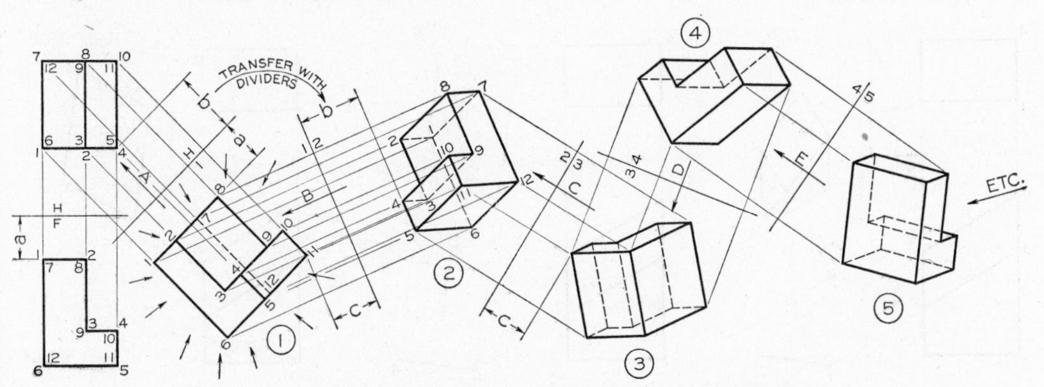

Fig. 490 "Chain Reaction"—Successive Auxiliary Views.

From primary auxiliary view 1, a *secondary auxiliary view* 2 can be drawn; then from it a third auxiliary view 3, and so on. An infinite number of such successive auxiliary views may be drawn, a process which may be likened to the "chain reaction" in nuclear explosions.

However, secondary auxiliary view 2 is not the only one that can be projected from primary auxiliary view 1 and thus start an independent "chain reaction." As shown by the arrows around view 1, an infinite number of secondary auxiliary views, with different lines of sight, may be projected. *Any auxiliary view projected from a primary auxiliary view is a secondary auxiliary view.* Furthermore, any succeeding auxiliary view may be used to project an infinite number of "chains" of views from it.

In this example, folding lines are more convenient than reference-plane lines.

In Auxiliary view 1, all numbered points of the object are the same distance from folding line H/1 as they are in the front view from folding line H/F. These distances, such as distance a, are transferred from the front view to the auxiliary view with the aid of dividers.

To draw the secondary auxiliary view 2, drop the front view from consideration, and center attention on the sequence of three views: the top view, view 1, and view 2. Draw arrow B toward view 1 in the direction desired for view 2, and draw light projection lines parallel to the arrow. Draw folding line 1/2 perpendicular to the projection lines and at any convenient distance from view 1. Locate all numbered points in view 2 from folding line 1/2 at the same distances they are in the top view from folding line H/1, using the dividers to transfer distances. For example, transfer distance b to locate points 4 and 5. Connect points with straight lines, and determine visibility. The corner nearest the observer (11) for view 2 will be visible, and the one farthest away (1) will be hidden, as shown.

To draw views 3, 4, etc., repeat the above procedure, remembering that each time we will be concerned only with a sequence of three views. In drawing any auxiliary view, the paper may be revolved so as to make the last two views line up as regular views.

281. To Find the True Size of an Oblique Surface. A typical requirement of a secondary auxiliary view is to show the true size and shape of an oblique surface, such as surface 1-2-3-4 in Fig. 491. In this case folding lines are used, but the same results can be obtained with reference lines. Proceed as follows:

I. Draw primary auxiliary view showing surface 1-2-3-4 as a line. As explained in §272, the edge view (EV) of a plane is found by getting the point view of a line

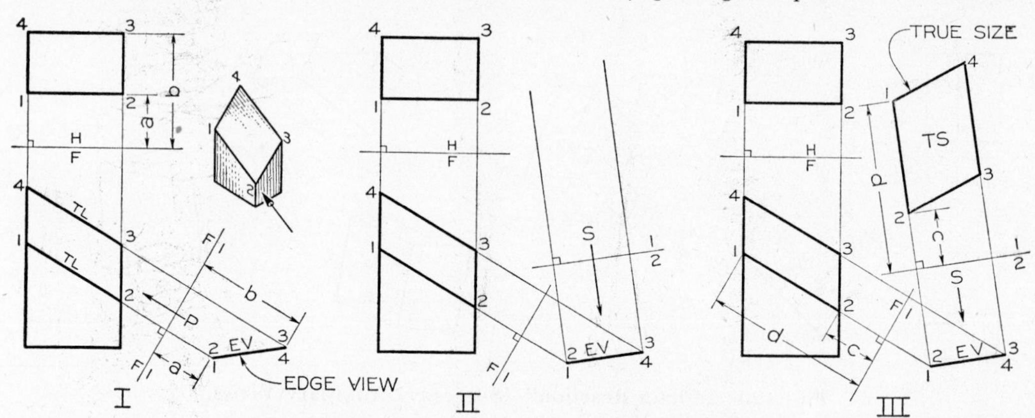

Fig. 491 Finding True Size of Oblique Surface.

in that plane. To get the point view of a line, the line of sight must be assumed parallel to the line. Therefore, draw arrow P parallel to lines 1-2 and 3-4, which are true length (TL) in the front view, and draw projection lines parallel to the arrow. Draw folding line H/F between the top and front views and F/1 between the front and auxiliary views, perpendicular to the respective projection lines. All points in the auxiliary view will be the same distance from the folding line F/1 as they are in the top view from folding line H/F. Lines 1-2 and 3-4 will appear as points in the auxiliary view, and plane 1-2-3-4 will therefore appear *edgewise*, or as a line.

II. Draw arrow S perpendicular to the edge view of plane 1-2-3-4 in the primary auxiliary view, and draw projection lines parallel to the arrow. Draw folding line 1/2 perpendicular to these projection lines and at a convenient distance from the primary auxiliary view.

III. Draw secondary auxiliary view. Locate each point (transfer with dividers) the same distance from the folding line 1/2 as it is in the front view to the folding line F/1, as for example, dimensions c and d. The true size (TS) of the surface 1-2-3-4 will be shown in the secondary auxiliary view, since the direction of sight, arrow S, was taken perpendicular to it.

282. Use of Reference Planes. In Fig. 492 (a), it is required to draw an auxiliary view in which triangular surface 1-2-3 will appear in true size and shape. In order for the true size of the surface to appear in the secondary auxiliary view, arrow S must be assumed perpendicular to the edge view of that surface; so it is necessary

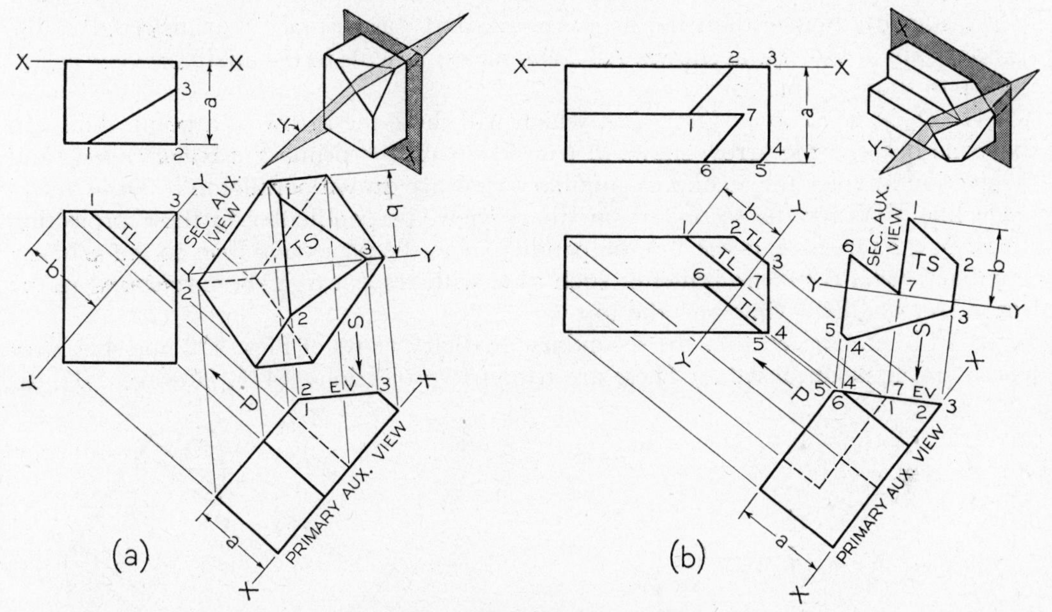

Fig. 492 Use of Reference Planes.

to have the edge view of surface 1-2-3 in the primary auxiliary view first. In order to do this, the direction of sight, arrow P, must be parallel to a line in surface 1-2-3 that appears true length (TL) in the front view. Hence, arrow P is drawn parallel to line 1-2 of the front view, line 1-2 will appear as a point in the primary auxiliary view, and surface 1-2-3 must therefore appear edgewise in that view.

In this case it is convenient to use reference lines and to assume the reference plane X (for drawing the primary auxiliary view) coinciding with the back surface of the object, as shown. For the primary auxiliary view, all depth measurements, as *a* in the figure, are transferred with dividers from the top view with respect to the reference line X-X.

For the secondary auxiliary view, reference plane Y is assumed cutting through the object for convenience in transferring measurements. All measurements perpendicular to Y-Y in the secondary auxiliary view are the same as between the reference plane and the corresponding points in the front view. Note that corresponding measurements must be *inside* (toward the central view in the sequence of three views) or *outside* (away from the central view). For example, dimension b is on the side of Y-Y *away* from the primary auxiliary view in both places.

In Fig. 492 (b) it is required to find the true size and shape of surface 1-2-3-4-5-6-7, and not to draw the complete secondary auxiliary view. The method is similar to that described above.

283. Oblique Direction of Sight Given. In Fig. 493 two views of a block are given, with two views of an arrow indicating the direction in which it is desired to look at the object to obtain a view. Proceed as follows:

I. *Draw primary auxiliary view of both the object and the assumed arrow*, which will show the true length *of the arrow.* In order to do this, assume a horizontal reference plane X-X in the front and auxiliary views, as shown. Then assume a direction of sight perpendicular to the given arrow. In the front view, the butt end of the arrow is a distance *a* higher than the arrow point, and this distance is transferred to the primary auxiliary view as shown. All *height* measurements in the auxiliary view correspond to those in the front view.

II. *Draw secondary auxiliary view*, which will show the arrow as a point. This can be done because the arrow shows in true length in the primary auxiliary view, and projection lines for the secondary auxiliary view are drawn parallel to it. Draw reference line Y-Y, for the secondary auxiliary view, perpendicular to these projection lines. In the top view, draw Y-Y perpendicular to the projection lines to the primary auxiliary view. All measurements, such as *b*, with respect to Y-Y, correspond in the secondary auxiliary view and the top view.

It will be observed that the secondary auxiliary views of Figs. 492 and 493 have considerable pictorial value. These are trimetric projections, §475. However, the di-

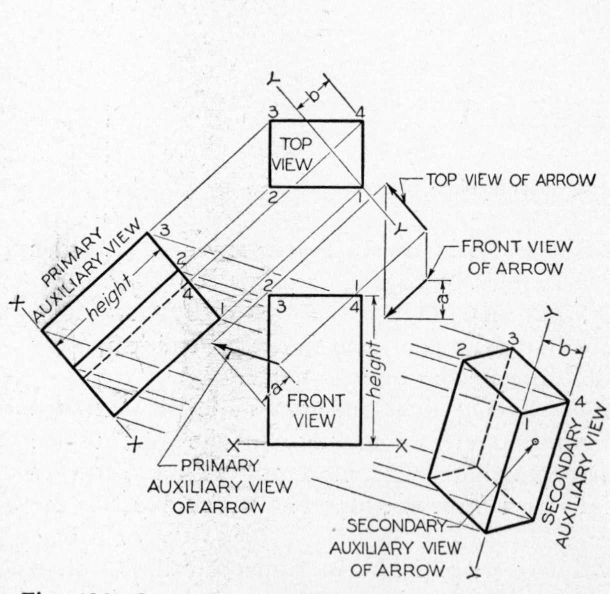

Fig. 493 Secondary Auxiliary View with Direction of Sight Given.

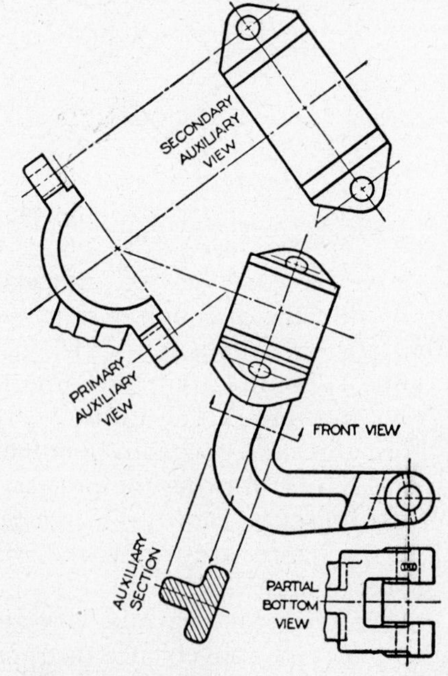

Fig. 494 Secondary Auxiliary View.

rection of sight could be assumed, in the manner of Fig. 493, to produce either isometric or dimetric projections. If the direction of sight is assumed parallel to the diagonal of a cube, the resulting view is an *isometric projection*, §449.

A typical application of a secondary auxiliary view in machine drawing is shown in Fig. 494. All views are partial views, except the front view. The partial secondary auxiliary view illustrates a case in which break lines are not needed. Note the use of an auxiliary section to show the true shape of the arm.

284. Ellipses. As shown in §235, if a circle is viewed obliquely, the result is an ellipse. This often occurs in successive auxiliary views, because of the variety of directions of sight. In Fig. 495 (a) the hole appears as a true circle in the top view. The circles appear as straight lines in the primary auxiliary view, and as ellipses in the secondary auxiliary view. In the latter, the major axis AB of the ellipse is parallel to the projection lines and equal in length to the true diameter of the circle as shown in the top view. The minor axis CD is perpendicular to the major axis, and its foreshortened length is projected from the primary auxiliary view.

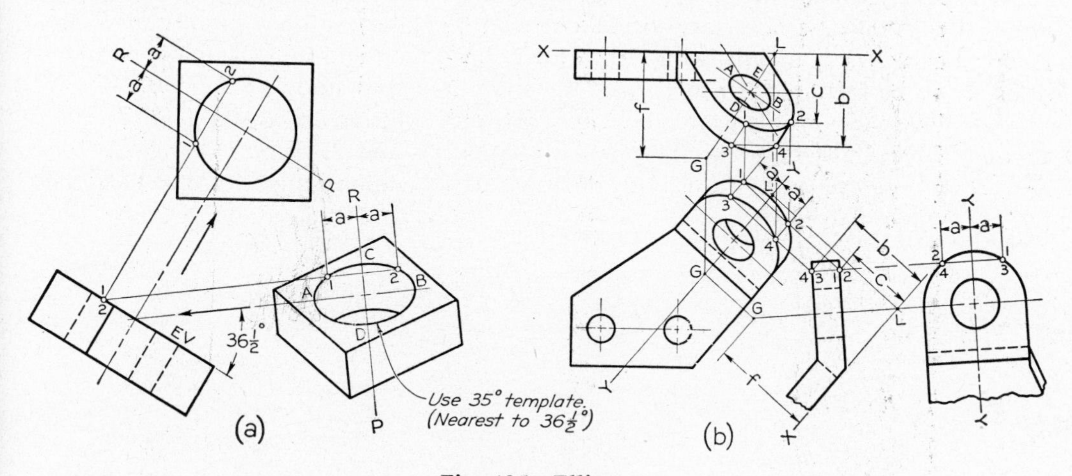

Fig. 495 Ellipses.

The ellipse can be completed by projecting points, such as 1 and 2, symmetrically located about CD, with distances *a* equal in the top and secondary auxiliary views as shown, and finally, after a sufficient number of points have been plotted, by applying the irregular curve, §66.

Since the major and minor axes are easily found, any of the "true" ellipse methods of Figs. 200-202 and 204 (a) may be used, or an approximate ellipse, Fig. 208, may be found sufficiently accurate for a particular drawing. Or the ellipses may be easily and rapidly drawn with the aid of an ellipse template, §163. The "angle" of ellipse to use is the one which most closely matches the angle between the direction of sight arrow and the plane (EV) containing the circle, as seen in this case in the primary auxiliary view. Here the angle is 36½°, so a 35° ellipse is selected.

At (b) are shown successive auxiliary views in which the true circular shapes appear in the secondary auxiliary view, and the elliptical projections in the front and top views. It is necessary to construct the circular shapes in the secondary auxiliary view, then to project plotted points back to the primary auxiliary view, the front view, and finally to the top view, as shown in the figure for points 1, 2, 3, and 4. The final curves are then drawn with the aid of the irregular curve.

If the major and minor axes are found, any of the "true" ellipse methods may be used; or better still, an ellipse template, §163, may be employed. The major and minor axes are easily established in the front view, but in the top view they are more difficult to find. The major axis AB is at right angles to the center line GL of the hole, and equal in length to the true diameter of the hole. The minor axis ED is at right angles to the major axis. Its length is found by plotting several points in the vicinity of one end of the minor axis, or by using descriptive geometry to find the angle between the line of sight and the inclined surface, and by this angle selecting the ellipse guide required.

285. Auxiliary-View Problems. The problems in Figs. 496-498 are to be drawn mechanically or freehand on $8\frac{1}{2}'' \times 11''$ paper, Layout A-2 or A-3. If partial auxiliary views are not assigned, the auxiliary views are to be complete views of the entire object, including all necessary hidden lines.

It is often difficult to space properly the views of an auxiliary view drawing. In some cases it may be necessary to make a trial blocking-out on a preliminary sheet before starting the actual drawing. Allowances for dimensions must be made if dimensions are to be included. In such case, the student should study §§327-350.

Auxiliary-view problems in convenient form for solution are available in *Technical Drawing Problems*, by Giesecke, Mitchell, and Spencer, and *Technical Drawing Problems, Series 2*, by Spencer and Grant, both designed to accompany this text, and published by The Macmillan Company.

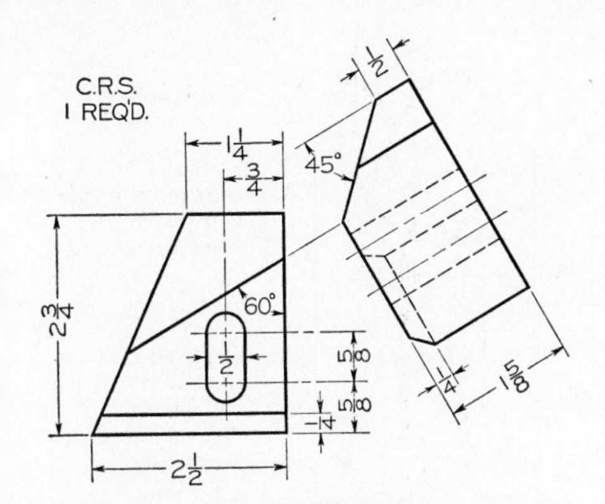

Fig. 496 R. H. Finger.
Given: Front and Aux. views.
Req'd: Complete Front, Aux., L. Side, and Top views (Layout A-3).

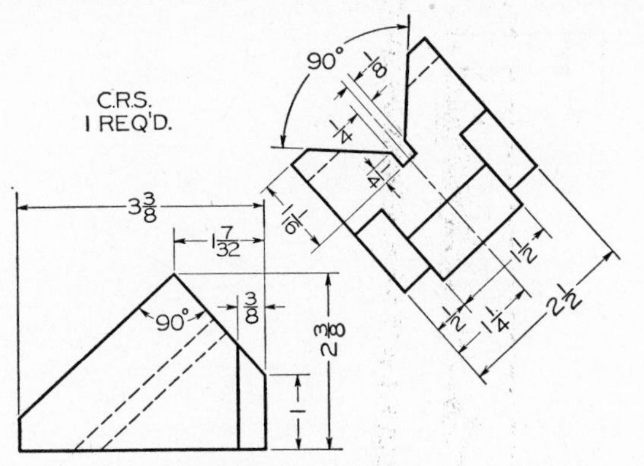

Fig. 497 V-Block.
Given: Front and Aux. views.
Req'd: Complete Front, Top, and Aux. views (Layout A-3).

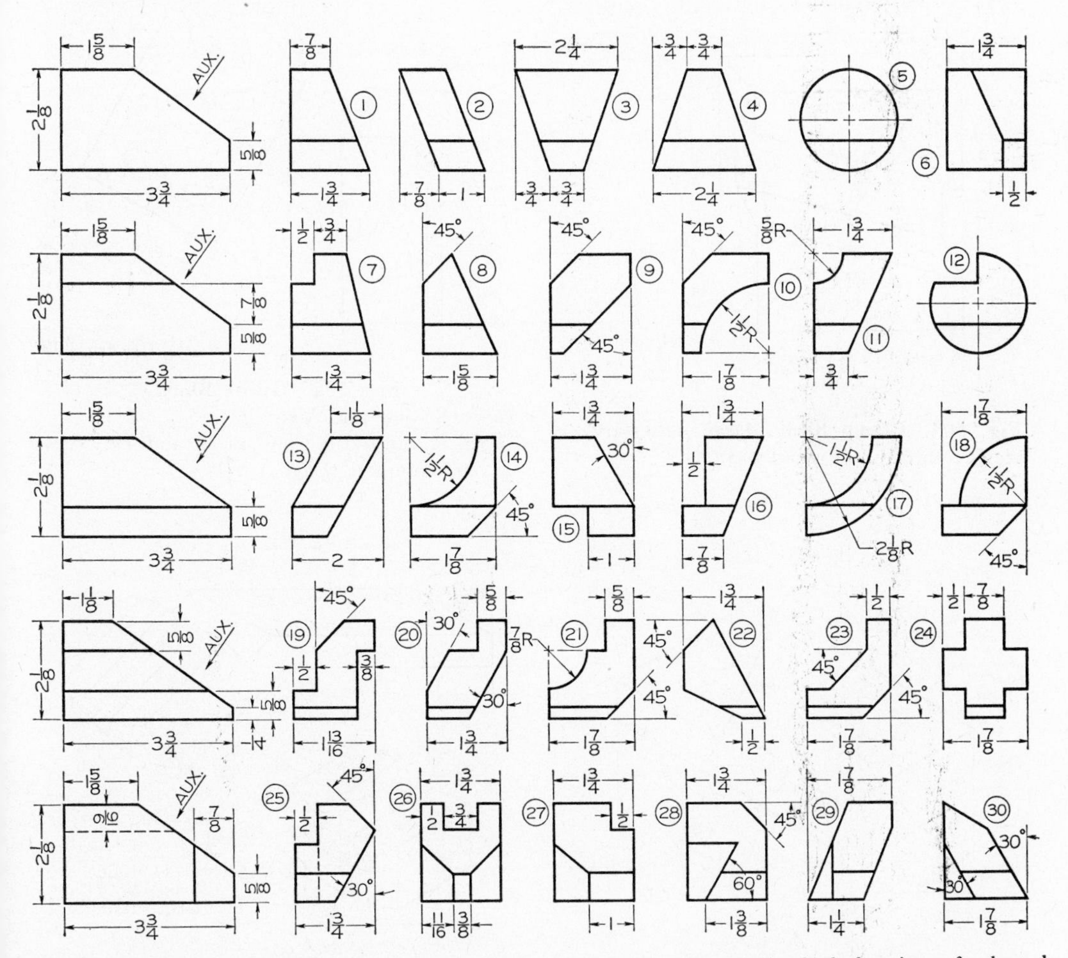

Fig. 498 Auxiliary-View Problems. Make freehand sketch or mechanical drawing of selected problem as assigned by instructor. Draw given Front and R. Side views, and add complete Auxiliary view, including all hidden lines (Layout A-3).

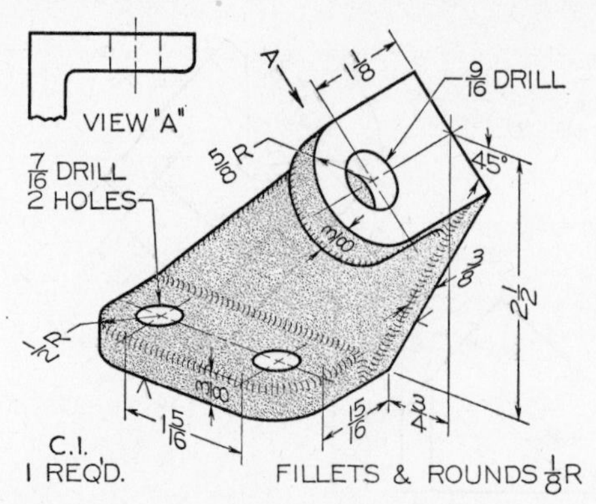

Fig. 499 Anchor Bracket. Draw necessary views or partial views (Layout A-3).

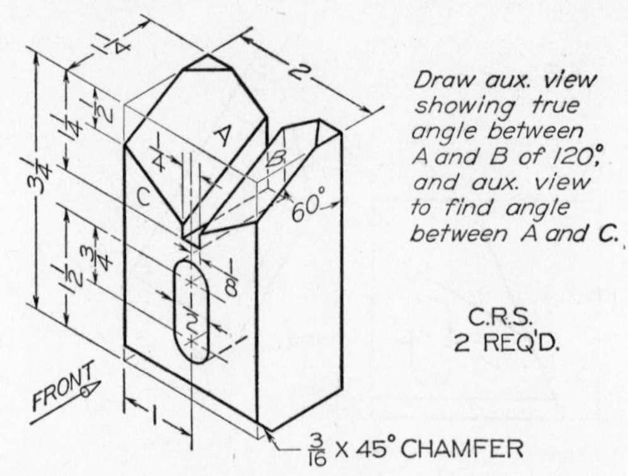

Fig. 500 Centering Block. Draw complete Front, Top, and R. Side views, plus indicated Auxiliary views (Layout B-3).

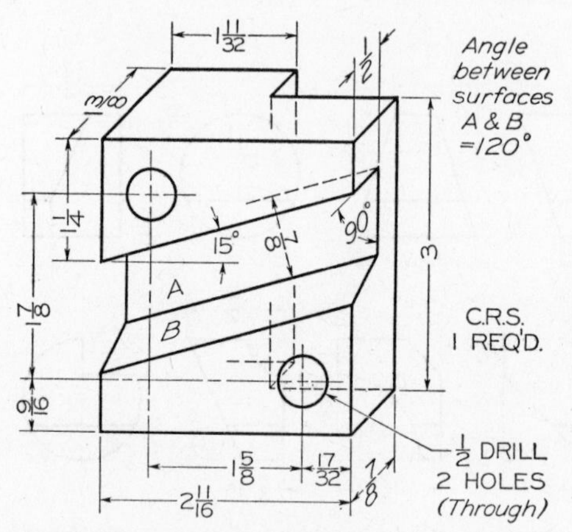

Fig. 501 Clamp Slide. Draw necessary views completely (Layout B-3).

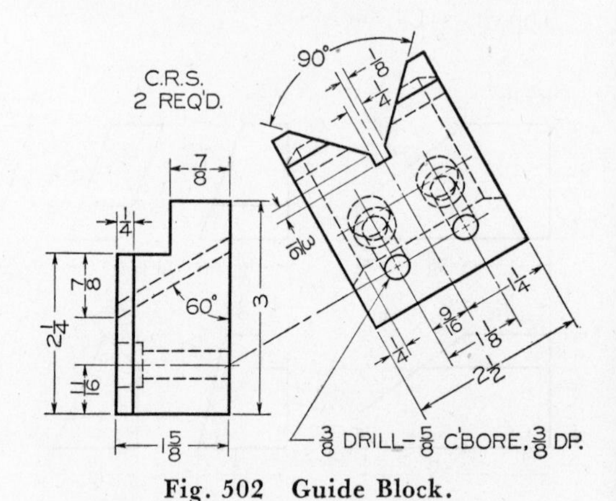

Fig. 502 Guide Block.
Given: R. Side and Aux. view.
Req'd: R. Side, Aux., plus Front and Top views —all complete (Layout B-3).

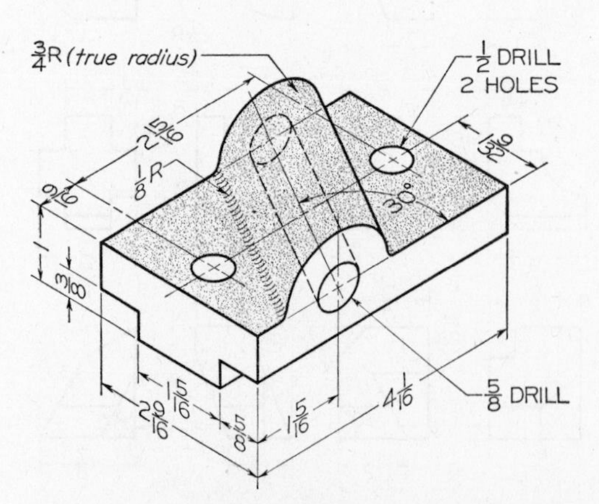

Fig. 503 Angle Bearing. Draw necessary views, including a complete Auxiliary view (Layout A-3).

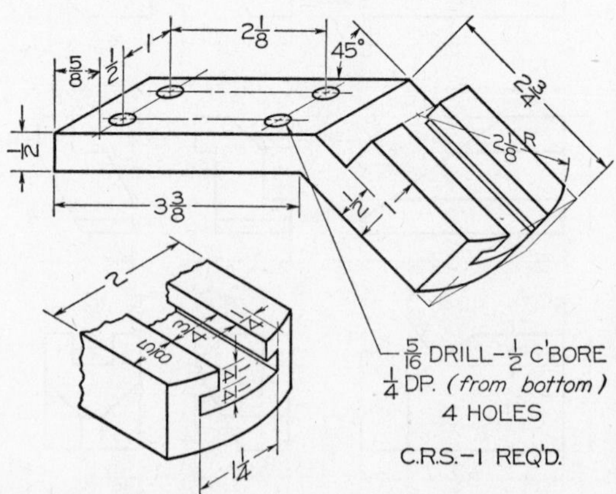

Fig. 504 Guide Bracket. Draw necessary views or partial views (Layout B-3).

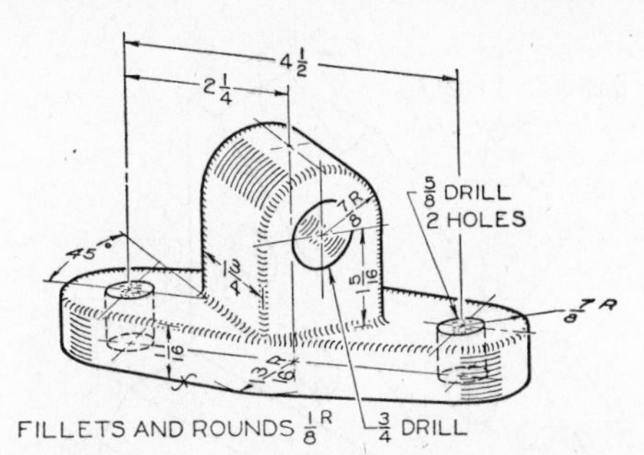

Fig. 505 Rod Guide. Draw necessary views, including complete Auxiliary view showing true shape of upper rounded portion (Layout B-4).

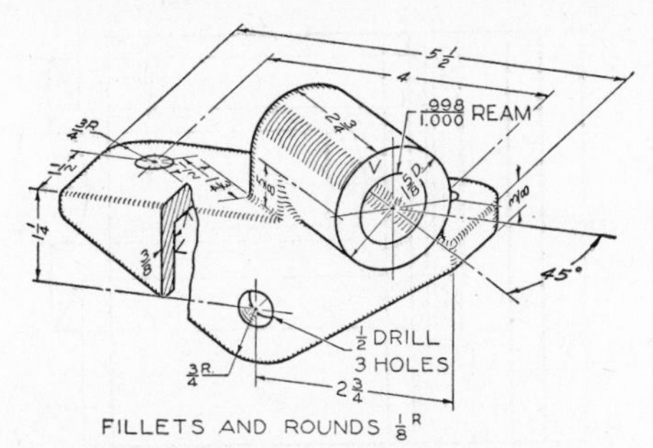

Fig. 506 Brace Anchor. Draw necessary views, including partial Auxiliary view showing true shape of cylindrical portion (Layout B-4).

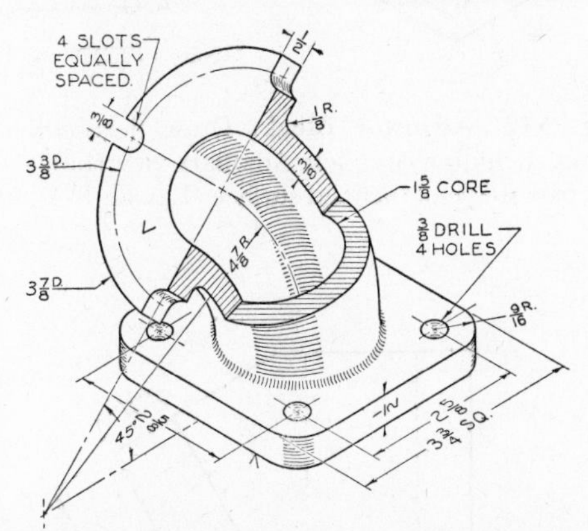

Fig. 507 45° Elbow. Draw necessary views, including a broken section and two half views of flanges (Layout B-4).

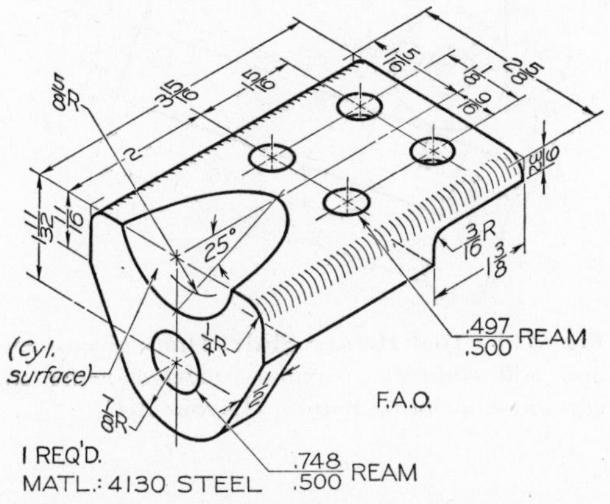

Fig. 508 Angle Guide. Draw necessary views, including a partial Auxiliary view of cylindrical recess (Layout B-4).

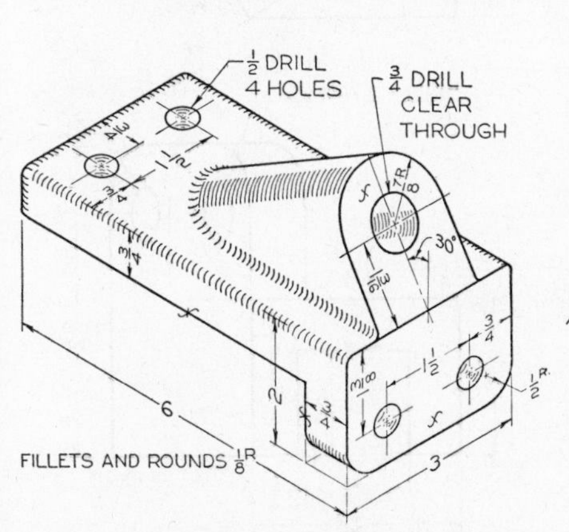

Fig. 509 Anchor Support. Draw necessary views, including partial Auxiliary view showing true shape of upper rounded portion, and partial views showing upper left end and lower right end (Layout B-4).

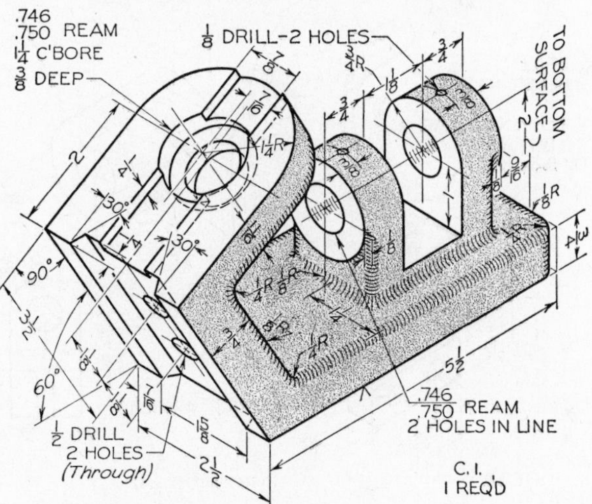

Fig. 510 Control Bracket. Draw necessary views, including partial Auxiliary views and regular views (Layout B-4).

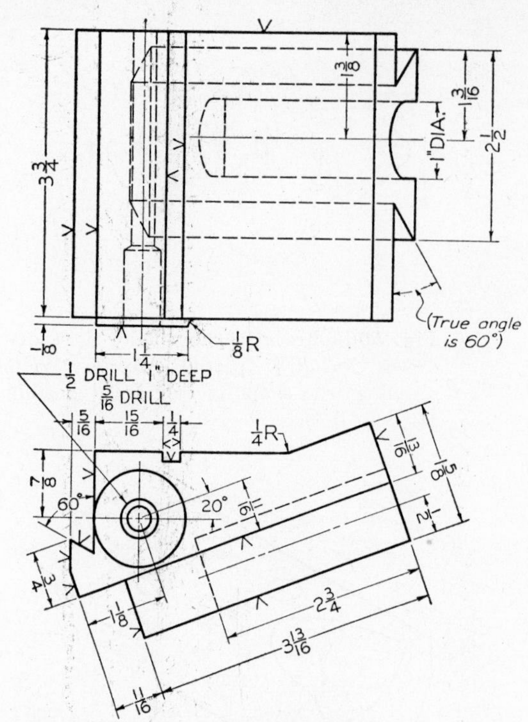

Fig. 511 Tool Holder Slide. Draw given views, and add complete Auxiliary view showing true curvature of slot on bottom (Layout B-4).

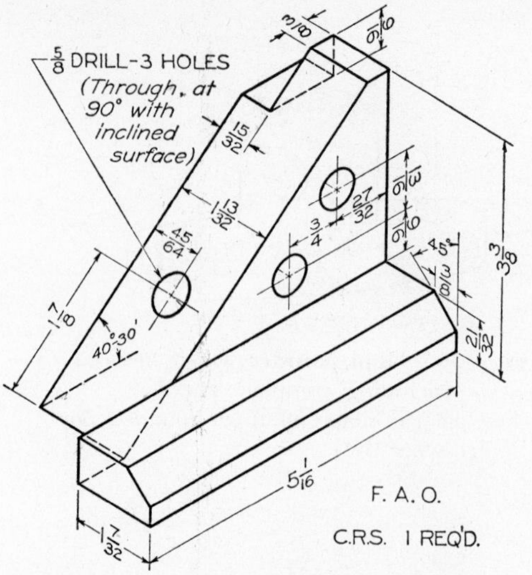

Fig. 512 Adjustor Block. Draw necessary views, including complete Auxiliary view showing true shape of inclined surface (Layout B-4).

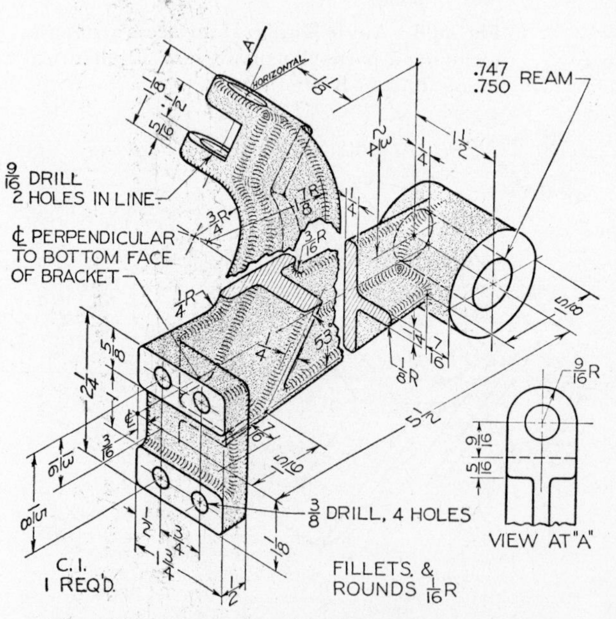

Fig. 513 Guide Bearing. Draw necessary views and partial views, including two Partial Auxiliary views (Layout C-4).

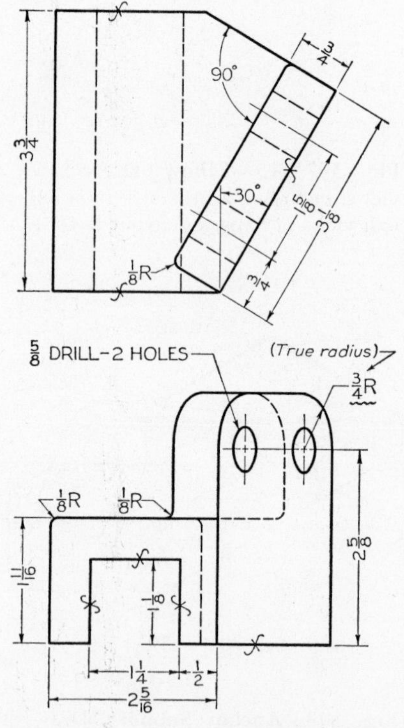

Fig. 514 Drill Press Bracket. Draw given views and add complete Auxiliary view showing true shape of inclined face (Layout B-4).

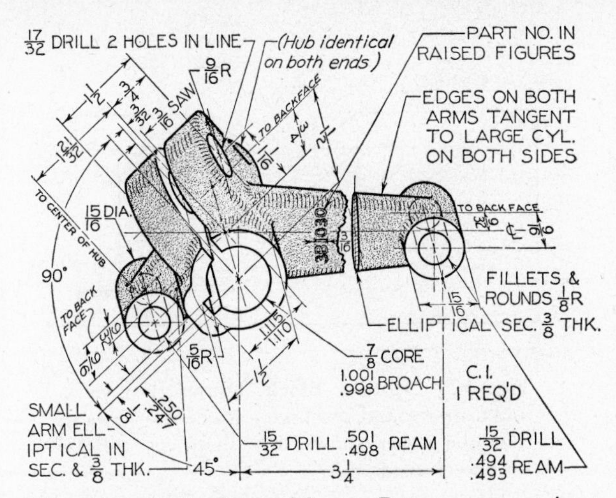

Fig. 515 Brake Control Lever. Draw necessary views and partial views (Layout B-4).

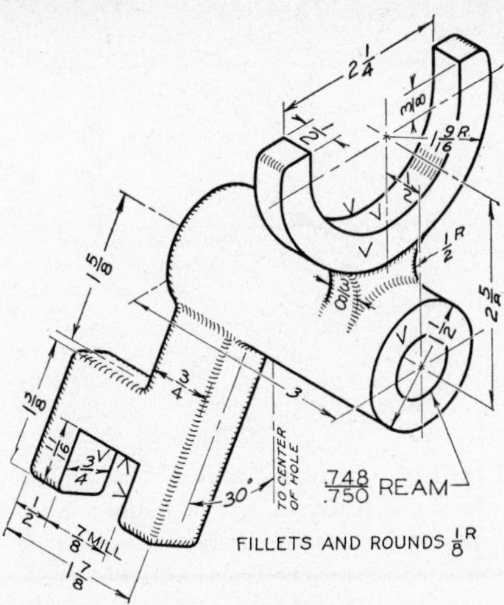

Fig. 516 Shifter Fork. Draw necessary views, including partial Auxiliary view showing true shape of inclined arm (Layout B-4).

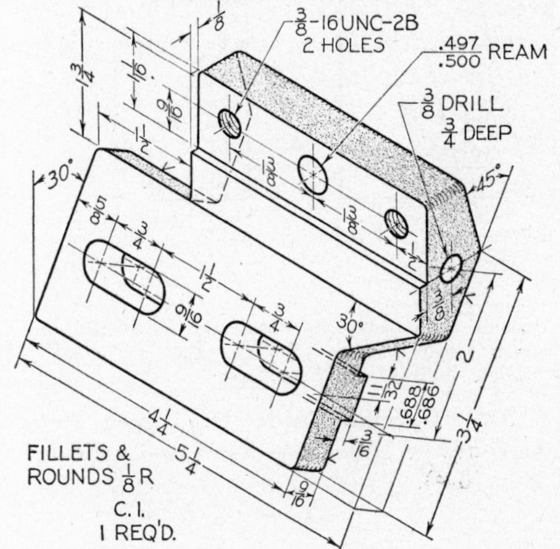

Fig. 517 Cam Bracket. Draw necessary views or partial views as needed. For threads, see §402 (Layout B-4).

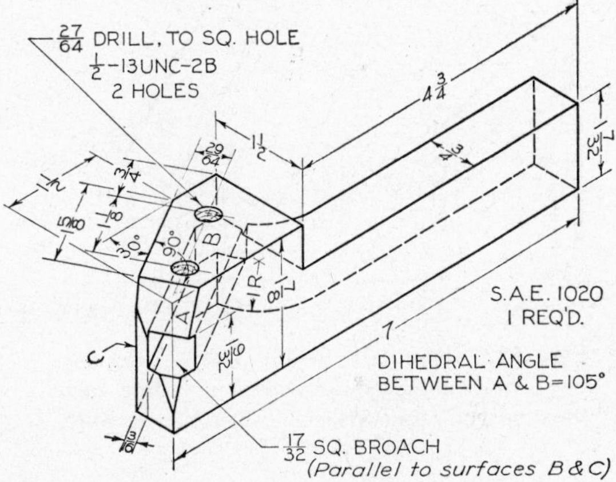

Fig. 518 R.H. Tool Holder. Draw necessary views, including partial Auxiliary views showing 105° angle and square hole true size. For threads, see §402 (Layout B-4).

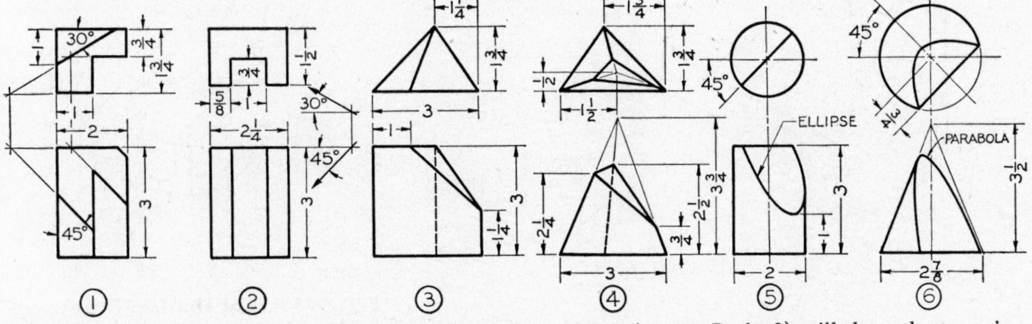

Fig. 519 Draw secondary Auxiliary views, complete, which (except Prob. 2) will show the true sizes of the inclined surfaces. In Prob. 2, draw secondary Auxiliary view as seen in direction of arrow (Layout B-3).

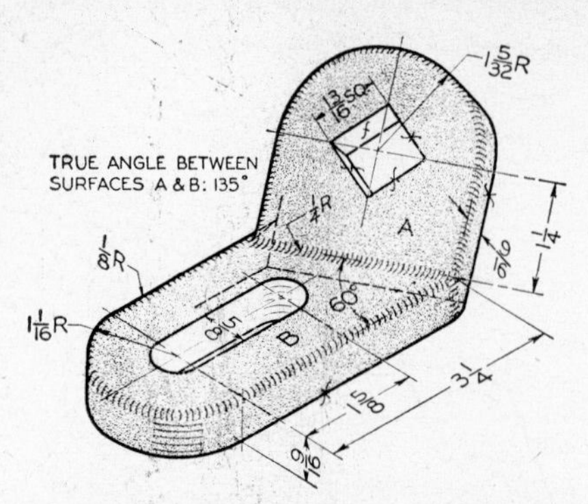

Fig. 520 Control Bracket. Draw given views and primary and secondary Auxiliary views so that the latter shows true shape of oblique surfaces (Layout B-4).

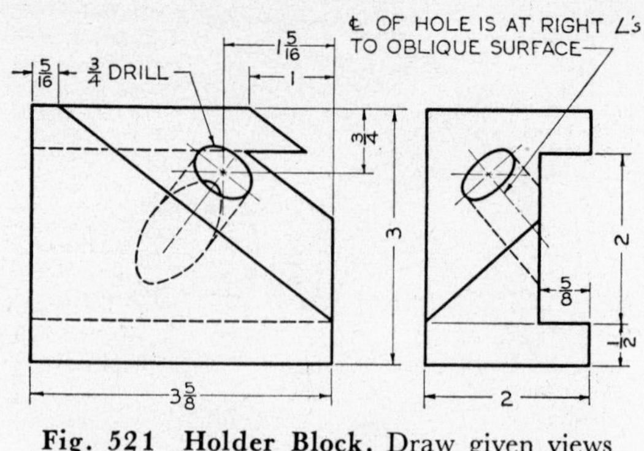

Fig. 521 Holder Block. Draw given views and primary and secondary Auxiliary views so that the latter shows true shape of oblique surface (Layout B-4).

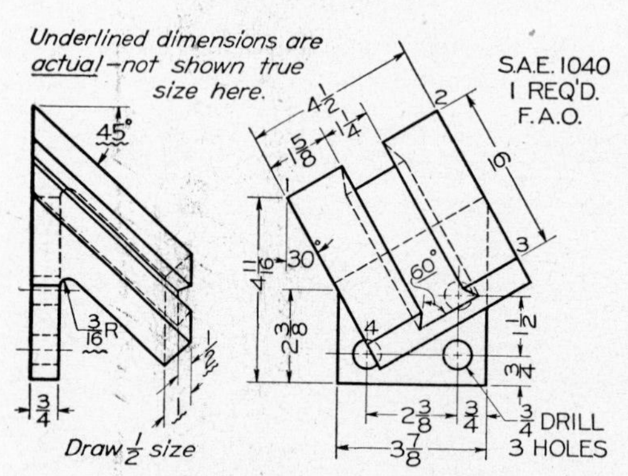

Fig. 522 Dovetail Slide. Draw complete given views and auxiliary views, including view showing true size of surface 1-2-3-4 (Layout B-4).

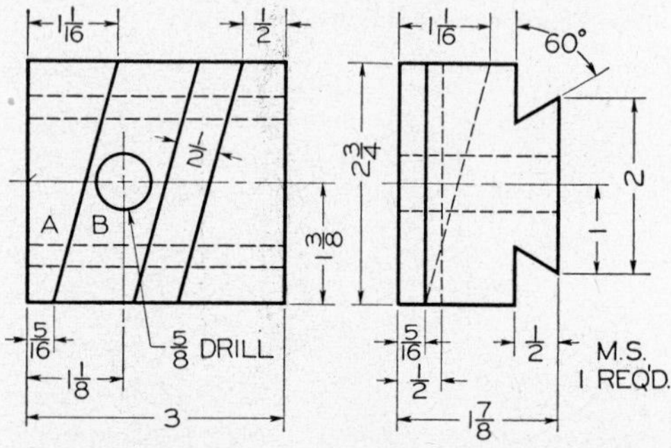

Fig. 523 Dovetail Guide. Draw given views plus complete Auxiliary views as indicated (Layout B-4).

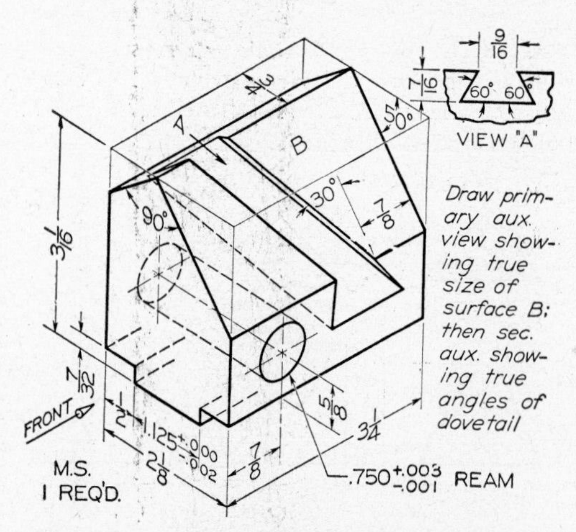

Fig. 524 Adjustable Stop. Draw complete Front and Auxiliary views plus partial R. Side view. Show all hidden lines (Layout C-4).

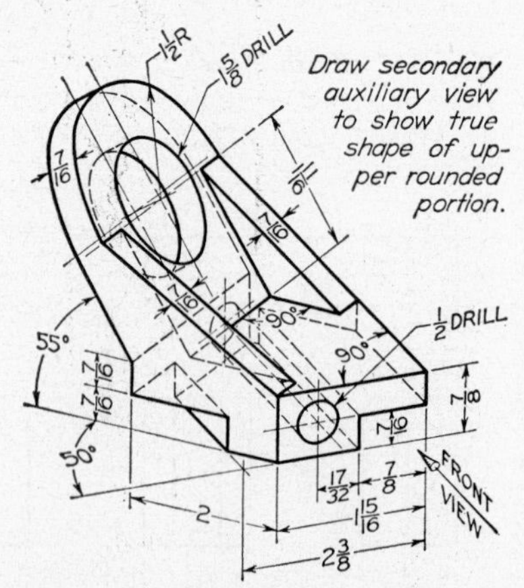

Fig. 525 Tool Holder. Draw complete Front view, and primary and secondary Auxiliary views as indicated (Layout B-4).

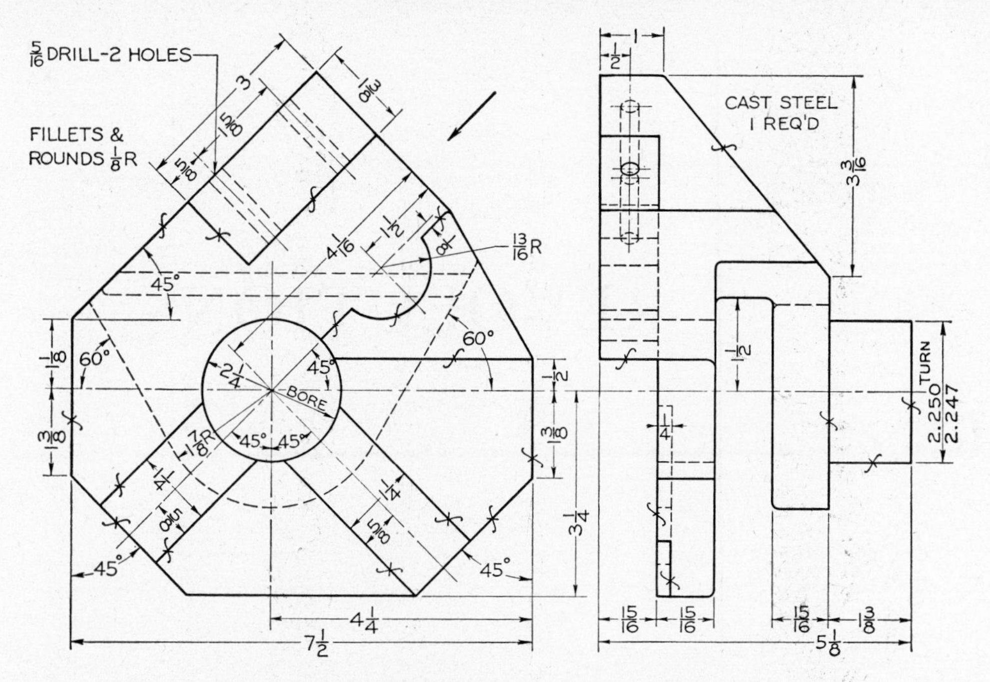

Fig. 526 Box Tool Holder for Turret Lathe.

Given: Front and R. Side views.

Req'd: Front and L. Side views, and complete Auxiliary view as indicated by arrow (Layout C-4).

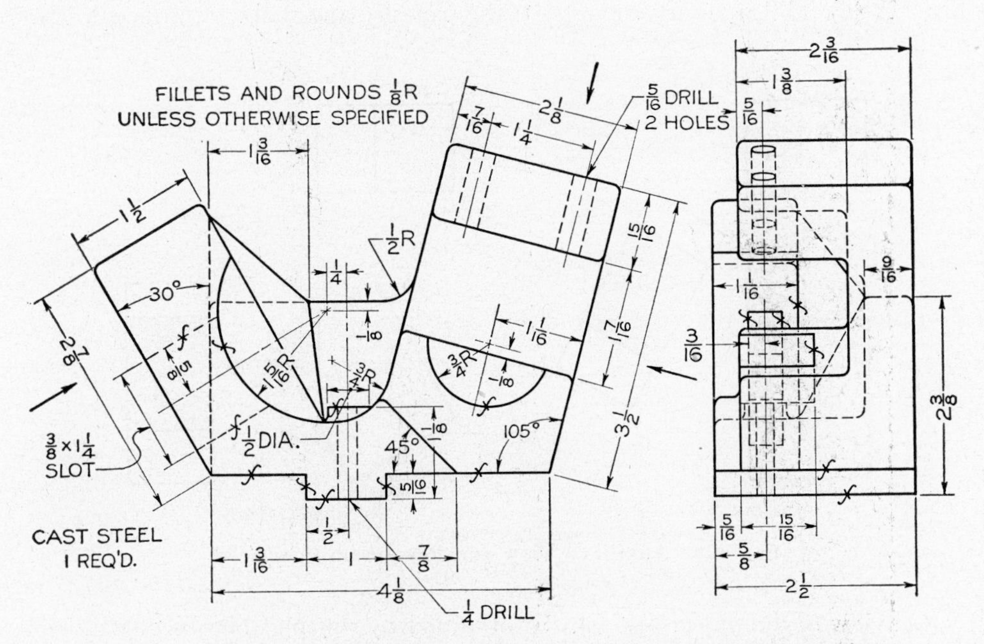

Fig. 527 Pointing Tool Holder for Automatic Screw Machine.

Given: Front and R. Side views.

Req'd: Front view and three partial Auxiliary views (Layout C-4).

CHAPTER 9

REVOLUTIONS

286. Revolutions Compared with Auxiliary Views. To obtain an auxiliary view, the observer shifts his position with respect to the object, as shown by the arrow in Fig. 528 (a). The auxiliary view shows the true size and shape of surface A. Exactly

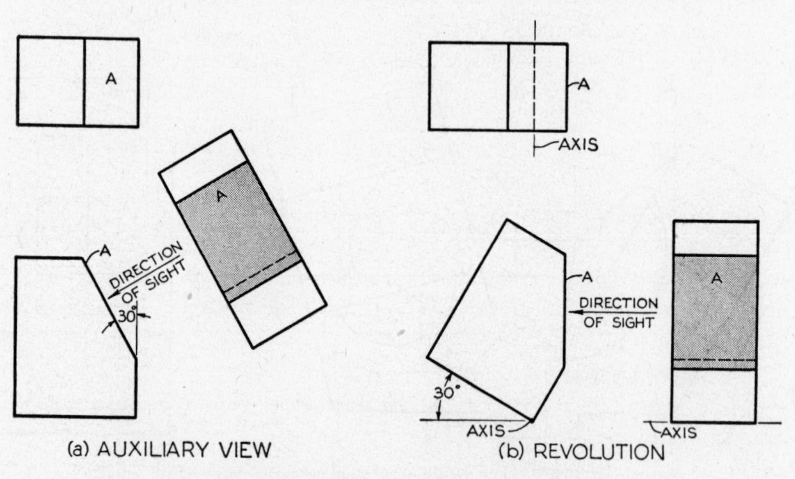

Fig. 528 Auxiliary View and Revolution Compared.

the same view of the object also can be obtained by shifting the object with respect to the observer, as shown at (b). Here the object is revolved until surface A appears in its true size and shape in the right-side view. The *axis of revolution* is assumed perpendicular to the frontal plane of projection, as shown.

Note that the view in which the axis of revolution appears as a point (in this case the front view) *revolves but does not change shape*, and that in the views in which the axis is shown as a line in true length the *dimensions of the object parallel to the axis do not change*.

To make a revolution drawing, the view on the plane of projection that is perpendicular to the axis of revolution is drawn first, since it is the only view that remains unchanged in size and shape. This view is drawn revolved either *clockwise* (as the hands of a clock move) or *counterclockwise* about a point that is the end view, or point view, of the axis of revolution. This point may be assumed at any convenient point on or outside the view. The other views are then projected from this view.

The axis of revolution is usually considered perpendicular to one of the three principal planes of projection. Thus an object may be revolved about an axis perpendicular to the horizontal, frontal, or profile planes of projection, and the views drawn in the new positions. Such a process is called a *primary revolution*. If this drawing is then used as a basis for another revolution, the operation is called a *successive revolution*. Obviously, this process may be continued indefinitely, which reminds us of the "chain reaction" in successive auxiliary views, Fig. 490.

287. Revolution about Axis Perpendicular to Frontal Plane. A primary revolution is illustrated in Fig. 529. An imaginary axis XY is assumed, about which the

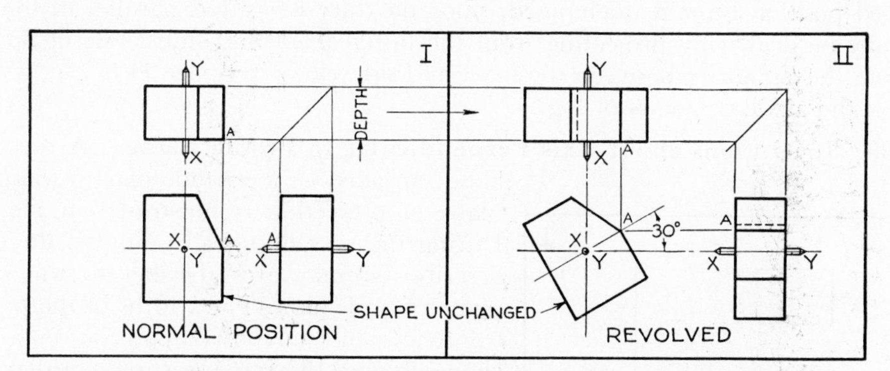

Fig. 529 Primary Revolution about Axis Perpendicular to Frontal Plane.

object is to revolve to the desired position. In this case the axis is selected perpendicular to the frontal plane of projection, and during the revolution all points of the object describe circular arcs parallel to that plane. The axis may pierce the object at any point or may be exterior to it. In Space II, the front view is drawn *revolved* (but not changed in shape) through the angle desired (30° in this case), and the top and side views are obtained by projecting from the front view. The *depth* of the top view and the side view is found by projecting from the top view of the first unrevolved position (Space I), *because the depth, since it is parallel to the axis, remains unchanged*. If the front view of the revolved position is drawn directly without first drawing the normal unrevolved position, the depth of the object, as shown in the revolved top and side views, may be drawn to known dimensions. No difficulty should be encountered by the student who understands how to obtain projections of points and lines, §212.

Note the similarity between the top and side views in Space II of Fig. 529 and some of the auxiliary views of Fig. 476 (c).

288. Revolution about Axis Perpendicular to Horizontal Plane. A revolution about an axis perpendicular to the horizontal plane of projection is shown in Fig. 530. An imaginary axis XY is assumed perpendicular to the top plane of pro-

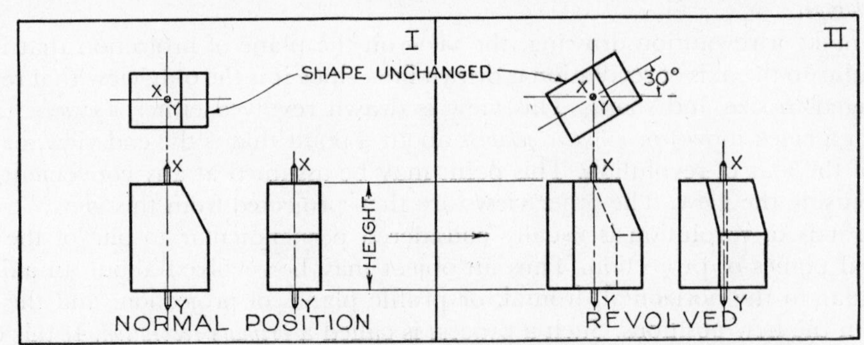

Fig. 530 Primary Revolution about Axis Perpendicular to Horizontal Plane.

jection, the top view is drawn revolved (but not changed in shape) to the desired position (30° in this case), and the other views are obtained by projecting from this view. During the revolution, all points of the object describe circular arcs parallel to the horizontal plane. The *heights* of all points in the front and side views in the revolved position remain unchanged, since they are measured parallel to the axis, and may be drawn by projecting from the initial front and side views of Space I.

Note the similarity between the front and side views in Space II of Fig. 530 and some of the auxiliary views of Fig. 477 (c).

289. Revolutions about Axis Perpendicular to Profile Plane. A revolution about an axis XY perpendicular to the profile plane of projection is illustrated in Fig. 531. During the revolution, all points of the object describe circular arcs parallel to the profile plane of projection. The widths of the top and front views in the revolved position remain unchanged, since they are measured parallel to the axis, and may be obtained by projection from the top and front views of Space I, or may be set off by direct measurement.

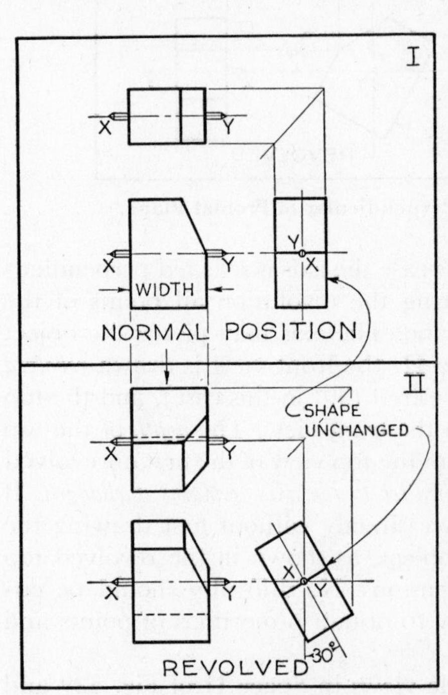

Fig. 531 Primary Revolution about Axis Perpendicular to Profile Plane.

Note the similarity between the top and front views in Space II of Fig. 531 and some of the auxiliary views of Fig. 478 (c).

290. Successive Revolutions. It is possible to draw an object in an infinite number of revolved positions by making successive revolutions. Such a procedure, Fig. 532, limited to three or four stages, offers excellent practice in multiview projection. While it is possible to make several revolutions of a simple object without the aid of a system of numbers, it is absolutely necessary in successive revolutions to assign a number or a letter to every corner of the object. See §212.

The numbering or lettering must be consistent in the various views of the several stages of revolution. Figure 532 shows four sets of multiview drawings numbered I, II, III, and IV, respectively. These represent the same object in different positions with reference to the planes of projection.

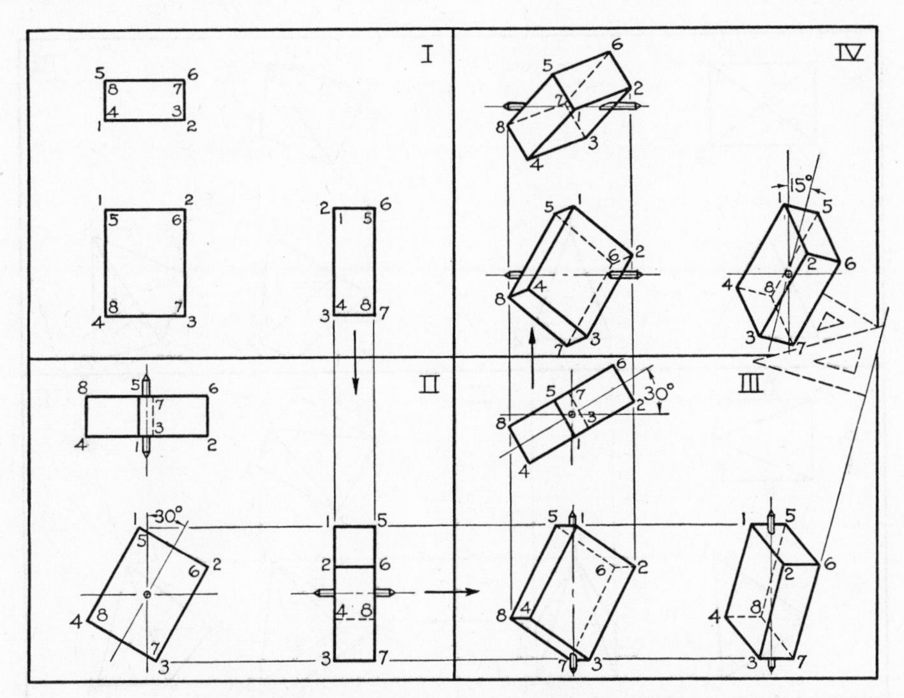

Fig. 532 Successive Revolutions of a Prism.

In Space I the object is represented in its normal position, with its faces parallel to the planes of projection. In Space II the object is represented after it has been revolved clockwise through an angle of 30° about an axis perpendicular to the frontal plane. The drawing in Space II is placed under Space I so that the side view, whose width remains unchanged, can be projected from Space I to Space II as shown.

During the revolution, all points of the object describe circular arcs parallel to the frontal plane of projection and remain at the same distance from that plane. The side view, therefore, may be projected from the side view of Space I and the front view of Space II. The top view may be projected in the usual manner from the front and side views of Space II.

In Space III the object is taken as represented in Space II and is revolved counter-clockwise through an angle of 30° about an axis perpendicular to the horizontal plane of projection. During the revolution, all points describe *horizontal* circular arcs and remain at the same distance from the horizontal plane of projection. The top view is copied from Space II but is revolved through 30°. The front and side views are obtained by projecting from the front and side views of Space II and from the top view of Space III.

In Space IV the object is taken as represented in Space III and revolved clock-wise through 15° about an axis perpendicular to the profile plane of projection. During the revolution, all points of the object describe circular arcs parallel to the profile plane of projection and remain at the same distance from that plane. The side view is copied, §135, from the side view of Space III but revolved through 15°. The front and top views are projected from the side view of Space IV and from the top and front views of Space III.

A similar successive revolution applied to a pyramid is shown in Fig. 533.

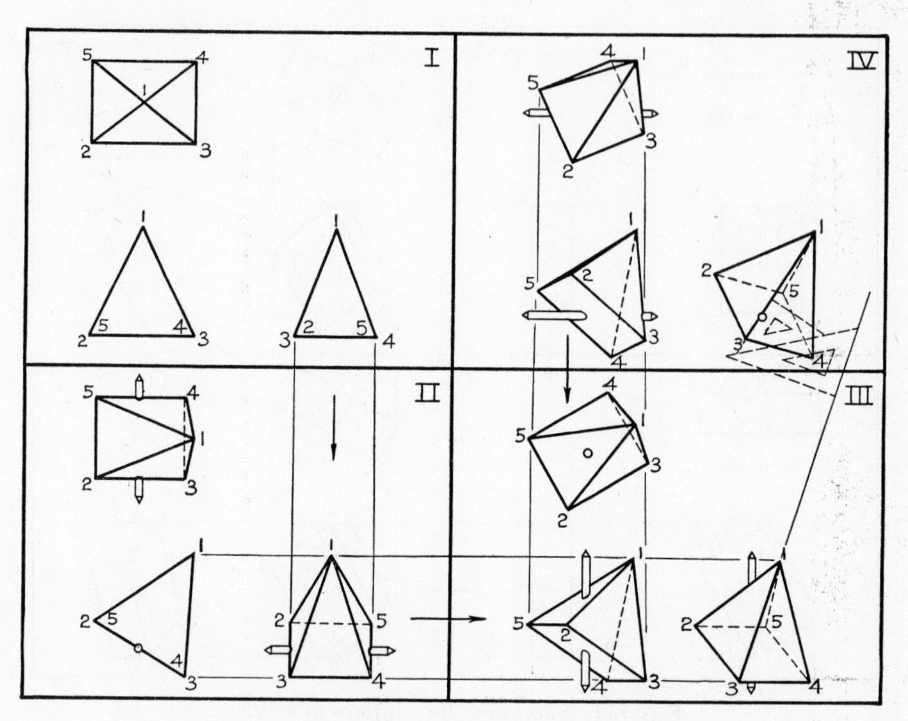

Fig. 533 Successive Revolutions of a Pyramid.

A convenient method of copying a view in a new revolved position is to use tracing paper as described in §136. Either a tracing can be made and transferred by rubbing, or the prick-points may be made and transferred, as shown.

In Spaces III and IV of Figs. 532 and 533, each view is an axonometric projection, §448. An isometric projection can be obtained by revolution, as shown in Fig. 824, and a dimetric projection, §472, can be constructed in a similar manner. If neither an isometric nor a dimetric projection is specifically sought, the successive revolution will produce a trimetric projection, §475, as shown in Figs. 532 and 533.

291. To Find the True Length of a Line. If a line is parallel to one of the planes of projection, its projection on that plane is equal in length to the line, Fig. 303. In Fig. 534 (a), the element AB of the cone is oblique to the planes of projection; hence its projections are foreshortened. If AB is revolved about the axis of the cone until it coincides with either of the contour elements, for example AB′, it will be shown in its true length in the front view because it will then be parallel to the frontal plane of projection.

Likewise, at (b), the edge of the pyramid CD is shown in its true length CD′ when it has been revolved about the axis of the pyramid until it is parallel to the frontal plane of projection. At (c), the line EF is shown in its true length when it has been revolved about a vertical axis until it is parallel to the frontal plane of projection.

The true length of a line may also be found by constructing a right triangle, as shown at (d), whose base is equal to the top view of the line, and whose altitude is the difference in elevation of the ends. The hypotenuse of the triangle is equal to the true length of the line.

In these cases the lines are revolved until parallel to a plane of projection. The true length of a line may also be found by leaving the line stationary but shifting the position of the observer; that is, the method of auxiliary views, §279.

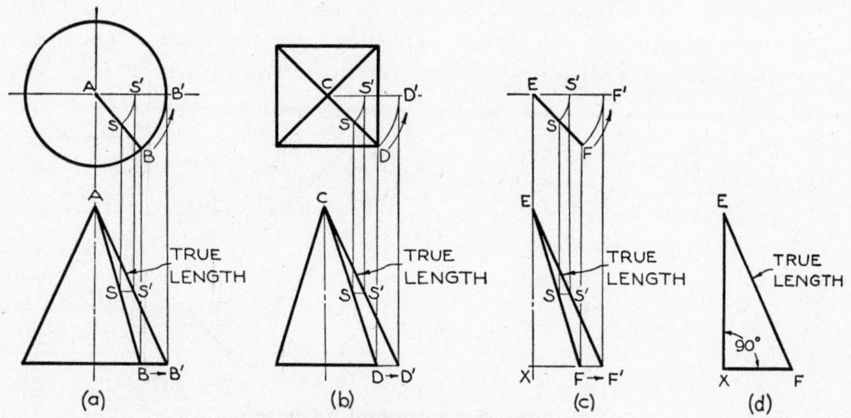

Fig. 534 True Length of a Line.

292. True Size of a Plane Surface. If a surface is parallel to one of the planes of projection, its projection on that plane is true size, Fig. 302. In Fig. 535 (a), the inclined surface 1-2-3-4 is foreshortened in the top and side views and appears as a line in the front view. Line 2-3 is taken as the axis of revolution, and the surface is revolved clockwise in the front view to the position 4_R-3 and projected to the side view at 4_R-1_R-2-3, which is the true size of the surface. In this case the surface was revolved until parallel to the profile plane of projection.

At (b), triangular surface 1-2-3 is revolved until parallel to the horizontal plane of projection so that the surface appears true size in the top view, as shown.

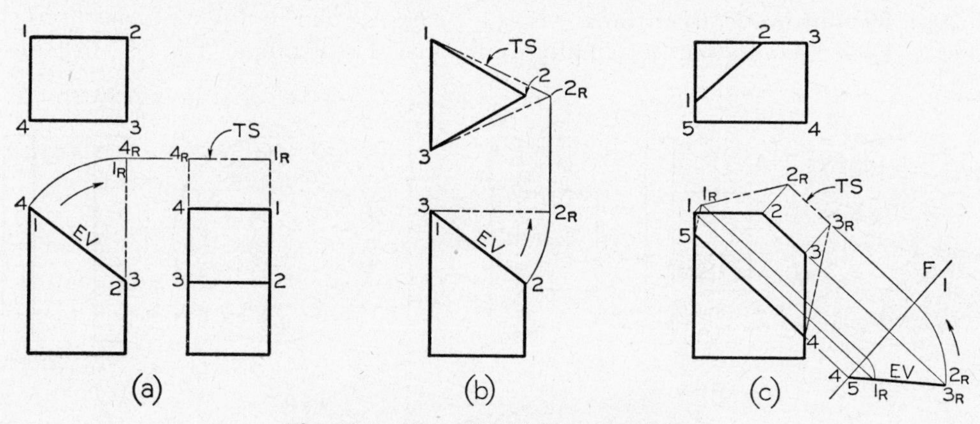

Fig. 535 True Size of Plane Surfaces.

At (c), the true size of the oblique surface cannot be found by a simple primary revolution. The true size can be found by two successive revolutions, or by a combination of an auxiliary view and a primary revolution. The latter is shown at (c). First, draw an auxiliary view that will show the edge view (EV) of the plane. See Fig. 491. Second, revolve the edge view of the surface until it coincides with the folding line F/1, as shown. All points in the front view, except those in the axis of revolution, will describe circular arcs parallel to the reference plane F/1. These arcs will appear in the front view as lines parallel to the folding line, such as $2\text{-}2_R$, $3\text{-}3_R$, etc. The true size of the surface is found by connecting the points with straight lines.

293. Revolution of Circles. As shown in Fig. 202 (a) to (c), a circle, when viewed obliquely, appears as an ellipse. In that case the coin is revolved by the fingers. The geometric construction of this revolution is shown in Fig. 536 (a). In

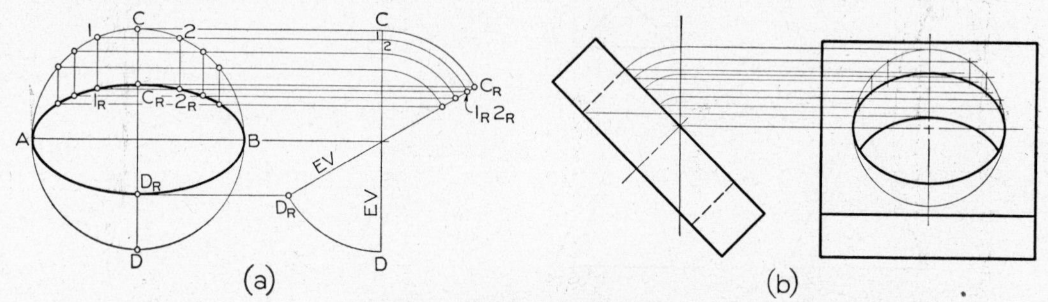

Fig. 536 Revolution of a Circle.

the front view the circle appears at ACBD, and in the side view as line CD, which is really the edge view of the plane containing the circle. In the side view, CD (the side view of the circle) is revolved through any desired acute angle to $C_R D_R$.

To find points on the ellipse, draw a series of horizontal lines across the circle in the front view. Each line will cut the circle at two points, as 1 and 2. Project these points across to the vertical line representing the unrevolved circle; then, revolve each point and project horizontally to the front view to establish points on the ellipse. Plot as many points as necessary to secure a smooth curve.

An application of this construction to the representation of a revolved object with a large hole is shown at (b).

294. Revolution Conventions. In some cases regular multiview projections are either awkward, confusing, or actually misleading. For example, Fig. 537 (a) shows

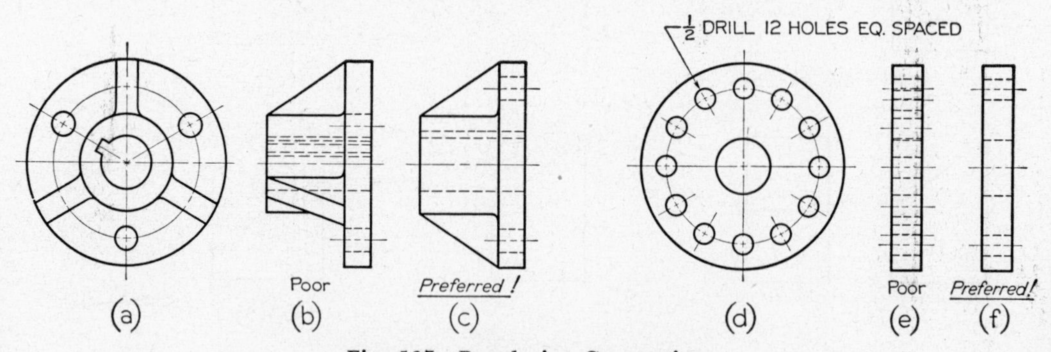

Fig. 537 Revolution Conventions.

an object that has three triangular ribs, three holes equally spaced in the base, and a keyway. The right-side view at (b) is a regular projection and is not recommended. The lower ribs appear in a foreshortened position, the holes do not appear in their true relation to the rim of the base, and the keyway is projected as a confusion of hidden lines.

The conventional method shown at (c) is preferred, not only because it is simpler to read, but requires less drafting time. Each of the features mentioned has been revolved in the front view to lie along the vertical center line from where it is projected to the correct side view at (c).

At (d) and (e) are shown regular views of a flange with many small holes. The hidden holes at (e) are confusing and take unnecessary time to draw. The preferred representation at (f) shows the holes revolved, and the drawing is clear.

Another example is shown in Fig. 538. As shown at (b), a regular projection results in a confusing foreshortening of an inclined arm. In order to preserve the

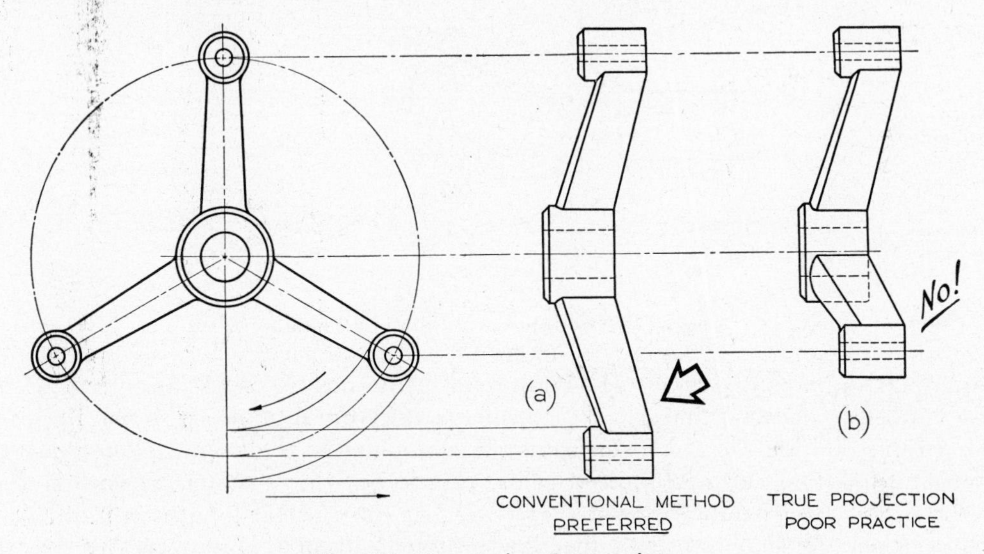

Fig. 538 Revolution Conventions.

appearance of symmetry about the common center, the lower arm is revolved to line up vertically in the front view so that it projects true length in the side view at (a).

Revolutions of the type discussed here are frequently used in connection with sectioning. Such sectional views are called *aligned sections*, §257.

295. Counterrevolution. The reverse procedure to revolution is *counterrevolution*. For example, if the three views of Space II in Fig. 529 are given, the object can be drawn in the unrevolved position of Space I by counterrevolution. The front view is simply counterrevolved back to its normal upright position, and the top and side views are drawn as shown. Similarly, in Fig. 532, the object may be counterrevolved from its position of Space IV to its unrevolved position of Space I by simply reversing the process.

In practice, it sometimes becomes necessary to draw the views of an object located on or parallel to a given oblique surface. In such an oblique position, it is very difficult to draw the views of the object, because of the foreshortening of lines. The work is greatly simplified by counterrevolving the oblique surface to a simple position, completing the drawing, and then revolving to the original given position.

An example is shown in Fig. 539. Assume that the oblique surface 8-4-3-7 (three views in Space I) is given, and that it is required to draw the three views of a prism $\frac{1}{2}''$ high, having the given oblique surface as its base. Revolve the surface about any horizontal axis X-X, perpendicular to the side view, until the edges 8-4 and 3-7 are

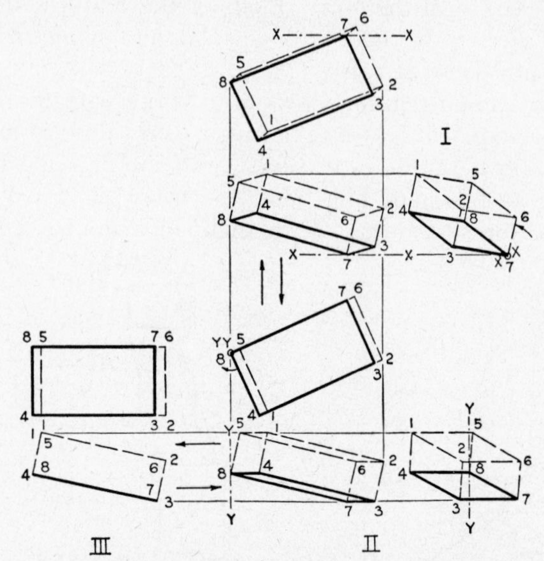

Fig. 539 Counterrevolution of a Prism.

horizontal, as shown in Space II. Then revolve the surface about any vertical axis Y-Y until the edges 8-7 and 4-3 are parallel to the frontal plane, as shown in Space III. In this position the given surface is perpendicular to the frontal plane, and the front and top views of the required prism can be drawn, as shown by dashed lines in the figure, because the edges 4-1 and 3-2, etc., are parallel to the frontal plane and are therefore shown in their true lengths, one-half inch. Having drawn the two views in Space III, counterrevolve the object from III to II and then from II to I to find the required views of the given object in Space I.

296. Revolution Problems. In Figs. 540-544 are problems covering primary revolutions, successive revolutions, and counterrevolutions. Additional problems, in convenient form for solution, are available in *Technical Drawing Problems*, by Giesecke, Mitchell, and Spencer, and *Technical Drawing Problems, Series 2*, by Spencer and Grant, both designed to accompany this text, and published by The Macmillan Company.

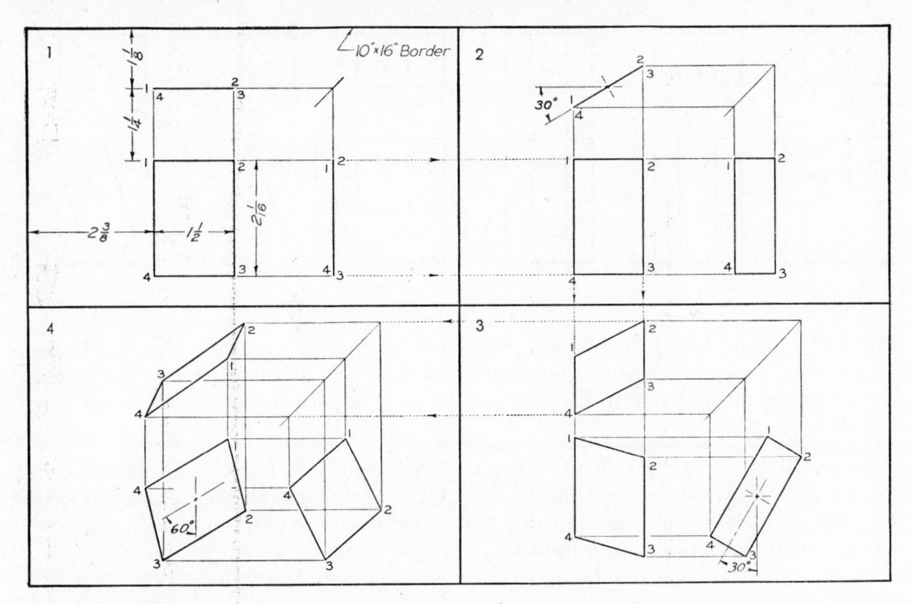

Fig. 540 Using Size B sheet, divide into four equal parts, as shown. Draw given views of rectangle, and then the primary revolution in Space 2, followed by successive revolutions in Spaces 3 and 4. Number points as shown. Omit dimensions. Letter date below border at left, and your name correspondingly at the right.

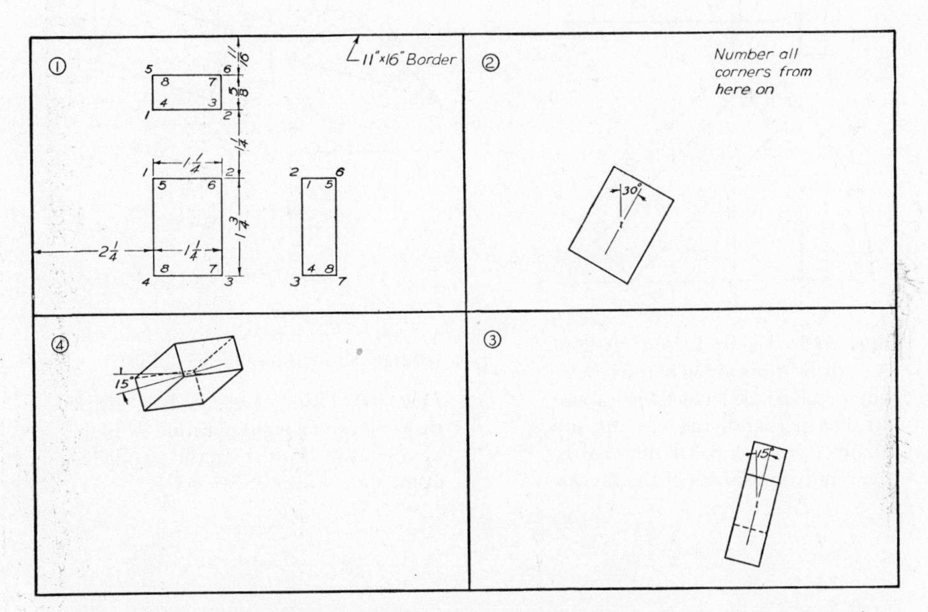

Fig. 541 Using Size B sheet, divide into four equal parts, as shown. Draw given views of prism as shown in Space 1; then draw three views of the revolved prism in each succeeding space, as indicated. Number all corners. Omit dimensions. Letter date below border at the left, and your name correspondingly at the right.

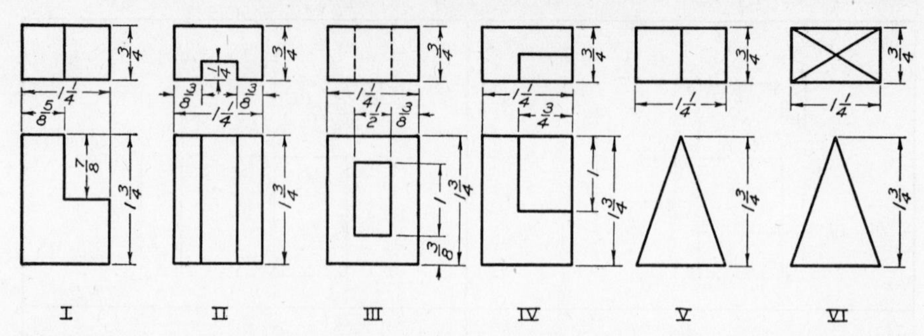

Fig. 542 *Prob. I:* Using Size B sheet, divide into four equal parts as in Fig. 540. In the upper two spaces, draw a simple revolution as in Fig. 529, and in the lower two spaces, draw a simple revolution as in Fig. 530; but for each problem use a block assigned from Fig. 542.

Prob. II: Using Size B sheet, divide into four equal parts as in Fig. 540. In the two left-hand spaces, draw a simple revolution as in Fig. 531, but use an object assigned from Fig. 542. In the two right-hand spaces, draw another simple revolution as in Fig. 531, but use a different object taken from Fig. 542, and revolve through 45° instead of 30°.

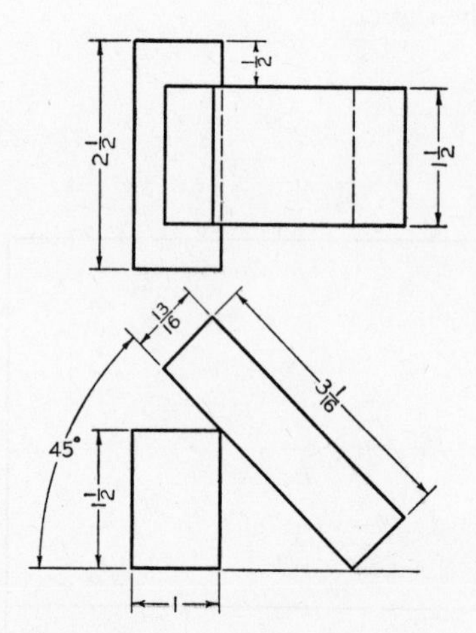

Fig. 543 Using Layout A-2 or A-3, draw three views of the blocks, but revolved 30° clockwise about an axis perpendicular to the top plane of projection. Do not change the relative positions of the blocks.

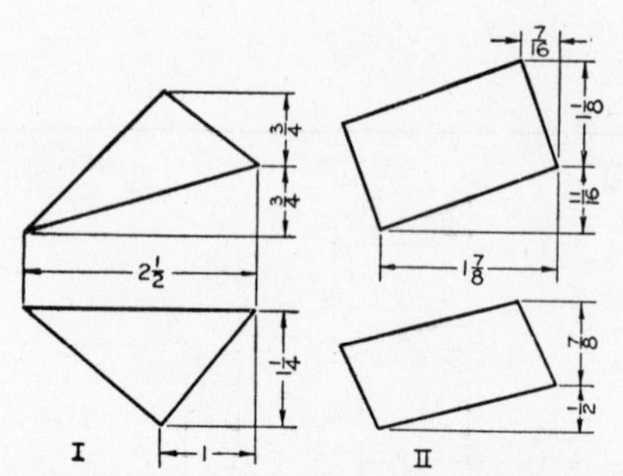

Fig. 544 *Prob. I:* Using Layout B-3, draw three views of a right prism $1\frac{1}{2}''$ high that has as its lower base the triangle shown above. See §295.

Prob. II: Using Layout B-3, draw three views of a right pyramid 2″ high, having as its lower base the parallelogram shown above. See §295.

CHAPTER 10

SHOP PROCESSES

by
*J. George H. Thompson** and
John Gilbert McGuire†

297. Introduction. The test of the usefulness of any working drawing is whether the object described can be satisfactorily produced without further information than that furnished on the drawing. The drawing must furnish information as to shape, size, material, and finish, and in general suggest the shop processes required to produce the desired object. It is the purpose of this chapter to provide the young engineer with some information about certain fundamental shop terms and processes and to assist him in using this information on his drawings. The student engineer should seize every opportunity to visit shops, in order that he may learn what can be done on various machines and thus be able to make practical specifications on his drawings. The draftsman in any organization must have a thorough knowledge of what can be done in his shop before he can properly indicate on his drawing the machining operations, heat-treatment, finish, and the accuracy that should be maintained on each part.

298. Shop Processes. The shop starts with what might be called *raw stock* and modifies this until it agrees with the detail drawing. The shape of the raw stock may have to be altered. For example, an automobile body shell may be pressed out of sheet steel on a massive press before it conforms with the drawing. The size may also have to be changed; for example, a 2″ diameter bar may have to be turned down on a lathe until it becomes 1.774″ diameter.

Changing the shape and size of the material of which a part is being made requires one or more of the following: (1) removing part of the original material, (2) adding more material, and (3) redistributing original material. Cutting, as turning on a lathe,

*Professor of Mechanical Engineering, Agricultural and Mechanical College of Texas.
†Professor of Engineering Drawing, Agricultural and Mechanical College of Texas.

or punching holes by means of a power press, removes material. Welding, brazing, soldering, metal spraying, and electro-chemical plating add material. Forging, pressing, drawing, extruding, and spinning redistribute material.

The shop is also concerned with the characteristics of the material of which the part is being produced. Frequently, the characteristics of the material may have to be altered to agree with the properties specified by the designer on the drawing. For example, the part may be required to have a Brinell hardness* of 400. Since steel is not ordinarily supplied in this degree of hardness, it would become necessary to harden such a part by heat treatment, §325.

299. Manufacturing Methods and the Drawing. Before the draftsman prepares a drawing for the production of a part, he should consider what manufacturing process is to be used, as this determines the representation of the detailed features of the part, the choice of dimensions, and the machining accuracy. Principal types are (1) *casting*, (2) *machining from standard stock*, (3) *welding*, (4) *forming from sheet stock*, and (5) *forging*. A knowledge of these processes, along with a thorough understanding of the intended use of the part, will help determine the basic manufacturing process. Drawings that reflect these manufacturing methods are shown in Fig. 545.

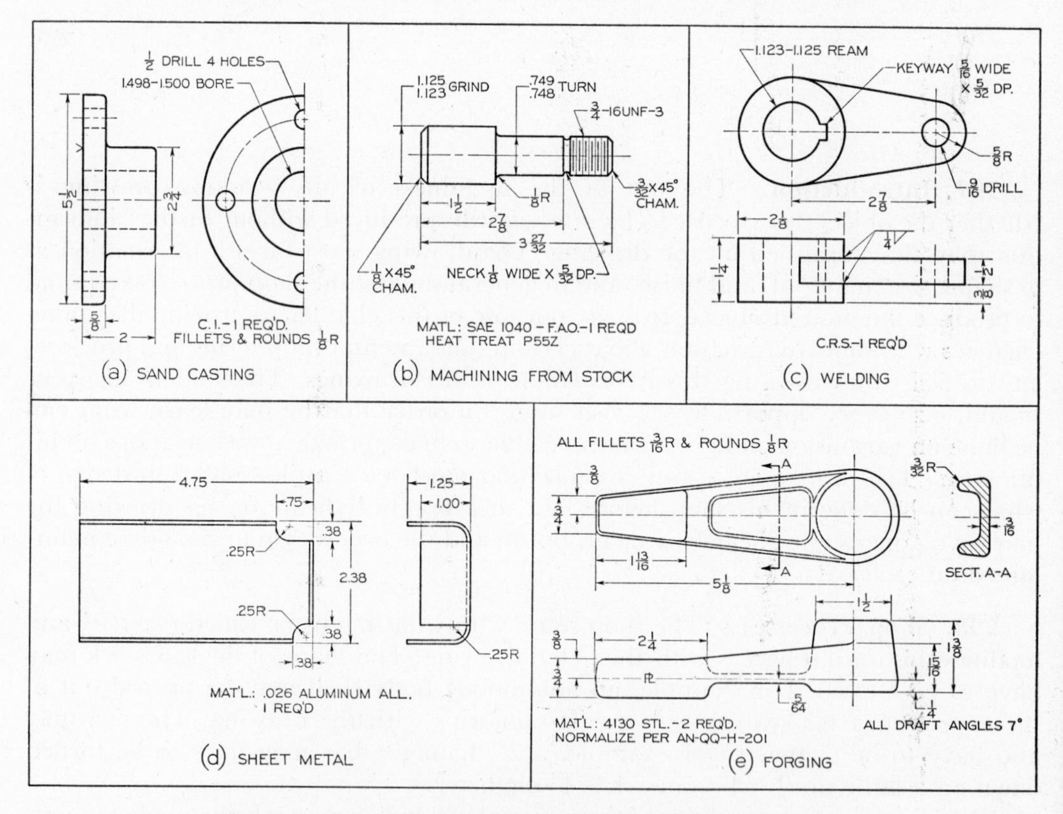

Fig. 545 Drawings for Different Manufacturing Processes.

*A Brinell number indicates hardness. The number is obtained by measuring the diameter of indentation produced by a standard steel ball loaded according to specified conditions. A steel of "150 Brinell" is moderately soft. "250 to 400 Brinell" is as hard as can ordinarily be cut with single-pointed steel cutting tools. Harder materials can be machined by grinding or by the use of non-ferrous cutting tools, such as Stellite or Tungsten-Carbide.

In sand casting, Fig. 545 (a), all unfinished surfaces remain rough, and all rough corners are filleted or rounded. Sharp corners appear where two surfaces intersect if at least one is a finished surface. Finish marks are shown on the edge views of all finished surfaces. See §§341 and 342. Pattern draft is usually not shown on the drawing. Dimensions are given for the patternmaker and the machinist on the same drawing, §352.

In drawings of parts machined from standard stock, (b), most surfaces are machined. In some cases, as on shafting, the surface existing on the raw stock is often accurate enough to be left unfinished. Corners are usually sharp, but fillets and rounds are machined when necessary. For example, an interior corner may be machined with a radius to provide greater strength.

On welding drawings, (c), the several pieces are cut to size, brought together and then welded. Welding symbols, Chapter 25, are used to indicate the welds required. Generally there are no fillets and rounds. Certain surfaces may be machined after welding, or in some cases before welding. Notice that lines are shown where the separate pieces are joined.

On sheet-metal drawings, (d), the thickness of material is uniform and is usually given in the material specification note. Bend radii and bend reliefs at corners are specified according to standard practice. See §712. For dimensions, whole numbers and common fractions may be used. In the aircraft and automotive industries especially, the complete decimal system, §368, may be used as shown in the figure. Allowances of extra material for joints may be required.

For forged parts, separate drawings are usually made for the diemaker and for the machinist. Thus, a forging drawing, (e), provides only the information to produce the forging, and the dimensions given are those needed by the diemaker. See §324. All corners are rounded and filleted, and are so shown on the drawing. The draft is drawn to scale, and is usually specified by degrees in a note, as shown.

300. Sand Casting. While there are a number of processes used to produce castings, sand molds are used far more extensively than any other types of molds.

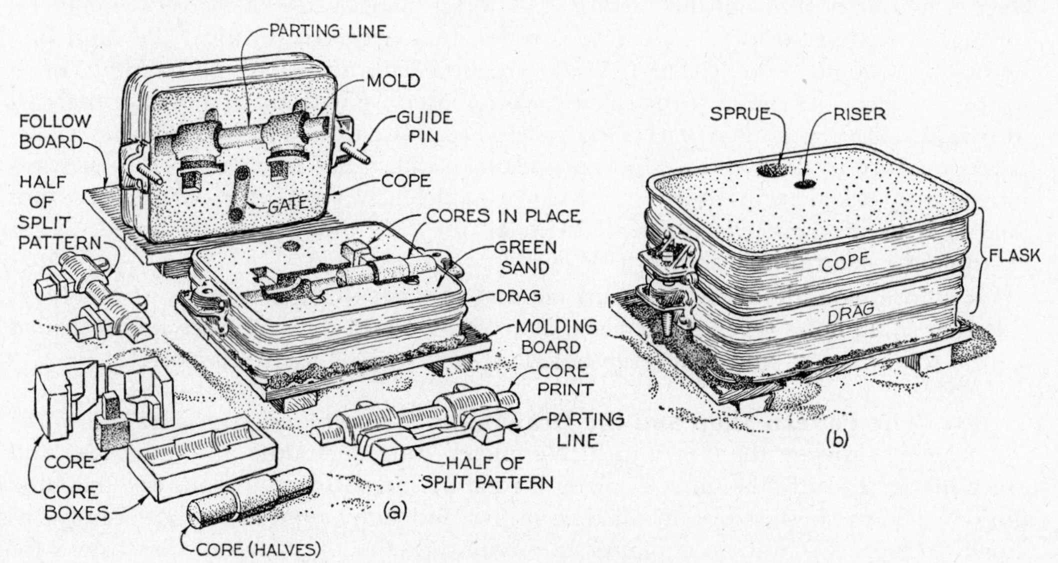

Fig. 546 Sand Molding (of Table Bracket, Fig. 737).

Sand molds are made by ramming sand around a wood "model," or pattern, and then carefully withdrawing the pattern, leaving a cavity to receive the molten metal, as shown in Fig. 546 (a). The sand is contained in a two-part box called a *flask*, (b), the upper part of which is called the *cope* and the lower the *drag*. For more complicated work, one or more intermediate boxes, called *cheeks*, may be introduced between the cope and the drag. The pattern must be of such shape that it will "pull away" from both the cope and the drag. The plane of separation of the two halves of the pattern marks the *parting line* on the pattern. On each side of the parting line the pattern must be tapered slightly to permit the withdrawal of the pattern from the sand, unless a segmented pattern is used. This taper is known as the *draft*. Although the draft is usually not shown, and dimensions are not given for it on the working drawing, the design must be such that the draft can be properly built into the pattern by the patternmaker.

A *sprue stick*, or round peg, is placed in position during the ramming process and then removed to leave a hole through which the metal may be poured. The part of the hole adjacent to the casting is called the *gate*, and the vertical part the *sprue*, Fig. 546. On some molds another hole, known as the *riser*, is provided to allow the gases to escape and to provide a reservoir of metal that feeds back to the casting as it cools and shrinks.

Since shrinkage occurs when metal cools, patterns are made slightly oversize. The patternmaker accomplishes this by using a *shrink rule* whose units are oversize according to the shrinkage characteristics of the metal used. For example, cast iron shrinks about $\frac{1}{8}''$ per linear foot as it cools, while steel shrinks about $\frac{3}{16}''$ per linear foot. Although allowance for shrinkage is not shown on the working drawing but is taken care of entirely by the pattern shop, the patternmaker must refer to the drawing for the kind of metal to be used in the casting.

A *core print* is a projection added to a pattern to form a cavity in a mold into which a corresponding portion of a *core* will rest, as shown in Fig. 546, thus forming an anchor to hold the core in place. Cores are made of sand and are used to provide certain hollow portions of the casting. The most common use of a core is to extend through a casting to form a *cored hole*. When it is necessary to form the sand into shapes that would ordinarily not permit the necessary adhesion and strength, or in instances where the shape of the casting would interfere with the removal of the pattern, a *dry sand core* is used. Dry sand cores, Fig. 546, are made by ramming a prepared mixture of sand and a binding substance into a *core box;* the core is then removed and baked in a *core oven* to make it sufficiently rigid. A *green sand core* (not baked) is used when it is practical to make the core along with the mold, as for example the central hole shown in Fig. 547.

It is from considerations of shrinkage and of draft that one arrives at a general rule that small holes (even if so placed as to draw out of the sand) are better drilled from a solid casting, and that large holes are better cored and then bored.

301. The Pattern Shop and the Drawing. The *pattern shop* receives the working drawing showing the object in its completed state, including all dimensions and finish marks. Usually the same drawing is used by the pattern shop and the machine shop; hence, it should contain all dimensions and notes needed for both shops, as shown in Fig. 547. Some companies follow the practice of giving all dimensions for the pattern shop or forge shop in pencil and those for the machine shop in ink. On the blueprint, the difference is easily distinguishable. If the part is large and complicated, two separate drawings are sometimes made, one showing the pattern dimensions, and the other the machine dimensions.

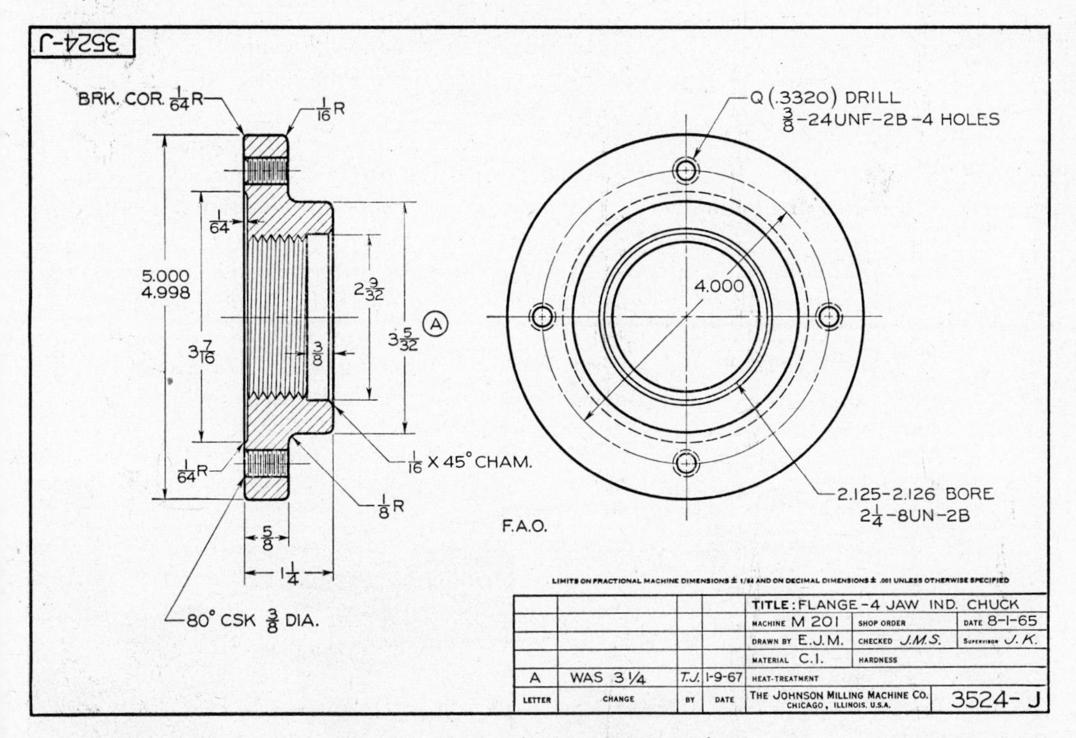

Fig. 547 A Detail Working Drawing.

Pattern dimensions, §352, which are used to make a pattern, should not be closer than $\frac{1}{16}''$, while dimensions for the machine shop must often reflect a tolerance of a few thousandths of an inch or less. Therefore, to dimension correctly, a careful study of a part must be made to determine which dimensions are to be used by the patternmaker, §303.

Finish marks, §342, are as important to the patternmaker as to the machinist because additional material on each surface to be machined must be provided. For small and medium-sized castings, $\frac{1}{16}''$ to $\frac{1}{8}''$ is usually sufficient; larger allowances are made if there is probability of distortion or warping. On the *Flange*, Fig. 547, it is necessary for the patternmaker to provide material for finish on all surfaces.

Sometimes the patternmaker works directly from the drawing; in other cases, he may find it desirable to make his own pattern layout. Except for simple objects, it is common practice for the pattern shop to prepare full-size (to shrink rule) pattern layout drawings on white pine boards. Wood is used instead of paper because it is more durable and because paper stretches and shrinks excessively. The pattern is then checked against this *pattern layout drawing* on which draft, shrinkage, coreprint dimensions, and other such information are clearly shown.

302. Fillets and Rounds. *Fillets* (inside rounded corners) and *rounds* (outside rounded corners) must be taken care of on the pattern, in order to provide for strength and appearance in the casting, Fig. 548. Crystals of cooling metal tend to arrange themselves perpendicular to the exterior surfaces as indicated at (b) and (c). If the corners of a casting are rounded as shown at (c), a much stronger casting results. It has been demonstrated that the load a part can safely carry may sometimes be doubled by simply increasing the radius of a fillet; therefore, a fillet should have as large a radius as possible.

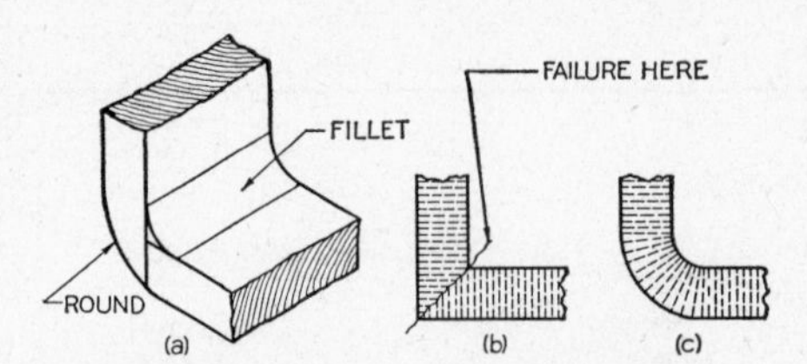

Fig. 548 Fillets and Rounds.

Rounds are constructed by rounding off the corners of the pattern (by sanding, planing, or turning) while fillets are constructed of leather, wax, or wood. Preformed leather, with a range of radii, is available for this purpose. A leather fillet of proper radius is fastened with glue and firmly pressed in place with the use of an unheated spherical tool, as shown in Fig. 549 (a). Wax fillets are applied in much the same way as leather but with the use of a hot spherical tool and without the use of glue, as shown at (b). For some cylindrical patterns it is possible to form (on the lathe) the proper radius for a fillet as an integral part of the wood pattern.

(a) Application of Leather Fillet to Pattern.

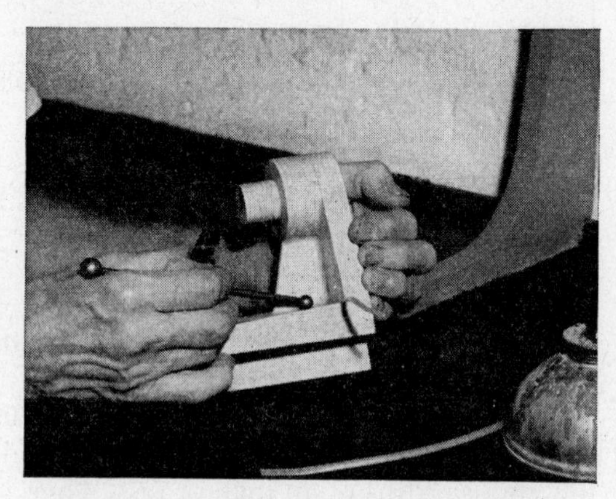

(b) Application of Wax Fillet to Pattern.

Fig. 549 Fillets.

All fillets and rounds should be shown on the drawing, drawn to scale with the use of a compass or with a circle guide, as in Fig. 547. For a general discussion of fillets and rounds from the standpoint of representation on the drawing, see §§239-241. For dimensioning of fillets and rounds, see §341.

303. Pattern Construction. After the patternmaker has examined the drawing, Fig. 547, and has determined such things as material and finish, he constructs the pattern, Fig. 550, of a durable wood (usually white pine or mahogany), or of plastic, using dimensions given on the working drawing or pattern layout drawing and making proper allowances for metal shrinkage, machining, and draft. At (a) a cylindrical block of wood is first mounted on the face plate of a wood lathe and then turned to the proper diameter (5″ plus allowance for shrinkage and machining). At (b) the hub diameter ($3\frac{5}{32}$″ on working drawing) is checked. It should be noted that the $\frac{1}{8}$″ fillet between the hub and the flange has been formed on the pattern. Also, it may be noted at (c) that the flange thickness $\frac{5}{8}$″ on working drawing) is checked with a shrink rule, and at (d) the center hole is being sanded on a spindle sander to provide a 1° draft taper to the hole previously turned on the lathe. If a quantity of molds were to be made from the pattern, the pattern stock would probably be glued up from several pieces of wood with the grain arranged to minimize warpage. For a large number of molds, a *master pattern* would be constructed of wood and a duplicate working pattern cast in aluminum for actual use in the foundry. In this way the master pattern would be protected from excessive wear.

(a) Turning to Required Diameter.

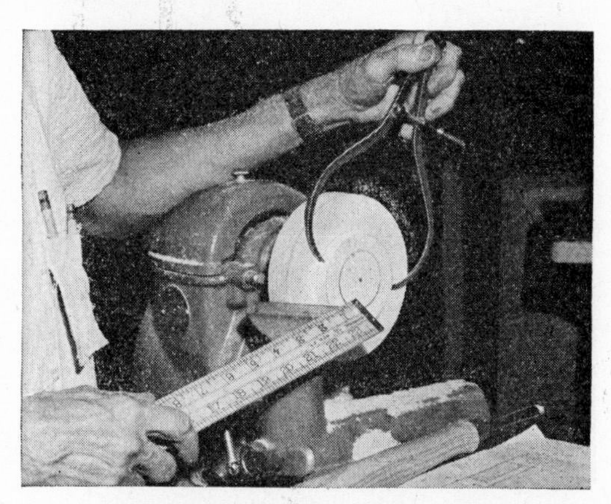

(b) Checking Diameter of Hub.

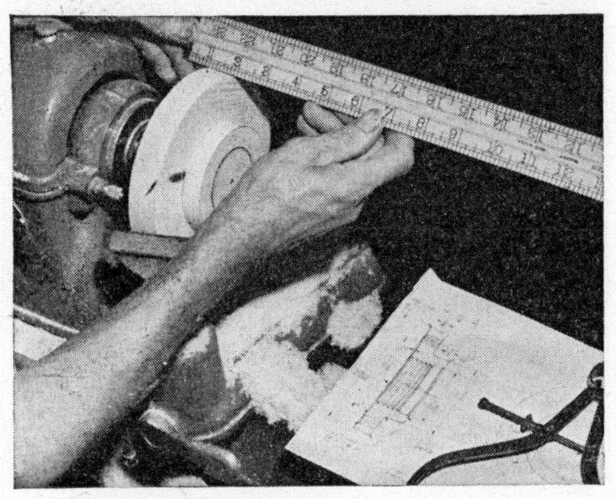

(c) Checking Thickness of Flange.

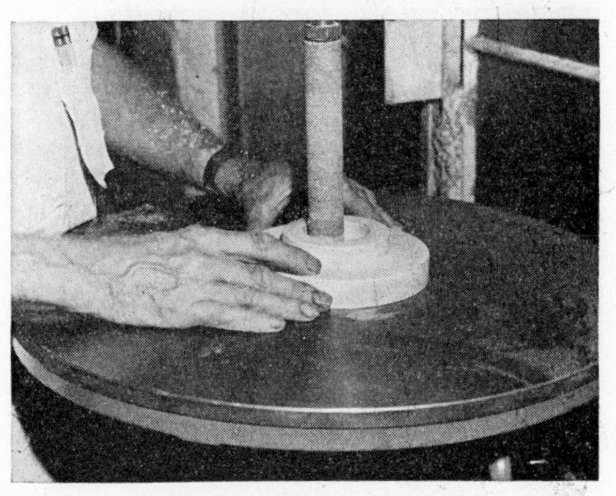

(d) Sanding Draft in Hole.

Fig. 550 Making a Pattern.

304. The Foundry. After the pattern has been completed, as described above, it is sent to the foundry for casting. The steps in preparing a sand mold for the *Flange*, Fig. 547, are illustrated in Fig. 551. At (a) sand has been rammed around the pattern, and then at (b) the pattern has been carefully removed, leaving a cavity for molten metal. After the cope has been placed in position on the drag at (c), molten metal is poured through the *sprue* (hole leading to cavity) into the mold. After sufficient cooling time, the casting is removed from the broken mold, as shown at (d).

Although sand molds* are the least expensive to make, sometimes *plaster of Paris*

*See *Materials and Processes*, 2nd Edition, edited by James F. Young of General Electric Company (New York-John Wiley 1954).

molds are used when it is desired to produce castings with smoother surfaces. Stainless steel, for example, can be successfully cast in plaster of Paris molds.

The lost wax process* produces castings of great dimensional accuracy and surface detail. In this process, a wax pattern is melted after the mold has been formed. Thus, none of the details of the mold are injured by the removal of the pattern, and for that reason shapes may be molded that would be impossible if a pattern had to be drawn out of the mold.

(a) Sand Rammed Around Pattern.

(b) Pattern Removed From Mold.

(c) Pouring of Molten Metal.

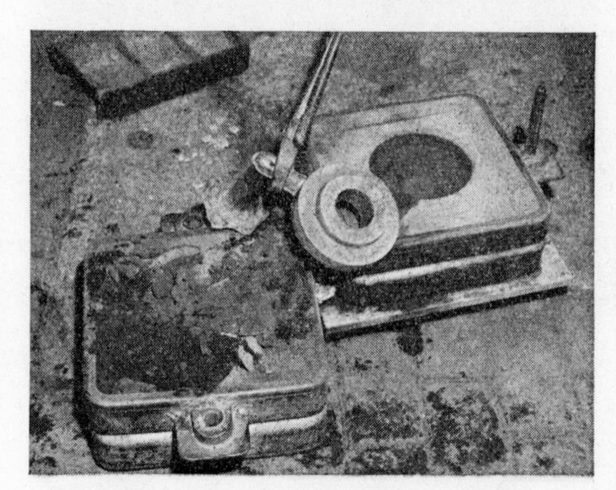

(d) Removal of Casting From Mold.

Fig. 551 Preparation of Mold and Pouring Molten Metal.

A casting process of growing importance is *centrifugal casting*. Molten metal is poured into a mold that is already rotating and that may continue to rotate until the metal has solidified. The effect of centrifugal force produces a less porous casting than that produced in a sand mold. The process is extensively used in the manufacture of cast iron pipe, and for steel gears and discs.

Die casting is the process of forcing molten metal into metallic dies by a force greater than atmospheric pressure. These castings are accurate in size and shape and possess surfaces superior in appearance and accuracy to those produced by other casting processes. Die castings, as a rule, require little or no machining. Although

Ibid.

die castings generally cannot be heat-treated or put into service at high temperatures, this process is the fastest of the casting processes, and consequently the least expensive for mass produced items, such as 'automobile carburetors, door handles, and radio and television chassis.

305. Powder Metallurgy. Powder metallurgy is an old process (dating back to 3000 B.C.) but is today an important technique of producing metallic parts from powders of a single metal, of two or more metals, or of a combination of metals and nonmetals. It consists of successive steps of mixing the powders, compressing them at high pressures (30 or 40 tons per sq. in.) into a preliminary shape, and heating them at high temperatures, but below the melting point of the principal metal. This process bonds the individual particles and produces a part with smooth surfaces that are quite accurate in dimension ($\pm$ .001 in. per in. in the direction perpendicular to the direction of the applied pressure and approximately $\pm$ .005 in. per in. in the direction of pressure).

Some of the advantages of this process include the elimination of scrap, elimination of machining, suitability to mass production, better control of composition and structure of a part, and the ability to produce parts made from a mixture of metals and nonmetals. Some of the disadvantages include high cost of dies, lower physical properties, higher cost of raw materials, and the limitation on design of a part. Powders when pressed do not flow around corners, nor do they transmit pressures as a liquid; therefore a part designed for casting will normally require redesigning in order to be made by powder metallurgy. Much specific information on this subject is available for the guidance of the designer.*

306. The Machine Shop and the Drawing. The casting is sent from the foundry to the machine shop for machining. The drawing is used in the machine shop to obtain information for the machining operations. In some cases, a special drawing is made for the machine shop, and in rare instances, a drawing is furnished for each machining operation. As a rule, however, the working drawings used in the preceding shops are also used in the machine shop to bring the product to completion.

Surfaces to be machined must be held to much greater accuracy than cast surfaces, and consequently the dimensions must reflect the desired accuracy for machining.

307. Machine Tools. Some of the more common *machine tools* are the *engine lathe*, *drill press*, *milling machine*, *shaper*, *grinding machine*, *planer*, and *boring mill*. A brief description of these machines will be given in the following paragraphs; a more detailed discussion of machines used in the machine shop may be found in the text, *Machine Shop Training Course*.†

308. Engine Lathe. The *engine lathe*, Fig. 552, is one of the most versatile machines used in the machine shop, and on it are performed such operations as *turning*, *boring*, *reaming*, *facing*, *threading*, and *knurling*. Figure 565 shows how a workpiece may be held in the lathe by means of a *chuck* (essentially a rotating vise). The *cutting tool* is fastened in the *tool holder* of the lathe and fed mechanically into the work as required. Several lathe operations will be illustrated in the following pages.

**Ibid.*, Chapter 16.
†Jones, Franklin D., *Machine Shop Training Course*, Vol. 1 and II (New York: The Industrial Press, 1940).

309. Drill Press. The *drill press*, Fig. 553, is one of the most-used machine tools in the shop. Some of the operations that may be performed on this machine are *drilling*, *reaming*, *boring*, *spot facing*, *counterboring*, and *countersinking*. The *sensitive drill press* is used for light work and is fed by hand. The *heavy-duty drill press* is used for

Fig. 552 Engine Lathe.

heavy work and is fed mechanically at the required speed. The *radial drill press*, shown in Fig. 553, with its adjustable head and spindle, is very versatile and is especially suitable for large work. A *multiple-spindle drill press* supports a number of spindles driven from the same shaft and is used in mass production. A table of twist drills is given in Appendix 14.

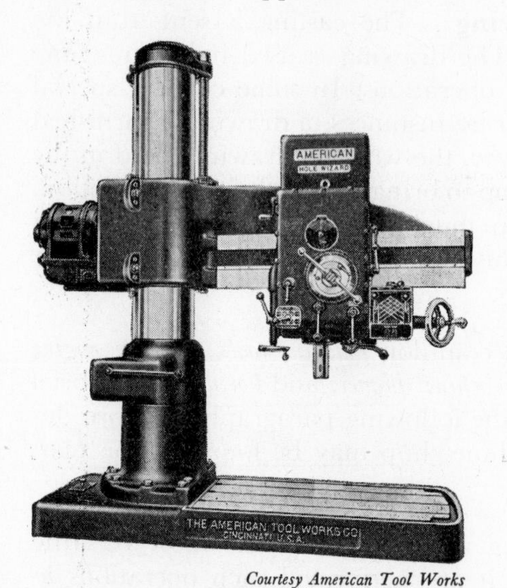

Courtesy American Tool Works

Fig. 553 Radial Drill Press.

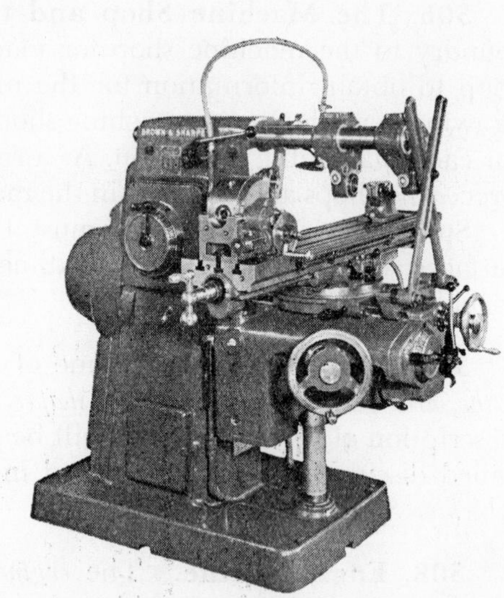

Courtesy Brown & Sharpe Mfg. Co.

Fig. 554 Milling Machine.

310. Milling Machine. On the *milling machine*, Fig. 554, cutting is accomplished by feeding the work into a rotating cutter. Gear cutting, Fig. 555, is one of the principal uses of the milling machine, although it is also used for cutting keyways, and with the many different cutters available, Fig. 556, it is possible to produce just about any surface. In addition, it is practical to drill, ream, and bore on the milling machine.

Fig. 555 Cutting Teeth on Gear in Milling Machine.

311. Shaper. On the *shaper*, Figs. 557 and 558, work is held in a vise while a single-pointed cutting tool mounted in a reciprocating head is forced to move in a straight line past the stationary work. Between succeeding strokes of the tool, the vise that holds the work is fed mechanically into the path of the tool for the next cut alongside the one just completed.

312. Planer. On the *planer*, Fig. 559, work is fixed to a table that is moved mechanically so that the cutting takes place between stationary tools and moving work. This machine, with its reciprocating bed, and a tool head that is adjustable both horizontally and vertically, is used principally for machining large plane surfaces, or surfaces on a large number of pieces, as shown in the figure. In the extremely large planers, the table is stationary and the tool actually is carried along a track past the work in order to cut.

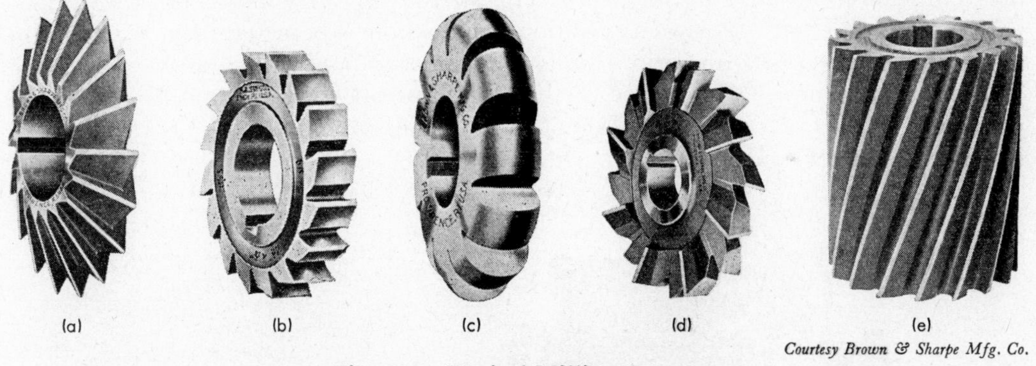

Courtesy Brown & Sharpe Mfg. Co.

Fig. 556 Typical Milling Cutters.

313. Grinding Machine. A *grinding machine*, Fig. 560, is used for removing a relatively small amount of material to bring the work to a very fine and accurate finish. In grinding, the work is fed mechanically against the grinding wheel, and the depth of cut may be varied from .001″ to .00025″. Attachments for grinding machines are available for grinding plane surfaces as well as both external and internal cylindrical and conical surfaces. Also, specially formed grinding wheels are often used to cut gear teeth, threads, and other shapes.

314. Boring Mill. The *vertical boring mill*, Fig. 561, is used for facing, turning, and boring on heavy work. Castings weighing up to twenty tons are commonly handled on this machine. The vertical boring mill has a large rotating table and a non-rotating cutting tool that moves mechanically into the work.

The *horizontal boring machine* and the *jig borer* are similar to the milling machine in that the cutting action is between a rotating tool and non-rotating work. The *horizontal boring machine* is suitable for accurate boring, reaming, facing, counter-boring, and milling of pieces larger than could be handled on the milling machine.

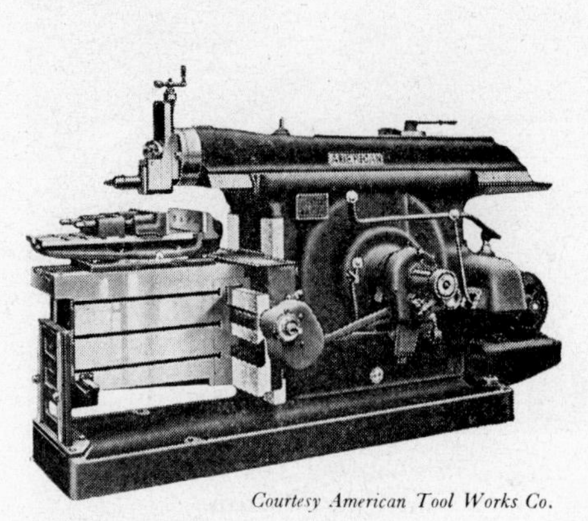

Courtesy American Tool Works Co.

Fig. 557 Shaper.

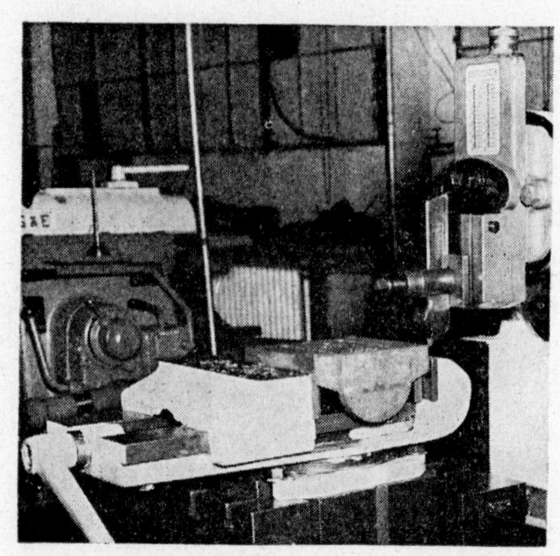

Fig. 558 Machining Plane Surface on Shaper.

The *jig borer* is a precision machine that somewhat resembles a drill press in its basic features of a rotating vertical spindle supporting a cutting tool and a stationary horizontal table for holding the work. The *precision jig borer*, however, is equipped with a table that can be locked in position while a hole is being cut, but that may be moved between cutting operations so as to locate one hole with respect to another. The work is not moved with respect to the table; instead, the table may be accurately positioned in two perpendicular directions by means of two accurate lead screws or by micrometer measuring bars.

Courtesy Cincinnati Planer Co.

Fig. 559 Planer.

Since conditions for boring vary, it is difficult to give a figure for accuracy of a bored hole. On S.I.P. (Swiss) jig-boring machines, holes are said to be bored within 0.00008″ of true size and are said to be located to within 0.00005″ accuracy. Diamond-boring machines* are said to produce holes to the following limits: 0.00001″ for

Courtesy Brown & Sharpe Mfg. Co.

Fig. 560 Grinding Machine.

Courtesy Cincinnati Planer Co.

Fig. 561 Vertical Boring Mill.

out-of-round, 0.00001″ for straightness, and 0.00001″ for size. Thus the boring operation at its best is capable of very accurate results. However, the tables in Appendices 5-11 are more representative of current practice.

315. Broaching. *Broaching* is illustrated in Fig. 562, in which a broach is shown in an arbor press cutting a keyway in a small gear. Broaching is basically similar to a single-stroke filing operation. As the broach is forced through the work, each succeeding tooth bites deeper and deeper into the metal, thus enlarging and forming the keyway as the broach passes through the hole. Another typical broach, together with the corresponding drawing calling for the use of this broach, is shown in Fig. 563.

The illustrations in Figs. 562 and 563 are examples of *internal broaching*. Square, cylindrical, hexagonal, and other shaped holes are produced by this process. An initial hole, produced by drilling, punching, or otherwise, is necessary in order to permit the broach to enter the work. Thus, internal broaching only enlarges and improves the shape of a hole that has already been produced by some other operation.

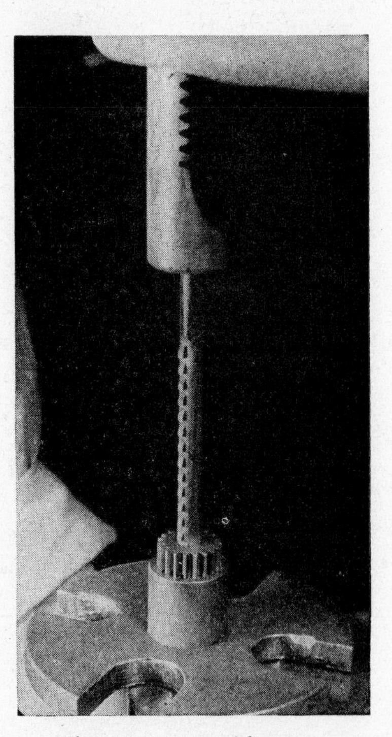

Fig. 562 Broaching.

*Diamond is the material used for the cutting tool.

In addition to internal broaching, *surface broaching* is used to produce flat surfaces on such parts as engine blocks and cylinder heads.

Broaching is distinctly a mass-production process in which there is usually involved a large expenditure of money both for the broaches themselves and for the special machines developed for their use in mass production. Broaching is one of our most accurate cutting processes, and work produced by it is of a high order of precision. It is common to hold holes to within .0005″ of the desired diameter with this process.

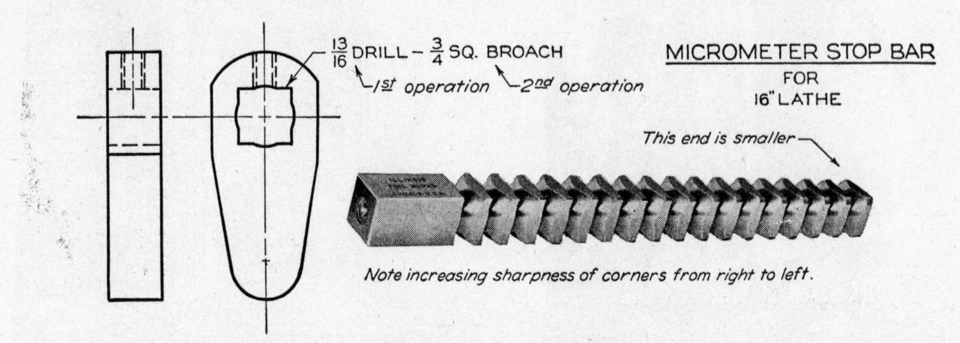

Fig. 563 Broaching.

316. Holes. Holes are produced in metal by *coring*, *piercing* (punching), *flame cutting*, or *drilling*. The first three of these methods are very rough. Drilling, Figs. 564 (a) and 565, while producing a hole superior in finish to coring, piercing, or flame cutting, does not produce a very accurate hole either in roundness, straightness,

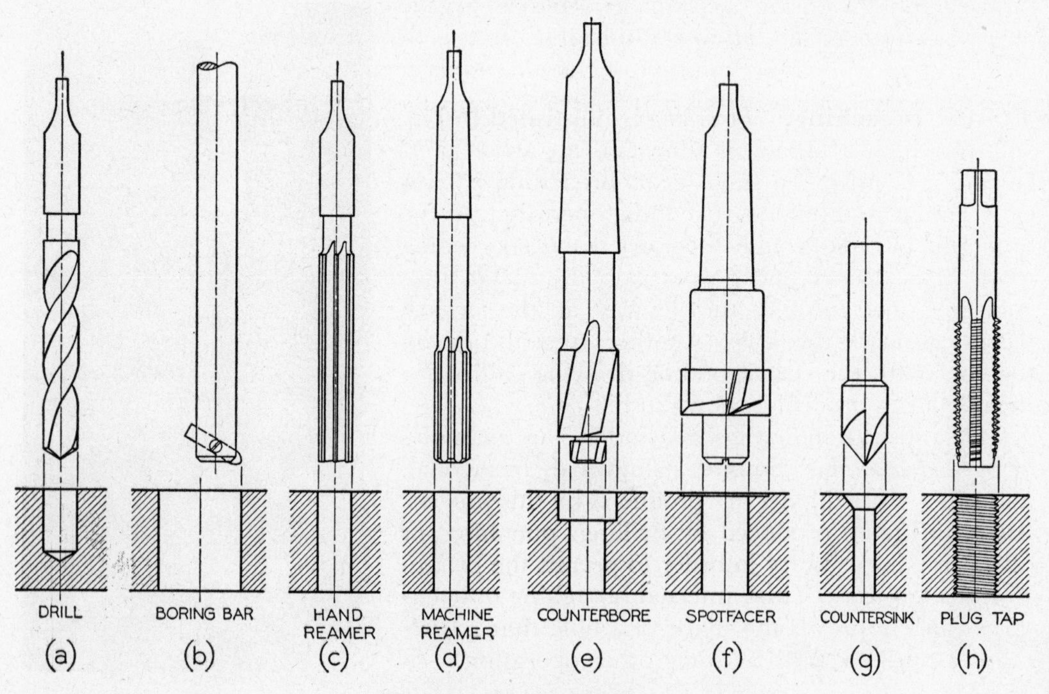

Fig. 564 Types of Holes.

or size. Also, under most conditions, drills cut holes slightly larger than their nominal size. A twist drill is somewhat flexible, which makes it tend to follow a path of least resistance. For work that demands greater accuracy, drilling is followed by *boring*, Fig. 564 (b), or by *reaming*, (c) or (d), and Fig. 566. When a drilled hole is to be finished by boring, it is drilled undersize and the boring tool, which is supported by a relatively rigid bar, generates a hole that is round and straight. Boring is also used

Fig. 565 Drilling $\frac{31}{64}''$ Hole in Gear Blank.

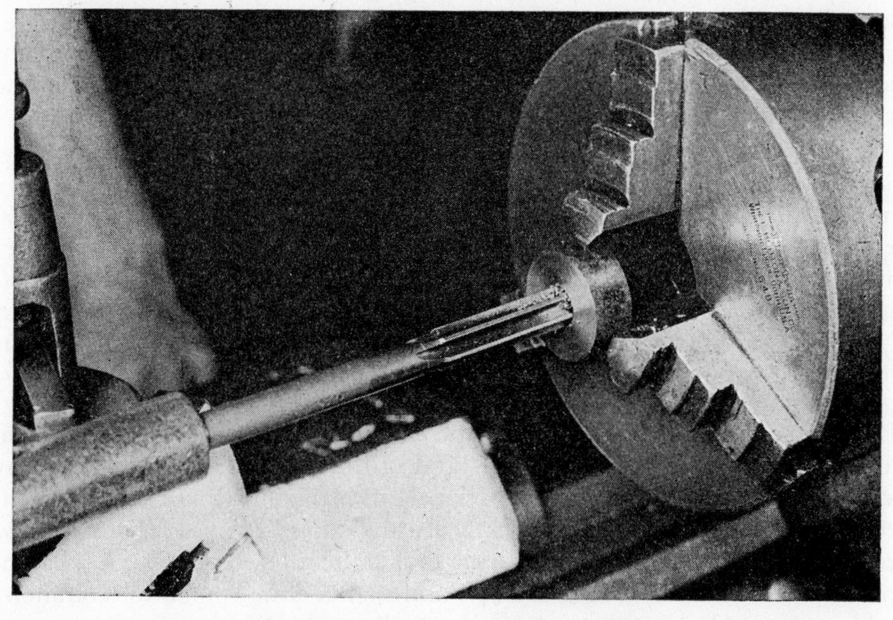

Fig. 566 Reaming to Finish $\frac{1}{2}''$ Hole.

in finishing a cored hole. Also, reaming is used for enlarging ($\frac{1}{64}''$ approx.) and improving the surface quality of a drilled or bored hole. Good practice is to drill, bore, and then ream to produce an accurate and finely finished hole. Standard reamers are available in $\frac{1}{64}''$ increments of diameter.

Counterboring, Fig. 564 (e), is the cutting of an enlarged cylindrical portion of a previously produced hole, usually to receive the head of a fillester-head or socket-head screw, Figs. 689 and 690.

Spotfacing is similar to counterboring, but quite shallow, usually about $\frac{1}{16}''$ deep or deep enough to clean a rough surface, Fig. 564 (f), or to finish the top of a boss to form a bearing surface. Although the depth of a spotface is commonly drawn $\frac{1}{16}''$ deep, the actual depth is usually left to the shop. It is also good practice to include a note, "spotface to clean." A spotface provides an accurate bearing surface for the underside of a bolt or screw head.

Countersinking, Fig. 564 (g), is the process of cutting a conical taper at one end of a hole, usually to receive the head of a flat-head screw, Figs. 689 and 690, or to provide a seat for a lathe center, as shown on part No. 9 of Fig. 788. Actually, this operation is an example of drilling and countersinking, which is accomplished in one operation with a tool called a *combined drill and countersink*, made especially for the purpose.

Tapping is the threading of small holes by the use of one or more taps, Fig. 564 (h). Before tapping may be accomplished, a hole must be previously drilled, as shown in Fig. 680.

The location of holes is a subject that is as important as the production of the holes themselves. Details are beyond the scope of this chapter, but it should be mentioned that the layout of the part to be machined is usually a necessary operation before cutting begins.

For the production of holes, the layout is customarily accomplished by first treating the surface with copper sulphate, Prussian blue, or other suitable substances so that scratches will be clearly visible. Sharp-pointed instruments (scribers) are used to scratch center lines on the work. The surface plate or layout table is often used for such layouts. Frequently the work is clamped to a toolmaker's angle and the center lines scratched on with a vernier height gage.

Of the several methods commonly used to locate work relative to the cutting tool, the use of toolmaker's buttons is probably the most common. Buttons are small hardened-steel rings which may be temporarily attached to the work so that the centers of the rings are precisely where the centers of the holes are to be. The work is then set up with the center of one of the buttons in the center of rotation of, say, the faceplate of an engine lathe. The button is then unscrewed and a hole drilled, bored, and reamed in proper location. By repeating this process, the required number of holes may be produced in their proper locations.

317. Measuring Devices in the Shop. Inasmuch as the machinist uses various measuring devices depending upon the kind of dimensions shown on the drawing, it is evident that to dimension correctly, the engineering draftsman must have at least a working knowledge of the common measuring tools. The *machinist's steel rule*, or *scale*, is the most commonly used measuring tool in the shop, Fig. 567 (a). The smallest divisions on one scale of this rule are $\frac{1}{64}''$, and such a scale is used for common fractional dimensions. Also, many machinist's rules have a decimal scale with the smallest division of 0.01″, which is used for dimensions given on the drawing by the decimal system, §368. For checking the nominal size of outside diameters, the *outside spring caliper* and steel scale are used as shown at (b) and (c). Likewise, the *inside spring caliper* is used for checking nominal dimensions, as shown at (d) and (e). Another use for the outside caliper, (f), is to check the nominal distance between holes (center-to-center). The *combination square* may be used for checking height, as shown at (g), and for a variety of other measurements.

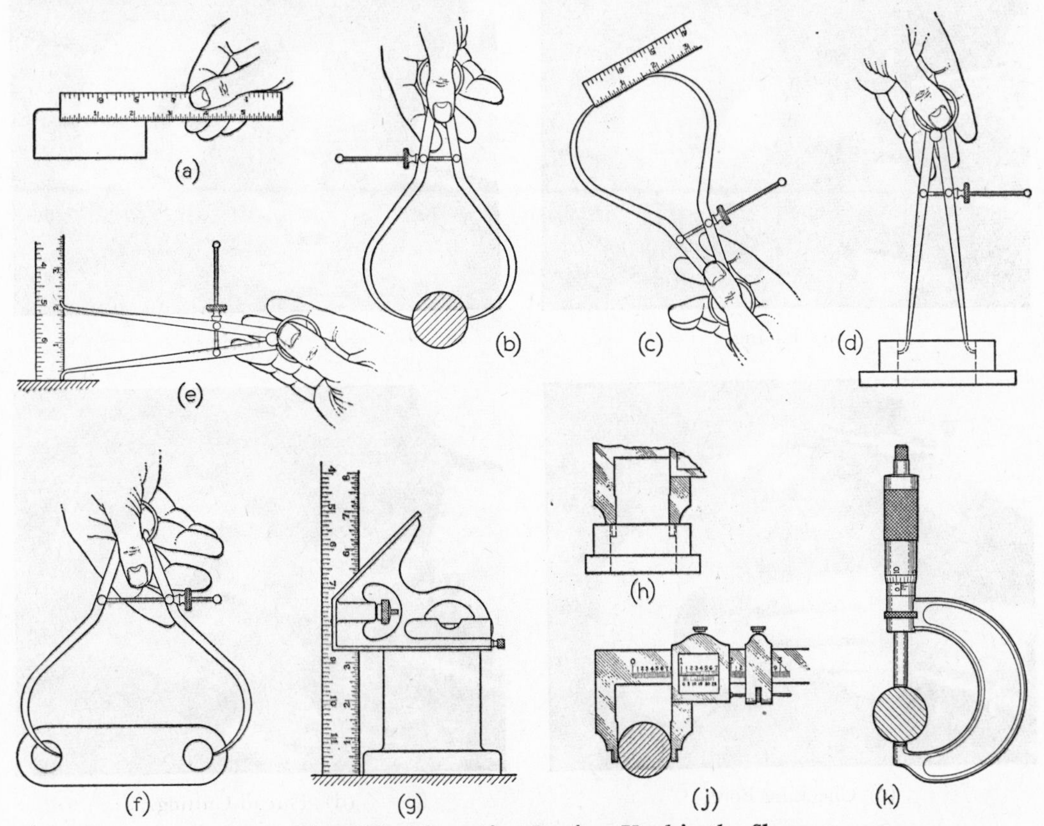

Fig. 567 Measuring Devices Used in the Shop.

For dimensions that require more precise measurements, the *vernier caliper*, (h) and (j), or the micrometer caliper, (k), may be used. It is common practice to check measurements to 0.001″ with these instruments and in some instances they are used to measure directly to 0.0001″.

Most measuring devices in the shop are adjustable so they can each be employed to measure any size within their range of designed usage. There is also a need for measuring devices designed to be used for only one particular dimension. These are called *fixed gages* because their setting is fixed and cannot be changed in the shop.

The subject of gages and gaging is a specialized field, and involves so many technical considerations that large companies employ highly trained men to attend to nothing but this one feature of their operations.

318. Machine Operations. To demonstrate how the machinist uses drawings, the operations called for on the working drawing in Fig. 547 will be shown in the following steps. This drawing was also used by the pattern shop and the foundry, Figs. 550 and 551.

STEP I. After the casting has been properly cleaned, it is then *chucked* in the lathe, as shown in Fig. 568 (a). The drawing specifies that the casting is to be finished

(a) Facing. (b) Boring.

(c) Checking Bore. (d) Thread Cutting.

(e) Checking with Thread Gage. (f) Checking Relief Depth.

Fig. 568 Machining and Checking.

on all surfaces; therefore the machinist must produce a smooth finish on the flat surface at the end of the cylinder. This is called *facing*. After this surface has been machined, it serves as a control surface in locating parallel surfaces.

Step II. The next machine operation on this casting is that of boring the cen-

(a) Checking Relief Diameter.

(b) Checking Chamfer Depth.

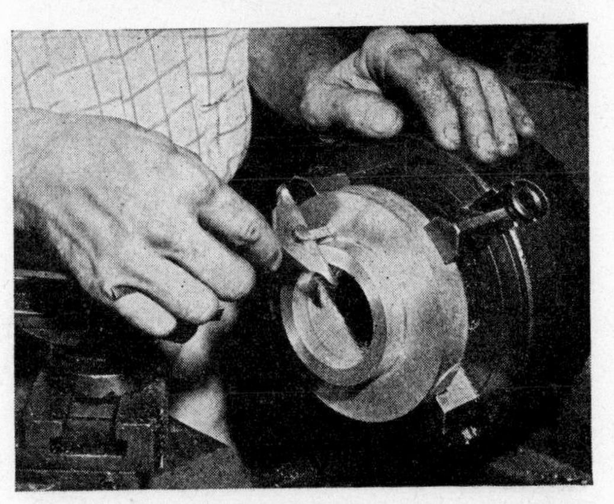

(c) Checking Hub Length.

(d) Checking Hub Diameter.

(e) Checking Round.

(f) Checking Flange Diameter.

Fig. 569 Checking.

tral hole, as shown at (b). The machinist reads 2.125-2.126 from the note 2.125-2.126 BORE $2\frac{1}{4}$-8UN-2B on the drawing to determine the required size of the bored hole. This note indicates that he is allowed a tolerance of .001″, and the finished hole diameter must be between 2.125″ and 2.126″. Although this hole is checked, as shown

(a) Facing Back Surface.

(b) Checking Relief Diameter.

(c) Laying Out Small Holes.

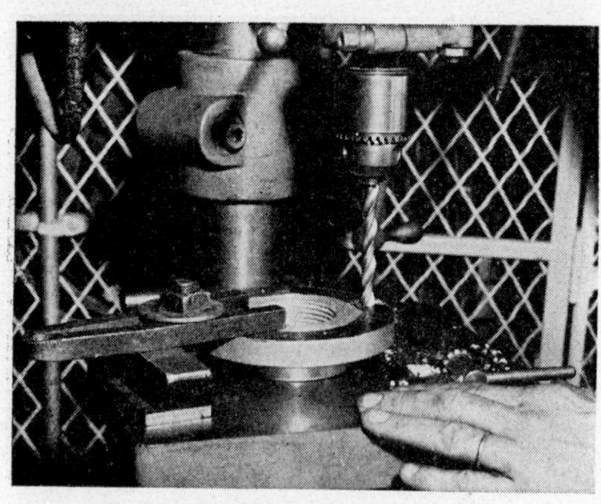

(d) Drilling Small Holes.

(e) Countersinking Small Holes.

(f) Tapping Small Holes.

Fig. 570 Machining and Checking.

at (c), using an inside spring caliper and an outside micrometer caliper, other procedures commonly used are either to check the hole directly with an inside micrometer caliper or to check it with a "go-no-go" gage.

STEP III. After boring is completed, the internal thread is cut with a 60° thread cutting tool, as shown at (d). The $2\frac{1}{4}$ in the thread note $2\frac{1}{4}$-8UN-2B, Fig. 547, indicates that the major diameter of the thread must be $2\frac{1}{4}''$; 8UN indicates 8 Unified National threads per inch; and 2B indicates a Class 2 internal thread. The finished thread is checked with a thread plug gage, as shown at (e).

STEP IV. The threads are now relieved by boring $2\frac{9}{32}''$ diameter by $\frac{3}{8}''$ deep, and checked as shown at (f) and 569 (a). These dimensions are checked with a steel scale, since the required tolerance is only $\pm\frac{1}{64}''$ on all fractional dimensions on this drawing, as shown in Fig. 547.

STEP V. The next step is to cut the $\frac{1}{16} \times 45°$ chamfer, and check as shown in Fig. 569 (b). The cutting tool is set at a 45° angle, and the depth of cut is checked with a depth gage set at $\frac{1}{16}''$.

STEP VI. The next associated group of machine operations includes facing the flange, turning the hub diameter, and turning the $\frac{1}{8}''$ radius between the hub and flange. The $\frac{5}{8}''$ length of the hub is checked with the depth gage, as shown at (c). The $3\frac{5}{32}''$ hub diameter is checked at (d), with a spring caliper and steel scale, because this dimension is not critical. The $\frac{1}{16}''$ radius on the hub is checked with a radius gage, as shown at (e).

After this group of operations is completed, the FLANGE is removed from the chuck of the lathe and screwed onto the lathe spindle.

STEP VII. As shown at (f), the outside diameter of the FLANGE is turned to size and checked with a micrometer caliper. The outside diameter dimension $\begin{smallmatrix}5.000\\4.998\end{smallmatrix}$ is critical, as it must fit in a mating part of the 4-JAW INDEPENDENT CHUCK.

STEP VIII. The back surface is now faced, Fig. 570 (a), to provide a smooth flat surface and to bring the over-all length of the FLANGE to $1\frac{1}{4}''$. Again, since this dimension is not critical, it is checked with the spring caliper and steel scale. The facing tool is used to break the back surface corner $\frac{1}{64}$ R and to round the forward corner $\frac{1}{16}$ R as dimensioned in Fig. 547.

STEP IX. After the back surface is faced, it is then relieved, Fig. 570 (b), to a diameter of $3\frac{7}{16}''$ and to a depth of $\frac{1}{64}''$, as dimensioned in Fig. 547. By relieving the back surface of the flange in this manner, a better seat is insured between the flange and its mating part in the 4-JAW INDEPENDENT CHUCK.

STEP X. The next machining operation is to locate and drill the four small holes, as shown in Fig. 570 (c) and (d). After locating the centers 90° apart on a 4″ diameter circle, the machinist selects a "Q" diameter (0.3320) drill, and drills the four holes.

STEP XI. The next step is to countersink, Fig. 570 (e), the four small holes with an 80° countersink to a diameter equal to the major diameter of the threads. The countersinking facilitates the tapping of the holes.

STEP XII. The last step in machining the FLANGE is to tap the four small holes as shown at (f). The $\frac{3}{8}$ in the drawing note $\frac{3}{8}$-24UNF-2B indicates that the major diameter of the tapped holes must be $\frac{3}{8}''$; 24UNF indicates 24 Unified National Fine threads per inch of thread length; and 2B indicates a Class 2 fit internal thread.

319. Automation. *Automation* is the term applied to systems of automatic machines and processes. As the demand for national defense materiel and consumer goods increases, industry is resorting increasingly to automation in an effort to offset scarcity and high cost of skilled labor. These automatic devices include machine tools that follow directions punched on a tape, computers that make thousands of intricate mathematical calculations in a matter of seconds or a fraction of a second, and gages measuring electrically in millionths of an inch.

Automation does not mean that there will be less thinking for the engineer, but more. Therefore, the engineering student is going to need even more intensive grounding in basic fundamentals than previously. In addition, the enterprising student will seize opportunities for keeping informed on current industrial developments.

320. Stock forms. Many standardized structural shapes are available in stock sizes for the fabrication of parts or structures. Among these are bars of various shapes, flat stock, rolled structural shapes, and extrusions, Fig. 571. The manufacturing

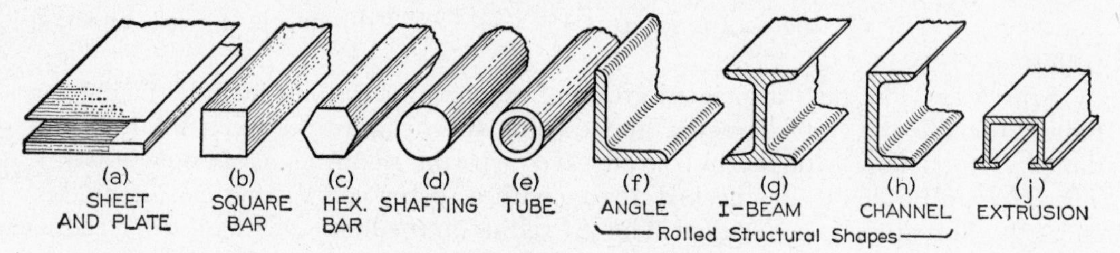

Fig. 571 Common Stock Forms.

processes of rolling, drawing, and extruding often add a great deal of toughness and strength to a metal, and these stock forms so processed are very useful in the manufacture of small machined parts such as screws, and for the fabrication of welded and riveted structures. See also §634.

321. Welding. *Welding* is a process of joining metals by fusion into a single homogeneous mass. *Arc welding*, *gas welding*, *resistance welding*, and *atomic hydrogen welding* are commonly used, as well as the old method of *forge welding*, which is still used to some extent. Welding drawings are discussed in Chapter 25.

Fig. 572 Use of Drilling Jig.

Welded structures are built up in most cases from stock forms, particularly plate, tubing, and angles. Often both heat-treating and machine shop operations must be performed on welded machine parts. Since welding frequently distorts a part or structure enough to alter permanently the dimensions to which it has been cut in the shop, it is usual practice not to weld accurate work after finish machining has been done unless the volume of the welded material is very small in comparison to the volume of the part.

322. Jigs and Fixture. A general-purpose machine tool may have its effectiveness on a specific job increased by means of *jigs* or *fixtures*. A *jig* is a device that holds the work and guides the tool; it is usually not rigidly fixed to a machine. A *drilling jig*, Fig. 572, is a common device by means of which holes on many duplicate parts may be drilled exactly alike. A *fixture* is rigidly attached to the machine, becoming in reality an extension of it, and holds the work in position for the cutting tools without acting as a guide for them. Drawings of one of the important fixtures used in the production of the Ford V-8 connecting rod, Figs. 725-727, are shown in Figs. 819-821. This fixture was built at considerable expense for the single purpose of holding the connecting rod in the exact position required for the efficient and speedy execution of a single operation.

Jigs and fixtures are usually designed in a tooling department by tool designers. Usually they are built by machinists of much better than average skill, using especially accurate equipment. Such tooling devices are commonly held to tolerances one-tenth of those applied to the parts to be produced on these jigs and fixtures.

Jigs and fixtures may be grouped into two general classes: *manufacturing tooling* and *assembly tooling*. Manufacturing tooling consists of devices used in producing individual parts. An example of this is a fixture for holding a connecting rod in a milling machine when the ends are being faced by straddle milling.

Assembly tooling consists of devices to hold work and guide tools as parts are being assembled. For example, the center section of the wing of the B-52 airplane was assembled in a special jig in which parts were held in place, drilled when together, and riveted. The assembly jig was so built that the component parts could fit together only when correctly located, and so that no measurement was ever made and no blueprint ever referred to by workers assembling the wing. Such assembly tooling enables a small group of highly skilled workers to produce jigs or fixtures by means of which precision products may be produced and assembled by cheaper and less-skilled labor than would otherwise be required.

323. Hot and Cold Working of Metals. Let us now review some of the ideas we have already discussed separately in several of the preceding sections. Recall that in each of the casting operations, §§300 and 304, the casting assumes its shape by filling a mold as a liquid and does not appreciably change its shape as it solidifies. Notice that in many manufacturing processes, however, the shape of the part is changed after the metal has solidified. If these processes involve slowly pressing the part (squeezing) or rapidly and repeatedly striking the part (hammering), then the term *mechanical working* is used to describe the operation.

The metallurgists have learned that the practical effects of mechanical working depend a great deal on the temperature at which the metal is squeezed or hammered. The terms *hot working* and *cold working* are used widely in engineering today, and the student will find in his metallurgy courses and machine design courses that he will devote a great deal of effort to the study of the *hot working* and *cold working* of metals.

324. Forging.* *Forging* is the process of shaping metal to a desired form by means of pressing or hammering. Generally, forging is *hot forging*, in which the metal is heated to a required temperature before forging. Some softer metals can be forged without heating, and this is *cold forging*.

When metal is worked under pressure, as in forging, the material is compressed and greatly strengthened. Forging is not suitable for producing parts with large in-

*ASA Y14.9 (*Tentative*).

teriors or with unsupported thin walls. Complicated shapes are extremely expensive because of the costly dies.

Forgings are made of steel, copper, brass, bronze, aluminum, magnesium, and more recently, the new light-weight titanium.

Closed-die forgings, or drop forgings, are formed between dies that are machined to the desired shapes. Open-die forgings, or hand forgings, are formed between flat dies and manipulated by hand during forging to obtain the required shape.

Usually the first step in the design of a forging is to locate the *parting plane*, which separates the upper and lower dies. As a rule, this surface follows the center line of the piece, as shown in Fig. 573 (a), and should pass through the largest portion of the object.

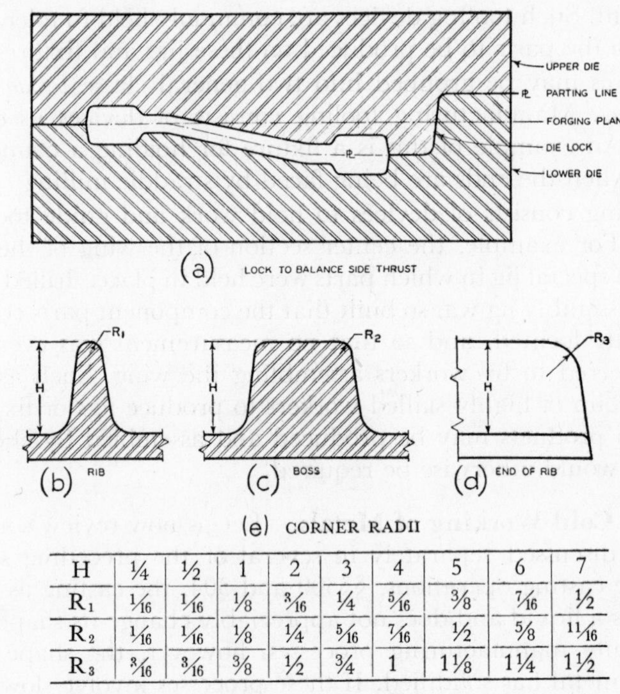

(e) CORNER RADII

H	1/4	1/2	1	2	3	4	5	6	7
R_1	1/16	1/16	1/8	3/16	1/4	5/16	3/8	7/16	1/2
R_2	1/16	1/16	1/8	1/4	5/16	7/16	1/2	5/8	11/16
R_3	3/16	3/16	3/8	1/2	3/4	1	1 1/8	1 1/4	1 1/2

Fig. 573 Forging Design (ASA Y14.9, *Tentative*).

Draft, or taper, must be provided on all forgings. Although draft is not usually shown on drawings for sand casting, it is always shown on forging drawings. A typical forging drawing is shown in Fig. 725. For outside surfaces, the standard draft angle is 7° for all materials. For inside surfaces in aluminum and magnesium, the draft is also 7°, but in steel the angle is usually 10°.

Small fillets and rounds in forgings reduce die life and decrease the quality of the forging. Minimum rounds or corner radii are shown in Fig. 573 (b) to (e), and minimum fillet radii in Fig. 574. Whenever possible, larger radii than those shown should be used.

The minimum thickness of a web depends on the area at the parting line. Minimum values are shown below. Ribs will increase in thickness as the height increases.

Area in Sq. In.	4	20	80	200	400	600	800	1000
Thickness	3/32	1/8	3/16	1/4	5/16	3/8	7/16	1/2

Minimum values conform to radius R in Fig. 573 (b).

The working drawing, as it comes to the shop, may show the completed or machined part, in which case the necessary forging and machining allowances are made by the diemaker. However, since dies are so expensive and the forging itself is so important, it is common practice to make separate forging and machining drawings, as shown in Figs. 725 and 726. It is standard practice to show visible lines for rounded corners as if the corners were sharp. See §241. Note that the parting plane is represented by a center line.

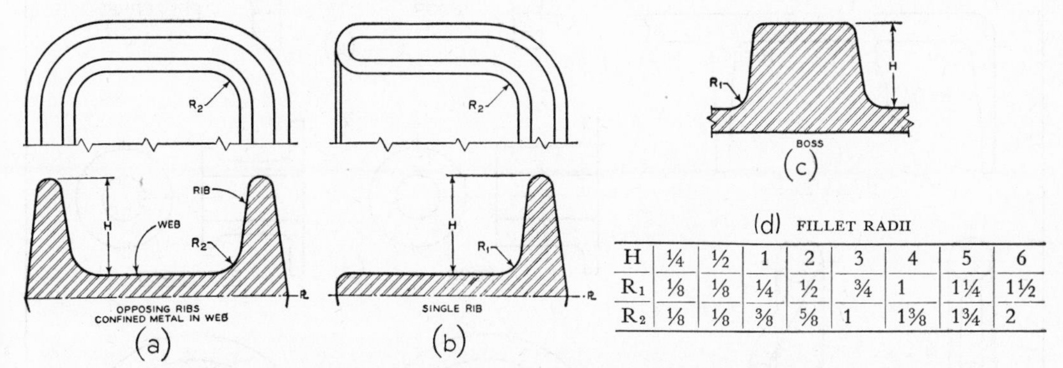

H	¼	½	1	2	3	4	5	6
R_1	⅛	⅛	¼	½	¾	1	1¼	1½
R_2	⅛	⅛	⅜	⅝	1	1⅜	1¾	2

Fig. 574 Minimum Radii for Forgings (ASA Y14.9, *Tentative*).

325. Heat Treating. We have seen above that heat, along with heavy forces, often plays an important role in the making of a part. Sometimes the heat itself, without any mechanical working, is enough to improve a piece significantly. When heat alone is so employed and when the object is to change the properties of the metal rather than merely to change its shape, we use the term *heat treating*.

Let us mention briefly some of the heat-treating terms the draftsman will see on his drawings. *Annealing* and *normalizing* soften the metal and improve its grain structure. They involve heating to the critical temperature range and then slowly cooling. *Hardening* requires heating to above the critical temperature followed by rapid cooling-quenching in oil, water, or brine, or in some instances in air (air-hardened steels). *Tempering* reduces internal stresses caused by hardening. Tempering also improves the toughness and ductility. *Surface Hardening* is a way of hardening the surface of a steel part while leaving the inside of the piece soft. Surface hardening is accomplished by *carburizing*, followed by heat treatment; by *cyaniding*, followed by heat treatment; by *nitriding;* by *induction hardening;* and by *flame hardening*.

326. Do's and Don'ts of Practical Design. In Figs. 575 and 576 are shown a number of examples in which a knowledge of shop processes and limitations is essential for good design.

Many difficulties in producing good castings result from abrupt changes in section or thickness. In Fig. 575 (a) rib thicknesses are uniform so that the metal will flow easily to all parts. Fillet radii are equal to the rib thickness—a good general rule to follow. When it is necessary to join a thin member to a thicker member, the thin member should be thickened as it approaches the intersection, as shown at (b).

At (c), (g), and (h) coring is used to produce walls of more uniform sections. At (d) an abrupt change in sections is avoided by making thinner walls and leaving a collar, as shown.

At (e) and (f) are shown examples in which the preferred design will tend to allow the castings to cool without introducing internal stresses. The less desirable

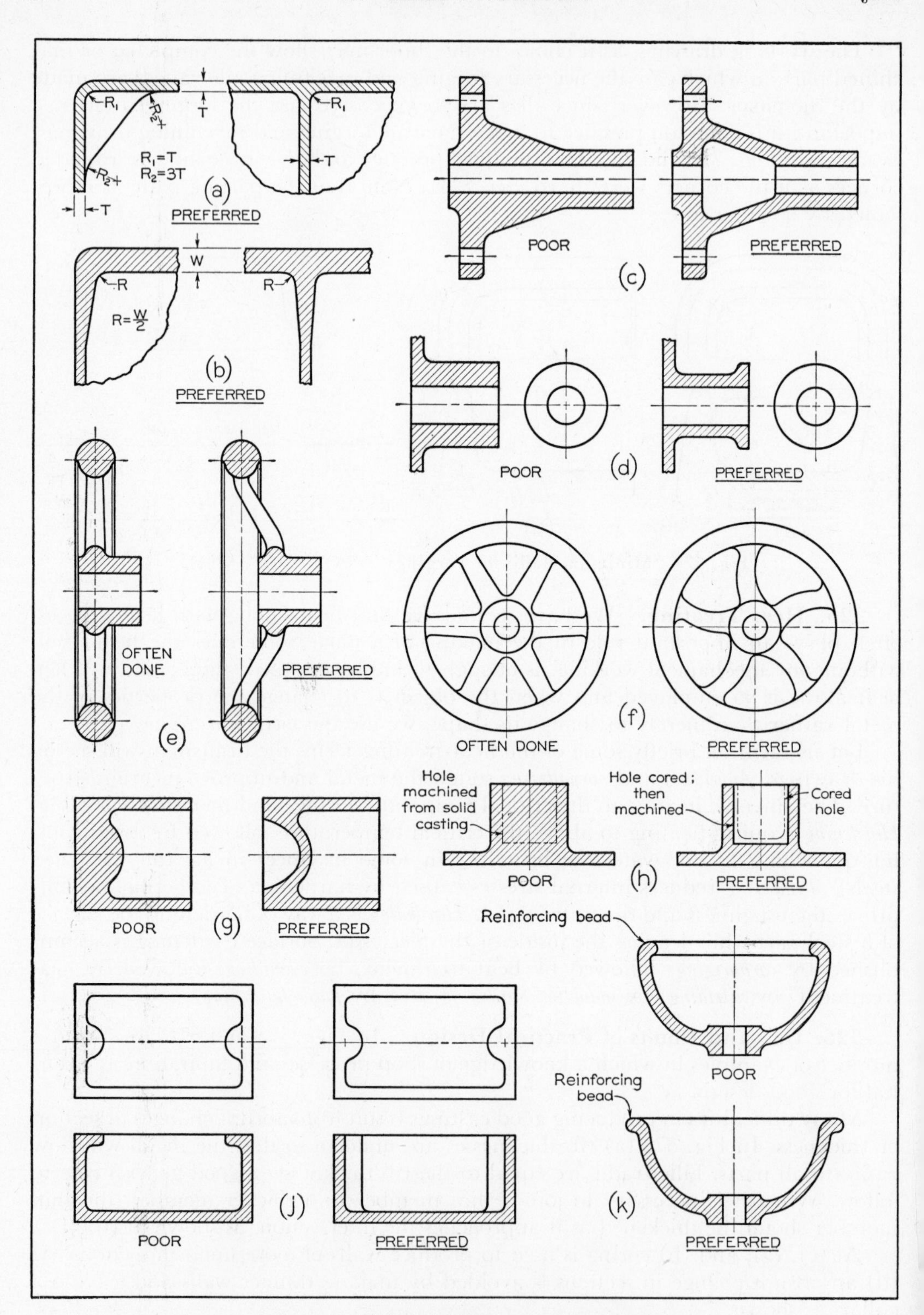

Fig. 575 Casting Design Do's and Don'ts.

design is more likely to crack as it cools, since there is no "give" in the design. Curved spokes are preferable to straight spokes, and an odd number of spokes is better than an even number because direct stresses along opposite spokes are avoided.

The design of a part may cause unnecessary trouble and expense for the pattern shop and foundry without any gain in the usefulness of the design. For example, in the "poor" designs at (j) and (k), one-piece patterns would not withdraw from the sand, and two-piece patterns would be necessary. In the "preferred" examples, the design is just as useful and is conducive to economical work in the pattern shop and foundry.

As shown in Fig. 576 (a), a narrower piece of stock sheet metal can be used for certain designs that can be linked or overlapped. In this case, the stampings may be overlapped if dimension W is increased slightly, as shown. By such an arrangement, great savings in scrap metal can often be effected.

The maximum hardness that can be obtained in the heat treatment of steel depends on the carbon content of the steel. To get this hardness, it is necessary to cool rapidly (quench) after heating to the temperature required. In practice it is often impossible to quench uniformly because of the design. In the design above at (b), the piece is solid and will harden well on the outside but will remain soft and relatively weak on the inside. However, as shown in the preferred example, a hollow piece can be quenched from both the outside and the inside. Thus it is possible for a hardened hollow shaft to be stronger than a hardened solid shaft of the same diameter.

As shown at (c), the addition of a rounded groove, called a *neck*, around a shaft next to a shoulder will eliminate a practical difficulty in precision grinding. It is not only more expensive to grind a sharp internal corner, but such sharp corners often lead to cracking and failure.

The design at the right at (d) eliminates a costly "reinforced" weld, which would be required by the design on the left. The strong virgin metal with a generous radius is present at the point at which the stress is likely to be most severe. It is possible to make the design on the left as strong as the design on the right, but it is more expensive and requires expert skill and special equipment.

It is difficult to drill into a slanting surface, as shown at the left at (e). The drilling will be greatly facilitated if a boss is provided, as shown at the right.

At (f), the design at the left requires accurately boring or reaming a blind hole all the way to a flat bottom, which is difficult and expensive. It is better to drill deeper than the hole is to be finished, as shown at the right, in order to provide room for tool clearance and for chips.

In the upper example at (g), the drill and counterbore cannot be used for the hole in the center piece because of the raised portion at the right end. In the approved example, the end is redesigned to provide access for the drill and counterbore.

In the design above at (h), the ends are not the same height. As a result, each flat surface must be machined separately. In the design below, the ends are the same height, the surfaces are in line horizontally, and only two machining operations are necessary. It is always good design to simplify and limit the machining as much as possible.

The design at the left at (j) requires that the housing be bored for the entire length in order to receive a pressed bushing. Machining time can be decreased if the cored recess is made, as shown. This assumes that average loads would be applied in use.

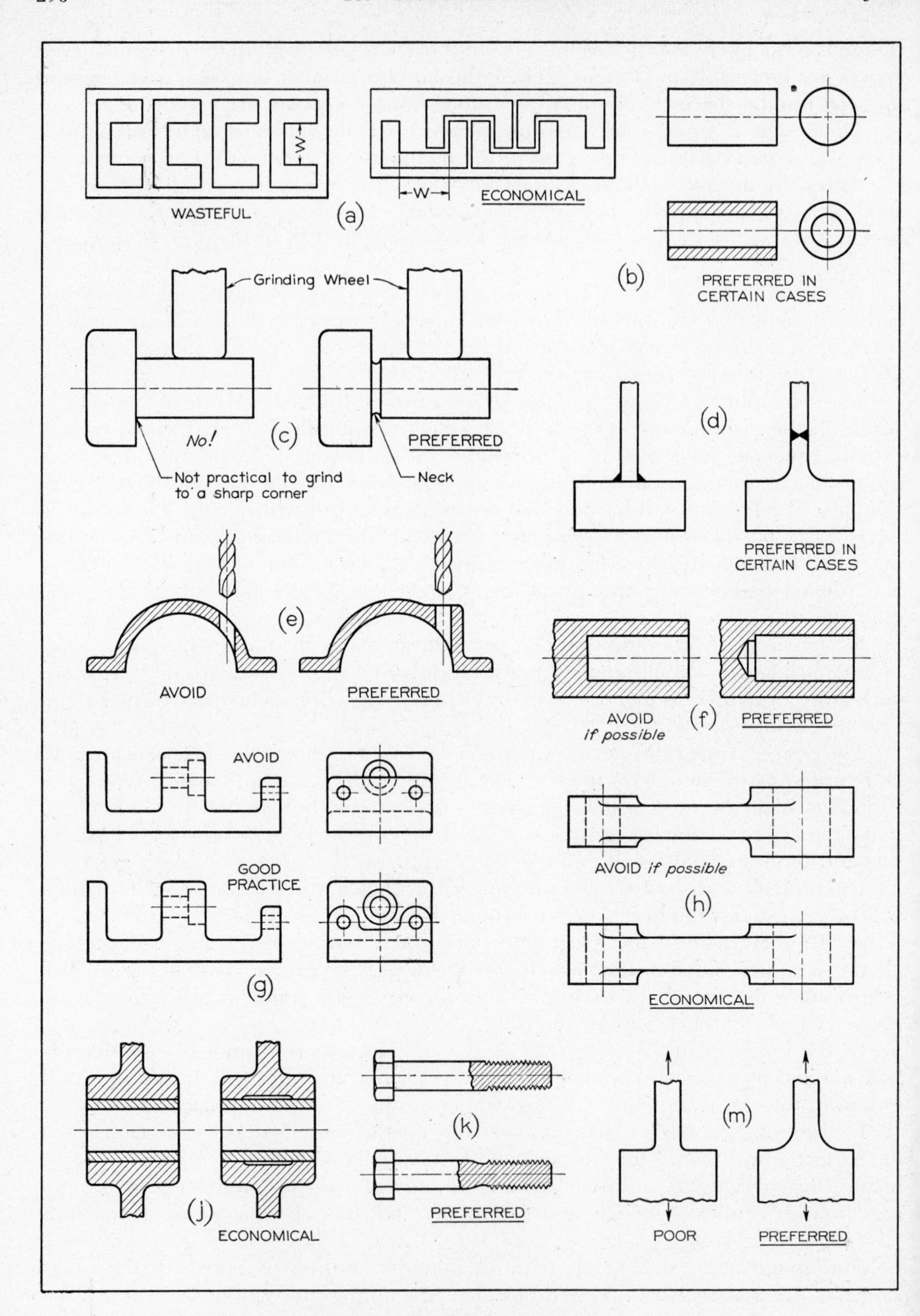

Fig. 576 Design Do's and Don'ts.

At (k) the lower bolt is encircled by a rounded groove no deeper than the root of the thread. This makes a gentle transition from the small diameter at the root of the threads and the large diameter of the body of the bolt, producing less stress concentration and a stronger bolt. In general, sharp internal corners should be avoided as points of stress concentration and possible failure.

At (m) is shown a $\frac{1}{4}''$ steel plate being pulled, as shown by the arrows. By increasing the radius from $\frac{1}{8}$R to $\frac{1}{2}$R, the strength of the plate is increased a great deal by decreasing stress concentration.

CHAPTER 11

DIMENSIONING*

327. Historical Measurements. We have all heard of "rule of thumb." Actually, at one time an inch was defined as the width of a thumb, and a foot was simply the length of a man's foot. In old England, an inch used to be "three barley corns, round and dry." In the time of Noah and the Ark, a *cubit* was the length of a man's forearm, or about 18 inches.

In 1793, when Napoleon was rising to power, France adopted the metric system, based on the *meter*, which is 39.37 inches. In the meantime England was setting up a more accurate determination of the *yard*, which was legally defined in 1824 by act of Parliament. A foot was one-third of a yard, and an inch was one-thirty-sixth of a yard. From these specifications, graduated rulers, scales, and many types of measuring devices have been developed to achieve even more accuracy of measurement and inspection.

Until this century, common fractions were considered adequate for dimensions; but as products became more complicated, and as it became necessary to have interchangeable parts in order to support mass production, more accurate specifications were required and it became necessary to turn to the decimal system. Today, decimals are widely used, and it is only a question of time until practically all dimensions will be expressed in decimals in this country. See §§335 and 368.

328. Size Description. In addition to a complete *shape description* of an object, as discussed in previous chapters, a drawing must also give a complete *size description;* that is, it must be *dimensioned.*

In the early years of machine manufacturing, the designing and production functions were closely allied under one roof. In many cases these processes were even

*See ASA Y14.5-1957.

carried out by the same individual. Design drawings, mostly of the assembly type, were scaled by the workmen to obtain the dimensions. It was up to the shop man to make the parts correctly and to see to it that they would fit together and operate properly. If any question arose, he could always consult the designer who would be nearby. Under these conditions it was not necessary for drawings to carry detailed dimensions and notes.

The need for *interchangeability* of parts is the basis for the development of modern methods of size description. Drawings today must be dimensioned so that workmen in widely separated places can make mating parts fit properly when brought together finally in the assembly shop, §351.

The increasing need for precision manufacturing and the necessity to control sizes for interchangeability has shifted responsibility for size control from the machinist to the designing engineer and the draftsman. The workman no longer exercises judgment in engineering matters, but only in the proper execution of instructions given on the drawings. It is therefore necessary for engineers and draftsmen to be familiar with materials and methods of construction, and with requirements of the shops. The engineering student or the draftsman should seize every opportunity to familiarize himself with the fundamental shop processes, especially *patternmaking*, *foundry*, *forging*, and *machine-shop practice*, discussed in the previous chapter.

The drawing should show the object in its completed condition, and should contain all necessary information to bring it to that final state. Therefore, in dimensioning a drawing, the draftsman should keep in mind the finished piece, the shop processes required, and above all the function of the part in assembly. Whenever possible—that is, when there is no conflict with functional dimensioning—dimensions should be given that are convenient for the workman. These dimensions should be given primarily for the machinist, who should not find it necessary to calculate, scale, or assume any dimensions. Do not give dimensions to points or surfaces that are not accessible to the shop man.

Dimensions should not be duplicated or superfluous, §355. Only those dimensions should be given that are needed to produce and inspect the part exactly as intended by the designer. The student often makes the mistake of giving the dimensions he used *to make the drawing*. These are not necessarily the dimensions required. There is much more to the theory of dimensioning, as we shall see.

329. Scale of Drawing. Drawings should be made to scale, and the scale should be indicated in the title strip. See §40 for a general discussion of scales.

A wavy line should be drawn under any dimension that is not to scale, Fig. 589, or the abbreviation NTS (Not to Scale) should be indicated. This may be necessary when a change is made in the drawing that is not important enough to justify making an entirely new drawing.

330. Learning to Dimension. Dimensions are given in the form of linear distances, angles, or notes.

First, the student must learn the *technique of dimensioning:* the character of the lines, the spacing of dimensions, the making of arrowheads, etc. A typical dimensioned drawing is shown in Fig. 577. Note the strong contrast between the visible lines of the object and the thin lines used for the dimensions.

Second, the student must learn the rules of *placement of dimensions* on the drawing. These practices assure a logical and practical arrangement with maximum legibility.

Third, the student should learn the *choice of dimensions*. Formerly, manufacturing processes were considered the governing factor in dimensioning. Now function is considered first and shop processes second. The proper procedure is to dimension

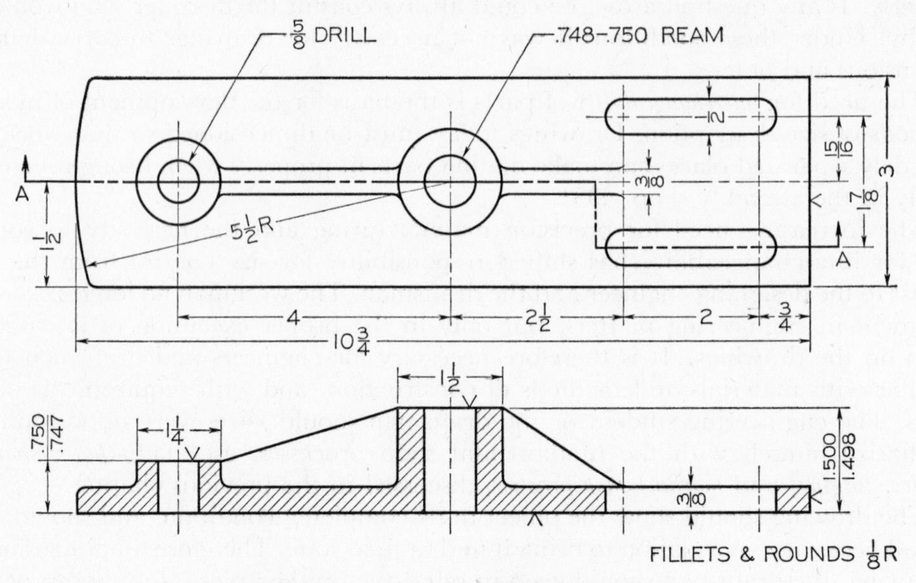

Fig. 577 Technique of Dimensioning.

tentatively for function, then review the dimensioning to see if any improvements from the standpoint of the shop can be made without adversely affecting the functional dimensioning. A "geometric breakdown," §345, will assist the beginner in selecting dimensions. In most cases dimensions thus determined will be functional, but this method should be accompanied by a logical analysis of the functional requirements.

331. Lines Used in Dimensioning. A *dimension line*, Fig. 578 (a), is a fine, dark, solid line terminated by arrowheads, which indicates the direction and extent

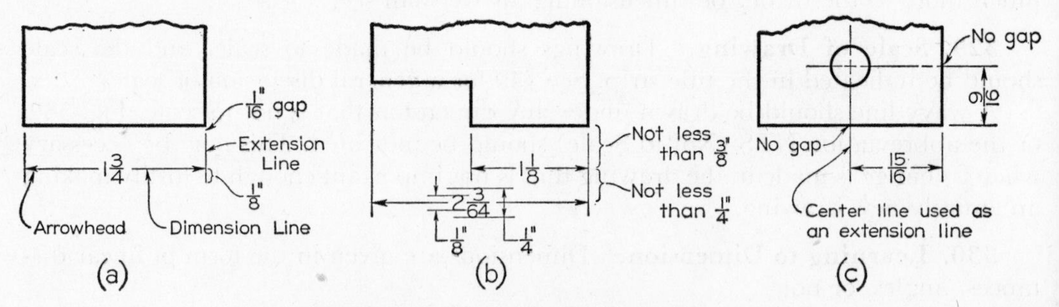

Fig. 578 Dimensioning Technique.

of a dimension. In machine drawing, in which common fractions are used, the line is broken, usually near the middle, to provide an open space for the dimension figure. In structural and architectural drawing, it is customary to place the dimension figure above an unbroken dimension line, Fig. 1079.

As shown at (b), the dimension line nearest the object outline should be spaced at least $\frac{3}{8}''$ away. All other parallel dimension lines should be at least $\frac{1}{4}''$ apart, and more if space is available. *The spacing of dimension lines should be uniform throughout the drawing.*

An *extension line*, (a), is a fine, dark, solid line that "extends" from a point on the drawing to which a dimension refers. The dimension line meets the extension lines at right angles except in special cases, as in Fig. 582 (a). A gap of about $\frac{1}{16}''$ should be left where the extension line would join the object outline. The extension line should extend about $\frac{1}{8}''$ beyond the outermost arrowhead, (a) and (b).

A *center line* is a fine, dark line composed of alternate long and short dashes, and is used to represent axes of symmetrical parts and to denote centers. As shown in Fig. 578 (c), center lines are commonly used as extension lines in locating holes and other features. When so used, the center line crosses over other lines of the drawing without gaps. A center line should always end in a long dash.

332. Placement of Dimension and Extension Lines. A correct example of the placement of dimension lines and extension lines is shown in Fig. 579 (a). The shorter dimensions are nearest to the object outline. Dimension lines should not cross extension lines as at (b), which results from placing the shorter dimensions outside.

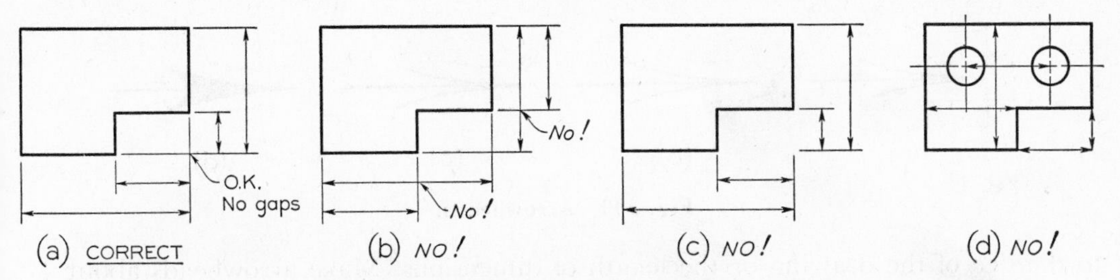

Fig. 579 Dimension and Extension Lines.

Note that it is perfectly satisfactory to cross extension lines, as shown at (a). They should never be shortened, as at (c). A dimension line should never coincide with, or form a continuation of, any line of the drawing, as shown at (d). Avoid crossing dimension lines wherever possible.

Dimensions should be lined up and grouped together as much as possible, as shown in Fig. 580 (a), and not as at (b).

In many cases, extension lines and center lines must cross visible lines of the object, Fig. 581 (a). When this occurs, gaps should not be left in the lines, as at (b).

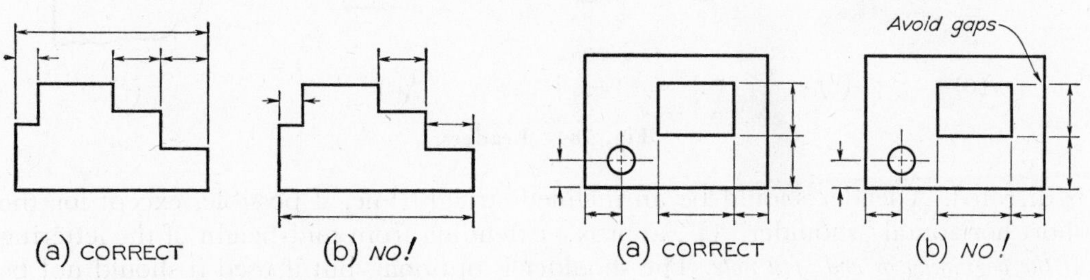

Fig. 580 Grouped Dimensions. **Fig. 581 Crossing Lines.**

Dimension lines are normally drawn at right angles to extension lines, but an exception may be made in the interest of clearness, as shown in Fig. 582 (a). In crowded conditions, gaps in extension lines near arrowheads may be left, in order

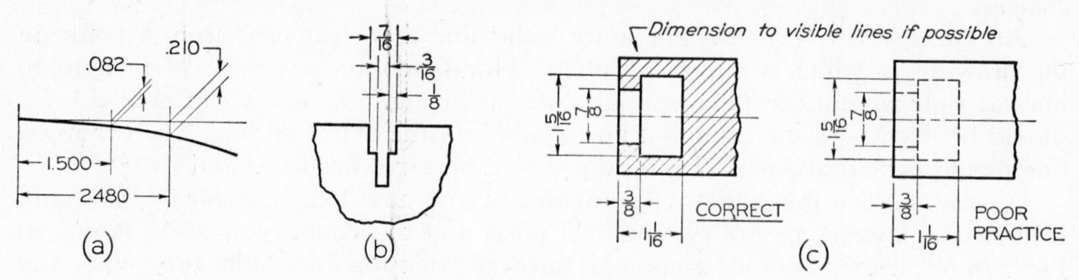

Fig. 582 Placement of Dimensions.

to clarify the dimensions, as shown at (b). In general, avoid dimensioning to hidden lines, as shown at (c).

333. Arrowheads. *Arrowheads*, Fig. 583, indicate the extent of dimensions. They should be uniform in size and style throughout the drawing, and not varied according

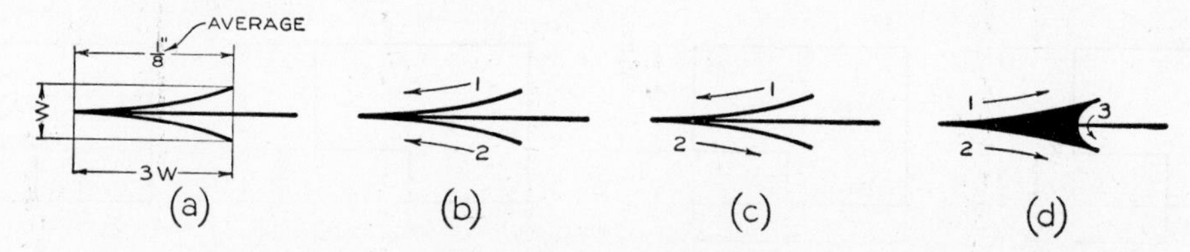

Fig. 583 Arrowheads.

to the size of the drawing or the length of dimensions. Make arrowheads about $\frac{1}{8}''$ long and very narrow, (a). Use strokes toward the point or away from the point as desired, (b) to (d). The method at (b) is easier when the strokes are drawn toward the draftsman. For best appearance, fill in the arrowhead as at (d). A suitable pen for inking arrowheads is the Gillott's 303.

334. Leaders. A *leader*, Fig. 584, is a thin solid line leading from a note or dimension, and terminated by an arrowhead touching the part to which attention

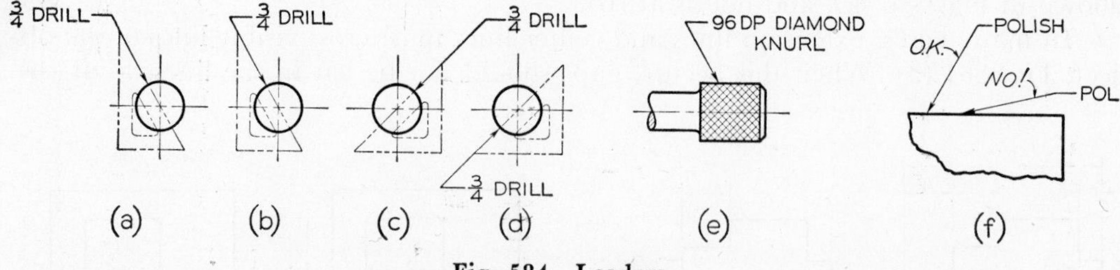

Fig. 584 Leaders.

is directed. A leader should be an inclined straight line, if possible, except for the short horizontal "shoulder" ($\frac{1}{4}''$, approx.) extending from mid-height of the lettering *at the beginning or end of a note*. The shoulder is optional, but if used it should not be drawn so as to underline the lettering.

A leader to a circle should be radial, so that if extended it would pass through the center. A drawing presents a more pleasing appearance if leaders near each other are drawn parallel. Leaders should cross as few lines as possible, and should never cross each other. They should not be drawn parallel to nearby lines of the drawing, allowed to pass through a corner of the view, drawn unnecessarily long, or drawn horizontally or vertically on the sheet. A leader should not be drawn at a small angle, as shown at (f).

335. Fractional and Decimal Dimensions. In the early days of machine manufacturing in this country, the workman would scale the undimensioned design drawing to obtain any needed dimensions, and it was his responsibility to see to it that the parts fitted together properly. When blueprinting came into use, workmen in widely separated localities could use the same drawings, and it became the practice to dimension the drawings, more or less completely, in inches and common fractions, such as $\frac{1}{4}''$, $\frac{1}{8}''$, $\frac{1}{16}''$, $\frac{1}{32}''$, and $\frac{1}{64}''$. The smallest dimension the workman was supposed to measure directly was $\frac{1}{64}''$, which was the smallest division on the ordinary machinist's scale.* (However, a good machinist could "split" sixty-fourths with ease.) When close fits were required, the drawing would carry a note, such as "running fit," or "drive fit," and the workman would make considerably finer adjustment of size than $\frac{1}{64}''$. Workmen were skilled, and it should not be thought that very accurate and excellent fits were not obtained. Hand-built machines were often beautiful examples of precision workmanship.

This system of units and common fractions is still used in architectural and structural work where close accuracy is relatively unimportant, and where the steel tape or framing square is used to set off measurements. Architectural and structural drawings are therefore dimensioned in this manner.

Also today there are many types of manufacturing in which units and common fractions are almost universally used, because extreme accuracy is not necessary. An example of this is railway car drawing, in which the structure is very large, and extremely fine measurements are generally not required. However, there are also many small articles manufactured today in which the ordinary machinist's scale is sufficiently accurate, and units and common fractions for dimensions on drawings are considered perfectly satisfactory and suitable for the purpose.

As industry has progressed, there has been greater and greater demand for more accurate specifications of the important functional dimensions—more accurate than the $\frac{1}{64}''$ permitted by the machinists scale. Since it was cumbersome to use still smaller fractions, such as $\frac{1}{128}$ or $\frac{1}{256}$, it became the practice to give decimal dimensions, such as 4.2340, 3.815, etc., for the dimensions requiring accuracy. Along with this, many of the dimensions, such as pattern dimensions, forging dimensions, and relatively unimportant machine dimensions, were still expressed in whole numbers and common fractions.

Thus, a given drawing today may be dimensioned entirely with whole numbers and common fractions, or entirely with decimals, or with a combination of the two. The last is the most common, especially with reference to machined castings or forgings, in which the rough dimensions usually need not be closer than $\pm \frac{1}{64}''$. However, practice is rapidly moving toward the adoption of the complete decimal system, §368, for all dimensions.

*Decimal scales, graduated in fiftieths of an inch, are coming into increasing use, Fig. 628.

On any drawing decimal dimensions should be used wherever the degree of accuracy required is closer than $\pm \frac{1}{64}''$. For decimal equivalents of common fractions, see the front endpaper of this book. For rounding off decimals, see §368.

336. Dimension Figures. The importance of good lettering of dimension figures cannot be overstated. The shop produces according to the directions on the drawing, and to save time and prevent costly mistakes, all lettering should be perfectly legible. A complete discussion of numerals is given in §§92-94.

As shown in Fig. 585 (a), the standard height for whole number is $\frac{1}{8}''$, and for

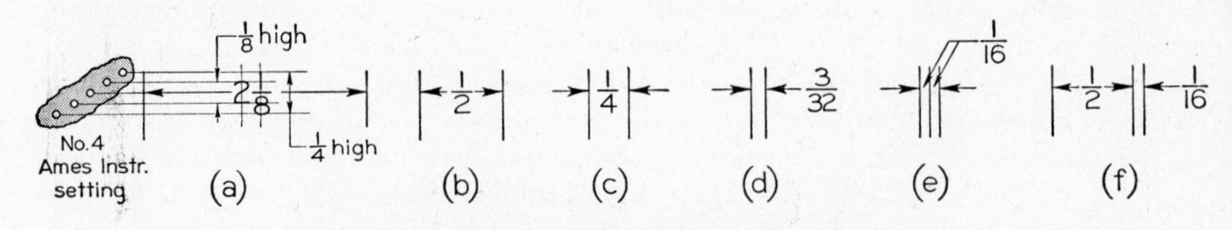

Fig. 585 Common-Fraction Dimension Figures.

fractions double that, or $\frac{1}{4}''$. Beginners should use guide lines as shown in Figs. 124-126. As shown in Fig. 127 (a), the numerator and denominator of a fraction should be clearly separated from the fraction bar. As shown at (c) in the same figure, the fraction bar should always be horizontal. An exception to this may be made in crowded places, as in parts lists, but never in dimensioning.

Legibility should never be sacrificed by crowding dimension figures into limited spaces. For every such case there is a practical and effective method, as shown in Fig. 585. At (a) and (b) there is enough space for the dimension line, the numeral, and the arrowheads. At (c) there is only enough room for the figure, and the arrowheads are placed outside. At (d) both the arrowheads and the figure are placed outside. Other methods are shown at (e) and (f).

If necessary, a removed partial view may be drawn to an enlarged scale to provide the space needed for clear dimensioning, Fig. 421.

Methods of lettering decimal dimension figures are shown in Fig. 586. All nu-

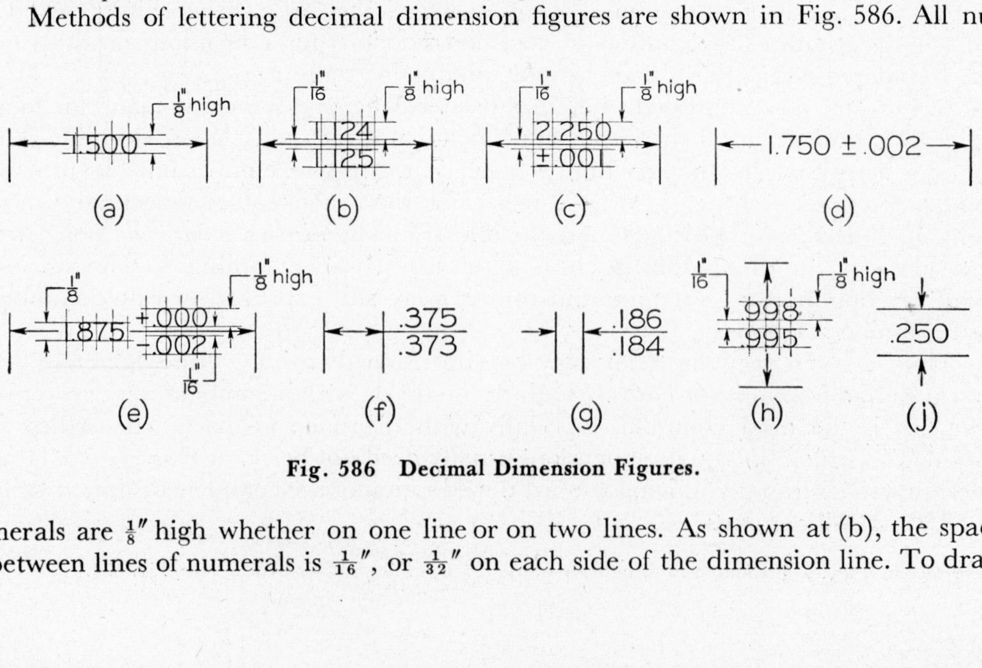

Fig. 586 Decimal Dimension Figures.

merals are $\frac{1}{8}''$ high whether on one line or on two lines. As shown at (b), the space between lines of numerals is $\frac{1}{16}''$, or $\frac{1}{32}''$ on each side of the dimension line. To draw guide lines with the Braddock-Rowe Triangle, Fig. 120, use the "fraction" holes at the left side of the triangle. For the Ames Lettering Instrument, Fig. 121, use the No. 4 setting and the center column of holes.

Make all decimal points bold, allowing ample space. Never letter a dimension figure over any line of the drawing. Place dimension figures outside a sectioned area if possible, Fig. 587 (a). When a dimension must be placed on a sectioned area, leave an opening in the section lining for the dimension figure, (b).

In a group of parallel dimension lines, the numerals should be staggered as in Fig. 588 (a), and not stacked up one above the other, as at (b).

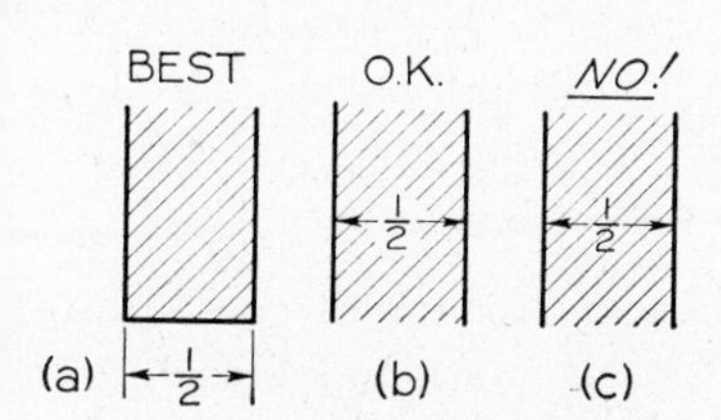

Fig. 587 Dimensions and Section Lines.

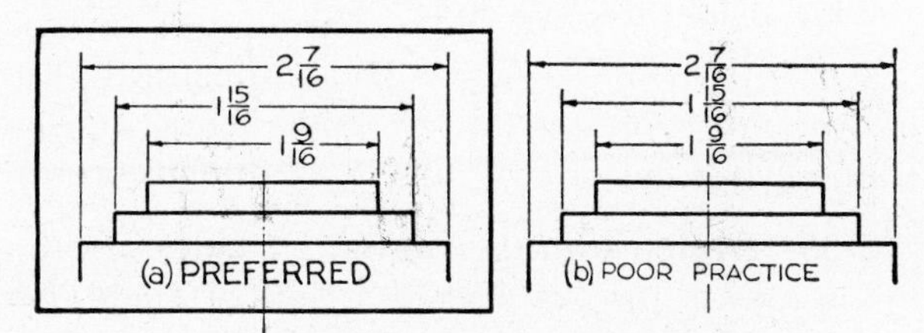

Fig. 588 Staggered Numerals.

337. Direction of Dimension Figures.

Two systems of reading direction for dimension figures are approved by the ASA. In the *aligned system*, Fig. 589 (a), all dimension figures are aligned with the dimension lines so that they may be read from the bottom or from the right side of the sheet. Dimension lines should not run in the directions included in the shaded area of Fig. 590, if avoidable. Notes should always be lettered horizontally on the sheet.

In the *unidirectional system*, Fig. 589 (b), all dimension figures and notes are lettered

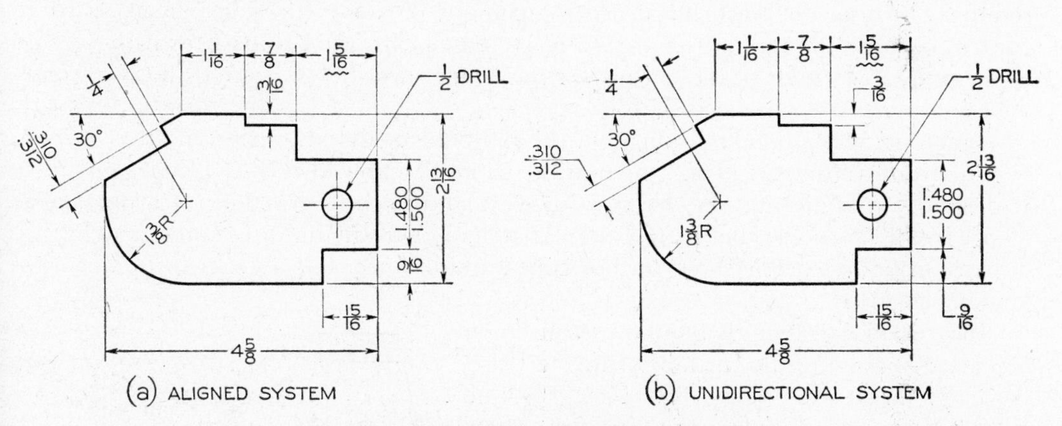

Fig. 589 Directions of Dimension Figures.

horizontally on the sheet. The unidirectional system has been extensively adopted in the aircraft, automotive, and other industries in which drawings are often very large and therefore hard to read from the side.

338. Feet and Inches.

Inches are indicated by the symbol ″ placed slightly above and to the right of the numeral, thus: $2\frac{1}{2}''$. *Feet* are indicated by the symbol ′

similarly placed, thus: 3′-0, 5′-6, 10′-0¼. It is customary in such expressions to omit the inch marks.

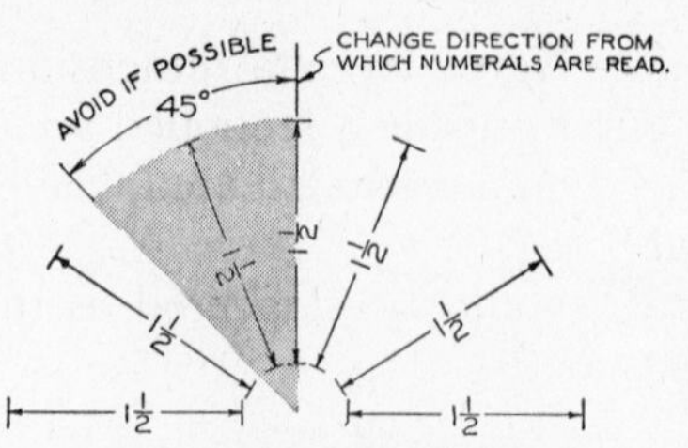

Fig. 590 Directions of Dimensions.

It is standard practice to omit inch marks when all dimensions on a drawing are in inches, except when there is a possibility of misunderstanding. Thus, 1 VALVE should be 1″ VALVE, and 1 DRILL should be 1″ DRILL, etc.

In some industries all dimensions, regardless of size, are given in inches; in others dimensions up to 72″ inclusive are given in inches, and those greater are given in feet and inches. In structural and architectural drafting, all dimensions of one foot or over are expressed in feet and inches. In locomotive, aircraft, and sheet-metal drafting, it is customary to give all dimensions in inches.

339. Dimensioning Angles. Angles are dimensioned by means of coordinate dimensions of the two legs of a right triangle, Fig. 591 (a), or by means of a linear

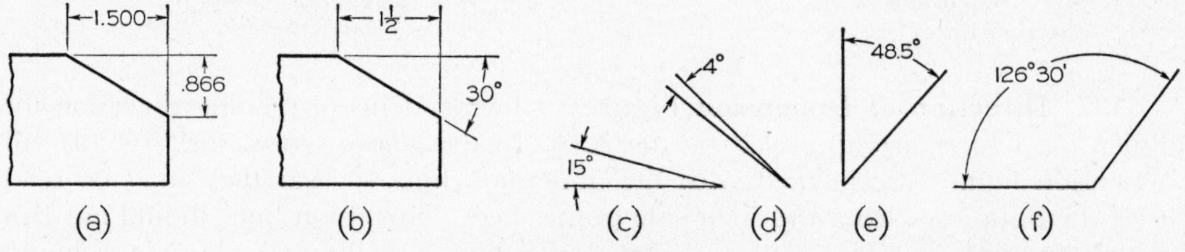

Fig. 591 Angles.

dimension and an angle in degrees, (b). The coordinate method is suitable for work requiring a high degree of accuracy. Variations of angle (in degrees) are hard to control because the amount of variation increases with the distance from the vertex of the angle. Methods of indicating various angles are shown from (c) to (f). Tolerances of angles are discussed in §387.

When degrees alone are indicated, the symbol ° or the abbreviation DEG is used. When minutes alone are given, the number should be preceded by 0°. *Example:* 0° 23′. If desired, an angle may be given in degrees and decimal fractions of a degree, as 49.5°. In all cases, whether in the aligned system or in the unidirectional system, the dimension figures are lettered on horizontal guide lines. For a general discussion of angles, see §109.

In structural drawings, angular measurements are made by giving the ratio of "run" to "rise," with the larger size being 12″, Figs. 1081 and 1085. These right triangles are referred to as *bevels*, §642. In civil engineering drawings, *slope* represents the angle with the horizontal, while *batter* is the angle referred to the vertical. Both are expressed by making one member of the ratio equal to 1, as shown in Fig. 592. *Grade*, as of a highway, is similar to slope but is expressed in percentage of rise per 100 ft. of run. Thus a 20 ft. rise in a 100 ft. run is a grade of 20 per cent.

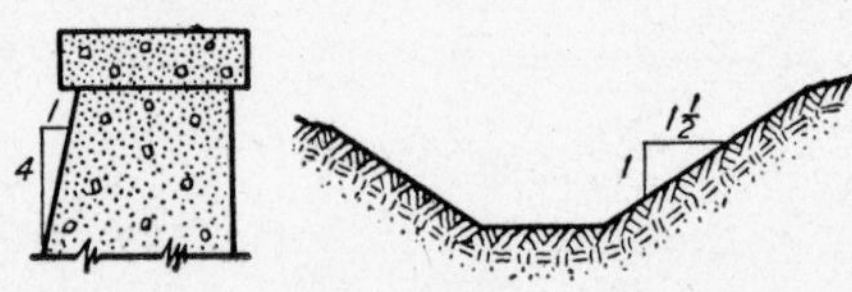

Fig. 592 Angles in Civil Engineering Projects.

340. Dimensioning Arcs. A circular arc is dimensioned in the view in which its true shape is shown, by giving the numeral denoting its radius, followed by the abbreviation R, as shown in Fig. 593. Centers are indicated by small crosses except

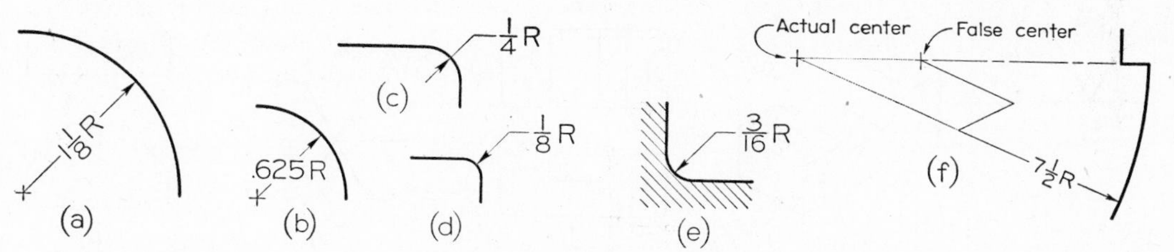

Fig. 593 Dimensioning Arcs.

for small or unimportant radii. Crosses should not be drawn for undimensioned arcs. As shown at (a) and (b), when there is room enough, both the numeral and the arrowhead are placed inside the arc. At (c) the arrowhead is left inside, but the numeral had to be moved outside. At (d) both the arrowhead and the numeral had to be moved outside. At (e) is shown an alternate method to (c) or (d) to be used when section lines or other lines are in the way. Note that in the unidirectional system, all of these numerals would be lettered horizontally on the sheet.

For a long radius, as shown at (f), when the center falls outside the available space, the dimension line is drawn toward the actual center; but a false center may be indicated and the dimension line "jogged" to it, as shown.

341. Fillets and Rounds. Individual fillets and rounds are dimensioned as any arc, as shown in Fig. 593 (b) to (e). If there are only a few and they are obviously the same size, as in Fig. 615 (5), one typical radius is sufficient. However, fillets and rounds are often quite numerous on a drawing and most of them are likely to be some standard size, as $\frac{1}{8}$"R or $\frac{1}{4}$"R. In such cases it is customary to give a note in the lower portion of the drawing to cover all such uniform fillets and rounds, thus: FILLETS $\frac{1}{4}$R AND ROUNDS $\frac{1}{8}$R UNLESS OTHERWISE SPECIFIED, or ALL CASTING RADII $\frac{1}{4}$R UNLESS NOTED, or simply ALL FILLETS AND ROUNDS $\frac{1}{8}$R.

For a discussion of fillets and rounds in the shop, see §302.

342. Finish Marks. A *finish mark* is used to indicate that a surface is to be machined, or finished, as on a rough casting or forging. To the patternmaker or diemaker, a finish mark means that allowance of extra metal in the rough workpiece must be provided for the machining. See §301. On drawings of parts to be machined from rolled stock, finish marks are generally unnecessary, as it is obvious that the surfaces are finished. Similarly, it is not necessary to show finish marks when a shop operation is specified in a note that indicates machining, as in drilling, reaming, boring, countersinking, counterboring, broaching, etc., or when the dimension implies finished surfaces, such as .250-.245 DIA.

Two styles of finish marks are approved by the ASA, the newer V symbol and the older *f* symbol. The first, Fig. 594 (a), is like a capital V, made about $\frac{1}{8}$" high in conformity with the dimension figures. For best results it should be drawn with the aid of the 30° × 60° triangle.

At (b) is shown a simple casting having several finished surfaces, and at (c) are shown two views of the same casting, showing how the finish marks are indicated on a drawing. The *finish mark is shown only on the edge view of a finished surface, and is*

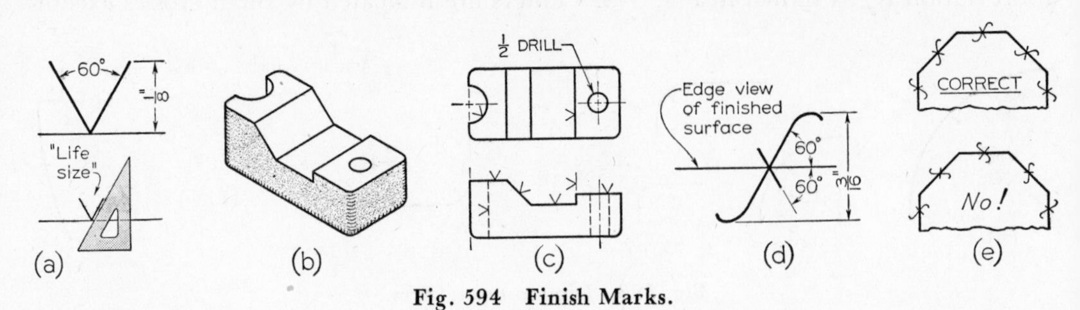

Fig. 594 Finish Marks.

repeated in any other view in which the surface appears as a line, even if the line is a hidden line. The point of the V should point inward toward the body of metal in a manner similar to that of a tool bit. When it is necessary to control surface roughness of finished surfaces, the V is used as a base for the more elaborate surface quality symbols, as discussed in §367.

The old symbol *f* is still more widely used, though it is executed in a variety of forms. It is approved by the ASA in the form shown in Fig. 594 (d) and (e), and will undoubtedly continue in use for many years. It is shown on the edge views of finished surfaces as described above for the V-type finish marks.

If a part is to be finished all over, finish marks should be omitted, and a general note should be lettered on the lower portion of the sheet, such as FINISH ALL OVER, or F.A.O.

343. Dimensions On or Off Views. *Dimensions should not be placed upon a view unless the clearness of the drawing is promoted thereby.* The ideal form is shown in Fig. 595 (a), in which all dimensions are placed outside the view. Compare this with the

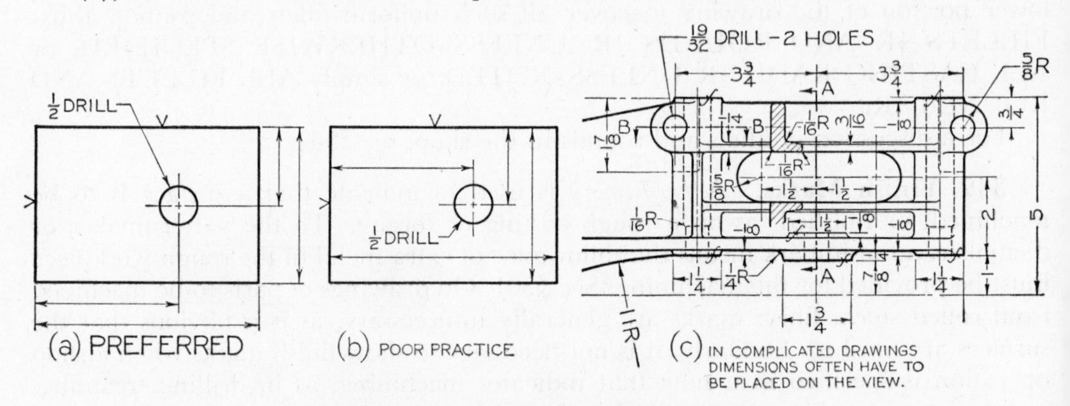

Fig. 595 Dimensions On or Off the Views.

evidently poor practice shown at (b). This is not to say that a dimension should never be placed on a view, for in many cases, particularly in complicated drawings, this is necessary, as shown at (c). Certain radii and other dimensions are given on the views, but in each case investigation will reveal a good reason for placing the dimension on the view. *Place dimensions outside of views, except where directness of applica-*

tion and clarity are gained by placing them on the views where they will be closer to the features dimensioned.

344. Contour Dimensioning. Views are drawn to describe the shapes of the various features of the object, and dimensions are given to define exact sizes and locations of those shapes. It follows that *dimensions should be given where the shapes are shown,* that is, in the views where the contours are delineated, as shown in Fig. 596 (a). Incorrect placement of the dimensions is shown at (b).

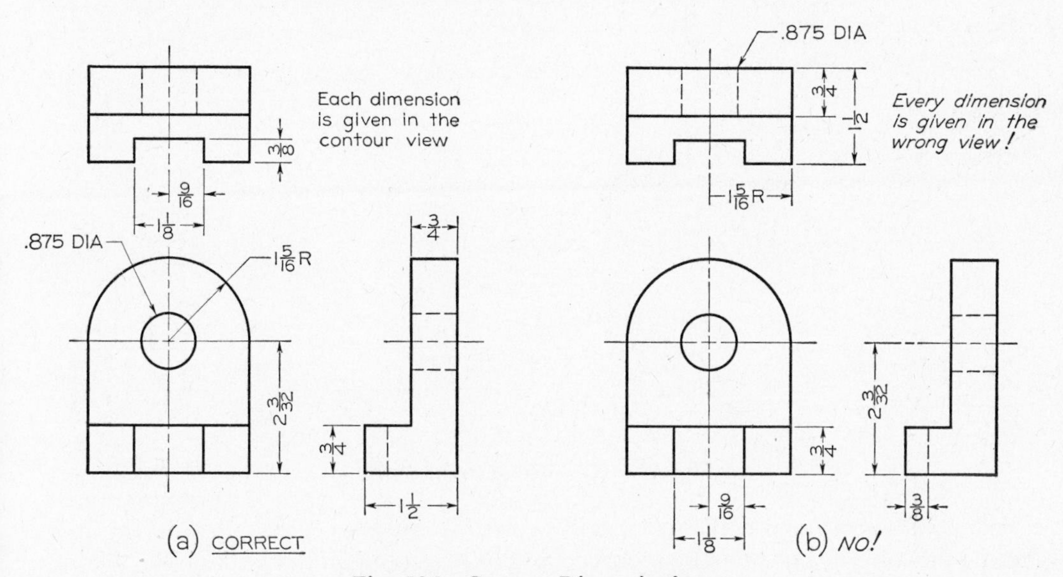

Fig. 596 Contour Dimensioning.

If individual dimensions are attached directly to the contours that show the shapes being dimensioned, it will automatically prevent the attachment of dimensions to hidden lines, as shown for the depth $\frac{3}{8}$ of the slot at (b). It will also prevent the attachment of dimensions to a visible line the meaning of which is not clear in a particular view, such as dimension $\frac{3}{4}$ for the height of the base at (b).

Although the placement of notes for holes follows the contour rule wherever possible, as shown at (a), the diameter of an external cylindrical shape is preferably given in the rectangular view where it can be readily found near the dimension for the length of the cylinder, as shown in Figs. 597 (b), 600, and 601.

345. Geometric Breakdown. Engineering structures are composed largely of simple geometric shapes, such as the prism, cylinder, pyramid, cone, sphere, etc., as shown in Fig. 597 (a). They may be exterior (positive) or interior (negative) forms. For example, a steel shaft is a positive cylinder, and a round hole is a negative cylinder.

These shapes result directly from the necessity in design to keep forms as simple as possible, and from the requirements of the fundamental shop operations. Forms having plane surfaces are produced by planing, shaping, milling, etc., while forms having cylindrical, conical, or spherical surfaces are produced by turning, drilling, reaming, boring, countersinking, and other rotary operations. See Chapter 10, "Shop Processes."

The dimensioning of engineering structures begins with two steps: **first,** giving the dimensions showing the *sizes* of the simple geometric shapes, called *size dimensions;* and **second,** giving the dimensions *locating* these elements with respect to each other, called *location dimensions*. This method of geometric analysis is very helpful in dimensioning any object, but must be modified when there is a conflict with either the function of the part in the assembly or with the production requirements in the shop.

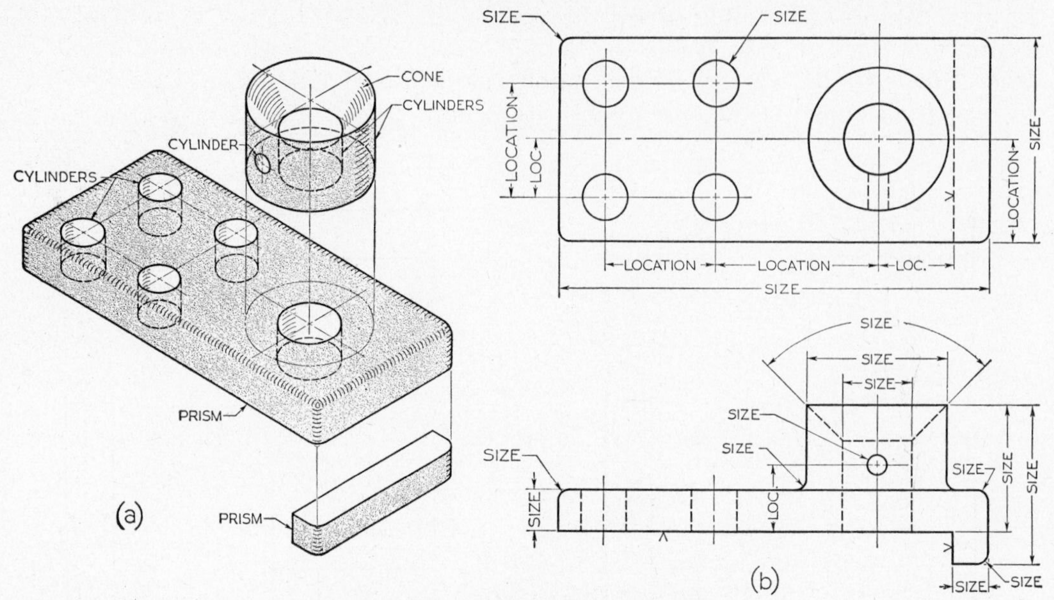

Fig. 597 Geometric Breakdown.

In Fig. 597 (b) is shown a multiview drawing of the object shown in isometric at (a). Here it will be seen that each geometric shape is dimensioned with size dimensions, and these shapes are then located with respect to each other with location dimensions. Note that a *location dimension locates a three-dimensional geometric element,* and not just a surface; otherwise, all dimensions would have to be classified as location dimensions.

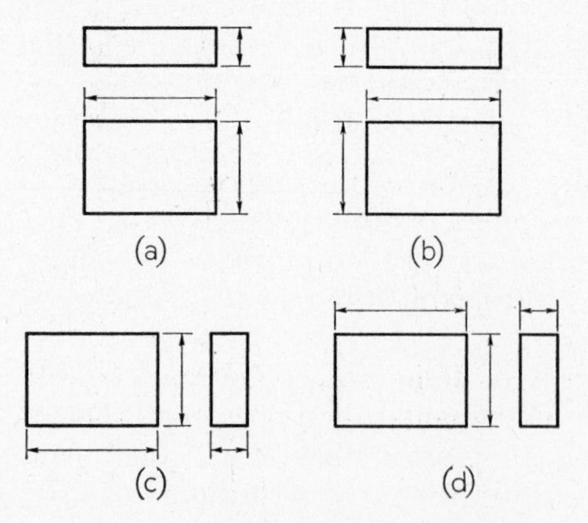

Fig. 598 Dimensioning Rectangular Prisms.

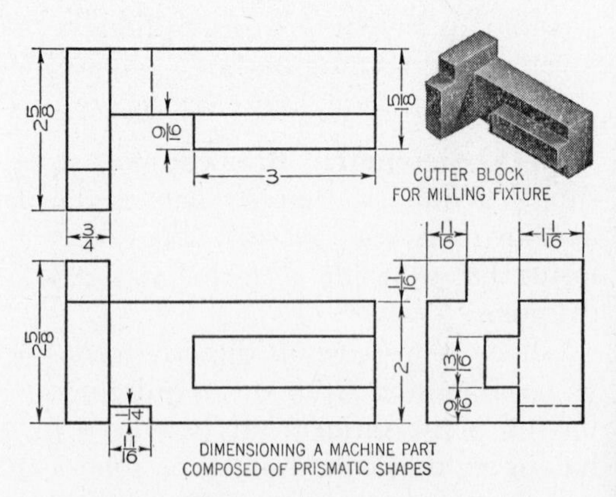

Fig. 599 Dimensioning Prismatic Shapes.

346. Size Dimensions—Prisms. The right rectangular prism, Fig. 159, is probably the most common geometric shape. Front and top views are dimensioned as shown in Fig. 598 (a) or (b). The height and width are given in the front view, and the depth in the top view. The vertical dimensions can be placed on the left or right provided both of them are placed in line. The horizontal dimension applies to both the front and top views, and should be placed between them, as shown, and not above the top or below the front view.

Front and side views should be dimensioned as at (c) or (d). The horizontal dimensions can be placed above or below the views, provided both are placed in line. The dimension between views applies to both views and should not be placed elsewhere without a special reason.

An application of size dimensions to a machine part composed entirely of rectangular prisms is shown in Fig. 599.

347. Size Dimensions—Cylinders. The right circular cylinder is the next most common geometric shape, and is commonly seen as a shaft or a hole. The general method of dimensioning a cylinder is to give both its diameter and its length in the rectangular view, Fig. 600. If the cylinder is drawn in a vertical position, the length

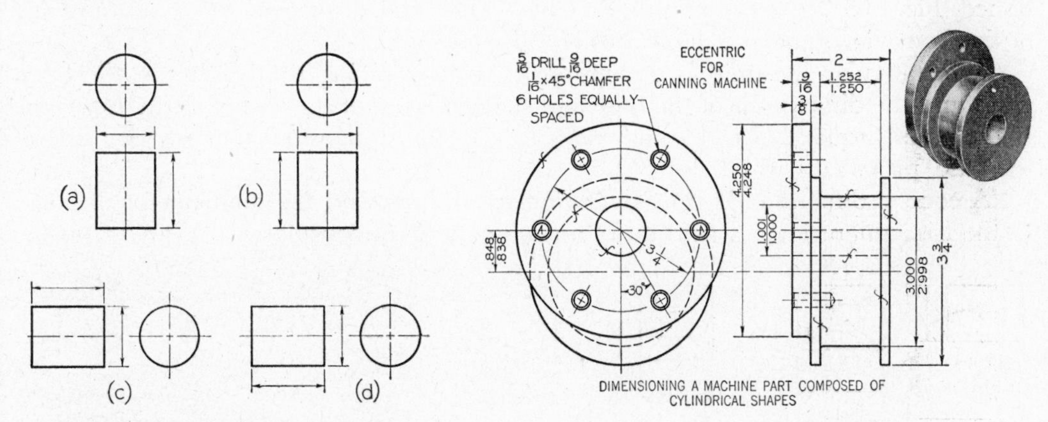

Fig. 600 Dimensioning Cylinders.

Fig. 601 Dimensioning Cylinders.

or altitude of the cylinder may be given at the right as at (a), or on the left as at (b). If the cylinder is drawn in a horizontal position, the length may be given above the rectangular view as at (c), or below as at (d). An application showing the dimensioning of cylindrical shapes is shown in Fig. 601.

The ASA approves the use of a diagonal diameter in the circular view, in addition to the method shown in Fig. 600, but the authors do not recommend this method except in special cases when clearness is gained thereby. The use of several diagonal diameters on the same center is definitely to be discouraged, since the result is usually confusing.

The radius of a cylinder should never be given, since measuring tools, such as the micrometer caliper, are designed to check diameters.

Small cylindrical holes, such as drilled, reamed, or bored holes, are usually dimensioned by means of notes specifying the diameter and the depth, together with required shop operations, Figs. 601 and 604.

When it is not clear from the views that a dimension indicates a diameter, the abbreviation DIA should be given after the dimension figure, Fig. 602 (a). In some cases, DIA may be used to eliminate the circular view, as shown at (b).

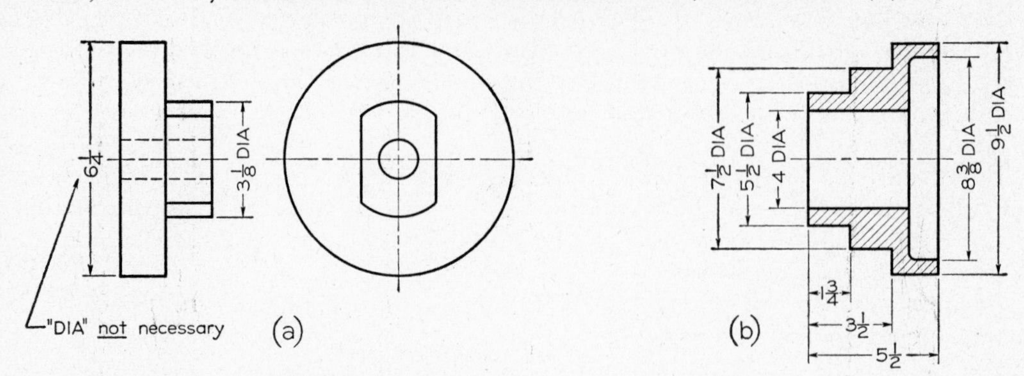

Fig. 602 Use of DIA in Dimensioning Cylinders.

348. Size Dimensions—Miscellaneous Shapes. A triangular prism is dimensioned, Fig. 603 (a), by giving the height, width, and displacement of the top edge in the front view, and the depth in the top view.

A rectangular pyramid is dimensioned, (b), by giving the heights in the front view, and the dimensions of the base and the centering of the vertex in the top view. If the base is square, (c), it is necessary to give the dimensions for only one side of the base, provided it is labeled SQ.

A cone is dimensioned, (d), by giving its altitude and the diameter of the base in the triangular view. A frustum of a cone may be dimensioned, (e), by giving the

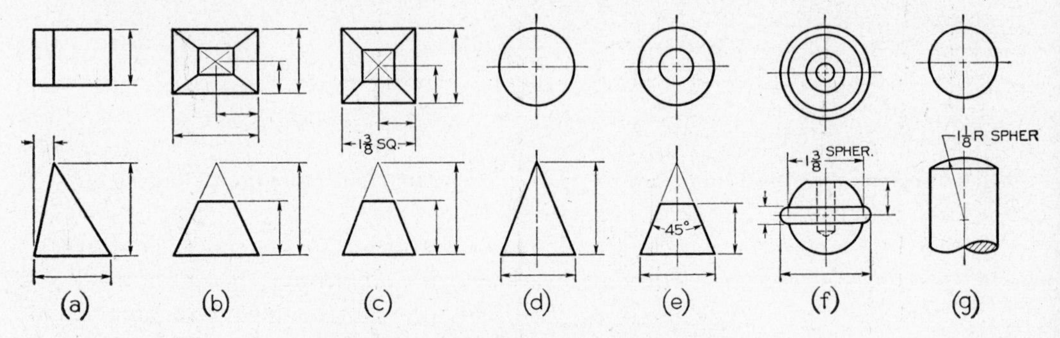

Fig. 603 Dimensioning Various Shapes.

vertical angle and the diameter of one of the bases. Another method is to give the length and the diameters of both ends in the front view. Still another is to give the diameter at one end and the amount of taper per foot in a note, §358.

At (f) is shown a two-view drawing of a plastic knob. The main body is spherical, and is dimensioned by giving its diameter, followed by the abbreviation SPHER. A bead around the knob is in the shape of a torus, Fig. 159, and it is dimensioned by giving the thickness of the ring and the outside diameter, as shown. At (g) a spherical end is dimensioned by giving its radius, followed by SPHER.

Internal shapes corresponding to the external shapes in Fig. 603 would be dimensioned in a similar manner.

349. Size Dimensioning of Holes. Holes that are to be drilled, bored, reamed, punched, cored, etc., are usually specified by standard notes, as shown in Figs. 323, 604 (a), and 616. The order of items in a note corresponds to the order of procedure in the shop in producing the hole. Two or more holes are dimensioned by a single note, the leader pointing to one of the holes, as shown at the top of Fig. 604 (a).

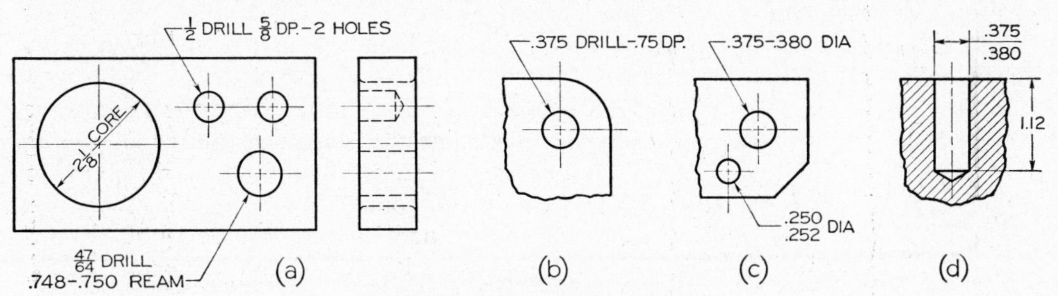

Fig. 604 Dimensioning Holes.

As shown in Figs. 323 and 604, the leader of a note should, as a rule, point to the circular view of the hole. It should point to the rectangular view only when clearness is promoted thereby. When the circular view of the hole has two or more concentric circles, as for counterbored, countersunk, or tapped holes, the arrowhead should touch the outer circle, Fig. 616 (c) to (j).

Notes should always be lettered horizontally on the paper, and guide lines should always be used.

The use of decimal fractions instead of common fractions to designate drill sizes has gained wide acceptance,* Fig. 604 (b). For numbered or letter-size drills, Appendix 14, it is recommended that the decimal size be given in this manner, or given in parenthesis, thus: #28 (.1405) DRILL, or "P" (.3230) DRILL.

On drawings of parts to be produced in large quantity for interchangeable assembly, dimensions and notes may be given without specification of the shop process to be used. Only the dimensions of the hole are given, without reference to whether the holes are to be drilled, reamed, punched, etc., Fig. 604 (c) and (d). It should be realized that even though shop operations are omitted from a note, the tolerances indicated would tend to dictate the shop processes required.

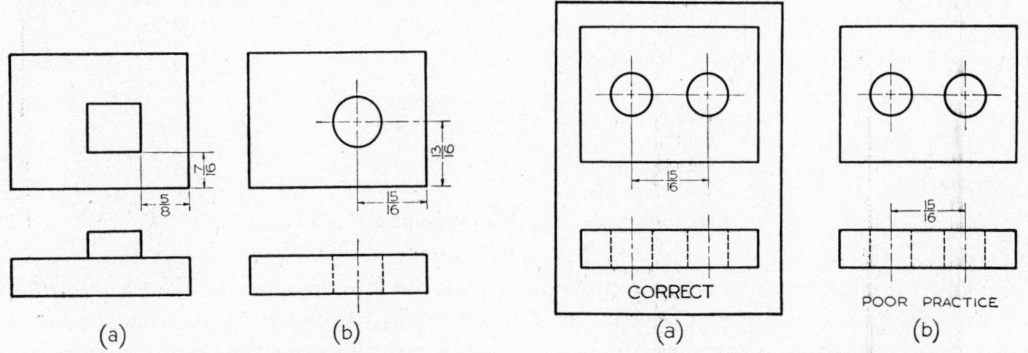

Fig. 605 Location Dimensions. **Fig. 606 Locating Holes.**

*Although drills are still usually listed fractionally in most manfacturers' catalogs, many companies have long since supplemented drill and wire sizes with a decimal value. In many cases the number, letter, or common fraction has been replaced by the decimal size.

350. Location Dimensions. After the geometric shapes composing a structure have been dimensioned for *size*, as discussed above, *location dimensions* must be given to show the relative positions of these geometric shapes, as shown in Fig. 597. As shown in Fig. 605 (a), rectangular shapes, whether in the form of solids or of recesses, are located with reference to their faces. As shown at (b), cylindrical or conical holes or bosses, or other symmetrical shapes, are located with reference to their center lines.

As shown in Fig. 606, location dimensions for holes are preferably given in the circular view of the holes.

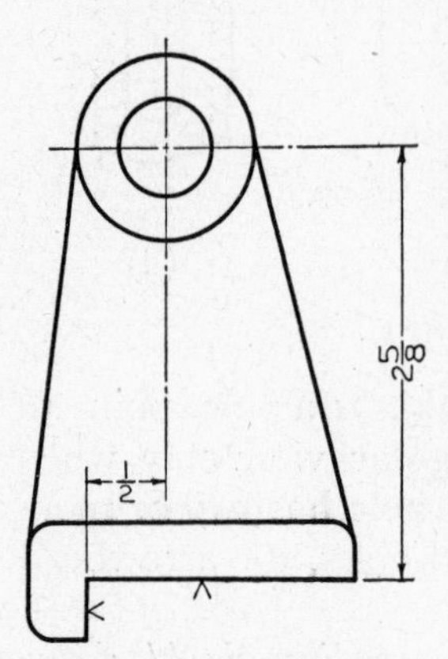

Fig. 607 Dimensions to Finished Surfaces.

Location dimensions should lead to finished surfaces wherever possible, Fig. 607, because rough castings and forgings vary in size, and unfinished surfaces cannot be relied upon for accurate measurements. Of course, the *starting dimension*, used in locating the first machined surface on a rough casting or forging, must necessarily lead from a rough surface, or from a center or a center line of the rough piece. See Figs. 813 and 814, parts 12 and 6.

In general, location dimensions should be built from a finished surface as a datum plane, or from an important center or center line.

When several cylindrical surfaces have the same center line, as in Fig. 602 (b), it is not necessary to locate them with respect to each other.

Holes equally spaced about a common center may be dimensioned, Fig. 608 (a), by giving the diameter (diagonally) of the *circle of centers*, or *bolt circle*, and specifying "equally spaced" in the note.

Holes unequally spaced, (b), are located by means of the bolt circle diameter plus angular measurements with reference to *only one* of the center lines, as shown.

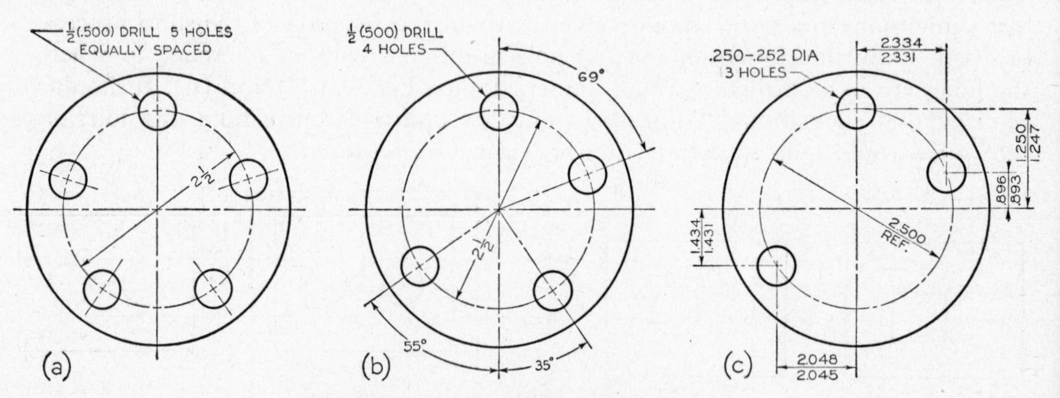

Fig. 608 Locating Holes about a Center.

Where greater accuracy is required, coordinate dimensions should be given, as at (c). In this case, the diameter of the bolt circle is marked REF to indicate that it is to be used only as a *reference dimension*. Reference dimensions are given for information only. They are not intended to be measured and do not govern the shop operaions. They represent calculated dimensions and are often useful in showing the intended design sizes. See Fig. 609 for other examples.

When several non-precision holes are located on a common arc, they are dimensioned, Fig. 609 (a), by giving the radius and the angular measurements from a *base line*, as shown. In this case, the base line is the horizontal center line.

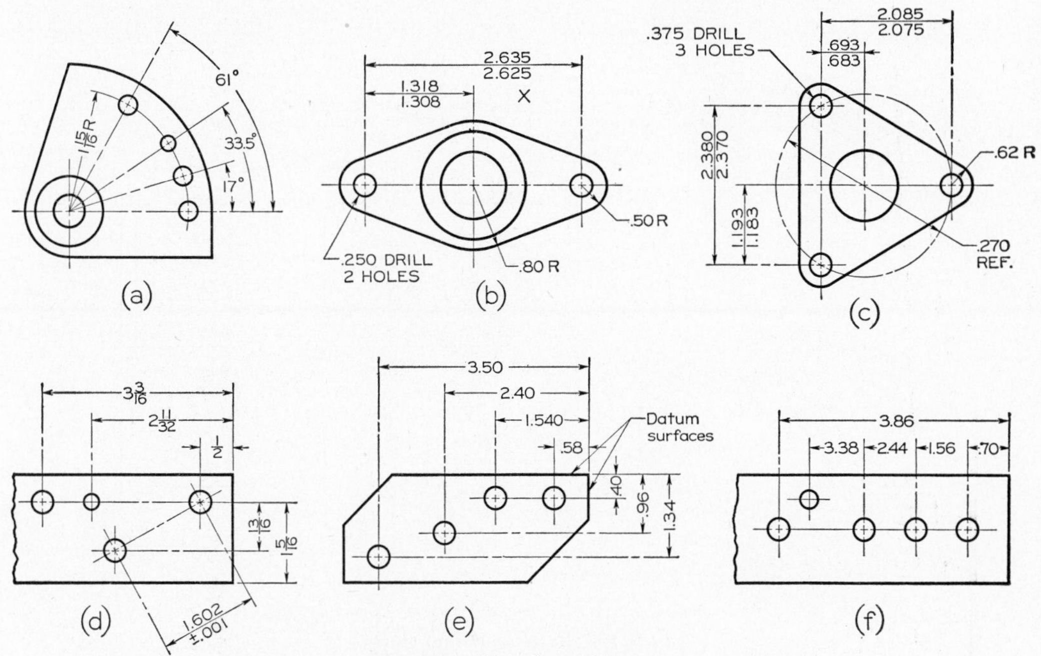

Fig. 609 Locating Holes.

At (b) the three holes are on a common center line. One dimension locates one small hole from the center; the other gives the distance between the small holes. Note the omission of a dimension at X. This method is used when (as is usually the case) the distance between the small holes is the important consideration. If the relation between the center hole and each of the small holes is more important, then include the distance at X, and mark the over-all dimension REF.

At (c) is another example of coordinate dimensioning. The three small holes are on a bolt circle whose diameter is marked REF, for reference purposes only. From the main center, the small holes are located in two mutually perpendicular directions.

At (d) is shown another example of locating holes by means of linear measurements. In this case, one such measurement is made at an angle to the coordinate dimensions because of the direct functional relationship of the two holes.

At (e) the holes are located from two *base lines* or *datums*. When all holes are located from a common datum, the sequence of measuring and machining operations is controlled, over-all tolerance accumulations are avoided, and proper functioning of the finished part is assured as intended by the designer. The datum surfaces selected must be more accurate than that required of any measurement made from them, and must be accessible during manufacture and arranged so as to facilitate tool and fixture design. Thus it may be necessary to specify accuracy of the datum surfaces in terms of straightness, roundness, flatness, etc. See §385.

At (f) is shown a method of giving, in a single line, all of the dimensions from a common datum. Each dimension except the first has a single arrowhead, and is accumulative in value. The final and longest dimension is separate and complete.

These methods of locating holes are equally applicable to locating pins or other symmetrical features.

351. Mating Dimensions. In dimensioning a single part, its relation to mating parts must be taken into consideration. For example, in Fig. 610 (a), a Guide Block

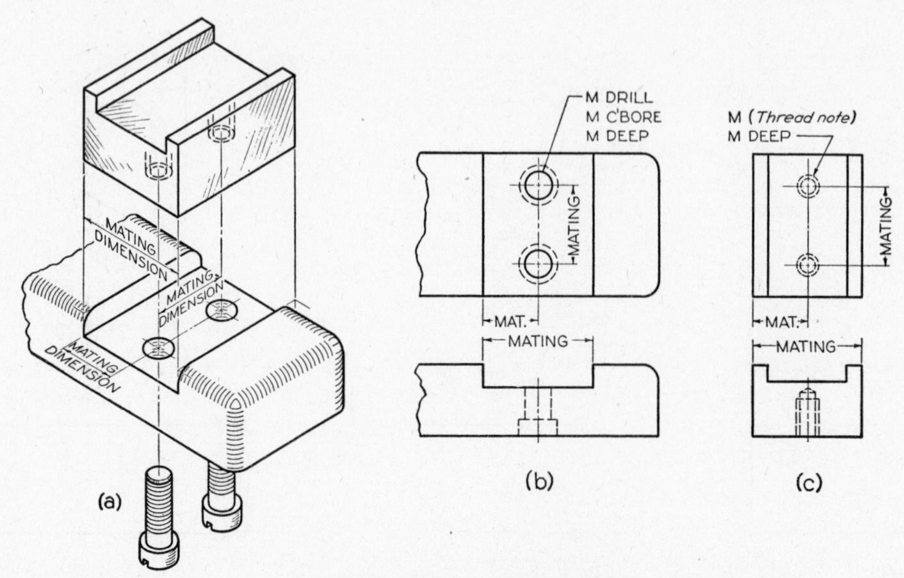

Fig. 610 Mating Dimensions.

fits into a slot in a Base. Those dimensions common to both parts are *mating dimensions*, as indicated.

These mating dimensions should be given on the multiview drawings in the corresponding locations, as shown at (b) and (c). Other dimensions are not mating dimensions since they do not control the accurate fitting together of the two parts. The actual *values* of two corresponding mating dimensions may not be exactly the same. For example, the width of the slot at (b) may be dimensioned $\frac{1}{32}''$ or several thousandths larger than the width of the Block at (c), but these are mating dimensions figured

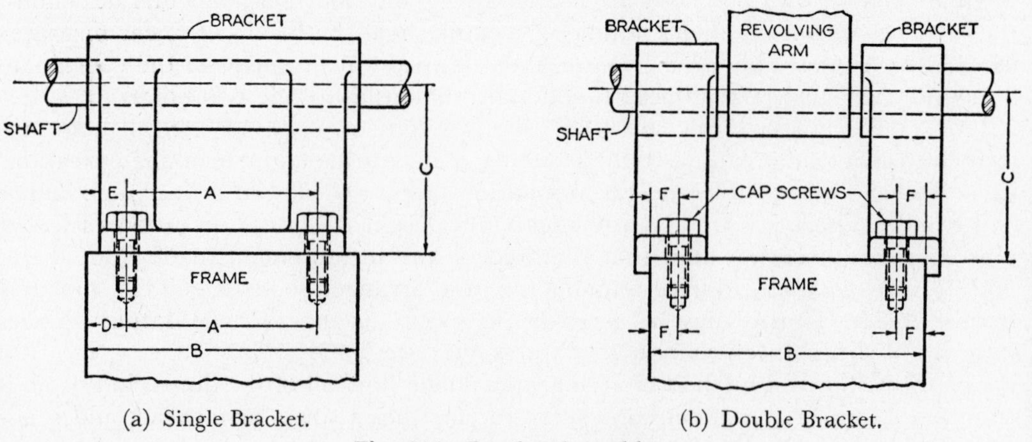

(a) Single Bracket. (b) Double Bracket.

Fig. 611 Bracket Assembly.

from a single basic width. It will be seen that the mating dimensions shown might have been arrived at from a geometric breakdown, §345. However, the mating dimensions need to be identified so that they can be specified in the corresponding locations on the two parts, and so that they can be given with the degree of accuracy commensurate with the proper fitting of the parts.

In Fig. 611 (a) the dimension A should appear on both the drawings of the Bracket and of the Frame, and is therefore a necessary mating dimension. At (b), which shows a redesign of the Bracket into two parts, dimension A is not used on either part, as it is not necessary to control closely the distance between the cap screws. But dimensions F are now essential mating dimensions and should appear correspondingly on the drawings of both parts. Other dimensions E, D, B, and C, at (a), are not mating dimensions, since they do not directly affect the mating of the parts.

352. Machine, Pattern, and Forging Dimensions. In Fig. 610 (a), the Base is machined from a rough casting; the patternmaker needs certain dimensions to make the pattern, and the machinist needs certain dimensions for the machining. In some cases one dimension will be used by both. Again, in most cases, these dimensions will be the same as those resulting from a geometric breakdown, §345, but it is important to identify them in order to assign values to them intelligently.

The same part is shown in Fig. 612, with the machine dimensions and pattern dimensions identified by the letters M and P. The patternmaker is interested only in the dimensions he needs to make the pattern, and the machinist, in general, is concerned only with the dimensions he needs to machine the part. It frequently occurs that a dimension that is convenient for the machinist is not convenient for the patternmaker, or vice-versa. Since the patternmaker uses the drawing only once while making the pattern and the machinist refers to it continuously, the dimensions should be given primarily for the convenience of the machinist.

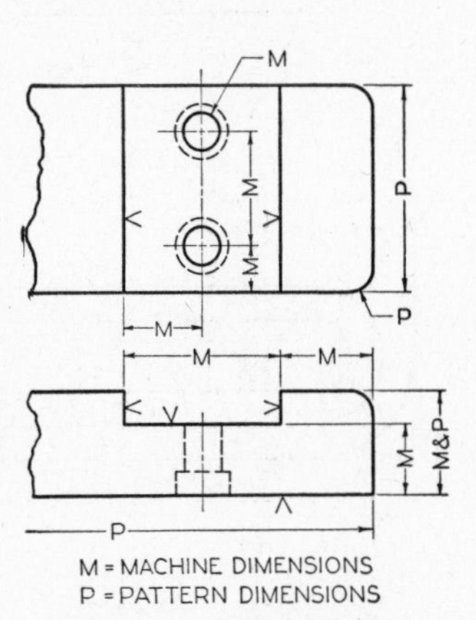

Fig. 612 Machine and Pattern Dimensions.

If the part is large and complicated, two separate drawings are sometimes made, one showing the pattern dimensions, and the other the machine dimensions. The usual practice, however, is to prepare one drawing for both the patternmaker and the machinist. See §301.

For forgings it is common practice to make separate forging drawings and machining drawings. A forging drawing of a connecting rod, showing only the dimensions needed in the forge shop, is shown in Fig. 725. A machining drawing of the same part, but containing only the dimensions needed in the machine shop, is shown in Fig. 726. See also Figs. 813 and 814.

Unless the complete decimal system is used, §368, the pattern dimensions are always nominal, usually to the nearest $\frac{1}{16}''$, and given in whole numbers and common fractions. If a machine dimension is given in whole numbers and common fractions, the machinist is usually allowed a tolerance (permissible variation in size) of $\frac{1}{64}''$, corresponding to his steel scale which has $\frac{1}{64}''$ divisions. Some companies specify a tolerance of .010″ on all common fractions. If greater accuracy is required, the dimensions are given in decimal form to three or more places, §§368 and 371.

353. Dimensioning Curves. Curved shapes may be dimensioned by giving a group of radii, as shown in Fig. 613 (a). Note that in dimensioning the $4\frac{15}{16}$R arc

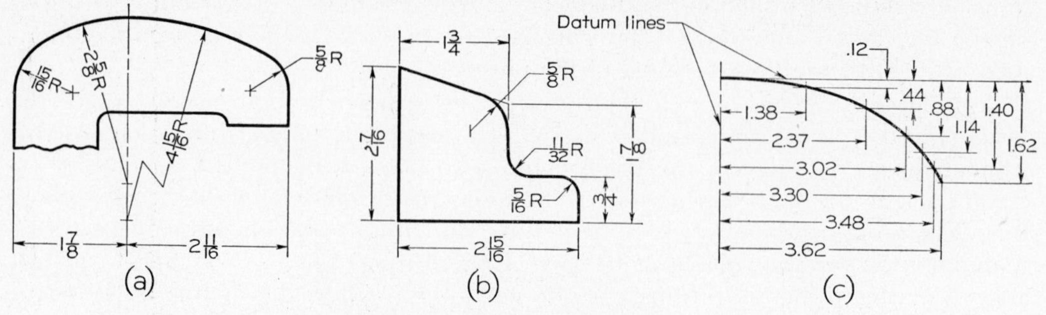

Fig. 613 Dimensioning Curves.

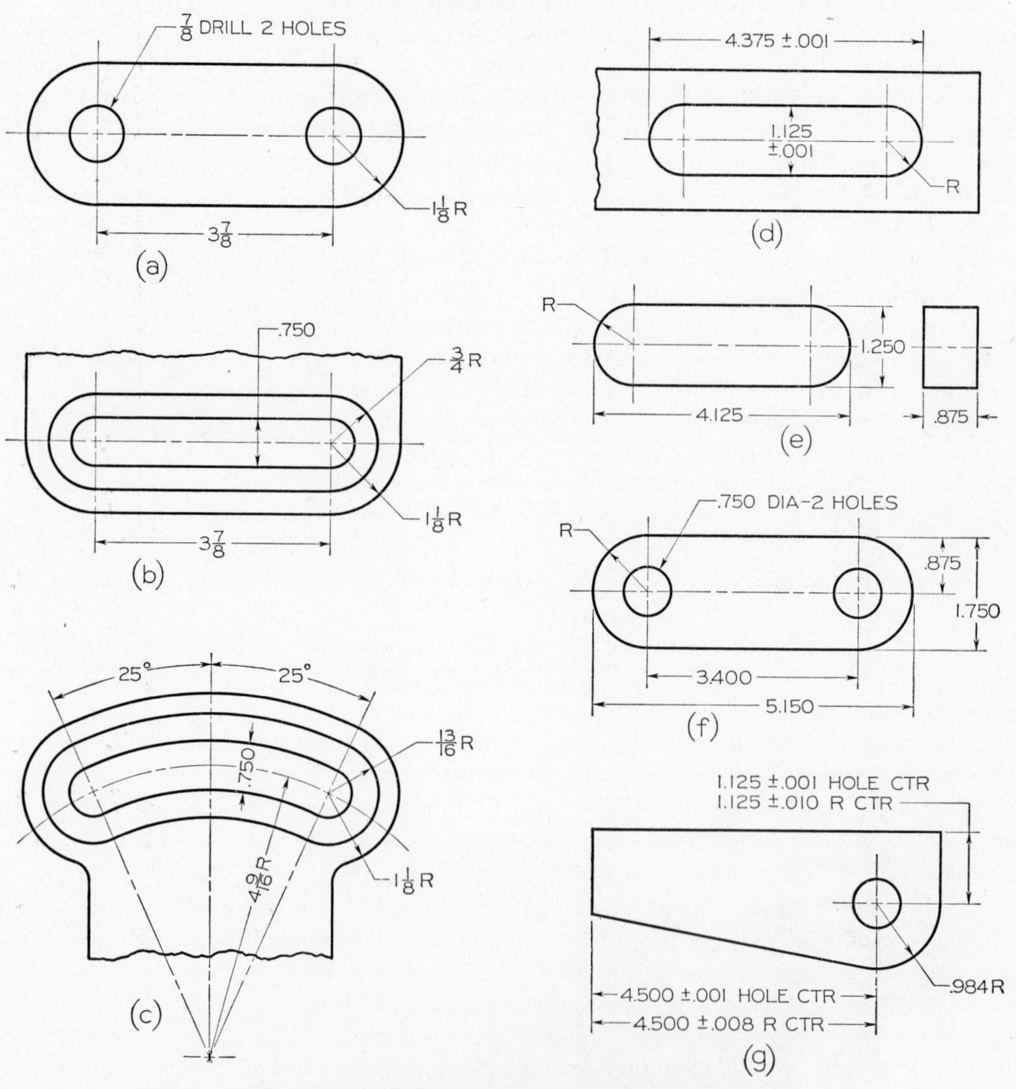

Fig. 614 Dimensioning Rounded-End Shapes.

whose center is inaccessible, the center may be moved in along a center line, and a jog made in the dimension line. See also Fig. 593 (f). Another method is to dimension the outline envelope of a curved shape so that the various radii are self-locating from "floating centers," as at (b). Either a circular or a non-circular curve may be dimensioned by means of coordinate dimensions referred to datums, as shown at (c).

354. Dimensioning Rounded-End Shapes. Methods of dimensioning rounded-end shapes depend upon the degree of accuracy required, Fig. 614. When precision is not necessary, the methods used are those which are convenient for the shop, as at (a), (b), and (c).

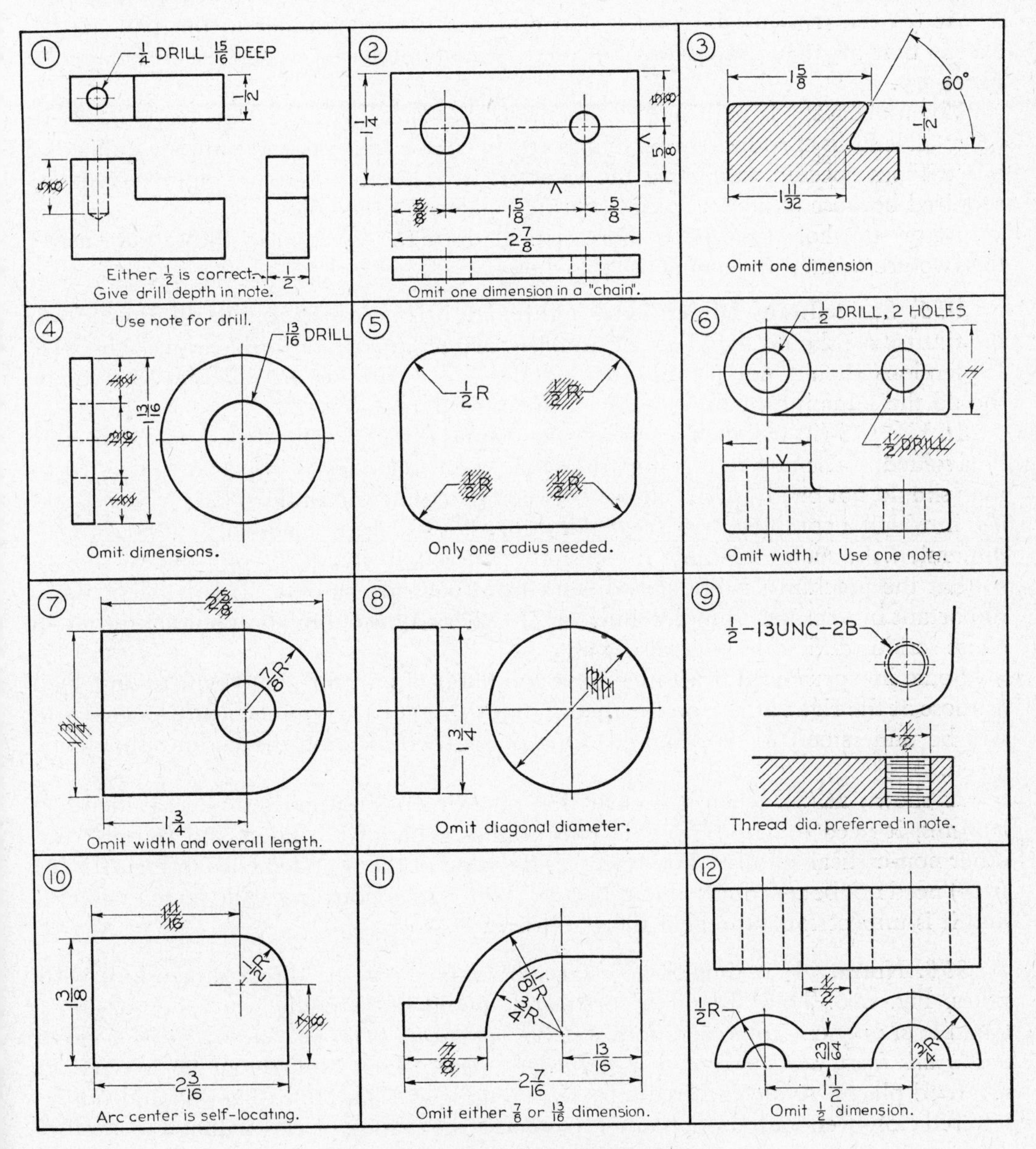

Fig. 615 Superfluous Dimensions.

At (a) the link, to be cast or to be cut from sheet metal or plate, is dimensioned as it would be laid out in the shop, by giving the center-to-center distance and the radii of the ends. Note that only one such radius dimension is necessary.

At (b) the pad on a casting, with a milled slot, is dimensioned from center-to-center for the convenience of both the patternmaker and the machinist in layout. An additional reason for the center-to-center distance is that it gives the total travel of the milling cutter, which can be easily controlled by the machinist. The width dimension indicates the diameter of the milling cutter; hence it is incorrect to give the radius of a machined slot. On the other hand, a cored slot (see §300) should be dimensioned by radius in conformity with the patternmakers' layout procedure.

At (c) the semicircular pad is laid out in a similar manner to the pad at (b), except that angular dimensions are used. Angular tolerances, §387, can be used if necessary.

When accuracy is required, the methods shown at (d) to (g) are recommended. Over-all lengths of rounded-end shapes are given in each case, and radii are indicated, but without specific values. In the example at (f), the center-to-center distance is required because of necessity for accurate location of the holes.

At (g) the hole location is more critical than the location of the radius; hence the two are located independently, as shown.

355. Superfluous Dimensions. Although it is necessary to give all dimensions, the draftsman should avoid giving unnecessary or superfluous dimensions, Fig. 615. Dimensions should not be repeated on the same view or on different views, nor should the same information be given in two different ways.

In Fig. 615 (2) is shown a type of superfluous dimensioning that should generally be avoided, especially in machine drawing where accuracy is important. The workman should not be allowed a choice between two dimensions. *Avoid "chain" dimensioning*, in which a complete series of detail dimensions is given, together with an over-all dimension. In such cases, one dimension of the chain should be omitted, as shown, so that the machinist is obliged to work from one surface only. This is particularly important in tolerance dimensioning, §371, where an accumulation of tolerances can cause serious difficulties. See also §381.

Some inexperienced draftsmen have the habit of omitting both dimensions, such as those at the right at (2), on the theory that the holes are symmetrically located and will be understood to be centered. One of the two location dimensions should be given.

As shown at (5), when it is clear that one dimension applies to several identical features, it need not be repeated. This applies generally to fillets and rounds, and other non-critical features. For example, the radii of the rounded ends in Fig. 614 (a) to (c) need not be repeated; and in Fig. 577 both ribs are obviously the same thickness, and it is unnecessary to repeat the $\frac{3}{8}''$ dimension.

356. Notes. It is usually necessary to supplement the direct dimensions with notes, Fig. 616. They should be brief and should be carefully worded so as to be capable of only one interpretation. *Notes should always be lettered horizontally on the sheet, with guide lines, and arranged in a systematic manner.* They should not be lettered in crowded places. Avoid placing notes between views, if possible. They should not be lettered closely enough to each other to confuse the reader, or close enough to another view or detail to suggest application to the wrong place. Leaders should be as short

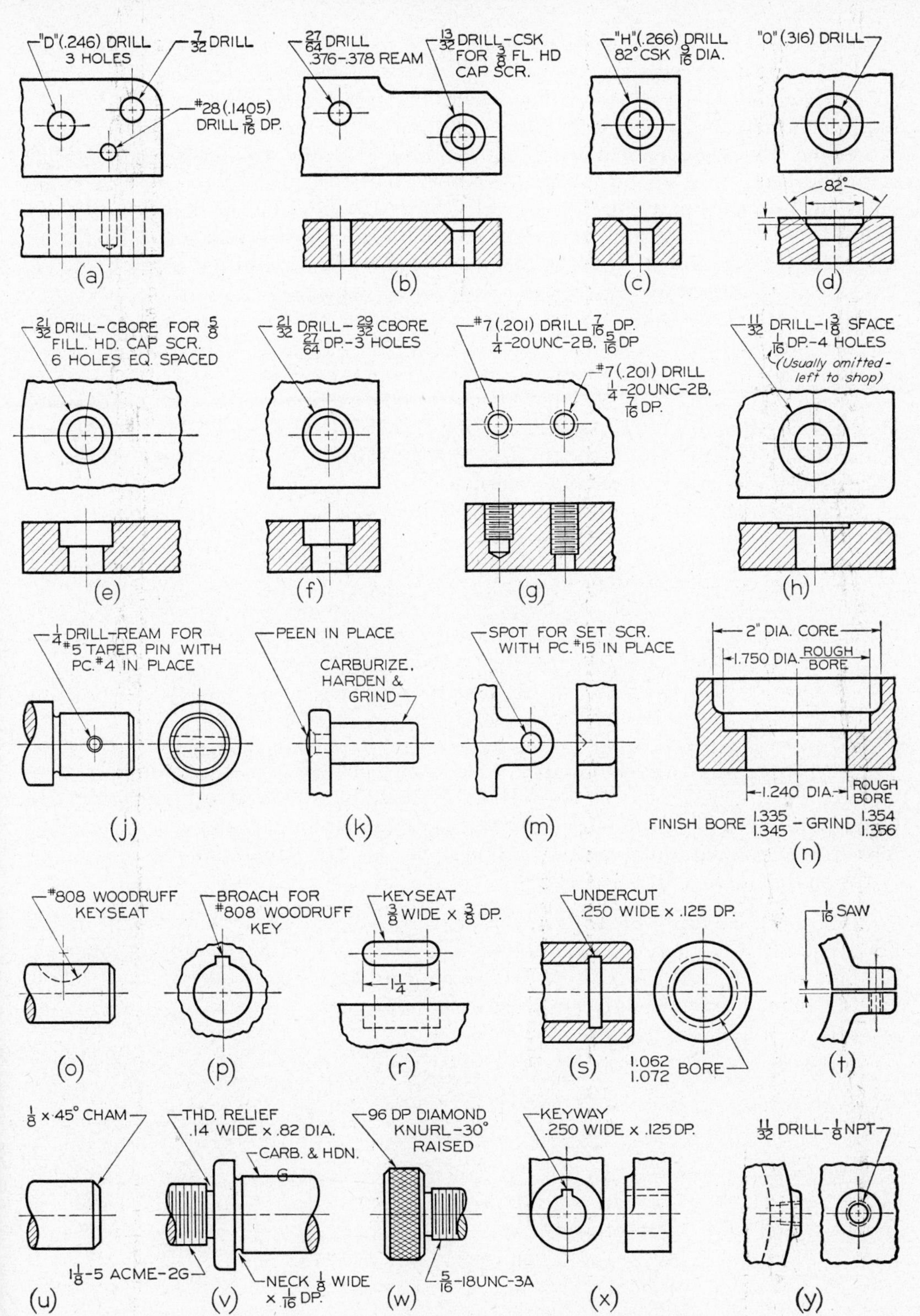

Fig. 616 Local Notes. See also Figs. 323 and 604.

as possible and cross as few lines as possible. They should never run through a corner of a view or through any specific points or intersections.

Notes are classified as *general notes* when they apply to an entire drawing, or *local notes* when they apply to specific items.

General Notes. General notes should be lettered in the lower right-hand corner of the drawing, above or to the left of the title block, or in a central position below the view to which they apply; e.g.: FINISH ALL OVER, or BREAK SHARP EDGES TO $\frac{1}{32}$ R; or SAE 3345-BRINELL 340-380; or ALL DRAFT ANGLES 3° UNLESS OTHERWISE SPECIFIED, or DIMENSIONS APPLY AFTER PLATING. In machine drawings, the title strip or title block will carry many general notes, including material, general tolerances, heat treatment, pattern information, etc. See Fig. 721.

Local Notes. Local notes apply to specific operations only, and are connected by a leader to the point at which such operations are performed; e.g.: $\frac{1}{4}$ DRILL-4 HOLES; or $\frac{1}{16} \times 45°$ CHAMFER; or 33p. DIAMOND KNURL, RAISED. The leader should be attached at the front of the first word of a note, or just after the last word, and not at any intermediate place.

For information on notes applied to holes, see §349.

Certain commonly used abbreviations may be used freely in notes, as THD, DIA, MAX, etc. The less common abbreviations should be avoided as much as possible. All abbreviations should conform to ASA Z32.13-1950. See Appendix 4 for American Standard abbreviations.

In general, leaders and notes should not be placed on the drawing until the dimensioning is substantially completed. If notes are lettered first, it will be found almost invariably that they will be in the way of necessary dimensions, and will have to be moved.

357. Dimensioning of Threads. Local notes are used to specify dimensions of threads. For tapped holes the notes should, if possible, be attached to the circular views of the holes, as shown in Fig. 616 (g). For external threads, the notes are usually placed in the longitudinal views where the threads are more easily recognized, as at (v) and (w). For a detailed discussion of thread notes, see §410.

358. Dimensioning of Tapers. A *taper* is a conical surface on a shaft or in a hole. The usual method of dimensioning a taper is to give the amount of taper per foot in a note as: TAPER 2″ PER FT (often with TO GAGE added), and then to give the diameter at one end, plus the length, or give the diameters at both ends and omit the length. "*Taper per foot*" *means the difference in diameter in one foot of length.*

Standard machine tapers are used on machine spindles, shanks of tools, pins,

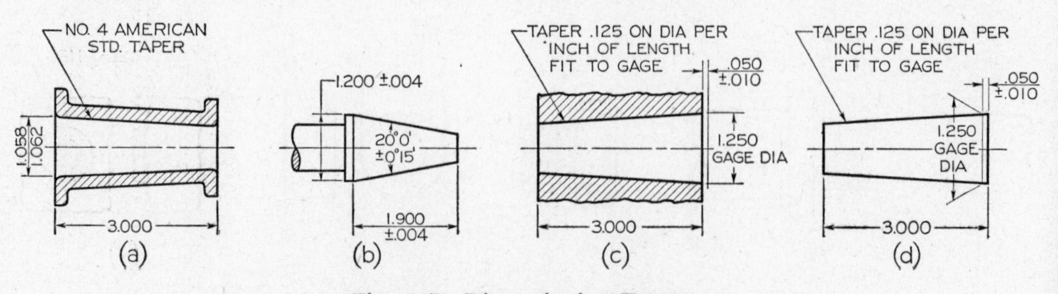

Fig. 617 Dimensioning Tapers.

etc., and are described in "Machine Tapers, Self-Holding, and Steep Taper Series," ASA B5.10-1953. Such standard tapers are dimensioned on a drawing by giving the diameter, usually at the large end, the length, and a note, such as NO. 4 AMERICAN STANDARD TAPER. See Fig. 617 (a).

For not-too-critical requirements, a taper may be dimensioned by giving the diameter at the large end, the length, and the included angle, all with proper tolerances, (b). Or the diameters of both ends, plus the length, may be given with necessary tolerances.

For close-fitting tapers, the amount of *taper per inch on diameter* is indicated as shown at (c) and (d). A gage line is selected and located by a comparatively generous tolerance, while other dimensions are given appropriate tolerances as required.

359. Dimensioning Chamfers. A *chamfer* is a beveled or sloping edge. When the angle is not 45°, it is dimensioned by giving the length and the angle, as in Fig. 618(a). A 45° chamfer also may be dimensioned in a manner similar to that shown at (a), but usually it is dimensioned by note as at (b).

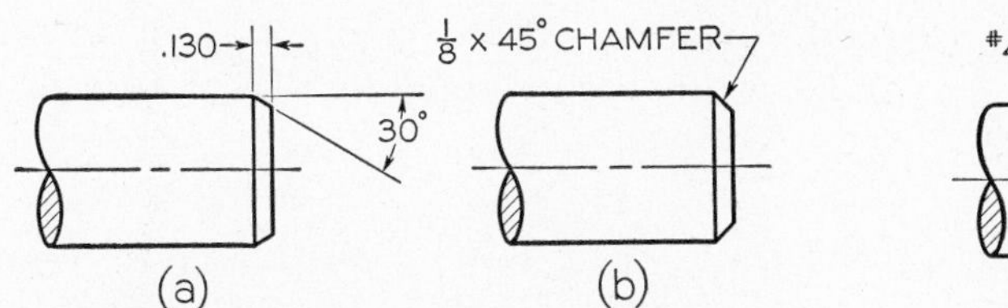

Fig. 618 Dimensioning Chamfers.

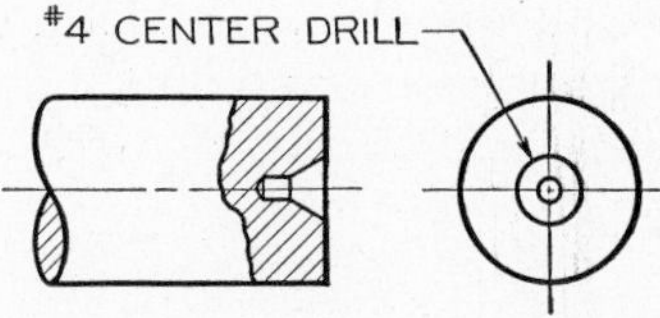

Fig. 619 Shaft Center.

360. Shaft Centers. Shaft centers are required on shafts, spindles, and other conical or cylindrical parts for turning, grinding, and other operations. Such a center may be dimensioned as shown in Fig. 619. Normally the centers are produced by a combined drill and countersink. See Appendix 42 for a table of shaft center sizes.

361. Dimensioning Keyways. Methods of dimensioning keyways for Woodruff keys and stock keys are shown in Fig. 620. Note, in both cases, the use of a dimension

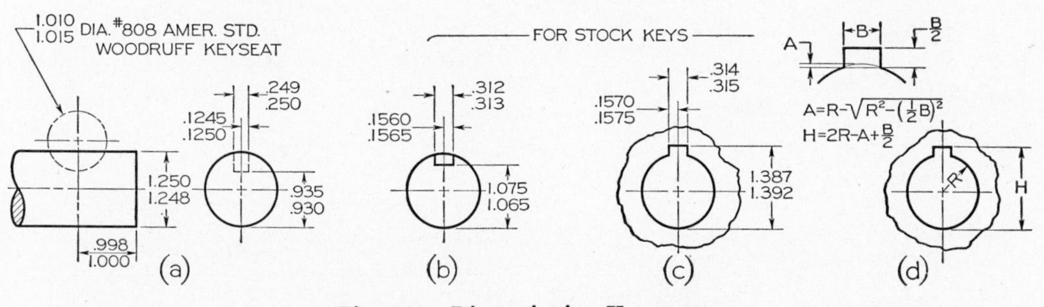

Fig. 620 Dimensioning Keyways.

to center the keyway in the shaft or collar. The preferred method of dimensioning the depth of a keyway is to give the dimension from the bottom of the keyway to the opposite side of the shaft or hole, as shown. The method of computing such a dimension is shown at (d). Values for A may be found in machinists' hand books.

For general information about keys and keyways, see §423.

362. Dimensioning Knurls. A *knurl* is a roughened surface to provide a better hand-grip or to be used for a press fit between two parts. For hand-gripping purposes, it is necessary only to give the pitch of the knurl, the type of knurling, and the length of the knurled area, Fig. 621 (a) and (b). To dimension a knurl for a

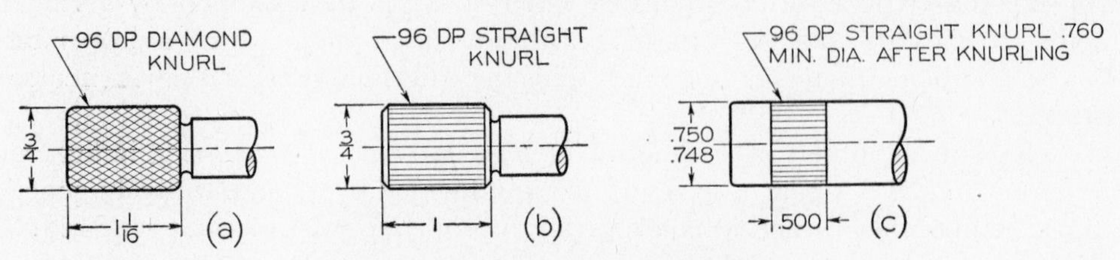

Fig. 621 Dimensioning Knurls.

press fit, the toleranced diameter before knurling should be given, (c). A note should be added giving the pitch and type of knurl and the minimum diameter after knurling. See ASA B5.30-1953.

363. Dimensioning Along Curved Surfaces. When angular measurements are unsatisfactory, chordal dimensions or linear dimensions upon the curved surfaces may be given, as shown in Fig. 622.

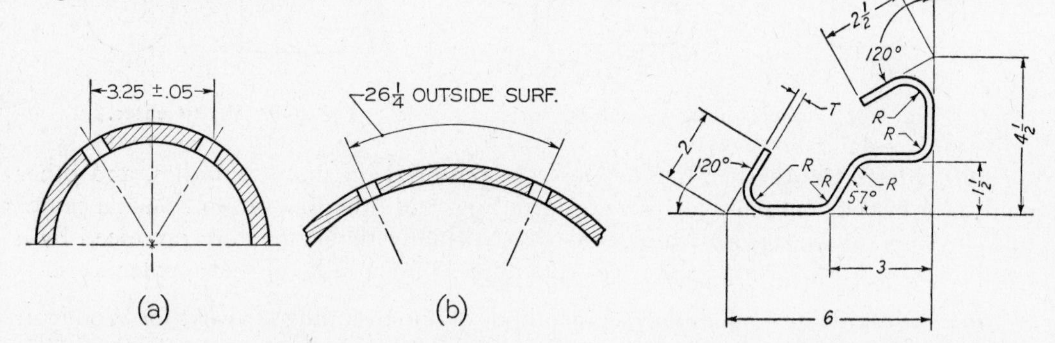

Fig. 622 Dimensioning Curved Surfaces. **Fig. 623 Profile Dimensioning.**

364. Sheet Metal Bends. In sheet metal dimensioning, allowance must be made for bends, as explained in §712. The intersection of the plane surfaces adjacent to a bend is called the *mold line*, and this line, rather than the center of the arc, is used to terminate dimensions, Fig. 623.

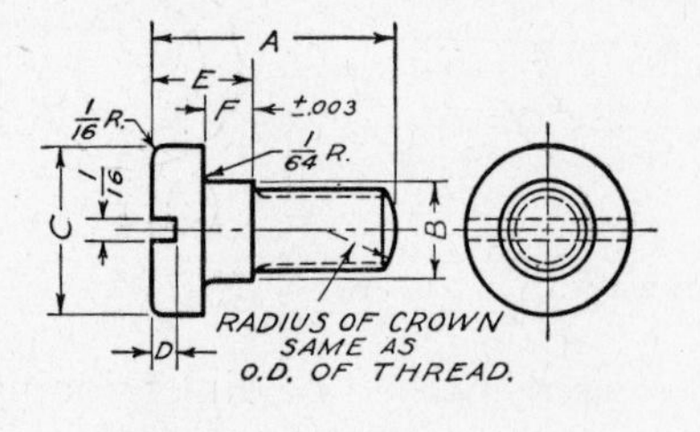

DETAIL	A	B	C	D	E	F	U.S.S.THD.	STOCK	LBS.
1	$\frac{5}{8}$	$\frac{3}{8}$	$\frac{5}{8}$	$\frac{1}{16}$	$\frac{1}{4}$	.135	$\frac{5}{16}$X18	$\frac{3}{4}$ DIA.	.09
2	$\frac{7}{8}$	$\frac{3}{8}$	$\frac{5}{8}$	$\frac{3}{32}$	$\frac{3}{8}$	.197	$\frac{5}{16}$X18	$\frac{3}{4}$ DIA.	.12
3	1	$\frac{7}{16}$	$\frac{3}{4}$	$\frac{1}{8}$	$\frac{3}{8}$	.197	$\frac{3}{8}$X16	$\frac{7}{8}$ DIA.	.19
4	$1\frac{1}{4}$	$\frac{1}{2}$	$\frac{7}{8}$	$\frac{1}{8}$	$\frac{1}{2}$	.260	$\frac{7}{16}$X14	1 DIA	.30
5	$1\frac{1}{2}$	$\frac{9}{16}$	1	$\frac{5}{32}$	$\frac{5}{8}$	.323	$\frac{1}{2}$X13	$1\frac{1}{8}$ DIA	.46

Fig. 624 Tabular Dimensioning.

365. Tabular Dimensions. A series of objects having like features but varying in dimensions may be represented by one drawing, Fig. 624. Letters are substituted for dimension figures on the drawing, and the varying dimensions are given in tabular form. The dimensions of many standard parts are given in this manner in the various catalogs and handbooks.

366. Standards. Dimensions should be given, wherever possible, to make use of readily available materials, tools, parts, and gages. The dimensions for many commonly used machine elements, such as bolts, screws, nails, keys, tapers, wire, pipes, sheet metal, chains, belts, ropes, pins, rolled metal shapes, etc., have been standardized, and the draftsman must obtain these sizes from company standards manuals, from published handbooks, from American Standards, or from manufacturers' catalogs. Tables of some of the more common items are given in the Appendix of this text.

Such standard parts are not delineated on detail drawings unless they are to be altered for use, but are drawn conventionally on assembly drawings and are listed in parts lists, §435. Common fractions are generally used to indicate the nominal sizes of standard parts or tools. If the decimal system is used exclusively, all such sizes ordinarily are expressed in decimals; for example: .250 DRILL instead of $\frac{1}{4}$ DRILL.

367. Surface Roughness, Waviness, and Lay. The modern demands of the automobile, the airplane, and other machines for parts that can stand heavier loads and higher speeds with less friction and wear have increased the need for accurate control of surface quality by the designer. Simple finish marks are no longer enough to specify surface finish on such parts.

The ASA* recommends a system of symbols for use on drawings that is now broadly used by American industry. These symbols define *roughness*, *waviness*, and *lay*. *Surface roughness* refers to the small peaks and valleys that will be found on any surface, however smooth in appearance. *Waviness* refers to the larger undulations of a surface upon which the roughness features are superimposed. *Lay* refers to the direction of the tool marks, scratches, or grains of a surface.

Surface finish is intimately related to the functioning of a surface, and proper specification of finish of such surfaces as bearings, seals, etc., is very important. If the finish of a surface is not important, it may not be necessary to specify the surface quality. It is to be used where needed only, since the cost of producing a surface becomes greater as the quality of surface called for is finer. Generally speaking, the ideal finish is the roughest one that will do the job satisfactorily.

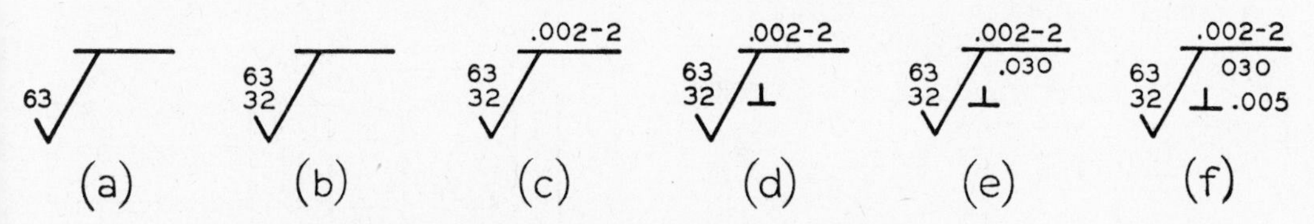

Fig. 625 Surface Quality Symbols (ASA B46.1-1955).

When surface quality is to be specified, the ASA V finish mark, §342, is used as a base, and the right side is extended upward as on a check mark, with a horizontal bar across the top, as shown in Fig. 625 (a). When it is necessary to specify only roughness height, and the width between ridges or direction of tool marks is unim-

*ASA B46.1-1955.

portant, the simple symbol at (a) is used, but the horizontal cross-bar may be omitted if desired. The roughness is indicated by the numeral in microinches (one microinch = one-millionth inch) representing the *arithmetic-average deviation* of the surface from the mean line in a profile. The arithmetic average is an expression of the average amount of deviation of peaks and valleys from a mean line. The higher the number of microinches, the rougher the surface. Surface roughness in terms of arithmetic-average can be easily measured with a surfagage, or other tracer-type electrical instrument. This method is superior to the method of comparison, visually and by touch, with sample surfaces having accurately measured surface irregularities, although the latter method has a definite usefulness. When it is desired to specify maximum and minimum average roughness height, the upper and lower numbers are placed as at (b).

When "waviness height" is to be specified, the numerical value, in inches, is placed above the horizontal bar as at (c). The maximum "waviness width," in inches, is given to the right of the waviness height number. In this example the waviness width is 2″.

If it is further desired to specify the lay, it will be indicated by the addition of a symbol at the right of the V as at (d). The lay symbols are:

= Parallel to the boundary line of the surface.
⊥ Perpendicular to the boundary line of the surface.
× Angular in both directions to the boundary line of the surface.
M Multidirectional.
C Approx. circular relative to the center of the surface.
R Approx. radial relative to the center of the surface.

When it is necessary to indicate the "roughness-width cut-off ratings," it is given in inches underneath the horizontal cross-bar as shown at (e). This number expresses the "maximum width in inches of surface irregularities to be included in the measurement of roughness height."* Standard values to be used are (in inches) 0.003, 0.010, 0.030, 0.100, 0.300, and 1.000.

"Roughness width" is the maximum allowed spacing, in inches, between repetitive units of the surface pattern. The numeral is placed under the right portion of the cross-bar, as shown at (f).

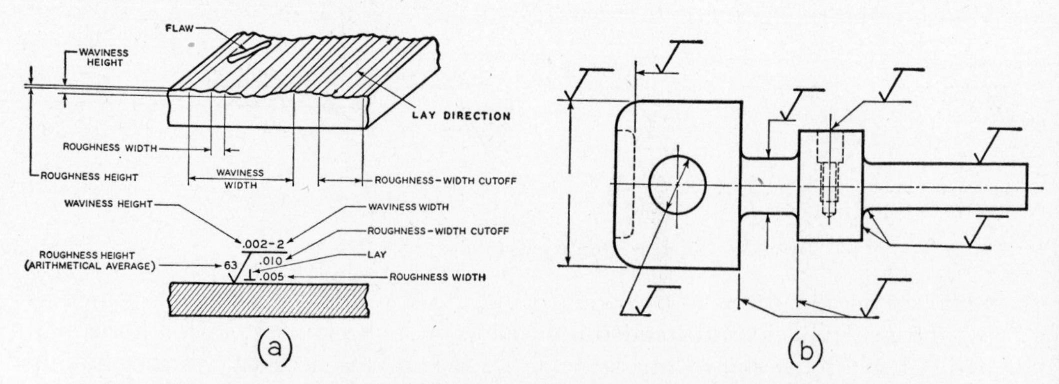

Fig. 626 Surface Characteristics and Applications of Symbols (ASA B46.1-1955).

*ASA B46.1-1955.

The relation of symbols to surface characteristics is illustrated in Fig. 626 (a). Only those numbers required to specify the surface adequately for the function should be included in the symbol.

The symbol is always made in the standard upright position—never at an angle or upside down, etc. Applications are shown at (b).

The typical range of surface roughness values that may be obtained from the various production methods is shown in Fig. 627. Preferred roughness-height values are shown at the top of the chart.

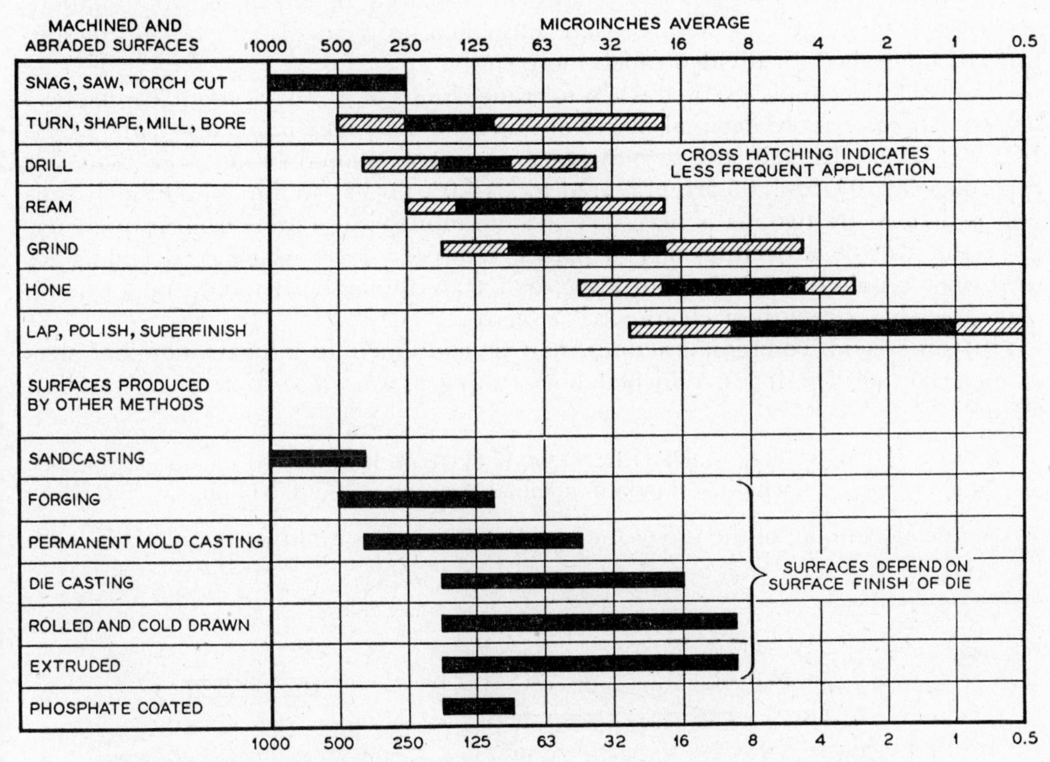

Fig. 627 Surface Roughness and Common Production Methods (ASA B46.1-1955).

368. Complete Decimal System. As shown in §335, the ever-increasing requirement for accuracy has brought greater use of decimals, while common fractions were continued for dimensions that did not require great accuracy. The use of both common fractions and decimals on the same drawings has caused a great deal of confusion, and there is today a definite trend toward the use of a complete decimal system for all dimensions. The metric system has many advantages, and undoubtedly would have been adopted if it had not been necessary to scrap all measuring devices in use and undergo the upheaval of making the change from the inch system.

It was found that a decimal system, based upon the inch as a unit of measure, could assure most of the advantages of the metric system, and could be adopted by American industries without too much difficulty.

In 1932, the Ford Motor Company adopted a complete decimal system. The shop scale adopted, Fig. 628, is divided on one edge into inches and tenths, and on the other edge into inches, tenths, and fiftieths. Thus, the smallest division is one-fiftieth, or .02″; two divisions are .04″, etc., so that when necessary to halve any measurement, the result will still be a two-place decimal. This scale has now been widely adopted, especially in the automotive and aircraft industries.

Fig. 628 Ford Special Rule.

The ASA recommends the complete decimal system* for optional use to replace common fractions. In this system, a two-place decimal is used when a common fraction is regarded as sufficiently accurate. When common fractions are replaced by decimals, computations are simplified, as decimals can be added, subtracted, multiplied, or divided much more easily.

Two-place decimals are used when tolerance limits of ± .01, or more, can be permitted. Three or more decimal places are used for tolerance limits less than ± .01. In a two-place decimal, the second place preferably should be an even digit (for example: .02, .04, and .06 are preferred to .01, .03, or .05) so that when the dimension is divided by two, as is necessary in determining the radius from a diameter, the result will be a decimal of two places. However, odd two-place decimals are used when required for design purposes, such as in dimensioning points on a smooth curve, or when strength or clearance is a factor.

In this system, common fractions may be continued to indicate nominal sizes of materials, drilled holes, punched holes, threads, keyways, and other standard features.

Examples: $\frac{1}{4}$-20UNC-2A; $\frac{5}{16}$ DRILL; STOCK $1\frac{1}{4} \times 1\frac{1}{2}$. If desired, decimals may be used for everything, including standard nominal sizes, as .250-20UNC-2B, or .750 HEX.

A typical example of the use of the complete decimal system is shown in Fig. 629.

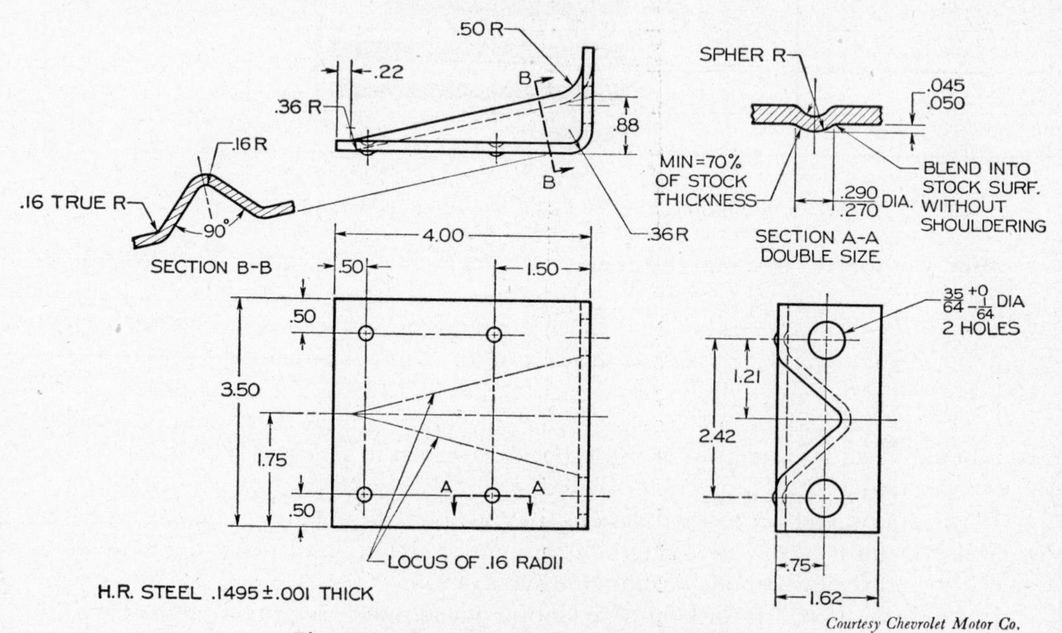

Courtesy Chevrolet Motor Co.

Fig. 629 Complete Decimal Dimensioning.

*ASA Y-14.5-1957.

When a decimal value is to be rounded off to a lesser number of places than that available, the method is as follows:

"When the figure beyond the last figure to be retained is less than 5, the last figure retained should not be changed.

Example: 3.46325, if cut off to three places, should be 3.463.

"When the figures beyond the last place to be retained amount to more than 5, the last figure retained should be increased by 1.

Example: 8.37652 if cut off to three places, should be 8.377.

"When the figure beyond the last place to be retained is exactly 5 with only zeros following, the preceding number, if even, should be unchanged; if odd, it should be increased by 1.

Example: 4.365 becomes 4.36 when cut off to two places. Also 4.355 becomes 4.36 when cut off to two places."*

Shop scales and drafting scales for use in the complete decimal system are standardized by the ASA.† The drafting scale is known as the *decimal scale*, and is discussed in §39.

The use of the complete decimal system means not only an expensive change-over of measuring equipment, but also a change-over in thinking on the part of draftsmen and designers. They must discontinue thinking in terms of units and common fractions, and think in terms of tenths, fiftieths, and hundredths of an inch. However, once the new system is installed, it is obvious that the advantages in computation, in checking, and in simplified dimensioning technique will be considerable. There is no question but that industry in general is moving toward the adoption of a complete decimal system.

369. Do's and Don'ts of Dimensioning. The following check list summarizes briefly most of the situations in which a beginning draftsman is likely to make a mistake in dimensioning. The student should check his drawing by this list before submitting it to his instructor.

1. Each dimension should be given clearly, so that it can be interpreted in only one way.
2. Dimensions should not be duplicated or the same information be given in two different ways, and no dimensions should be given except those needed to produce or inspect the part.
3. Dimensions should be given between points or surfaces which have a functional relation to each other or which control the location of mating parts.
4. Dimensions should be given to finished surfaces or important center lines in preference to rough surfaces wherever possible.
5. Dimensions should be so given that it will not be neccssary for the machinist to calculate, scale, or assume any dimension.
6. Dimensions should be attached to the view where the shape is best shown (contour rule).
7. Dimensions should be placed in the views where the features dimensioned are shown true size.
8. Avoid dimensioning to hidden lines wherever possible.
9. Dimensions should not be placed upon a view unless clearness is promoted thereby.

*ASA Z25.1-1947.
†ASA Z75.1-1955.

10. Dimensions applying to two adjacent views should be placed between views, unless clearness is promoted by placing them outside.
11. The longer dimensions should be placed outside all intermediate dimensions, so that dimension lines will not cross extension lines.
12. In machine drawing, omit all inch marks, except where necessary for clearness; e.g.: 1″ Valve.
13. Do not expect the workman to assume a feature is centered (as a hole on a plate), but give a location dimension from one side. However, if a hole is to be centered on a symmetrical rough casting, mark the center line ℄, and omit the locating dimension from the center line.
14. A dimension should be attached to only one view (extension lines not connecting two views).
15. Detail dimensions should "line up" in chain fashion.
16. Avoid a complete chain of detail dimensions; better omit one, otherwise one detail dimension or the over-all dimension should be marked REF.
17. A dimension line should never be drawn through a dimension figure. A figure should never be lettered over any line of the drawing.
18. Dimension lines should be spaced uniformly throughout the drawing. They should be at least $\frac{3}{8}''$ from the object outline and $\frac{1}{4}''$ apart.
19. No line of the drawing should be used as a dimension line or coincide with a dimension line.
20. A dimension line should never be joined end-to-end (chain fashion) with any line of the drawing.
21. Dimension lines should not cross, if avoidable.
22. Dimension lines and extension lines should not cross, if avoidable. Extension lines may cross each other.
23. When extension lines cross extension lines or visible lines, no break in either line should be made.
24. A center line may be extended and used as an extension line, in which case it is still made like a center line.
25. Center lines should generally not extend from view to view.
26. Leaders for notes should be straight, not curved, and pointing to the circular views of holes wherever possible.
27. Leaders should slope at 45°, 30°, or 60° with horizontal but may be made at any odd angle except vertical or horizontal.
28. Leaders should extend from the beginning or end of a note, the horizontal "shoulder" extending from the mid-height of the lettering.
29. Dimension figures should be approximately centered between the arrowheads, except that in a "stack" of dimensions, the figures should be "staggered."
30. Dimension figures should be about $\frac{1}{8}''$ high for whole numbers and $\frac{1}{4}''$ high for fractions.
31. Dimension figures should never be crowded or in any way made difficult to read.
32. Dimension figures should not be lettered over sectioned areas unless necessary, in which case a clear space should be left for the dimension figures.
33. Dimension figures for angles should generally be lettered horizontally.
34. Fraction bars should always be parallel to the dimension lines, never inclined.
35. The numerator and denominator of a fraction should never touch the fraction bar.
36. Notes should always be lettered horizontally on the sheet.
37. Notes should be brief and clear, and the wording should be standard in form, Fig. 616.
38. Finish marks should be placed on the edge views of all finished surfaces, including hidden edges and the contour and circular views of cylindrical surfaces.
39. Finish marks should be omitted on holes or other features where a note specifies a machining operation.

40. Finish marks should be omitted on parts made from rolled stock.
41. If a part is finished all over, omit all finish marks, and use the general note: FINISH ALL OVER, or F.A.O.
42. A cylinder is dimensioned by giving both its diameter and length in the rectangular view, except when notes are used for holes. A diagonal diameter in the circular view may be used in cases where clearness is gained thereby.
43. Holes to be bored, drilled, reamed, etc., are size-dimensioned by notes in which the leaders preferably point toward the circular views of the holes. Indications of shop processes may be omitted from notes.
44. Drill sizes are preferably expressed in decimals. Particularly for drills designated by number or letter, the decimal size must also be given.
45. In general, a circle is dimensioned by its diameter, an arc by its radius.
46. Avoid diagonal diameters, except for very large holes and for circles of centers. They may be used on positive cylinders when clearness is gained thereby.
47. A diameter dimension figure should be followed by DIA except when it is obviously a diameter.
48. A radius dimension figure should always be followed by the letter R. The radial dimension line should have only one arrowhead, and it should touch the arc.
49. Cylinders should be located by their center lines.
50. Cylinders should be located in the circular views, if possible.
51. Cylinders should be located by coordinate dimensions in preference to angular dimensions where accuracy is important.
52. When there are several rough non-critical features obviously the same size (fillets, rounds, ribs, etc.), it is necessary to give only typical dimensions, or to use a note.
53. When a dimension is not to scale, it should be underscored with a wavy line or marked, NTS or NOT TO SCALE
54. Mating dimensions should be given correspondingly on drawings of mating parts.
55. Pattern dimensions should be given in common whole numbers and fractions to the nearest $\frac{1}{16}''$.
56. Decimal dimensions should be used when greater accuracy than $\frac{1}{64}''$ is required on a machine dimension.
57. Avoid cumulative tolerances, especially in limit dimensioning, described in §381.

370. Dimensioning Problems. It is expected that most of the students' practice in dimensioning will be in connection with working drawings assigned from other chapters. However, a limited number of special dimensioning problems are available here in Figs. 630 and 631. The problems are designed for Layout A-3 ($8\frac{1}{2}'' \times 11''$ paper), and are to be drawn with instruments and dimensioned to a full-size scale.

Dimensioning problems in convenient form for solution may be found in *Technical Drawing Problems*, by Giesecke, Mitchell, and Spencer, and in *Technical Drawing Problems, Series 2*, by Spencer and Grant, both designed to accompany this text and published by The Macmillan Company.

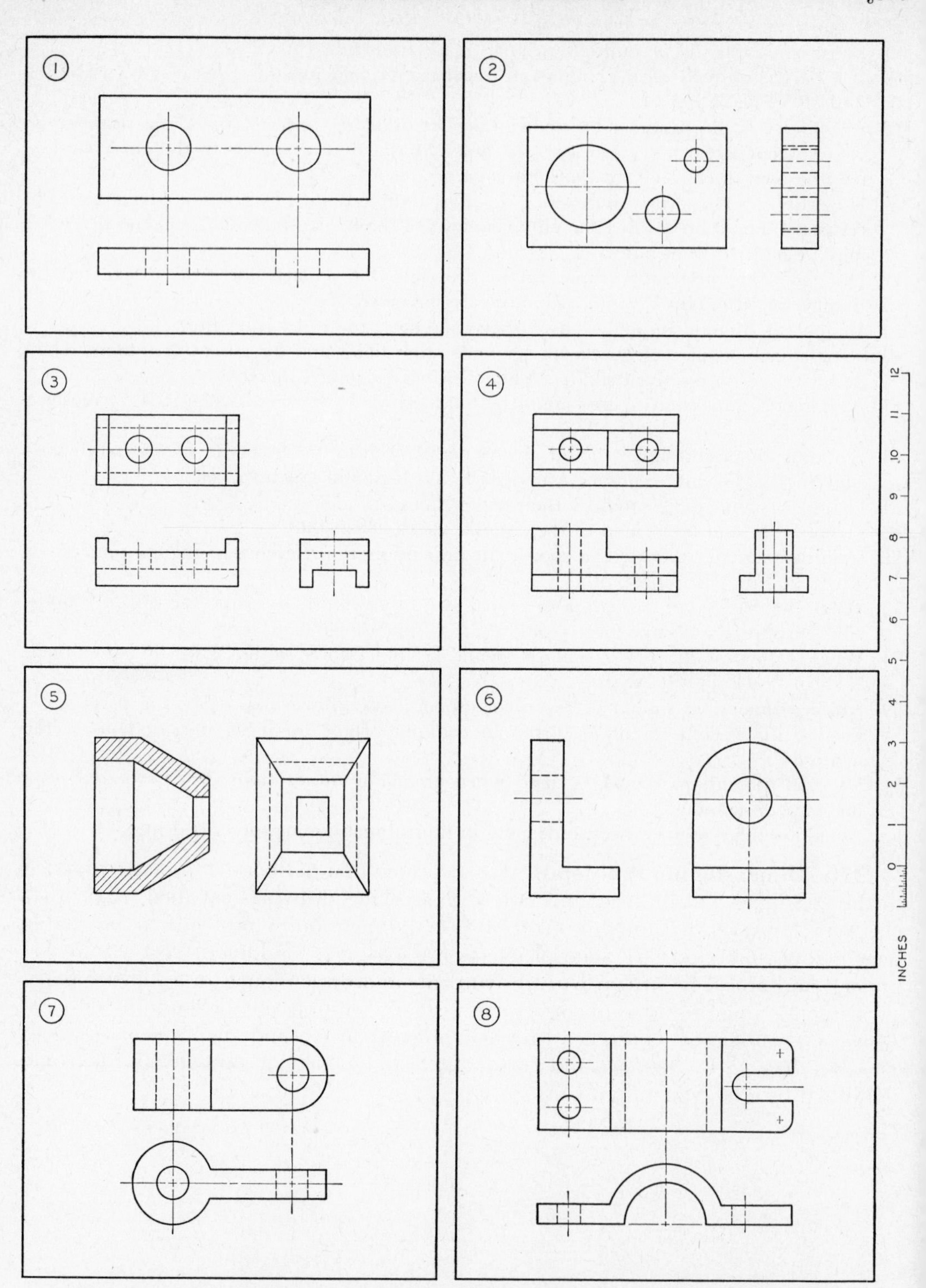

Fig. 630 Using Layout A-3, draw assigned problem with instruments. To obtain sizes, place bow dividers on the views on this page and transfer to scale at the side to obtain values in inches (nearest $\frac{1}{16}''$). Dimension drawing completely, full size.

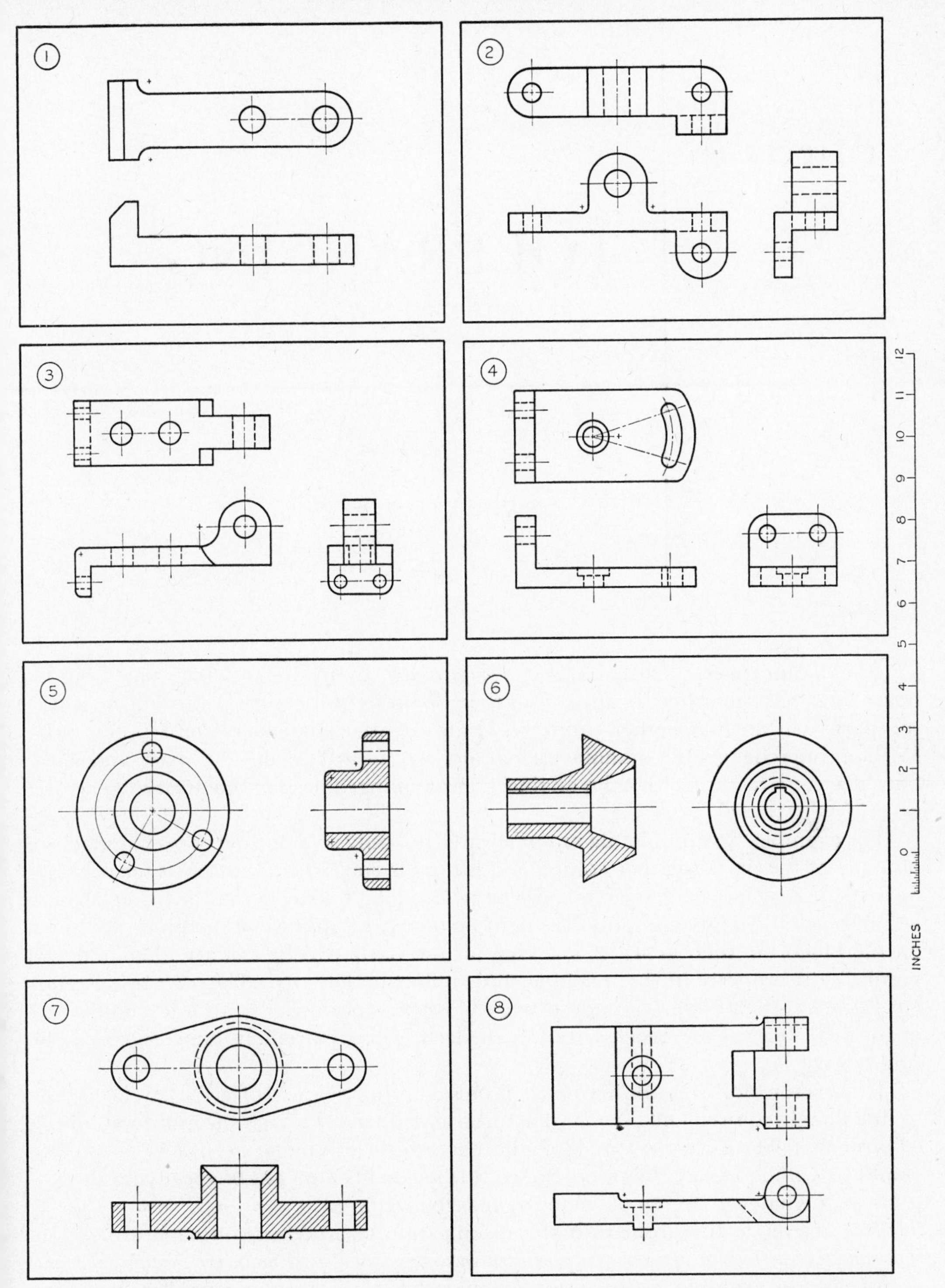

Fig. 631 Using Layout A-3, draw assigned problem with instruments. To obtain sizes, place bow dividers on the views on this page and transfer to scale at the side to obtain values in inches (nearest $\frac{1}{16}''$). Dimension drawing completely, full size.

CHAPTER 12

TOLERANCING*

371. Tolerances. *Interchangeable manufacturing*, by means of which parts can be made in widely separated localities and then be brought together for assembly, where the parts will all fit together properly, is an essential element of mass production. Without interchangeable manufacturing, modern industry could not exist, and without effective size-control by the engineer, interchangeable manufacturing could not be achieved.

For example, an automobile manufacturer not only subcontracts the manufacture of many of the parts to other companies, but he is concerned with parts for replacement. All parts in each category must be near enough alike so that any one of them will fit properly in any assembly. It might be thought that if the dimensions are given on the blueprint, such as $4\frac{5}{8}$, 2.362, etc., all the parts will be exactly alike and will naturally fit properly in the machine. But, unfortunately, *it is impossible to make anything to exact size.* It can be made to very close dimensions, even to a few millionths of an inch (e.g.: gage blocks), but such accuracy is extremely expensive or even prohibitive.

However, exact sizes are not needed, only varying degrees of accuracy according to functional requirements. A manufacturer of children's tricycles would soon go out of business if he insisted on making the parts with jet-engine accuracy, as no one would be willing to pay the price. So what is needed is a means of specifying dimensions with whatever degree of accuracy may be required.

The answer to the problem is the specification of a *tolerance* on each dimension. *Tolerance is the amount of variation permitted in the size of a part or in the location of points or surfaces.* For example, a dimension given as 1.625″ ± .002″ means that it may be (on the manufactured part) 1.627″ or 1.623″, or anywhere between these dimensions.

*See ASA Y14.5-1957.

The tolerance, or total amount of variation "tolerated," is .004″. Thus, it becomes the function of the draftsman or designer to specify the allowable error that may be tolerated for a given dimension and still permit the satisfactory functioning of the part. Since greater accuracy costs more money, he would not specify the closest tolerance, but instead would specify as generous a tolerance as possible.

The old method of indicating dimensions of two mating parts was to give the nominal dimensions of the two parts in common whole numbers and fractions and to indicate by a note the kind of fit that was desired, as shown in Fig. 632 (a), and

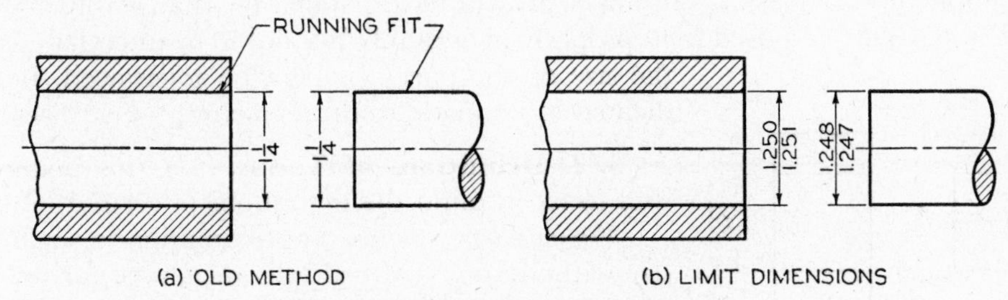

Fig. 632 Fits between Mating Parts.

then to depend upon the workman to produce the parts so that they would fit together and function properly. Other types of fit included "drive fit," "sliding fit," "tunking fit," "force fit," etc.

In the example shown at (a), the machinist would make the hole close to $1\frac{1}{4}$″ diameter and would then make the shaft, say, .003″ less in diameter. It would not matter if the hole were several thousands more or less than 1.250″; he could make the shaft about .003″ less and obtain the desired fit. But this method would not work in quantity production, since the sizes would vary considerably and would not be interchangeable; that is, any given shaft would not fit properly in any hole.

In order to control the dimensions of quantities of the two parts so that any two mating parts would be interchangeable, it became necessary to assign tolerances to the dimensions of the parts, as shown at (b). The diameter of the hole may be machined not less than 1.250″ and not more than 1.251″, these two figures representing the *limits* and the difference between them, .001″, being the *tolerance*. Likewise, the shaft must be produced between the limits of 1.248″ and 1.247″, the tolerance on the shaft being the difference between these, or .001″.

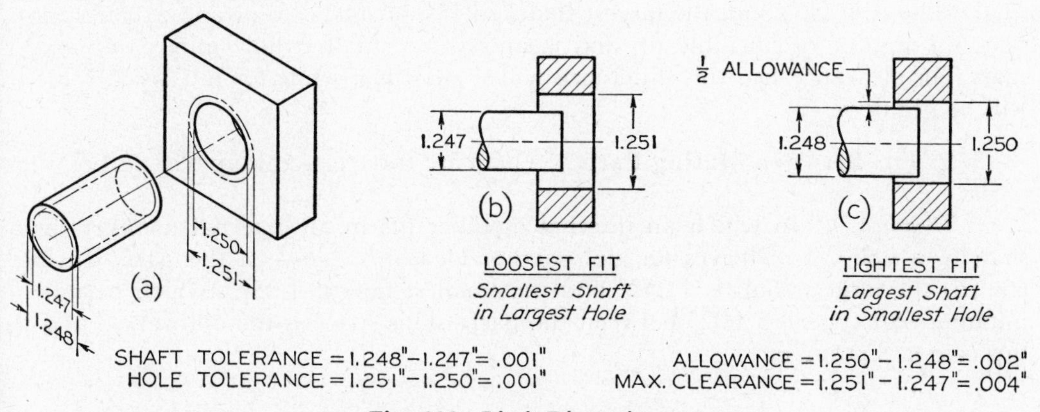

Fig. 633 Limit Dimensions.

A pictorial illustration of the dimensions in Fig. 632 (b) is shown in Fig. 633 (a). The maximum shaft is shown solid, and the minimum shaft is shown in phantom. The difference in diameters, .001″, is the tolerance on the shaft. Similarly, the tolerance on the hole is the difference between the two limits shown, or .001″. The loosest fit, or maximum clearance, occurs when the smallest shaft is in the largest hole, as shown at (b). The tightest fit, or minimum clearance, occurs when the largest shaft is in the smallest hole, as shown at (c). The difference between these, .002″, is the *allowance*. The average clearance is .003″, which is the same difference as allowed in the example of Fig. 632 (a), but any shaft will fit any hole interchangeably.

When parts are required to fit properly in assembly but not to be interchangeable, the size of one part need not be toleranced, but indicated to be made to fit at assembly, Fig. 634.

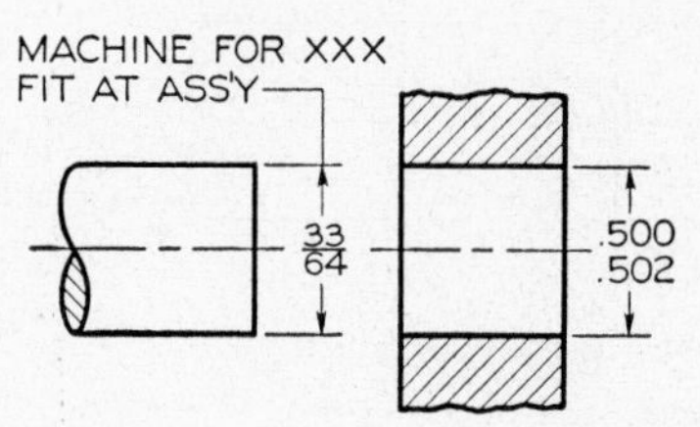

Fig. 634 Non-Interchangeable Fit.

372. Definitions of Terms. At this point it is well to fix in mind the definitions of certain terms:

Nominal Size—a close approximation to a standard size without any specified tolerance, used for the purpose of general identification. In Fig. 632 (b), the nominal size of both hole and shaft is $1\frac{1}{4}''$.

Basic Size—the "size from which limits of size are derived by the application of allowances and tolerances."* It is the exact theoretical size from which limits are figured. In Fig. 632 (b) the basic size is the decimal equivalent of the nominal size $1\frac{1}{4}''$, or 1.250″.

Design Size—the "size from which the limits of size are derived by the application of tolerances. When there is no allowance, the design size is the same as the basic size."† In Fig. 632 (b) the design size of the hole is 1.250″ (smallest hole), and of the shaft is 1.248″ (largest shaft).

Tolerance—"the total permissible variation of a size, or the difference between the limits. In Fig. 633 (a) the tolerance on either the shaft or hole is the difference between the limits, or .001″.

Limits—the maximum and minimum sizes indicated by a toleranced dimension. In Fig. 633 (a) the limits for the hole are 1.250″ and 1.251″, and for the shaft are 1.248″ and 1.247″.

Allowance—the minimum clearance space (or maximum interference) intended between mating parts. In Fig. 633 (c) the allowance is the difference between the smallest hole, 1.250″, and the largest shaft, 1.248″, or .002″. Allowance, then, represents the tightest permissible fit, and is simply the smallest hole minus the largest shaft. For clearance fits, this difference will be positive, while for interference fits it will be negative.

373. Fits Between Mating Parts. There are three general types of fits between parts:

1. *Clearance fit*, in which an internal member fits in an external member (as a shaft in a hole), and leaves an air space or clearance between the parts. In Fig. 632 (b) the largest shaft is 1.248″ and the smallest hole is 1.250″, which permits a minimum air space of .002″ between the parts. This space is the allowance, and in this case it is positive.

*ASA Y14.5-1957.
†*Ibid.*

2. *Interference fit*, in which the internal member is larger than the external member such that there is an actual interference of metal. In Fig. 635 (a) the smallest shaft is 1.2513″, and the largest hole is 1.2506″, so that there is an actual interference of metal amounting to at least .0007″.

3. *Transition fit*, in which the fit might be either a clearance fit or an interference fit. In Fig. 635 (b) the smallest shaft, 1.2503″, will fit in the largest hole, 1.2506″, with .0003″ to spare. But the largest shaft, 1.2509″, will have to be forced into the smallest hole, 1.2500″, with an interference of metal (negative allowance) of .0009″.

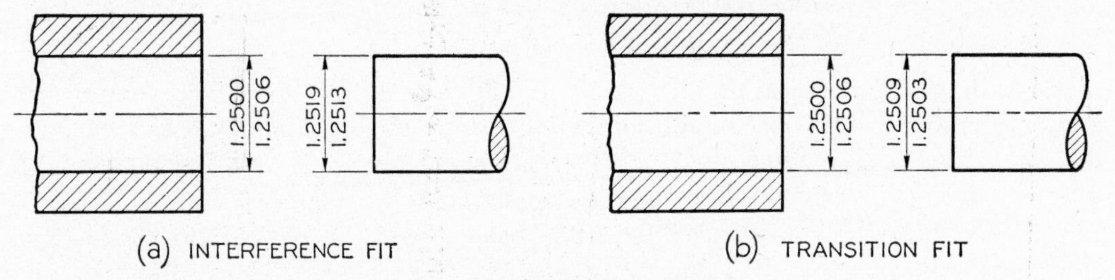

Fig. 635 Fits Between Parts.

374. Selective Assembly. If allowances and tolerances are properly given, mating parts can be completely interchangeable. But for close fits, it is necessary to specify very small allowances and tolerances, and the cost may be very high. In order to avoid this expense, *selective assembly* is often used. In selective assembly, all parts are inspected and classified into several grades according to actual sizes, so that "small" shafts can be matched with "small" holes, "medium" shafts with "medium" holes, etc. In this way, very close fits often may be obtained at much less expense than by machining all mating parts to very accurate dimensions. Since a transition fit may or may not represent an interference of metal, interchangeable assembly generally is not as satisfactory as selective assembly.

375. Basic Hole System. Standard reamers, broaches, and other standard tools are often used to produce holes, and standard plug gages are used to check the actual sizes. On the other hand, shafting can easily be machined to any size desired. Therefore, toleranced dimensions are most commonly figured on the so-called *basic hole system*. In this system, the *minimum hole is taken as the basic size*, an allowance is assigned, and tolerances are applied on both sides of, and away from, this allowance.

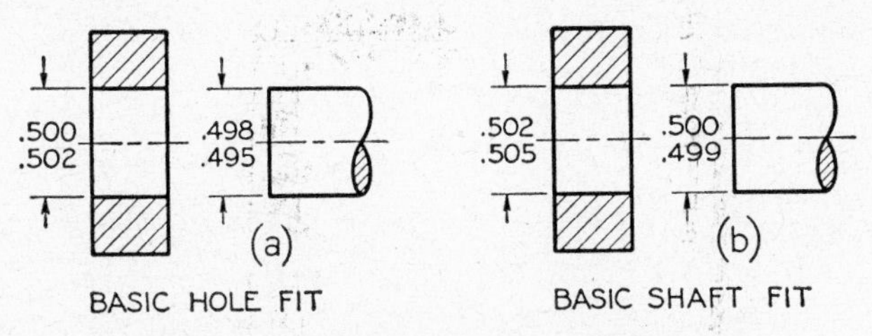

Fig. 636 Basic Hole and Basic Shaft Systems.

In Fig. 636 (a) the minimum size of the hole, .500″, is taken as the basic size. An allowance of .002″ is decided upon and subtracted from the basic hole size, giving the maximum shaft .498″. Tolerances of .002″ and .003″, respectively, are applied to the hole and shaft to obtain the maximum hole of .502″ and the minimum shaft of

.495″. Thus the minimum clearance between the parts is .500″ − .498″ = .002″ (smallest hole minus largest shaft), and the maximum clearance is .502″ − .495″ = .007″ (largest hole minus smallest shaft).

In the case of an interference fit, the maximum shaft size would be found by *adding the desired allowance* (maximum interference) to the basic hole size. In Fig. 635 (a), the basic size is 1.2500″. The maximum interference decided upon was .0019″, which added to the basic size gives 1.2519″, the largest shaft size.

The basic hole size can be changed to the basic shaft size by subtracting the allowance for a clearance fit, or adding it for an interference fit. The result is the largest shaft size, which is the new basic size.

376. Basic Shaft System. In some branches of industry, such as textile machinery manufacturing, in which use is made of a great deal of cold-finished shafting, the *basic shaft system* is often used. This system should be used only when there is a reason for it. For example, it is advantageous when several parts having different fits, but one nominal size, are required on a single shaft. In this system *the maximum shaft is taken as the basic size*, an allowance for each mating part is assigned, and tolerances are applied on both sides of, and away from, this allowance.

In Fig. 636 (b) the maximum size of the shaft, .500″, is taken as the basic size. An allowance of .002″ is decided upon and added to the basic shaft size, giving the minimum hole .502″. Tolerances of .003″ and .001″, respectively, are applied to the hole and shaft to obtain the maximum hole .505″ and the minimum shaft .499″. Thus the minimum clearance between the parts is .502″ − .500″ = .002″ (smallest hole minus largest shaft), and the maximum clearance is .505″ − .499″ = .006″ (largest hole minus smallest shaft).

In the case of an interference fit, the minimum hole size would be found by *subtracting the desired allowance from the basic shaft size.*

The basic shaft size may be changed to the basic hole size by adding the allowance for a clearance fit or subtracting it for an interference fit. The result is the smallest hole size, which is the new basic size.

377. Unilateral Tolerances. "A unilateral tolerance allows variations in only one direction from a design size."* This method is advantageous when a critical size is approached as material is removed during manufacture, as in the case of close-fitting holes and shafts.

In Fig. 637 (a) the design size is 1.878″. The tolerance .002″ is all in one direction—toward a smaller size. If this is a shaft diameter, the design size 1.878″ is the

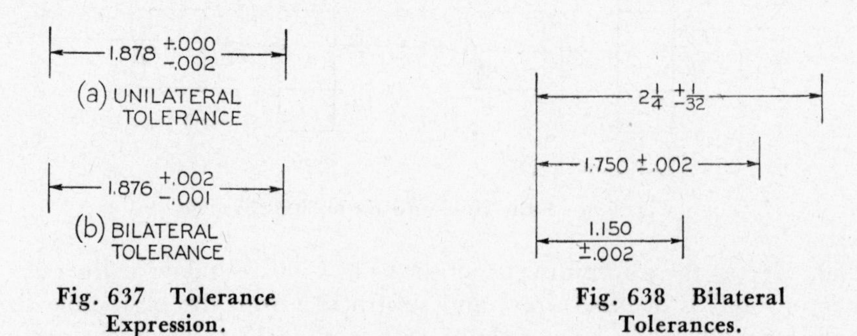

Fig. 637 Tolerance Expression.

Fig. 638 Bilateral Tolerances.

*ASA Y14.5-1957.

size nearest the critical size because it is nearest to the tolerance zone; hence, the tolerance is taken *away* from the critical size. A unilateral tolerance is always all plus or all minus; that is, either the plus or the minus value must be zero. However, the zeros should not be omitted.

378. Bilateral Tolerances. "A bilateral system of tolerances allows variations in both directions from a design size."† Bilateral tolerances are usually given with location dimensions or with any dimensions that can be allowed to vary in either direction.

In Fig. 637 (b) the design size is 1.876″, and the actual size may be larger by .002″ or smaller by .001″. If it is desired to specify an equal variation in both directions, the combined plus or minus symbol (±) is used with a single value, as shown in Fig. 638.

379. Specification of Tolerances. For fractional dimensions, the worker is not expected to work closer than he can be expected to measure with a steel rule. It is customary to indicate an over-all tolerance for all common-fraction dimensions by means of a printed note in or just above the title block, such as ALL FRACTIONAL DIMENSIONS ± .010 UNLESS OTHERWISE SPECIFIED; HOLD FRACTIONAL DIMENSIONS TO ± $\frac{1}{64}$ UNLESS OTHERWISE NOTED. See Figs. 721 and 725. General angular tolerances also may be given, as: ANGULAR TOLERANCE 1°.

General tolerances on decimal dimensions in which tolerances are not given may also be covered in a general printed note, such as DECIMAL DIMENSIONS TO BE HELD TO ± .001. Thus if a dimension 3.250 is given, the worker machines between the limits 3.249 and 3.251. See Fig. 716. Every dimension on a drawing should have a tolerance, either direct or by general tolerance note, except that commerical material is often assumed to have the tolerances set by commercial standards.

A tolerance of a decimal dimension should be given in decimal form to the same number of places. A tolerance of a dimension given in common-fraction form should be given as a common fraction. See Fig. 638.

Several methods of expressing tolerances in dimensions are approved by ASA†, as follows:

1. *Plus and Minus.* Two unequal tolerance numbers are given, one plus and one minus, with the plus above the minus, Fig. 637. One of the numbers may be zero, if desired.

2. *Plus or Minus.* A single tolerance value is given, preceded by the plus or minus

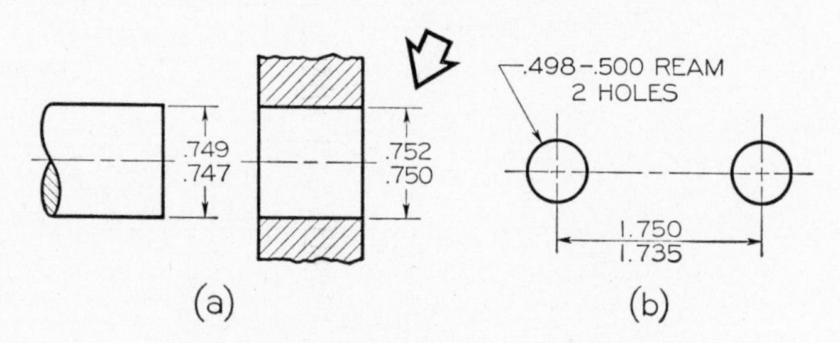

Fig. 639 Method of Giving Limits.

*ASA Y14.5-1957.
†Ibid.

symbol (±), Fig. 638. This method should be used when the plus and minus values are equal.

3. *Limits.* The maximum and minimum limits of size are specified, as shown in Fig. 639. The high limit is placed above the low limit, thus: $\frac{.749}{.747}$ DIA. In note form, the low limit is given first, thus: .498-.500 REAM.

The old method of placing limits is also approved in the new ASA standard. For a shaft the high limit is placed above the low limit, since a shaft is machined from "large to small." This agrees with the rule above. However, for a hole dimensioned directly (see arrow), the low limit is placed above, since a hole is machined from "small to large," thus: $\frac{.750}{.752}$. This is the only difference between the methods. However the method adopted should be used consistently throughout a drawing.

A typical example of limit dimensioning is given in Fig. 640.

4. MIN or MAX is often placed after a number to indicate minimum or maximum dimensions desired. For example, a thread length may be dimensioned thus: |←1.500→| MIN FULL THD or a radius dimensioned: .05 R MAX↘ Other applications include depths of holes and chamfers.

5. Angular tolerances are usually bilateral and in terms of degrees, minutes, and seconds.

Examples: 25° ± 1°, 25° 0′ ± 0° 15′, or 25° ± .25°. See also §387.

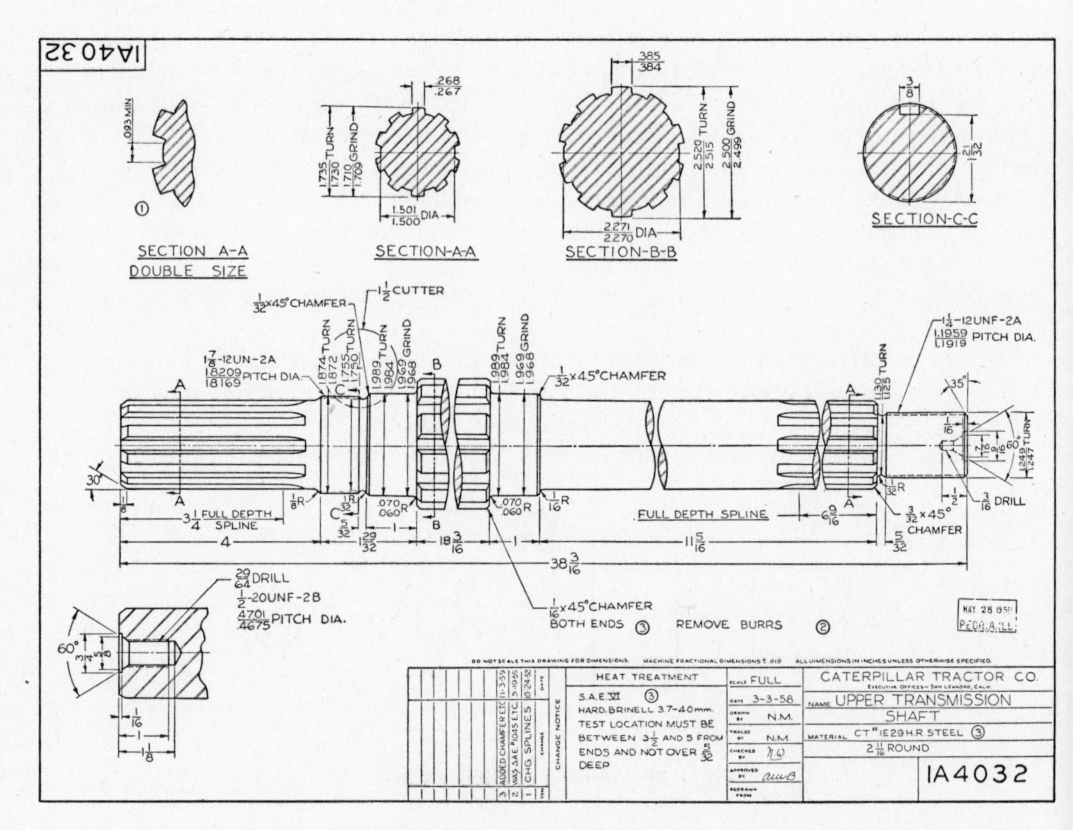

Fig. 640 Limit Dimensions.

380. American Standard Limits and Fits. For many years the American Standard ASA B4a-1925 "Tolerances, Allowances and Gages for Metal Fits" has been used in varying degrees. This standard was superseded in 1955 by ASA B4.1-1955 "Preferred Limits and Fits for Cylindrical Parts." It defines terms and recommends preferred standard sizes, allowances, tolerances, and fits.

The standard includes Appendix tables 5–11, which give "a series of standard types and classes of fits on a unilateral hole basis such that the fit produced by mating parts in any one class will produce approximately similar performance throughout the range of sizes. These tables prescribe the fit for any given size, or type of fit; they also prescribe the standard limits for the mating parts which will produce the fit."*

Letter symbols to identify the five types of fit are:

RC Running or Sliding Fit
LC Locational Clearance Fit
LT Transition Fit
LN Locational Interference Fit
FN Force or Shrink Fit

These letter symbols, plus a number indicating the class of fit within each type, are used to indicate a complete fit. Thus FN 4 means a Class 4 Force Fit. The fits are described as follows:†

RUNNING AND SLIDING FITS

Running and sliding fits, for which limits of clearance are given in Appendix tables 5 and 6, are intended to provide a similar running performance, with suitable lubrication allowance, throughout the range of sizes. The clearances for the first two classes, used chiefly as slide fits, increase more slowly with diameter than the other classes, so that accurate location is maintained even at the expense of free relative motion.

These fits may be described briefly as follows:

RC 1 *Close sliding fits* are intended for the accurate location of parts which must assemble without perceptible play.

RC 2 *Sliding fits* are intended for accurate location, but with greater maximum clearance than class RC 1. Parts made to this fit move and turn easily but are not intended to run freely, and in the larger sizes may seize with small temperature changes.

RC 3 *Precision running fits* are about the closest fits which can be expected to run freely, and are intended for precision work at slow speeds and light journal pressures, but are not suitable where appreciable temperature differences are likely to be encountered.

RC 4 *Close running fits* are intended chiefly for running fits on accurate machinery with moderate surface speeds and journal pressures, where accurate location and minimum play is desired.

RC 5, RC 6 *Medium running fits* are intended for higher running speeds, or heavy journal pressures, or both.

RC 7 *Free running fits* are intended for use where accuracy is not essential, or where large temperature variations are likely to be encountered, or under both these conditions.

RC 8, RC 9 *Loose running fits* are intended for use where materials such as cold-rolled shafting and tubing, made to commercial tolerances, are involved.

*ASA B4.1-1955.
†Ibid.

LOCATIONAL FITS

Locational fits, Appendix tables 7–10, are fits intended to determine only the location of the mating parts; they may provide rigid or accurate location, as with interference fits, or provide some freedom of location, as with clearance fits. Accordingly they are divided into three groups: clearance fits, transition fits, and interference fits.

These are more fully described as follows:

LC *Locational clearance fits* are intended for parts which are normally stationary, but which can be freely assembled or disassembled. They run from snug fits for parts requiring accuracy of location, through the medium clearance fits for parts such as spigots, to the looser fastener fits where freedom of assembly is of prime importance.

LT *Transition fits* are a compromise between clearance and interference fits, for application where accuracy of location is important, but either a small amount of clearance or interference is permissible.

LN *Locational interference fits* are used where accuracy of location is of prime importance, and for parts requiring rigidity and alignment with no special requirements for bore pressure. Such fits are not intended for parts designed to transmit frictional loads from one part to another by virtue of the tightness of fit, as these conditions are covered by force fits.

FORCE FITS

Force or shrink fits, Appendix table 11, constitute a special type of interference fit, normally characterized by maintenance of constant bore pressures throughout the range of sizes. The interference therefore varies almost directly with diameter, and the difference between its minimum and maximum value is small, to maintain the resulting pressures within reasonable limits.

These fits may be described briefly as follows:

FN 1 *Light drive fits* are those requiring light assembly pressures, and produce more or less permanent assemblies. They are suitable for thin sections or long fits, or in cast-iron external members.

FN 2 *Medium drive fits* are suitable for ordinary steel parts, or for shrink fits on light sections. They are about the tightest fits that can be used with high-grade cast-iron external members.

FN 3 *Heavy drive fits* are suitable for heavier steel parts or for shrink fits in medium sections.

FN 4, FN 5 *Force fits* are suitable for parts which can be highly stressed, or for shrink fits where the heavy pressing forces required are impractical.*

In the tables for each class of fit, the range of nominal sizes of shafts or holes is given in inches. To simplify the tables and reduce the space required to present them, the other values are given in thousandths of an inch. Minimum and maximum limits of clearance are given, the top number being the least clearance, or the allowance, and the lower number the maximum clearance, or greatest looseness of fit. Then, under the heading "Standard Limits" are given the limits for the hole and for the shaft that are to be applied algebraically to the basic size to obtain the limits of size for the parts.

For example, take a 2.0000″ basic diameter with a Class RC 1 fit. This fit is given in Appendix table 5. In the column headed "Nominal Size Range, Inches," find 1.97-3.15 which embraces the 2.0000″ basic size. Reading to the right we find under "Limits of Clearance" the values 0.4 and 1.2, representing the maximum and minimum clearance between the parts *in thousandths of an inch*. To get these values in

*ASA B4.1-1955.

inches, simply multiply by one thousandth, thus: $\frac{4}{10} \times \frac{1}{1000} = .0004$. Or, to convert 0.4 thousandths to inches, simply move the decimal point three places to the left, thus: .0004″. Therefore, for this 2.0000″ diameter, with a Class RC 1 fit, the minimum clearance, or allowance, is .0004″, and the maximum clearance, representing the greatest looseness, is .0012″.

Reading farther to the right we find under "Standard Limits" the value + 0.5, which converted to inches is .0005″. Add this to the basic size thus: 2.0000″ + .0005″ = 2.0005″, the upper limit of the hole. Since the other value given for the hole is zero, the lower limit of the hole is the basic size of the hole, or 2.0000″. The hole would then be dimensioned as follows:

$$\begin{matrix} 2.0000 \\ 2.0005 \end{matrix} \quad \text{or} \quad 2.0000 \begin{matrix} +.0005 \\ -.0000 \end{matrix}$$

The limits for the shaft are read as − .0004″ and − .0007″. To get the limits of the shaft, subtract these values from the basic size, thus:

2.0000″ − .0004″ = 1.9996″, the upper limit of the shaft.
2.0000″ − .0007″ = 1.9993″, the lower limit of the shaft.

The shaft would then be dimensioned as follows:

$$\begin{matrix} 1.9996 \\ 1.9993 \end{matrix} \quad \text{or} \quad 1.9996 \begin{matrix} +.0000 \\ -.0003 \end{matrix}$$

381. Accumulation of Tolerances. In limit dimensioning, it is very important to consider the effect of one tolerance on another. When the location of a surface in a given direction is affected by more than one tolerance figure, the tolerances are *cumulative*. For example, in Fig. 641 (a), if dimension Z is omitted, surface A would

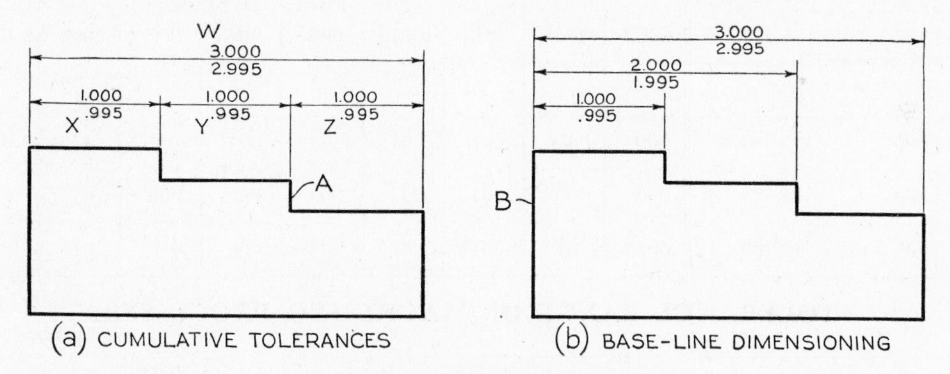

Fig. 641 Cumulative Tolerances.

be controlled by both dimensions X and Y, and there could be a total variation of .010 instead of the variation of .005 permitted by dimension Y, which is the dimension directly applied to surface A. Further, if the part is made to all the minimum tolerances of X, Y, and Z, the total variation in the length of the part will be .015, and the part can be as short as 2.985. However, the tolerance on the over-all dimension W is only .005, permitting the part to be only as short as 2.995. The part is superfluously dimensioned.

In some cases, for functional reasons it may be desired to hold all three small dimensions X, Y, and Z closely without regard to the over-all length. In such a case the over-all dimension should be marked REF. In other cases it may be desired to hold two small dimensions X and Y and the over-all closely without regard to dimension Z. In that case, dimension Z should be omitted, or marked REF.

As a rule, it is best to dimension each surface so that it is affected by only one dimension. This can be done by referring all dimensions to a single datum surface, such as B, as shown at (b). See also Fig. 609 (d) to (f).

382. Tolerances and Shop Processes. As has been repeatedly stated in this chapter, tolerances should be as coarse as possible and still permit satisfactory use of the part. If this is done, great savings can be effected as a result of the use of less expensive tools, lower labor and inspection costs, and reduced scrapping of material.

In Fig. 642 is shown a chart of tolerances and shop processes that may be used as a guide by the draftsman in selecting tolerances. This chart appears in MIL-STD-8A and in the SAE Dimensioning Standard, March, 1955. See also Chapter 10 for detailed information on shop processes.

Range of Sizes From	To & Incl.	Tolerances								
.000	.599	.00015	.0002	.0003	.0005	.0008	.0012	.002	.003	.005
.600	.999	.00015	.00025	.0004	.0006	.001	.0015	.0025	.004	.006
1.000	1.499	.0002	.0003	.0005	.0008	.0012	.002	.003	.005	.008
1.500	2.799	.00025	.0004	.0006	.001	.0015	.0025	.004	.006	.010
2.800	4.499	.0003	.0005	.0008	.0012	.002	.003	.005	.008	.012
4.500	7.799	.0004	.0006	.001	.0015	.0025	.004	.006	.010	.015
7.800	13.599	.0005	.0008	.0012	.002	.003	.005	.008	.012	.020
13.600	20.999	.0006	.001	.0015	.0025	.004	.006	.010	.015	.025

TOLERANCE RANGE OF MACHINING PROCESSES

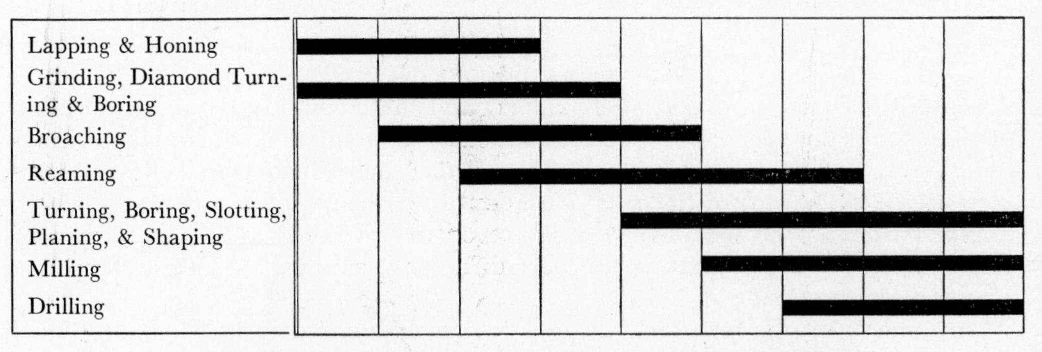

Fig. 642 Tolerances Related to Shop Processes (from MIL-STD-8A).

383. Positional Tolerances. In §350 are shown a number of examples of the traditional methods of locating holes—that is, by means of rectangular coordinates or angular dimensions. Each dimension has a tolerance, either given directly or indicated by a general note.

For example, in Fig. 643 (a) is shown a hole located from two surfaces at right angles to each other. As shown at (b), the center may lie anywhere within a square tolerance zone the sides of which are equal to the tolerances. Thus the total variations along either diagonal of the square will be greater than the indicated tolerance.

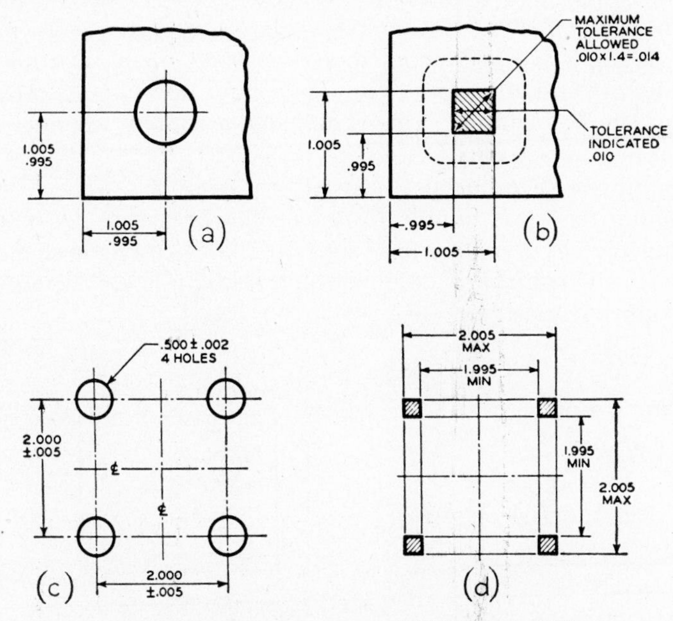

Fig. 643 Tolerance Zones (ASA Y14.5-1957).

If four holes are dimensioned with rectangular coordinates as at (c), the square tolerance zones for the holes would be as shown at (d). The locational tolerances are actually greater than indicated by the dimensions.

In Fig. 644 I, hole A is selected as a datum and the other three are located from it. The square tolerance zone for hole A results from the tolerances on the two

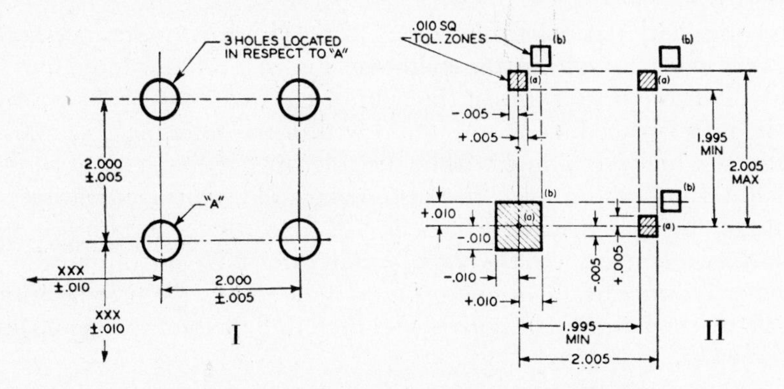

Fig. 644 Tolerance Zones (ASA Y14-5.1957).

rectangular coordinate dimensions locating hole A. The sizes of the tolerance zones for the other three holes result from the tolerances between the holes, while their locations will vary according to the actual location of the datum hole A. Two of the many possible zone patterns are shown at II. The locations marked (a) are those when datum hole A is at its theoretically correct position, and the locations marked (b) are those when datum hole A is at the extreme upper right corner of its tolerance zone.

Thus, with the dimensions shown at I it is difficult to say whether the resulting parts will actually fit the mating parts satisfactorily even though they conform to the tolerances shown on the drawing.

These disadvantages are overcome by giving exact theoretical locations by untoleranced dimensions and then specifying by a note how far actual positions may be displaced from these locations. This is called *true-position dimensioning*. It will be seen that the tolerance zone for each hole will be a circle, the size of the circle depending upon the amount of variation permitted from "true position."

The ASA committee* was unable to agree on whether to designate the radius or the diameter of the circular tolerance zone; it therefore recommended both. The two methods are illustrated in Fig. 645, in which notes indicate the amount of variation permitted from true position.

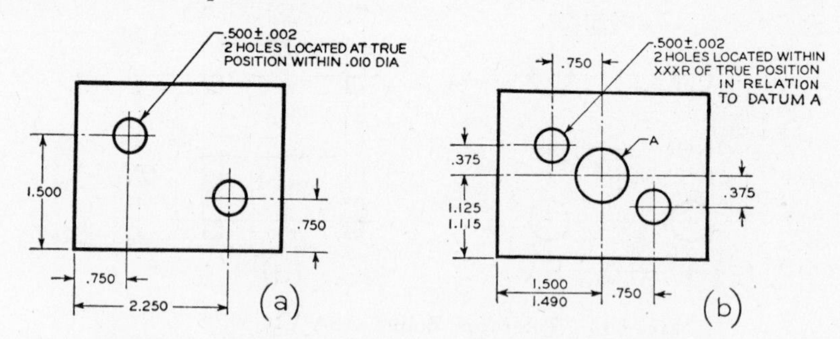

Fig. 645 True-Position Dimensioning (ASA Y14.5-1957).

To prevent misunderstanding, true position should always be established with respect to a datum. In simple arrangements, the choice of datum is obvious and it does not require identification, Fig. 645 (a). When necessary for clearness, a phrase should be added to the true-position note, such as IN RELATION TO DATUM A, as shown at (b).

Actually, the "circular tolerance zone" is a cylindrical tolerance zone, and the axis of the hole must be within the cylinder, Fig. 646. The center line of the hole may coincide with the center line of the cylindrical tolerance zone, (a), or it may be parallel to it but displaced so as to remain within the tolerance cylinder, (b), or it may be inclined while remaining within the tolerance cylinder, (c). In this last case we see that the true position tolerance also defines the limits of squareness variation.

In terms of the cylindrical surface of the hole, the true-position specification indicates that all elements on the hole surface must be on or outside a cylinder whose diameter is equal to the diameter of the hole minus the true position tolerance (diameter, or twice the radius), with the center line of the cylinder located at true position, Fig. 647.

*ASA Y14.5-1957.

The use of basic untoleranced dimensions to locate features at true position avoids one of the chief difficulties in tolerancing—the accumulation of tolerances, §381, even in a chain of dimensions, Fig. 648.

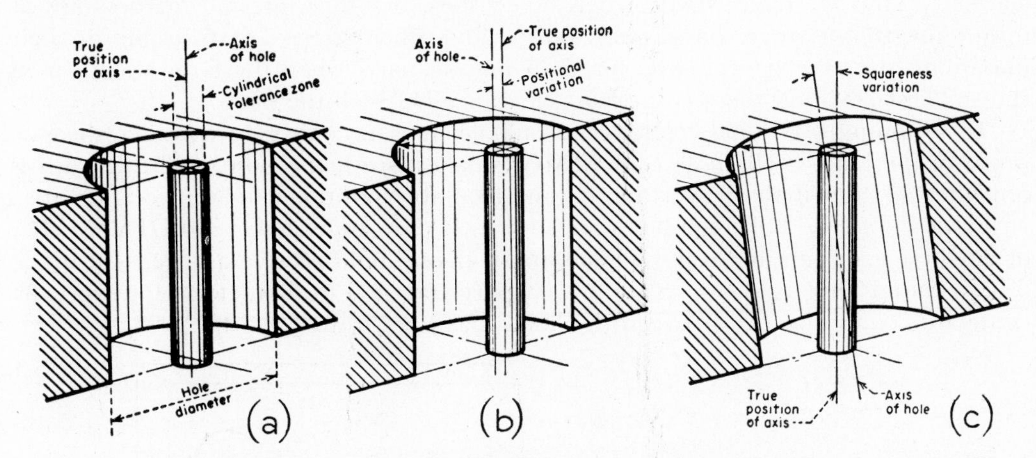

Fig. 646 Cylindrical Tolerance Zone (ASA Y14.5-1957).

While features such as holes and bosses may vary in any direction from the true-position axis, other features, such as slots, may vary on either side of a true-position plane. The note may be worded in either of two ways:

1. 6 SLOTS LOCATED AT TRUE POSITION WITHIN .010 WIDE ZONE.
2. 6 SLOTS LOCATED WITHIN .005 EITHER SIDE OF TRUE POSITION.

It has been quite common in some industries to use the note: LOCATED WITHIN .005 OF TRUE POSITION, which means the same as 2 above.

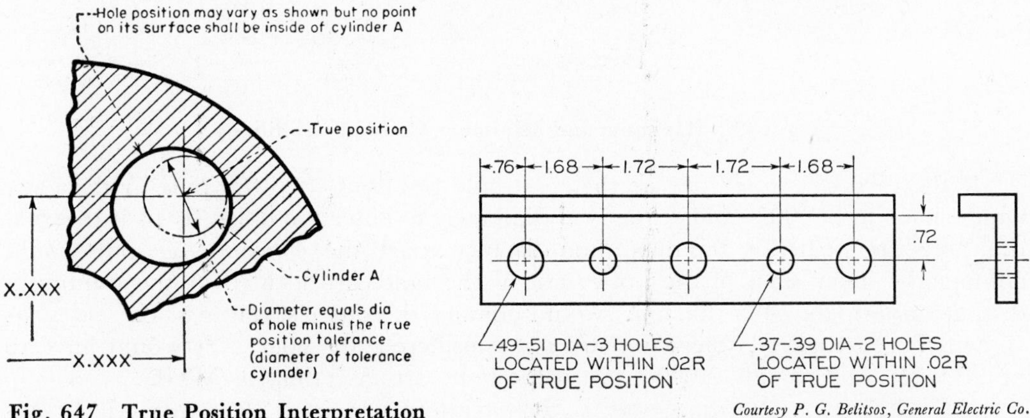

Fig. 647 True Position Interpretation (ASA Y14.5-1957).

Courtesy P. G. Belitsos, General Electric Co.

Fig. 648 No Tolerance Accumulation.

Since the exact locations of the true positions of the tolerances are given by untoleranced dimensions, it is important to prevent the application of general tolerances to these. A note should be added to the drawing such as: GENERAL TOLERANCES DO NOT APPLY TO BASIC TRUE-POSITION DIMENSIONS.

384. Maximum Material Condition. *Maximum material condition*, usually abbreviated to MMC, means "a feature of a finished product contains the maximum amount of material permitted by the toleranced size dimensions shown for that feature." Thus we have MMC when holes, slots, or other internal features are at minimum size, or when shafts, pads, bosses, and other external features are at their maximum size. We have MMC for both mating parts when the largest shaft is in the smallest hole and there is the least clearance between the parts.

In assigning positional tolerance to a hole, it is necessary to consider the size limits of the hole. If the hole is at MMC (smallest size), the positional tolerance is not affected; but if the hole is larger, the available positional tolerance is greater. In Fig. 649 (a) two half-inch holes are shown. If they are exactly .500 DIA (MMC, or smallest size) and are exactly 2.000 apart, they should receive a gage, (b), made of two round pins fixed in a plate 2.000 apart and .500 in diameter. However, the center-to-center distance between the holes may vary from 1.995 to 2.005.

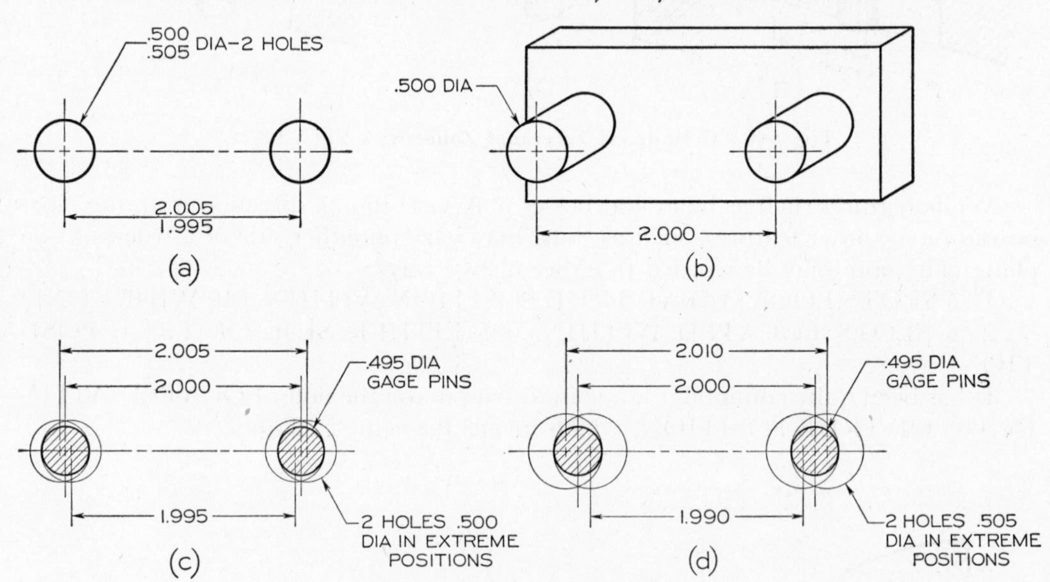

Fig. 649 Maximum and Minimum Material Conditions.

If the .500 dia. holes are at their extreme positions, (c), the pins in the gage would have to be .005 smaller, or .495 diameter, to enter the holes. Thus if the .500 dia. holes are located at the maximum distance apart, the .495 dia. gage pins would contact the inner sides of the holes; and if the holes are located at the minimum distance apart, the .495 dia. pins would contact the outer surfaces of the holes, as shown. If gagemakers' tolerances are not considered, the gage pins would have to be .495 dia. and exactly 2.000 apart if the holes are .500 dia., or MMC.

If the holes are .505 dia.—that is, at maximum size, as at (d)—they will be accepted by the same .495 dia. gage pins at 2.000 apart if the inner sides of the holes contact the inner sides of the gage pins, and the outer sides of the holes contact the outer sides of the gage pins, as shown. Thus the holes may be 2.010 apart, which is beyond the tolerance permitted for the center-to-center distance between the holes. Similarly, the holes may be as close together as 1.990 from center-to-center, which again is outside the specified positional tolerance.

Thus when the holes are not at MMC, i.e., when they are at maximum size, a greater positional tolerance becomes available. Since all features may vary in size, it is necessary to be made clear on the drawing at what condition of size the true position tolerance applies. In all but a few exceptional cases, the additional positional tolerance available when holes are larger than minimum size is acceptable and desirable. Parts thus accepted can be freely assembled whether the holes or other features are within the specified positional tolerance or not. This practice has been recognized and used in manufacturing for years, as is evident from the use of fixed-pin gages, which have been commonly used to inspect parts and control the least favorable condition of assembly. Thus it has become common practice for both manufacturing and inspection to assume that positional tolerance applies to MMC, and that greater positional tolerance becomes permissible when the part is not at MMC.

However, this practice should be followed with caution. The designer should be conscious of the additional tolerance and should permit it when it is practical and desirable. The ASA* recommends that "Where the requirements of the maximum material condition (MMC) apply, it shall be stated:

1. by use of a general note; or
2. by adding the abbreviation "MMC" to each applicable size specification; or
3. by suitable coverage in a specification which is referenced on the drawing.

"Where maximum material condition is not used, the true position tolerance applies regardless of feature size. The following shall be added to the drawing note: REGARDLESS OF FEATURE SIZE.

"For example:

1. 6 HOLES LOCATED AT TRUE POSITION WITHIN .010 DIA REGARDLESS OF HOLE SIZE.
2. 6 HOLES LOCATED WITHIN .005 R OF TRUE POSITION REGARDLESS OF HOLE SIZE."†

385. Geometric Tolerances. *Geometric tolerance*, or "tolerance of form" specifies "how far actual surfaces are permitted to vary from the perfect geometry implied by drawings."* The term "geometric" refers to the various geometric forms, as a plane, a cylinder, a cone, a square, or a hexagon. Theoretically these are perfect forms, but since it is impossible to produce perfect forms, it may be necessary to specify the amount of variation permitted. Geometric tolerances define conditions of straightness, flatness, parallelism, squareness, angularity, symmetry, concentricity, and roundness.

Methods of indicating geometric tolerances by means of notes, as recommended by ASA†, are shown in Figs. 650 and 651. At the right of each example, the meaning of the tolerance note or dimension is illustrated. In these examples, tolerances of form are considered individually, not taking into account the effects of combinations with other tolerances of position, size, or form.

An example of combined expressions is shown in Fig. 651 (j), in which a hole is to be held more closely parallel to a flat surface than its positional tolerance alone would indicate. As shown at the right, the requirement that the center line be anywhere within the .010 DIA positional tolerance cylinder is further restricted by the

*ASA Y14.5-1957.
†*Ibid.*

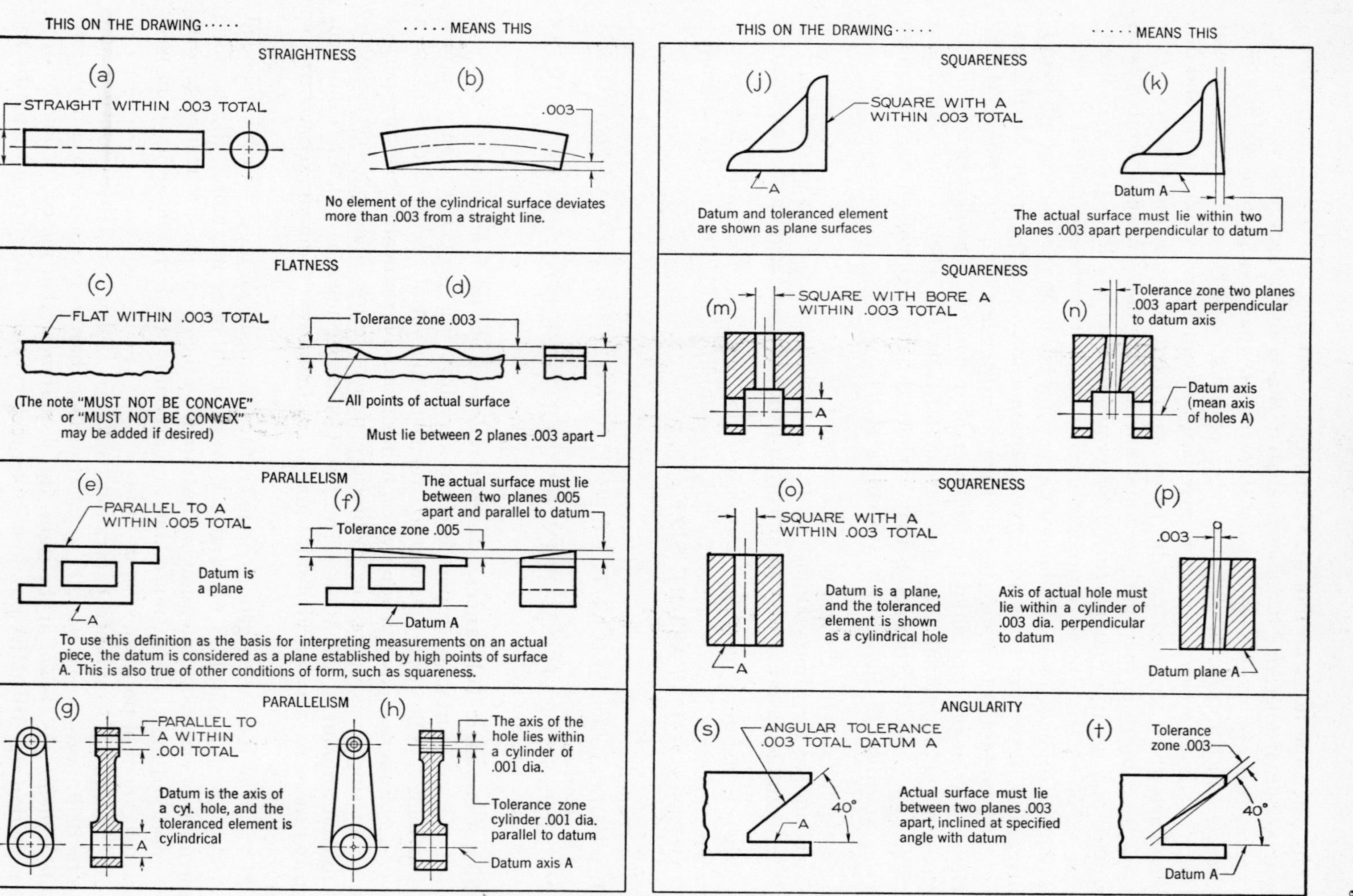

Fig. 650 Tolerances of Form (ASA Y14.5-1957).

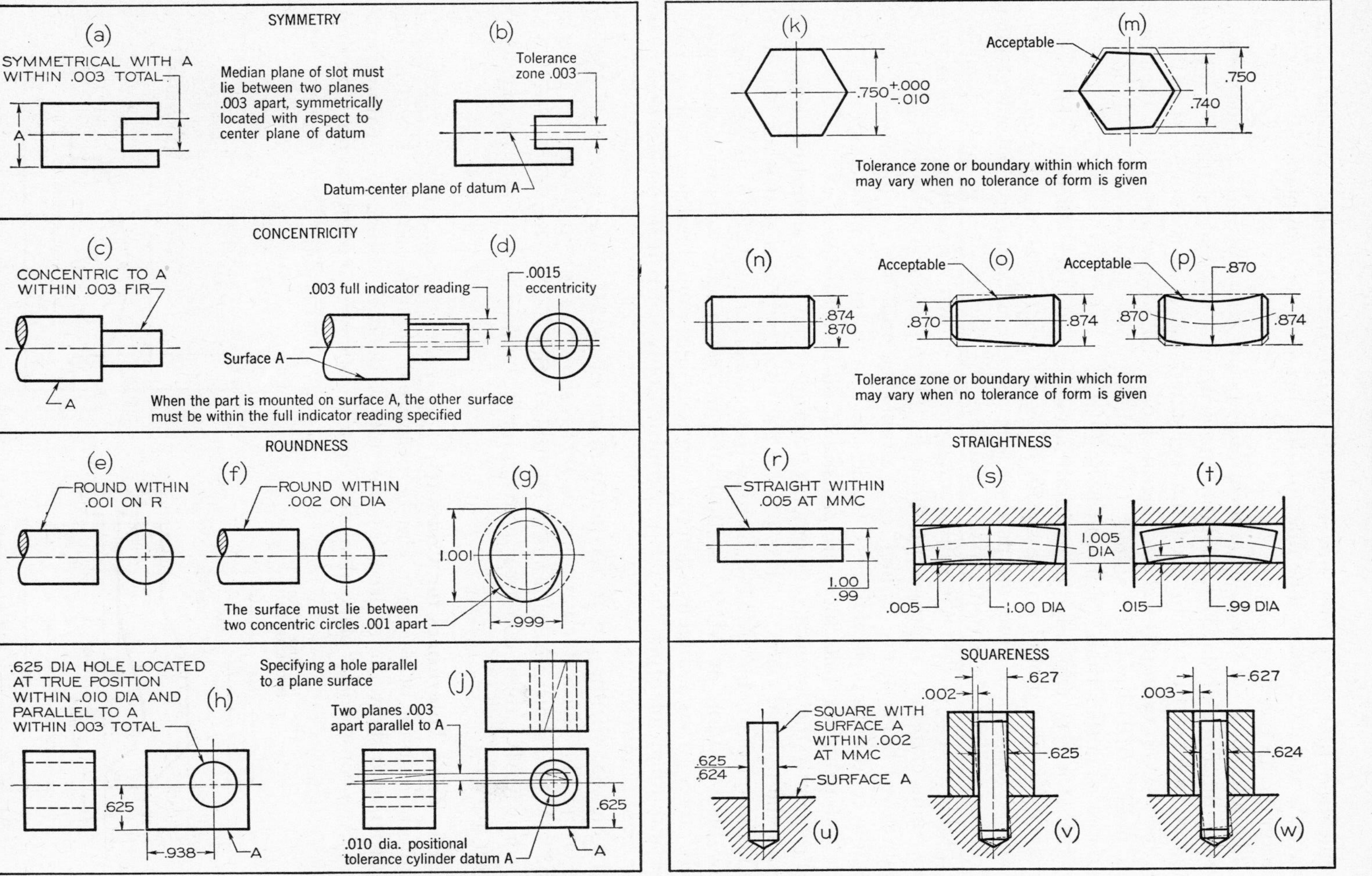

Fig. 651 Tolerances of Form (ASA Y14.5-1957).

specification that the center line must lie between two planes parallel to datum surface A, and .003 apart. The diagonal line between the two planes is one possible position in this case.

The tolerances on size and on form and specifications of surface roughness may overlap. But for a given part, variation in form must not exceed the size tolerance; therefore, tolerance of form need not be stated unless it is smaller than the size tolerance. As an example, if a size tolerance of ± .005 is permitted, any tolerance of form under .010 should be stated. Or if the flatness tolerance is .001, a surface roughness smoother than 1000√ should be indicated.

Since it is very expensive to maintain accurate geometric tolerances in the shop, tolerances of form should not be specified except when established shop procedures cannot be depended upon to produce the necessary accuracy. When geometric tolerances are not indicated on the drawing, the actual part is understood to be acceptable if it is within the dimensional limits shown, regardless of variations in form. Some extreme variations are shown in Fig. 651 (k to p). However, in the case of fabricated bars, sheets, tubing, etc., established industry standards prescribe acceptable conditions of straightness, flatness, etc., and these standards are understood to hold if geometric tolerances are not shown on the drawing.

Frequently the amount of geometric tolerance can vary according to the actual sizes of the finished parts. In Fig. 651 (r) the note indicates that when the shaft is at maximum diameter (MMC), it may have an error of .005 in straightness. This allowable variation at MMC is illustrated at (s). If .005 variation is allowed at MMC, then .015 is available when the part is at minimum diameter, as shown at (t). This additional tolerance should be permitted only when it does not interfere with functional requirements.

In Fig. 651 (u) the pin at maximum diameter (MMC) may have a .002 squareness error. This allowable variation of .002 is shown at (v), in which the gage hole is .627 (MMC shaft plus squareness tolerance). But if .002 squareness error is permitted at MMC, then .003 error is available if the part is at minimum diameter as shown at (w).

386. Zone Tolerances for Contours. "A 'zone' tolerance may be given when a uniform amount of variation can be permitted along a contour. The drawing is constructed to show the desired contour fully defined by dimensions without tolerances. At a conspicuous place along the contour, one or two phantom lines are drawn

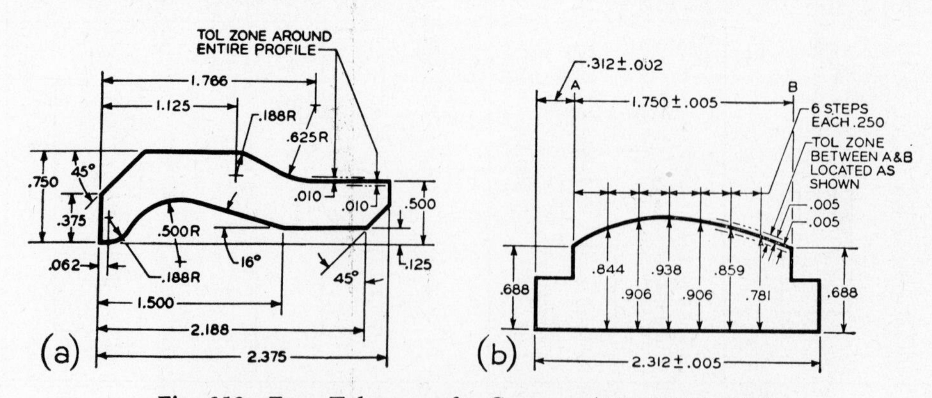

Fig. 652 Zone Tolerances for Contours (ASA Y14.5-1957).

with dimension lines and arrowheads to indicate the location of the tolerance zone. The value of the tolerance is given by a note."* See Fig. 652 (a). The distance between the contour and phantom lines is usually exaggerated for clarity. In cases where some limits on a drawing are given by zone tolerance while others are given by a general tolerance, it is necessary to indicate the extent of the zone tolerance, as at (b). The tolerance zone may be symmetrical about the contour line (bilateral), or it may be all on either side (unilateral).

387. Tolerances of Angles. Bilateral tolerances are usually given on angular dimensions where accuracy must be maintained, as illustrated in Fig. 653. In figuring

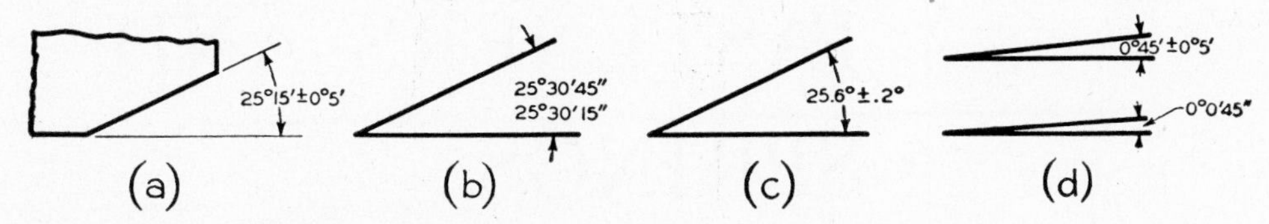

Fig. 653 Tolerances of Angles (ASA Y14.5-1957).

tolerances on angles, it should be kept in mind that the width of the tolerance zone increases as the distance from the vertex of the angle increases. The tolerance should be figured after considering the total allowable displacement at the point farthest from the vertex of the angle, and a tolerance specified that will not exceed this. The use of angular tolerances may be avoided by using gages. Taper turning is often handled by machining to fit a gage or by fitting to the mating part.

If an angular surface is located by a linear and an angular dimension, Fig. 654 (a), the surface must lie within a tolerance zone as shown at (b). The angular zone will be wider as the distance from the vertex increases. In order to avoid the accumulation

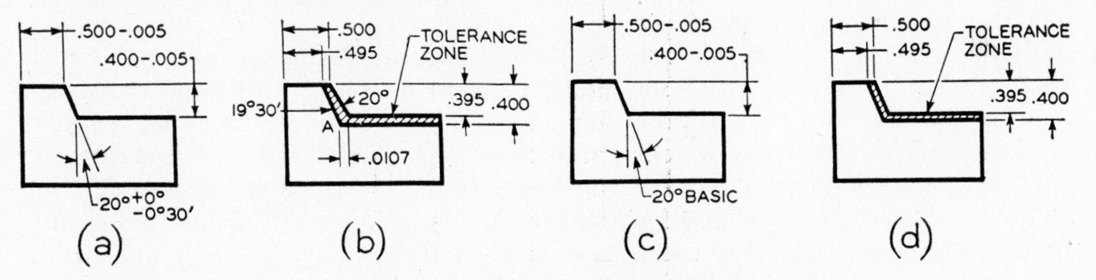

Fig. 654 Angular Tolerance Zones (ASA Y14.5-1957).

of tolerances, i.e., to decrease the tolerance zone, the *basic angle* tolerancing method of (c) is recommended by the ASA†. The angle is marked BASIC and no angular tolerance is specified. The tolerance zone is now defined by two parallel planes, resulting in improved angular control, (d). See also Fig. 650 (s) and (t) for tolerancing angularity as tolerance of form.

*ASA Y14.5-1957.
†*Ibid.*

CHAPTER 13

THREADS, FASTENERS, AND SPRINGS

388. Screw Threads.* The concept of the screw thread seems to have occurred first to Archimedes, the third-century B.C. mathematician, who wrote briefly on spirals and invented several simple devices applying the screw principle. By the first century B.C. the screw was a familiar element, but was crudely cut from wood or filed by hand on a metal shaft. Nothing more was heard of the screw thread in Europe until the fifteenth century, though the Greeks and the Arabs had preserved their knowledge of it. Leonardo da Vinci understood the screw principle, and he has left sketches showing how to cut screw threads by machine. In the sixteenth century, screws appeared in German watches, and screws were used to fasten suits of armour. In 1569 the screw-cutting lathe was invented by the Frenchman, Besson, but the method did not take hold, and for another century and a half nuts and bolts continued to be made largely by hand. In the eighteenth century, screw manufacturing got started in England during the Industrial Revolution.

In these early times, there was no such thing as standardization. The nuts of one manufacturer would not fit the bolts of another. In 1841 Sir Joseph Whitworth started crusading for a standard screw thread, and soon the Whitworth thread was accepted throughout England. In 1864 in the United States, a committee named by the Franklin Institute adopted a thread proposed by William Sellers of Philadelphia; but the Sellers' nuts would not screw on a Whitworth bolt, or vice versa, the thread angles being different.

In 1935 the American Standard Thread, with the same 60° V form of the old Sellers' thread, was adopted in this country. Still there was no standardization between countries. In peace-time it was a nuisance; in World War I it was a serious

*See ASA Y14.6-1957.

inconvenience, and finally in World War II the obstacle was so great that it was decided to do something about it, and talks began between the Americans, British, and Canadians (the so-called ABC countries). Finally in 1948 agreement was reached on unification of American and British screw threads. The new thread was called the *Unified Screw Thread*, and it represents a compromise between the American Standard and Whitworth systems, allowing complete interchangeability of threads in the three countries.

Today screw threads are vital to our industrial life. They are used for hundreds of different purposes, the three basic applications being: (1) to *hold parts* together, (2) to *adjust parts* with reference to each other, and (3) to *transmit power*.

389. Definitions of Terms. The following definitions apply to screw threads generally, Fig. 655:

Screw Thread. A ridge of uniform section in the form of a helix, §170, on the external or internal surface of a cylinder.

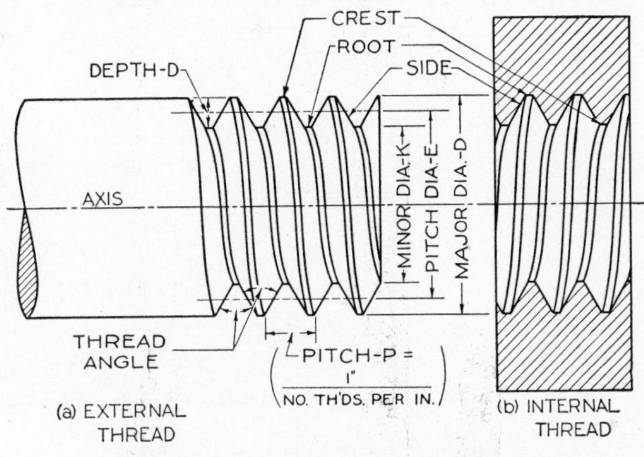

Fig. 655 Screw-Thread Nomenclature.

External Thread. A thread on the outside of a member, as on a shaft.

Internal Thread. A thread on the inside of a member, as in a hole.

Major Diameter. The largest diameter of a screw thread (applies to both internal and external threads).

Minor Diameter. The smallest diameter of a screw thread (applies to both internal and external threads).

Pitch. The distance from a point on a screw thread to a corresponding point on the next thread measured parallel to the axis. The pitch P is equal to 1″ divided by the number of threads per inch.

Pitch Diameter. The diameter of an imaginary cylinder passing through the threads so as to make equal the widths of the threads and the widths of the spaces cut by the cylinder.

Lead. The distance a screw thread advances axially in one turn.

Angle of Thread. The angle included between the sides of the thread measured in a plane through the axis of the screw.

Crest. The top surface joining the two sides of a thread.

Root. The bottom surface joining the sides of two adjacent threads.

Side. The surface of the thread that connects the crest with the root.

Axis of Screw. The longitudinal center line through the screw.

Depth of Thread. The distance between the crest and the root of the thread measured normal to the axis.

Form of Thread. The cross section of thread cut by a plane containing the axis.

Series of Thread. Standard number of threads per inch for various diameters.

390. Screw-Thread Forms. Various *forms* of threads are in use to meet the general functions listed above, Fig. 656. For holding parts together, the American

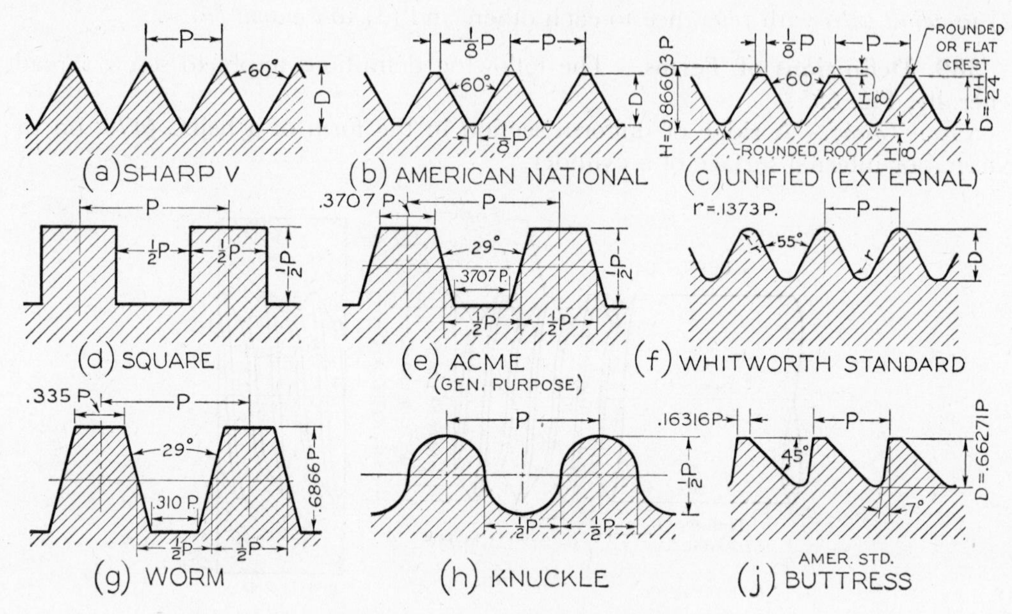

Fig. 656 Screw-Thread Forms.

National Thread is used most in the United States, it having superseded the old 60° *Sharp-V thread.* The flattened roots and crests make the former the stronger thread. This thread was originally called the United States Standard, or the Sellers thread. For purposes of certain adjustments, the Sharp-V thread is still useful because of the friction resulting from the increased area of the thread-face. It is also used in brass pipe work.

The *Unified Thread** is the new standard thread agreed upon by the United States, Canada, and Great Britain in 1948. The crest of the external thread may be flat or rounded, and the root is rounded; otherwise the thread form is essentially the same as the American National. The Unified Thread is the new American Standard, and industries are gradually changing over to it.

The *Square Thread* is theoretically the ideal thread for power transmission, since its face is nearly at right angles to the axis; but owing to the difficulty of cutting it with dies and because of other inherent disadvantages, such as the fact that split nuts will not readily disengage, the Square thread has been displaced to a large extent by the Acme thread. The Square thread is not standardized.

The *Acme Thread* is a modification of the Square thread, and has largely replaced

*ASA B1.1-1949.

it. It is stronger than the Square thread, is easier to cut, and has the advantage of easy disengagement from a split nut, as on the lead screw of a lathe.

The *Whitworth Thread* has been the British Standard, and is being replaced by the Unified Thread. Its uses correspond to those of the American National Standard.

The *Standard Worm Thread* is similar to the Acme thread, but is deeper. It is used on shafts to carry power to worm wheels.

The *Knuckle Thread* is usually rolled from sheet metal but is sometimes cast, and is used in modified forms in electric bulbs and sockets, bottle tops, etc.

The *Buttress Thread* is designed to transmit power in one direction only, and is used in the breech-locks of large guns, in jacks, in airplane propeller hubs, and in other mechanisms of similar requirements.

391. Thread Pitch. The *pitch* of a thread, of whatever form, is the distance parallel to the axis between corresponding points on adjacent threads, Figs. 655 (a) and 656. The pitch is simply one inch divided by the number of threads per inch. The thread tables give the number of threads per inch for each standard diameter. Thus a Unified Coarse thread, Appendix table 12, of 1″ diameter, has eight threads per inch, and the pitch P equals $\frac{1}{8}$″.

As shown in Fig. 657 (a), if a thread has only four threads per inch, the pitch and the threads themselves are quite large. If there are, say, sixteen threads per inch, the pitch is only $\frac{1}{16}$″, and the threads are relatively small, as shown at (b).

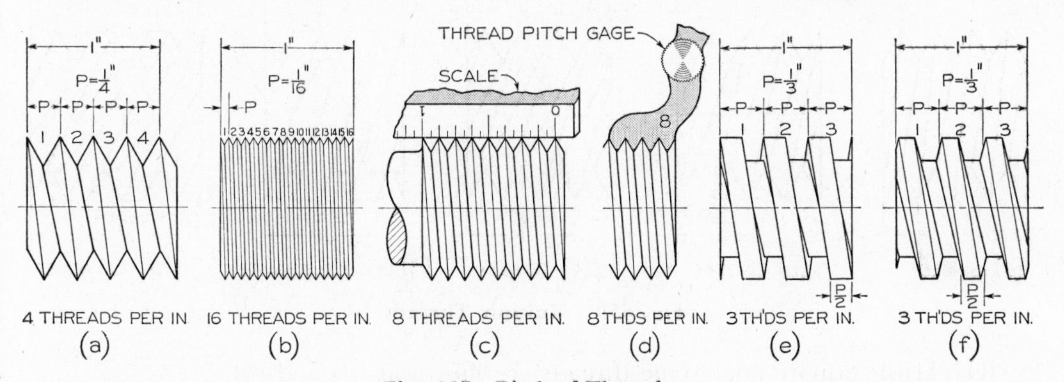

Fig. 657 Pitch of Threads.

The pitch or number of threads per inch can easily be measured with an ordinary scale, (c), or with a *thread pitch gage*, (d).

It will be seen at (e) and (f) that the pitch for Square and Acme threads includes a thread ridge and a space.

392. Right-Hand and Left-Hand Threads. A right-hand thread, when viewed toward an end, winds in a clockwise and receding direction. Thus a right-hand thread is one which advances into the nut when turned clockwise, and a left-hand thread is one which advances into the nut when turned counterclockwise, Fig. 658. A thread is always considered to be right-hand (RH) unless otherwise specified. A left-hand thread is always marked LH on a drawing. See Fig. 674 (a).

393. Single and Multiple Threads. A *single* thread, as the name implies, is composed of one ridge; in this thread the lead is equal to the pitch. *Multiple* threads are composed of two or more ridges running side by side. As shown in Fig. 659 (a) to (c), the *slope line* is the hypotenuse of a right triangle whose short side equals $\frac{1}{2}$P for single threads, P for double threads, $1\frac{1}{2}$P for triple threads, etc. This applies to all forms of threads. In *double* threads, the lead is twice the pitch; in *triple* threads the

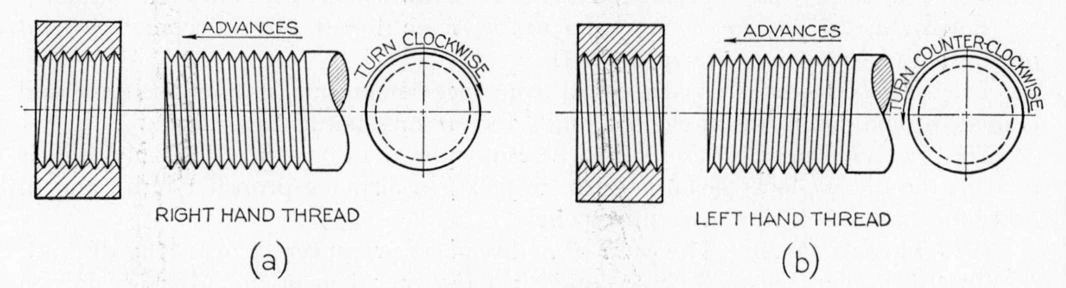

Fig. 658 Right-Hand and Left-Hand Threads.

lead is three times the pitch, etc. On a drawing of a single or triple thread, a root is opposite a crest; in the case of a double or quadruple thread, a root is drawn opposite a root. Therefore, in one turn, a double thread advances twice as far as a single thread; a triple thread advances three times as far, etc.

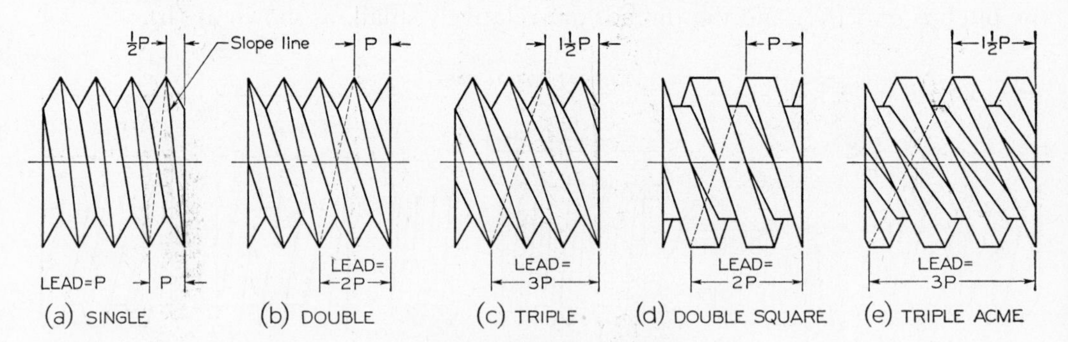

Fig. 659 Multiple Threads.

RH Triple Square and Acme threads are shown at (d) and (e).

Multiple threads are used wherever quick motion, but not great power, is desired, as on fountain pens, toothpaste caps, valve stems, etc. The threads on a valve stem are frequently multiple threads, to impart quick action in opening and closing the

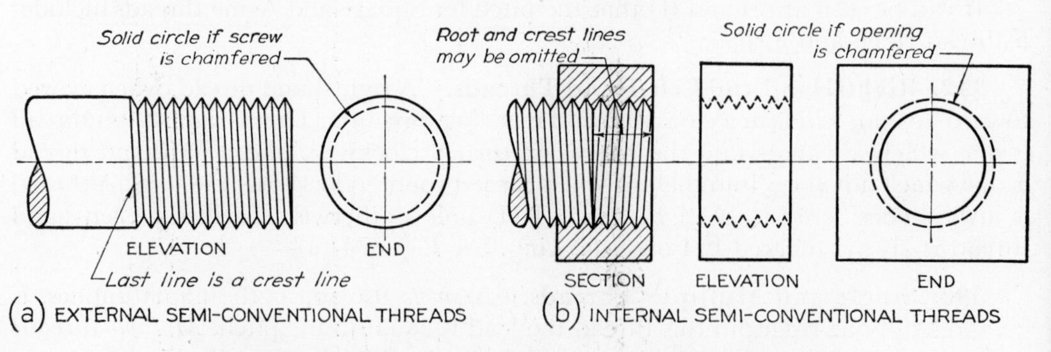

Fig. 660 Semi-Conventional American Standard Threads.

valve. Multiple threads on a shaft can be recognized and counted by observing the number of thread endings on the end of the screw.

394. Detailed Representation of Threads. The true projection of the helical curves of a screw thread, Fig. 655, presents a pleasing appearance, but this does not compensate for the laborious task of plotting the helices, §170. Consequently, the true projection is never used in practice.

When the diameter of the thread on the drawing is over one inch approximately, a pleasing drawing may be made by the *detailed representation* method, in which the true profiles of the threads (any form of thread) are drawn; but the helical curves are replaced by straight lines, as shown in Fig. 660 for Unified Threads.* Whether the crests or roots are flat or rounded, they are represented by single lines and not double lines as in Fig. 655; consequently, Sharp-V and American National threads are drawn in exactly the same way.

395. Detailed Representation—"V" Threads. The detailed representation for Sharp-V, American National, and Unified threads is the same, since the flats, if any, are disregarded. The steps in drawing are, Fig. 661:

I. Draw center line and lay out length and major diameter.

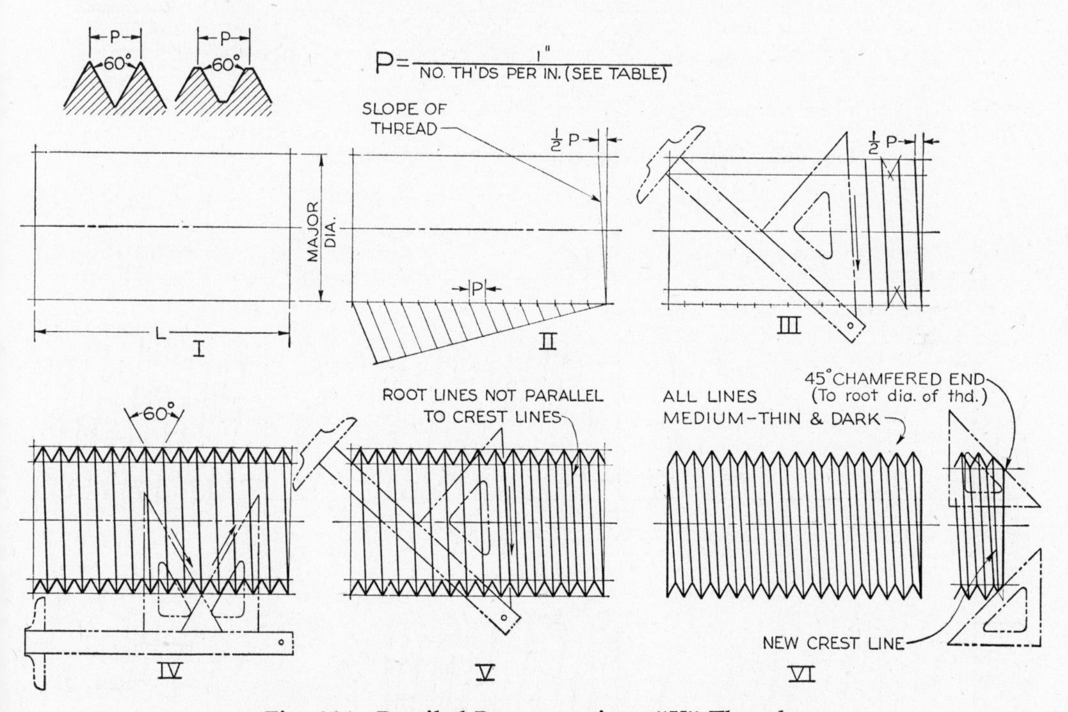

Fig. 661 Detailed Representation—"V" Threads.

II. Find the number of threads per inch in Appendix table 12. This number depends upon the major diameter of the thread, whether the thread is internal or external. Find P (pitch) by dividing 1″ by the number of threads per inch, §391. Establish the slope of the thread by offsetting the slope line ½P for single threads,

*A thread 1½″ dia., if drawn half size, would be less than 1″ dia. on the drawing, and hence would be too small for this method of representation.

P for double threads, $1\frac{1}{2}$P for triple threads, etc.* For right-hand threads, the slope line slopes upward to the left; for left-hand threads, the slope line slopes upward to the right. If the number of threads per inch conforms to the scale, the pitch can be set off directly. For example, eight threads per inch can easily be set off with the architects scale, and ten threads per inch with the engineers scale. Otherwise, use the bow dividers or use the parallel-line method shown in Fig. 661 (II).

III. From the pitch-points, draw crest lines parallel to slope line. These should be dark medium-sharp lines. Slide triangle on T-square (or another triangle) to make lines parallel. Draw two V's to establish depth of thread, and draw guide lines for root of thread, as shown.

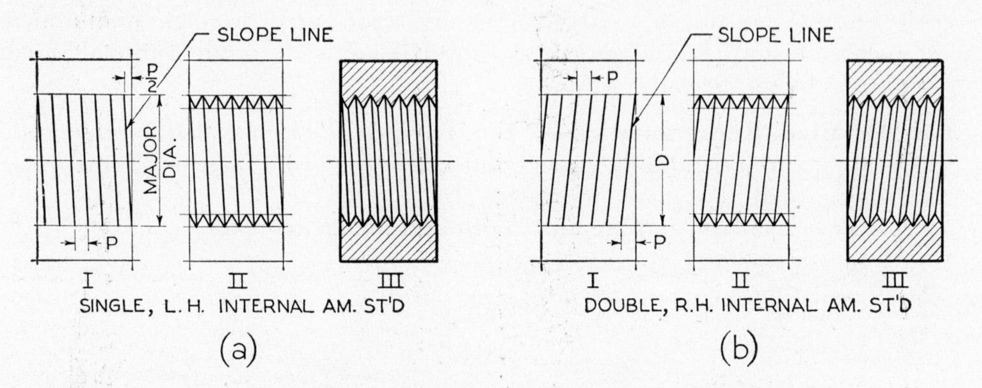

Fig. 662 Detailed Representation for Internal "V" Threads.

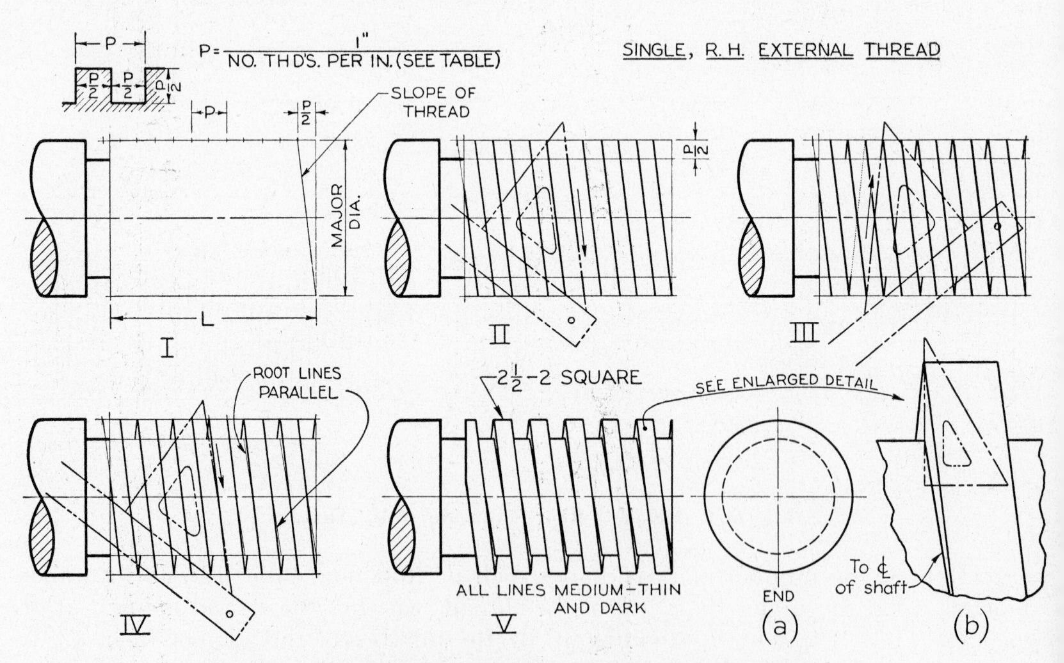

Fig. 663 Detailed Representation—External Square Threads.

*These offsets are the same in terms of P for any form of thread.

IV. Draw 60° V's finished weight. These V's should stand vertically; i.e., they should not "lean" with the thread.

V and VI. Draw root lines dark at once. Root lines will *not* be parallel to crest lines. Slide triangle on straightedge to make root lines parallel. In the final drawing, all thread lines should be approximately the same weight: medium-thin, but dark.

When the end is chamfered (usually 45° with end of shaft, sometimes 30°), the chamfer extends to the thread depth. The chamfer creates a new crest line, which is then drawn between the two new crest points. It is not parallel to the other crest lines.

The corresponding internal detailed threads, in section, are drawn as shown in Fig. 662. Notice that for LH threads the lines slope upward to the left, as shown at (a), while for RH threads the lines slope upward to the right, as at (b). Make all final thread lines medium-thin, but dark.

396. Detailed Representation—Square Threads. The steps in drawing the detailed representation of an external Square thread when the major diameter is over 1″ (approx.) on the drawing are shown in Fig. 663.

I. Draw center line, and lay out length and major diameter of thread. Determine P by dividing 1″ by the number of threads per inch. See Appendix table 20. For a single RH thread, the lines slope upward to the left, and the slope line is offset $\frac{P}{2}$ *as for all single threads of any form.* On the upper line, set off spaces equal to $\frac{P}{2}$, as shown, using a scale if possible; otherwise use the bow dividers or the parallel-line method to space the points.

II. From the $\frac{P}{2}$ points on the upper line, draw crest lines parallel to slope line, dark and fairly thin. Draw guide lines for root of thread, making the depth $\frac{P}{2}$, as shown.

III. Draw parallel visible back edges of threads.

IV. Draw parallel visible root lines. Note enlarged detail at (b).

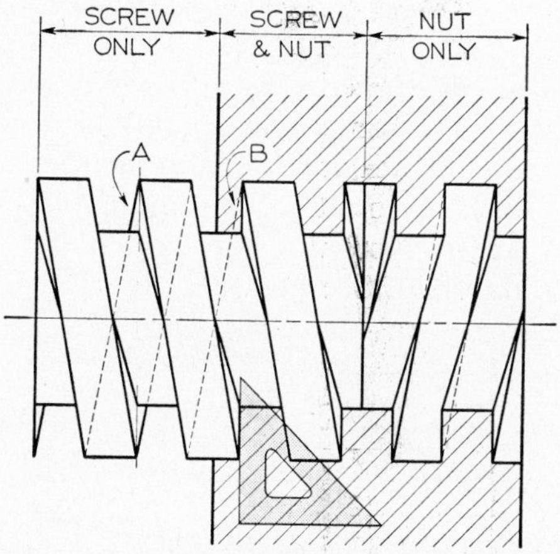

V. Accent the lines. All lines should be medium-thin, and dark.

Fig. 664 Square Threads in Assembly.

Note the end view of the shaft at (a). The root circle is hidden, and no attempt is made to show the true projection. If the end is chamfered, a solid circle would be drawn instead of the hidden circle.

An assembly drawing, showing an external Square thread partly screwed into the nut, is shown in Fig. 664.

The detail of the Square thread at A is the same as shown in Fig. 663. But when the external and internal threads are assembled, the thread in the nut overlaps and covers up half of the "V," as shown at B.

The internal thread construction is the same as in Fig. 665. Note that the thread lines representing the back half of the internal threads (since the thread is in section), slope in the opposite direction from those on the front side of the screw.

Steps in drawing a single internal Square thread in section are shown in Fig. 665.

Note in Step II that a crest is drawn opposite a root. This is the case for both single and triple threads. For double or quadruple threads, a crest is opposite a crest. Thus the construction in Step I is the same for any multiple of thread. The differences

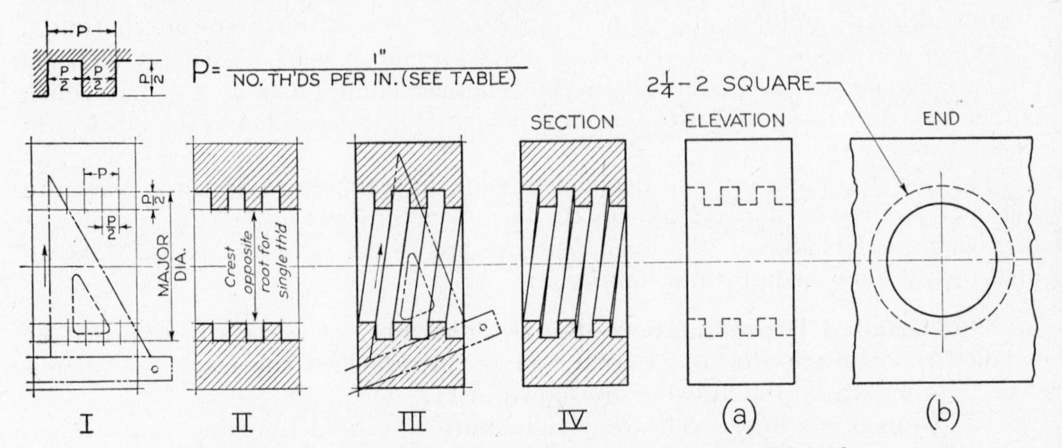

Fig. 665 Detailed Representation—Internal Square Threads.

are developed in Step II where the threads and spaces are distinguished and outlined.

The same internal thread is shown in elevation in Fig. 665 (a). The profiles of the threads are drawn in their normal position, but with hidden lines; and the sloping lines are omitted for simplicity. The end view of the same internal thread is shown at (b). Note that the hidden and solid circles are opposite those for the end view of the shaft. See Fig. 663 (a).

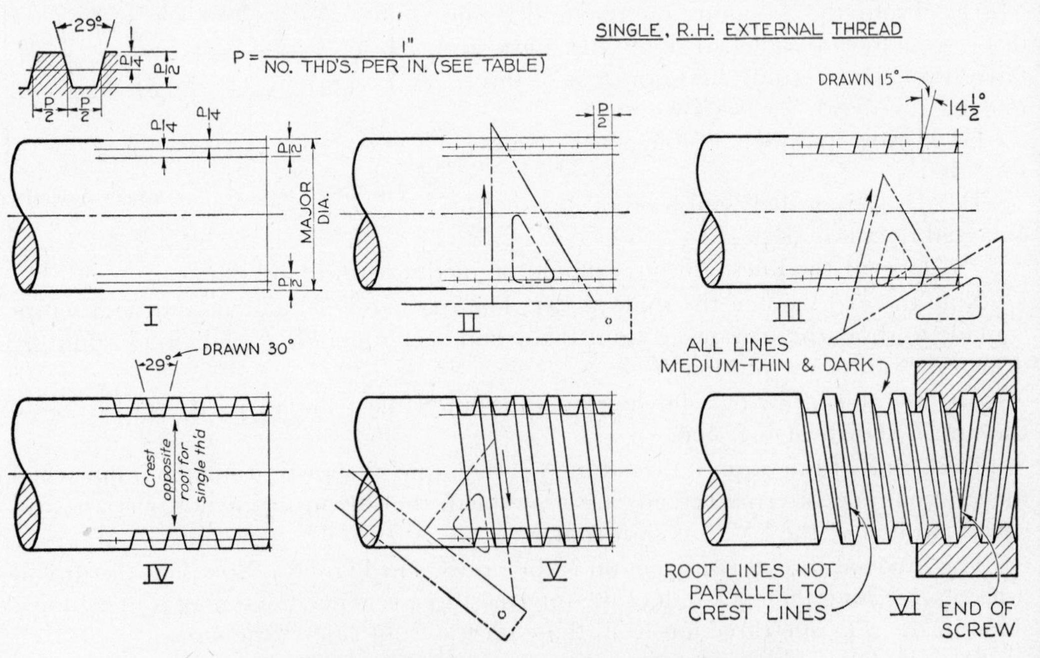

Fig. 666 Detailed Representation—Acme Threads.

397. Detailed Representation—Acme Thead. The steps in drawing the detailed representation of Acme threads when the major diameter is over 1″ (approx.) on the drawing are shown in Fig. 666.

I. Draw center line, and lay out length and major diameter of thread. Determine P by dividing 1″ by the number of threads per inch. See Appendix table 20. Draw construction lines for the root diameter, making the thread depth $\frac{P}{2}$. Draw construction lines halfway between the crest and root guide lines.

II. On the intermediate construction lines, lay off $\frac{P}{2}$ spaces, as shown. Set off spaces directly with scale if possible (for example, if $\frac{P}{2} = \frac{1}{10}$″, use the engineers scale); otherwise, use bow dividers or parallel-line method.

III. Through alternate points, draw construction lines for sides of threads (draw 15° instead of 14½°).

IV. Draw construction lines for other sides of threads. Note that for single and triple threads, a crest is opposite a root, while for double and quadruple threads, a crest is opposite a crest. Heavy-in tops and bottoms of threads.

V. Draw parallel crest lines, final weight at once.

VI. Draw parallel root lines, final weight at once, and heavy-in the thread profiles. All lines should be medium-thin and dark. Note that the internal threads in the back of the nut slope in the opposite direction to the external threads on the front side of the screw.

End views of Acme threaded shafts and holes are drawn exactly like those for the Square thread, Figs. 663 and 665.

The ASA has approved Acme* and Stub Acme† threads, and provides two types of Acme threads, "general-purpose" and "centralizing" threads. The general-purpose threads, shown in Fig. 666, have three classes, 2G, 3G, and 4G. These provide generous fits allowing free movement. The centralizing threads have a limited clearance on major diameters so as to maintain approximate alignment of the thread axis and prevent wedging on the flanks of the threads. They have five classes, 2C, 3C, 4C, 5C, and 6C.

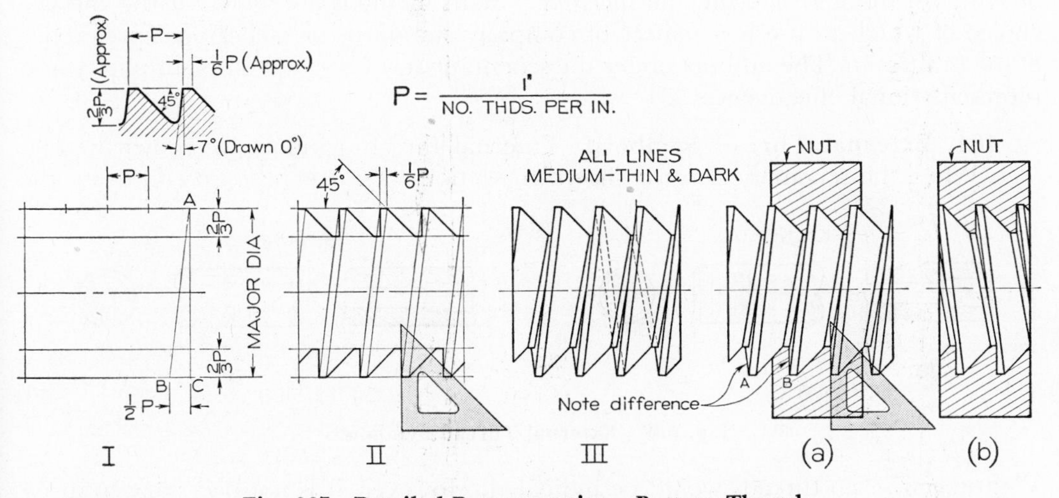

Fig. 667 Detailed Representation—Buttress Threads.

*ASA B1.5-1952.
†ASA B1.8-1952.

398. Detailed Representation—Buttress Thread. The steps in drawing an American Standard LH Buttress Thread are shown in Fig. 667:

I. Lay out major and minor thread diameters, making the depth of thread $\frac{2}{3}$P, which is very close to the formula depth .66271P. Set off BC = $\frac{1}{2}$P to establish slope line AB of the LH thread.

II. Construct profiles of threads, ignoring the 7° for the left side of the threads, and drawing it perpendicular to the centerline of the shaft. Although the root is rounded, draw flat like the crest. Make the crest $\frac{1}{6}$P approximately, which is close to the formula width .16316P.

III. Heavy-in crest lines, draw root lines, and darken the profiles. Make all lines medium-thin and dark.

The external and internal assembled threads are shown at (a). Note the differences in the threads at A and B. The internal thread in section is shown at (b).

399. Use of Phantom Lines. In representing objects having a series of identical features, time may be saved and satisfactory representation effected by the use of phantom lines, as in Fig. 668. Threaded shafts thus represented may be shortened

Fig. 668 Use of Phantom Lines.

without the use of conventional breaks, but must be correctly dimensioned. The same methods may be applied to springs, §426. This use of phantom lines is limited almost entirely to detail drawings.

400. Thread Symbols. *Thread symbols* are used to represent threads of small diameter, under approximately 1″ dia. on the drawing. The symbols are the same for all forms of thread, as Unified, Square, Acme, etc. The ASA approves two sets of symbols, the *schematic* and the *simplified*.* Both methods are widely used, and the choice of which to use is a matter of company standards or of personal preference of the draftsman. The authors prefer the schematic form because of its simplicity and representational effectiveness.

401. External Thread Symbols. External thread symbols are shown in Fig. 669. When the schematic form is shown in section, (a), it is necessary to draw the

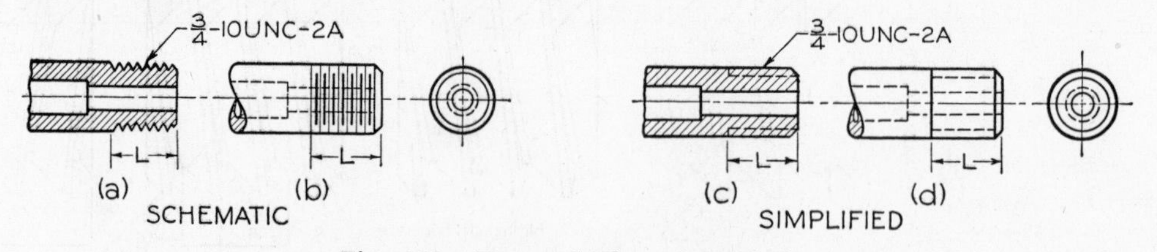

Fig. 669 External Thread Symbols.

V's; otherwise no threads would be indicated. However, it is not necessary to draw the V's to scale or according to the actual slope of the crest lines. To draw the V's,

*ASA Y14.6-1957.

use the schematic thread depth, Fig. 671 (a), and let the pitch be determined by the V's.

Schematic threads in elevation, Fig. 669 (b), are indicated by alternate long and short lines at right angles to the center line, the root lines being preferably thicker than the crest lines. Although theoretically the crest lines would be spaced according to actual pitch, the lines would often be very crowded and tedious to draw, thus defeating the purpose of the symbol, to save drafting time. Therefore, in practice the experienced draftsman spaces the crest lines carefully by eye, as he does section lines, and then adds the heavy root lines spaced by eye half-way between the crest lines. In general, the spacing should be proportionate for all diameters. For convenience in drawing, proportions for the schematic symbol are given in Fig. 671.

Simplified external thread symbols are shown in Fig. 669 (c) and (d). The threaded portions are indicated by hidden lines parallel to the axis at the approximate depth of the thread, whether in section or in elevation. To draw these lines, use the schematic depth of thread as given in the table in Fig. 671.

402. Internal Thread Symbols. Internal thread symbols are shown in Fig. 670. The schematic thread in section is exactly the same as the external symbol in Fig.

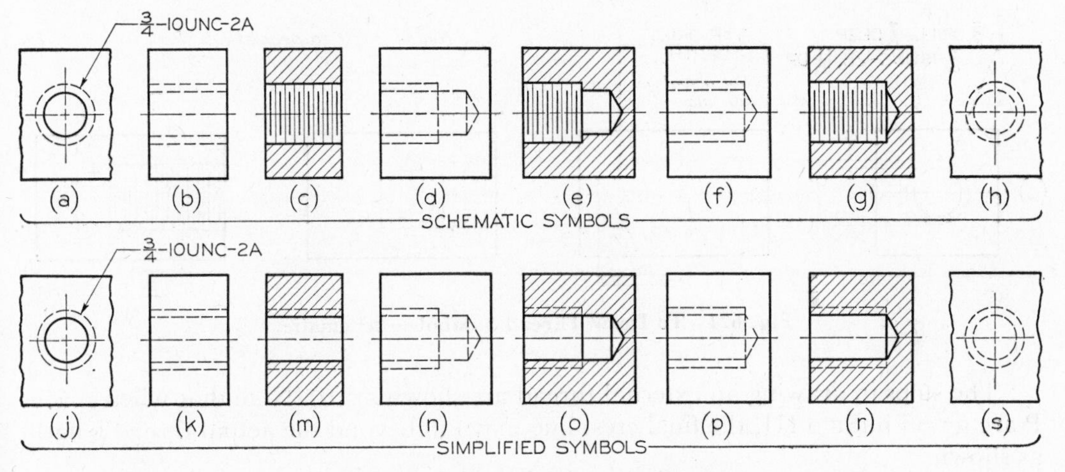

Fig. 670 Internal Thread Symbols.

669 (b). Hidden threads, by either method, are represented by pairs of hidden lines. The hidden dashes should be staggered, as shown. Note that the only differences between the schematic and simplified internal thread symbols occur in the sectional views.

In the case of blind tapped holes, the drill depth normally is drawn at least three schematic pitches beyond the thread length, as shown in Fig. 670 (d), (e), (n), and (o). The representations at (f) and (p) are used to represent the use of a bottoming tap, when the length of thread is the same as the depth of drill. See also §413.

403. To Draw Schematic Thread Symbols. In Fig. 671 (a) is shown a table of values of depth and pitch to use in drawing schematic thread symbols. These values are selected to produce a well-proportioned symbol and to be convenient to set off with the scale. The experienced draftsman will carefully space the lines by eye, but the student should use the scale. Note that the values of D and P are for the diameter *on the drawing*. Thus a $1\frac{1}{2}''$ dia. thread at half scale would be $\frac{3}{4}''$ dia. on the drawing, and values of D and P for a $\frac{3}{4}''$ major diameter would be used.

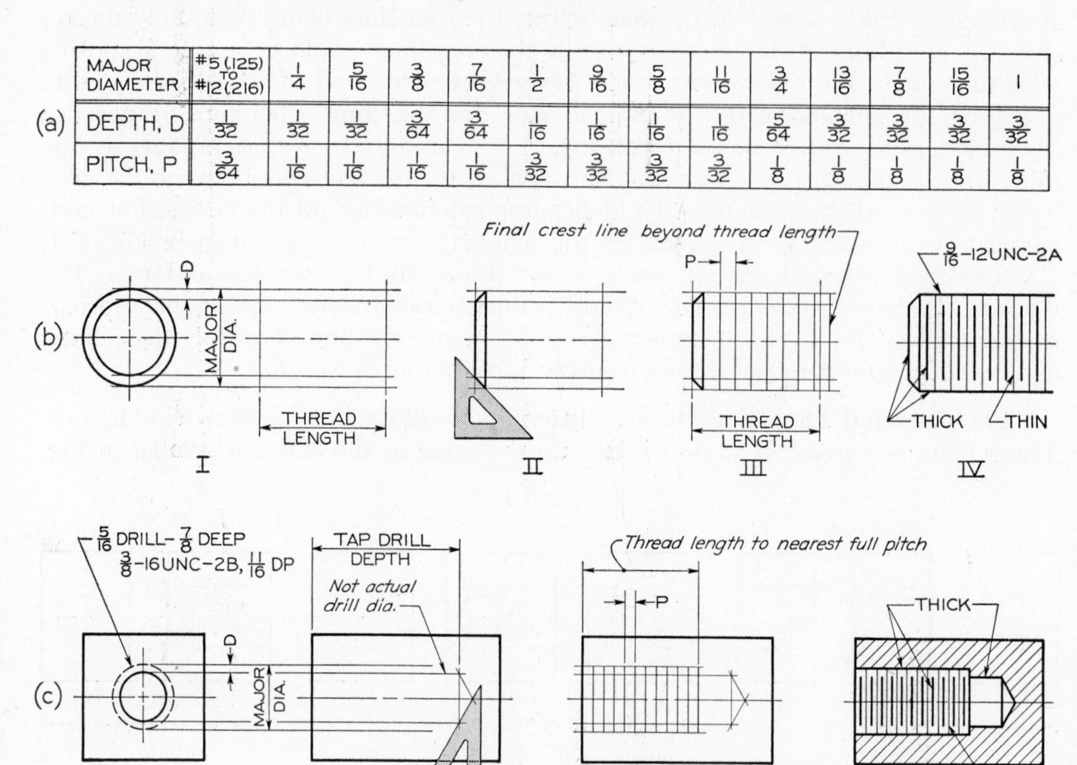

(a)

MAJOR DIAMETER	#5 (.125) TO #12 (.216)	1/4	5/16	3/8	7/16	1/2	9/16	5/8	11/16	3/4	13/16	7/8	15/16	1
DEPTH, D	1/32	1/32	1/32	3/64	3/64	1/16	1/16	1/16	1/16	5/64	3/32	3/32	3/32	3/32
PITCH, P	3/64	1/16	1/16	1/16	1/16	3/32	3/32	3/32	3/32	1/8	1/8	1/8	1/8	1/8

Fig. 671 To Draw Thread Symbols—Schematic.

The steps in drawing an external thread are shown at (b). Note that when spaces P are set off in Step III, the final crest line may fall beyond the actual thread length, as shown.

To draw the external thread in section, Fig. 669 (a), use the schematic thread depth, and let the pitch be determined by the V's.

The steps in drawing an internal thread are shown at (c). Here again the symbol thread length may be slightly longer than the actual given thread length. If the tap drill depth is known or given, the drill is drawn to that depth, as shown. If the thread note omits this information, as is often done in practice, the draftsman merely draws the hole about three thread pitches (schematic) beyond the thread length. The tap drill diameter is represented approximately, as shown, and not to actual size.

404. Threads in Section. Detailed representations of large threads in section are shown in Figs. 660, 662, and 664-667. As indicated by the note in Fig. 660 (b), the root lines and crest lines may be omitted in internal sectional views, if desired.

External thread symbols are shown in section in Fig. 669. Note that in the schematic symbol, the V's must be drawn. Internal thread symbols in section are shown in Fig. 670.

Threads in an assembly drawing are shown in Fig. 672. It is customary not to section a stud or a nut, or any solid part, unless necessary to show some internal shapes. See §441. Note that when external and internal threads are sectioned in assembly, the V's are required to show the threaded connection.

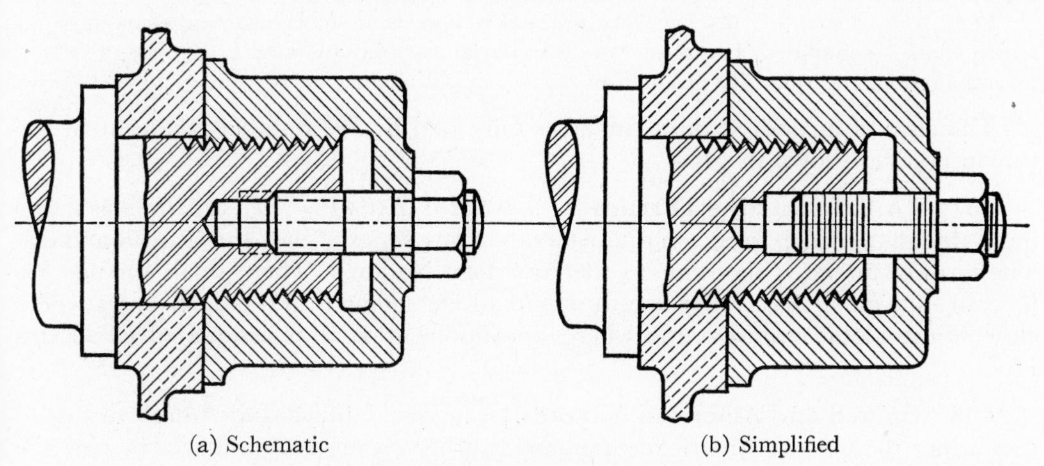

(a) Schematic (b) Simplified

Fig. 672 Threads in Assembly (ASA Z14.1-1946).

405. American National Thread. The old American (National) Standard thread was adopted in 1935. The *form*, or profile, Fig. 656 (b), is the same as the old Sellers' profile, or U.S. Standard, and is known as the *National Form*. The methods of representation are the same as for the Unified thread. American National threads are being replaced by the Unified threads, but the old threads will be frequently encountered on drawings for a long time.

Five *series* of threads were embraced in the old standard,* as follows:

1. *Coarse Thread:* A general-purpose thread for holding purposes. Designated NC (National Coarse.)
2. *Fine Thread:* A greater number of threads per inch; used extensively in automotive and aircraft construction. Designated NF (National Fine).
3. *8-Pitch Thread:* All diameters have 8 threads per inch. Used on bolts for high-pressure pipe flanges, cylinder-head studs, and similar fasteners. Designated 8N (National Form, 8 threads per inch).
4. *12-Pitch Thread:* All diameters have 12 threads per inch; used in boiler work and for thin nuts on shafts and sleeves in machine construction. Designated 12N (National Form, 12 threads per inch).
5. *16-Pitch Thread:* All diameters have sixteen threads per inch; used where necessary to have a fine thread regardless of diameter, as on adjusting collars and bearing retaining nuts. Designated 16N (National Form, sixteen threads per inch).

406. American National Thread Fits. The old standard* also established for general use four classes of screw thread *fits* between mating threads (as between bolt and nut). These fits are produced by the application of tolerances listed in the standards, and are described as follows:

Class 1 Fit. Recommended only for screw thread work where clearance between mating parts is essential for rapid assembly and where shake or play is not objectionable.

*ASA B1.1935.

Class 2 Fit. Represents a high quality of commercial thread product, and is recommended for the great bulk of interchangeable screw thread work.

Class 3 Fit. Represents an exceptionally high quality of commercially threaded product and is recommended only in cases where the high cost of precision tools and continual checking are warranted.

Class 4 Fit. Intended to meet very unusual requirements more exacting than those for which Class 3 is intended. It is a selective fit if initial assembly by hand is required. It is not, as yet, adaptable to quantity production.

The class of fit desired on a thread is indicated by the number of the fit in the thread note, as shown in §410.

407. S.A.E. Extra Fine Threads.* *The S.A.E. Extra Fine Thread Series* has many more threads per inch for given diameters than any series of the American Standard. The form of thread is the same as the American National. These small threads are used in thin metal where the length of thread engagement is small, in cases where close adjustment is required, and where vibration is great. It is designated EF (Extra Fine).

408. Unified and American Threads.† The new Unified and American thread constitutes the present American Standard. Earlier American Standards in part are continued in the new standard. The parts carried over, but not "unified," are clearly identified in Appendix table 12, by being in lighter type. The Unified thread form is shown in Fig. 673. The standard lists six different series of numbers of threads per

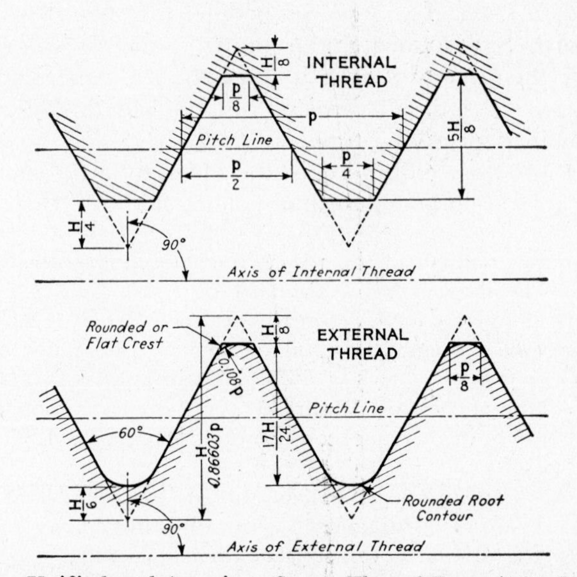

Fig. 673 Unified and American Screw Thread Form (ASA B1.1-1949).

inch for the various standard diameters, and selected combinations of special diameters and pitches.

The six series are the *Coarse Thread Series* (UNC or NC) recommended for general use corresponding to the old National Coarse thread; the *Fine Thread Series* (UNF

*ASA B1. 1935.
†ASA B1.1-1949.

or NF), for general use in automotive and aircraft work and in applications where a finer thread is required; the *Extra-Fine Series* (UNEF or NEF), which is the same as the SAE Extra Fine Series, used particularly in aircraft and aeronautical equipment and generally for threads in thin walls; and the *8-*, *12-*, and *16-Pitch Thread Series* (8UN or 8N, 12UN or 12N, and 16UN or 16N), recommended for the uses corresponding to the old 8-, 12-, and 16-Pitch American National threads. In addition, there are three special thread series: UNS, NS, and UN, which involve special combinations of diameter, pitch, and length of engagement.

409. Unified and American Thread Fits. The tables of the Standard* specify tolerances and allowances defining the several classes of fit (degree of looseness or tightness) between mating threads. In the symbols for fit, the letter A refers to external threads, and B to internal threads. In the new standard, Classes 1A and 1B take the place of old Class 1, and have generous tolerances facilitating rapid assembly and disassembly; Classes 2A and 2B are used in the normal production of screws, bolts, and nuts, as well as a variety of general applications; and Classes 3A and 3B are newly toleranced classes for highly accurate and close-fitting requirements. Class 4 of the old standard has been dropped because of its infrequent and specialized use. Classes 2 and 3, because of their long and widespread use, are continued as American National, but are not unified. They will, in time, be superseded by the Unified classes.

410. Thread Notes. Thread notes for American National threads are shown in Fig. 674. These same symbols are used in correspondence, on shop and store-

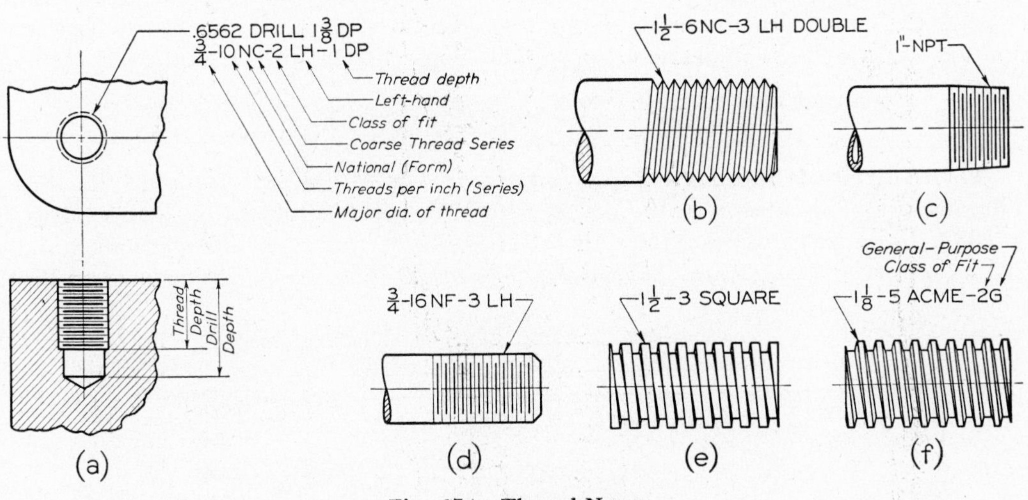

Fig. 674 Thread Notes.

room cards, and in specifications for parts, taps, dies, tools, and gages. A thread note for a blind tapped hole is shown at (a). In a complete note, the tap drill and depth should be given, though in practice they are often omitted and left to the shop. For tap drill sizes, see Appendix table 12. If the LH symbol is omitted, the thread is understood to be RH. If the thread is a multiple thread, the word DOUBLE, TRIPLE, or QUADRUPLE should precede the thread depth; otherwise the thread

*ASA B1.1-1949.

is understood to be single. Thread notes for holes are preferably attached to the circular views of the holes, as shown.

Thread notes for external threads are preferably given in the longitudinal view of the threaded shaft, as shown from (b) to (f). Examples of 8-, 12-, and 16-Pitch threads, not shown in the figure, are 2-8N-2, 2-12N-2, and 2-16N-2. An example of a special thread designation is: $1\frac{1}{2}$-7N-LH.

General-purpose Acme threads are indicated by the letter G, and centralizing Acme threads by the letter C. Typical thread notes are: $1\frac{3}{4}$-4 ACME-2G or $1\frac{3}{4}$-6 ACME-4C.

Thread notes for Unified and American threads are shown in Fig. 675. Unified

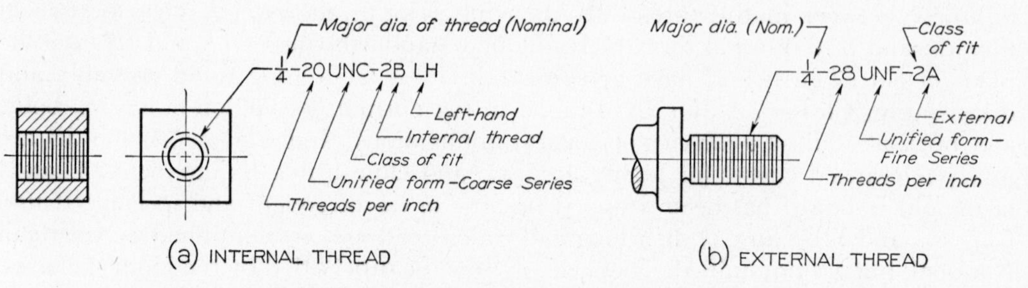

Fig. 675 Unified and American Thread Notes.

threads are distinguished from American National threads by the insertion of the letter U before the series letters, and by the letters A or B (for external or internal, respectively) after the numeral designating the class of fit. If the letters LH are omitted, the thread is understood to be RH. Some typical thread notes are:

$\frac{1}{4}$-20UNC-2A TRIPLE $\qquad$ $1\frac{3}{4}$-16UN-2A
$\frac{9}{16}$-18UNF-2B $\qquad$ $1\frac{5}{8}$-10NS-2 PD 1.5600-1.5532

411. American Standard Pipe Threads. The American Standard for Pipe Threads, originally known as the Briggs Standard, was formulated by Robert Briggs in 1882. Two general types of pipe threads have been approved as American Standard: *Taper Pipe Threads* and *Straight Pipe Threads.** The profile of the Taper Pipe

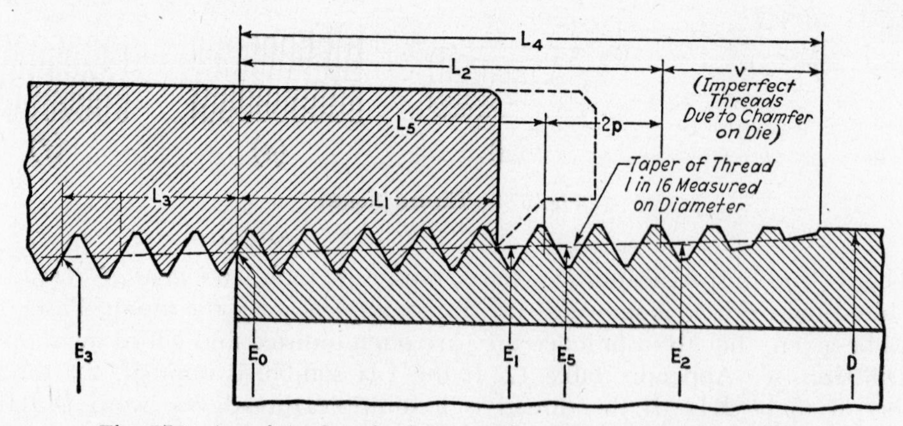

Fig. 676 American Standard Taper Pipe Thread (ASA B2.1-1945).

*ASA B2.1-1945.

thread is illustrated in Fig. 676. See Appendix table 36. The taper of the thread is 1 in 16 or 0.75″ per foot measured on the diameter and along the axis. The angle between the sides of the thread is 60°. The depth of the sharp V is 0.8660p, and the basic maximum depth of the truncated thread is 0.800p, where p = pitch. The basic pitch diameters E_0 and E_1, and the basic length of the effective external taper thread L_2, are determined by the formulas:

$$E_0 = D - (0.050D + 1.1)\frac{1}{n}$$

$$E_1 = E_0 + 0.0625L_1$$

$$L_2 = (0.80D + 6.8)\frac{1}{n}$$

where D = Basic O.D. of pipe

E_0 = Pitch dia. of thd. at end of pipe.
E_1 = Pitch dia. of thd. at large end of internal thd.
L_1 = Normal engagement by hand.
n = No. of threads per in.

The ASA† has also recommended two modified taper pipe threads for: (1) Dryseal Pressure-Tight Joints and (2) Rail Fitting Joints. The former is used to provide a metal-to-metal joint eliminating the need for a sealer, and is used in refrigeration, marine, automotive, aircraft, and ordnance work. The latter is used to provide a rigid mechanical thread joint as required in rail fitting joints.

While Taper Pipe threads are recommended for general use, there are certain types of joints where straight pipe threads are used to advantage. The number of threads per inch, the angle, and the depth of thread are the same as on the Taper Pipe thread, but the threads are cut parallel to the axis. Straight Pipe threads are used for pressure-tight joints for pipe couplings, fuel and oil line fittings, drain plugs, free-fitting mechanical joints for fixtures, loose-fitting mechanical joints for locknuts, and loose-fitting mechanical joints for hose couplings.

Pipe threads are represented by detailed or symbolic methods in a manner similar to the representation of Unified and American threads. The symbolic representation (schematic or simplified) is recommended for general use regardless of diameter, Fig.

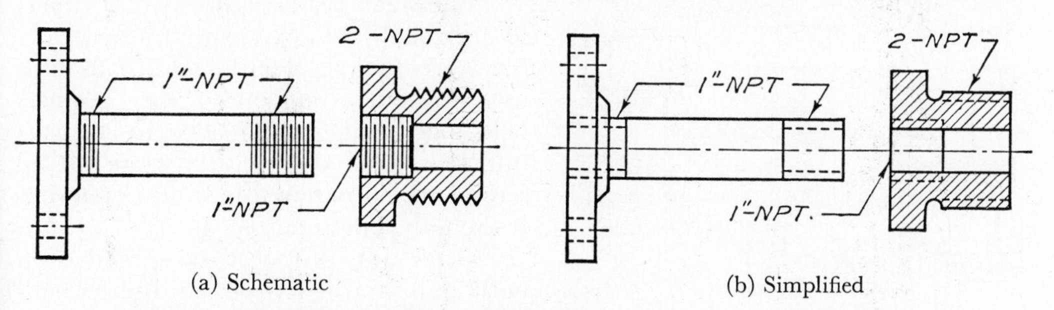

Fig. 677 Conventional Representation of Pipe Threads (ASA Z14.1-1946).

677, the detailed method being approved only when the threads are large and when it is desired to show the profile of the thread as, for example, in a sectional view of an assembly. See Fig. 1112.

As shown in Fig. 677, it is not necessary to draw the taper on the threads unless there is some reason to emphasize it, since the thread note indicates whether the thread is straight or tapered. If it is desired to show the taper, it should be exag-

†*Ibid.*

gerated, as shown in Fig. 678, where the taper is drawn $\frac{1}{16}''$ per 1″ *on radius*, instead of the actual taper of $\frac{1}{16}''$ *on diameter*. American Standard Taper Pipe threads are indicated by a note giving the nominal diameter followed by the letters NPT (National Pipe Taper), as shown in Fig. 677. When Straight Pipe threads are specified, the letters NPS (National Pipe Straight) are used. In practice, the tap drill size is normally not given in the thread note.

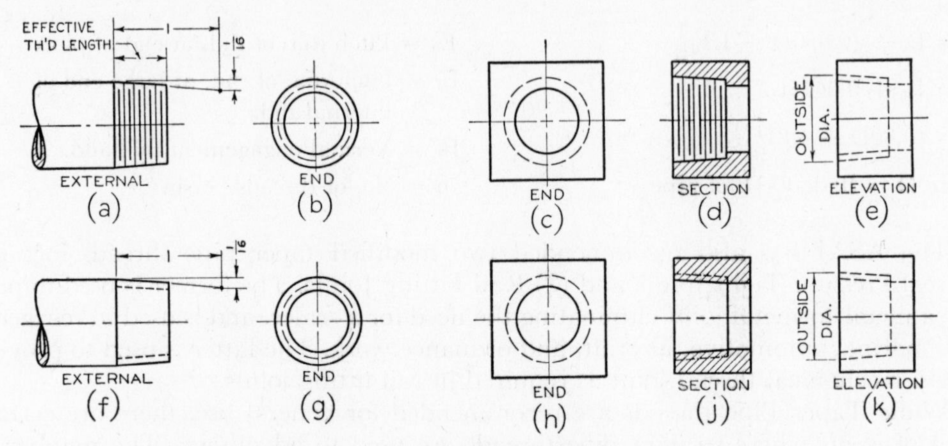

Fig. 678 Conventional Pipe Thread Representation.

For numbers of threads per inch and other data on pipe threads, see Appendix table 36. For a general discussion of piping drawings, see Chapter 24.

412. Bolts, Studs, and Screws. The term *bolt* is generally used to denote a "through bolt" which has a head on one end and is passed through clearance holes in two or more aligned parts and is threaded on the other end to receive a nut to tighten and hold the parts together. See Fig. 679 (a), and §§414 and 415.

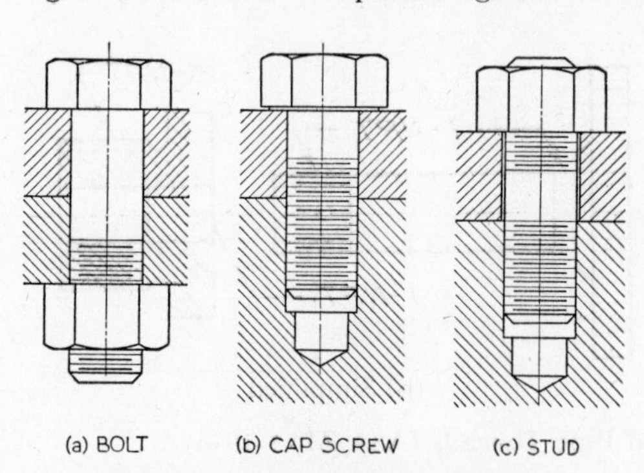

Fig. 679 Bolt, Screw, and Stud.

A hexagon head *cap screw*, (b), is similar to a through bolt, except that it generally has a greater length of thread because it is used without a nut, one of the members held together being threaded to act as a nut. It is screwed on with a wrench. Cap screws are not screwed into thin materials if strength is desired. See §418.

A *stud*, (c), is a steel rod threaded on both ends, and is screwed into place with a pipe wrench or, preferably, with a stud driver. As a rule, a stud is passed through a clearance hole in one member and is screwed into another member, a nut being used on the free end, as shown.

A *machine screw*, Fig. 690, is similar to the slotted-head cap screws, but is in general smaller. It may be used with or without a nut.

A *set screw*, Fig. 691, is a screw with or without a head which is screwed through one member and whose special point is forced against another member to prevent relative motion between the two parts.

It is customary not to section bolts, nuts, screws, and similar parts when drawn in assembly, as shown in Figs. 679 and 689, because they do not themselves require sectioning for clearness. See §441.

413. Tapped Holes. The bottom of a drilled hole is conical in shape, as formed by the point of the twist drill, Fig. 680 (a) and (b). When an ordinary drill is used

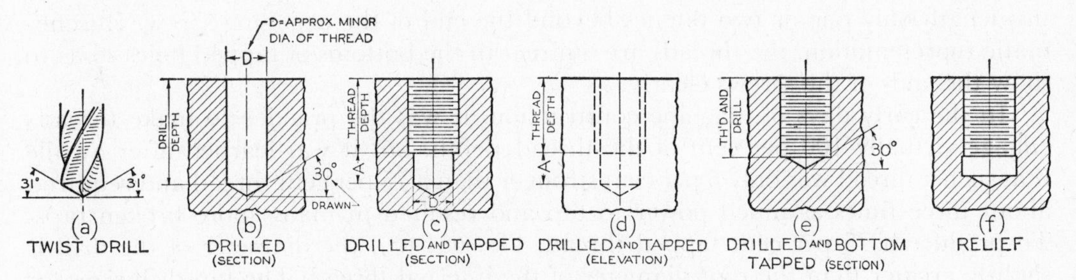

Fig. 680 Drilled and Tapped Holes.

in connection with tapping, it is referred to as a *tap drill.* On drawings, an angle of 30° is used to approximate the actual 31°.

The thread length is the length of full or perfect threads. The tap drill depth is the depth of the cylindrical portion of the hole and does not include the cone point, (b). The portion A of the drill depth shown beyond the threads at (c) and (d) includes the several imperfect threads produced by the chamfered end of the tap. This distance A varies according to drill size and whether a plug tap, Fig. 564 (h), or a bottoming tap is used to finish the hole. For drawing purposes, when the tap drill depth is not specified, the distance A may be drawn equal to three or four schematic thread pitches, Fig. 671.

A tapped hole finished with a bottoming tap is drawn as shown at (e). Blind bottoming holes should be avoided wherever possible. A better procedure is to cut a relief with its diameter slightly greater than the major diameter of the thread, (f).

One of the chief causes of tap breakage is insufficient tap drill depth, in which the tap is forced against a bed of chips in the bottom of the hole. Therefore, the draftsman should never draw a blind hole when a through hole of not much greater length can be used; and when a blind hole is necessary, he should provide generous tap drill depth. Tap drill sizes for Unified and American threads are given in Appendix table 12.

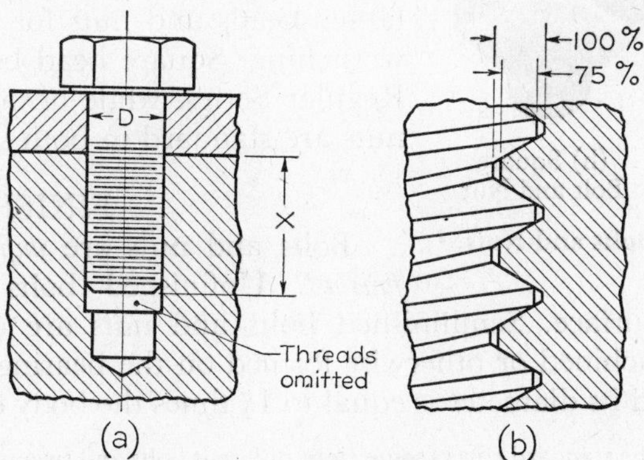

Fig. 681 Tapped Holes.

The thread length in a tapped hole depends upon the major diameter and the material being tapped. In Fig. 681 (a), the minimum engagement length X, when both parts are steel, is equal to the diameter D of the thread. When a steel screw is screwed into cast iron, brass, or bronze, $X = 1\frac{1}{2}D$; and when screwed into aluminum, zinc, or plastic, $X = 2D$.

Since the tapped thread length contains only full threads, it is necessary to make this length only one or two pitches beyond the end of the engaging screw. In schematic representation, the threads are omitted in the bottoms of tapped holes so as to show the ends of the screws clearly.

In the early days of machine construction, it was the practice to make threads engage virtually 100 per cent of the thread depth, but it was learned after awhile that such a thread was only 5 per cent stronger than a 75 per cent thread and required about three times as much power to tap and resulted in many more broken taps. To produce a 75 per cent tapped thread, Fig. 681 (b), the diameter of the drill is slightly greater than the root diameter of the internal thread. The tap drill sizes for Unified and American threads given in Appendix table 12 are for 75 per cent threads. It is good practice to give the tap drill size in the thread note, §410.

When a bolt or a screw is passed through a clearance hole in one member, the hole may be drilled $\frac{1}{32}''$ larger than the screw up to $\frac{3}{8}''$ diameter, and $\frac{1}{16}''$ larger for larger diameters. For more precise work, the clearance hole may be only $\frac{1}{64}''$ larger than the screw up to $\frac{3}{8}''$ diameter, and $\frac{1}{32}''$ larger for larger diameters. Closer fits may be specified for special conditions. The clearance spaces on each side of a screw or bolt need not be shown on a drawing unless it is necessary to show that there is no thread engagement, in which case the clearance spaces are drawn about $\frac{1}{32}''$ wide for clarity.

(a) Hexagon Bolt and Nut

(b) Square Bolt and Nut

Fig. 682 Standard Bolts and Nuts.

414. American Standard Bolts and Nuts. American Standard bolts and nuts* are produced in two forms: square and hexagon, Fig. 682. Square heads and nuts are chamfered at 30°, and hexagon heads and nuts are chamfered at 25°. Both are drawn at 30° for simplicity.

BOLT SERIES

Two *series* are standardized: the *Regular Series* for general use and the *Heavy Series*, which have larger heads and nuts for heavier use or easier wrenching. Square head bolts come only in the Regular Series, while hexagon bolts and square nuts are standard in both series.

FINISH

Bolts and nuts are *unfinished*, *semifinished*, or *finished*. Unfinished bolts and nuts are not machined on any surface. Semifinished bolts and nuts are unfinished except for a "washer face" machined or otherwise formed on the bearing surface. The washer face is $\frac{1}{64}''$ thick, and its diameter is equal to $1\frac{1}{2}$ times the body diameter D. For nuts,

*ASA B18.2-1955. The standard covers some eight different bolts and twenty nuts. For complete details, see the standard.

the bearing surface may be a circular surface produced by chamfering. Finished bolts and nuts have closer tolerances and a more finished appearance, but are not completely machined. Square bolts and nuts are always unfinished, while hexagon bolts and nuts may be unfinished, semifinished, or finished.

PROPORTIONS

Sizes based on diameter D of the bolt body, which are either exact formula proportions or close approximations for drawing purposes are:

Regular Series: $W = 1\frac{1}{2}D$ $H = \frac{2}{3}D$ $T = \frac{7}{8}D$
Heavy Series: $W = 1\frac{1}{2}D + \frac{1}{8}''$ $H = \frac{3}{4}D$ $T = D$
Where W = width across flats, H = head height, and T = nut height.

The washer face is always included in the head or nut height for finished and semifinished hexagon bolt heads and nuts.

THREADS

Unfinished and semifinished bolts have Coarse threads, Class 2A, while finished bolts have Coarse, Fine, or 8-Pitch threads, Class 2A. Unfinished nuts have Coarse threads, Class 2B. Finished and semifinished nuts have Coarse or Fine threads, Class 2B, while certain of these also may have 8-Pitch threads.

THREAD LENGTHS

For bolts up to 6″ in length, thread length $= 2D + \frac{1}{4}''$.
For bolts over 6″ in length, thread length $= 2D + \frac{1}{2}''$.

Bolts too short for these formulas are threaded as close to the head as practicable. For drawing purposes, this may be taken as three thread pitches, approximately. The threaded end of a bolt may be rounded or chamfered, but is usually drawn with a 45° chamfer from the thread depth, Fig. 683.

BOLT LENGTHS

Bolt lengths have not been standardized because of the endless variety required by industry. The following increments (differences in successive lengths) are compiled from manufacturers' catalogs. These increments apply to stock sizes of cut-thread bolts. Long bolts of small diameter or short bolts of large diameter would have to be ordered "special."

Square Head Bolts
Lengths $\frac{1}{2}''$ to $\frac{3}{4}''$ = $\frac{1}{8}''$ increments
Lengths $\frac{3}{4}''$ to 5″ = $\frac{1}{4}''$ increments
Lengths 5″ to 12″ = $\frac{1}{2}''$ increments
Lengths 12″ to 30″ = 1″ increments

Hexagon Head Bolts
Lengths $\frac{3}{4}''$ to 8″ = $\frac{1}{4}''$ increments
Lengths 8″ to 20″ = $\frac{1}{2}''$ increments
Lengths 20″ to 30″ = 1″ increments

For dimensions of standard bolts and nuts, see Appendix table 16.

415. To Draw American Standard Bolts. In practice, standard bolts and nuts are not shown on detail drawings unless they are to be altered, but they appear so frequently on assembly drawings that a suitable but rapid method of drawing them must be used. They may be drawn from exact dimensions taken from tables* if accuracy is important, as in figuring clearances; but in the great majority of cases

*See Appendix table 16.

the conventional representation, in which proportions based upon diameter are used, will be sufficient, and a considerable amount of time may be saved. Three typical

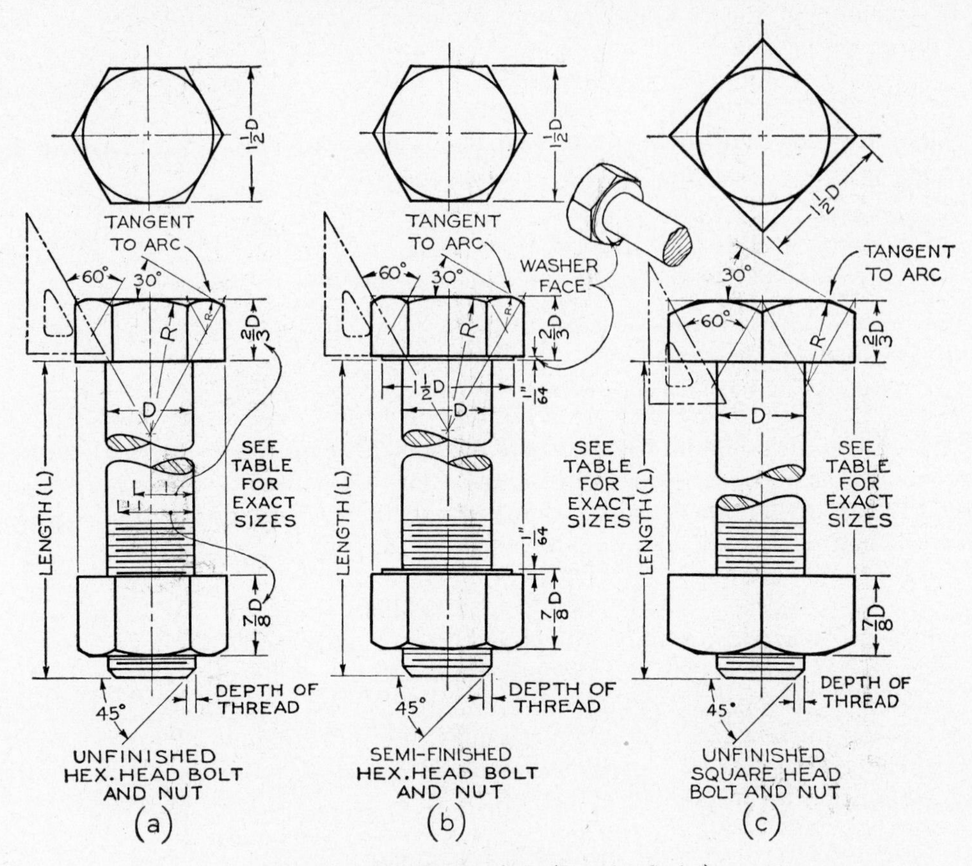

Fig. 683 Bolt Proportions (Regular Series).

bolts illustrating the use of these proportions for the Regular Series bolts are shown in Fig. 683.

As shown in Fig. 955 (c), the curves produced by the chamfer on the bolt heads and nuts are hyperbolas. However, in actual practice these curves are always represented approximately by means of circular arcs, as shown in Fig. 683.

Generally, bolt heads and nuts should be drawn "across corners" in all views, regardless of projection, as shown in Figs. 776-778. This conventional violation of projection is used to prevent confusion between the square and hexagon heads and nuts and to show actual clearances. Only when there is a special reason should bolt heads and nuts be drawn across flats. In such cases, the conventional proportions shown in Fig. 684 are used.

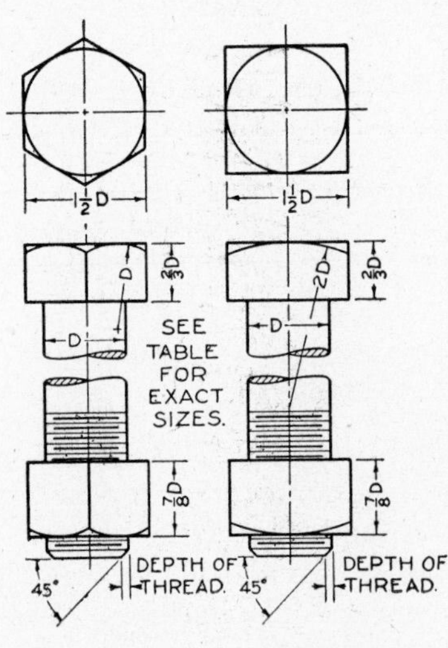

Fig. 684 Bolts "Across Flats."

Steps in drawing hexagon bolts and nuts are illustrated in Fig. 685, and those for square bolts and nuts in Fig. 686. Before starting, the diameter of the bolt, the length from the under side of the bearing surface to the tip, the type of head (square or hexagon), and the series (Regular or Heavy), as well as the type of finish, must be known.

If only the longitudinal view of a bolt is needed, it is necessary to draw only the lower half of the top views in Figs. 685 and 686 *with light construction lines* in order to project the corners of the hexagon or square to the front view. These construction lines may then be erased if desired.

The head and nut heights can be spaced off with the dividers on the shaft diameter and then transferred as shown in both figures, or the scale may be used as in Fig. 167. The heights should not be determined by arithmetic.

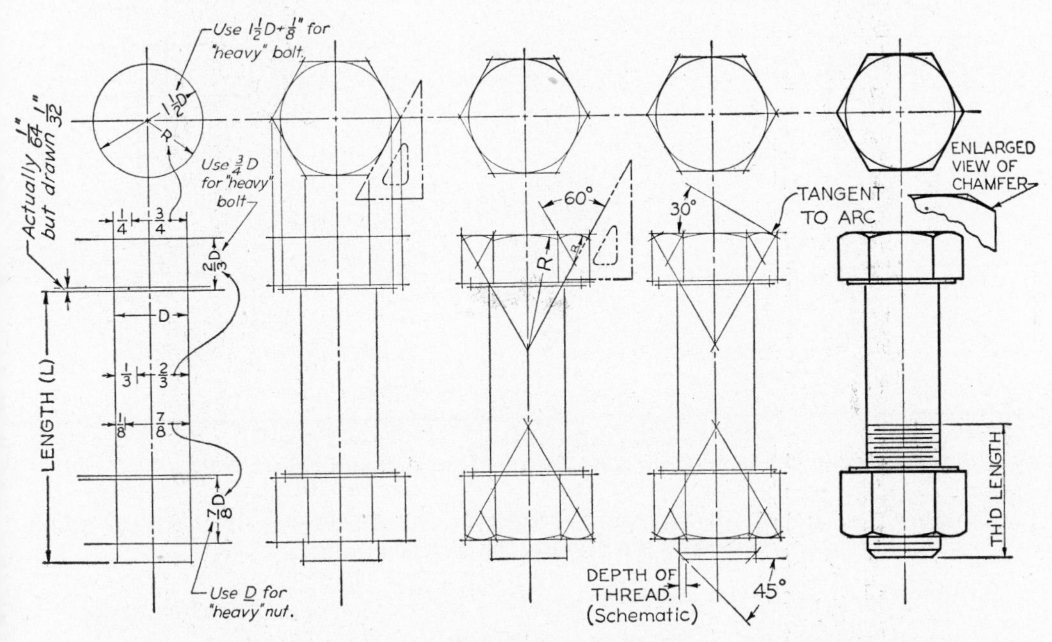

Fig. 685 Steps in Drawing Semifinished Hexagon-Head Bolt and Hexagon Nut.

The $\frac{1}{64}''$ washer face has a diameter equal to the distance across flats of the bolt head or nut. It appears only on the semifinished or finished hexagon bolts or nuts, the $\frac{1}{64}''$ thickness being drawn at $\frac{1}{32}''$ for clearness. The $\frac{1}{32}''$ is included in the head or nut height.

Threads should be drawn schematically for bolt diameters of 1″ or less on the drawing, Fig. 671 (b), and by detailed representation for larger diameters, §§394 and 395. The threaded end of the bolt should be chamfered at 45° from the schematic thread depth, Fig. 671 (a).

On drawings of small bolts or nuts (under $\frac{1}{2}''$ dia. approx.) where the chamfer is hardly noticeable, the chamfer on the head or nut may be omitted in the longitudinal view.

Many styles of templates are available for saving time in drawing bolt heads and nuts. One of these is the *Draftsquare*, Fig. 90 (b).

416. Specifications for Bolts and Nuts. In specifying bolts in parts lists, in correspondence, or elsewhere, the following information must be covered in order:

1. Diameter of bolt shank
2. Thread specification (see §410)
3. Length of bolt
4. Finish of bolt
5. Type of head
6. Name

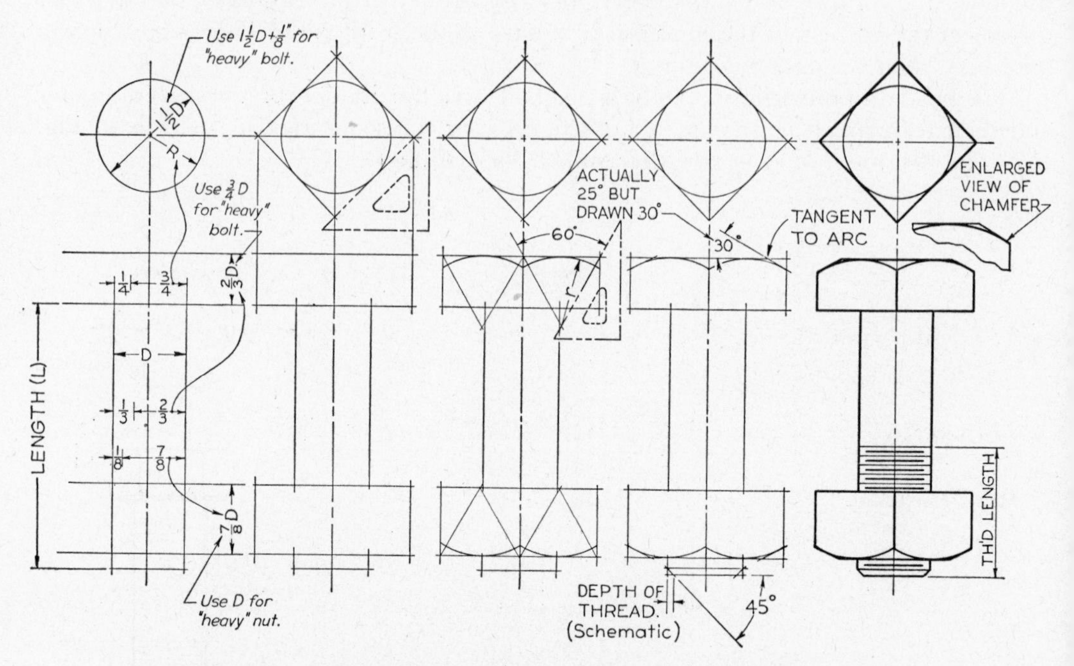

Fig. 686 Steps in Drawing Square-Head Bolt and Square Nut.

Example (Complete): $\frac{3}{4}$-10UNC-2A × $2\frac{1}{2}$ SEMIFINISHED HEXAGON HEAD BOLT
Example (Abbreviated): $\frac{3}{4}$ × $2\frac{1}{2}$ SEMIFIN HEX HD BOLT

Nuts may be specified as follows:

Example (Complete): $\frac{5}{8}$-11UNC-2B SQUARE NUT
Example (Abbreviated): $\frac{5}{8}$ SQ NUT

For either bolts or nuts, the word REGULAR is assumed if left out of the specification. If the Heavy Series is intended, the word HEAVY should appear as the first word in the name of the fastener. Similarly, finish need not be mentioned if the bolt or nut is unfinished.

417. Locknuts and Locking Devices. Many types of special nuts and devices to prevent nuts from unscrewing are available, some of the most common of which are illustrated in Fig. 687. The American Standard Jam Nuts*, (a) and (b), are the same as the Regular or Heavy Hexagon Nuts, except that they are thinner. The application at (b), where the larger nut is on top and is screwed on more tightly, is recommended. They are the same distance across flats as the corresponding Regular or Heavy hexagon nuts ($1\frac{1}{2}$D or $1\frac{1}{2}$D + $\frac{1}{8}$"). They are slightly over $\frac{1}{2}$D in thickness, but are drawn $\frac{1}{2}$D for simplicity. They are available finished, unfinished, and semi-

*ASA B18.2-1955.

finished in the Regular and Heavy series. The tops of all are flat and chamfered at 30°, and the semifinished and finished forms have either a washer face or a chamfered bearing surface.

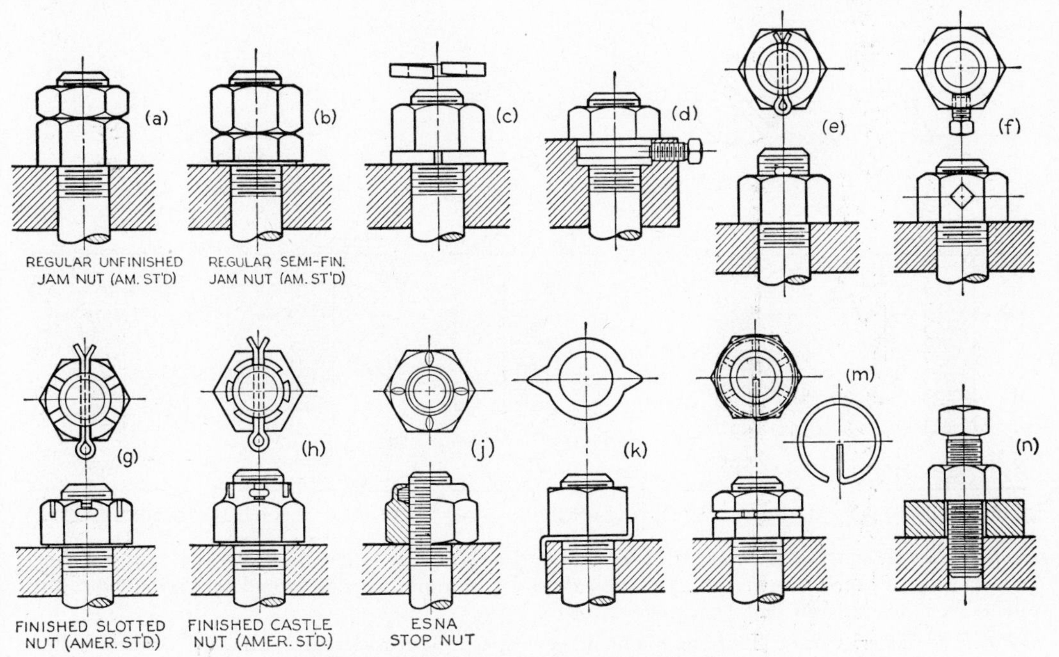

Fig. 687 Locknuts and Locking Devices.

The lock washer shown at (c), and the cotter pin, (e), (g), and (h) are very common. See Appendix tables 25 and 28. The set screw, (f), is often made to press against a plug of softer material, such as brass, which in turn presses against the threads without deforming them.

For use with cotter pins (see Appendix table 28), the ASA* recommends a Finished Slotted Nut, (g), and a Finished Castle Nut, (h), as well as Regular and Heavy Semifinished Slotted Nuts and a Finished Thick Slotted Nut.

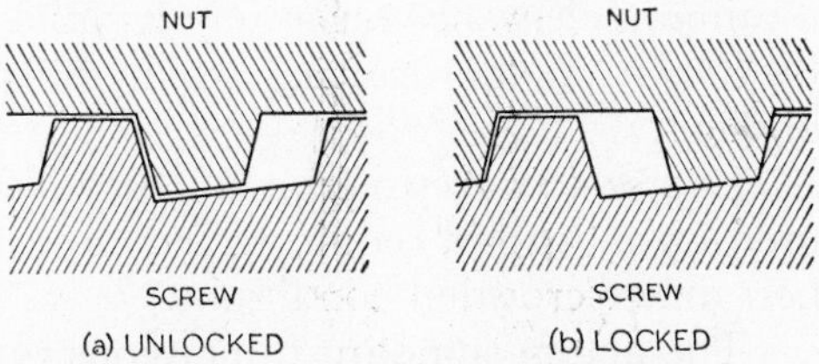

The Dardelet thread is self-locking, and is illustrated in Fig. 688. The form of the thread is similar to that of the Acme, but the crest of the thread in the nut and the root of the thread on the screw are tapered at about 6° with the axis of the screw, so that when the nut is tightened the crest of the thread of the nut is wedged against the root of the thread on the screw, which locks the thread firmly.

NUT NUT
SCREW SCREW
(a) UNLOCKED (b) LOCKED

Fig. 688 Dardelet Self-Locking Thread.

418. American Standard Cap Screws.† The five types of American Standard Cap Screws are shown in Fig. 689. The first four have these standard heads, while

**Ibid.*
†ASA B18.6.2-1956 and ASA B18.3-1954.

the Socket Head Cap Screws, (e), have several different shapes of round heads and sockets. Cap screws are regularly produced in finished form and are used on machine tools and other machines, for which accuracy and appearance are important. The ranges of sizes and exact dimensions are given in Appendix tables 16 and 17.

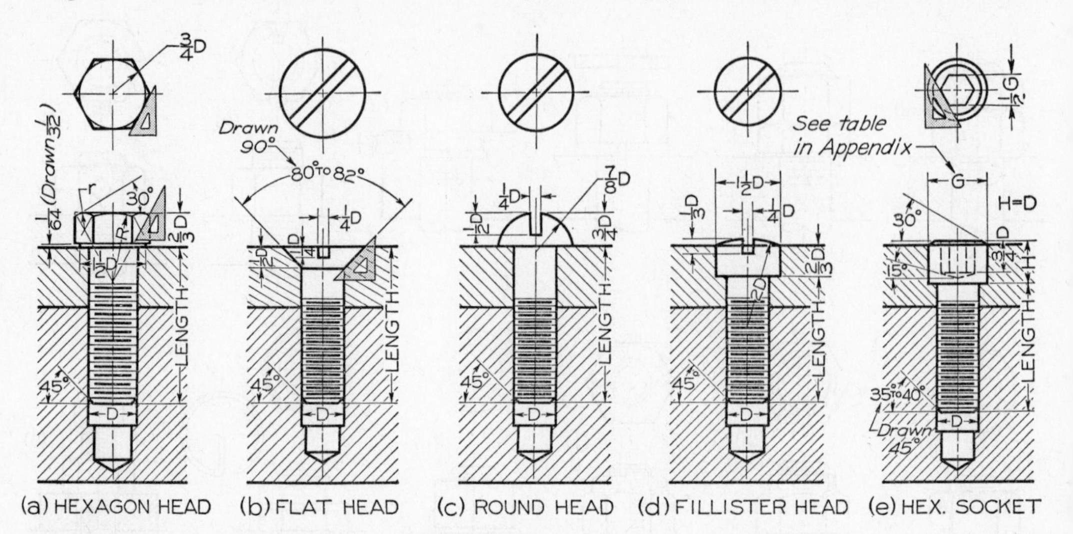

Hexagon Head Screws: Coarse, Fine, or 8-Thread Series, 2A. Thread length = 2D + $\frac{1}{4}''$ up to 6″ long, and 2D + $\frac{1}{2}''$ if over 6″ long. For screws too short for formula, threads extend to approximate three pitches from head. Screw lengths not standardized.

Slotted Head Screws: Coarse, Fine, or 8-Thread Series, 2A. Thread length = 2D + $\frac{1}{4}''$. Screw lengths not standardized.

Hexagon Socket Screws: Coarse or Fine threads, 3A. Coarse thd. length = 2D + $\frac{1}{2}''$ where this would be over $\frac{1}{2}$L; otherwise thd. length = $\frac{1}{2}$L. Fine thd. length = $1\frac{1}{2}$D + $\frac{1}{2}''$ where this would be over $\frac{3}{8}$L; otherwise thd. length = $\frac{3}{8}$L. Increments in screw lengths = $\frac{1}{8}''$ for screws $\frac{1}{8}''$ to 1″ long, $\frac{1}{4}''$ for screws 1″ to 3″ long, and $\frac{1}{2}''$ for screws $3\frac{1}{2}''$ to 6″ long.

Fig. 689 American Standard Cap Screws (See Appendix tables 16 and 17).

Cap screws ordinarily pass through a clearance hole in one member and screw into another. The hole is drilled slightly larger than the screw in order to permit minimum clearance for assembly, as explained in §413. The clearance hole need not be shown on the drawing when the presence of the unthreaded clearance hole is obvious.

Cap screws are inferior to studs if frequent removal is necessary; hence they are used on machines requiring few adjustments. The slotted or socket-type heads are best under crowded conditions.

The actual standard dimensions may be used in drawing the cap screws whenever exact sizes are necessary, but this is seldom the case. In Fig. 689 the dimensions are given in terms of body diameter D, and they closely conform to the actual dimensions. The resulting drawings are almost exact reproductions and are easy to draw. The Hexagonal Head Cap Screw is drawn in the same manner as the semi-finished bolt shown in Fig. 685. The points are drawn chamfered at 45° from the schematic thread depth.

For correct representation of tapped holes, see §413. For information on drilled, countersunk, or counterbored holes, see §§238 and 316.

In an assembly section, it is customary not to section screws, bolts, shafts, or other solid parts whose center lines lie in the cutting plane. Such parts in themselves do not require sectioning, and are therefore shown "in the round," Fig. 689 and §441.

Note that screwdriver slots are drawn at 45° in the circular views of the heads, without regard to true projection, and that threads in the bottom of the tapped holes are omitted so that the ends of the screws may be clearly seen. A typical cap screw note is as follows:

Example (Complete): $\frac{3}{8}$-16UNC-2A × $2\frac{1}{2}$ HEXAGON HEAD CAP SCREW
Example (Abbreviated): $\frac{3}{8}$ × $2\frac{1}{2}$ HEX HD CAP SCR

419. American Standard Machine Screws. Machine screws are similar to cap screws but are in general smaller (0.060 to 0.750 dia.). The ASA* has approved eight forms of heads, as shown in Appendix table 18. The hexagon head may be slotted if desired. All others are available in either slotted or recessed-head forms. American

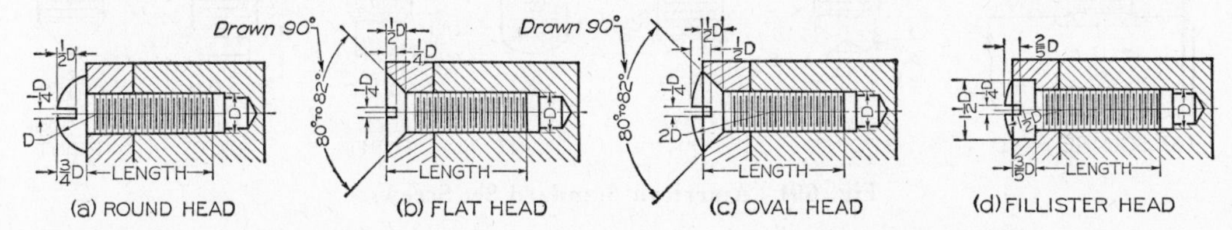

Threads: National Coarse or Fine, Class 2 fit. On screws 2″ long or less, threads extend to within 2 thds. of head; on longer screws thd. length = $1\frac{3}{4}$″. Screw lengths not standardized.

Fig. 690 American Standard Machine Screws. (See Appendix table 18.)

Standard Machine Screws are regularly produced with a naturally bright finish, not heat-treated, and are regularly supplied with plain-sheared ends, not chamfered.

Machine screws are particularly adapted to screwing into thin materials, and all the smaller-numbered screws are threaded nearly to the head. They are used extensively in firearms, jigs, fixtures, and dies. Machine screw nuts are used mainly on the round head and flat head types, and are hexagonal in form.

Exact dimensions of machine screws are given in Appendix 18, but they are seldom needed for drawing purposes. The four most common types of machine screws are shown in Fig. 690, where proportions based on diameter D conform closely to the actual dimensions and produce almost exact drawings. Clearance holes and counterbores should be made slightly larger than the screws, as explained in §413.

Note that the threads in the bottom of the tapped holes are omitted so that the ends of the screws will be clearly seen. Observe also that it is conventional practice to draw the screwdriver slots at 45° in the circular view without regard to true projection.

A typical machine screw note is as follows:

Example (Complete): No. 10 (.1900)-32NF-3 × $\frac{5}{8}$ FILLISTER HEAD MACHINE SCREW
Example (Abbreviated): No. 10 (.1900) × $\frac{5}{8}$ FILL HD MACH SCR

420. American Standard Set Screws. The function of set screws, Fig. 691 (a), is to prevent relative motion, usually rotary, between two parts, such as the move-

*ASA B18.6-1947.

ment of the hub of a pulley on a shaft. A set screw is screwed into one part so that its point bears firmly against another part. If the point of the set screw is cupped, (e), or if a flat is milled on the shaft, (a), the screw will hold much more firmly. Obviously

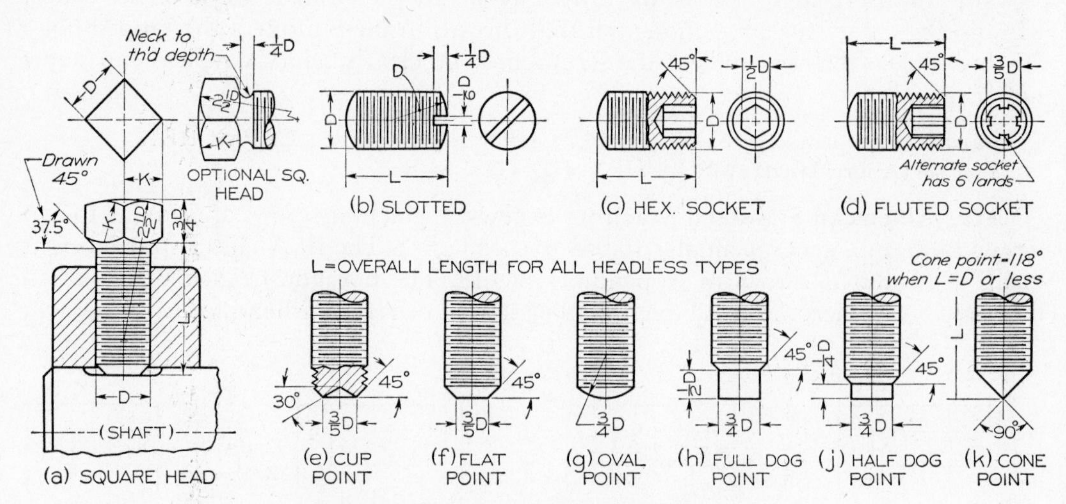

Fig. 691 American Standard Set Screws.

set screws are not efficient when the load is heavy or is suddenly applied. Usually they are manufactured of steel, and case hardened.

The American Standard Square Head Set Screw* is shown in Fig. 691 (a), and the American Standard Slotted Headless Set Screw* at (b). Two American Standard Socket Set Screws† are illustrated at (c) and (d). American Standard set screw points are shown from (e) to (k). The headless set screws have come into greater use because the projecting head of headed set screws has caused many industrial casualties; this has resulted in legislation prohibiting their use in many states.

Most of the dimensions in Fig. 691 are American Standard formula dimensions, and the resulting drawings are almost exact representations.

Square head set screws have Coarse, Fine, or 8-Pitch threads, Class 2A, but are usually furnished with Coarse threads, since the square head set screw is generally used on the rougher grades of work. Slotted headless and socket set screws have Coarse or Fine threads, Class 3A.

Nominal diameters of socket set screws are No's. 0 to 10, 12, $\frac{1}{4}$, $\frac{5}{16}$, $\frac{3}{8}$, $\frac{7}{16}$, $\frac{1}{2}$, $\frac{9}{16}$, $\frac{5}{8}$, $\frac{3}{4}$, $\frac{7}{8}$, 1, $1\frac{1}{8}$, $1\frac{1}{4}$, $1\frac{3}{8}$, $1\frac{1}{2}$, $1\frac{3}{4}$, and 2. Square head set screws are No. 10 to $1\frac{1}{2}$ only, while slotted headless set screws are No. 5 to $\frac{3}{4}$ only, of this series of diameters.

Socket set screw lengths are standardized† as follows:

Lengths $\frac{1}{8}''$ to $\frac{5}{8}''$, increments = $\frac{1}{16}''$.
Lengths $\frac{5}{8}''$ to 1″, increments = $\frac{1}{8}''$.
Lengths 1″ to 4″, increments = $\frac{1}{4}''$.
Lengths 4″ to 6″, increments = $\frac{1}{2}''$.

Square head set screw lengths are not standardized, but manufacturers list increments as follows:

Lengths $\frac{3}{8}''$ to 1″, increments = $\frac{1}{8}''$.
Lengths 1″ to 4″, increments = $\frac{1}{4}''$.
Lengths 4″ to 5″, increments = $\frac{1}{2}''$.
Lengths 5″ and 6″, increment = 1″.

*ASA B18.6.2-1956.
†ASA B18.3-1954.

Slotted headless set screw lengths are not standardized, but manufacturers list increments as follows:

Lengths $\frac{1}{8}''$ to $\frac{3}{8}''$, increments = $\frac{1}{16}''$.
Lengths $\frac{3}{8}''$ to $1''$, increments = $\frac{1}{8}''$.
Lengths $1''$ to $2''$, increments = $\frac{1}{4}''$.

Set screws are specified as follows:

Example (Complete): $\frac{3}{8}$-16UNC-2A × $\frac{3}{4}$ SQUARE HEAD FLAT POINT SET SCREW
Example (Abbreviated): $\frac{3}{8}$ × $1\frac{1}{4}$ SQ HD FLAT PT SET SCR
Example (Abbreviated): $\frac{7}{16}$ × $\frac{3}{4}$ HEX SOCK CUP PT SET SCR

421. American Standard Wood Screws. Wood screws with three types of head have been standardized,* Fig. 692. The dimensions shown closely approximate the actual dimensions and are more than sufficiently accurate for use on drawings.

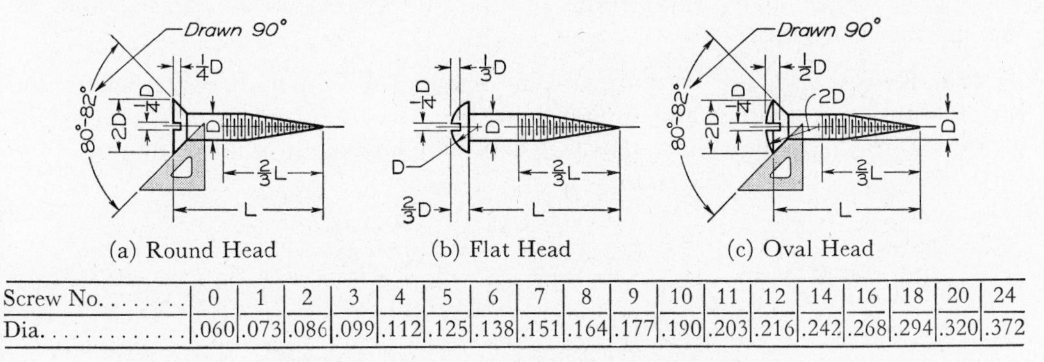

Screw No.	0	1	2	3	4	5	6	7	8	9	10	11	12	14	16	18	20	24
Dia.	.060	.073	.086	.099	.112	.125	.138	.151	.164	.177	.190	.203	.216	.242	.268	.294	.320	.372

Fig. 692 American Standard Wood Screws.

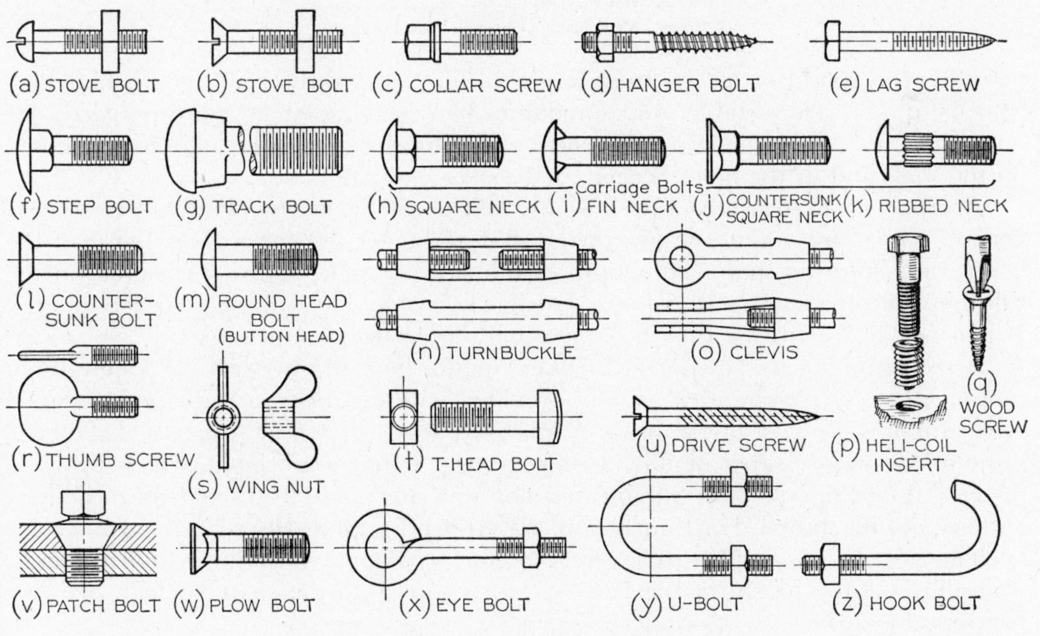

Fig. 693 Miscellaneous Bolts and Screws.

*ASA B18.6.1-1956.

Instead of the screwdriver slot, the Phillips recessed head is becoming more popular. Two styles of cross recesses have been standardized by the ASA.* Many examples may be seen on the automobile. A special screwdriver is used, as shown in Fig. 693 (q), and results in rapid assembly without damage to the head.

422. Miscellaneous Fasteners. Many other types of fasteners have been devised for specialized uses. Some of the more common types are shown in Fig. 693. A number of these are American Standard Round Head Bolts,† including Carriage, Button Head, Step, and Countersunk Bolts.

Aero-Thread Inserts, or Heli-Coil Inserts, as shown at (p), are shaped like a spring except that the cross section of the wire conforms to threads in the screw and in the hole. These are made of phosphor bronze or stainless steel, and they provide a hard, smooth protective lining for tapped threads in soft metals and in plastics. These inserts have many applications in aircraft engines and accessories, and are coming into wider use.

423. Keys. *Keys* are used to prevent relative movement between shafts and wheels, couplings, cranks, and similar machine parts attached to or supported by shafts, Fig. 694. For light duty, that is, when the tendency for relative motion is not

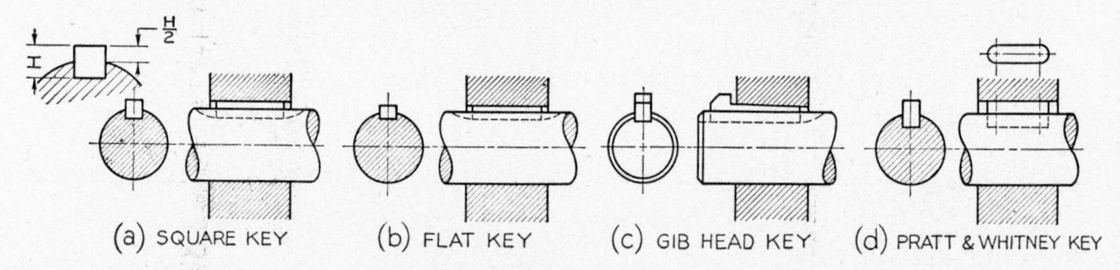

Fig. 694 Square and Flat Keys.

very great, a round or *pin key* may be used. For heavy duty, only rectangular keys (flat or square) are suitable, and sometimes two rectangular keys are necessary for one connection. For even stronger connections, interlocking *splines* may be machined on the shaft and in the hole. See Fig. 640.

A *square key* is shown in Fig. 694 (a), and a *flat key* at (b). The widths of keys generally used are about one-fourth the shaft diameter. In either case, one-half the key is sunk into the shaft. The depth of the keyway or keyseat is measured on the side—not in the center, (a). Square and flat keys may have the top surface tapered $\frac{1}{8}''$ per foot, in which case they become square taper or flat taper keys.

A rectangular key that prevents rotary motion but permits relative longitudinal motion is a *feather key*, and is usually provided with *gib heads*, or otherwise fastened so it cannot slip out of the keyway. A gib head key is shown at (c). It is exactly the same as the square taper or flat taper key, except that a gib head, which provides for easy removal, is added. Square and flat keys are made from cold-finished stock and are not machined. For dimensions, see Appendix table 19.

The *Pratt & Whitney Key*, (d), is rectangular in shape, with semi-cylindrical ends. Two-thirds of the height of the P & W key is sunk into the shaft keyseat. See Appendix table 23.

Ibid.
†ASA B18.5-1952.

The American Standard* *Woodruff Key* is semicircular in shape, Fig. 695. The key fits into a semicircular key slot cut with a Woodruff cutter, as shown, and the top of the key fits into a plain rectangular keyway. Sizes of keys for given shaft diameters are not standardized, but for average conditions it will be found satisfactory to

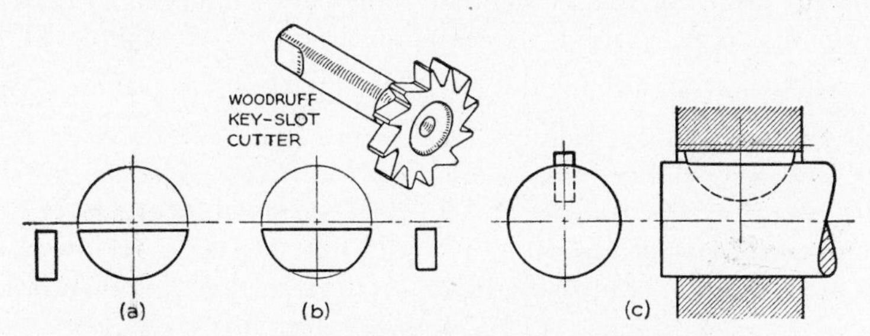

Fig. 695 Woodruff Keys and Key-Slot Cutter.

select a key whose diameter is approximately equal to the shaft diameter. For dimensions, see Appendix table 21.

A *keyseat* is in a shaft; a *keyway* is in the hub or surrounding part.

Typical specifications for keys are:

Example: $\frac{1}{4}$" × $1\frac{1}{2}$" SQ KEY
Example: No. 204 WOODRUFF KEY
Example: $\frac{1}{4}$" × $\frac{3}{16}$" × $1\frac{1}{2}$" FLAT KEY
Example: No. 10 P & W KEY

Notes for nominal specifications of keyways and keyseats are shown in Fig. 616 (o), (p), (r) and (x). For production work, keyways and keyseats should be dimensioned as shown in Fig. 620.

424. Machine Pins. American Standard Machine Pins† include *taper pins*, *straight pins*, *dowel pins*, *clevis pins*, and *cotter pins*. For light work, the taper pin is effective for fastening hubs or collars to shafts, as shown in Fig. 696, in which the hole through the collar and shaft is drilled and reamed when the parts are assembled. For slightly heavier duty, the taper pin may be used parallel to the shaft as for square keys, as shown in Fig. 694 (a). See Appendix table 27.

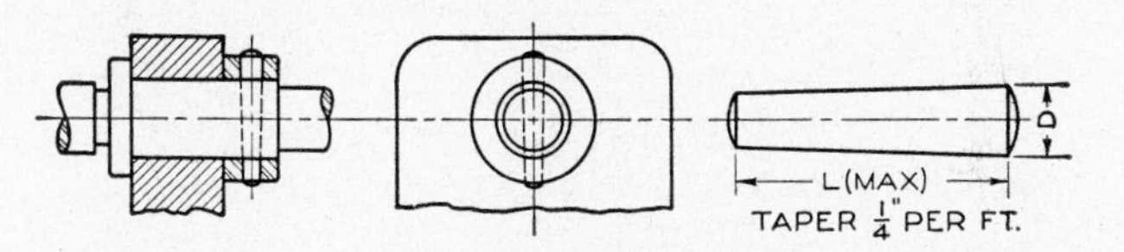

Fig. 696 Taper Pin.

Dowel pins are cylindrical (American Std.) or conical in shape, and are used for a variety of purposes, chief of which is to keep two parts in a fixed position or to preserve alignment. The taper dowel pin is most commonly used, and is recommended where accurate alignment is essential. Dowel pins are usually made of machinery steel and are hardened and ground in a centerless grinder.

The clevis pin is used in a clevis and is held in place by a cotter pin. For the latter, see Appendix table 28.

*ASA B17f-1930 (R-1955).
†ASA B5.20-1954.

425. Rivets. *Rivets* are regarded as permanent fastenings as distinguished from removable fastenings, such as bolts and screws. They are generally used to hold sheet metal or rolled steel shapes together, and are made of wrought iron, soft steel, or copper, or occasionally other metals.

To fasten two pieces of metal together, holes are punched, drilled, or punched and then reamed, slightly larger in diameter than the shank of the rivet. Rivet diameters in practice are made from $d = 1.2\sqrt{t}$ to $d = 1.4\sqrt{t}$, where d is the rivet diameter and t is the metal thickness. The larger size is used for steel and single-riveted joints, and the smaller may be used for multiple-riveted joints. In structural work it is common practice to make the hole $\frac{1}{16}''$ larger than the rivet.

When the red-hot rivet is inserted, a "dolly bar," having a depression the shape of the driven head, is held against the head. A riveting machine is then used to drive

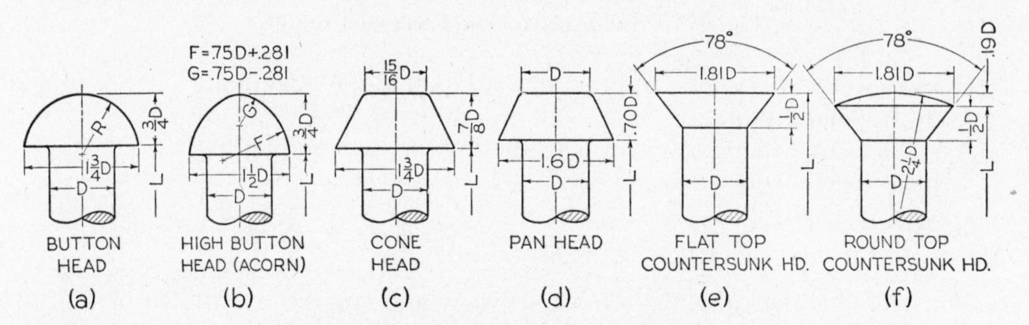

Fig. 697 American Standard Large Rivets.

the rivet and to form the head on the plain end. This action causes the rivet to swell and fill the hole tightly. See §638.

American Standard Large Rivets are used in structural work of bridges, buildings, and in ship and boiler construction, and are shown in their exact formular proportions in Fig. 697. The button head and countersunk head types, (a) and (e), are the rivets

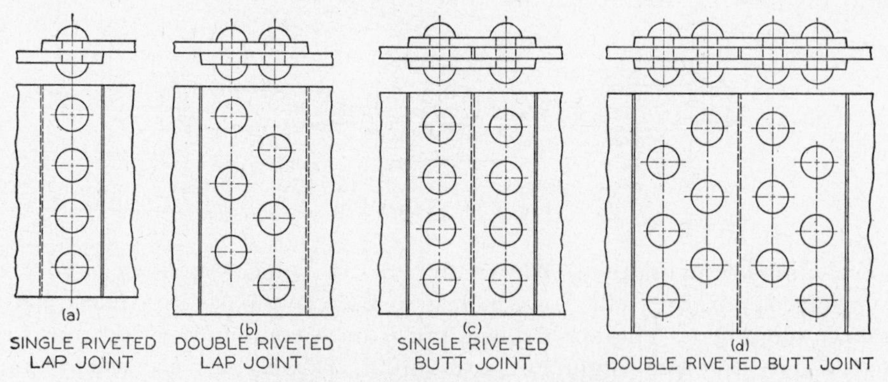

Fig. 698 Common Riveted Joints.

most commonly used in structural work. The button head and cone head are commonly used in tank and boiler construction.

Typical riveted joints are illustrated in Fig. 698. Notice that the longitudinal view of each rivet shows the shank of the rivet with both heads made with circular arcs, and the circular view of each rivet is represented by only the outside circle of the head. In structural drafting, where there may be many such circles to draw, the drop spring bow, Fig. 65, is a popular instrument.

Since many engineering structures are too large to be built in the shop, they are built in the largest units possible and then transported to the desired location. Trusses are common examples of this. The rivets driven in the shop are called *shop rivets*, and those driven on the job are called *field rivets*. Solid black circles are used to represent field rivets, and other standard symbols are used to show other features, as shown in Fig. 699.

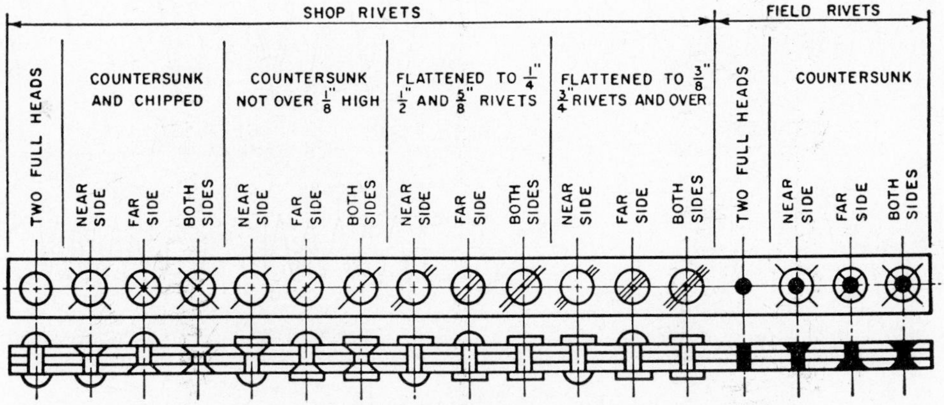

Fig. 699 Conventional Rivet Symbols (ASA Y14.14-Tentative).

For light work, small rivets are used. American Standard Small Solid Rivets are illustrated with dimensions showing their standard proportions in Fig. 700. Included in the same Standard* are Tinners', Coopers', and Belt Rivets.

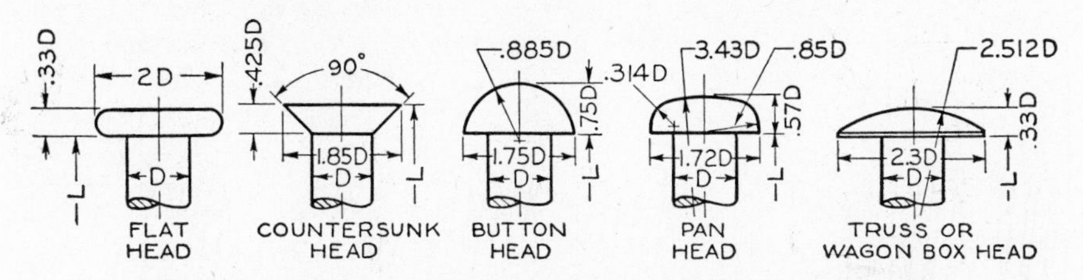

Fig. 700 American Standard Small Solid Rivet Proportions.

426. Springs. "A *spring* is a mechanical devise designed to store energy when deflected and to return the equivalent amount of energy when released."† Springs are classified as *helical springs*, Fig. 701, or *flat springs*, Fig. 706. Helical springs may be cylindrical or conical, but are usually the former.

There are three types of helical springs: *compression springs*, which offer resistance to a compressive force, Fig. 701 (a) to (e), *extension springs*, which offer resistance to a pulling force, Fig. 704, and *torsion springs*, which offer resistance to a torque load or twisting force, Fig. 705.

*ASA B18.1-1955.
†ASA Y14.13 (*Proposed*).

On working drawings, true projections of helical springs are never drawn because of the labor involved. As in the drawing of screw threads, the detailed and schematic methods, employing straight lines in place of helical curves, are used as shown in Fig. 701.

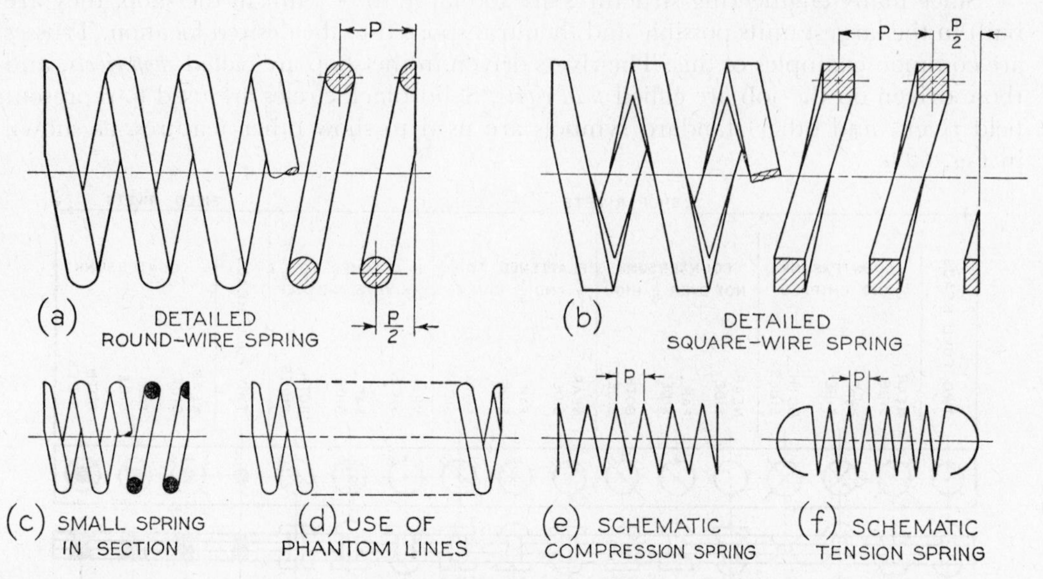

Fig. 701 Helical Springs.

The elevation view of the square-wire spring is similar to the square thread with the core of the shaft removed, Fig. 663. Standard section lining is used if the areas in section are large, as shown in Fig. 701 (a) and (b). If these areas are small, the sectioned areas may be made solid black, (c). In cases where a complete picture of

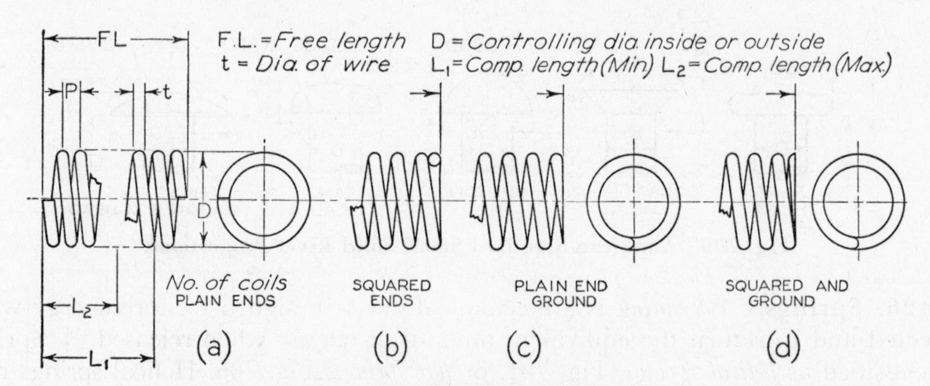

Fig. 702 Compression Springs.

the spring is not necessary, phantom lines may be used to save time in drawing the coils, (d). If the drawing of the spring is too small to be represented by the outlines of the wire, it may be drawn by the schematic method, in which single lines are used, (e) and (f).

Compression springs have *plain ends*, Fig. 702 (a), or *squared* (*closed*) *ends*, (b). The ends may be *ground* as at (c), or both *squared and ground* as at (d). Required dimensions are indicated in the figure.

In a detail drawing of a compression spring, the coils are not drawn, Fig. 703. The spring is symbolically shown by a rectangle and diagonals, and necessary specifications are included as dimensions and notes. Either the I.D. or the O.D. is given, depending upon whether the spring works on a rod or in a hole.

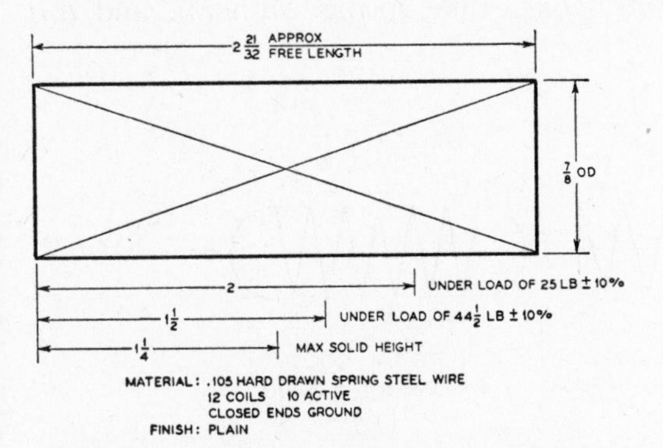

Fig. 703 Compression Spring Drawing (ASA Y14.13—Proposed).

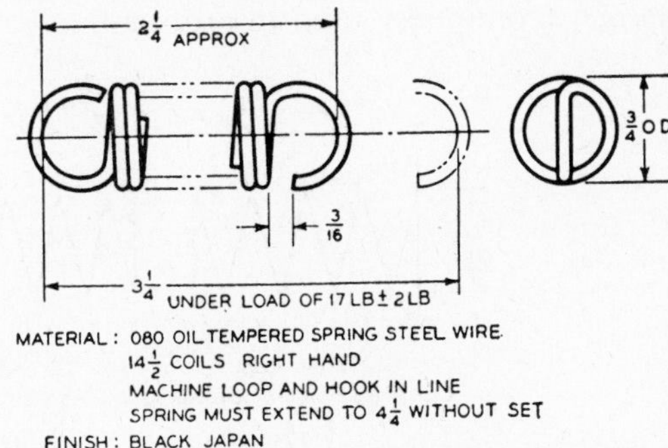

Fig. 704 Extension Spring Drawing (ASA Y14.13—Proposed).

Many companies use standard printed spring drawings with a printed form to be filled in by the draftsman, providing the necessary information as indicated in Fig. 703, plus load at a specified deflected length, the load rate, finish, type of service, and other data.

An extension spring may have any one of many types of ends, and it is therefore necessary to draw the spring or at least the ends and a few adjacent coils, Fig. 704.

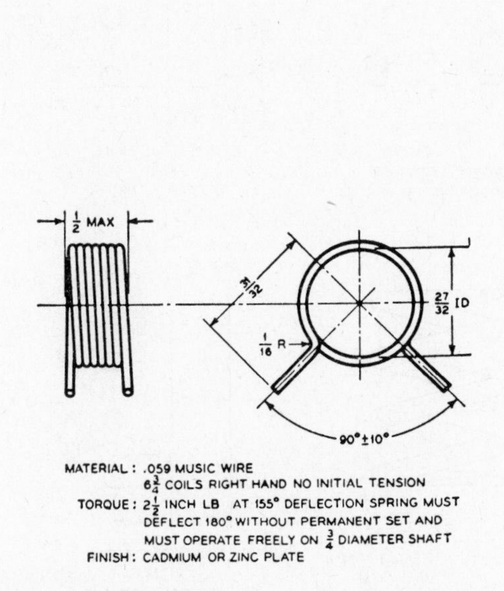

Fig. 705 Torsion Spring Drawing (ASA Y14.13—Proposed).

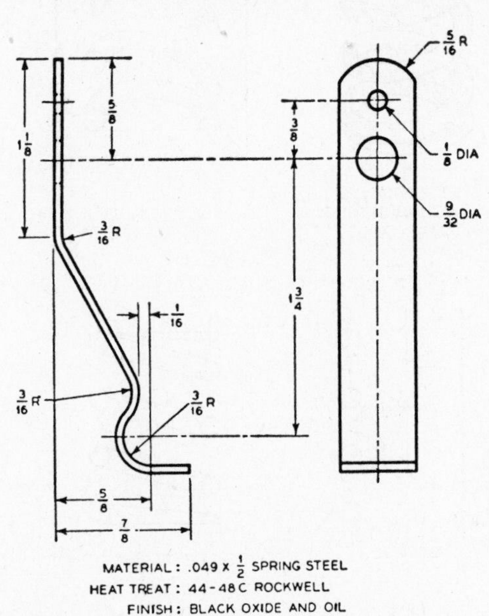

Fig. 706 Flat Spring (ASA Y14.13—Proposed).

Note the use of phantom lines to show the continuity of coils. Printed forms are used when a given form of spring is produced with differences in verbal specification only.

A typical torsion spring drawing is shown in Fig. 705. Here also printed forms are used when there is sufficient uniformity in product to permit a common representation.

A typical flat spring drawing is shown in Fig. 706. Other types of flat springs are *power springs* (or flat coil springs), *Belleville springs* (like spring washers), and *leaf springs* (commonly used in automobiles).

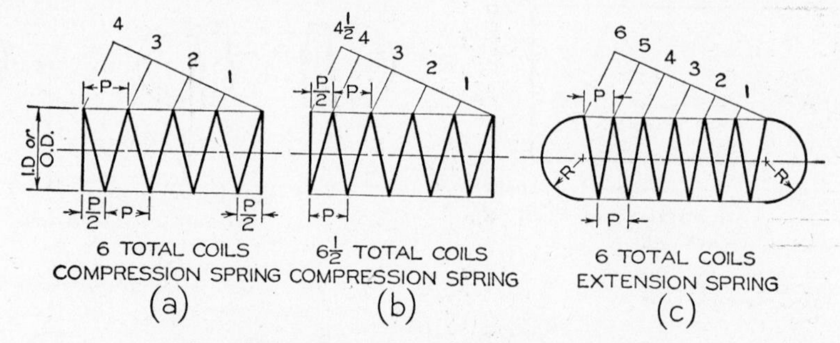

Fig. 707 Schematic Spring Representation.

427. To Draw Helical Springs.

The construction for a schematic elevation view of a compression spring having six total coils is shown in Fig. 707 (a). Since

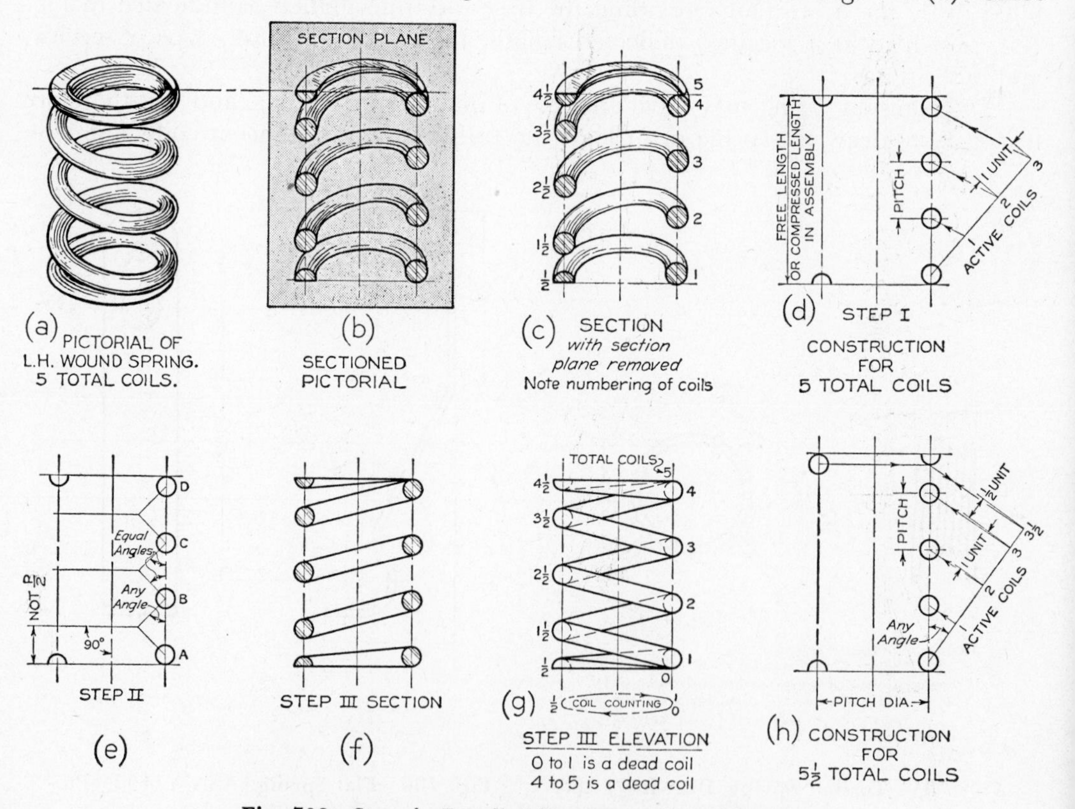

Fig. 708 Steps in Detailed Representation of Spring.

the ends are closed, or squared, two of the 6 coils are "dead" coils, leaving only four full pitches to be set off along the top of the spring, as shown.

If there are 6½ total coils, as at (b), the $\frac{P}{2}$ spacings will be on opposite sides of the spring. The construction of an extension spring with 6 active coils and loop ends is shown at (c).

In Fig. 708 are shown the steps in drawing a sectional view and an elevation view of a compression spring by detailed representation. The given spring is shown pictorially at (a). At (b) a cutting plane has passed through the center line of the spring, and the front portion of the spring has been removed. At (c) the cutting plane has been removed. Steps in constructing the spring through several stages to obtain the sectional view are shown at (d) to (f). The corresponding elevation view is shown at (g).

If there is a fractional number of coils, such as 5½ coils at (h), note that the half-rounds of sectional wire are placed on opposite sides of the spring.

428. Thread and Fastener Problems. It is expected that the student will make use of the information in this chapter in connection with working drawings at the end of the next chapter, where many different kinds of threads and fasteners are required. However, several problems are included here for specific assignment in this area. All are to be drawn on tracing paper or detail paper, size 11″ × 17″.

Thread and fastener problems in convenient form for solution may be found in *Technical Drawing Problems*, by Giesecke, Mitchell, and Spencer, and in *Technical Drawing Problems, Series 2*, by Spencer and Grant, both designed to accompany this text, and published by The Macmillan Company.

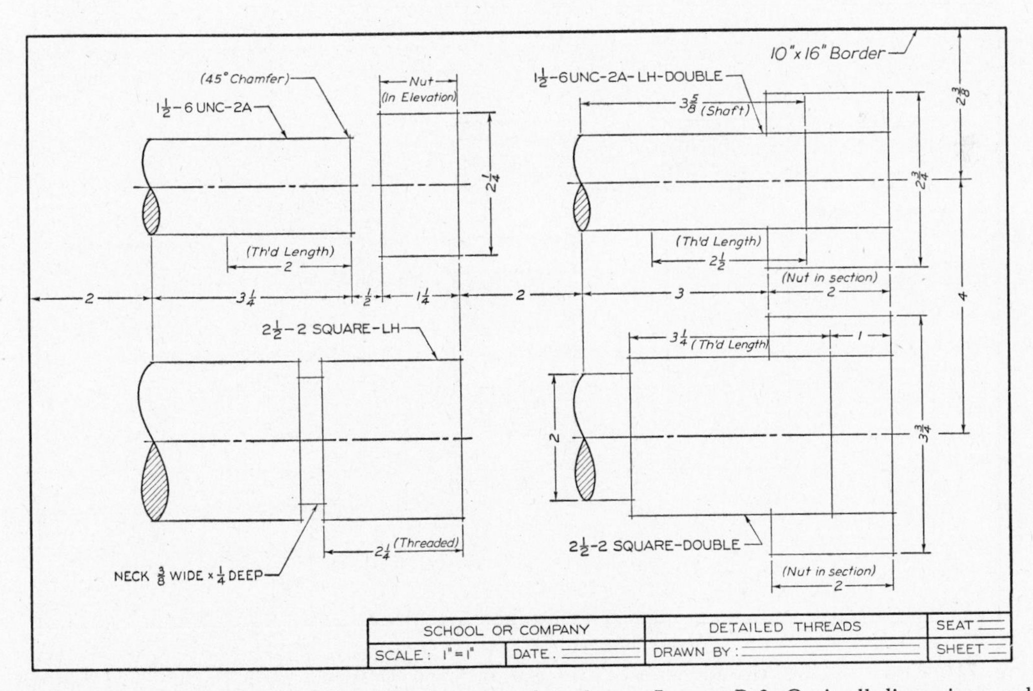

Fig. 709 Draw specified detailed threads arranged as shown. Layout B-3. Omit all dimensions and notes given in inclined letters. Letter only the thread notes and the title strip.

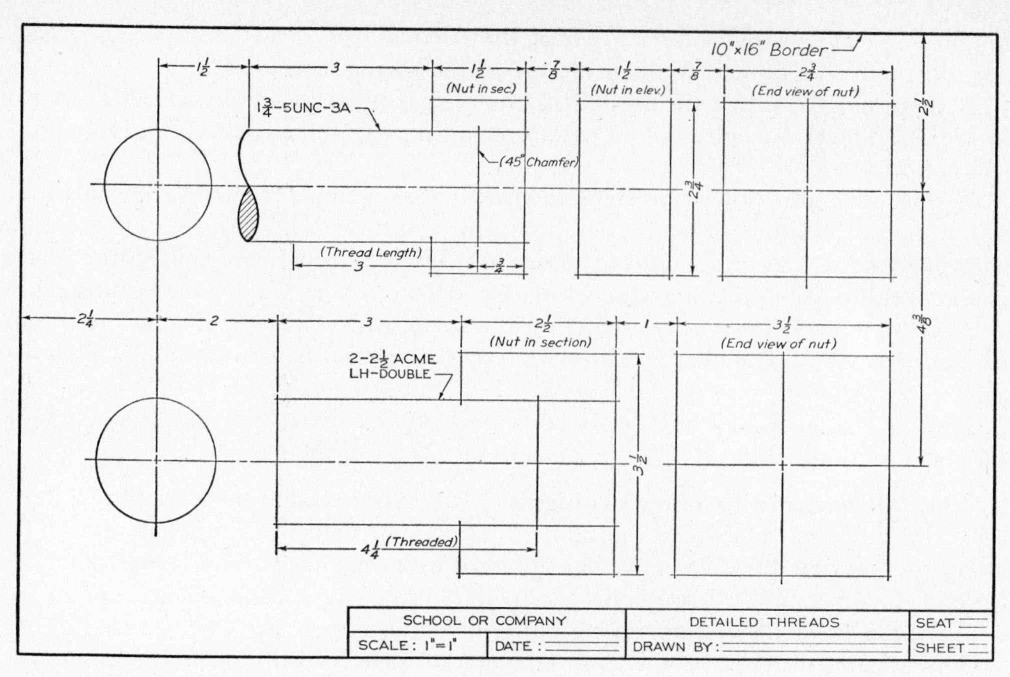

Fig. 710 Draw specified detailed threads, arranged as shown. Layout B-3. Omit all dimensions and notes given in inclined letters. Letter only the thread notes and the title strip.

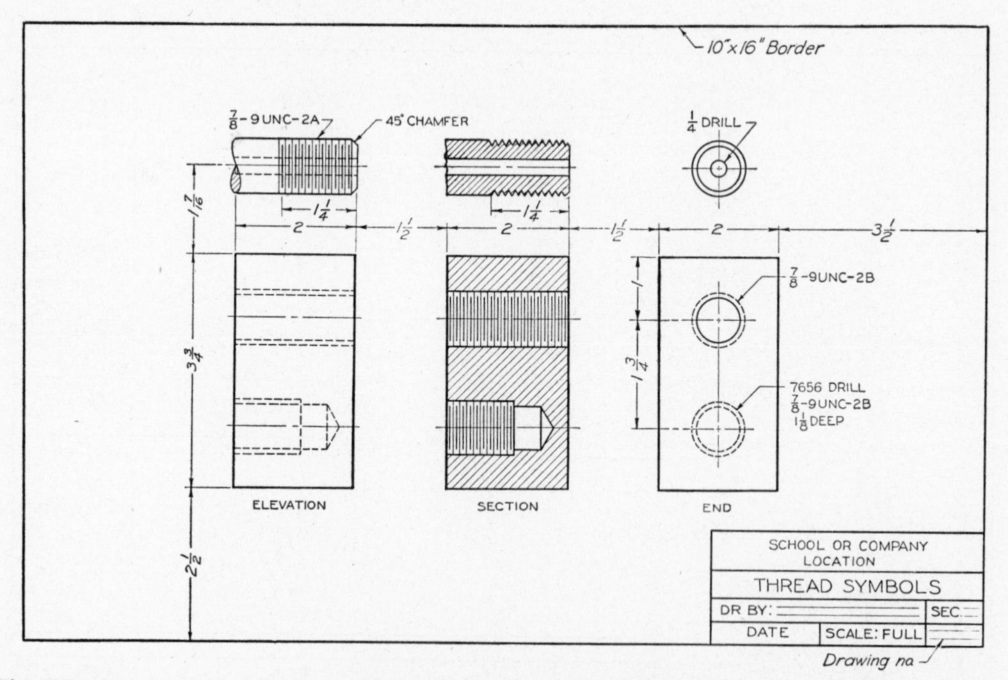

Fig. 711 Draw specified thread symbols, arranged as shown. Draw schematic or simplified symbols, as assigned by instructor, Layout B-5. Omit all dimensions and notes given in inclined letters. Letter only the drill and thread notes, the titles of the views, and the title strip.

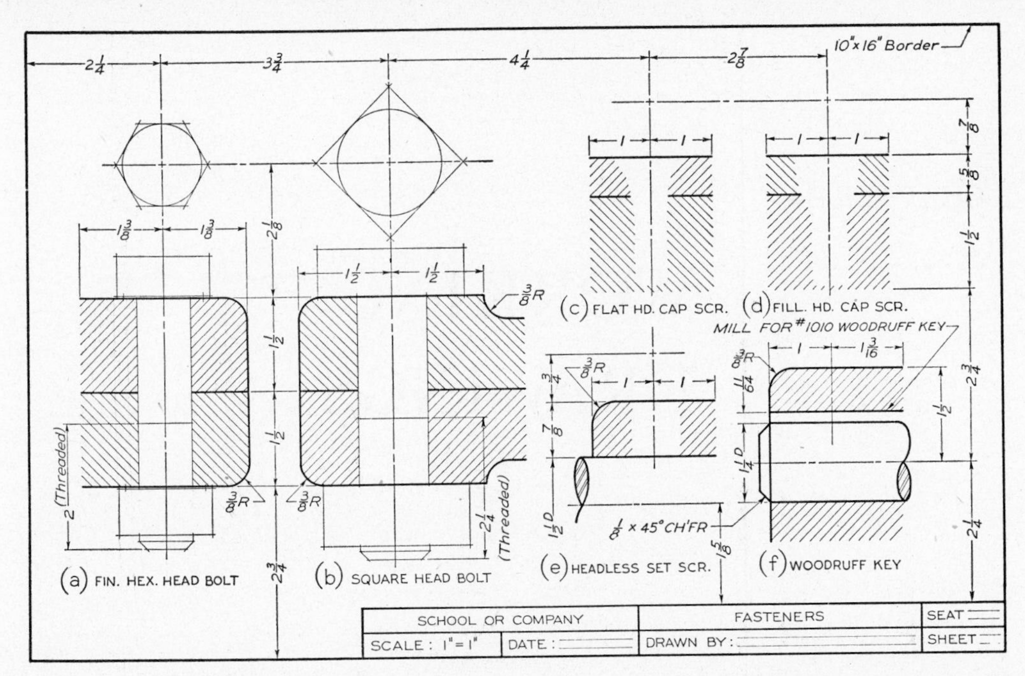

Fig. 712 Draw fasteners, arranged as shown, Layout B-3. At (a) draw $\frac{7}{8}$-9UNC-2A $\times$ 4 Fin. Hex. Hd. Bolt. At (b) draw $1\frac{1}{8}$-7UNC-2A $\times$ 4 Sq. Hd. Bolt. At (c) draw $\frac{3}{8}$-24UNC-2A $\times$ $1\frac{1}{2}$ Flat Hd. Cap Screw. At (d) draw $\frac{7}{16}$-14UNC-2A $\times$ 1 Fill. Hd. Cap Screw. At (e) draw $\frac{1}{2}$ $\times$ 1 Headless Slotted Set Screw. At (f) draw front view of No. 1010 Woodruff Key. Letter titles under each figure as shown.

CHAPTER 14

WORKING DRAWINGS

429. The Young Engineer. A typical engineering department is shown in Fig. 713. Many of the men have considerable training and experience; others are young graduates who are gaining experience. Many engineering graduates do not realize how much there is to learn after they get out of school, and that it will be necessary for them to start at the bottom and work up. Very much to the point is the following statement by the chief engineer of a large corporation:*

Many of the engineering students whom we interview have the impression that if they go to work at the drafting board, they will be only draftsmen doing routine work. This impression is completely erroneous, because all of our engineers work at the board at least occasionally. Actually, drawing is only one phase of responsibility which includes site evaluations, engineering calculations, cost estimates, preliminary layouts, engineering specifications, equipment selection, complete drawings (with the help of draftsmen), and follow-up on construction and installation.

Our policy is to promote from within, and it is our normal practice to hire engineers at the time they finish school, and to give them the opportunity for growth and development by diversified experience. These newly-hired engineers without experience are assigned to productive work at a level which their education and experience qualify them to handle successfully. The immediate requirement is for the young engineer to obtain practical engineering experience, and to learn our equipment and processes. In design work, these initial assignments are on engineering details in any one of several fields of engineering study (structural, mechanical, electrical, etc.). Our experience has shown that it is not wise to give a newly-graduated engineer without experience a problem in advanced engineering, such as creative design, on the assumption that he can make quick sketches or layouts and then have them detailed by someone else. Rather than start a young engineer at an advanced responsibility level where he may fail or make costly mistakes, we assign him initially to work which requires

*C. G. R. Johnson, Kimberly-Clark Corp.

him to make complete and accurate detail drawings, and his assignments become increasingly complex as he demonstrates the ability to do work of an advanced engineering calibre. If he demonstrates the capacity to assume responsibility, he is given direction of other engineers with less experience who in turn do detailed engineering for him.

Courtesy George Gorton Machine Co.

Fig. 713 Section of an Engineering Department.

430. Evolution of Design. Nothing could be truer than the old saying, "necessity is the mother of invention." A new mechanical idea is the result of a need. If a new device, machine, or gadget is really needed, it may be assumed that people will buy it, if it does not cost too much. The questions arise: Is there a wide potential market? Can this device be made at a price that people will be willing to pay? If these questions can be answered satisfactorily, then the inventor or the officials of a company may elect to go ahead and develop the new device.

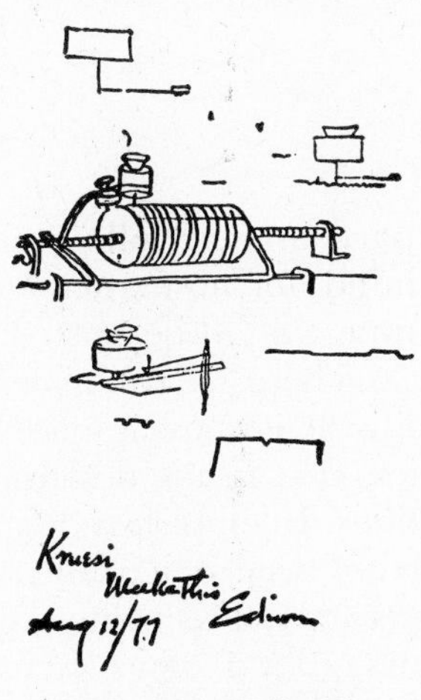

Fig. 714 Edison's Phonograph.*

A new machine or structure, or an improvement thereof, must exist in the mind of the engineer before it can exist in reality. This original conception is usually placed on paper in the form of a freehand "idea sketch," Figs. 232 and 714*, and is then followed by other sketches

*Original sketch of Thomas A. Edison's first conception of the phonograph; reproduced by special permission of Mrs. Edison.

developing more fully the idea in the mind of the engineer, designer, or inventor.

The sketches are followed by a study of the *kinematic* problems involved, if any. What source of power is to be used—manual operation or electric motor? What type of motion is needed? Is it necessary to translate rotary motion into linear motion? These problems are solved graphically by means of a kinematic drawing in which the various parts are shown in skeleton form. A pulley is represented by a circle, meshing gears by tangent pitch circles, an arm by a single line, paths of motion by center lines, etc. At this stage certain basic calculations may be made, such as those related to velocity, acceleration, etc.

These preliminary studies are followed by the *design layout*, or simply the *layout*, which is usually full size so the designer can clearly visualize the actual sizes and proportions, and is executed accurately with instruments, Fig. 715. At this time all

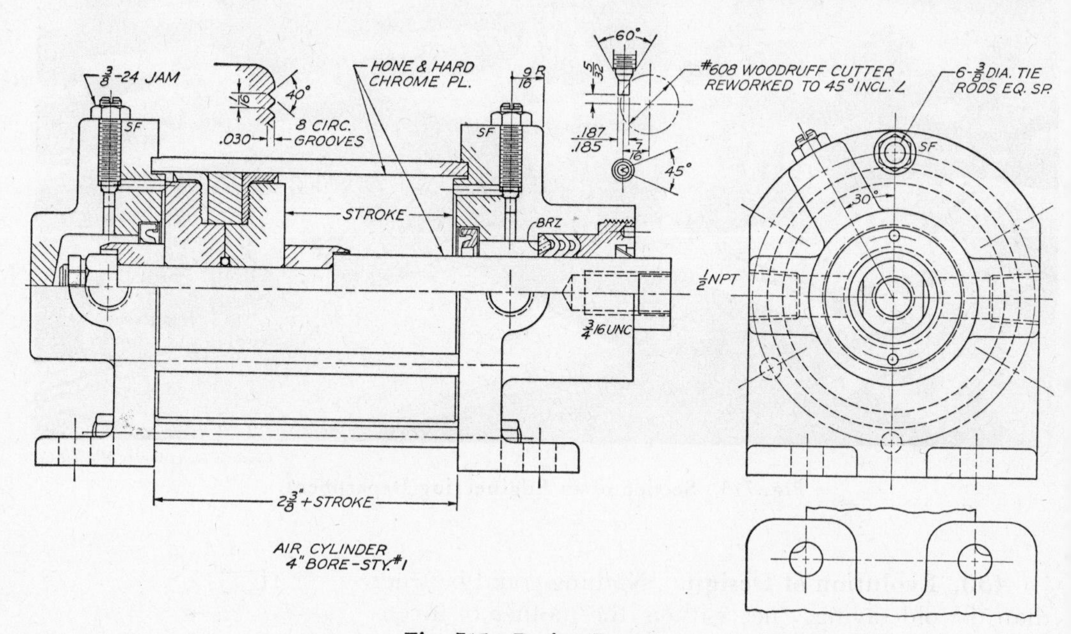

Fig. 715 Design Layout.

parts are carefully designed for strength and function. Costs are kept constantly in mind, for no matter how well the machine performs, it must be built to sell, or the time will all be wasted.

There are two general types of design: *scientific design* and *empirical design*. In scientific design, use is made of the principles of physics, mathematics, chemistry, mechanics, etc., in the design of new structures intended to function under assigned conditions. In empirical design, use is made of data compiled in handbooks which, in turn, have been determined with the aid of the sciences, or have been learned by experience. Practically all ordinary design is a combination of scientific and empirical design. A designer should therefore be equipped with adequate engineering and scientific knowledge, and should have access to handbooks relating to his field.

During the layout stage, the designer will rely a great deal upon what has gone before. He is nearly always concerned with the improvement of an existing mechanism, or with the redesign of a machine from a different approach in which nearly all the details will be similar to others he has used before. His experience has given him a sense of proportion and size which enables him to design the non-critical or more standard features by eye or with the aid of empirical data. Stress analysis and detailed computation will be necessary in connection with high speeds or heavy loads or other special requirements or conditions.

As shown in Fig. 715, the layout is an assembly drawing showing how the parts fit together and the basic proportions of the various parts. Auxiliary views or sections are used if necessary. Section lining may be omitted to save time. All lines are very sharp and the drawing is made as accurately as possible, since all dimensions are omitted except a few key dimensions that the designer wishes the draftsman to use. Any notes or other information needed by the detailers will be given on the layout.

Special attention is given to clearances of moving parts, to ease of assembly, and to serviceability. Standard parts are used wherever necessary. Most companies maintain some form of *engineering standards manual* containing much of the empirical data and detailed information that is regarded as "company standard." Materials and costs are carefully considered. Although functional considerations must come first, the designer will keep constantly in mind the problems of manufacturing. A slight change in material or in the shape of some part may, in some cases, be made without any loss of effectiveness and yet may save hundreds or thousands of dollars. The ideal design is the one that will do the job required at the lowest possible cost.

431. Working Drawings. After the layout has been approved by the chief engineer or others delegated by him, it is turned over to the draftsmen to make the

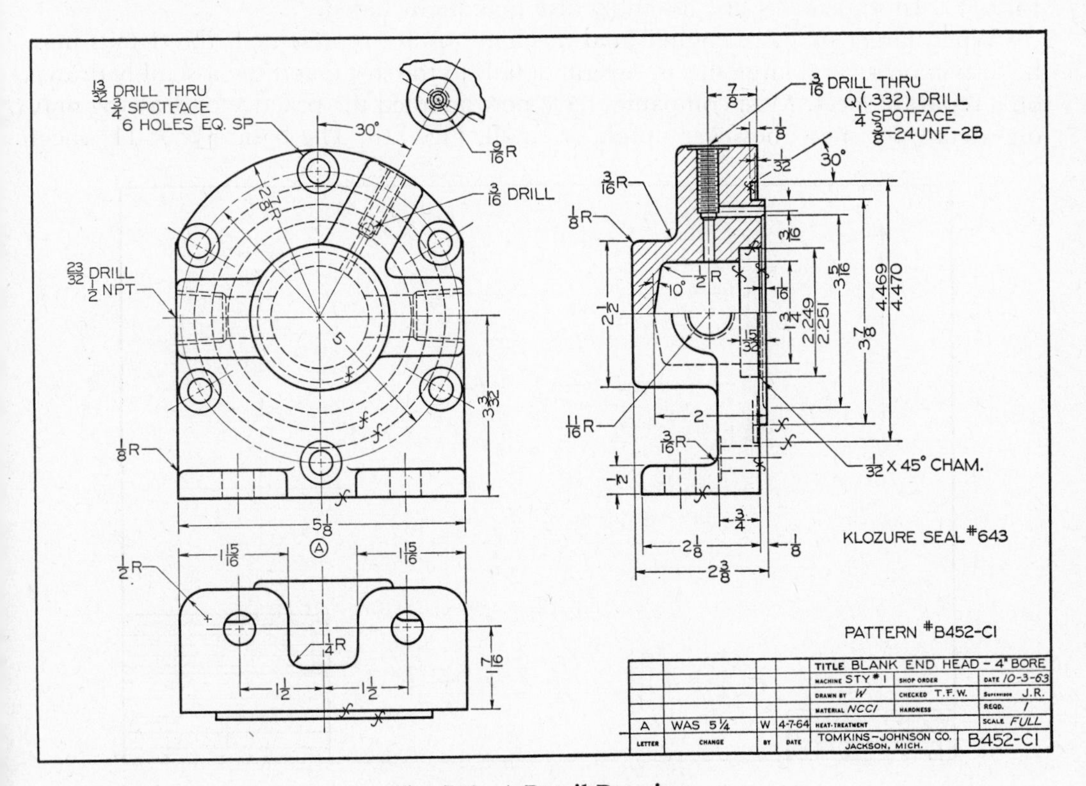

Fig. 716 A Detail Drawing.

production drawings. The draftsmen, or detailers, "pick off" the details from the layout with the aid of the scale or with dividers. The necessary views, §189, are drawn of each part, and complete dimensions and notes, Chapter 11, are added so that the drawings will describe the parts completely. These working drawings of the individual parts are also referred to as *detail drawings*. The parts may be represented individually on separate sheets, or if the project is small, the details may be grouped on a single sheet, §432.

A detail drawing of one of the parts from the design layout of Fig. 715 is shown in Fig. 716.

After the parts have been detailed, an *assembly drawing* is made, showing how all the parts fit in the completed machine or structure. The assembly may be made by tracing the various details in place directly from the detail drawings, or the assembly may be traced from the original design layout; but if either is done, the value of the assembly for checking purposes, §444, will be largely lost. The various types of assemblies are discussed in §§439-444.

Finally, in order to protect the manufacturer, a *patent drawing*, which is often a form of assembly, is prepared and filed in the U.S. Patent Office. Patent drawings are line-shaded, lettered in script, and otherwise follow the rigid rules of the Patent Office, §720.

432. Number of Details per Sheet. Two general methods are followed in industry regarding the grouping of details on sheets. If the machine or structure is small or composed of few parts, all the details may be shown on one large sheet, Fig. 717. In some cases, the assembly also may be included.

When larger or more complicated mechanisms are represented, the details may be drawn on several large sheets, several details to the sheet, and the assembly drawn on a separate sheet. Most companies have now adopted the practice of drawing only one detail per sheet, however simple or small, Fig. 718. The basic $8\frac{1}{2}'' \times 11''$ sheet

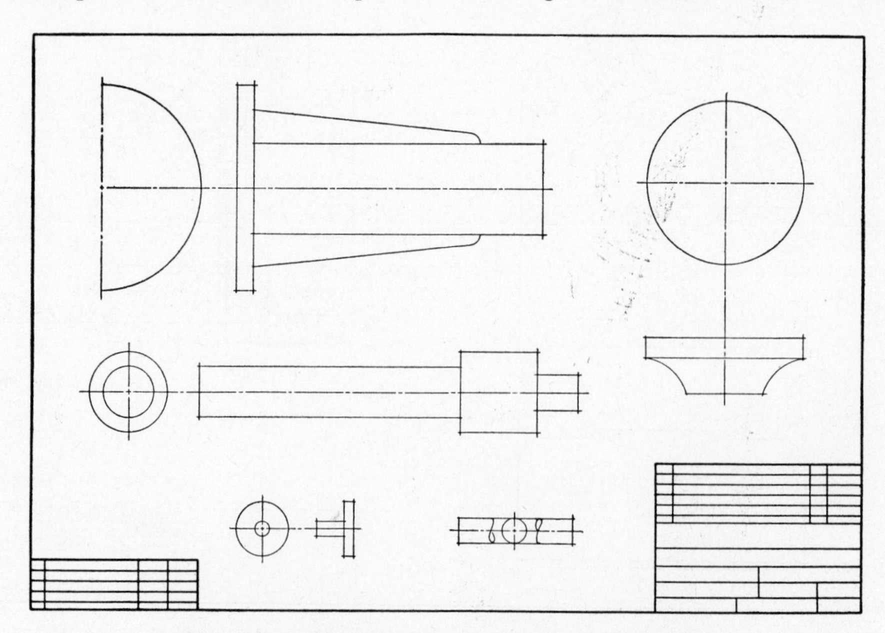

Fig. 717 "Blocking-In" the Views. (See Fig. 772)

is most commonly used for details, multiples of this size being used for larger details or the assembly. For American Standard sheet sizes, see §73.

When several details are drawn on one sheet, careful consideration must be given to spacing. The draftsman should determine the necessary views for each detail, and

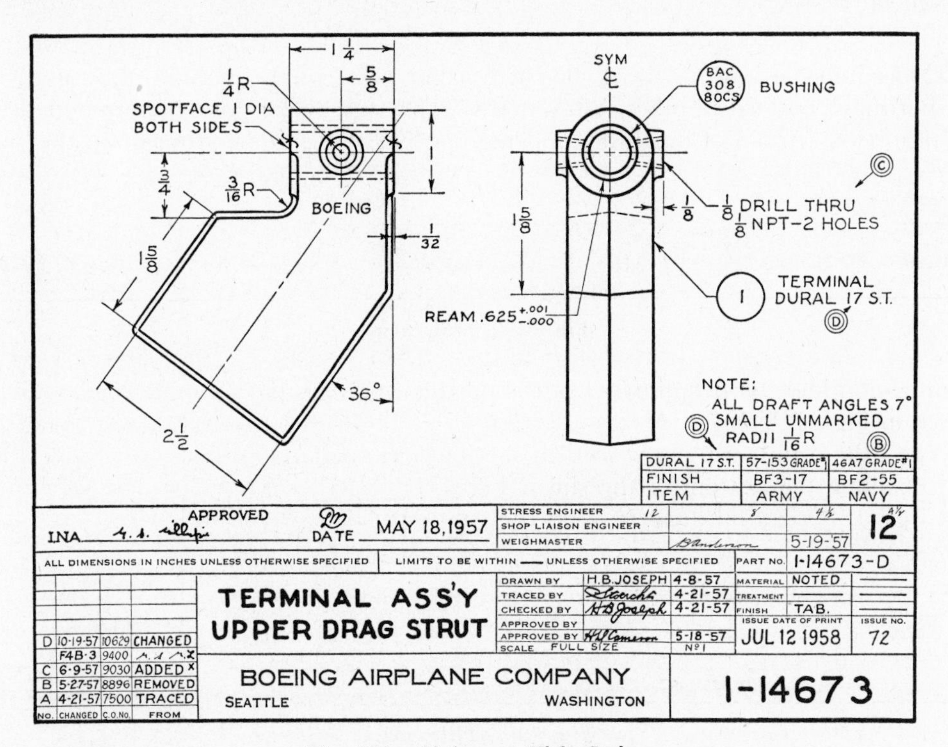

Fig. 718 Elaborate Title Strip.

block-in all views lightly before beginning to draw any view, as shown in Fig. 717. Ample space should be allowed for dimensions and notes. A simple method to space the views is to cut out rectangular scraps of paper roughly equal to the sizes of the views, and to move these around on the sheet until a suitable spacing is decided upon. The corner locations are then marked on the sheet, and the scraps of paper are discarded.

The same scale should be used for all details on a single sheet, if possible. When this is not possible, the scales for the dissimilar details should be clearly noted under each.

433. Title and Record Strips. The function of the title and record strip is to show, in an organized manner, all necessary information not given directly on the drawing with its dimensions and notes, as in Fig. 718. Obviously, the type of title used depends upon the filing system in use, the processes of manufacture, and the requirements of the product. The following information should generally be given in the title form:

1. Name of the object represented.
2. Name and address of the manufacturer.
3. Name and address of the purchasing company, if any.

4. Signature of the draftsman who made the drawing, and the date of completion.
5. Signature of the tracer, if any, and the date of completion.
6. Signature of the checker, and the date of completion.
7. Signature of the chief draftsman, chief engineer, or other official, and the date of approval.
8. Scale of the drawing.
9. Number of the drawing.

Other information may be given, such as material, quantity, heat treatment, finish, hardness, pattern number, estimated weight, superseding and superseded drawing numbers, symbol of machine, and many other items, depending upon the plant

SUPERSEDES DWG. No. 17913 -18233
SUPERSEDED BY DWG. No.

PROPERTY OF HUGHES TOOL CO. HOUSTON, TEXAS	DRAWN BY M^cK.	REVISIONS					FILE No.	SCALE —
	TRACED BY M^cK.	A 10-15-58 M^cK	F 1-14-59 B.R.A.	L	Q	V	F/D	18305
	CHECKED BY R.P.S.	B 10-17-58 M^cK	G 1-31-59 B.R.A.	M	R	W		
	DATE BEGUN 10-11-58	C 10-19-58 B.R.A.	H 2-2-59 R.P.S.	N	S	X		
	DATE FINISHED 10-11-58	D 10-21-58 R.P.S.	J	O	T	Y		
		E 10-24-58 M^cK.	K	P	U	Z		

Fig. 719 Title Strip.

organization and the peculiarities of the product. Some typical commercial titles are shown in Figs. 718-721.

The title form is usually placed in the lower right-hand corner of the sheet, Fig. 722, or along the bottom of the sheet, Figs. 718-721, because drawings are usually filed in flat, horizontal drawers, and the title must be easily found. However, many

REPORT ALL ERRORS TO FOREMAN

ALTERATIONS	DATE OF CHG.	NO. REQUIRED	MATERIAL	HEAT TREATMENT	PART NAME	DRAWN BY	
		1	SAE 3115	SEE NOTE	FEED WORM SHAFT	H.F.	UNIT 3134
		REPLACED BY	REPLACES	OLD PART NO.	DRAWN FOR SIMPLEX & DUPLEX (1200)	TRACED BY E.E.Z.	ALSO USED ON ABOVE MACHINES
				563-310	ENGINEERING DEPARTMENT	CHECKED BY C.STB.	FIRST USED ON LOT / LAST USED ON LOT
		JUN 25 1957		SCALE FULL SIZE	KEARNEY & TRECKER CORPORATION MILWAUKEE, WISCONSIN, U. S. A.	APPROVED BY / DATE 7-10-56	17840 B

Fig. 720 Title Strip.

filing systems are in use, and the location of the title form is completely governed by the system employed.

Lettering should be single-stroke vertical or inclined Gothic capitals, Figs. 122 and 123. The items in the title form should be lettered in accordance with their relative importance. The drawing number should receive greatest emphasis, closely followed by name of the object and the name of the company. The date and scale,

DO NOT SCALE THIS DRAWING FOR DIMENSIONS. MACHINE FRACTIONAL DIMENSIONS ± .010. ALL DIMENSIONS IN INCHES UNLESS OTHERWISE SPECIFIED.

SYM	CHANGE	DATE	CHANGE NOTICE	HEAT TREATMENT		CATERPILLAR TRACTOR CO. EXECUTIVE OFFICES — SAN LEANDRO, CALIF.
2	CHGD. MATL. ETC.	10-22-59		S.A.E. VIII.	SCALE FULL	NAME FIRST, FOURTH & THIRD SLIDING PINION
1	WAS #2345-ETC.	5-21-58		HARD. ROCKWELL C-50-56	DATE 6-26-58	MATERIAL C.T. #1E36 STEEL ② ①
				NOTE 3 TEST LOCATIONS	DRAWN BY S.G.	UPSET FORGING 3 $\frac{7}{8}$ ROUND MAX.
					TRACED BY L.R.	
					CHECKED BY n.w.	
					APPROVED BY amB.	1A4045
					REDRAWN FROM	

Fig. 721 Title Strip.

and the draftsmen's and checkers' names are important, but they do not deserve prominence. Greater importance of items is indicated by heavier lettering, larger lettering, wider spacing of letters, or by a combination of these methods.

Most companies have adopted standard title forms and have them printed on standard sheets of tracing cloth or paper, so that the draftsmen need merely fill in the blank spaces, as shown in Figs. 718-721.

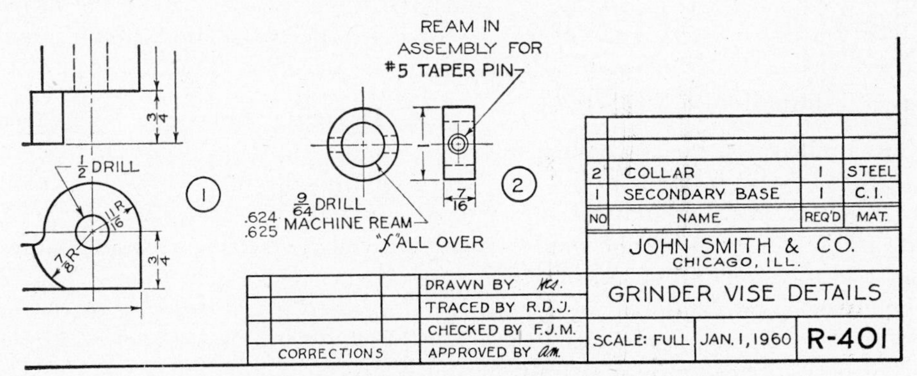

Fig. 722 Identification of Details with Parts List.

Drawings constitute important and valuable information regarding the products of a manufacturer. Hence, carefully designed, well-kept, systematic files are generally maintained for the filing of drawings.

434. Drawing Numbers. Every drawing should be numbered. Some companies use serial numbers, such as 60412, or a number with a prefix or suffix letter to indicate the sheet size, as A60412 or 60412-A. A size A sheet would probably be the standard $8\frac{1}{2}'' \times 11''$ or $9'' \times 12''$, and the B size a multiple thereof. Many different numbering schemes are in use in which various parts of the drawing number indicate different things, such as model number of the machine, the general nature or use of the part, etc. In general it is best to use a simple numbering system and not to load the number with too many indications.

The drawing number should be lettered at least $\frac{1}{4}''$ high in the lower right and upper left corners of the sheet, Fig. 729. If the drawing is rendered in ink, the Leroy, Wrico, or Speedball pens, §§85 and 99, may be used.

435. Parts Lists. A bill of material, or *parts list*, consists of an itemized list of the several parts of a structure shown on a detail drawing or an assembly drawing. This list is often given on a separate sheet, Fig. 812, but is frequently lettered directly on the drawing, Fig. 728. The title strip alone is sufficient on detail drawings of only one part, Fig. 718, but a parts list is necessary on detail drawings of several parts, Fig. 722.

Parts lists on machine drawings contain the part numbers or symbols, a descriptive title of each part, the number required, the material specified, and frequently other information, such as pattern numbers, stock sizes of materials, weights of parts, etc.

Parts are listed in general order of size or importance. The main castings or forgings are listed first, parts cut from cold-rolled stock second, and standard parts such as fasteners, bushings, roller bearings, etc., third. If the parts list rests on top of the title box or strip, the order of items should be from the bottom upward, Figs. 722 and 728, so that new items can be added later, if necessary. If the parts list is placed in the upper right corner, the items should read downward.

Each detail on the drawing may be identified with the parts list by the use of a small circle containing the part number, placed adjacent to the detail, as in Fig. 722. One of the sizes in Fig. 723 will be found suitable, depending on the size of the drawing.

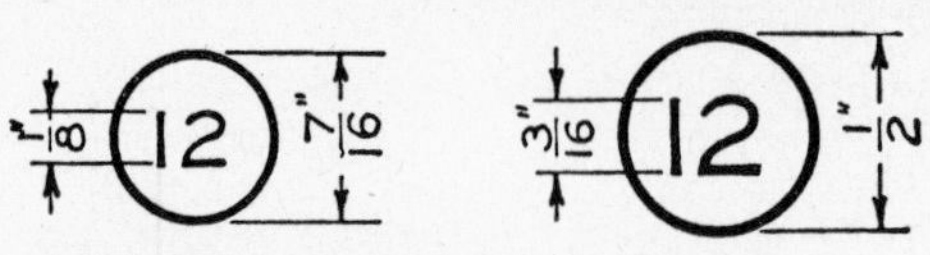

Fig. 723 Identification Numbers.

Standard parts are not drawn, but are listed in the parts list, §366.

436. Drawing Revisions. Changes on drawings are necessitated by changes in design, changes in tools, desires of customers, or by errors in design or in production. In order that the sources of all changes of information on drawings may be understood, verified, and accessible, an accurate record of all changes should be made on the drawings. The record should show the character of the change, by whom, when, and why made.

The changes are made by erasures directly on the original drawing or by means of erasure fluid on a reproduction print. See §734. Additions are simply drawn in on the original. The removal of information by crossing out is not recommended. If a dimension is not noticeably affected by a change, it may be underlined with a wavy line as shown in Fig. 589 to indicate that it is not to scale. In any case, prints of each issue are kept on file to show how the drawing appeared before the revision. New prints are issued to supersede old ones each time a change is made.

If considerable change on a drawing is necessary, a new drawing may be made and the old one then stamped OBSOLETE and placed in the "obsolete" file. In the title block of the old drawing, under "SUPERSEDED BY . . . ," or "REPLACED BY . . . ," (see Figs. 719 and 720), the number of the new drawing is entered. On the new drawing, under "SUPERSEDES . . . ," or "REPLACES . . . ," the number of the old drawing is entered.

Various methods are used to reference the area on a drawing where the change is made, with the entry in the revision block. The most common is to place numbers or letters in small circles near the places where the changes were made and to use the same numbers or letters in the revision block, Fig. 724. On zoned drawings the zone of the correction would be shown in the revision block. In addition, a brief description of the change should be made, and the date and the initials of the person making the change should be given.

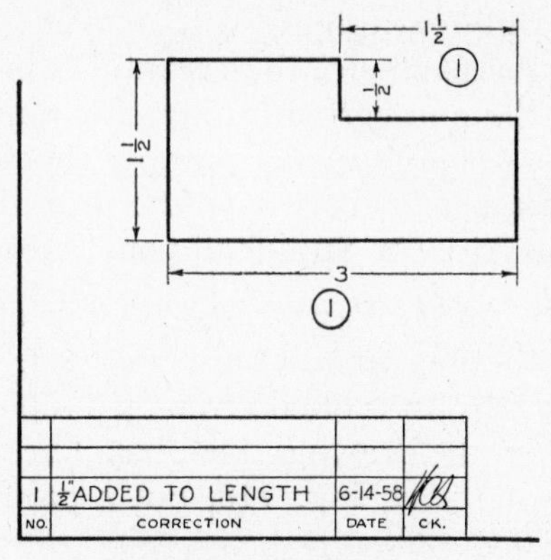

Fig. 724 Revisions.

437. Checking. The importance of accuracy in technical drawing cannot be overestimated. In commercial offices, errors sometimes cause tremendous unnecessary expenditures. *The draftsman's signature on a drawing identifies him, and he is held responsible for the accuracy of his work.*

In small offices, checking is usually done by the designer or by one of the draftsmen. In large offices, experienced engineers are employed who devote their entire time to checking.

The pencil drawing, upon completion, is carefully checked and signed by the draftsman who made it. The drawing is then checked by the designer for function, economy, practicability, etc. Corrections, if any, are then made by the original draftsman. If the drawing is to be traced, the tracing is checked against the original drawing. If it is correct, it is signed by the tracer.

The final checker should be able to discover all remaining errors. If his work is to be effective, he must proceed in a systematic way, studying the drawing with particular attention to the following points:

1. Soundness of design, with reference to function, strength, materials, economy, manufacturability, serviceability, ease of assembly and repair, lubrication, etc.
2. Choice of views, partial views, auxiliary views, sections, line work, lettering, etc.
3. Dimensions, with special reference to repetition, ambiguity, legibility, omissions, errors, and finish marks. Special attention should be given to tolerances.
4. Standard parts. In the interest of economy, as many parts as possible should be standard.
5. Notes, with special reference to clear wording and legibility.
6. Clearances. Moving parts should be checked in all possible positions to assure freedom of movement.
7. Title form information.

438. Simplified Drafting. In the old days, drafting technique was somewhat of an end in itself. Drawings were inked in with carefully-graduated line weights, usually on white paper; and shading and even colors were extensively used to produce an artistic effect. All parts were drawn in minute detail—every bolt or rivet head, valve, or fitting—nothing was left out. Lettering was ornate, like the signs on storefronts and on Studebaker wagons of that day. Time apparently meant nothing to the draftsman or his employer.

As manufacturing for a competitive mass market developed, it was realized that drafting time was a considerable element of cost. Colors and line-shading were dropped. Lettering was simplified, §79, into single-stroke Gothic. Inking was dropped (except for certain highly-finished work) in favor of the rapidly-made dark pencil drawing directly on tracing paper or pencil cloth. Improved reproduction processes produced prints equally as clear as ink tracings.

In 1935 the American Standards Association issued the first American Standard, entitled *Drawings and Drafting Room Practice*. A revised and improved revision was issued in 1946. This standard is now in process of revision and great expansion, so that when completed there will be some seventeen sections or booklets, §7. These publications incorporate the best and most representative practices today in this country, and the authors are in full accord with them. These standards advocate simplification in many ways; for example: partial views, half views, thread symbols, piping symbols, single-line spring drawings, etc.

However, in the past few years some industries have felt that drafting practice should be much further simplified.* The drastic changes suggested and the poor drafting often used as examples in company manuals have tended to alienate support from many who are really sympathetic to the basic idea. A summary of practices advocated in simplified drafting is as follows:

1. Use word description in place of drawing wherever practicable.

2. Never draw an unnecessary view. Often a view can be eliminated by using abbreviations or symbols, as HEX, SQ, DIA, ℄, etc.

3. Draw partial views instead of full views wherever possible. Draw half views of symmetrical parts.

*For a complete treatment of this subject, see *Simplified Drafting Practice*, by W. L. Healy and A. H. Rau (New York, John Wiley & Sons, 1953).

4. Avoid elaborate, pictorial, or repetitive detail as much as possible. Use phantom lines to avoid drawing repeated features, §399.

5. On assemblies, represent bolts, nuts, and other hardware by simple center lines, crosses, or boxes.

6. Omit unnecessary hidden lines. See §192.

7. Omit section lining, or use only partial section lining wherever it can be done without loss of clarity.

8. Represent holes by means of crosses and center lines wherever practicable.

9. Use symbolic representation wherever possible, as piping symbols, thread symbols, etc.

10. Draw entirely freehand, or mechanically plus freehand, wherever practicable.

11. Avoid lettering as much as possible. For example, parts lists should be typed on a separate sheet.

12. Use labor-saving devices wherever practicable, such as templates, plastic overlays, etc.

It will be noted that most of these practices have long been approved by ASA. Some elements of simplified drafting that are not now recognized by the majority of industry will undoubtedly in time find their way into the ASA standards. Until they do, the student should follow the ASA standards as exemplified throughout this book. Fundamentals should come first—shortcuts perhaps later.

A list of the American Drafting Standards is given in Appendix 1.

439. Assembly Drawings. An assembly drawing shows the assembled machine or structure, with all detail parts in their functional positions. Assembly drawings

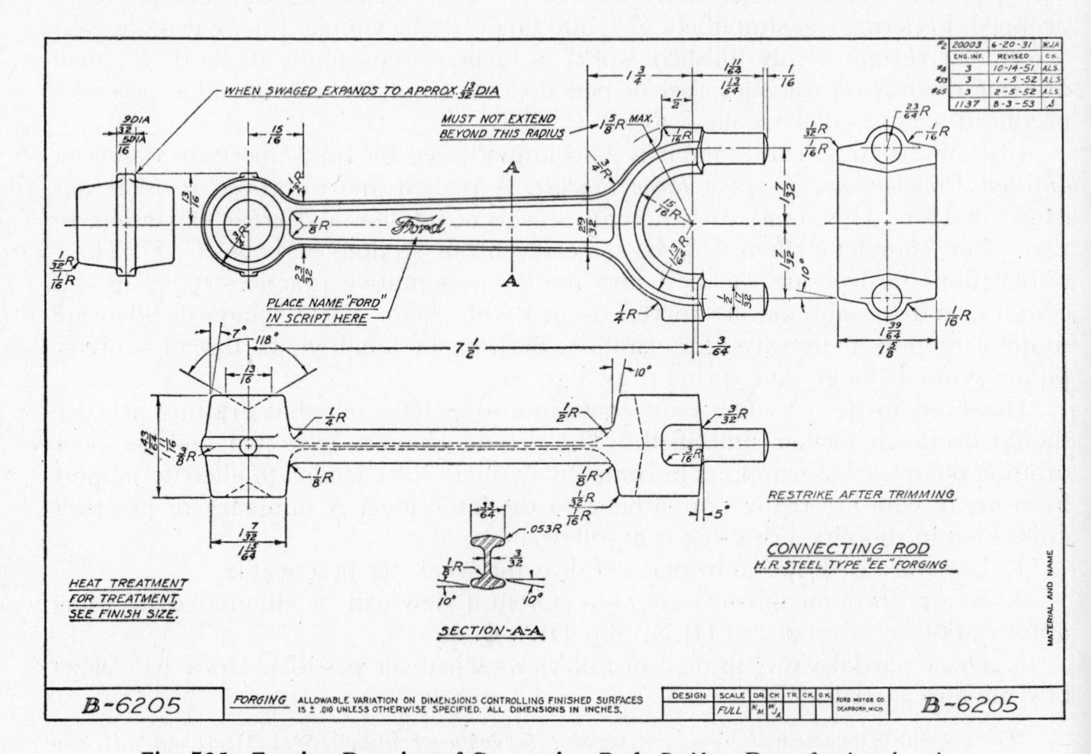

Fig. 725 Forging Drawing of Connecting Rod. *Problem:* Draw forging drawing.

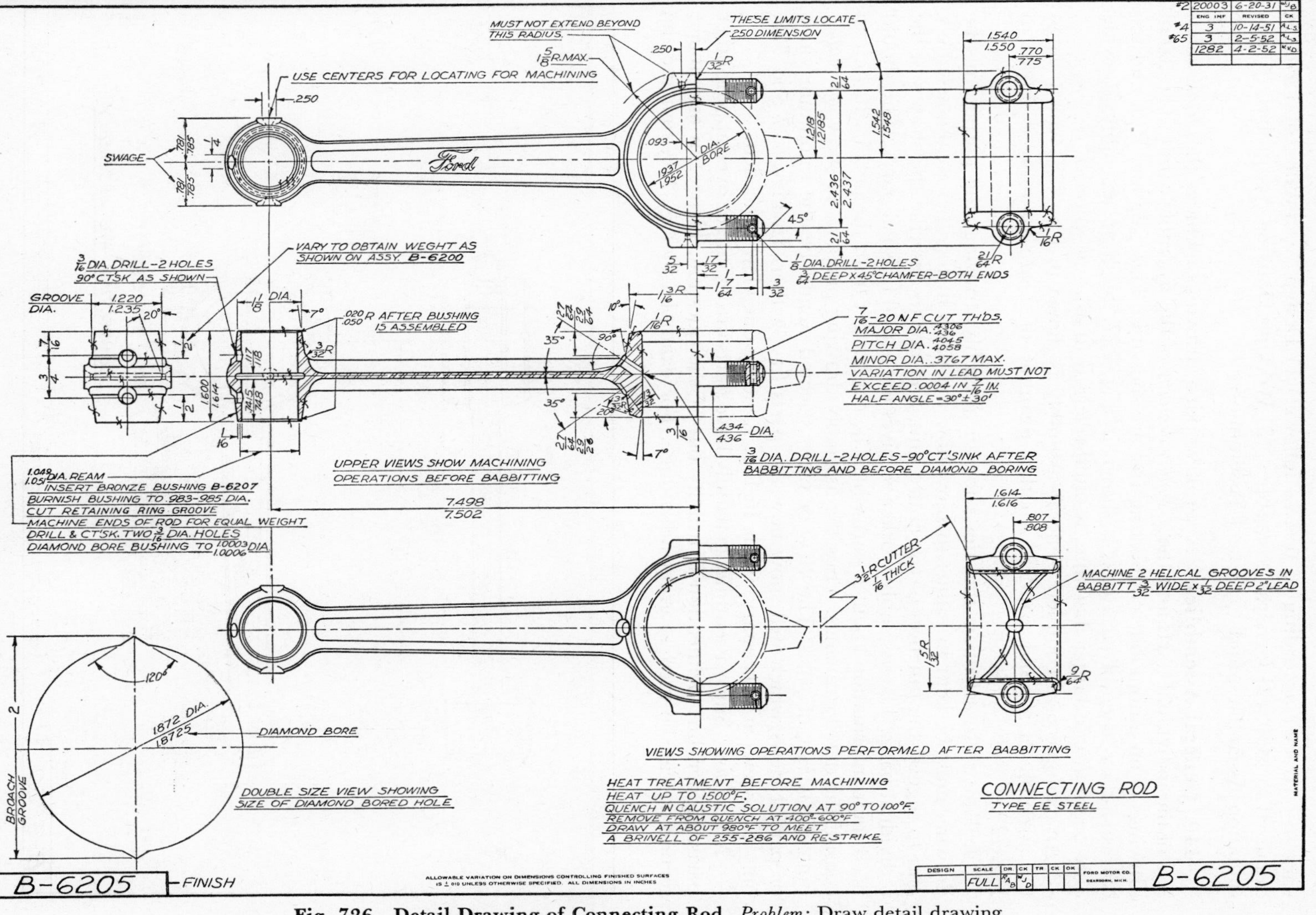

Fig. 726 Detail Drawing of Connecting Rod. *Problem:* Draw detail drawing.

vary in character according to use, as follows: (1) Design Assemblies, or Layouts, discussed in §430, (2) General Assemblies, (3) Working Drawing Assemblies, (4) Installation Assemblies, and (5) Check Assemblies.

440. General Assemblies. A set of working drawings includes the *detail drawings* of the individual parts and the *assembly drawing* of the assembled unit. The detail drawings of an automobile connecting rod are shown in Figs. 725 and 726, and the corresponding assembly drawing is shown in Fig. 727. Such an assembly, showing only one unit of a larger machine, is often referred to as a *sub-assembly*.

An example of a complete general assembly is shown in Fig. 728, which shows the assembly of a hand grinder. Another example of a sub-assembly is shown in Fig. 729.

1. Views. In selecting the views for an assembly drawing, the purpose of the drawing must be kept in mind: to show how the parts fit together in the assembly and to suggest the function of the entire unit, and not to describe the shapes of the individual parts. The assembly worker receives the actual finished parts. If he should need some information about a part which he cannot get from the part itself, he can consult the detail drawings. Thus the assembly drawing purports to show *not shapes*, but *relationships*, of parts. The view or views selected should be the minimum views or partial views which will show how the parts fit together. In Figs. 727 and 728, only two views are necessary, while in Fig. 729 only one view is needed.

2. Sections. Since assemblies generally have parts fitting into or overlapping other parts, hidden-line delineation is usually out of the question. Hence, in assemblies, sectioning can be used to great advantage. For example, in Fig. 728, imagine the

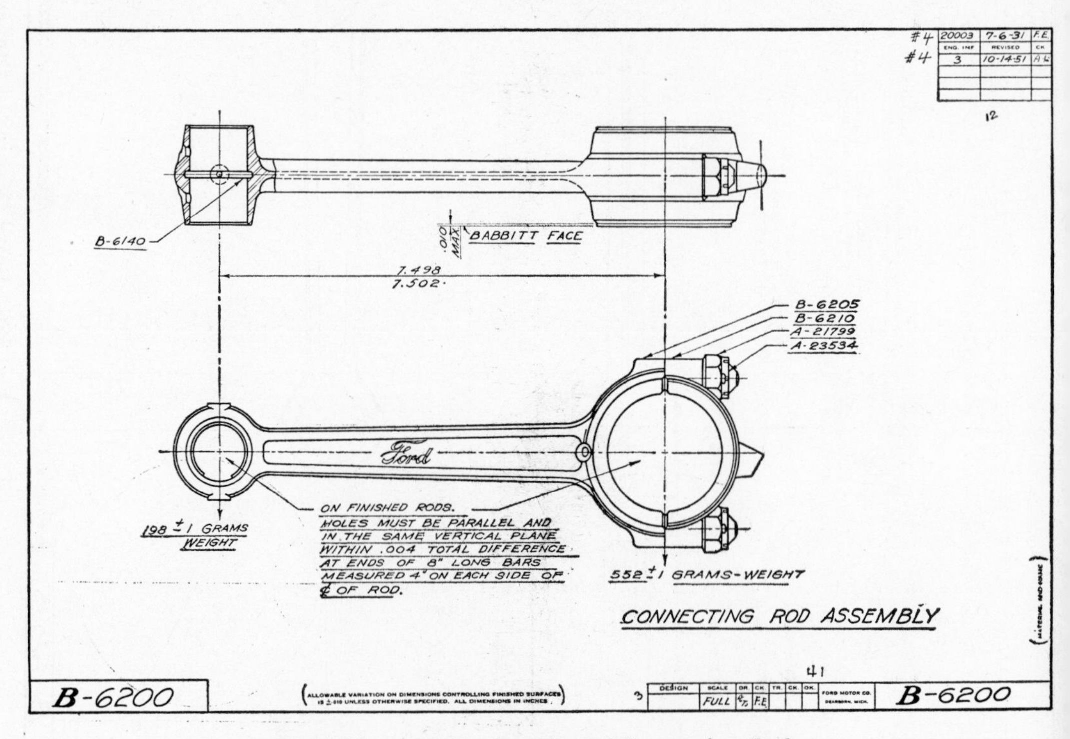

Fig. 727 Assembly Drawing of Connecting Rod.

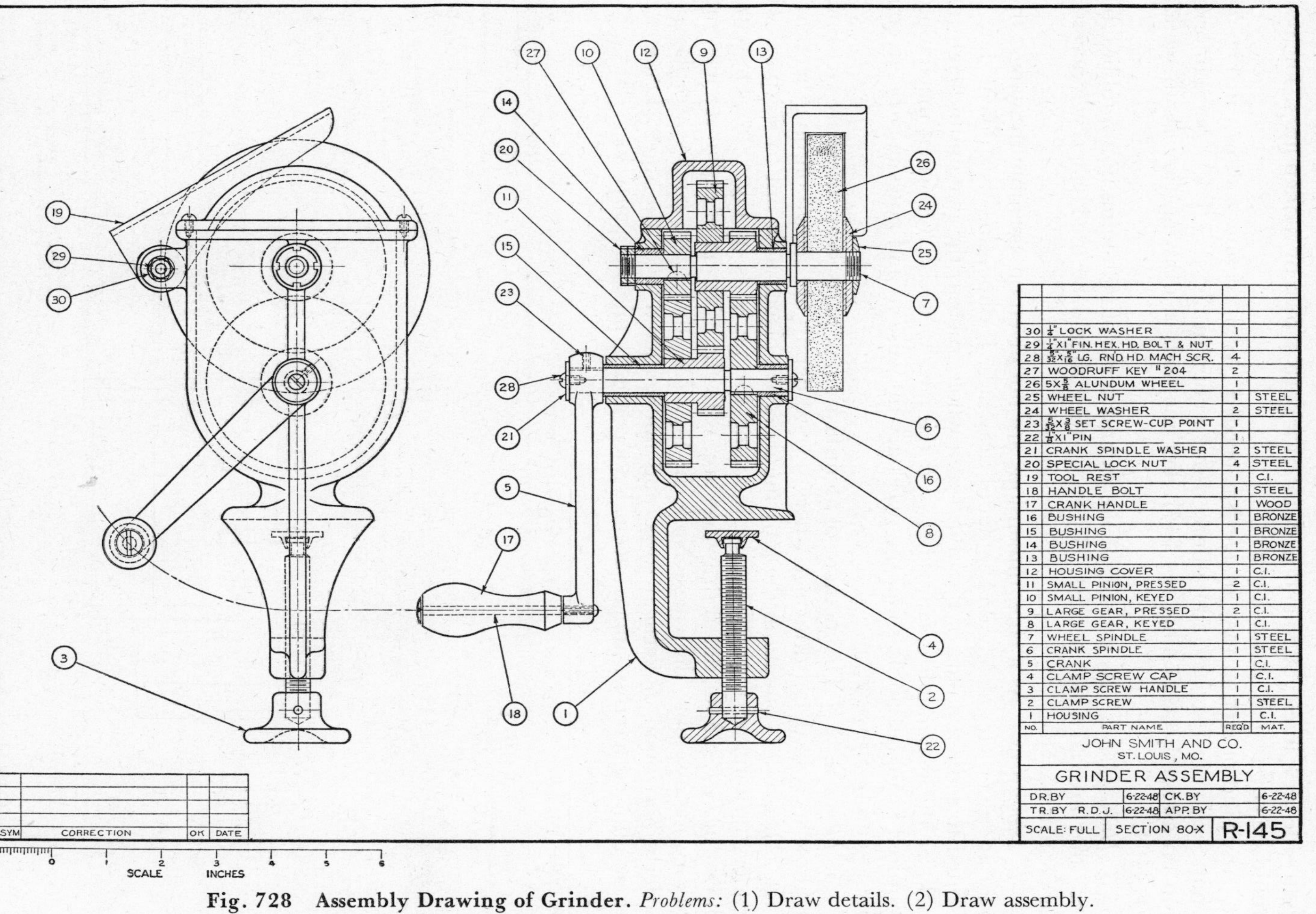

30 1/4" LOCK WASHER 1
29 1/4" X 1" FIN. HEX. HD. BOLT & NUT 1
28 5/32" X 5/16" LG. RND. HD. MACH SCR. 4
27 WOODRUFF KEY #204 2
26 5 X 5/8 ALUNDUM WHEEL 1
25 WHEEL NUT 1 STEEL
24 WHEEL WASHER 2 STEEL
23 5/32 X 3/8 SET SCREW-CUP POINT 1
22 1/8" X 1" PIN 1
21 CRANK SPINDLE WASHER 2 STEEL
20 SPECIAL LOCK NUT 4 STEEL
19 TOOL REST 1 C.I.
18 HANDLE BOLT 1 STEEL
17 CRANK HANDLE 1 WOOD
16 BUSHING 1 BRONZE
15 BUSHING 1 BRONZE
14 BUSHING 1 BRONZE
13 BUSHING 1 BRONZE
12 HOUSING COVER 1 C.I.
11 SMALL PINION, PRESSED 2 C.I.
10 SMALL PINION, KEYED 1 C.I.
9 LARGE GEAR, PRESSED 2 C.I.
8 LARGE GEAR, KEYED 1 C.I.
7 WHEEL SPINDLE 1 STEEL
6 CRANK SPINDLE 1 STEEL
5 CRANK 1 C.I.
4 CLAMP SCREW CAP 1 C.I.
3 CLAMP SCREW HANDLE 1 C.I.
2 CLAMP SCREW 1 STEEL
1 HOUSING 1 C.I.
NO. PART NAME REQ'D MAT.
JOHN SMITH AND CO.
ST. LOUIS, MO.
GRINDER ASSEMBLY
DR. BY 6-22-48 CK. BY 6-22-48
TR. BY R. D. J. 6-22-48 APP. BY 6-22-48
SCALE: FULL SECTION 80-X R-145
SYM CORRECTION OK DATE
SCALE INCHES

Fig. 728 Assembly Drawing of Grinder. *Problems:* (1) Draw details. (2) Draw assembly.

right-side view in elevation and with interior parts shown by hidden lines. The result would be completely unintelligible.

Any kind of section may be used as needed. A full section is shown in Fig. 728, a half section in Fig. 729, and a broken-out section in Fig. 727. For general information on assembly sectioning, see §441. For methods of drawing threads in sections, see §404.

3. Hidden Lines. As a result of the extensive use of sectioning in assemblies, hidden lines are often not needed. However, they should be used wherever necessary for clearness.

4. Dimensions. As a rule, dimensions are not given on assembly drawings, since they are given completely on the detail drawings. If dimensions are given, they are limited to some function of the object as a whole, such as the maximum height of a jack, or the maximum opening between the jaws of a vise. Or when machining is required in the assembly shop, the necessary dimensions and notes may be given on the assembly drawing.

5. Identification. The methods of identification of parts in an assembly are similar to those used in detail drawings where several details are shown on one sheet, as in Fig. 722. Circles containing the part numbers are placed adjacent to the parts with leaders terminated by arrowheads touching the parts as in Fig. 728. The circles shown in Fig. 723 for detail drawings are, with the addition of radial leaders, satisfactory for assembly drawings. Note, in Fig. 728, that these circles are placed in orderly horizontal or vertical rows, and not scattered over the sheet. Leaders are never allowed to cross, and adjacent leaders are parallel or nearly so.

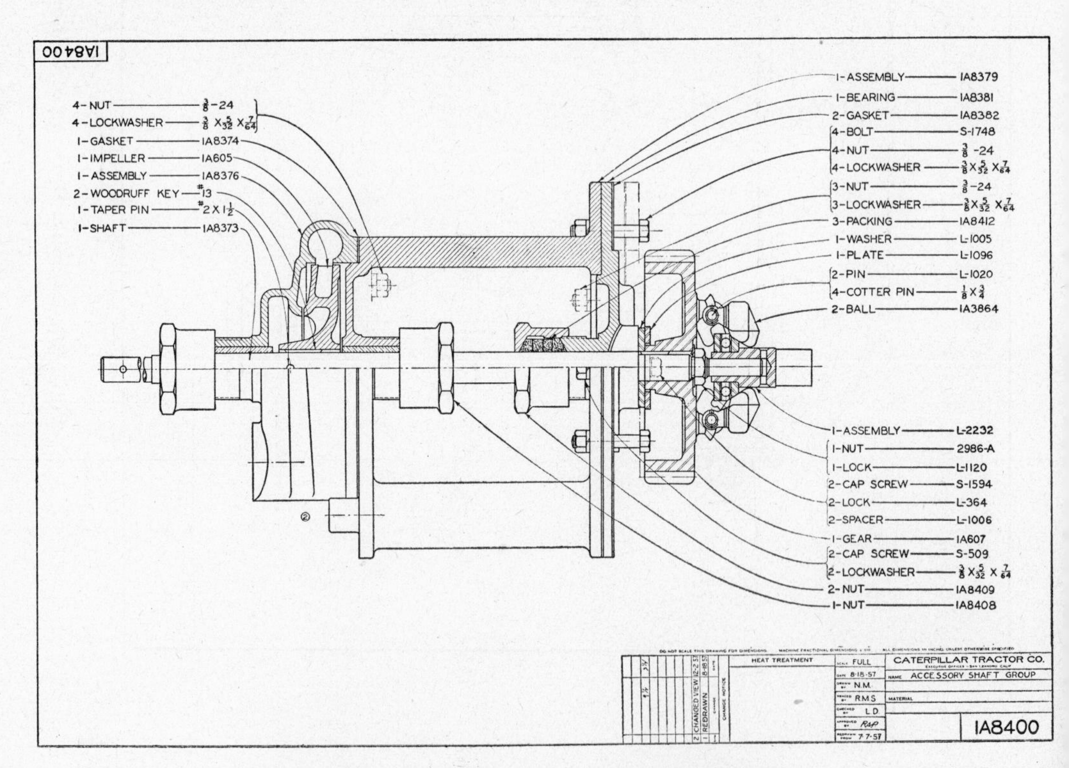

Fig. 729 Sub-Assembly of Accessory Shaft Group.

The parts list includes the part numbers or symbols, a descriptive title of each part, the number required per machine or unit, the material specified, and frequently other information, such as pattern numbers, stock sizes, weights, etc. Frequently the parts list is lettered or typed on a separate sheet, as shown in Fig. 812.

Another method of identification is to letter the part names, numbers required, and part numbers, at the end of leaders as shown in Fig. 729. More commonly, however, only the part numbers are given, together with straight-line leaders.

6. Drawing Revisions. Methods of recording changes are the same as those for detail drawings. See §436 and Figs. 718-721.

441. Assembly Sectioning. In assembly sections it is necessary not only to show the cut surfaces but to distinguish between adjacent parts. This is done by drawing the section lines in opposing directions, as shown in Fig. 730. The first large area, (a), is section-lined at 45°. The next large area, (b), is section-lined at 45° in the opposite direction. Additional areas are then section-lined at other angles, as 30° or 60° with horizontal, as shown at (c). If necessary, "odd" angles may be used. Note at (c) that in small areas it is necessary to space the section lines closer together. The section lines in adjacent areas should not meet at the visible lines separating the areas.

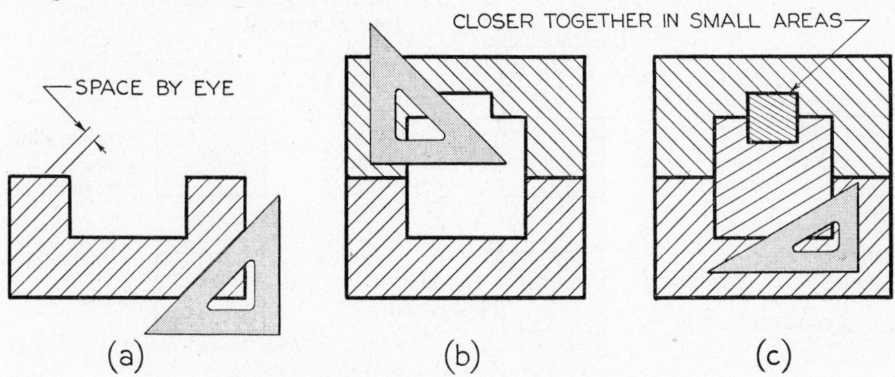

Fig. 730 Section Lining (Full Size).

For general use, the cast iron general-purpose section lining is recommended for assemblies. Wherever it is desired to give a general indication of the materials used,

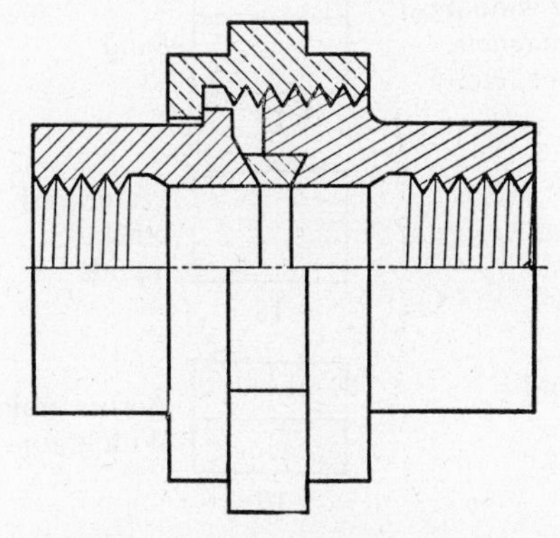
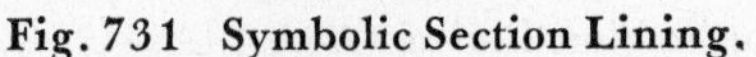

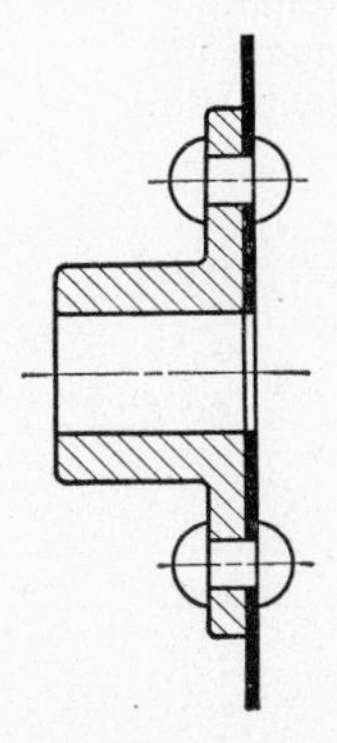

Fig. 731 Symbolic Section Lining. **Fig. 732 Sectioning Thin Parts.**

symbolic section lining may be used, as in Fig. 731. The American Standard symbols for section lining are shown in Fig. 733.

In sectioning relatively thin parts in assembly, such as gaskets, sheet metal parts, etc., section lining is ineffective, and such parts should be shown solid black, Fig. 732.

Often solid objects, or parts which themselves do not require sectioning, lie in the path of the cutting plane. It is customary and standard practice to show such parts unsectioned, or "in the round." These include bolts, nuts, shafts, keys, screws, pins, ball or roller bearings, gear teeth, spokes, ribs, etc. Many of these are shown in Fig. 734. See how many you can find. See similar examples in Figs. 728 and 729.

442. Working-Drawing Assembly. A *working-drawing assembly*, Fig. 735, is a combined detail and assembly drawing. Such drawings are often used in place of separate detail and assembly drawings when the assembly is simple enough for all of its parts to be shown clearly in the single drawing. In some cases, all but one or two parts can be drawn and dimensioned clearly in the assembly drawing, in which event these parts are detailed separately on the same sheet. This type of drawing is

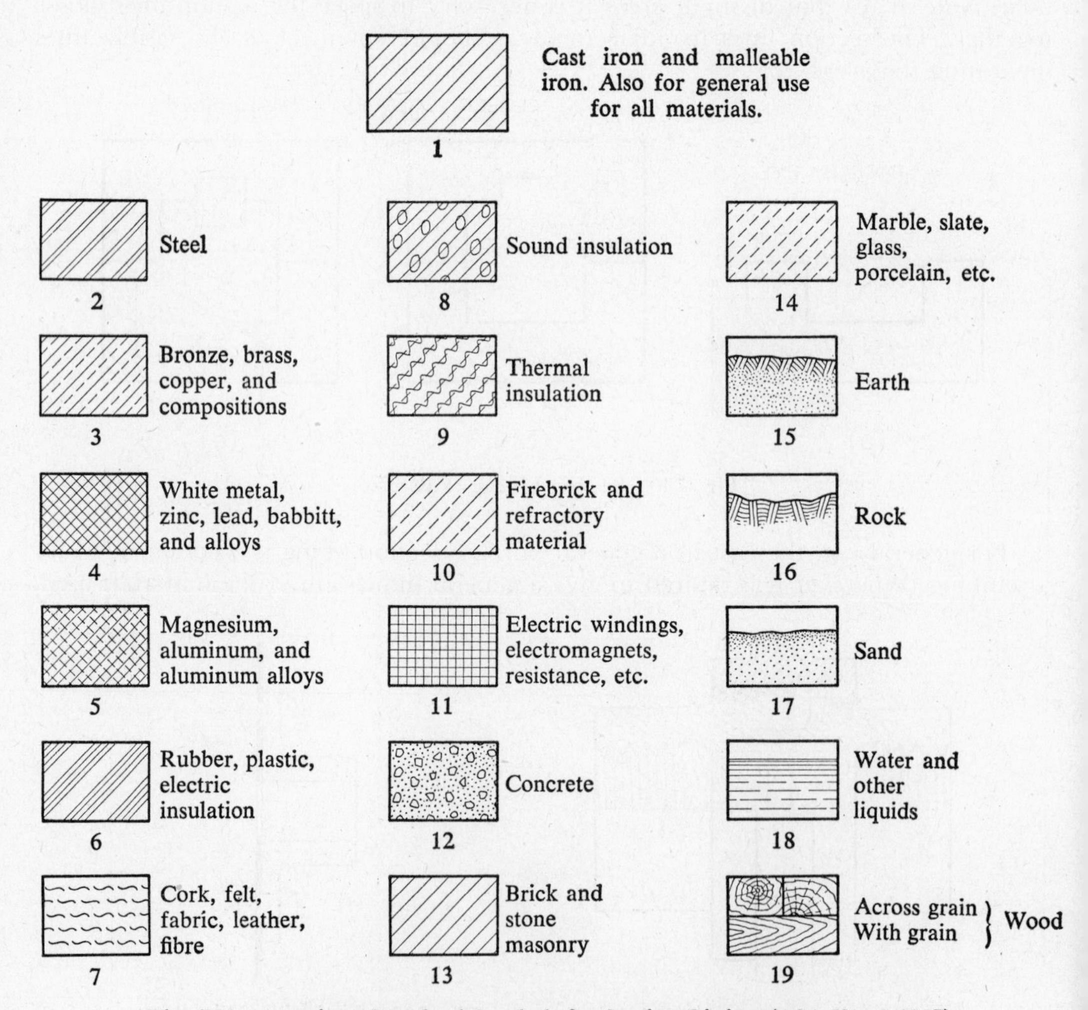

Fig. 733 American Standard Symbols for Section Lining (ASA Y14.2-1957).

common in valve drawings, locomotive sub-assemblies, and in drawings of jigs and fixtures.

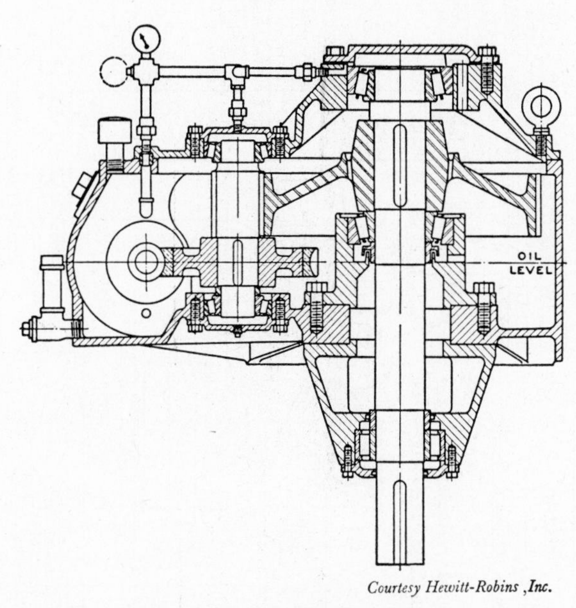

Courtesy Hewitt-Robins, Inc.

Fig. 734 Assembly Section.

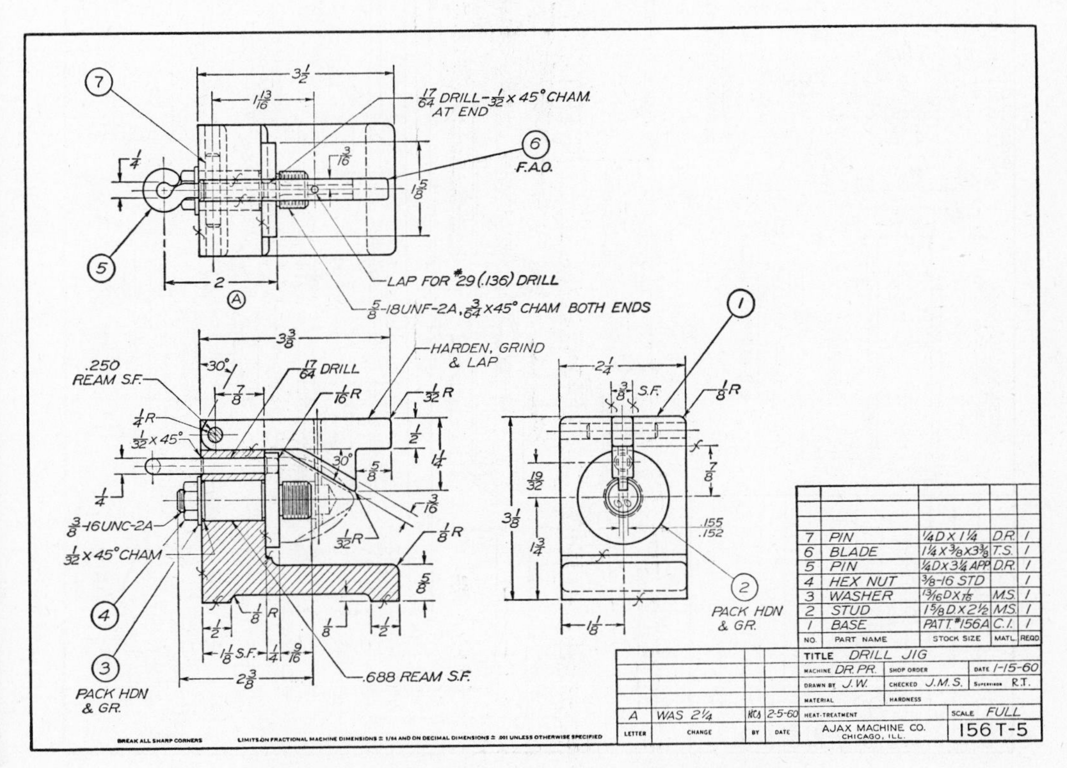

Fig. 735 Working Drawing Assembly of Drill Jig.

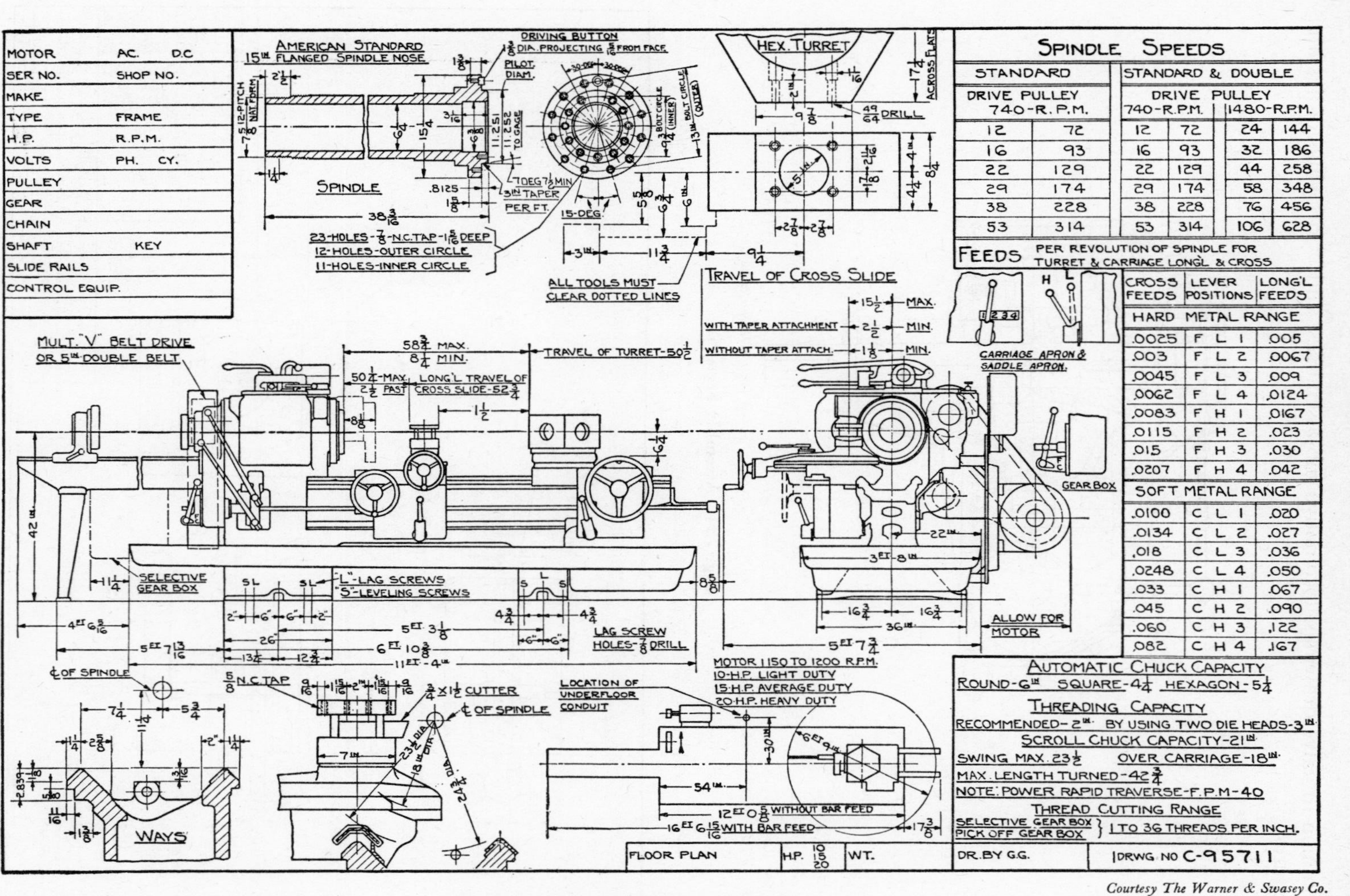

Fig. 736 Installation Assembly.

Courtesy The Warner & Swasey Co.

443. Installation Assemblies. An assembly made specifically to show how to install or erect a machine or structure is an *installation assembly.* This type of drawing is also often called an *outline assembly,* because it shows only the outlines and the relationships of exterior surfaces. A typical installation assembly is shown in Fig. 736. In aircraft drafting, an installation drawing (assembly) gives complete information for placing details or sub-assemblies in their final positions in the airplane. An example is shown in Fig. 1155.

444. Check Assemblies. After all detail drawings of a unit have been made, it may be necessary to make a *check assembly,* especially if a number of changes were made in the details. Such an assembly is drawn accurately to scale in order to check graphically the correctness of the details and their relationship in assembly. After the check assembly has served its purpose, it may be converted into a general assembly drawing.

445. Working Drawing Problems. These problems are presented to provide practice in making regular working drawings of the type used in industry. Owing to the variations in sizes and in scales which may be used, the student is to select his own sheet sizes and scales when these are not specified, subject to the approval of the instructor. Standard sheet layouts are shown on the rear end paper of this book.

The statements for each problem are intentionally brief, so that the instructor may amplify or vary the requirements when making assignments. Some tracings in pencil and in ink should be assigned, but this also is left to the instructor.

The student should clearly understand that in problems presented in pictorial form, the placement of dimensions and finish marks cannot always be followed in the drawing. *The dimensions given are in most cases those needed to make the parts, but due to the limitations of pictorial drawing they are not in all cases the dimensions which should be shown on the drawing.* In the pictorial problems the rough and finished surfaces are shown, but finish marks are usually omitted. The student should add all necessary finish marks. Also in Figs. 746-762, dimensions are not necessarily given in their best locations because of the brevity of the views. The student should place all dimensions in the preferred places in the final drawings.

Each problem should be preceded by a thumbnail sketch or a complete technical sketch, fully dimensioned. Any of the title blocks shown on the rear endpaper of this book may be used, or the student may design the title block if so assigned by the instructor.

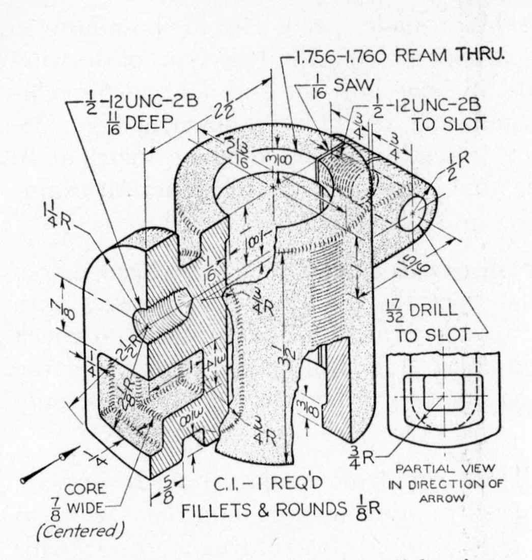

Fig. 737 **Table Bracket.** Make detail drawing. Use Size B sheet.

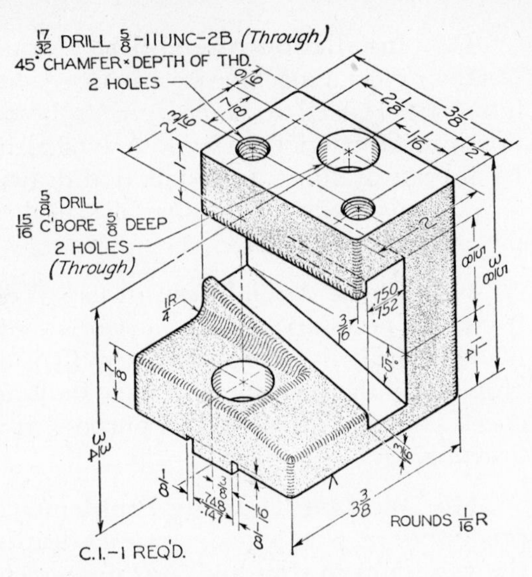

Fig. 738 **R.H. Tool Post.** Make detail drawing. Use Size B sheet.

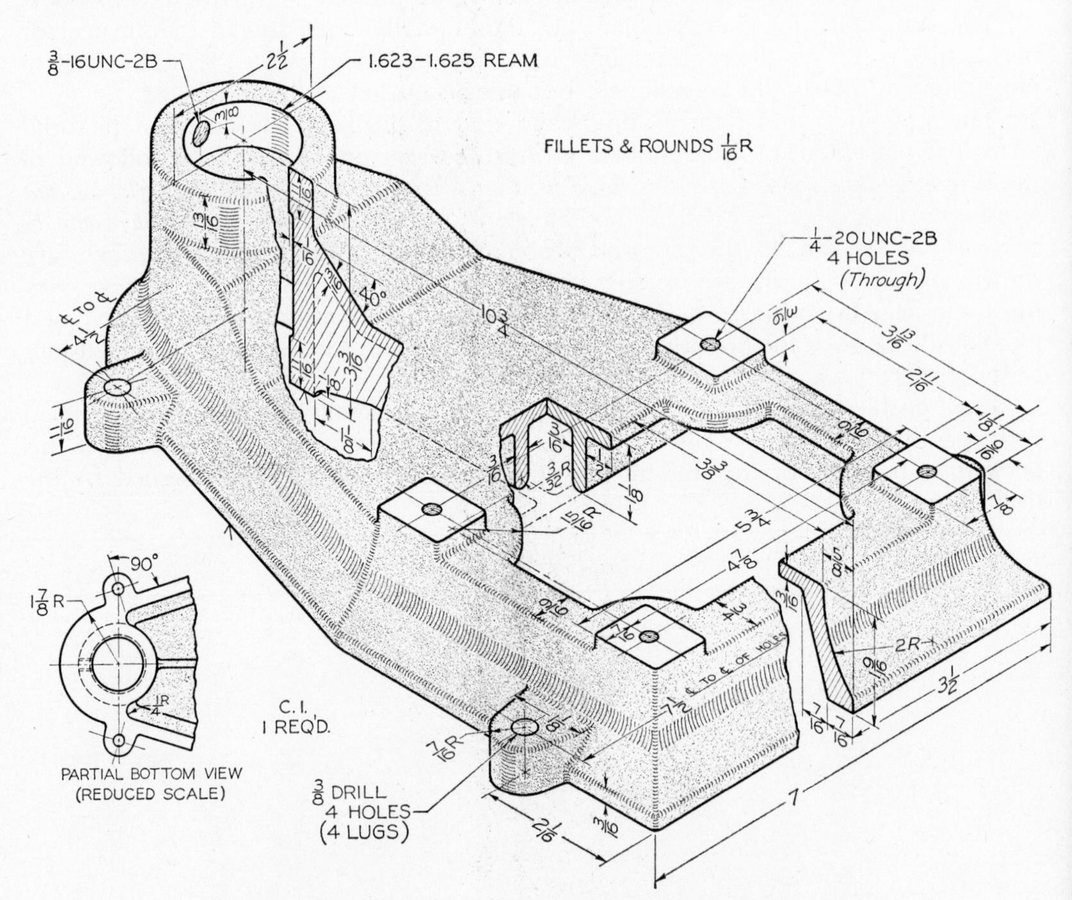

Fig. 739 **Drill Press Base.** Make detail drawing. Use Size C sheet.

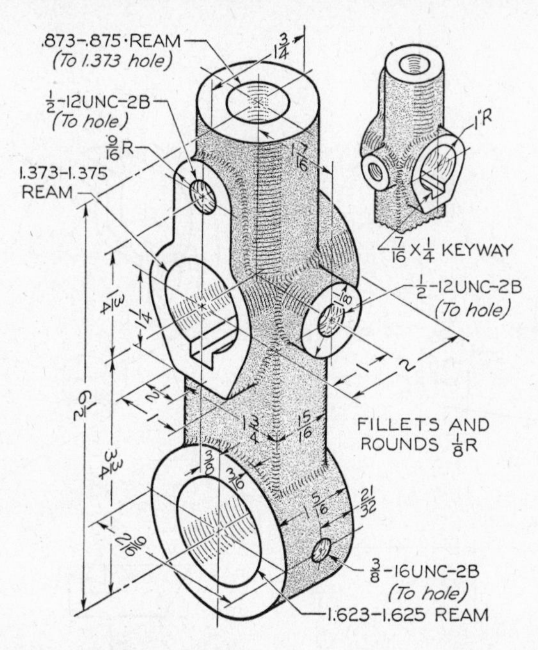

Fig. 740 Idler Arm. Make detail drawing. Use Size B sheet.

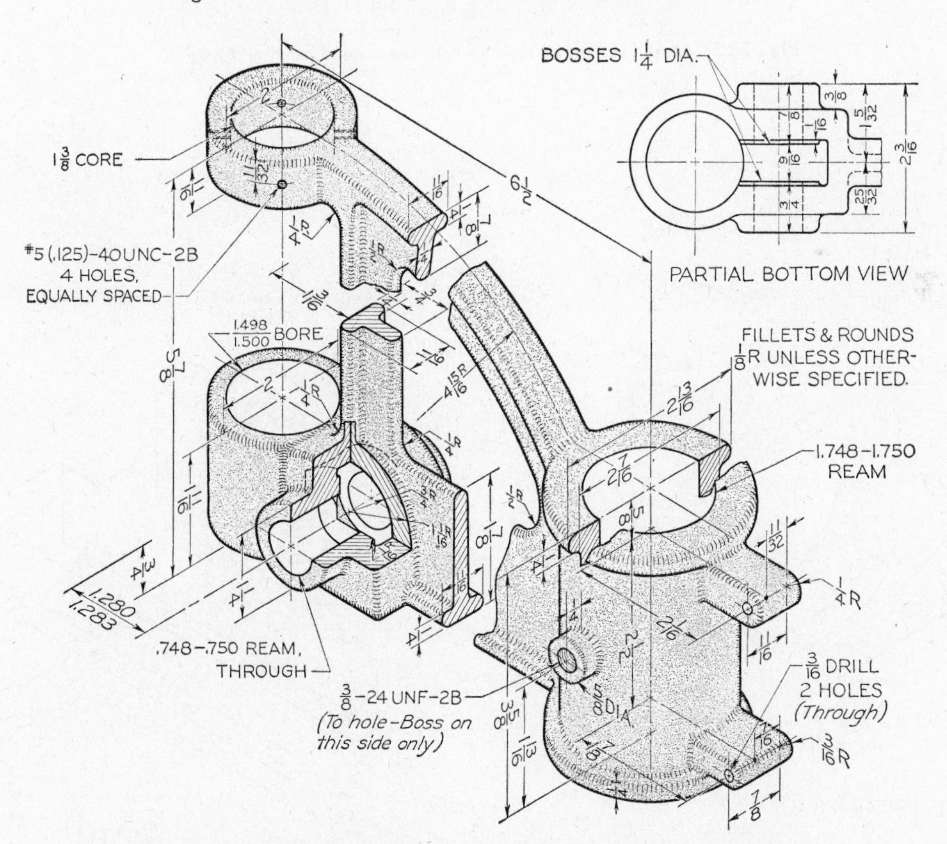

Fig. 741 Drill Press Bracket. Make detail drawing. Use Size C sheet.

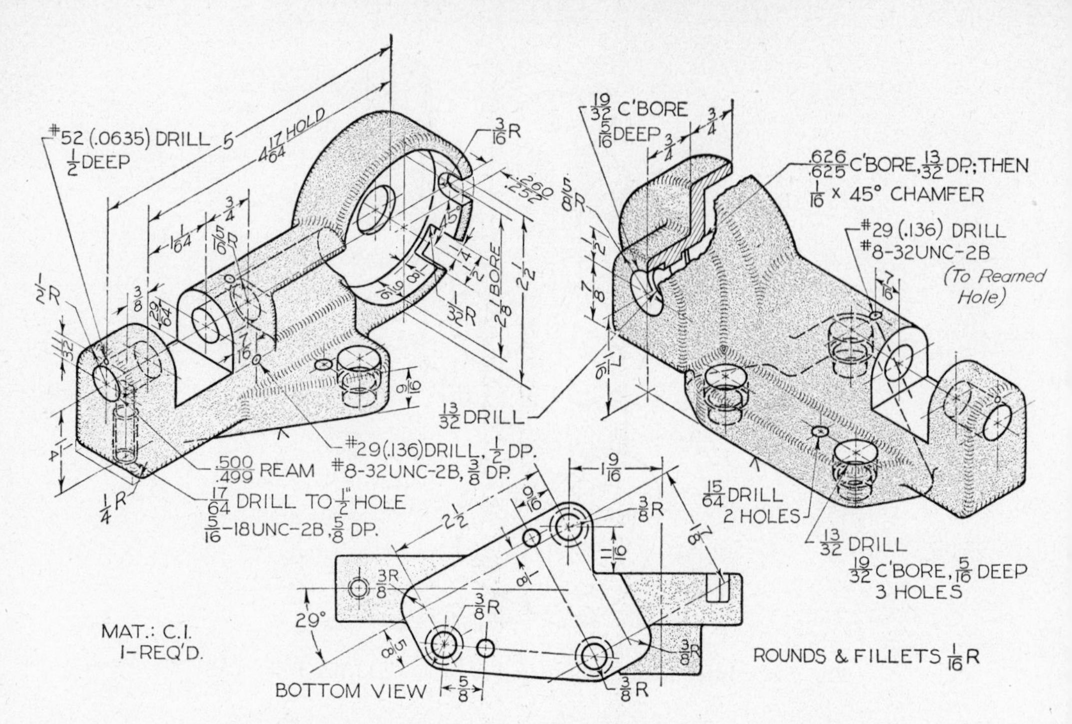

Fig. 742 Dial Holder. Make detail drawing. Use Size C sheet.

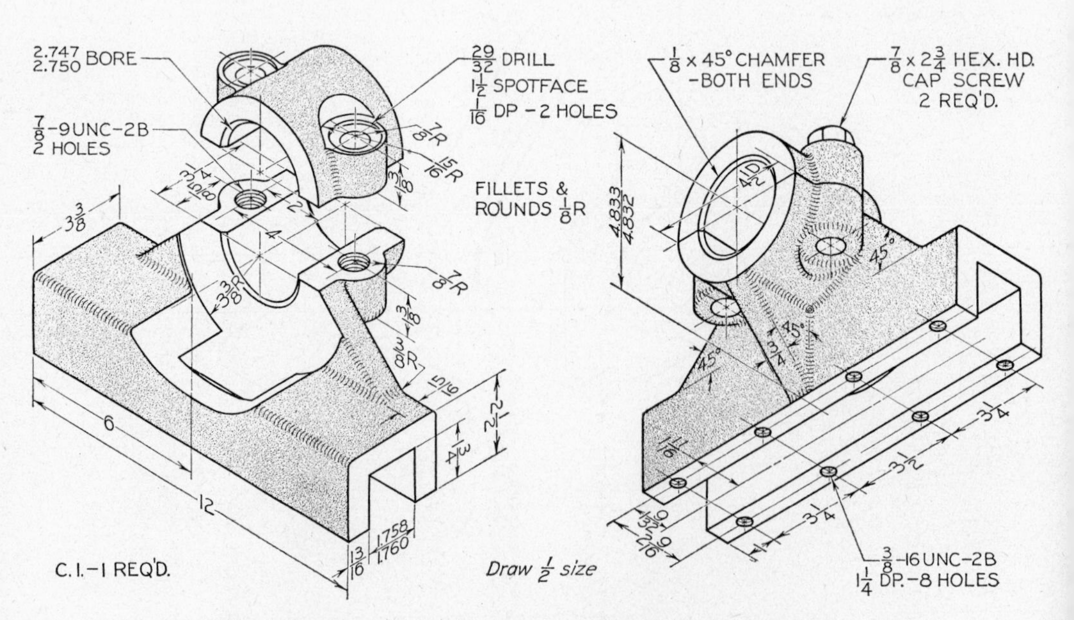

Fig. 743 Rack Slide. Make detail drawings. Use Size B sheet.

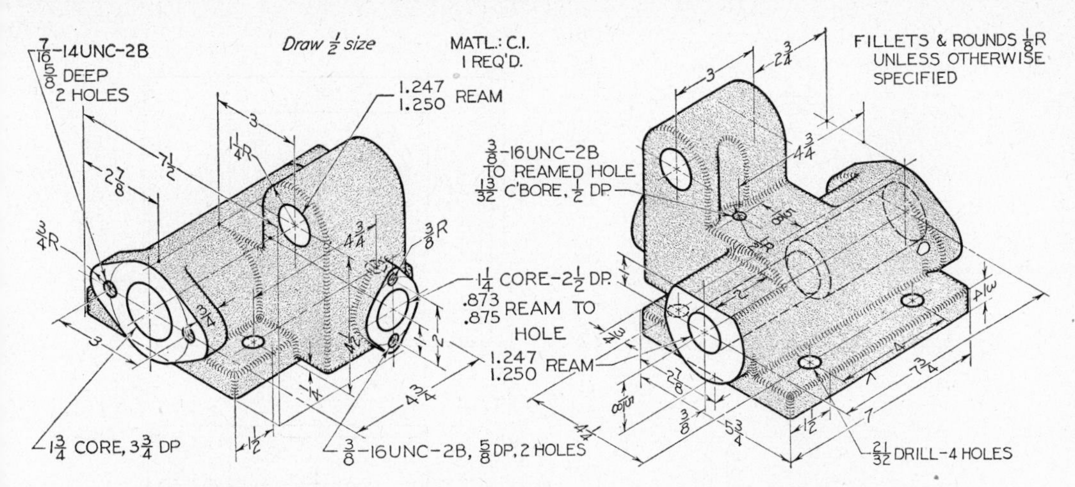

Fig. 744 Automatic Stop Box. Make detail drawing. Use Size B sheet.

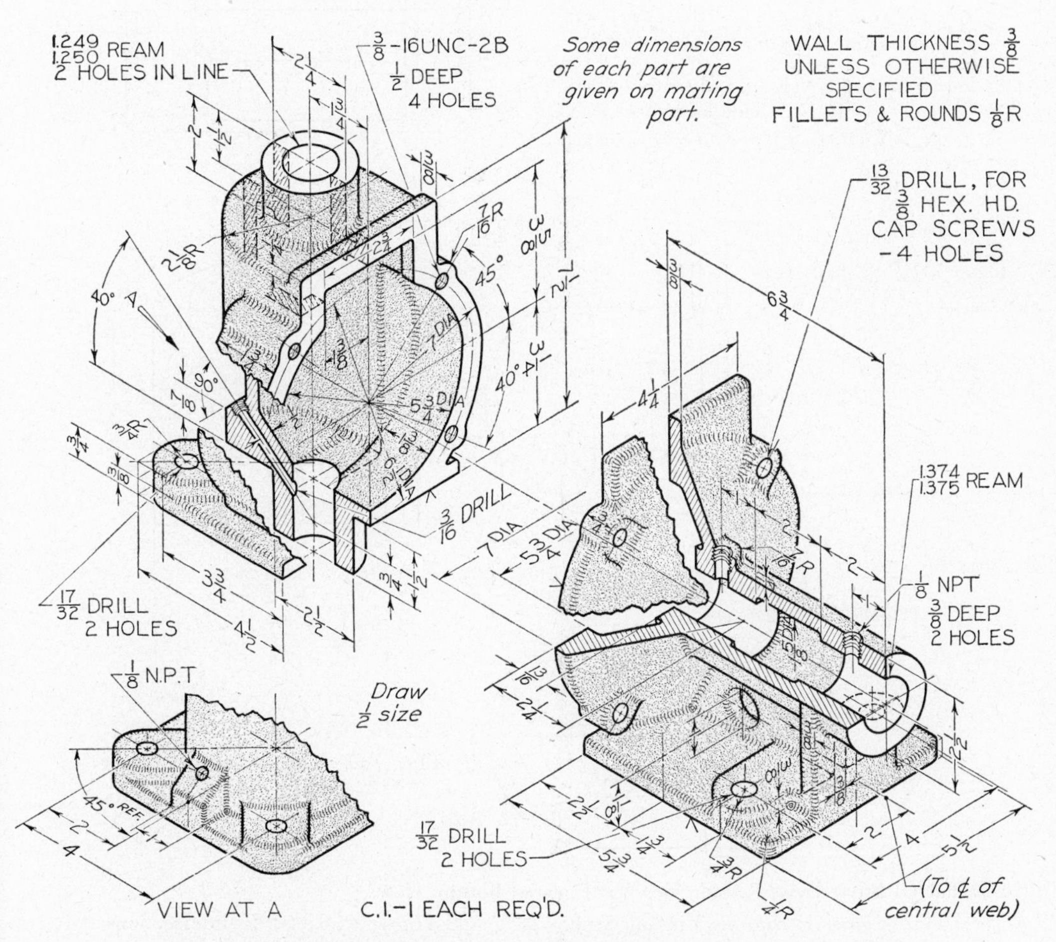

Fig. 745 Conveyor Housing. Make detail drawings. Use Size C sheets.

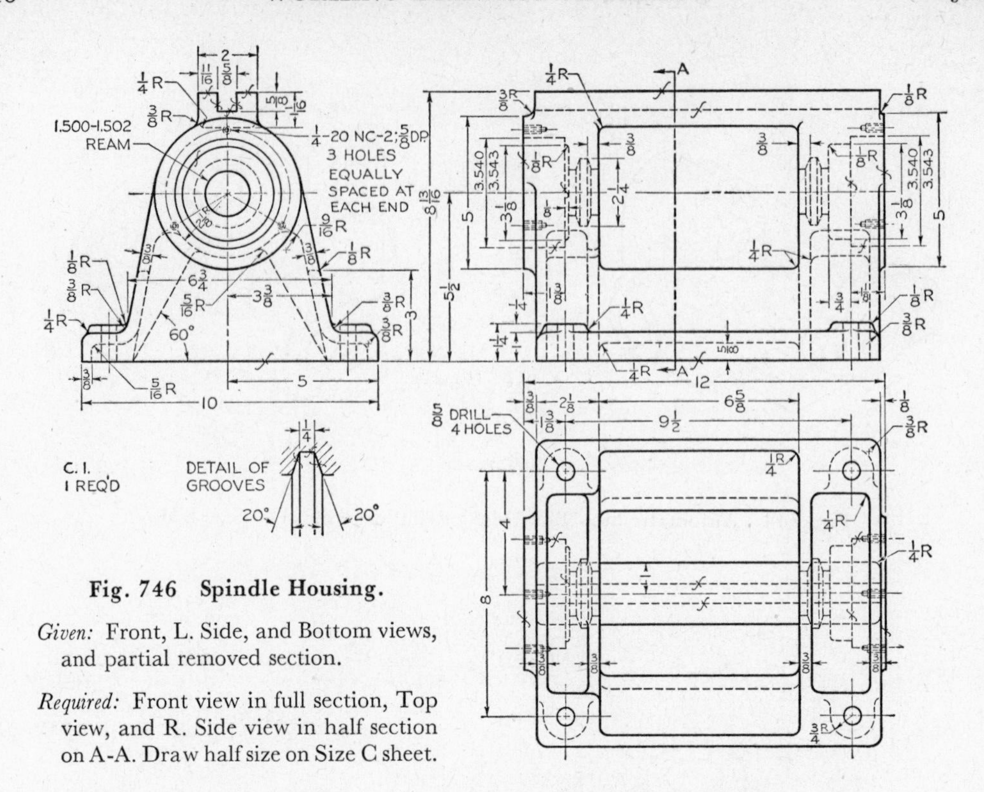

Fig. 746 Spindle Housing.

Given: Front, L. Side, and Bottom views, and partial removed section.

Required: Front view in full section, Top view, and R. Side view in half section on A-A. Draw half size on Size C sheet.

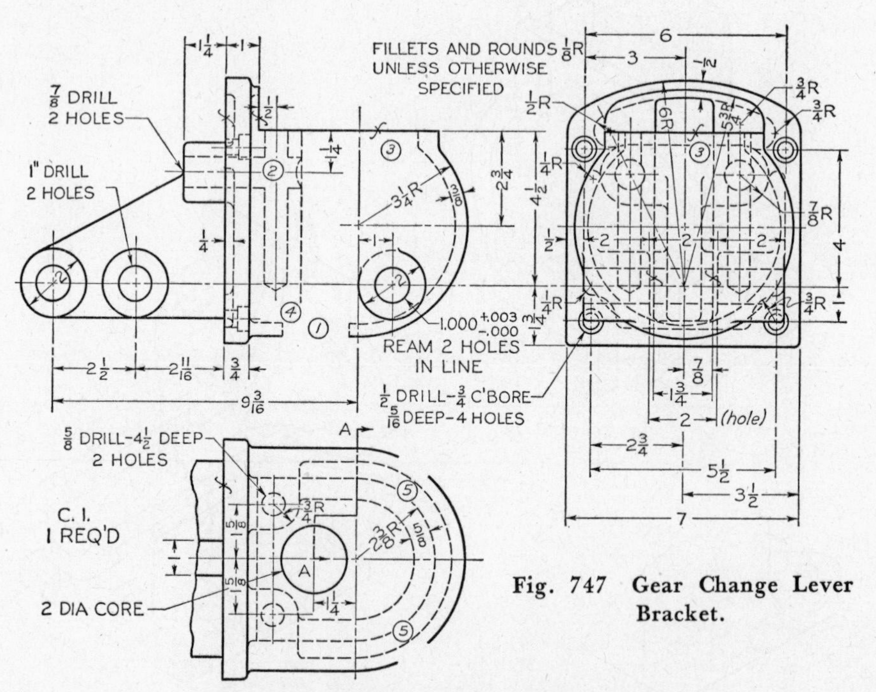

Fig. 747 Gear Change Lever Bracket.

Given: Partial Front view, R. Side view, and partial Bottom view.

Required: Front view, L. Side view in half Section A-A, and Top view. Construct intersections at 1, 2, 3, and 4. Complete the visible lines for bosses at 5. Draw half size on Size C sheet.

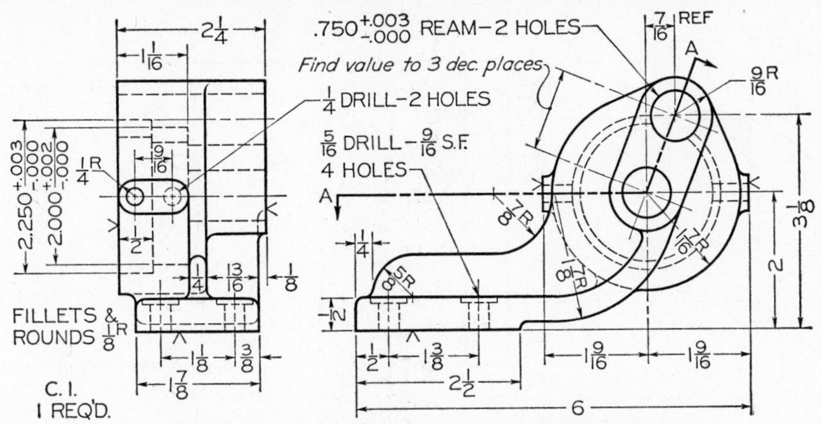

Fig. 748 Pump Bracket for a Thread Milling Machine.

Given: Front and L. Side views.

Required: Front and R. side views, and Top view in section on A-A. Draw full size on Size B sheet.

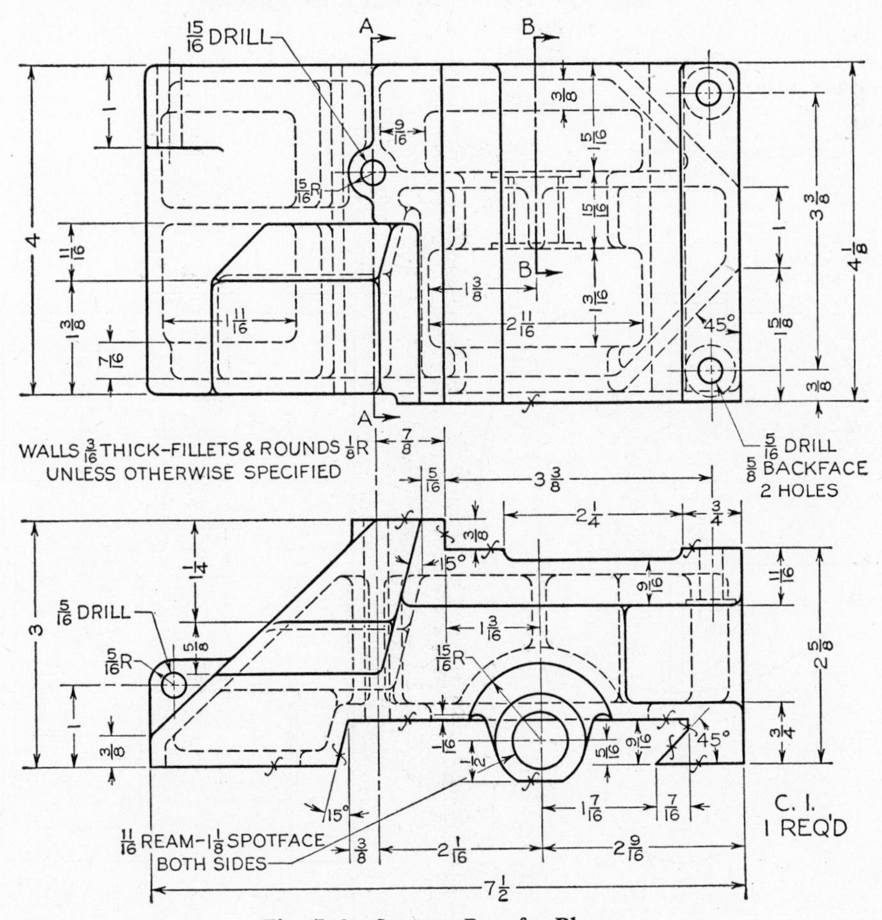

Fig. 749 Support Base for Planer.

Given: Front and Top views.

Required: Front, Top, and L. Side view full Section A-A, and removed Section B-B. Draw full size on Size C sheet.

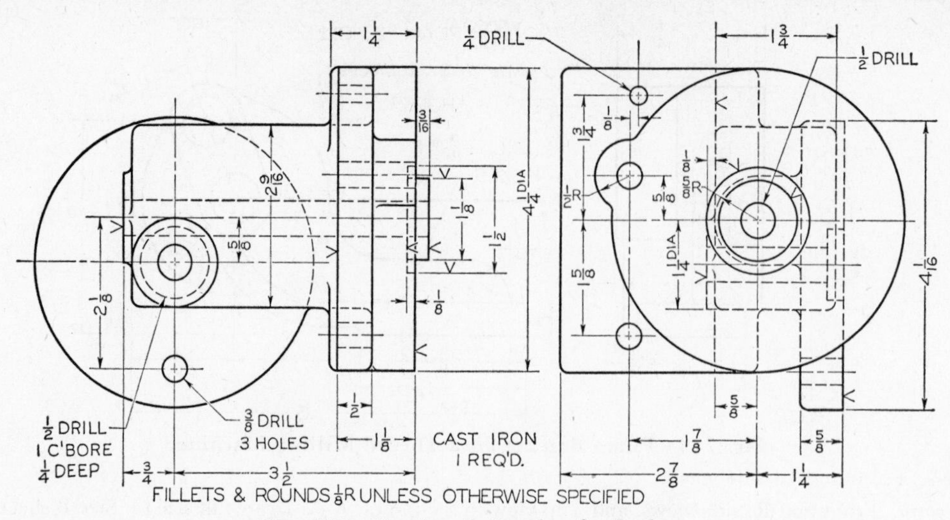

Fig. 750 Headstock Bracket for Lathe.

Given: Front and R. side views.

Required: Take present Front view as new Top view; then add new Front and L. Side views. Draw full size on Size C sheet.

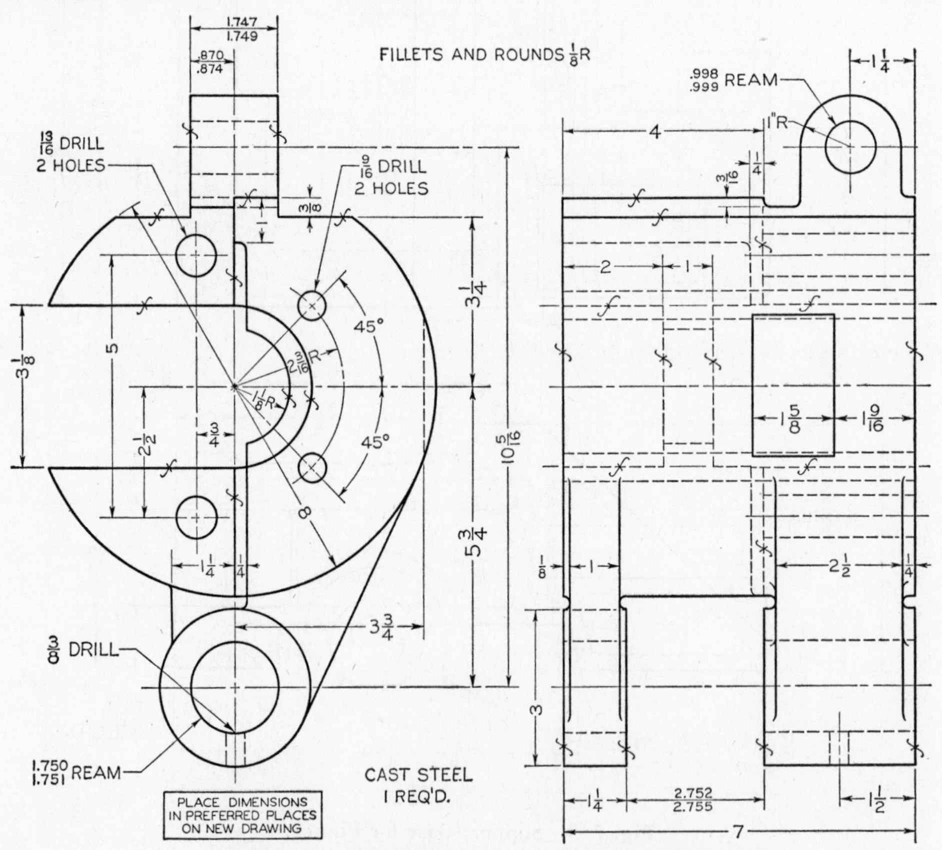

Fig. 751 Fixture Base for 60-Ton Vertical Press.

Given: Front and R. Side views.

Required: Revolve Front view 90° clockwise; then add Top and L. Side Views. Draw half size on Size C sheet.

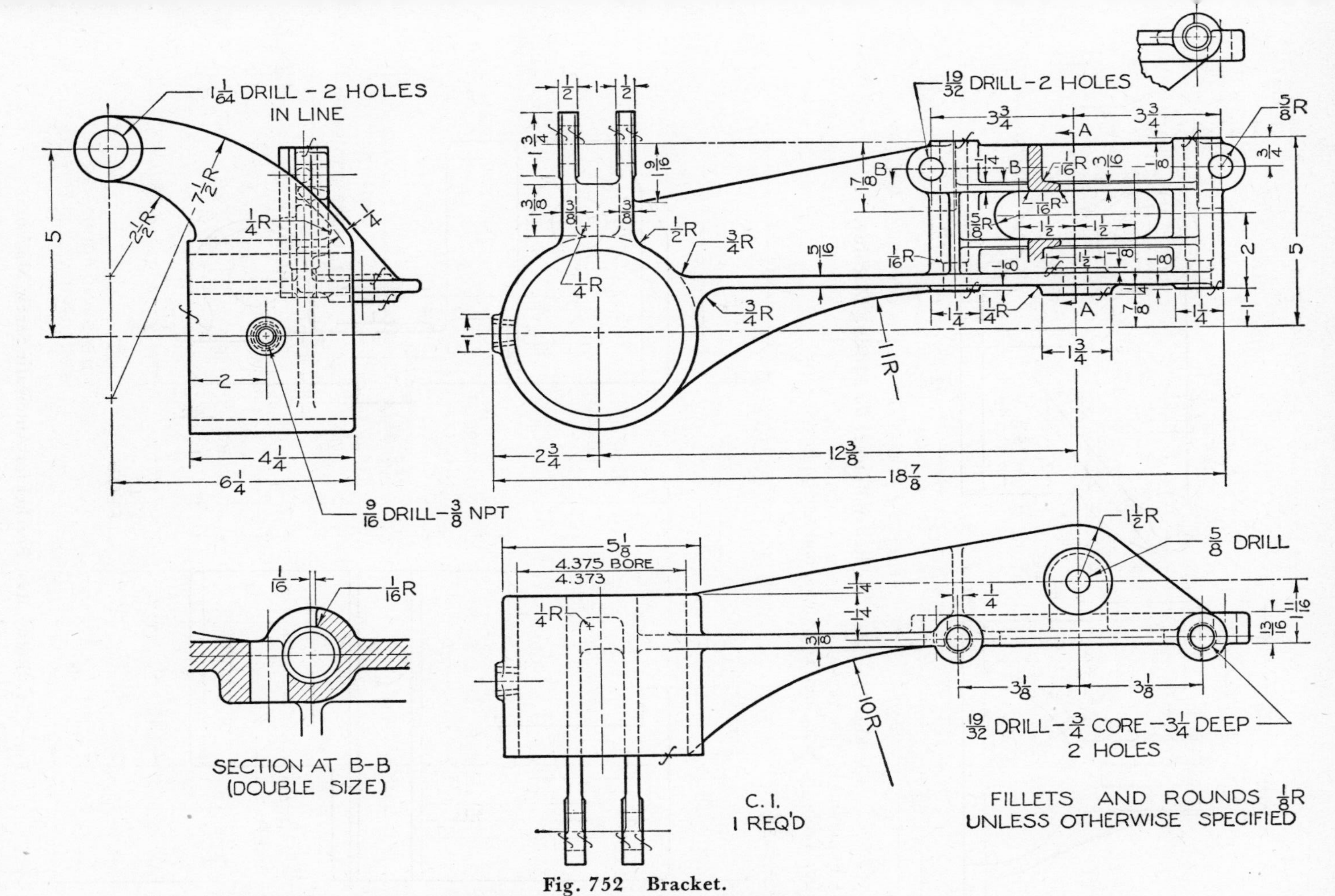

Fig. 752 Bracket.

Given: Front, L. Side, and Bottom views, and partial removed section.
Required: Make detail drawing. Draw Front, Top, and R. Side views, and removed Sections A-A and B-B. Draw half size on Size C sheet. Draw Sec. B-B full size.

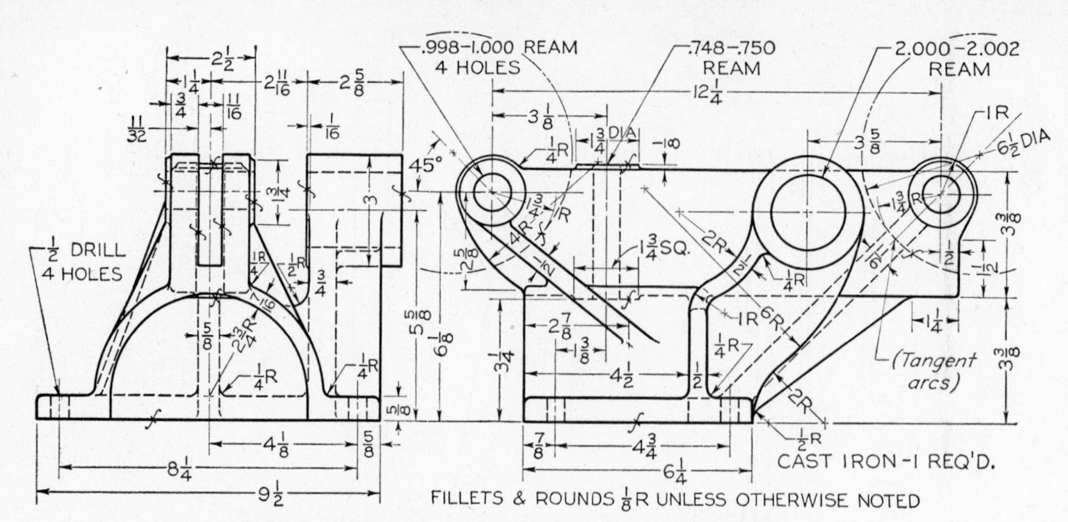

Fig. 753 Belt Tightener Bracket for Milling Machine.

Given: Front and L. Side views.

Required: Front, R. Side, and Top views, half size on Size C sheet.

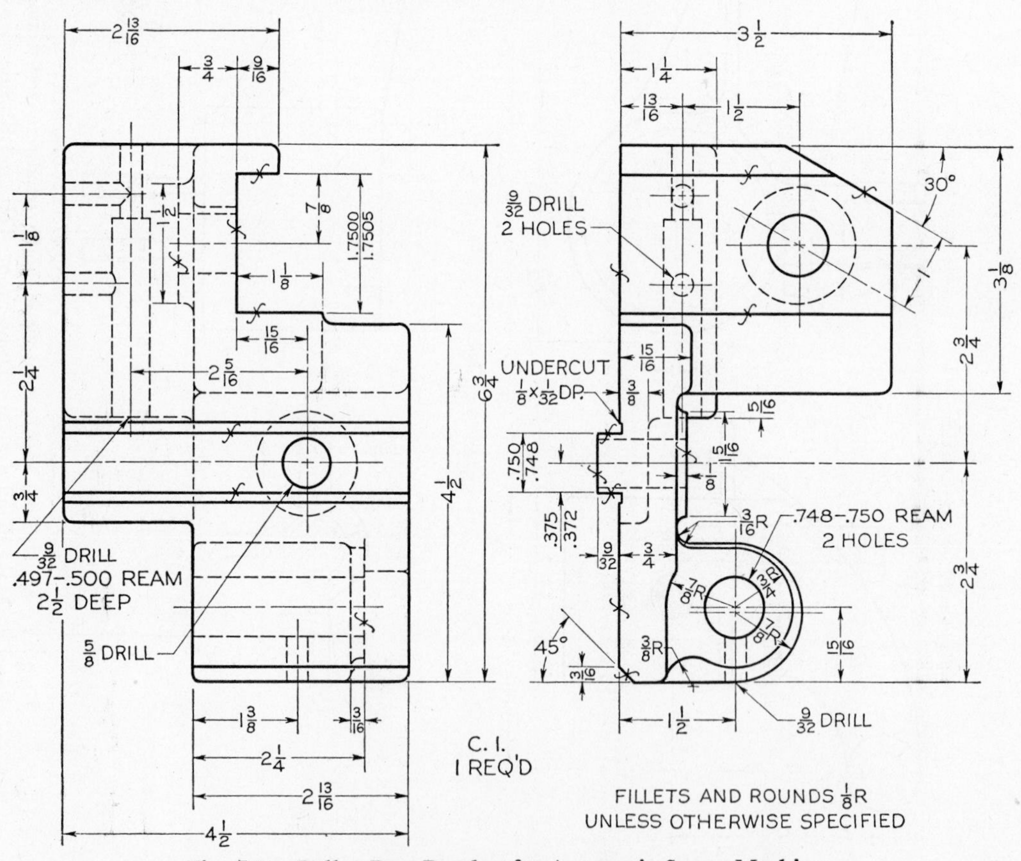

Fig. 754 Roller Rest Bracket for Automatic Screw Machine.

Given: Front and L. Side views.

Required: Revolve Front view 90° counterclockwise; then add Top and L. Side views. Draw full size on Size C sheet.

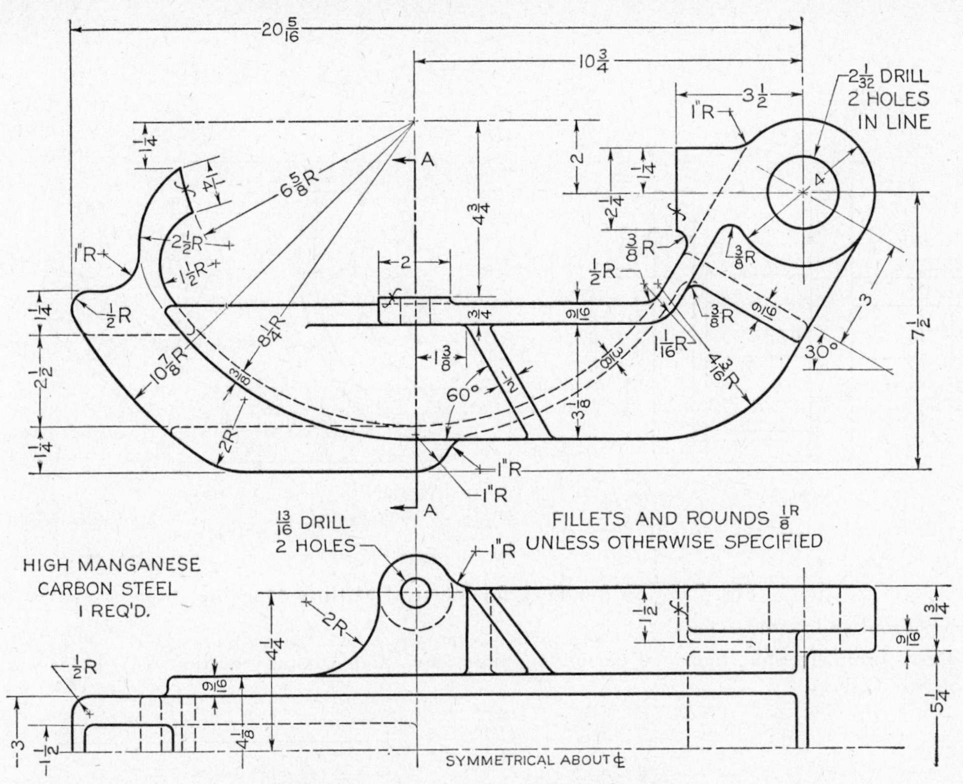

Fig. 755 Sheave Guard for Skimmer.

Given: Front view and half Bottom view.

Required: Front view, complete Top view, L. Side view, and removed Section A-A. Draw half size on Size C sheet.

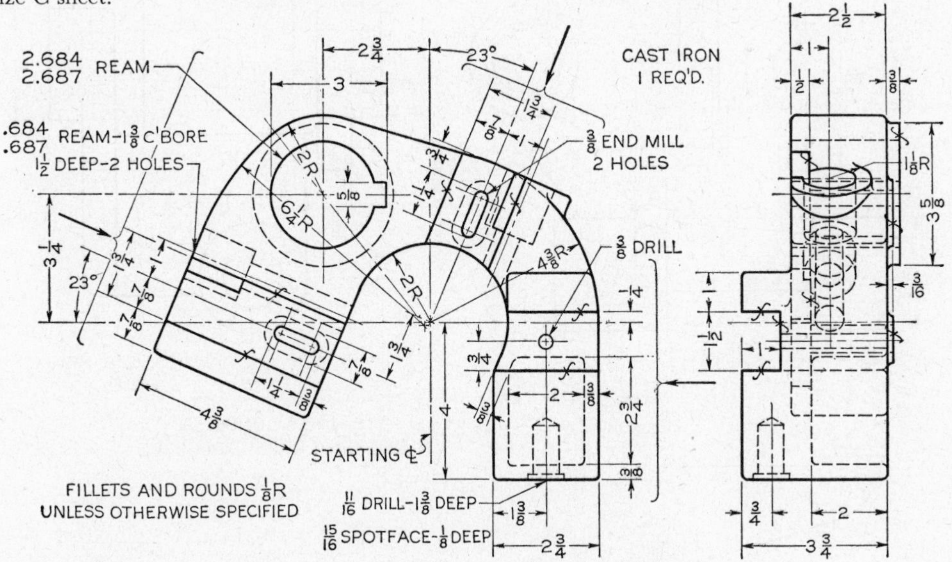

Fig. 756 Guide Bracket for Gear Shaper.

Given: Front and R. Side views.

Required: Front view and two partial Auxiliary views, and a partial R. Side view taken in direction of arrows. Draw half size on Size C sheet.

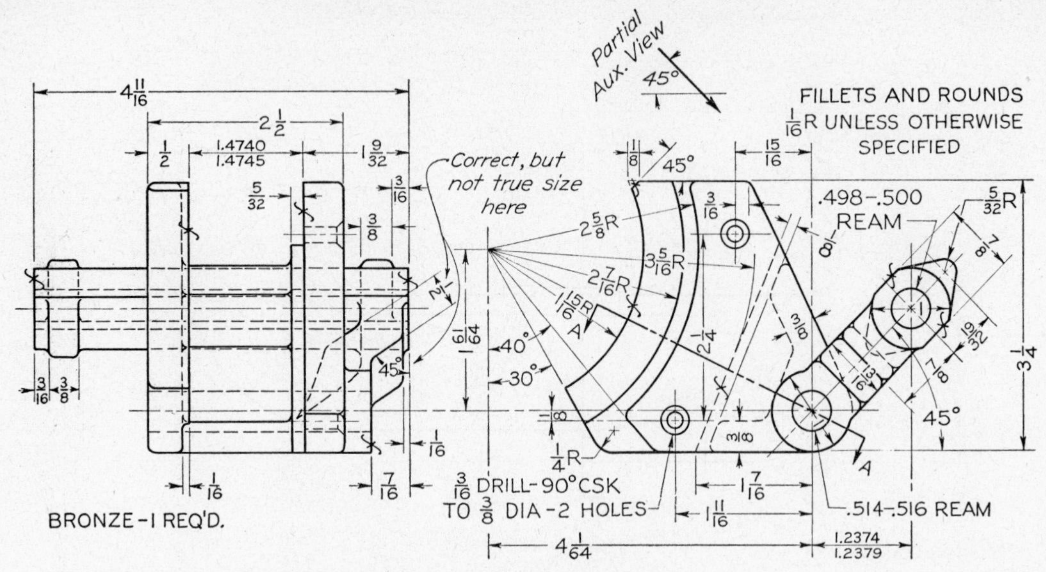

Fig. 757 Shifter Fork for Vertical Milling Machine.

Given: Front and L. Side views.

Required: Front, L. Side, indicated partial Auxiliary view, and Auxiliary Section A-A. Draw full size on Size C sheet.

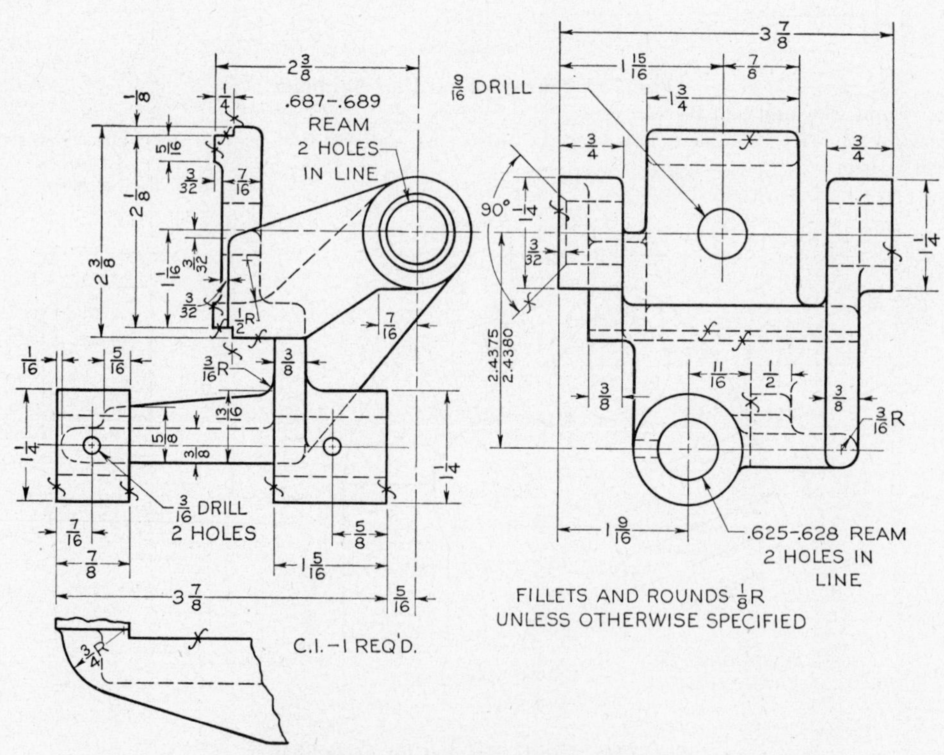

Fig. 758 Bearing for a Worm Gear.

Given: Front and R. Side views.

Required: Front, Top, and L. Side views. Draw full size on Size C sheet.

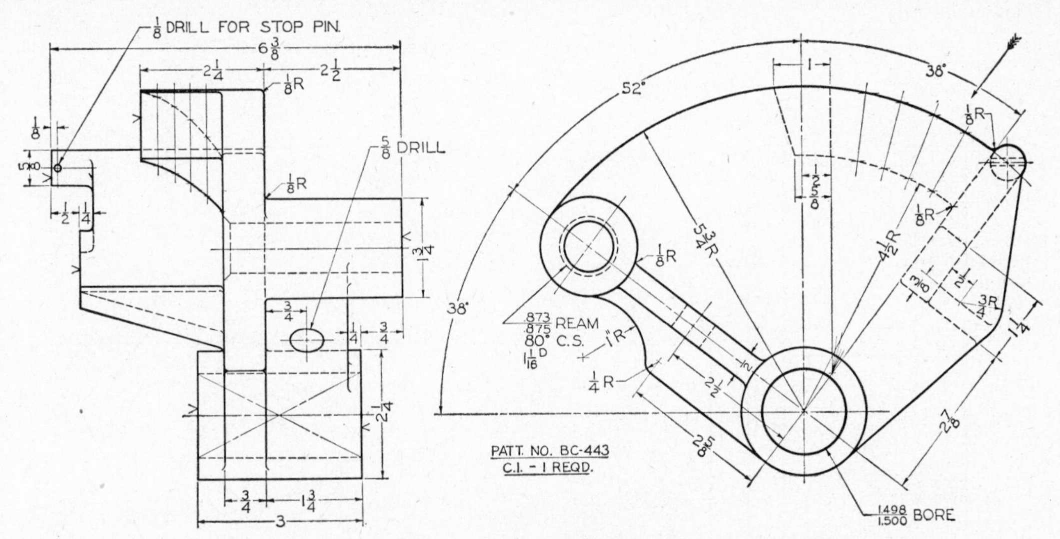

Fig. 759 Unloader Slide Lever.

Given: Front and L. side views.
Required: Front, R. side, and complete Auxiliary view as indicated by arrow. Draw full size on Size C sheet.

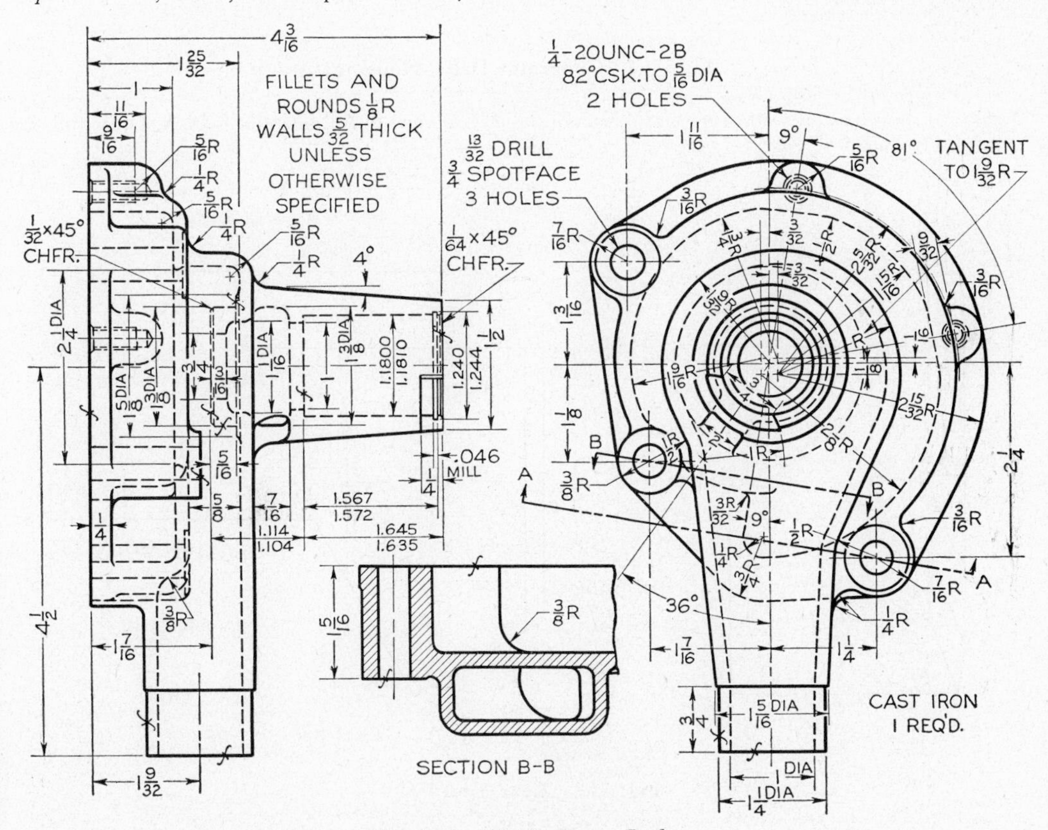

Fig. 760 Water Pump Body.

Given: Front and L. side views, with detail Section B-B.
Required: Front view, R. side in full section, and Auxiliary Section A-A. Include in Section A-A all visible lines behind the cutting plane. Omit Section B-B. Draw full size on Size C sheet.

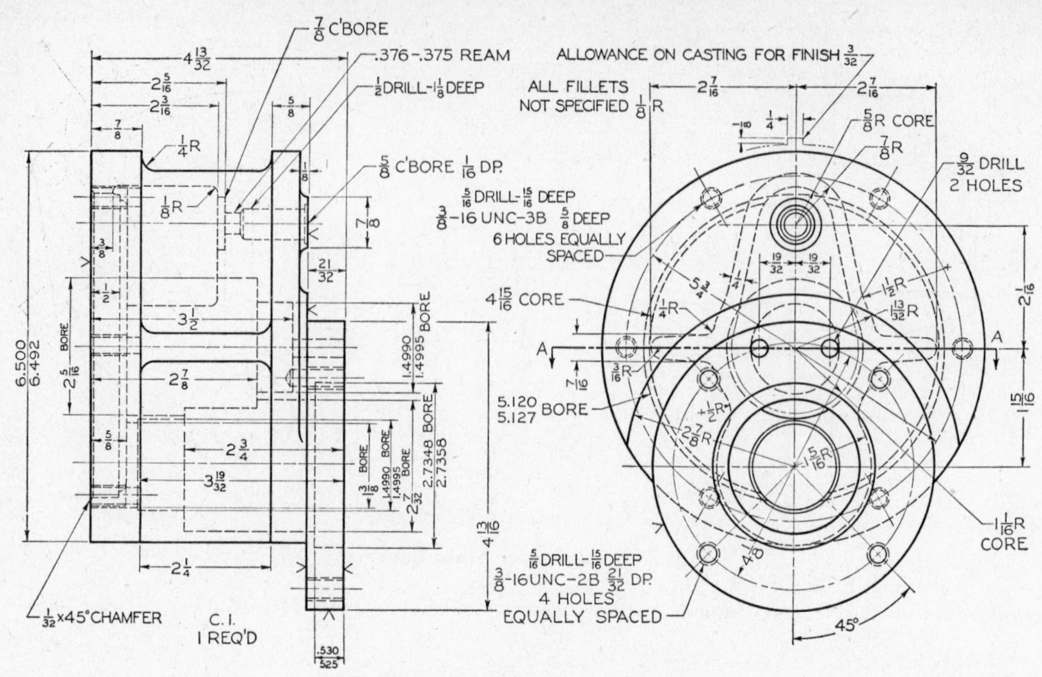

Fig. 761 Generator Drive Housing.

Given: Front and L. Side views.

Required: Front view, R. Side view in full section, and Top view in full Section on A-A. Draw full size on Size C sheet.

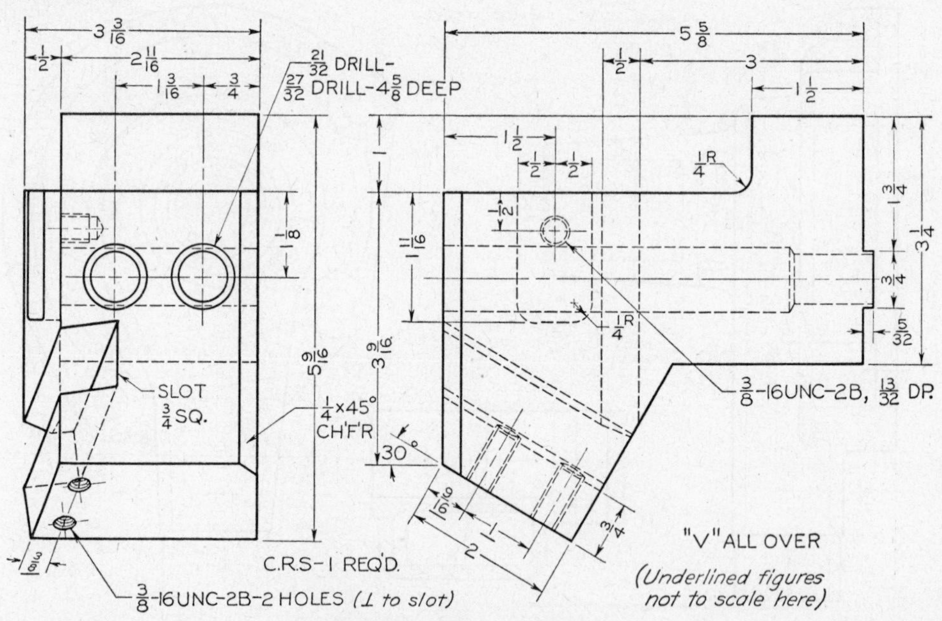

Fig. 762 Rear Tool Post.

Given: Front and L. Side views.

Required: Take L. Side view as Top view in new drawing; then add Front and L. Side views, and a Primary Auxiliary view followed by a Secondary Auxiliary view taken so as to show true end view of $\frac{3}{4}''$ slot. Draw Front and L. Side views approx. $8\frac{1}{2}''$ apart. Complete all views, except show only the necessary hidden lines in Auxiliary views. Draw full size on Size C sheet.

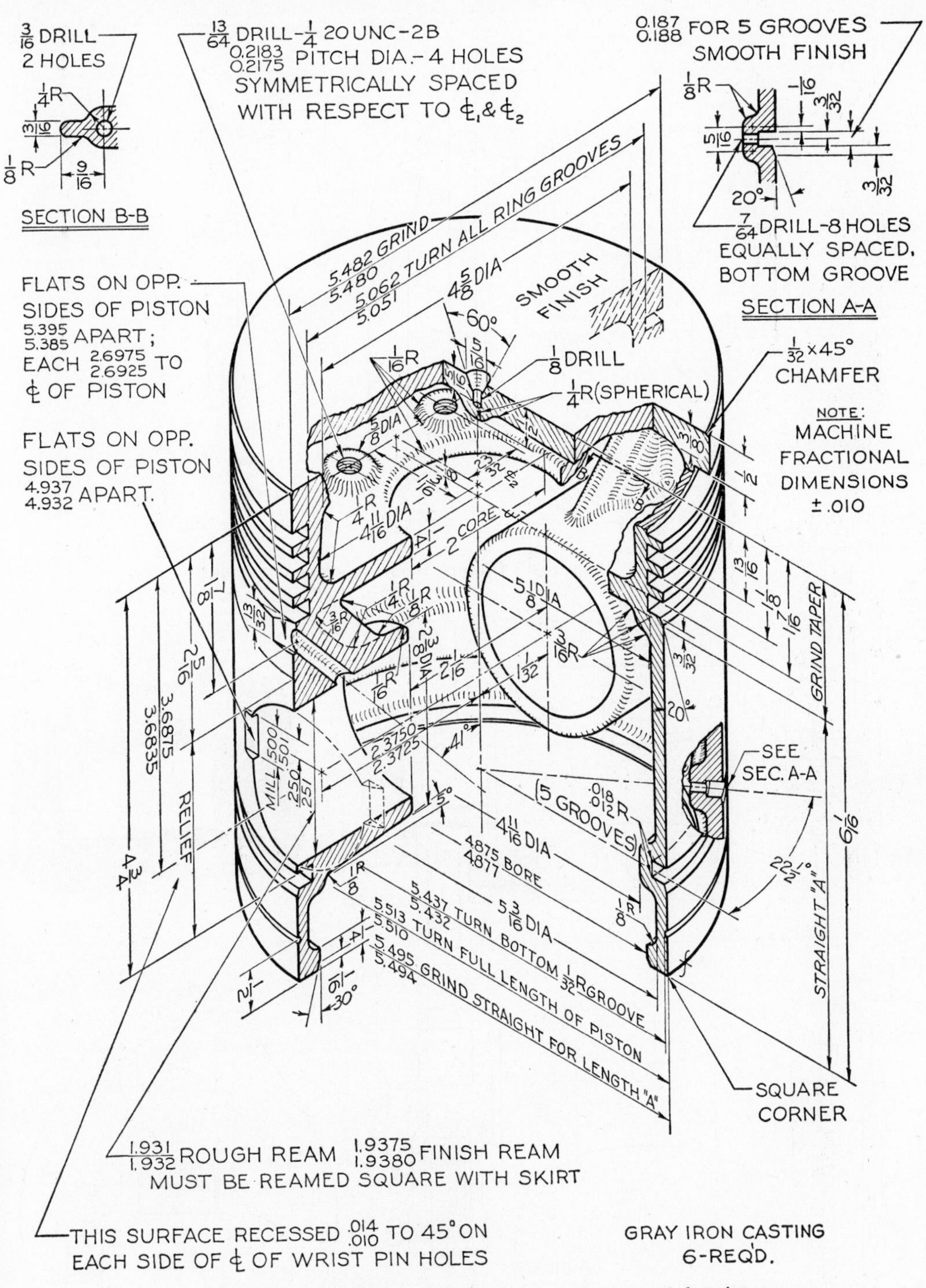

Fig. 763 Caterpillar Tractor Piston. Make detail drawing.

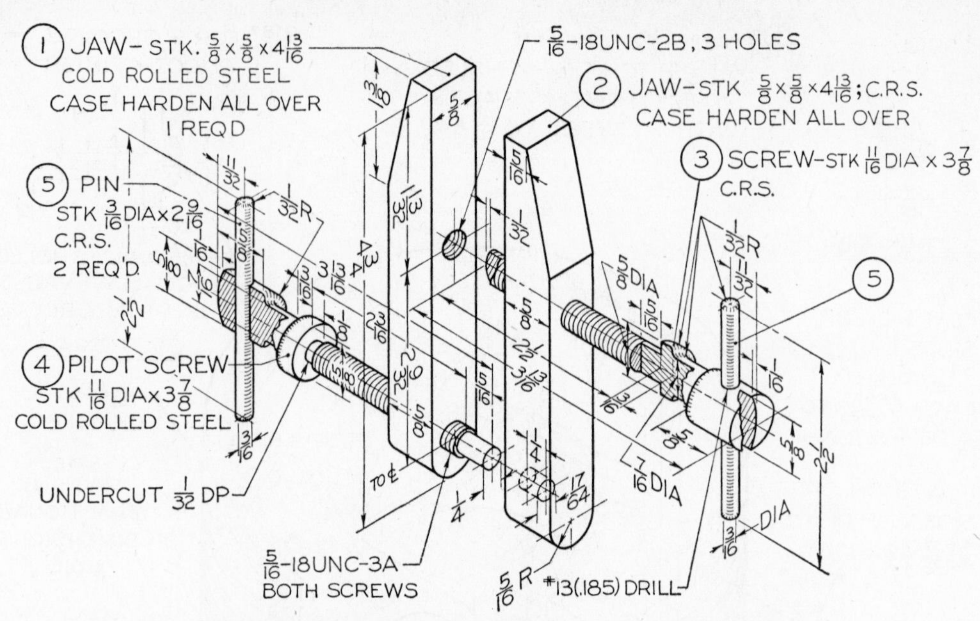

Fig. 764 Machinist's Clamp. Draw details and assembly.

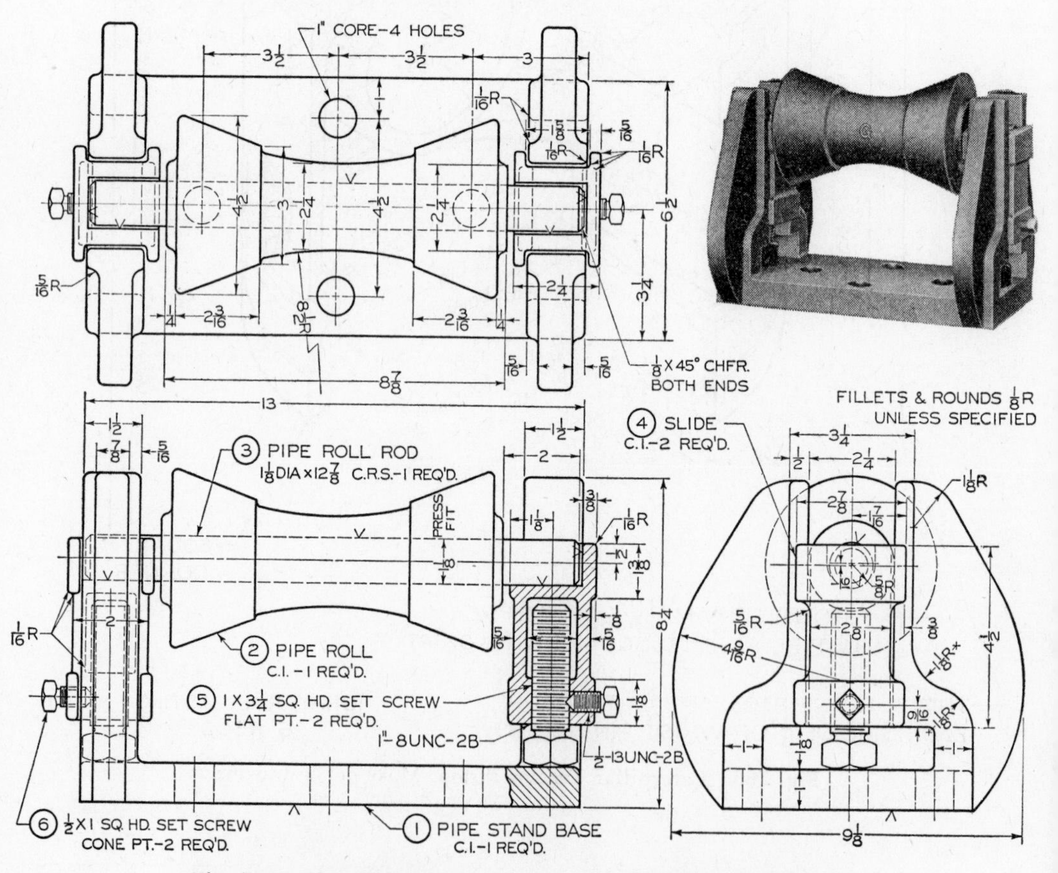

Fig. 765 Pipe Roll Stand. (1) Draw details. (2) Draw assembly.

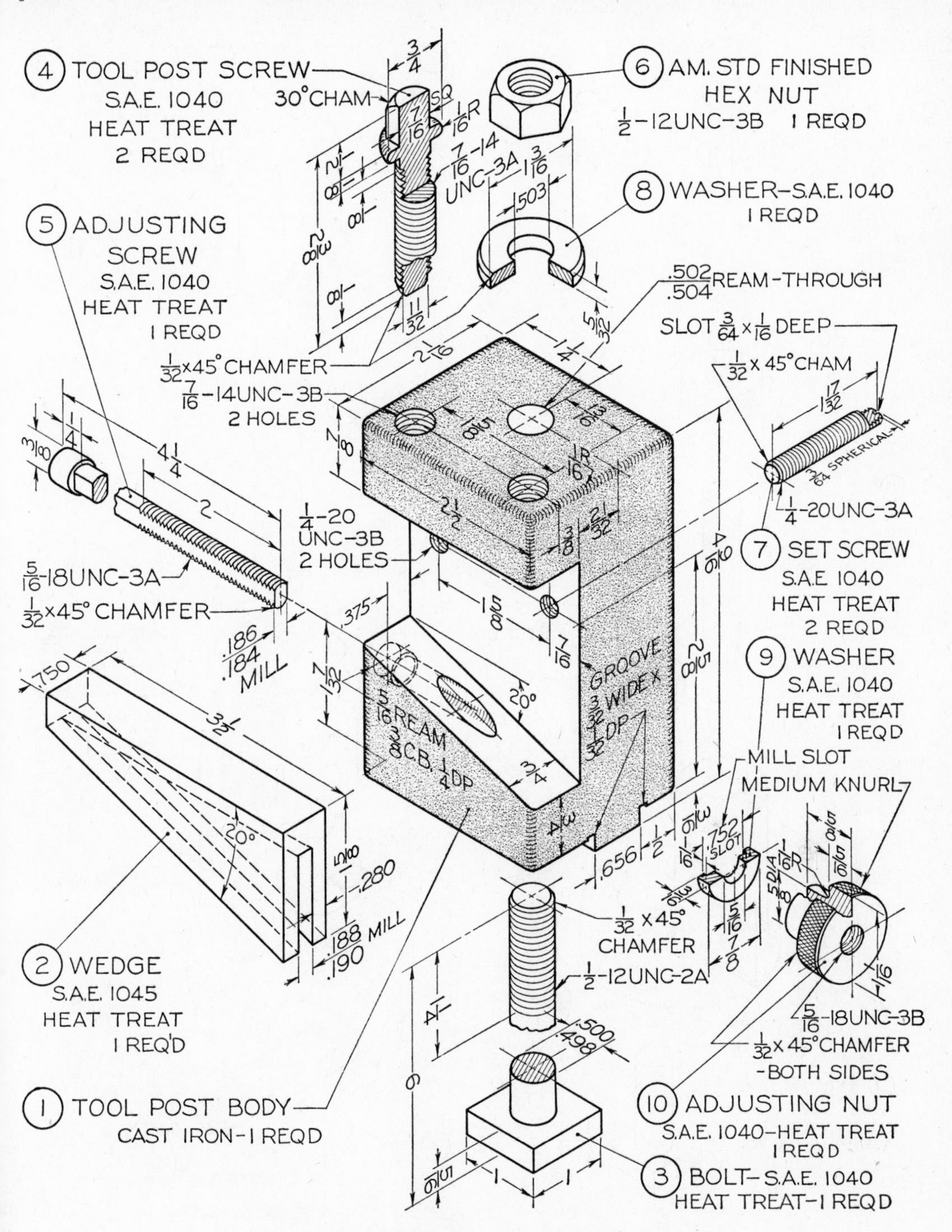

Fig. 766 Tool Post. (1) Draw details. (2) Draw assembly.

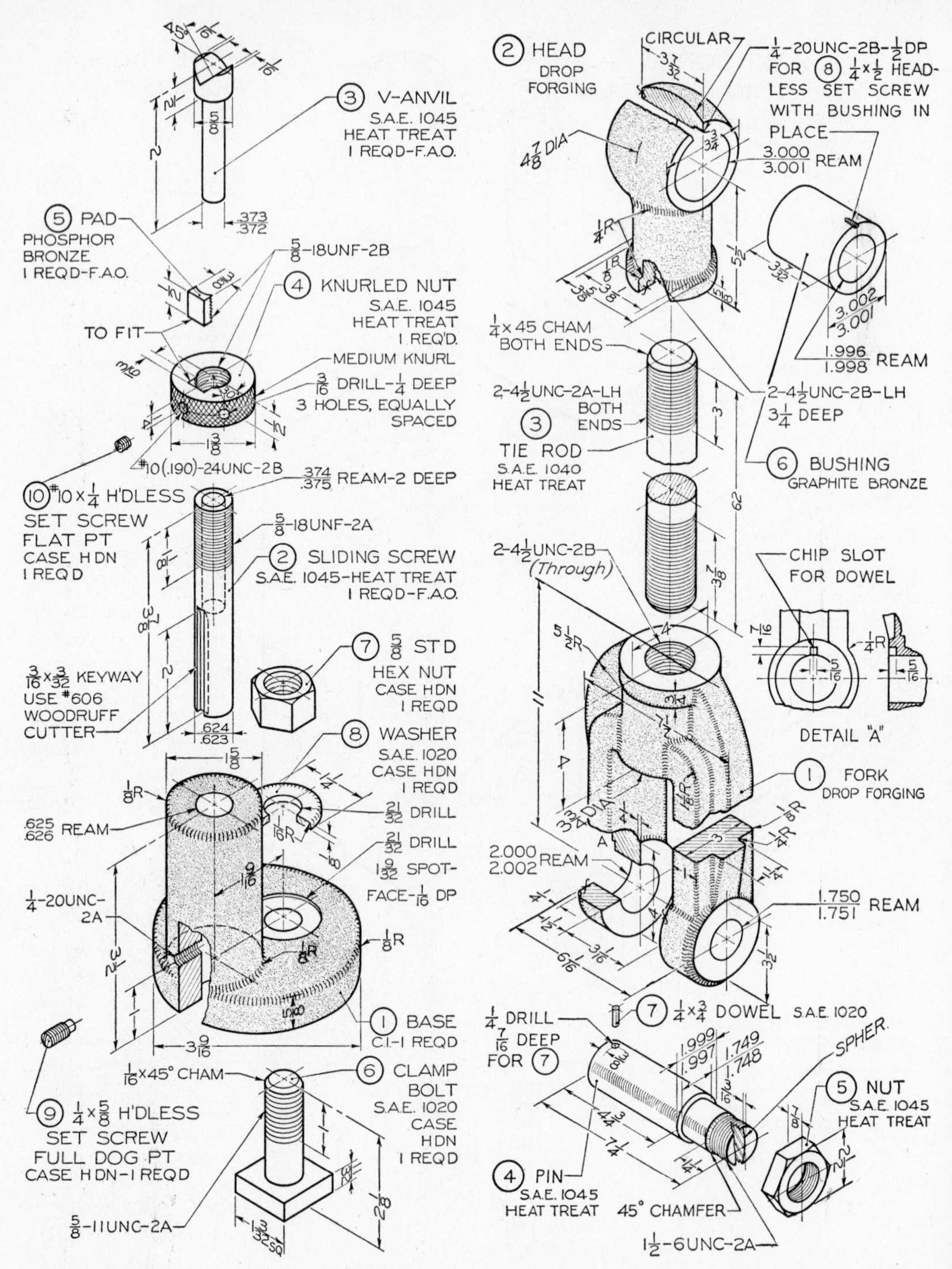

Fig. 767 Milling Jack. (1) Draw details. (2) Draw assembly.

Fig. 768 Connecting Bar. (1) Draw details. (2) Draw assembly.

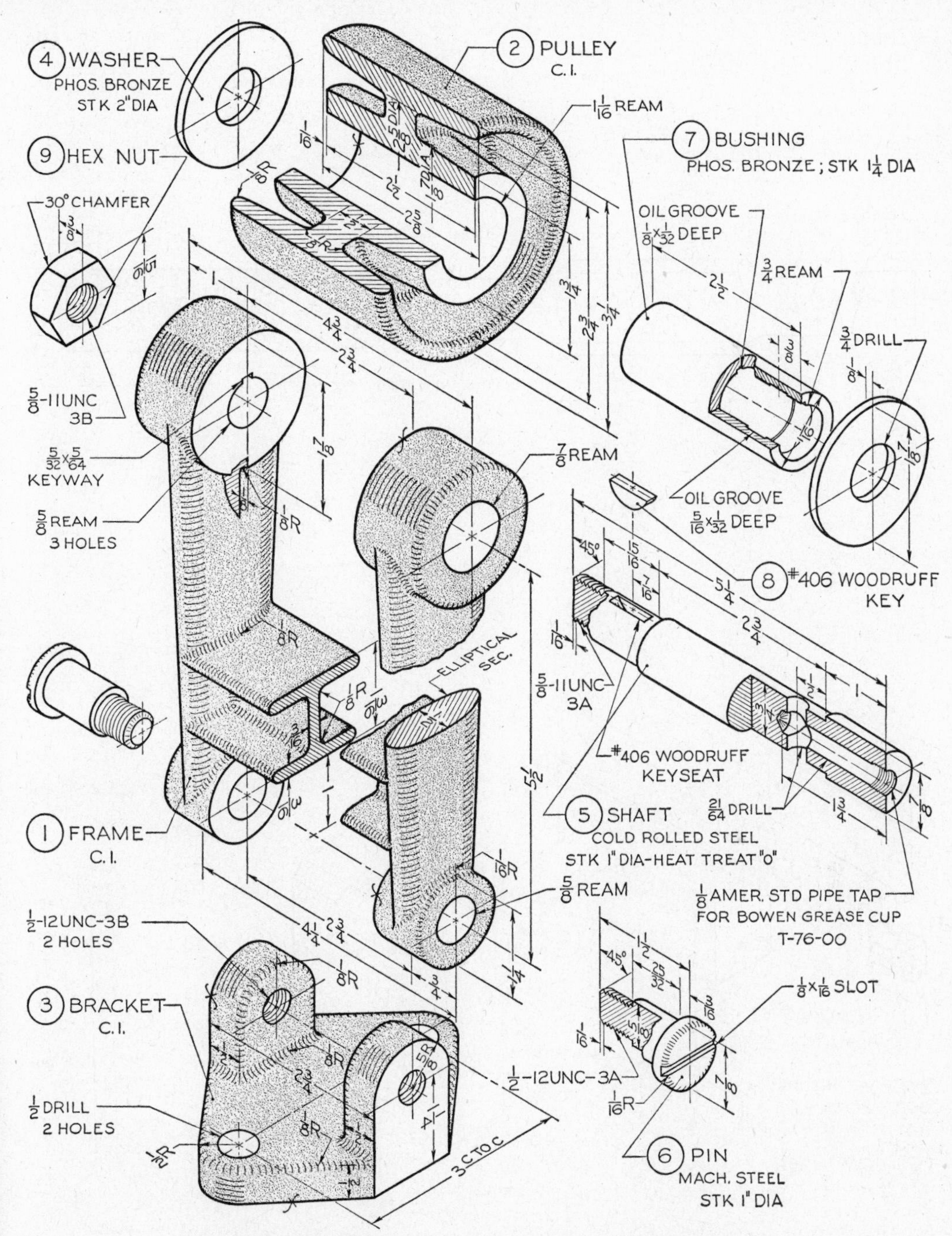

Fig. 769 Belt Tightener. (1) Draw details. (2) Draw assembly.

It is assumed that the parts are to be made in quantity and they are to be dimensioned for interchangeability on the detail drawings. Using Tables of Limits in Appendix tables 5–11, give dimensions as follows:

1. Bushing fit in pulley: Class LN 2 fit.
2. Shaft fit in bushing: Class RC 5 fit.
3. Shaft fits in frame: Class RC 2 fit.
4. Pin fit in frame: Class RC 5 fit.
5. Pulley hub length plus washers fit in frame: Allowance .005 and tolerances .004.
6. Make bushing .010″ shorter than pulley hub.
7. Bracket fit in frame: same as 5 above.

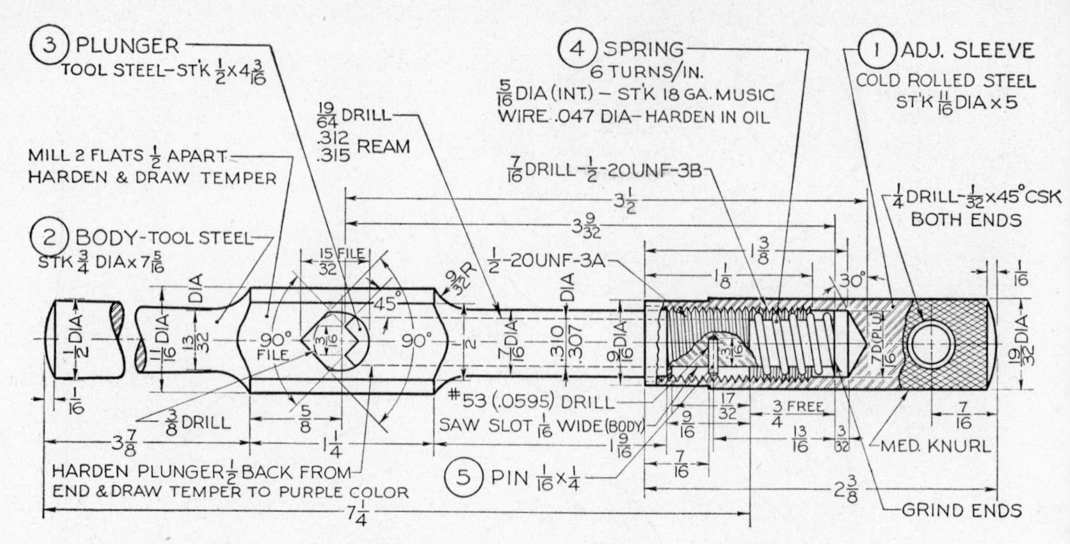

Fig. 770 Tap Wrench. (1) Draw details. (2) Draw assembly.

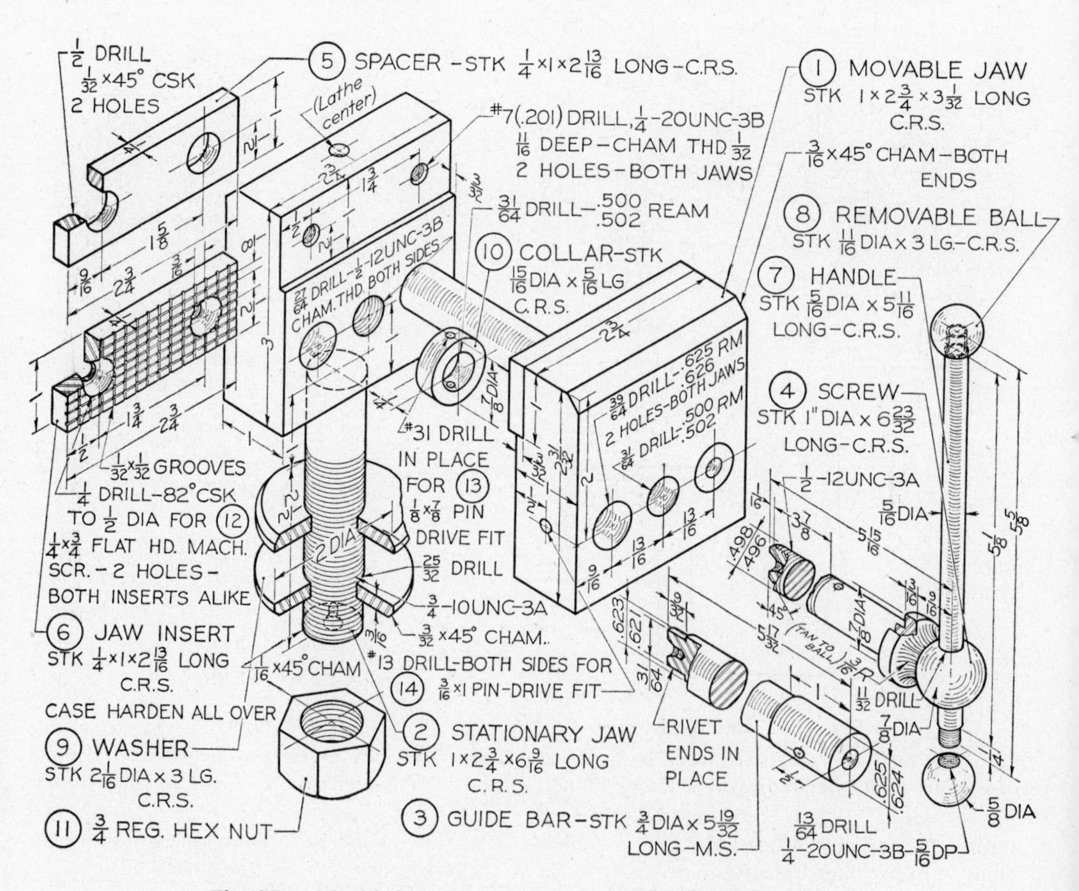

Fig. 771 Machinist's Vise. (1) Draw details. (2) Draw assembly.

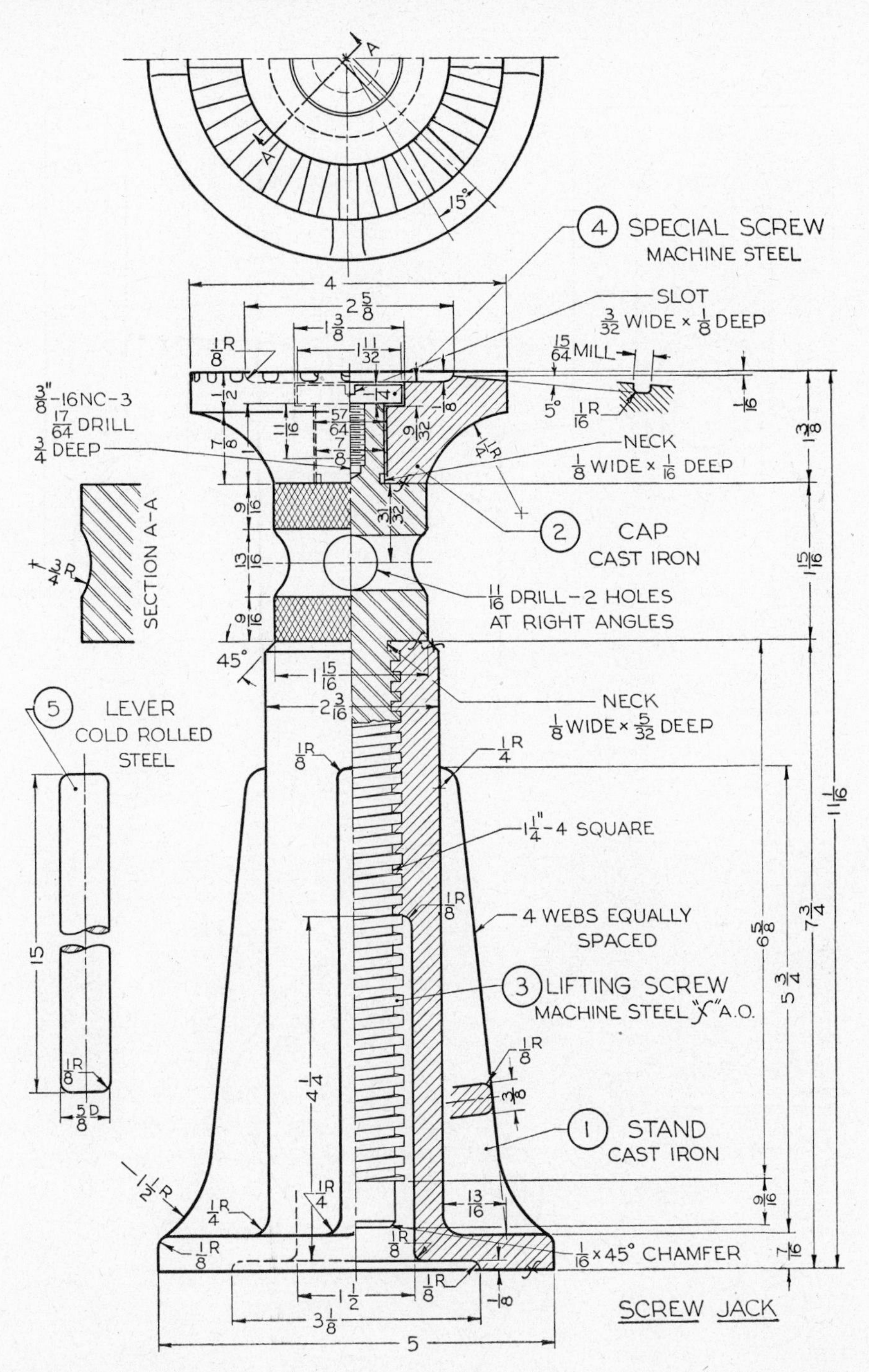

Fig. 772 Screw Jack. (1) Draw details (see Fig. 717, showing "blocked-in" views on Sheet Layout C-678). (2) Draw assembly.

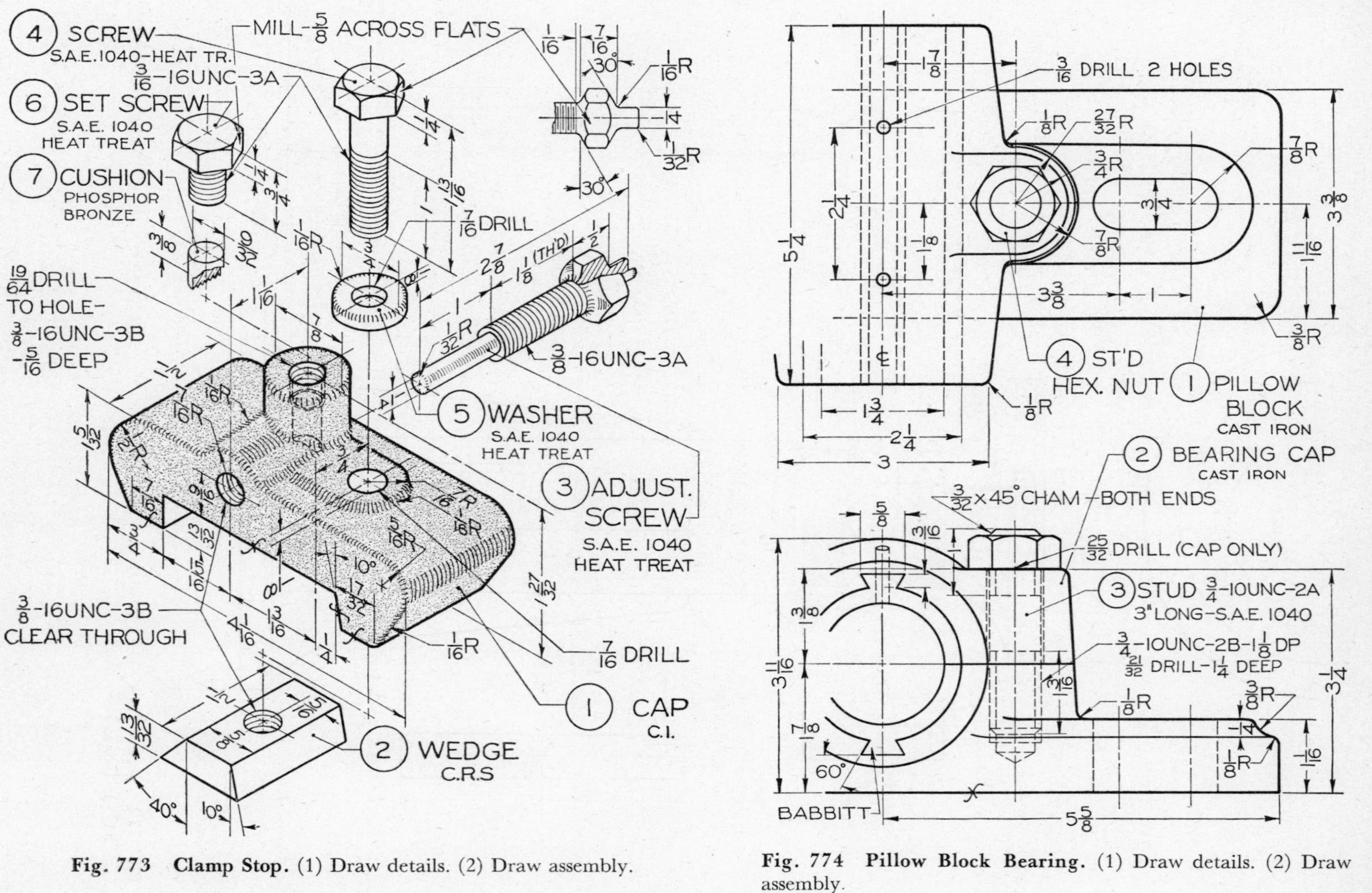

Fig. 773 **Clamp Stop.** (1) Draw details. (2) Draw assembly.

Fig. 774 **Pillow Block Bearing.** (1) Draw details. (2) Draw assembly.

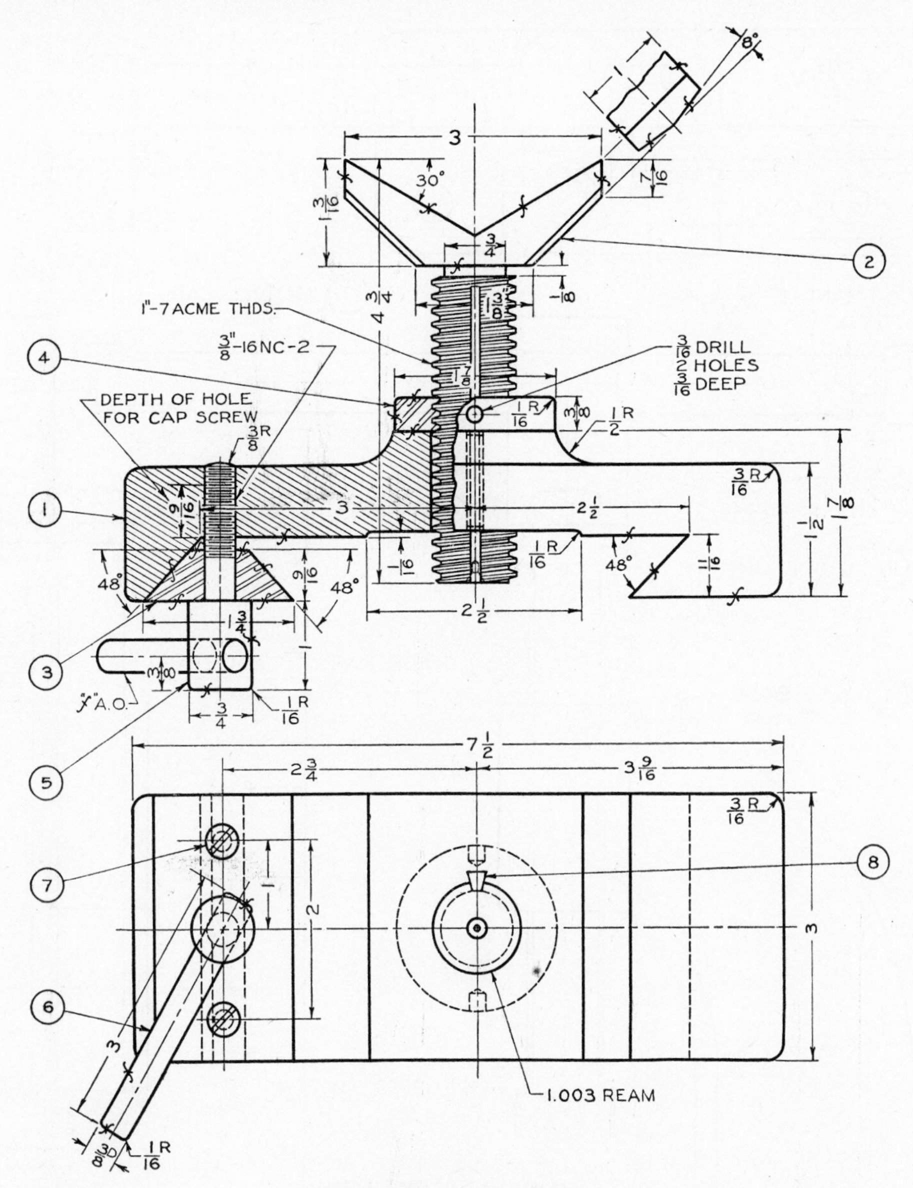

PARTS LIST							
NO.	PART NAME	MAT.	REQ'D.	NO.	PART NAME	MAT.	REQ'D.
1	BASE	C.I.	1	5	CLAMP SCREW	S.A.E. 1020	1
2	REST	S.A.E. 1020	1	6	CLAMP HANDLE	S.A.E. 1020	1
3	CLAMP	S.A.E. 1020	1	7	$\frac{1}{4}$ X 1 FILL. HD. CAP SCREW		2
4	ADJUSTING NUT	S.A.E. 1020	1	8	$\frac{7}{32}$ X $\frac{7}{32}$ X $\frac{1}{8}$ - 1 LONG - KEY	S.A.E. 1030	1

Fig. 775 Centering Rest. (1) Draw details. (2) Draw assembly.

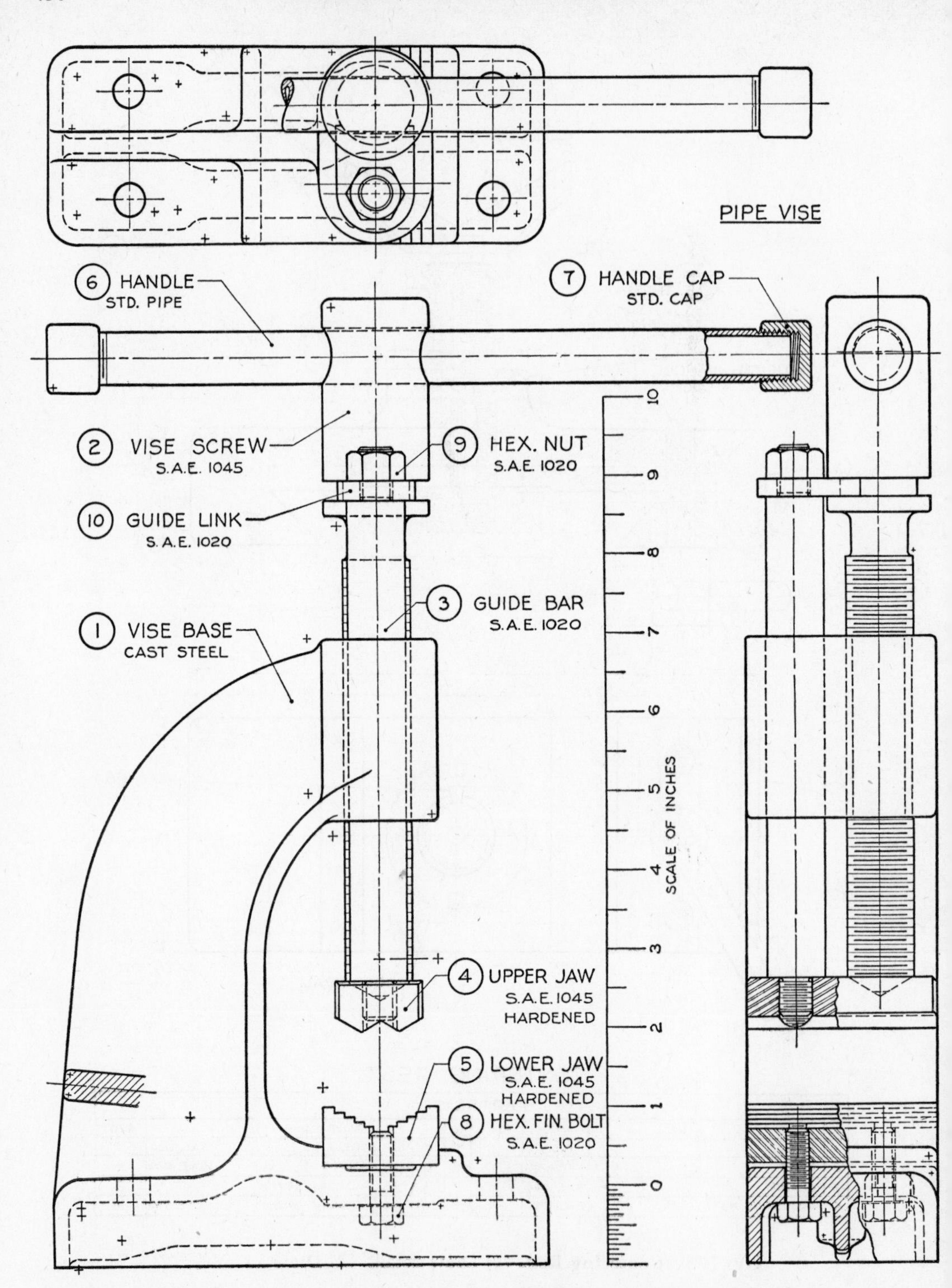

Fig. 776 Pipe Vise. (1) Draw details. (2) Draw assembly. To obtain dimensions, take distances directly from figure with dividers; then set dividers on printed scale and read measurements in inches. All threads are Unified and American Coarse Threads except the American Standard Pipe Threads on Handle and Handle Caps.

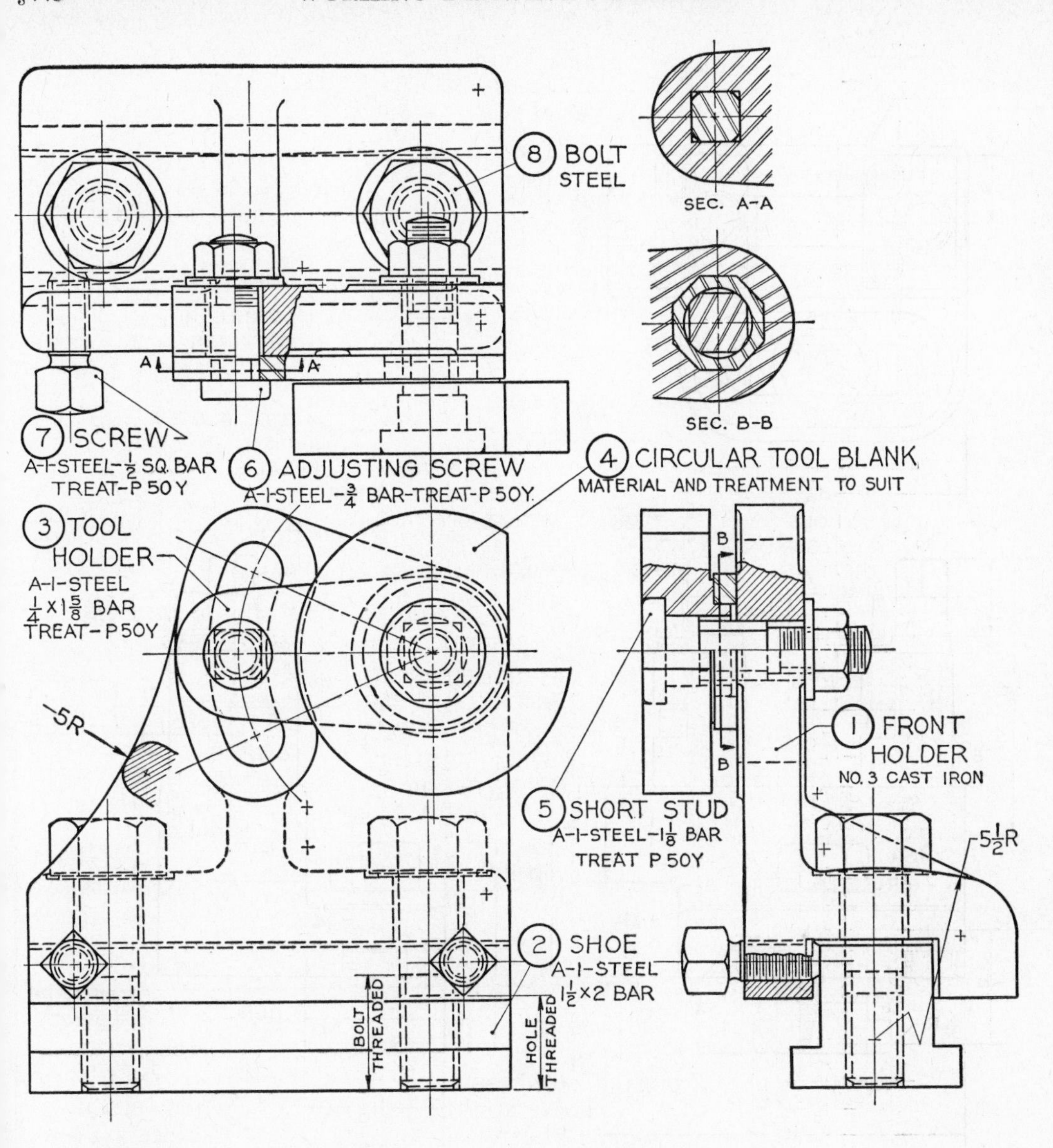

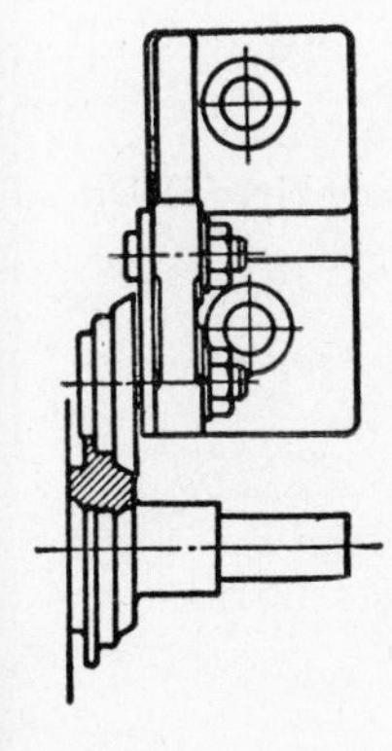

Fig. 777 Front Circular Forming Cutter Holder. (1) Draw details. (2) Draw assembly. Above layout is half size. To obtain dimensions, take distances directly from figure with dividers, and double them.

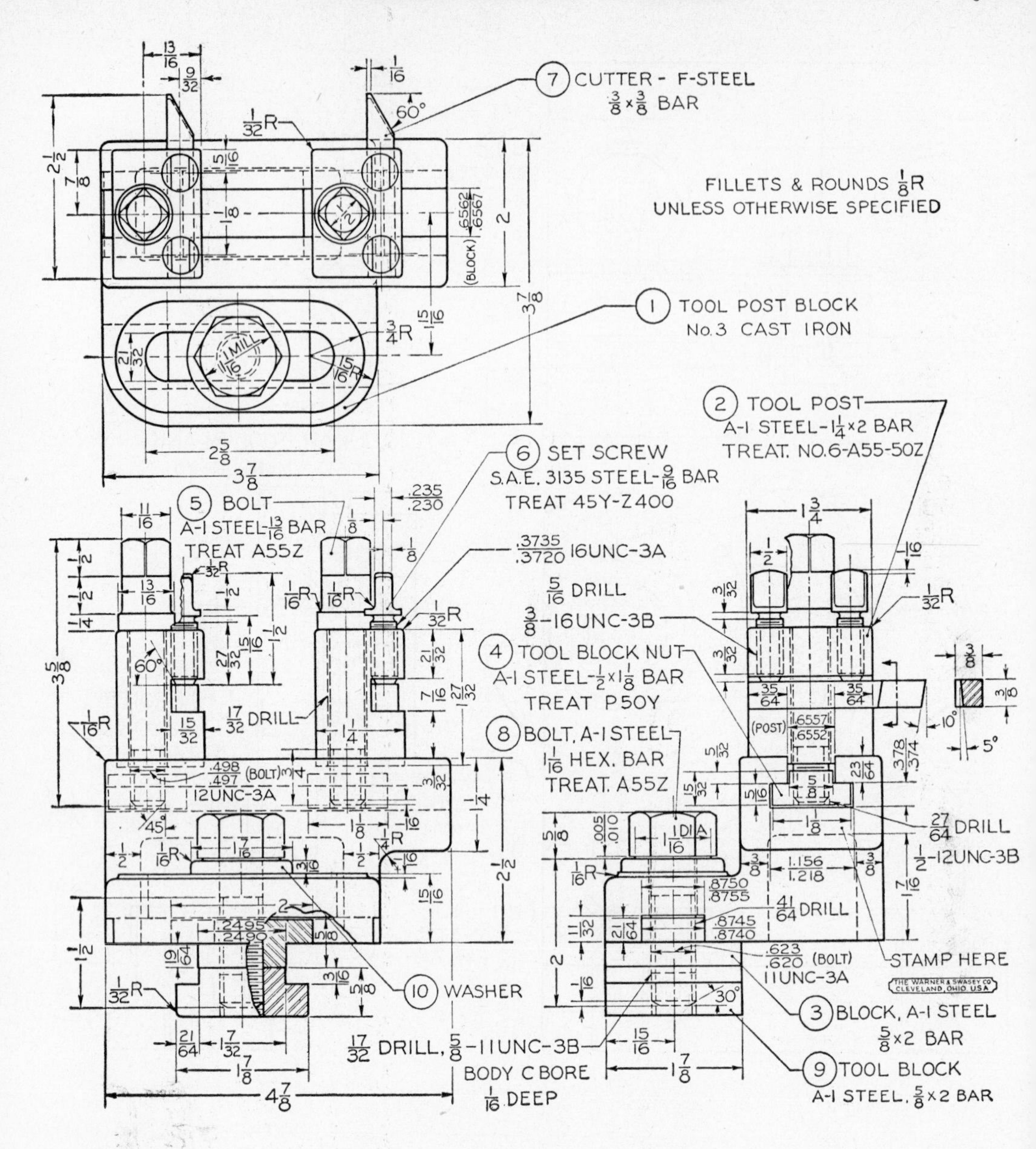

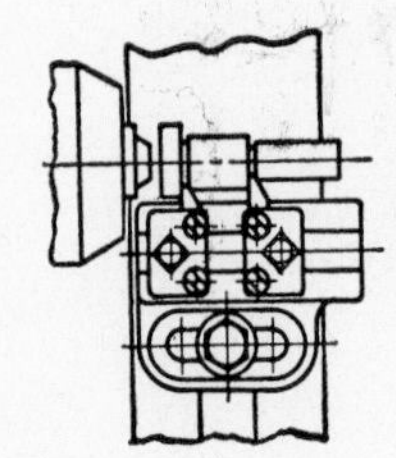

Fig. 778 Necking Tool Block. (1) Draw details. (2) Draw assembly. (3) Draw isometric exploded assembly.

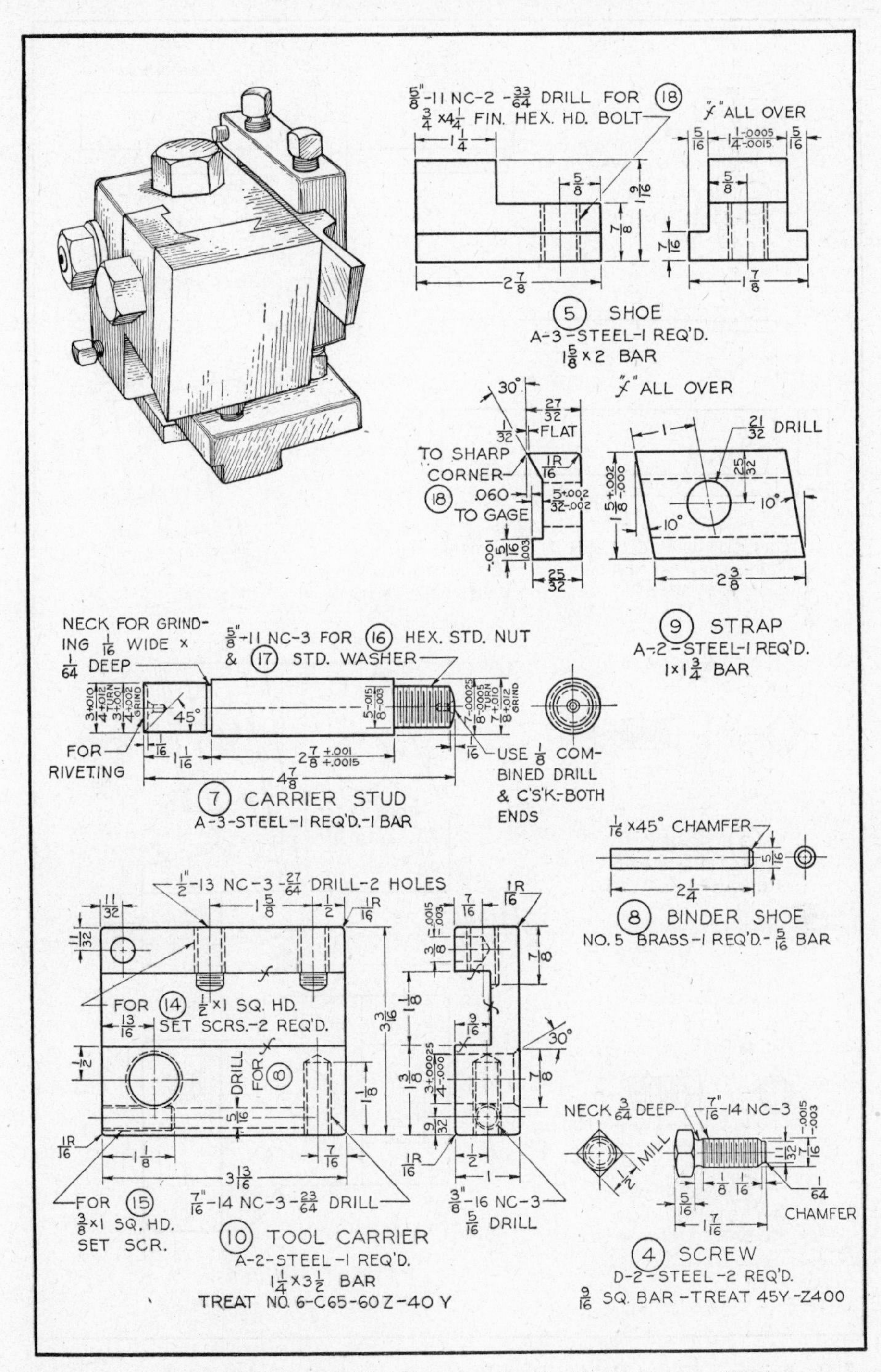

Fig. 779 Forming Cutter Holder Details. (1) Draw details. (2) Draw assembly (see also Fig. 780).

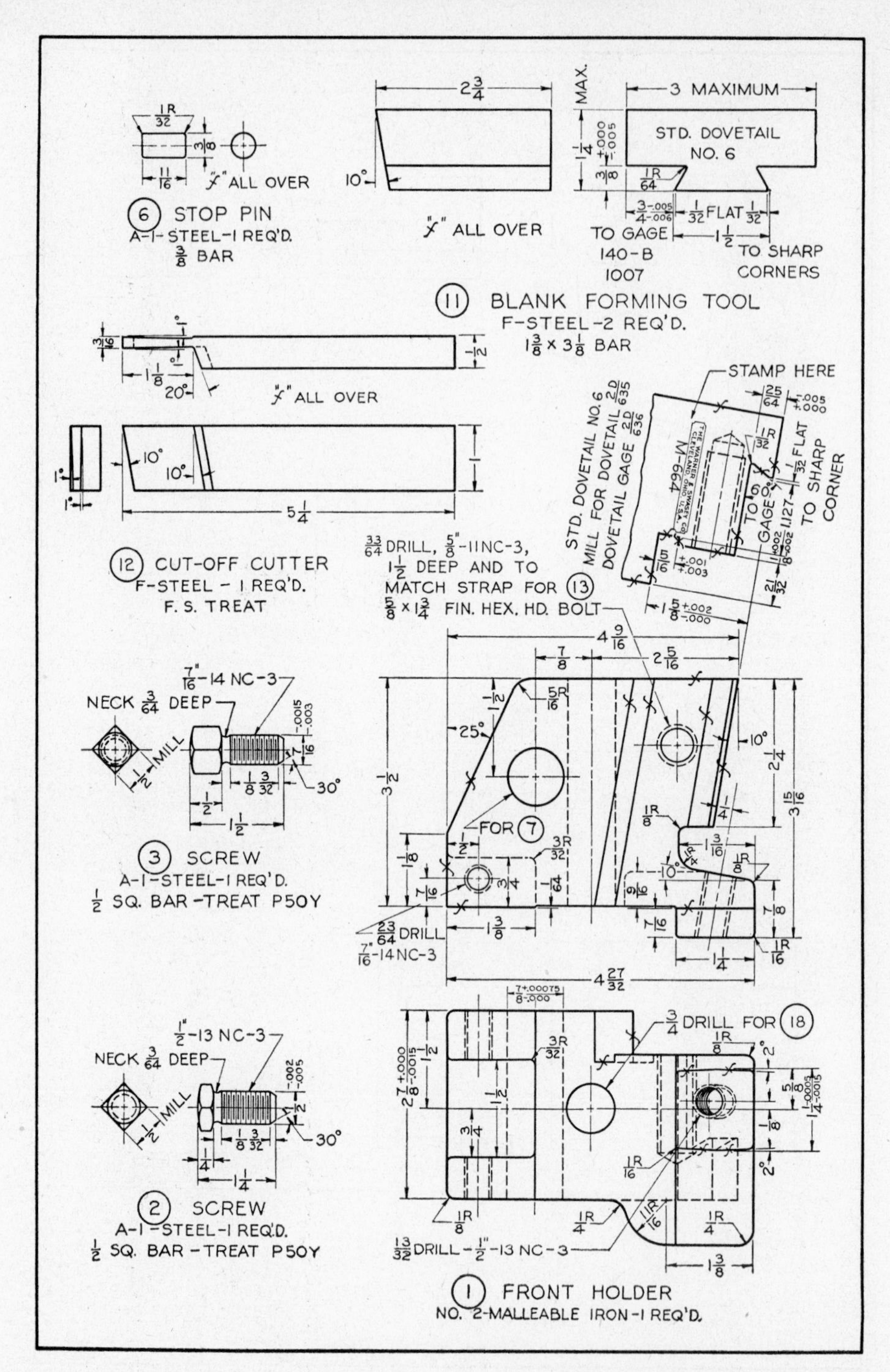

Fig. 780 Forming Cutter Holder Details (*Continued*). (1) Draw details. (2) Draw assembly.

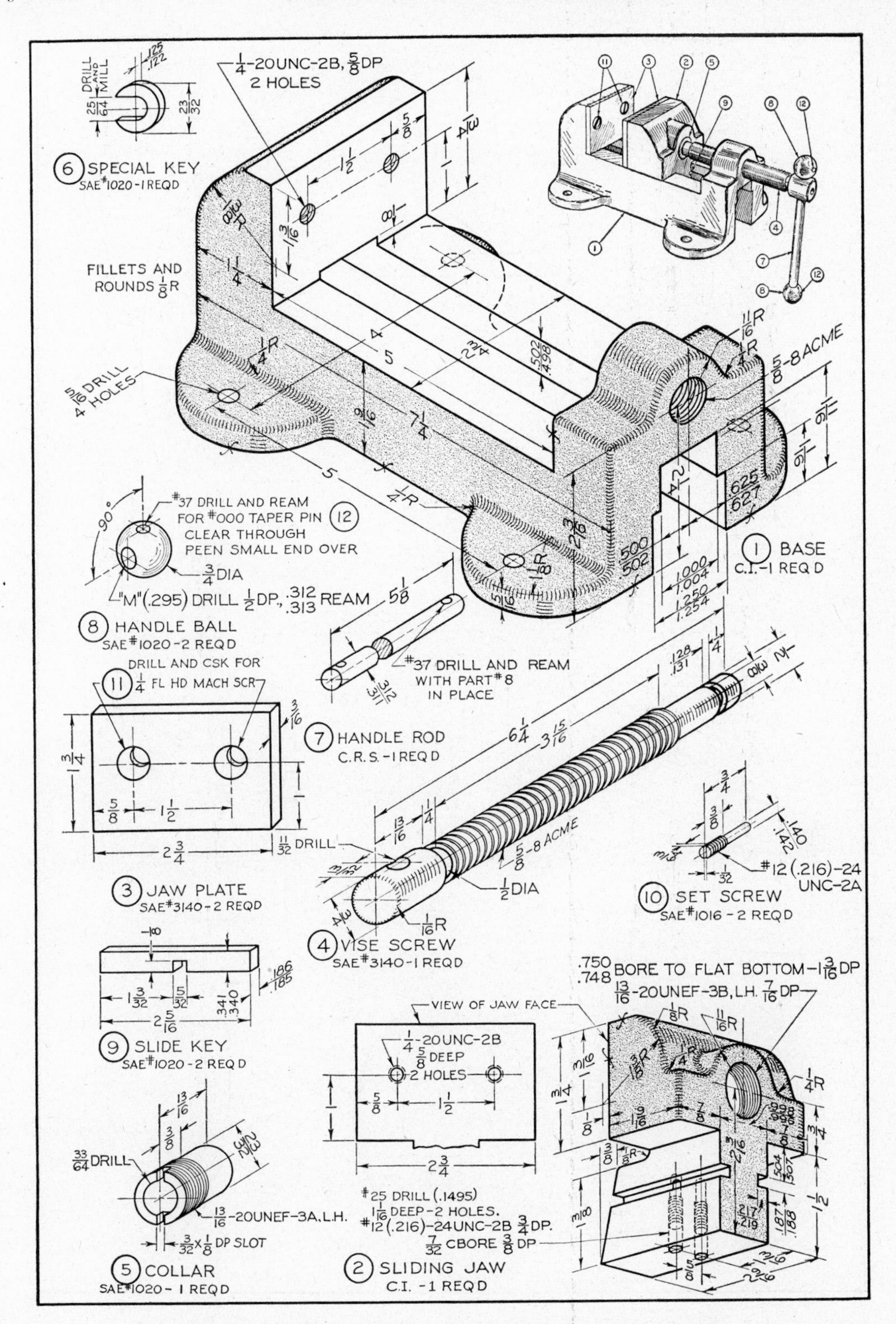

Fig. 781 Machine Vise. (1) Draw details. (2) Draw assembly.

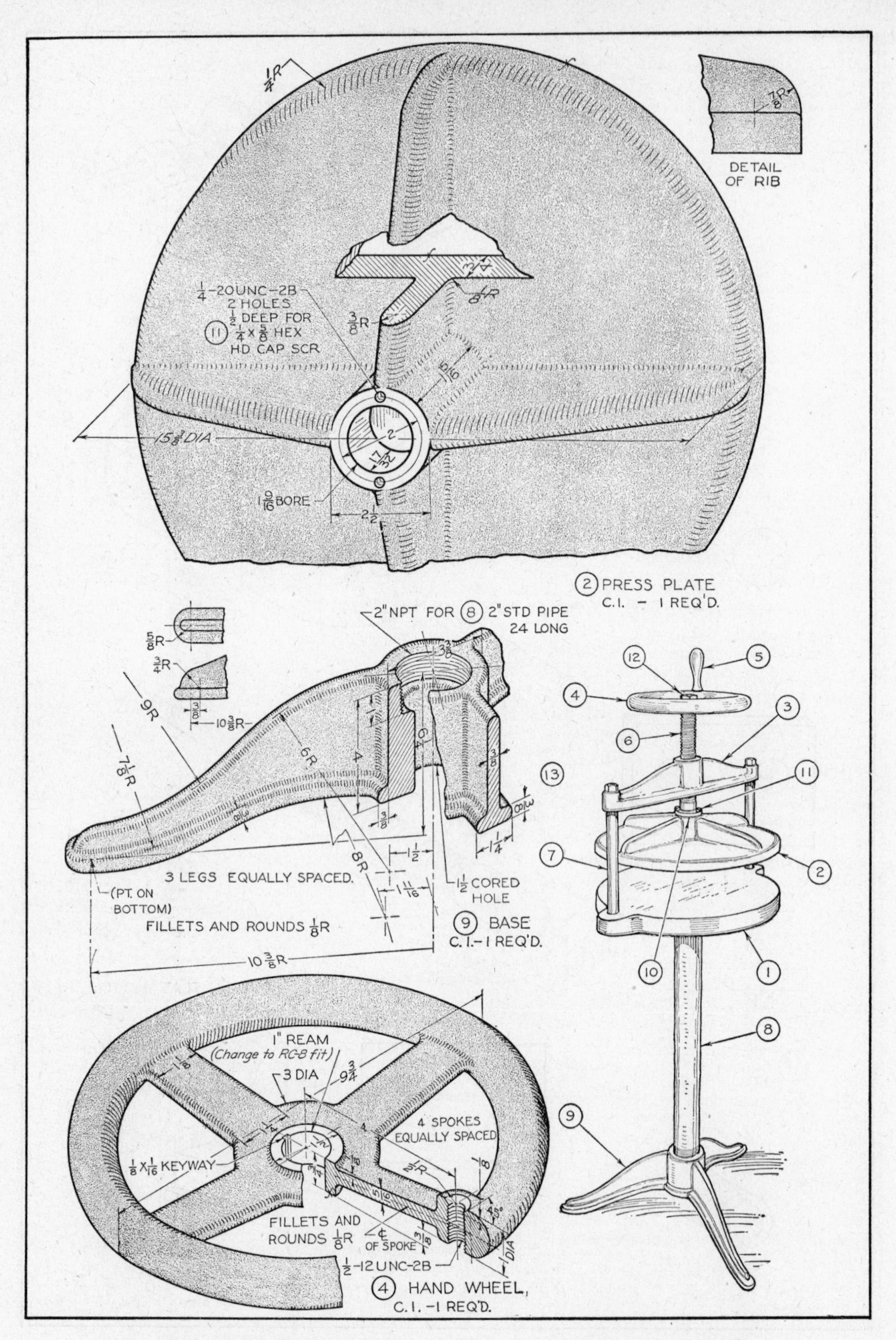

Fig. 782 Press. (1) Draw details. (2) Draw assembly (see also Fig. 783).

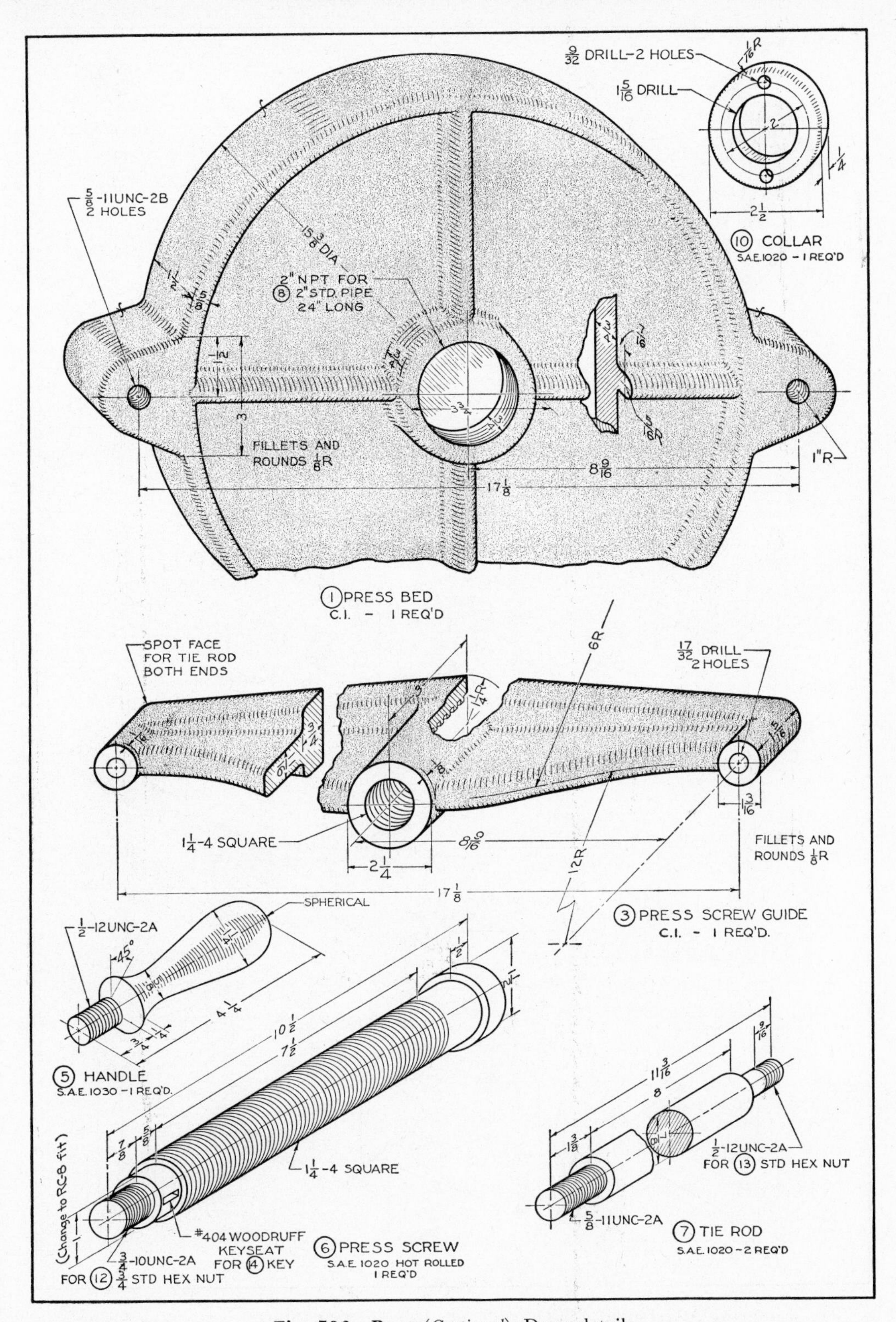

Fig. 783 **Press** (*Continued*). Draw details.

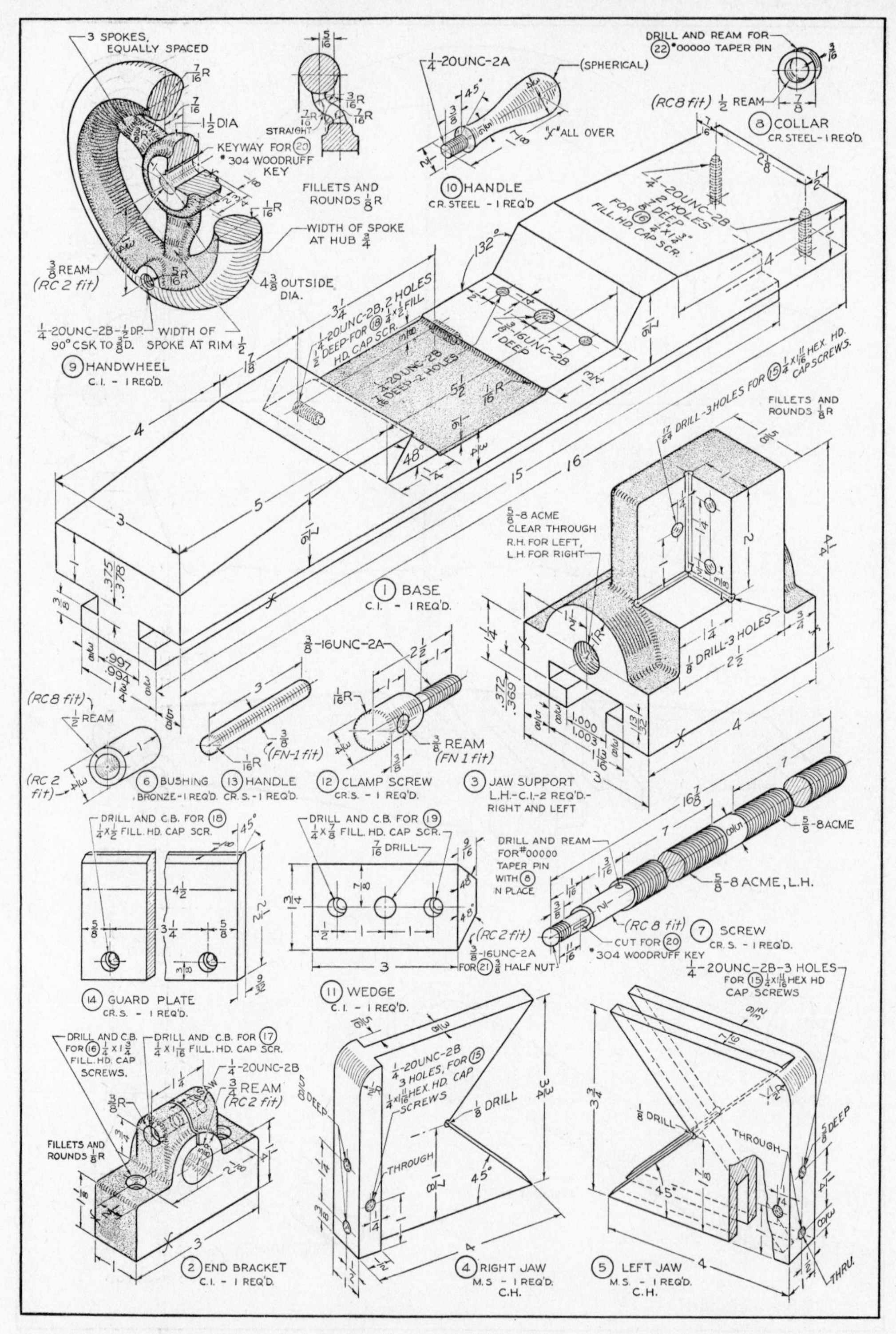

Fig. 784 Centering Attachment. (1) Draw details. Use American Standard tables for fits indicated. See Appendix tables 5–11. (2) Draw assembly (see Fig. 785).

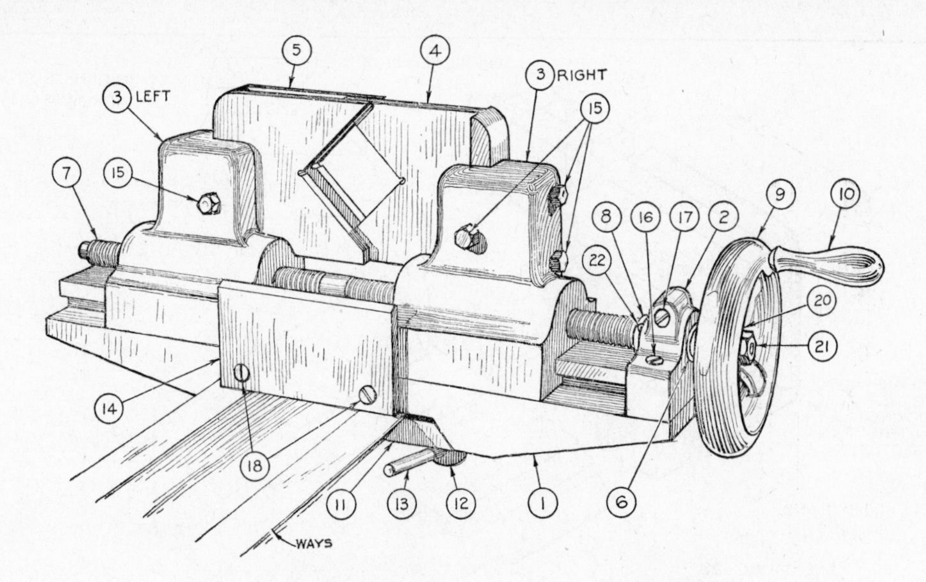

Fig. 785 **Centering Attachment.** (See Fig. 784.)

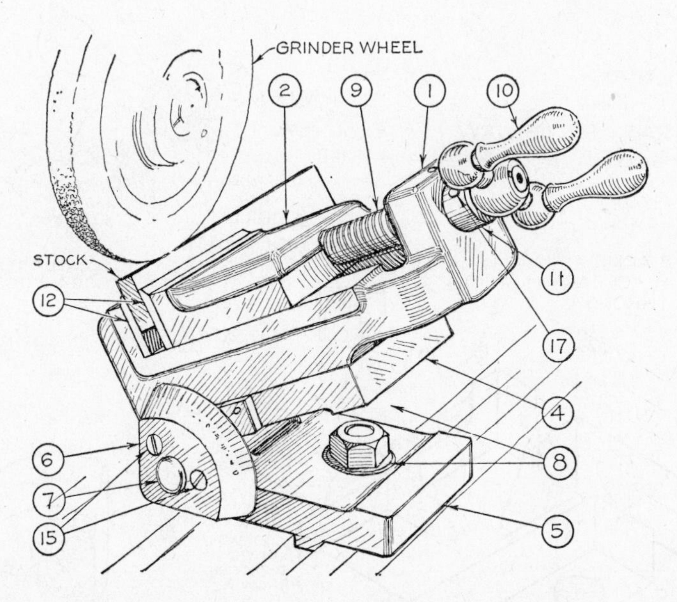

Fig. 786 **Grinder Vise.** (See Figs. 787 and 788.)

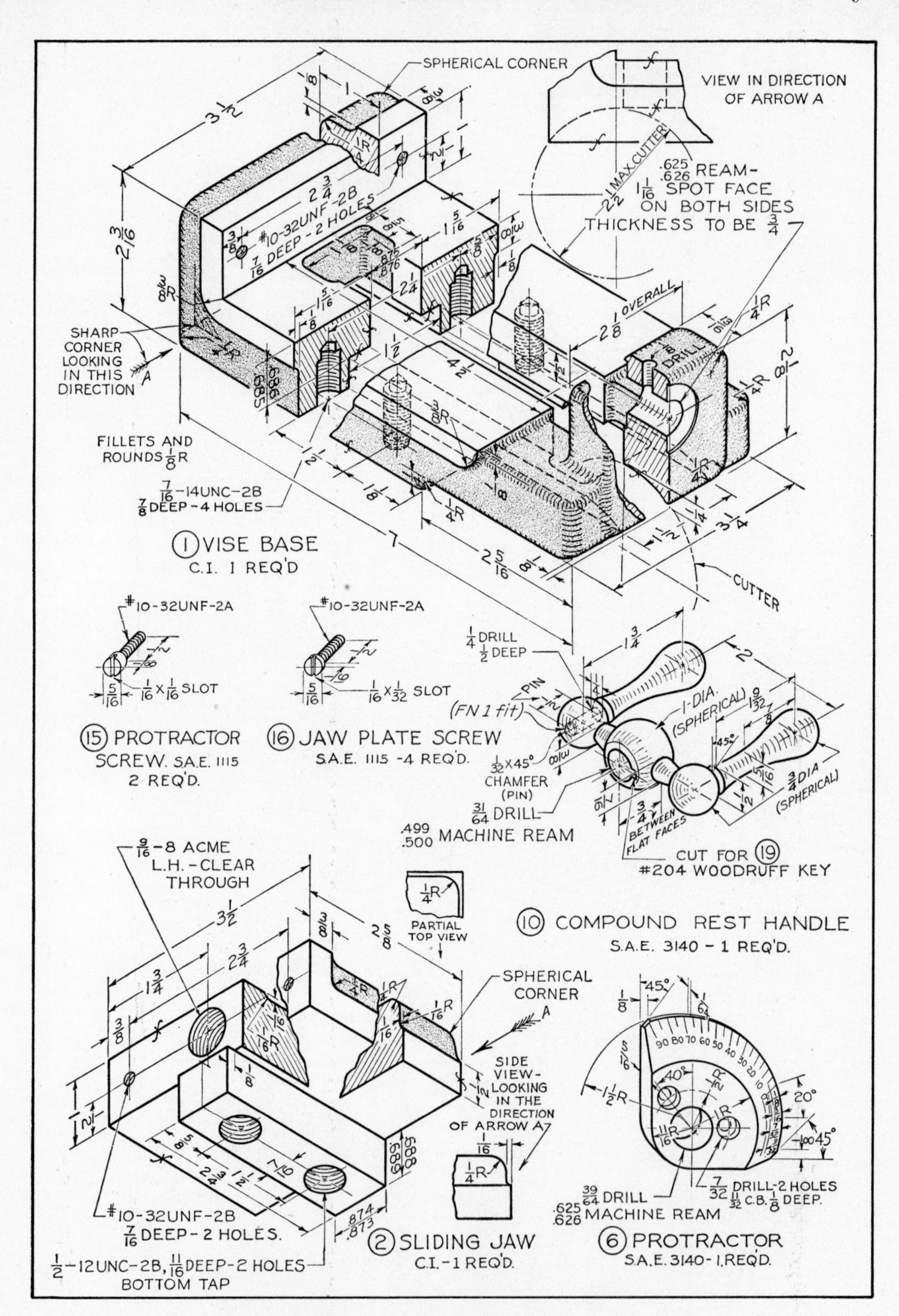

Fig. 787 Grinder Vise. (1) Draw details. (2) Draw assembly (see Figs. 786 and 788).

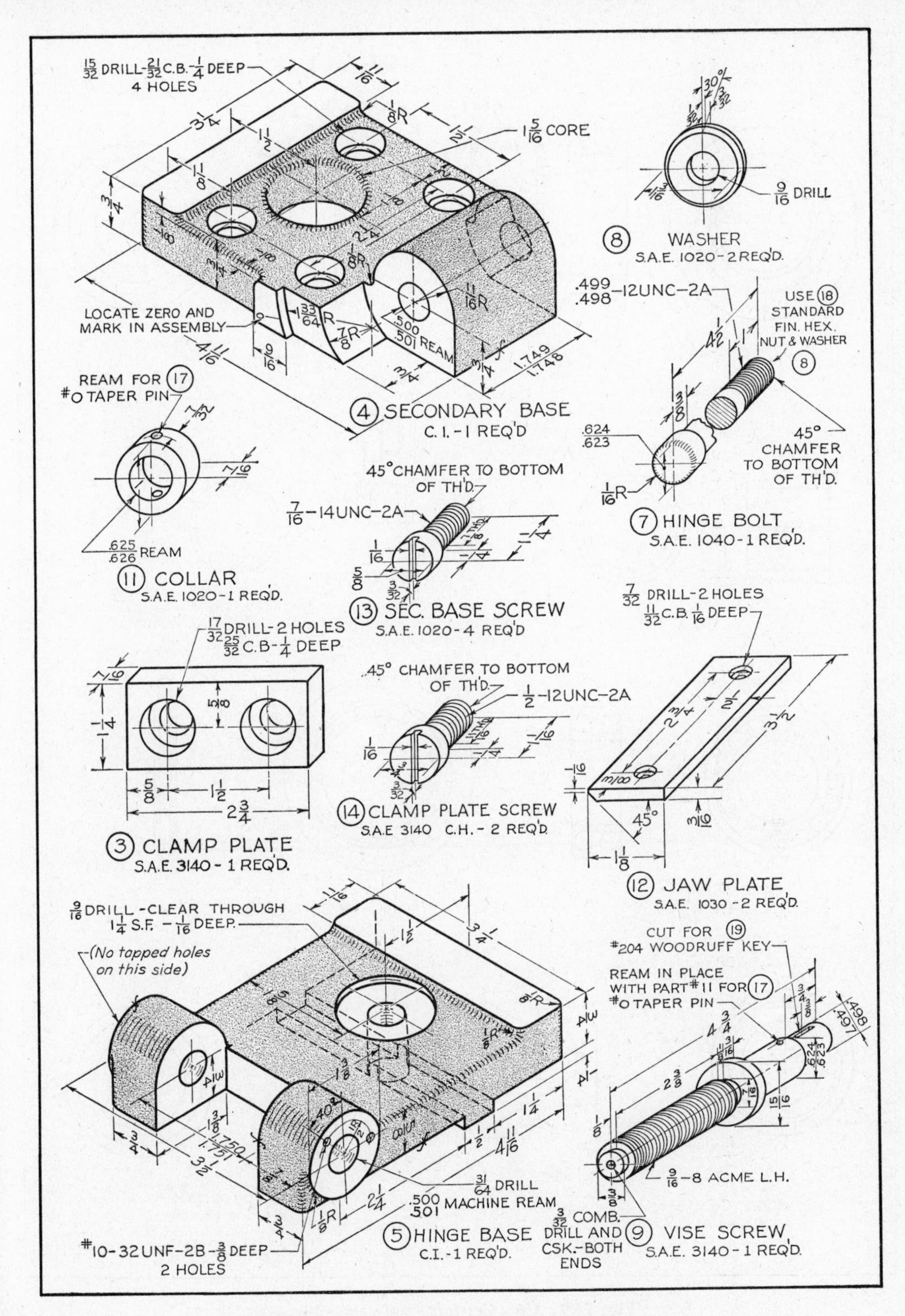

Fig. 788 Grinder Vise (*Continued*). Draw details.

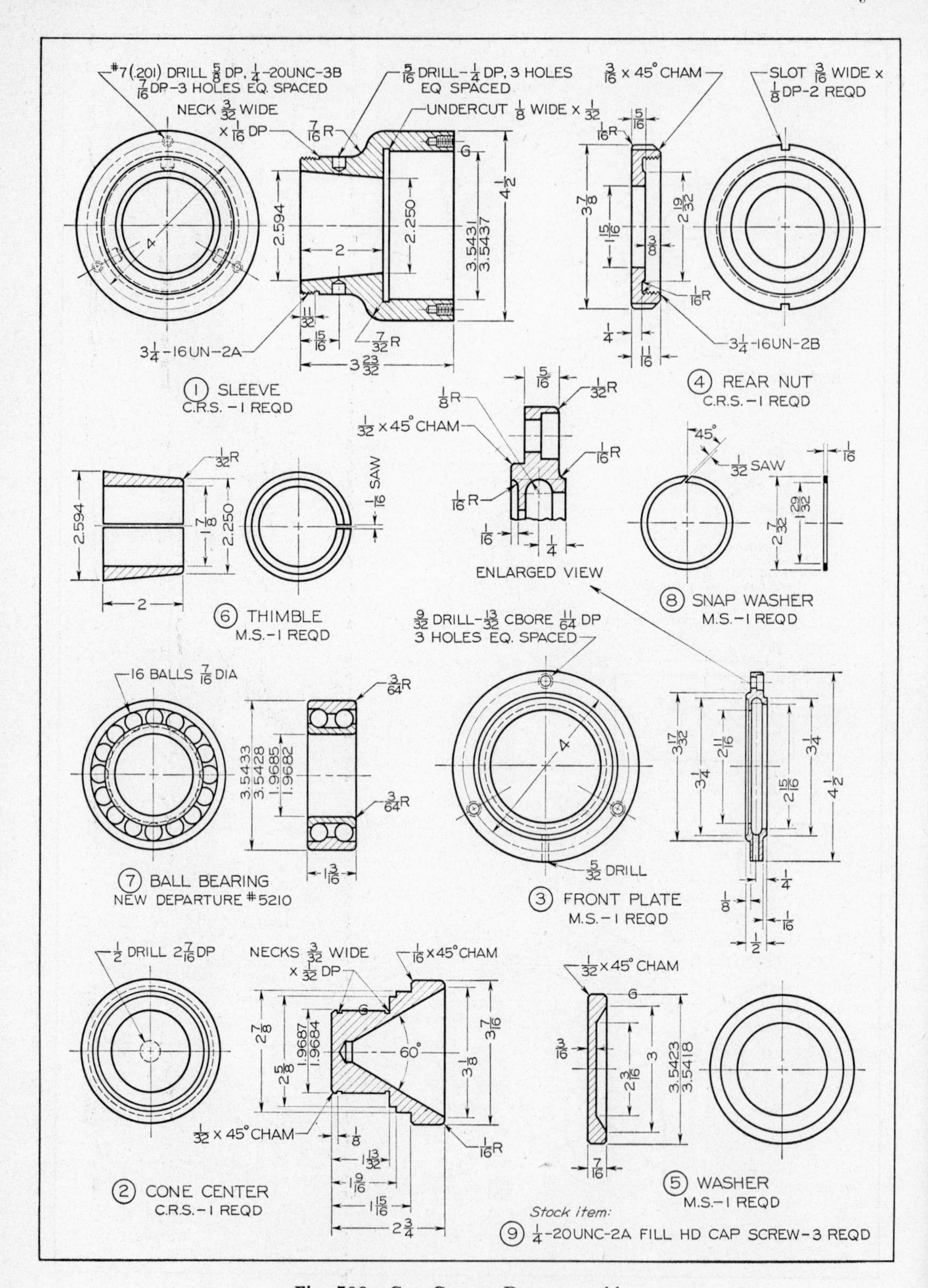

Fig. 789 Cup Center. Draw assembly.

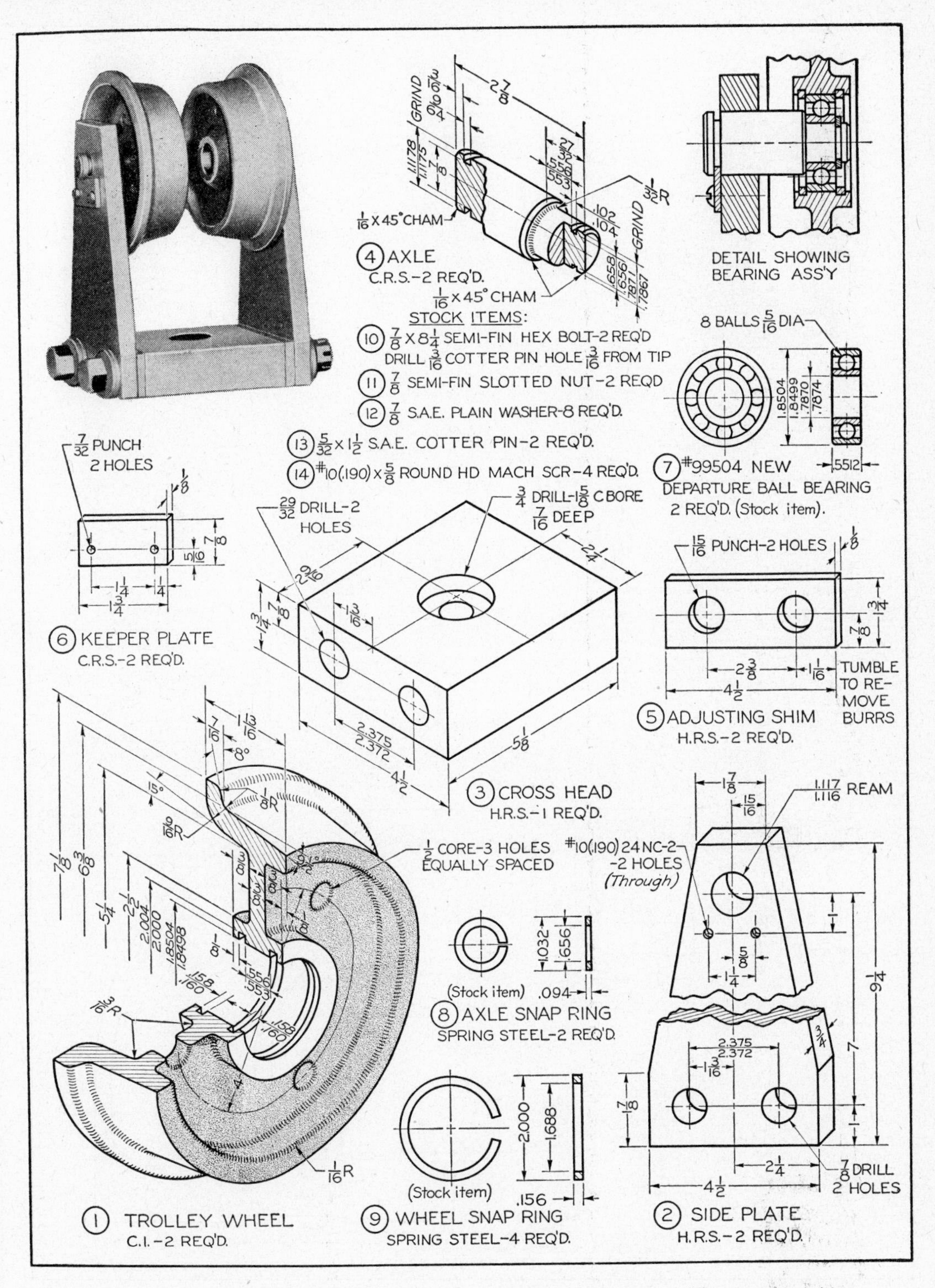

Fig. 790 Trolley. (1) Draw details, omitting parts 7-14 inclusive. (2) Draw assembly.

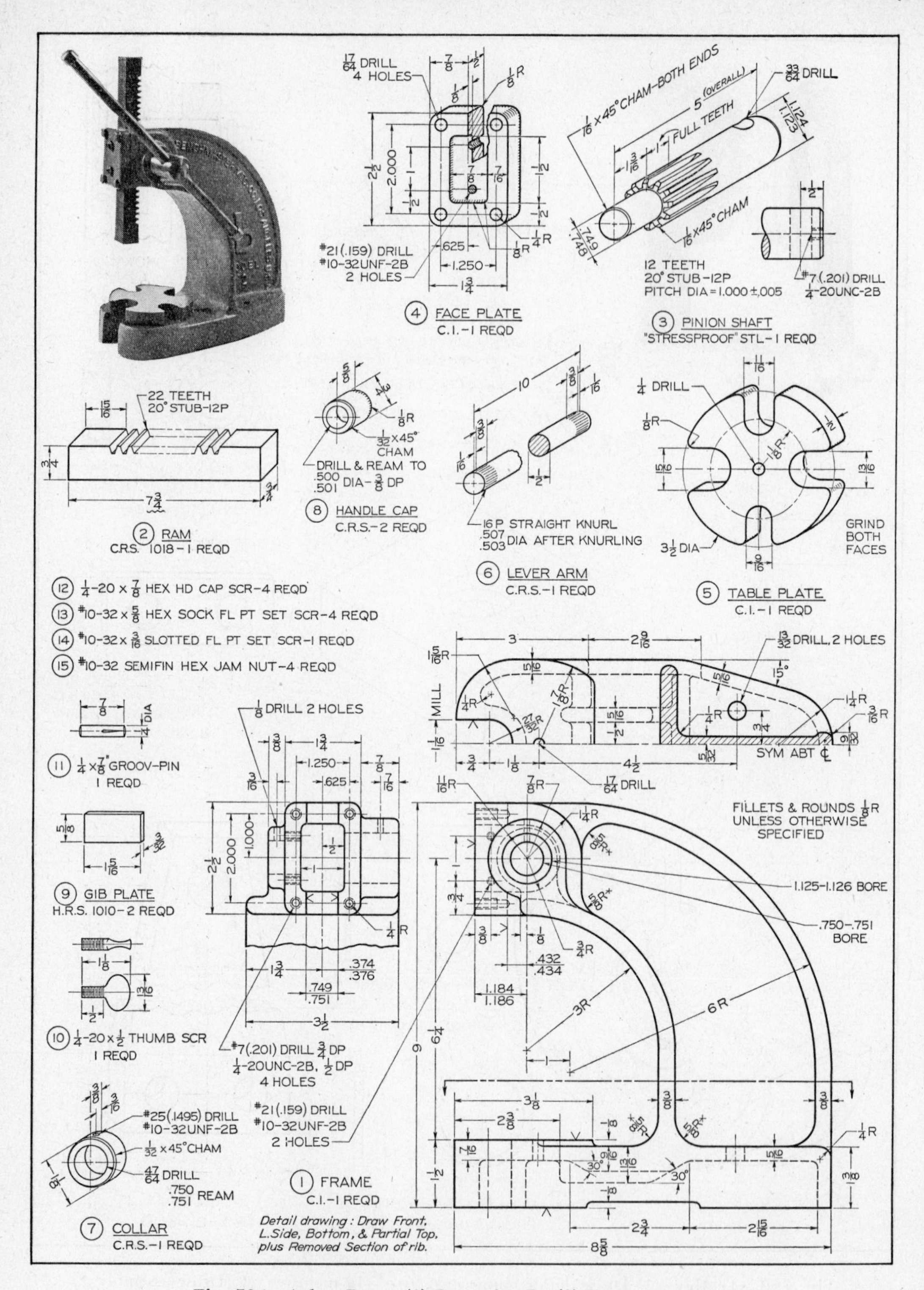

Fig. 791 Arbor Press. (1) Draw details. (2) Draw assembly.

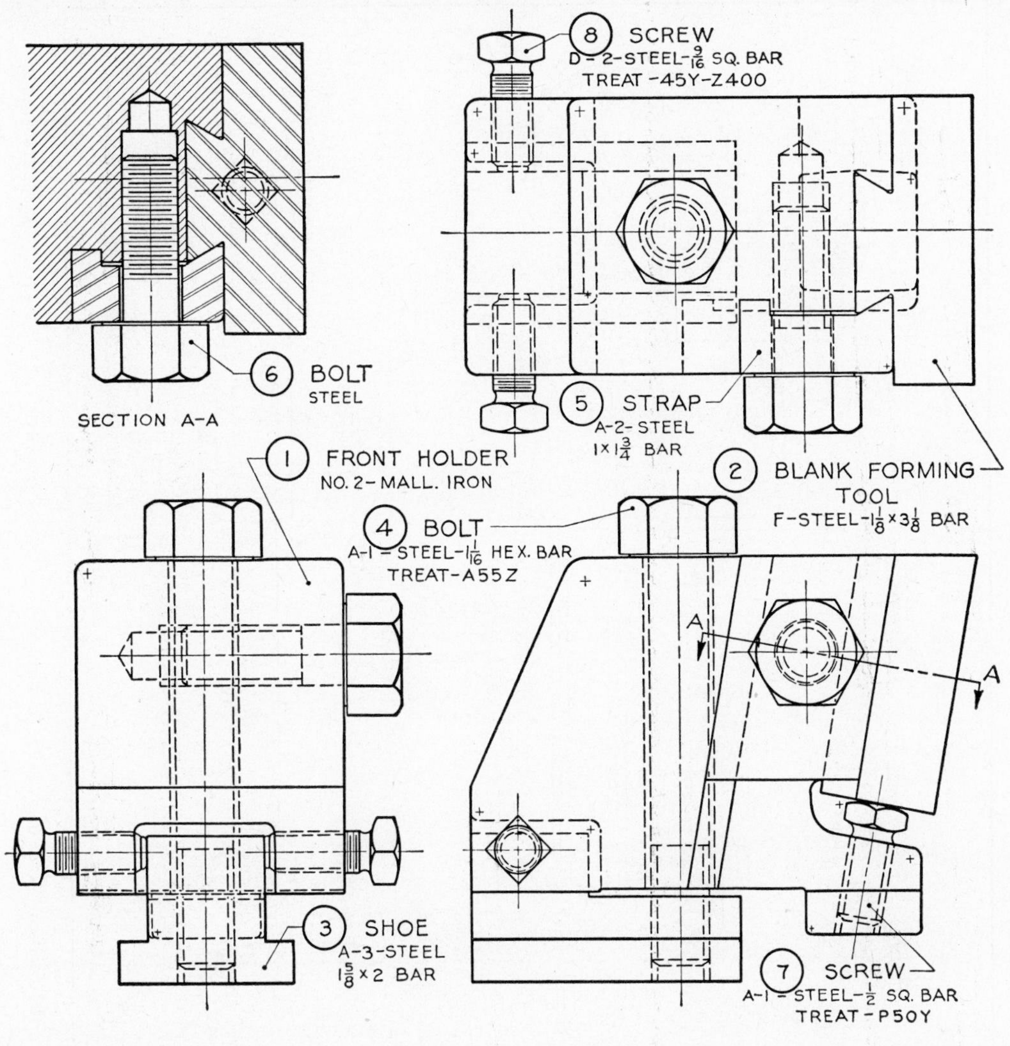

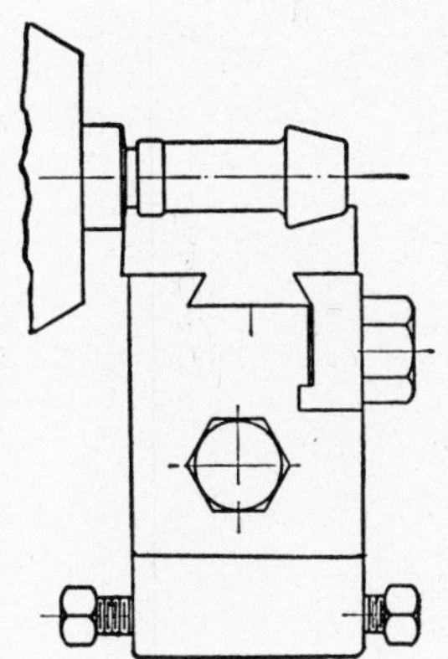

Fig. 792 F. S. Forming Cutter Holder. (1) Draw details. (2) Draw assembly. Above layout is half size. To obtain dimensions, take distances directly from figure with dividers, and double them.

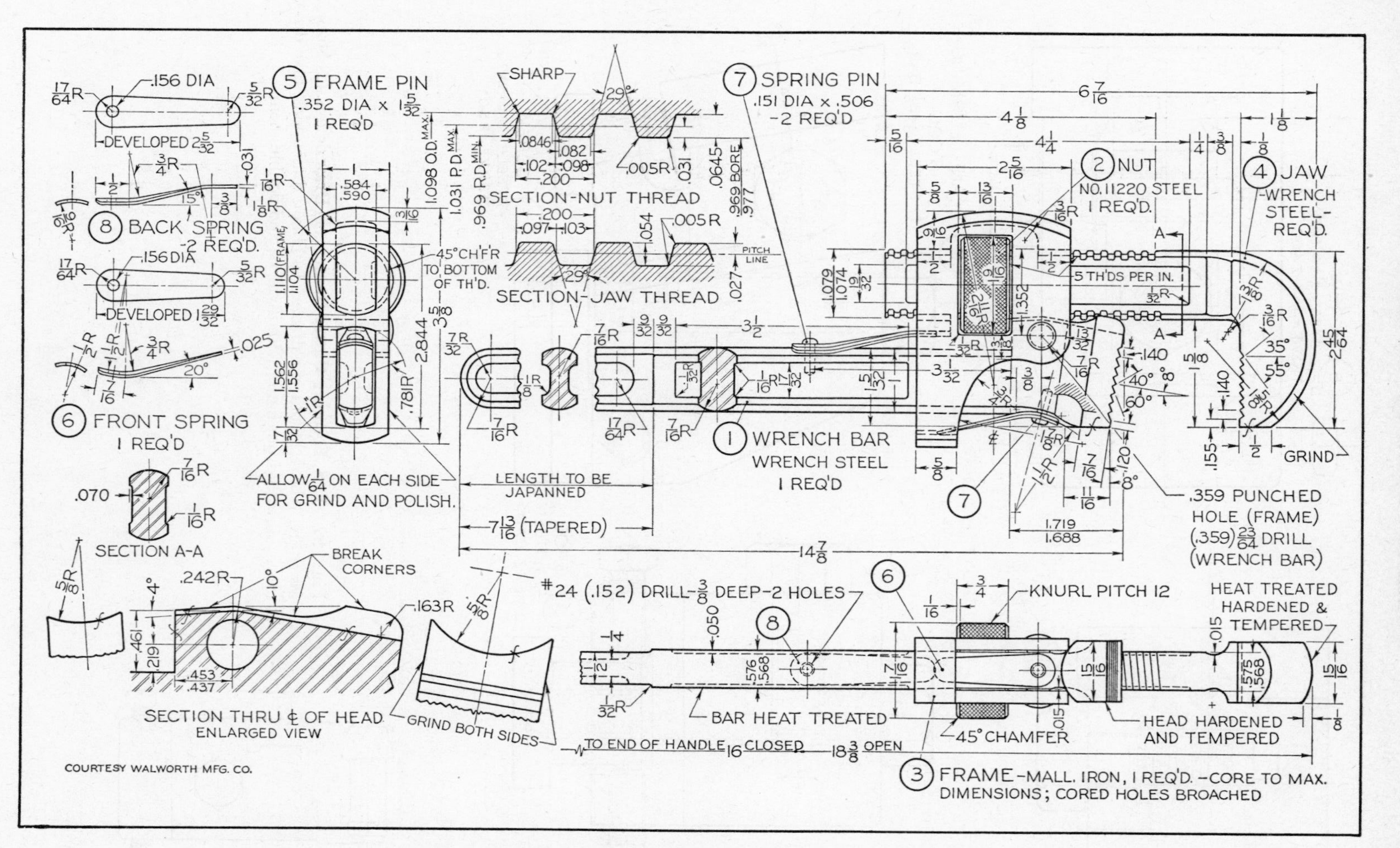

Fig. 793 18" Stillson Wrench. (1) Draw details. (2) Draw assembly.

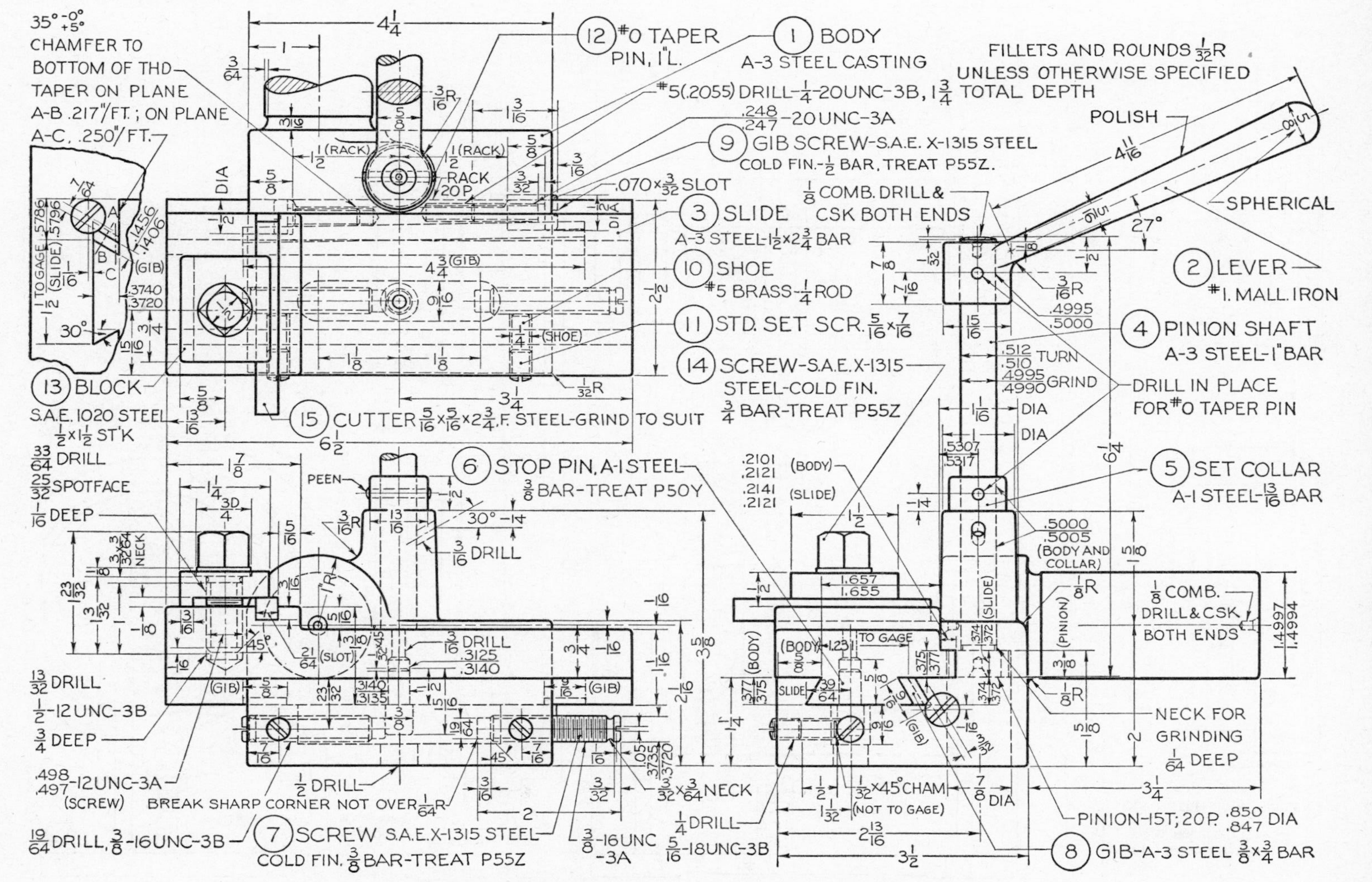

Fig. 794 Plain Rack Tool. (1) Draw details. (2) Draw assembly.

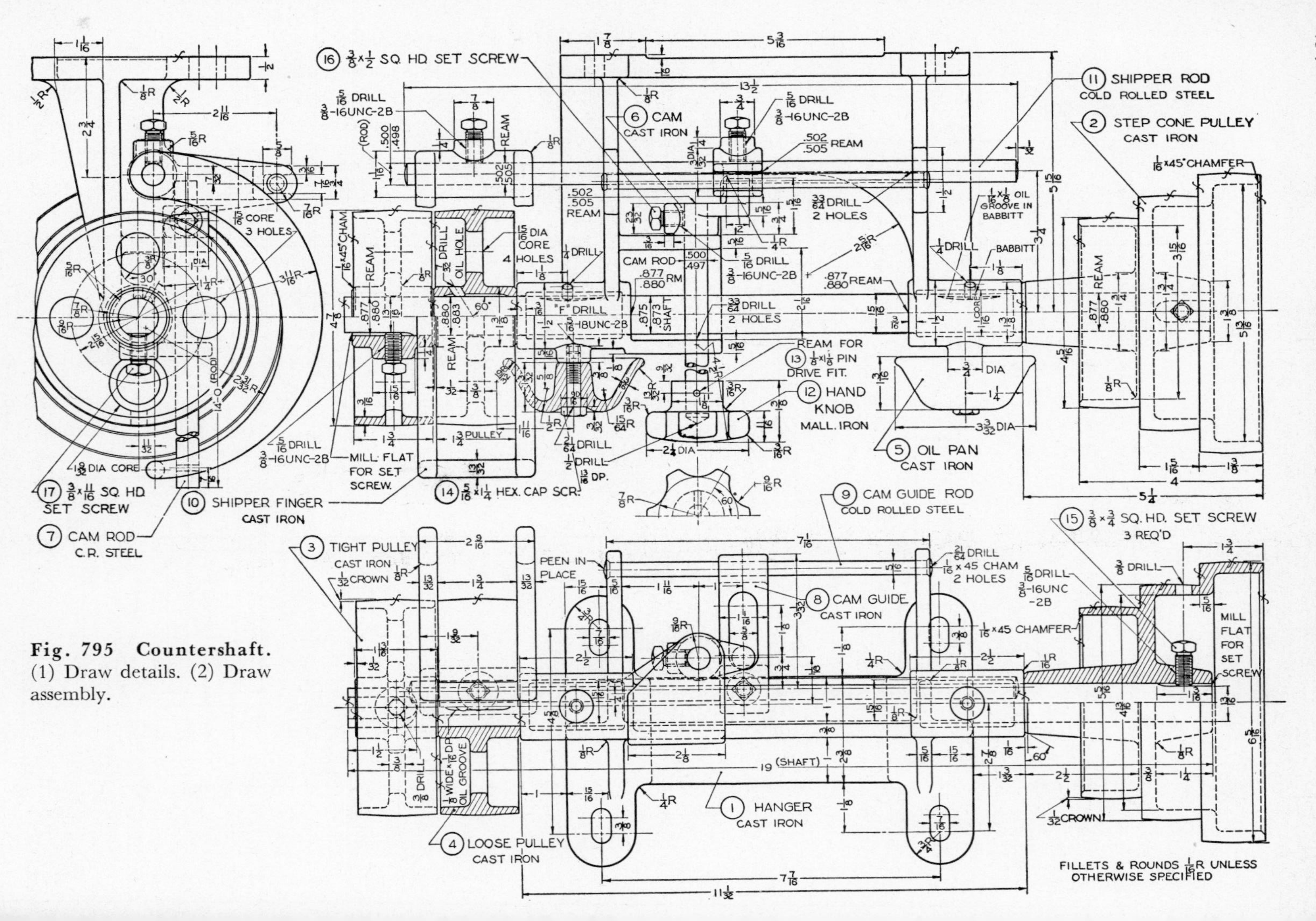

Fig. 795 Countershaft. (1) Draw details. (2) Draw assembly.

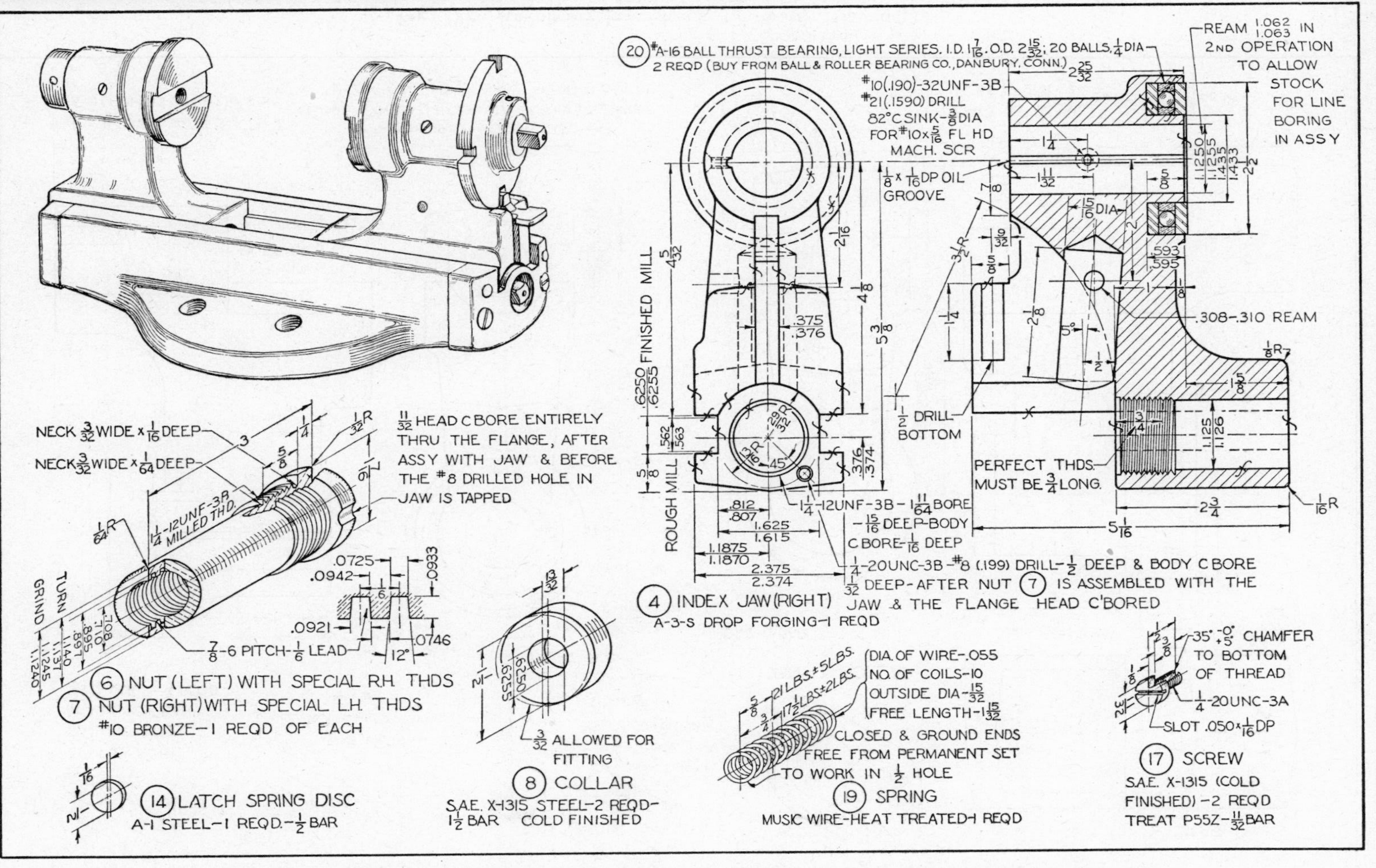

Fig. 796 Revolving Jaw Chuck. (1) Draw details. (2) Draw assembly (see Figs. 797-799).

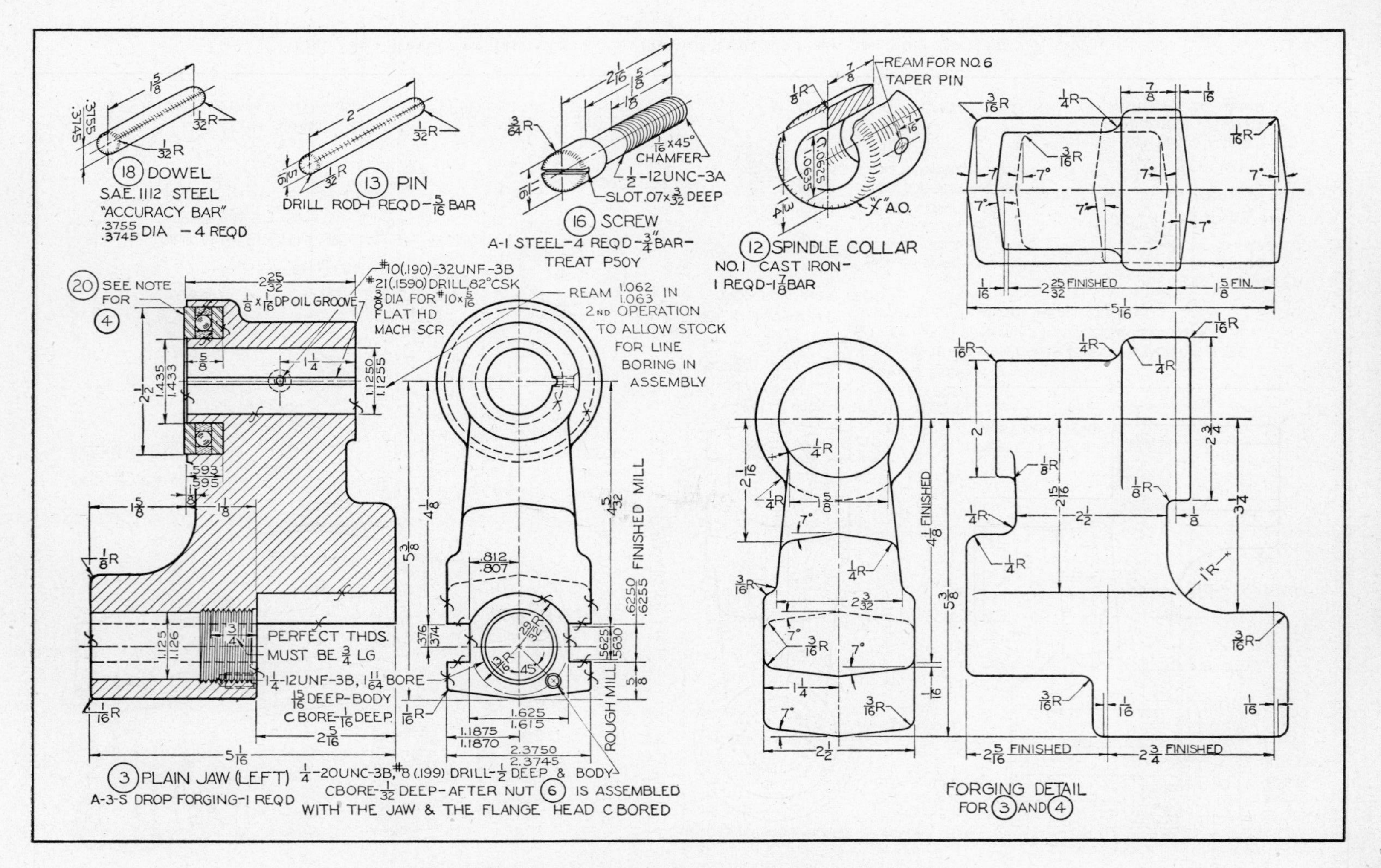

Fig. 797 Revolving Jaw Chuck *(Continued)*. Draw details.

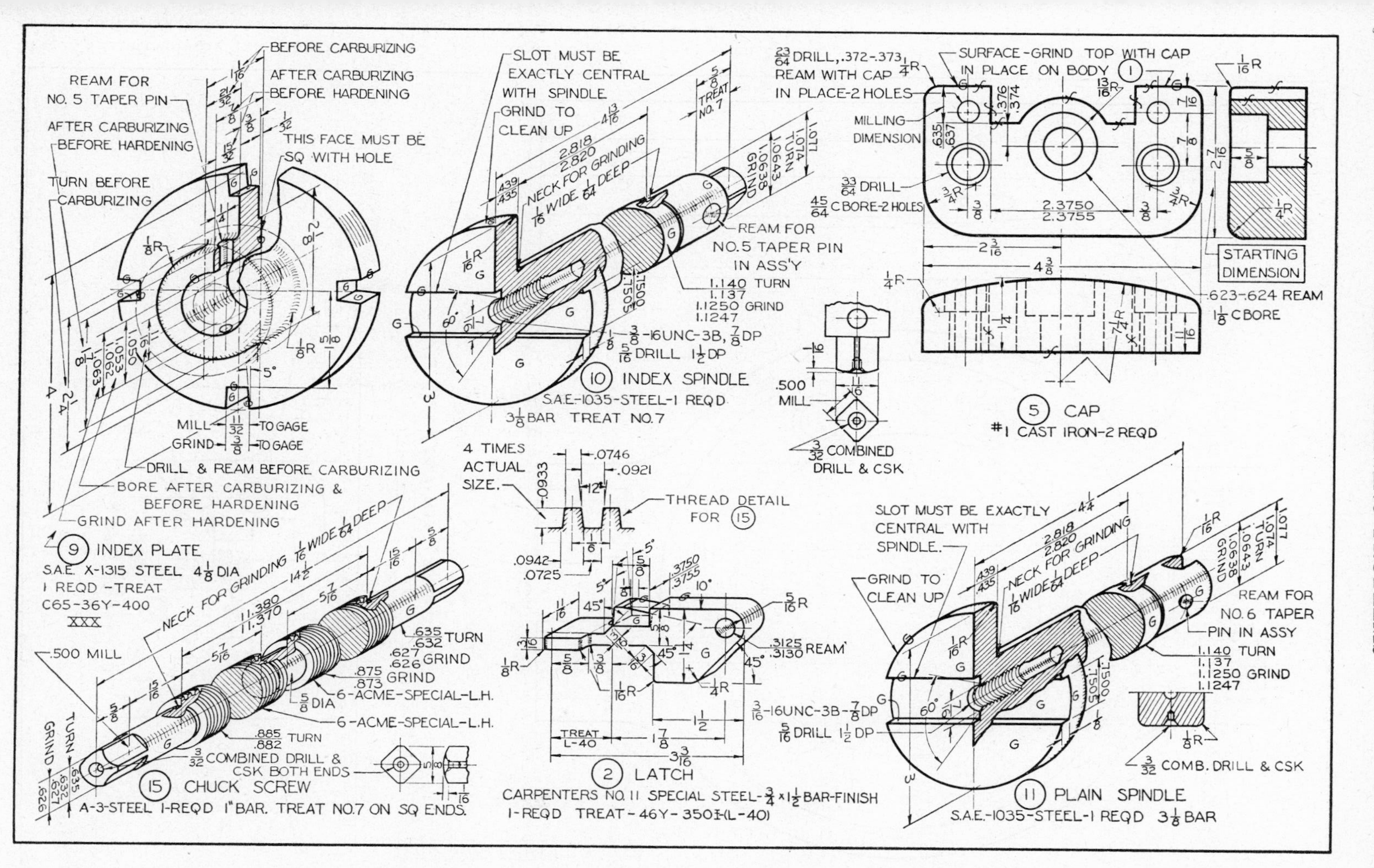

Fig. 798 Revolving Jaw Chuck (*Continued*). Draw details. For Part No. 5: Revolve given Front view 180°; then add R. Side and Bottom views.

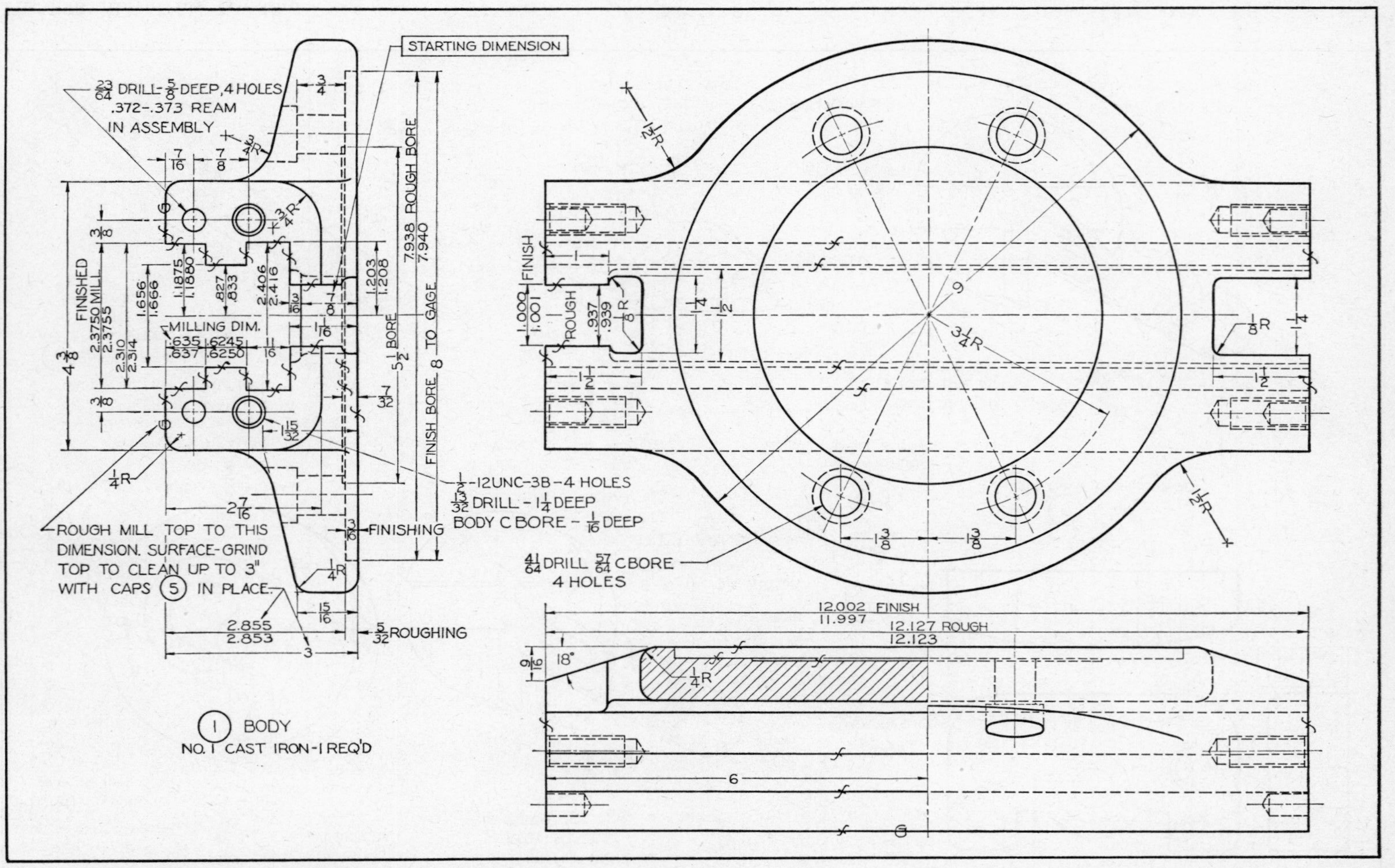

Fig. 799 Revolving Jaw Chuck (*Continued*). Draw detail drawing. Take given L. Side view as the R. Side view in the new drawing; then add Front view, and Bottom view in half section.

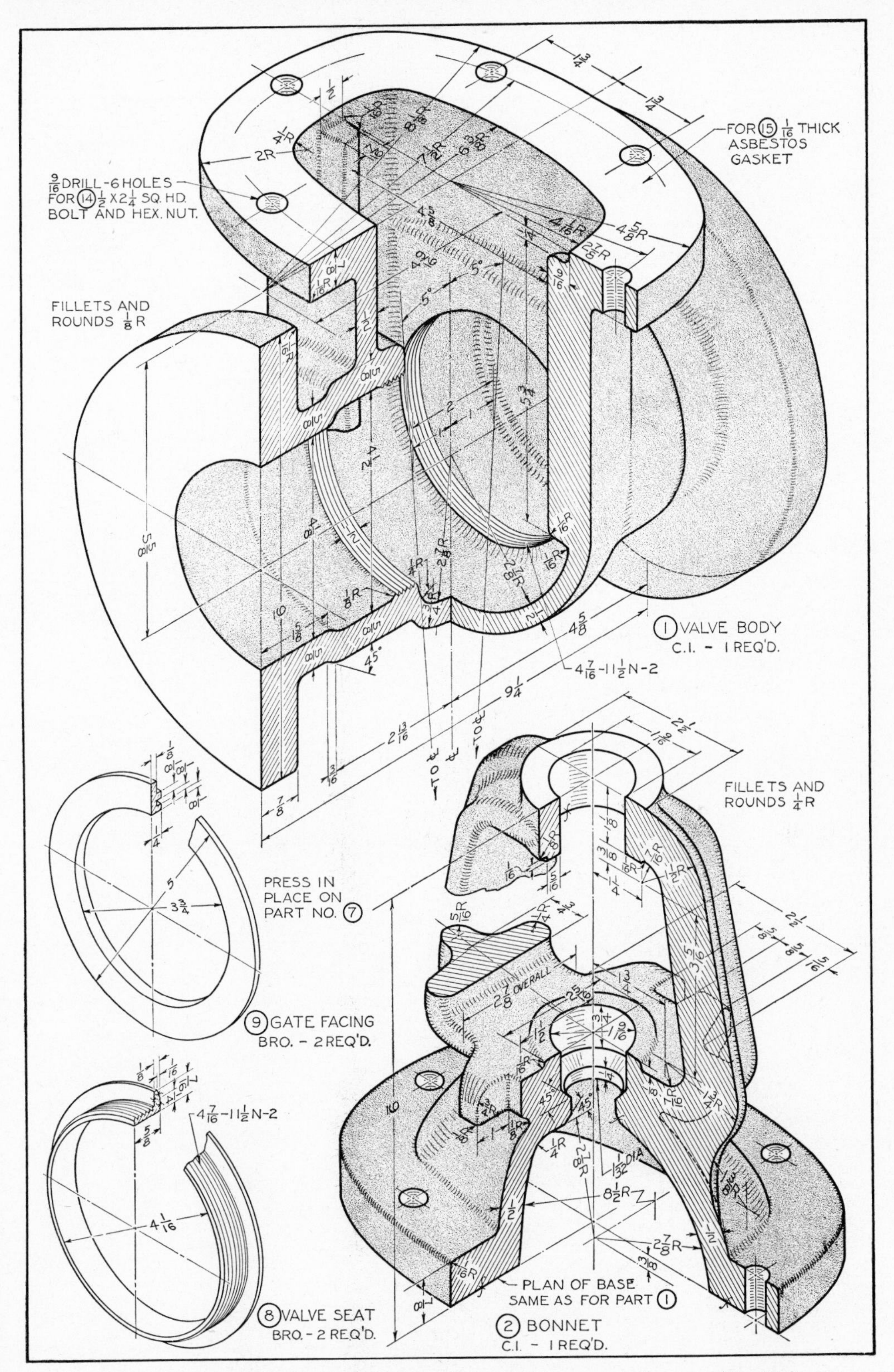

Fig. 800 Gate Valve. (1) Draw details. (2) Draw assembly (see also Fig. 801).

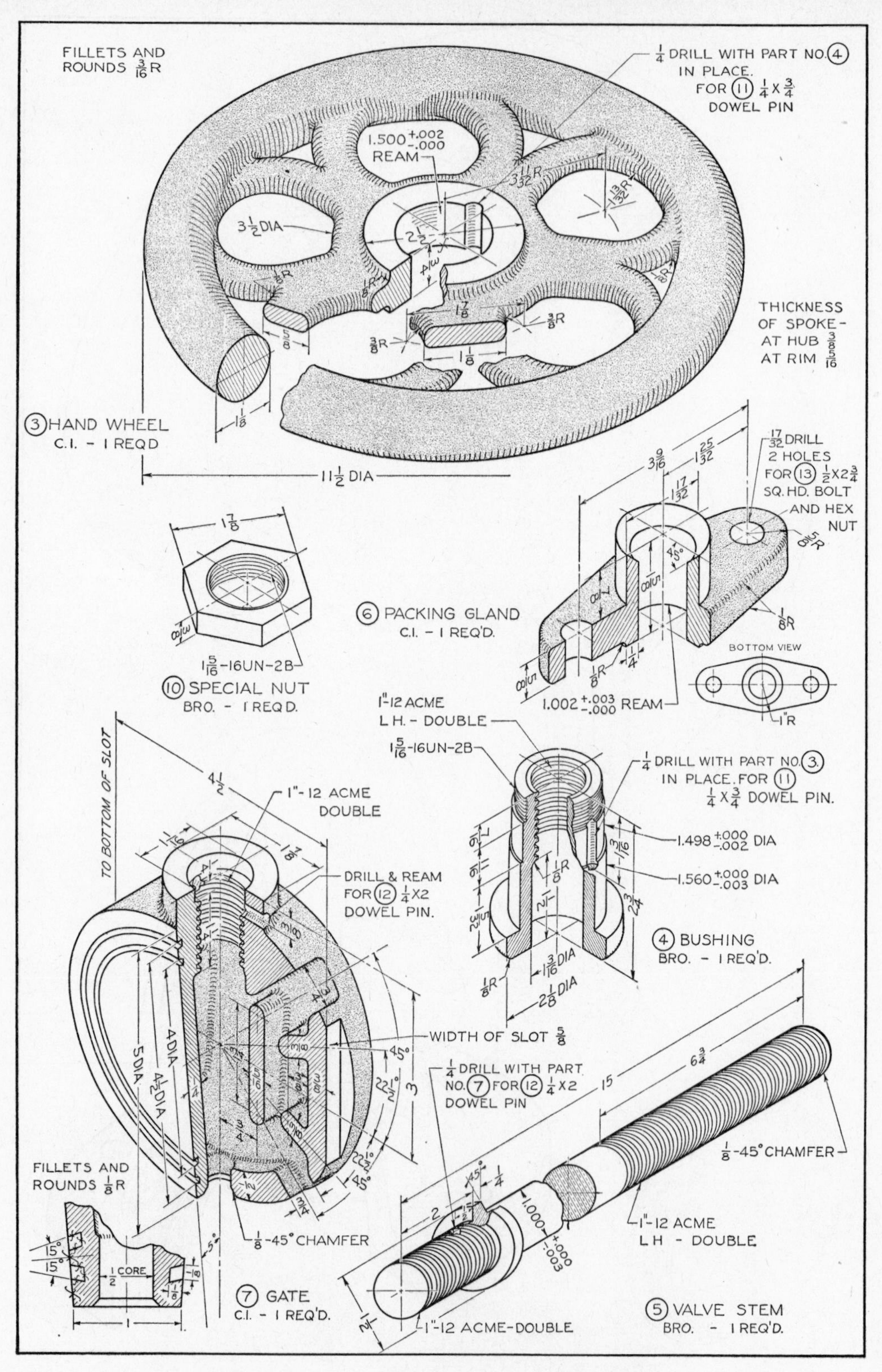

FILLETS AND ROUNDS $\frac{3}{16}$R
$\frac{1}{4}$ DRILL WITH PART NO. (4) IN PLACE. FOR (11) $\frac{1}{4}$ X $\frac{3}{4}$ DOWEL PIN
1.500 $^{+.002}_{-.000}$ REAM
$3\frac{11}{32}$R
$\frac{13}{32}$R
$3\frac{1}{2}$DIA
$2\frac{1}{2}$
$\frac{3}{4}$
$\frac{1}{8}$R
$1\frac{7}{8}$
$\frac{3}{8}$R
$\frac{5}{8}$
$1\frac{1}{8}$
THICKNESS OF SPOKE - AT HUB $\frac{3}{8}$ AT RIM $\frac{5}{16}$
(3) HAND WHEEL
C.I. - 1 REQ'D
$11\frac{1}{2}$ DIA
$3\frac{9}{16}$
$1\frac{25}{32}$
$1\frac{17}{32}$
$\frac{17}{32}$ DRILL 2 HOLES FOR (13) $\frac{1}{2}$X$2\frac{3}{4}$ SQ. HD. BOLT AND HEX NUT
$\frac{5}{8}$R
45°
$1\frac{7}{8}$
$\frac{3}{8}$
$1\frac{5}{16}$-16UN-2B
(10) SPECIAL NUT
BRO. - 1 REQ'D.
(6) PACKING GLAND
C.I. - 1 REQ'D.
$1\frac{1}{8}$
$\frac{5}{8}$
$\frac{1}{8}$R
$\frac{1}{8}$R
$\frac{1}{4}$
1.002 $^{+.003}_{-.000}$ REAM
BOTTOM VIEW
1"R
1"-12 ACME L.H. - DOUBLE
$1\frac{5}{16}$-16UN-2B
$\frac{1}{4}$ DRILL WITH PART NO. (3) IN PLACE. FOR (11) $\frac{1}{4}$ X $\frac{3}{4}$ DOWEL PIN.
1.498 $^{+.000}_{-.002}$ DIA
1.560 $^{+.000}_{-.003}$ DIA
$\frac{3}{16}$
$\frac{3}{16}$
$\frac{1}{8}$R
$1\frac{1}{16}$
$\frac{1}{2}$
$2\frac{3}{4}$
$2\frac{3}{8}$
$1\frac{3}{16}$DIA
$2\frac{1}{8}$DIA
$\frac{1}{8}$R
(4) BUSHING
BRO. - 1 REQ'D.
TO BOTTOM OF SLOT
$4\frac{1}{2}$
1"-12 ACME DOUBLE
$1\frac{1}{16}$
$1\frac{7}{8}$
DRILL & REAM FOR (12) $\frac{1}{4}$X2 DOWEL PIN.
$\frac{3}{8}$
WIDTH OF SLOT $\frac{5}{8}$
45°
$22\frac{1}{2}$°
$22\frac{1}{2}$°
45°
3
4DIA
$4\frac{1}{2}$DIA
5DIA
$\frac{3}{4}$
$\frac{1}{2}$
$\frac{3}{4}$
$\frac{1}{8}$-45° CHAMFER
FILLETS AND ROUNDS $\frac{1}{8}$R
15°
15°
$\frac{1}{2}$ CORE
1
5°
(7) GATE
C.I. - 1 REQ'D.
$\frac{1}{4}$ DRILL WITH PART NO. (7) FOR (12) $\frac{1}{4}$ X 2 DOWEL PIN
15
$6\frac{3}{4}$
$\frac{1}{8}$-45° CHAMFER
45°
$\frac{1}{4}$
2
1.000 $^{+.000}_{-.003}$
1"-12 ACME LH - DOUBLE
$1\frac{1}{2}$
1"-12 ACME-DOUBLE
(5) VALVE STEM
BRO. - 1 REQ'D.

Fig. 801 Gate Valve (*Continued*). Draw details.

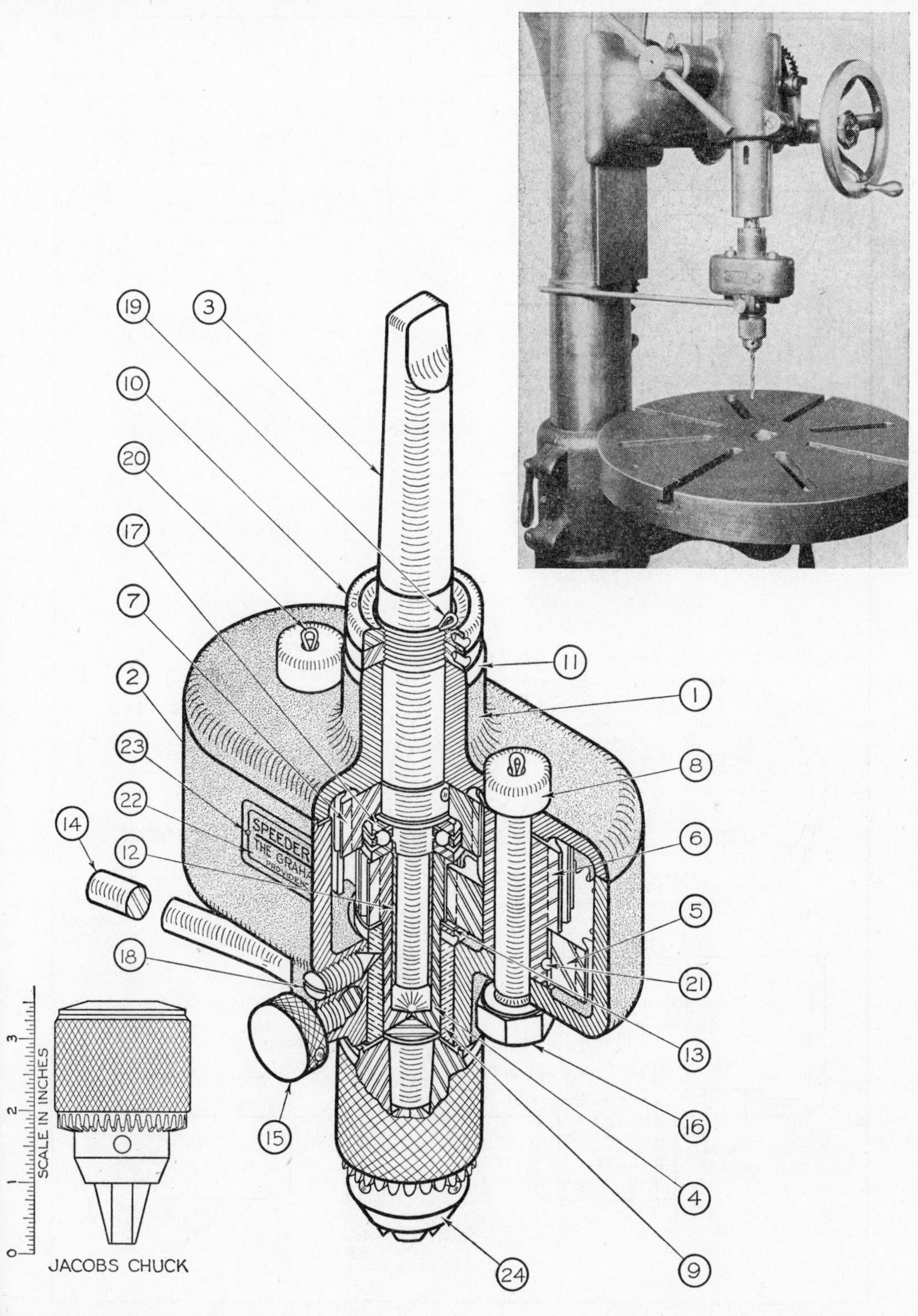

Fig. 802 Drill Speeder. (See Figs. 803 and 804.) Part 17 is a Thrust Bearing $1\frac{11}{32}$ O.D. × .625 I.D. × $\frac{9}{16}$.

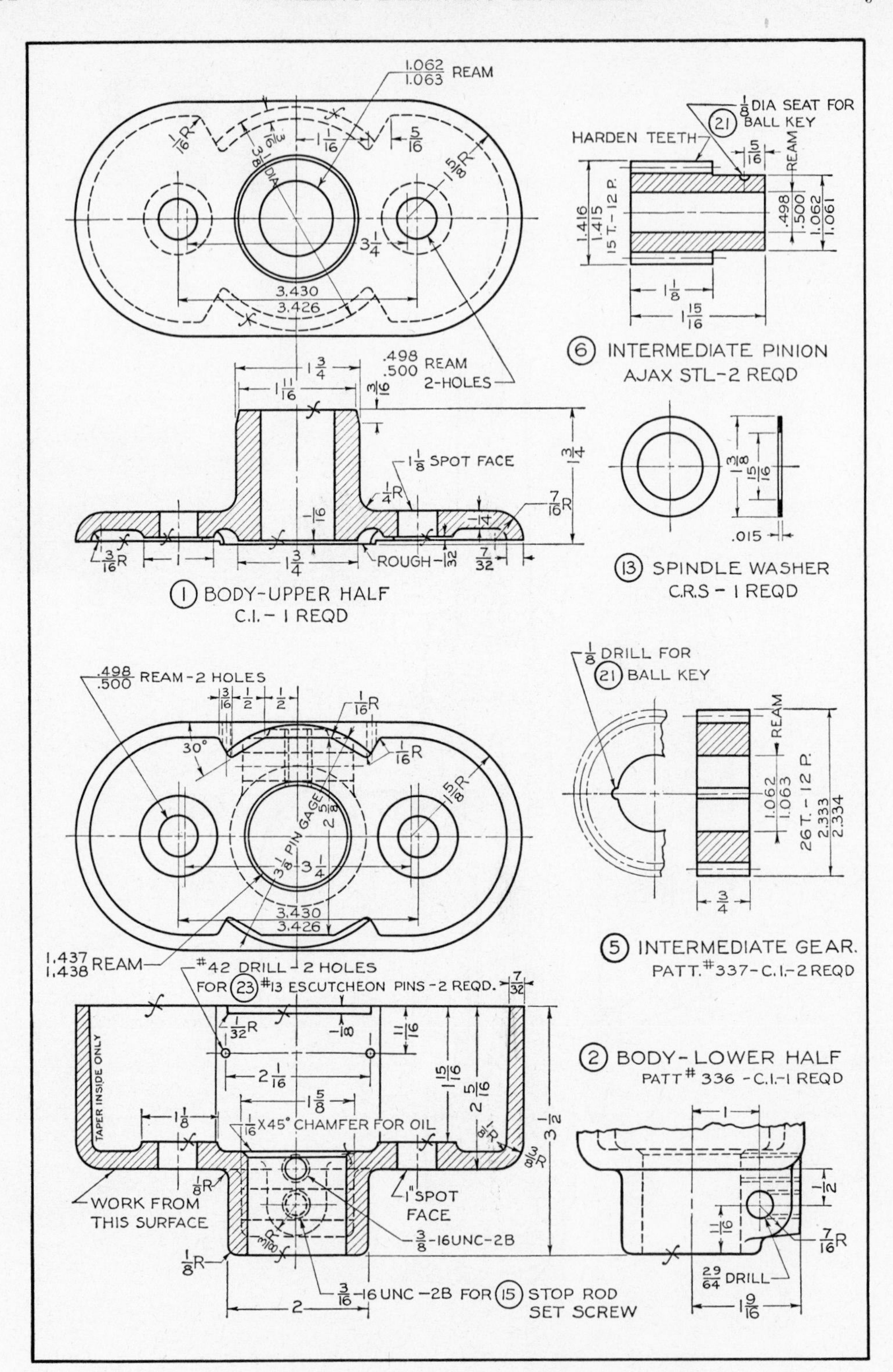

Fig. 803 Drill Speeder (*Continued*). (1) Draw details. (2) Draw assembly (see Fig. 802).

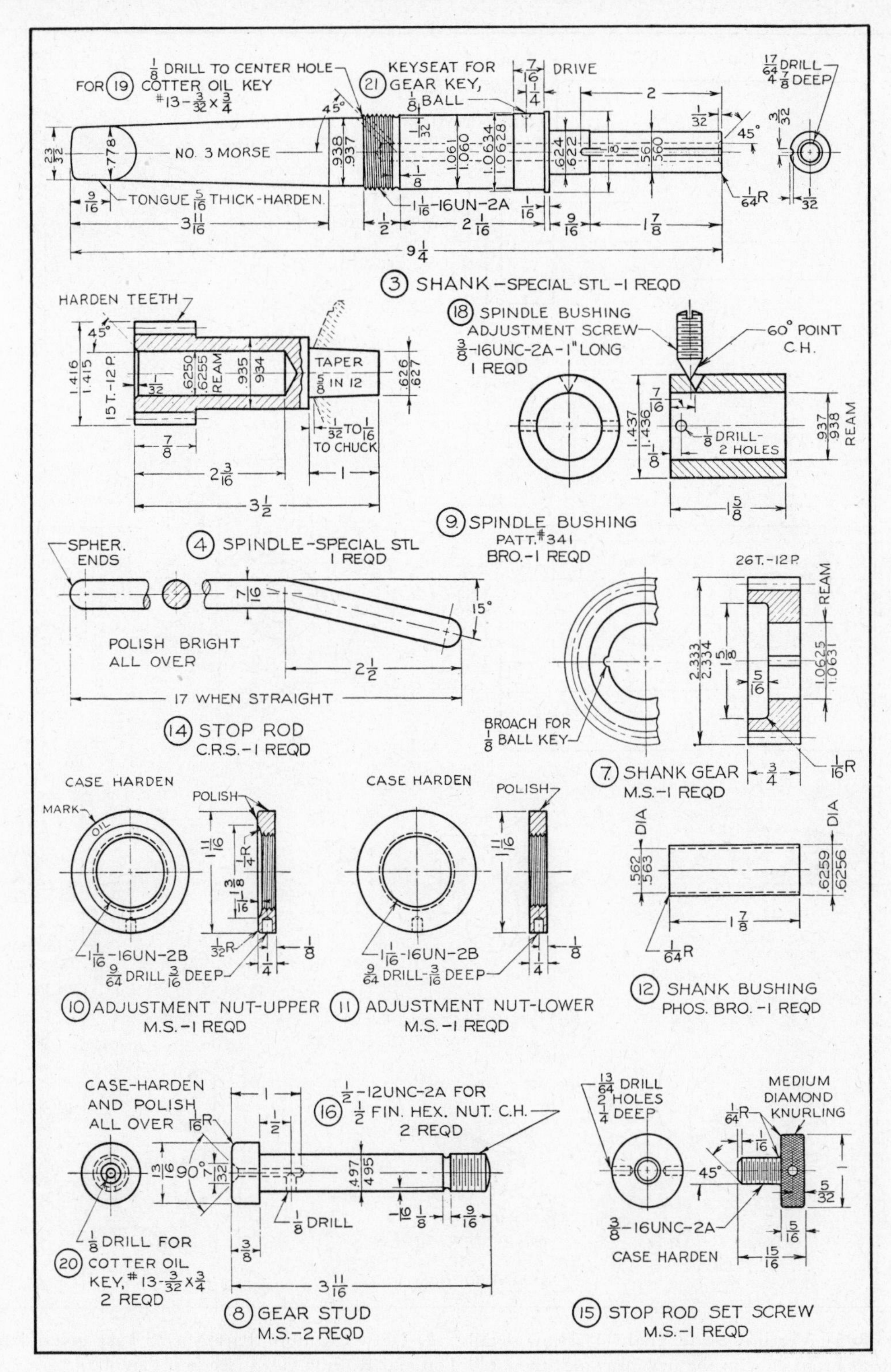

Fig. 804 Drill Speeder (*Continued*). Draw details.

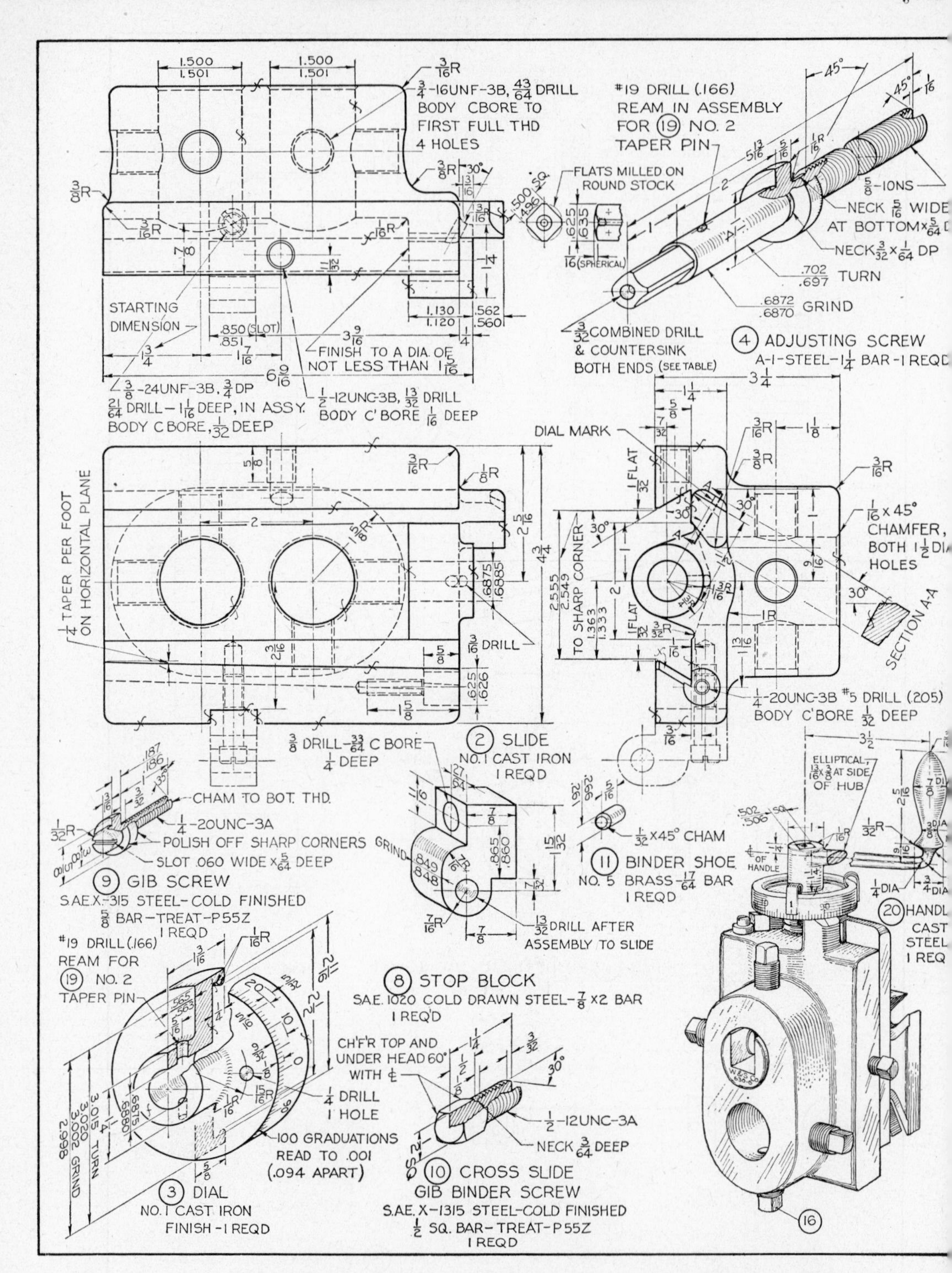

Fig. 805 Vertical Slide Tool. (1) Draw details. (2) Draw assembly. Part No. 2: Take given Top view as Front view in the new drawing; then add Top and R. Side views (see also Fig. 806).

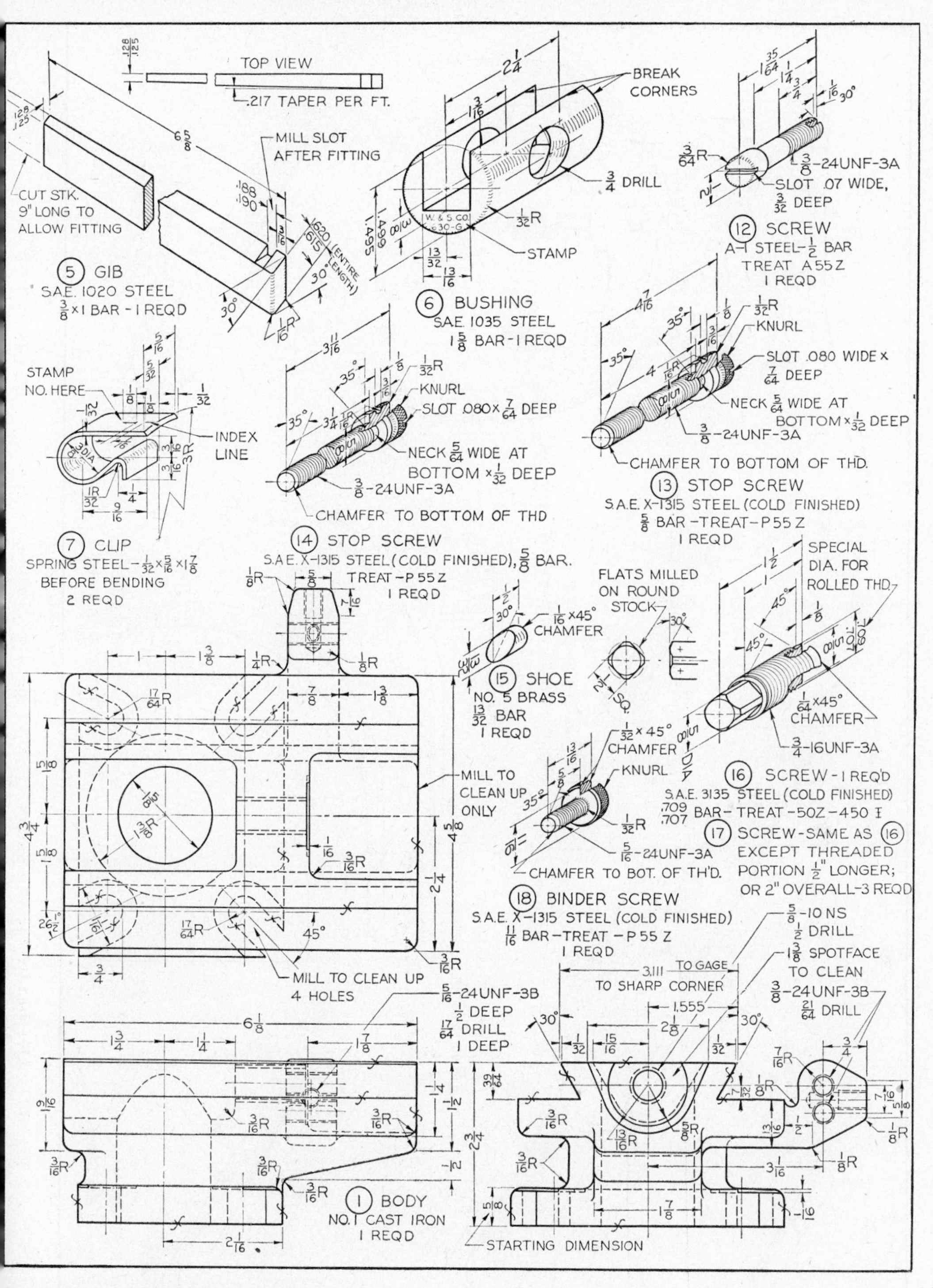

Fig. 806 Vertical Slide Tool (*Continued*). Part No. 1: Take Top view as Front view in the new drawing; then add Top and R. Side views.

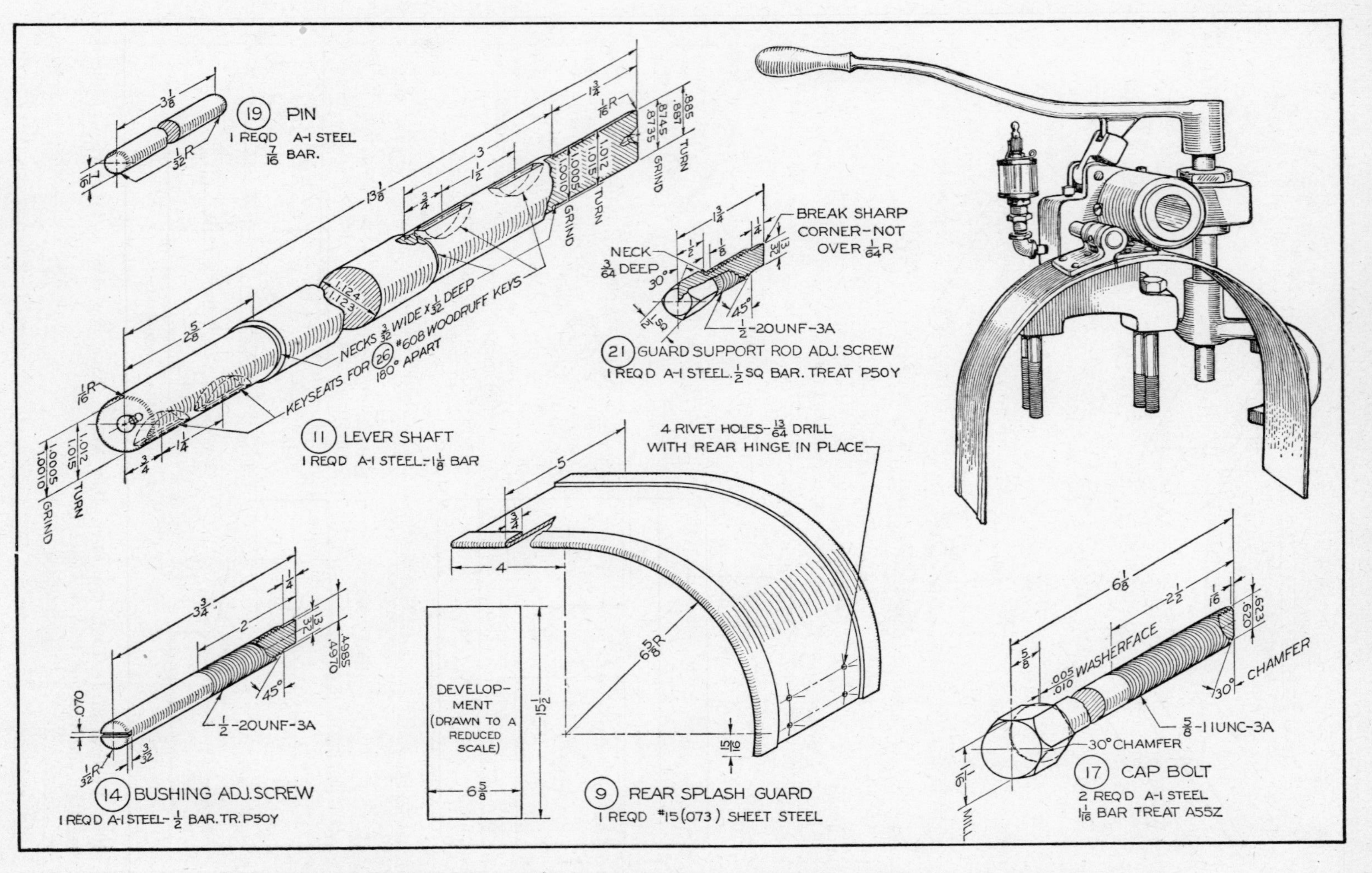

Fig. 807 Overhead Pilot Attachment. (1) Draw details. (2) Draw assembly (see also Figs. 808-810).

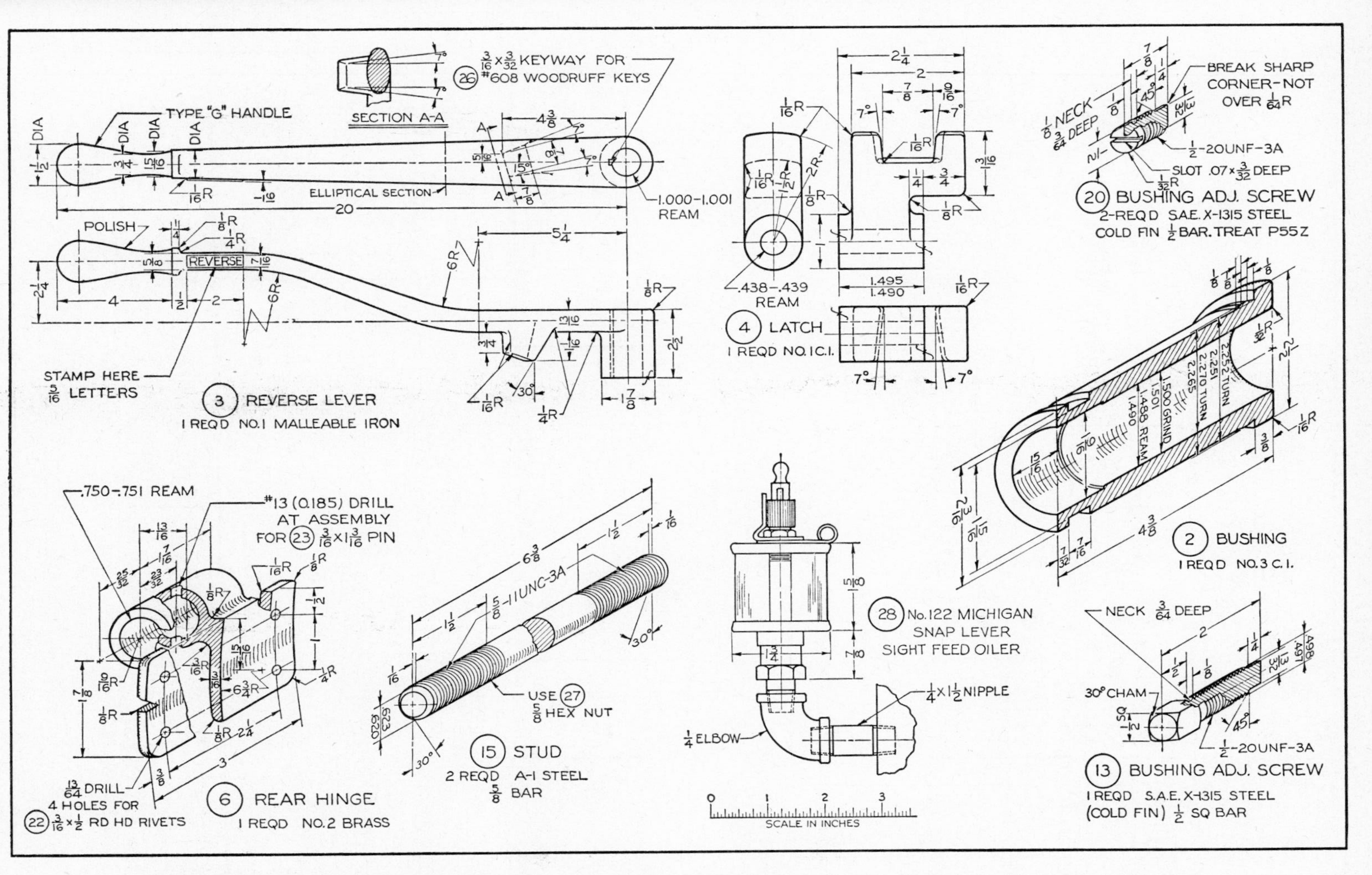

Fig. 808 Overhead Pilot Attachment (*Continued*). Draw details. Part No. 4: Draw Front, Top, and R. Side views.

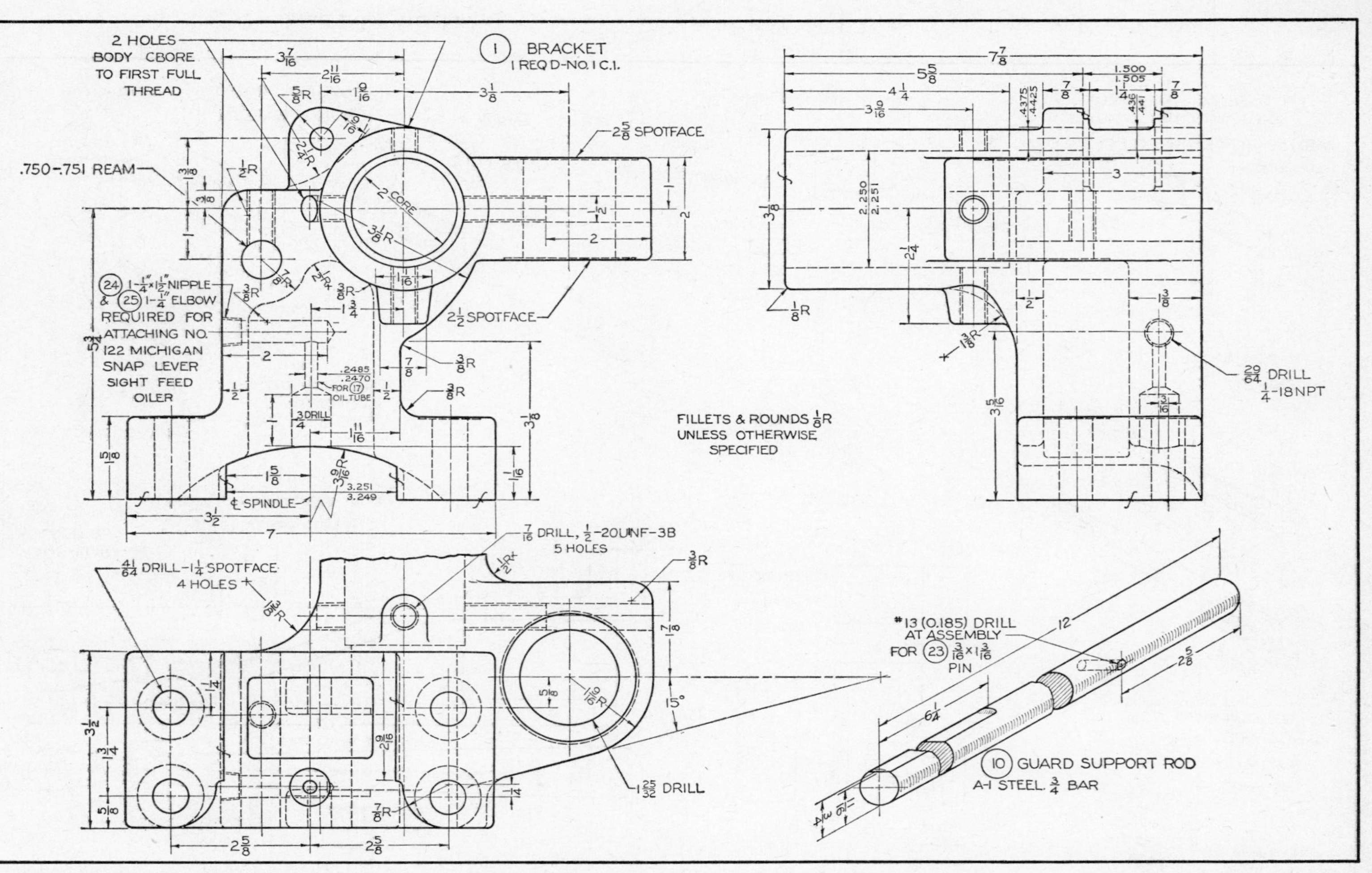

Fig. 809 Overhead Pilot Attachment (*Continued*). Part No. 1: Draw present Front view and add Top and L. Side views, omitting present R. Side view and partial Bottom view.

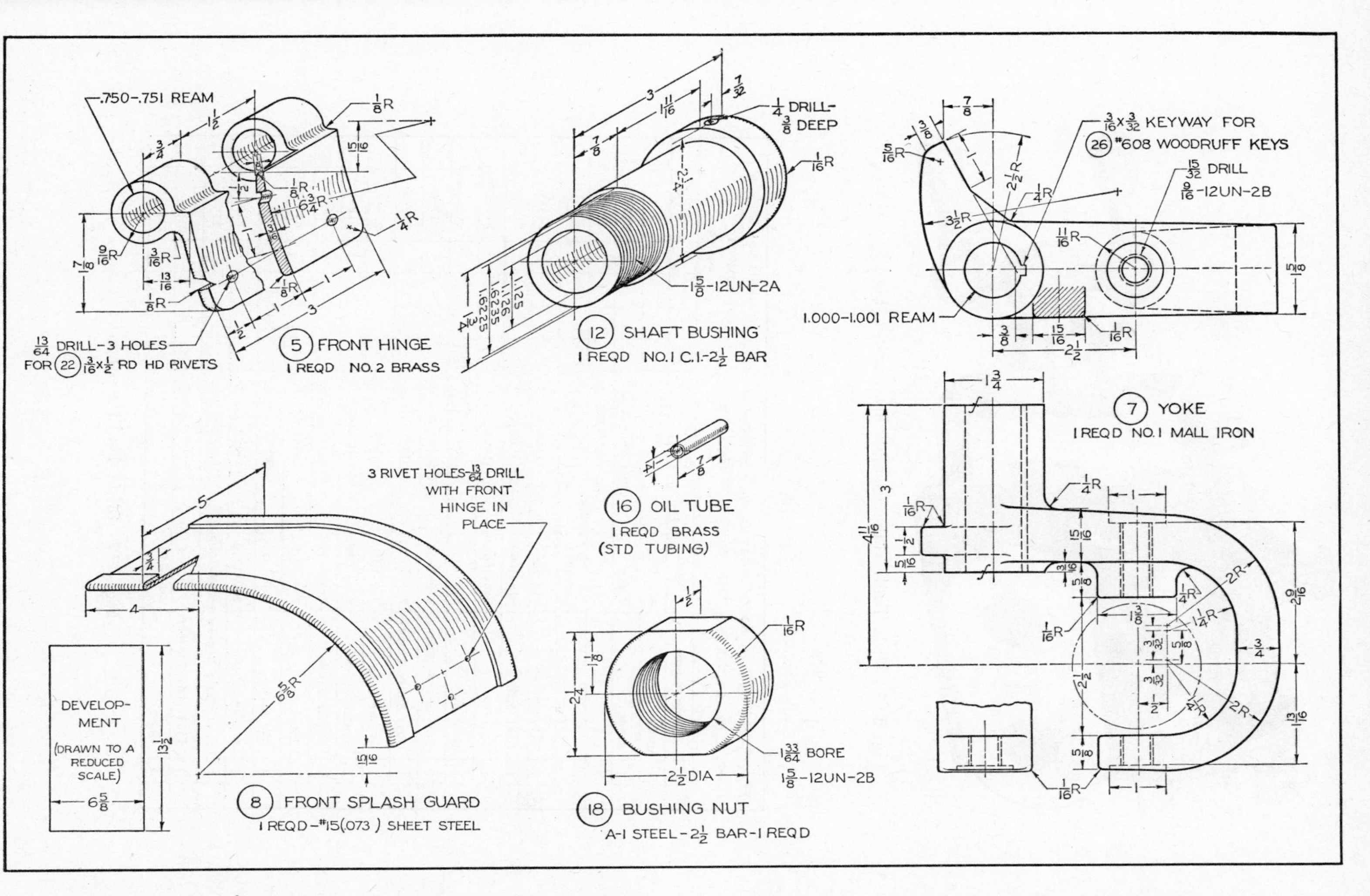

Fig. 810 Overhead Pilot Attachment *(Continued)*. Part No. 7: Revolve given Top view 180°; then add Front and R. Side views.

Fig. 811 Slide Tool. Make assembly drawing (see Figs. 813-815).

PARTS LIST NO. OF SHEETS 2 SHEET NO. 1 MACHINE NO. M-219

NAME NO. 4 SLIDE TOOL (SPECIFY SIZE OF SHANK REQ'D.) LOT NUMBER

NO. OF PIECES

TOTAL ON MACH.	NO. PCS.	NAME OF PART	PART NO.	CAST FROM PART NO.	TRACING NO.	MATERIAL	ROUGH WEIGHT PER PC.	DIA.	LENGTH	MILL	PART USED ON	NO. REQ. FINISH
	1	Body	219-12		D-17417	A-3-S D.F.						
	1	Slide	219-6		D-19255	A-3-S D.F.					219-12	
	1	Nut	219-9		E-19256	#10 BZ.					219-6	
	1	Gib	219-1001		C-11129	S.A.E. 1020					219-6	
	1	Slide Screw	219-1002		C-11129	A-3-S					219-12	
	1	Dial Bush	219-1003		C-11129	A-1-S					219-1002	
	1	Dial Nut	219-1004		C-11129	A-1-S					219-1002	
	1	Handle	219-1011		E-18270	(Buy from Cincinnati Ball Crank Co.)					219-1002	
	1	Stop Screw (Short)	219-1012		E-51950	A-1-S					219-6	
	1	Stop Screw (Long)	219-1013		E-51951	A-1-S					219-6	
	1	Binder Shoe	219-1015		E-51952	#5 Brass					219-6	
	1	Handle Screw	219-1016		E-62322	X-1315 C.F.					219-1011	
	1	Binder Screw	219-1017		E-63927	A-1-S					219-6	
	1	Dial	219-1018		E-39461	A-1-S					219-1002	
	2	Gib Screw	219-1019		E-52777	A-1-S		$\frac{1}{4}$-20	1		219-6	
	1	Binder Screw	280-1010		E-24962	A-1-S					219-1018	
	2	Tool Clamp Screws	683-F-1002		E-19110	D-2-S					219-6	
	1	Fill. Hd. Cap Scr.	1-A			A-1-S		$\frac{3}{8}$	$1\frac{3}{8}$		219-6 219-9	
	1	Key	No.404 Woodruff								219-1002	

Fig. 812 Slide Tool Parts List.

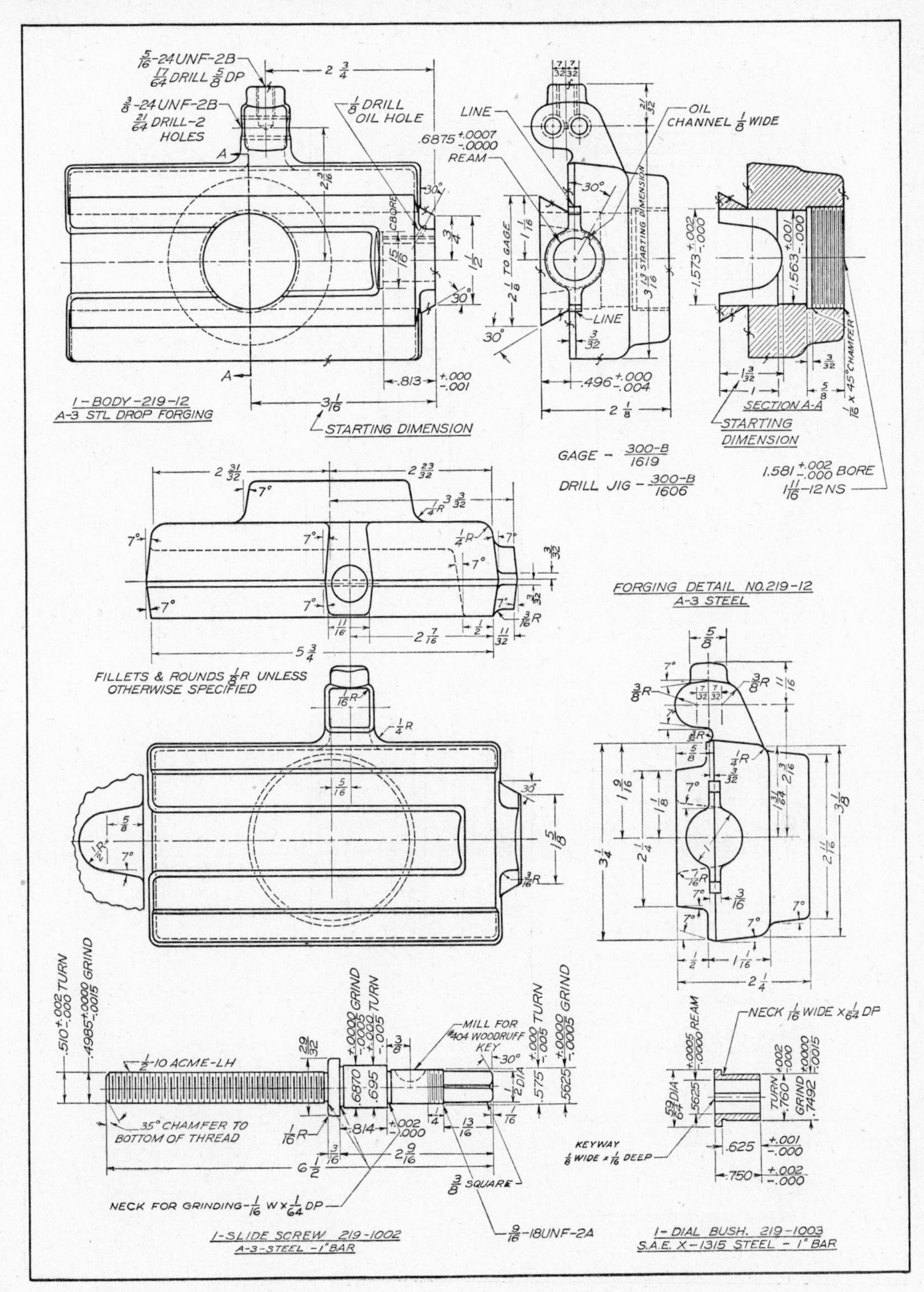

Fig. 813 Slide Tool (*Continued*). (1) Draw details. (2) Draw assembly (see Fig. 811).

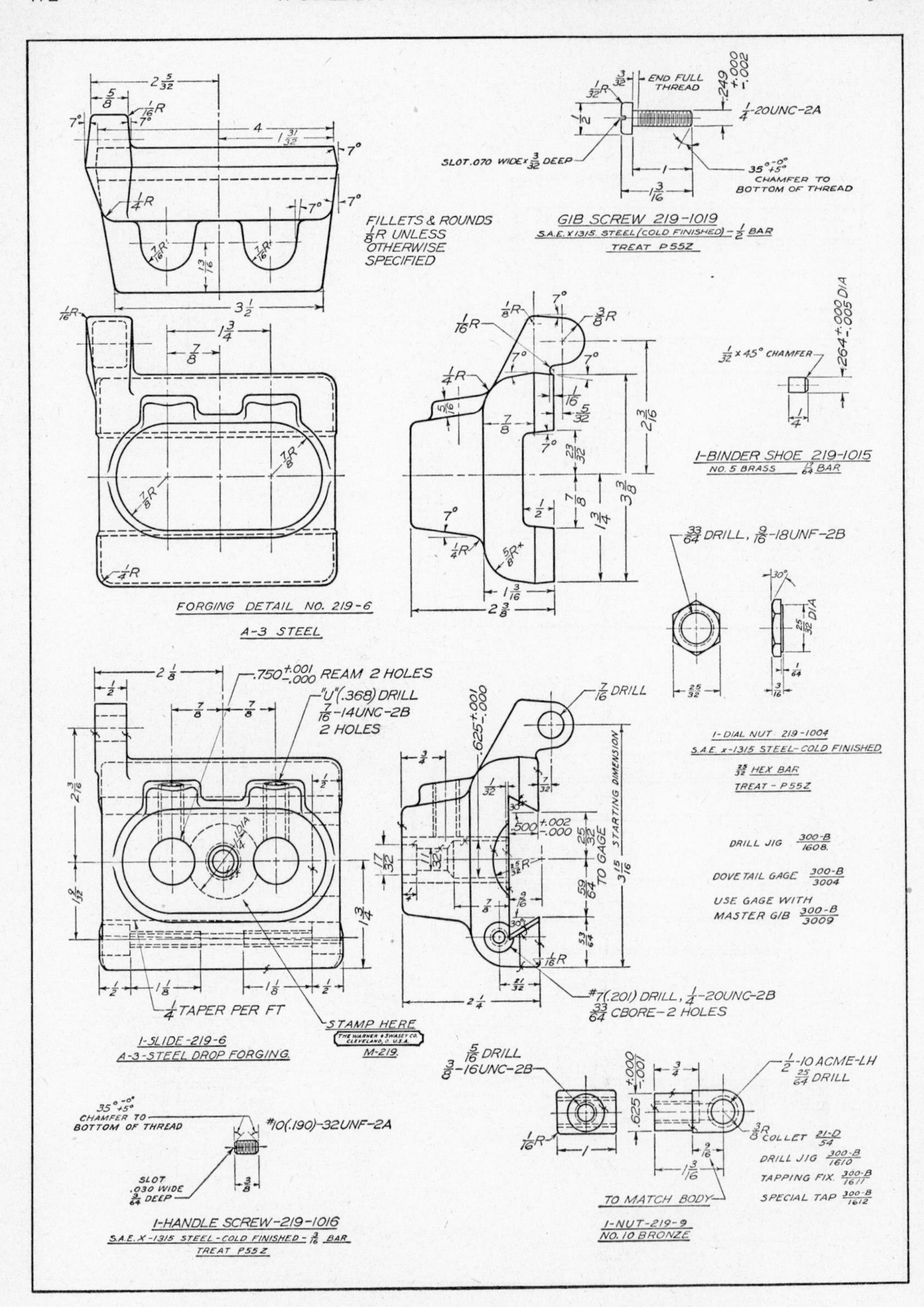

Fig. 814 Slide Tool (*Continued*). (1) Draw details. (2) Draw assembly.

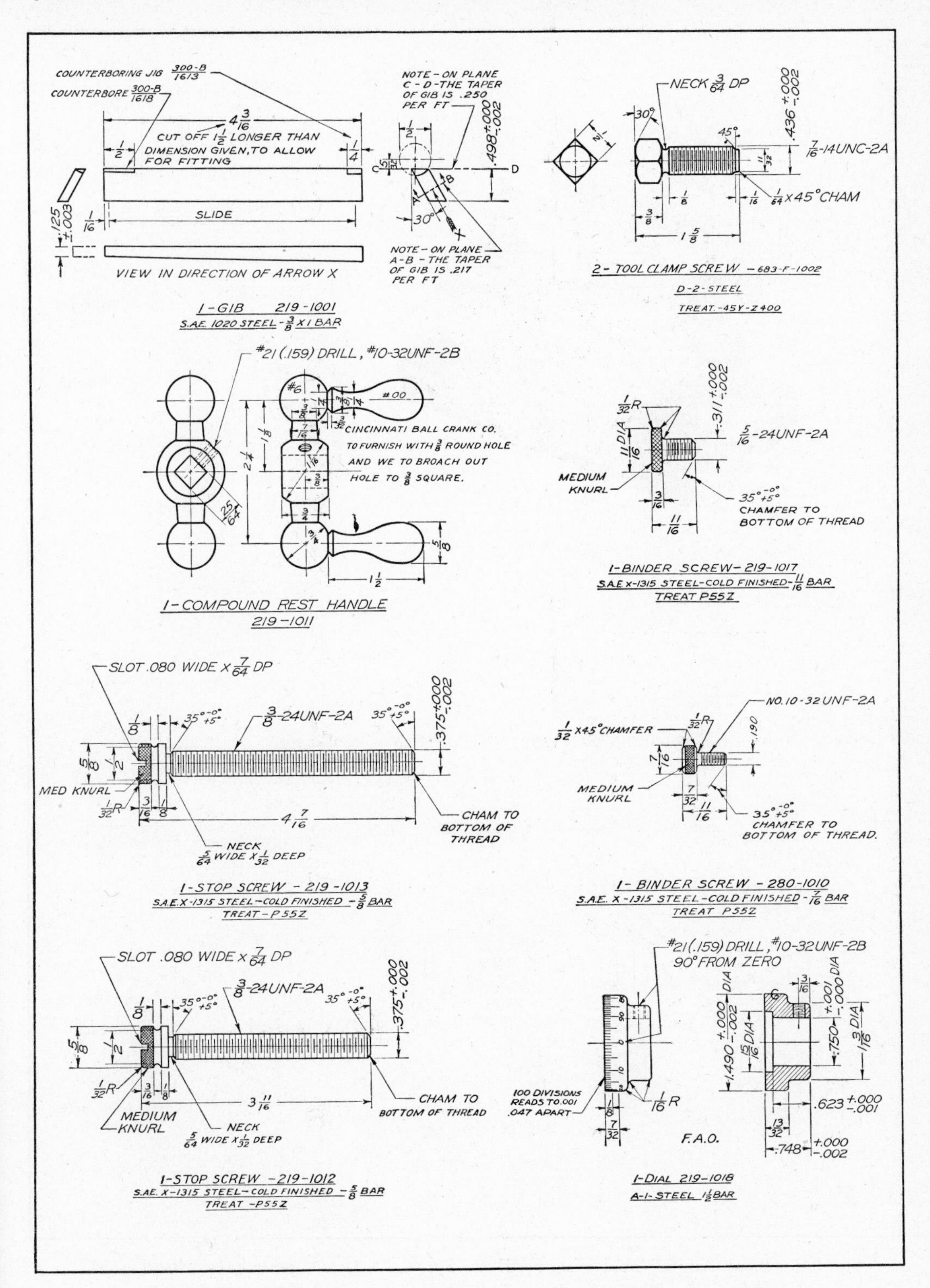

Fig. 815 Slide Tool (*Continued*). (1) Draw details. (2) Draw assembly.

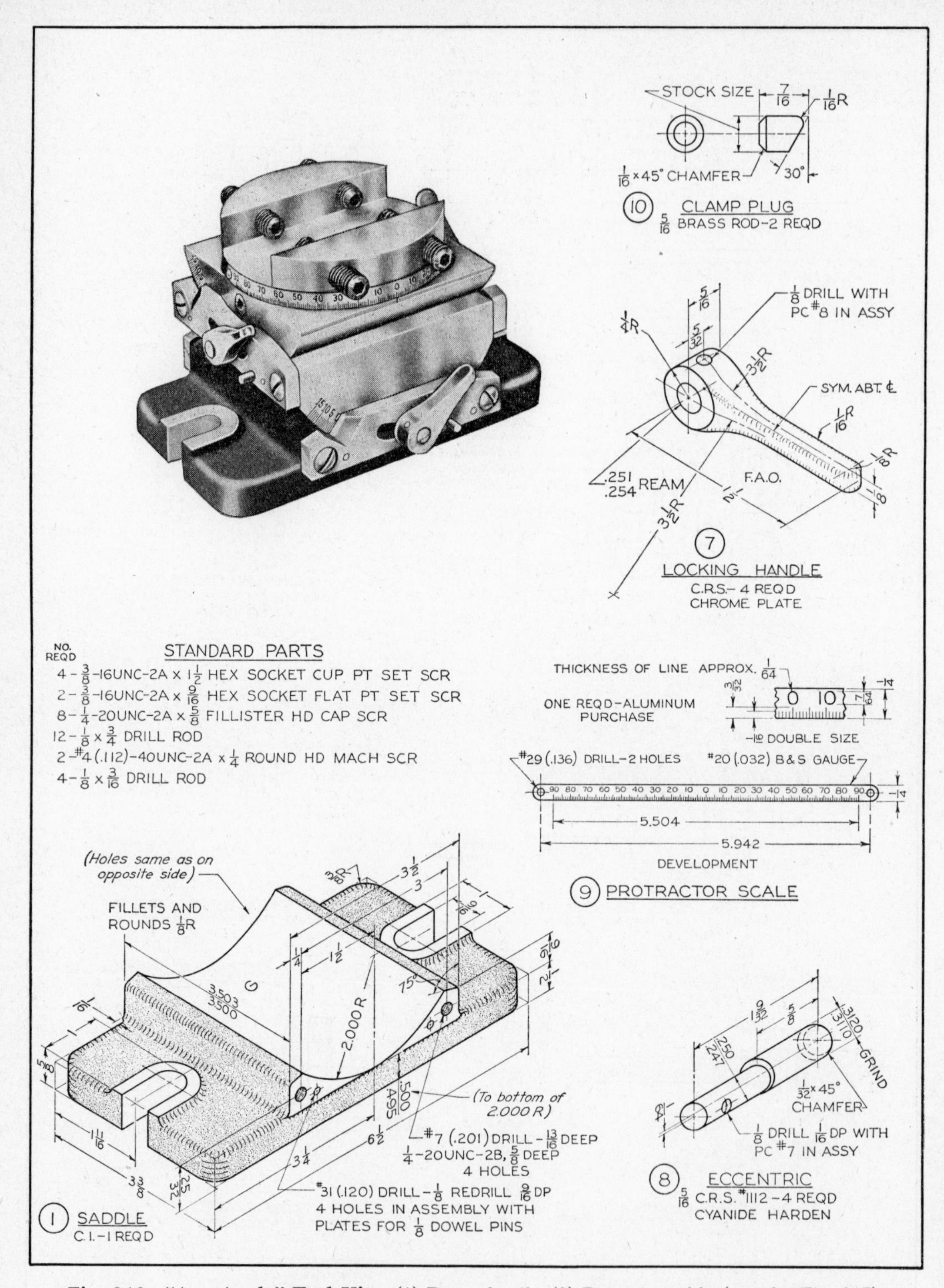

Fig. 816 "Any-Angle" Tool Vise. (1) Draw details. (2) Draw assembly (see also Fig. 817).

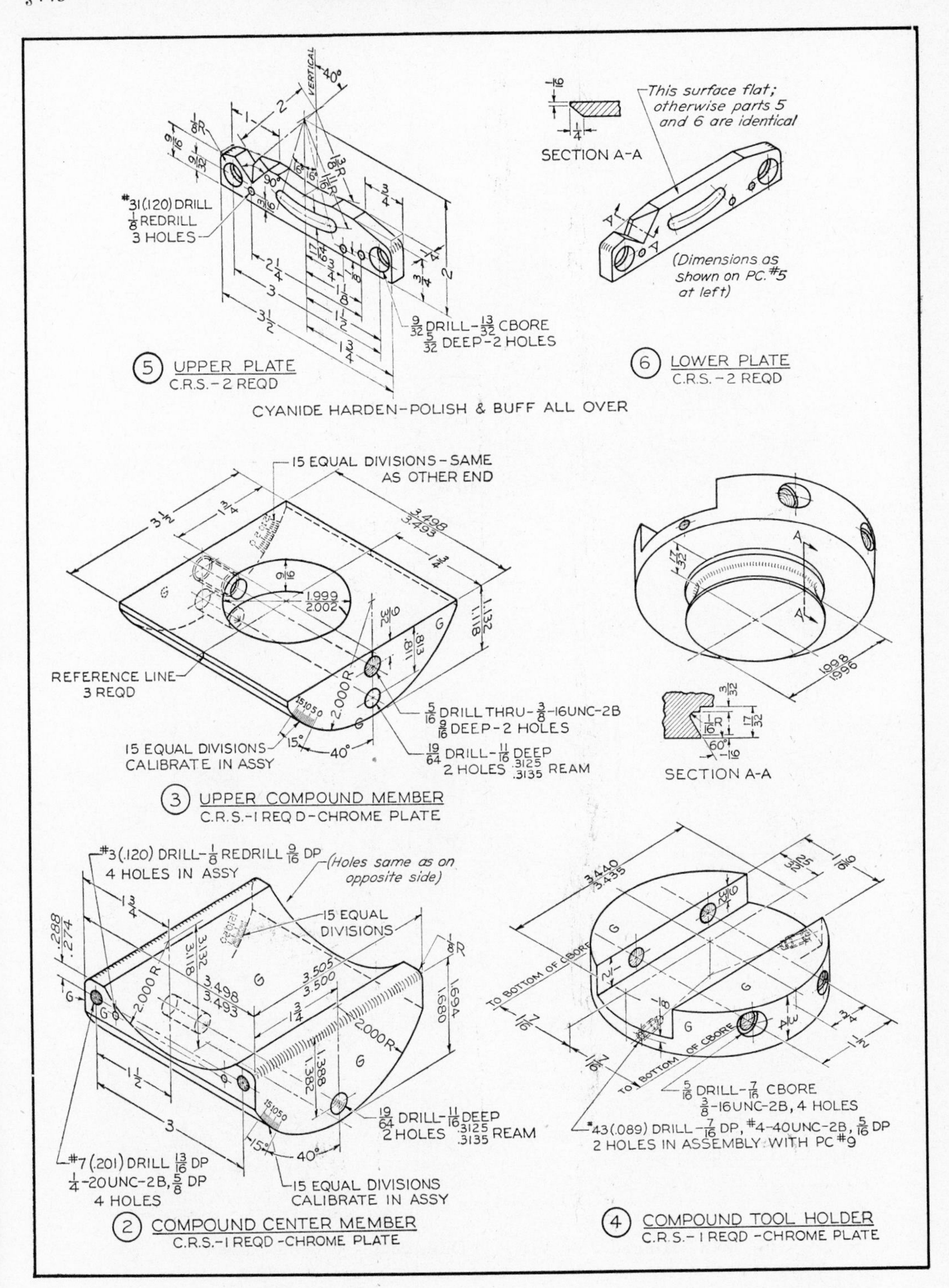

Fig. 817 "Any-Angle" Tool Vise (*Continued*). (1) Draw details. (2) Draw assembly.

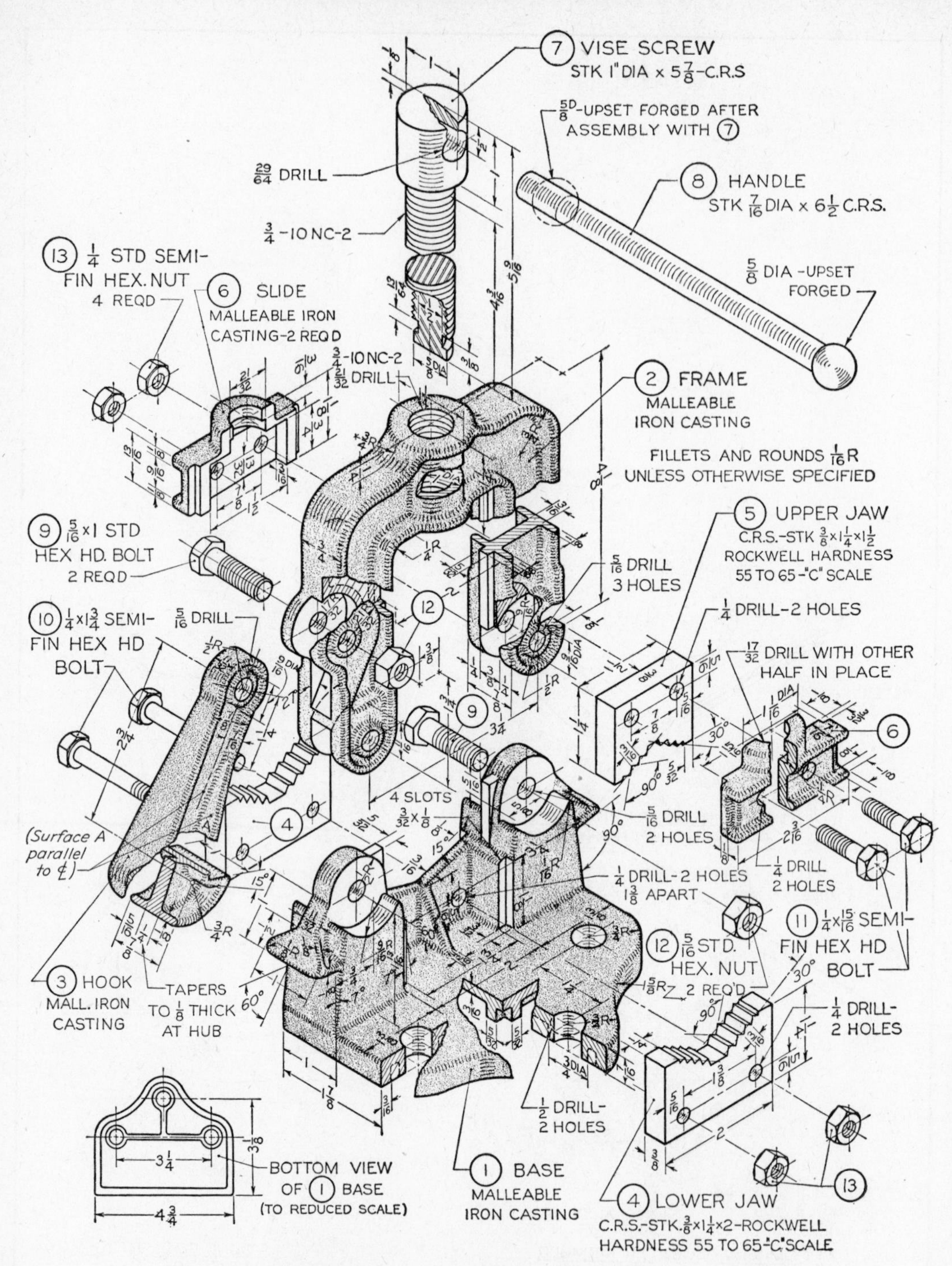

Fig. 818 Hinged Pipe Vise. (1) Draw details. (2) Draw assembly.

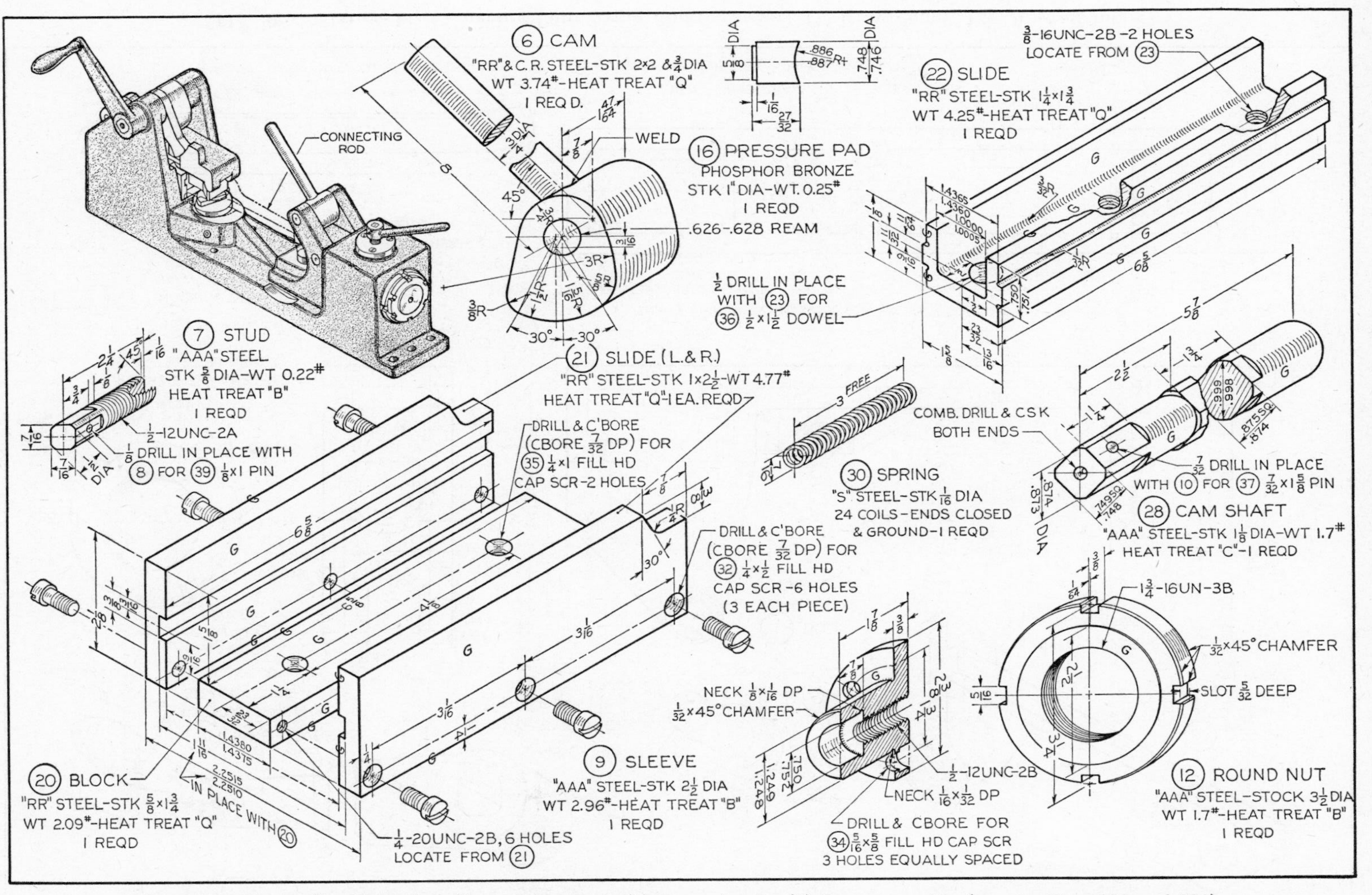

Fig. 819 Fixture for Centering Connecting Rod. (1) Draw details. (2) Draw assembly (see also Figs. 820 and 821).

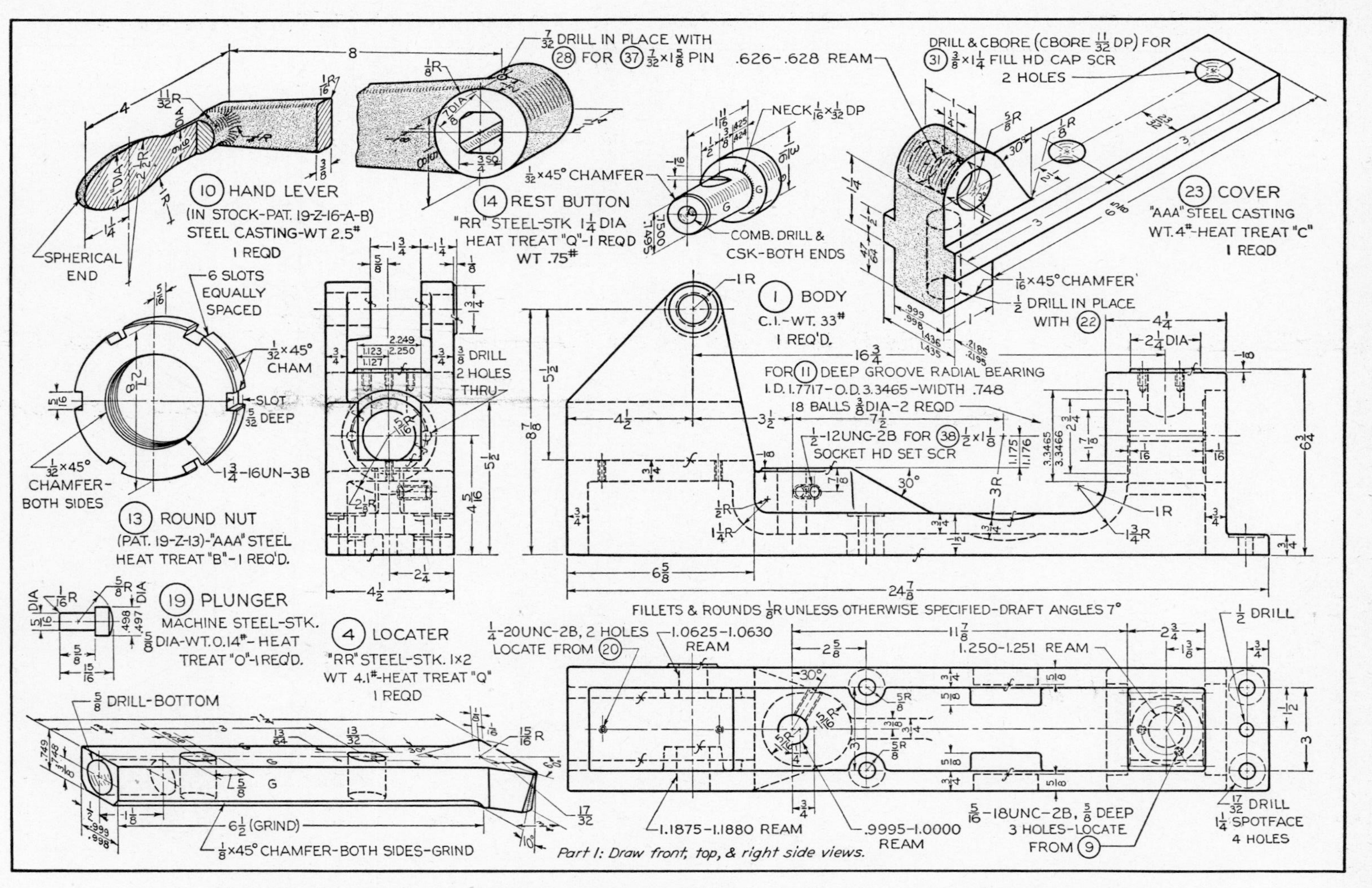

Fig. 820 Fixture for Centering Connecting Rod (*Continued*). (1) Draw details. (2) Draw assembly.

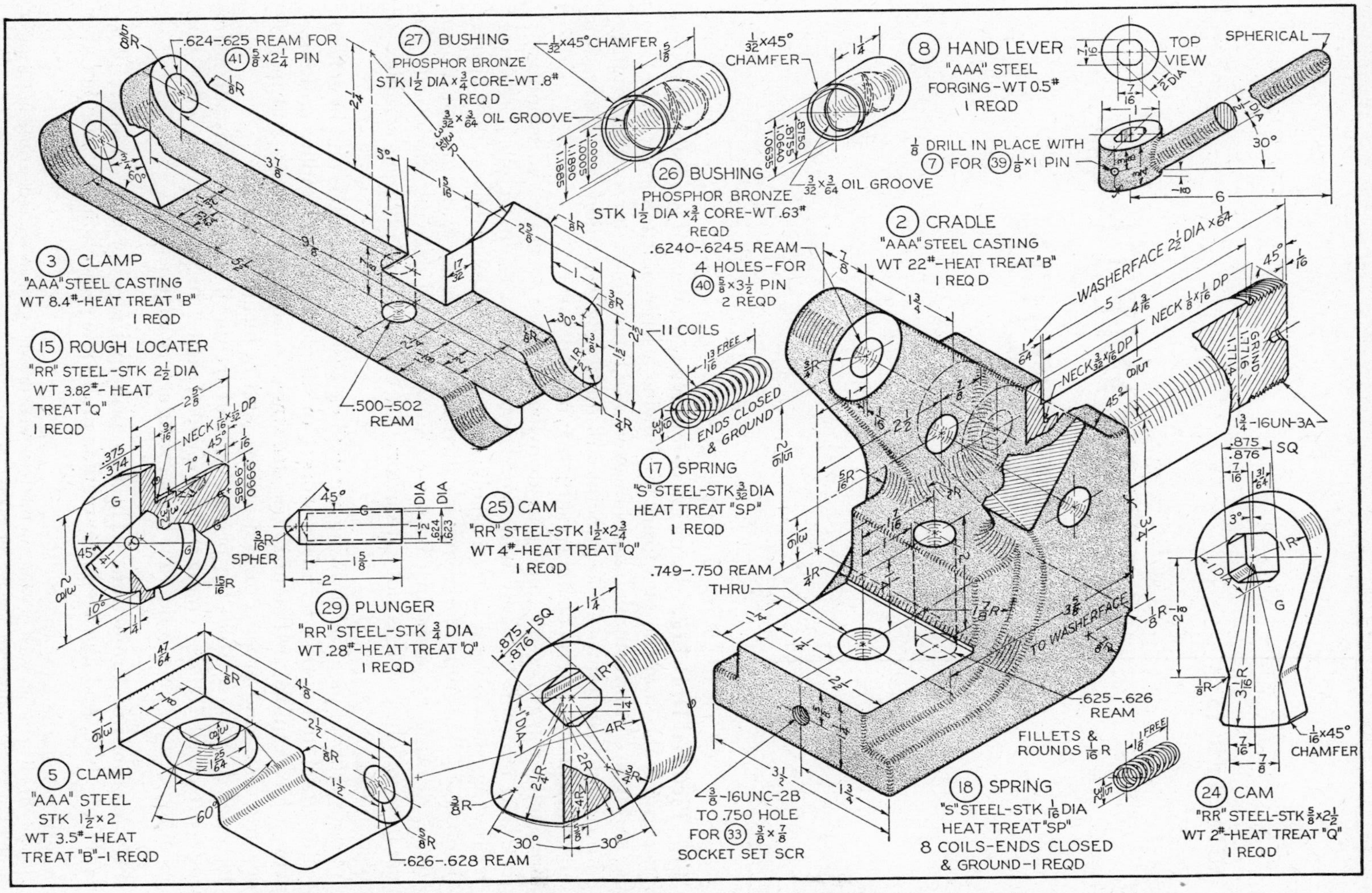

Fig. 821 Fixture for Centering Connecting Rod (*Continued*). (1) Draw details. (2) Draw assembly.

CHAPTER 15

AXONOMETRIC PROJECTION*

446. Pictorial Drawing. By means of multiview drawing, it is possible to represent accurately the most complex forms by showing a series of exterior views and sections. This type of representation has, however, two limitations: its execution requires a thorough understanding of the principles of multiview projection, and its reading requires a definite exercise of the constructive imagination.

Frequently it is necessary to prepare drawings that are accurate and scientifically correct, and that can be easily understood by persons without technical training. Such drawings show several faces of an object at once, approximately as they appear to the observer. This type of drawing is called *pictorial drawing*, to distinguish it from multiview drawing discussed in Chapter 6. Since pictorial drawing shows only the appearances of objects, it is not satisfactory for completely describing complex or detailed forms.

Various types of pictorial drawing are used extensively in catalogs and in general sales literature, and also in technical work,† to supplement and amplify multiview drawings; for example, in Patent Office drawings, in piping diagrams, and in machine, structural, and architectural drawings, and in furniture design.

Pictorial drawing enables the person without technical training to visualize the object represented. It also enables the trained designer to visualize the successive stages of the design, and to develop it in a satisfactory manner.

447. Methods of Projection. The four principal types of projection are illustrated in Fig. 822, and all except the regular multiview projection, (a), are pictorial

*See §10. See also ASA Y14.4-1957.

†Practically all of the pictorial drawings in this book were drawn by the methods described in Chapters 15, 16, and 17. See especially Figs. 336-395 for examples.

types since they show several sides of the object in a single view. In all cases the views, or projections, are formed by the piercing points in the plane of projection of an infinite number of visual rays or projectors.

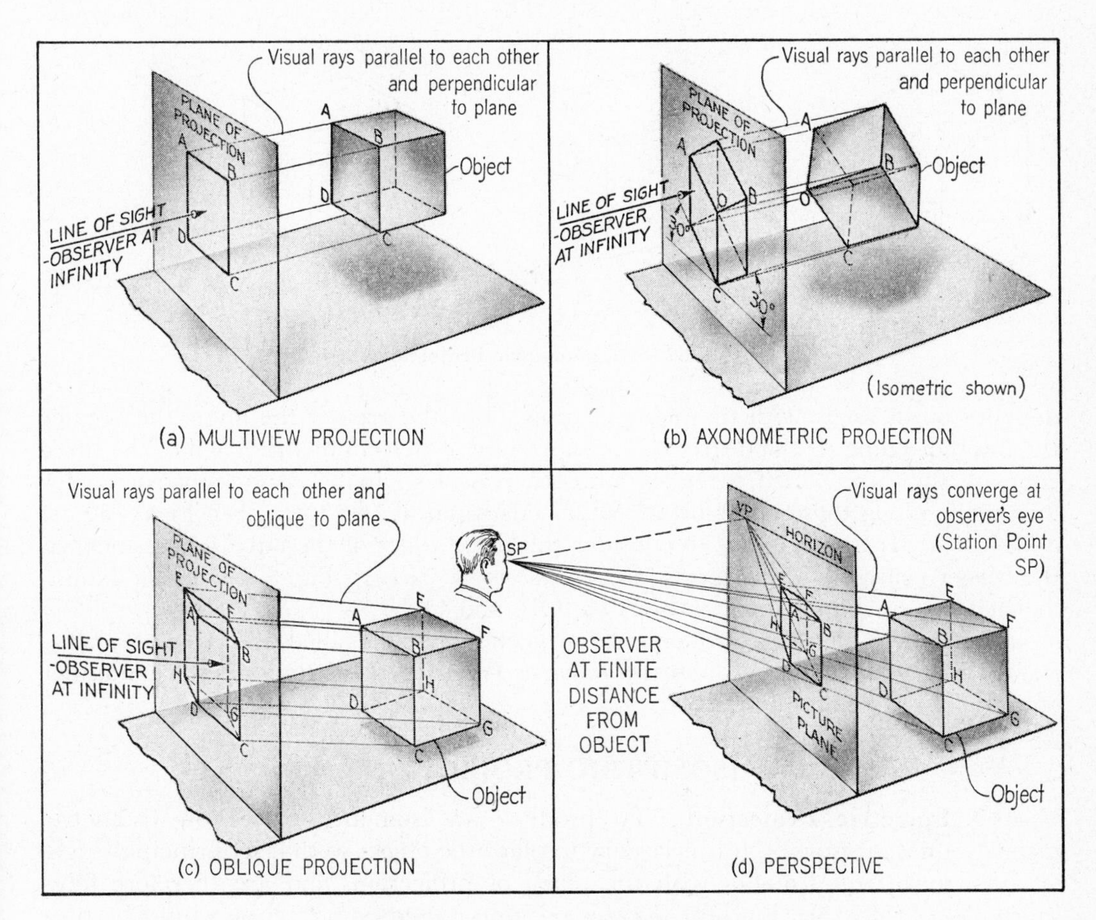

Fig. 822 Four Types of Projection.

In both multiview projection, (a), and *axonometric projection*, (b), the observer is considered to be at infinity, and the visual rays are perpendicular to the plane of projection. Therefore, both are classified as *orthographic projections*, §10.

In *oblique projection*, (c), the observer is considered to be at infinity, and the visual rays are parallel to each other but oblique to the plane of projection. See Chapter 16.

In *perspective*, (d), the observer is considered to be at a finite distance from the object, and the visual rays extend from the observer's eye, or the Station Point (SP), to all points of the object to form a so-called "cone of rays." See Chapter 17.

448. Types of Axonometric Projection. The distinguishing feature of axonometric projection, as compared to multiview projection, is the inclined position of the object with respect to the plane of projection. Since the principal edges and surfaces of the object are inclined to the plane of projection, the lengths of the lines, the sizes of the angles, and the general proportions of the object vary with the infinite number of possible positions in which the object may be placed with respect to the plane of projection. Three of these are shown in Fig. 823.

In these cases the edges of the cube are inclined to the plane of projection, and are therefore foreshortened. See Fig. 303. The degree of foreshortening of any line

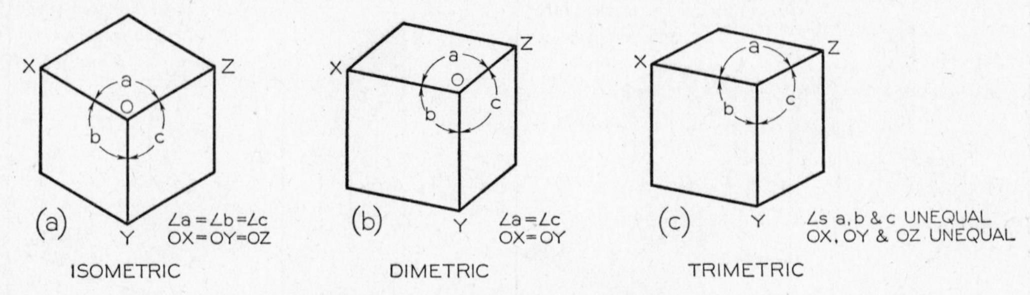

Fig. 823 Axonometric Projections.

depends on its angle with the plane of projection; the greater the angle the greater the foreshortening. If the degree of foreshortening is determined for each of the three edges of the cube which meet at one corner, scales can be easily constructed for measuring along these edges or any other edges parallel to them. See Figs. 860 (a) and 865 (a). It is customary to consider the three edges of the cube which meet at the corner nearest to the observer as the *axonometric axes.* In Fig. 822 (b), the axonometric axes, or simply the *axes,* are OA, OB, and OC.

As shown in Fig. 823, axonometric projections are classified as (a) *Isometric Projection,* (b) *Dimetric Projection,* and (c) *Trimetric Projection,* depending upon the number of scales of reduction required.

ISOMETRIC PROJECTION

449. Isometric Projection. To produce an isometric projection (isometric means "equal measure"), it is necessary to place the object so that its principal edges or axes, make equal angles with the plane of projection, and are therefore foreshortened equally. See Fig. 271. In this position the edges of a cube would be pro-

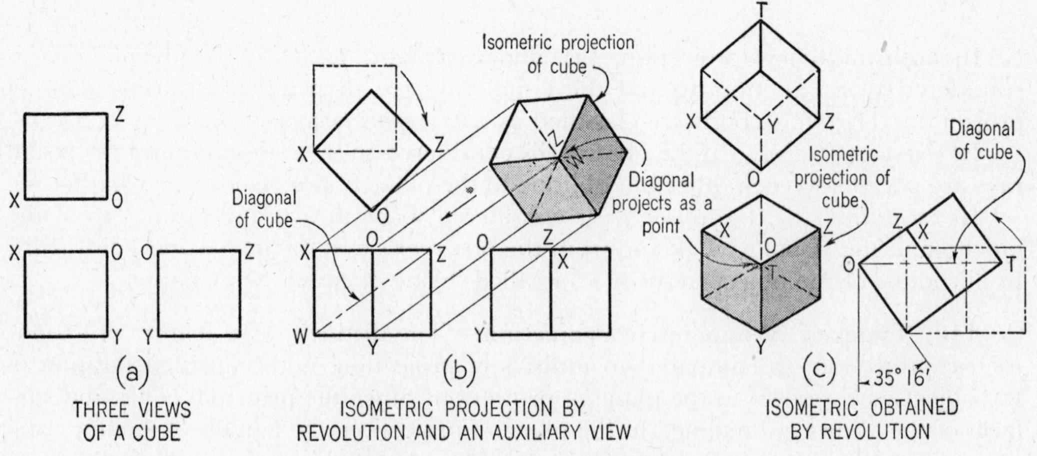

Fig. 824 Isometric Projection.

jected equally and would make equal angles with each other (120°), as shown in Fig. 823 (a).

In Fig. 824 (a) is shown a multiview drawing of a cube. At (b) the cube is shown revolved through 45° about an imaginary vertical axis. Now an auxiliary view in the direction of the arrow will show the cube diagonal ZW as a point, and the cube appears as a true isometric projection. However, instead of the auxiliary view at (b) being drawn, the cube may be further revolved as shown at (c), this time the cube being tilted forward about an imaginary horizontal axis until the three edges OX, OY, and OZ make equal angles with the frontal plane of projection and are, therefore, foreshortened equally. Here again, a diagonal of the cube, in this case OT, appears as a point in the isometric view. The front view thus obtained is a true isometric projection. In this projection the twelve edges of the cube make angles of about 35° 16′ with the frontal plane of projection. The lengths of their projections are equal to the lengths of the edges multiplied by $\sqrt{\frac{2}{3}}$, or by 0.816, approximately. Thus the projected lengths are about 80 per cent of the true lengths, or still more roughly, about three-fourths of the true lengths. The projections of the axes OX, OY, and OZ make angles of 120° with each other, and are called the *isometric axes.* Any line parallel to one of these is called an *isometric line;* a line which is not parallel is called a *non-isometric line.* It should be noted that the angles in the isometric projection of the cube are either 120° or 60° and that all are projections of 90° angles. In an isometric projection of a cube, the faces of the cube, or any planes parallel to them, are called *isometric planes.*

450. The Isometric Scale. A correct isometric projection may be drawn with the use of a special *isometric scale,* prepared on a strip of paper or cardboard, Fig. 825. All distances in the isometric scale are $\sqrt{\frac{2}{3}}$ times true size, or approximately 80 per cent of true size. The use of the isometric scale is illustrated in Fig. 826 (a). A scale of 9″ = 1′-0, or $\frac{3}{4}$-size scale, could be used to approximate the isometric scale.

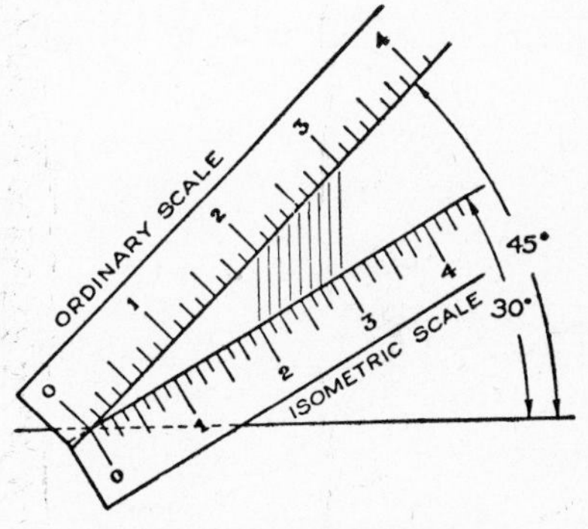

Fig. 825 Isometric Scale.

451. Isometric Drawing. When a drawing is prepared with an isometric scale, or otherwise as the object is actually *projected* on a plane of projection, it is an *isometric projection,* as illustrated in Fig. 826 (a). When it is prepared with an ordinary scale, it is an *isometric drawing,* illustrated in (b). The isometric drawing (b) is about 22½ per cent larger than the isometric projection (a), but the pictorial value is obviously the same in both.

Since the isometric projection is foreshortened and an isometric drawing is full size, it is customary to make an isometric drawing rather than an isometric projection, because it is so much easier to execute and, for all practical purposes, is just as satisfactory as the isometric projection.

452. Isometric Drawing. The steps in constructing an isometric drawing of an object composed only of normal surfaces, §224, are illustrated in Fig. 827. Notice that all measurements are made parallel to the main edges of the enclosing box, that is, parallel to the isometric axes. No measurement along a diagonal (non-

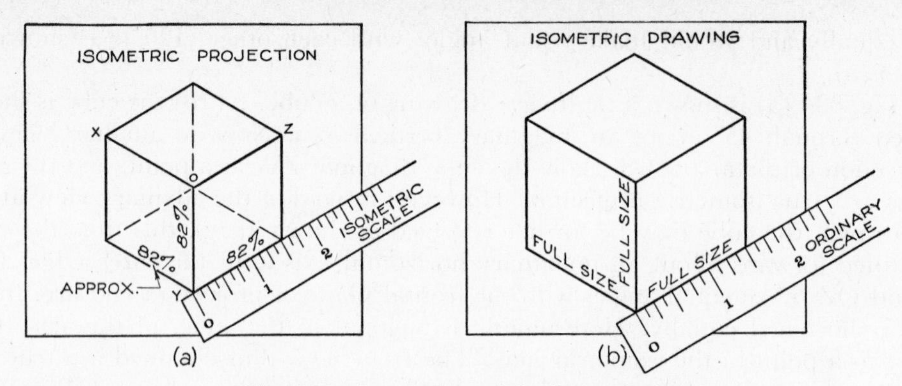

Fig. 826 Isometric and Ordinary Scales.

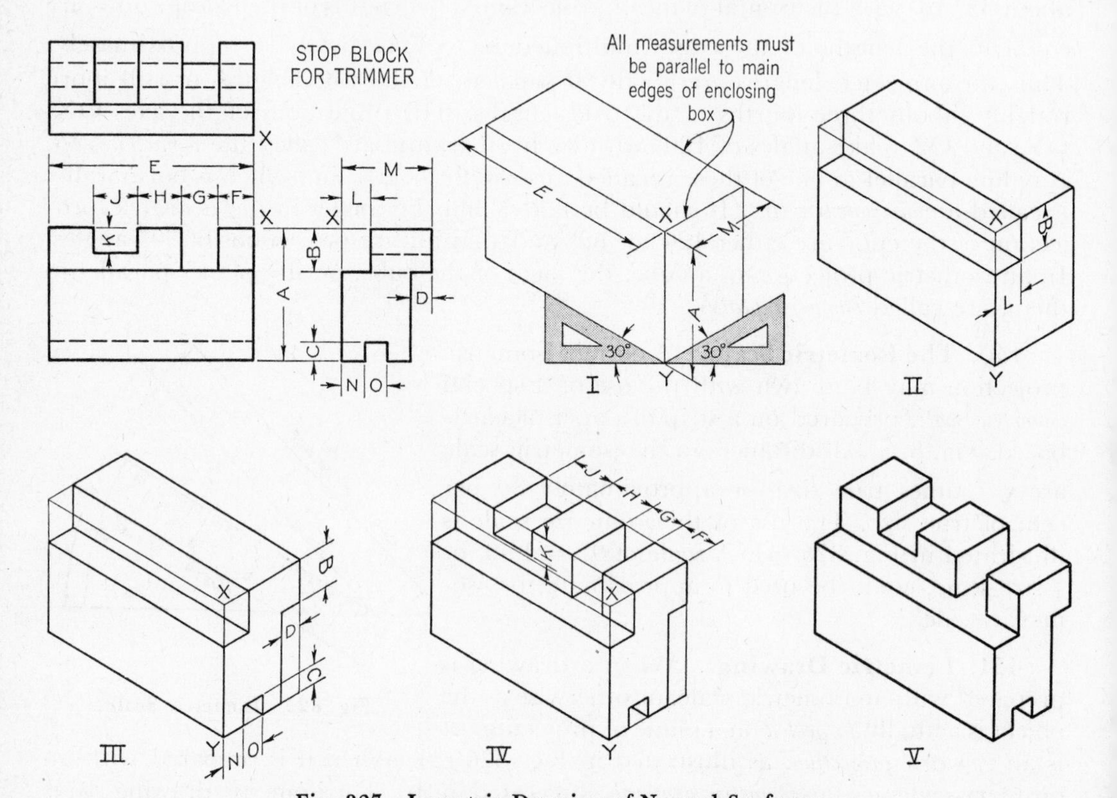

Fig. 827 Isometric Drawing of Normal Surfaces.

isometric line) on any surface or through the object can be set off directly with the scale. The object may be drawn in the same position by beginning at the corner Y, or any other corner, instead of at the corner X.

The method of constructing an isometric drawing of an object composed partly of inclined surfaces (and oblique edges) is shown in Fig. 828. Notice that inclined surfaces are located by *offset measurements* along isometric lines. For example, dimensions E and F are set off to locate the inclined surface M, and dimensions A and B are used to locate surface N.

For sketching in isometric, see §§199-202.

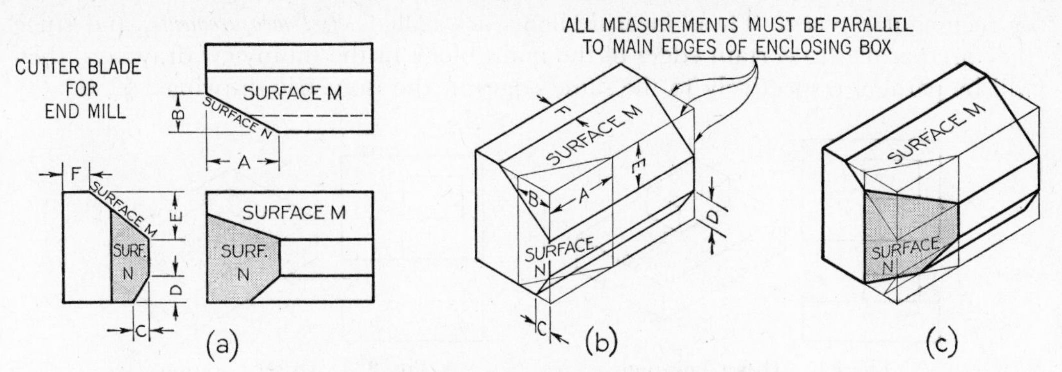

Fig. 828 Inclined Surfaces in Isometric.

453. Other Positions of the Isometric Axes. The isometric axes may be placed in any desired position according to the requirements of the problem, as shown in Fig. 829, but the angle between the axes must remain 120°. The choice of the direc-

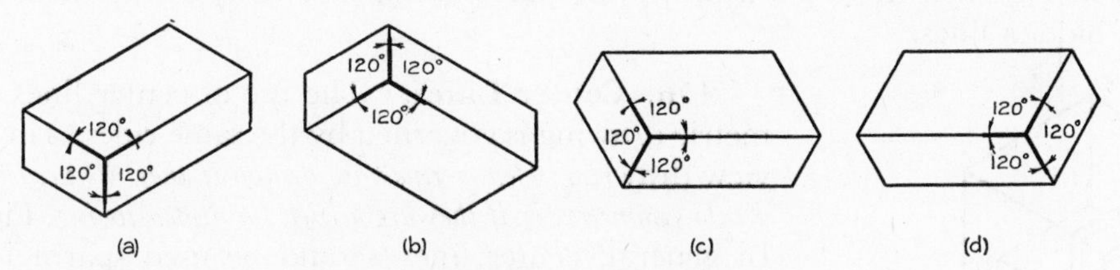

Fig. 829 Positions of Isometric Axes.

tions of the axes is determined by the position from which the object is usually viewed, Fig. 830, or by the position which best describes the shape of the object. If possible, both requirements should be met.

If the object is characterized by considerable length, the long axis may be placed horizontally for best effect, as shown in Fig. 831.

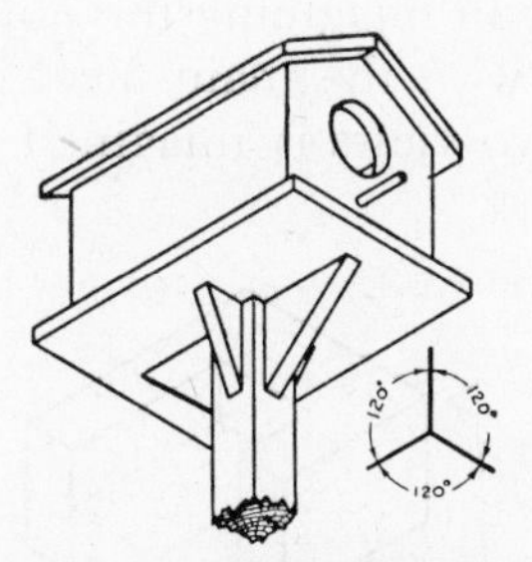

Fig. 830 An Object Naturally Viewed from Below.

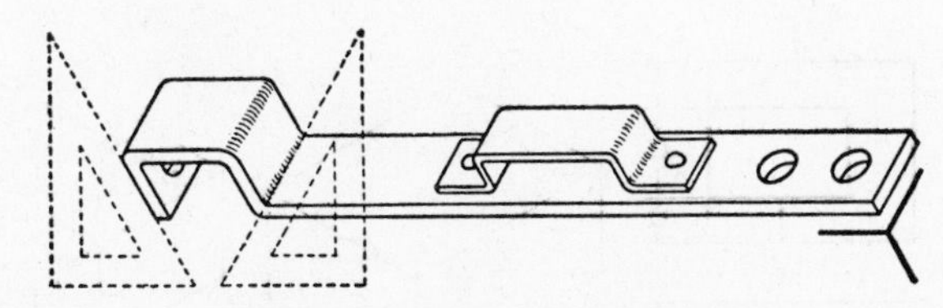

Fig. 831 Long Axis Horizontal.

454. Offset Location Measurements. The method of locating one part with respect to another is illustrated in Figs. 832 and 833. In each case, after the main block has been drawn, the offset lines CA and BA in the multiview drawing are drawn full size in the isometric drawing, thus locating corner A of the small block or rectangular recess. These measurements are called *offset measurements*, and since they are parallel to certain edges of the main block in the multiview drawings, they will be parallel respectively to the same edges in the isometric drawings, §230.

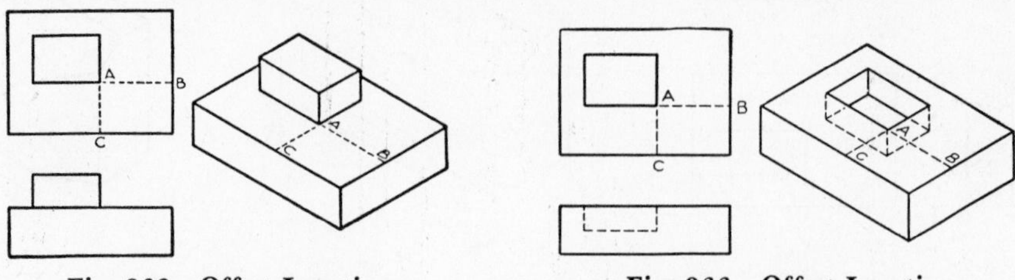

Fig. 832 Offset Location Measurements. **Fig. 833 Offset Location Measurements.**

455. Hidden Lines. The use of hidden lines in isometric drawing is governed by the same rules as in all other types of projection: *Hidden lines are omitted unless they are needed to make the drawing clear.* A case in which hidden lines are needed is illustrated in Fig. 834, in which a projecting part cannot be clearly shown without the use of hidden lines.

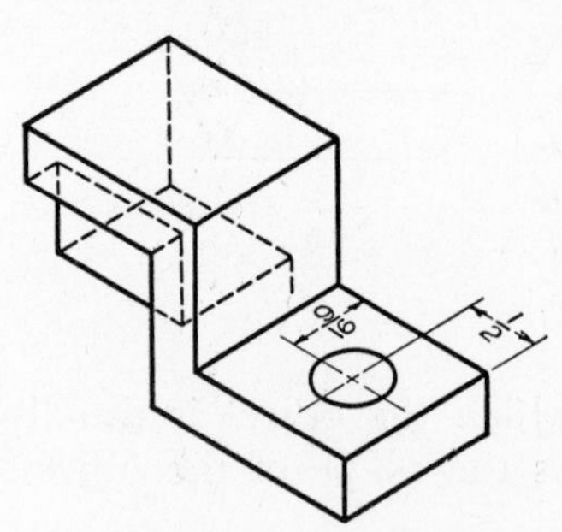

Fig. 834 Use of Hidden Lines.

456. Center Lines. The use of center lines in isometric drawing is governed by the same rules as in multiview drawing: *Center lines are drawn if they are needed to indicate symmetry, or if they are needed for dimensioning*, Fig. 834. In general, center lines should be used sparingly, and omitted in cases of doubt. The use of too many center lines may produce a confusion of lines, which diminishes the clearness of the drawing. Examples in which center lines are not needed are shown in Figs. 830 and 831. Examples in which they are needed are seen in Figs. 858 (a), 737, 738, etc.

457. Box Construction. Objects of rectangular shape may be more easily drawn by means of *box construction*, which consists simply in imagining the object to be enclosed in a rectangular box whose sides coincide with the main faces of the object. For example, in Fig. 835 the object shown in two views is imagined to be

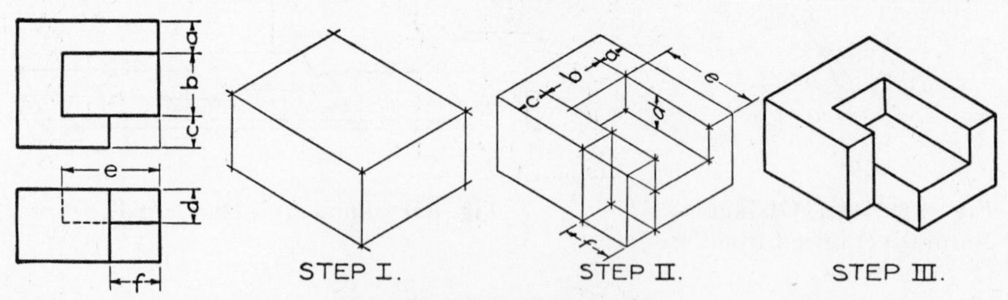

Fig. 835 Box Construction.

enclosed in a construction box. This box is then drawn lightly with construction lines, I, the irregular features are then constructed, II, and finally, III, the required lines are made heavy.

458. Non-Isometric Lines. Since the only lines of an object that are drawn true length in an isometric drawing are the isometric axes or lines parallel to them, *non-isometric* lines cannot be set off directly with the scale. For example, in Fig. 836 (a),

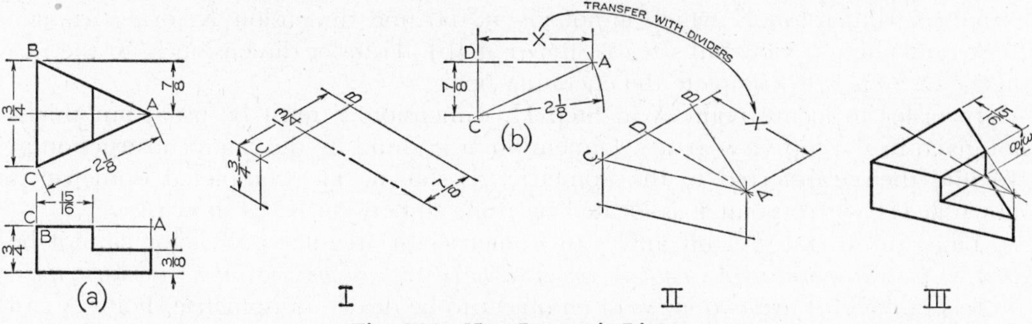

Fig. 836 Non-Isometric Lines.

the inclined lines BA and CA are shown in their true lengths ($2\frac{1}{8}''$) in the top view, but since they are not parallel to the isometric axes, they will not be true length in the isometric. Such lines are drawn in isometric by means of box construction and offset measurements. First, as shown at I, the measurements $1\frac{3}{4}''$, $\frac{3}{4}''$, and $\frac{7}{8}''$ can be set off directly since they are made along isometric lines. The non-isometric $2\frac{1}{8}''$ dimension cannot be set off directly, but if one half of the given top view is constructed full size to scale as shown at (b), the dimension X can be determined. This dimension is parallel to an isometric axis and can be transferred with dividers to the isometric at II. The dimensions $\frac{15}{16}''$ and $\frac{3}{8}''$ are parallel to isometric lines and can be set off directly, as shown in Step III.

To realize the fact that non-isometric lines will not be true length in the isometric drawing, set your dividers on BA of Step II and then compare with BA on the given top view. Do the same for line CA. It will be seen that BA is shorter and CA is longer in the isometric than the corresponding lines in the given views.

459. Angles in Isometric. As shown in §231, angles project true size only when the plane of the angle is parallel to the plane of projection. An angle may project larger or smaller than true size, depending upon its position. Since in isometric the various surfaces of the object are usually inclined to the plane of projection, it follows that angles generally will not be projected true size. For example, in the multiview drawing in Fig. 837 (a), none of the three 60° angles will be 60° in the isometric

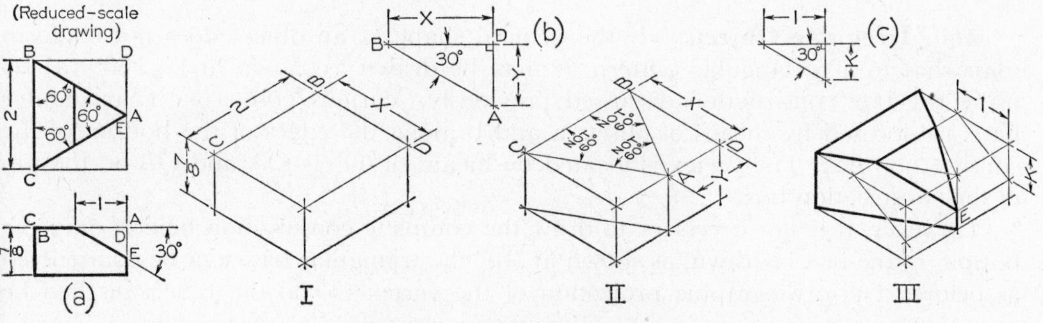

Fig. 837 Angles in Isometric.

drawing. To realize this fact, measure each angle in the isometric of Step II with your protractor and note the number of degrees compared to the true 60°. No two angles are the same; two are smaller and one larger than 60°.

As shown in Step I, the enclosing box can be drawn from the given dimensions, except for dimension X which is not given. To find dimension X, draw triangle BDA from the top view full size, as shown at (b). Transfer dimension X to the isometric in Step I, to complete the enclosing box.

In order to locate point A in Step II, dimension Y must be used, but this is not given in the top view at (a). Dimension Y is found by the same construction at (b) and then transferred to the isometric, as shown. The completed isometric is shown at III where point E is located by using dimension K, as shown.

Thus, in order to set off angles in isometric, the regular protractor cannot be used.* *Angular measurements must be converted to linear measurements along isometric lines.*

In Fig. 838 (a) are two views of an object to be drawn in isometric. Point A can easily be located in the isometric, Step I, by measuring $\frac{7}{8}''$ down from point O.

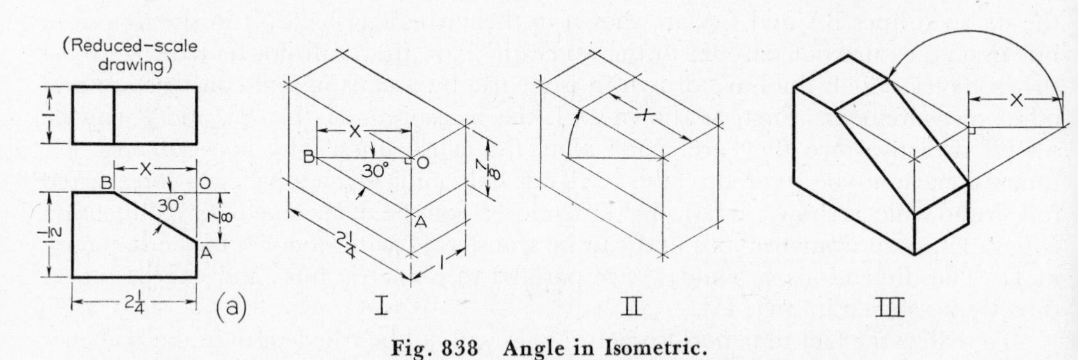

Fig. 838 Angle in Isometric.

However, in the given drawing at (a) the location of point B depends upon the 30° angle, and to locate B in the isometric the linear dimension X must be known. This distance can be found graphically by drawing the right triangle BOA attached to the isometric, as shown. The distance X is then transferred to the isometric with the compass or dividers, as shown at II. Actually, the triangle could be attached in several different positions. One of these is shown in Step III.

When angles are given in degrees, it is necessary to convert the angular measurements into linear measurements. This is best done by drawing a right triangle separately, as in Fig. 837 (b), or attached to the isometric, as in Fig. 838.

460. Irregular Objects. If the general shape of an object does not conform somewhat to a rectangular pattern, it may be drawn as shown in Fig. 839 (a) by using the box construction discussed previously. Various points of the triangular base are located by means of offsets a and b along the edges of the bottom of the construction box. The vertex is located by means of offsets OA and OB on the top of the construction box.

However, it is not necessary to draw the complete construction box. If only the bottom of the box is drawn, as shown at (b), the triangular base can be constructed as before. The orthographic projection of the vertex O′ on the base can then be

*Special protractors for setting off angles on isometric surfaces are available from dealers.

located by offsets O′A and O′B, as shown, and from this point the vertical center line O′O can be erected, using measurement C.

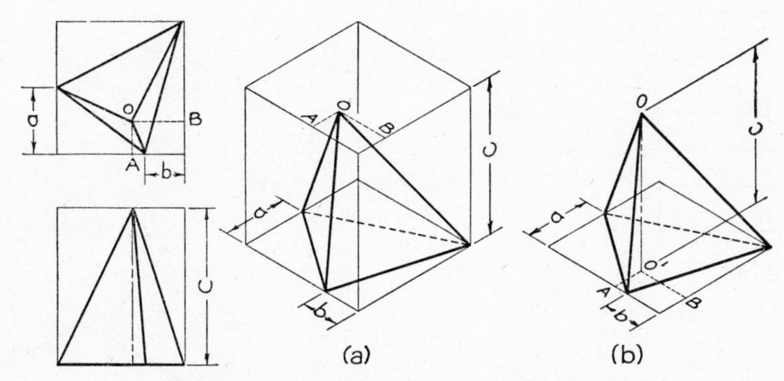

Fig. 839 Irregular Object in Isometric.

Sometimes an irregular object may be drawn by means of a series of sections, as illustrated in Fig. 840. The edge views of a series of imaginary cutting planes are shown in the top and front views of the multiview drawing at (a). At (b) the various

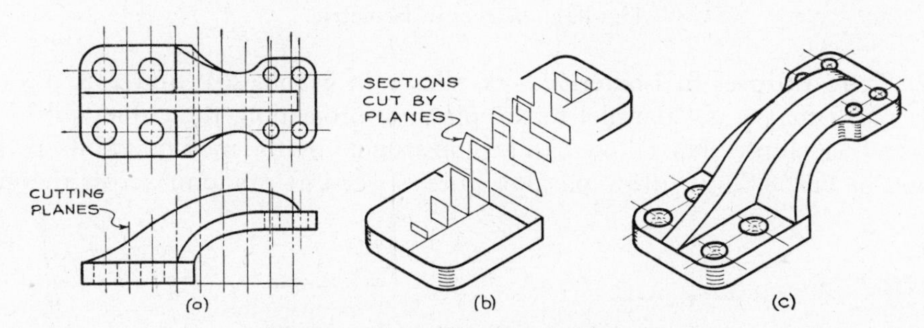

Fig. 840 Use of Sections in Isometric.

sections are constructed in isometric, and at (c) the object is completed by drawing lines through the corners of the sections. In the isometric at (b), all height dimensions are taken from the front view at (a), and all depth dimensions from the top view.

461. Curves in Isometric. Curves may be drawn in isometric by means of a series of offset measurements similar to those discussed in §454. In Fig. 841 any desired number of points, such as A, B, C, etc., are selected at random along the curve in the given top view at (a). Enough points should be chosen to fix accurately the path of the curve; the more points used, the greater the accuracy. Offset grid lines are then drawn from each point parallel to the isometric axes.

As shown at I, offset measurements a and b are laid off in the isometric to locate point A on the curve. Points B, C, and D are located in a similar manner, as shown at II. A light freehand curve is sketched smoothly through the points as shown at III. Points A′, B′, C′, and D′ are located directly under points A, B, C, and D, as shown at IV, by drawing vertical lines downward, making all equal to dimension C, the height of the block. A light freehand curve is then drawn through the points.

The final curve is heavied in with the aid of the irregular curve, §66, and all straight lines are darkened to complete the isometric at V.

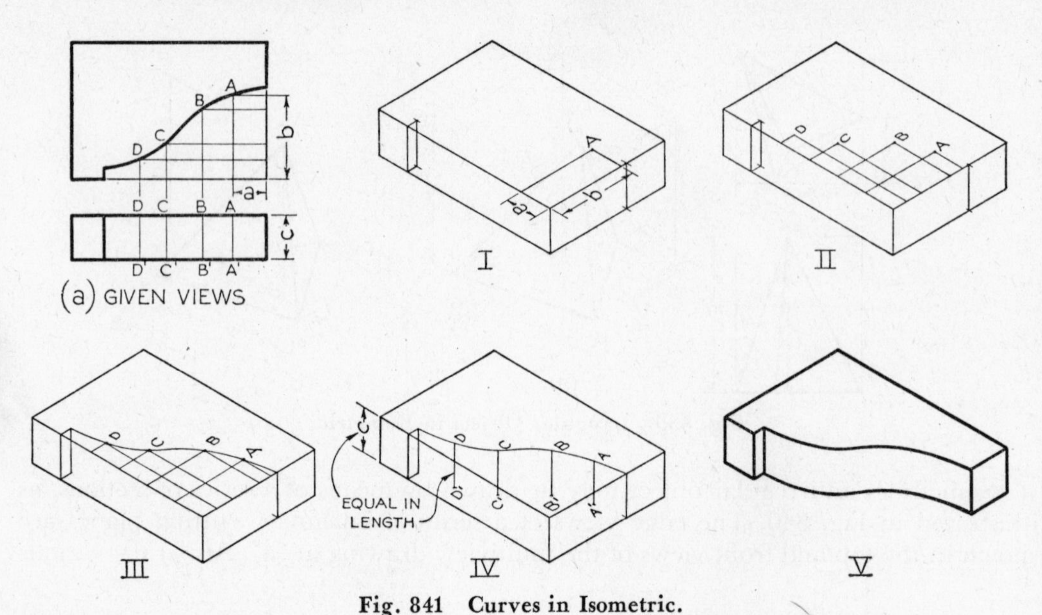

Fig. 841 Curves in Isometric.

462. True Ellipses in Isometric. As shown in §§158, 201, and 235, if a circle lies in a plane that is not parallel to the plane of projection, the circle will be projected as a true ellipse. The ellipse can be constructed by the method of offsets, §461. As shown in Fig. 842 (a), draw parallel lines, spaced at random, across the circle;

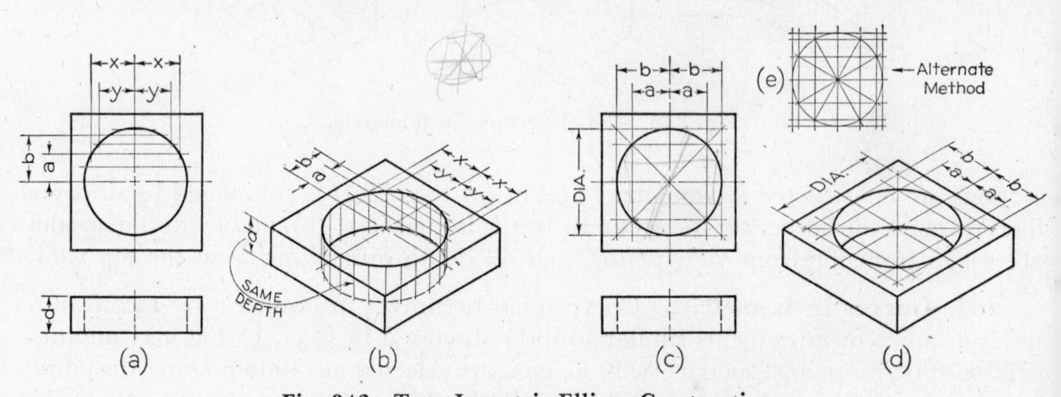

Fig. 842 True Isometric Ellipse Construction.

then transfer these lines to the isometric as shown at (b), with the aid of the dividers. To locate points in the lower ellipse, drop points of the upper ellipse down a distance equal to the height d of the block and draw the ellipse, part of which will be hidden, through these points. Draw the final ellipses with the aid of the irregular curve, §66.

A variation of the method of offsets, which provides eight points on the ellipse, is illustrated at (c) and (d). If more points are desired, parallel lines, as at (a), can be added. As shown at (c), draw a square around the given circle, and draw diagonals. Through the points of intersection of the diagonals and the circle, draw another square, as shown. Draw this construction in the isometric, as shown at (d), transferring distances a and b with the dividers.

A similar method that provides twelve points on the ellipse is shown at (e). The given circle is divided into twelve equal parts, using the 30° × 60° triangle, Fig. 34. Lines parallel to the sides of the square are drawn through these points. The entire construction is then drawn in isometric, and the ellipse is drawn through the points of intersection.

When the center lines shown in the top view at (a) are drawn in isometric, (b), they become the *conjugate diameters* of the ellipse. The ellipse can then be constructed on the conjugate diameters by the methods of Figs. 203 and 204 (b).

When the 45° diagonals at (c) are drawn in isometric at (d), they coincide with the major and minor axes of the ellipse, respectively. Note that the minor axis is equal in length to the sides of the inscribed square at (c). The ellipse can be constructed upon the major and minor axes by any of the methods in §§155-160.

Remember the rule: *The major axis of the ellipse is always at right angles to the center line of the cylinder, and the minor axis is at right angles to the major axis and coincides with the center line.*

Accurate ellipses may be drawn with the aid of ellipse guides, §§163 and 465, or with a special *ellipsograph*, or with the *Circular Drawing Machine*, Fig. 870.

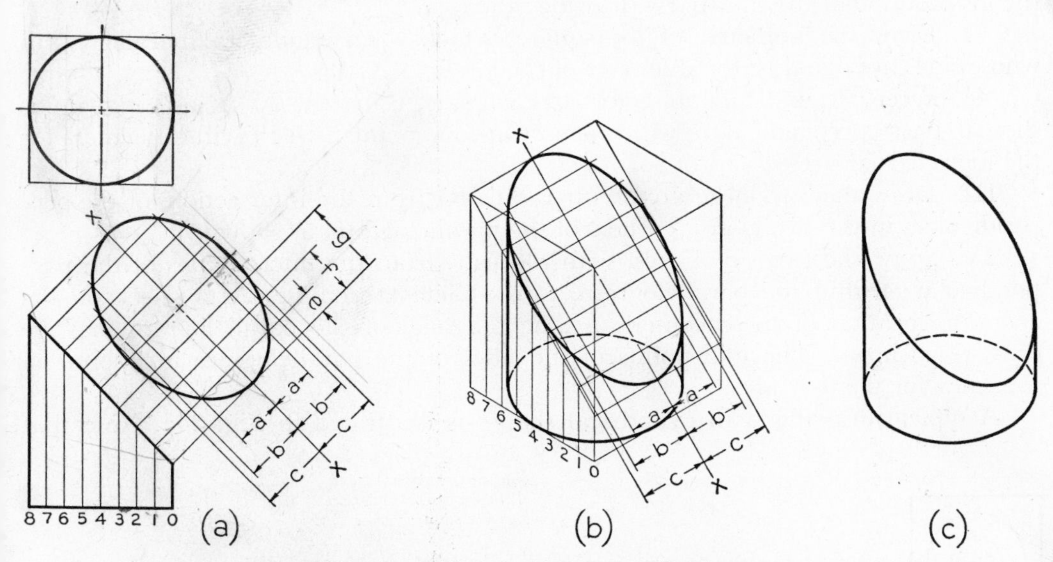

Fig. 843 Ellipse in Non-Isometric Plane.

If the curve lies in a non-isometric plane, all offset measurements cannot be applied directly. For example, in Fig. 843 (a) the elliptical face shown in the auxiliary view lies in a non-isometric plane. The cylinder is enclosed in a construction box, and the box is then drawn in isometric, as shown at (b). The base is drawn by the method of offsets, as shown in Fig. 842. The inclined ellipse is constructed by locating a number of points on the ellipse in the isometric, and drawing the final curve by means of the irregular curve, §66.

Measurements a, b, c, etc., are parallel to an isometric axis, and can be set off in the isometric at (b) on each side of the center line X-X, as shown. Measurements e, f, g, etc., are not parallel to any isometric axis, and cannot be set off directly in isometric. However, when these measurements are projected to the front view and down to the base, as shown at (a), they can then be set off along the lower edge of the construction box, as shown at (b). The completed isometric is shown at (c).

463. Approximate Four-Center Ellipse. An approximate ellipse is sufficiently accurate for nearly all isometric drawings. The method commonly used, called the *four-center ellipse*, is illustrated in Figs. 844-846. It can be used only for ellipses in isometric planes.

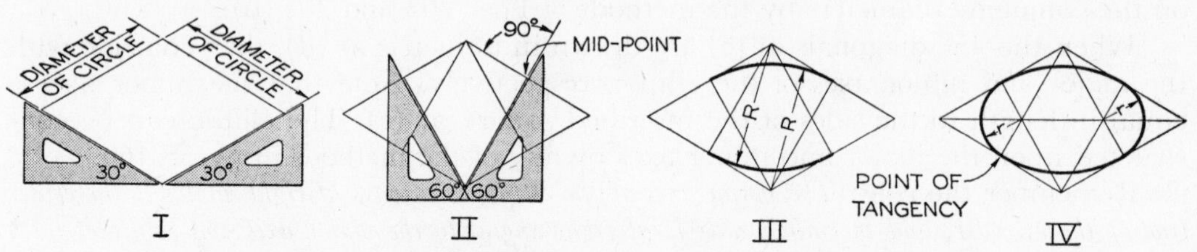

Fig. 844 Steps in Drawing Four-Center Ellipse.

To apply this method, Fig. 844, draw, or conceive to be drawn, a square around the given circle in the multiview drawing; then:

I. Draw the isometric of the square, which is an equilateral parallelogram whose sides are equal to the diameter of the circle.

II. Erect perpendicular bisectors to each side, using the 30° × 60° triangle as shown. These perpendiculars will intersect at four points, which will be centers for the four circular arcs.

III. Draw the two large arcs, with radius R, from the intersections of the perpendiculars in the two closest corners of the parallelogram, as shown.

IV. Draw the two small arcs, with radius r, from the intersections of the perpendiculars within the parallelogram, to complete the ellipse. As a check on the accurate location of these centers, a long diagonal of the parallelogram may be drawn, as shown. The mid-points of the sides of the parallelogram are points of tangency for the four arcs.

A typical drawing with cylindrical shapes is illustrated in Fig. 845. Note that

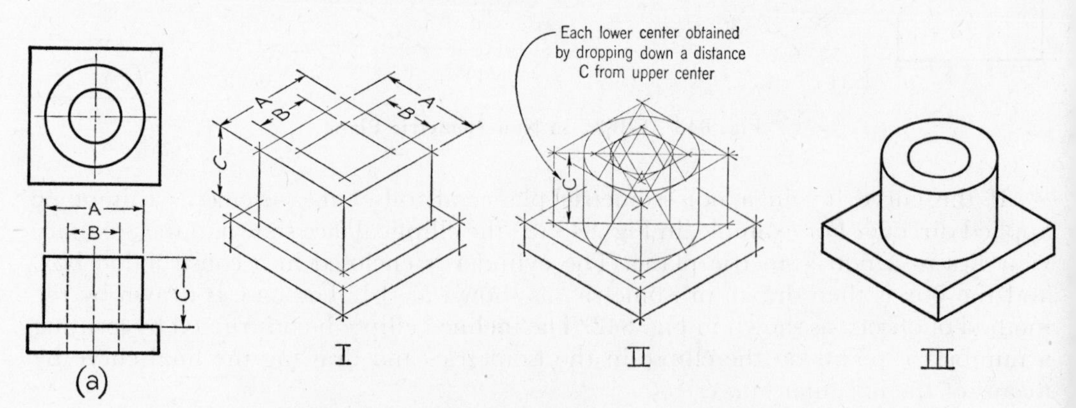

Fig. 845 Isometric Drawing of a Bearing.

the centers of the larger ellipse cannot be used for the smaller ellipse, though the ellipses represent concentric circles. Each ellipse has its own parallelogram and its own centers. Observe also that the centers of the lower ellipse are obtained by projecting the centers of the upper large ellipse down a distance equal to the height of the cylinder.

The construction of the four-center ellipse upon the three visible faces of a cube is shown in Fig. 846, a study of which shows that all diagonals are horizontal or 60° with horizontal; hence the entire construction is made with the T-square and 30° × 60° triangle.

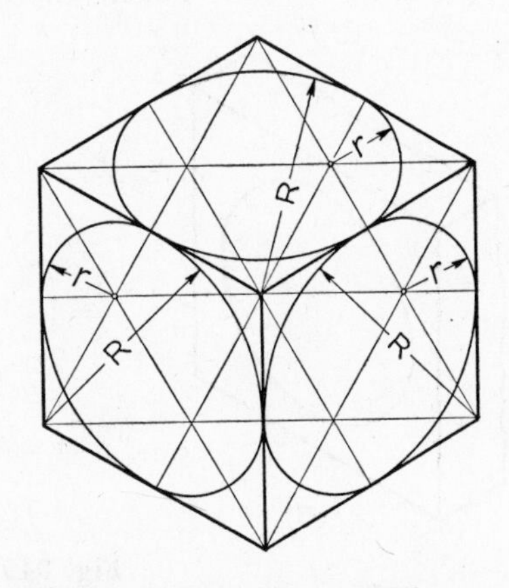

Fig. 846 Four-Center Ellipses.

Actually the four-center ellipse deviates considerably from the true ellipse. As shown in Fig. 847 (a), the four-center ellipse is somewhat shorter and "fatter" than the true ellipse. In constructions where tangencies or intersections with the four-center ellipse occur in the zones of error, the four-center ellipse is unsatisfactory, as shown at (b) and (c).

For a much closer approximation to the true ellipse, the Orth four-center ellipse, Fig. 848, which requires only one more step than the regular four-center ellipse, will be found sufficiently accurate for almost any problem.

When it is more convenient to start with the isometric center lines of a hole or

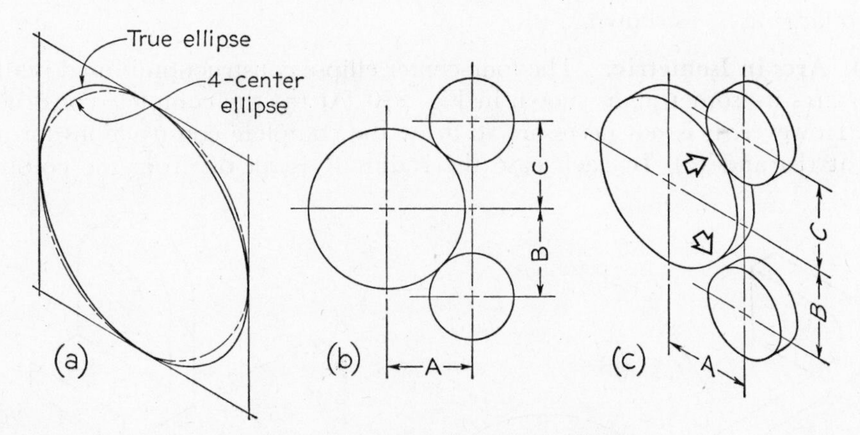

Fig. 847 Faults of Four-Center Ellipse.

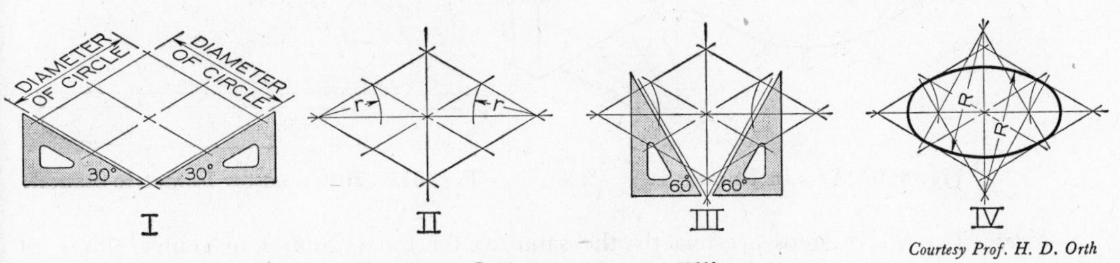

Courtesy Prof. H. D. Orth

Fig. 848 Orth Four-Center Ellipse.

cylinder in drawing the ellipse, rather than the enclosing parallelogram, the *alternate four-center ellipse* is recommended, Fig. 849. A completely constructed ellipse is shown at (a), and the steps followed are shown at the right in the figure.

I. Draw the isometric center lines. From the center, draw a construction circle equal to the actual diameter of the hole or cylinder. The circle will intersect the center lines at four points A, B, C, and D.

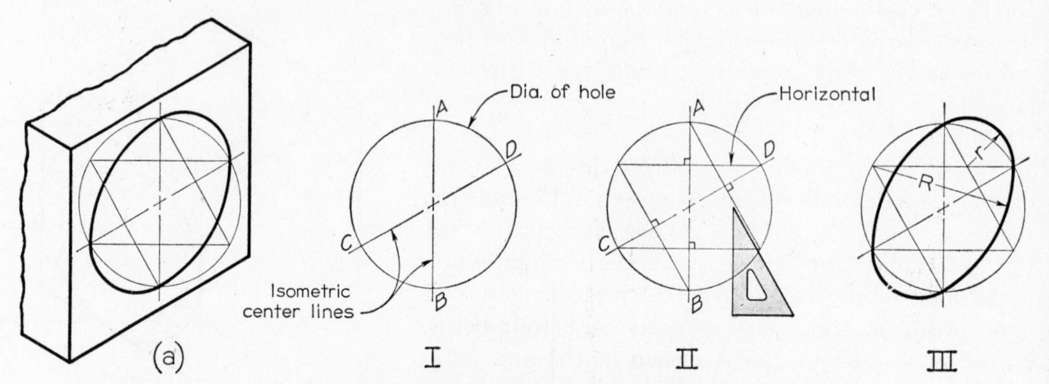

Fig. 849 Alternate Four-Center Ellipse.

II. From the two intersection points on one center line, erect perpendiculars to the other center line; then from the two intersection points on the other center line, erect perpendiculars to the first center line.

III. With the intersections of the perpendiculars as centers, draw two small arcs and two large arcs, as shown.

464. Arcs in Isometric. The four-center ellipse construction is used in drawing circular arcs in isometric, as shown in Fig. 850. At (a) the complete construction is shown. However, it is not necessary to draw the complete constructions for arcs, as shown at (b) and (c). In each case the radius R is set off from the construction

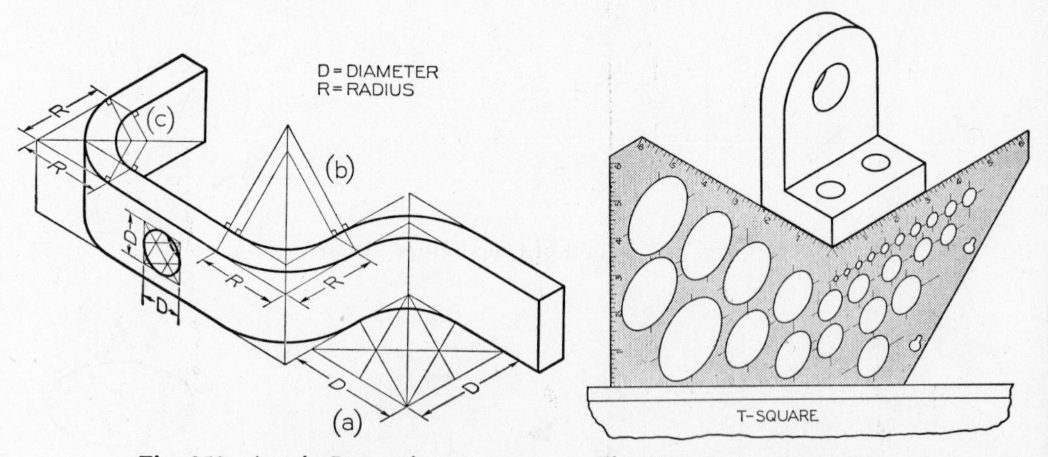

Fig. 850 Arcs in Isometric. **Fig. 851 Instrumaster Isometric Stencil.**

Note: The above steps are exactly the same as for the regular four-center ellipse of Fig. 844 except for the use of the isometric center lines instead of the enclosing parallelogram.

corner; then at each point, perpendiculars to the lines are erected, their intersection being the center of the arc. Note that the R distances are equal in both cases (b) and (c), but that the actual radii used are quite different.

If a truer elliptic arc is required, the Orth construction, Fig. 848, can be used. Or a true elliptic arc may be drawn by the method of offsets, §462, or with the aid of an ellipse guide, §465.

465. Ellipse Guides. One of the principal time-consuming elements in pictorial drawing is the construction of ellipses. A wide variety of ellipse guides, or templates, is available for ellipses of various sizes and proportions. See §163. They are not available in every possible size, of course, and it may be necessary to "use the fudge factor," such as leaning the pencil or pen when inscribing the ellipse, or shifting the template slightly for drawing each quadrant of the ellipse.

The *Instrumaster Isometric Stencil*, *Fig. 851, combines the angles, scales, and ellipses on the same instrument. The ellipses are provided with markings to coincide with the isometric center lines of the holes—a convenient feature in isometric drawing.

466. Intersections. To draw the elliptical intersection of a cylindrical hole in an oblique plane, Fig. 852, draw the ellipse in the isometric plane on top of the con-

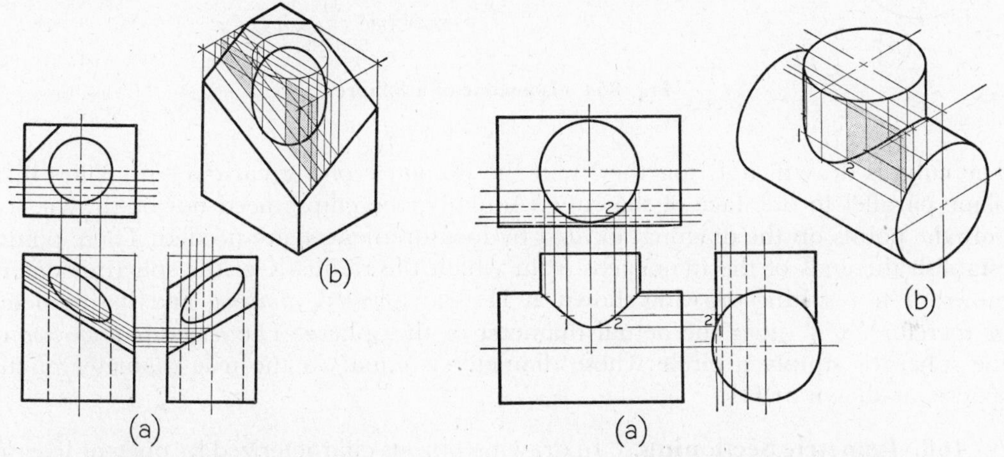

Fig. 852 Oblique Plane and Cylinder. **Fig. 853 Intersection of Cylinders.**

struction box, (b); then project points down to the oblique plane as shown. It will be seen that the construction for each point forms a trapezoid, which is produced by a slicing plane parallel to a lateral surface of the block.

To draw the curve of intersection between two cylinders, Fig. 853, pass a series of imaginary cutting planes through the cylinders parallel to their axes, as shown. Each plane will cut elements on both cylinders that intersect at points on the curve of intersection, as shown at (b). As many points should be plotted as necessary to assure a smooth curve. For most accurate work, the ends of the cylinders should be drawn by the Orth construction, or with ellipse guides, or by one of the true-ellipse constructions.

467. The Sphere in Isometric. The isometric drawing of any curved surface is evidently the envelope of all lines which can be drawn on that surface. For the

*John R. Cassell Co., Inc., New York, N. Y.

sphere, the great circles (circles cut by any plane through the center) may be selected as the lines on the surface. Since all great circles, except those which are perpendicular or parallel to the plane of projection, are shown as ellipses having equal major axes, it follows that their envelope is a circle whose diameter is the major axis of the ellipses.

In Fig. 854 (a) two views of a sphere enclosed in a construction cube are shown.

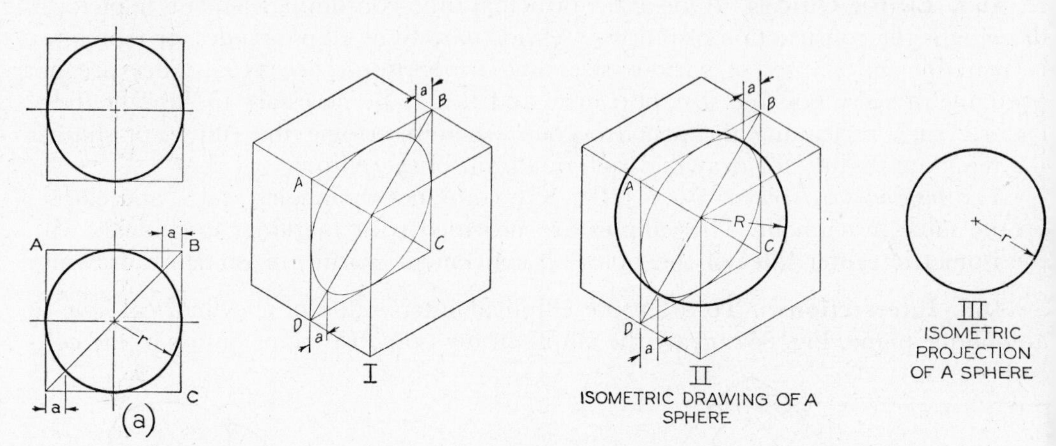

Fig. 854 Isometric of a Sphere.

The cube is drawn at I, together with the isometric of a great circle that lies in a plane parallel to one face of the cube. Actually, the ellipse need not be drawn, for only the points on the diagonal located by measurements *a* are needed. These points establish the ends of the major axis from which the radius R of the sphere is determined. The resulting drawing shown at II is an *isometric drawing*, and its diameter is, therefore, $\sqrt{\frac{3}{2}}$ times the actual diameter of the sphere. The *isometric projection* of the sphere is simply a circle whose diameter is equal to the true diameter of the sphere, as shown at III.

468. Isometric Sectioning. In drawing objects characterized by open or irregular interior shapes, isometric sectioning is as appropriate as in multiview drawing. An *isometric full section* is shown in Fig. 855. In such cases it is usually best to draw the cut surface first, and then to draw the portion of the object that lies behind the

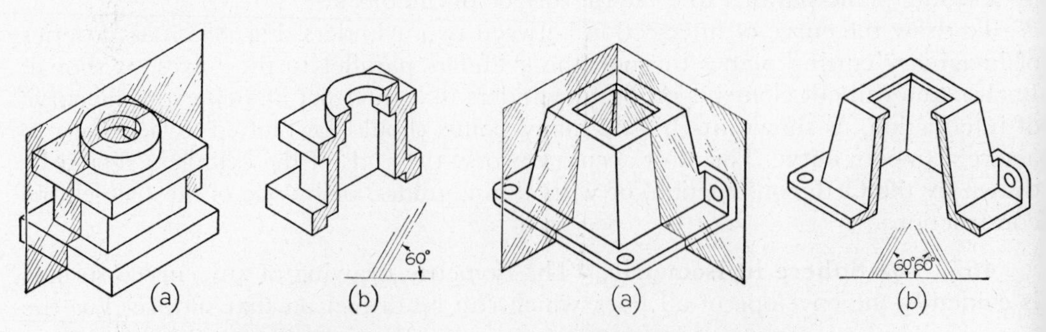

Fig. 855 Isometric Full Section. **Fig. 856 Isometric Half Section.**

cutting plane. Other examples of isometric full sections are shown in Figs. 373, 410, and 411.

An *isometric half section* is shown in Fig. 856. The simplest procedure in this case is to make an isometric drawing of the entire object and then the cut surfaces. Since only a quarter of the object is removed in a half section, the resulting pictorial drawing is more useful than full sections in describing both exterior and interior shapes together. Other typical isometric half sections are shown in Figs. 412, 440, 441, etc.

Isometric broken-out sections are also sometimes used. Examples are shown in Figs. 380, 450, and 737.

Section lining in isometric drawing is similar to that in multiview drawing. Section lining at an angle of 60° with horizontal, Figs. 855 and 856, is recommended, but the direction should be changed if at this angle the lines would be parallel to a prominent visible line bounding the cut surface, or to other adjacent lines of the drawing.

469. Isometric Dimensioning. Isometric dimensions are similar to ordinary dimensions, but are expressed in pictorial form. The method of drawing numerals and arrowheads in isometric is shown in Fig. 857 (a). For the $2\frac{1}{2}''$ dimension, the

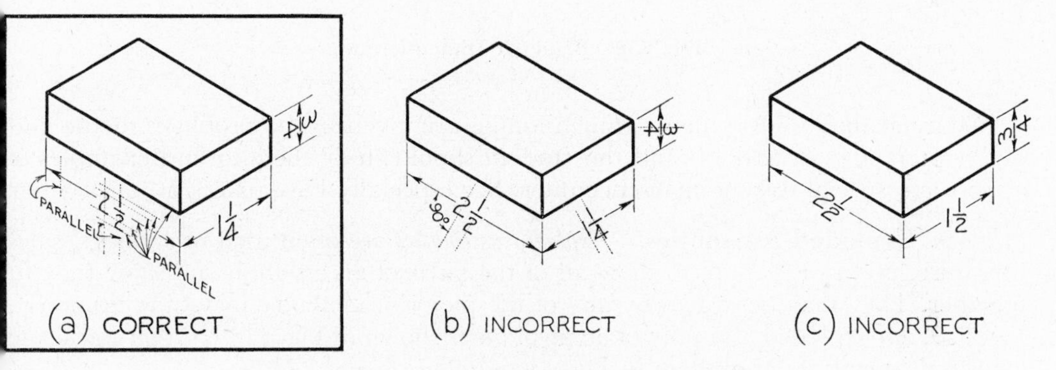

Fig. 857 Numerals and Arrowheads in Isometric.

two extension lines, the dimension line, and the numerals are all drawn in the isometric plane of one face of the object. The "horizontal" guide lines for the numerals are drawn parallel to the dimension line, and the vertical guide lines are drawn parallel to the extension lines. The rear ends of the arrowheads should line up parallel to the extension lines. Note that the lettering is *vertical lettering drawn in isometric*. Inclined lettering should not be used in pictorial dimensioning.

As shown at (b), the guide lines used to keep the letters vertical should not be perpendicular to the dimension lines. The example at (c) is incorrect because the $2\frac{1}{2}''$ and $1\frac{1}{2}''$ dimensions are not lettered in the plane of the corresponding dimension lines and extension lines, and the $\frac{3}{4}''$ dimension is awkward to read because of its reversed position.

Correct and incorrect practices in isometric dimensioning are illustrated in Fig. 858. At (b) the $3\frac{1}{8}''$ dimension runs to a wrong extension line at the right, and consequently the dimension does not lie in an isometric plane. Near the left side, a number of lines cross one another unnecessarily, and terminate on the wrong lines. The upper

½″ drill hole is located from the edge of the cylinder when it should be dimensioned from its center line. Study these two drawings carefully to discover additional mistakes at (b).

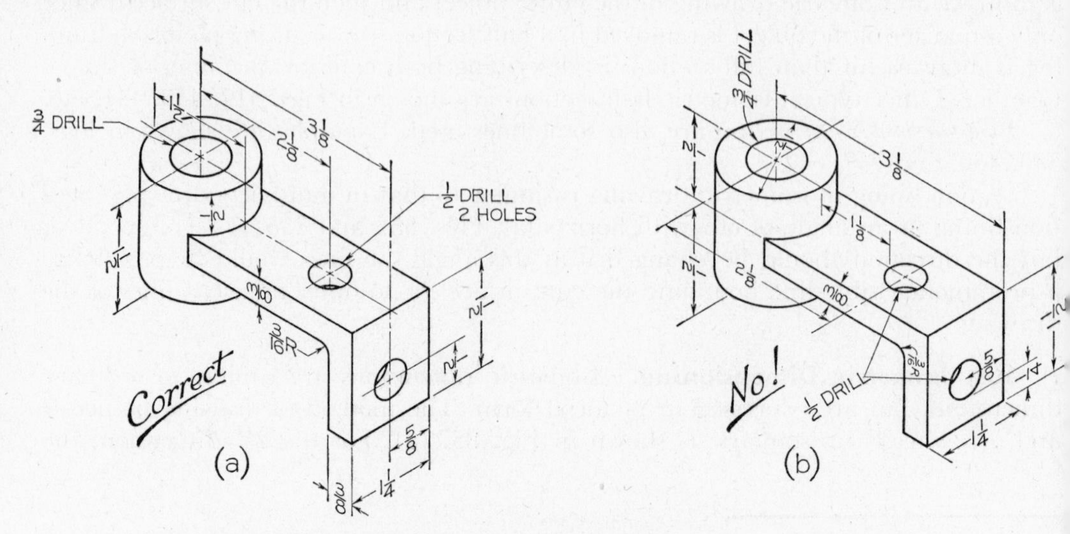

Fig. 858 Isometric Dimensioning.

Many examples of isometric dimensioning are given in the problems at the end of Chapters 6, 7, 8, and 14, and the student should study these to find examples of almost any special case he may encounter. See especially Figs. 336-375.

470. Exploded Assemblies. *Exploded assemblies* are often used in catalogs, sales literature, and in the shop, to show all of the parts of an assembly and how they fit together. They may be drawn by any of the pictorial methods, including isometric, Fig. 859. An exploded assembly of an airplane is shown in Fig. 1149. Other isometric exploded assemblies are shown in Figs. 766-769, 771, and 773.

471. Piping Diagrams. Isometric and oblique drawings are well adapted for representation of piping layouts, as illustrated in Figs. 1116 and 1121, as well as for all other structural work to be represented pictorially.

DIMETRIC PROJECTION

472. Method of Projection. A *dimetric projection* is an axonometric projection of an object so placed that two of its axes make equal angles with the plane of projection, and the third axis makes either a smaller or a greater angle. Hence, the two axes making equal angles with the plane of projection are foreshortened equally, while the third axis is foreshortened in a different ratio.

Generally the object is so placed that one axis will be projected in a vertical position. However, if the relative positions of the axes have been determined, the projection may be drawn in any revolved position, as in isometric drawing. See §453.

The angles between the *projections of the axes* must not be confused with the angles the *axes themselves* make with the plane of projection.

473. Dimetric Drawing. The positions of the axes may be assumed such that any two angles between the axes are equal and over 90°, and the scales determined graphically, as shown in Fig. 860 (a), in which OP, OL, and OS are the projections of the axes or converging edges of a cube. In this case, angle POS = angle LOS.

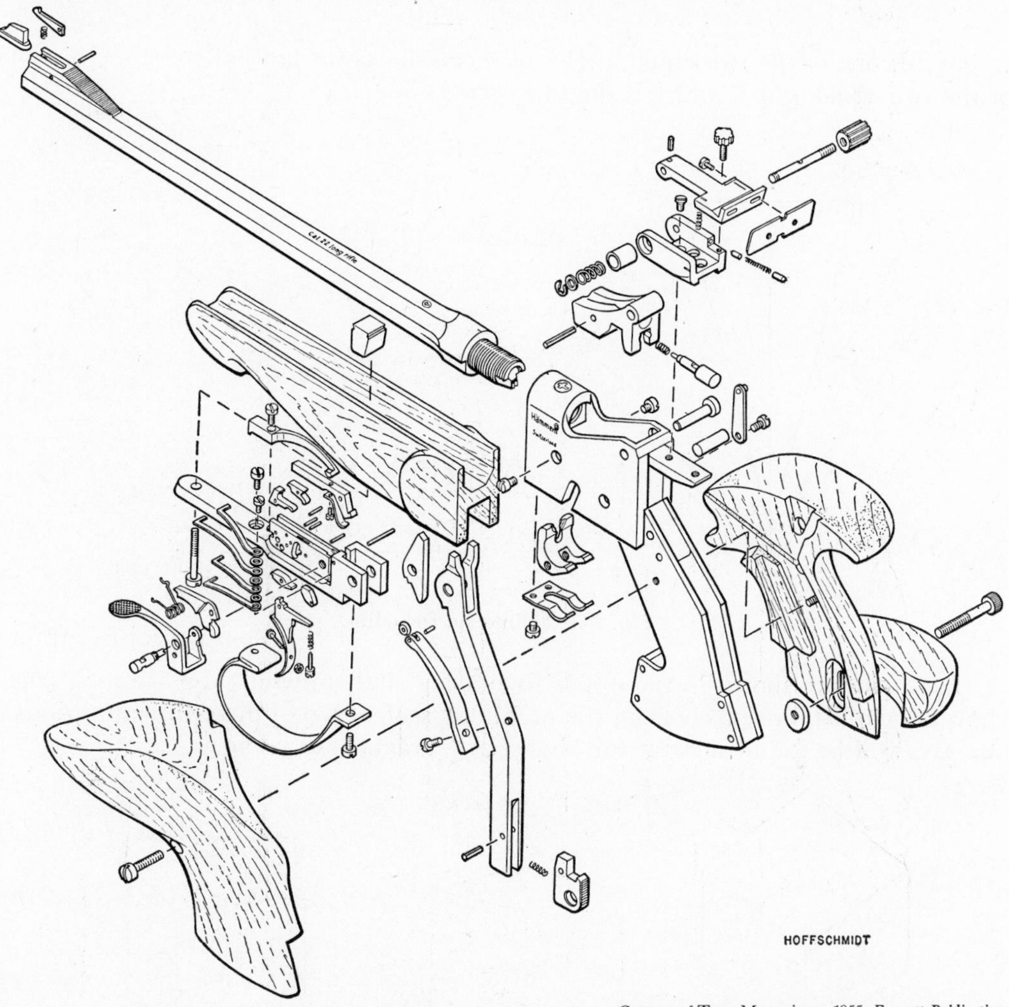

Courtesy of True Magazine *c 1955, Fawcett Publications.*

Fig. 859 Isometric Exploded Assembly of Hammerli Match Pistol.

Lines PL, LS, and SP are the lines of intersection of the plane of projection with the three visible faces of the cube. From descriptive geometry we know that since line LO is perpendicular to the plane POS, in space, its projection LO is perpendicular to PS, the intersection of the plane POS and the plane of projection. Similarly, OP is perpendicular to SL, and OS is perpendicular to PL.

If the triangle POS is revolved about the line PS as an axis into the plane of projection, it will be shown in its true size and shape as PO′S. If regular full-size scales are marked along the lines O′P and O′S, and the triangle is counterrevolved to its original position, the dimetric scales are found on the axes OP and OS, as shown.

In order to avoid the preparation of special scales, use can be made of available scales on the architects scale by assuming the scales and calculating the positions of the axes, as follows:

$$\cos a = -\frac{\sqrt{2h^2v^2 - v^4}}{2hv}$$

where a is one of the two equal angles between the projections of the axes, h is one of the two equal scales, and v is the third scale.

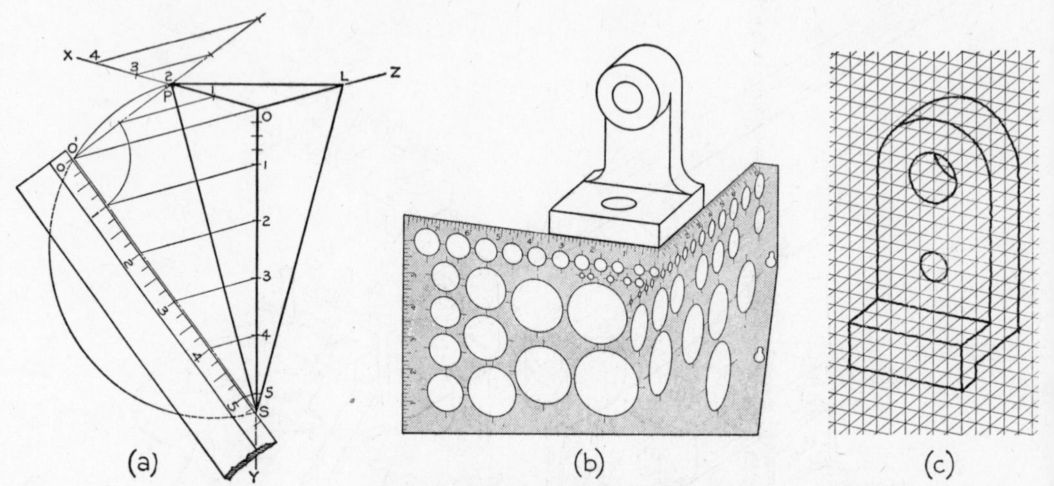

Fig. 860 Dimetric Drawing.

Examples are shown in the upper row of Fig. 861, in which the assumed scales, shown encircled, are taken from the architects scale. One of these three positions of the axes will be found suitable for almost any practical drawing.

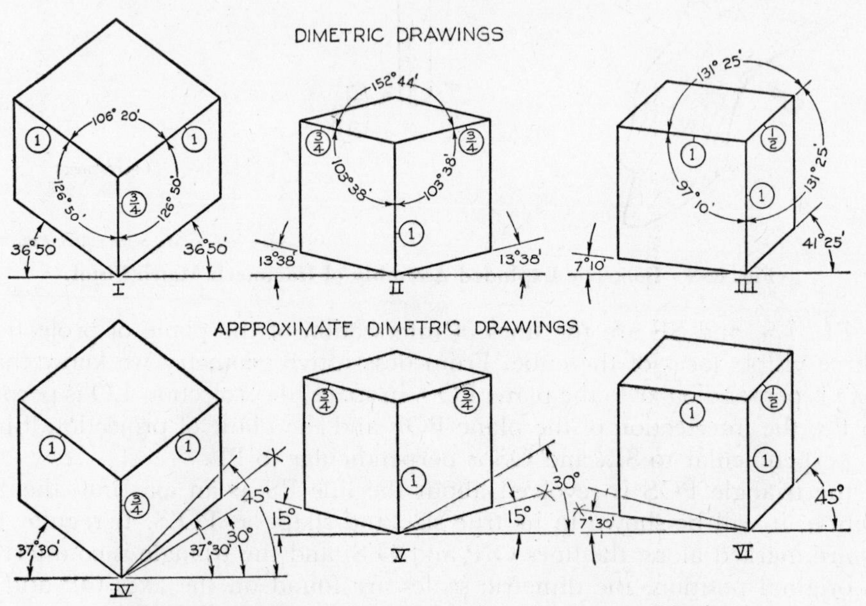

Fig. 861 Angles of Axes Determined by Assumed Scales.

The Instrumaster Dimetric Stencil,* Fig. 860 (b), has angles of approximately 11° and 39° with horizontal, which provides a picture similar to that in Fig. 861 III. In addition, the stencil has ellipses corresponding to the axes, and accurate scales along the edges.

The Instrumaster Dimetric Graph paper,* Fig. 860 (c), can be used to sketch in dimetric as easily as to sketch isometrics on isometric paper. The grid lines slope in conformity to the angles on the Dimetric Stencil at (b), and are printed on vellum. The grid lines do not reproduce on prints.

474. Approximate Dimetric Drawing. Approximate dimetric drawings, which closely resemble true dimetrics, can be constructed by substituting for the true angles shown in the upper half of Fig. 861, angles that can be obtained with the ordinary triangles and compass, as shown in the lower half of the figure. The resulting drawings will be sufficiently accurate for all practical purposes.

The procedure in preparing an approximate dimetric drawing, using the position of VI in Fig. 861, is shown in Fig. 862. The offset method of drawing a curve is shown in the figure. Other methods for drawing ellipses are the same as in trimetric drawing, §477.

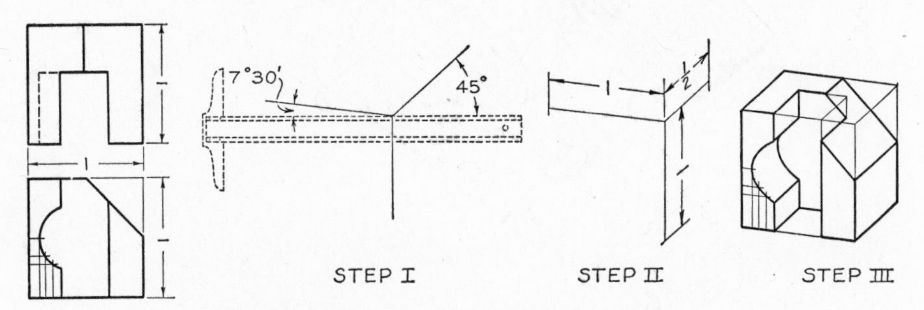

Fig. 862 Steps in Dimetric Drawing.

The steps in making a dimetric sketch, using a position similar to that in Fig. 861 V, are shown in Fig. 863. The two angles are equal and about 20° with horizontal for the most pleasing effect.

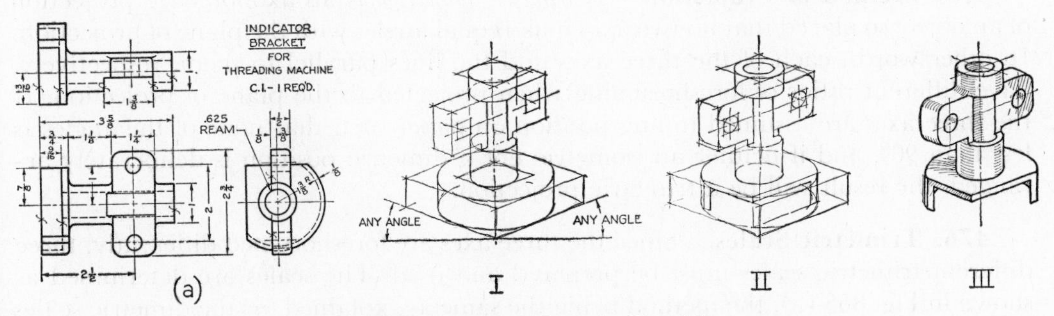

Fig. 863 Steps in Dimetric Sketching.

*John R. Cassell Co., Inc., New York, N.Y.

An exploded approximate dimetric drawing of an adding machine is shown in Fig. 864. The dimetric axes used are those in Fig. 861 IV.

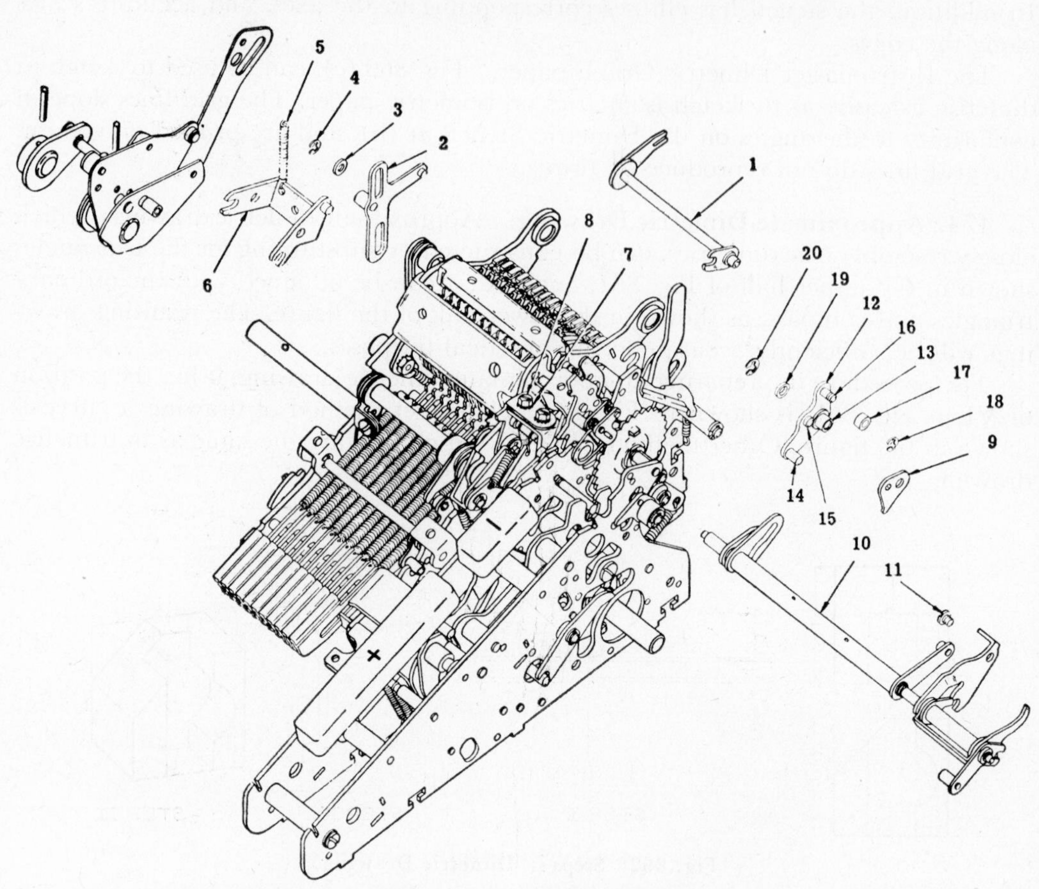

Courtesy Victor Adding Machine Co.

Fig. 864 Exploded Dimetric of an Adding Machine.

TRIMETRIC PROJECTION

475. Method of Projection. A *trimetric projection* is an axonometric projection of an object so placed that no two axes make equal angles with the plane of projection. In other words, each of the three axes and the lines parallel to them, respectively, have different ratios of foreshortening when projected to the plane of projection. If the three axes are assumed in any position on paper such that none of the angles is less than 90°, and if neither an isometric nor a dimetric position is deliberately arranged, the result will be a trimetric projection.

476. Trimetric Scales. Since the three axes are foreshortened differently, three different trimetric scales must be prepared and used. The scales are determined as shown in Fig. 865 (a), the method being the same as explained for the dimetric scales in §473. As shown at (a), any two of the three triangular faces can be revolved into the plane of projection to show the true lengths of the three axes. In the revolved position, the regular scale is used to set off inches or fractions thereof. When the axes have been counterrevolved to their original positions, the scales will be correctly foreshortened, as shown. These dimensions should be transferred to the edges of three thin cards and marked OX, OZ, and OY for easy reference.

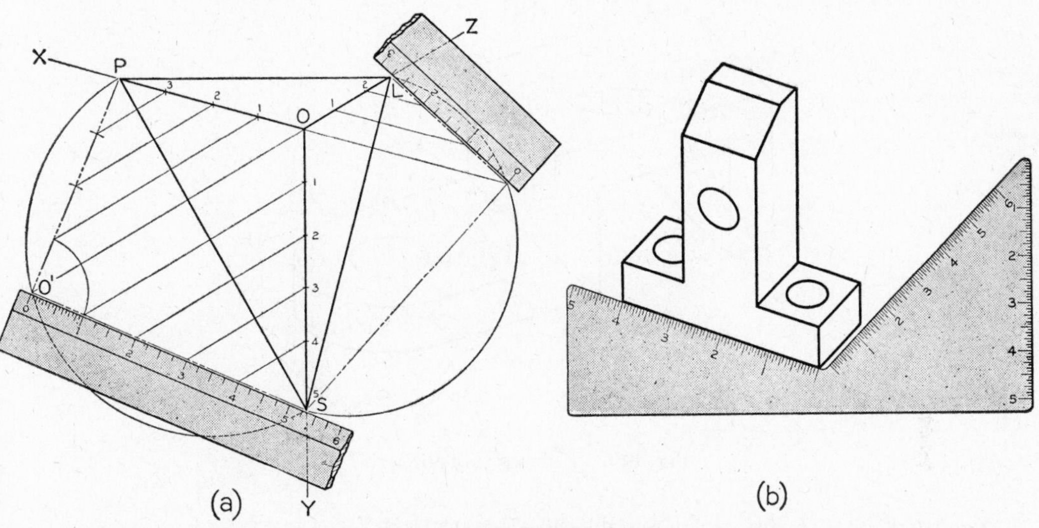

Fig. 865 Trimetric Scales.

A special trimetric angle may be prepared from Bristol Board or plastic, as shown at (b). Perhaps six or seven such guides, using angles for a variety of positions of the axes, would be sufficient for all practical requirements.*

477. Trimetric Ellipses. The trimetric center lines of a hole, or on the end of a cylinder, become the conjugate diameters of the ellipse when drawn in trimetric. The ellipse may be drawn upon the conjugate diameters by the methods of Fig. 203 or 204 (b). Or the major and minor axes may be determined from the conjugate diameters, Fig. 205 (b), and the ellipse constructed upon them by any of the methods of Figs. 199-202, and 204 (a), or with the aid of an ellipse guide, Fig. 207.

One of the advantages of trimetric is the infinite number of positions of the object available. The angles and scales can be handled without too much difficulty, as shown in §476. However, the infinite variety of ellipses has been a discouraging factor.

In drawing any axonometric ellipse, keep the following in mind:

1. On the drawing, the major axis is always perpendicular to the center line, or axis, of the cylinder.

2. The minor axis is always perpendicular to the major axis; that is, on the paper it coincides with the axis of the cylinder.

3. The length of the major axis is equal to the actual diameter of the cylinder.

Thus we know at once the directions of both the major and minor axes, and the length of the major axis. *We do not know the length of the minor axis.* If we can find it, we can easily construct the ellipse with the aid of an ellipse guide or any of a number of ellipse constructions mentioned above.

*Commercial templates of this type are available from Charles Bruning Co., Inc., Chicago, Ill.

In Fig. 866 (a), center O is located as desired, and horizontal and vertical construction lines that will contain the major and minor axes are drawn through O. Note that the major axis will be on the horizontal line perpendicular to the axis of the hole, and the minor axis will be perpendicular to it, or vertical.

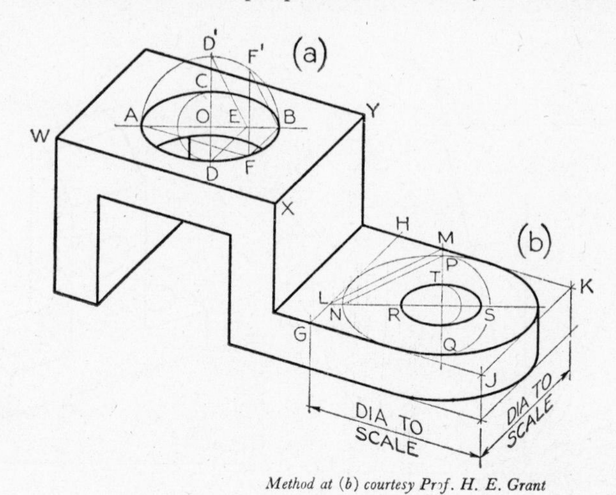

Method at (b) courtesy Prof. H. E. Grant

Fig. 866 Ellipses in Trimetric.

Set the compass for the actual radius of the hole and draw the semicircle, as shown, to establish the ends A and B of the major axis. Draw AF and BF parallel to the axonometric edges WX and YX, respectively, to locate F which lies on the ellipse. Through F draw a vertical line to intersect the semicircle at F′, and join F′ to B as shown. From D′ where the minor axis, extended, intersects the semicircle, draw D′E and ED parallel to F′B and BF, respectively. Point D is one end of the minor axis. From center O, strike arc DC to locate C, the other end of the minor axis. Upon these axes, a true ellipse can be constructed, or drawn with the aid of an ellipse guide. A simple method for finding the "angle" of ellipse guide to use is shown in Fig. 207 (c). If an ellipse guide is not available, an approximate four-center ellipse, Fig. 208, will be found satisfactory in most cases.

In constructions where the enclosing parallelogram for an ellipse is available or easily constructed, the major and minor axes can be readily determined as shown at (b). The directions of both axes, and the length of the major axis, are known. Extend the axes to intersect the sides of the parallelogram at L and M, and join the points with a straight line. From one end N of the major axis, draw a line NP parallel to LM. The point P is one end of the minor axis. To find one end T of the minor axis of the smaller ellipse, it is only necessary to draw RT parallel to LM or NP.

The method of constructing an ellipse on an oblique plane in trimetric is similar to that shown for isometric in Fig. 852.

478. Axonometric Projection by the Method of Intersections. Instead of constructing axonometric projections with the aid of specially prepared scales, as explained in the preceding paragraphs, an axonometric projection can be obtained directly by projection from two orthographic views of the object. This method is called the *method of intersections*; it was developed by Prof. L. Eckhart of the Vienna College of Engineering, and published by him in 1937.

To understand this method, let us assume, Fig. 867, that the axonometric projection of a rectangular object is given, and it is required to find its three orthographic projections: the top view, front view, and side view.

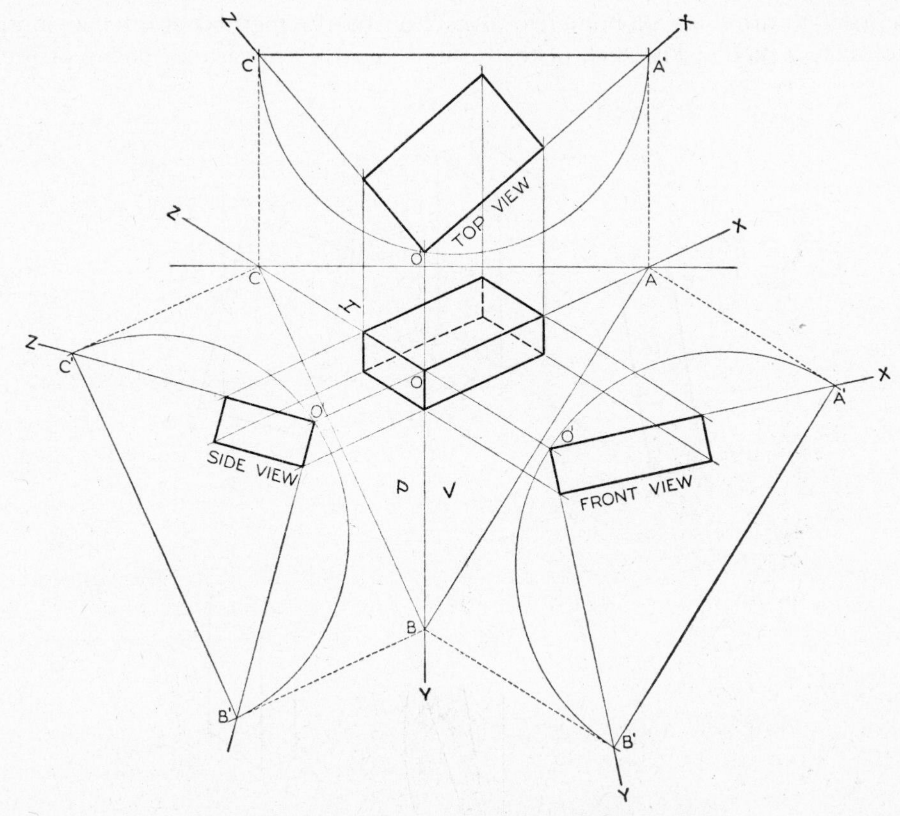

Fig. 867 Views from an Axonometric Projection.

Assume that the object is placed so that its principal edges coincide with the coordinate axes, and assume that the plane of projection (the plane upon which the axonometric projection is drawn) intersects the three coordinate planes in the triangle ABC. From descriptive geometry we know that lines BC, CA, and AB will be perpendicular, respectively, to axes OX, OY, and OZ. Any one of the three points A, B, or C may be assumed at pleasure on one of the axes, and the triangle ABC drawn.

To find the true size and shape of the top view, revolve the triangular portion of the horizontal plane AOC, which is in front of the plane of projection, about its base CA, into the plane of projection. In this case, the triangle is revolved *inward* to the plane of projection through the smallest angle made with it. The triangle will then be shown in its true size and shape, and the top view of the object can be drawn in the triangle by projection from the axonometric projection, as shown, since all width dimensions remain the same. In the figure, the base CA of the triangle has been moved upward to C′A′ so that the revolved position of the triangle will not overlap its projection.

In the same manner, the true sizes and shapes of the front view and side view can be found, as shown.

It is evident that if the three orthographic projections, or in most cases any two of them, are given in their relative positions, as shown in Fig. 867, the directions of the projections could be reversed so that the intersections of the projecting lines would determine the required axonometric projection.

In order to draw an axonometric projection by the method of intersections, it is well to make a sketch, Fig. 868, of the desired general appearance of the projection.

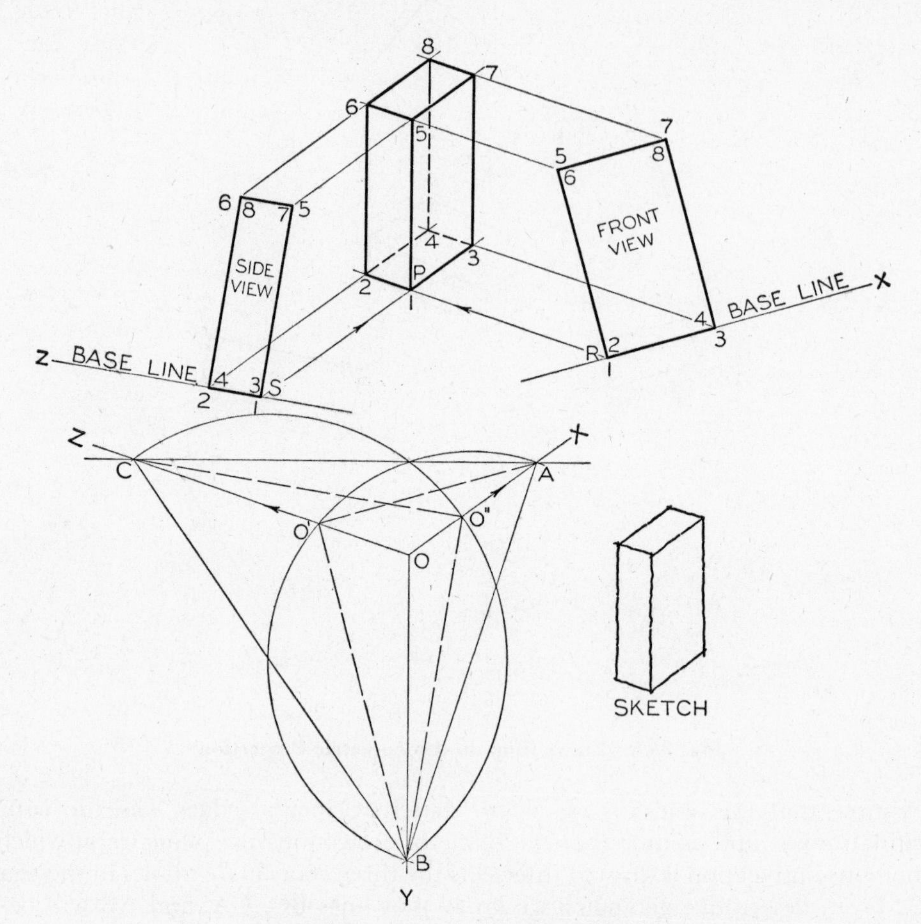

Fig. 868 Axonometric Projection.

Even if the object is a complicated one, this sketch need not be complete, but may be only a sketch of an enclosing box. Draw the projections of the coordinate axes OX, OY, and OZ, parallel to the principal edges of the object as shown in the sketch, and the triangle ABC to represent the intersection of the three coordinate planes with the plane of projection.

Revolve the triangle ABO about its base AB as the axis into the plane of projection. Line OA will revolve to O′A, and this line, or one parallel to it, must be used as the base line of the front view of the object. The projecting lines from the front view to the axonometric must be drawn parallel to the projection of the unrevolved Z-axis, as indicated in the figure.

Similarly, revolve the triangle COB about its base CB as the axis into the plane of projection. Line CO will revolve to CO″, and this line, or one parallel to it, must be used as the base line of the side view. The direction of the projecting lines must be parallel to the projection of the unrevolved X-axis, as shown.

Draw the front-view base line at pleasure, but parallel to O′X, and with it as the base, draw the front view of the object. Draw the side-view base line at pleasure, but parallel to O″C, and with it as the base, draw the side view of the object, as shown. From the corners of the front view, draw projecting lines parallel to OZ, and from the corners of the side view, draw projecting lines parallel to OX. The intersections of these two sets of projecting lines determine the desired axonometric projection. It will be an isometric, a dimetric, or a trimetric projection, depending upon the form of the sketch used as the basis for the projections, §448. If the sketch is drawn so that the three angles formed by the three coordinate axes are equal, the resulting projection will be an isometric projection; if two of the three angles are equal, the resulting projection will be a dimetric projection; and if no two of the three angles are equal, the resulting projection will be a trimetric projection.

In order to place the desired projection in a specific location on the drawing,

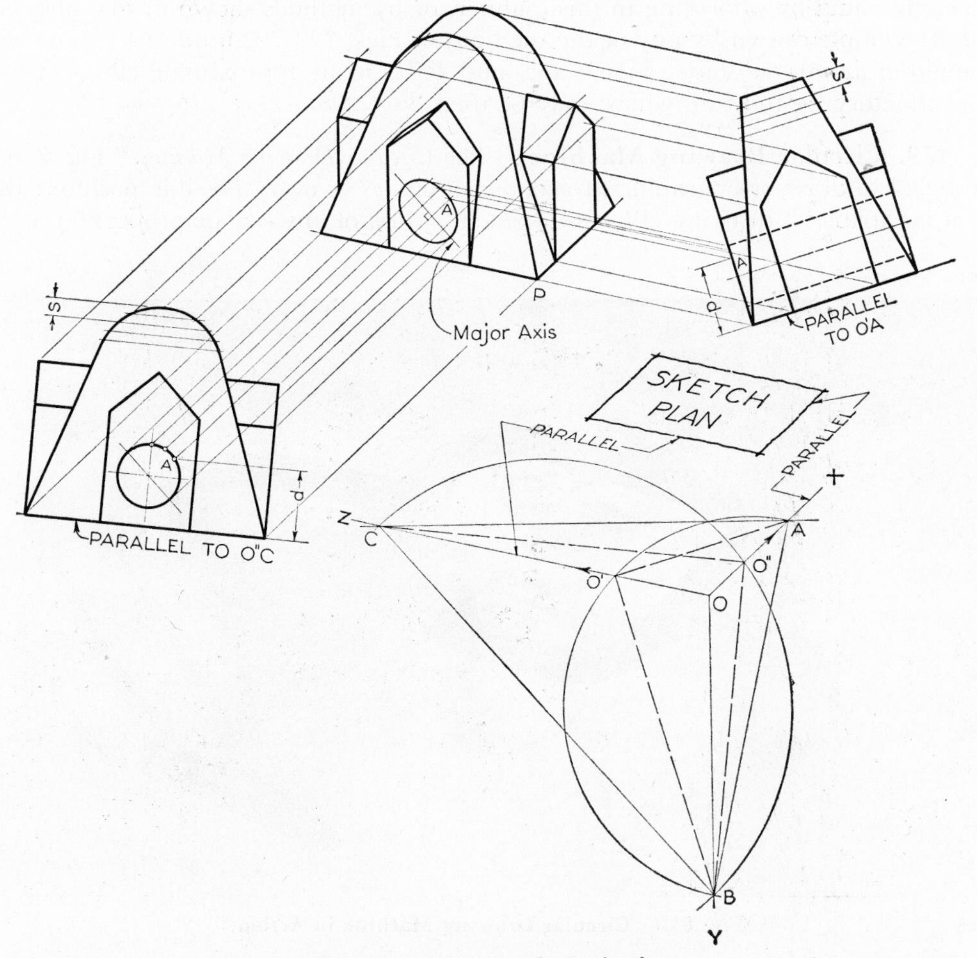

Fig. 869 Axonometric Projection.

Fig. 868, select the desired projection P of the point 1, for example, and draw two projecting lines PR and PS to intersect the two base lines and thereby to determine the locations of the two views on their base lines.

Another example of this method of axonometric projection is shown in Fig. 869. In this case, it was deemed necessary only to draw a sketch of the plan or base of the object in the desired position, as shown. The axes are then drawn with OX and OZ parallel respectively to the sides of the sketch plan, and the remaining axis OY is assumed in a vertical position. The triangles COB and AOB are revolved, and the two base lines drawn parallel to O″C and O′A as shown. Point P, the lower front corner of the axonometric drawing, was then chosen at pleasure, and projecting lines drawn toward the base lines parallel to axes OX and OZ to locate the positions of the views on the base lines. The views are drawn upon the base lines, or cut apart from another drawing and fastened in place with drafting tape or thumbtacks.

To draw the elliptical projection of the circle, assume any points, such as A, on the circle in both front and side views. Note that point A is the same altitude d above the base line in both views. The axonometric projection of point A is found simply by drawing the projecting lines from the two views. The major and minor axes may be easily found by projecting in this manner, or by methods shown in Fig. 866, and the true ellipse drawn by any of the methods of Figs. 199-202, and 204 (a), or with the aid of an ellipse guide, §§163, 465, and 477. Or an approximate ellipse, which is satisfactory for most drawings, may be used, Fig. 208.

479. Circular Drawing Machine. The *Circular Drawing Machine*,* Fig. 870, is designed to draw orthographic projections of circles in every possible position; that is, it is capable of drawing ellipses in pencil or ink, of any size or proportion, up to

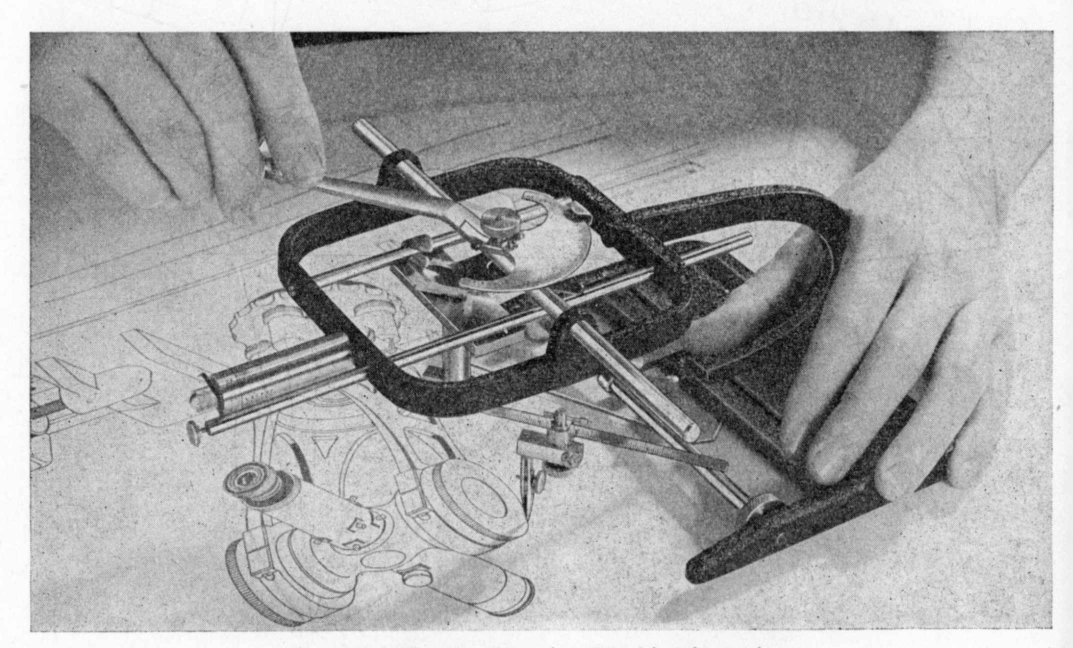

Fig. 870 Circular Drawing Machine in Action.

*Charles Bruning Co. Inc., Chicago, Ill.

a size having a 6″ major axis. Actually, the machine could more properly be called an axonometric machine. Some of the unusual features include a rapid means of locating the instrument on the axes over the center of the ellipse to be drawn. The base is made to roll on the paper so that repeated or related ellipses on the same axis can be drawn quickly. A built-in protractor provides a rapid means of setting off angles, and this feature, combined with the elliptical functions of the instrument, make it possible to perform geometrical constructions, such as bisecting lines, erecting perpendiculars, drawing tangent arcs, etc., *directly in trimetric.*

The machine is used in conjunction with a set of trimetric angles similar to the one shown in Fig. 865 (b). All straight trimetric lines are easily drawn to scale along the edges of the angles, and the machine is then mounted upon the axes of cylindrical shapes, and all necessary ellipses quickly drawn.

480. Axonometric Problems. A large number of problems to be drawn axonometrically are given in Figs. 871-874. The earlier isometric sketches may be drawn on isometric paper, §202; later sketches should be made on plain drawing paper. On drawings to be executed with instruments, show all construction lines required in the solutions.

For additional problems, see Figs. 333-335, 896-898, and 928.

Axonometric problems in convenient form for solution may be found in *Technical Drawing Problems*, by Giesecke, Mitchell, and Spencer, and in *Technical Drawing Problems, Series 2*, by Spencer and Grant, both designed to accompany this text, and published by The Macmillan Company.

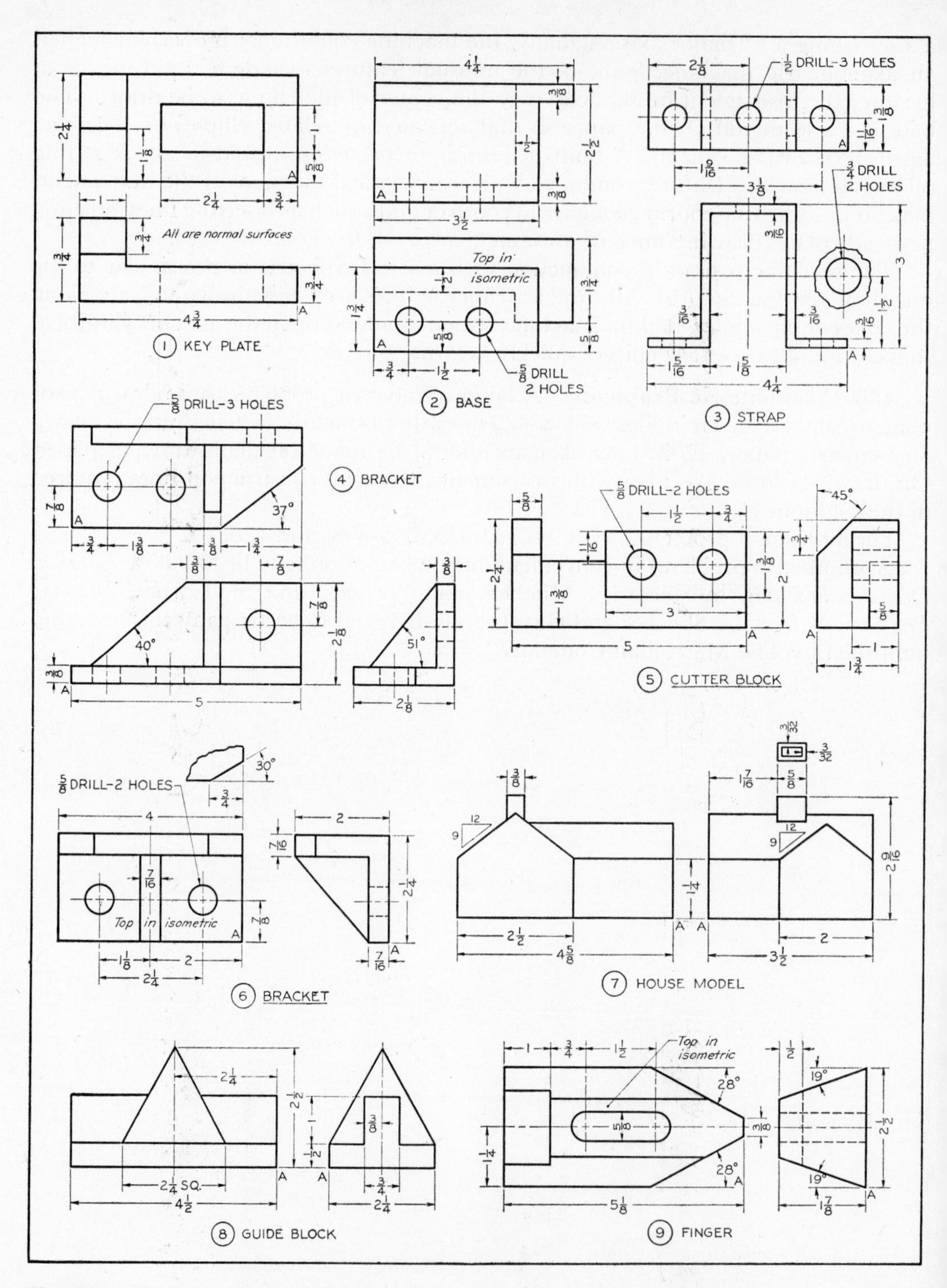

Fig. 871 (1) Make freehand isometric sketches. (2) Make isometric drawings with instruments on Layout A-2. (3) Make dimetric drawings with instruments, using Layout A-2 and position assigned from Fig. 861. (4) Make trimetric drawings, using instruments, with axes chosen to show the objects to best advantage. If dimensions are required, study §469.

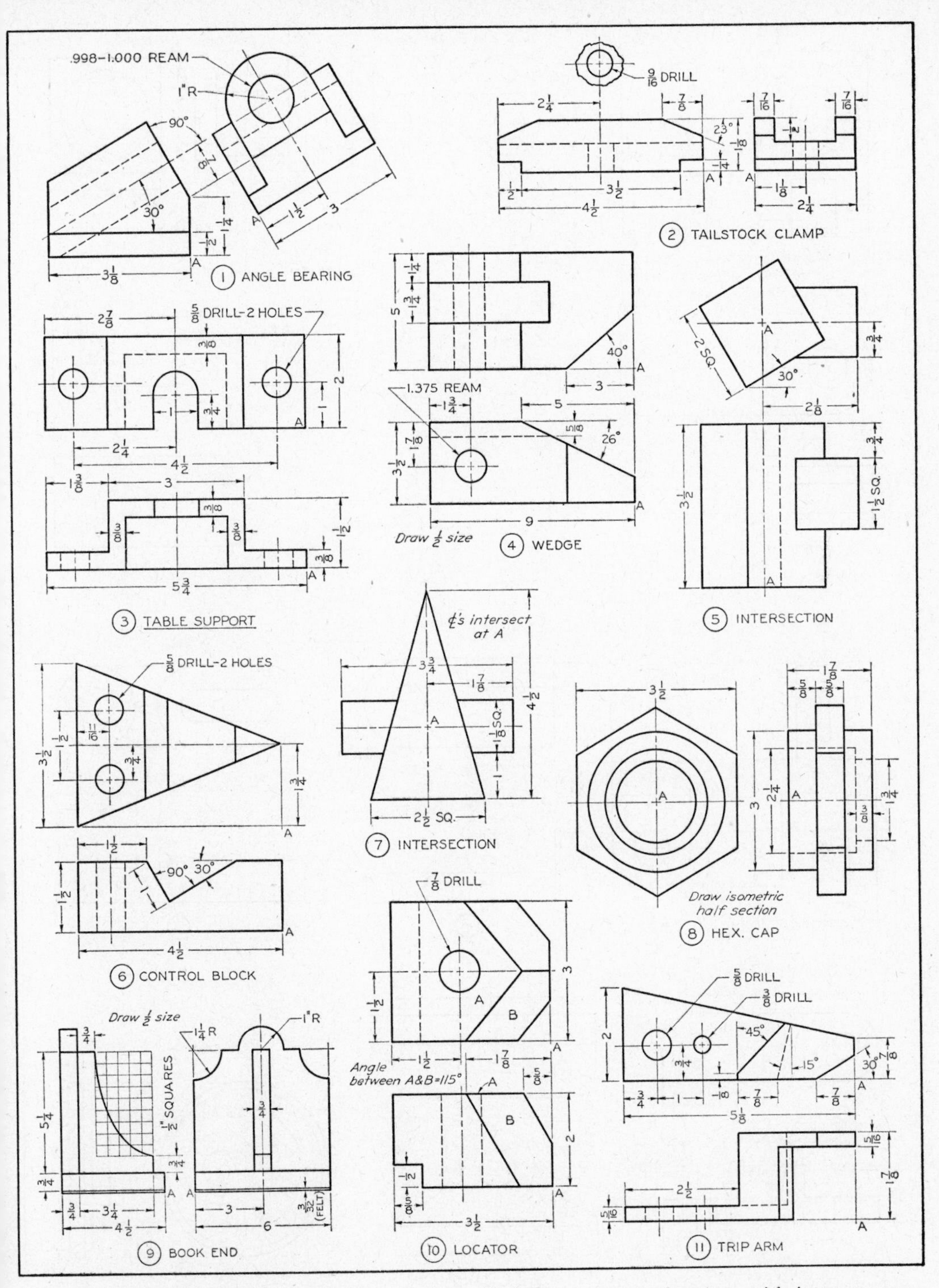

Fig. 872 (1) Make freehand isometric sketches. (2) Make isometric drawings with instruments on Layout A-2. (3) Make dimetric drawings with instruments, using Layout A-2 and position assigned from Fig. 861. (4) Make trimetric drawings, using instruments, with axes chosen to show the objects to best advantage. If dimensions are required, study §469.

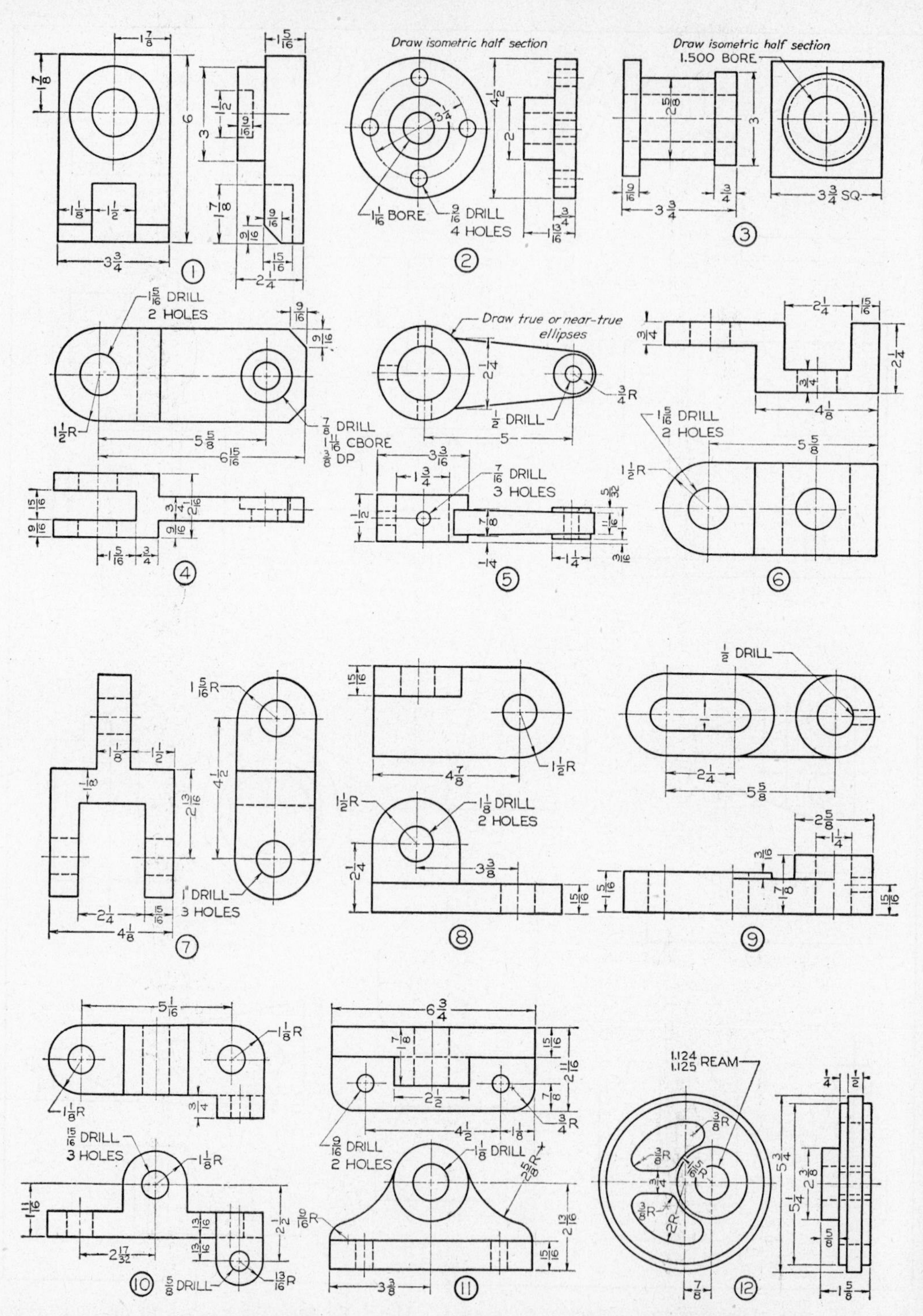

Fig. 873 (1) Make isometric freehand sketches. (2) Make isometric drawings with instruments, using Size A or Size B sheet, as assigned. (3) Make dimetric drawings with instruments, using Size A or Size B sheet, as assigned, and position assigned from Fig. 861. (4) Make trimetric drawings, using instruments, with axes chosen to show the objects to best advantage. If dimensions are required, study §469.

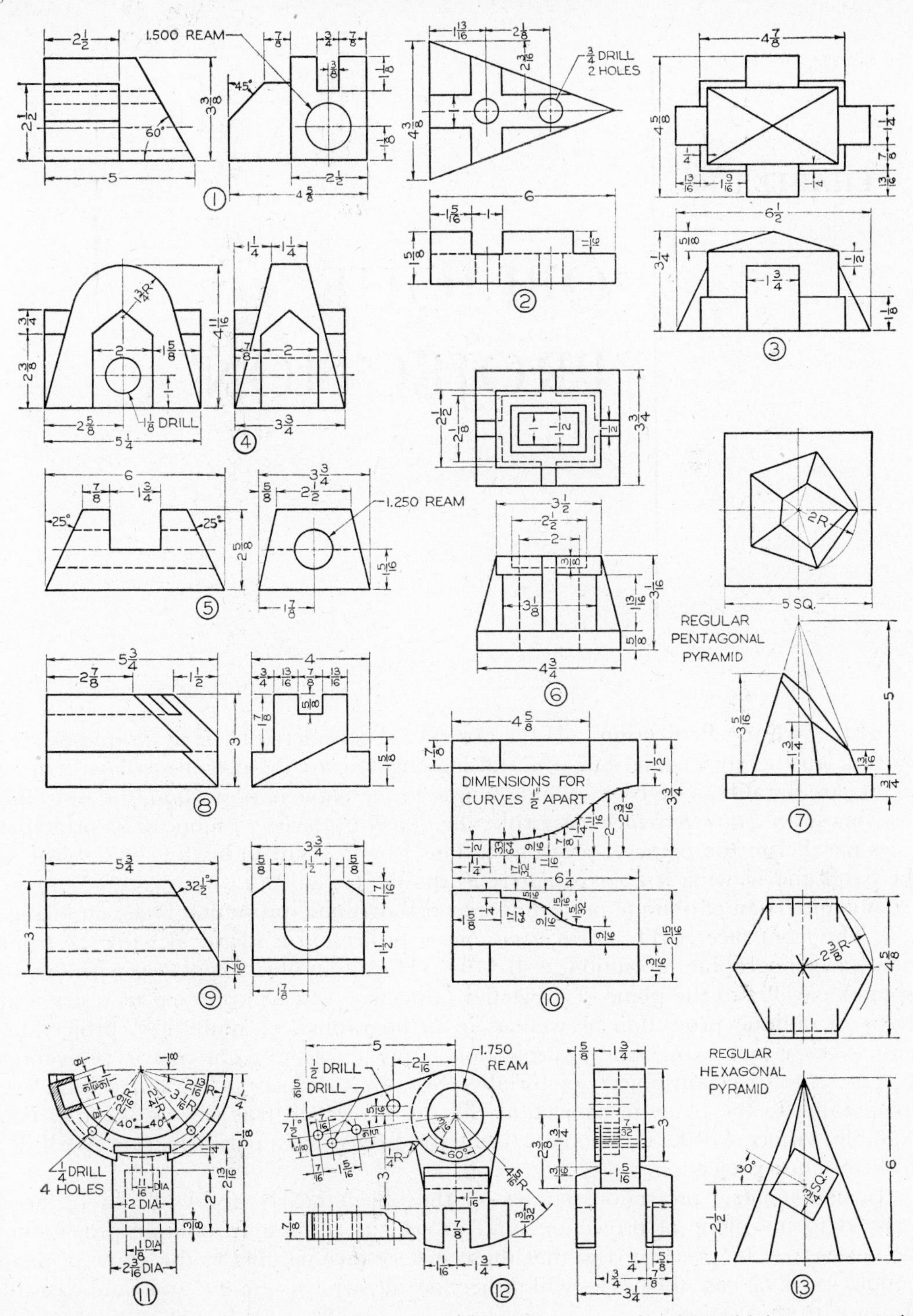

Fig. 874 (1) Make isometric freehand sketches. (2) Make isometric drawings with instruments, using Size A or Size B sheet, as assigned. (3) Make dimetric drawings with instruments, using Size A or Size B sheet, as assigned, and position assigned from Fig. 861. (4) Make trimetric drawings, using instruments, with axes chosen to show the objects to best advantage. If dimensions are required, study §469. For additional problems, assignments may be made from any of the problems in Figs. 896-898 and 928.

CHAPTER 16

OBLIQUE PROJECTION*

481. Oblique Projection. If the observer is considered to be at an infinite distance from the object, Fig. 822 (c), and looking toward the object so that the projectors are parallel to each other and oblique to the plane of projection, the resulting drawing is an *oblique projection.* As a rule, the object is placed with one of its principal faces parallel to the plane of projection. This is equivalent to holding the object in the hand and viewing it approximately as shown in Fig. 277.

A comparison of oblique projection and orthographic projection is shown in Fig. 875. The front face A′B′C′D′ in the oblique projection is identical with the front view, or orthographic projection, $A^vB^vC^vD^v$. Thus if an object is placed with one of its faces parallel to the plane of projection, that face will be projected true size and shape in oblique projection as well as in orthographic, or multiview, projection. This is the reason why oblique projection is preferable to axonometric projection in representing certain objects pictorially. Note that surfaces of the object that are not parallel to the plane of projection will not project in true size and shape. For example, surface ABFE on the object (a square) projects as a parallelogram A′B′F′E′ in the oblique projection.

In axonometric projection, circles on the object nearly always lie in surfaces inclined to the plane of projection, and project as ellipses. In oblique projection, the object may be positioned so that those surfaces are parallel to the plane of projection, in which case the circles will project as full-size true circles, and can be easily drawn with the compass.

A comparison of the oblique and orthographic projections of a cylindrical object is shown in Fig. 876. In both cases, the circular shapes project as true circles. Note

*See ASA Y14.4-1957.

that although the observer, looking in the direction of the oblique arrow, does see these shapes as ellipses, the drawing, or projection, represents not what he sees, but what is projected upon the plane of projection. This curious situation is peculiar to oblique projection.

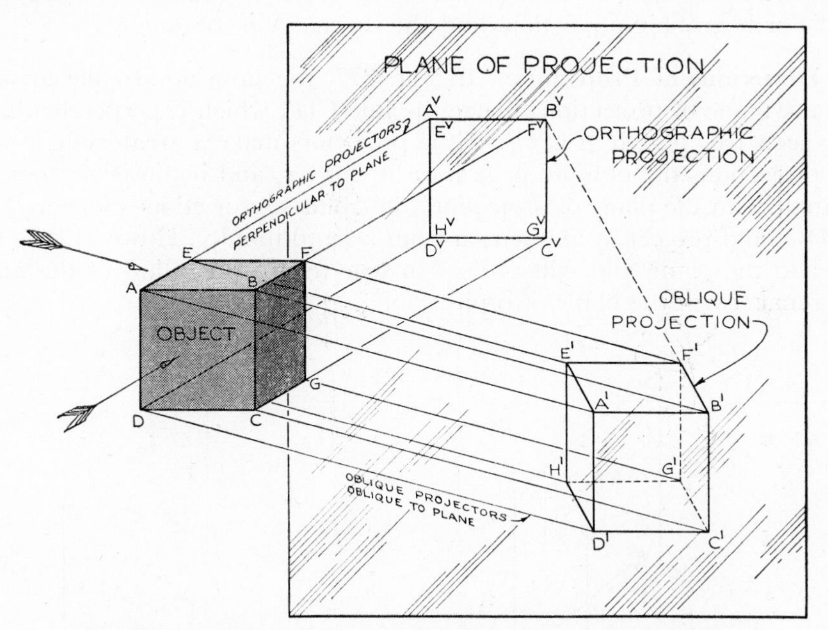

Fig. 875 Comparison of Oblique and Orthographic Projections.

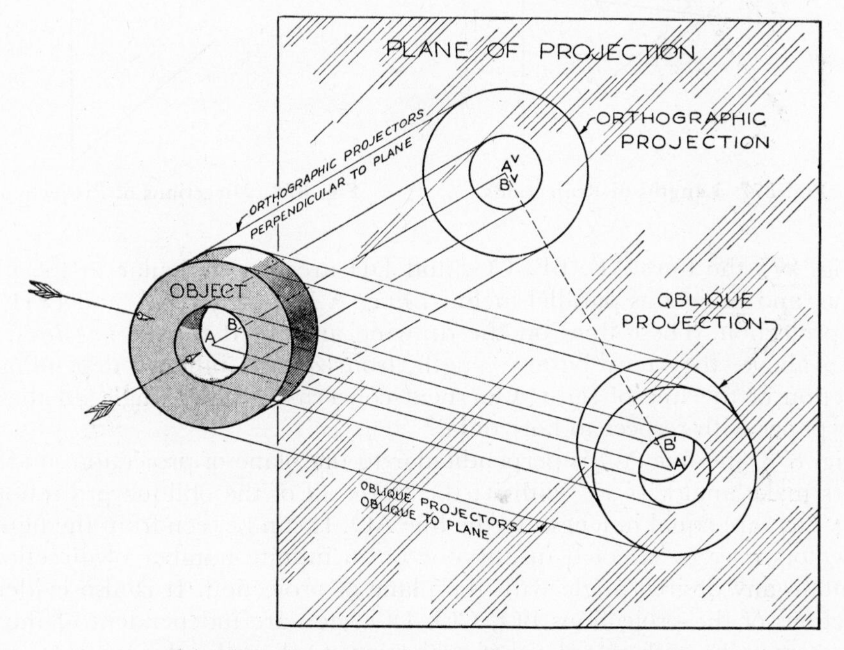

Fig. 876 Circles Parallel to Plane of Projection.

Observe that the axis AB of the cylinder projects as a point A^VB^V in the orthographic projection, since the line of sight is parallel to AB. But in the oblique projection, the axis projects as a line A′B′. The more nearly the direction of sight approaches the perpendicular with respect to the plane of projection, the larger the angle between the projectors and the plane, the closer the oblique projection moves toward the orthographic projection, and the shorter A′B′ becomes.

482. Directions of Projectors. In Fig. 877, the projectors make an angle of 45° with the plane of projection; hence the line CD′, which is perpendicular to the plane, projects true length at C′D′. If the projectors make a greater angle with the plane of projection, the oblique projection is shorter, and if the projectors make a smaller angle with the plane of projection, the oblique projection is longer. Theoretically, CD′ could project in any length from zero to infinity. However, the line AB is parallel to the plane and will project in true length regardless of the angle the projectors make with the plane of projection.

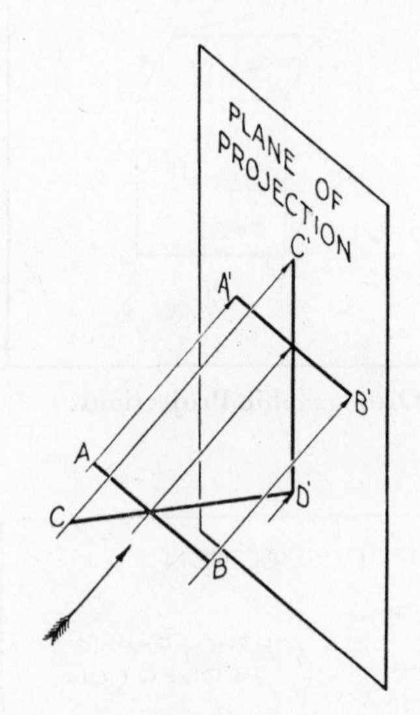

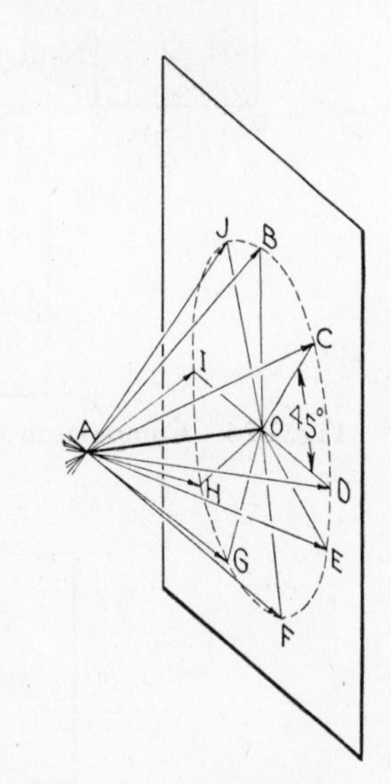

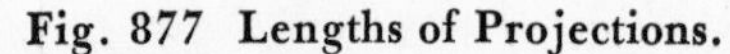

Fig. 877 Lengths of Projections.

Fig. 878 Directions of Projectors.

In Fig. 875 the lines AE, BF, CG, and DH are perpendicular to the plane of projection, and project as parallel inclined lines A′E′, B′F′, C′G′, and D′H′ in the oblique projection. These lines on the drawing are called the *receding lines*. As we have seen above, they may be any length, from zero to infinity, depending upon the direction of the line of sight. Our next concern is: What angle do these lines make on paper with respect to horizontal?

In Fig. 878, the line AO is perpendicular to the plane of projection, and all the projectors make angles of 45° with it; therefore, all of the oblique projections BO, CO, DO, etc., are equal in length to the line AO. It can be seen from the figure that the projectors may be selected in any one of an infinite number of directions and yet maintain any desired angle with the plane of projection. It is also evident that the directions of the projections BO, CO, DO, etc., are independent of the angles the projectors make with the plane of projection. Ordinarily, this inclination of the projection is 45° (CO in the figure), 30°, or 60° with horizontal, since these angles may be easily drawn with the triangles.

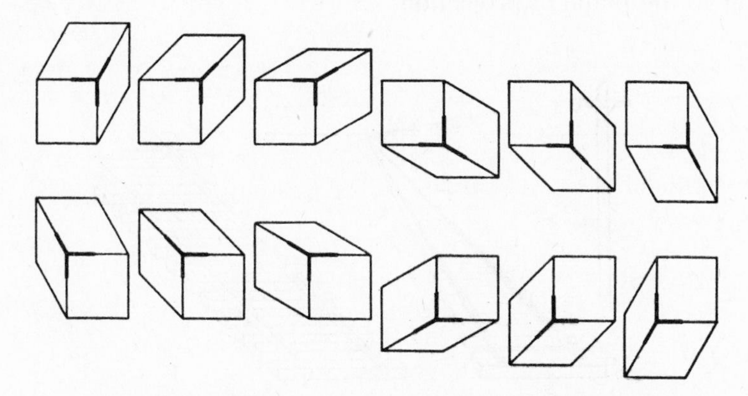

Fig. 879 Variation in Direction of Receding Axis.

483. Angles of Receding Lines. As shown above, the receding lines may be drawn at any convenient angle. Some typical drawings with the receding lines in various directions are shown in Fig. 879. The angle that should be used in an oblique drawing depends upon the shape of the object and the location of its significant features. For example, in Fig. 880 (a) a large angle was used in order to obtain a better view of the rectangular recess on the top, while at (b) a small angle was chosen to show a similar feature on the side.

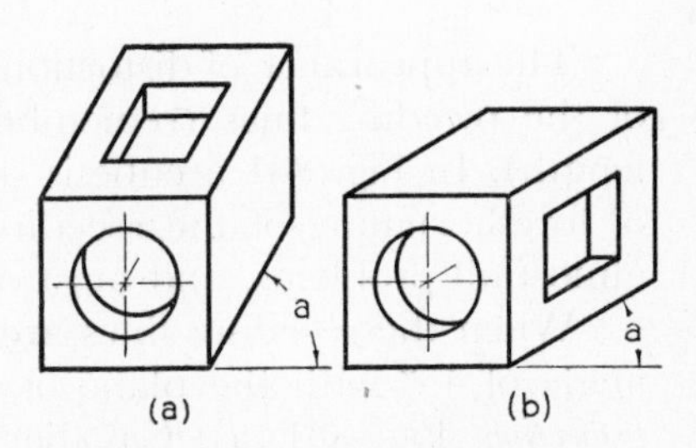

Fig. 880 Angle of Receding Axis.

484. Length of Receding Lines. Since the eye is accustomed to seeing objects with all receding parallel lines appearing to converge, an oblique projection presents an unnatural appearance, the seriousness of the distortion depending upon the object shown. For example, the object shown in Fig. 881 (a) is a cube, the receding lines being full length; but the receding lines appear to be too long and to diverge at the

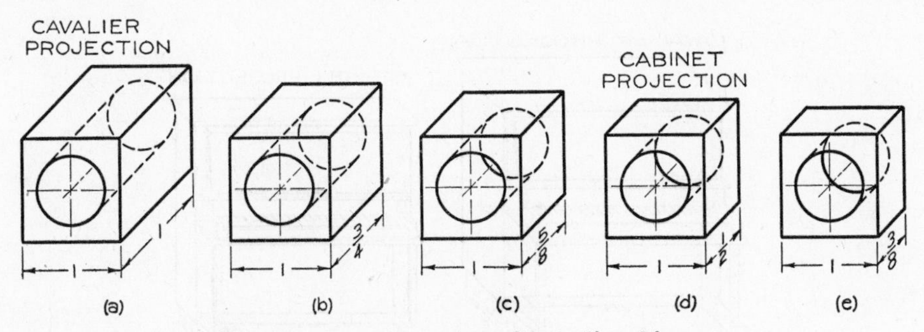

Fig. 881 Foreshortening of Receding Lines.

rear of the block. A striking example of the unnatural appearance of an oblique drawing when compared with the natural appearance of a perspective is shown in Fig.

882. This example points up one of the chief limitations of oblique projection: objects characterized by great length should not be drawn in oblique with the long dimension perpendicular to the plane of projection.

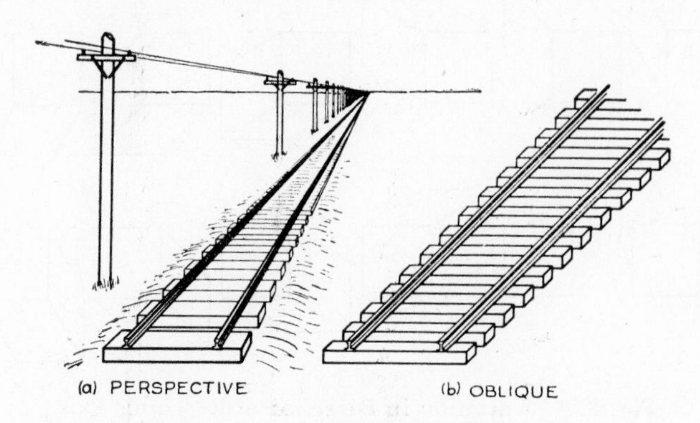

Fig. 882 Unnatural Appearance of Oblique Drawing.

The appearance of distortion may be materially lessened by decreasing the length of the receding lines (remember, we established in §482 that they could be any length). In Fig. 881 a cube is shown in five oblique drawings with varying degrees of foreshortening of the receding lines. The range of scales chosen is sufficient for almost all problems, and most of the scales are available on the architects scale.

When the receding lines are true length—that is, when the projectors make an angle of 45° with the plane of projection—the oblique drawing is called a *cavalier projection*, Fig. 881 (a). Cavalier projections originated in the drawing of medieval fortifications, and were made upon horizontal planes of projection. On these fortifications the central portion was higher than the rest, and was called *cavalier* because of its dominating and commanding position.

When the receding lines are drawn to half size, as at (d), the drawing is commonly known as a *cabinet projection*. The term is attributed to the early use of this type of oblique drawing in the furniture industries. A comparison of cavalier projection and cabinet projection is shown in Fig. 883.

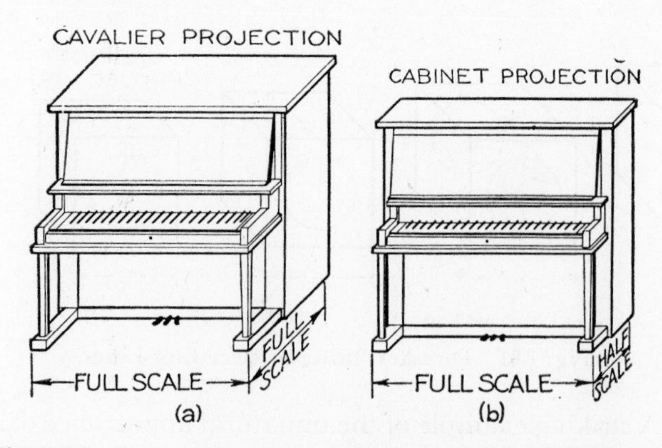

Fig. 883 Comparison of Cavalier and Cabinet Projections.

485. Choice of Position. The face of an object showing the essential contours should generally be placed parallel to the plane of projection, Fig. 884. If this is done, distortion will be kept at a minimum and labor reduced. For example, at (a) and (c) the circles and circular arcs are shown in their true sizes and shapes and may be quickly drawn with the compass, while at (b) and (d) these curves are not shown in their true sizes and shapes, and must be plotted as free curves or in the form of ellipses.

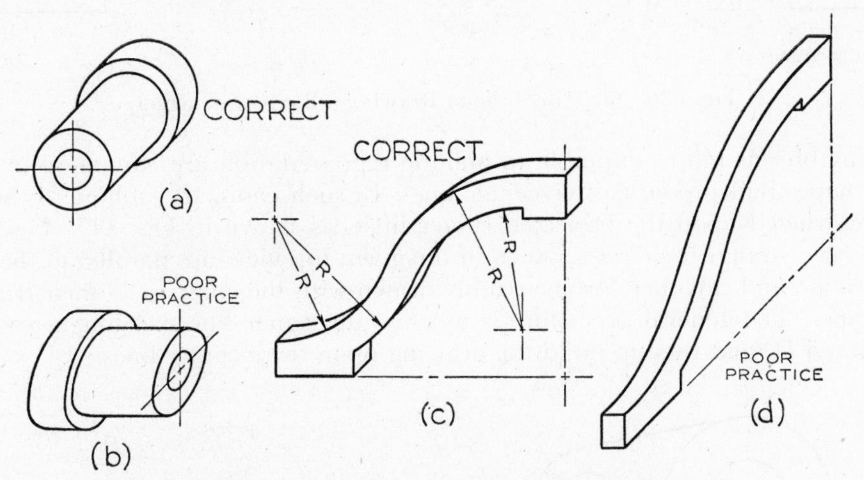

Fig. 884 Essential Contours Parallel to Plane of Projection.

The longest dimension of an object should generally be placed parallel to the plane of projection, as shown in Fig. 885.

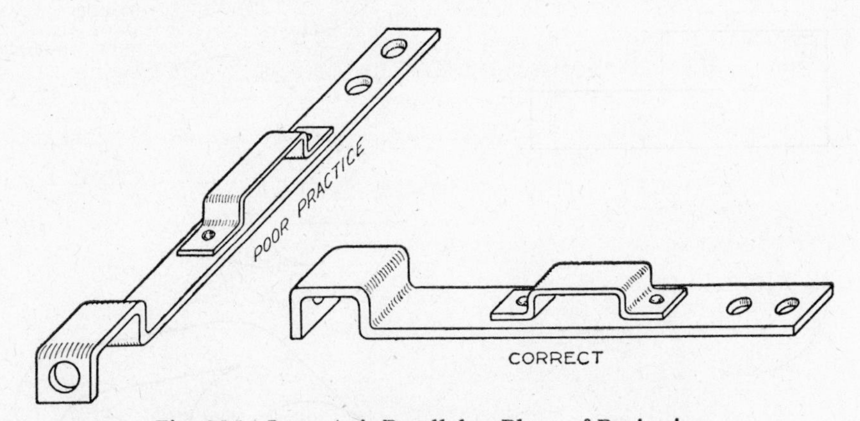

Fig. 885 Long Axis Parallel to Plane of Projection.

486. Steps in Oblique Drawing. The steps in drawing a cavalier drawing of a rectangular object is shown in Fig. 886. As shown in Step I, draw the axes OX and OY perpendicular to each other, and the receding axis OZ at any desired angle with horizontal. Upon these axes, construct an enclosing box, using the over-all dimensions of the object.

As shown at II, block out the various shapes in detail, and as shown at III, heavy-in all final lines.

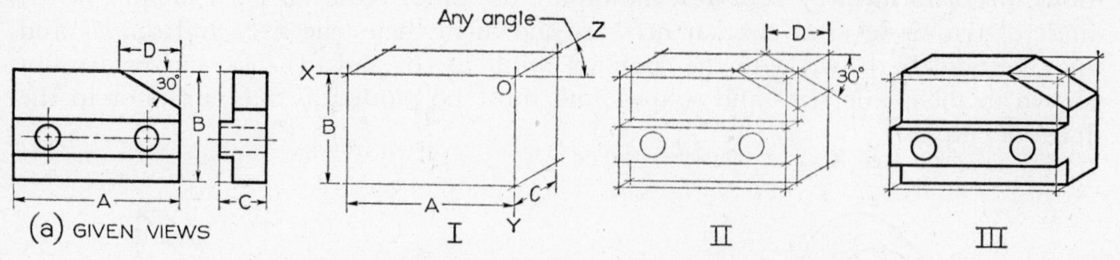

Fig. 886 Steps in Oblique Drawing—Box Construction.

Many objects most adaptable to oblique representation are composed of cylindrical shapes built upon axes or center lines. In such cases, the oblique drawing is best constructed upon the projected center lines, as shown in Fig. 887. The object is positioned so that the circles shown in the given top view are parallel to the plane of projection and can therefore be readily drawn with the compass in their true sizes and shapes. The general procedure is to draw the center-line skeleton, as shown in Steps I and II, and then to build the drawing upon these center lines.

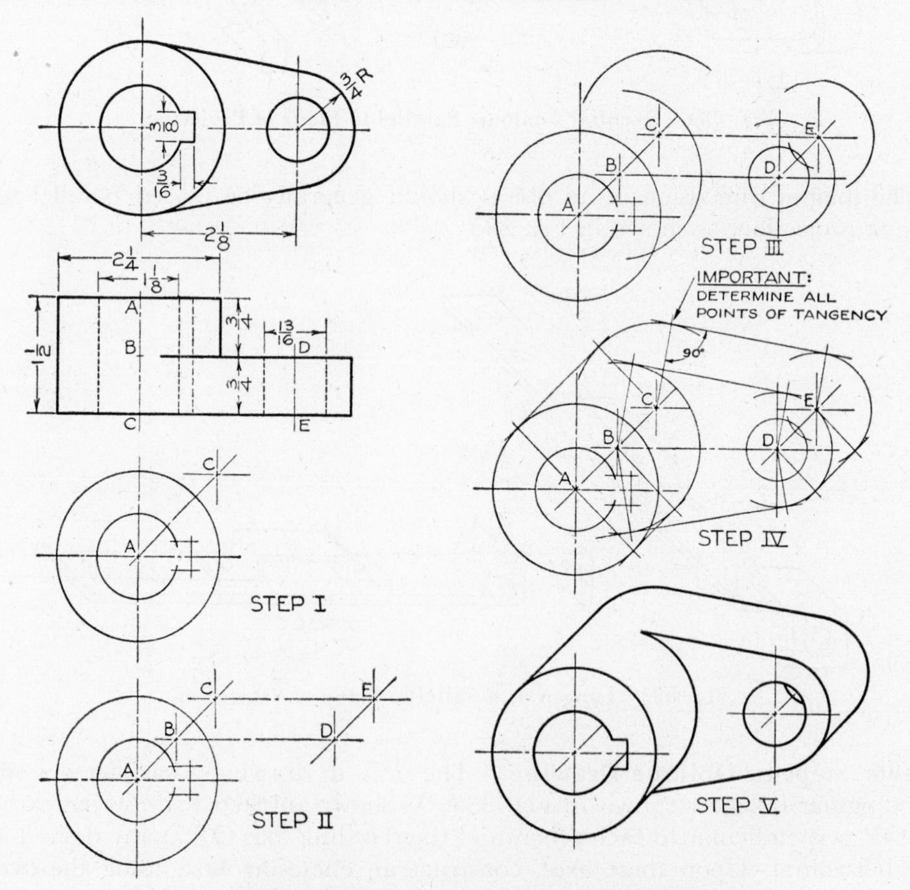

Fig. 887 Steps in Oblique Drawing—Skeleton Construction.

It is very important to construct all points of tangency, as shown in Step IV, especially if the drawing is to be inked. For a review of tangencies, see §§140-148. The final cavalier drawing is shown in Step V.

For applications of oblique drawing to piping, see Figs. 1116 and 1117.

487. Four-Center Ellipse. It is not always possible to place an object so that all of its significant contours are parallel to the plane of projection. For example, the object shown in Fig. 888 (a) has two sets of circular contours in different planes, and both cannot be placed parallel to the plane of projection.

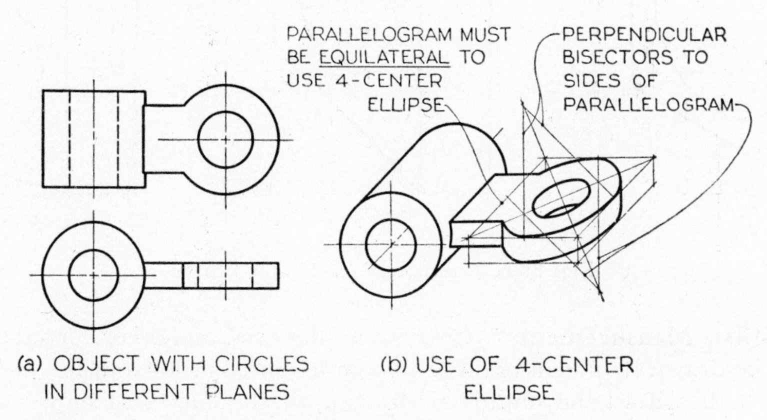

Fig. 888 Circles and Arcs Not Parallel to Plane of Projection.

In the oblique drawing at (b), the regular four-center method of Fig. 844 was used to construct ellipses representing circular curves not parallel to the plane of projection. This method can be used only in cavalier drawing in which case the enclosing parallelogram is equilateral—that is, the receding axis is drawn to full scale. The method is the same as in isometric: erect perpendicular bisectors to the four sides of the parallelogram; their intersections will be centers for the four circular arcs. If the angle of the receding lines is other than 30° with horizontal, as in this case, the centers of the two large arcs will not fall in the corners of the parallelogram.

The regular four-center method described above is not convenient in oblique unless the receding lines make 30° with horizontal so that the perpendicular bisectors may be drawn easily with the 30° × 60° triangle and the T-square without the necessity of first finding the mid-points of the sides. A more convenient method is the alternate four-center ellipse drawn upon the two center lines, as shown in Fig. 889. This is the same method as used in isometric, Fig. 849, but in oblique drawing it varies slightly in appearance according to the different angles of the receding lines.

First, draw the two center lines. Then from the center, draw a construction circle equal in diameter to the actual hole or cylinder. The circle will intersect each center line at two points. From the two points on one center line, erect perpendiculars to the other center line; then from the two points on the other center line, erect perpendiculars to the first center line. From the intersections of the perpendiculars, draw four circular arcs, as shown.

It must be remembered that the four-center ellipse can be inscribed only in an equilateral parallelogram, and hence cannot be used in any oblique drawing in which the receding axis is foreshortened. Its use is limited, therefore, to cavalier drawing.

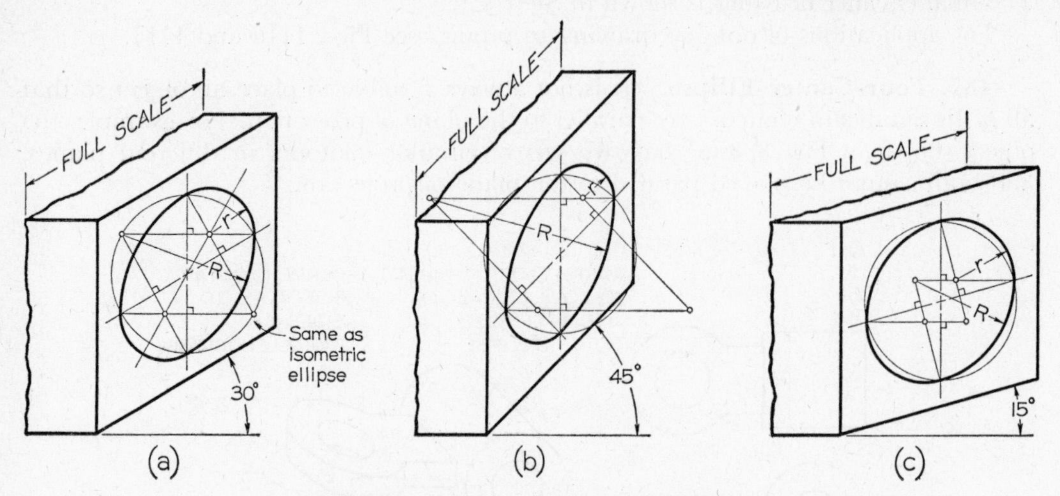

Fig. 889 Alternate Four-Center Ellipse.

488. Offset Measurements. Circles, circular arcs, and other curved or irregular lines may be drawn by means of offset measurements, as shown in Fig. 890. The offsets are first drawn on the multiview drawing of the curve, as shown at (a), and

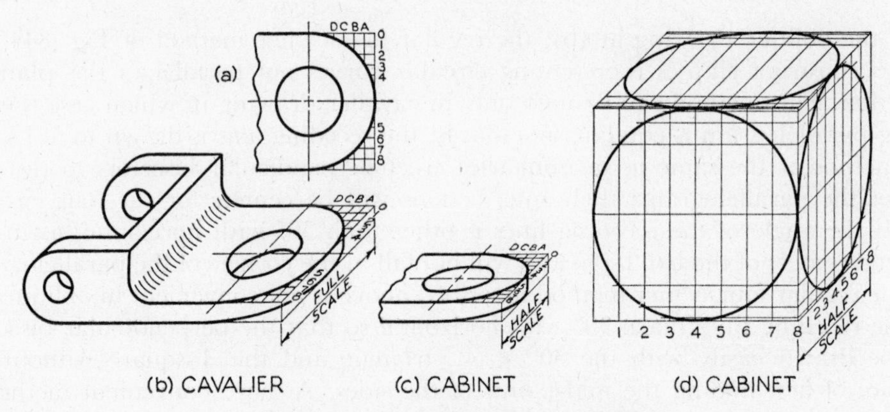

Fig. 890 Use of Offset Measurements.

these are transferred to the oblique drawing, as shown at (b). In this case, the receding axis is full scale, and therefore all offsets can be drawn full scale. The four-center ellipse could be used, but the method here is more accurate. The final curve is drawn with the aid of the irregular curve, §66.

If the oblique drawing is a cabinet drawing, as shown at (c), or any oblique drawing in which the receding axis is drawn to a reduced scale, the offset measurements parallel to the receding axis must be drawn to the same reduced scale. In this case, there is no choice of methods, since the four-center ellipse could not be used. A method of drawing ellipses in a cabinet drawing of a cube is shown at (d).

As shown in Fig. 891, a free curve may be drawn in oblique by means of offset measurements. This figure also illustrates a case in which hidden lines are used to make the drawing clearer.

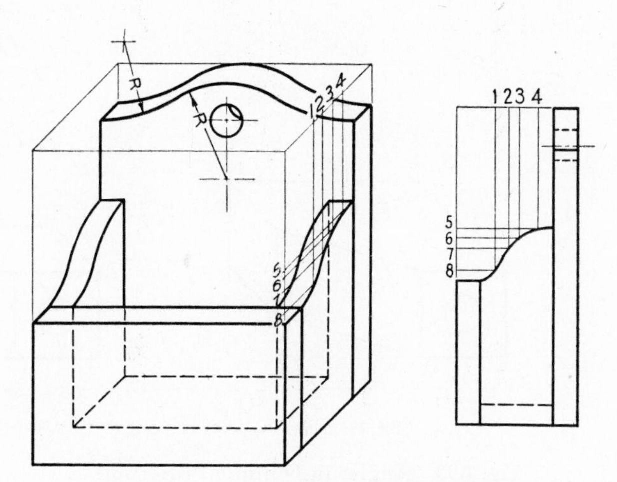

Fig. 891 Use of Offset Measurements.

The use of offset measurements in drawing an ellipse in a plane inclined to the plane of projection is shown in Fig. 892. At (a) a number of parallel lines are drawn to represent imaginary slicing planes. Each plane will cut a rectangular surface between the front end of the cylinder and the inclined surface. These rectangles are drawn in oblique, as shown at (b), and the curve is drawn through corner points, as indicated. The final cavalier drawing is shown at (c).

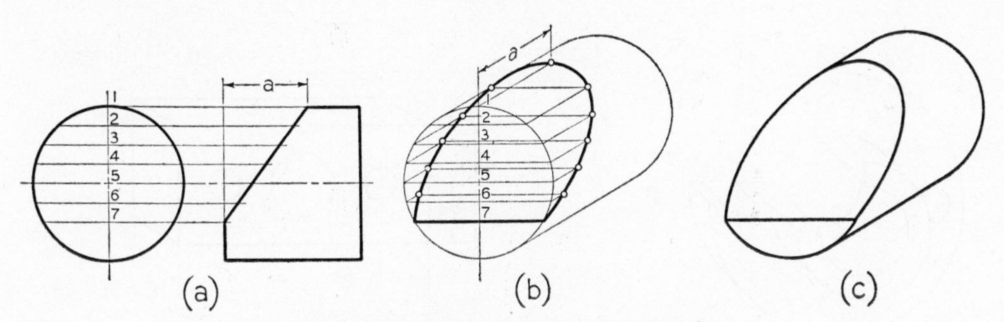

Fig. 892 Use of Offset Measurements.

489. Angles in Oblique Projection. If an angle that is specified in degrees lies in a receding plane, it is necessary to convert the angle into linear measurements in order to draw the angle in oblique. For example, in Fig. 893 (a) an angle of 30° is given. In order to draw the angle in oblique, we need to know dimensions AB and BC. The distance AB is given as $1\frac{1}{4}''$, and can be set off directly in the cavalier drawing, as shown at (b). Distance BC is not known, but can easily be found by constructing the right triangle ABC at (c) from the given dimensions in the top view at (a). The length BC is then transferred with the dividers to the cavalier drawing, as shown.

In cabinet drawing, it must be remembered that *all receding dimensions* must be reduced to half size. Thus, in the cabinet drawing at (d), the distance BC must be half the side BC of the right triangle at (e), as shown.

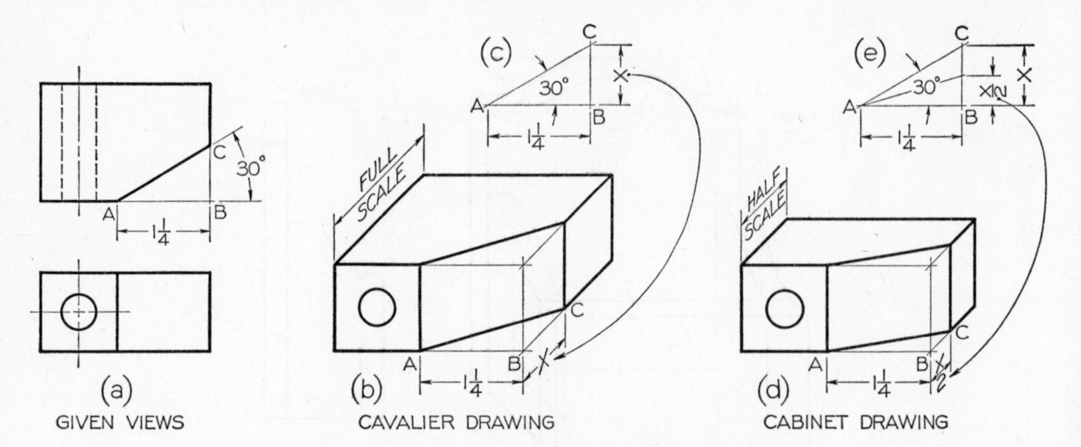

Fig. 893 Angles in Oblique Projection.

490. Oblique Sections. Sections are often useful in oblique drawing, especially in the representation of interior shapes. An *oblique half section* is shown in Fig. 894. Other examples are shown in Figs. 440, 441, 444, and following figures. *Oblique full sections*, in which the plane passes completely through the object, are seldom used because they do not show enough of the exterior shapes. In general, all the types of sections discussed in §468 for isometric drawing may be applied equally to oblique drawing.

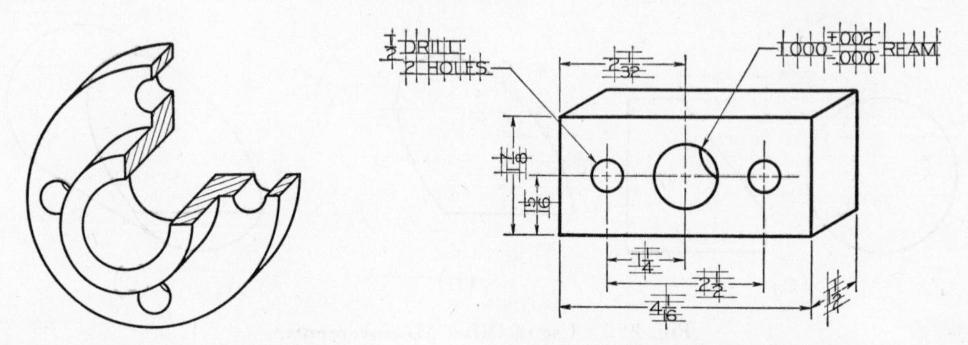

Fig. 894 Oblique Half Section. **Fig. 895 Oblique Dimensioning.**

491. Oblique Dimensioning. An oblique drawing may be dimensioned in a similar manner to that described in §469 for isometric drawing, as shown in Fig. 895. The general principles of dimensioning, as outlined in Chapter 11, must be followed. As shown in the figure, all dimensions must lie in the planes of the object to which they apply. Dimensions should be placed outside the outlines of the drawing except when greater clearness or directness of application results from placing the dimensions directly on the view. For many other examples of oblique dimensioning, see Figs. 342, 346, 347, and others on following pages.

492. Oblique Sketching. Methods of sketching in oblique on plain paper are illustrated in Fig. 277. Ordinary rectangular cross-section paper is very useful in oblique sketching, Fig. 278. The height and width proportions can be easily controlled by simply counting the squares. A very pleasing depth proportion can be obtained by sketching the receding lines at 45° diagonally through the squares, and through half as many squares as the actual depth would indicate.

493. Oblique Drawing Problems. A large number of problems to be drawn in oblique—either Cavalier or Cabinet—are given in Figs. 896-898. They may be drawn freehand, §203, using cross-section paper or plain drawing paper as assigned by the instructor, or they may be drawn with instruments. In the latter case, all construction lines should be shown on the completed drawing.

Many additional problems suitable for oblique projection will be found in Figs. 333-335, 871-874, and 928.

Oblique problems in convenient form for solution may be found in *Technical Drawing Problems*, by Giesecke, Mitchell, and Spencer, and in *Technical Drawing Problems, Series 2*, by Spencer and Grant, both designed to accompany this text, and published by The Macmillan Company.

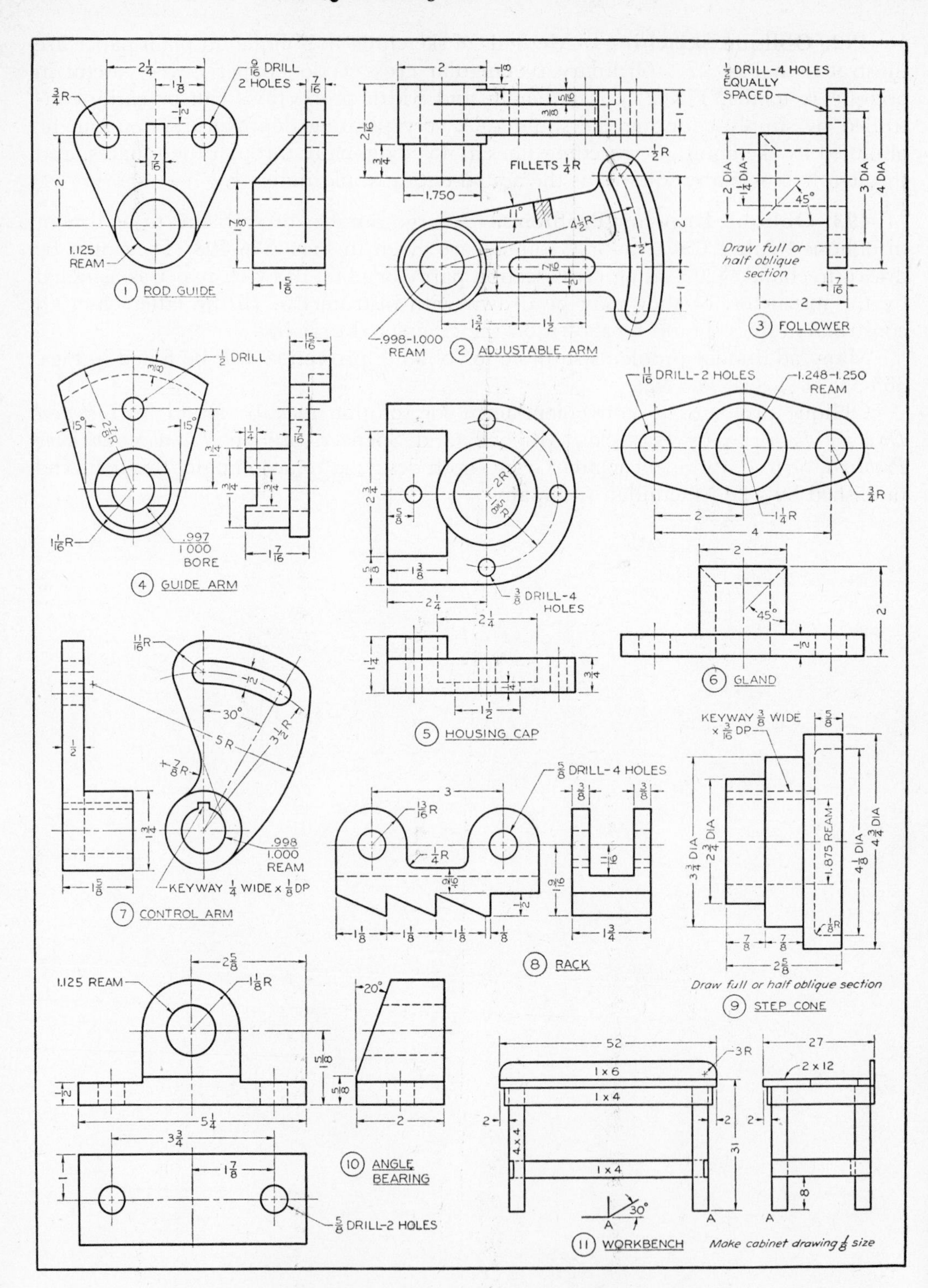

Fig. 896 (1) Make freehand oblique sketches. (2) Make oblique drawings with instruments, using Size A or Size B sheet, as assigned. If dimensions are required, study §491.

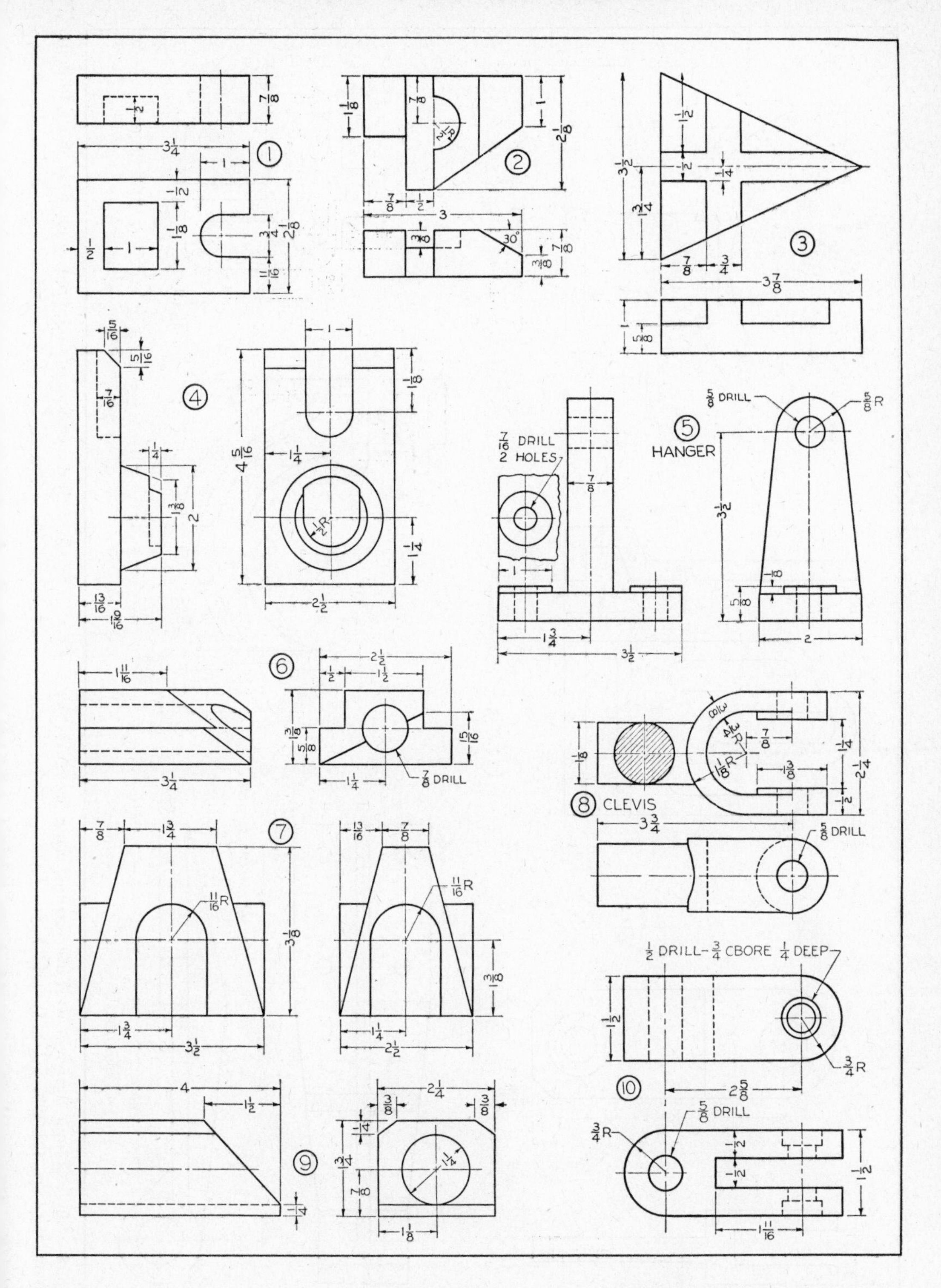

Fig. 897 Make oblique drawings with instruments, using Size A or Size B sheet, as assigned. If dimensions are required, study §491. For additional problems, see Figs. 496-498, 871-874, and 928.

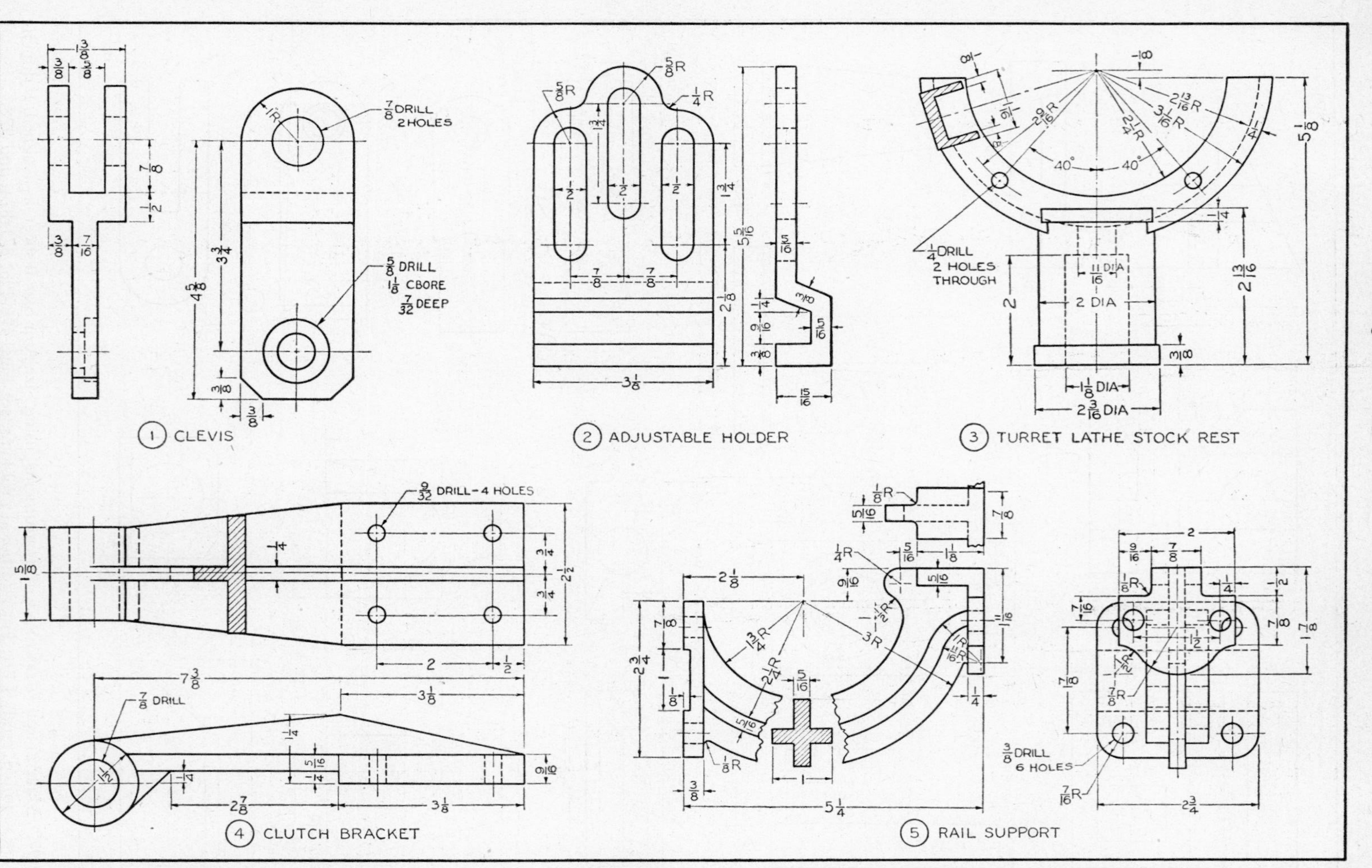

Fig. 898 Make oblique drawings with instruments, using Size A or Size B sheet, as assigned. If dimensions are required, study §491. For additional problems, see Figs. 496-498, 871-874, and 928.

CHAPTER 17

PERSPECTIVE*

494. General Principles. *Perspective*, or *central projection*, excels all other types of projection in the pictorial representation of objects because it more closely approximates the view obtained by the human eye, Fig. 899. Geometrically, an ordinary photograph is a perspective. While perspective is of major importance to the architect,

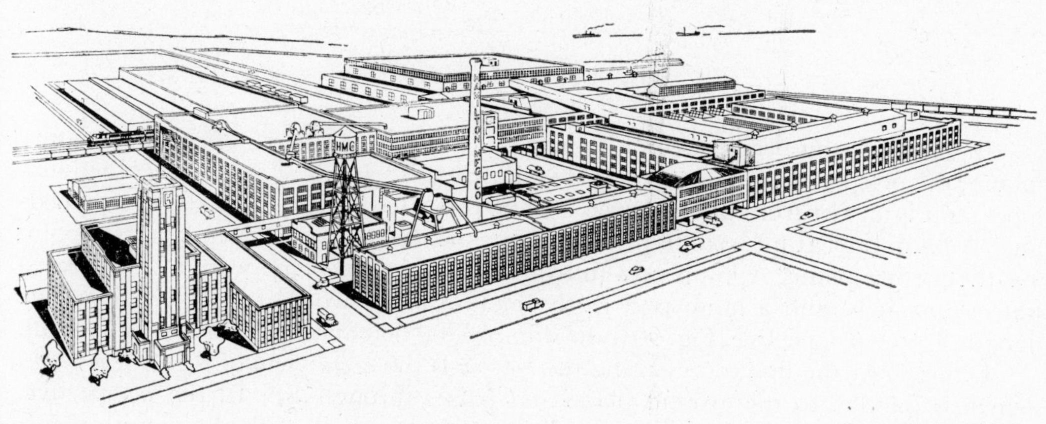

Courtesy Hamilton Mfg. Co.

Fig. 899 Perspective of a Factory.

industrial designer, or illustrator, the engineer at one time or another is apt to be concerned with the pictorial representation of objects, and should understand the basic principles.

*See ASA Y14.4-1957.

As explained in §10, a perspective involves four main elements: (1) the observer's eye, (2) the object being viewed, (3) the plane of projection, and (4) the projectors from the observer's eye to all points on the object. The plane of projection is placed between the observer and the object,† as shown in Fig. 8, and the collective piercing points in the plane of projection of all of the projectors produce the perspective.

In Fig. 900, the observer is shown looking through an imaginary plane of projection along a boulevard. This plane is called the *picture plane*, or simply PP. The position of the observer's eye is called the *station point*, or simply SP. The lines from

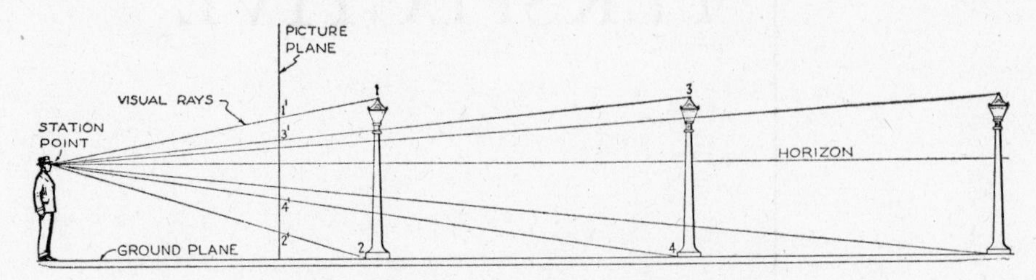

Fig. 900 Looking through the Picture Plane.

SP to the various points in the scene are the projectors, or more properly in perspective, *visual rays*. The points where the visual rays pierce PP are the *perspectives* of the respective points. Collectively, these piercing points form the perspective of the object or the scene as viewed by the observer. The perspective thus obtained is shown in Fig. 901.

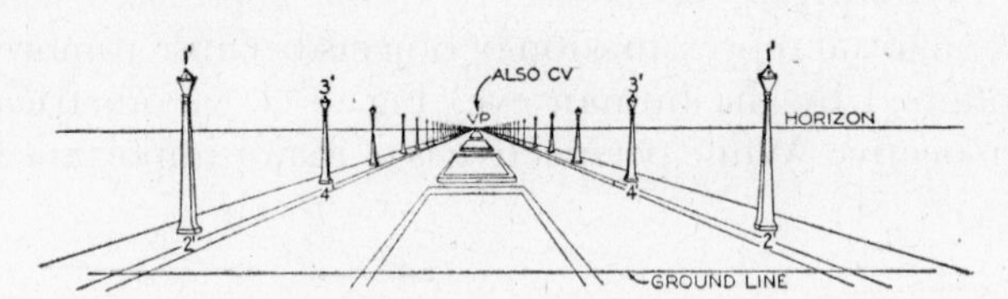

Fig. 901 A Perspective.

In Fig. 900, the perspective of lamp post 1–2 is shown at 1′–2′, on the picture plane; the perspective of lamp post 3–4 is shown at 3′–4′, etc. Each succeeding lamp post, as it is farther from the observer, will be projected smaller than the one preceding. A lamp post at an infinite distance from the observer would appear as a point on the picture plane. A lamp post in front of the picture plane would be projected taller than it is, and a lamp post in the picture plane would be projected in true length. In the perspective, Fig. 901, the diminishing heights of the posts are apparent.

In Fig. 900, the line representing the *horizon* is the edge view of the *horizon plane*, which is parallel to the ground plane and passes through SP. In the perspective, Fig. 901, the horizon is the line of intersection of this plane with the picture plane, and represents the eye level of the observer, or SP. Also, in Fig. 900, the *ground plane* is the edge view of the ground upon which the object usually rests. In Fig. 901, the *ground line*, or GL, is the intersection of the ground plane with the picture plane.

In Fig. 901, it will be seen that lines that are parallel to each other but not parallel to the picture plane, such as curb lines, sidewalk lines, and lines along the tops and

†Except as explained in §499.

bottoms of the lamp posts, all converge toward a single point on the horizon. This point is called the *vanishing point*, or VP, of the lines. Thus, the first rule to learn in perspective is: *All parallel lines that are not parallel to PP vanish at a single vanishing point, and if these lines are parallel to the ground, the vanishing point will be on the horizon.* Parallel lines that are also parallel to PP, such as the lamp posts, remain parallel and do not converge toward a vanishing point.

495. Multiview Perspective. A perspective can be drawn by the ordinary methods of multiview projection, as shown in Fig. 902. In the upper portion of the drawing are shown the top views of the station point, the picture plane, the object, and the visual rays. In the right-hand portion of the drawing are shown the right-

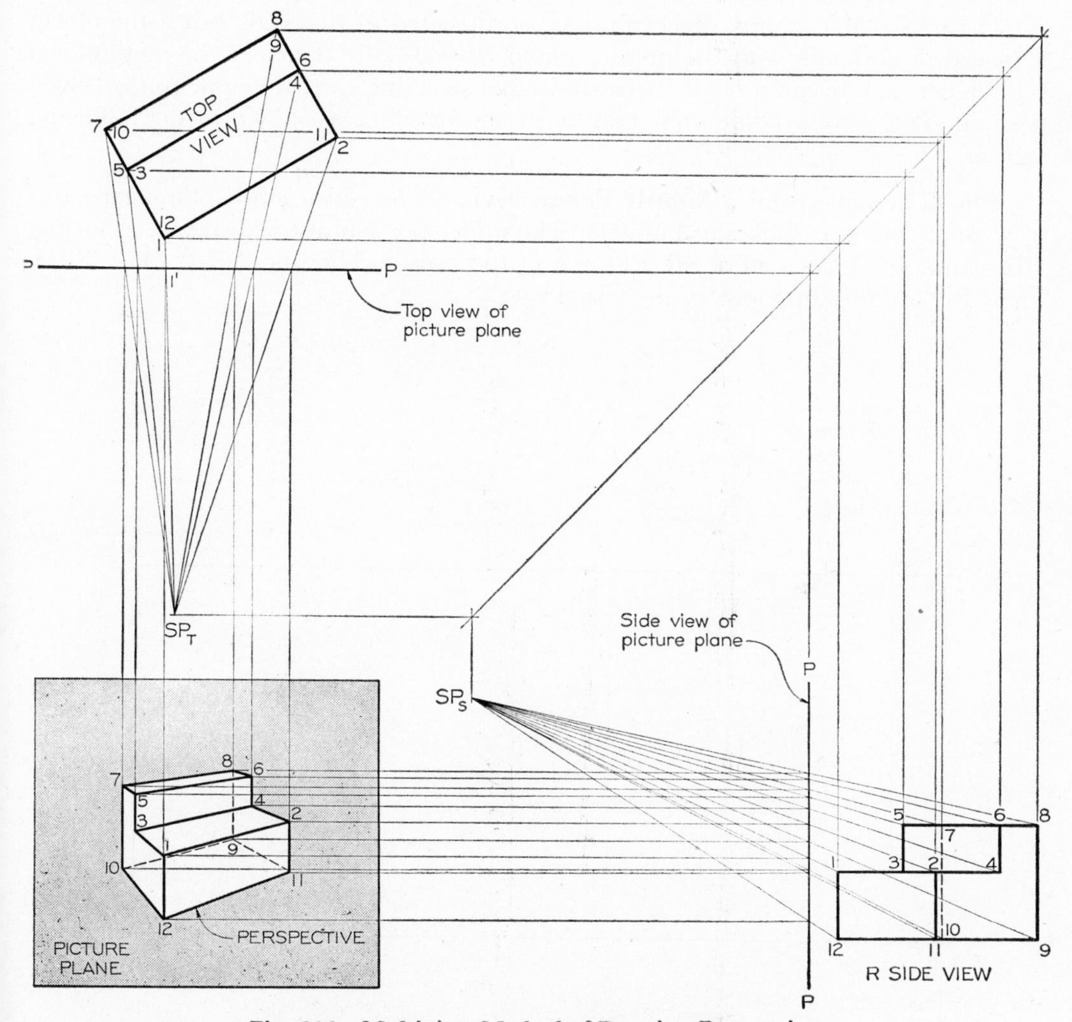

Fig. 902 Multiview Method of Drawing Perspective.

side views of the same station point, picture plane, object, and visual rays. In the front view, the picture plane coincides with the plane of the paper, and the perspective is drawn upon it. Note the method of projecting from the top view to the side view, which conforms to the usual multiview methods shown in Fig. 292.

To obtain the perspective of point 1, a visual ray is drawn in the top view from SP_T to point 1 on the object. From the intersection 1′ of this ray with the picture plane, a projection line is drawn downward till it meets a similar projection line from the side view. This intersection is the perspective of point 1, and the perspectives of all other points are found in a similar manner.

Observe that all parallel lines that are also parallel to the picture plane (the vertical lines) remain parallel and do not converge, whereas the other two sets of parallel lines converge toward vanishing points. However, the vanishing points are not needed in the multiview construction of Fig. 902, and are therefore not shown; but if the converging lines should be extended, it will be found that they meet at two vanishing points (one for each set of parallel lines).

The perspective of any object may be constructed in this way, but if the object is placed at an angle with the picture plane, as is usually the case, the method is a bit cumbersome because of the necessity of constructing the side view in a revolved position. The revolved side view can be dispensed with, as shown in the following section.

496. The Set-up for a Simple Perspective. The construction of a perspective of a simple prism is shown in Fig. 903. The upper portion of the drawing, as in Fig. 902, shows the top views of SP, PP, and of the object. The lines SP-1, SP-2, SP-3, and SP-4 are the top views of the visual rays.

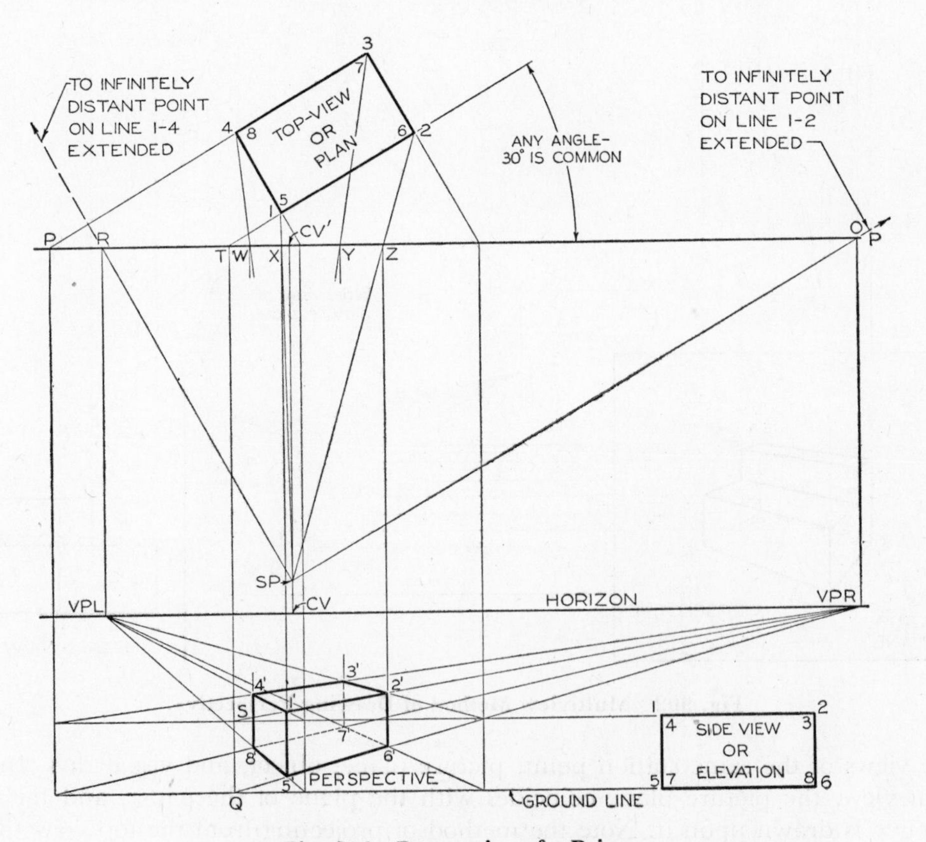

Fig. 903 Perspective of a Prism.

In the side view, a departure from Fig. 902 is made, in that a *revolved* side view is not required. All that is needed is any elevation view that will provide the necessary elevation or height measurements. If these dimensions are known, no view at one side is required.

The perspective itself is drawn in the front-view position, the picture plane being considered as the plane of the paper upon which the perspective is drawn. The ground line is the edge view of the ground plane or the intersection of the ground plane with the picture plane. The horizon is a horizontal line in the picture plane that is the line of intersection of the horizon plane with the picture plane. Since the horizon plane passes through the observer's eye, or SP, the horizon is drawn at the level of the eye; that is, at the distance above the ground line representing, to scale, the altitude of the eye above the ground.

The *center of vision*, or CV, is the orthographic projection (or front view) of SP on the picture plane, and since the horizon is at eye level, CV will always be on the horizon.* In Fig. 903, the top view of CV is CV′, found by dropping a perpendicular from SP to PP. The front view CV is found by projecting downward from CV′ to the horizon.

497. To Draw an Angular Perspective. Since objects are defined principally by edges that are straight lines, the drawing of a perspective resolves itself into drawing the *perspective of a line*. If a draftsman can draw the perspective of a line, he can draw the perspective of any object, no matter how complex.

To draw the perspective of any horizontal straight line not parallel to PP, for example the line 1–2 in Fig. 903, proceed as follows:

I. *Find the piercing point in PP of the line.* In the top view, extend line 1–2 until it pierces PP at T; then project downward to the level of the line 1–2 projected horizontally from the side view. The point S is the piercing point of the line.

II. *Find the vanishing point of the line.* The vanishing point of a line is the piercing point in PP of a line drawn through SP parallel to that line. Hence, the vanishing point VPR of the line 1–2 is found by drawing a line from SP parallel to that line and finding the top view of its piercing point O, and then projecting downward to the horizon. The line SP-O is actually a visual ray drawn toward the infinitely distant point on line 1–2 of the object, extended, and the vanishing point is the intersection of this visual ray with the picture plane. The vanishing point is, then, the perspective of the infinitely distant point on the line extended.

III. *Join the piercing point and the vanishing point with a straight line.* The line S-VPR is the line joining these two points, and it is the perspective of a line of infinite length containing the required perspective of the line 1–2.

IV. *Locate the end points of the perspective of the line.* The end points 1′ and 2′ can be found by projecting down from the piercing points of the visual rays in PP, or by simply drawing the perspectives of the remaining horizontal edges of the prism. In practice, it is best to use both methods as a check on the accuracy of the construction. To locate the end points by projecting from the piercing points, draw visual rays from SP to the points 1 and 2 on the object in the top view. The top views of the piercing points are X and Z. Since the perspectives of points 1 and 2 must lie on the line S-VPR, project downward from X and Z to locate points 1′ and 2′.

After the perspectives of the horizontal edges have been drawn, the vertical edges

*Except in three-point perspective, §504.

can be drawn, as shown, to complete the perspective of the prism. Note that *vertical heights can be measured only in the picture plane*. If the front vertical edge 1–5 of the object were actually in PP—that is, if the object were situated with the front edge in PP—the vertical height could be set off directly full size. If the vertical edge is behind PP, a plane of the object, such as surface 1–2–5–6 can be extended forward until it intersects PP in line TQ. The line TQ is called a *measuring line*, and the true height SQ of line 1–5 can be set off with a scale or projected from the side view as shown.

If a large drawing board is not available, one vanishing point, such as VPR, may fall off the board. By using one vanishing point VPL, and projecting down from the piercing points in PP, vanishing point VPR may be eliminated. However, a valuable means of checking the accuracy of the construction will be lost.

498. Position of the Station Point. The center line of the cone of visual rays should be directed toward the approximate center, or center of interest, of the object. In a perspective of the type shown in Fig. 903,* the location of SP in the plan view should be slightly to the left and not directly in front of the center of the object, and at such a distance from it that the object can be viewed at a glance without turning the head. This is accomplished if a cone of rays with its vertex at SP and a vertical angle of about 30° entirely encloses the object, as shown in Fig. 904.

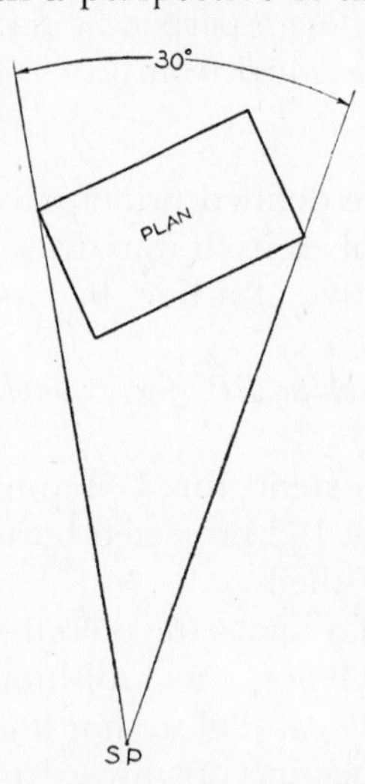

Fig. 904 Distance from SP to Object.

In the perspective portion of Fig. 903, SP does not appear, because SP is in front of the picture plane. However, the orthographic projection CV of SP in the picture plane does show the height of SP with respect to the ground plane. Since the horizon is at eye level, it also shows the altitude of SP. Therefore, in the perspective portion of the drawing, the horizon is drawn a distance above the ground line at which it is desired to assume SP. For most small and medium-size objects, such as machine parts or furniture, SP is best assumed slightly above the top of the object. Large objects, such as buildings, are usually viewed from a station point about the altitude of the eye above the ground, or about 5′-6.

499. Location of the Picture Plane. In general, the picture plane is placed in front of the object, as in Fig. 905 (b) and (c). However, it

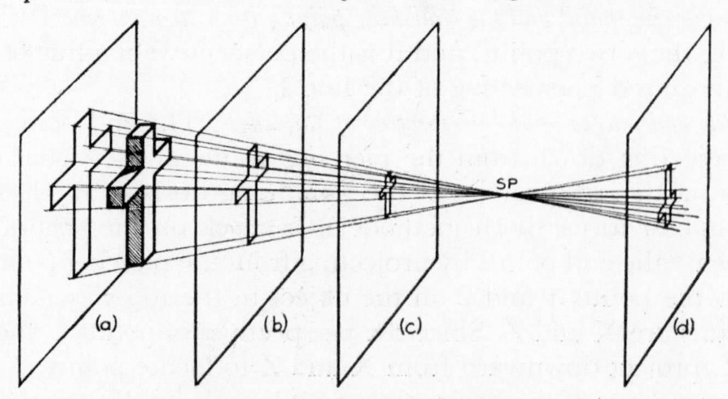

Fig. 905 Location of PP.

*Two-point perspective, §503.

may be placed behind the object, as shown at (a), and it may even be placed behind SP, as shown at (d), in which event the perspective is reversed, as in the case of a camera. Of course the usual position of the picture plane is between SP and the object. The perspectives in Fig. 905 differ in size but not in proportion. As shown at (b) and (c), the farther the picture plane is from the object, the smaller the perspective will be. This distance may be assumed, therefore, with the thought of controlling the scale of the perspective. In practice, however, the object is usually assumed with the front corner in the picture plane to facilitate vertical measurements. See Fig. 910.

500. Position of the Object with Respect to the Horizon. To compare the elevation of the object with that of the horizon is equivalent to referring it to the level of the eye or SP, because the horizon is on a level with the eye.* The differences in effect produced by placing the object on, above, or below the horizon are shown in Fig. 906.

If the object is placed above the horizon, it is above the level of the eye, or above SP, and will appear as seen from below. Likewise, if the object is below the horizon, it will appear as seen from above.

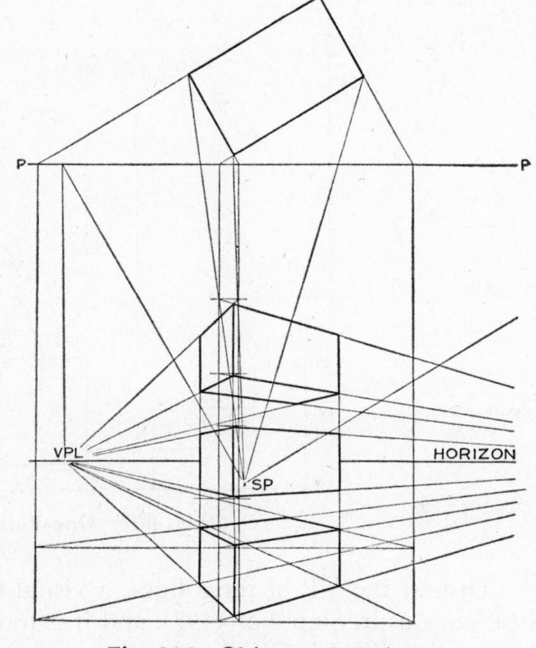

Fig. 906 Object and Horizon.

501. The Three Types of Perspectives. Perspectives are classified according to the number of vanishing points required, which in turn depends upon the position of the object with respect to the picture plane.

If the object is situated with one face parallel to the plane of projection, only one vanishing point is required, and the result is a *one-point perspective* or *parallel perspective*, §502.

If the object is situated at an angle with the picture plane but with vertical edges parallel to the picture plane, two vanishing points are required, and the result is a *two-point perspective* or an *angular perspective*. This is the most common type of perspective, and is the one described in §497. See also §503.

If the object is situated so that no system of parallel edges is parallel to the picture plane, three vanishing points are necessary, and the result is a *three-point perspective*, §504.

502. One-Point Perspective. In one-point perspective, Fig. 822 (d), the object is placed so that two sets of its principal edges are parallel to PP, and the third set is perpendicular to PP. This third set of parallel lines will converge toward a single vanishing point in perspective, as shown.

In Fig. 907, the plan view shows the object with one face parallel to the picture

*Except in three-point perspective, §504.

plane. If desired, this face could be placed *in* the picture plane. The piercing points of the eight edges perpendicular to PP are found by extending them to PP and then projecting downward to the level of the lines as projected across from the elevation view.

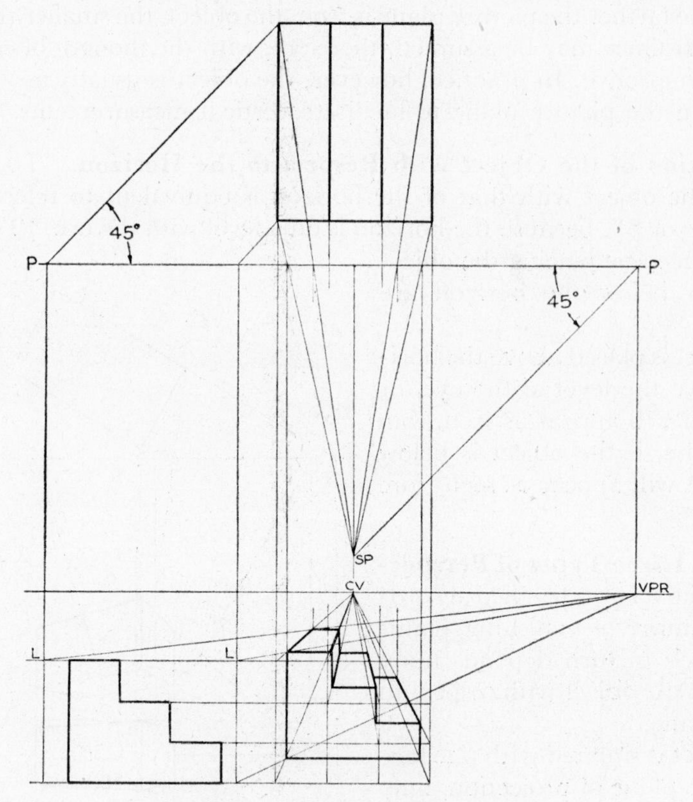

Fig. 907 One-Point Perspective.

To find the VP of these lines, a visual ray is drawn from SP parallel to them (the same as in Step 2 of §497), and it is found that the *vanishing point of all lines perpendicular to PP is in CV.* By connecting the eight piercing points with the vanishing point CV, the indefinite perspectives of the eight edges are obtained.

To cut off on these lines the definite lengths of the edges of the object, horizontal lines are drawn from the ends of one of the edges in the top view and at any desired angle with PP, 45° for example, as shown. The piercing points and the vanishing point VPR of these lines are found, and the perspectives of the lines drawn. The intersections of these with the perspectives of the corresponding edges of the object determine the lengths of the receding edges. The perspective of the object may then be completed as shown.

One of the most common uses of parallel perspective is in the representation of interiors of buildings, as illustrated in Fig. 908.

An adaptation of one-point perspective, which is simple and effective in representing machine parts, is shown in Fig. 909. The front surface of the cylinder is placed in PP, and all circular shapes are parallel to PP; hence these shapes will be projected as circles and circular arcs in the perspective. SP is located in front and to one side of the object, and the horizon is placed well above the ground line. The single vanishing point is on the horizon in CV.

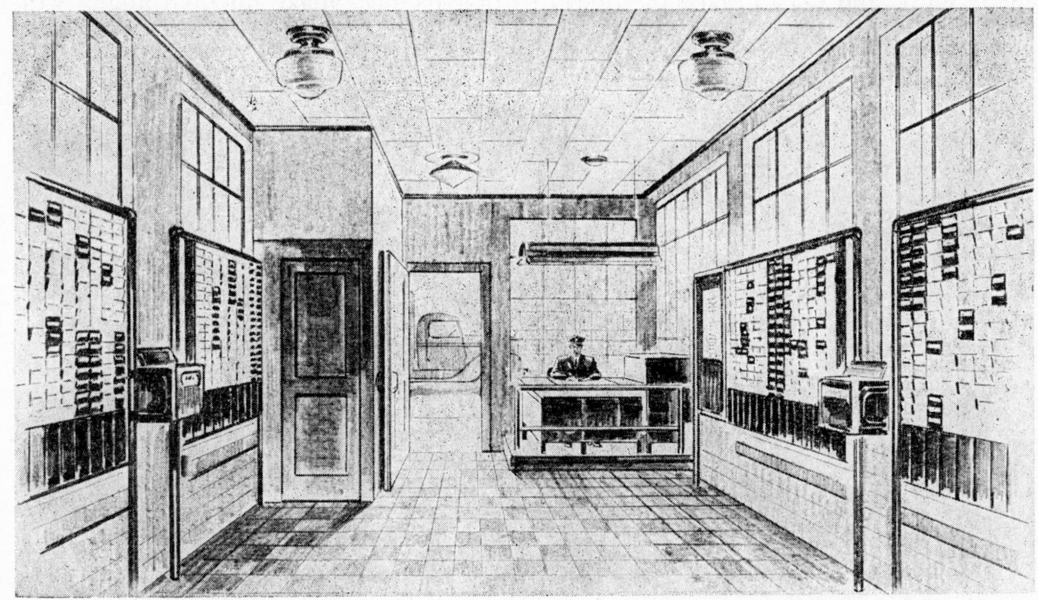

Courtesy Eaton Manufacturing Company

Fig. 908 One-Point Perspective.

The two circles and the keyway in the front surface of the object will be drawn true size because they lie in PP. The circles are drawn with the compass on center O′. To locate R′, the perspective center of the large arc, draw visual ray SP-R; then from its intersection X with PP, project down to the center line of the large cylinder, as shown.

To find the radius T′W′ at the right end of the perspective, draw visual rays SP-T and SP-W, and from their intersections with PP, project down to T′ and W′ on the horizontal center line of the hole.

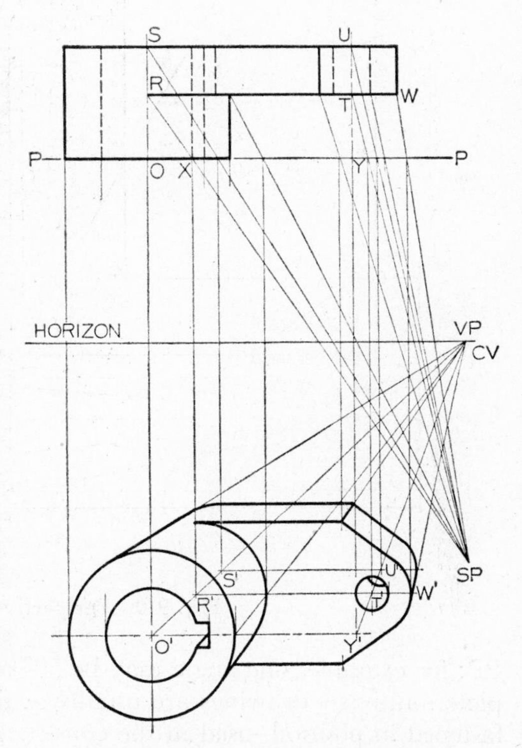

Fig. 909 One-Point Perspective.

503. Two-Point Perspective. In two-point perspective, the object is placed so that one set of parallel edges is vertical and has no vanishing point, while the two other sets each have vanishing points. This is the most common type and is the method discussed in §497. It is suitable especially for representing buildings in architectural drawing, or large structures in civil engineering, such as dams or bridges.

The perspective of a small building is shown in Fig. 910. It is common practice (1) to assume a vertical edge of an object in PP so that direct measurements may be made on it, and (2) to place the object so that its faces make unequal angles with

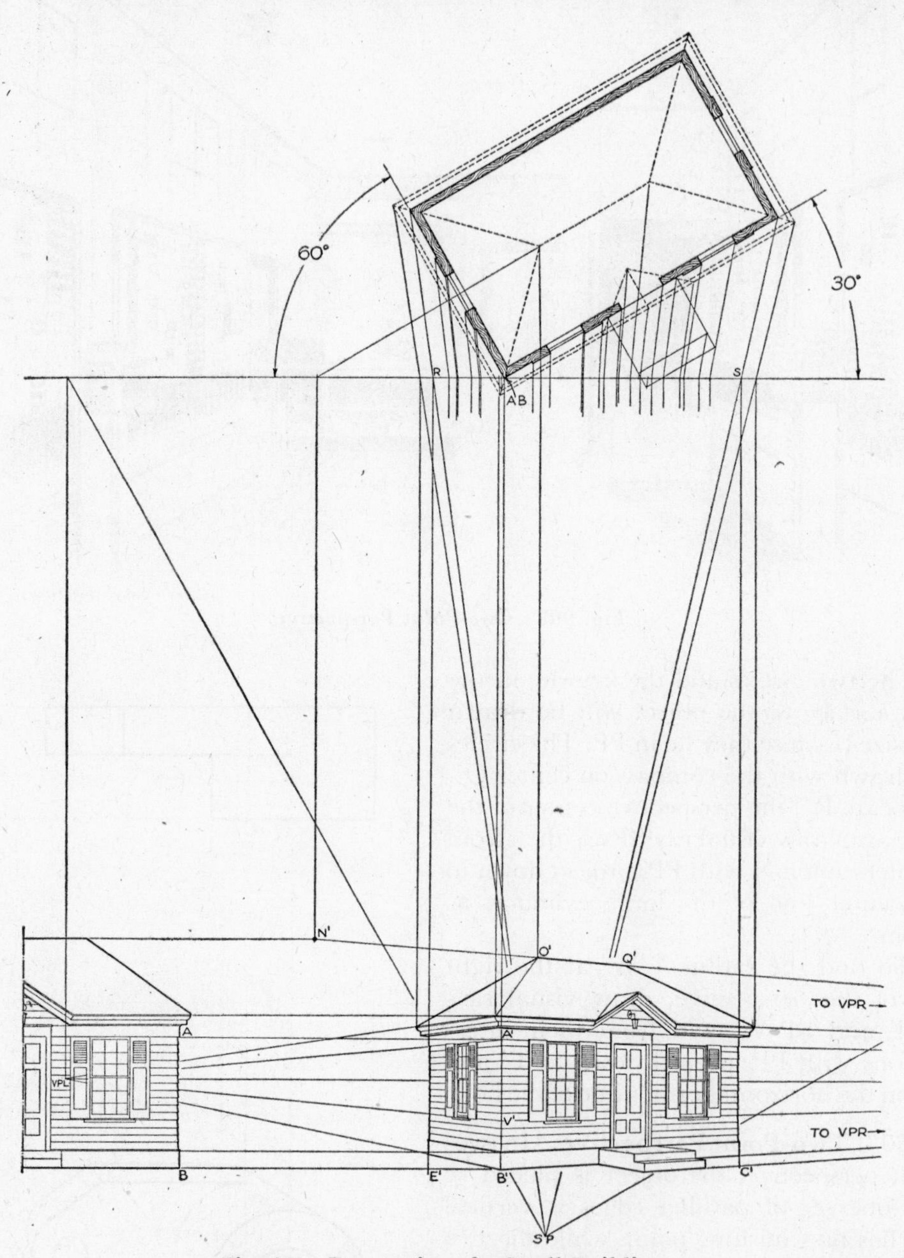

Fig. 910 Perspective of a Small Building.

PP; for example, one angle may be 30° and the other 60°. In practical work, complete multiview drawings are usually available, and the plan and elevation may be fastened in position, used in the construction of the perspective, and later removed.

Since the front corner AB lies in PP, its perspective A′B′ may be drawn full size

by projecting downward from the plan and across from the elevation. The lengths of the receding lines from this corner are cut off by vertical lines SC′ and RE′ drawn from the intersections S and R, respectively, of the visual rays to these points of the

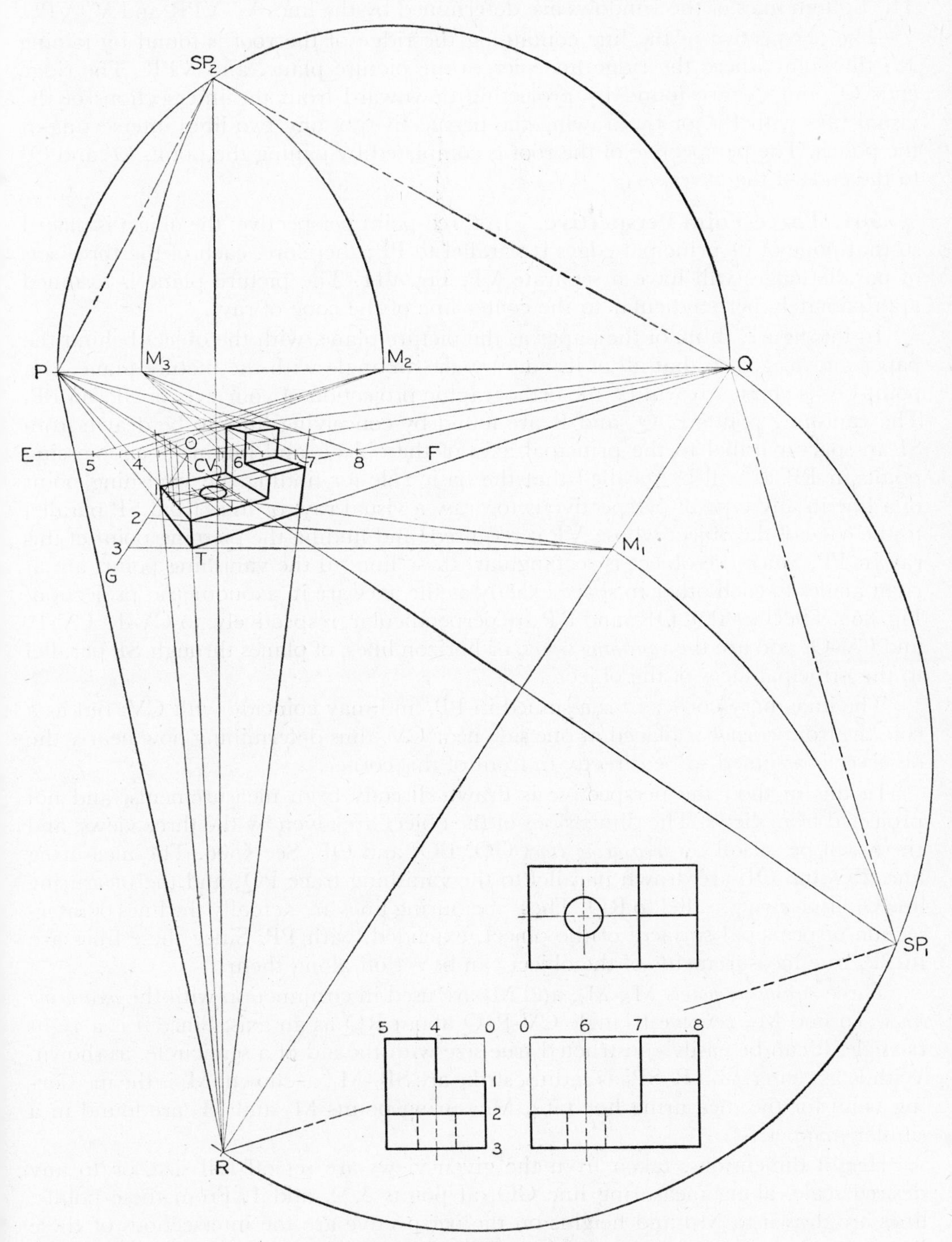

Fig. 911 Three-Point Perspective.

object. The perspectives of the tops of the windows and the door are determined by the lines A′-VPR and A′-VPL, and their widths and lateral spacings are determined by projecting downward from the intersections with PP of the respective visual rays. The bottom lines of the windows are determined by the lines V′-VPR and V′-VPL.

The perspective of the line containing the ridge of the roof is found by joining N′, the point where the ridge line pierces the picture plane, and VPR. The ridge ends O′ and Q′ are found by projecting downward from the intersections of the visual rays with PP, or by drawing the perspectives of any two lines intersecting at the points. The perspective of the roof is completed by joining the points O′ and Q′ to the ends of the eaves.

504. Three-Point Perspective. In three-point perspective, the object is placed so that none of its principal edges is parallel to PP; therefore, each of the three sets of parallel edges will have a separate VP, Fig. 911. The picture plane is assumed approximately perpendicular to the center line of the cone of rays.

In this figure, think of the paper as the picture plane, with the object behind the paper and placed so that all of its edges make an angle with the picture plane. If a point CV is chosen, it will be the orthographic projection of your eye, or SP, on PP. The vanishing points P, Q, and R are found by conceiving lines to be drawn from SP in space parallel to the principal axes of the object, and finding their piercing points in PP. It will be recalled that the basic rule for finding the vanishing point of a line in any type of perspective is to draw a visual ray, or line, from SP parallel to the edge of the object whose VP is required, and finding the piercing point of this ray in PP. Since the object is rectangular, these lines to the vanishing points are at right angles to each other in space exactly as the axes are in axonometric projection, Fig. 865. The lines PQ, QR, and RP are perpendicular, respectively, to CV-R, CV-P, and CV-Q, and are the *vanishing traces*, or horizon lines, of planes through SP parallel to the principal faces of the object.

The imaginary corner O is assumed in PP, and may coincide with CV; but as a rule the front corner is placed at one side near CV, thus determining how nearly the observer is assumed to be directly in front of this corner.

In this method the perspective is drawn directly from measurements, and not projected from views. The dimensions of the object are given by the three views, and these will be set off on *measuring lines* GO, EO, and OF. See §506. The measuring lines EO and OF are drawn parallel to the vanishing trace PQ, and the measuring line GO is drawn parallel to RQ. These measuring lines are actually the lines of intersection of principal surfaces of the object, extended, with PP. Since these lines are in PP, true measurements of the object can be set off along them.

Three *measuring points* M_1, M_2, and M_3 are used in conjunction with the *measuring lines*. To find M_1, revolve triangle CV-R-Q about RQ as an axis. Since it is a right triangle, it can be easily constructed true size with the aid of a semicircle, as shown. With R as center, and R-SP_1 as radius, strike arc SP_1-M_1, as shown. M_1 is the measuring point for the measuring line GO. Measuring points M_2 and M_3 are found in a similar manner.

Height dimensions, taken from the given views are set off full size, or to any desired scale, along measuring line GO, at points 3, 2, and 1. From these points, lines are drawn to M_1, and heights on the perspective are the intersections of these lines with the perspective front corner OT of the object. Similarly, the true depth of the object is set off on measuring line EO from O to 5, and the true width is set off on measuring line OF from O to 8. Intermediate points can be constructed in a similar manner.

505. The Perspective Linead and Template. *1. Perspective Linead.* The perspective linead, Fig. 912 (a), consists of three straight-edged blades which can be clamped to each other at any desired angles. This instrument is convenient in drawing lines toward a vanishing point outside the limits of the drawing.

Before starting such a drawing, a small-scale diagram should be made, as indicated at (b), in which the relative positions of the object, PP, and SP are assumed, and the distances of the vanishing points from CV determined. Draw any line LL

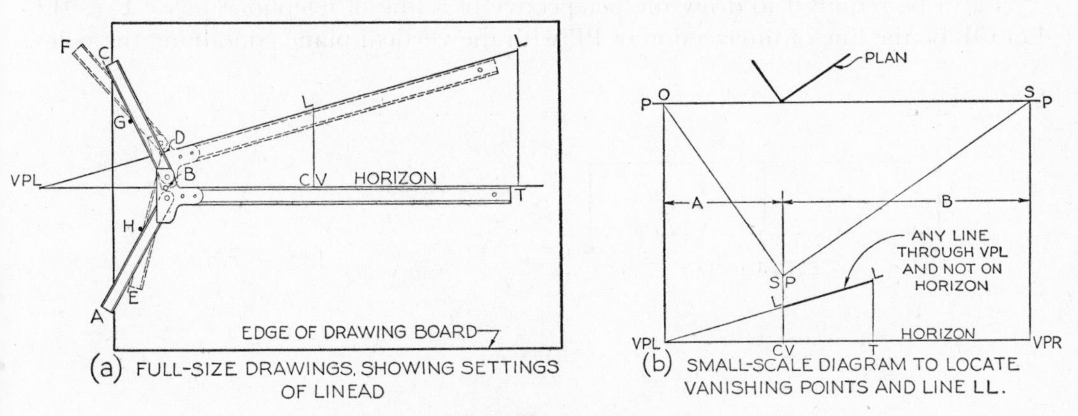

Fig. 912 Perspective Linead.

through a vanishing point as shown; then on the full-size drawing, assume CV and locate LL, as shown at (a).

To set the linead, clamp the blades in any convenient position; set the edge of the long blade along the horizon, and draw the lines BA and BC along the short blades. Then set the edge of the long blade along the line LL, and draw the lines DE and DF to intersect the lines first drawn at points G and H. Set pins at these points. If the linead is moved so that the short blades touch the pins, all lines drawn

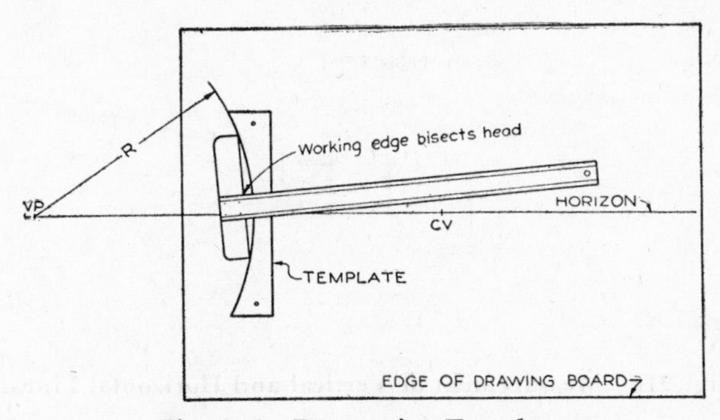

Fig. 913 Perspective Template.

along the edge of the long blade will pass through VPL. This method is based on the principle that *an angle inscribed in a circle is measured by half the arc it subtends.*

2. Templates. Fig. 913. A template of thin wood or heavy cardboard, cut in the form of a circular arc, may be used instead of a perspective linead. If the template is attached to the drawing board so that the inaccessible VP is at the center of the circular arc, and the T-square is moved so that the head remains in contact with the template, lines drawn along the edge of the blade will, if extended, pass through the inaccessible VP.

If the edge of the blade does not pass through the center of the head, the lines drawn will be tangent to a circle whose center is at VP and whose radius is equal to the distance from the center of the head to the edge of the blade.

506. Measurements in Perspective. As explained in §494, all lines in PP are shown in their true lengths, and all lines behind PP are foreshortened.

Let it be required to draw the perspective of a line of telephone poles, Fig. 914. Let OB be the line of intersection of PP with the vertical plane containing the poles.

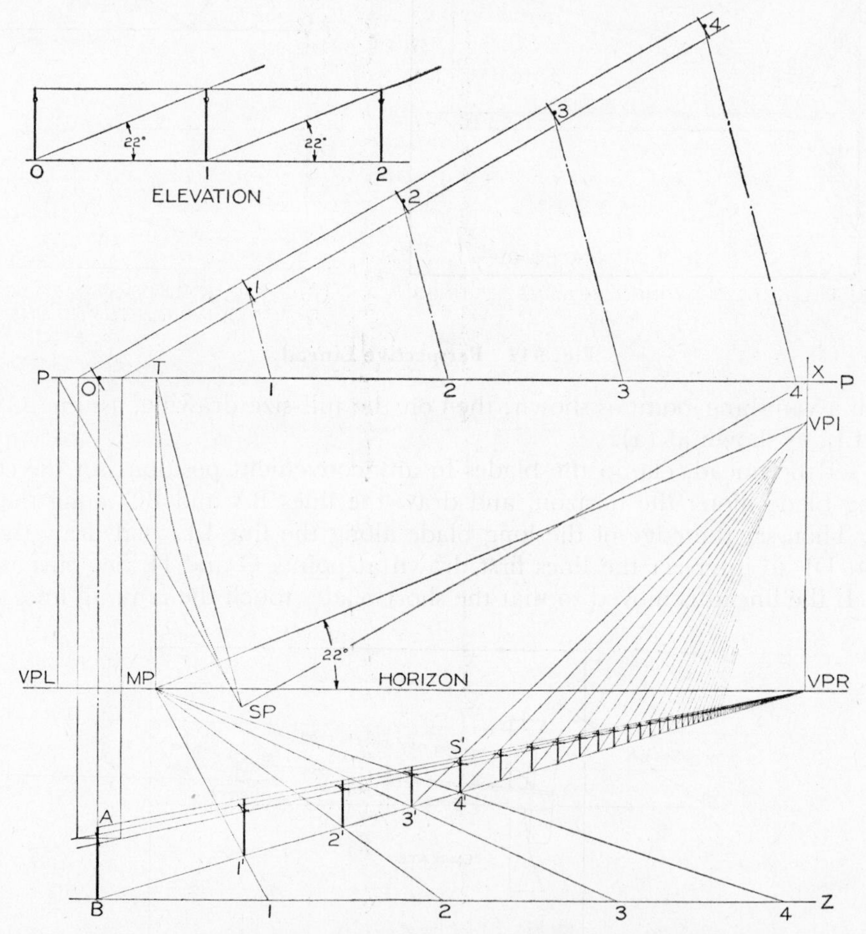

Fig. 914 Measurement of Vertical and Horizontal Lines.

In this line, the height AB of a pole is set off directly to the scale desired, and the heights of the perspectives of all poles are determined by drawing lines from A and B to VPR.

To locate the bottoms of the poles along the line B-VPR, set off along PP the distances 0-1, 1-2, 2-3, etc., equal to the distance from pole to pole; draw the lines 1-1, 2-2, 3-3, etc., forming a series of isosceles triangles 0-1-1, 0-2-2, 0-3-3, etc. The lines 1-1, 2-2, 3-3, etc., are parallel to each other, and therefore have a common vanishing point MP, which is found in the usual manner by drawing from SP a line SP-T parallel to the lines 1-1, 2-2, 3-3, etc., and finding its piercing point MP (*measuring point*) in PP.

Since the line SP-X is parallel to the line of poles 1-2-3, etc., the triangle SP-X-T is an isosceles triangle, and T is the top view of MP. The point T may be determined by setting off the distance X-T equal to SP-X or simply by drawing the arc SP-T with center at X and radius SP-X.

Having the measuring point MP, find the piercing points in PP of the lines 1-1, 2-2, 3-3, etc., and draw their perspectives as shown. Since these lines are horizontal lines, their piercing points fall in a horizontal line BZ in PP, at the bottom of the drawing. Along BZ the true distances between the poles are set off; hence BZ is called a *measuring line*. The intersections 1′, 2′, 3′, etc., of the perspectives of the lines 1-1, 2-2, 3-3, etc., with the line B-VPR determine the spacing of the poles. It will be seen that only a few measurements may be made along the measuring line BZ within the limits of the drawing. For additional measurements, the *diagonal method* of spacing may be employed, as shown. Since all diagonals from the bottom of each pole to the top of the succeeding pole are parallel, they have a common vanishing point VPI, which may be found as explained in §507. Evidently, the diagonal method may be used exclusively in the solution of this problem.

The method of direct measurements may be applied also to lines inclined to PP and to the ground plane, as illustrated in Fig. 915 for the line XE, which pierces PP at X. If the end of the house is conceived to be revolved about a vertical axis

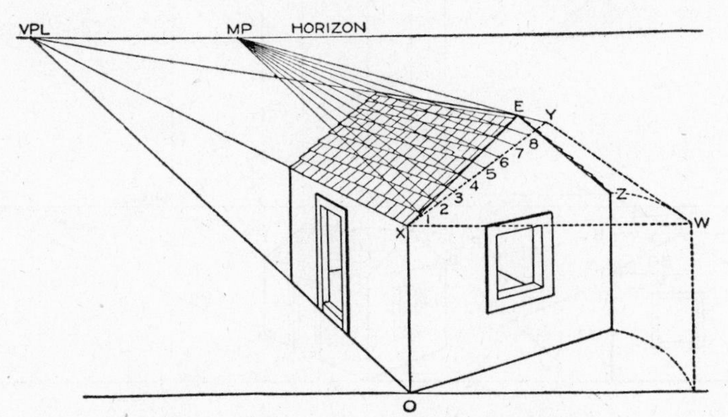

Fig. 915 Measurement of Inclined Lines.

XO into PP, the line XE would be shown in its true length and inclination at XY. This line XY may be used as the measuring line for XE; it remains only to find the corresponding measuring point MP. The line YE is the horizontal base of an isosceles triangle having its vertex at X, and a line drawn parallel to it through SP will determine MP, as described for Fig. 914.

507. Vanishing Points of Inclined Lines. The vanishing point of an inclined line is determined, as for all other lines, by finding the piercing point in PP of a line drawn from SP parallel to the given line.

In Fig. 916 is shown the perspective of a small building. The vanishing point of the inclined roof line C′E′ can be determined as follows: If a plane is conceived to be passed through the station point and parallel to the end of the house (plan view),

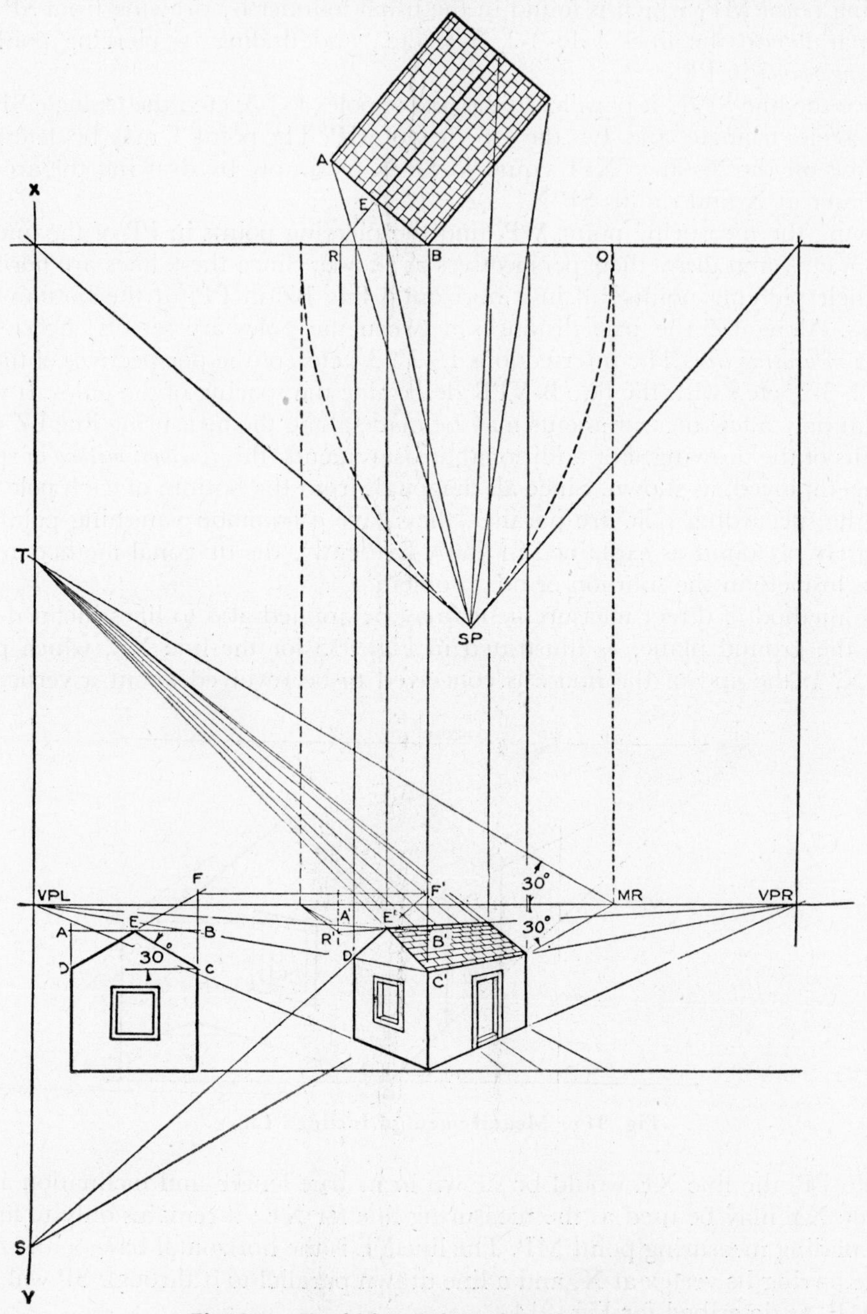

Fig. 916 Vanishing Points of Inclined Lines.

it would intersect PP in the line XY, through VPL, and perpendicular to the horizon. Since the line drawn from SP parallel to C′E′ (in space) is in the plane SP-X-Y, it will pierce PP at some point T in XY. To find the point T, conceive the plane SP-X-Y revolved about the line XY as an axis into PP. The point SP will then fall on the horizon at a point shown by O in the top view, and by MR in the front view. From the point MR draw the revolved position of the line SP-T (now MR-T) making an angle of 30° with the horizon and thus determining the point T, which is the vanishing point of the line C′E′ and of all lines parallel to that line. The vanishing point S of the line D′E′ is evidently in the line XY because D′E′ is in the same vertical plane as the line C′E′. The vanishing point S is as far below the horizon as T is above the horizon, because the line E′D′ slopes downward at the same angle at which the line C′E′ slopes upward.

The perspectives of inclined lines can generally be found without finding the vanishing points, by finding the perspectives of the end points and joining them. The *perspective of any point may be determined by finding the perspectives of any two lines intersecting at the point.* Obviously, it would be best to use horizontal lines, parallel respectively to systems of lines whose vanishing points are already available. For example, in Fig. 916, to find the perspective of the inclined line EC, the point E′ is the intersection of the horizontal lines R′-VPR and B′-VPL. The point C′ is already established, since it is in PP; but if it were not in PP, it could be easily found in the same manner. The perspective of the inclined line EC is, therefore, the line joining the perspectives of the end points E′ and C′.

508. Curves and Circles in Perspective. If a circle is parallel to PP, its perspective is a circle. If the circle is inclined to PP, its perspective may be any one of the conic sections, in which the base of the cone is the given circle, the vertex is SP, and the cutting plane is PP. But since the center line of the cone of rays should be

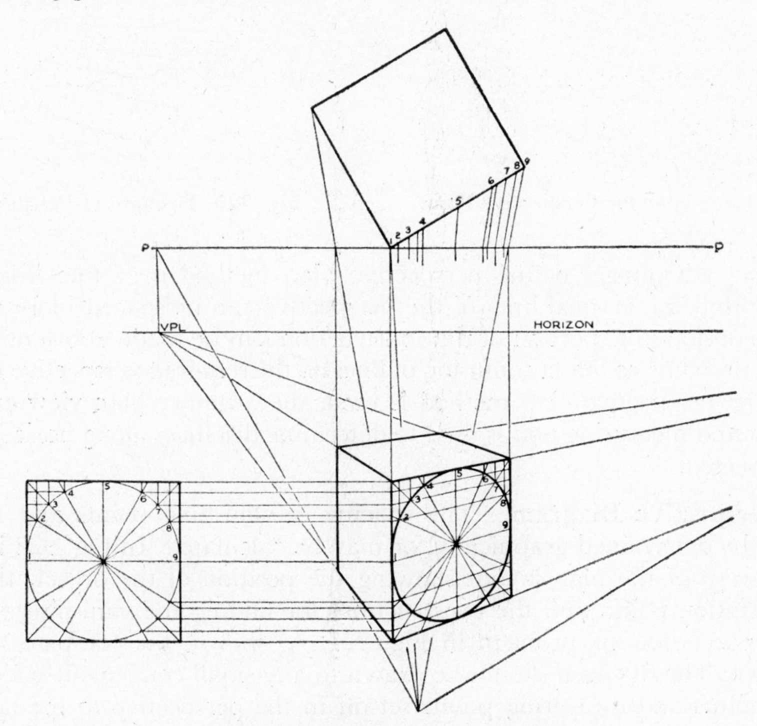

Fig. 917 Circles in Perspective.

approximately perpendicular to the picture plane, the perspective will generally be an ellipse. The ellipse may be constructed by means of lines intersecting the circle, as shown in Fig. 917. The radial lines in the elevation view at the left can be easily drawn with the 45° and 30° × 60° triangles. A convenient method for determining the perspective of any plane curve is shown in Fig. 918.

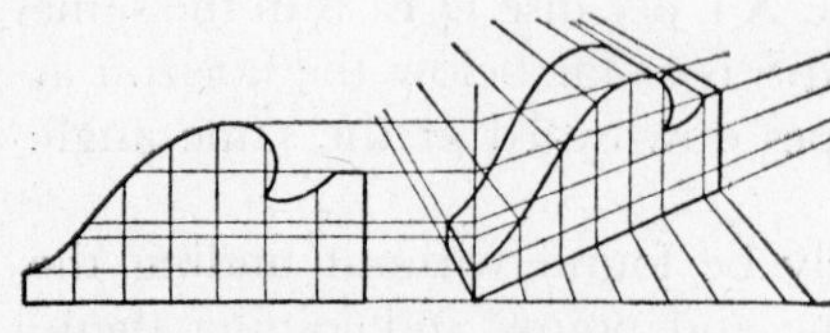

Fig. 918 Curves in Perspective.

509. The Perspective Plan Method. A perspective may be drawn by drawing first the perspective of the plan of the object, as shown in Fig. 919 (a), then the vertical lines, (b), and finally, the connecting lines, (c). However, in drawing intricate structures, the superimposition of the perspective upon the perspective plan causes a confusion of lines. For this reason, the perspective of the plan from which the location of vertical lines is determined is drawn either above or below its normal location. A suggestion of the range of possible positions of the perspective plan is given in Fig. 920. The use of the perspective plan below the perspective is illustrated in Fig. 921.

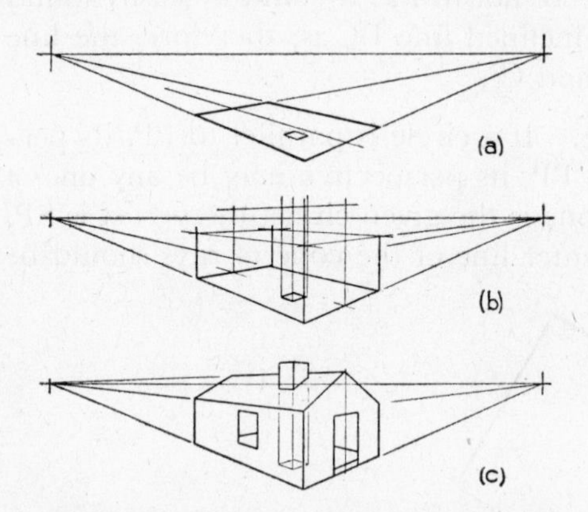

Fig. 919 Building upon the Perspective Plan.

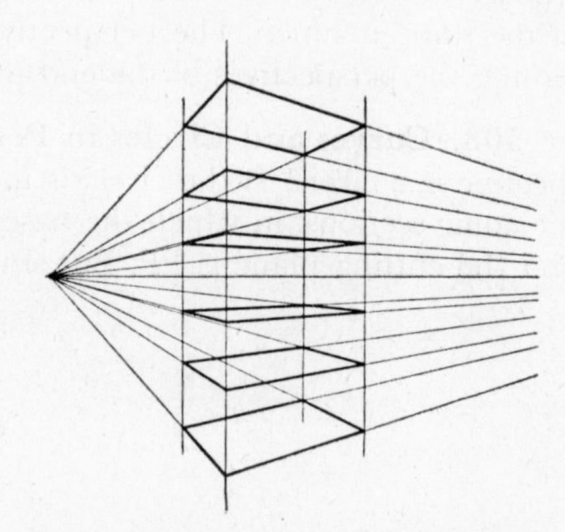

Fig. 920 Positions of Perspective Plan.

The chief advantages of the perspective plan method over the ordinary plan method are that the vertical lines of the perspective can be spaced more accurately and that a considerable portion of the construction can be made above or below the perspective drawing, so that a confusion of lines on the required perspective is avoided.

When the perspective plan method is used, the ordinary plan view can be dispensed with and measuring points used to determine distances along horizontal edges in the perspective.

510. Perspective Diagram. The spacing of vanishing points and measuring points may be determined graphically, or may be calculated. In Fig. 922 is shown a simple diagram of the plan layout showing the position of the object, the picture plane, the station point, and the constructions for finding the vanishing points and measuring points for the problem in Fig. 921. As shown, the complete plan need not be drawn. The diagram should be drawn to any small convenient scale, and the vanishing points and measuring points set off in the perspective to the larger scale desired.

Since in practice, structures are usually considered in one of a limited number of simple positions with reference to the picture plane, such as 30° × 60°, 45° × 45°, 20° × 70°, etc., a table of measurements for locating vanishing points and measuring points may be easily prepared, and may be used to avoid the necessity of a special construction for each drawing.

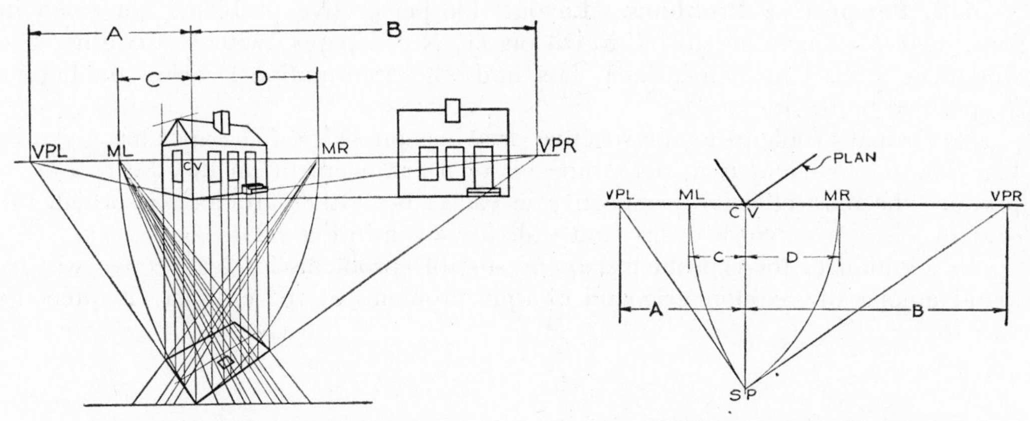

Fig. 921 Perspective Plan Method.

Fig. 922 Perspective Diagram.

511. Perspective Sketching. Methods of sketching in one-point and two-point perspective on plain paper are shown in Figs. 279 and 280. Obviously the best results in perspective sketching will be obtained by the person with art training or aptitude, or both; but the average draftsman can learn to make very creditable sketches if he applies himself to it.

An excellent aid in perspective sketching is the *perspective grid*, Fig. 923, which is a printed sheet of grid lines arranged for the most commonly used positions.

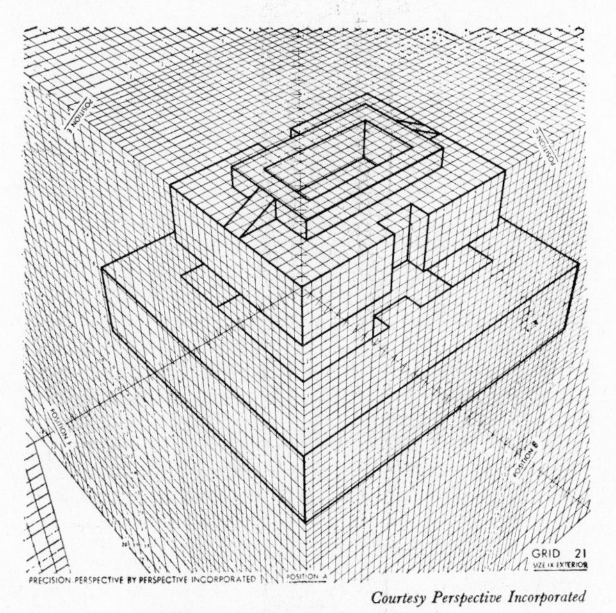

Courtesy Perspective Incorporated

Fig. 923 Perspective Sketches.

Usually such a grid is furnished as a master sheet along with a pad of tracing paper the same size as the grid sheet. The grid is simply placed under a sheet of tracing paper, and the sketch is easily made by following the grid lines. Although the grid is of great assistance to the inexperienced sketcher, its usefulness actually increases in proportion to the ability of the draftsman.

512. Perspective Problems. Layouts for perspective problems are given in Figs. 924-927. These are to be drawn on 11″ × 17″ paper, with $\frac{1}{2}$″ margins. The student is to letter his name, date, class, and other information $\frac{1}{8}$″ below the border as specified by the instructor.

Additional problems for perspective drawings on 11″ × 17″ paper are given in Fig. 928. The student is to determine his own arrangement on the sheet so as to produce the most effective perspective in each case. Other problems in which the student selects his own sheet size and scale are given in Fig. 929.

In addition to these problems, many suitable problems for perspective will be found among the axonometric and oblique problems at the ends of Chapters 15 and 16.

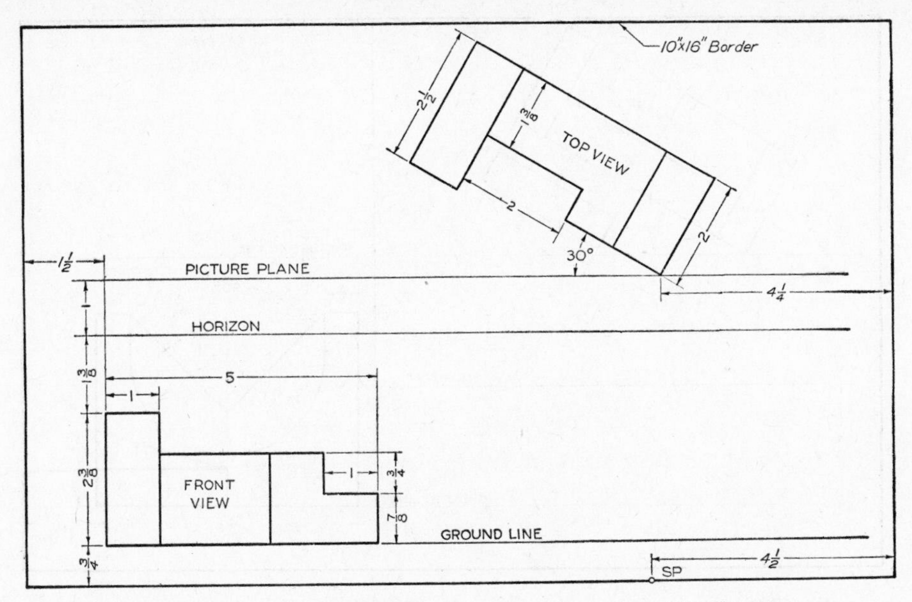

Fig. 924 Draw views and perspective. Omit dimensions. Use Size B sheet.

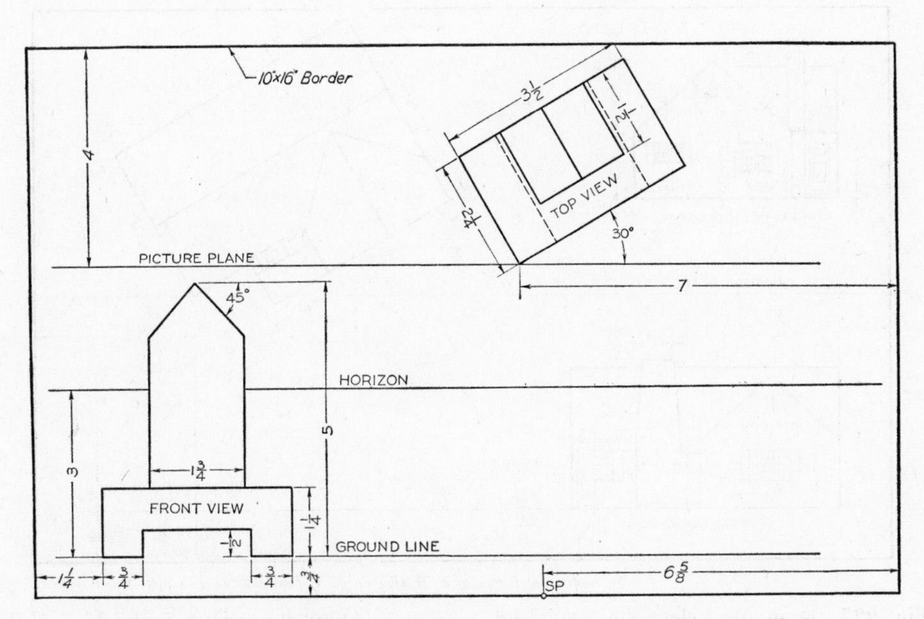

Fig. 925 Draw views and perspective. Omit dimensions. Use Size B sheet.

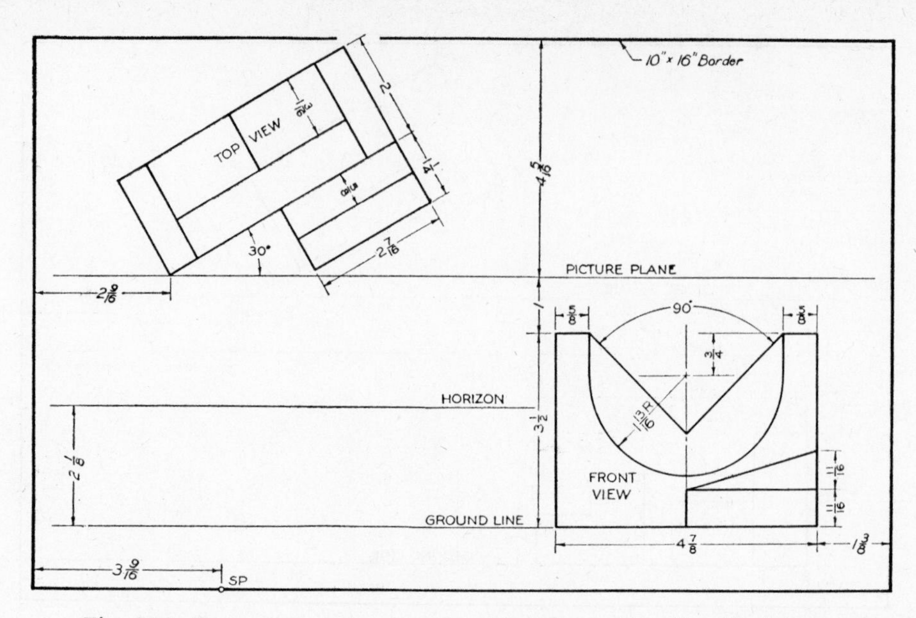

Fig. 926 Draw views and perspective. Omit dimensions. Use Size B sheet.

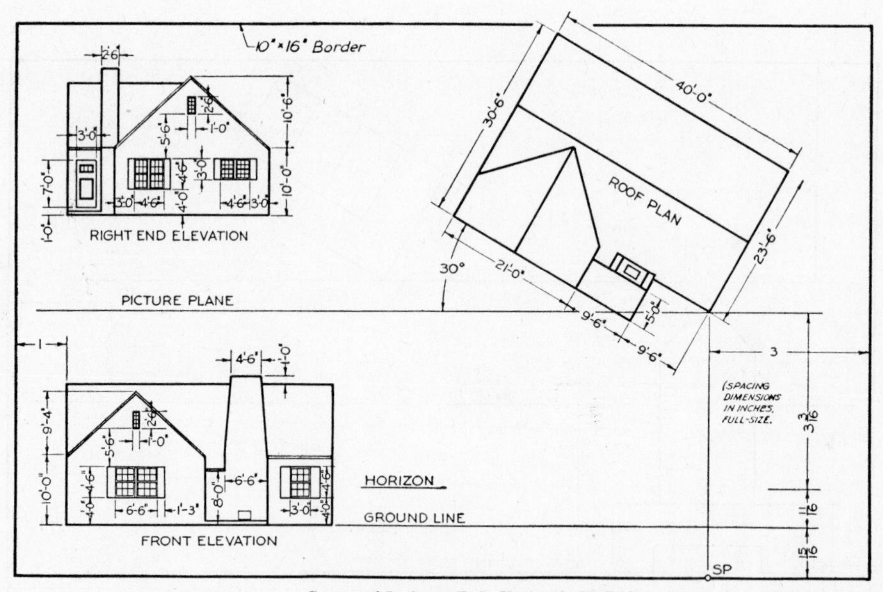

Courtesy of Professors F. R. Hughes, J. N. Eckle, and D. F. Grant, Yale University.

Fig. 927 Draw front elevation, plan, and perspective. Omit dimensions. Scale: $\frac{1}{8}'' = 1'\text{-}0$. Use Size B sheet.

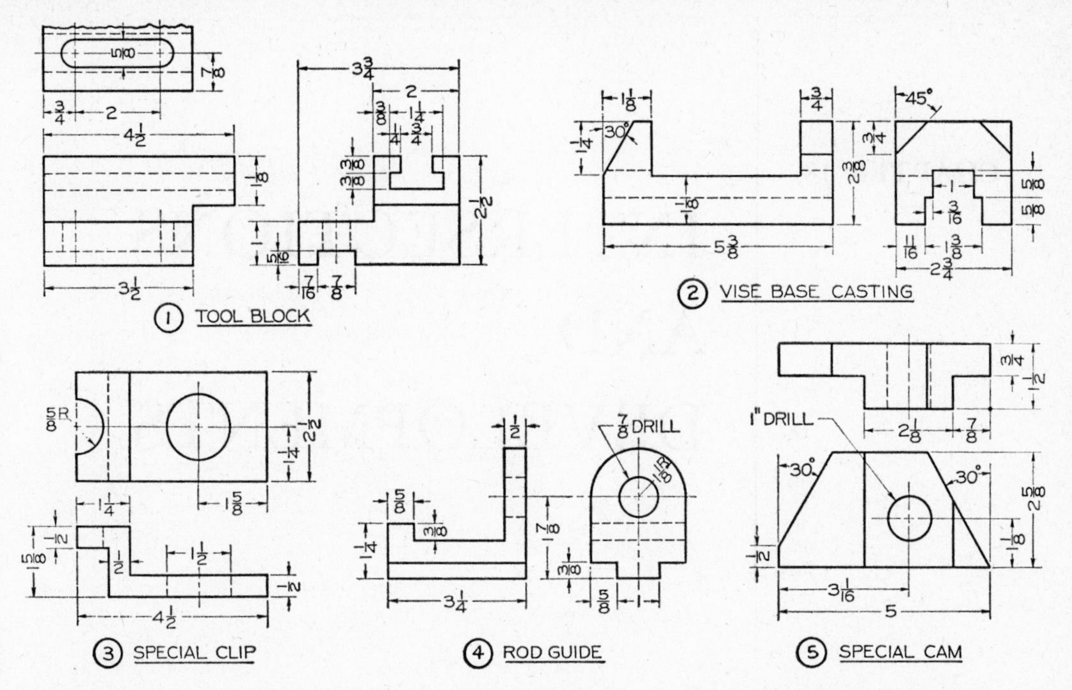

Fig. 928 Draw side or front elevation, plan, and perspective of assigned problem. Omit dimensions. Use Size B sheet.

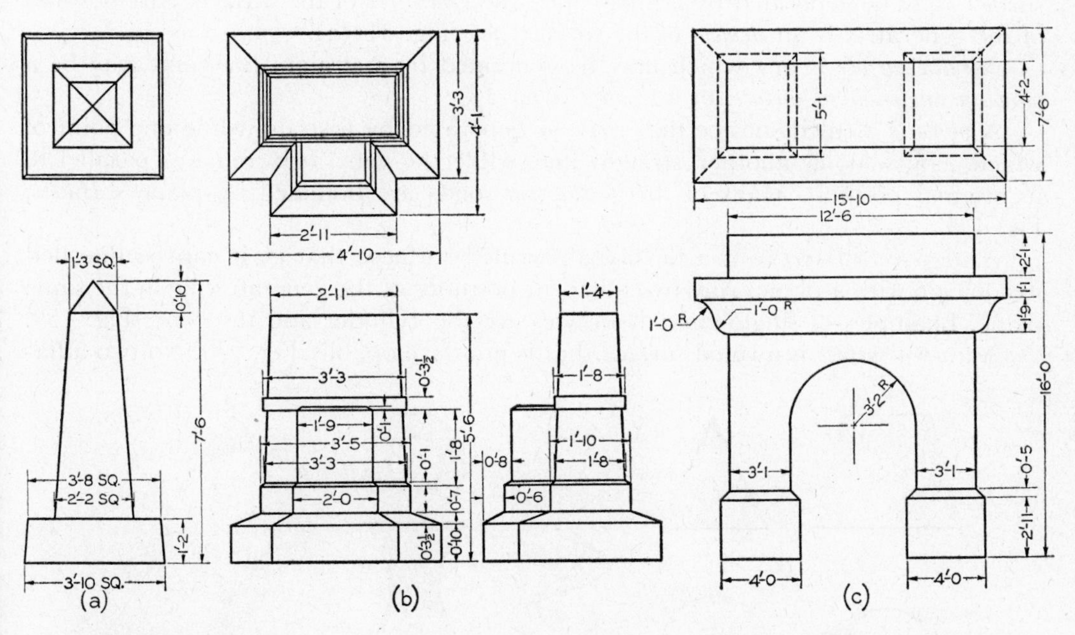

Fig. 929 Draw side or front elevation, plan, and perspective of assigned problem. Omit dimensions. Select sheet size and scale.

CHAPTER 18

INTERSECTIONS AND DEVELOPMENTS

513. Surfaces. A *surface* is a geometric magnitude having two dimensions. A surface may be generated by a line, called the *generatrix* of the surface. Any position of the generatrix is an *element* of the surface. See Fig. 314(a).

A *ruled surface* is one which may be generated by a straight line, and may be a *plane*, a *single-curved surface*, or a *warped surface*.

A *plane* is a ruled surface that may be generated by a straight line one point of which moves along another straight line, while the generatrix remains parallel to its original position. Many of the geometric solids are bounded by plane surfaces, Fig. 159.

A *single-curved surface* is a developable ruled surface; that is, it can be unrolled to coincide with a plane. Any two adjacent positions of the generatrix lie in the same plane. Examples of single-curved surfaces are the cylinder and the cone, Fig. 159.

A *warped surface* is a ruled surface that is not developable, Fig. 930. No two adja-

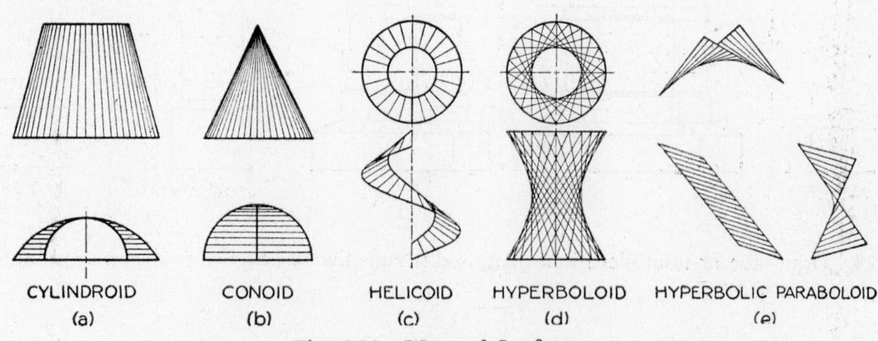

Fig. 930 Warped Surfaces.

cent positions of the generatrix lie in the same plane. Many exterior surfaces on an airplane or automobile are warped surfaces.

A double-curved surface may be generated only by a curved line, and has no straight-line elements. Such a surface, generated by revolving a curved line about a straight line in the plane of the curve, is called a *double-curved surface of revolution.* Common examples are the *sphere*, *torus*, and *ellipsoid*, Fig. 159, and the *hyperboloid*, Fig. 930 (d).

A *developable surface* is one which may be unfolded or unrolled so as to coincide with a plane, §516. Surfaces composed of single-curved surfaces, or of planes, or of combinations of these types, are developable. Warped surfaces and double-curved surfaces are not developable. They may be developed approximately by dividing them into sections and substituting for each section a developable surface; that is, a plane or a single-curved surface. If the material used is sufficiently pliable, the flat sheets may be stretched, pressed, stamped, spun, or otherwise forced to assume the desired shape. Non-developable surfaces are often produced by a combination of developable surfaces which are then formed slightly to produce the required shape.

514. Solids. *Solids* bounded by plane surfaces are *polyhedra*, the most common of which are the pyramid and prism, Fig. 159. Convex solids whose faces are all equal regular polygons are *regular polyhedra.* The simple regular polyhedra are the *tetrahedron*, *cube*, *octahedron*, *dodecahedron*, and *icosahedron*, known as the five *Platonic solids.*

Plane surfaces that bound polyhedra are *faces* of the solids. Lines of intersection of faces are *edges* of the solids.

A solid generated by revolving a plane figure about an axis in the plane of the figure is a *solid of revolution.*

Solids bounded by warped surfaces have no group name. The most common example of such solids is the screw thread.

INTERSECTIONS OF PLANES AND SOLIDS

515. Principles of Intersections. The principles involved in intersections of planes and solids have their practical application in the cutting of openings in roof surfaces for flues and stacks, in wall surfaces for pipes, chutes, etc., and in the building of sheet-metal structures like tanks, boilers, etc.

In such cases, the problem is generally one of determining the true size and shape of the intersection of a plane and one of the more common geometric solids. The intersection of a plane and a solid is the locus of the points of intersection of the elements of the solid with the plane. For solids bounded by plane surfaces, it is necessary only to find the points of intersection of the edges of the solid with the plane, and to join these points, in consecutive order, with straight lines. For solids bounded by curved surfaces, it is necessary to find the points of intersection of several elements of the solid with the plane and to trace a smooth curve through these points. The curve of intersection of a plane and a circular cone is a *conic section.* The various conic sections are illustrated and defined in Fig. 198.

516. Developments. The *development* of a surface is that surface laid out on a plane, Fig. 931. Practical applications of developments occur in sheet-metal work, stone cutting, and pattern making.

Single-curved surfaces and the surfaces of polyhedra can be developed. Warped surfaces and double-curved surfaces can be developed only approximately. See §513.

In sheet-metal layout, extra material must be provided for laps or seams. If the material is heavy, the thickness may be a factor, and the crowding of metal in bends

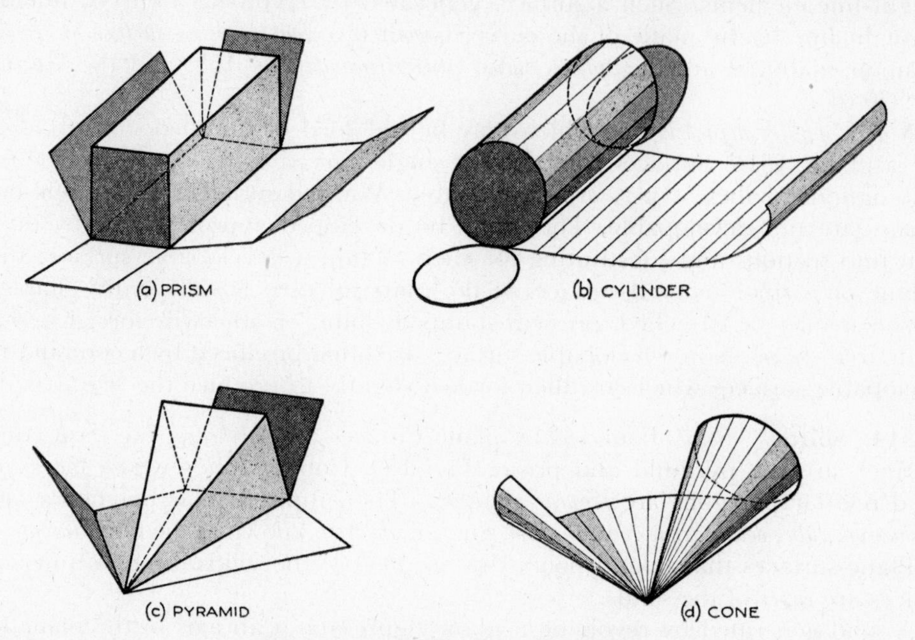

Fig. 931 Development of Surfaces.

must be considered. See §712. The draftsman must also take stock sizes into account, and should make his layouts so as to economize in the use of material and of labor. In preparing developments, it is best to put the seam at the shortest edge and to

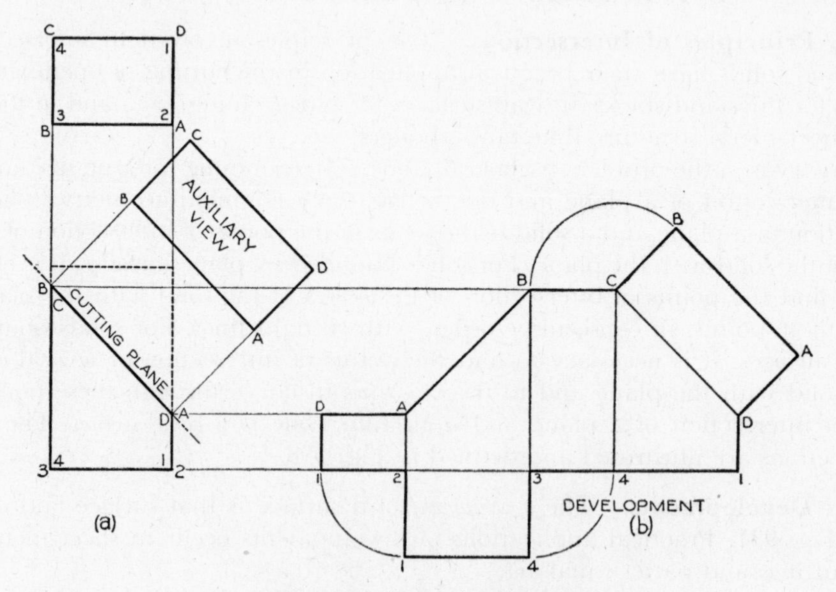

Fig. 932 Plane and Prism.

attach the bases at edges where they match, so as to economize in soldering, welding, or riveting.

It is common practice to draw development layouts with the inside surfaces up. In this way, all fold lines and other markings are related directly to inside measurements, which are the important dimensions in all ducts, pipes, tanks, and other vessels; and in this position they are also convenient for use in the fabricating shop.

517. To Find the Intersection of a Plane and a Prism and the Development of the Prism. Fig. 932.

(a) *Intersection.* The true size and shape of the intersection is shown in the auxiliary view. See Chapter 8. The length AB is the same as AB in the front view, and the width AD is the same as AD in the top view.

(b) *Development.* On the straight line 1-1, called the *stretch-out line*, set off the widths of the faces 1-2, 2-3, etc., taken from the top view. At the division points, erect perpendiculars to 1-1, and set off on each the length of the respective edge, taken from the front view. The lengths can be projected across from the front view, as shown. Join the points thus found by straight lines to complete the development of the lateral surface. Attach to this development the lower base and the upper base, or auxiliary view, to obtain the development of the entire surface of the frustum of the prism.

518. To Find the Intersection of a Plane and a Cylinder and the Development of the Cylinder. Fig. 933.

(a) *Intersection.* The intersection is an ellipse whose points are the piercing points

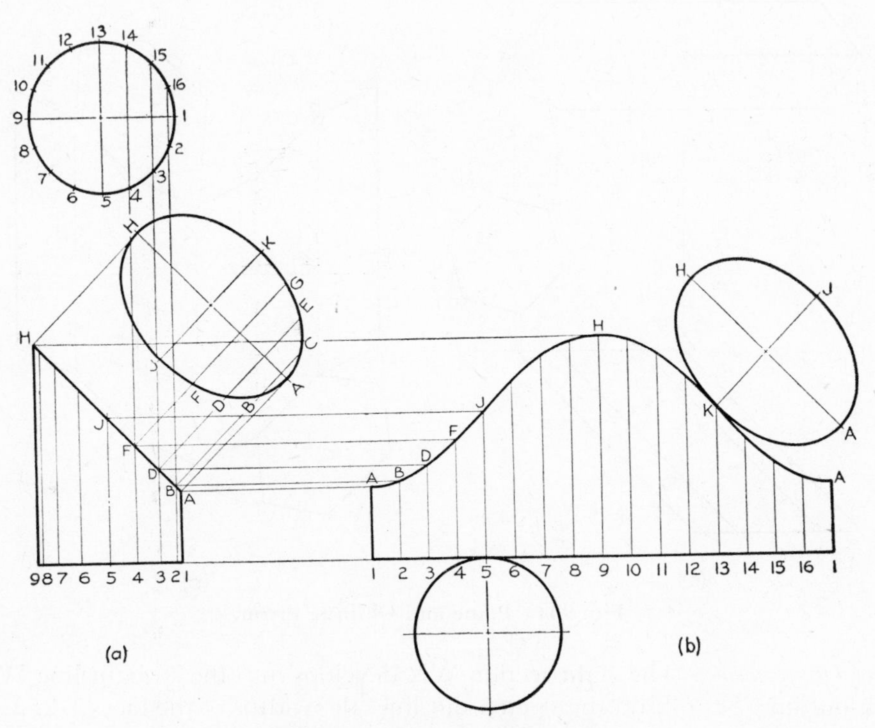

Fig. 933 Plane and Cylinder.

in the secant plane of the elements of the cylinder. In spacing the elements, it is best, though not necessary, to divide the circumference of the base into *equal* parts, and to draw an element at each division point. In the auxiliary view, the widths BC, DE, etc., are taken from the top view at 2-16, 3-15, etc., respectively, and the curve is traced through the points thus determined, with the aid of the irregular curve, §66.

The major axis AH and the minor axis JK are shown true length in the front view and the top view, respectively; therefore, the ellipse may also be constructed as explained in §§155-158 or with the aid of an ellipse template, §163.

(b) *Development.* The base of the cylinder develops into a straight line 1-1, the stretch-out line, equal to the circumference of the base, whose length may be determined by calculation (πd), by setting off with the bow dividers, or by rectifying the arcs of the base 1-2, 2-3, etc., §152. Divide the stretch-out line into the same number of equal parts as the circumference of the base, and draw an element through each division perpendicular to the line. Set off on each element its length, projected from the front view, as shown; then trace a smooth curve through the points A, B, D, etc., §66, and attach the bases.

519. To Find the Intersection of a Plane and an Oblique Prism and the Development of the Prism. Fig. 934.

(a) *Intersection.* The right section cut by the plane WX is a regular hexagon, as shown in the auxiliary view; the oblique section, cut by the horizontal plane YZ, is shown in the top view.

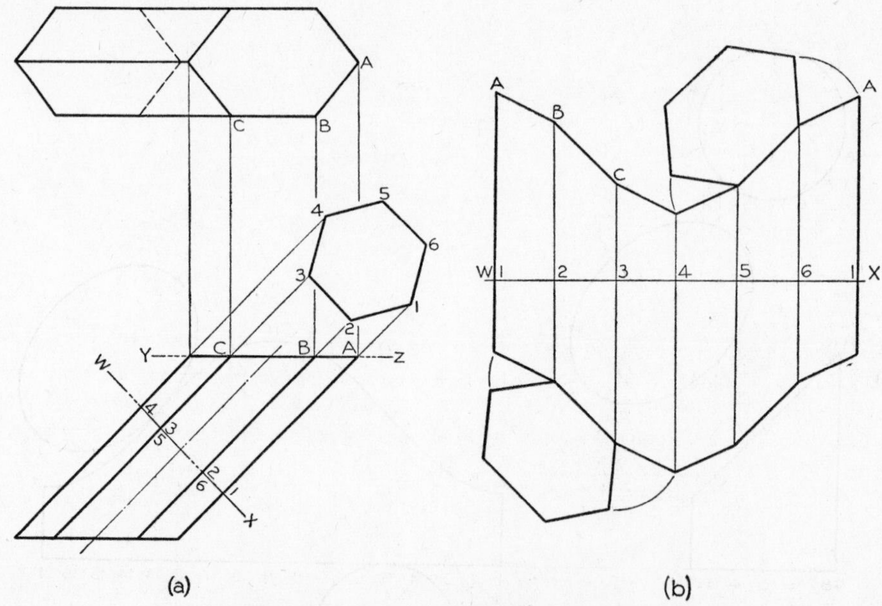

Fig. 934 Plane and Oblique Prism.

(b) *Development.* The right section WX develops into the straight line WX, the stretch-out line. Set off, on the stretch-out line, the widths of the faces 1-2, 2-3, etc., taken from the auxiliary view, and draw a line through each division perpendicular to the line. Set off, from the stretch-out line, the lengths of the respective edges measured from WX in the front view. Join the points A, B, C, etc., with straight lines, and attach the bases, which are shown in their true sizes in the top view.

520. To Find the Intersection of a Plane and an Oblique Cylinder. Fig. 935.

(a) *Intersection.* The right section cut by the plane WX is a circle, shown in the auxiliary view. The intersection of the horizontal plane YZ with the cylinder is an ellipse shown in the top view, whose points are found as explained for the auxiliary

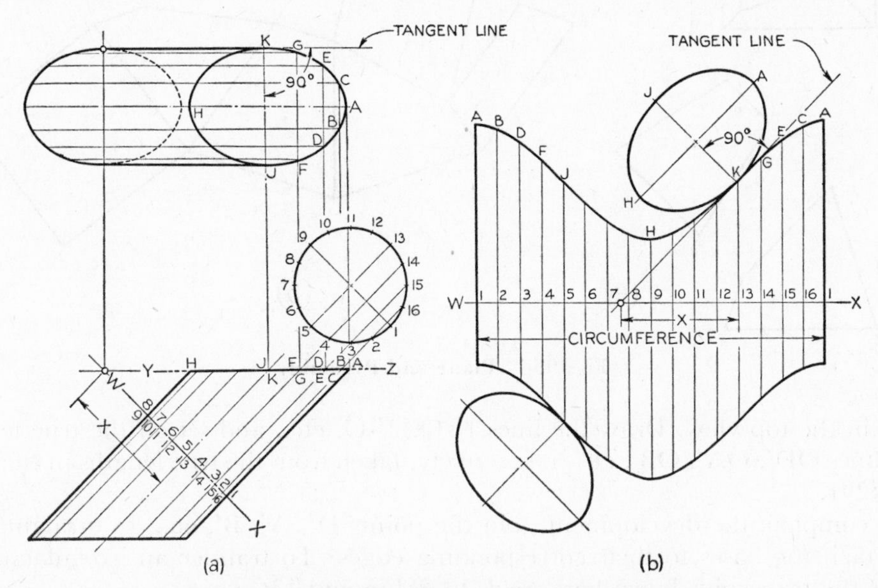

Fig. 935 Plane and Oblique Circular Cylinder.

view in §518 (a). The major axis AH is shown true length in the top view, and the minor axis JK is equal to the diameter of the cylinder; therefore, the ellipse may be constructed as explained in §§155-158, or with the aid of an ellipse template, §163.

(b) *Development.* The cylinder may be considered as a prism having an infinite number of edges; therefore, the development is found in a manner similar to that of the oblique prism shown in Fig. 934.

The circle of the right section cut by plane WX develops into a straight line 1-1, the stretch-out line, equal in length to the circumference of the circle (πd). Divide the stretch-out line into the same number of equal parts as the circumference of the circle as shown in the auxiliary view, and draw elements through these points perpendicular to the line. Set off on each element its length, taken from the front view with dividers, as shown; then trace a smooth curve through the points A, B, D, etc., §66, and attach the bases.

521. To Find the Intersection of a Plane and a Pyramid and to Develop the Resulting Truncated Pyramid. Fig. 936.

(a) *Intersection.* The intersection is a trapezoid whose vertices are the points in which the edges of the pyramid pierce the secant plane. In the auxiliary view, the altitude of the trapezoid is projected from the front view, and the widths AD and BC are transferred from the top view with dividers.

(b) *Development.* With O in the development as center and O-1′ in the front view (the true length of one of the edges) as radius, draw the arc 1′, 2′, 3′, etc. Inscribe the cords 1′-2′, 2′-3′, etc., equal respectively to the sides of the base, as

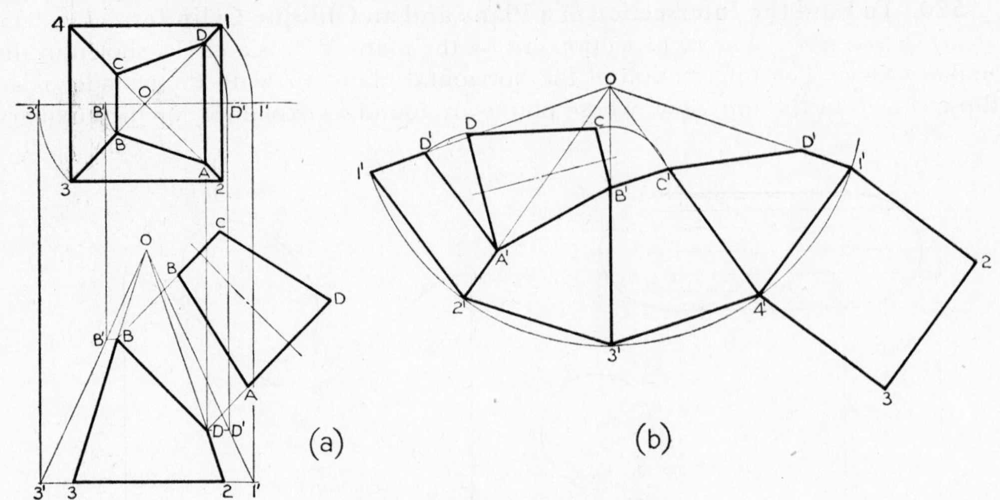

Fig. 936 Plane and Pyramid.

shown in the top view. Draw the lines 1′-O, 2′-O, etc., and set off the true lengths of the lines OD′, OA′, OB′, etc., respectively, taken from the true lengths in the front view, §291.

To complete the development, join the points D′, A′, B′, etc., by straight lines, and attach the bases to their corresponding edges. To transfer an irregular figure, such as the trapezoid shown here, refer to §§135 and 136.

522. To Find the Intersection of a Plane and a Cone and to Develop the Lateral Surface of the Cone. Fig. 937.

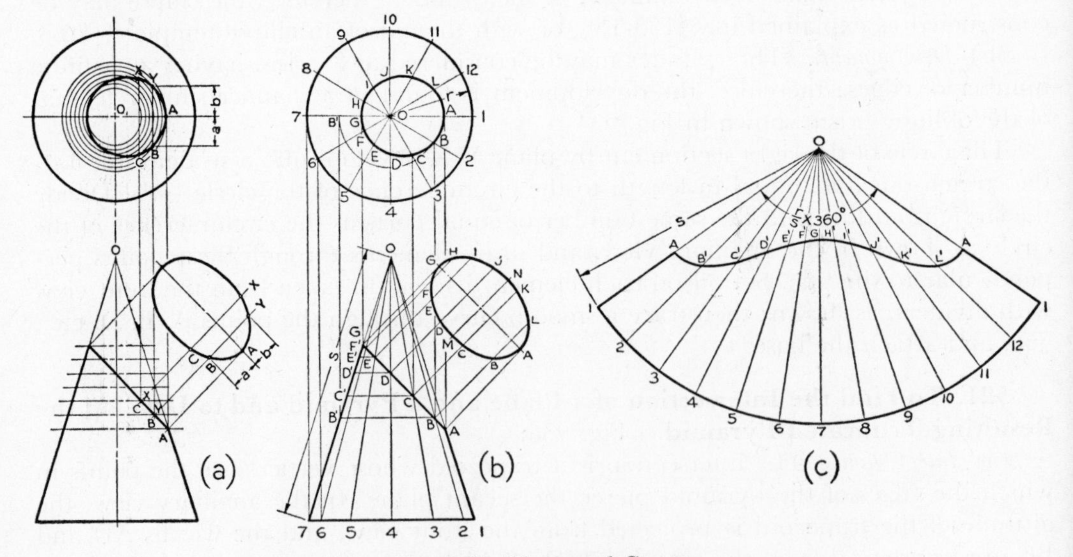

Fig. 937 Plane and Cone.

(a) *Intersection.* The intersection is an ellipse. If a series of horizontal cutting planes is passed perpendicular to the axis, as shown, each plane will cut a circle from the cone that will show in true size and shape in the top view. Points in which these circles intersect the original secant plane are points on the ellipse. Since the secant plane is shown edgewise in the front view, all of these piercing points may be found in that view and projected to the others, as shown.

(b) *Intersection.* This method is most suitable when a development also is required, since it utilizes elements that are also needed in the development. The piercing points of these elements in the secant plane are points on the intersection. Divide the base into any number of equal parts, and draw an element at each division point. These elements pierce the secant plane in points A, B, C, etc. The top views of these points are found by projecting upward from the front view, as shown. In the auxiliary view, the widths BL, CK, etc., are taken from the top view. The ellipse is then drawn with the aid of the irregular curve, §66.

The major axis of the ellipse, shown in the auxiliary view, is equal to AG in the front view. The minor axis MN bisects the major axis, and is equal to the minor axis of the ellipse in the top view. With these axes, the ellipse may also be constructed, as explained in §§155-158, or with the aid of an ellipse template, §163.

(c) *Development.* The cone may be considered as a pyramid having an infinite number of edges; hence the development is found in a manner similar to that explained for the pyramid in §521. The base of the cone develops into a circular arc,

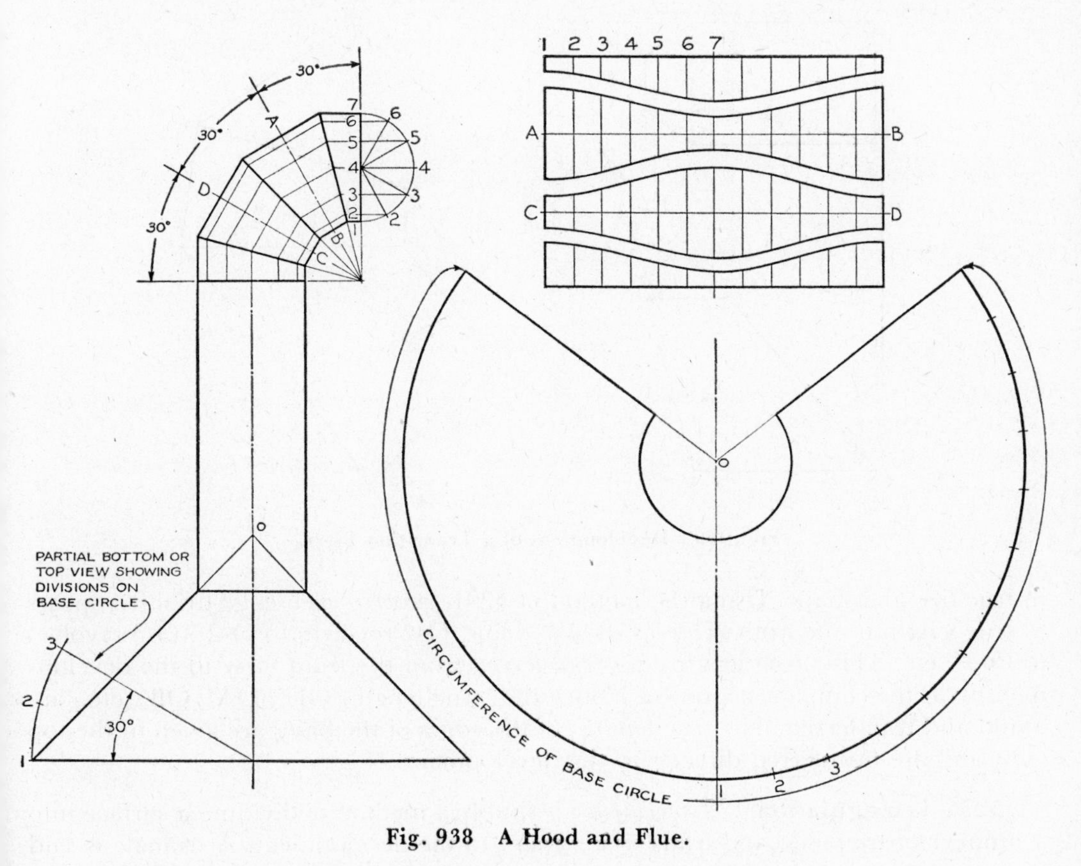

Fig. 938 A Hood and Flue.

with the slant height of the cone as its radius and the circumference of the base as its length, §152. The lengths of the elements in the development are taken from the element O-7 or O-1 in the front view, (b). Instead of our finding the true circumference of the base, the vertical angle 1-O-1, in the development, can be set off equal to $\frac{r}{s}360°$ (where r is the radius of the base, and s the slant height of the cone).

523. To Find the Development of a Hood and Flue. Fig. 938. Since the hood is a conical surface, it may be developed as described in §522. The two end sections of the elbow are cylindrical surfaces, and may be developed as described in §518. The two middle sections of the elbow are cylindrical surfaces, but since their bases are not perpendicular to the axes, they will not develop into straight lines. They will be developed in a manner similar to that for an oblique cylinder, §520 (b). If the auxiliary planes AB and DC are passed perpendicular to the axes, they will cut right sections from the cylinders, which will develop into the straight lines AB and CD in the developments.

If the developments are arranged as shown, the elbow can be constructed from a rectangular sheet of metal without wasting material. The patterns are shown separated after cutting. Before cutting, the adjacent curves coincided.

524. To Find the Development of a Truncated Oblique Rectangular Pyramid. Fig. 939. None of the four lateral surfaces is shown in the multiview drawing

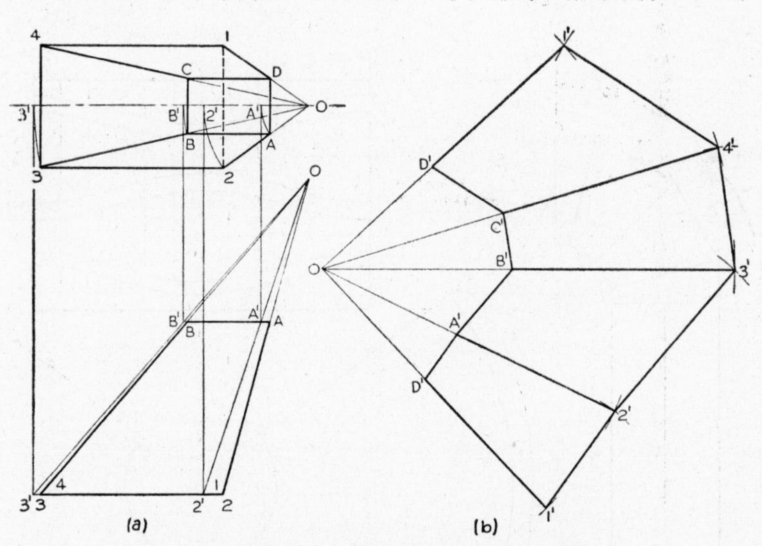

Fig. 939 Development of a Transition Piece.

in true size and shape. Using the method of §291, revolve each edge until it appears in true length in the front view, as shown. Thus, O-2 revolves to O-2′, O-3 revolves to O-3′, etc. These true lengths are transferred from the front view to the development with the compass, as shown. Notice that true lengths OD′, OA′, OB′, etc., are found and transferred. The true lengths of the edges of the bases are given in the top view and are transferred directly to the development.

525. Triangulation. *Triangulation* is simply a method of dividing a surface into a number of triangles, and transferring them to the development. A triangle is said to be "indestructible," because if its sides are of given lengths, it can be only one shape. A triangle can be easily transferred by transferring the sides with the aid of the compass, §135.

526. To Find the Development of an Oblique Cone by Triangulation. Fig. 940. Divide the base, in the top view, into any number of equal parts, and draw an element at each division point. Find the true length of each element, §291. If the divisions of the base are comparatively small, the lengths of the chords may be set

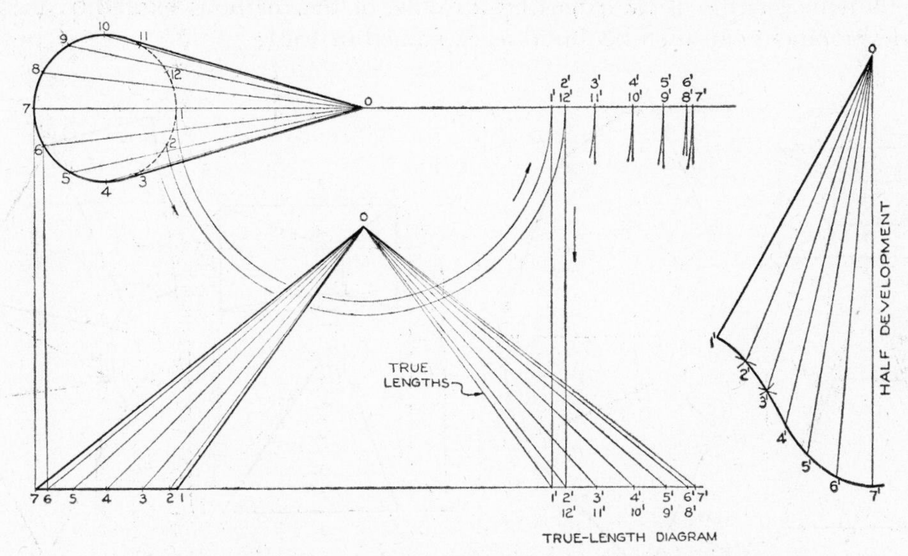

Fig. 940 Development of an Oblique Cone by Triangulation.

off in the development as representing the lengths of the respective subtending arcs. In the development, set off O-1′ equal to O-1 in the front view where it is shown true length. With 1′ in the development as center, and the chord 1-2 taken from the top view as radius, strike an arc at 2′. With O as center, and O-2′, the true length of the element O-2 from the "true-length" diagram, as radius, draw the arc at 2′. The intersection of these arcs is a point on the development of the base of the cone. The points 3′, 4′, etc., in the curve are found in a similar manner, and the curve is traced through these points with the aid of the irregular curve, §66.

Since the development is symmetrical about element O-7′, it is necessary to lay out only half the development, as shown.

527. Transition Pieces. A *transition piece* is one that connects two differently shaped, differently sized, or skewed-position openings, Fig. 941. In most cases, transi-

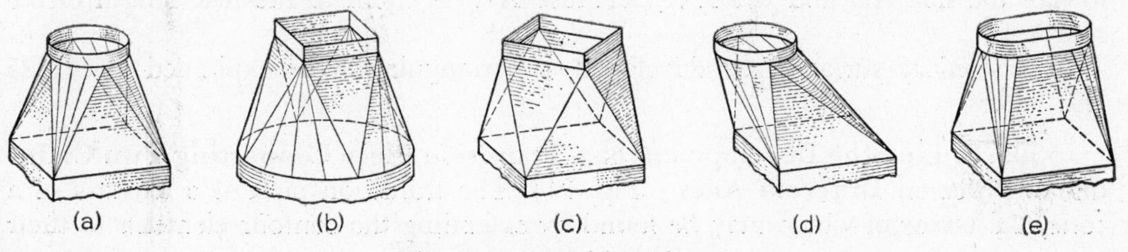

Fig. 941 Transition Pieces.

tion pieces are composed of pláne surfaces and conical surfaces, the latter being developed by triangulation. Triangulation can also be used to develop, approximately, certain warped surfaces. Transition pieces are extensively used in air-conditioning, heating, ventilating, and similar works.

528. To Find the Development of a Transition Piece Connecting Rectangular Pipes on the Same Axis. Fig. 942 (a). The transition piece is a frustum of a pyramid. Find the vertex O of the pyramid by extending its edges to their intersection. Find the true lengths of the edges by any one of the methods explained in §291. The development can then be found as explained in §521.

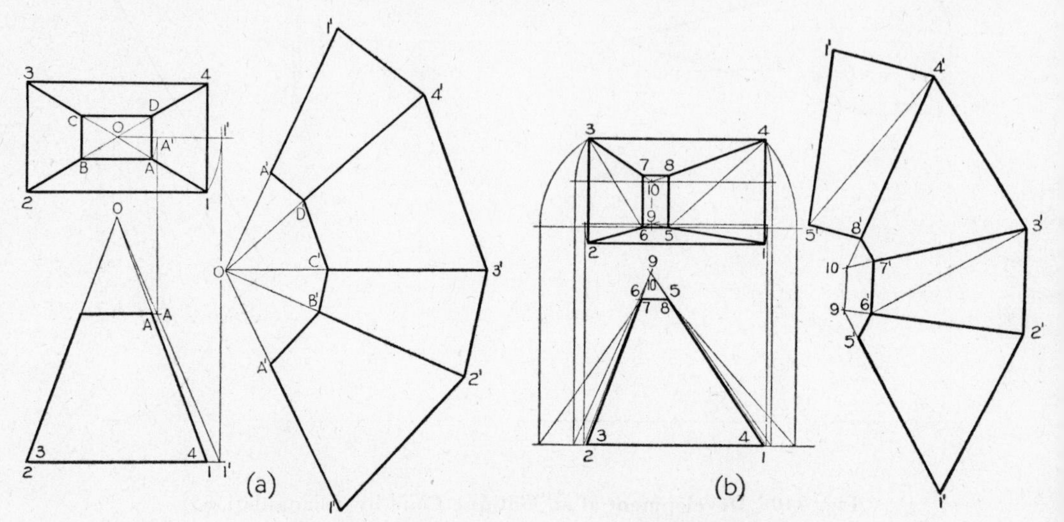

Fig. 942 Development of a Transition Piece.

If the transition piece is not a frustum of a pyramid, as in Fig. 942 (b), it can best be developed by triangulation, §525, as shown for the faces 1-5-8-4 and 2-6-7-3, or by extending the sides to form triangles, as shown for faces 1-2-6-5 and 3-4-8-7, and then finding the true lengths of the sides of the triangles, §291, and setting them off as shown.

As a check on the development, lines parallel on the surface must also be parallel on the development; for example, 8′-5′ must be parallel to 4′-1′ on the development.

529. To Find the Development of a Transition Piece Connecting a Circular Pipe and a Rectangular Pipe on the Same Axis. Fig. 943. The transition piece is composed of four isosceles triangles and four conical surfaces. Begin the development on the line 1′-S, and draw the right triangle 1′-S-A, whose base SA is equal to half the side AD and whose hypotenuse A-1′ is equal to the true length of the side A-1.

The conical surfaces are developed by triangulation as explained in §§525 and 526.

530. To Find the Development of a Transition Piece Connecting Two Cylindrical Pipes on Different Axes. Fig. 944. The transition piece is a frustum of a cone, the vertex of which may be found by extending the contour elements to their intersection A.

The development can be found by triangulation, as explained in §§525 and 526. The sides of each triangle are the true lengths of two adjacent elements of the cone, and the base is the true length of the curve of the base of the cone between the two

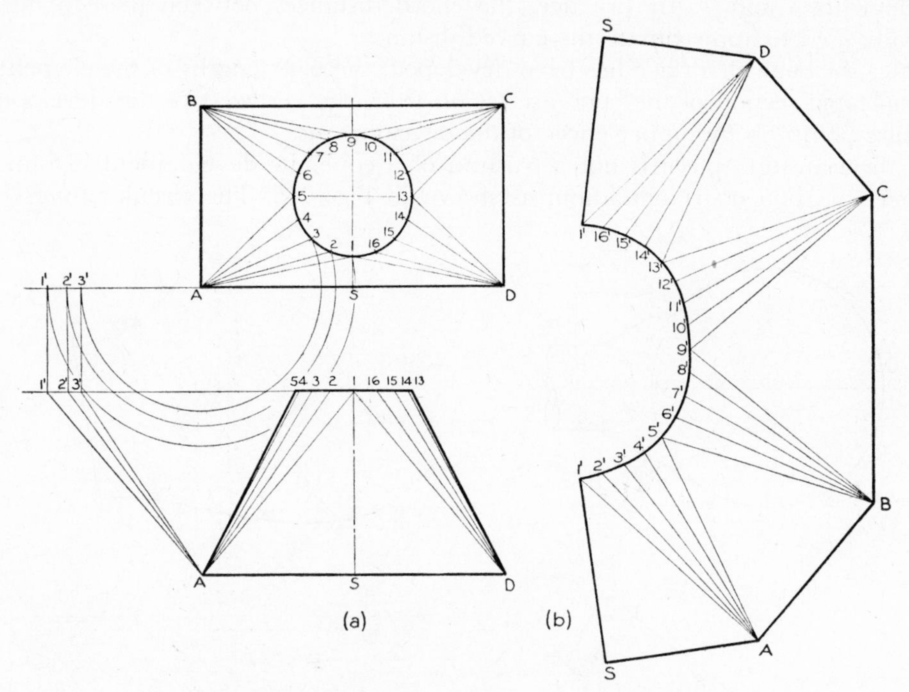

Fig. 943 Development of a Transition Piece.

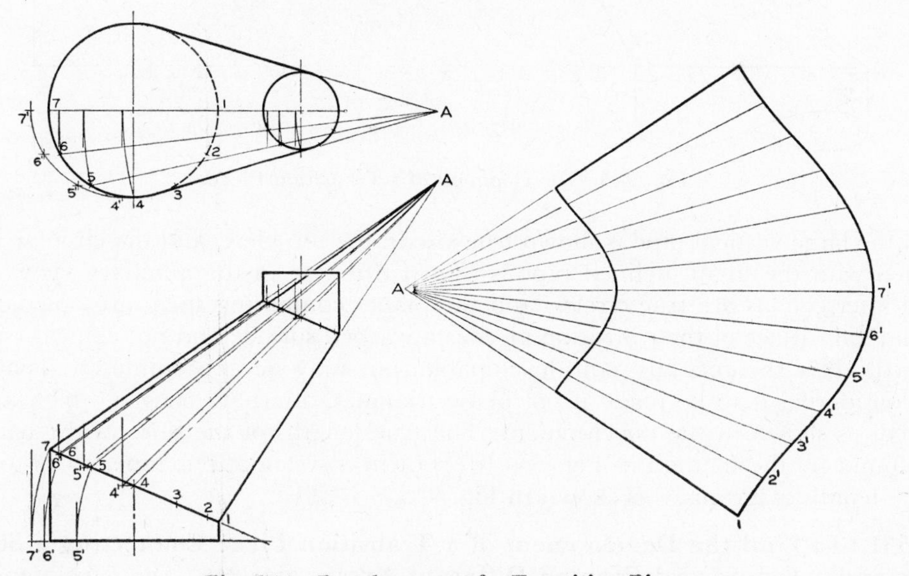

Fig. 944 Development of a Transition Piece.

elements. This curve is not shown in its true length in either view, and the plane of the base of the frustum must therefore be revolved until it is horizontal in order to find the distance from the foot of one element to the foot of the next. When the plane of the base is thus revolved, the foot of any element, such as 7, revolves to 7′, and the curve 6′-7′ (top view) is the true length of the curve of the base between the elements 6 and 7. In practice, the chord distances between these points are generally used to approximate the curved distances.

After the conical surface has been developed, the true lengths of the elements on the truncated section of the cone are set off from the vertex A of the development to secure points on the upper curve of the development.

If the transition piece is not a frustum of a cone, its development is found by another variation of triangulation, as shown in Fig. 945. The circular intersection

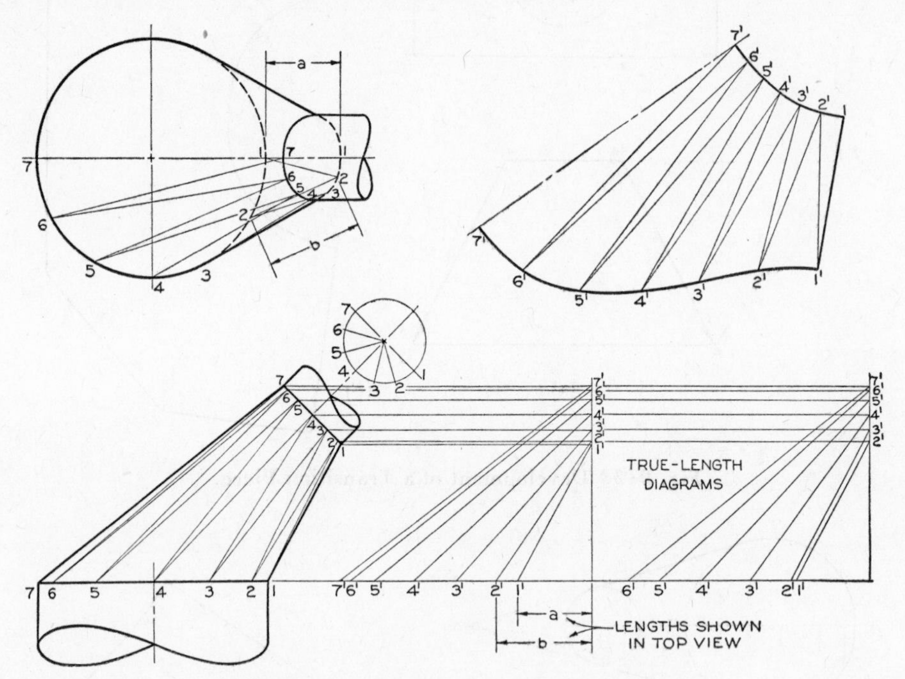

Fig. 945 Development of a Transition Piece.

with the large vertical pipe is shown true size in the top view, and the circular intersection with the small inclined pipe is shown true size in the auxiliary view. Since both intersections are true circles, and the planes containing them are not parallel, the lateral surface of the transition piece is a warped surface and not conical (single-curved). It is theoretically non-developable, but may be approximately developed by considering it to be made up of plane triangles, alternate ones of which are inverted, as shown in the development. The true lengths of the sides of the triangles are found by the method of Fig. 534 (d), but in a systematic manner so as to form "true-length diagrams," as shown in Fig. 945.

531. To Find the Development of a Transition Piece Connecting a Square Pipe and a Cylindrical Pipe on Different Axes. Fig. 946. The development of the transition piece is made up of four plane triangular surfaces and four triangular conical surfaces similar to those in Fig. 943. The development is made in a similar manner to those described in §§526 and 529.

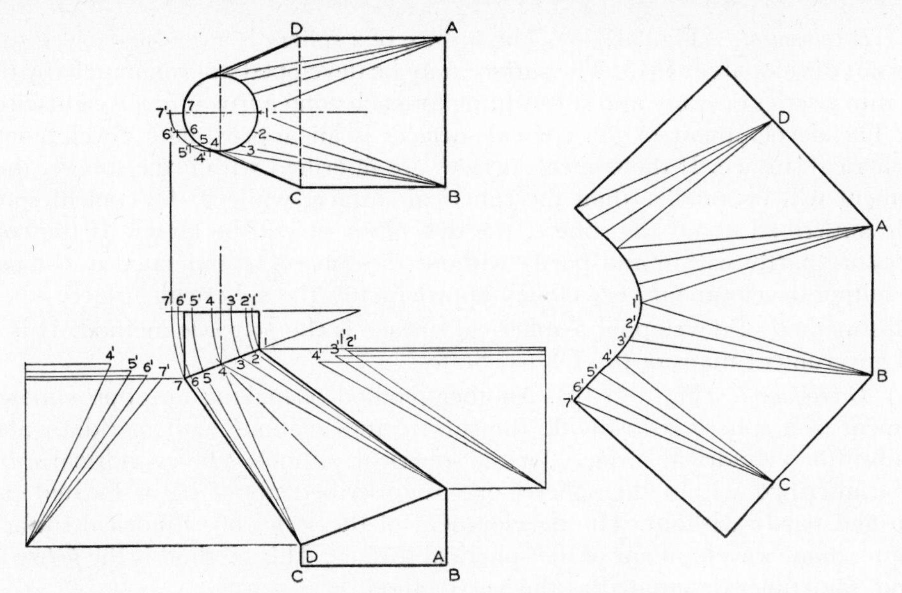

Fig. 946 Development of a Transition Piece.

532. To Find the Intersection of a Plane and a Sphere, and to Find the Approximate Development of the Sphere. Fig. 947.

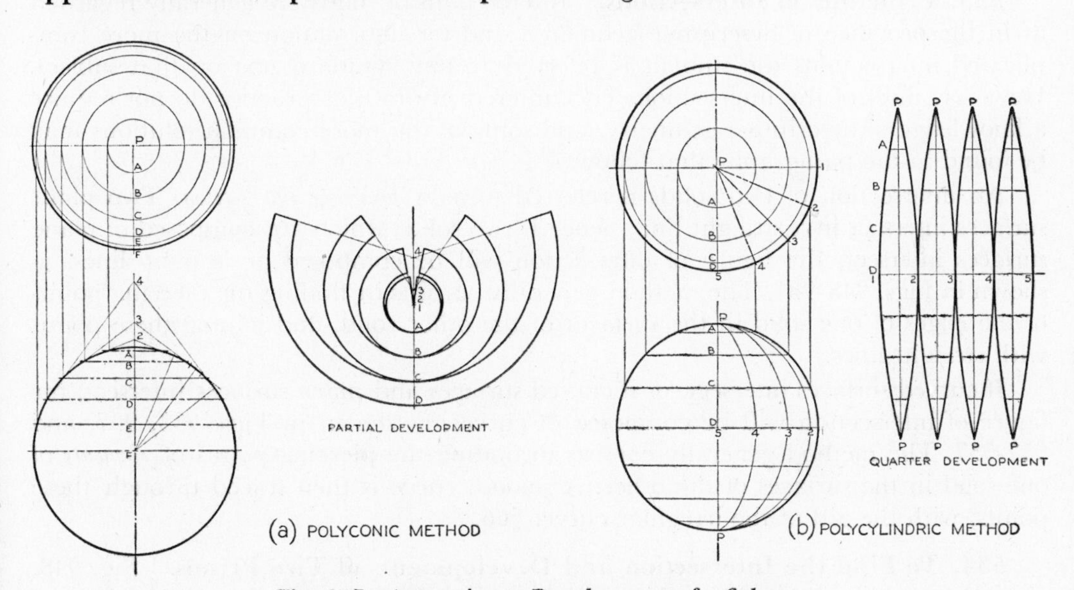

Fig. 947 Approximate Development of a Sphere.

(a) *Intersection.* The intersection of a plane and a sphere is a circle, as shown in the top views in Fig. 947, the diameter of the circle depending upon where the plane is passed. Any circle cut by a plane through the center of the sphere is called a *great circle.* If a plane passes through the center, and perpendicular to the axis, the resulting great circle is called the *equator.* If a plane contains the axis, it will cut a great circle called a *meridian.*

(a) *Development.* Fig. 947 (a). The surface of a sphere is a double-curved surface and is not developable, §513. The surface may be developed approximately by dividing it into a series of zones and substituting for each zone a frustum of a right-circular cone. The development of the conical surfaces is an approximate development of the spherical surface. If the conical surfaces are inscribed within the sphere, the development will be smaller than the spherical surface, while if the conical surfaces are circumscribed about the sphere, the development will be larger. If the conical surfaces are partly within and partly without the sphere, as indicated in the figure, the resulting development very closely approximates the spherical surface.

This method of developing a spherical surface is the *polyconic* method. It is used on all government maps of the United States.

(b) *Development.* Fig. 947 (b). Another method of making an approximate development of a sphere is to divide the surface into sections with meridian planes, and substitute cylindrical surfaces for the spherical sections. The cylindrical surfaces may be inscribed within the sphere, or circumscribed about it, or located partly within and partly without. The development of the series of cylindrical surfaces is an approximate development of the spherical surface. This method is the *polycylindric* method, sometimes designated as the *gore* method.

INTERSECTIONS OF SOLIDS

533. Principles of Intersections. Intersections of solids are generally regarded as in the province of descriptive geometry, and for information on the more complicated intersections the student is referred to any standard text on that subject. However, most of the intersections encountered in drafting practice do not require a knowledge of descriptive geometry, and some of the more common solutions may be found in the paragraphs that follow.

An intersection of two solids is referred to as a *figure of intersection.* Two plane surfaces intersect in a straight line; hence if two solids which are composed of plane surfaces intersect, the figure of intersection will be composed of straight lines, as shown in Figs. 948-951. The method generally consists in finding the piercing points of the edges of one solid in the surfaces of the other solid, and joining these points with straight lines.

If curved surfaces intersect, or if curved surfaces and plane surfaces intersect, the figure of intersection will be composed of curves, as shown in Figs. 933, 937, and 952-957. The method generally consists in finding the piercing points of *elements* of one solid in the surfaces of the other. A smooth curve is then traced through these points, with the aid of the irregular curve, §66.

534. To Find the Intersection and Developments of Two Prisms. Fig. 948.

(a) *Intersection.* The points in which the edges A, B, C, and D of the horizontal prism pierce the vertical prism are vertices of the intersection. The edges D and B of the horizontal prism intersect the edges 3 and 7 of the vertical prism at the points E, F, L, and M. The edges A and C of the horizontal prism intersect the faces of the vertical prism at the points G, H, J, and K. The intersection is completed by joining these points in order by straight lines.

(b) *Developments.* To develop the lateral surface of the horizontal prism, set off on the vertical stretch-out line A-A the widths of the faces AB, BC, etc., taken from the end view, and draw the edges through these points, as shown. Set off, from the stretch-out line, the lengths of the edges AG, BL, etc., taken from the front view or from the top view, and join the points G, L, J, etc., by straight lines.

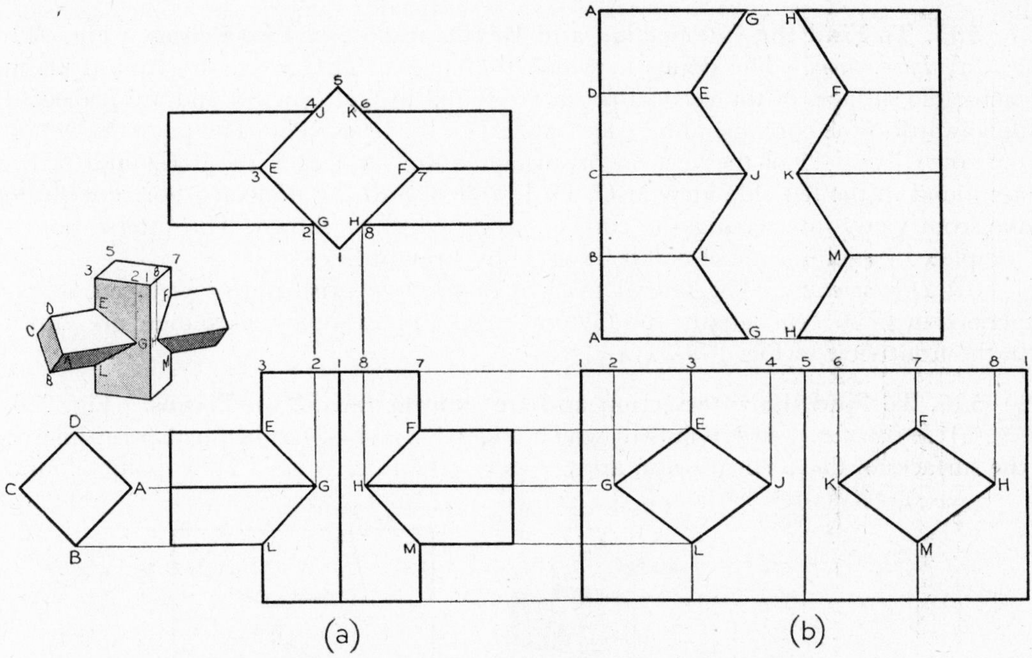

Fig. 948 Two Prisms at Right Angles to Each Other.

To develop the lateral surface of the vertical prism, set off on the stretch-out line 1-1 the widths of the faces 1-3, 3-5, etc., taken from the top view, and draw the edges through these points, as shown. Set off on the stretch-out line, the distances

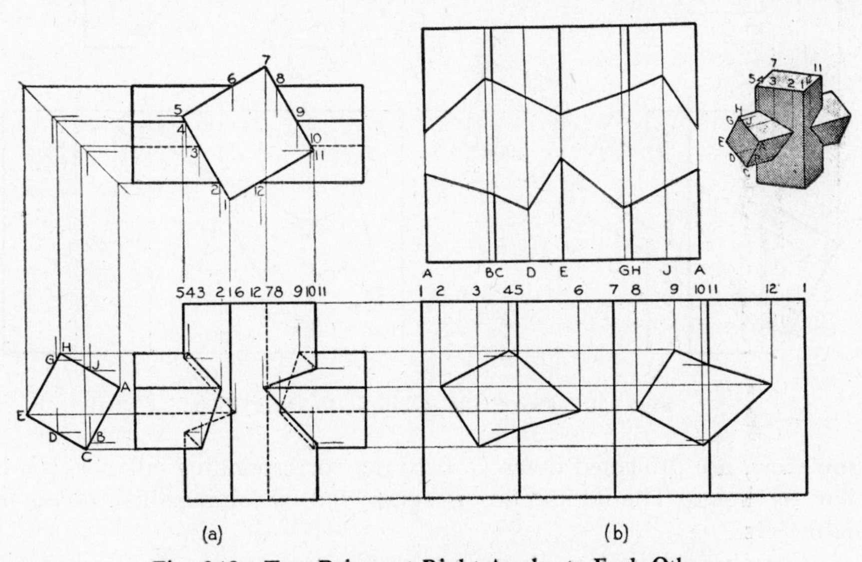

Fig. 949 Two Prisms at Right Angles to Each Other.

1-2, 5-4, 5-6, and 1-8, taken from the top view, and draw the intermediate elements parallel to the principal edges. Take the lengths of the principal edges and of the intermediate elements from the front view, and join the points E, G, L, etc., in order with straight lines, to complete the development.

535. To Find the Intersection and Developments of Two Prisms. Fig. 949.

(a) *Intersection.* The points in which the edges ACEH of the horizontal prism pierce the surfaces of the vertical prism are found in the top view and are projected downward to the corresponding edges ACEH in the front view. The points in which the edges 5 and 11 of the vertical prism pierce the surfaces of the horizontal prism are found in the left side view at G, D, J, and B, and are projected horizontally to the front view, intersecting the corresponding edges as shown. The intersection is completed by joining these points in order by straight lines.

(b) *Developments.* The lateral surfaces of the two prisms are developed as explained in §534. True lengths of all lateral edges and lines parallel to them are shown in the front view of Fig. 949 at (a).

536. To Find the Intersection and Developments of Two Prisms. Fig. 950.

(a) *Intersection.* The points in which edges 1-2-3-4 of the inclined prism pierce the surfaces of the vertical prism are vertices of the intersection. These points, found

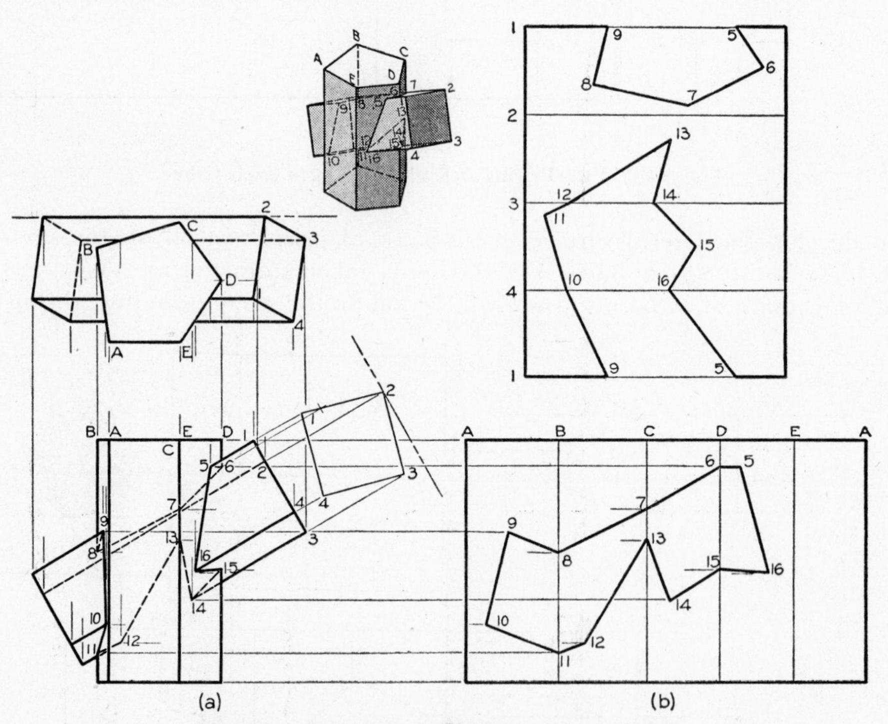

Fig. 950 Two Prisms Oblique to Each Other.

in the top view, are projected downward to the corresponding edges 1-2-3-4 in the front view, as shown. The intersection is completed by joining these points in order by straight lines.

(b) *Developments.* The lateral surfaces of the two prisms are developed as explained in §534. True lengths of all edges of both prisms are shown in the front view of Fig. 950 at (a).

537. To Find the Intersection and the Developments of Two Prisms. Fig. 951. In this case the edges of the inclined prism are oblique to the planes of projection, and in the front and top views none of the edges is shown true length, §229, and none of the faces is shown true size, §228. Furthermore, none of the angles, including the

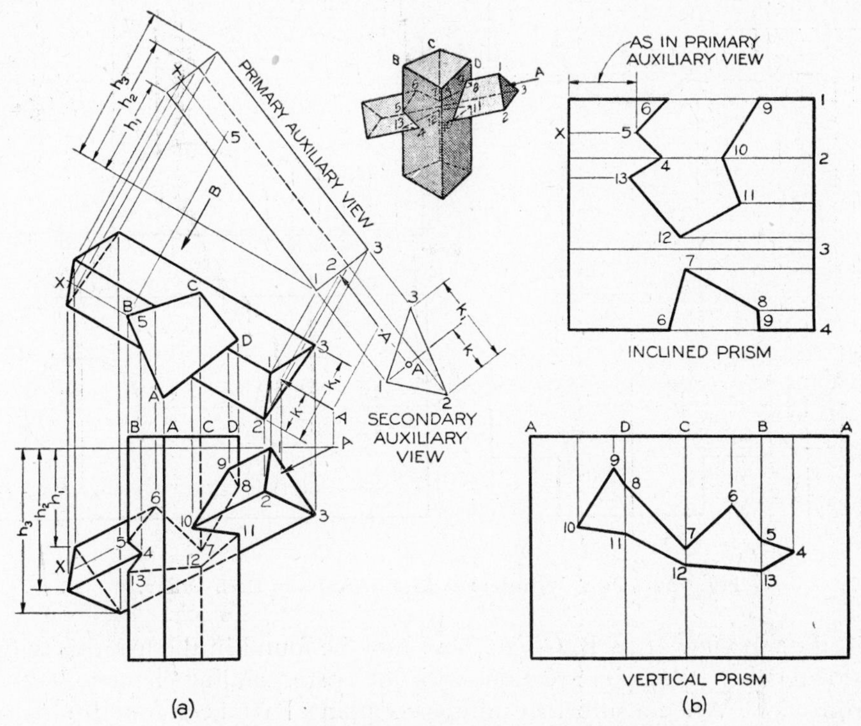

Fig. 951 Two Prisms Oblique to Each Other.

angle of inclination, is shown true size, §231. Therefore, it is necessary to draw a secondary auxiliary view, §280, to obtain the true size and shape of the right section of the inclined prism.

The direction of sight, indicated by arrow A, is assumed perpendicular to the end face 1-2-3, that is, parallel to the principal edges of the prism. The primary auxiliary view, taken in the direction of arrow B, shows the true lengths of the edges, the true inclination of the prism with respect to the horizontal and, incidentally, the true length and inclination of arrow A. In the secondary auxiliary view, arrow A is shown as a point, and the end face 1-2-3 is shown in its true size.

(a) *Intersection.* The points in which the edges 1-2-3 of the inclined prism pierce the surfaces of the vertical prism are vertices of the intersection, found first in the top view and then projected downward to the front view.

(b) *Developments.* The lateral surfaces of the two prisms are developed as explained in §534. True lengths of the edges of the vertical prism are shown in the front view. True lengths of the edges of the inclined prism can be shown in the primary auxiliary view; true lengths to the vertices of the intersection may be found in this view, as shown for line X-5.

538. To Find the Intersection and Developments of Two Cylinders. Fig. 952.

(a) *Intersection.* Assume a series of elements (preferably equally spaced) on the horizontal cylinder, numbered 1-2-3, etc., in the side view, and draw their top and front views. Their points of intersection with the surface of the vertical cylinder are

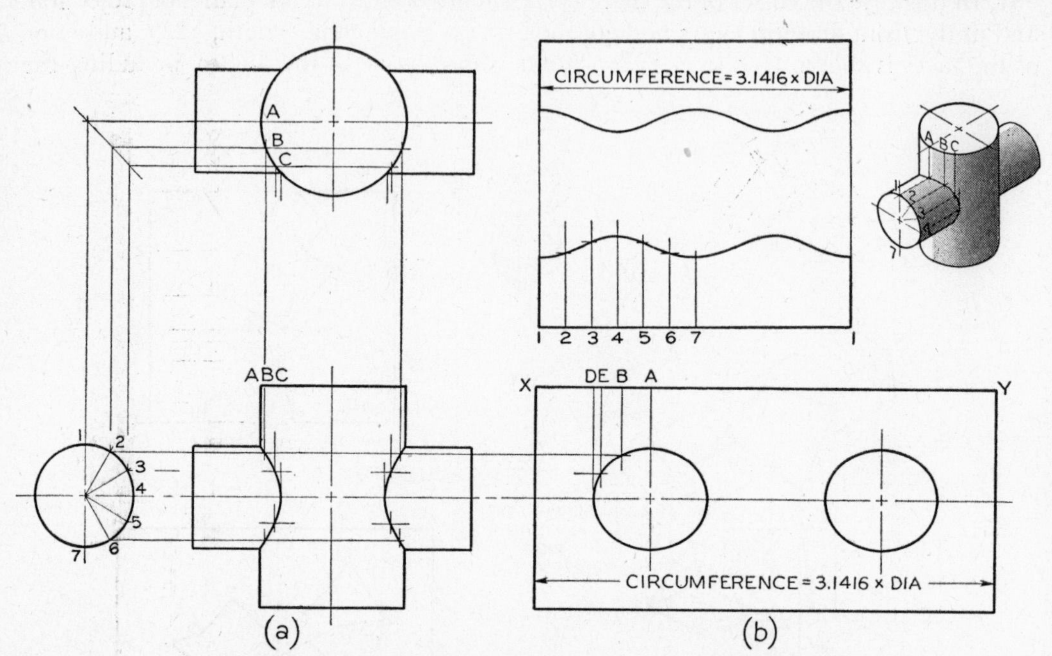

Fig. 952 Two Cylinders at Right Angles to Each Other.

shown in the top view at A, B, C, etc., and may be found in the front view by projecting downward to their intersections with the corresponding elements 1-2-3, etc., in the front view. When a sufficient number of points have been found to determine the intersection, the curve is traced through the points with the aid of the irregular curve, §66. See also Fig. 321(c).

(b) *Developments.* The lateral surfaces of the two cylinders are developed as explained in §518. True lengths of all elements of both cylinders are shown in the front view. Since both cylinders have bases at right angles to the center lines, the circles will develop as straight lines, and the developments will be rectangular, as shown. The length XY of the development of the vertical cylinder is equal to the circumference of the cylinder, or πd, and the length 1-1 of the development of the horizontal cylinder is determined in the same way. Those elements of the large cylinder which pierce the small cylinder can be identified in the top view as elements A, B, C, etc. When these are drawn in the development, the points of intersection are found at their intersections with the corresponding elements of the horizontal cylinder taken from the front view, thus determining one of the figures of intersection, as shown in Fig. 952 (b).

539. To Find the Intersections and Developments of Two Cylinders. Fig. 953.

(a) *Intersection.* A revolved right section of the inclined cylinder is divided into a number of equal parts 1-2-3, etc., and an element is drawn at each of the division points. The points of intersection of these elements with the surface of the vertical cylinder are shown in the top view at B, C, D, etc., and are found in the front view by projecting downward to intersect the corresponding elements 1-2-3, etc. The curve is traced through these points with the aid of the irregular curve, §66.

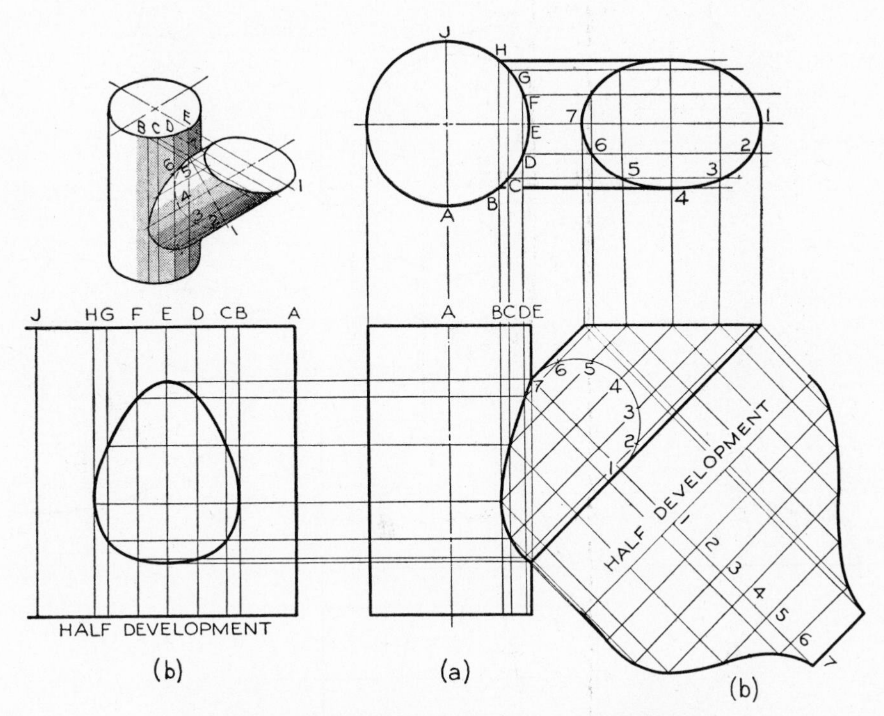

Fig. 953 Two Cylinders Oblique to Each Other.

(b) *Developments.* The lateral surfaces of the two cylinders are developed as explained in §§518 and 520. True lengths of all elements of both cylinders are shown in the front view.

540. To Find the Intersection and Developments of a Prism and a Cone. Fig. 954.

(a) *Intersection.* Points in which the edges of the prism intersect the surface of the cone are shown in the side view at A, C, and F. Intermediate points such as B, D, E, and G are piercing points of any lines on the lateral surface of the prism parallel to the edges. Through all of the piercing points in the side view, elements of the cone are drawn, and then drawn in the top and front views. The intersections of the elements of the cone with the edges of the prism (and lines along the prism drawn parallel thereto) are points of the intersections. The figures of intersection are traced through these points with the aid of the irregular curve, §66.

The elements 6, 5, 4, etc., in the side view of the cone may be regarded as the edge views of cutting planes which cut these elements on the cone and edges or elements on the prism. The intersection of corresponding edges or elements on the two solids are points on the figure of intersection.

Another method of finding the figure of intersection is to pass a series of horizontal parallel planes through the solids in the manner of Fig. 937 (a). The plane will cut circles on the cone and straight lines on the prism, and their intersections will be points on the figure of intersection. See also Fig. 955 (b).

(b) *Developments.* The lateral surface of the prism is developed as explained in §534. True lengths of all edges and lines parallel thereto are shown in both the front

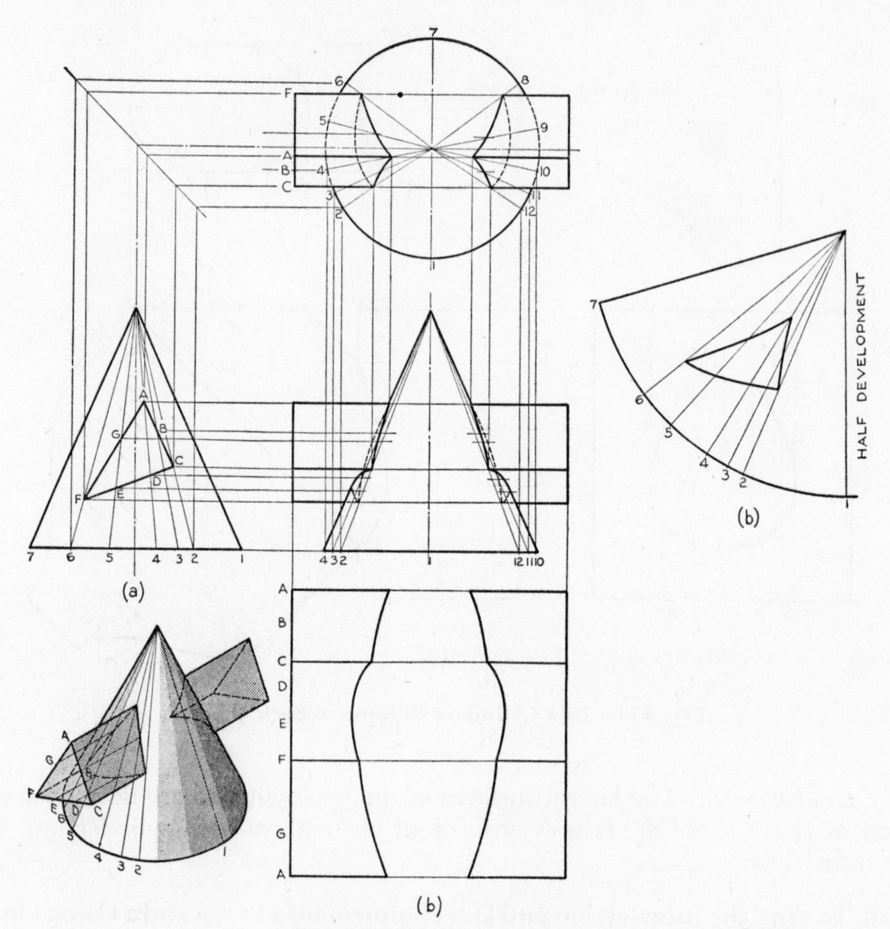

Fig. 954 Prism and Cone.

and top views. The lateral surface of the cone is developed as explained in §522. True lengths of elements from the vertex to points on the intersections are found as shown in Fig. 534 (a).

541. To Find the Intersection of a Prism and a Cone with Edges of Prism Parallel to Axis of Cone.

Fig. 955 (a). Since the lateral surfaces of the prism are parallel to the axis of the cone, the figure of intersection will be composed of a series of hyperbolas, §§154 and 167. If a series of planes is assumed containing the axis of the cone, each plane will contain edges of the prism or will cut lines parallel to them along the prism, and will cut elements on the cone that intersect these at points of the figure of intersection.

Fig. 955 (b). The intersection is the same as at (a), but found in a different manner. Here a series of parallel planes perpendicular to the axis of the cone cut circles of varying diameters on the cone. These circles are shown true size in the top view, where also are shown the piercing points of these circles in the vertical plane surfaces of the prism. The front views of these piercing points are found by projecting downward to the corresponding cutting-plane lines.

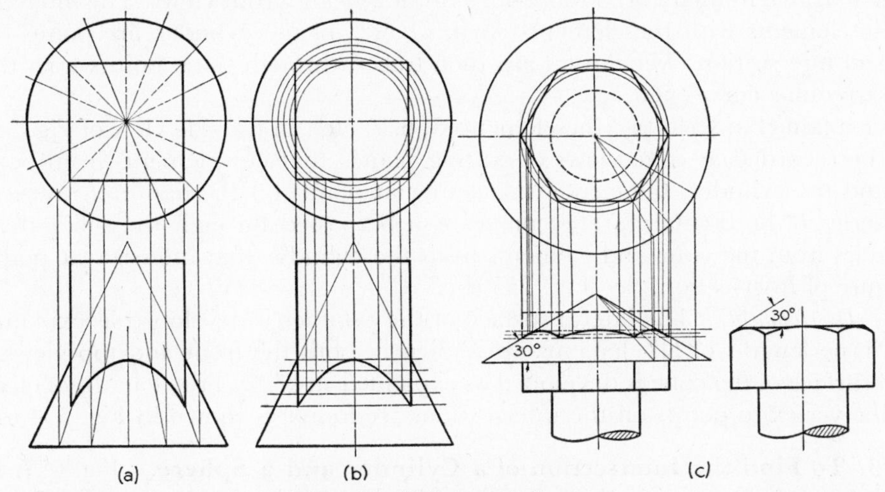

Fig. 955 Prisms and Cones.

Fig. 955 (c). The chamfer of an ordinary hexagon bolt head or hexagon nut is actually a conical surface that intersects the six vertical sides of a hexagonal prism to form hyperbolas. At (c) the methods of both (a) and (b) are shown to illustrate how points may be found by either method.

In machine drawings of bolts and nuts, these hyperbolic curves are approximated by means of circular arcs, as shown in Fig. 683.

542. To Find the Intersections and the Developments of a Cylinder and a Cone. Fig. 956.

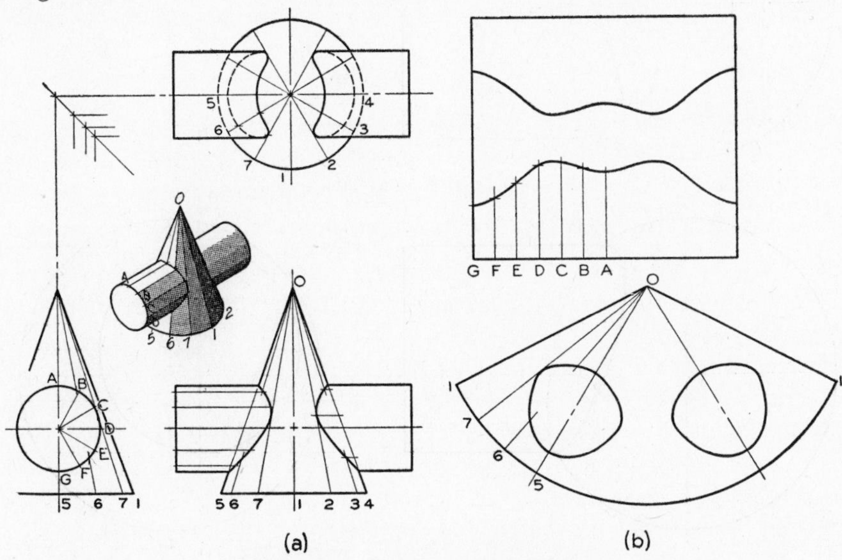

Fig. 956 Cone and Cylinder.

(a) *Intersections.* Points in which elements of the cylinder (preferably equally spaced to facilitate the development) intersect the surface of the cone are shown in the side view at A, B, C, etc. The elements of the cylinder are here shown as points. Elements of the cone are then drawn from the vertex through each of these points, and then drawn in their correct locations in the top and front views. The intersections of these elements with the elements A, B, C, etc., of the cylinder are points on the figures of intersection. The curves are then traced through these points with the aid of the irregular curve, §66.

As explained in §540 (a), the elements 5, 6, 7, etc., in the side view of Fig. 956 (a) could be regarded as edge views of cutting planes that cut elements from both the cone and the cylinder, the elements meeting at points on the figure of intersection. Or a series of horizontal parallel planes can be passed through the solids that will cut circles from the cone and elements from the cylinder that intersect at points on the figure of intersection. See Fig. 955 (b).

(b) *Development.* The lateral surface of the cylinder is developed as explained in §538. True lengths of all elements are shown in both the front and top views. The lateral surface of the cone is developed as explained in §522. True lengths of elements from the vertex to points on the intersections are found as shown in Fig. 534 (a).

543. To Find the Intersection of a Cylinder and a Sphere. Fig. 957. Horizontal planes 1, 2, 3, etc., which appear edgewise in the front and side views, cut elements A, B, C, etc., from the cylinder and circular arcs 1′, 2′, 3′, etc., from the sphere. The intersections of the elements with the arcs produced by the corresponding

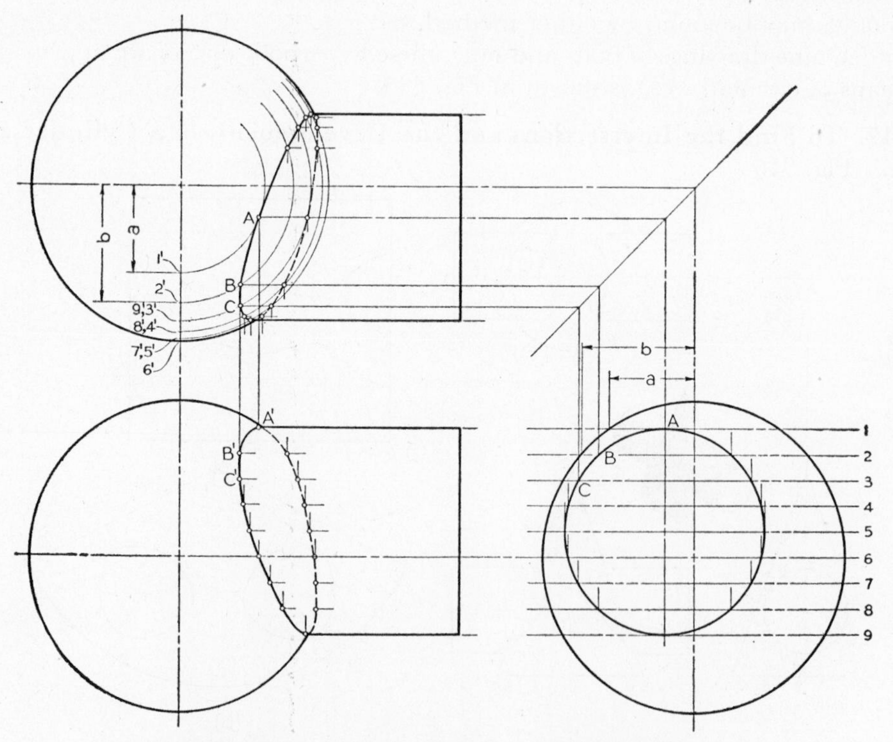

Fig. 957 Intersection of Sphere and Cylinder.

planes are points on the figure of intersection. Join the points with a smooth curve, §66.

544. Intersection and Development Problems. A wide selection of intersection and development problems is provided in Figs. 958-965. These problems are designed to fit on 11″ × 17″ sheets. Dimensions should be included on the given views. The student is cautioned to take special pains to obtain accuracy in these drawings, and to draw smooth curves.

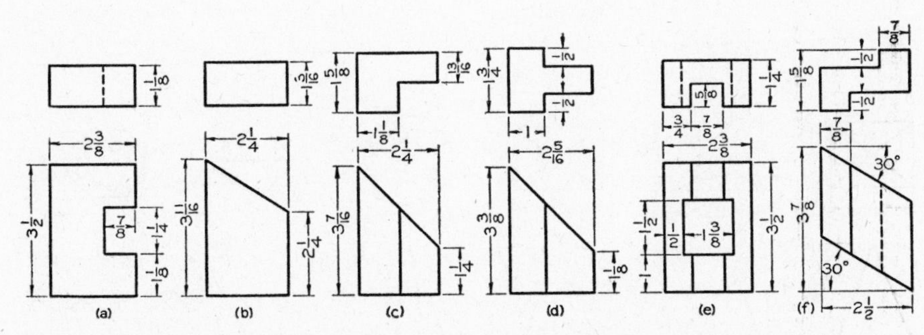

Fig. 958 Draw given views and develop lateral surface. Layout B-3.

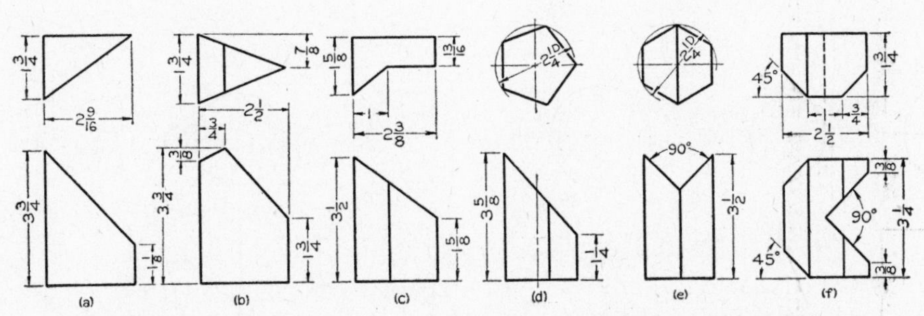

Fig. 959 Draw given views and develop lateral surface. Layout B-3.

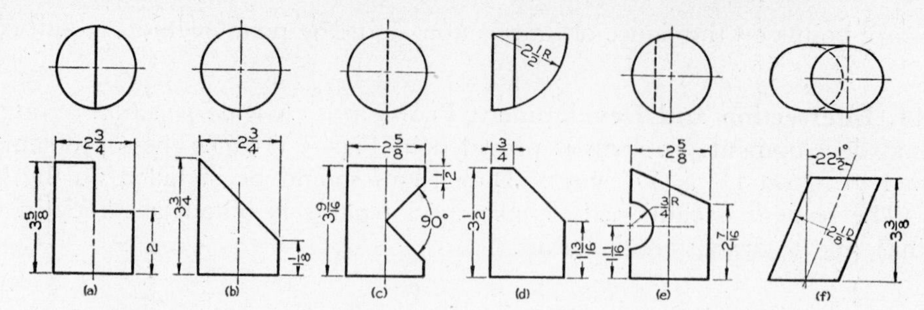

Fig. 960 Draw given views and develop lateral surface. Layout B-3.

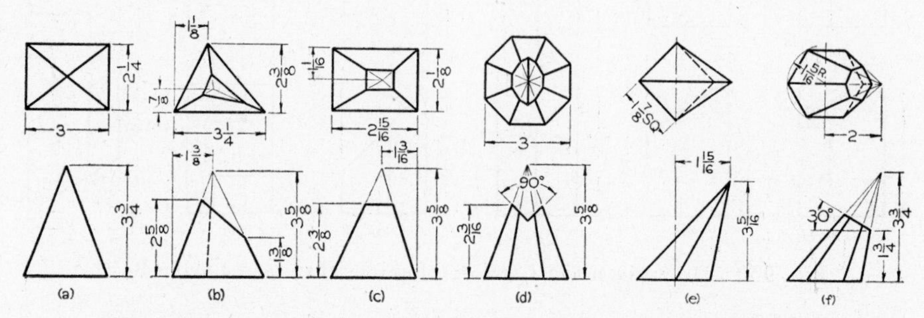

Fig. 961 Draw given views and develop lateral surface. Layout B-3.

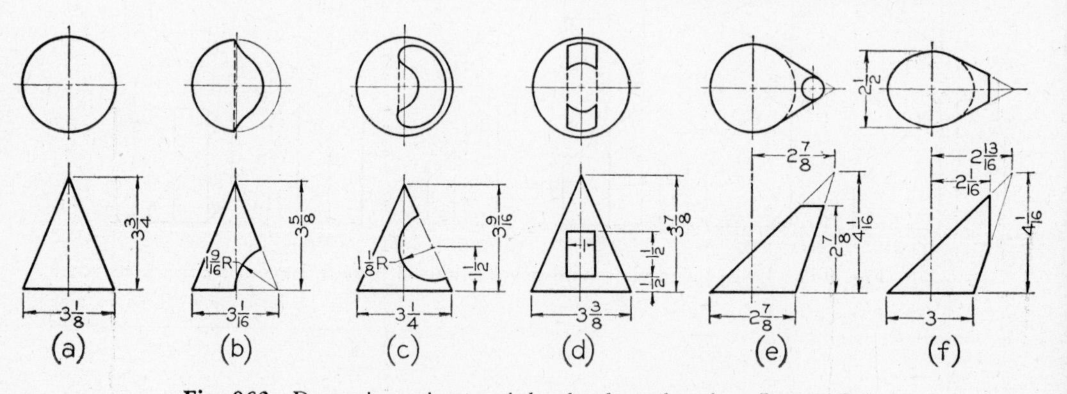

Fig. 962 Draw given views and develop lateral surface. Layout B-3.

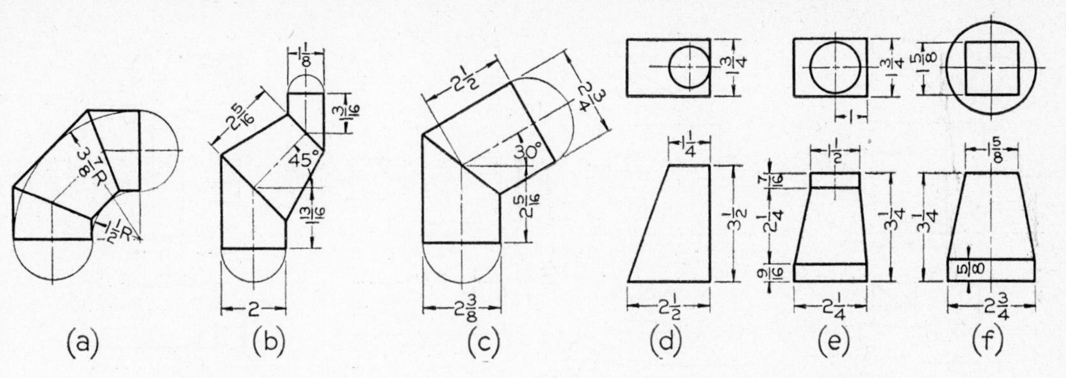

Fig. 963 Draw given views and develop lateral surface. Layout B-3.

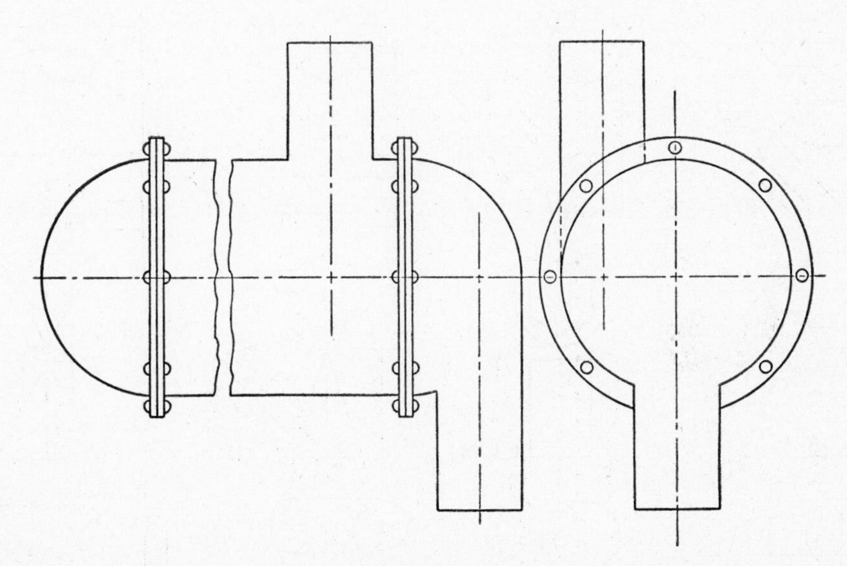

Fig. 964 Draw the two views of the CONDENSER, as shown. Transfer all measurements with dividers, making your drawing three times the size shown. Find the intersections of the small cylindrical pipes with the main portion. The ends are spherical. Layout B-3.

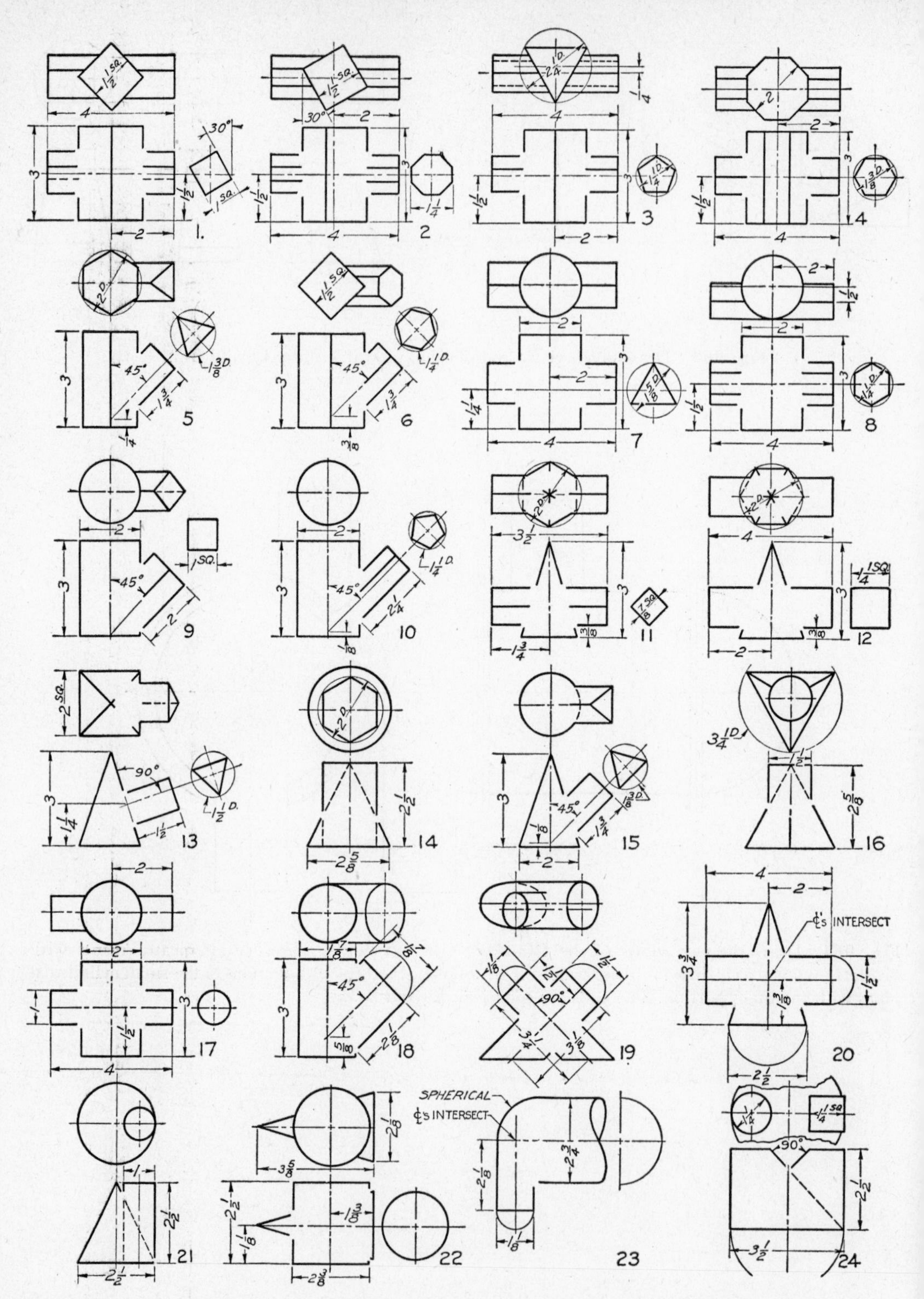

Fig. 965 Draw the given views of assigned object, and complete the intersection. Then develop lateral surfaces. Layout B-3.

CHAPTER 19

GEARING AND CAMS

*By B. Leighton Wellman**

545. Gears. *Gears*, Fig. 966, are used to transmit motion, rotating or reciprocating, from one machine part to another. They may be classified according to the posi-

Fig. 966 An Assortment of Gears. *Courtesy Charles Bond Co.*

*Professor of Mechanical Engineering and Head of Division of Engineering Drawing, Worcester Polytechnic Institute.

tion of the shafts that they connect. Parallel shafts, for example, may be connected by *spur gears*, *helical gears*, or *herringbone gears*. Intersection shafts may be connected by *bevel gears* having either straight, skew, or spiral teeth. Non-parallel, non-intersecting shafts may be connected by crossed *helical gears*, *hypoid gears*, or a *worm* and *worm gear*. A spur gear meshed with a *rack* will convert rotary motion to reciprocating motion.

Because the design of gears is a complex problem involving involute geometry, strength and wear characteristics, precision of manufacture, and inspection control, the scope of this chapter must be restricted to the drafting of straight-tooth spur and bevel gears, and worm gearing.

546. Spur Gears. The *friction wheels* shown in Fig. 967 (a) will transmit motion and power from one shaft to another parallel shaft. However, friction gears are subject to slipping, and excessive pressure is required between the wheels to obtain

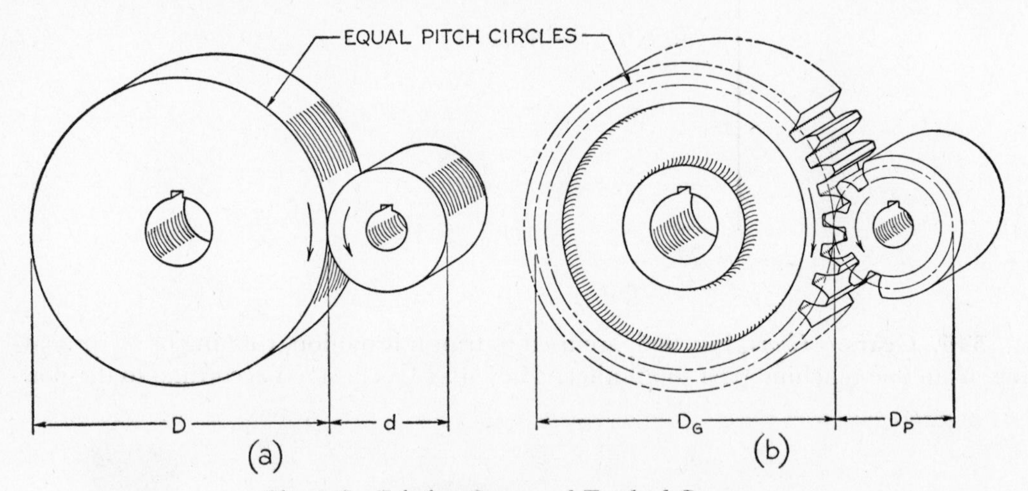

Fig. 967 Friction Gears and Toothed Gears.

the necessary frictional force. If teeth of the proper shape are provided on the cylindrical surfaces, the resulting spur gears, Fig. 967 (b), will transmit the same motion and power without slipping and with greatly reduced bearing pressures.

If a friction wheel of diameter D turns at n rpm, the linear velocity of a point on its periphery will be πDn. But the pitch circles of a pair of mating spur gears correspond exactly to the outside diameters of the friction wheels, and since the gears turn in contact without slipping, they must have the same linear velocity at the pitch line. Therefore,

$$\pi D_G n_G = \pi D_P n_P \quad \text{or} \quad \frac{D_G}{D_P} = \frac{n_P}{n_G} = m_G$$

where D_G and D_P are the pitch diameters of the larger gear (called the *gear*) and the smaller gear (called the *pinion*); n_G and n_P are the rpm of the gears; m_G is the gear ratio, expressed as the ratio of larger gear to smaller.

The teeth on mating gears must be of equal width and spacing; hence the number of teeth on each gear, N, is directly proportional to its pitch diameter, or

$$\frac{N_G}{N_P} = \frac{D_G}{D_P} = \frac{n_P}{n_G} = m_G$$

547. Spur Gear Definitions and Formulas.* Proportions and shapes of gear teeth are well standardized, and the terms defined below and in Fig. 968 are common

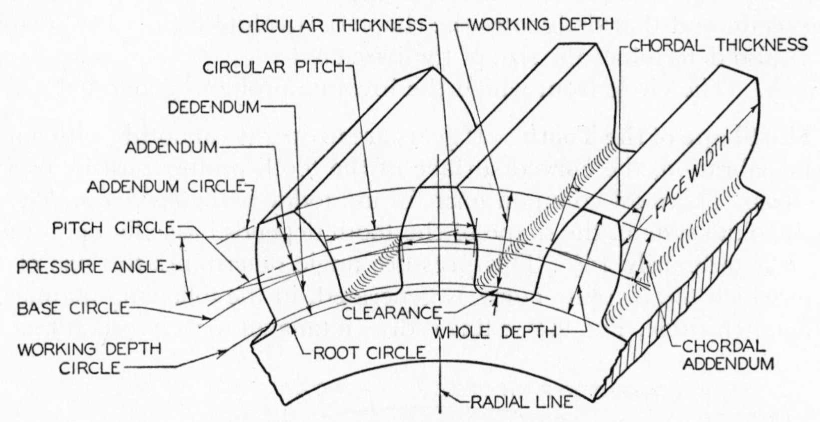

Fig. 968 Gear Tooth Nomenclature.

to all spur gears. The dimensions relating to tooth height are for full-depth $14\frac{1}{2}°$ or 20° involute teeth.

Pitch Circle. An imaginary circle that corresponds to the circumference of the friction gear from which the spur gear is derived.

Pitch Diameter (D_G or D_P). The diameter of the pitch circle of gear or pinion.

Number of Teeth (N_G or N_P). The number of teeth on the gear or pinion.

Diametral Pitch (P). A ratio equal to the number of teeth on the gear per inch of pitch diameter. $P = N/D$.

Circular Pitch (p). The distance measured along the pitch circle from a point on one tooth to the corresponding point on the adjacent tooth. It thus includes one tooth and one space. $p = \pi D/N$. It is useful to note that $p \times P = \pi$.

Addendum (a). The radial distance from the pitch circle to the top of the tooth. $a = 1/P$.

Dedendum (b). The radial distance from the pitch circle to the bottom of the tooth space. $b = 1.157/P$.

Outside Diameter (D_O). The diameter of the addendum circle. It is equal to the pitch diameter plus twice the addendum. $D_O = D + 2a = (N + 2)/P$.

Root Diameter (D_R). The diameter of the root circle. It is equal to the pitch diameter minus twice the dedendum. $D_R = D - 2b = (N - 2.314)/P$.

Whole Depth (h_t). The total height of the tooth. It is equal to the addendum plus the dedendum. $h_t = a + b = 2.157/P$.

Working Depth (h_k). The distance that a tooth projects into the mating space. It is equal to twice the addendum. $h_k = 2a = 2/P$.

Clearance (c). The distance between the top of a tooth and the bottom of the mating space. It is equal to the dedendum minus the addendum. $c = b - a = 0.157/P$.

Circular Thickness (t). The thickness of a tooth measured along the pitch circle. It is equal to one-half the circular pitch. $t = p/2 = \pi/2P$.

Chordal Thickness (t_c). The thickness of a tooth measured along a chord of the pitch circle. $t_c = D \sin(90°/N)$.

*Letter symbols conform to ASA B6.5-1949.

Chordal Addendum (a_c). The radial distance from the top of a tooth to the chord of the pitch circle. $a_c = a + \frac{1}{2}D\ [1 - \cos(90°/N)]$.

Pressure Angle (ϕ). The angle that determines the direction of pressure between contacting teeth, and that designates the shape of involute teeth—$14\frac{1}{2}°$ involute, for example. It also determines the size of the base circle.

Base Circle. The circle from which the involute profile is generated.

548. The Shape of the Tooth. If gears are to operate smoothly with a minimum of noise and vibration, the curved surface of the tooth profile must be of a definite geometric form. The most common form in use today is the *involute profile*.

In the involute system, the shape of the tooth depends basically upon the *pressure angle*, which is ordinarily $14\frac{1}{2}°$. This pressure angle determines the size of the base circle, from which the involute curve is generated, in the following manner: at any point on the pitch circle, Fig. 969, a line is drawn tangent to it; a second line is drawn

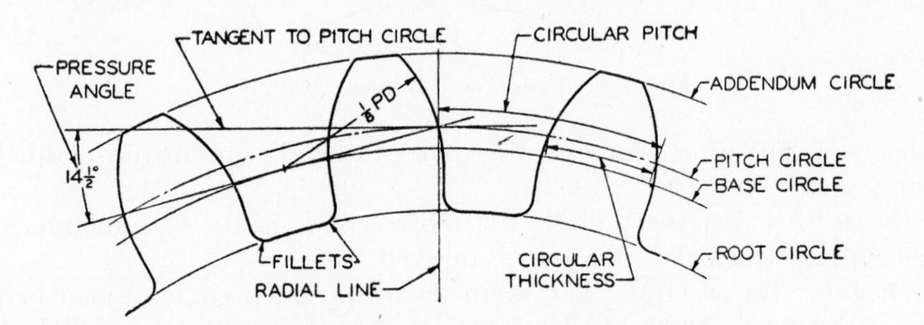

Fig. 969 Approximate Representation of Involute Spur Gear Teeth.

through the point of tangency at an angle of $14\frac{1}{2}°$ (frequently approximated as 15° on the drawing) with the tangent; the base circle is then drawn tangent to the $14\frac{1}{2}°$ line.

If the exact shape of the tooth is desired, the portion of the profile from the base circle to the addendum circle can be drawn as the involute of the base circle. The method of construction is shown in Fig. 216 (d) and (e). That part of the profile below the base circle is drawn as a radial line which terminates in the fillet at the root circle. The fillet should be equal in radius to one and one-half times the clearance.

For display drawings and more rapid construction, the involute curve can be approximated with one or two circular arcs. The construction shown in Fig. 969 employs a single arc. The base circle is drawn as described above, and the spacing of the teeth is set off along the pitch circle. With a radius equal to one-eighth the pitch diameter, and with centers on the base circle, circular arcs are drawn through the spaced points on the pitch circle, and are extended from the addendum circle to slightly below the base circle. Below the base circle, a radial line and fillet complete the profile.

A closer approximation is that known as *Grant's Odontograph*, in which the involute is replaced by two circular arcs as shown in Fig. 970. The base circle and tooth spacing are established as previously described. The face of the tooth from P to A is drawn with the face radius R, and the portion of the flank from P to O is drawn with the flank radius r. Both arcs are drawn from centers located on the base circle. The table gives the correct face and flank radii for gears of *one* diametral pitch. For other pitches, the figures in the table must be divided by the diametral pitch. Below

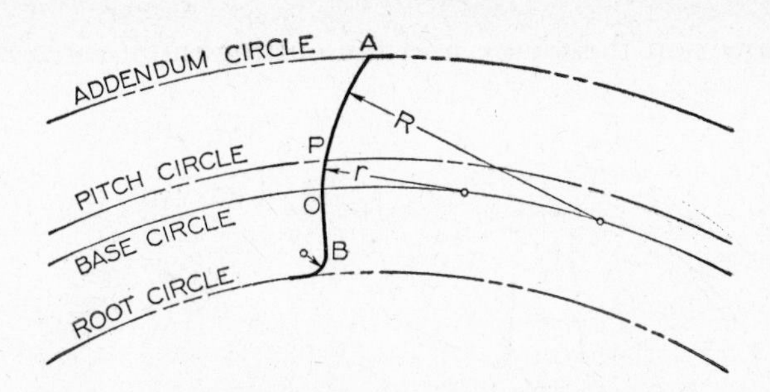

Divide radii given in table by Diametral Pitch.

No. of Teeth	Face Radius R	Flank Radius r	No. of Teeth	Face Radius R	Flank Radius r	No. of Teeth	Face Radius R	Flank Radius r
12	2.51	0.96	24	3.64	2.24	35	4.39	3.16
13	2.62	1.09	25	3.71	2.33	36	4.45	3.23
14	2.72	1.22	26	3.78	2.42	37–40	4.20	4.20
15	2.82	1.34	27	3.85	2.50	41–45	4.63	4.63
16	2.92	1.46	28	3.92	2.59	46–51	5.06	5.06
17	3.02	1.58	29	3.99	2.67	52–60	5.74	5.74
18	3.12	1.69	30	4.06	2.76	61–70	6.52	6.52
19	3.22	1.79	31	4.13	2.85	71–90	7.72	7.72
20	3.32	1.89	32	4.20	2.93	91–120	9.78	9.78
21	3.41	1.98	33	4.27	3.01	121–180	13.38	13.38
22	3.49	2.06	34	4.33	3.09	181–360	21.62	21.62
23	3.57	2.15						

Fig. 970 Grant's $14\frac{1}{2}°$ Involute Odontograph.

the base circle, the flank of the tooth is completed with a radial line OB and a fillet.

If the pressure angle is increased to 20° and the height of the tooth is reduced, the teeth are called *stub* teeth. Such teeth are drawn in the same manner except that $a = 0.8/P$, $b = 1/P$, and the pressure angle is made 20°. The major advantage of stub teeth is that they are stronger than the $14\frac{1}{2}°$ standard involute teeth.

When the gear teeth are formed on a flat surface, the result is a *rack*, Fig. 971. In the involute system, the sides of rack teeth are straight, and are inclined at an angle equal to the pressure angle. To mesh with a gear, it is obvious that the linear pitch of the rack must be the same as the circular pitch of the gear, and the rack teeth must have the same height proportions as the gear teeth.

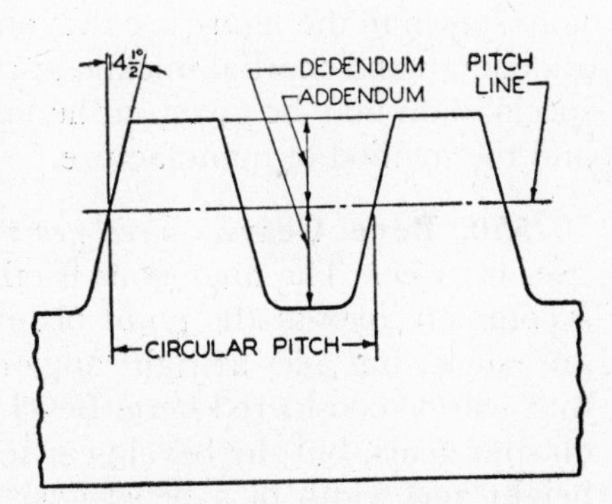

Fig. 971 Involute Rack Teeth.

549. Working Drawings of Spur Gears. A typical working drawing of a spur gear is shown in Fig. 972. Since the teeth are cut to standard shape with special cutters, it is not necessary to show individual teeth on the drawing. Instead, the addendum and root circles are drawn as phantom lines, Fig. 24, and the pitch circle as a center line. Thus the drawing actually shows only a gear blank—a gear complete except for teeth. Since the machining of the blank and the cutting of the teeth are separate operations in the shop, the necessary dimensions are arranged in two groups:

the blank dimensions are shown on the views, and the cutting data are given in a note or table.

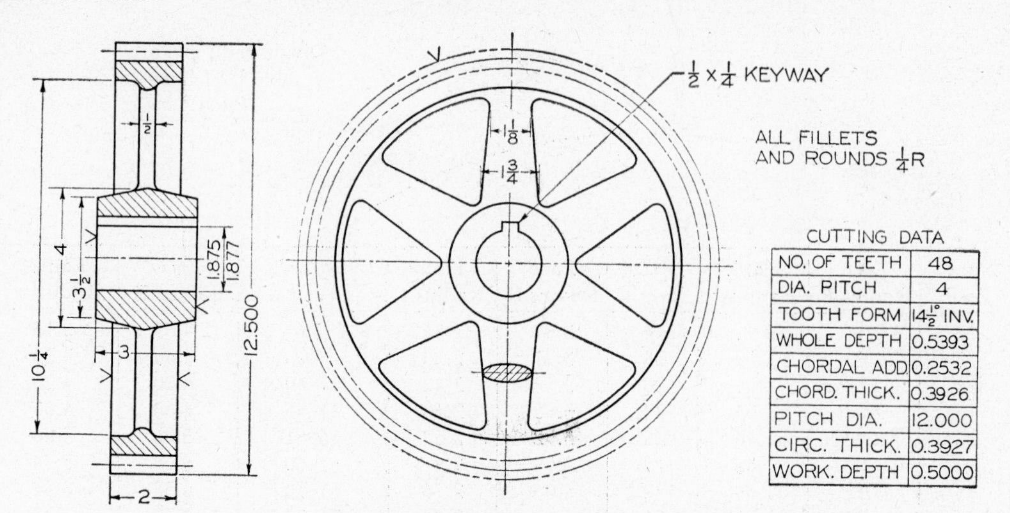

CUTTING DATA	
NO. OF TEETH	48
DIA. PITCH	4
TOOTH FORM	$14\frac{1}{2}°$ INV.
WHOLE DEPTH	0.5393
CHORDAL ADD	0.2532
CHORD. THICK.	0.3926
PITCH DIA.	12.000
CIRC. THICK.	0.3927
WORK. DEPTH	0.5000

Fig. 972 Working Drawing of a Spur Gear.

Before laying out the working drawing, the draftsman must calculate the gear dimensions. For example, if the gear must have 48 teeth of 4 diametral pitch, with $14\frac{1}{2}°$ full-depth involute profile, as in Fig. 972, then the following items should be calculated in this order: pitch diameter, addendum, dedendum, outside diameter, root diameter, whole depth, chordal thickness, and chordal addendum. The dimensions shown in the figure are the minimum requirements for a spur gear. The chordal addendum and chordal thickness are given to aid in checking the finished gear. Other special data may be given in the table, according to the degree of precision required and the method of manufacture.

550. Bevel Gears. *Bevel gears* are used to transmit power between shafts whose axes intersect. The analogous friction drive would consist of a pair of cones having a common apex at the point of intersection of the axes. The axes may intersect at any angle, but axes at right angles occur most frequently, and this is the only case that will be considered here. Bevel gear teeth have the same involute shape as teeth on spur gears, but the bevel gear teeth are tapered toward the cone apex; hence the height and width of a bevel gear tooth vary as the distance from the cone apex. Whereas spur gears are interchangeable—a spur gear of given pitch will run properly with any other spur gear of the same pitch and tooth form—this is not true of bevel gears, which must be designed in pairs and will run only with each other.

The speed ratio of bevel gears can be calculated from the same formulas given for spur gears in §546.

551. Bevel Gear Definitions and Formulas. The design of bevel gears is very similar to that of spur gears; therefore, many of the spur gear terms are applied with slight modification to bevel gears. The *pitch diameter* D of a bevel gear is the diameter of the base of the pitch cone, and the *circular pitch* p of the teeth is measured along this circle. The *diametral pitch* P is also based on this circle; hence the relationship of these three items is the same as for spur gears.

The important dimensions and angles of a bevel gear are illustrated in Fig. 973. The *pitch cone* is shown as the triangle OAB. Examination of Fig. 974 will reveal the pitch cones for the mating gear and pinion shown there. Evidently, the pitch angle of each gear depends upon the relative diameters of the gears. Therefore, the pitch angles, Γ(*gamma*), are determined from the following equations:

$$\tan \Gamma_G = \frac{D_G}{D_P} = \frac{N_G}{N_P}, \quad \tan \Gamma_P = \frac{D_P}{D_G} = \frac{N_P}{N_G}.$$

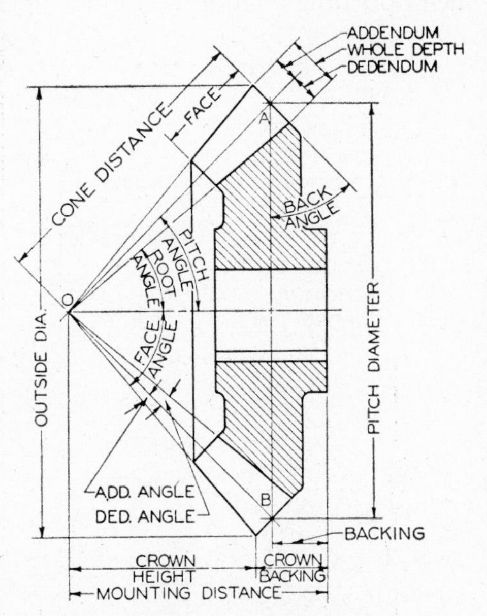

Fig. 973 Bevel Gear Nomenclature.

Other terms are defined as:

Pitch Diameter (D_G or D_P). The diameter of the base of the pitch cone. $D_G = N_G/P$ and $D_P = N_P/P$.

Cone Distance (A). The slant height of the pitch cone; hence the same for both gear and pinion. $A = D/2 \sin \Gamma$.

Addendum (a). The same as for a spur gear of the same diametral pitch. It is measured at the large end of the tooth. $a = 1/P$.

Dedendum (b). The same as for a spur gear of the same diametral pitch. It is measured at the large end of the tooth. $b = 1.157/P$.

Addendum Angle (α). The angle subtended by the addendum. It is the same for both gear and pinion. $\tan \alpha = a/A$.

Dedendum Angle (δ). The angle subtended by the dedendum. It is the same for both gear and pinion. $\tan \delta = b/A$.

Face Angle (Γ_O). The angle between the top of the teeth and the gear axis. $\Gamma_O = \Gamma + \alpha$.

Root Angle (Γ_R). The angle between the root of the teeth and the gear axis. $\Gamma_R = \Gamma - \delta$.

Back Angle. This angle is usually equal to the pitch angle.

Outside Diameter (D_O). The diameter of the outside or crown circle of the gear. $D_O = D + 2a \cos \Gamma$.

Crown Height (X). The distance parallel to the gear axis from the cone apex to the crown of the gear. $X = \frac{1}{2}D_O/\tan \Gamma_O$.

Backing (Y). The distance from the base of the pitch cone to the rear of the hub.

Crown Backing (Z). For shop use, the crown backing is more practical than the backing; hence dimension Z is given on drawings instead of Y. $Z = Y + a \sin \Gamma$.

Mounting Distance (M). This dimension is used primarily for inspection and assembly purposes. $M = Y + \frac{1}{2}D/\tan \Gamma$.

Face Width (F). The face width should not exceed $\frac{1}{3}A$.

Equivalent Number of Teeth $(N)_e$. This information is needed in selecting the proper cutter to form the teeth. $N_e = N/\cos \Gamma$.

552. Working Drawings of Bevel Gears. As in the case of spur gears, a working drawing of a bevel gear gives only the dimensions of the gear blank. The necessary data for cutting the teeth are given in a note or table. A single sectional view, Fig.

974, usually will provide all necessary information. If a second view is required, only the gear blank is drawn, and the tooth profiles are omitted. Two gears are shown in their operating relationship. On detail drawings, each gear is usually drawn separate-

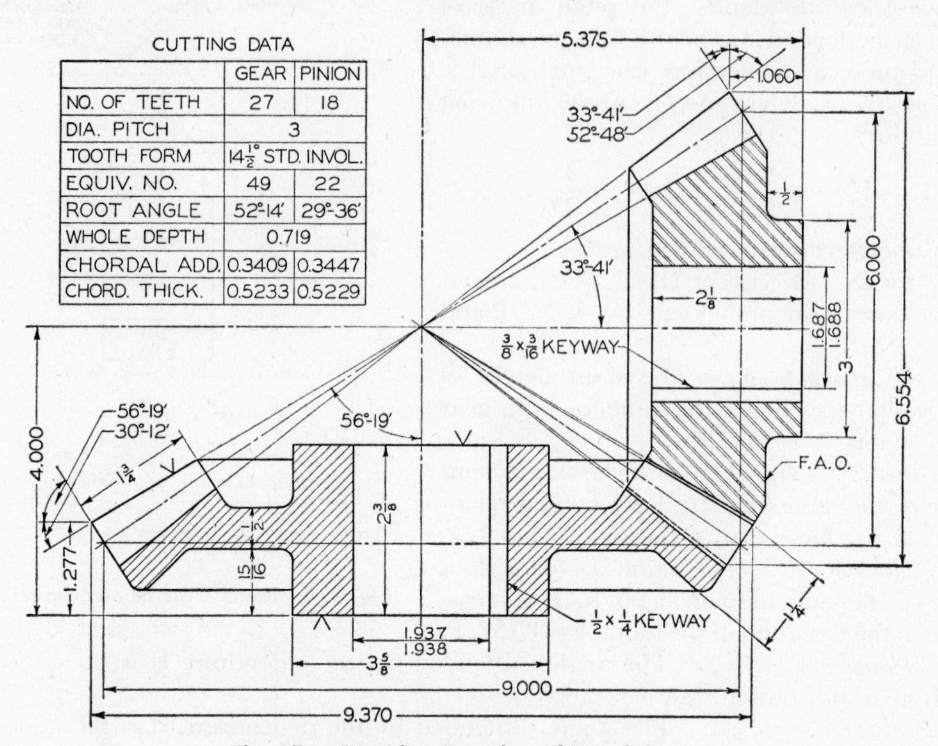

CUTTING DATA

	GEAR	PINION
NO. OF TEETH	27	18
DIA. PITCH	3	
TOOTH FORM	$14\frac{1}{2}°$ STD. INVOL.	
EQUIV. NO.	49	22
ROOT ANGLE	52°-14′	29°-36′
WHOLE DEPTH	0.719	
CHORDAL ADD.	0.3409	0.3447
CHORD. THICK.	0.5233	0.5229

Fig. 974 Working Drawing of Bevel Gears.

ly, Fig. 973, and fully dimensioned. Proper placing of the gear-blank dimensions is largely dependent upon the shop methods used in producing the gear, but the scheme shown is commonly followed.

553. Worm Gears. Worm gears are used to transmit power between non-intersecting shafts that are at right angles to each other. The *worm* is a screw having a thread of the same shape as a rack tooth. The *worm wheel* is similar to a spur gear except that the teeth have been twisted and curved to conform to the shape of the worm. A large speed ratio is obtainable with worm gearing, since a *single-thread worm* in one revolution advances the worm wheel only one tooth and space.

In Fig. 975, a worm and worm wheel are shown engaged. The section taken through the center of the worm and perpendicular to the axis of the worm wheel shows that the worm section is identical with a rack, and that the wheel section is identical with a spur gear. Consequently, in this plane the height proportions of thread and gear teeth are the same as for a spur gear of corresponding pitch.

Pitch (p). The axial pitch of the worm is the distance from a point on one thread to the corresponding point on the next thread measured parallel to the worm axis. The pitch of the worm must exactly equal the circular pitch of the gear.

Lead (L). The lead is the distance that the thread advances axially in one turn. The lead is always a multiple of the pitch. Thus for a single-thread worm, the lead equals the pitch; for a double-thread worm, the lead is twice the pitch, etc.

Lead Angle (λ). The angle between a tangent to the helix at the pitch diameter

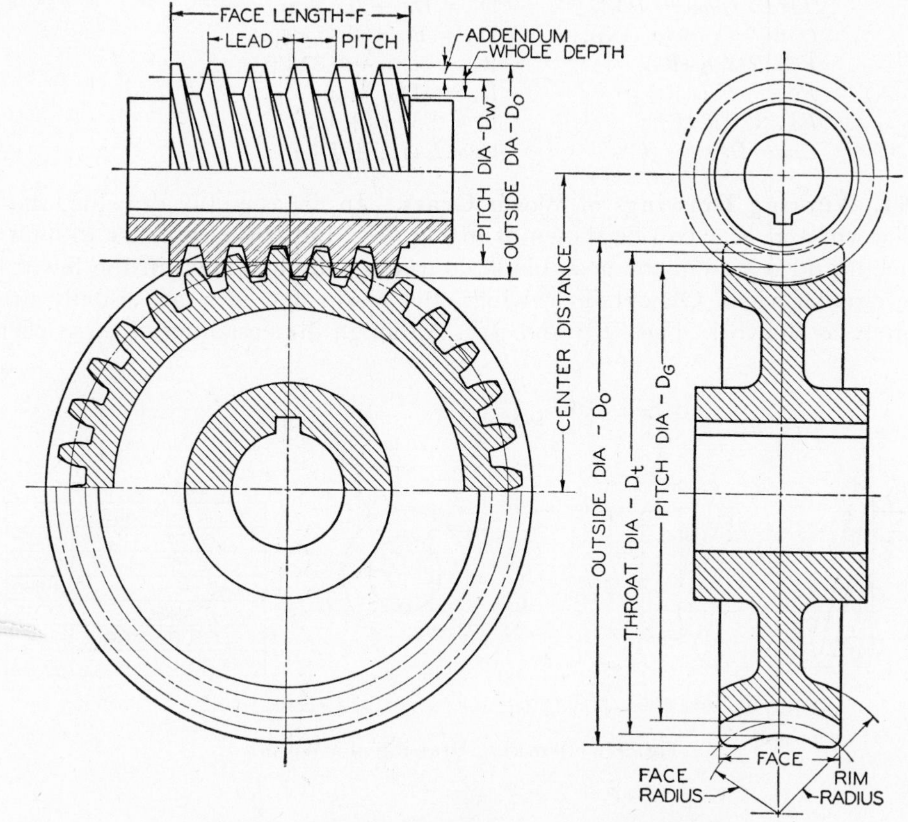

Fig. 975 Double Thread Worm and Worm Gear.

and a plane perpendicular to the axis of the worm. The lead angle can be calculated from the formula

$$\tan \lambda = \frac{L}{\pi D_W}$$

where D_W is the pitch diameter of the worm.

The speed ratio of worm gears depends only upon the *number of threads* on the worm and the *number of teeth* on the gear. Therefore $m_G = N_G/N_W$ where N_G is the number of teeth on the gear, and N_W is the number of threads on the worm.

For $14\frac{1}{2}°$ standard involute teeth and single-thread or double-thread worms, the following proportions are the recommended practice of the A.G.M.A. (American Gear Manufacturers' Association). All formulas are expressed in terms of circular pitch p instead of diametral pitch. It is easier to machine the worm and the hob used to cut the gear if the circular pitch has an even rational value such as $\frac{5}{8}''$.

For the worm:

Pitch Diameter (D_W).	$D_W = 2.4\ p + 1.1$ (recommended value, but may be varied).
Whole Depth (h_t).	$h_t = 0.686\ p$
Outside Diameter (D_O).	$D_O = D_W + 0.636\ p$
Face Length (F).	$F = p\ (4.5 + N_G/50)$

For the gear:

Pitch Diameter (D_G).	$D_G = p\ N_G/\pi$
Throat Diameter (D_t).	$D_t = D_G + 0.636\ p$
Outside Diameter (D_O).	$D_O = D_t + 0.4775\ p$
Face Radius (R_f).	$R_f = \frac{1}{2}D_W - 0.318\ p$
Rim Radius (R_r).	$R_r = \frac{1}{2}D_W + p$
Face Width (F).	$F = 2.38\ p + 0.25$
Center Distance (C).	$C = \frac{1}{2}\ (D_G + D_W)$

554. Working Drawings of Worm Gears. In an assembly drawing, the engaged worm and gear can be shown as in Fig. 975, but it is customary to omit the gear teeth and represent the gear blank conventionally as shown in the lower half of the circular view. On detail drawings, the worm and gear are usually drawn separately as shown in Figs. 976 and 977. Although dimensioning of these parts is

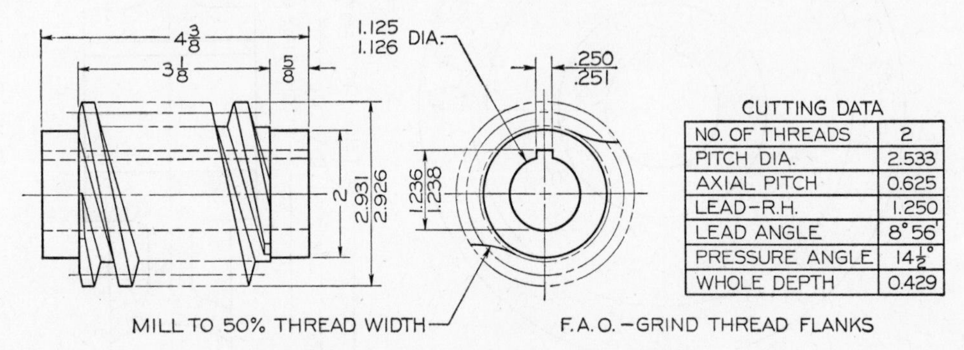

Fig. 976 Working Drawing of a Worm.

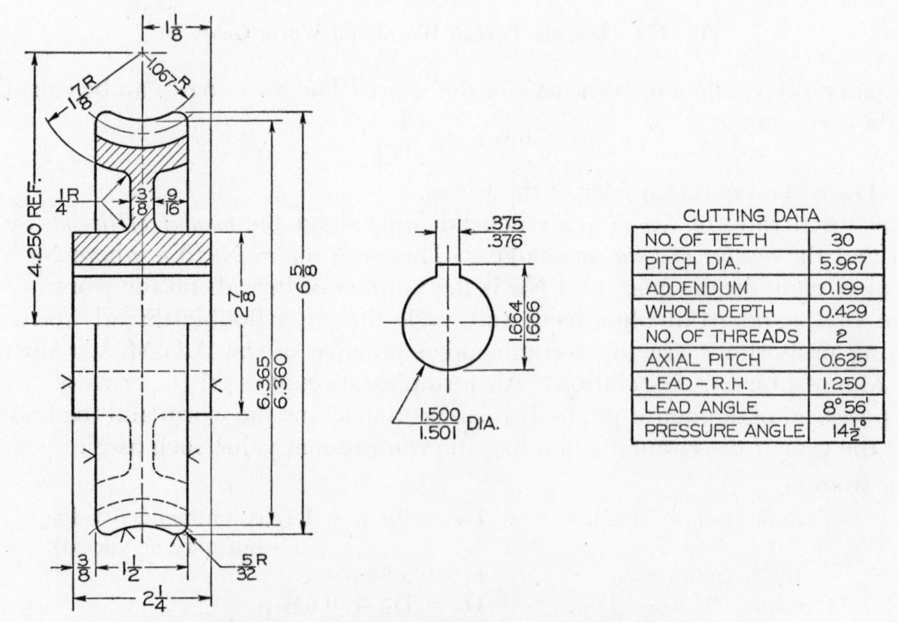

Fig. 977 Working Drawing of a Worm Gear.

again largely dependent upon the method of production, it is standard practice to dimension the blanks on the views, and to give the cutting data in tabular form, as shown. Note that those dimensions which closely affect the engagement of the gear and worm have been given as decimal or limit dimensions; other dimensions, such as rim radius, face lengths, and gear outside diameter have been rounded to convenient common fractional values.

555. Cams. *Cams* provide a simple means for obtaining unusual and irregular motions that would be difficult to produce otherwise. Figure 978 (a) illustrates the basic principle of the cam. A shaft rotating at uniform speed carries an irregularly

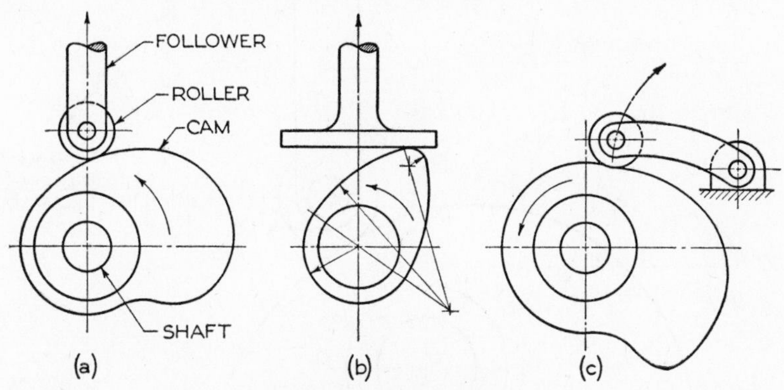

Fig. 978 Disk Cams.

shaped disk called the *cam*; a reciprocating plunger, called the *follower*, presses a small roller against the curved surface of the cam. Rotation of the cam thus causes the follower to reciprocate with a definite cyclic motion according to the shape of the cam profile. The roller is held in contact with the cam by gravity or a spring. The problem of the draftsman is to construct the cam profile necessary to obtain the desired motion of the follower.

An automobile valve cam that operates a flat-faced follower is shown at (b). The profile of this cam is composed of circular arcs for ease in manufacture. At (c) is shown a disk cam with the roller follower attached to a swinging arm.

556. Displacement Diagrams. Since the motion of the follower is of primary importance, its rate of speed and its various positions should be carefully planned in a *displacement diagram* before the cam profile is constructed. A displacement diagram, Fig. 979, is a curve showing the displacement of the follower as ordinates erected

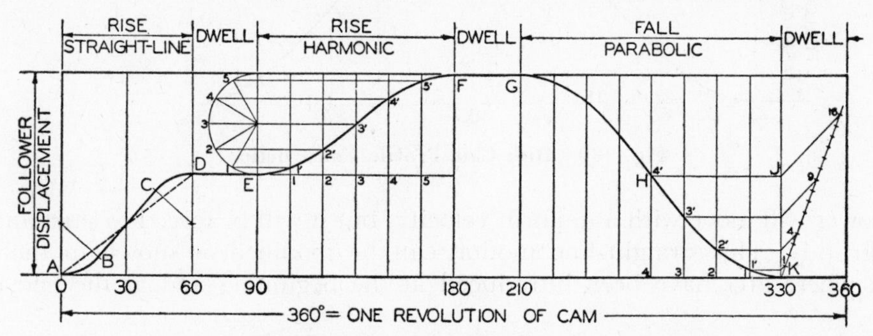

Fig. 979 Displacement Diagram with Typical Curves.

on a base line that represents one revolution of the cam. The follower displacement should be drawn to scale, but any convenient length can be used to represent the 360° of cam rotation.

The motion of the follower as it rises or falls depends upon the shape of the curves in the displacement diagram. In this diagram, four commonly employed types of curves are shown. If a straight line is used, such as the dashed line AD in the figure,

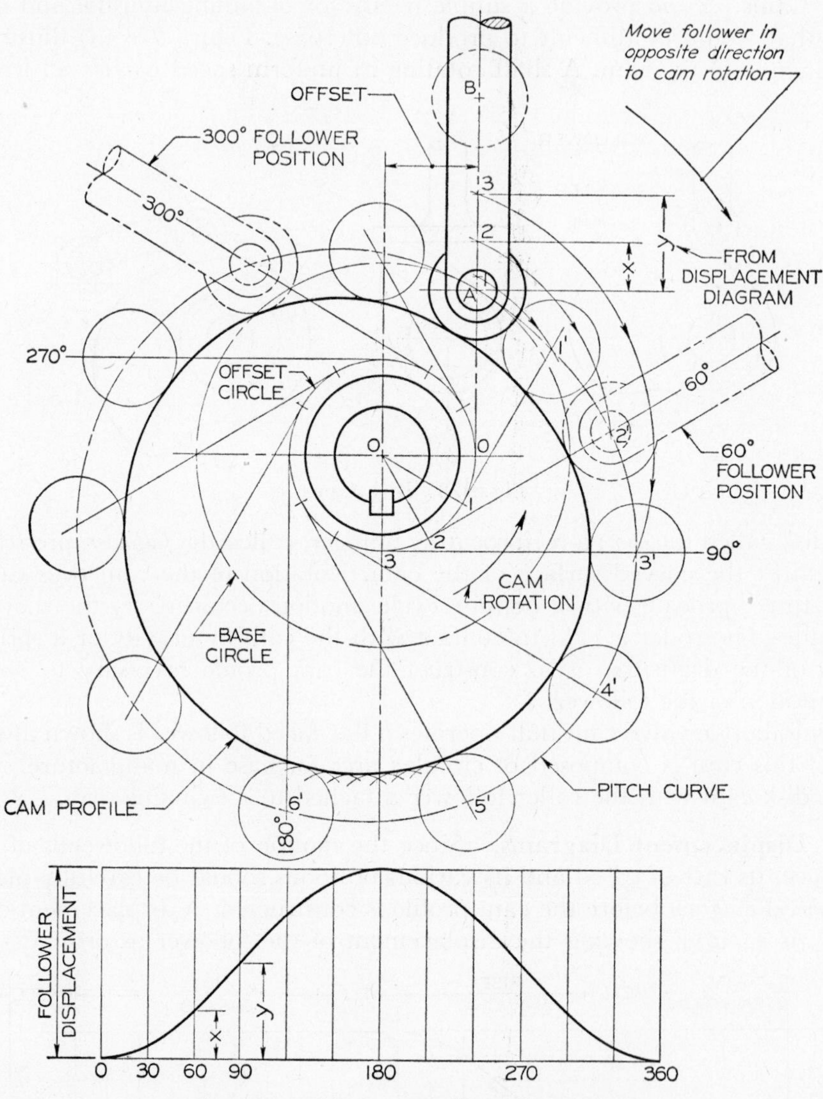

Fig. 980 Disk Cam Profile Construction.

the follower will move with a uniform velocity, but it will be forced to start and stop very abruptly. This straight-line motion can be modified as shown in the curve ABCD, where arcs have been introduced at the beginning and at the end of the period.

The curve shown at EF is one which gives harmonic motion to the follower. To construct this curve, a semicircle is drawn whose diameter is equal to the desired rise. The circumference of the semicircle is divided into equal arcs, the number of divisions being the same as the number of horizontal divisions. Points on the curve are then found by projecting horizontally from the divisions on the semicircle to the corresponding ordinates.

The parabolic curve shown at GHK gives the follower constantly accelerated and decelerated motion. This motion is analogous to that of a falling body. The half of the curve from G to H is exactly the reverse of the half from H to K. To construct the curve HK, the vertical height from K to J is divided into distances proportional to 1^2, 2^2, 3^2, etc., or 1, 4, 9, etc., the number of such divisions being the same as the number of horizontal divisions. See §§123 and 165. Points on the curve are found by projecting horizontally from the divisions on the line JK to the corresponding ordinates.

557. The Cam Profile. The general procedure in constructing a cam profile is shown in Fig. 980. The disk cam rotating counterclockwise on its shaft raises and lowers the roller follower along the straight line AB. The axis of the follower is *offset* from the center line of the cam. The displacement diagram at the bottom of the figure shows the desired follower motion.

With the follower in its lowest or initial position, the center of the roller is at A, and OA is the radius of the *base circle*. With the same center O, an *offset circle* is drawn with radius equal to the offset. As the cam turns, the extended center line of the follower will always be tangent to this offset circle. But since the cam must remain stationary while it is being drawn, an equivalent rotative effect must be obtained by imagining that the cam stands still while the follower rotates about the cam in the opposite direction. Therefore, the offset circle is divided into twelve equal angular divisions corresponding to the divisions used in the displacement diagram. These divisions begin at zero and are numbered in an opposite direction to the cam rotation. Tangent lines are then drawn from each point on the offset circle as shown.

The points 1, 2, 3, etc., on the follower axis AB indicate successive positions of the center of the roller, and are located by transferring ordinates such as x and y from the displacement diagram. Thus when the cam has rotated 60°, the follower roller must rise a distance x to position 2, and after 90° of rotation, a distance y to position 3, etc.

It should now be observed that while the center of the roller moved from its initial position A to position 2, for example, the cam rotated 60° *counterclockwise*. Point 2 must therefore be revolved *clockwise* about the cam center O to the corresponding 60° tangent line to establish point 2′. In this position, the complete follower would appear as shown by the phantom outline. Points 1′, 2′, 3′, etc., represent consecutive positions of the roller center, and a smooth curve drawn through these points is called the *pitch curve*. To obtain the actual cam profile, the roller must be drawn in a number of positions, and the cam profile drawn tangent to the roller circles as shown. The best results are obtained by first drawing the pitch curve very carefully, and then drawing a great many closely spaced roller circles with centers on the pitch curve as shown between points 5′ and 6′.

If the roller is on a pivoted arm, as shown in Fig. 981 (a), then the displacement of the roller center is along the circular arc AB. The height of the displacement diagram (not shown) should be made equal to the rectified length of arc AB. Ordinates from the diagram are then transferred to arc AB to locate the roller positions 1, 2, 3, etc. As the follower is revolved about the cam, pivot point C moves in a circular path of radius OC to the consecutive positions C_1, C_2, etc. Length AC is constant for

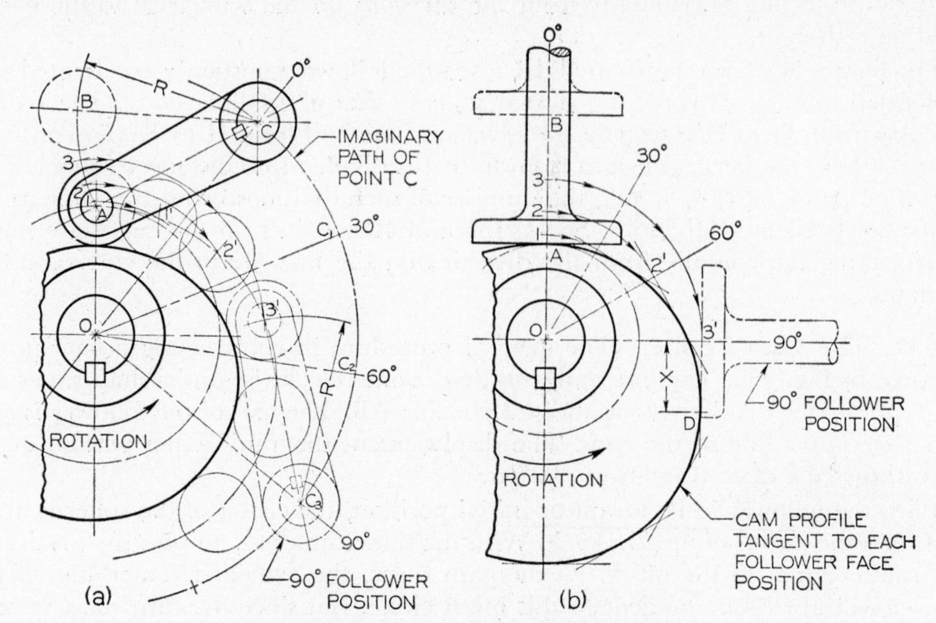

Fig. 981 Pivoted and Flat-Faced Followers.

all follower positions; hence, from each new position of point C the follower arc of radius R is drawn as shown at the 90° position. The roller centers 1, 2, 3, etc., are now revolved about the cam center O to intersect the follower arcs at 1′, 2′, 3′, etc. After the pitch curve is completed, the cam profile is drawn tangent to the roller circles.

The construction for a flat-faced follower is shown at (b). The initial point of contact is at A, and points 1, 2, 3, etc., represent consecutive positions of the follower face. Then for the 90° position, point 3 must be revolved 90°, as shown, to position 3′, and the flat face of the follower is drawn through point 3′ at right angles to the cam radius. When this procedure has been repeated for each position, the cam profile will be enveloped by a series of straight lines, and the cam profile is drawn inside and tangent to these lines. Note that the point of contact, initially at A, changes as the follower rises: at 90°, for example, contact is at D, a distance X to the right of the follower axis.

558. Cylindrical Cams. When the follower movement is in a plane parallel to the cam shaft, some form of cylindrical cam must be employed. In Fig. 982, for example, the follower rod moves vertically parallel to the cam axis as the attached roller follows the groove in the rotating cam cylinder.

A cylindrical cam of diameter D is required to lift the follower rod a distance AB with harmonic motion in 180° of cam motion, and to return the rod in the remaining 180° with the same motion. The displacement diagram is drawn first, and conveniently placed directly opposite the front view. The 360° length of the diagram must be made equal to πD so that the resulting curves will be a true development of the outer surface of the cam cylinder. The pitch curve is drawn to represent the required motion, and a series of roller circles are then drawn to establish the sides of the groove tangent to these circles. This completes the development of the outer cylinder, and actually provides all information needed for making the cam; hence it is not uncommon to omit the curves in the front view.

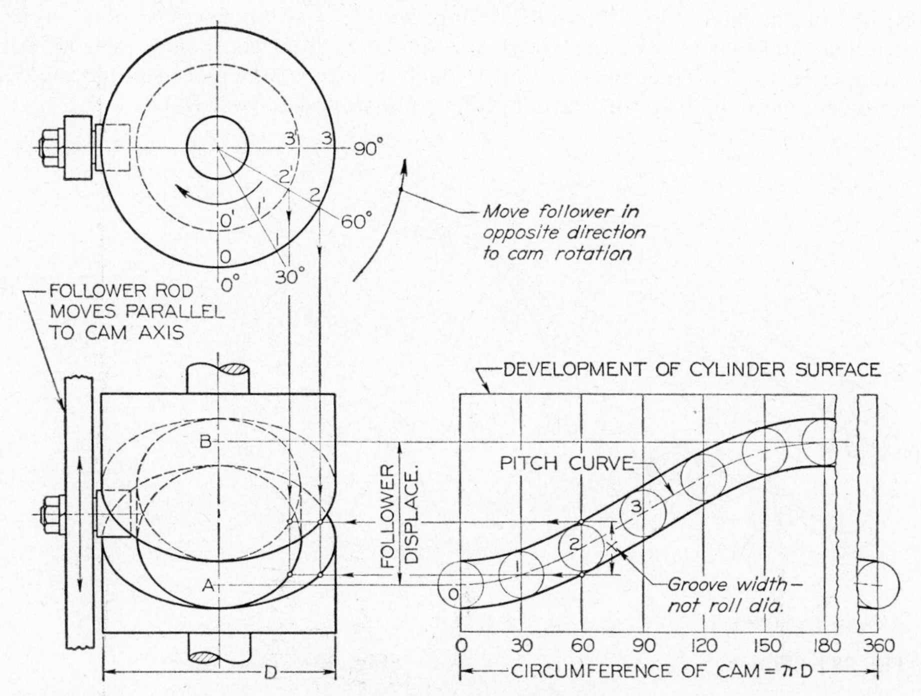

Fig. 982 Cylindrical Cam.

To complete the front view, points on the curves are projected horizontally from the development. For example, at 60° in the development, the width of the groove measured parallel to the cam axis is X, a distance somewhat greater than the actual roller diameter. This width X is projected to the front view directly below point 2, the corresponding 60° position in the top view, to establish two points on the outer curves. The inner curves for the bottom of the groove can be established in the same manner except that the groove width X is located below point 2′, which lies on the inner diameter. The inner curves are only approximate because the width at the bottom of the groove is actually slightly greater than X, but the exact bottom width can only be determined by drawing a second development for the inner cylinder.

559. Problems on Gearing and Cams. The following problems are given to provide practice in laying out and making working drawings of the common types of gears and cams. Where paper sizes are not given, the student is to select his own scale and sheet layout.

GEARING

Prob. 1. A 12-tooth 1 DP pinion engages a 15-tooth gear. Make a full-size drawing of a segment of each gear showing how the teeth mesh. Construct the $14\frac{1}{2}°$ involute teeth exactly, noting any points where the teeth appear to interfere.

Prob. 2. The same as Prob. 1, but use 20° stub teeth.

Prob. 3. The same as Prob. 1, but use a rack in place of the 12-tooth pinion.

Prob. 4. Make a display drawing of the pinion in Fig. 983. Show two views, drawing the teeth by the approximate method shown in Fig. 969. Draw double size.

Prob. 5. The same as Prob. 4, but use Grant's Odontograph to draw the teeth.

Prob. 6. A spur gear has 60 teeth of 5 diametral pitch. The face width is $1\frac{1}{2}''$. The shaft is $1\frac{3}{16}''$ in diameter. Make the hub $2''$ long, and $2\frac{1}{4}''$ in diameter. Calculate accurately all dimensions, and make a working drawing of the gear. Show six spokes, each $1\frac{1}{8}''$ wide at the hub, tapering to $\frac{3}{4}''$ wide at the rim, and $\frac{1}{2}''$ thick. Use your own judgment for any dimensions not given. Draw half size on Layout A-3, or full size on Layout B-3.

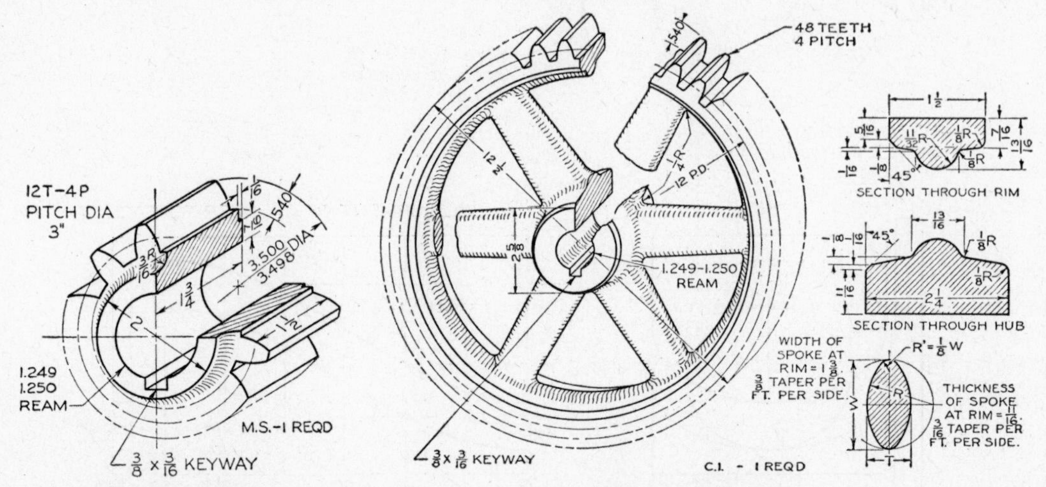

Fig. 983 Pinion. **Fig. 984 Spur Gear.**

Prob. 7. Make a working drawing of the intermediate pinion shown in Fig. 803 (Part 6). Check the gear dimensions by calculation.

Prob. 8. The same as Prob. 7, but use the intermediate gear shown in Fig. 803 (Part 5).

Prob. 9. The same as Prob. 7, but use the pinion shown in Fig. 983.

Prob. 10. The same as Prob. 7, but use the spur gear shown in Fig. 984.

Prob. 11. A pair of bevel gears have teeth of 4 diametral pitch. The pinion has 13 teeth, the gear 25 teeth. The face width is $1\frac{1}{8}''$. The pinion shaft is $\frac{15}{16}''$ in diameter, and the gear shaft is $1\frac{3}{16}''$ in diameter. Calculate accurately all dimensions, and make a working drawing showing the gears engaged as in Fig. 974. Make the hub diameters approximately twice the shaft diameters. Select key sizes from Appendix table 19. The backing for the pinion must be $\frac{5}{8}''$; for the gear $1\frac{1}{4}''$. Use your own judgment for any dimensions not given.

Prob. 12. Make a working drawing of the pinion in Prob. 11.

Prob. 13. Make a working drawing of the gear in Prob. 11.

Prob. 14. The same as Prob. 12, but show two views of the pinion.

Prob. 15. The same as Prob. 11, but use 5 diametral pitch, 30 and 15 teeth. The face is $1''$. Shafts: pinion, $1''$ diam.; gear, $1\frac{1}{2}''$ diam. Backing: pinion, $\frac{1}{2}''$, gear, $1''$.

Prob. 16. The same as Prob. 11, but use 4 diametral pitch, both gears 20 teeth. Select the correct face width. Shafts: $1\frac{3}{8}''$ diam. Backing: $\frac{3}{4}''$.

Prob. 17. Fig. 985 (a) shows a countershaft end used on a trough conveyor, and (b) shows the layout for an assembly drawing of a similar unit. Using a scale of half size, make an assembly drawing of the complete unit (given layout dimensions are full size). The following full-size dimensions are sufficient to establish the position and general outline of the parts; the minor detail dimensions, web and rib shapes, and fillets and rounds should be designed by the student with the help of the photograph.

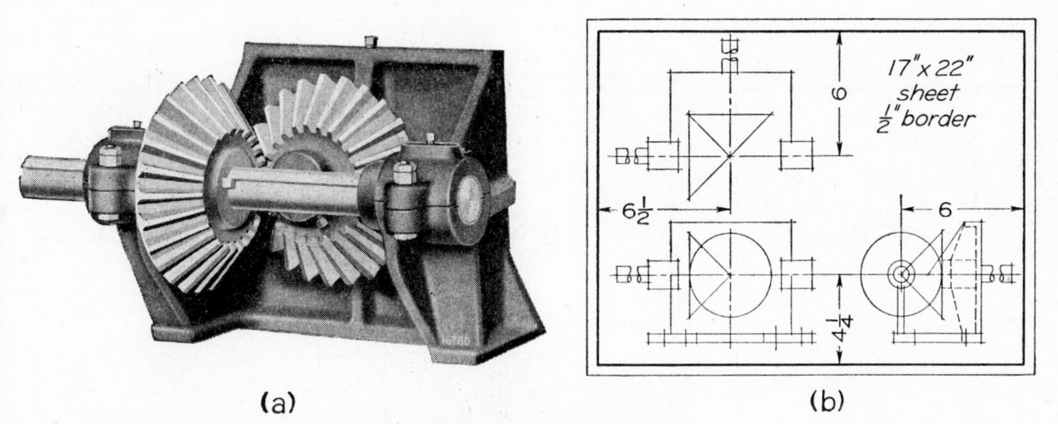

Fig. 985 Countershaft End. *Photo courtesy Link-Belt Co.*

The gears are identical in size and are similar in shape to the one in Fig. 973. There are 24 teeth of 3 diametral pitch in each gear. Face width, 2″. Shafts, $1\frac{1}{2}''$ diam., extending 7″ beyond left bearing, 4″ beyond rear (break shafts as shown). Hub diameters, $2\frac{3}{4}''$. Backing, 1″. Hub lengths, $3\frac{1}{4}''$. Front gear is held by a square gib key (see Appendix table 19); back gear by a square key and a $\frac{3}{8}''$ set screw. Collar on front shaft next to right bearing is $\frac{3}{4}''$ thick, $2\frac{1}{4}''$ outside diameter, with a $\frac{3}{8}''$ set screw.

On the main casting, the split bearings are $2\frac{3}{4}''$ diam., 3″ long, and 10″ apart. Each bearing cap is held by two $\frac{1}{2}''$ bolts, $3\frac{1}{2}''$ apart, center to center. Oil holes have a $\frac{1}{4}''$ pipe tap for plug or grease cup. Shaft center lines are $6\frac{1}{2}''$ above bottom surface, 8″ from rear surface of casting. Main casting is $11\frac{1}{2}''$ high, and its base is 16″ long, $8\frac{3}{4}''$ wide, and $\frac{3}{4}''$ thick. All webs, ribs, and walls are uniformly $\frac{1}{2}''$ thick. In the base are eight holes (not shown in the photograph) for $\frac{1}{2}''$ bolts, two outside each end, and two inside each end.

Proceed by first blocking in the gears in each view, and then block in the principal main casting dimensions. Fill in details only after principal dimensions are clearly established.

Prob. 18. The worm and worm gear shown in Fig. 975 have a circular pitch of $\frac{5}{8}''$, and the gear has 32 teeth of $14\frac{1}{2}°$ involute form. The worm is double threaded. Make an assembly drawing similar to Fig. 975. The teeth on the gear may be drawn by either method of §548. Calculate dimensions accurately, and use A.G.M.A. proportions. Shafts: worm, $1\frac{1}{8}''$ diam.; gear, $1\frac{5}{8}''$ diam.

Prob. 19. Make a working drawing of the worm in Prob. 18.

Prob. 20. Make a working drawing of the gear in Prob. 18.

Prob. 21. The same as Prob. 19, but the worm is single-thread.

Prob. 22. A single-thread worm has a lead of $\frac{3}{4}''$. The worm gear has 28 teeth of standard form. Make a working drawing of the worm. The shaft is $1\frac{1}{4}''$ in diameter.

Prob. 23. Make a working drawing of the gear in Prob. 22. The shaft is $1\frac{1}{2}''$ in diameter.

CAMS

Prob. 24. Fig. 986 (a). Draw the displacement diagram, and determine the cam profile that will give the radial roller follower this motion: up $1\frac{1}{2}''$ in 120°, dwell 60°, down in 90°, dwell 90°. Motions are to be unmodified straight line and of uniform velocity. The roller is $\frac{3}{4}''$ in diameter, and the base circle is 3″ in diameter. Note that the follower has zero offset. The cam rotates clockwise.

Prob. 25. The same as Prob. 24, except that the straight-line motions are to be modified by arcs whose radii are equal to one-half the rise of the follower.

Prob. 26. The same as Prob. 24, except that the upward motion is to be harmonic, and the downward motion parabolic.

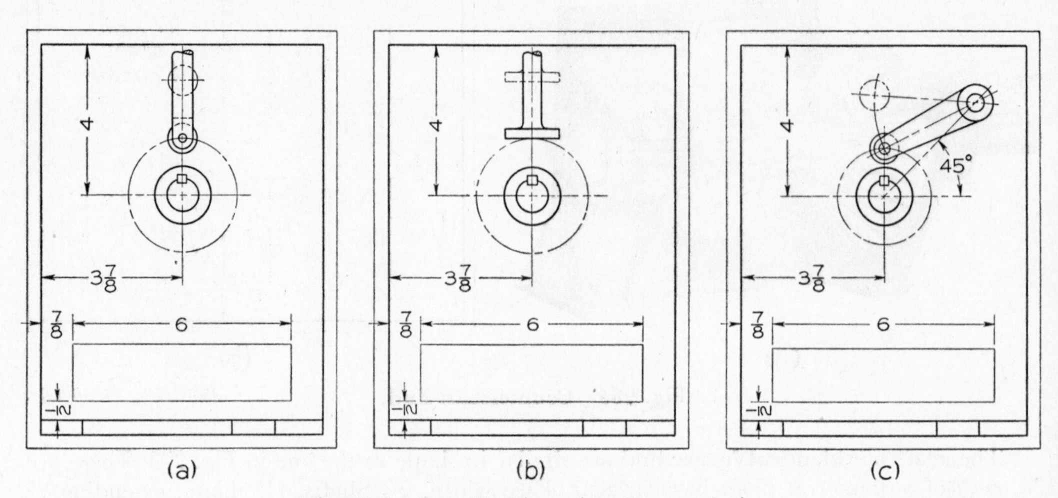

Fig. 986 Cam Problems (use Layout A-1).

Prob. 27. The same as Prob. 26, except that the follower is offset 1″ to the left of the cam center line.

Prob. 28. Fig. 986 (b). Draw the displacement diagram, and determine the cam profile that will give the flat-faced follower this motion: dwell 30°, up $1\frac{1}{2}$″ on a parabolic curve in 180°, dwell 30°, down with harmonic motion in 120°. The base circle is 3″ in diameter. After completing the cam profile, determine the necessary width of face of the follower by finding the position of the follower where the point of contact with the cam is farthest from the follower axis. The cam rotates counterclockwise.

Prob. 29. Fig. 986 (c). Draw the displacement diagram, and determine the cam profile that will swing the pivoted follower through an angle of 30° with the same motion as prescribed in Prob. 28. The radius of the follower arm is $2\frac{3}{4}$″, and in its lowest position the center of the roller is directly over the center of the cam. The base circle is $2\frac{1}{2}$″ in diameter, and the roller is $\frac{3}{4}$″ in diameter. The cam rotates counterclockwise.

Prob. 30. The same as Prob. 29, except that the motion is to be that given in Probs. 24 and 25.

Prob. 31. The same as Prob. 29, except that the motion is to be that given in Probs. 24 and 26.

Prob. 32 (Layout A-3). Using an arrangement like Fig. 982, construct a *half* development, and complete the front view for the following cylindrical cam: Cam, $2\frac{1}{2}$″ diam., $2\frac{1}{2}$″ high. Roller, $\frac{1}{2}$″ diam.; cam groove, $\frac{3}{8}$″ deep. Cam shaft, $\frac{5}{8}$″ diam.; follower rod, $\frac{5}{8}$″ wide, $\frac{5}{16}$″ thick. Motion: up $1\frac{1}{2}$″ with harmonic motion in 180°, down with same motion in remaining 180°. Cam rotates counterclockwise. Assume lowest position of the follower at the front center of cam.

Prob. 33 (Layout B-3). The same as Prob. 32, but construct a *full* development for the following motion: up $1\frac{1}{2}$″ with parabolic motion in 120°, dwell 90°, down with harmonic motion in 150°.

CHAPTER 20

GRAPHS

560. Graphical Representation. In previous chapters we have seen how graphical representation is used instead of word-descriptions to describe the size, shape, material, and fabrication methods for the manufacture of actual objects. Graphical representation is also used extensively to represent engineering facts, statistics, and laws of phenomena. A pictorial or graphical description is much more impressive and easier understood than a numerical tabulation or a verbal description. These graphical descriptions are synonymously termed *graphs* or *charts* or *diagrams*.

Tabulated data in Fig. 987 (a) showing the increase in automobile horsepower for

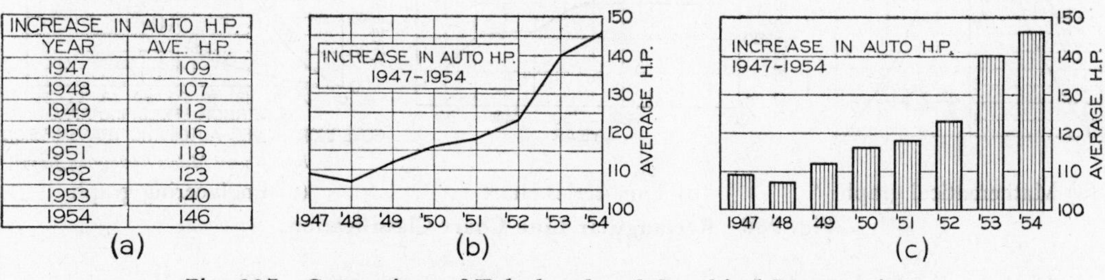

INCREASE IN AUTO H.P.	
YEAR	AVE. H.P.
1947	109
1948	107
1949	112
1950	116
1951	118
1952	123
1953	140
1954	146

Fig. 987 Comparison of Tabulated and Graphical Presentations.

the years 1947 to 1954, inclusive, are presented as a *line graph* at (b) and a *bar graph* at (c). The greater effectiveness of graphical representation is evident.

Graphical constructions are used in three general ways: (1) graphical presentation of data, (2) graphical analysis, and (3) graphical computation. Graphic presentations and analysis are discussed in this chapter, and graphical computations are discussed in Chapter 21.

The type of graphical presentation used depends upon any of the following factors:

1. The type of reader to be reached.
2. The most efficient type of graph to help the reader visualize the significant features.
3. General purpose of chart.
4. Features of a relationship that are considered significant.
5. Occasion for its use.
6. Nature or amount of data.

561. Rectangular Coordinate Line Charts. The *rectangular coordinate line chart* is the type in which values of two related variables are plotted on coordinate paper, and the points, joined together successively, form a continuous line or "curve."

The following are some of the advantages of a line chart:

1. Comparison of a large number of plotted values in a compact space.
2. Comparison of the relative movements (trends) of several curves on the same chart. There should not be more than two or three curves on the same chart, and there should be some definite relationship between them.
3. Interpolation of intermediate values.
4. Representation of movement or over-all trend (relative change) of a series of values, rather than the difference between values (absolute amounts).

Line charts are *not* particularly suited for:

1. Presenting relatively few plotted values in a series.
2. Emphasizing changes or difference in absolute amounts.
3. Showing extreme or irregular movement of data.

Rectangular line charts may be classified as (1) *mathematical graphs*, (2) *time series charts*, or (3) *engineering charts*. Any of these may have one or more curves on the same chart.

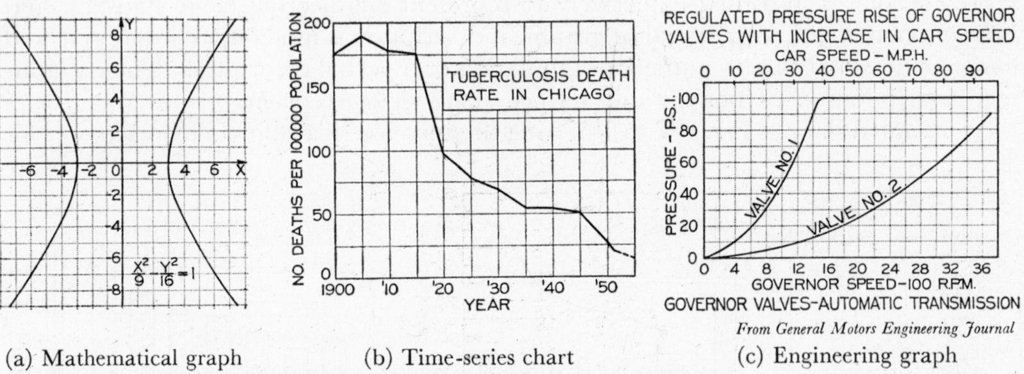

From General Motors Engineering Journal

(a) Mathematical graph (b) Time-series chart (c) Engineering graph

Fig. 988 Rectangular Line Chart Classification.

If the values plotted along the axes are pure numbers (positive and negative), showing the relationship of an equation, the plot is commonly called a mathematical graph, Fig. 988 (a). When one of the variables is any unit of time, the chart is known as a time-series chart, (b). This is one of the most common forms, since time is frequently one of the variables. Line, bar, or surface chart forms may be used for time-series charts, line charts being the most widely used in engineering practice. The plotting of values of any two related physical variables on a rectangular coordinate grid is referred to as an engineering chart, graph, or diagram, (c).

562. Design and Layout of Rectangular Coordinate Line Charts. The steps in drawing a typical co-ordinate line chart are shown in Fig. 989:

I. Select the type of co-ordinate paper, §563. Locate axes, determine the variables for each, and choose appropriate scale or scales. Letter the unit values along the axes, §564.

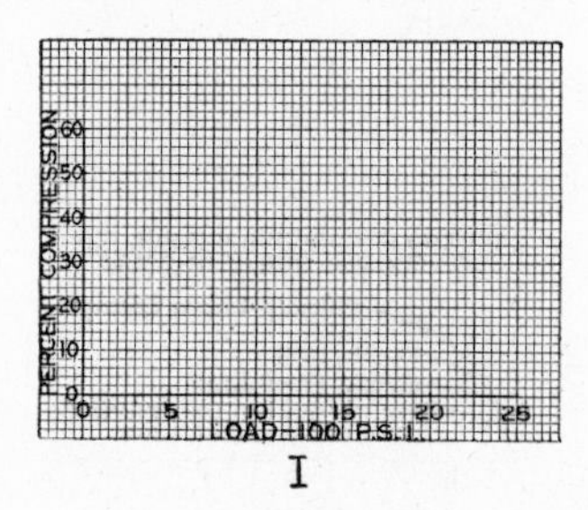

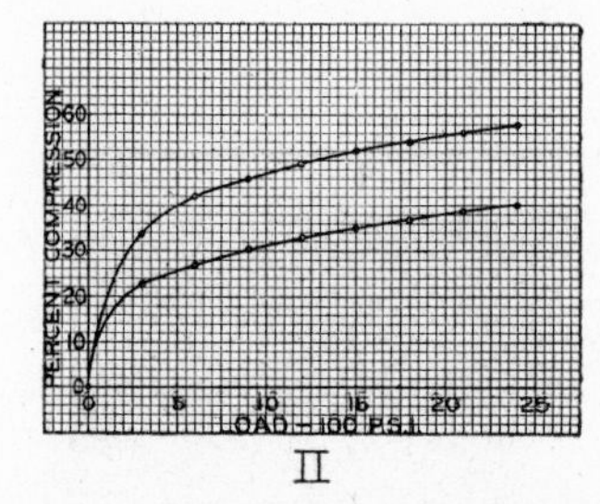

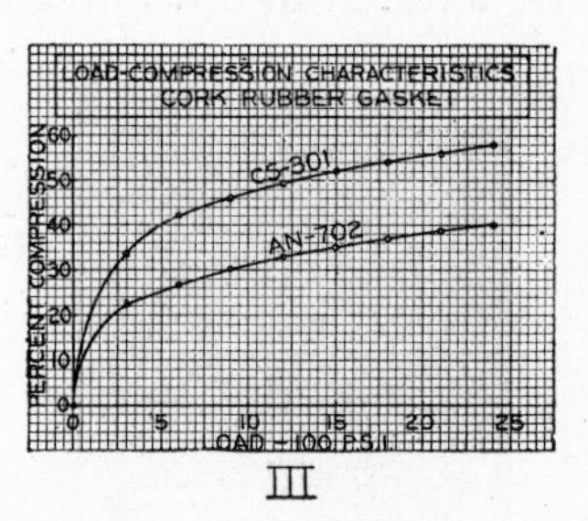

Fig. 989 Steps in Drawing a Graph.

II. Plot the points representing the data and draw the curve or curves, §565.

III. Identify the curves by lettering names or symbols, §565. Letter the title, §566. Ink in the graph if desired.

Much of the above procedure is applicable also to the other forms of charts, graphs, and diagrams to be discussed in subsequent sections.

563. Grids and Composition. To simplify the plotting of values along the perpendicular axes and to eliminate the use of a special scale to locate them, *coordinate paper*, or "cross section paper," ruled with grids, is generally used and can be purchased already printed; or the grids can be drawn on blank paper if desired.

Printed coordinate papers are available in various sizes and spacings of grids, $8\frac{1}{2}'' \times 11''$ being the most common paper size. The spacing of the grid lines may be $\frac{1}{10}''$, $\frac{1}{20}''$, or multiples of $\frac{1}{16}''$. A spacing of $\frac{1}{8}''$ to $\frac{1}{4}''$ is preferred. Closely spaced coordinate ruling is generally avoided for publications, charts reduced in size, and charts used for lantern slides. Much of engineering graphical analysis, however, requires (1) close study, (2) interpolation, and (3) only one copy with possibly a few prints, which can be readily prepared with little effort. These are therefore usually plotted on the closely spaced, printed coordinate paper. Printed papers can be obtained in several colors of lines and in various weights and grades. A thin translucent paper is used when prints are required. A special non-reproducible grid coordinate paper is available in which reproductions will not show the grids on the prints.

Since printed papers do not have sufficient margins to accommodate the axes and nomenclature, the axes should be placed far enough inside the grid area to permit sufficient space for axes and lettering, as shown in Fig. 990 (a) and (b). As much of the remaining grid space as possible should be used for the curve—that is, the scale should be such as to spread the curve out over the available space. A title block (and tabular data, if any) should be placed in an open space on the chart, as shown. If only one copy of the chart is required, tabular data may be placed on the back of the graph or on a separate sheet.

Charts prepared for printed publications, conferences, or projection (lantern slides) generally do not require accurate or detailed interpolation and should emphasize the major facts presented. Coordinate grids drawn on blank paper for these graphs have definite advantages when compared to printed paper. The charts at (b) and (c) show the same information plotted on printed coordinate paper and plain paper, respectively. The specially prepared sheet should have as few grid rulings as necessary, or none, as at (c), to allow a clear interpretation of the curve. Lettering is not placed upon grid lines, which improves the ease of reading. The title and other data can be lettered in open areas, completely free of grid lines.

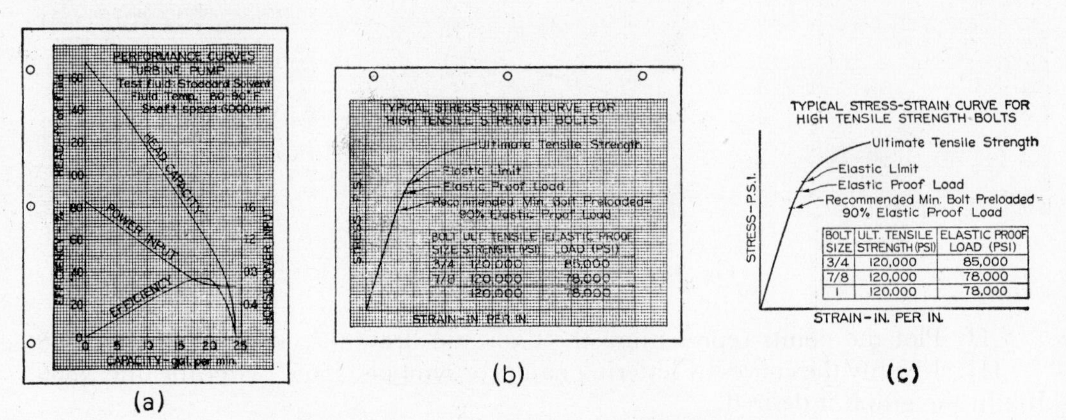

Fig. 990 Printed and Prepared Coordinate Paper.

The layout of specially prepared grids is restricted by the over-all paper size required, or space limitations for lantern slides and other considerations. Space is first provided for margins and for axes nomenclature; the remaining space is then divided into a sufficient number of grid spaces needed for the range of values to be plotted. Another important consideration of composition that affects the spacing of grids is the slope or trend of the curve, as discussed in §564.

Since independent variable values are generally placed along the horizontal axes, especially in time-series charts, vertical rulings can be made for each value plotted, if uniformly spaced, Fig. 991 (a). If there are many values to plot, intermediate values

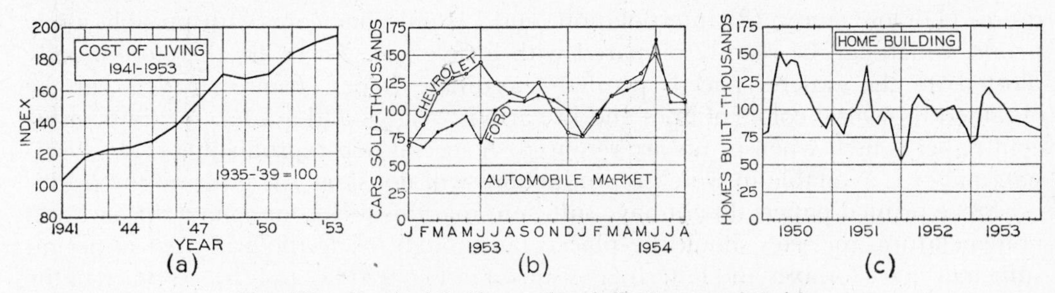

Fig. 991 Vertical Rulings—Specially Prepared Grids.

can be designated by points on the curves, (b), or *ticks* along the axis, (c), with the grid rulings omitted. The horizontal lines are generally spaced according to the available space and the range of values.

The weight of the grid rulings should be thick enough to guide the eye in reading the values, but thin enough to provide contrast and emphasize the curve. The thickness of the lines generally should decrease as the number of rulings increase. As a general rule, as few rulings should be used as possible, but if a large number of rulings are necessary, major divisions should be drawn heavier than the sub-division rulings, for ease of reading.

564. Scales and Scale Designation. The choice of scale is the most important factor of composition and curve significance. Rectangular coordinate line graphs have values of the two related variables plotted with reference to two mutually perpendicular coordinate axes, meeting at a zero point or origin, Fig. 992 (a). The horizontal

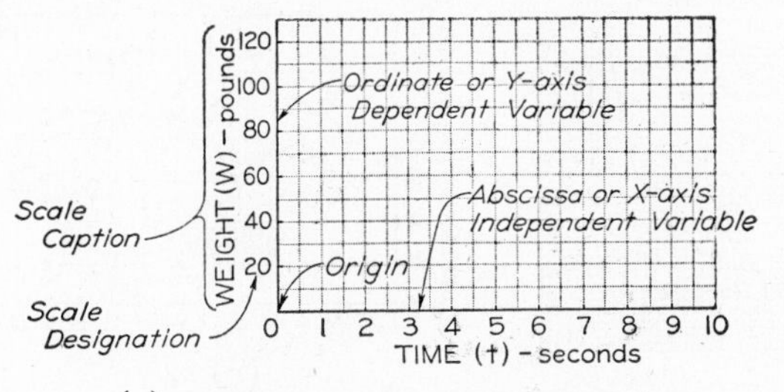

(a) COORDINATE AXES DESIGNATION

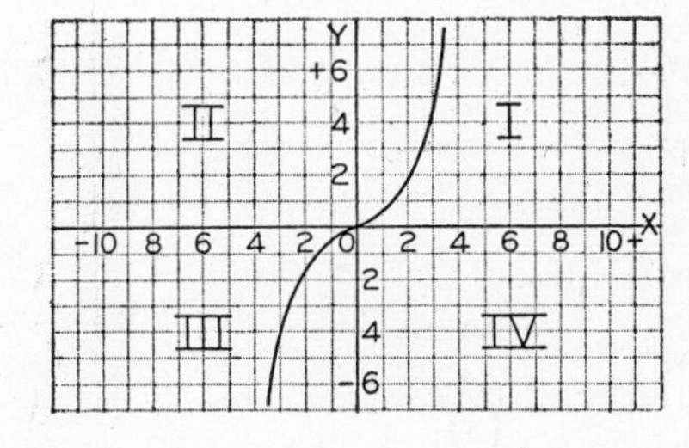

(b) MATHEMATICAL GRAPH AXES DESIGNATION

Fig. 992 Axes Designation.

axis, normally designated as an X-axis, is called the *abscissa*. The vertical axis is denoted as a Y-axis and is called the *ordinate*. It is common practice to place independent values along the abscissa and the dependent values along the ordinate. For example, if in an experiment at certain time intervals, related values are observed, recorded, or determined, the amount of these values is dependent upon the time intervals (independent or controlled) chosen. The values increase from the point of origin toward the right on the X-axis and upward on the Y-axis.

Mathematical graphs, (b), quite often contain positive and negative values, which necessitates the division of the coordinate field into four quadrants, numbered counterclockwise as shown. Positive values increase toward the right on the X-axis and upward on the Y-axis, from the origin. Negative values increase (negatively) to the left on the X-axis and downward on the Y-axis.

Generally a full range of values is desirable, beginning at zero and extending slightly beyond the largest value, to avoid crowding. The available coordinate area should be used as completely as possible. However, certain circumstances require special consideration, to avoid wasted space. For example, if the values to be plotted along one of the axes do not range near zero, a "break" in the grid may be shown,

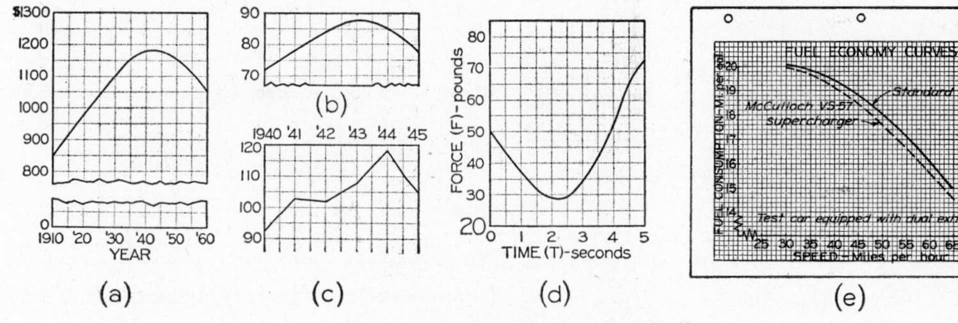

Fig. 993 Axes Scale "Breaks."

as in Fig. 993. However, when relative amount of change is required, Fig. 990 (a), the axes or grid should not be broken or the zero line omitted. If the absolute amount is the important consideration, the zero line may be omitted, as in Fig. 993 (b) to (d). Time designations of years naturally are fixed, and have no relation to zero.

If a few given values to be plotted are widely separated in amount from the others, the total range may be very great, and when this is compressed to fit on the sheet, the resulting curve will tend to be "flat," as shown in Fig. 994 (a). In such cases it is

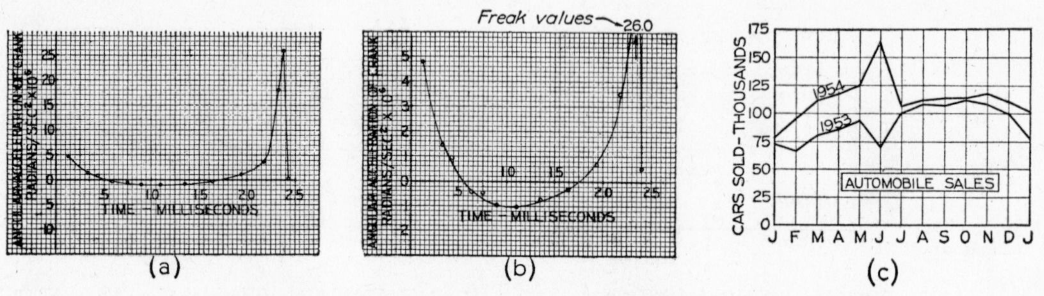

Fig. 994 "Freak" Values and Combined Curves.

best to arrange for such values to fall off the sheet and to indicate them as "freak" values. The curve may then be drawn much more satisfactorily, as shown at (b).

A convenient manner in which to show related curves having the same units along the abscissa, but different ordinate units, is to place one or more sets of ordinate units along the left margin and another set of ordinate values along the right margin, as shown in Fig. 990 (a), using the same rulings. Multiple scales are also sometimes established along the abscissa, such as for time units of months covering multiple years, as in Fig. 991(b). A more compact arrangement for one of the curves is shown in Fig. 994 (c). In this case, the purpose is to compare the 1953 and 1954 sales of one car rather than compare sales of two different cars.

The choice of scales deserves careful consideration, since it has a controlling influence on the depicted rate of change of the dependent variable. The *slope* of the curve (trend) should be chosen to represent a true picture of the data or a correct impression of the trend.

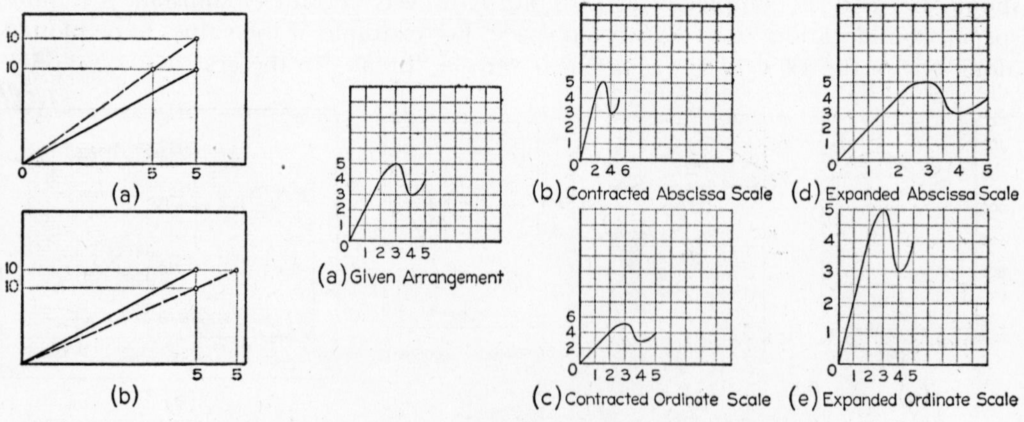

Fig. 995 Slopes.

Fig. 996 Effects of Scale Designation.

The slope of a curve is affected by the spacing of the rulings and its designations. A slope or trend can be made to appear "steeper" by increasing the ordinate scale or decreasing the abscissa scale, Fig. 995 (a), and "flatter" by increasing the abscissa scale or decreasing the ordinate scale, (b). As shown in Fig. 996, a variety of slopes or shapes can be obtained by expanding or contracting the scales. A deciding factor is the impression desired to be conveyed graphically.

Normally an angle greater than 40° with the horizontal gives an impression of a significant rise or increase of ordinate values, while an angle of 10° or less suggests an insignificant trend, Fig. 997 (a) and (b). *The slope chosen should emphasize the significance of the data plotted.* Some relationships are customarily presented in a conventional shape, as shown at (c) and (d). In this case, an expanded abscissa scale, as shown at (d), should be avoided.

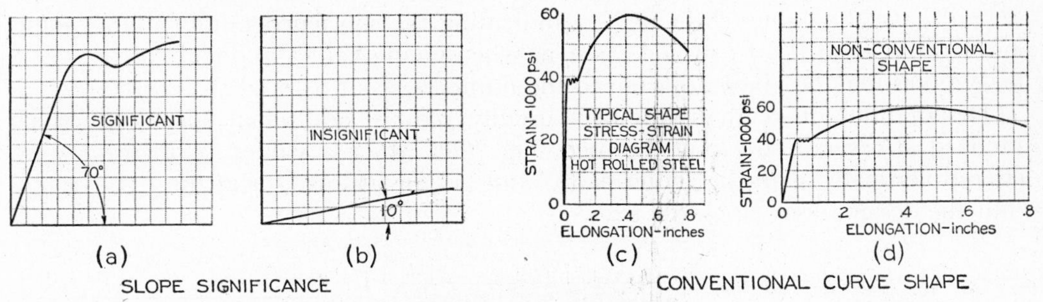

Fig. 997 Curve Shapes.

Scale designations should be placed outside the axes, where they can be shown clearly. Abscissa nomenclature is placed along the axis so that it can be read from the bottom of the graph. Ordinate values are generally lettered so that they can also be read from the bottom; but ordinate captions, if lengthy, are lettered to be read from the right. The values can be shown on both the right and left sides of the graph, or along the top and bottom, if necessary for clearness, as when the graph is exceptionally wide or tall or when the rulings are close together and hard to follow. When the major interest (e.g., maximum or minimum values) is situated at the right, the ordinate designations may be placed along the right, Fig. 987 (b) and (c). This arrangement also encourages reading the chart first and then the scale magnitudes.

When grid rulings are specially prepared on blank paper, every major division ruling should have its value designated, Fig. 998 (a). The labeled divisions should

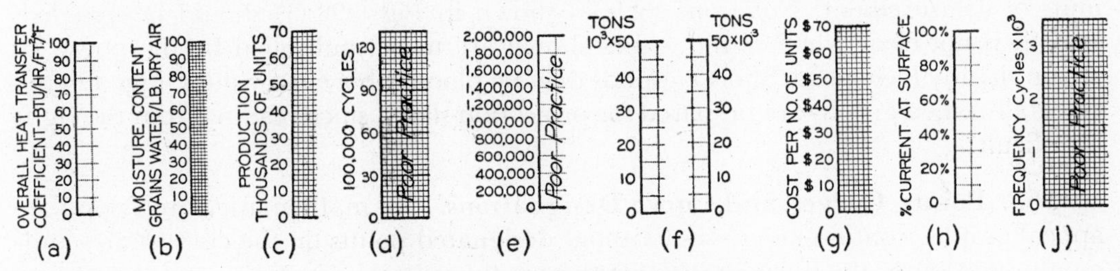

Fig. 998 Scale Designations.

not be closer than 0.25″ and rarely more than 1″ apart. Intermediate values (rulings or ticks) should not be identified and should be spaced no closer than 0.05″. If the rulings are numerous and close together, as on printed graph paper, only the major values are noted, as at (b) and (c). The assigned values should be consistent with the minor divisions. For example, major divisions designated as 0, 5, 10, etc., should not have 2 or 4 minor intervals, since resulting values of 1.25, 2.5, 3.75, etc., are undesirable. Similarly, odd numbered major divisions of 3, 5, 7, etc., or multiples of odd numbers with an even number of minor divisions, should be avoided as shown at (d). The numbers, if three digits or smaller, can be fully given. If the numbers are larger than three digits, (e), dropping the ciphers is recommended, if the omission is indicated in the scale caption, as at (c). Values are shortened to even hundreds, thousands, or millions, in preference to tens of thousands, etc. Graphs for technical use can have the values shortened by indicating the shortened number times some power of 10, as at (f). In special cases, as when giving values in dollars or per cent, the symbols may be given adjacent to the numbers, as at (g) and (h).

Designations other than numbers usually require additional space; therefore standard abbreviations should be used, Fig. 999. Abscissa values for these may be lettered vertically as in the center at (a) and (b), or inclined, as at the right in (c), to fit the designations along the axes.

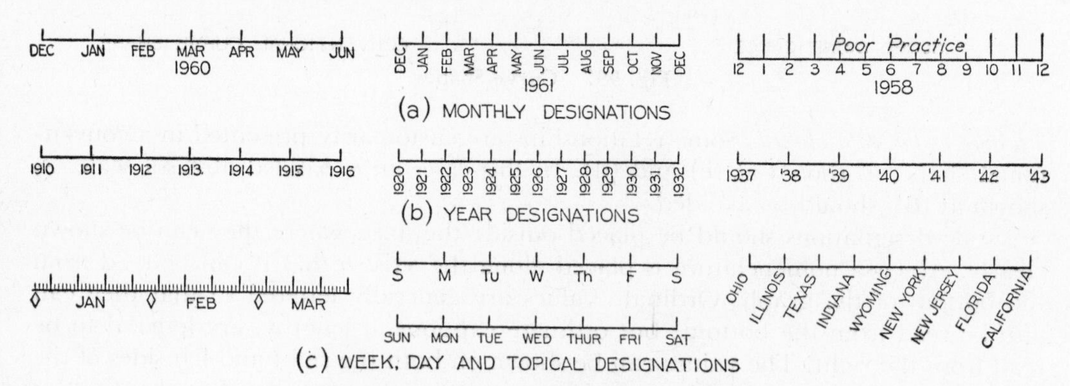

Fig. 999 Non-numerical Designations.

Scale captions (or titles) should be placed along the scales so that they can be read from the bottom for the abscissa, and from the right for the ordinate. Captions include the name of the variable, symbol (if any), units of measurement, and any explanation of digit omission in the values. If space permits, the designations are lettered completely, but if necessary, standard abbreviations may be used for the units of measurement. Notations such as shown in Fig. 998 (j) should be avoided, since it is not clear whether the values shown are to be multiplied by the power of ten or already have been. Short captions may be placed above the values, Fig. 998 (f), especially when graphs are prepared for projection slides, since reading from the right is difficult.

565. Points, Curves, and Curve Designations. In mathematical and popular-appeal graphs, continuous curves (without designated points on the curves) are commonly used, since the purpose is to emphasize the general significance of the curves. On graphs prepared from observed data, as in laboratory experiments, points are usually designated by various symbols, Fig. 1000. If more than one curve is plotted on the same grid, a combination of these symbols may be used, one type for each curve. The use of open-point symbols is recommended, except in cases of "scatter" diagrams, Fig. 1023, where the filled-in points are more visible. In general, filled-in symbols should be used only when more than three curves are plotted on the same graph, and a different identification is required for each curve.

Line curves are generally presented for any of three types of relationships: (1) Observed relationships, usually plotted with observed data points connected by straight lines, Fig. 1001 (a). (2) Empirical relationships, (b), normally reflecting the author's interpretation of his series of observations. They are smooth curves or straight lines fitted to the data by eye or by formulas chosen empirically. (3) Theoretical relationships, (c), in which the curves are without point designations, though observed values may be plotted to compare them with a theoretical curve, as shown. The curve thus drawn is based on theoretical considerations only, in which a theoretical function (equation) is used to compute values for the curve.

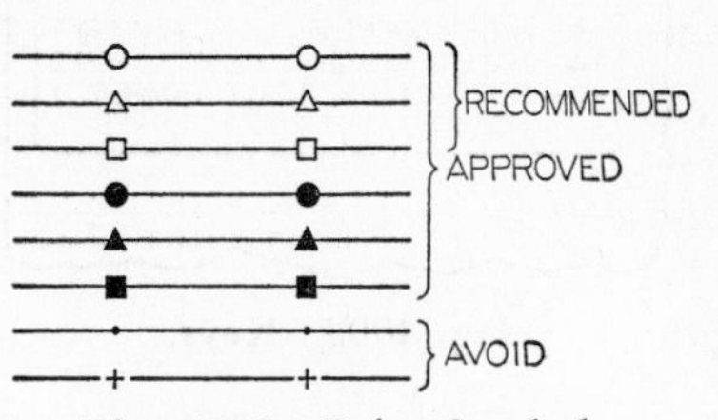

Fig. 1000 Point Symbols.

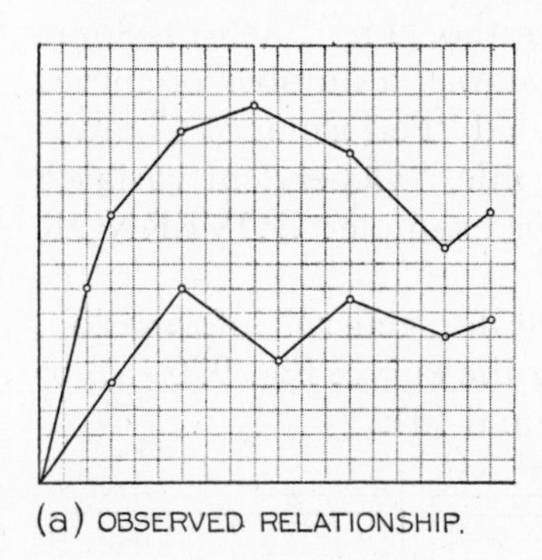

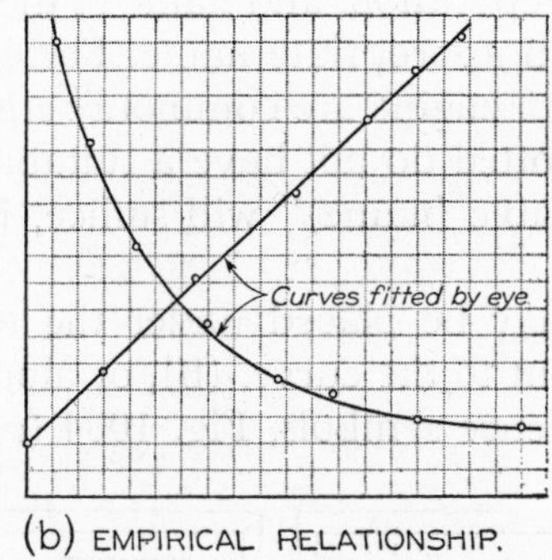

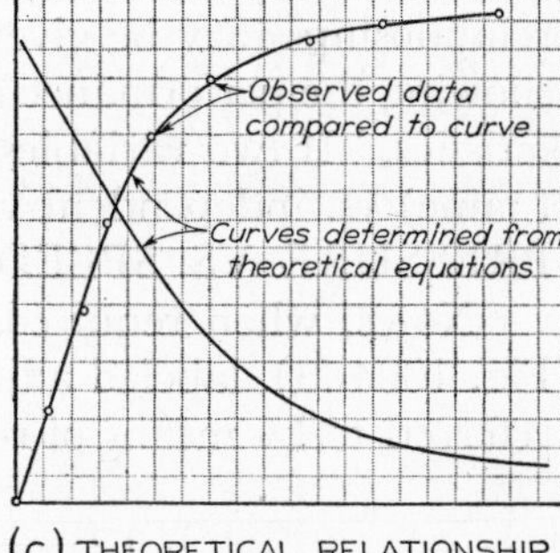

Fig. 1001 Curve Fitting.

When a number of curves are to be plotted on the same grid, they can be distinguished by the use of various types of lines, Fig. 1002 (a). However, solid lines are used for the curves wherever possible, while the dashed line is commonly used for

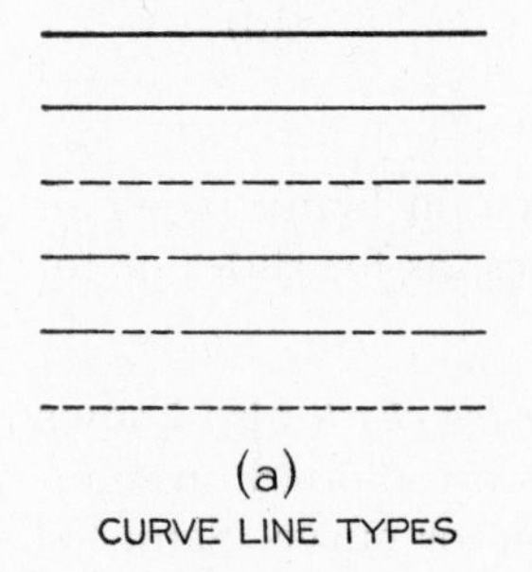

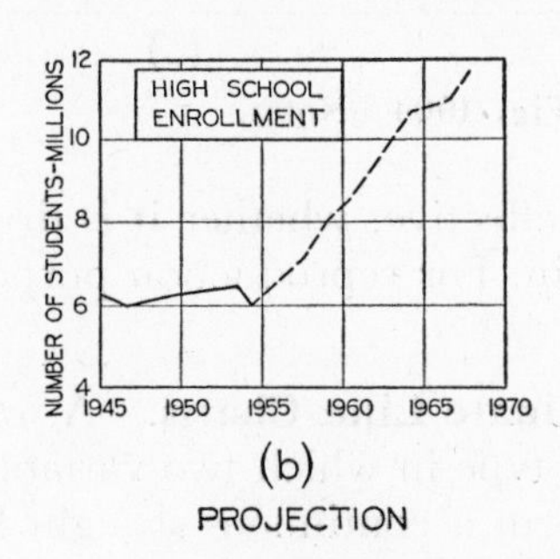

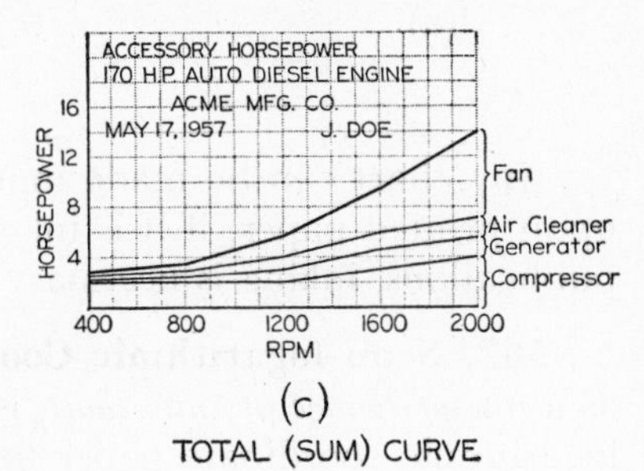

Fig. 1002 Curve Lines.

projections (estimated values, such as future expectations), as shown at (b). The curve should be heavier in weight than the grid rulings, but a difference in weight can also be made between various curves to emphasize a preferred curve or a total value curve (sum of two or more curves), as shown at (c). A *key*, or *legend*, should be placed in an isolated portion of the grid, preferably enclosed by a border, to denote point symbols or line types that are used for the curves. If the grid lines are drawn on blank paper, a space should be left vacant for this information, Fig. 1003. Keys may be placed off the grids below the title, if space permits. Curves are also designated by means of notes, Fig. 988 (c), in which no key is necessary. Colored lines are very effective for distinguishing the various curves on a grid, but they cannot be used for multiple copies.

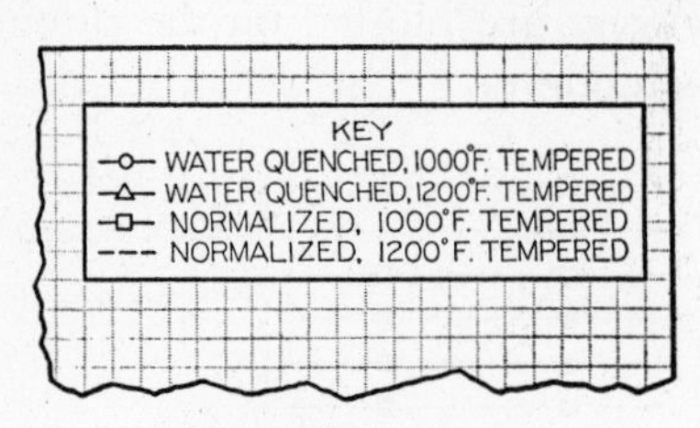

Fig. 1003 Keys.

566. Titles. Titles for a graph may be placed on or off the grid surface. If placed on the grid, white space should be left for the title block, but if printed coordinate paper is used, a heavy border should enclose the title block. If further emphasis is desired, the title may be underlined. The content of title blocks varies according to method of presentation. Typical title blocks include title, sub-title, institution or company, date of preparation, and name of the author, Fig. 1002 (c). Some relationships may be given an appropriate name. For example, a number of curves showing the performance of an engine are commonly entitled "Performance Characteristics." If two variables plotted do not have a suitable title, "Dependent variable (name) vs. independent variable (name)" will suffice. For example: GOVERNOR PRESSURE VS. SPEED.

Notes, when required, may be placed under the title for general information, Fig. 1004 (a), labeled adjacent to the curve, (b), or along the curve, Fig. 990 (a), or referred to by means of reference symbols, Fig. 1004 (c) and (d).

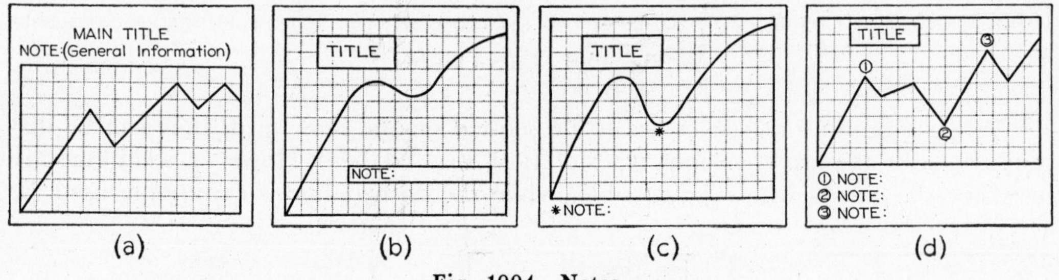

Fig. 1004 Notes.

Any chart can be made more effective, whether it is drawn on blank paper or upon printed paper, if it is inked in. For reproduction purposes, as for slides or for publications, inking is necessary.

567. Semi-logarithmic Coordinate Line Charts. A *semi-log chart*, also known as a *rate-of-change* or *ratio chart*, is a type in which two variables are plotted on semi-logarithmic coordinate paper to form a continuous straight line or curve. Semi-log paper contains uniformly divided vertical rulings and logarithmically divided horizontal rulings.

Semi-log charts have the same advantages as rectangular coordinate line charts, (arithmetic charts), §561. When rectangular coordinate line charts give a false impression of the trend of a curve, the semi-log charts would be more effective in revealing whether the rate of change is increasing, decreasing, or constant. Semi-log charts are also useful in the derivation of empirical equations, §627.

Semi-log charts, as rectangular coordinate line graphs, are not recommended for presenting only a few plotted values in a series, for emphasizing change in absolute amounts, or for showing extreme or irregular movement or trend of data.

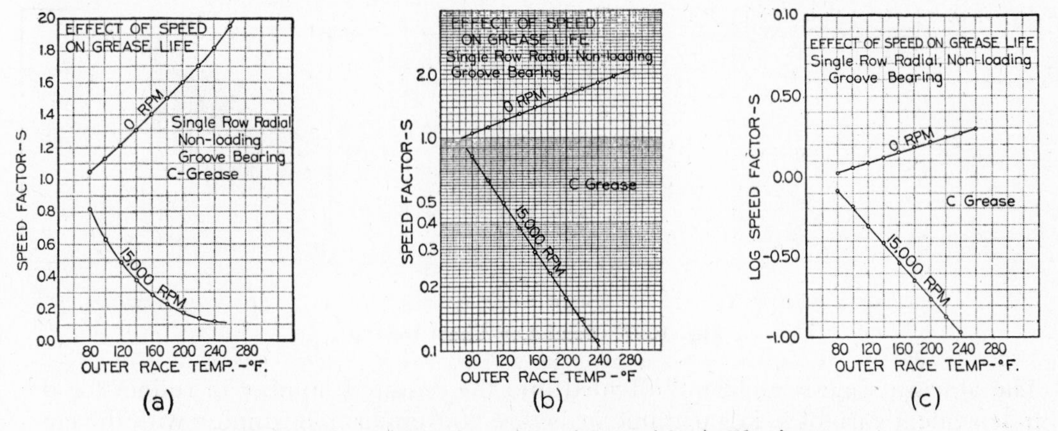

Fig. 1005 Arithmetic and Semi-Logarithmic Plottings.

In Fig. 1005 (a) and (b), data are plotted on rectangular coordinate grids (arithmetic) and on semi-logarithmic coordinate grids, respectively. The same data, which produce curves on the arithmetic graph, produce straight lines on the semi-log grid. The straight lines permit an easier analysis of the trend or movements of the variables. If the logarithms of the ordinate values are plotted on a rectangular coordinate grid, instead of the actual values, straight lines will result on the arithmetic graph, as shown at (c). The straight lines produced on semi-log grid provide a simple means of deriving empirical equations, as discussed in §627.

Straight lines are not necessarily obtained on a semi-log grid, but if they do occur, it means that the rate of change is constant, Fig. 1006 (a). Irregular curves can be

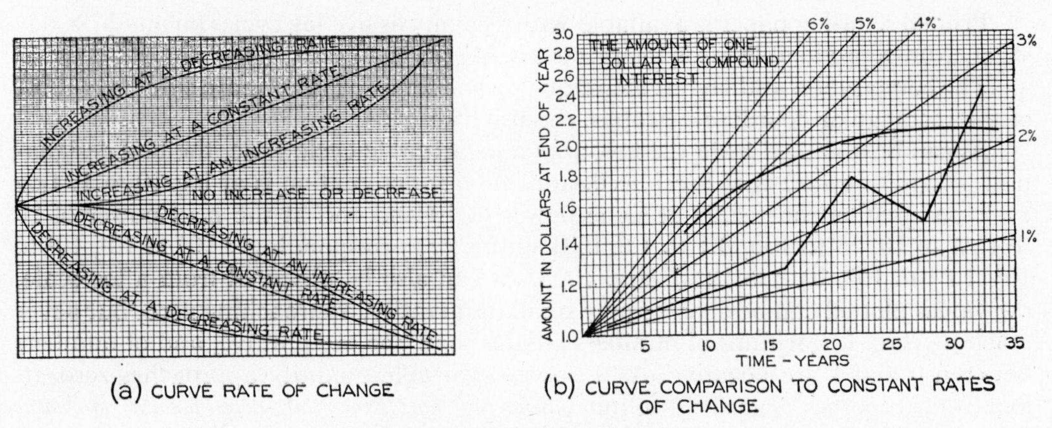

Fig. 1006 Rates of Change.

compared to constant rate scales individually or between a series of curves, as shown at (b).

568. Design and Layout of Semi-Log Charts. Semi-log graphs are usually prepared on printed semi-log coordinate paper. Graphs for publication, however, requiring fewer grid rulings, are plotted on specially prepared grid scales. As illustrated in Fig. 1007 (a), the logarithmically divided scale is generally placed along the ordinate.

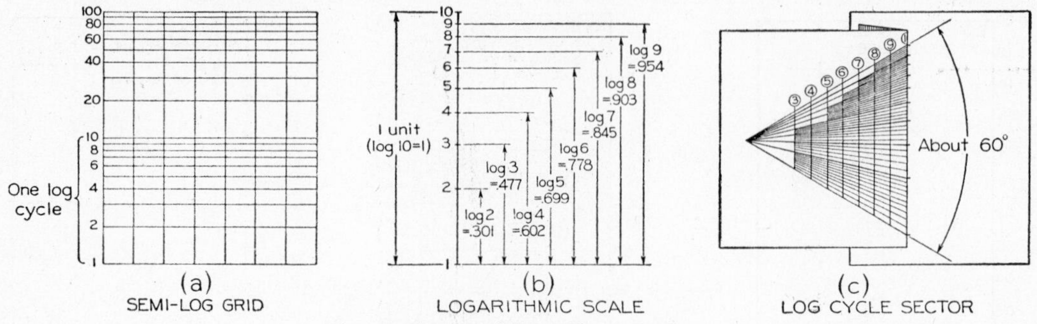

Fig. 1007 Semi-Log Chart Design.

The abscissa scale is uniformly divided into the required number of rulings for the independent variables. Logarithmic scales are non-uniform, beginning with the largest space at the bottom and decreasing in size for ten divisions, which is known as one *log cycle*. The locations of major divisions and sub-divisions for the semi-log grid are determined by taking the logarithm to the base 10 (common logarithm) of the value to be plotted. As shown at (b), the major divisions of one log cycle, from 1 to 10, ranges in value (logarithmically) from 0 to 1. If this logarithmic proportion is maintained, the log cycle can be laid out to any height, but the remaining divisions depend upon the height chosen for one unit of log cycle. For example, if a length of 10″ for one log cycle is suitable, the number two division is at 3.01″ = (0.301 × 10″) from the origin of the cycle (number one division); number three is at 4.77″ from the origin, etc. A convenient device for laying out logarithmic scales is a *log cycle sector* prepared on a separate sheet of paper, as shown at (c). The sector is folded along the length of log cycle desired, and divisions are marked on the scale being prepared adjacent to the folded edge.

Printed semi-log paper is available with as many as five log cycles on an $8\frac{1}{2}'' \times 11''$ sheet. Non-reproducible grids for prints are also available.

The composition and layout techniques of the semi-log graphs are similar to those of the rectangular coordinate graphs, discussed in §564.

As can be observed from the layout of a logarithmic cycle scale, the first designation (origin of cycle) is at log 1 = 0, and the last designation is at log 10 = 1. Log cycle designation must start at some power of 10 and end at the next power of 10 (e.g., 1 to 10, 10 to 100, 0.1 to 1.0, 0.01 to 0.1, or correspondingly 10^0 to 10^1, 10^1 to 10^2, 10^{-1} to 10^0, 10^{-2} to 10^{-1}), Fig. 1008. As can also be observed from a log scale, the values plotted can be decreasingly small in quantity, approaching zero, but never reaching zero. Logarithms of numbers greater than one are positive, and of numbers between 0 and 1 are negative. As a positive variable quantity approaches zero, its logarithm becomes negatively infinite. *Semi-log charts, therefore, never have a zero value on the logarithmic scale.* The cycle designation determines the number of cycles required as a result of observing the range of values to be plotted. Cycle designation also permits the plotting of an extensive range of values, since each cycle accommodates an entire range of one power of 10.

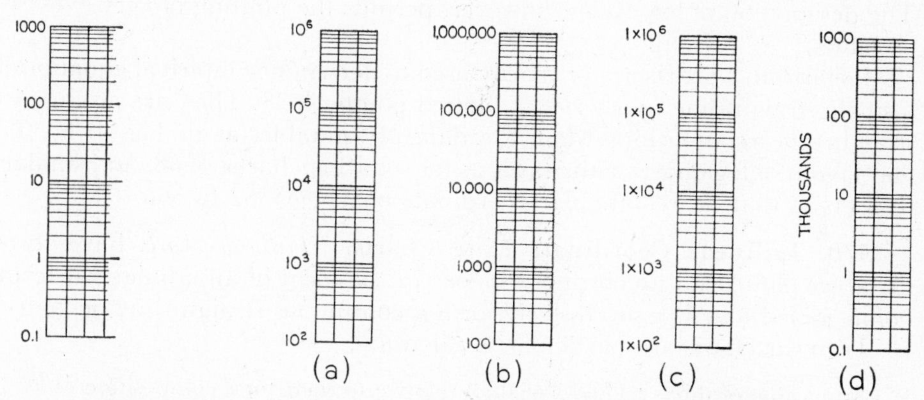

Fig. 1008 Cycle Designation. **Fig. 1009 Log Scale Designation.**

The techniques of scale and chart designation and captions are similar to those discussed for rectangular coordinate line graphs. Values shown along the logarithmic axis, however, can be designated in a number of ways, as shown in Fig. 1009. The scale shown at (a) is preferred, since it is the simplest and will be understood by any person who makes frequent use of semi-log charts.

569. Logarithmic Coordinate Line Charts. *Logarithmic charts* have two variables plotted on a logarithmic coordinate grid to form a continuous line or a "curve." Printed logarithmic paper contains logarithmically spaced horizontal and vertical rulings. As in the case of semi-log charts, paper containing as many as five cycles on an axis can be purchased.

Log charts are applicable for the comparison of a large number of plotted values

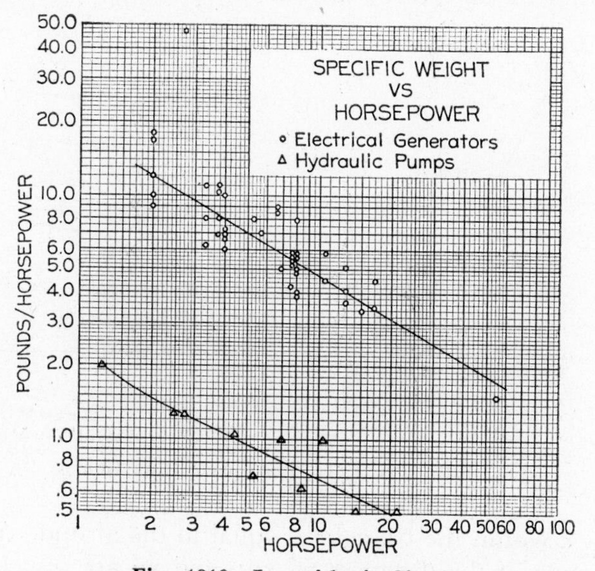

Fig. 1010 Logarithmic Chart.

in a compact space, and for the comparison of the relative trends of several curves on the same chart. This form of graph is not the best form or presentation of relatively few plotted values in a series, or for emphasizing change in absolute amounts. The designation of log cycles, however, permits the plotting of very extensive ranges of values.

Logarithmic charts are primarily used to determine empirical equations by fitting a single straight line to a series of plotted points, §628. They are also used to obtain straight-line relationships when the data are suitable, as in Fig. 1010. The design and layout of log charts is the same as for semi-log charts, §568, and similar in many respects to that for rectangular coordinate charts, §§562 to 566.

570. Trilinear Coordinate Line Charts. *Trilinear charts* have three related variables plotted on a coordinate paper in the form of an equilateral triangle. The points joined together successively form a continuous straight line or "curve."

Trilinear charts are particularly suited for:

1. Comparison of three related variables relative to their total composition (100%).
2. Analysis of the composition structure by a combination of curves; e.g., metallic microstructure of trinary alloys, Fig. 1012 (a).
3. Emphasizing change in amount or differences between values.

The trilinear chart is *not* recommended for:

1. Emphasizing movement or trend of data.
2. Comparison of three related but dissimilar physical quantities; e.g., force, mass, acceleration, time, etc.

Trilinear charts are most frequently applied in the metallurgical and chemical fields, because of the frequency of three variables in metallurgical and chemical composition. The basis of application is the geometric principle that the sum of the perpendiculars to the three sides from any point within an equilateral triangle is equal to the altitude of the triangle.

In an equilateral triangle ABC, Fig. 1011 (a), the sum of the distances xf, xm, and

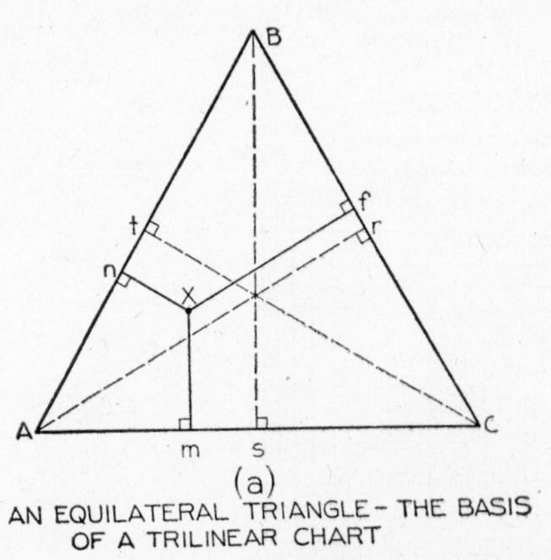

(a)
AN EQUILATERAL TRIANGLE - THE BASIS OF A TRILINEAR CHART

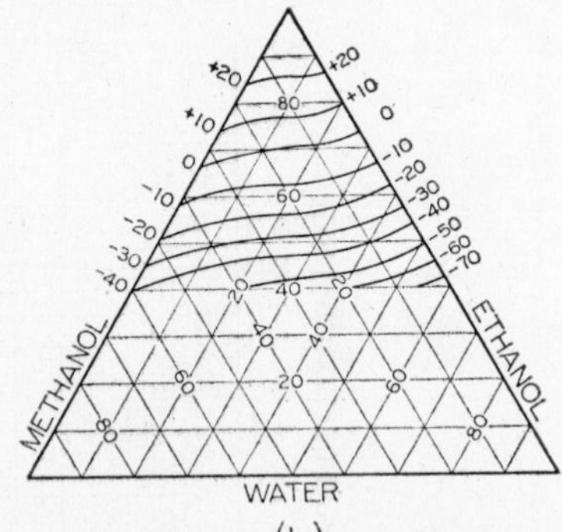

(b)
FREEZING TEMPERATURES OF MIXTURES OF WATER, METHANOL AND ETHANOL

Fig. 1011 Trilinear Coordinate Line Charts.

xn from the point x within the triangle, is equal to the altitude Ar, Bs, or Ct of the triangle. For example, if the distances xf, xm, and xn are respectively 50, 30, and

20 units, the altitude of the triangle is 100 units, and the point x will represent the function composed of, or resulting from, 50 parts of A, 30 parts of B, and 20 parts of C. At (b) is shown a chart for various freezing temperatures, with the mixture proportions by volume of water, methanol, and ethanol required. For example, a freezing temperature of $-40°F$ can be established by mixing 50 parts of water with 10 parts of ethanol and 40 parts of methanol.

571. Design and Layout of Trilinear Charts. Trilinear charts are also usually drawn on printed grid paper. Charts for publications, requiring maximum clarity, can be plotted on grids drawn on plain paper or traced from an original plotting on a printed sheet. The trilinear coordinate grid is an equilateral triangle, which must have the same space between rulings along each axis. Normally each side of the triangle is divided into ten divisions for per cent plots, with as many subdivisions as required, Fig. 1012. If only a portion of a coordinate field is to be used, only a portion of the grid need be shown, Fig. 1013.

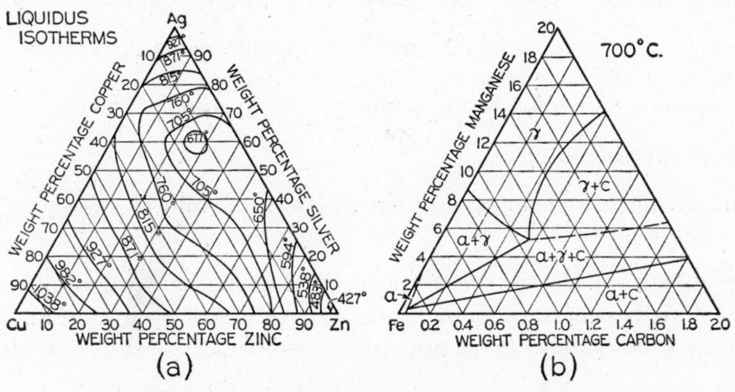

Fig. 1012 Metallurgical Trilinear Charts.

The scale designations can be placed to be read from the bottom of the sheet or can be tilted as shown in Fig. 1011 (b), but for ease of reading, all scale notations should be placed outside the grid area. Other techniques of scale designations are the same as for rectangular coordinate line graphs, §564.

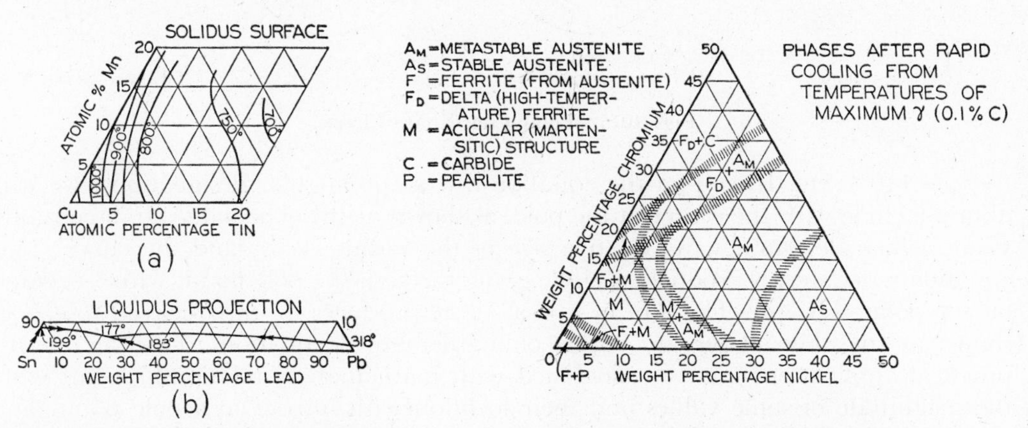

Fig. 1013 Partial Trilinear Charts. **Fig. 1014 Shaded Band Curves.**

Since individual points are generally not shown on trilinear charts, the curves are drawn as continuous lines.

Curves are frequently single-weight solid lines, but dashed lines are used for projected (anticipated) information, Fig. 1012 (b). When the curve-fitting information varies within a range, shaded bands are used, Fig. 1014.

When the chart indicates the composition of various parts or different units, the symbols are placed within the appropriate areas, e.g., α, β, $\alpha + \gamma$, which are microstructure symbols, Fig. 1012 (b). Curve designations are placed along the curves or in breaks in the curve; e.g., the temperature values shown in Fig. 1012 (a).

Since most of the grid area is utilized by the curves, the title, keys, and notes are placed off the grid, as shown in Fig. 1014.

572. Polar Coordinate Line Charts. *Polar charts* have two variables, one a *linear* magnitude and the other an angular quantity, plotted on a polar coordinate grid with respect to a pole (origin) to form a continuous line or "curve."

Polar charts are particularly adaptable to the following applications:

1. Comparison of two related variables, one being a linear magnitude (called a *radius vector*), and the second an angular value.
2. Indicating movement or trend or location with respect to a pole point.

Polar charts are *not* suited for:

1. Emphasizing changes in amounts or differences between values.
2. Interpolating intermediate values.

As shown in Fig. 1015 (a), the zero degree line is the horizontal right axis. To locate a point P, it is necessary to know the radius vector r, and an angle θ, (e.g., 5, 70°). The point P could also be denoted as (5, 430°), (5, −290°), and (−5, 250°),

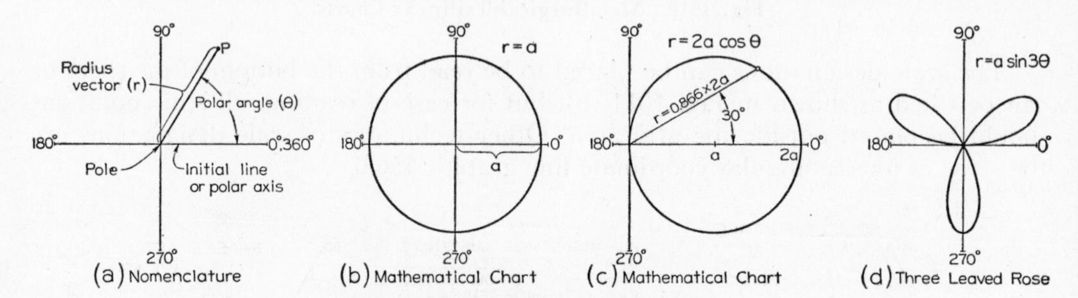

Fig. 1015 Polar Coordinate Charts.

(−5, −110°), etc. If we plot the equation r = a (no angular designation), we will obtain a circle with the center at the pole, as shown at (b). The value a is a constant value, which determines the relative size of the radius vector and the curve. The equation $r = 2a \cos \theta$ produces a circle going through the pole point, with its center on the polar axis, (c). The plot of $r = a \sin 3\theta$ produces a "three-leaved rose," as shown at (d). The above are charts of mathematical equations; however, many practical applications are not concerned with mathematical relationships but with the magnitude of some values and their location with respect to a pole point. For example, Fig. 1016 (a) and (b) illustrate stress charts from experimentation and for stress visualization, respectively. At (c) is shown a polar chart of a bearing load diagram, and the graph at (d) indicates the noise distribution from a jet engine.

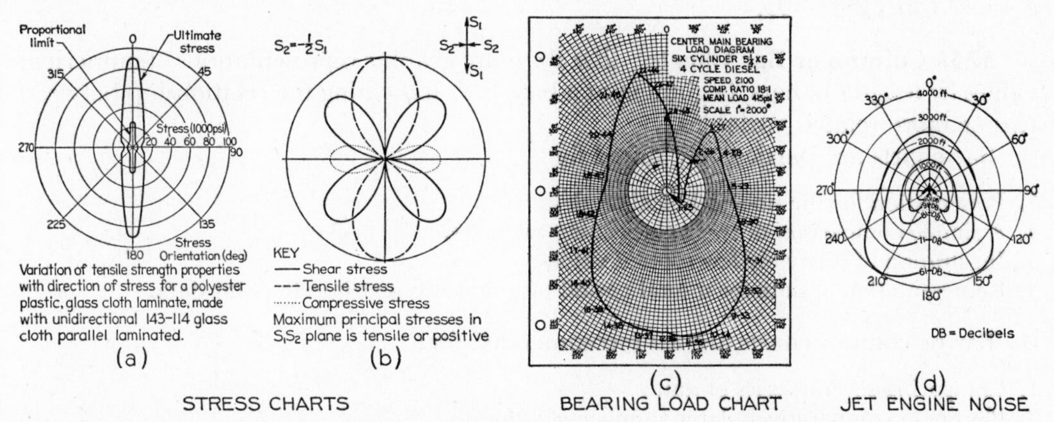

Fig. 1016 Polar Coordinate Charts. *(a) and (b) adapted from charts by Robert L. Stedfeld and F. W. Kinsman, respectively, with permission of* Machine Design. *(c) adapted from R. R. Slaymaker,* Bearing Lubrication Analysis, *copyright 1955 by John Wiley and Sons, with permission of the publisher and Clevite Corporation. (d) adapted from chart by G. S. Schairer, with permission of author and Society of Automotive Engineers.*

573. Design and Layout of Polar Charts. Polar charts are usually prepared on printed polar grids on $8\frac{1}{2}'' \times 11''$ paper, Fig. 1016 (c). Charts for publication may be plotted on a grid drawn upon blank paper, which involves the required number of subdivisions of 360° and a sufficient number of concentric circles, equally spaced, to denote the values of radius vectors. See Fig. 1016 (d).

The initial line, or polar axis, is normally the horizontal line extending from the pole towards the right. Upon many polar charts and printed grids, however, the upper vertical line is used as a polar axis. The sense of direction from 0° to 360° can be either clockwise or counterclockwise. A proper sense of direction may at times be determined by a physical sense of rotation, as for the load diagram in Fig. 1016 (c). Mathematical diagrams, in which the standard four quadrants are used, follow a counterclockwise sense of direction and a horizontal polar axis.

Angular designations are placed along the outskirts of the polar grid, while radius vector values are lettered along either the horizontal or vertical axis, wherever space permits. As in the case of rectangular coordinate graphs, numbers no larger than three digits should be used, with proper notations of actual values, Fig. 1016 (a). All captions should be placed to be read from the bottom of the graph.

Points are normally not designated on polar charts. As in other forms of charts, curves may be distinguished by various types of lines, Fig. 1016 (b), when more than one curve is required. Curves may be designated with a leader and arrow to a note, or with a legend.

Since in many polar charts the grid area is more or less filled with the curve, the scale and curve designations, titles, keys, and additional notes are placed off the grid. If a printed grid is used, the title information may be placed on white paper and fastened onto the grid to facilitate ease of reading, as at (c).

574. Nomographs or Alignment Charts. *Nomographs* or *alignment charts* consist of straight or curved scales, arranged in various configurations so that a straight line drawn across the scales intersects them at values satisfying the equation represented.

Alignment charts can be used for analysis, but the predominant application is for computation. The design and layout of some common forms of nomographs are discussed in §§587-605.

575. Column or Bar Charts. *Bar charts* are graphic representations of numerical values by lengths of bars, beginning at a base line, indicating the relationship between two or more related variables.

Bar charts are particularly suited for:

1. Presentation for the non-technical reader.
2. A simple comparison of two values along two axes.
3. Illustration of relatively few plotted values.
4. Representation of data for a total period of time in comparison to point data.

Bar or column charts are *not* recommended for:

1. Comparing several series of data.
2. Plotting a comparatively large number of values.

The bar chart is effective for popular or nontechnical use because it is most easily read and understood. It is therefore used extensively by newspapers, magazines, and similar publications. The bars may be placed horizontally or vertically; when they are placed vertically, the presentation is often called a *column chart.*

Bar charts are plotted with reference to two mutually perpendicular coordinate axes, similar to those in rectangular coordinate line charts. The charts may be a simple bar chart with two related variables, Fig. 1017 (a), a grouped bar chart (three or more related variables), (b), or a combined bar chart (three or more related variables), (c). Another example of bar graphs for three or more related variables employs pictorial symbols, composed to form bars, as shown at (d). Bar charts can also effectively indicate a "deviation" or difference between values, (e).

576. Design and Layout of Bar Charts. If only a few values are to be represented by vertical bars, the chart should be higher than wide, Fig. 1017 (a). When a relatively large number of values are plotted, a chart wider than high is preferred, (b) and (c). Composition is dictated by the number of bars used, whether they are to be vertical or horizontal, and the available space.

A convenient method of spacing bars is to divide the available space into twice as many equal spaces as bars are required, Fig. 1018 (a). Center the bars on every other division mark, beginning with the first division at each end, as shown at (b). When the series of data is incomplete, the missing bars should be indicated by the use of ticks, (c), indicating the lack of data. The bars should be spaced uniformly when the data used are uniformly distributed. When irregularities in data exist, the bars should be spaced accordingly, as shown.

Bar charts may be drawn on printed coordinate paper; however, clarity for popular use is promoted by the use of blank paper and the designation of only the major rulings perpendicular to the bars. If the values of bars are individually noted, the perpendicular rulings may be omitted completely.

Since bar charts are used extensively to show differences in values for given periods of time, the values or amounts should be proportionate to the heights or lengths of the bars. The zero or principal line of reference should never be omitted. Normally, the full length of the bars should be shown to the scale chosen. When a few exceptionally large values exist, the columns may be broken as shown in Fig. 1017 (e), with a notation included indicating the full value of the bar.

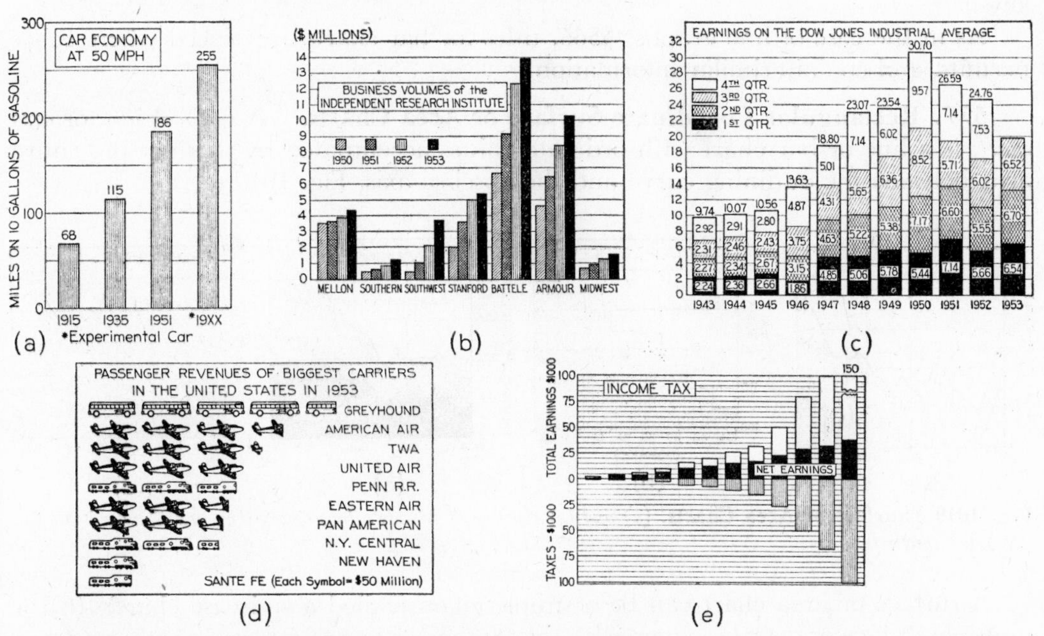

Fig. 1017 Bar or Column Charts. *(b) adapted from* Chemical Week, *copyright 1953, with permission of the publisher. (c) adapted from* Barron's *with permission of the publisher. (d) adapted with permission from chart copyright 1954 by the* Chicago Tribune.

Scale designations are normally placed along the base line of the bars and adjacent to the rulings.

Standard abbreviations may be used for bar designations. The techniques of scale designations used for line curves, §564, are also applicable to bar charts. The values of the bars may be designated above the bars for simple and grouped bar charts, Fig. 1017 (a) and (c). Spaces between bars should be wider than the bars when there are relatively few bars. Spaces between bars should be narrower than the bars when there are many bars, as at (b) and (c).

Bars are normally emphasized by shading or by solid filling-in. Some of the common forms of shading are shown in the figure. Weight and spacing of shading is dependent upon the amount of area to be shaded and the final size of the chart, and is therefore a matter of judgment.

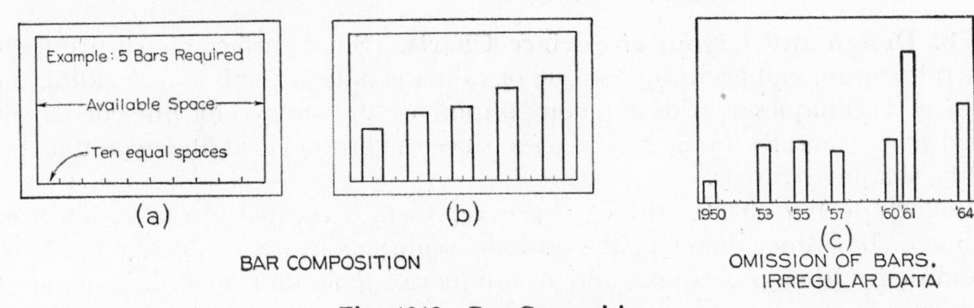

Fig. 1018 Bar Composition.

Bar designations may be placed across several bars when possible, as shown at (e). Other methods include the use of notes with leaders and arrowheads, or the use of keys.

As in the case of line graphs, §566, titles for bar charts are placed where space permits, and contain similar information.

577. Rectangular Coordinate Surface or Area Charts. A *surface chart*, or *area chart*, is a line or bar chart with ordinate values accentuated by shading the entire area between the outlining curve and the abscissa axis, Fig. 1019 (a).

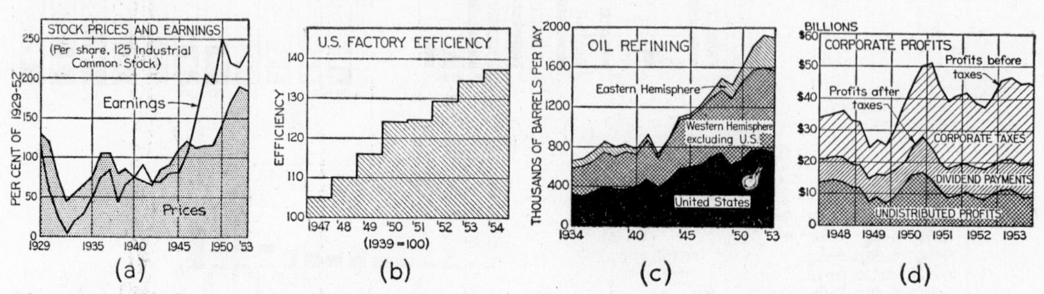

Fig. 1019 Surface or Area Charts. *(a) and (c) courtesy of Moody's Investors Service and Standard Oil Co. (N.J.), respectively.*

A surface or area chart can be a simple chart (called a staircase chart), (b), a multiple surface or "strata" chart with one surface or layer on top of another, (c), or a combined surface chart indicating the distribution of components in relation to their total, (d). A surface and line chart combined to compare earnings with prices is shown at (a).

Surface charts are used effectively to:

1. Accentuate or emphasize data that appear weak as a line chart.
2. Emphasize amount or ordinate values.
3. Represent the components of a total, usually expressed as a per cent of a total, or compared to 100%
4. Present a general picture.

Surface charts are *not* recommended for:

1. Emphasizing accurate reading of values of charts containing more than one curve.
2. Showing line curves that intersect or cross one another.

A map of terrestrial or geographic features is a form of area chart, or surface chart, representing a graphic picture of areas, surfaces, or the relationship of their component parts to the total configuration.

578. Design and Layout of Surface Charts. Since surface charts are used as a general picture, and accurate reading of values is difficult, only major rulings need be shown. Techniques of grids and composition are the same as for line curves, §563. Printed grids generally should not be used, since the printed grid lines would interfere with the shading of areas.

Similarly to bar charts, surface charts represent a comparison of values from a zero line or base line; therefore the ordinate scale should not be broken or the zero line omitted. Other procedures and techniques of scale and scale designations are the same as for line curves.

The shading of surfaces is accomplished by the use of black (solid) and white areas, lines, and dots, Fig. 1019. The darker shading tones should be used at the bottom of the chart, and progressively lighter tones should be used as each strata is shaded proceeding upward, as shown at (c). The weight and spacing of lines and dots are dependent upon the final size of the chart, and are a matter of judgment on the part of the draftsman. Projections, or extensions of a curve beyond present available data, can be distinguished by dashed outlines and lighter line shading.

Surfaces should be designated by placing the labels entirely on the surface, where possible, as at (c). Small surfaces can be denoted by a label with a leader and arrowhead, (d). The area of the labels should be clear of shading for ease in reading. Legends should not be used as a means of designation if direct labeling is possible.

The methods of chart designations are the same as for rectangular coordinate line graphs.

579. Pie Area Charts. *Pie charts*, or *sector charts*, are used to compare component parts in relation to their total by the use of circular areas.

Pie charts are an effective method for:

1. Representing data on a per cent basis.
2. Popular presentation of a general picture.
3. Showing relatively few plotted values.
4. Emphasizing amounts rather than the trend of data.

A pie chart is normally presented as a true circular area, Fig. 1020 (a), or in pictorial form, (b). Since most applications are concerned with monetary values, a disk or "coin" is commonly used for the circular area.

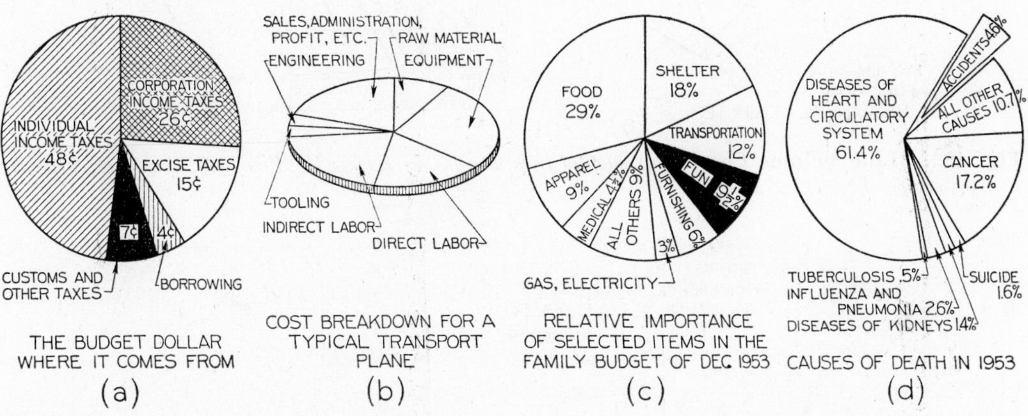

Fig. 1020 Pie, or Sector, Charts. (*b*) *adapted from drawing by E. A. Green with permission of* Machine Design.

580. Design and Layout of Pie Charts. Grids are not used for pie charts. The circular area is drawn to a desired size, within a permissible space. The determination of the various sizes of the sectors is based upon 360° being equivalent to 100%. Therefore the following relationship exists:

$$\frac{100\%}{360^\circ} = \frac{x\%}{\theta^\circ} \text{ or } \theta^\circ = \frac{x\%\,(360^\circ)}{100\%} = 3.6\ (x\%)$$

Therefore, 1 per cent is represented by 3.6°, 2 per cent by 7.2°, etc.

Sector values (per cent and amount) are placed within the sectors where possible. If a sector is small, a note with a leader and arrowhead will suffice. Labels should be clear of any shading and lettered to read from the bottom of the chart, where possible, Fig. 1020 (c) and (d). Another technique of sector designation is to shade the areas, (a). If one of the parts is to be emphasized and compared with the other parts, it can be shaded a different tone from the others, (c), or separated from the circular area, as shown at (d).

Titles should be placed above or below the figure.

581. Volume Charts. A *volume chart* is the graphic representation of three related variables with respect to three mutually perpendicular axes in space. Volume charts are generally difficult to prepare, and thus are not often used. The method of construction is not discussed in detail, but some of the forms will be considered in the following paragraphs.

Figure 1021 illustrates line volume charts plotted with respect to three axes (Cartesian coordinates), shown in isometric and oblique projection at (a) and (b), respectively. Bar graphs can be similarly presented in three dimensions on an isometric grid.

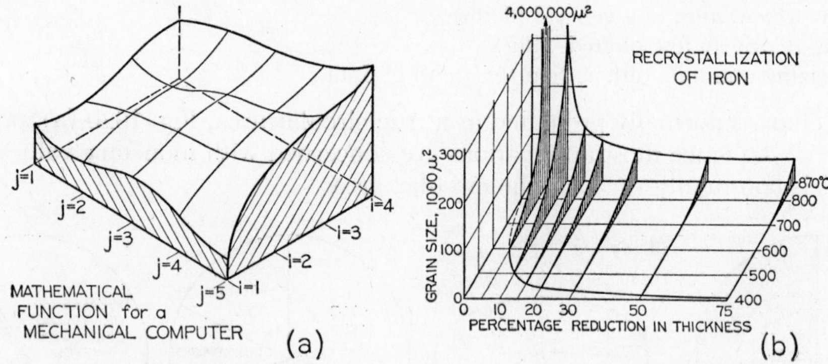

Fig. 1021 Line Volume Charts. (*a*) *adapted from drawing by Eugene W. Pike and Thomas R. Silverberg with permission of* Machine Design.

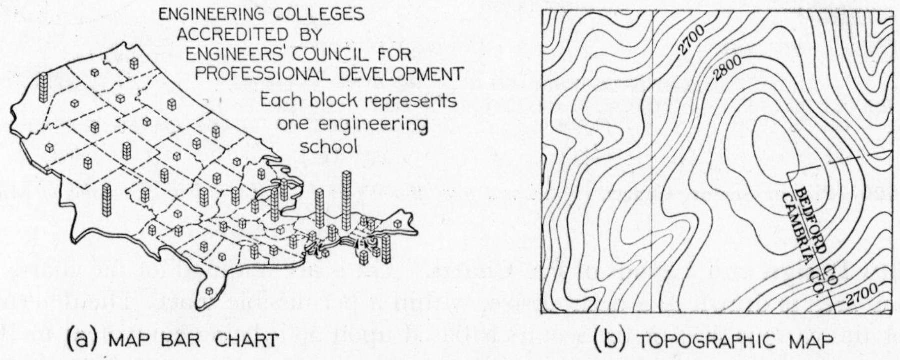

Fig. 1022 Map Charts. (*a*) *adapted from* Engineering: A Creative Profession, *3rd Ed., copyright 1956 by Engineers' Council for Professional Development, with permission of the publisher.*

A combination of bars and maps may be used to represent three related variables, as shown in Fig. 1022 (a). Topograph map construction also is a graphic representation of three related variables, (two dimensions in a horizontal plane and one dimension in a vertical direction) all drawn in one plane of the drawing paper, as shown at (b).

582. Rectangular Coordinate Distribution Charts. When the data observed or obtained vary greatly, the data can be plotted on a rectangular coordinate grid for the purpose of observing the distribution or areas of major concentration, with no attempt to fit a curve, Fig. 1023. Charts of this nature are commonly called *scatter diagrams*.

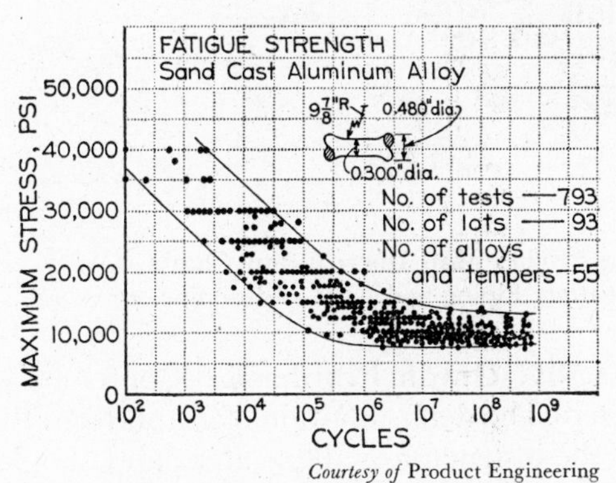

Courtesy of Product Engineering

Fig. 1023 "Scatter" Diagram.

583. Flow Charts. *Flow charts* are predominantly schematic representations of the flow of a process; for example, manufacturing production processes, electric or hydraulic circuits, etc., Fig. 1024. Pictorial forms may be used as shown at (a). Schematic symbols are also used, if applicable, (b) and (c). Blocks with captions are used in the simplest form of flow chart, as illustrated at (d).

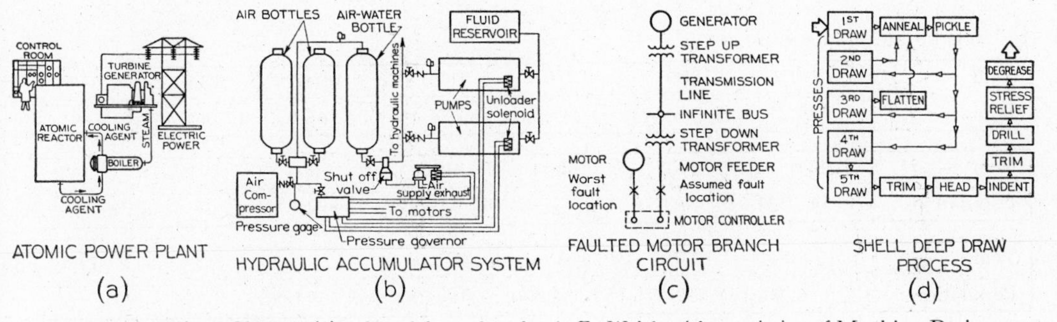

Fig. 1024 Flow Charts. (*b*) *adapted from chart by A. F. Welsh with permission of* Machine Design.

Organization charts are similar to flow charts, except that they are usually representations of the arrangement of personnel and physical items of a definite organization, Fig. 1025.

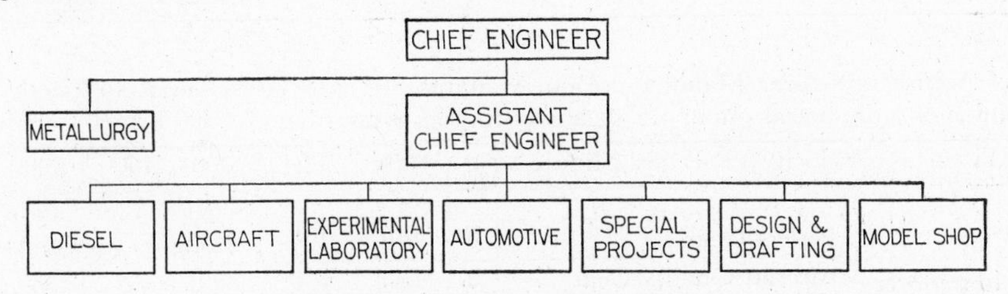

Fig. 1025 Organization Chart.

584. Map Distribution Charts. When it is desired to present data according to geographical distribution, maps are commonly used. Locations and emphasis of data can be shown by dots of various sizes, Fig. 1026 (a), by the use of symbols, (b), by shading of areas, (c), and by the use of numbers or colors.

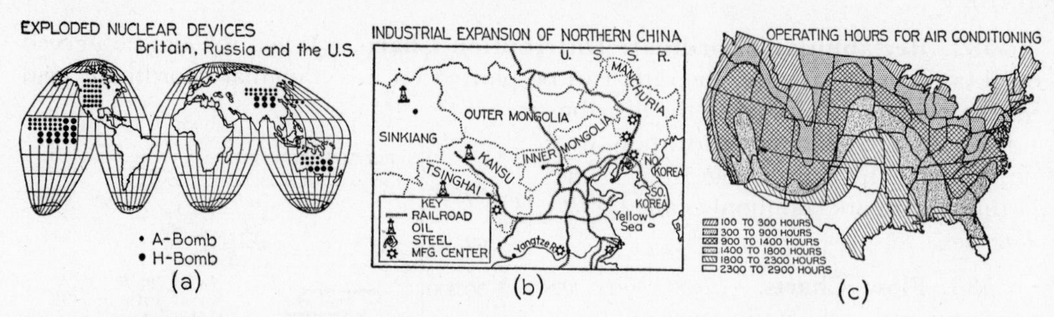

Fig. 1026 Map Distribution Charts. *(a), (b), and (c) courtesy of* Look *Magazine,* Life *Magazine, and* Heating, Piping and Air Conditioning, *respectively.*

585. Graph Problems. Construct an appropriate form of graph for the data listed. The determination of graph form (line, bar, surface, etc.) is left to the discretion of the instructor or the student and should be based on the nature of the data or the form of presentation desired. In the tables below, the data for the two related variables are separated by a double line. In some cases, more than one curve or more than one series of bars is required.

1. United States Population.

Year	1790	1800	1810	1820	1830	1840	1850
Population	3,929,214	5,308,483	7,239,881	9,638,453	12,866,020	17,069,453	23,191,876

Year	1860	1870	1880	1890	1900	1910	1920
Population	31,443,321	38,558,371	56,155,783	62,947,714	75,994,575	96,977,266	105,710,620

Year	1930	1940	1950	1953	1954	1955	1956
Population	122,775,046	131,669,275	151,132,000	160,000,000	165,000,000	....	....

2. College-Age Youth Attending College.

College Enrollment Per Cent of College-Age Youth	4.25	4.75	8.0	11.5	16.5	30.5	38.0	48.0
Year	1900	1910	1920	1930	1940	1950	1960	1970

3. Effect of Accuracy of Gear Manufacture on Available Strength Horsepower. (60 teeth gear and 30 teeth pinion of 6 diametral pitch, 1.5″ face width, $14\frac{1}{2}°$ pressure angle; 500 Brinell Case hardness).

Pitch Line Velocity—Feet Per Minute		0	1000	2000	3000	4000	5000
Horsepower	Perfect Gear	0	140	290	Straight Line Curve through Two Points		
	Aircraft Quality Gear	0	100	180	250	320	380
	Accurate Quality Gear	0	75	105	120	130	140
	Commercial Quality Gear	0	50	60	70	72	72

Noise Limit: 2750 FPM—Accurate Quality Gear
1400 FPM—Commercial Quality Gear

4. Average Tax Rates.

Year		1929	1939	1942	1945	1949	1950	1953	1954	1955
Tax Rate	Earnings Less than $5000	0.1	1.2	8.1	9.8	7.8	7.3	9.4	8.5	6.9
Per Cent	Earnings $5000 or More	6.1	10.5	32.5	29.0	15.9	17.0	20.3	18.9	17.6

5. "Depreciation Band" for Fifteen Makes of American Cars (sedans of vintage 1941 through 1952).

Year		1941	1942	War Years	1946	1947	1948	1949	1950	1951	1952
Per cent of	Maximum	22	23		38	41.5	47.5	57	73	79	98
Original Value	Minimum	7	7		14	16.5	21.5	33.5	42	55	64.5

6. Comparison Curves of Horsepower at the Rear Wheels (as shown by dynamometer tests).

Engine RPM		2000	2200	2400	2600	2800	3000	3200	3400	3600
HP at Rear Wheels	McCulloch Supercharged with Dual Exhausts	77.0	82.0	88.0	95.5	105.0	112.5	117.0	119.0	118.5
	McCulloch Supercharged	73.0	77.5	83.5	91.5	99.0	105.5	109.5	111.5	111.5
	Unsupercharged-Dual Exhausts	64.5	70.5	75.0	79.0	82.5	84.5	85.5	83.5	
	Unsupercharged	59.5	65.0	69.0	73.5	76.5	78.5	79.5	77.0	

7. Psychological Analysis of Work Efficiency and Fatigue.

Hours of Work		9-10 AM	10-11 AM	11-12 AM	12-1 PM	Lunch	2-3 PM	3-4 PM	4-5 PM	After 8 Hour Day 5-6 PM	6-7 PM	7-8 PM
Relative Production Index	Heavy Work	100	108	104	98		103	99	98	91	86	68
	Light Work	96	104	104	103		100	102	101	94	93	83

8. Metal Hardness Comparison (Mohs' hardness scale).

Metal	Lead	Tin	Cadmium	Zinc	Gold	Silver	Aluminum	Copper
Comparative Degree of Hardness. Diamond = 10	1.5	1.8	2.0	2.5	2.5	2.7	2.9	3.0

Metal	Nickel	Platinum	Iron	Cobalt	Tungsten	Chromium	Diamond
Comparative Degree of Hardness. Diamond = 10	3.5	4.3	4.5	5.5	7.5	9.0	10.0

9. 1953 Crime Figures for Chicago and New York (as supplied by the F.B.I.).

Crime		Murders	Robberies	Burglaries	Auto Thefts	Aggravated Assaults
No. of	Chicago	291	6,980	13,279	7,688	4,352
Crimes	New York	321	8,890	44,948	13,153	9,028

10. Relative Strength and Weight of Aircraft Structural Forging.

Metal	Steel		Titanium	
	Alloy Steel 4340 MIL-S-5000 (HT 180 kps)	Stainless Steel 18-8 AMS 5640 SAE 30303F	Titanium Alloy AMS 4925 (annealed)	Titanium Comm. Pure AMS 4921 (annealed)
Ultimate Strength (Min.)	100%	41.5%	82%	44.5%
Strength Per Unit of Weight	73%	30.5%	100%	57%
Weight	100%		57.5%	

Metal	Aluminum		Magnesium
	Aluminum 75S-T6 AMS 4139 QQ-A-367	Aluminum 14S-T6 AMS 4135 QQ-A-367	Magnesium AZ-61A AMS 4358 AN-M-20
Ultimate Strength (Min.)	41.5%	36%	23.3%
Strength per Unit of Weight	86%	74.5%	75.5%
Weight	35%		23%

11. Essential Qualities of a Successful Engineer (average estimate based on 1500 inquiries from practicing engineers).

Character	Judgment	Efficiency	Understanding Human Nature	Technical Knowledge	Total
41%	$17\frac{1}{2}$%	$14\frac{1}{2}$%	14%	13%	100%

12. Automobile Accident Analysis.

Type of Accident	Cross Traffic (Grade crossing, highway, railway)	Same Direction	Head-on	Fixed Object	Pedestrian	Misc.	Total
per cent	21	30	21	11	10	7	100

13. Plot the following data on rectangular coordinate paper and on 2-cycle × 1-cycle log-log paper. Loss of Head for Water Flowing in Iron Pipes.

Velocity Feet per Second		0	1	2	3	4	5	6	7
Loss of Head Ft./1000 ft.	1″ pipe	0	6.0	23.5	50				
	2″ pipe	0	2.9	9.5	20	34	51		
	4″ pipe	0	1.6	4.2	8.2	13.5	20	28	37
	6″ pipe	0	0.7	2.4	4.7	7.7	11.4	15.5	20

14. Plot the following data on polar coordinate paper. Use the upper vertical axis as the zero axis. The data for 0°-180° is identical for both sides of a 360° plot. The data represents light distribution in a vertical plane for a bulb suspended from the ceiling with the filament at the origin of the polar chart.

Orientation—Degrees	0	10	20	30	40	50	60	70	80	90
Candle Power	140	210	310	320	310	310	300	290	280	250

Orientation—Degrees	100	110	120	130	140	150	160	170	180
Candle Power	270	290	295	300	315	330	340	350	340

CHAPTER 21

ENGINEERING GRAPHICS

586. Introduction. The engineer, in performing his design work, uses mathematics for deriving equations, solving particular solutions of equations, or for calculations using equations. In general, mathematical problems may be performed *algebraically* (using numerical and mathematical symbols) or *graphically* (using drawing techniques). The algebraic method is predominantly a verbal approach in comparison to the visual methods of graphics; therefore errors and discrepancies are more evident and subject to detection in a graphical presentation. The advantages of graphics are quite evident in any mathematics text, since most writers of mathematics supplement their algebraic notations with graphical illustrations to illuminate their writings and improve the visualization of the solutions.

Naturally the question of accuracy arises, but it must be remembered that measuring instruments are themselves graphical devices. The data obtained from an instrument reading or from measurements of physical quantities, in many cases are recorded to three significant figures. Such a degree of accuracy is practical and readily substantiated by a graphical method of computation.

The graphical methods cannot be used exclusively, nor can the algebraic methods be used to the fullest degree of effectiveness without the use of graphics. The engineer should be familiar with both methods in order to convey a clearer understanding of his analysis and designs.

Graphical computations are performed by the use of alignment charts and coordinate axes graphs, and are discussed in the subsequent articles on the basis of these two methods.

587. Nomographs or Alignment Charts. A *nomograph* is a diagram or a combination of diagrams representing a mathematical law. The Greek roots *nomos* (law) and *graphien* (to write) suggest this definition. The rectangular coordinate graphs discussed in the preceding chapter are the most common examples, showing graphically the relationship of two or more variables and their function. The term "nomograph," however, is more popularly applied to a combination of scales, arranged properly to represent mathematical laws (equations) for computational purposes. Some of the more common forms of nomographs are shown in Fig. 1027.

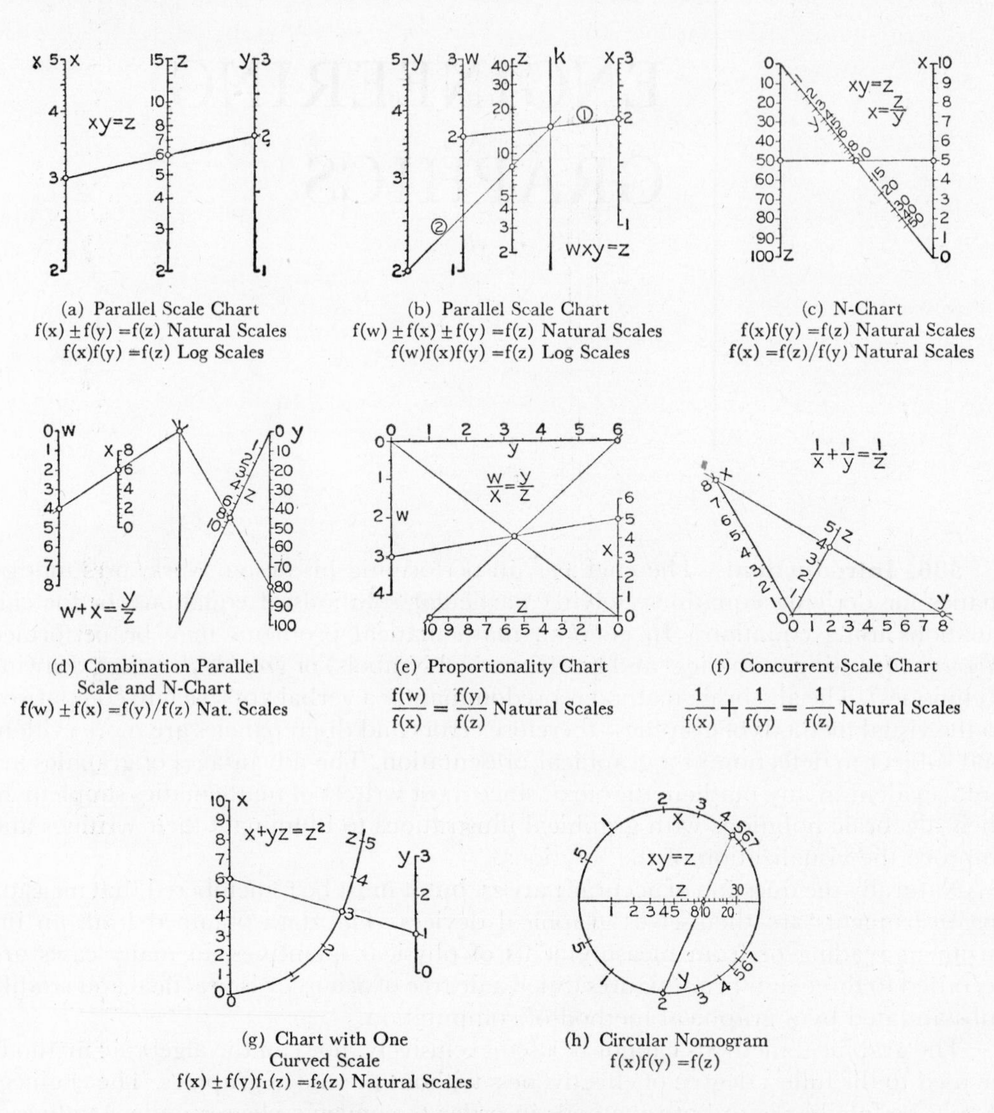

(a) Parallel Scale Chart
$f(x) \pm f(y) = f(z)$ Natural Scales
$f(x)f(y) = f(z)$ Log Scales

(b) Parallel Scale Chart
$f(w) \pm f(x) \pm f(y) = f(z)$ Natural Scales
$f(w)f(x)f(y) = f(z)$ Log Scales

(c) N-Chart
$f(x)f(y) = f(z)$ Natural Scales
$f(x) = f(z)/f(y)$ Natural Scales

(d) Combination Parallel Scale and N-Chart
$f(w) \pm f(x) = f(y)/f(z)$ Nat. Scales

(e) Proportionality Chart
$\frac{f(w)}{f(x)} = \frac{f(y)}{f(z)}$ Natural Scales

(f) Concurrent Scale Chart
$\frac{1}{f(x)} + \frac{1}{f(y)} = \frac{1}{f(z)}$ Natural Scales

(g) Chart with One Curved Scale
$f(x) \pm f(y)f_1(z) = f_2(z)$ Natural Scales

(h) Circular Nomogram
$f(x)f(y) = f(z)$

Fig. 1027 Common Forms of Nomographs or Alignment Charts.

Basically, a nomograph is used to solve a three-variable equation. A straight line joining known or given values of two of the variables intersects the scale of the third variable at a value that satisfies the equation represented, as at (a), (c), (f), (g), and (h). For this reason they are also called *alignment charts*. Two or more such charts can sometimes be combined to solve an equation containing more than three variables, as at (b), (d), and (e). The forms shown have fixed scales and a movable alignment line. However, *movable-scale nomographs* can be designed with a fixed direction alignment line, the slide rule being an example of this form.

588. Functional Scales. A mathematical equation employs a group of *variables* and *constants* to express their relationship. For example, the equation $y = x^2 + 2x + 3$ contains two variables, x and y, which are letter symbols designating quantities that may have several values in the equation. The variable x is expressed as the *function* $x^2 + 2x$; that is, the function of x, $f(x) = x^2 + 2x$. Other typical expressions of functions of a variable are $\frac{1}{x}$, log x, sin x, $x^2 + \frac{3}{x^3}$, $x-1$, etc. In the equation above, $f(y) = y$. A *constant* is any quantity that always has the same value, such as the number 3 in the equation. The number 2 of 2x is a *constant coefficient*.

The common forms of nomographs are composed of scales, each representing only one variable; hence, scales are designed for each individual variable. To draw a scale to a certain length L, representing the function of a variable between definite limits, the difference between the extreme values is multiplied by a proportionality factor called the *functional modulus*, m.

$L = m\,[f(x_{max.}) - f(x_{min.})] = m\,[f(x_2) - f(x_1)]$, where x_2 and x_1 represent the maximum and minimum values of the variables respectively, Fig. 1028. In this rela-

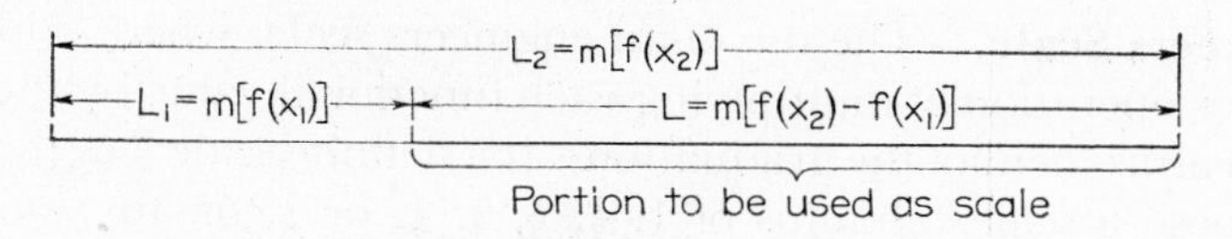

Fig. 1028 Functional Scale Relationship.

tionship, the range of values represents the first data chosen. Obviously a definite length could be chosen if desired and the functional modulus determined; or a convenient modulus could be chosen to simplify construction of the scale and consequently determine the length.

589. Scale Layout. As an example of functional scale construction, the equation of §588 is used, and the typical procedure for scale layout is illustrated in Fig.

STEP 1. Assume or determine values of the variable and the limits of the scale.

STEP 2. Compute the values of the function, substituting the values of (1).

STEP 3. Choose a convenient functional modulus and multiply it by the function values. Determine length of scale.

STEP 4. Lay off the distances obtained along a chosen line.

STEP 5. Denote the values of the variable (not the function) at the corresponding points determined.

x	0	1	2	3	4	5
f(x)	3	6	11	18	27	38
If $m = \frac{1}{8}$	$\frac{3}{8}$	$\frac{6}{8}$	$\frac{11}{8}$	$\frac{18}{8}$	$\frac{27}{8}$	$\frac{38}{8}$
L = 4.375″	0.375	0.750	1.375	2.250	3.375	4.750

0.750
.375
0 1 2 3 4 5
Measuring point
Scale of $f(x) = x^2 + 2x$

Fig. 1029 Functional Scale Layout and Design

1029 for the variable x, $f(x) = x^2 + 2x + 3$. The consequence of including the constant in the function of the variable is to add $\frac{3}{8}''$ to the scale, which is not used. Therefore, the resulting scale in effect is for the terms with variables only, $f(x) = x^2 + 2x$.

590. Scale Modulus. When each term of a variable in a function contains a constant coefficient, it may be more convenient to plot values from the function after combining the constant with the modulus. For example, if $f(x) = 2x^2 + 2x$, with x ranging in value from 0.5 to 5.0, $L = m\,[f(x_2) - f(x_1)] = m\,[2(25) + 2(2)] - [2(.25) + 2(.5)] = m\,[54 - 1.5] = 53.5m$. If the length $L = 10''$ is chosen, then $10'' = 53.5m$, and $m = 0.187$. The scale could be prepared from tabulated values for $2x^2 + 2x$. A more convenient method is to use the equation as $L = 2m\,[(x_2{}^2 + x_2) - (x_1{}^2 + x_1)]$. A common factor has been withdrawn from the coefficients and combined with the functional modulus. The product of the functional modulus and the constant coefficient is called a *scale modulus* and is commonly designated by a capital letter M. The scale modulus would be $M = 2m = 0.374$, and the function of the variable used would be $x^2 + x$.

A further advantage in preparing this scale would be to choose a more convenient scale modulus, resulting in a length close to the original length of 10″ desired. If $M = 0.333 = \frac{1}{3}$ is chosen, then $L = 0.333\,[(25 + 5) - (.25 + .5)] = 9.91''$. Since $M = \frac{1}{3} = \frac{10}{30}$, the 30 scale on an engineers scale could be used, §591.

591. Engineers Scale. The use of an engineers scale, where appropriate, eliminates the tedious operation of multiplying each functional value by the scale modulus to obtain the measurement of the graduations. If a definite scale length is not required, it is best to choose a scale modulus of 1, $\frac{1}{2}$, $\frac{1}{3}$, $\frac{1}{4}$, $\frac{1}{5}$, or $\frac{1}{6}$, or any multiple of these, as 3, $\frac{1}{30}$, $\frac{10}{30}$, etc., which would permit the use of the 10, 20, 30, 40, 50, or 60 scales on the engineers scale, §37.

The numbers 10, 20, 30, 40, etc., represent the number of subdivisions per inch on the scale. For example, the 30 scale, Fig. 1030 (a), has one inch divided into 30

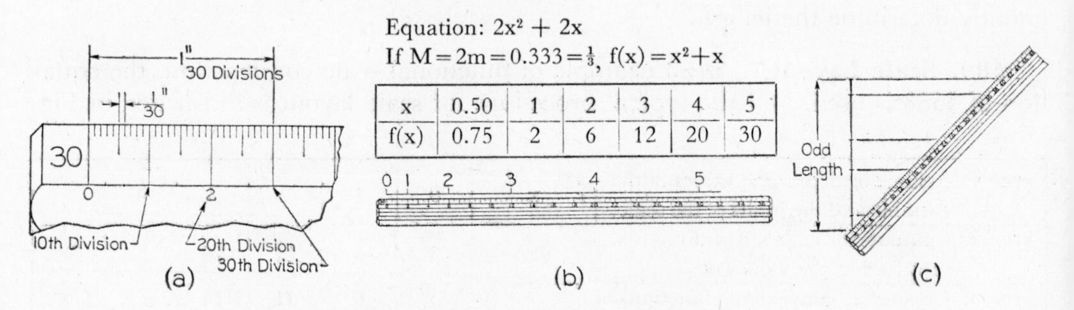

Fig. 1030 Use of Engineers Scale.

parts. The number 3 can be taken to represent number 0.3, 3, 30, 300, etc., depending upon the scale modulus used, 3, $\frac{1}{3}$, $\frac{1}{30}$, or $\frac{1}{300}$, respectively. At (b) is shown the equation of §590. Since $M = \frac{1}{3}$, the figure at one inch represents 3 as a value of the function $(x^2 + x)$. Number 5 (a value of the variable x) would be at 30 on the engineers scale; number 2 would be at 6 on the engineers scale, etc. When an odd scale modulus or length is desired, the odd length can be laid off by the parallel-line method, (c). The functional moduli or scale moduli are not considered.

592. Approximate Subdivision of Non-uniform Spaces. After the major graduations are located along the scale, the scale should be subdivided into smaller divisions. If the spaces between the major divisions are non-uniform, an approximate method of subdivision is to use a *sector*, constructed on a separate sheet of translucent paper or tracing cloth, as shown in Fig. 1031. In use, the sector is fitted to three previously located points on the scale, such as 3, 3.5, and 4 in the example. The subdivisions are located on the scale by piercing through the sector paper with a compass or divider needle point.

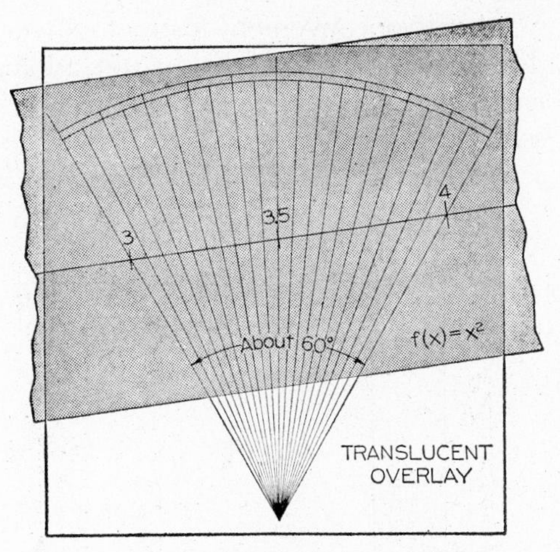

Fig. 1031 Sector for Scale Subdivision.

593. Conversion Scales. Two properly related scales placed adjacent to one another are known as *conversion scales*. The procedure for designing each scale is the same as discussed in §§589 and 590, with one exception: *the functional modulus must be the same for both scales*. As an example, the classical problem of a temperature conversion scale is illustrated in Fig. 1032. The equation for converting °C to °F is °F = $\frac{9}{5}$° C + 32 = 1.8° C + 32. The first step is to separate the two variables, ex-

If $M = m = \frac{1}{50}$, $L_{°C} = \frac{1}{50}[212 - (-40)] = \frac{1}{50}(252) = 5.04''$

Variable	°C	−40	−30	−20	−10	0	10	20	30	40	50	60	70	80	90	100
Function	1.8°C + 32	−40	−22	−4	14	32	50	68	86	104	122	140	158	176	194	212
M x Function	$\frac{1}{50}(1.8°C + 32)$	$-\frac{40}{50}$	$-\frac{22}{50}$	$-\frac{4}{50}$	$\frac{14}{50}$	$\frac{32}{50}$	$\frac{50}{50}$	$\frac{68}{50}$	$\frac{86}{50}$	$\frac{104}{50}$	$\frac{122}{50}$	$\frac{140}{50}$	$\frac{158}{50}$	$\frac{176}{50}$	$\frac{194}{50}$	$\frac{212}{50}$

When °C = −40°, °F = 1.8(−40) + 32 = −40°F; when °C = 100°, °F = 1.8(100) + 32 = 212°F

Variable	°F	−40	−20	0	20	40	60	80	100	120	140	160	180	200	220
Function	°F	−40	−20	0	20	40	60	80	100	120	140	160	180	200	220
M x Function	$\frac{1}{50}$(°F)	$-\frac{40}{50}$	$-\frac{20}{50}$	$-\frac{0}{50}$	$\frac{20}{50}$	$\frac{40}{50}$	$\frac{60}{50}$	$\frac{80}{50}$	$\frac{100}{50}$	$\frac{120}{50}$	$\frac{140}{50}$	$\frac{160}{50}$	$\frac{180}{50}$	$\frac{200}{50}$	$\frac{220}{50}$

Measuring point DEGREES CENTIGRADE
-40 -30 -20 -10 0 10 20 30 40 50 60 70 80 90 100
-40 -30 -20 -10 0 10 20 30 40 50 60 70 80 90 100 110 120 130 140 150 160 170 180 190 200 210
DEGREES FAHRENHEIT

Fig. 1032 Design and Layout of Temperature Conversion Chart.

pressing their functions individually. Thus, f (°F) = °F, f(°C) = 1.8°C + 32. If °C is to range from − 40°C to 100°C, the range of °F must be determined from the equation.

After tabulation of an appropriate distribution of intermediate values and the corresponding computed function values, the scale modulus multiplied by the function need not be determined if a convenient modulus is chosen for the use of the engineers scale. *Both scales are laid out from the same origin,* although it may not appear on the scale. The origin is at the zero value of the function, not the variable; in this case at 0°F, − 17.8°C.

Another convenient form of conversion scale is the series illustrated in Fig. 1033. The scales are parallel, but not adjacent, and a horizontal line must be drawn at right angles to the scales to determine the various forms of x. Thus, if from 6 on the scale x, a horizontal line is drawn across to scale x^2, the value 36 is found. These scales are called *natural (functional) scales.* If the function is the log of a variable, it is

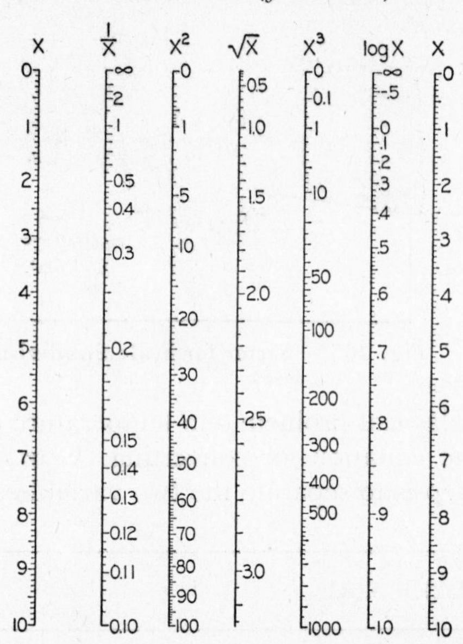

Fig. 1033 Conversion Scales.

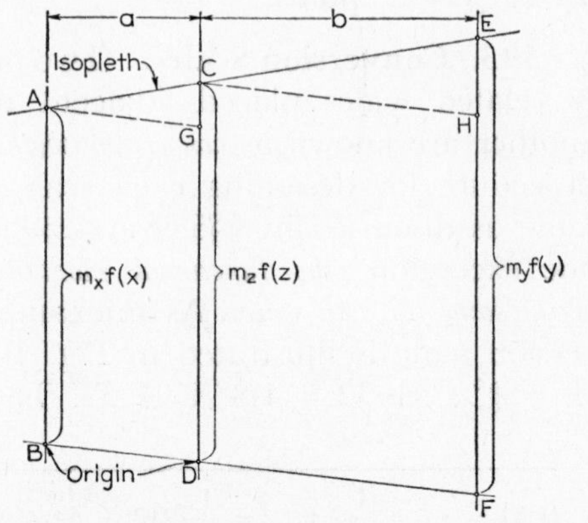

Fig. 1034 Parallel Scale Chart Geometric Derivation.

a natural scale, but if it is a function laid out logarithmically, it is called a log scale, §596. Only the first scale is uniformly divided; the remaining scales are non-uniform.

594. Natural, Parallel-Scale Nomographs, f(x) ± f(y) = f(z). The construction of nomographs for equations with three variables, involving the sum or difference of two variables, is based upon the following geometric derivation.

In Fig. 1034, AB, CD, and EF represent three parallel functional scales. The origins are aligned at any angle, and an *isopleth* (index line) crosses the three scales in any direction. Lines AG and CH are drawn parallel to BF.

From similar triangles,
$$\frac{CD - GD}{a} = \frac{EF - HF}{b}$$

Correspondingly,
$$\frac{m_z f(z) - m_x f(x)}{a} = \frac{m_y f(y) - m_z f(z)}{b}$$

Collecting terms,
$$m_x f(x) + \frac{a}{b} m_y f(y) = \left(1 + \frac{a}{b}\right) m_z f(z)$$

If $f(x) \pm f(y) = f(z)$, the coefficients must be equal;

therefore,
$$m_x = \frac{a}{b} m_y = \left(1 + \frac{a}{b}\right) m_z$$

or
$$\frac{m_x}{m_y} = \frac{a}{b} \qquad \text{Eq. (1)}$$

Substituting Eq. (1) into the coefficient of f(z),

$$m_x = \left(1 + \frac{m_x}{m_y}\right)m_z \quad \text{or} \quad m_z = \frac{m_x\, m_y}{m_x + m_y} \qquad \text{Eq. (2)}$$

Equation (1) is the relationship for spacing the outside scales after choosing moduli for these scales. Equation (2) is used to determine the modulus of the middle scale. These two relationships are used on all parallel-scale nomographs. Note that the two functions together on one side of the equality sign are the outside scales.

A few principles applicable to all parallel scale charts are illustrated in Fig. 1035. When positive values are laid out in one direction, negative values must be laid out

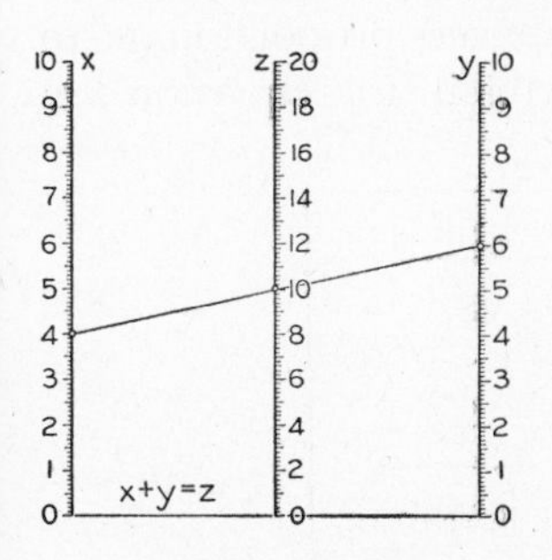

(a) Original form of equation

(b) A scale with a minus sign is inverted

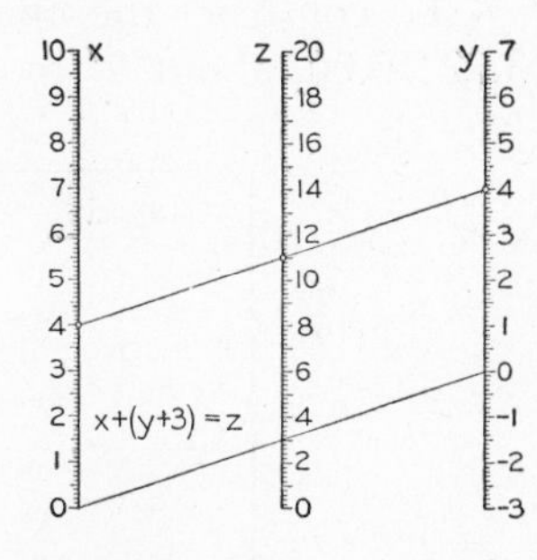

(c) A constant in a function shifts its scale up or down

Fig. 1035 Parallel Scale Variations.

in the opposite direction, (b). Note that a function with a negative sign has an inverted scale. Another approach would be to rearrange this relationship to be $x = z + y$, x being the middle scale. In this case, no scale would be inverted.

As mentioned previously, changing the alignment of origins to any angle including horizontal does not affect the equation.

595. Design and Layout of Three-Variable Parallel-Scale Nomographs. The example to be used is the equation for the velocity of a free falling body, $v_f^2 = v_o^2 + 2gs$, where v_f = final velocity, ft./sec.; v_o = original velocity, 10-50 ft./sec.; g = constant acceleration due to gravity, 32.2 ft./sec.^{2}; and s = displacement of body, 5-50 ft.

The first step is to separate the three variables and determine an appropriate length (for $8\frac{1}{2}'' \times 11''$ paper size) and functional moduli for the two outside scales, v_o and s. See upper portion of Fig. 1036.

If
$$m_{v_o} = \frac{1}{300}, \; L_{v_o} = \frac{1}{300}(2500 - 100) = 8''$$

If
$$m_s = \frac{1}{322}, \; L_s = \frac{1}{322}(64.4)(50 - 5) = 9''$$

Correspondingly, the spacing of the outside scales is

$$\frac{a}{b} = \frac{m_{v_o}}{m_s} = \frac{1/300}{1/322} = \frac{322}{300} = \frac{3.22}{3.00}$$

The functional modulus of the middle scale is

$$m_{v_f} = \frac{m_{v_o} m_s}{m_{v_o} + m_s} = \frac{(1/300)\,(1/322)}{(1/300) - (1/322)} = \frac{1}{622}$$

The range of v_f is found by substituting the maximum and minimum values into the equation.

$$L_{v_f} = \frac{1}{622}\,(5720 - 422) = 8.518''$$

The scale modulus is found by multiplying the functional modulus by the constant coefficient of each variable function.

Note that functional moduli are used to determine spacing and length of scales. A convenient scale modulus, however, aids in the layout of the scale.

It is not necessary to compute the length of the middle scale, v_f, but its computation is desirable, since it serves as a check on the preceding work.

To construct the chart, three parallel lines are drawn, spaced proportionate to 3.22::3.00, about 9″ to 10″ long, Fig. 1036. Since all the terms in the equation are

Variable	Function	m	M	Length—in.
v_o	v_o^2	$\frac{1}{300}$	$1 \times \frac{1}{300}$	8″
s	2gs	$\frac{1}{322}$	$\frac{62.2}{322} = \frac{2}{10} = \frac{10}{50}$	9″
v_f	v_f^2	$\frac{1}{622}$	$1 \times \frac{1}{622}$	8.518″

Spacing: $\frac{a}{b} = \frac{3.22}{3.00}$

(a) Scale design

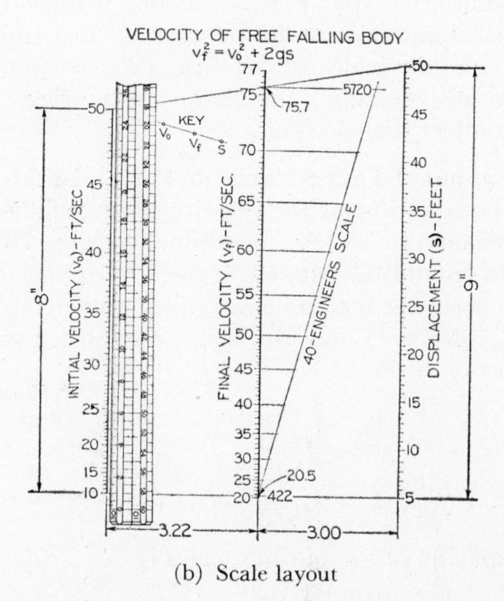

(b) Scale layout

Fig. 1036 Four-Variable Parallel-Scale Nomograph.

positive, the three scales increase in value in the same direction, normally upward, as shown here. In order to align the scale origins, draw a base line across the three scales. Usually the base line is drawn in a horizontal direction for convenience. With the "30" engineers scale, lay out the v_o scale, with one inch representing 300. The major values of 10, 15, 20, 25, etc., are located by squaring these values (since $f(v_o) = v_o^2$) and measuring them at 100, 225, 400, 625, etc., respectively. The use of the engineers scale eliminates the need for multiplying each functional value by the functional modulus. The s scale is laid out in a similar manner, by using the 50 scale. Since the constant coefficient was accounted for in the functional modulus (conveniently chosen), it is not necessary to multiply each value by 2g or the functional modulus.

After the two outside scales are laid out for their entire range, a line joining their top ends determines the correct height of the middle scale. The major scale divisions of the middle scale could be located by multiplying each functional value by the functional modulus and measuring this distance with a regular scale from the origin point. Since the modulus is odd, however, a more convenient method is to use a slightly larger engineers scale (40 or 50) and locate the major divisions by the parallel-line method as shown. From the equation, the values of v_f range from 20.5 to 75.7, and their functional values range from 422 to 5720 ($f(v_f) = v_f^2$). With the 40 scale placed with 422 at the origin (1″ = 400), the squares of the major values are set off with the engineers scale. The values of the limits should not be exactly 20.5 and 75.7. The scale should be extended to round figures of 20.0 and 77 ft./sec., but the values of 20.5 and 75.7 must lie on the base line and top isopleth, respectively.

After the scales have been properly arranged, the subdivisions can be located with the aid of a sector chart, Fig. 1031; and addition of title, legend, and scale captions completes the chart.

596. Logarithmic Parallel-Scale Nomographs, f(x)f(y) = f(z). Most engineering equations appear in the form of the product of variables. This form can be transformed into the sum or difference of variables by taking logarithms of both sides of the equation:

$$\log f(x) + \log f(y) = \log f(z)$$

Parallel-scale charts can be constructed for this form of equation by the use of logarithmic scales instead of natural scales. A convenient device for the construction of logarithmic scales is a log sector chart, Fig. 1007 (c). The use of a logarithmic sector eliminates the necessity of multiplying the logarithm of each value by the modulus to construct the scale.

597. Design and Layout of Logarithmic Parallel-Scale Nomographs. The primary design of log scale nomographs is similar to natural scale nomographs; the majority of the calculations are tabulated in Fig. 1037 (a).

After proper arrangement of the equation, it is converted to logarithmic form. The determination of scale and functional modulus, scale lengths, and spacing is similar to the form discussed in §595. Note that the scale modulus is the length of one complete log cycle used. A portion of a cycle, or more than one cycle, may be used. Cycles longer than those on the sector can be constructed by doubling or tripling a smaller cycle length; e.g., the 20″ cycle for f can be constructed by doubling lengths measured on the 10″ cycle, only a portion of a cycle being used (from 19.75 to 44.2).

Odd-length cycles can be conveniently determined by folding the sector at the appropriate subdivision between two even-length cycles (e.g., the 4.44″ cycle for F_o is between the 4″ and 5″ cycles, and is laid out almost twice since F_o ranges from 2 to 10 and from 10 to 100).

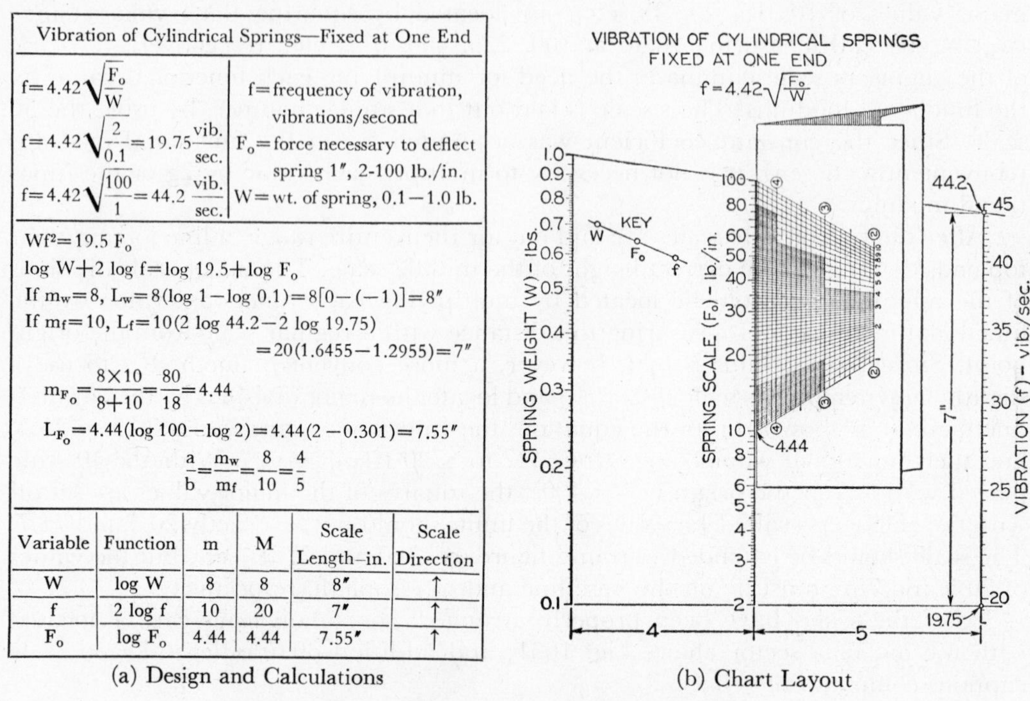

Variable	Function	m	M	Scale Length–in.	Scale Direction
W	log W	8	8	8″	↑
f	2 log f	10	20	7″	↑
F_o	log F_o	4.44	4.44	7.55″	↑

(a) Design and Calculations (b) Chart Layout

Fig. 1037 Logarithmic Parallel-Scale Nomograph.

Whether the form of equation is $x + y = z +$ constant or $xy = z \times$ constant, the constant is accounted for by determining the range of values for all variables from the equation with the constant. The primary result is to place the last scale in the proper position to accommodate the constant. This is automatically accomplished by placing the maximum and minimum values chosen or determined on the upper isopleth and base line respectively.

598. Parallel-Scale Four-Variable Nomographs, $f(w) \pm f(x) \pm f(y) = f(z)$ or $f(w)\,f(x)\,f(y) = f(z)$ or $f(w)\,f(x) = f(z)\,f(y)$. Parallel-scale nomographs may be constructed for any of the four-variable equation forms shown.

The design of four-variable nomographs requires transforming the equation into two groups of two variables each, and equating these two groups to a third term, f(k).

$$f(w) \pm f(x) = f(z) \pm f(y) = f(k)$$

and

$$\log f(w) \pm \log f(x) = \log f(z) \pm \log f(y) = \log f(k)$$

From the transformed equation, two separate three-variable parallel-scale nomographs are constructed. Both nomographs include the new term f(k). The f(k) is used to determine a scale and functional modulus, which *must be the same for both nomographs*. The scale for f(k) is never calibrated, but is used as a *pivot scale* and must have the same direction as the other scales.

Several scale arrangements for four-variable nomographs are shown in Fig. 1038.

599. Design and Layout of Four-Variable Parallel-Scale Nomographs. Since the method of design and layout is similar to the other forms discussed, the informa-

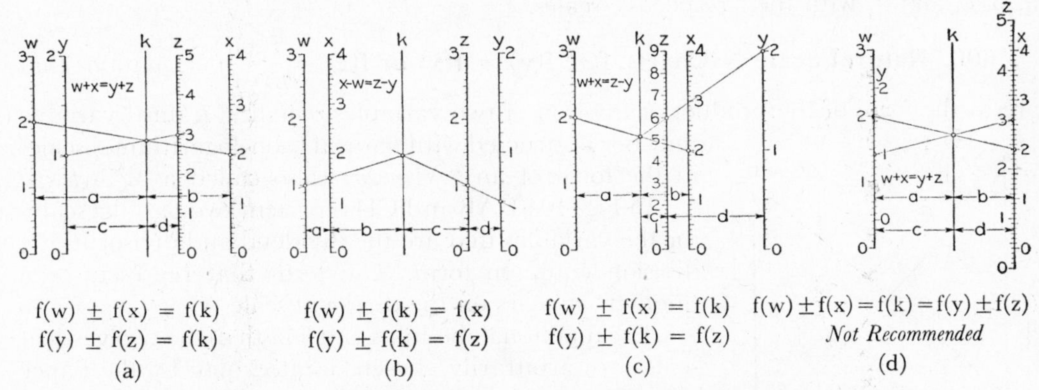

$f(w) \pm f(x) = f(k)$, $f(y) \pm f(z) = f(k)$ (a) $f(w) \pm f(k) = f(x)$, $f(y) \pm f(k) = f(z)$ (b) $f(w) \pm f(x) = f(k)$, $f(y) \pm f(k) = f(z)$ (c) $f(w) \pm f(x) = f(k) = f(y) \pm f(z)$ *Not Recommended* (d)

Fig. 1038 Arrangements of Four-Variable Parallel-Scale Nomographs.

tion for an example problem is tabulated and illustrated in Fig. 1039 (a) and (b). As shown in the upper portion of the figure at (a), equation (1) could be used, but would require an inverted scale; therefore equation (2) is preferred. The constant

JOULE'S HEATING LAW FOR ELECTRICAL CONDUCTORS

$H = 0.239\ RI^2t$	H = heat, calories
$\frac{H}{I^2} = 0.239\ Rt$	R = resistance, 1–10 ohms
	I = current, 5–25 amperes
(1) $\log H - 2\log I = \log k = \log 2.39 + \log R + \log t$	t = time, 1–60 seconds
	$H = 0.239(1)(25)(1) = 5.975$ cal.
(2) $\log H = \log k + 2\log I$; $\log k = \log 2.39 + \log R + \log t$	$H = 0.239(10)(625)(60) = 89{,}600$ cal.

If $m_H = 2$, $L_H = 2(\log 89{,}600 - \log 5.975) = 2(4.9525 - 0.7765) = 8.352''$

If $m_I = 6$, $L_I = 6(2\log 25 - 2\log 5) = 6(2)(1.398 - 0.699) = 8.388''$

Since f(H) is the middle scale, $m_H = \frac{m_I m_k}{m_I + m_k}$; $2 = \frac{6m_k}{6 + m_k}$; $m_k = 3$

$$\frac{a}{b} = \frac{m_I}{m_k} = \frac{6}{3} = \frac{2}{1}$$

If $m_R = 8$, $L_R = 8(\log 10 - \log 1) = 8(1 - 0) = 8''$

Since f(k) is the middle scale, $m_k = \frac{m_R m_t}{m_R + m_t}$; $3 = \frac{8m_t}{8 + m_t}$; $m_t = 4.8$

$$L_t = 4.8(\log 60 - \log 1) = 4.8(1.778 - 0) = 8.53''$$

$$\frac{c}{d} = \frac{m_R}{m_t} = \frac{8}{4.8} = \frac{2}{1.2}$$

Variable	Function	m	M	Length—in.	Direction
H	log H	2	2	8.352	↑
I	2 log I	6	12	8.388	↑
k	log k	3	3	—	↑
R	log R	8	8	8.000	↑
t	log t	4.8	4.8	8.530	↑

(a) Design

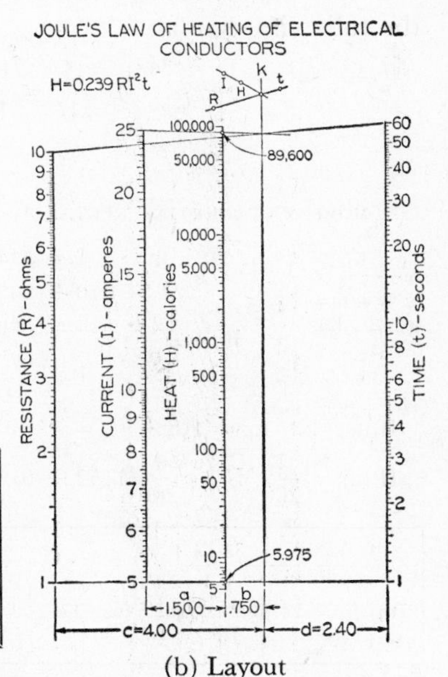

(b) Layout

Fig. 1039 Four-Variable Parallel-Scale Nomograph.

value is disregarded and automatically accounted for by placing maximum and minimum values on the isopleths and base lines, respectively. Note that the functional modulus of the middle scale does not have to be the last modulus determined. A common error, to be especially avoided, is incorrect association of the scale spacings a, b, c, and d, with their respective scales.

600. Natural Scale N-Charts, $f(z)\,f(y) = f(x)$ or $f(z) = \frac{f(x)}{f(y)}$. Equations that are in the form of the product or division of two variables equaling a third variable may be constructed with natural (non-logarithmic) scales in the form of an "*N*"*-chart* (also called a "*Z*"*-chart*).

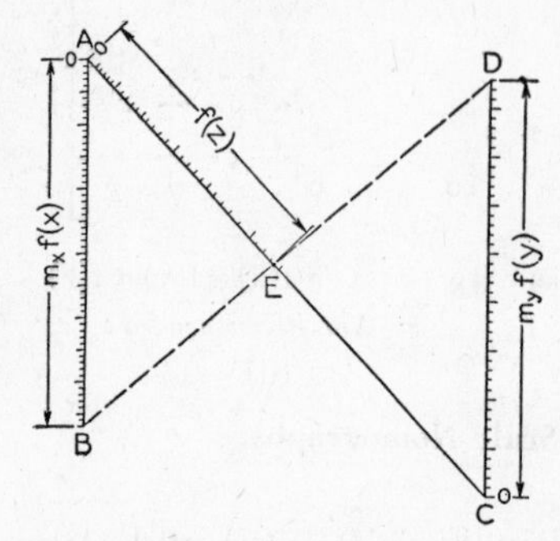

Fig. 1040 N-Chart Geometry.

In Fig. 1040, AB and CD represent two parallel scales for the variables that are the dividend and divisor in the division equation form. The term that represents the quotient appears on the diagonal scale AE.

The functional modulus and length of the two vertical scales are arbitrarily chosen, limited only by the paper size. *These scales must begin at zero* and extend in opposite directions. The diagonal scale connects the two zero points from A to C, and is graduated from A to the maximum value desired. The distance between the vertical scales is any convenient distance.

The outside scales are graduated in the same manner as previously described in §§589-591. The diagonal scale may be calibrated analytically or graphically. The analytical method may be found in the texts listed in the bibliography, p. 777. A simpler graphical method is discussed in §601.

601. Design and Layout of an N-Chart. The calculations and layout of an example problem are shown in Fig. 1041. The equation may be arranged in any of the following forms:

$$\frac{I}{b} = \frac{h^3}{12}; \quad \frac{I}{h^3} = \frac{b}{12}; \quad \text{or} \quad \frac{12\,I}{b} = h^3$$

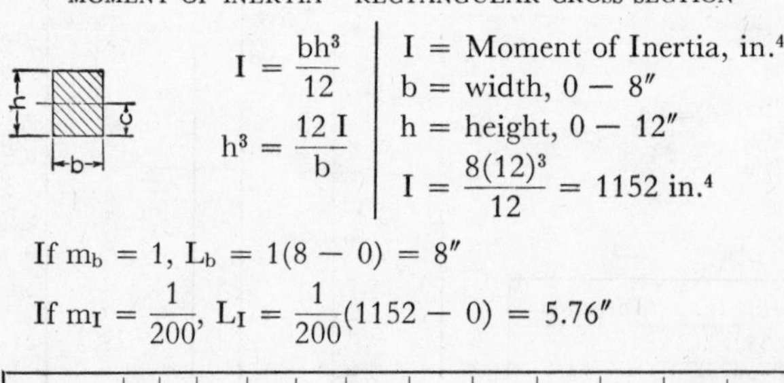

h	1	2	3	4	5	6	7	8	9	10	11	12
Values of I with b=6″ $I = h^3/2$	0.5	4.0	13.5	32.0	62.5	108.0	171.5	256.0	364.5	500.0	665.5	864.0

(a) Design

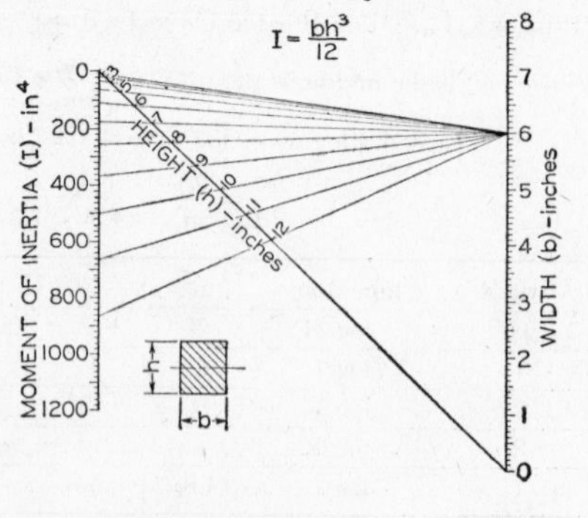

(b) Layout

Fig. 1041 N-Chart.

If the variable to a power were one of the vertical scales, it would be a non-uniform functional scale, requiring calibration and computation. Since the diagonal scale must be calibrated, $f(h) = h^3$ was chosen for this scale.

After reasonable scale lengths have been determined for f(I) and f(b), construct these scales parallel to each other and any convenient distance apart. Proper selection of moduli permits direct use of the engineers scale. A base line for alignment is not required.

Draw a line connecting the zero values of the two vertical scales. To calibrate the diagonal scale for f(h), select one convenient value of b (or I). This value should be so selected as to simplify the form of the equation for computation of the other values, when substituted into the equation. A value of $b = 6$ simplifies the equation to $I = \frac{h^3}{2}$. With this expression, determine the values of I for various values of h. Draw lines from 6 on the b scale to the values of I on the I scale to locate the corresponding values of h on the diagonal scale.

602. Large Value N-Charts. When the range of values for an N-chart are large numbers and do not extend to zero, the chart may be constructed with an additional computation. As shown in Fig. 1042, the scales of the chart are non-intersecting within the area of the paper. The conditions for an N-chart, however, are met, since the scales extended to their zero values would intersect. The lengths of the two vertical scales are determined for their large-number range of values. The two lengths (L) must be the same.

After computing new scale lengths from the zeros to the maximum values, the necessary horizontal distances b and c can be easily determined, since by similar triangles,

$$\frac{a}{d_1} = \frac{b}{d_2} = \frac{c}{d_3}.$$

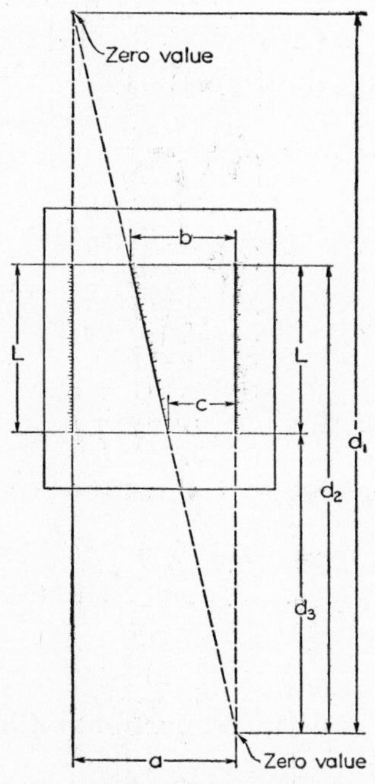

Fig. 1042 Non-intersecting N-chart.

603. Combination Parallel-Scale and N-Chart,

$$\mathbf{f(w) \pm f(x) = \frac{f(y)}{f(z)} \;[or = f(v)\, f(z)].}$$

The arrangement of variables within an equation for a nomograph makes it necessary that all the scales have the same type of graduations, either logarithmic or natural. An equation with addition or subtraction of variables has natural scales, while an equation with a multiplication or division of variables suggests logarithmic scales. If an equation of four variables contains any combination of addition or subtraction with multiplication or division, the use of an N-chart (natural scales) for the multiplication or division terms in combination with a natural parallel-scale chart for the addition or subtraction terms, can properly represent the equation, as shown in Fig. 1043. A fifth term f(k), equated to the equation is used as an uncalibrated vertical outside scale for the parallel-scale nomograph, and as a vertical scale for the

N-chart, since $f(w) \pm f(x) = f(k) = \frac{f(y)}{f(z)}$ or $= [f(y)\, f(z)]$.

604. Design and Layout of a Four-Variable Combination Nomograph. The method of design and layout is similar to the other forms discussed; the given data for an example problem are tabulated and illustrated in Fig. 1044.

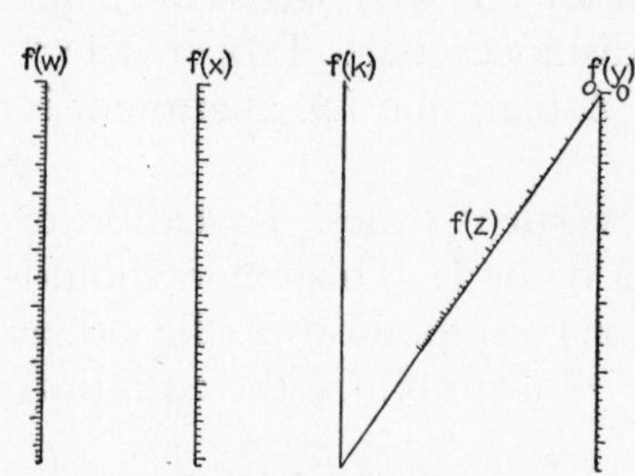

Fig. 1043 Combination Chart.

The equations are arranged as shown in step (2), since the f(k) scale is the outside parallel scale in combination as a vertical scale for the N-chart. If the equation contains a product of two variables, either variable can be placed on the diagonal scale of the N-chart. If the equation contains a division of two variables, the variable in the numerator must be placed on the diagonal scale. The lengths of only the three vertical scales must be determined. The resulting functional moduli for the parallel-scale chart are used for spacing the scales, while the N-chart spacing is not definite and is arranged simply with a convenient spacing.

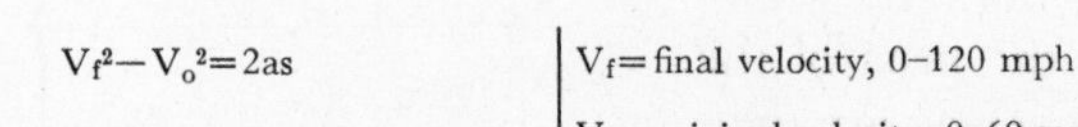

LINEAR MOTION—CONSTANT ACCELERATION

$V_f^2 - V_o^2 = 2as$

- V_f = final velocity, 0–120 mph
- V_o = original velocity, 0–60 mph
- a = acceleration, 0–2.2 ft./sec.^{2}
- s = displacement, 0–1 mile

1) $f(V_f) - f(V_o) = f(k) = f(a)f(s)$

2) $V_f^2 = k + V_o^2; \quad a = \dfrac{k}{s}$

3) $L_{V_f} = m_{V_f}(V_{f_2}^2 - V_{f_1}^2) = m_{V_f}(14400 - 0)$; if $m_{V_f} = 0.0005$, $L_{V_f} = 7.20''$

4) $L_{V_o} = m_{V_o}(V_{o_2}^2 - V_{o_1}^2) = m_{V_o}(3600 - 0)$; if $m_{V_o} = 0.001$, $L_{V_o} = 3.60''$

5) $m_{V_f} = \dfrac{m_{V_o} m_k}{m_{V_o} + m_k}$; $.0005 = \dfrac{.001\, m_k}{.001 + m_k}$; $m_k = .001$; $\dfrac{a}{b} = \dfrac{m_{V_o}}{m_k} = \dfrac{1}{1}$

6) $L_s = m_s(5280 - 0)$; if $m_s = .001$, $L_s = 5.28''$

(a) Design

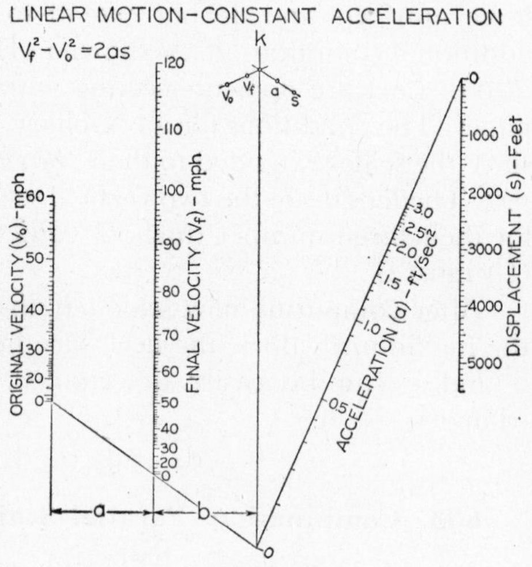

(b) Layout

Fig. 1044 Combination Parallel Scale and N-Chart.

605. Proportional Charts, $\dfrac{f(w)}{f(x)} = \dfrac{f(y)}{f(z)}$. Equations with four variables in the form of a proportion, or in the form $f(w)f(z) = f(y)f(x)$, can be represented by a nomograph with natural scales, called a *proportional chart*. The geometry of its construction is shown in Fig. 1045 (a).

From similar triangles, $\dfrac{AB}{EF} = \dfrac{a}{b} = \dfrac{AC}{DF}$

If $AB = L_w = m_w f(w)$ and $EF = L_x = m_x f(x)$, etc.,

$$\frac{m_w f(w)}{m_x f(x)} = \frac{m_y f(y)}{m_z f(z)}, \text{ and since } \frac{f(w)}{f(x)} = \frac{f(y)}{f(z)}, \text{ then } \frac{m_w}{m_x} = \frac{m_y}{m_z}$$

The proportionality of the functional moduli is the only relationship necessary for the construction of the chart. After three moduli have been determined for the scale lengths desired, the fourth modulus must be in proportion to the other values.

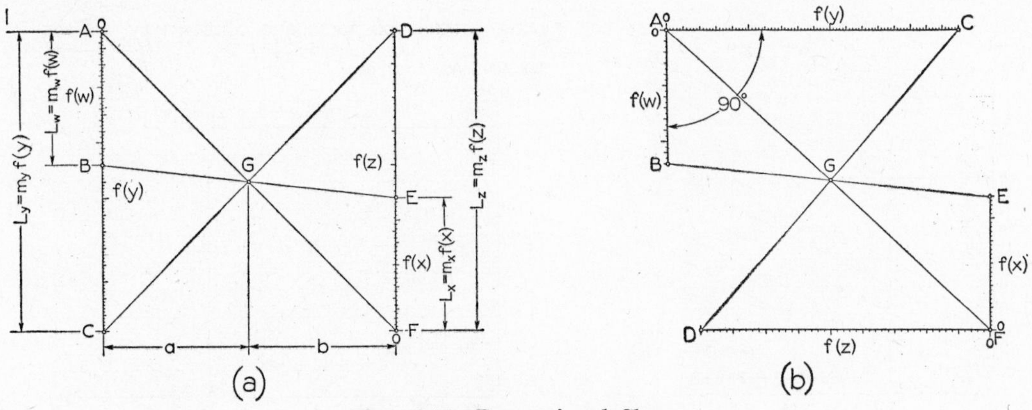

Fig. 1045 Proportional Chart.

Many arrangements of scale are possible; the most convenient form is shown at (b). The uncalibrated diagonal line connects the zero values of the four scales and is used as an index or pivot line. The relationship of the reversed directions for the vertical and horizontal scales, as based on the geometry employed, should be noted.

The design calculations and layout of an illustrative example of a proportional chart are shown in Fig. 1046.

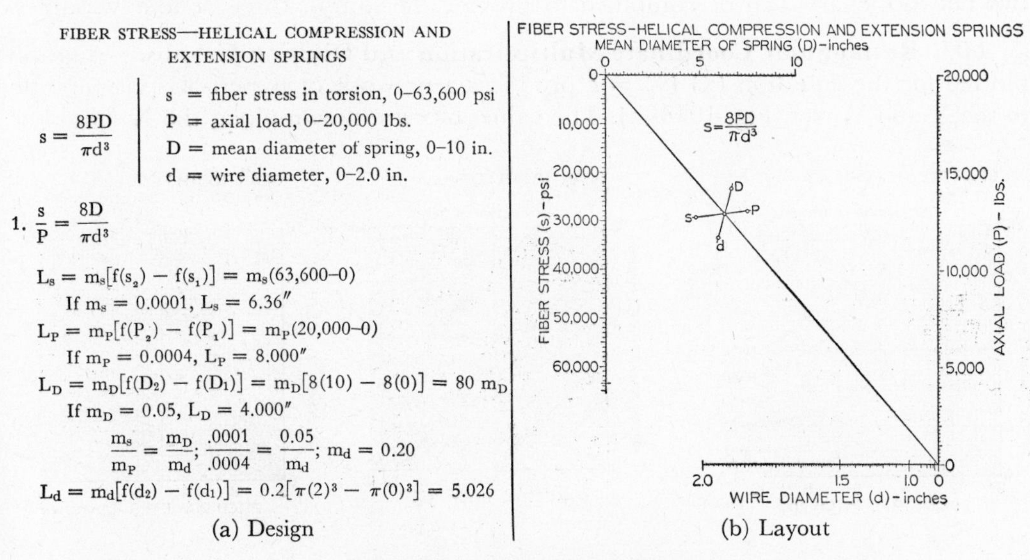

FIBER STRESS—HELICAL COMPRESSION AND EXTENSION SPRINGS

$$s = \frac{8PD}{\pi d^3}$$

s = fiber stress in torsion, 0–63,600 psi
P = axial load, 0–20,000 lbs.
D = mean diameter of spring, 0–10 in.
d = wire diameter, 0–2.0 in.

1. $\frac{s}{P} = \frac{8D}{\pi d^3}$

$L_s = m_s[f(s_2) - f(s_1)] = m_s(63{,}600-0)$

If $m_s = 0.0001$, $L_s = 6.36''$

$L_P = m_P[f(P_2) - f(P_1)] = m_P(20{,}000-0)$

If $m_P = 0.0004$, $L_P = 8.000''$

$L_D = m_D[f(D_2) - f(D_1)] = m_D[8(10) - 8(0)] = 80\ m_D$

If $m_D = 0.05$, $L_D = 4.000''$

$\frac{m_s}{m_P} = \frac{m_D}{m_d}$; $\frac{.0001}{.0004} = \frac{0.05}{m_d}$; $m_d = 0.20$

$L_d = m_d[f(d_2) - f(d_1)] = 0.2[\pi(2)^3 - \pi(0)^3] = 5.026$

(a) Design

(b) Layout

Fig. 1046 Proportional Chart.

606. Rectangular Coordinate Addition and Subtraction Graphs. When the equation for a straight line on rectangular coordinate grid, y = mx + b, is plotted at a slope value of m = 1, the resulting equation is y = x + b. This form of equation provides a chart with two variables x and y and a *network* of constant-value straight lines which can be used for addition or subtraction, as shown in Fig. 1047 (a). Such charts are called *network charts.* The constant value of the lines equals the intercept value of b in the equation $y = mx + b$, which is the y value when $x = 0$. The

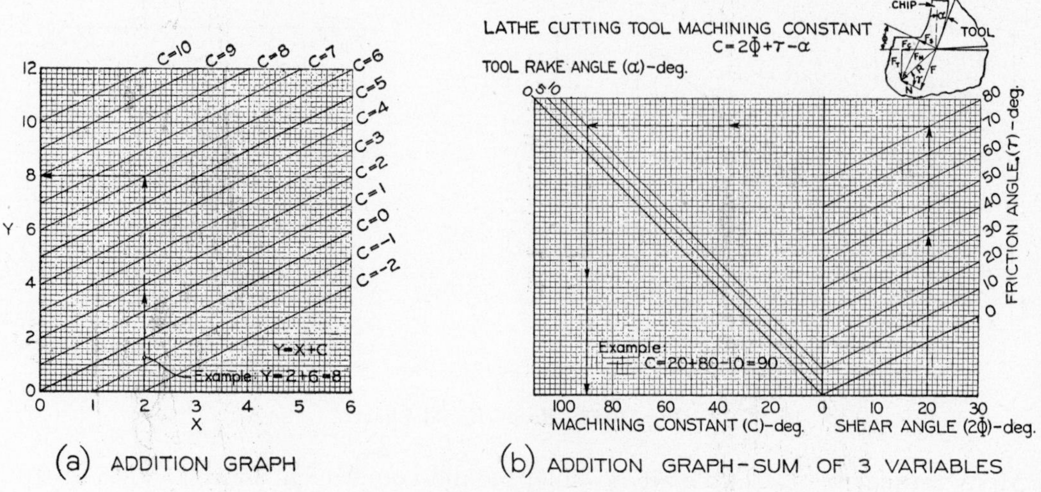

Fig. 1047 Addition Graphs.

slope (m) equals one, although the angle of the lines does not equal 45°, because one unit length on the X-axis was not made equal to one unit length on the Y-axis. The designated values along the axes are used to determine the slope. As shown at (b), two network charts can be combined to provide the sum of three or four variables.

607. Rectangular Coordinate Multiplication and Division Graphs. A graph plotted for the equation $(x)(y) = c$ produces a network of hyperbolas, asymptotic to the X and Y axes, Fig. 1048 (a). The values of x are shown along the X-axis. The

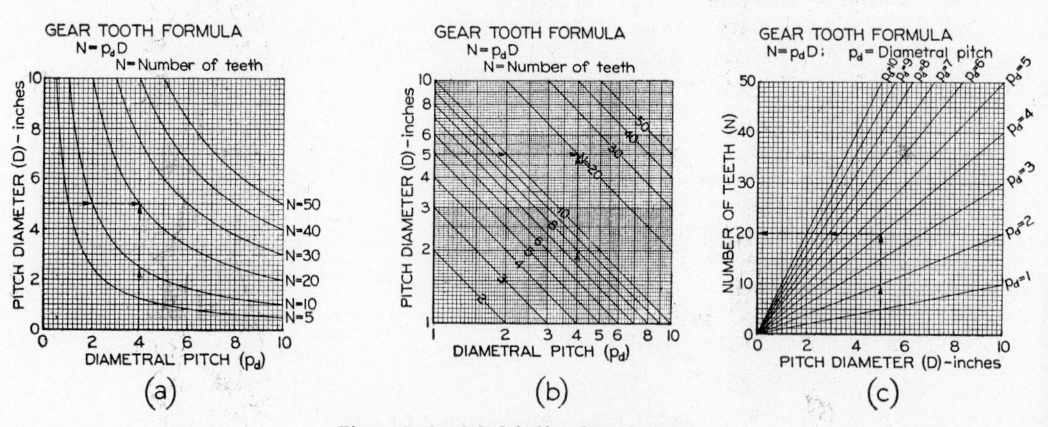

Fig. 1048 Multiplication Graphs.

corresponding y values are located on the Y-axis for a constant value of c. This network chart can be used for multiplication and division operations, but interpolation for intermediate values of c is inconvenient and curve plotting is time consuming.

A more convenient plot of the equation $(x)(y) = c$ results if the variables x and y are plotted on a logarithmic grid for constant values of c, as shown at (b). The basis for the graph is the equation $\log x + \log y = \log c$ and letting $\bar{x} = \log x$, $\bar{y} = \log y$ and $k = \log c$. Then $\bar{x} + \bar{y} = k$. The resulting network chart of straight lines has a slope of -1.

A third form of multiplication or division chart is obtained from the equation $y = mx + b$. When the intercept value $b = 0$, the resulting equation is $y = mx$. Values for x and y are plotted on the X and Y axes with various constant values for the slope m, as shown at (c). This form of chart is readily plotted and can be combined with another multiplication chart for an equation as (x) (c) = (y) (d), Fig. 1049.

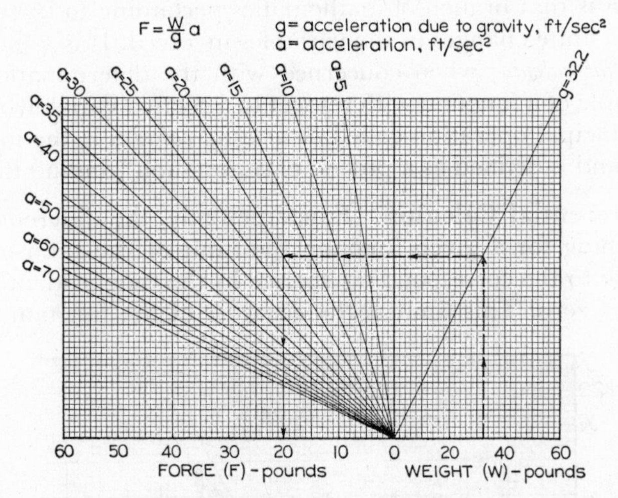

Fig. 1049 Combination Division-and-Multiplication Graph.

608. Rectangular Coordinate Graph Algebraic Solutions. Rectangular coordinate paper can be effectively used to visualize algebraic solutions. A common application is to solve two simultaneous polynomial equations with two unknowns. A graph of two linear equations is shown in Fig. 1050 (a), indicating the solution as the intersection of the two lines.

Another application is the determination of the roots of quadratic equations. A graph of a quadratic equation is shown at (b), indicating the intersection of the curve

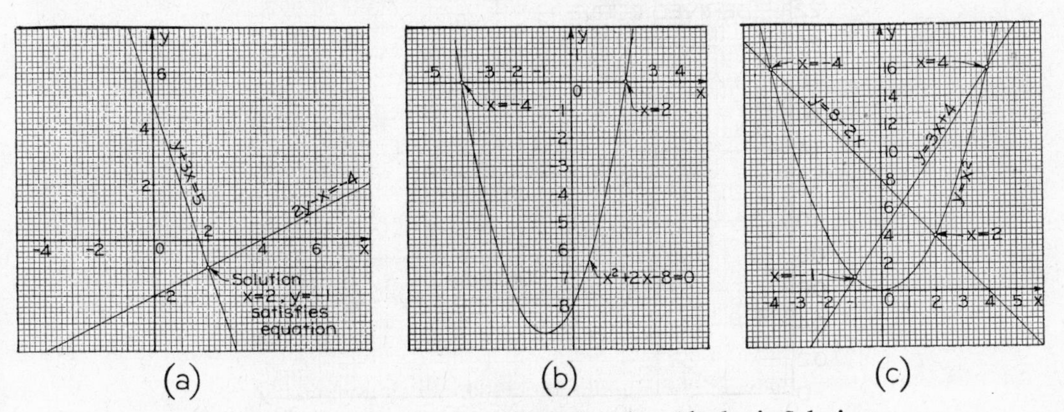

Fig. 1050 Rectangular Coordinate Graphs—Algebraic Solutions.

and the X-axis as the two roots of the equation. A more convenient procedure is to separate the equation as $x^2 = 8 - 2x = y$ and plot two separate curves, as shown at (c). The equation $y = x^2$ results in a parabola. The remaining portion of the quadratic equation $y = 8 - 2x$ is a linear equation. The intersection of the parabola and the straight line provides the same solution. The parabolic curve can be used with other linear portions of a quadratic equation to provide the roots, as shown at (c).

609. The Graphical Calculus. If two variables are so related that the value of one of them depends on the value assigned the other, then the first variable is said to be a function of the second. For example, the area of a circle is a function of the radius. *The calculus* is that branch of mathematics pertaining to the change of values of functions due to finite changes in the variables involved. It is a method of analysis called the *differential calculus* when concerned with the determination of the *rate of change* of one variable of a function with respect to a related variable of the same function. A second principal operation called *the integral calculus* is the inverse of the differential calculus and is defined as a process of summation (finding the total change).

610. The Differential Calculus. Fundamentally, the differential calculus is a means of determining for a given function the limit of the ratio of change in the dependent variable to the corresponding change in the independent variable, as this change approaches zero. This limit is the *derivative* of the function with respect to

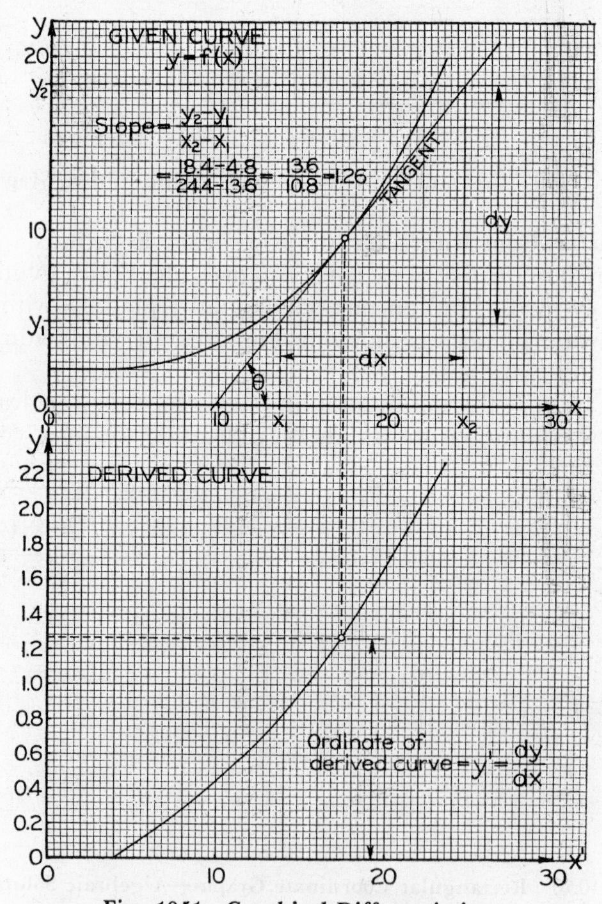

Fig. 1051 Graphical Differentiation.

the independent variable. The derivative of a function $y = f(x)$ may be denoted by $\frac{dy}{dx}$, $f'(x)$, y' or $D_x\, y$.

*The value of the derivative at any point of a curve is equal to the slope of the tangent line to the curve at that point.**

Accordingly, graphical differentiation is a process of drawing tangents or the equivalent (chords parallel to tangents) to a curve at any point and determining the value of a differential (value of the slope of the tangent line to the curve or to a parallel chord, at the point selected). The value of the slope is the corresponding ordinate value on the derived curve.

611. Graphical Differentiation—Slope Law. The *slope* of the curve at any point is the tangent of the angle (θ) between the X-axis of the graph and a tangent line to the curve at the selected point, Fig. 1051. The slope is also the rise or fall of the tangent line in the Y-direction per one unit of travel in the X-direction, the value of the slope being calculated in terms of the scale units of the X and Y axes.

The *Slope Law* as applied to differentiation may be stated as follows:

The slope at any point on a given curve is equal to the ordinate of the corresponding point on the next lower derived curve. The Slope Law is the graphical equivalent of differentiation of the calculus. The applications of the principles of the Slope Law are illustrated in the following examples.

In Fig. 1052 (a), the slope of OA and CD are constant and positive, indicated by the straight lines increasing in value upward (positively). Curve AB has a constant zero slope; curve BC has a constant negative slope as shown by the straight line de-

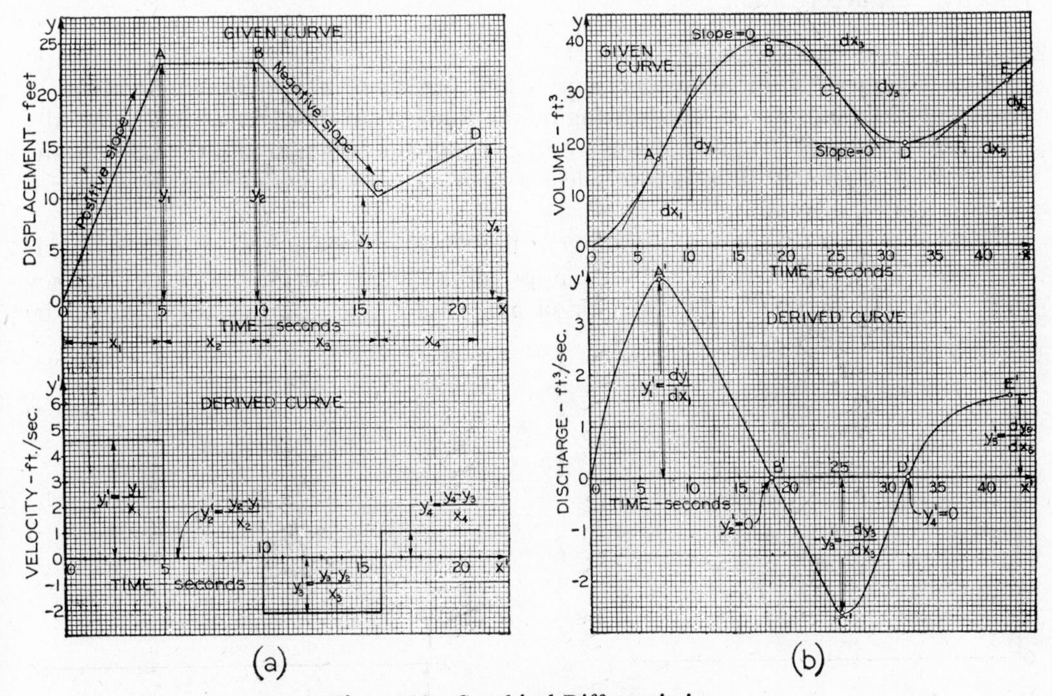

Fig. 1052 Graphical Differentiation.

*Most calculus texts include a geometric development of the principle upon which this theorem is based (Geometric Interpretation of the Derivative).

creasing in value downward (negatively). The constant slopes produce level derivative curves, which are derived in a definite order and graphed in the same order in projection below the given curve. Units assigned to the axes are at any convenient scale.

At (b), any point on the curve has a slope equal to $\frac{dy}{dx}$. The corresponding point on the derived curve has an ordinate y′ equal to the numerical value of $\frac{dy}{dx}$. Negative slopes have negative ordinate values on the derived curve. Zero slopes at B and D have zero values at B′ and D′ and are on the X-axis on the derived curve.

Since the differential calculus is concerned with the determination of a rate of change, problems involving the change of a physical quantity as a second related quantity varies, are applicable. The example at (a) involves the displacement of a body with respect to time, and the derived curve represents the velocity of the body at any instant. The second example, at (b), is a graph of the quantity of water discharged from a pipe plotted versus time. The ordinates of the derived curve represent $\frac{\text{quantity}}{\text{time}}$, which is the velocity of the water in the pipe at any given instant. The derived curve is one degree lower than the given curve. For example, if $y = x^3$, $\frac{dy}{dx} = 3x^2$. It is to be noted that the exponent of x in the derivative is one degree less or lower. If the derived curve $\frac{\text{velocity}}{\text{time}}$ were differentiated, an $\frac{\text{acceleration}}{\text{time}}$ curve would be derived, completing a family of curves, which includes the $\frac{\text{displacement}}{\text{time}}$, $\frac{\text{velocity}}{\text{time}}$, and $\frac{\text{acceleration}}{\text{time}}$ curves.

612. Tangent Line Constructions. The principal manipulation involved in graphical differentiation is the drawing of lines tangent to the given curve. Generally it is more convenient to draw a tangent line in a given direction and then determine the point of tangency, than it is to select a point of tangency and then determine a tangent line at this point.

Under certain conditions, the tangent line can be constructed geometrically. If a portion of the curve is assumed to be approximately circular, the point of tangency will lie on the perpendicular bisector of any chord, and the tangent line will be parallel to the chord, Fig. 1053 (a).

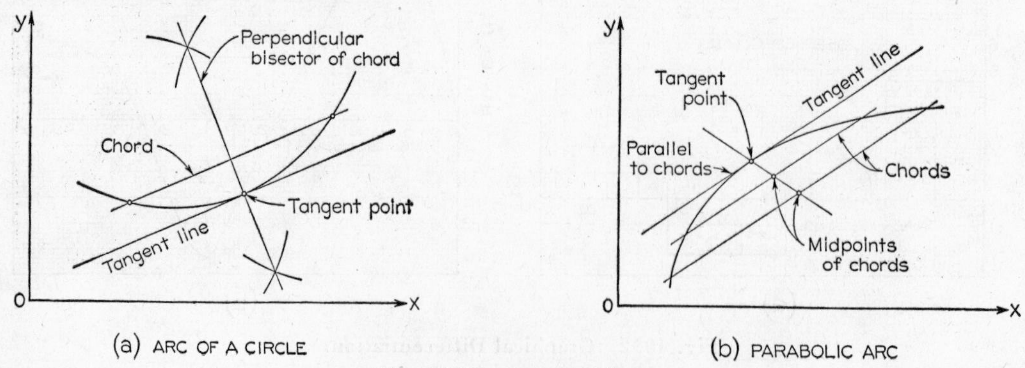

Fig. 1053 Tangent Lines.

If the curve approximates a parabola, ellipse, hyperbola, or circle, the tangent point is located at the intersection of the curve and a line drawn through the midpoint of two parallel chords, as shown at (b). The tangent line is drawn parallel to the chords.

If the curve is of unknown analytical form, difficulties arise in drawing the tangent. One method of drawing tangents is to mark a transparent strip at the edge with two dots no more than 2 mm. apart. If the dots lie on the curve, the edge of the strip approximately coincides with the tangent, the point of tangency being located midway between the dots, Fig. 1054 (a).

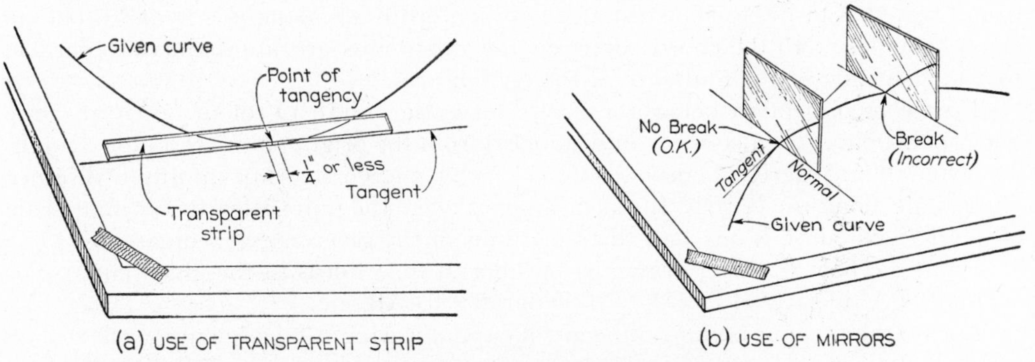

Fig. 1054 Tangent Line Determination.

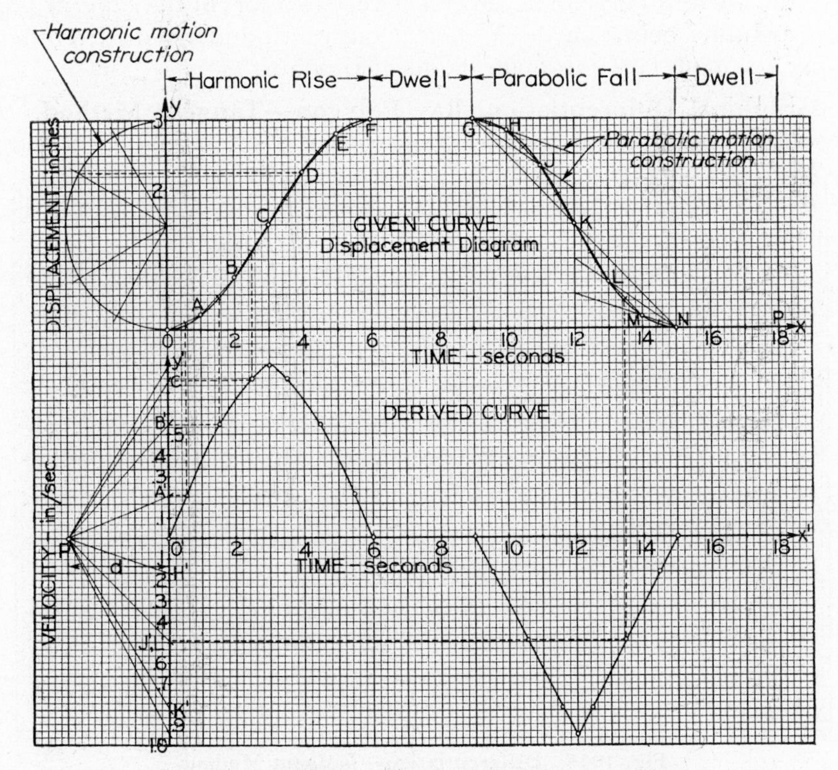

Fig. 1055 Differentiation—Ray Polygon Method.

A second method requires the use of a stainless steel or glass mirror, as shown at (b). The mirror is placed perpendicular to the graph surface and across the curve. The mirror is then moved until the curve and image coincide to form an unbroken line, as shown at the left at (b). A line drawn along the front edge of the stainless steel or the back edge of the glass mirror determines a normal to the tangent, as shown.

613. Graphical Differentiation, Ray Polygon—Chord Method. In Fig. 1055, a harmonic-rise and parabolic-fall motion for a cam is determined in the displacement diagram (given curve), and a derived curve is required. The given curve is divided into a number of short arcs, with equally spaced ordinates. Draw chords from O to A, A to B, etc. The distance between ordinates along the X-axis is chosen small enough so that the chords between the two points are approximately parallel to a tangent line at the midpoint of the chosen arc.

Locate in projection below the given curve the origin O′ of the derived curve and a pole point P at a convenient distance d from the origin point O′ on the X-axis. To avoid "flat" derived curves, d is generally chosen at some multiple distance of the scale along the X-axis. If this distance is twice the unit scale, the ordinate scale of the derived curve is one-half the scale units of the given curve ordinate.

From the pole point P, draw *rays* parallel to the chords in the given curve. For example, PA′ is parallel to OA; PB′ is parallel to AB, etc.

Locate the tangent points on the given curve for each subtended arc, as described in §612. Project each tangent point from the given curve downward to the derived curve until it intersects a horizontal line drawn from the appropriate points on the Y′-axis of the derived curve. The process is repeated for all the tangent points and their corresponding points on the Y′-axis. Construct the derived (derivative) curve by drawing a smooth curve through the located points.

614. Graphical Differentiation, Ray Polygon—Tangent Method. Another method similar to the chord method consists in the determination of the tangent

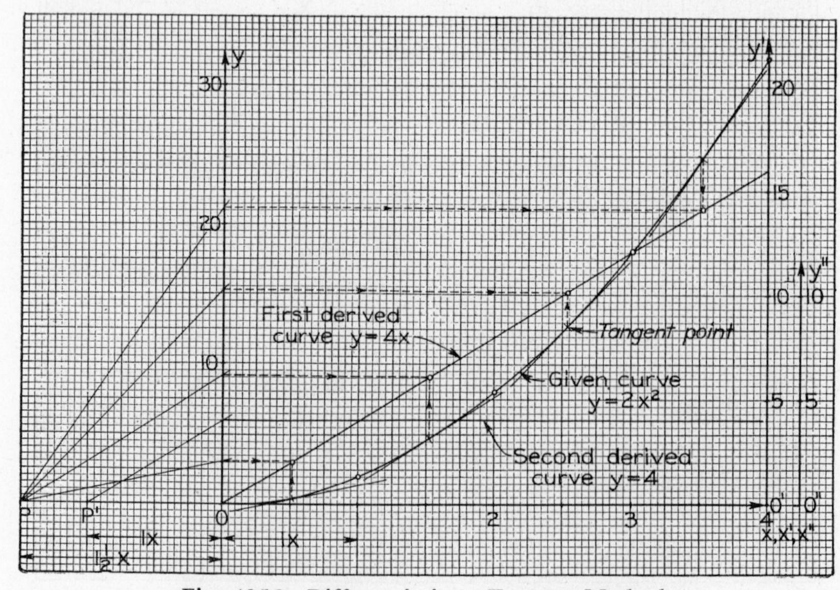

Fig. 1056 Differentiation—Tangent Method.

points and lines, without concern for the cords. In Fig. 1056, a derived curve is obtained from a given curve $y = 2x^2$ by the tangent method. A ray is drawn from the pole point parallel to each tangent line, and the derived curve points are located by projecting the tangent points vertically from the given curve until the lines intersect the appropriate horizontal projection lines from the Y′-axis of the derived curve.

The pole point P in this case is at a distance equal to $1\frac{1}{2}$times the unit scale. Therefore the ordinate scale for the first derived curve (y′) is $\frac{2}{3}$ times the ordinate scale of the given curve.

The differentiating of the derived curve (successive differentiation) results in a second derived curve. The pole point selected, P′, is at a distance equal to the unit scale of the X′-axis of the derived curve; therefore the ordinate scale for the second derived curve (y″) is equal to the ordinate scale of the first derived curve (y′).

If clarity permits, the successively derived curves can be conveniently placed one on top of the other, since the abscissa scale is the same for all the curves. Each curve, however, requires a different ordinate scale, since a "flat" curve is to be avoided.

In the example used, the first derived curve is an inclined straight line, which is expected, since if $y = 2x^2$,

$$\text{1st derivative: } \frac{dy}{dx} = 4x$$

The second derived curve appears as a line parallel to the X-axis, with a value of y = 4, which is also expected, since if y = 4x

$$\text{2nd differential: } \frac{d^2 y}{dx^2} = 4$$

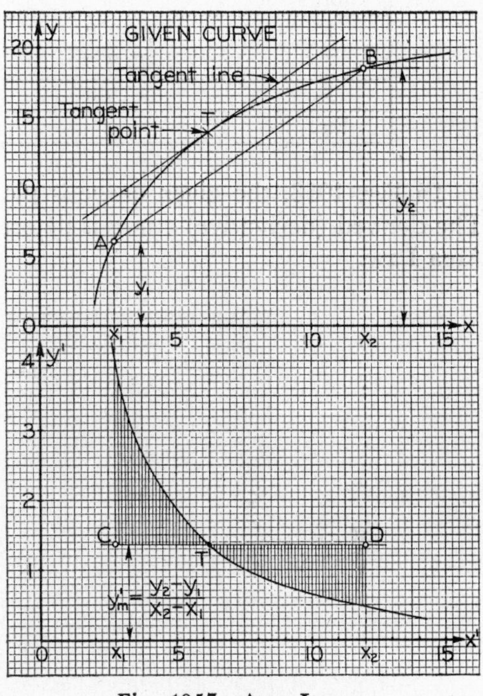

Fig. 1057 Area Law.

615. Area Law. Given a curve, Fig. 1057 (upper portion), a tangent to the arc is constructed at T. A length of arc AB is chosen so that T is at the midpoint of the length of arc. Through A and B a chord is constructed that is parallel to the tangent line. The coordinates of A are x_1, y_1, and of B are x_2, y_2. Since the chord and tangent are parallel, their slopes are equal. The slope of the tangent line at T is equal to the mean ordinate y'_m of T′ in the derived curve

$$y'_m = \frac{y_2 - y_1}{x_2 - x_1} \text{ and } y_2 - y_1 = y'_m (x_2 - x_1)$$

but $\quad y'_m (x_2 - x_1) = \text{Area of rectangle } CDx_2x_1$

Therefore $\quad y_2 - y_1 = \text{Area of rectangle } CDx_2x_1$

The law derived from this analysis, the *Area Law*, may be stated as follows:

The difference in the length of any two ordinates to a continuous curve equals the total net area between the corresponding ordinates of the next lower curve.

The application of the Area Law as stated, which provides for dividing the given curve into short arcs, permits the determination of the derivative curve for the given curve. The law is also applicable to the process of integration, as discussed in §618.

616. Graphical Differentiation—Area Law. As an example of the application of the Area Law, a practical problem in rate of interest may be used, as follows: An office building being erected costs $60,000 for the first story, $62,500 for the second, $65,000 for the third, etc. A basic cost of $400,000 is for the lot, plans, basement, etc. The net annual income is $4,000 per story. What number of stories will give the greatest rate of interest on the investment?

Initial computations are the following:

Stories	(1) Cost	(2) Income	Interest (2) ÷ (1)
1	$460,000	$ 4000	.87%
2	522,500	8000	1.53%
3	587,500	12,000	2.04%
Etc.			

In Fig. 1058, the given curve is plotted using the computed coordinate values. The derived (derivative) curve is determined from the given curve by applying the Area Law.

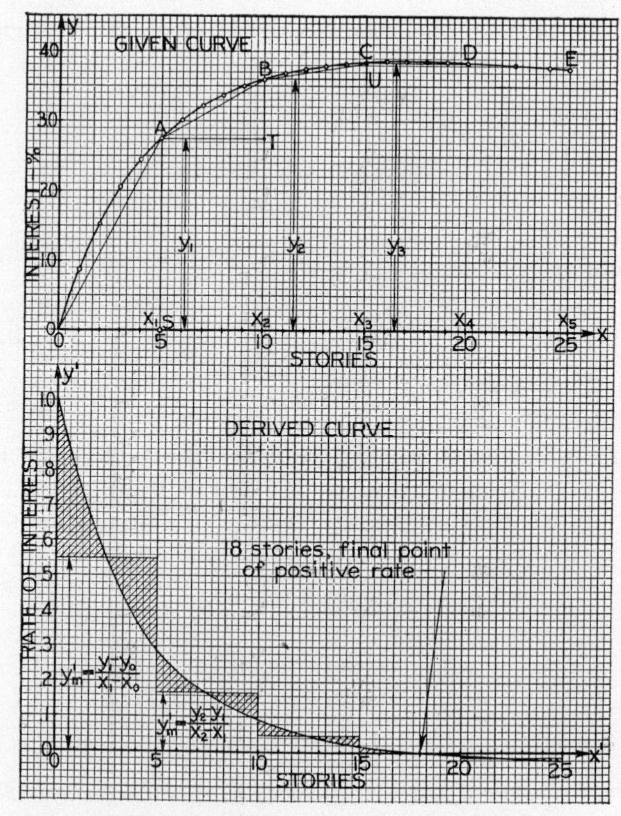

Fig. 1058 Differentiation—Area Law.

Divide the given curve into a number of short arcs, with equally spaced ordinates y_1, y_2, etc. The intervals between the ordinates along the X-axis should be chosen so that the chord intercepting the two points on the arcs OA, AB, BC, etc., will be approximately parallel to the tangent lines at the midpoints of the subtended arcs.

Determine the length of each mean ordinate for the appropriate interval by dividing the difference in length of the ordinates by the x distance between them.

$$y'_m = \frac{y_2 - y_1}{x_2 - x_1}, \text{ etc.}$$

The solution by this procedure is semi-graphical.

If the intervals between ordinates are made equal, the length of the mean ordinate for each interval of the derived curve will be proportional to SA, TB, UC, etc., which eliminates the calculations involved in the semi-graphical method.

Many of the derived curves determined are apt to appear "flat." To avoid a flat curve, the scale for the ordinates of the derived curve may be twice the scale of the ordinates of the given curve, $y'_m = 2\left(\frac{y_2 - y_1}{x_2 - x_1}\right)$. The distance on the ordinate of the derived curve can be rapidly determined by applying the dividers to the given curve ordinate differences SA, TB, UC, etc., and placing twice these distances on the corresponding ordinates of the derived curve.

Horizontal lines for each interval, and a smooth curve through the horizontal lines, are then drawn so that the triangular areas above and below the curve appear by eye to be approximately equal in area.

617. The Integral Calculus. *Integration* is a process of summation, the integral calculus being devised for the purpose of calculating the area bounded by curves. If the given area is assumed to be divided into an infinite number of infinitesimal parts called *elements*, the sum of all these elements is the total area required. The integral sign $\int$, the long S, was used by early writers and is still used in calculations to indicate "sum."

One of the most important applications of the integral calculus is the determination of a function from a given derivative, the inverse of differentiation. The process of determining such a function is *integration*, and the resulting function is called the *integral* of the given derivative or *integrand*. In many cases, however, the process of integration is used to determine the area bounded by curves, the expression of the function not being necessary.

Graphical solutions may be classified into three general groups: graphical, semi-graphical, and mechanical methods. Each of these methods will be discussed.

618. Graphical Solution—Area Law. Since integration is the inverse of differentiation, the Area Law as analyzed for differentiation, §615, is applicable to integration, but in reverse. The Area Law as applied to integration may be stated as follows: *The area between any two ordinates of a given curve is equal to the difference in length between the corresponding ordinates of the next higher curve.* If an $\frac{\text{acceleration}}{\text{time}}$ curve were given, the integral of this derivative curve would be the $\frac{\text{velocity}}{\text{time}}$ curve. If the $\frac{\text{velocity}}{\text{time}}$ curve is integrated as in Fig. 1059, the $\frac{\text{displacement}}{\text{time}}$ curve is obtained (the next higher curve).

In Fig. 1059, on the given derivative curve, uniformly (equally) spaced ordinates are established, and a mean ordinate is approximated between two consecutive ordinates by drawing a horizontal line which cuts the arc between the two ordinates, forming a pair of equal "triangles," as shown in the lower part of the figure.

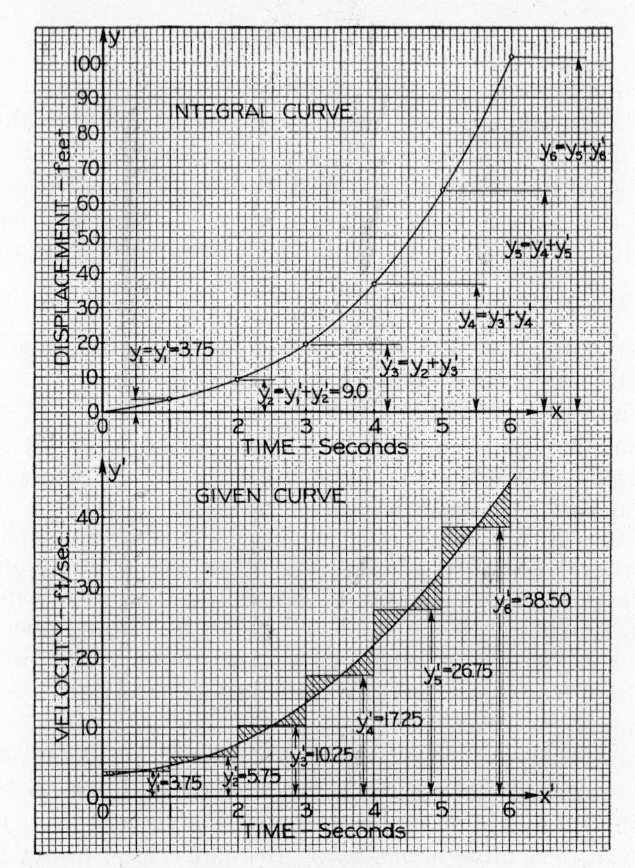

Fig. 1059 Integration—Area Law.

Draw the integral curve directly above the given curve, using the same scale along the X-axis as that on the X′-axis. The scale for the Y-axis must provide for the maximum value expected, and its length can be estimated. The value is equal to the total area under the given curve. For this example, the mean ordinate of the given curve is about 17; therefore the total area is approximately 102 square units.

The points on the integral curve are obtained directly from observation of the given curve. The first ordinate value (y_1) is at the end of the first one-second interval and is equal to the mean ordinate (y_1') obtained from the given curve. The second ordinate value (y_2) is for the end of the 2nd second interval, and is equal to the first mean ordinate value plus the second mean ordinate value ($y_2 = y'_1 + y'_2$), etc. If the scale of the ordinates for the given and integral curve are equal or proportional, these additions can be made graphically with the dividers.

619. Graphical Solution—Funicular Polygon Method. Another graphical method of integration is the *Funicular* or *String Polygon* method. If the lower degree curve is given, Fig. 1060, and the area between the curve ABCDEF and the X-axis is required (integration), the area obviously equals the sum of the rectangles ABKG, KLDC, and EFML.

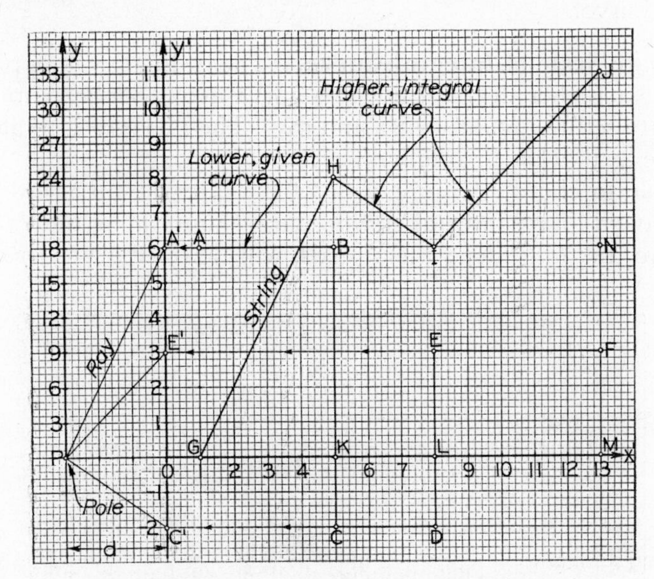

Fig. 1060 Integration—Funicular Polygon.

Locate a pole point P at any distance d to the left of the origin O on the extended X-axis. Project the lines AB, CD, and EF to the Y-axis to determine the points A′, C′, and E′, respectively. Draw rays from the pole point to points A′, C′, and E′. From point G draw a straight line (string) parallel to ray PA′ until it intersects the ordinate CKB, extended to the point of intersection at H. From point H draw a string parallel to ray PC′ until it intersects the ordinate DLE, extended to the point of intersection at I. From point I draw a string parallel to ray PE′ until it intersects the ordinate MF, extended to the intersection at J.

The curve GHIJ is the higher degree integral curve. Ordinate KH represents the area under line AB, ordinate IL represents the area under line ABCD, and ordinate JM represents the area under line ABCDEF.

Referring to Fig. 1060,

$$\text{Slope of ray } PA' = \frac{OA'}{PO} = \frac{OA'}{d}$$

$$\text{Slope of string } GH = \frac{KH}{GK}$$

since these slopes were constructed equal,

$$\frac{OA'}{d} = \frac{KH}{GK} \text{ or } OA' = d\left(\frac{KH}{GK}\right)$$

The ordinate OA′ of the lower curve, therefore, equals the slope of the string or upper curve multiplied by a constant d, the pole distance.

It will be seen that this analysis is a verification of the Slope Law, §611. A re-arrangement of the slope relationship verifies the Area Law, since $OA' = d\left(\frac{KH}{GK}\right)$, $OA'\ (GK) = d\ (KH) =$ Area of rectangle ABGK. KH is the ordinate of the higher degree integral curve.

620. Ordinate Scale of Integral Curve. The ordinate values placed along the Y-axis of the next higher degree (integral) curve are dependent upon the location of the pole point used to obtain strings and rays. The abscissa scale along the X-axis is kept the same as that of the original curve.

It is desirable to choose the unit scale of the integral curve ordinates a proper length so as to avoid "flat" integral curves. Larger units along the Y-axis, Fig. 1061, in comparison to the units along the Y′-axis, will produce a steeper curve. Another

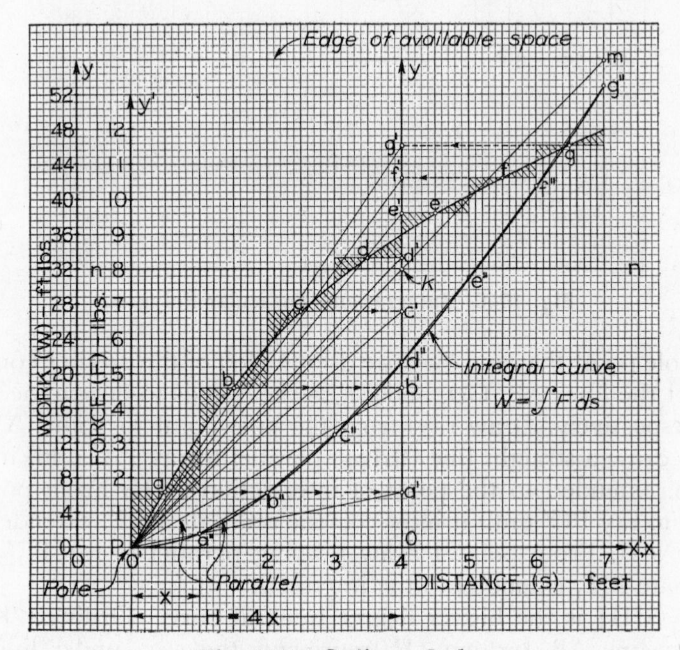

Fig. 1061 Ordinate Scale.

factor to consider is a choice of a pole distance (H). An even whole number multiple of the unit scale length along the abscissa will simplify the layout of the Y-axis (ordinate) scale units.

In the figure, H was chosen equal to four times the unit length of the abscissa scale, to the left of the origin. The vertical (ordinate) scale units of the integral curve are, therefore, four times the unit scale of the ordinate of the given curve (based on areas under the curves).

The problem in Fig. 1061 was arranged to obtain an integral curve with the best use of the available space, in order to obtain better accuracy. After plotting the given curve within the available space, a line Om is drawn from the origin to the desired end point of the integral curve. An average ordinate n-n is estimated so that the space above the given curve equals approximately the space below this curve.

An auxiliary ordinate axis for the integral curve is erected at the intersection of line n-n and Om. Divide the given curve into small increments (Area Law Method) and establish an average ordinate for each increment, a, b, c, etc. Project these ordinates to the auxiliary ordinate axis, a to a′, b to b′, etc. From the pole point, a ray is drawn to each of these points, Pa′, Pb′, Pc′, etc. Draw a segment O′a″ across the first increment from the origin parallel to the corresponding ray Pa′. From the end of this segment, a second segment a″b″ is drawn across the second increment parallel to the corresponding second ray Pb′. This process is continued for the remaining increments. A smooth curve is drawn through the end points of the segments.

621. Constants of Integration. The integral curves in many cases begin at zero, but it cannot be assumed that this is always the case.

In Fig. 1062, since the area under the given curve between any pair of ordinates gives only the difference between the lengths of the corresponding ordinates of the integral curve, the integration of the given curve gives only the shape of the integral curve. Its position with respect to the X-axis is determined by initial conditions that must be known or assumed. Therefore, when a curve is integrated by the calculus and limits are not stated, a quantity C, called the *constant of integration*, must be added to the resulting equation to provide for initial conditions, i.e., to fix the curve with respect to the coordinate axes.

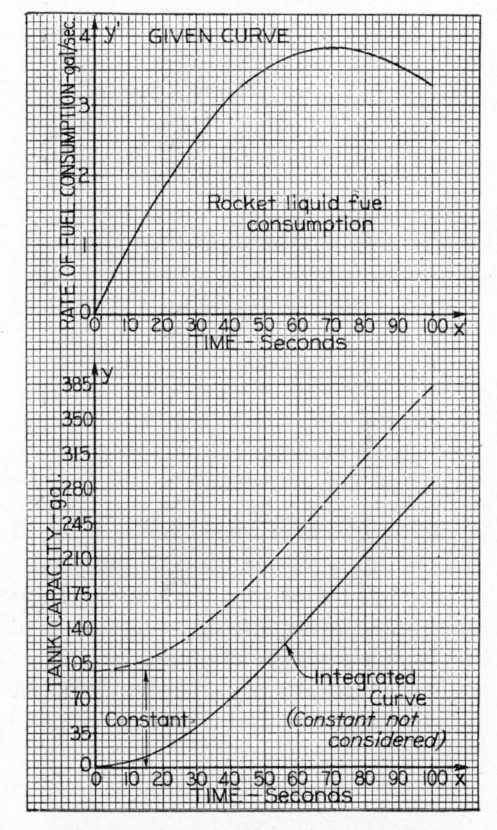

Fig. 1062 Constants of Integration.

622. Semi-Graphical Integration. When it becomes necessary to determine only an area and not an integral curve, many semi-graphical methods are applicable, some of which are the Rectangular Rule, Trapezoid Rule, Simpson's Rule, Durand's Rule, Weddle's Rule, Method of Gauss, and the Method of Parabolas. All of these rules and methods may be found in the references listed in the bibliography of this text on page 777. Simpson's Rule, which is one of the more accurate methods, is described in §623. The solutions from these methods become partly graphical, since the data used are measured from the graphical plot.

If we consider the curve ABC, Fig. 1063, as part of a parabola, and analyze the area OABCD on the same premise, then the configuration bounded by the curve, the X-axis, and the ordinates $x = x_0$ and $x = x_8$, has an area as follows:

$$A_s = \tfrac{1}{3}w\left[(y_0 + y_n) + 4\,(y_1 + y_3 + y_5 + \cdots + y_{n-1}) + 2\,(y_2 + y_4 + y_6 \cdots + y_{n-2})\right]$$

To apply Simpson's Rule, the given curve *must be divided into an even number of intervals*, and the required area is approximately equal to the sum of the extreme ordinates, plus four times the sum of the ordinates with odd subscripts, plus twice the sum of the ordinates with even subscripts, all multiplied by one-third the common distance between the ordinates.

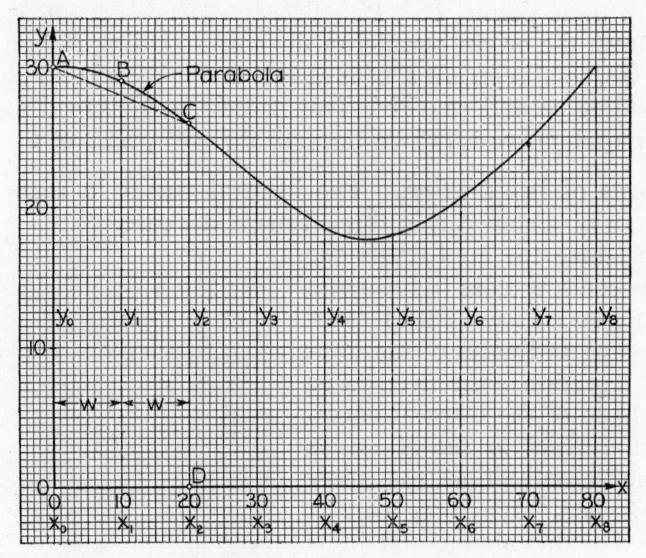

Fig. 1063 Simpson's Rule.

In the example shown in Fig. 1063,

$$A_s = \tfrac{1}{3}w\,[(y_0 + y_n) + 4(y_1 + y_3 + y_5 + y_7) + 2(y_2 + y_4 + y_6)]$$
$$= \tfrac{1}{3}\,(10)\,[(30 + 30) + 4(29 + 22 + 18 + 24.8) + 2(26 + 18.5 + 20.5)]$$
$$= 3.33\,[60 + 375 + 130] = 1883 \text{ square units per axes scale, not actual area.}$$

623. Integration—Mechanical Methods. When the requirement for integration is the determination of the area, mechanical integrators called *planimeters* may be used. The instrument is used manually to trace the outline of the area, and the instrument automatically records on a dial the area circumnavigated.

A common type of planimeter is the polar planimeter, Fig. 1064 (b). As illustrated at (a), by means of the polar arm OM, one end M of the tracer arm is caused to move in a circle, and the other end N is guided around a closed curve bounding the

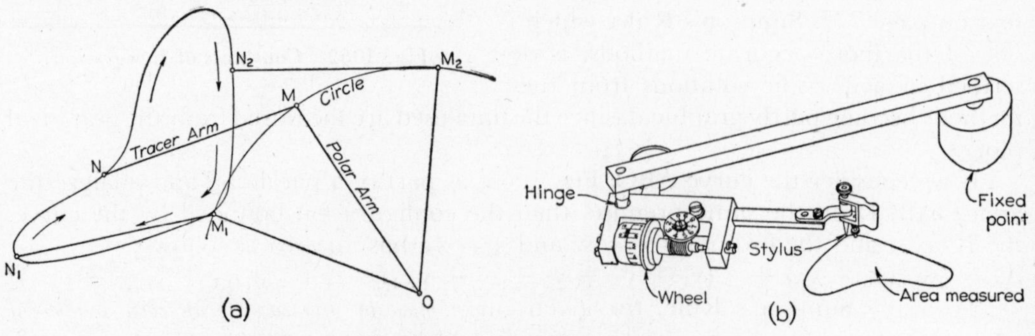

Fig. 1064 Polar Planimeter.

area measured. The area $M_1 N_1 NN_2 M_2 MM_1$ is "swept out" twice, but in opposite directions. The resulting displacement reading (difference in the two sweeps) on the integrating wheel indicates the amount of the area. The circumference of the wheel is graduated so that one revolution corresponds to a definite number of square units of area.

The ordinary planimeter used to measure indicator diagrams has a length $L = 4''$, and a wheel circumference of $2.5''$, so that one revolution of the wheel is $4 \times 2.5'' = 10$ sq. in. The wheel is graduated into ten parts, each part being further subdivided into ten parts, and a vernier scale facilitates a further subdivision into ten parts, enabling a reading to the nearest hundredth of a square inch.

624. Empirical Equations. Empirical equations by defintion are equations derived from experimental data or experience, as distinguished from equations derived from logical reasoning or hypothesis. At times the graphical analysis of plotted data is inadequate, and an equation for the data is required. The derivations of equations are varied in methods. A common method is to plot the data on rectangular, semi-logarithmic or logarithmic coordinate graph paper, with an attempt to obtain a straight line. If the plot turns out to be reasonably straight on one of these papers, an approximate (empirical) equation can be derived by geometric methods. Three of the more common methods of deriving the equation of a straight line are discussed in the following sections. The reader is referred to the references in the bibliography, p. 777, for additional methods of empirical equation derivation.

625. Empirical Equations—Rectangular Coordinate $Y = mX + a$. In Fig. 1065, the data presented at (a), plotted on rectangular coordinate paper, allows a straight line to be fitted to the points, as shown at (b). The equation for a straight line on rectangular coordinate paper is $Y = mX + a$. The procedure for equation derivation requires the determination of the slope of the line (m) and the Y-intercept (a) when $X = 0$.

Feed Rate (t) inches per revolution	.00055	.0011	.00235	.00295	.00370	.00620
Coefficient of friction (μ)	1.175	1.075	1.050	1.025	0.975	0.850

(a) Tabulated data

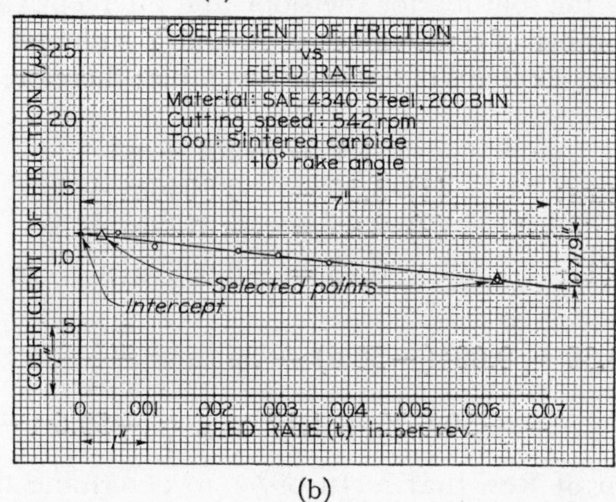

(b)

Fig. 1065 Rectangular Coordinate Solution.

Slope-Intercept Method. The slope-intercept method requires the use of a graphical plot. Since in a majority of cases the scale modulus on the Y-axis does not equal the scale modulus on the X-axis, the trigonometric slope of the line s must be divided by the ratio of the axes scale moduli r to determine the slope of the line.

$$m = \frac{s}{r} = \frac{\text{Trig. slope} = \dfrac{\text{y distance}}{\text{x distance}}}{\dfrac{\text{inches per unit y-axis}}{\text{inches per unit x-axis}}} = \frac{\dfrac{0.719''}{7''}}{\dfrac{2.00}{1000}} = \frac{-0.103}{0.002} = -51.5$$

The slope of the line may also be obtained as follows:

$$m = \frac{y_2 - v_1}{x_2 - x_1} = \frac{0.850 - 1.15}{0.0062 - 0.0003} = \frac{-0.30}{0.0059} = -50.9$$

By observation, the intercept values (Y value, when X = 0) is 1.17. The derived equation is $Y = -51.5\,X + 1.17$ or $\mu = -51.5\,t + 1.17$.

Selected Points Method. Since two unknowns must be determined, m and a, two equations solved simultaneously provides another method of equation derivation. Two points are selected; then data values on the line or points are established on the line. The X and Y values for these two points, substituted into the equation $Y = mX + a$ will provide a solution.

Selected points: (0.0003, 1.15) and (0.0062, 0.850)

$Y = mX + a$: (1) $1.15 = 0.0003m + a$
(2) $0.85 = 0.0062m + a$

Subtracting (2) from (1): $.30 = -0.0059m$; $m = -50.9$

Substituting in (1): $1.15 = 0.0003(-50.9) + a$; $a = 1.15 + 0.02 = 1.17$

Equation is $Y = -50.0X + 1.16$ or $\mu = -50.9\,t + 1.17$

Method of Averages. If a definite number of data values of Y were added together, and an equal number of computed values of Y (from a derived equation) were also added, a difference of zero between the two sums ($\Sigma Y_{data} - \Sigma Y_{computed} = 0$) would indicate close agreement. On this premise, if the equation $\Sigma Y = m\Sigma X + na$ is applied to two equal groups of data values of X and Y, solution of two simultaneous equations will provide the solution for the slope and intercept values. The term n in the equation is the number of data values of X and Y summed.

In the example used, six items of data are available; therefore two groups of three are added together.

$$\Sigma Y = m\Sigma X + na$$

Sum of first 3 data items: (1) $3.30 = 0.0040m + 3a$

Sum of second 3 data items: (2) $2.85 = 0.0129m + 3a$

Subtracting (2) from (1): $0.45 = -0.0089m$; $m = -50.6$

Substituting in (1): $3.30 = 0.004(-50.6) + 3a$; $a = \dfrac{3.502}{3} = 1.17$

Equation is $Y = -50.6\,X + 1.17$ or $\mu = -50.6\,t + 1.17$

626. Computation of Residuals. In order to determine the best equation of the three derived, *residuals* should be computed. With the use of the derived empirical equations, the *dependent variable* is computed using the independent variable data values. A comparison of computed values with observed data values reveals a plus or minus difference (residual), Fig. 1066. If the observed data are smaller than the computed value, the residual is commonly denoted as minus. The algebraic sum of the residuals for each equation suggests which is the best equation (the smallest sum, plus or minus). For the example problem used in §625, the equation derived by the slope-intercept method seems to be the best approximation.

COMPUTATION FOR RESIDUALS

Observed Data		Slope-Intercept $\mu = 51.5t + 1.17$		Selected Points $\mu = -50.9t + 1.17$		Method of Averages $\mu = -50.6t + 1.17$	
t	μ	μ_{c_1}	Residual	μ_{c_2}	Residual	μ_{c_3}	Residual
0.00005	1.175	1.142	+.033	1.142	+.033	1.142	+.033
0.0011	1.075	1.114	−.039	1.114	−.039	1.114	−.039
0.00235	1.050	1.049	+.001	1.050	0.000	1.051	−.001
0.00295	1.025	1.018	+.007	1.020	+.005	1.021	+.004
0.00370	0.975	0.979	−.004	0.982	−.007	0.983	−.008
0.00620	0.850	0.851	−.001	0.854	−.004	0.856	−.006
		Sum = −.003		Sum = −.012		Sum = −.017	

Fig. 1066 Residuals.

627. Empirical Equation—Semi-Log Coordinates $Y = A(10)^{mX}$. Data plotted on rectangular coordinate paper, which does not result in a straight line, may rectify to an approximate straight line graph on semi-logarithmic coordinate grid, if

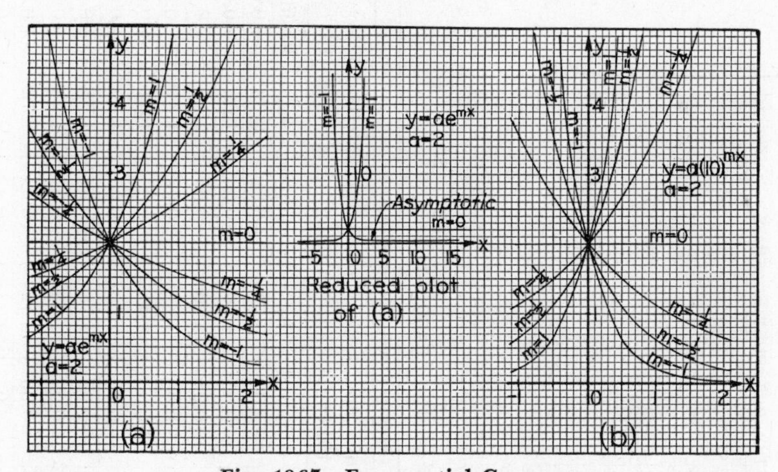

Fig. 1067 Exponential Curves.

the rectangular coordinate curve resembles an exponential curve, as shown in Fig. 1067 (a) or (b). The base of the exponent may be either e (e = 2.718) or 10.

Since semi-log paper has logarithmic divisions along the Y-axis, the equation for a straight line on this type of graph paper is

$$\log Y = mX + a$$

or

$$Y = 10^{mX}\,10^{a}$$

Friction Coeff. factor-k	1.00	1.05	1.15	1.30	1.45	1.65	1.80	2.00	2.15	2.35	2.60	2.80	3.00
log k	—	.021	.061	.114	.161	.218	.255	.301	.332	.371	.415	.447	.477
Ratio l/b	0	0.25	0.50	0.75	1.00	1.25	1.50	1.75	2.00	2.25	2.50	2.75	3.00

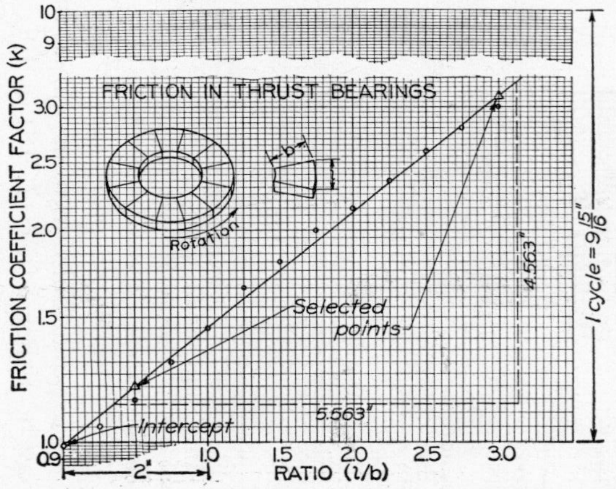

COMPUTATION FOR RESIDUALS

Observed Data		Slope intercept $k = 0.980(10)^{.165(l/b)}$		Selected Points $k = 0.993(10)^{.165(l/b)}$		Method of Averages $k = 0.980(10)^{.168(l/b)}$	
l/b	k	k_{c_1}	Residual	k_{c_2}	Residual	k_{c_3}	Residual
0.00	1.00	0.98	+0.02	0.99	+0.01	0.98	+0.02
0.25	1.05	1.08	−0.03	1.09	−0.04	1.08	−0.03
0.50	1.15	1.19	−0.04	1.20	−0.05	1.19	−0.04
0.75	1.30	1.30	0.00	1.32	−0.02	1.31	−0.01
1.00	1.45	1.42	+0.03	1.45	0.00	1.44	+0.01
1.25	1.65	1.58	+0.07	1.60	+0.05	1.59	+0.06
1.50	1.80	1.73	+0.07	1.76	+0.04	1.75	+0.05
1.75	2.00	1.91	+0.09	1.93	+0.07	1.93	+0.07
2.00	2.15	2.09	+0.06	2.12	+0.03	2.12	+0.03
2.25	2.35	2.31	+0.04	2.33	+0.02	2.34	+0.01
2.50	2.60	2.53	+0.07	2.57	+0.03	2.58	+0.02
2.75	2.80	2.79	+0.01	2.83	−0.03	2.83	−0.03
3.00	3.00	3.07	−0.07	3.11	−0.11	3.13	−0.13
		Sum =	+0.32	Sum =	0.00	Sum =	+0.03

SLOPE-INTERCEPT METHOD

Intercept = 0.980

$$\text{Slope (m)} = \frac{s}{r} = \frac{4.563''/5.563''}{9\tfrac{15}{16}'' \text{ per unit cycle}/2'' \text{ per unit}} = \frac{0.820}{4.97} = 0.165$$

Alternate:

$$\text{Slope (m)} = \frac{\log y_2 - \log y_1}{x_2 - x_1} = \frac{\log 3.1 - \log 1.2}{3.00 - 0.5} = \frac{0.4125}{2.50} = 0.165$$

$$Y = 0.980(10)^{0.165X}$$

$$\boxed{k = 0.980(10)^{0.165(l/b)}}$$

SELECTED POINTS METHOD

Selected points: (0.50, 1.2) and (3.00, 3.10)

$Y = A(10)^{mX}$; $\log Y = \log A + mX \log 10$

$\log 3.10 = \log A + m(3.00)(1)$; $0.492 = \log A + 3.00m$ (1)

$\log 1.20 = \log A + m(0.50)(1)$; $0.079 = \log A + 0.50m$ (2)

Subtracting: $0.413 = 2.50m$

$m = 0.165$

Substitute into (1): $0.492 = \log A + 3.00(0.165)$

$\log A = 0.492 - 0.495 = -0.003 = 9.997 - 10$

$A = 0.993$

$$\boxed{k = 0.993(10)^{0.165(l/b)}}$$

METHOD OF AVERAGES

$$\sum \log Y = n \log A + m \sum X \log 10$$

Sum of last 6 terms: $2.343 = 6 \log A + m(14.25)(1)$ (1)

Sum of first 6 terms: $0.830 = 6 \log A + m(5.25)(1)$ (2)

Subtracting: $1.513 = m(9.00)$; $m = 0.168$

Substitute into (2): $0.830 = 6 \log A + 5.25(0.168)(1)$

$$\log A = \frac{0.830 - 0.883}{6} = -0.009 = 9.991 - 10$$

$A = 0.980$

$$\boxed{k = 0.980(10)^{0.168(l/b)}}$$

Since 10^a will be the intercept on the Y-axis, $Y = A\ 10^{mX}$. The derivation of the empirical equation requires the solution of the values A (Y-axis intercept, when X = 0) and m(the slope of the straight line). The same three methods used for rectangular coordinate solutions, §625, are applicable. The graphical plot and essential calculations are presented in Fig. 1068 for an example problem.

628. Empirical Equations—Logarithmic Coordinates $Y = aX^m$. Curves that plot as a parabola through the origin, or a hyperbola asymptotic to the X and Y axes, are known as power curves, Fig. 1069, and can be rectified to a straight line by plotting the same data on logarithmic coordinate paper. The equation for a straight line on logarithmic paper is $\log Y = m \log X + \log A$ or $Y = aX^m$. The derivation of an empirical equation requires the determination of the values for the slope of the line m, and the intercept value (Y intercept when X = 1, or X intercept when Y = 1). Two solutions are possible, depending upon the axis intercept chosen. A preferred solution is to determine the intercept on the axis with the dependent variable. The three methods of equation derivation used for rectangular and semi-log coordinate solutions, §§625 and 627, are applicable to logarithmic coordinate plottings. The graphical plot and essential calculations are presented in Fig. 1070 as an example problem. Note that point A was shifted on the graph to point B to determine the Y intercept, when X = 1, for the slope-intercept method only.

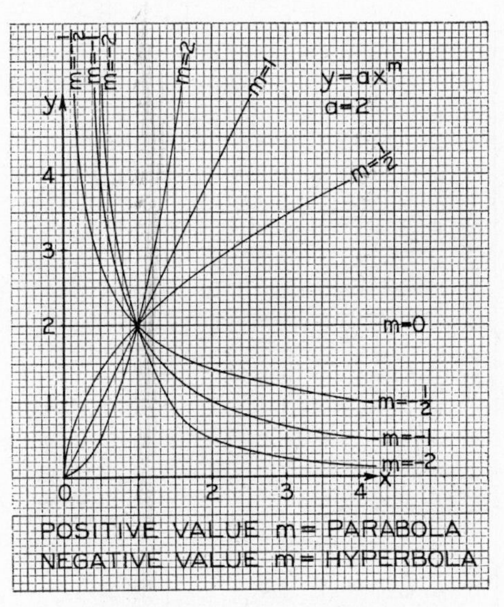

Fig. 1069 Power Curves.

INFLUENCE OF TOLERANCE ON COST OF MACHINING

Tolerance (T) inches	0.001	0.002	0.003	0.004	0.005	0.006	0.007
Relative Cost of Machining (C)	5.00	2.60	1.70	1.25	1.00	0.90	0.80
Log T	−3.000	−2.699	−2.523	−2.398	−2.301	−2.222	−2.155
Log C	0.699	0.415	0.231	0.097	0.000	−0.046	−0.097

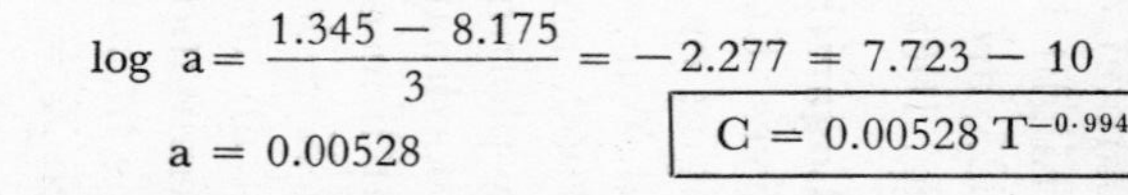

SLOPE-INTERCEPT METHOD

Intercept (a) = 0.0062, when x = 1.00

$$\text{Slope (m)} = \frac{-2.625''}{2.688''} = -0.976$$

$Y = 0.0062X^{-0.976}$

$C = 0.0062\ T^{-0.976}$

SELECTED POINTS METHOD

Selected points: (0.001, 5.00) & (0.009, 0.60)

$Y = aX^m$: $\log Y = \log a + m \log X$

(1): $0.699 = \log a + m(-3.00)$

(2): $(-0.222) = \log a + m\,(-2.046)$

$0.921 = -0.954m;\ m = -0.965$

Substitute in (1): $0.699 = \log a + (-0.965)(-3.000)$

$\log a = 0.699 - 2.895 = -2.196 = 7.804 - 10$

$a = 0.00637$

$C = 0.00637\ T^{-0.965}$

METHOD OF AVERAGES

$Y = aX^m$; $\log Y = \log a + m \log X$; $\sum \log Y = n \log a + m \sum \log X$

Sum of 1st three: (1) $1.345 = 3 \log a + m(-8.222)$

Sum of 2nd three: (2) $0.051 = 3 \log a + m(-6.921)$

$1.294 = m(-1.301);\ m = -.994$

Substitute in (1): $1.345 = 3 \log a + (-.994)(-8.222)$

$$\log a = \frac{1.345 - 8.175}{3} = -2.277 = 7.723 - 10$$

$a = 0.00528$

$C = 0.00528\ T^{-0.994}$

COMPUTATION FOR RESIDUALS

Observed Data		Slope-Intercept $C = 0.0062\ T^{-0.976}$		Selected Points $C = 0.00637T^{-0.965}$		Method of Averages $C = 0.00528T^{-0.994}$	
T	C	C_{C_1}	Residual	C_{C_2}	Residual	C_{C_3}	Residual
.001	5.00	5.21	−0.21	4.97	+0.03	5.02	−0.02
.002	2.60	2.67	−0.07	2.55	+0.05	2.54	+0.06
.003	1.70	1.80	−0.10	1.73	−0.03	1.69	+0.01
.004	1.25	1.35	−0.10	1.31	−0.06	1.28	−0.03
.005	1.00	1.09	−0.09	1.06	−0.06	1.02	−0.02
.006	0.90	0.91	−0.01	0.89	+0.01	0.86	+0.04
.007	0.80	0.79	+0.01	0.77	+0.03	0.73	+0.07
		Sum = −0.57		Sum = −0.03		Sum = +0.11	

Fig. 1070 Logarithmic Empirical Equation Derivation.

629. Engineering Graphics Problems. The following problems provide laboratory experience in applying the methods described in this chapter, and in addition introduce many typical engineering problems to the student. For convenience in assignment, the problems are classified according to type.

NOMOGRAPHY

Scale Layout Construct scales for the following:

1. $f(d) = \pi d$ — d = diameter of circle, 0 to 6 in.
2. $f(x) = 2x - 3$ — x = 0 to 15.

Two Variables—Conversion Scales. Construct two-variable conversion nomographs for the following:

3. $A = \dfrac{\pi d^2}{4}$ — A = area of a circle, in.^{2}
d = diameter of circle, 0 to 6 in.

4. $r = \dfrac{N^\circ}{57.3}$ — r = radians.
N° = number of degrees, 0° to 360°.
(1 revolution, 360° = 2 π radians; 1 radian = 57.3°)

Three Variables. Construct parallel-scale nomographs or N-charts for the following:

5. Velocity of free falling body.
$v_f = v_0 + gt$ — v_f = final velocity, ft./sec.
v_0 = initial velocity, 0 to 50 ft./sec.
t = time, 0 to 180 sec.
g = acceleration due to gravity, constant = 32.2 ft./sec.

6. Rectangular moment of inertia for circular hollow cylinder.
$I = \dfrac{\pi (D^4 - d^4)}{64}$ — I = rectangular moment of inertia, in.^{4}
D = outside diameter, 0.5 to 5.0 in.
d = inside diameter, 0.375 to 4.75 in.

7. Ohm's Law.
$V = I R$ — V = potential (voltage), volts.
I = electric current, 0.1 to 10 amperes.
R = resistance, 1 to 100 ohms.

8. Coefficient of friction.
$\mu = \dfrac{f}{N}$ — μ = coefficient of friction, 0.2 to 0.6
f = frictional force, lbs.
N = normal force (to plane of movement), 5 to 150 lbs.

Material	μ
wood on wood	.25 to .50
metal on wood	.20 to .60
metal on metal	.15 to .20

9. Kinetic Energy.
$E_k = \dfrac{Wv^2}{2g}$ — E_k = kinetic energy, ft.-lbs.
W = weight of body in motion, 10 to 200 lbs.
v = velocity of body, 5 to 100 ft./sec.
g = acceleration due to gravity = 32.2 ft./sec.^{2}

10. Torque–rpm–hp relationship.

$$T = \frac{63025\ hp}{N}$$

T = torque, 1000 to 100,000 in.-lb.
hp = horsepower, 0.02 to 3000 hp.
N = shaft speed, rpm.

Four Variables. Construct the appropriate four-variable nomograph (parallel-scale, combination N-chart and parallel-scale nomograph, or proportional chart) for the following:

11. Shunt electric motors.

$$I_a = \frac{V_t - E_c}{R_a}$$

I_a = armature current, amperes.
V_t = potential difference at the terminals, 220 to 440 volts.
E_c = counter e.m.f., 180 to 380 volts.
R_a = armature resistance, 0.10 to 0.50 ohms.

12. Critical speed of end-supported bare steel shafts.

$$N_c = \frac{46.886(10)^5\sqrt{D_1^2+D_2^2}}{L^2}$$

N_c = critical speed, rpm.
L = shaft length, 20 to 60 in.
D_1 = outside diameter, 1 to 4 in.
D_2 = inside diameter, 0.50 to 3.0 in.

For aluminum, multiply N_c by 1.0026.
For magnesium, multiply N_c by 0.9879.

13. Automotive gear ratios.

$$R\ S_c = D\ S_e$$

R = reduction ratios, 1 to 16.
S_c = car speed, 0 to 120 mi./hr.
D = wheel diameter, in.
S_e = engine speed, 0 to 3500 rpm.

14. Electromagnetic field intensity.

$$H = \frac{2\ \pi\ N\ I}{10r}$$

H = field intensity, Oersteds.
N = coil turns, 5 to 1000 turns.
I = current, 1 to 20 amperes.
r = coil radius, 1 to 10 cm.

15. Weight of steel plate.

$$W_p = 0.281wlt$$

W_p = weight of steel plate, lbs.
w = width of plate, 1 to 120 in.
l = length of plate, 1 to 120 in.
t = thickness of plate, $\frac{1}{4}$ to 3 in.

16. Vibration frequency of spring.

$$n = \frac{761{,}500d}{ND^2}$$

n = vibration per minute of spring.
d = wire diameter, 0 to 0.1875 in.
D = mean diameter of spring, 0 to 1.0 in.
N = number of active coils, 4 to 50.

RECTANGULAR COORDINATE GRAPHS

Arithmetic Charts. Construct the following multiplication graphs:

17. Plot Prob. 7 as a rectangular coordinate multiplication chart.
18. Plot Prob. 8 as a rectangular coordinate multiplication chart.

Algebra Solutions

19. A manufacturer of machine parts can sell x parts per week at a price p = 200 − 1.5x = (price per part). The cost of production $c = 0.5x^2 + 20x + 600$. Determine the total selling price and production cost for increments of five parts, for a total of 50 parts. Plot each curve, total selling price *vs* number of parts and production cost *vs* number of parts on the same sheet of graph paper.

Determine the "break-even" point, where selling price equals production costs. Determine the number of parts to be manufactured with the largest profit (profit = selling price minus production cost).

THE CALCULUS

In the following problems, plot the given data and perform the graphical differentiation or integration by any of the methods included in the text.

Differential Calculus

20. Determine the acceleration (mi./hr./sec.) *vs* time curve for the following tests performed on an 8-cylinder automobile engine.

Time (T)—Seconds		0	5	10	15	20	25	30	35	47	57
Velocity (v) Mi./Hr.	With Supercharger	0	35	55	67	75	81	86	90	97.1	
	Without Supercharger	0	28	45	54	61	67	71	75		86.7

21. From the data of tank capacity *vs* time for a rocket, determine the rate of fuel consumption, gal./sec.

Time (T)—Seconds	0	5	10	15	20	25	30	35	40	45	50	55	60
Liquid Fuel Tank Capacity (C)—Gallons	940	900	810	660	430	290	195	130	75	40	20	5	0

22. Determine the power curve for the following data. Power is the rate of doing work. 1 hp. = 33,000 ft.-lb./min.

Work (W)—Ft.-Lbs.	100,000	70,000	50,000	37,000	29,500	24,500	21,500	19,500
Time (T)—Minutes	1	2	3	4	5	6	7	8

Integral Calculus

23. For the data of Prob. 20, determine the displacement *vs* time curve.

24. Determine the work curve for the following data.

Pressure (p)—Lb./In.^{2}—psi	100	57	41	33	27.5	24	22	21	20	20
Volume (V)—In.^{3}	0	10	20	30	40	50	60	70	80	90

Plot volume as abscissa and pressure as ordinates. The integral curve is the work curve, in.-lb. *vs* in.^{3}, and is equivalent to the area under the given curve. The units for work are inch-pounds.

25. Plot the x and y coordinates given. The resulting shape is one-fourth the cross sectional area of an aircraft fuel tank. Compute the volume of the tank, if the length of the tank is 3 feet, after integrating the given curve to determine the area. Lay out a line EF to an appropriate scale representing the vertical height of the tank, and calibrate it as a measuring stick to the nearest 10 gallons. One gal. = 231 cu. in.

X—Feet	0.00	0.25	0.50	0.75	1.00	1.25	1.40	1.50
Y—Feet	2.0	1.975	1.890	1.740	1.500	1.120	0.750	0.00

EMPIRICAL EQUATIONS

Plot the data for the following problems on the appropriate form of graph paper. Determine the equation by three methods included in the text, and analyze with residuals the best equation.

Rectangular Coordinate Solution

26. Thermal conductivity of insulating materials.

Mean Temperature (T)—°F		40	80	120	160	200	240
Thermal Conductivity (C)	Celotex	0.35	0.355	0.36	0.365	0.372	0.380
B.T.U./Ft.² /Hr./°F/In.	Rockwool Blanket	0.24	0.27	0.30	0.32	0.35	0.38

27. Power at different altitudes for compound compression of air.

Altitude (A)—Feet Above Sea Level		1000	2000	3000	4000	5000	6000	7000	8000	9000	10,000
Power (P)—Indicated hp	80 lbs. gage	13.55	13.30	13.05	12.8	12.55	12.30	12.05	11.8	11.55	11.30
Per 100 Ft.³ /Min.	125 lbs. gage	16.83	16.50	16.17	15.83	15.50	15.17	14.85	14.50	14.17	13.82

Semi-Log Coordinate Solution

28. Allowable working stresses for compression springs; chrome vanadium wire, ASTM-A231, SAE 6150.

Wire Diameter (D)—Inches		.03	.05	.07	.09	.11	.13	.15	.17	.19	.21	.23	.25
Torsional Stress S_t 1000 psi	Severe Service	91	84	79	76	73	71	69	67	66	64	63	62
	Average Service	121	112	105	101	97	94	92	90	87	86	84	83

29. Electrical resistance of 95% alumina ceramic.

Temperature (T)—°C	400	450	500	550	600	650	700	750	800	850
Electrical Resistance (R)—megohms	240	120	60	31	15	7.5	3.7	2.0	0.9	0.47

Log-Log Coordinate Solution

30. Head-discharge curve, standard 1.00″ diameter circular orifice.

Discharge (D)—Ft³. /Sec.	4	5	6	7	8	9	10	15	20	25	30
Head (H)—Ft.	1.15	1.75	2.50	3.4	4.5	5.5	6.9	15.5	27	43	63

31. Power requirements for rotary drilling operations; 300 hydraulic hp required to pump 80 lb. mud through a $4\frac{1}{2}$ in. full hole string.

Mud Circulation (G)—Gal./Min.	300	400	500	600	700	800	900	1000
Stand Pipe Pressure (P)—1000 psi	17.2	12.8	10.2	8.5	7.3	6.35	5.7	5.1

32. Gas carburizing heat treatment of low carbon steel at 1700°F.

Carburizing Time (T)—Hrs.	4	8	12	16	20	24	28
Case Depth (D)—Inches	.047	.067	.084	.096	.110	.124	.136

CHAPTER 22

STRUCTURAL DRAWING

by
*E. I. Fiesenheiser**

630. Structural Drawing. *Structural drawing* consists of the preparation of design and working drawings for buildings, bridges, tanks, towers, and other structures, and comprises a very large field for the draftsman, as distinguished from the making of drawings for machines and machine parts. Although the basic principles used are the same for both types of work, certain of the methods of representation are different in structural drafting. The structural draftsman should be somewhat familiar with structural design principles. He must know a great deal about materials and the methods of fastening various members together in structures. It is also important that he be able to design connections of adequate strength to transmit the forces in a member to the other members to which it is joined.

Ordinarily the designing civil engineer determines the form and shape of a structure as well as the sizes of the main members to be used. The draftsman then makes the detail drawings, frequently under the engineer's supervision. Thus the draftsman is the engineer's first assistant and has the opportunity to learn something about design. In fact, in many cases, the draftsman's position is regarded as a steppingstone to one of greater responsibility.

The materials most commonly used in construction are wood, mild steel, concrete (plain, reinforced, and prestressed), structural clay products, and stone masonry. This chapter will be limited to discussions of the uses of these materials.

For structural drafting standards applied to structures used in connection with mechanical or electrical equipment, see ASA Y14.14, Structural Drafting.

631. Wood Construction. Many different types of wood are used as structural timber, among which are ash, birch, cedar, cypress, Douglas fir, elm, oak, pine,

*Director, Civil Engineering Dept., Illinois Institute of Technology.

poplar, redwood, and spruce. Authentic information concerning the strength properties of the various types and grades can be obtained from the *Wood Handbook No. 72* prepared by the U.S. Forest Products Laboratory in 1955.* For allowable stresses and design of connections, the student should refer to the *National Design Specification for Stress-Grade Lumber and its Fastenings* as recommended by the National Lumber Manufacturers Association, Washington, D.C.

Wood is in common use in the construction of homes and other buildings in the form of sills, columns, studs, floors, roof rafters, purlins, trusses, and roof sheathing. Typical drawings of wood structures and construction details may be obtained from the various publications of the National Lumber Manufacturers Association. Common methods of fastening timber members together involve the use of nails, screws, lag screws, drift bolts, bolts, steel plates, and various special timber connectors. Ordinarily a structural timber is cut so that the wood fibers, or grain, run parallel to the length. The strength resistance of wood is not the same in a direction perpendicular to the grain as it is parallel to the grain. Therefore, in designing connections, the direction of the force to be transmitted must be taken into account. Also, a proper spacing, edge distance, and end distance must be maintained for screws, bolts, and other connectors.

Typical bolted joints are shown in Fig. 1071. To transmit the forces, either wood or steel splice plates may be used, as shown at (a) and (b). A detail without splice plates is shown at (c). It should be realized that each type of connection requires a different design.

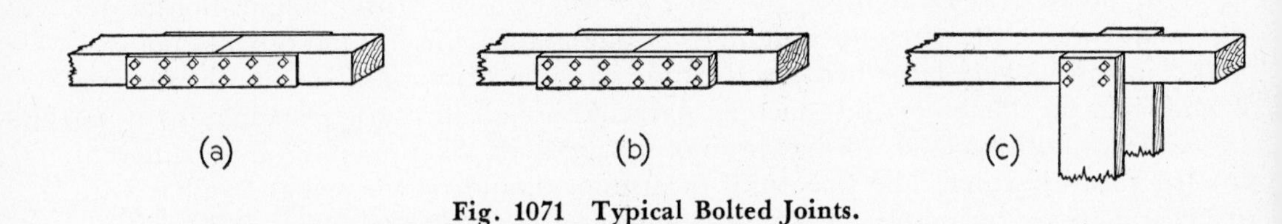

Fig. 1071 Typical Bolted Joints.

The use of *split-ring metal connectors*, §632, is now common. In Fig. 1072 is shown a drawing of the left half of a timber roof truss in which this type of connector is used. The left half only is drawn, since the truss is symmetrical about the centerline. By referring to the views of the top and bottom chords, the student may visualize the relative positions of the connecting members of the structure. In drawings of trusses, the view of the top chord is simply an auxiliary view. However, it is customary to show the lower cord by means of a section taken just above the lower chord with the *line of sight downward*. The lower chord is therefore shown in first-angle projection, §243.

Timber is used to some extent in the construction of highway truss bridges, as shown in Fig. 1073. Only the left half of the symmetrical structure is shown in the elevation view. The cross section, drawn for one-half the roadway width, and the bottom sectional view, indicate the framing and the manner in which the floor is supported.

As shown by this drawing, it is customary not to break the dimension lines, but

*Superintendent of Documents, U.S. Government Printing Office, Washington 25, D.C.

to letter the dimension above the line. This rule is followed quite generally in structural drafting.

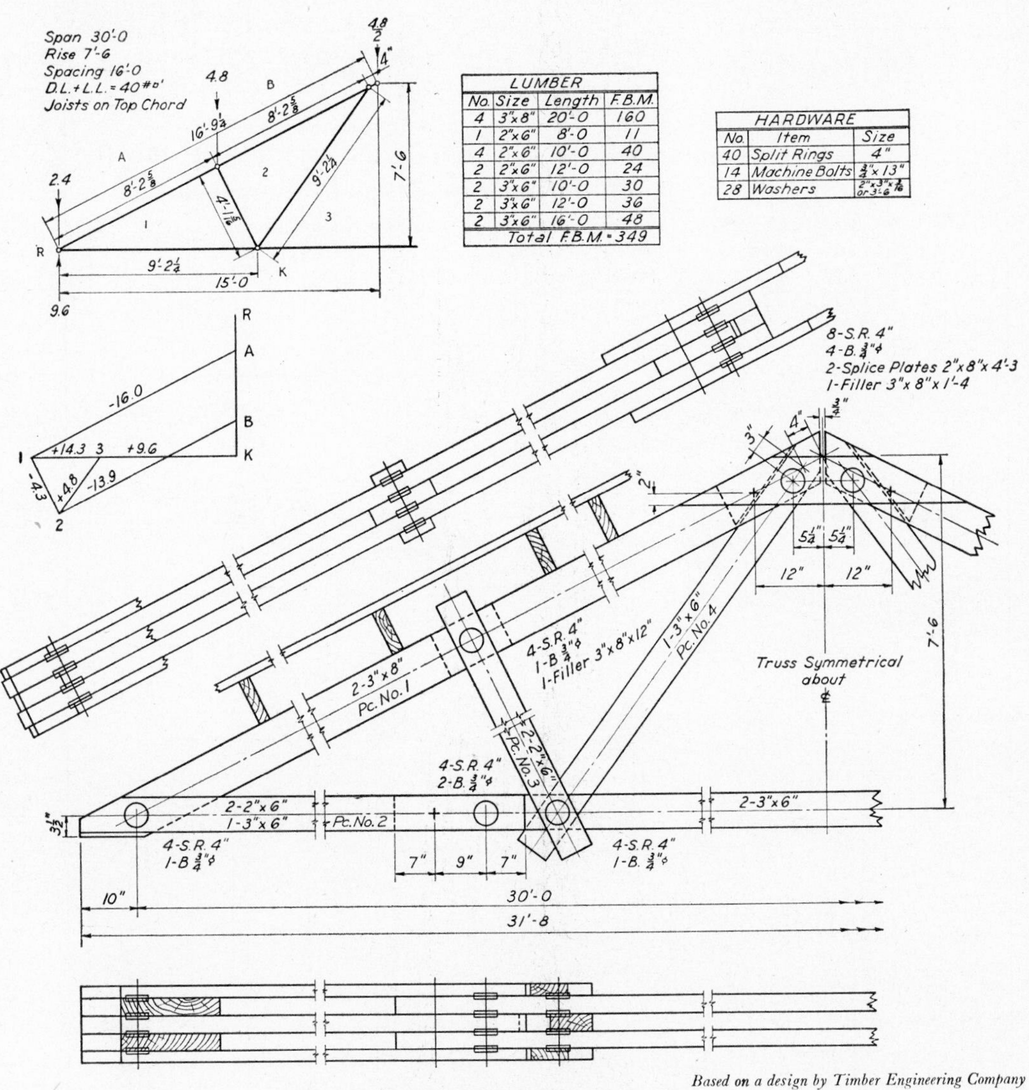

LUMBER			
No.	Size	Length	F.B.M.
4	3"x8"	20'-0	160
1	2"x6"	8'-0	11
4	2"x6"	10'-0	40
2	2"x6"	12'-0	24
2	3"x6"	10'-0	30
2	3"x6"	12'-0	36
2	3"x6"	16'-0	48
Total F.B.M.=349			

HARDWARE		
No.	Item	Size
40	Split Rings	4"
14	Machine Bolts	¾"x 13"
28	Washers	2"x3"x$\frac{3}{16}$ or 3'-6

Based on a design by Timber Engineering Company

Fig. 1072 A Roof Truss.

The actual size of a structural timber is ordinarily not the nominal size listed on the drawing. Surfacing or dressing reduces the nominal dimension $\frac{3}{8}''$ when it is 4″ or less and $\frac{1}{2}''$ when it is over 4″. For example, a 4″ × 8″ nominal-size timber would have the actual dimensions $3\frac{5}{8}'' \times 7\frac{1}{2}''$ when both sides and edges are surfaced. Usually the dressed size is understood and is not listed on drawings. If it is desired, however, to specify definitely that a piece is to be surfaced, the symbol S2S is used to indicate surfacing both sides, and S2E for surfacing both edges. If both sides and edges are to be surfaced, the symbol S4S may be used. When the nominal size given is to be adhered to, the term "full sawn" is generally used.

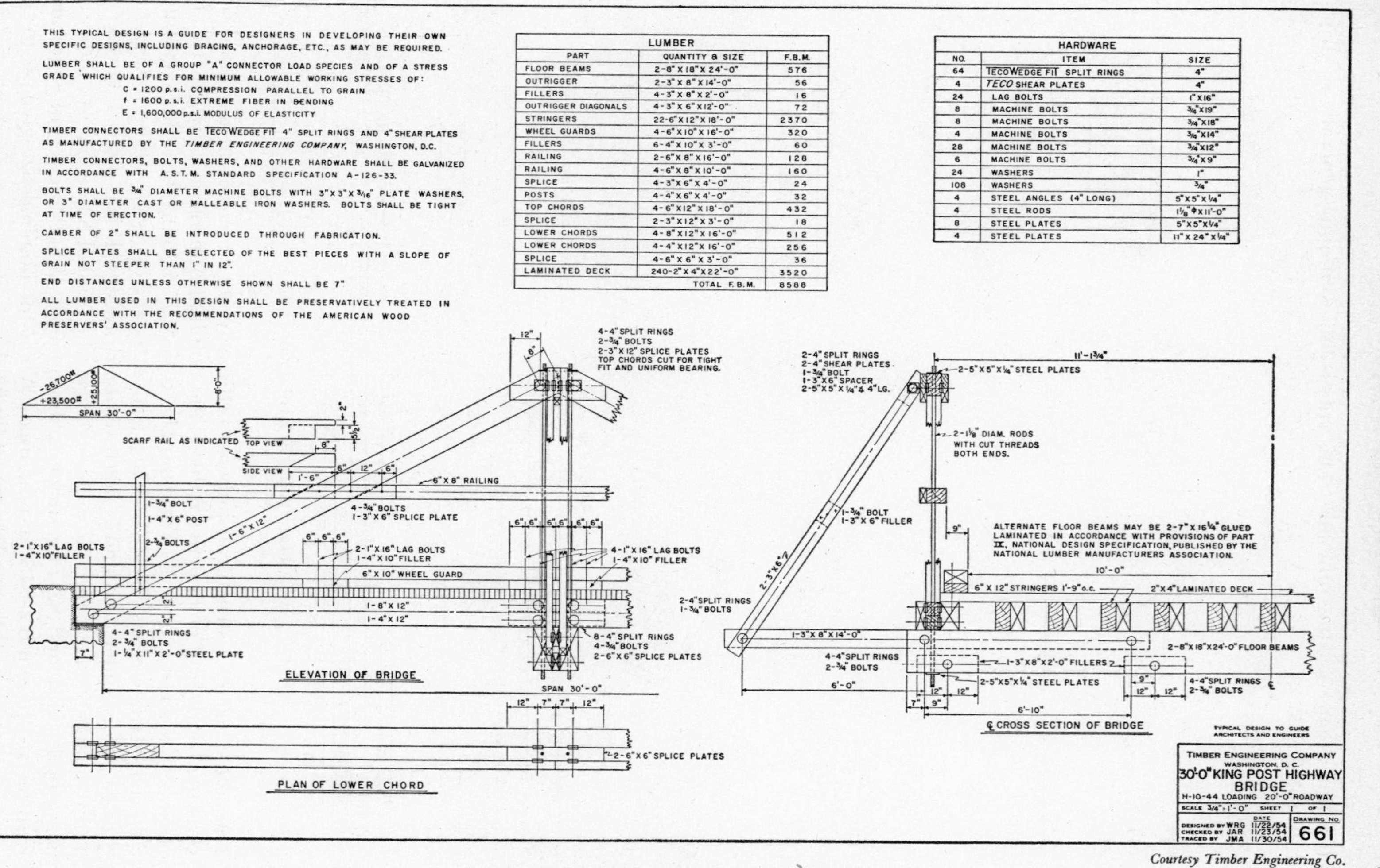

Courtesy Timber Engineering Co.

Fig. 1073 King Post Timber Bridge.

632. Metal Ring Connectors. If properly installed by skilled workmen in wood of proper grade and moisture content, the *metal ring connector* is a very satisfactory and useful device. The method consists in using either a toothed ring, called an *alligator*,

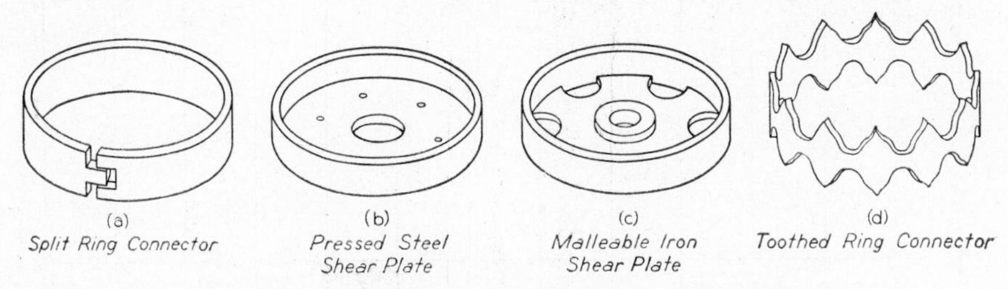

Fig. 1074 Metal Connectors for Timber Structures.

or a *split ring*, Fig. 1074. If the toothed ring, (d), is used, it is placed between the two members to be connected, and these are drawn together by tightening the bolt so that the teeth of the ring are forced into the two members, and the ring thus assists in transmitting stress from one member to the other. If the split ring is used, a groove is cut into each of the two members to be connected, the ring is placed in the grooves, and the *two members are held together by means of a bolt*, as shown in Fig. 1075. The open joint of the ring should be in a direction at right angles to that of the stress, so that as the stress is applied the ring is deformed slightly and transmits the pressure to the wood within the ring as well as to that without. With this connection, the tensile and shearing strengths of wood are more developed than by other methods of connection, and it is possible to use timber in tension much more economically.

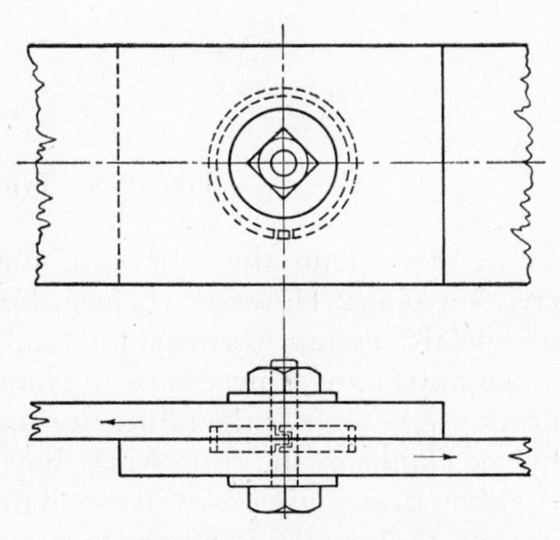

Fig. 1075 Method of Installing Split-Ring Connectors.

It is standard practice to show ring connectors by solid lines in order to save time, as shown in Figs. 1072 and 1073.

633. Structural Steel Drafting.* Structural steel drawings are ordinarily of two types: *engineering design drawings* made in the design engineer's office, and *shop drawings* usually made in the office of the steel fabricator.

Design drawings are concerned primarily with showing clearly the over-all dimensions of the structure, such as the locations of columns, beams, angles, and other structural shapes, and the listing of the sizes of these members. It is necessary to show a certain amount of detail, usually in the form of typical cross sections, special connections required, various notes, etc. In the case of a building floor, for example, a *floor plan* is drawn, showing the steel columns in cross section and the beam or girder framing by the use of single heavy lines, as shown in Fig. 1076. Members

*For a complete treatment of this subject, see *Structural Shop Drafting* (Vols. 1, 2, and 3), published by the American Institute of Steel Construction, 101 Park Ave., New York 17, N. Y.

framing from column to column, providing end support for other beams, are called *girders*, while smaller beams framing between girders are called *filler beams*. The designer's plans are sent to the steel fabricator who is to furnish the steel for the job.

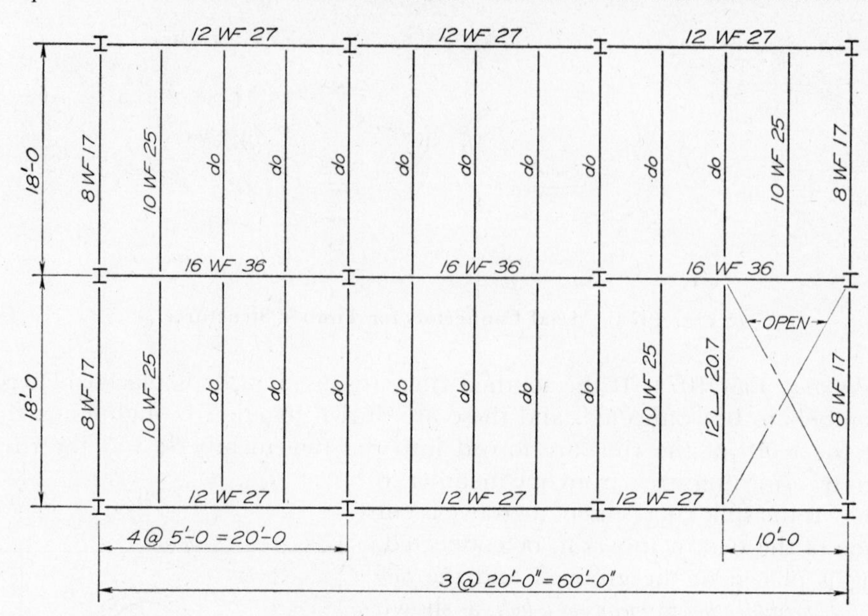

Fig. 1076 Typical Steel Floor Design Plan.

From these plans the fabricator makes the necessary detailed shop drawings and erection plans. However, before shop work is begun, the fabricator's drawings are sent to the design engineer for final checking, as the engineer has the authority to make any changes necessary to conform to the required strength and safety of connections. As soon as the fabricator has received the approved shop drawings returned by the engineer, the shop work may be carried out.

Shop drawings consist in detail drawings of all parts of the entire structure, showing exactly how the parts are to be made. Essentially, such drawings show all dimensions necessary for fabrication calculated to the nearest $\frac{1}{16}''$, the location of all holes for connections, details of connection parts, and the required sizes of all material. In addition, fabrication or construction methods may be specified by appropriate notes on the detail drawings, whenever such items are not covered by separate specifications. Obviously, fabrication and shop methods, as well as suitable field construction methods, must be fully understood by the detailer. The design of details and connections is an important part of the engineering of the structure and should not be neglected, as the connections of the various members must be adequate to transmit the forces in these members. Connection details should be drawn to a scale sufficiently large to show them clearly without crowding, although over-all lengths of members need not be drawn to any scale. All dimensions should be shown, since detail drawings should never be scaled in the shop or in the field by the workmen making the piece. An adequate system of piece marking should be employed. Each piece that is separately handled should have its own piece mark, and this piece mark should be shown wherever the member appears on the drawings. This mark also is painted on the

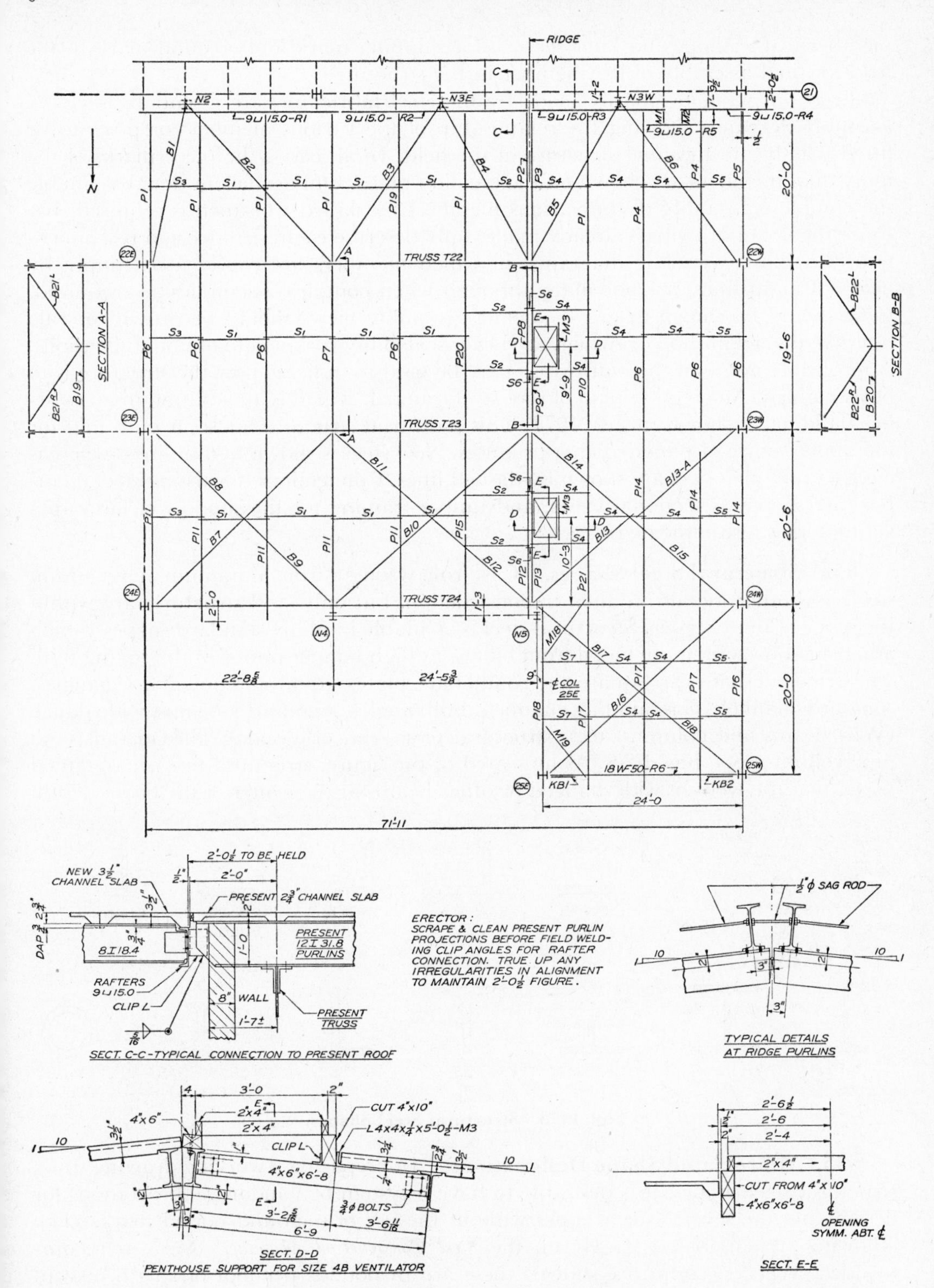

Fig. 1077 Roof Steel Erection Plan.

member in the shop, and later serves as a shipping mark and erection mark in the field for final assembly of the member in the structure.

Erection plans, ordinarily made by the steel fabricator, are essentially skeleton assembly drawings showing the relationship of the various members or parts to be fitted together in the final structure in the field. In all cases, the piece marks of the individual members are shown on the erection plans. Only sufficient detail to enable the complete assembly of the various members by skilled workmen is required, because the detail drawings, already made, fully describe each member and its connections. In most cases line diagrams, in which members are represented simply by heavy straight lines, are adequate although when complex assemblies are required, these should be shown in greater detail. Assembly views should be drawn to scale but, like the detail shop drawings, are never scaled by workmen to obtain dimensions. Appropriate notes on these drawings may be used to indicate how the structure is to be assembled. An erection plan of roof steel framing, which is an addition to an existing building, is shown in Fig 1077. This industrial structure houses a pulp mill for the manufacture of paper roofing products. New steel is shown by full lines, whereas existing roof members are shown by dashed lines. Connections to existing steel members are shown as sectional views, and timber framing for the support of large roof ventilators is also indicated.

634. Structural Steel Shapes. Cast iron, wrought iron, aluminum, magnesium, steel, and other metals are used in construction, but only steel structures are within the scope of this chapter. Structural steel is available in many standard shapes, which are formed by rolling the metal from billets, at high temperatures, in the rolling mill. The principal shapes are square and round bars, plates, equal-and unequal-leg angles, standard channels, car-building channels, bulb angles, standard I-beams, wide flange (WF) beams and columns, tees, structural tees, zees, crane and railroad rails, and pipe columns. See Fig. 1078. As indicated in the figure, structural tees are obtained by cutting the webs of wide-flanged or other beams at the center of the beam depth.

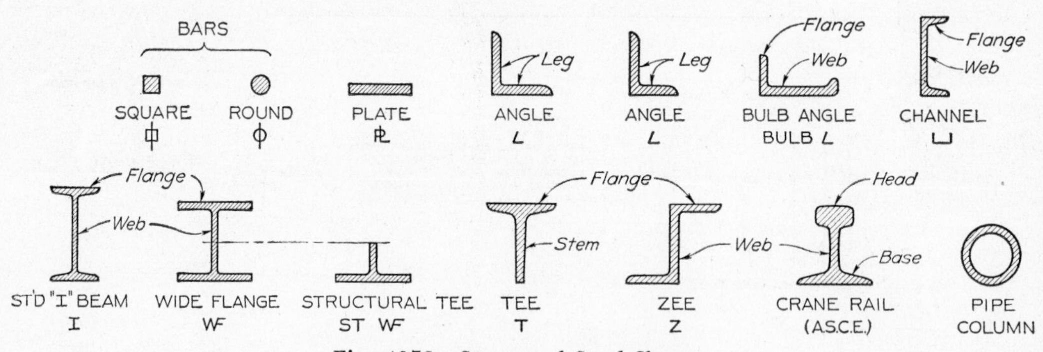

Fig. 1078 Structural Steel Shapes.

635. Abbreviated Shape Designations. In preparing a working drawing using structural steel shapes, it is desirable to have a system of standard abbreviations for designating the several steel shapes without the use of inch and pound marks. The following system, based largely on the *Steel Construction Manual*,* §637, is recommended. Where weights are shown, these are in pounds per foot of length, except in the case of rails.

*Published by the American Institute of Steel Construction, 101 Park Ave., New York 17, N. Y.

A System of Abbreviated Shape Designations

Square Bar	Size, symbol, length	¾ ⌧ 6′−4
Round Bar	Size, symbol, length	⅞ ⌀ 8′−6
Plate	Symbol, width, thickness, length	Pl 6 × ¼ × 9
Equal Leg Angle	Symbol, leg, leg, thickness, length	L 3 × 3 × ⅜ × 12′−0
Unequal Leg Angle	Symbol, long leg, short leg, thickness, length	L 2 × 1½ × ¼ × 9′−1
Bulb Angle	Symbol, web, flange, weight, length	Bulb L 5 × 2½ × 7.3 × 8′−3
Channel	Height, symbol, weight, length	6 ⊔ 10.5 × 15′−4
Standard Beam	Height, symbol, weight, length	12 I 31.8 × 10′−2
Wide Flange Shape	Height, symbol, weight, length	8 WF 17 × 30′−6⅞
Structural Tee	Symbol, height, weight, length	ST4 WF 8.5 × 30′−6⅞
Tee	Symbol, flange, stem, weight, length	T 5 × 3 × 11.5 × 11′−5
Zee	Symbol, web, flange, weight, length	Z 5 × 3¼ × 14.0 × 7′−11
Rail	Weight per yard, name	80 lb. rail
Pipe	Nominal diameter, name	4″ extra strong pipe

636. Scales for Detailing. Details should be drawn to a scale of ¾″ = 1′-0, or 1″ = 1′-0, using the architects scale, §36. Over-all lengths of members, however, need not be drawn to scale.

637. Specifications. The *Steel Construction Manual** gives detailed information concerning dimensions, weights, and properties of rolled steel structural shapes, rolling mill practice, and miscellaneous data for designing and estimating. The information contained in this manual is essential to structural steel drafting.

638. Riveting. One of the commonest methods of connecting steel members is by *riveting*. Structural rivets are made of soft carbon steel and are available in diameters ranging from ½″ to 1¼″. Rivets driven in the shop are called *shop rivets*, and those driven in the field (at the construction site) are called *field rivets*. Rivets are usually of the button-head type, Fig. 697 (a), and are driven hot, into holes $\frac{1}{16}$″ larger than the rivet diameter. The length of a rivet is the thickness (grip) of the parts being connected, plus the length of the shank necessary to form the driven head and to fill the hole. Excess shank length will produce capped heads, whereas lengths too short will not permit the formation of a full head. Shop rivets are ordinarily driven by large riveting machines that are part of the permanent shop equipment.

Field rivets are usually heated in a coal-burning forge with the use of a hand bellows, and when properly heated the shank will have a light cherry-red color and the head a dull red color. Riveting crews consist of four workmen: (1) the *heater*, who passes the hot rivets to the *sticker*; (2) the sticker, who receives the rivets and enters them into the holes; (3) the *bucker*, who holds the rivet firmly in the hole against the force of the rivet gun by use of a dolly bar; and (4) the *riveter*, who forms the head with a pneumatic hammer, forcing the shank to fill the hole completely. If the forge is at some distance from the work, rivets are tossed to the sticker, who catches them in a special can. After catching, the sticker strikes the head sharply against metal to remove cinders and scale before entering the rivet into the hole. The riveter holds the pneumatic hammer or rivet gun against the rivet with considerable force, and during driving rotates it slightly to assist in forming a round smooth head.

Several types of riveted joints are shown in Fig. 698.

**Ibid.*

639. Standard Riveted Beam Connections. Because of their common usage, the American Institute of Steel Construction recommends certain standard connections for attaching beams to other members. These connections are ordinarily adequate to transmit the end forces that beams carry. However, the draftsman should know the strength of these connections, and use them only when they are sufficient. For complete information, see the *Steel Construction Manual*. The "A," "H," and "HH" series are designed for the use of $\frac{7}{8}''$ rivets, and the "B," "K," and "KK" series are for $\frac{3}{4}''$ diameter rivets. See Appendix tables 29 and 30.

The use of angles B4 of the series in a typical detail drawing of a floor beam is shown in Fig. 1079. This drawing illustrates several important features: Shop rivets are shown as open circles on shop drawings, whereas holes for field rivets or bolts are

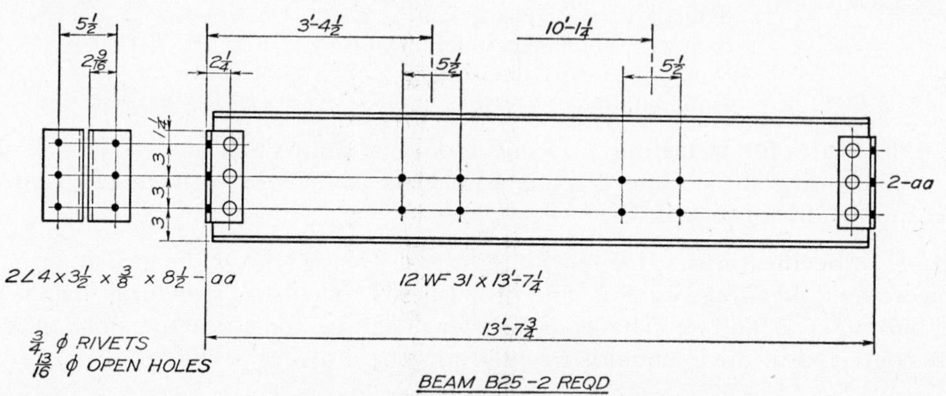

Fig. 1079 Floor Beam Shop Drawing.

blacked-in solid, Fig. 699; *gage lines* (lines passing through rivets or holes like center lines) are always shown, and it is desirable to line up holes or rivets on these lines where possible, rather than to "break the gage." It is necessary to locate the gage line of an angle for each leg in all cases, unless it has already been shown for the identical angle elsewhere on the drawing. The edge distance, from end rivet or hole to the end of the angle, must be given at one end, the billed length of the piece being worked out to provide the necessary edge distance at the other end. It is not necessary that the beam extend the full length of the distance back-to-back of end angles. In this case, as is customary, it is shown "set back" at both ends, the length of beam called for being $\frac{1}{2}''$ less than the $13'$-$7\frac{3}{4}$ distance. Below the sketch, the mark B25 is the piece or shipping mark which appears on the erection plan, and is to be painted on the member in the shop for identification. The end connection angles are fully detailed at the left end of the beam and are given the assembly mark aa. Therefore, at the right end where these same angles are again used, only the assembly mark 2-aa, to indicate two angles, is given. The figure $10'$-$1\frac{1}{4}$ is called an *extension figure*, as this is the distance from the back of the left-end angles to the center of the group of four holes. Note that this dimension is on the same horizontal line as the $3'$-$4\frac{1}{2}$ figure just to the left. It is customary to give the foot mark (′) after dimensions in feet, but not to give the (″) mark, designating inches.

A shop drawing for a riveted column is shown in Fig. 1080. Here the faces A, B, and C are marked for reference. It is customary to draw face A first at the left, then face B, then C. When the column is viewed in plan from above as shown in the section below view B, the faces are located progressively counterclockwise. Only occasionally is it necessary to draw face D (when the framing is very complex). The lower sectional view is called a *bottom section* in structural drafting, but it is always a section looking down, rather than a bottom view looking up. This sectional view shows the base detail with the 1″ thick plate, which will rest directly on the concrete foundation.

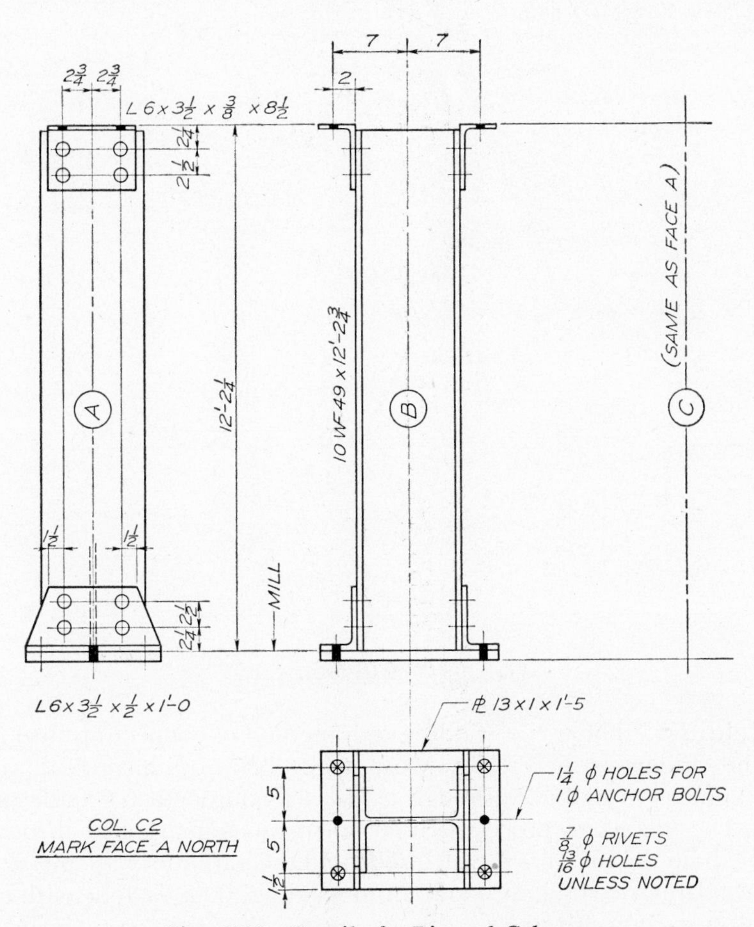

Fig. 1080 Detail of a Riveted Column.

In this view, rivets "countersunk far side" are shown, to tie the base plate to the connection angles. The bottom end of the column shaft is to be milled, so that it will have full contact area over the bearing plate. Milling is common practice when compression members bear against one another. The piece mark of the column is C2, which will be painted on the member before it leaves the shop.

A shop drawing for a riveted roof truss is shown in Fig. 1081. Only the left half is drawn, since the truss is symmetrical about the center line. The use of the gage lines of the members should be noted. Gage lines should be located as closely as possible to the centroidal axes of the members; and at the joints where members intersect, these gage lines should intersect at a single point to avoid unnecessary moment stresses due to eccentricities. It is noted that the locations of the open holes indicate where the field splices are to be, namely at the center line of the truss (℄) for the top chord, and at $5'\text{-}2\frac{5}{16}$ from the center line for the bottom chord.

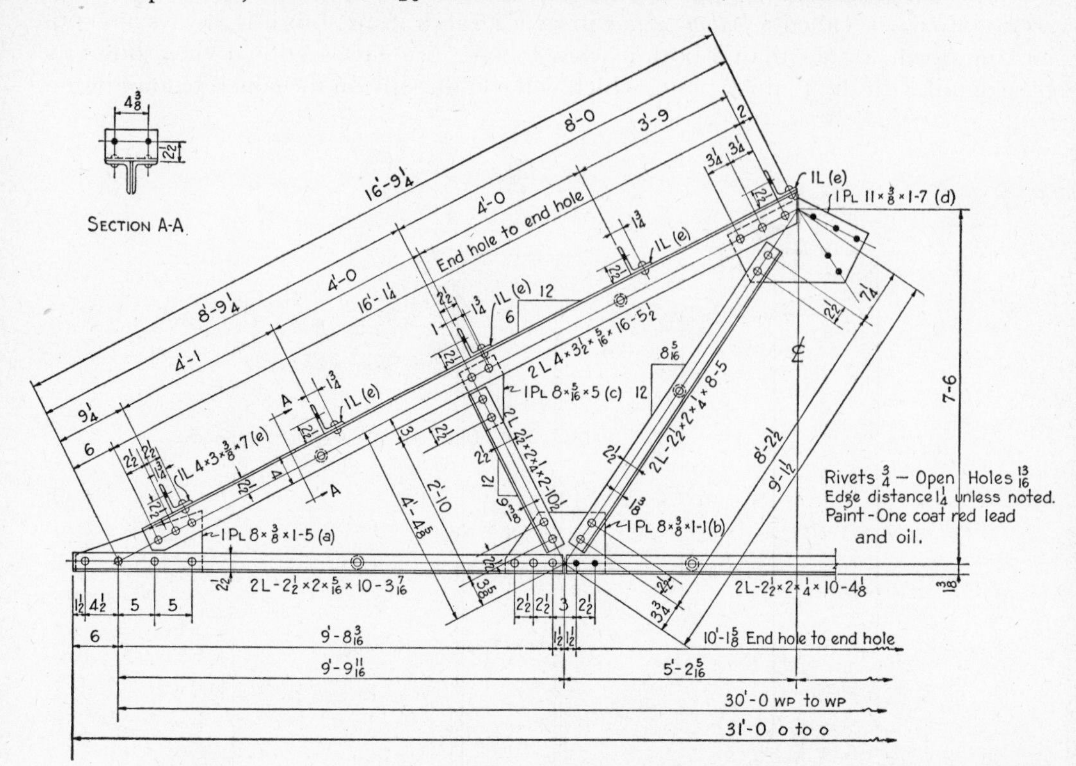

Based on a design by Fort Worth Structural Steel Company.

Fig. 1081 A Riveted Truss.

640. Welding. The use of welding as a method of connecting steel members of buildings and bridges is increasing. Most steel fabricators have both riveting and welding equipment available, although a few are equipped to handle only welded fabrication. The metal-arc process is used, energy being supplied through an electrode to unite both the metal of the electrode and the parent or base metal. Electrodes may be either bare or coated, although most welding today is done with coated electrodes. Of all types of welds, the *fillet weld* is most common in structural steel fabrication. For additional information, the student should refer to Chapter 25 and to the specifications of the *Steel Construction Manual* of the AISC. When a structure is to be welded, it should be designed throughout for this method of fabrication, as it is not possible to obtain the maximum economy of construction by merely substituting welding for riveting. The designations of welds by the use of standard symbols, Fig. 1124, have greatly simplified the making of shop drawings.

A beam with end connection angles shop welded to the beam web is shown in Fig. 1082. The outstanding legs of the angles are to be welded to the connecting columns in the field, as shown in the end view. This view pertains only to field erection. Open holes in the outstanding legs are for bolts, to facilitate positioning.

A shop drawing of diagonal bracing between two columns is shown in Fig. 1083. Here the diagonal angle members are shop welded to gusset plates which are to be bolted to the column flanges as a permanent installation in the field.

641. High Strength Steel Bolting. High tensile steel bolts, as a method of connecting the various members of steel structures, have recently come into fairly wide usage. These bolts are considered fully as satisfactory as rivets and in fact, for

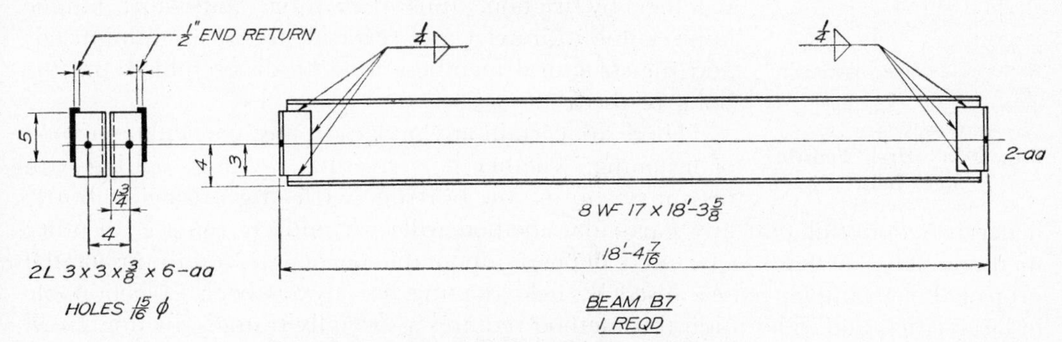

Fig. 1082 Detail of Welded Beam.

equal diameters, specifications allow equal values for either rivets or bolts, provided such bolts are used according to prescribed methods. Hardened steel washers should be used under both head and nut. Proper tensioning of the bolts may be done by the use of either a manual torque wrench or an air impact wrench, calibrated to deliver the necessary torque. It has also been discovered that if the steel parts are well drawn together, the bolts can be tightened sufficiently from a finger-tight position simply by turning the nut one full turn around with an ordinary manual wrench.

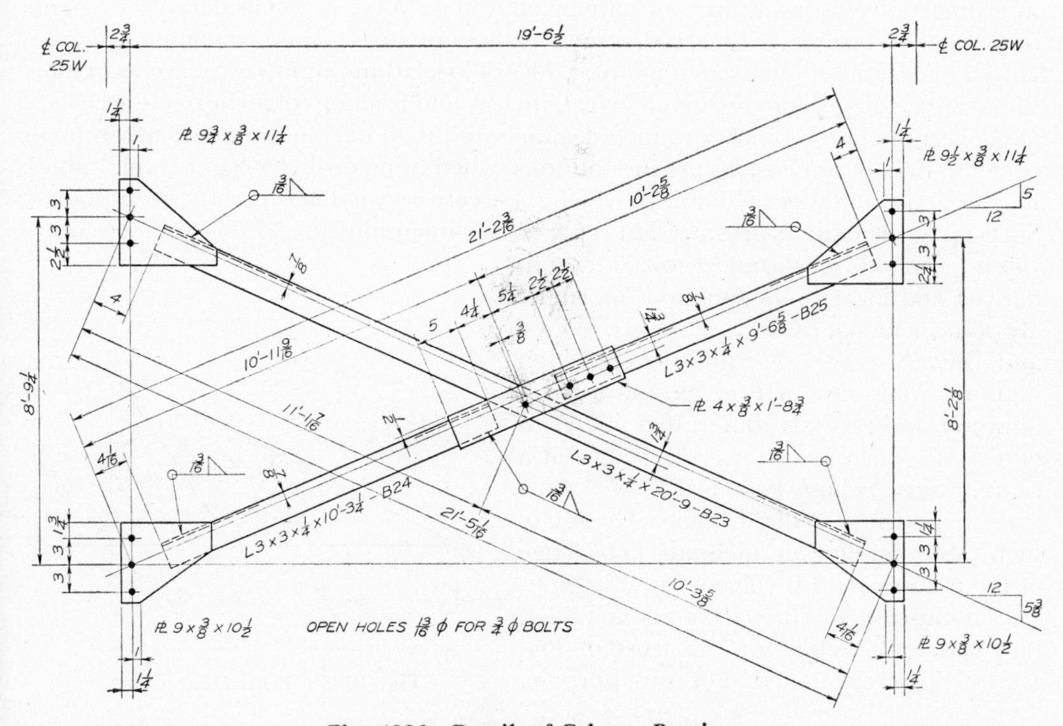

Fig. 1083 Details of Column Bracing.

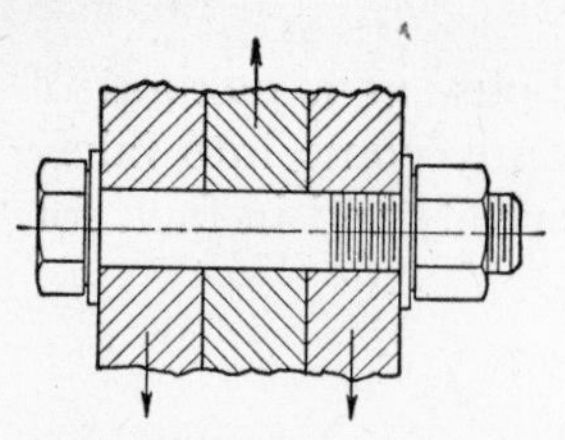

REGULAR SEMIFIN. HEX. BOLT
HEAVY SEMIFIN. HEX. NUT
HEAVY PLAIN CARBURIZED
WASHERS

Fig. 1084 High Tensile Steel Bolt.

The use of a high tensile steel bolt to transmit the force from the center plate into the two outside plates is shown in Fig. 1084. Among the advantages of this method of fastening are the following: the connected parts are held together by friction, thus preventing joint-slip; fatigue failures due to impact and reversal stresses are minimized; and the structural members may be disassembled and the bolts re-used.

There are certain advantages to any particular method of fastening, whether it is riveting, welding, bolting with ordinary bolts, or bolting with high tensile bolts; hence it is doubtful that any particular method will ever entirely replace all other methods. The method to be used depends upon the particular conditions of the project. For example, noise caused by field riveting has always been objectionable in large cities, and in addition this method requires a specially trained riveting crew. Although welding is noiseless, it requires highly skilled workmen and careful and difficult inspection to insure good workmanship. High tensile steel bolting requires less skilled workmen and has come to be accepted as an excellent method for connecting steel members in both shop and field.

642. The Calculation of Dimensions. Perhaps the most important part of the structural draftsman's work is the accurate calculation of dimensions. If there are incorrect dimensions on the drawings, they will result in serious errors and misfits when members are assembled in the field. Correction of such errors not only entails considerable expense, but often results in delaying the completion of the work. Since in ordinary steel work, dimensions are given to the closest $\frac{1}{16}''$, considerable precision is demanded. For skewed members, such as those in a truss, some trigonometry is involved. To facilitate such calculations, tables of logarithms and squares are extremely useful. Since dimensions are given in feet, inches, and fractions of inches, it is necessary to convert fractions of inches to inch decimals and then to convert inches to decimals of feet so that all dimensions are in foot units, when using ordinary logarithmic tables. Fortunately, tables are available in which these conversions have already been made.* Such tables include logarithms and squares for dimensions to $\frac{1}{16}''$ for distances up to 100 ft., and to $\frac{1}{8}''$ for distances to 200 ft. Both natural and logarithmic functions of angles are also included, as well as bevels, slopes, and rises. See §339. Ordinarily angles on shop drawings are not given in degrees, minutes, and seconds, but rather in terms of a *bevel*, which is the rise, or height of a right triangle whose base is 12. The slope is the hypotenuse of the triangle. The use of such tables implies an adequate knowledge of the trigonometry involved.

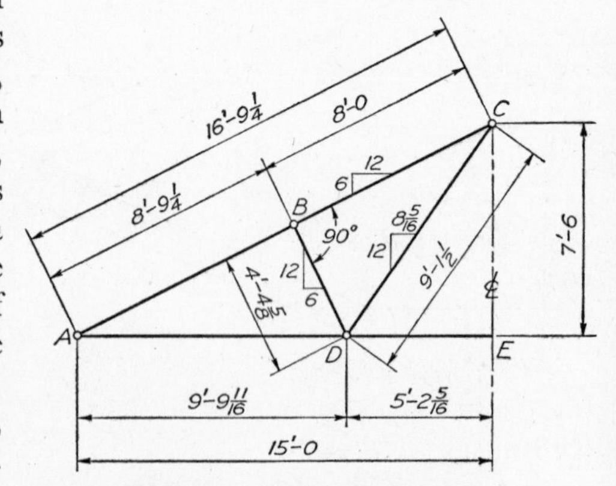

Fig. 1085 Truss Dimensions.

To illustrate the use of the above tables, the main dimensions for the truss of Fig. 1081 will be calculated. For this purpose,

*For example, *Smoley's New Combined Tables*, C. K. Smoley & Sons, Publishers, Scranton, Pa.

the line diagram of Fig. 1085, which shows the gage lines of the truss members, will be used. Length AE, one-half the truss span and CE, the height of truss, are the known dimensions. Point B is located at 8′-9¼ from A, as shown. From these dimensions, the other lengths and the bevels are to be determined. In a Fink truss such as this, the web member BD is perpendicular to the top chord AC. An arrangement of computations in tabular form is desirable, as in the following table:

Length	*Logarithm*	*Square*
CE = 7′-6	10.87506—10	56.25
AE = 15′-0	1.17609	225.00
	Log Tan A = 9.69897—10	281.25 = (AC)²
	Bevel AB = 6 in 12	AC = 16′-9¼
AB = 8′-9¼	0.94304	76.9275
	Log BD = 0.64201	
BD = 4′-4⅝		19.2319
		96.1594 = (AD)²
		AD = 9′-9¹¹⁄₁₆
DE = 5′-2⁵⁄₁₆	10.71539—10	26.96
CE = 7′-6	0.87506	56.25
	Log Tan DCE = 9.84033—10	83.21 = (DC)²
	Bevel DC = 8⁵⁄₁₆ in 12	DC = 9′-1½

643. Concrete Construction. Concrete is a building material made by mixing sand and gravel or other fine and coarse aggregate with *Portland cement* and water. The strength of concrete varies with the quality and relative quantities of the materials, with the manner of mixing, placing, and curing, and with the age of the concrete. The compressive strength of concrete depends on the mix design, but has been manufactured to develop an ultimate strength at 28 days as high as 7,000 lb. per sq. in. The tensile strength of the material is limited to about one-tenth the compressive strength. Portland cement is a controlled, manufactured product as compared to natural cements found in some localities. It derived its name from its color, which resembles that of a famous building stone found on the island of Portland in southern England.

Since the tensile strength of concrete is very limited, the usefulness of concrete as a building material can be materially improved by embedding steel reinforcing bars in it in such a way that the steel resists the tension, and the concrete mainly the compression. In this way the two materials act together in resisting forces and flexure. Concrete, combined in this way with steel, is called *reinforced concrete;* without the addition of steel rods or wires, it is called *plain concrete*. When the steel is pre-tensioned before the application of the superimposed load, thus producing an interior force within the member, the material is called *prestressed concrete*.

644. Reinforced Concrete Drawings. The design of the reinforcing for a reinforced concrete structure and the preparation of the corresponding drawings are complicated. In order to simplify this work and to secure uniformity in the many engineering offices, the American Concrete Institute, with the cooperation of the Concrete Reinforcing Steel Institute, has prepared a *Manual of Standard Practice for Detailing Reinforced Concrete Structures* which has been approved as A.C.I. Standard.

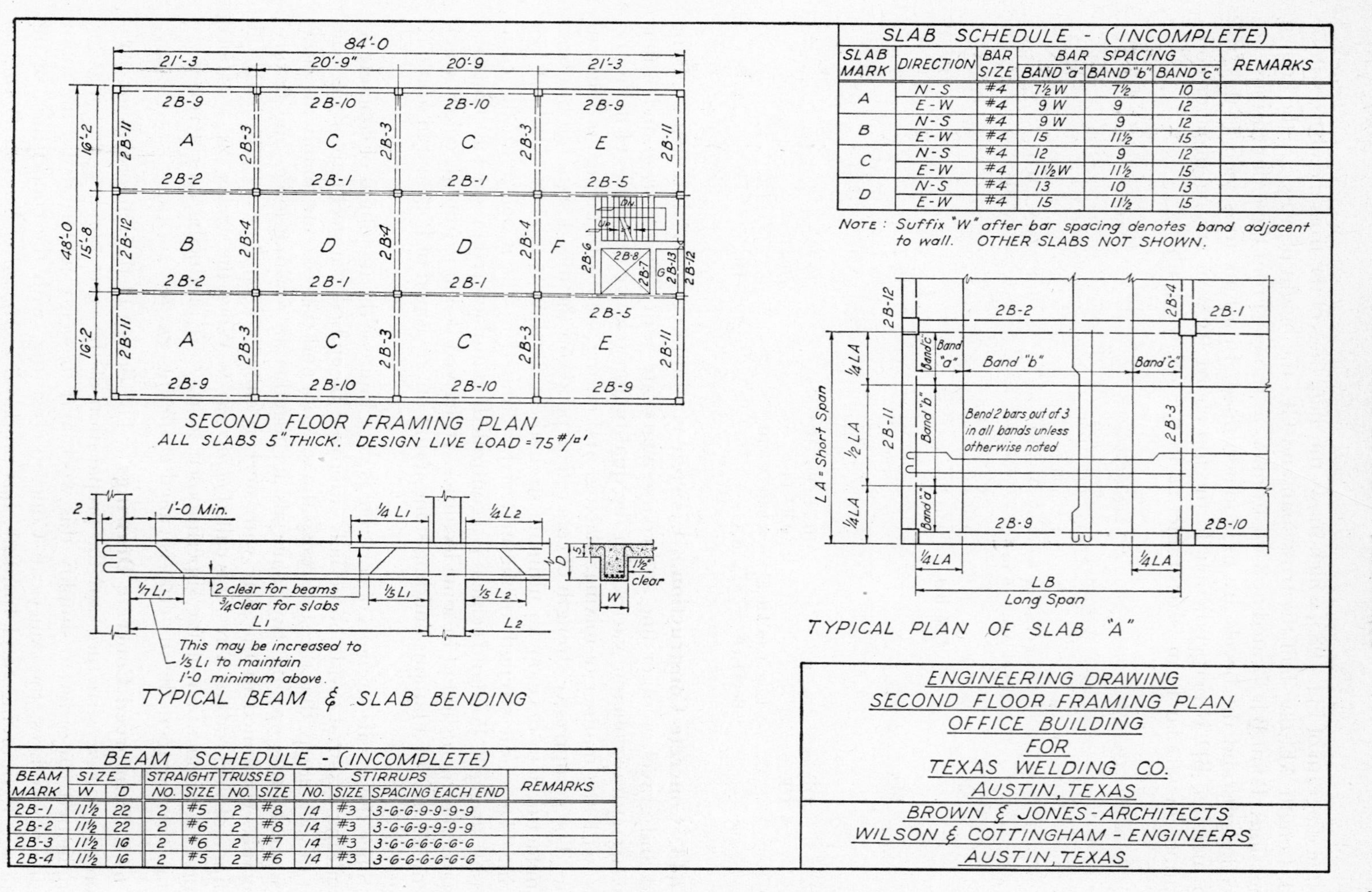

SLAB SCHEDULE - (INCOMPLETE)

SLAB MARK	DIRECTION	BAR SIZE	BAR SPACING BAND "a"	BAND "b"	BAND "c"	REMARKS
A	N-S	#4	7½ W	7½	10	
	E-W	#4	9 W	9	12	
B	N-S	#4	9 W	9	12	
	E-W	#4	15	11½	15	
C	N-S	#4	12	9	12	
	E-W	#4	11½ W	11½	15	
D	N-S	#4	13	10	13	
	E-W	#4	15	11½	15	

BEAM SCHEDULE - (INCOMPLETE)

BEAM MARK	SIZE W	D	STRAIGHT NO.	SIZE	TRUSSED NO.	SIZE	STIRRUPS NO.	SIZE	SPACING EACH END	REMARKS
2B-1	11½	22	2	#5	2	#8	14	#3	3-6-6-9-9-9-9	
2B-2	11½	22	2	#6	2	#8	14	#3	3-6-6-9-9-9-9	
2B-3	11½	16	2	#6	2	#7	14	#3	3-6-6-6-6-6-6	
2B-4	11½	16	2	#5	2	#6	14	#3	3-6-6-6-6-6-6	

Fig. 1086 Engineering Drawing for a Two-Way Slab and Beam Floor.

Also, a *Manual of Standard Practice for Detailing Reinforced Concrete Highway Structures*, copyrighted by the A.C.I., has now been published.*

It is recommended in the manuals that two sets of drawings be prepared, an *engineering drawing* and a *placing drawing*. The engineering drawing is to be prepared by the engineer who designs the structure, and the placing drawing by the manu-

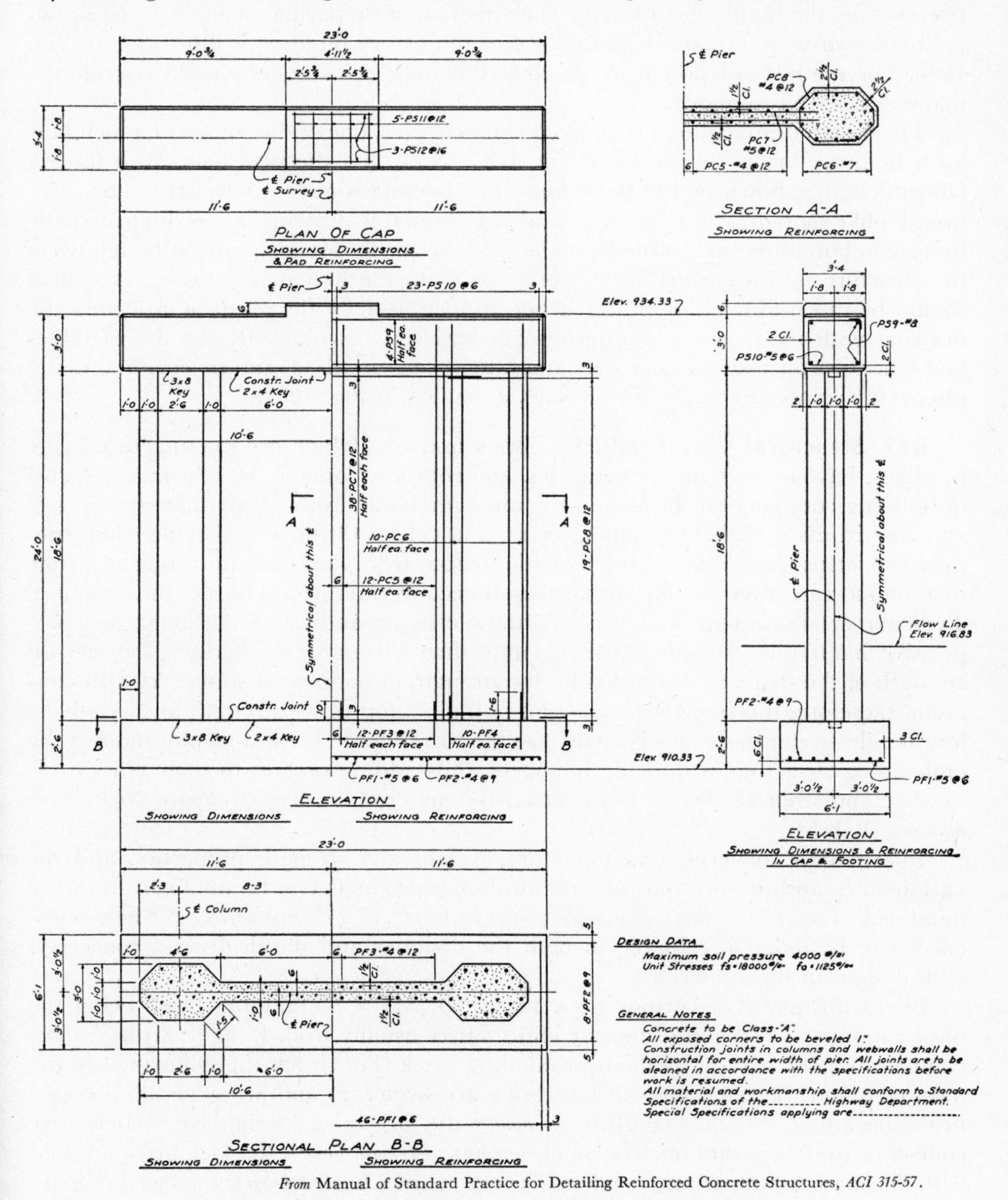

From Manual of Standard Practice for Detailing Reinforced Concrete Structures, *ACI 315-57*.

Fig. 1087 Reinforced Concrete Pier for Deck Girder.

*American Concrete Institute, P. O. Box 4754, Redford Station, Dearborn 19, Mich.

facturer who fabricates the reinforcing steel. The engineering drawing is to show the general arrangement of the structure, the sizes and reinforcements of the several members, and such other information as may be necessary for the correct interpretation of the designer's ideas. The placing drawing is to show the sizes and shapes of the several rods, stirrups, hoops, ties, etc., and to arrange them in tabular forms for ready reference by the building contractor. The method of preparing an engineering drawing for a two-way slab and beam floor of a multistory building is illustrated in Fig. 1086. For methods of preparing placing drawings, the student should consult the manuals referred to above.

The design drawing for a reinforced concrete pier, one of the supporting members for a highway bridge, is shown in Fig. 1087. Note that the steel bars, even though embedded, are shown by full lines, and that concrete is always stippled in cross section. Unlike shop drawings for structural steel, concrete drawings are ordinarily made to scale in both directions. Usually a scale of $\frac{1}{4}''$ to the foot is adequate, although when the structure is complicated, scales of $\frac{3}{8}''$ or $\frac{1}{2}''$ to the foot may be used. An effort should be made to avoid a cluttered appearance, usually the result of crowding the drawing with many notes. Cluttering can largely be avoided by the use of tables and schedules for listing of bar sizes and other necessary data, and by covering many important points in a single set of notes, as shown in Fig. 1087.

645. Structural Clay Products. Brick and tile, which are manufactured clay products, have been in use for centuries, and comprise some of the best-known forms of building construction. Brick and tile units are made from many different types of clay and in many different shapes, forms, and colors. Ordinarily they are built into masonry forms by the skilled brick or tile mason, who places the units one at a time in a soft mortar. After the mortar hardens, it becomes an integral part of the structure. Typical mortars contain sand, lime, Portland cement, and water. Although the compressive and tensile strengths of the clay units themselves are considerable, the over-all strength of the structure is limited by the strength of the mortar joints. As with concrete, therefore, the result is a structure of high compressive strength and relatively low tensile strength. Similarly, it is possible to reinforce brick and tile masonry by embedding steel rods within the members, thus adding greatly to their tension resistance and strength. When this is done, the material is called *reinforced brick* or *tile masonry* (R.B.M.).

Information concerning manufacture, weight and strength properties, and the various uses and applications of structural clay products can be obtained from the handbooks *Principles of Brick Engineering* and *Principles of Tile Engineering.** These references will be found invaluable to both the designer and the draftsman concerned with designs in this material.

Bricks are made of various sizes, the $2\frac{1}{4}'' \times 3\frac{3}{4}'' \times 8''$ building brick being the most common. Thickness of mortar joints varies usually from $\frac{1}{4}''$ to $\frac{3}{4}''$, with $\frac{1}{2}''$ most common. Of the several methods of bonding brick, Fig. 1088, the following are the most common: *running bond*—all face brick are stretchers and are generally bonded to the backing by metal ties; *American bond*—the face brick are laid alternately, five courses of stretchers and one course of headers; *Flemish bond*—the face brick are laid with alternate stretchers and headers in every course; *English bond*—the face brick are laid alternately, one course of stretchers and one course of headers. In modern work,

*Published by the Structural Clay Products Institute, Washington, D. C.

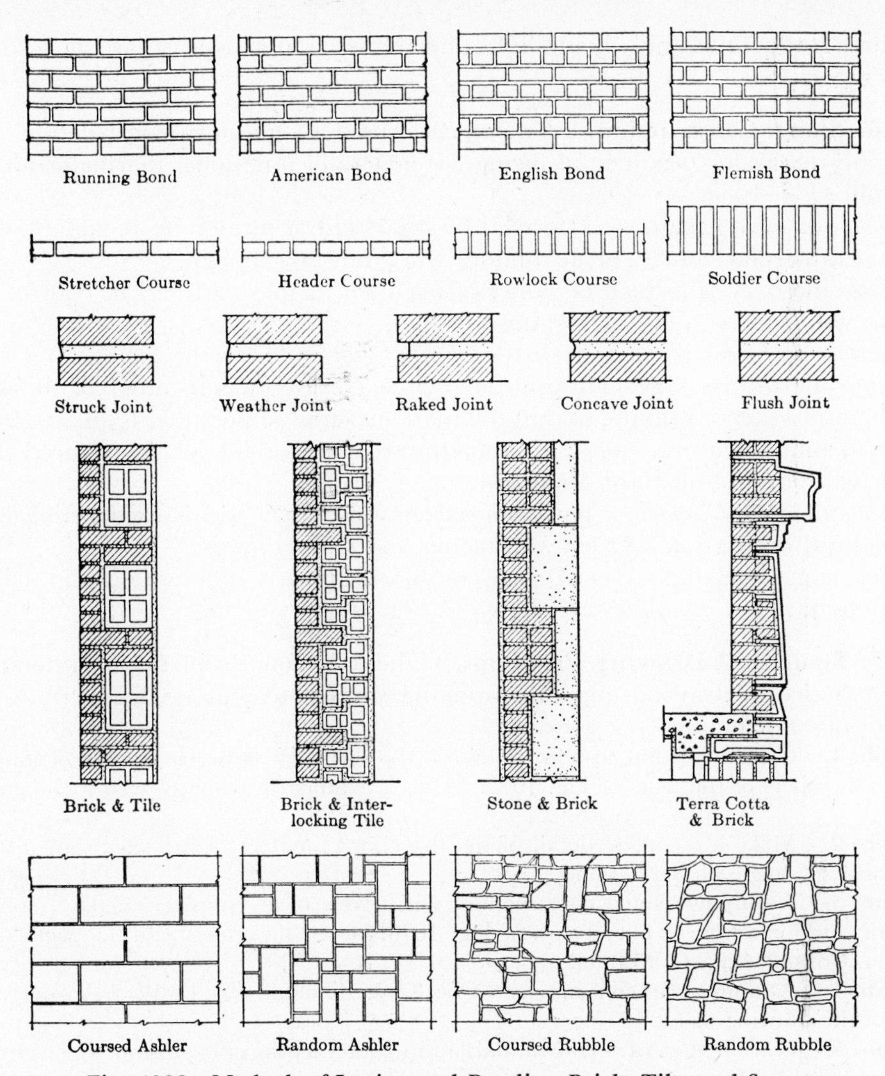

Fig. 1088 Methods of Laying and Bonding Brick, Tile, and Stone.

these standard methods of bonding are frequently modified to produce various artistic effects. Typical brick lintel arches are shown in Fig. 1089.

In addition to its use as a basic building material, tile is extensively used in fireproofing of structural steel members. Most building codes require that the steel members be enclosed in concrete or masonry so that fire will not cause collapse of the

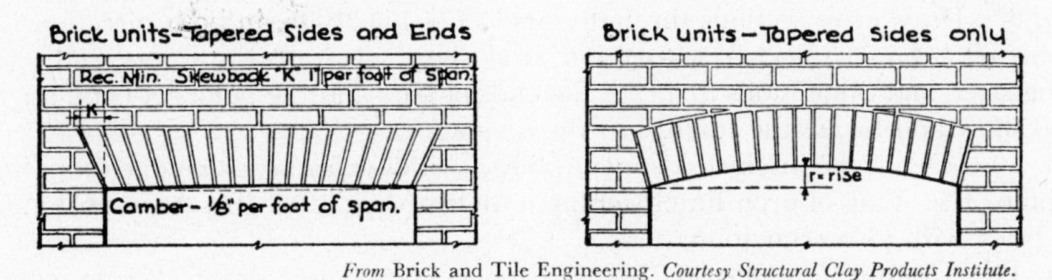

From Brick and Tile Engineering. *Courtesy Structural Clay Products Institute.*

Fig. 1089 Typical Lintels.

structure. Hollow tile units, being light and relatively inexpensive, are well adapted to this usage.

646. Stone Construction. *Natural stone*, used in masonry construction—most commonly today for ornamental facing—is generally limestone, marble, sandstone, or granite, Fig. 1088.

Ashlar (or *ashler*) masonry is formed of stones cut accurately to rectangular faces and laid in regular courses or at random with thin mortar joints.

Rubble masonry is formed of stones of irregular shapes and laid in courses or at random with mortar joints of varying thickness.

Manufactured stone is concrete made of fine aggregate for the facing, and coarse aggregate for the backing. The fine aggregate consists of screenings of limestone, marble, sandstone, or granite, so that the manufactured stone presents an appearance similar to that of natural stone. Manufactured stone is made of any desired shape, with or without architectural ornament.

Architectural terra cotta is a hard burned clay product, and is used primarily for architectural decoration and for wall facing and wall coping.

Brick, stone, tile, and terra cotta are combined in many different ways in masonry construction. A few examples are shown in Fig. 1088.

647. Structural Drawing Problems. The following problems are intended to afford practice in drawing and dimensioning simple structures, and in illustrating methods of construction.

Prob. 1. Calculate point-to-point lengths (the distances between centers of joints) of the web members of the truss of Fig. 1072. Make a detailed drawing of web member piece No. 4.

Prob. 2. Make a complete detail of the top chord member of the truss shown in Fig. 1072, for a 30° angle of inclination.

Prob. 3. Assuming riveted construction, with rivets of $\frac{3}{4}''$ diameter, make a complete shop drawing for a typical filler beam of Fig. 1076. Detail the same beam for welded construction. Consult A.I.S.C. Manual.

Prob. 4. Assuming the column size to be 8 WF 31, detail the 16 WF 36 girder at the center of the drawing of Fig. 1076.

Prob. 5. Consult the A.I.S.C. Manual, and list the depths of beams on which the connection angles of Appendix table 30 are normally used.

Prob. 6. From the unit weights given in the Manual, calculate the total weight for the beam and column assemblies of Figs. 1079 and 1080, and for the entire truss of Fig. 1081.

Prob. 7. Referring to Fig. 1081, make a complete detail for a truss of the same length, but change the height from 7′-6 to 8′-6. Use angle members of the same cross section sizes as those shown, but of different lengths, as needed.

Prob. 8. Referring to Fig. 1083, make a similar bracing detail, changing the distance between column centers from 20′-0 to 18′-6.

Prob. 9. Draw cross sections through panel D of Fig. 1086, in both directions. Include the supporting beams in each cross section, and show all dimensions, size and spacing of reinforcing steel, and dimensions to locate the ends of bars and the points of bend for the bent bars. Also show and locate the stirrups in these views.

Prob. 10. Detail the brickwork surrounding a window frame for an opening 4′-0 $\frac{7}{8}$ wide by 6′-9 high. Use type of arch lintel similar to that of Fig. 1089 (b). Assume standard-size building brick with $\frac{1}{2}''$ mortar joints.

Prob. 11. Consult *Principles of Tile Engineering*, §645, and draw a cross section view through a 12″ wall of composite brick and tile construction.

CHAPTER 23

TOPOGRAPHIC DRAWING AND MAPPING

by
*E. I. Fiesenheiser**

648. Introduction. Up to this point we have been concerned with the methods and techniques used in drawing man-made objects. *Topographic drawing* and *mapping* have to do with the representation of portions of the earth's surface, mainly its natural features, to a convenient scale. On such drawings the relative positions of natural features, with respect to certain definitely located points, are shown. Since the shape of the earth is spherical, any representation on a plane, such as a piece of paper, is necessarily somewhat distorted. See §532. In drawing large areas, therefore, some method of projection must be used which results in a minimum of distortion. In such work, the positions of the control or reference points are usually defined by spherical coordinates of latitude and longitude (meridians and parallels), which are shown as reference lines on the drawing. In drawing small areas to a relatively large scale, the distortion due to earth-curvature is so slight as to be unnoticeable, and may therefore be entirely neglected. Orthographic projection, as used in technical drawings, in which the line of sight is assumed to be perpendicular to the plane of the map, is therefore the method most commonly used.

649. Definitions. The purpose for which a map is to be used determines what features should be represented, what scale should be used, and what detail should be included. Certain commonly-used symbols are employed in representing natural features and man-made objects. Also, certain terms and various types of maps are common. The following definitions are therefore essential to an understanding of the subject.

1. A *plat* is a map, usually of a small area, plotted directly from a land survey. It does not ordinarily show relative ground elevations, and is drawn for some specific purpose, such

*Director, Civil Engineering Dept., Illinois Institute of Technology.

as the calculation of areas, the location of property lines, or the location of a building project. Usually it contains a *traverse*. See Fig. 1090.

2. A *traverse* consists of a series of intersecting straight lines of accurately measured lengths. At the points of intersection, the deflection angles between adjacent lines are measured and recorded. Starting at one point, rectangular coordinates of the other intersection points can therefore be calculated by trigonometry. A *closed traverse* thus becomes a closed polygon, and provides a method for checking the accuracy of the work. The land survey plat of Fig. 1090 illustrates a closed traverse.

3. *Elevations* are vertical distances above a common *datum* or reference plane. The elevation of a point on the surface of the ground is usually determined by differential leveling from some other point of known elevation. Commonly, elevations are referenced to mean sea level datum.

4. A *profile* is a line contained in a vertical plane, and depicts the relative elevations of various points along the line. Thus if a vertical section were to be cut into the earth, the top line of this section would represent the ground profile.

5. *Contours* are lines drawn on a map to locate, in the plan view, points of equal ground elevation. On a single contour line, therefore, all points have the same elevation.

6. *Hatchures* are short, parallel, or slightly divergent lines drawn in the direction of the slope. They are closely spaced on steep slopes and converge toward the tops of ridges and hills. Hatchures are shade lines to show relief. See Appendix 31.

7. *Monuments* are special installations of stone or concrete to mark the locations of points accurately determined by precise surveying. It is intended that such monuments be permanent or nearly so, and they are usually tied in by references to near-by natural features, such as trees, large boulders, etc.

8. *Cartography* is the science or art of map making.

9. *Topographic* maps depict (1) water, including seas, lakes, ponds, rivers, streams, canals, swamps, etc.; (2) *relief* or elevations of mountains, hills, valleys, cliffs, etc.; (3) *culture*, or the works of man, such as towns, cities, roads, railroads, boundaries, etc. See Figs. 1091 and 1100.

10. *Hydrographic* maps convey information concerning bodies of water, such as shoreline locations, relative elevations of points of stream, lake, or ocean beds, sounding depths, etc.

11. *Cadastral* maps are accurately-drawn maps of cities and towns, showing property lines and other features that control property ownership.

12. *Military* maps contain information of military importance in the area represented.

13. *Nautical* maps and charts show navigation features and aids, such as locations of buoys, shoals, lighthouses and beacons, and sounding depths.

14. *Aeronautical* maps and charts show prominent landmarks, towers, beacons, and elevations for the use of air navigators.

15. *Engineering* maps are made for special projects as an aid to location and construction. See Figs. 1094, 1097, 1098, and 1099.

16. *Landscape* maps are used in planning installations of trees, shrubbery, drives, etc., in the artistic design of area improvements. See Fig. 1096.

650. Sources of Information. The basis of all maps and topographic drawing is the *survey*. Surveying is the actual measurement of distances and elevations on the earth's surface. Hence all maps are plotted from field data provided by the surveyor. It will be appropriate to discuss briefly the surveying methods upon which map information is based.

Distances are ordinarily measured by steel tape, with stakes being driven to mark the points between field measurements. Distances may also be determined by measuring aerial photographs, when the scale of the photograph is known. An instrumental method, known as the *stadia* method, is also widely used in map making. The *stadia transit* is an optical instrument used in conjunction with a special *stadia rod*. By sighting

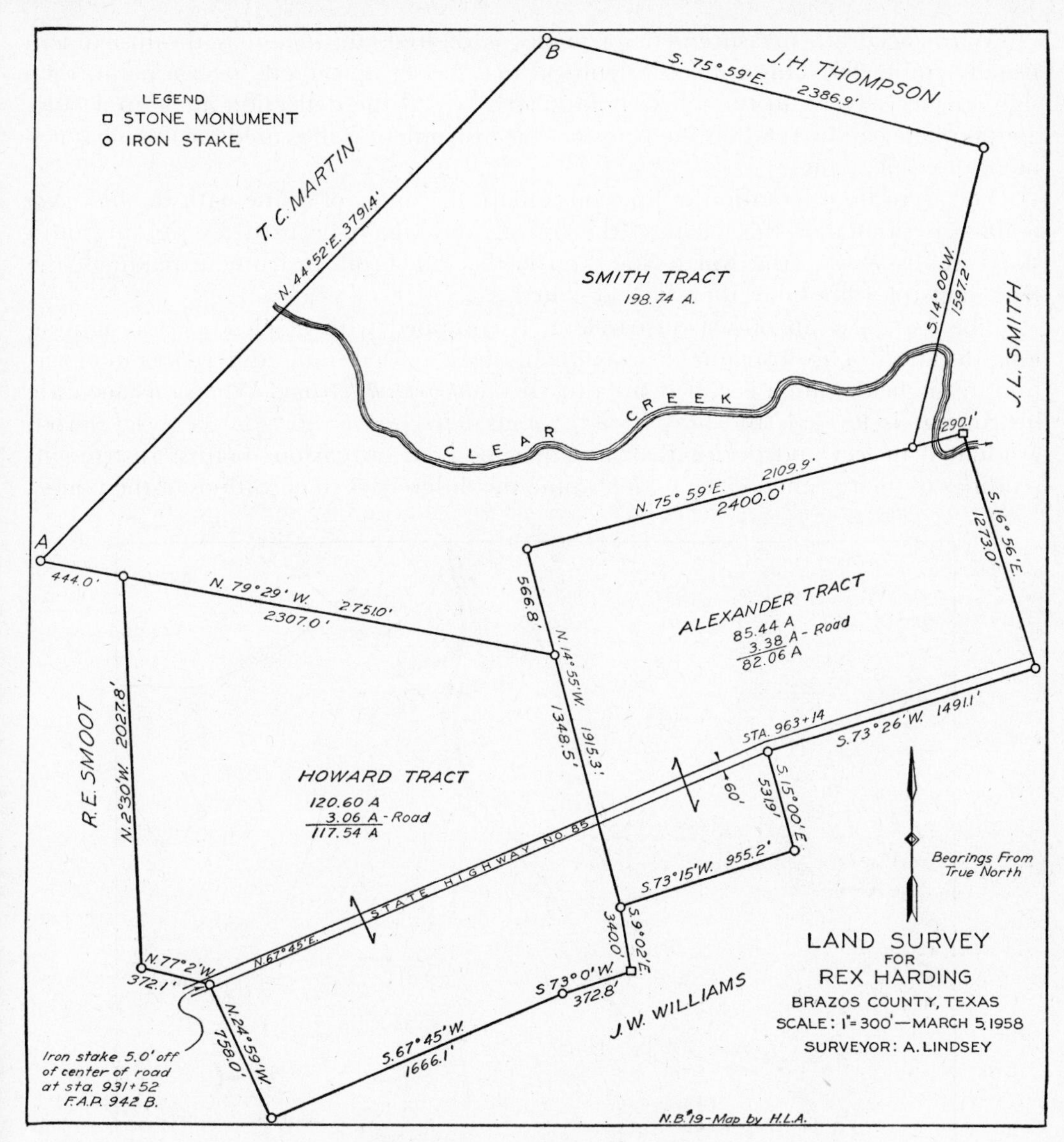

Fig. 1090 Land Survey Plat.

on the rod and using the necessary conversion factor, the instrument reading can easily be converted to distance.

By use of a *compass*, the bearing of a line, which is the angle between the line and the *magnetic north*, may be read on the compass. When all bearings of the lines of a *traverse* have been determined, the angles between the lines are easily computed by addition or subtraction. The correct method of listing bearings requires that they be referenced either to the north or to the south. A bearing is therefore either north, and east or west; or it is south, and east or west. See "N. 44°52′E." for traverse line AB of Fig. 1090. Compass readings are not to be regarded as of high accuracy, since magnetic north and true north are not quite the same; also local magnetism may affect the position of the compass needle.

When accurate measurement of angles is desired, the *transit* is the instrument usually employed. This optical instrument may be set up directly over a point, then sighted successively on two other points, after which the deflection angle in a horizontal plane can be read on the transit. This instrument is also used for the measurement of vertical angles.

For accurate orientation of lines on a map, the angle of a line with the *true north* is often desired. For this purpose the transit may again be used for sighting on a star (usually *Polaris*, the North Star), or on the sun. From instrument readings, the true angle of a line may then be calculated.

The *level*, also an optical instrument, is equipped with a telescope for sighting long distances. This instrument is commonly used to determine differences in elevation in the field, which is called the process of *differential leveling*. When this accurate instrument is leveled, the line of sight of its telescope is horizontal. A level rod, graduated in feet and decimals of feet may be held on various points. Instrument readings of the rod then serve to determine the differences in elevations of the points.

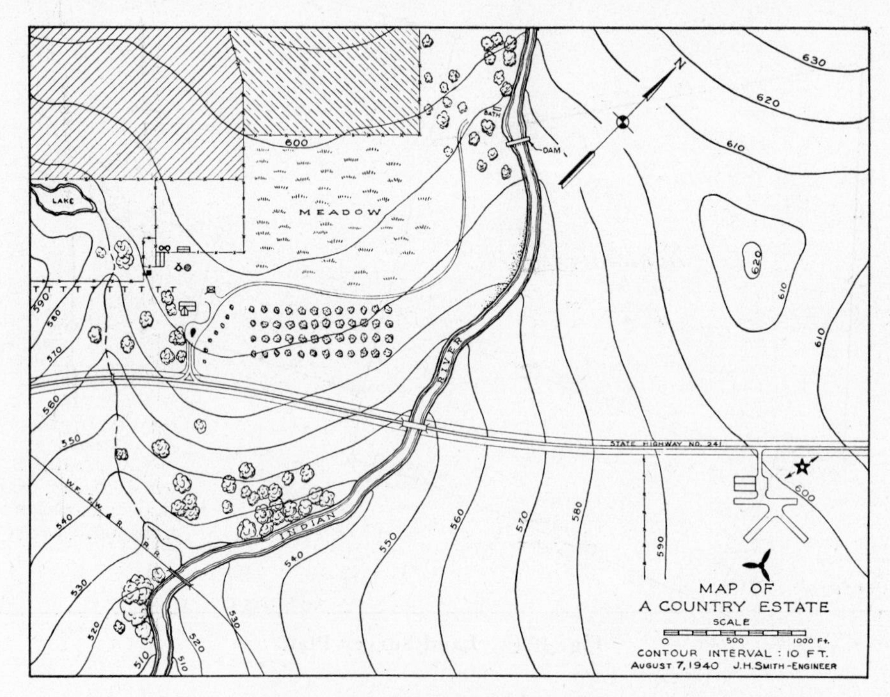

Fig. 1091 A Topographic Map.

Photogrammetry is now widely used for map surveying. This method utilizes actual photographs of the earth's surface and of artificial objects on the earth. Originally aerial photogrammetry was used mainly in mapping enemy territory during wartime. Now this method is used for governmental and commercial surveying, explorations, property valuation, etc. It has the great advantage of being easy to use in difficult terrain having steep slopes, where ground surveying would be difficult or nearly impossible. The utilization of aerial photographs is called *aerial photogrammetry*, whereas that utilizing photographs taken from ground stations with the axis of the camera lens nearly horizontal, is known as *terrestrial photogrammetry*. By combining the results of both types of observations, it is possible not only to determine the relative positions of objects in a horizontal plane, but it is also possible to determine relative elevations. Thus the science of photogrammetry can be the basis of contour mapping as well as plan mapping. Generally, aerial photographs are used in matched groups to form a *mosaic*. To form a satisfactory mosaic map, the group photographs must overlap slightly. A distinct advantage of photogrammetry is that a large area can be mapped from a single clear photograph. The method may be used in connection with ground surveying by photographing *control points* already located on the ground by precise surveying.

For additional information, which is beyond the scope of this chapter, the student may refer to the following texts: *Advanced Surveying and Mapping*, and *Elements of Photogrammetry*, by George D. Whitmore, International Textbook Company; *Elementary Topography and Map Reading*, by Samuel L. Greitzer, McGraw-Hill Book Company; and *Manual of Surveying Instructions for the Survey of the Public Lands of the United States*, prepared and published by the Bureau of Land Management, U.S. Government Printing Office, Washington, D.C.

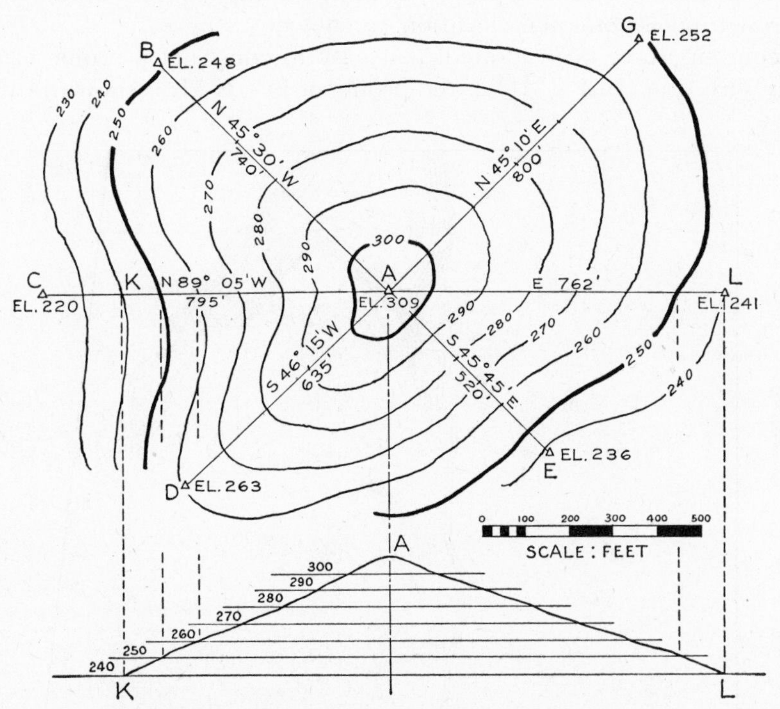

Fig. 1092 **Contours Determined from Control Points.**

651. Contours. Although contours have already been defined, their uses and characteristics and the methods of plotting them require further discussion.

A *contour interval* is the vertical distance between horizontal planes passing through successive contours. For example, in Fig. 1092 the contour interval is 10 ft. The contour interval should not change on any one map. It is customary to show every fifth contour by a line heavier than those representing intermediate contours.

If extended far enough, every contour line will close. At streams, contours form

V's pointing upstream. Should successive contours be evenly spaced, this means that the ground slopes uniformly, whereas closely-spaced contours indicate steep slopes.

Locations of points on contour lines are determined by *interpolation.* In Fig. 1092, the locations and elevations of seven control points are determined, and contour lines will be drawn on the assumption that the slope of the surface of the ground is uniform between station A and the six adjacent stations. To draw the contour lines, a contour interval of 10 ft. was adopted, and the locations of the points of intersection of the contour lines with the straight lines, joining the point A and the six adjacent points, was calculated as follows:

The horizontal distance between stations A and B is 740 ft. The difference in elevation of those stations is 61 ft. The difference in elevation of station A and contour 300 is 9 ft.; therefore, contour 300 crosses the line AB at a distance from station A of $\frac{9}{61}$ of 740, or 109.1 ft. Contour 290 crosses the line AB at a distance from contour 300 of $\frac{10}{61}$ of 740, or 121.3 ft. This 121.3 ft. distance between contour lines is constant along the line AB and can be set off without further calculation.

In the same way, points in which the contours cross the other lines of the survey can be interpolated. After this process is finished, the several contour lines can be drawn through points of equal elevation, as shown.

After contours have been plotted, it is easy to construct a profile of the ground line in any direction. In Fig. 1092, the profile of line KAL is shown in the lower or

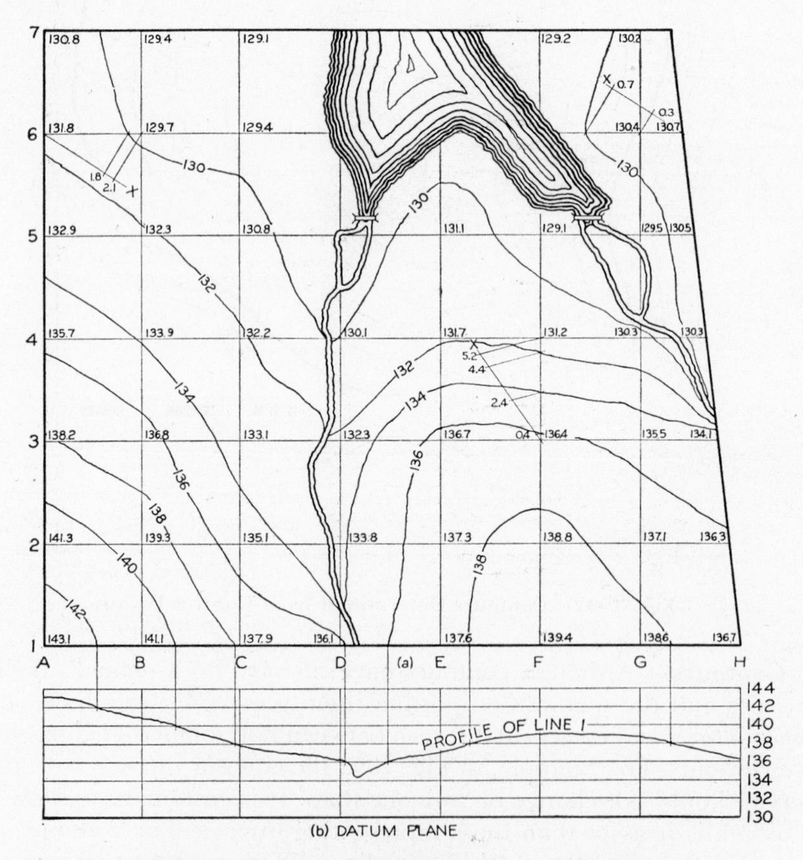

Fig. 1093 Contours Determined from Readings at Regular Intervals.

front view. It is customary, as shown here, to draw the profile to a larger vertical scale than that of the plan in order to emphasize the varying slopes.

Contour lines may also be plotted by use of the recorded elevations of points on the ground surface, as in Fig. 1093. This figure illustrates a *checker-board survey*, in which lines are drawn at right angles to each other dividing the survey into 100 ft. squares, and where elevations have been determined at the corners of the squares. The contour interval is taken as 2 ft., and the slope of the ground between adjacent stations is assumed to be uniform.

The points where the contour lines cross the survey lines can be located approximately by inspection, and accurately by the graphical method shown in Fig. 166, or by the numerical method explained above for Fig. 1092.

The points of intersection of contour lines with survey lines may also be found by constructing a profile of each line of the survey, as shown for line 1 in Fig. 1093 (b). Horizontal lines are drawn at elevations at which it is desired to show contours. The points in which the profile line intersects these horizontal lines indicate the elevations of points in which corresponding contour lines cross the survey line 1, and therefore can be projected upward, as shown, to locate these points.

It is obvious that the profile of any line can be constructed from the contour map by the converse of the process just described.

652. Symbols. Various natural and man-made features are designated by spe-

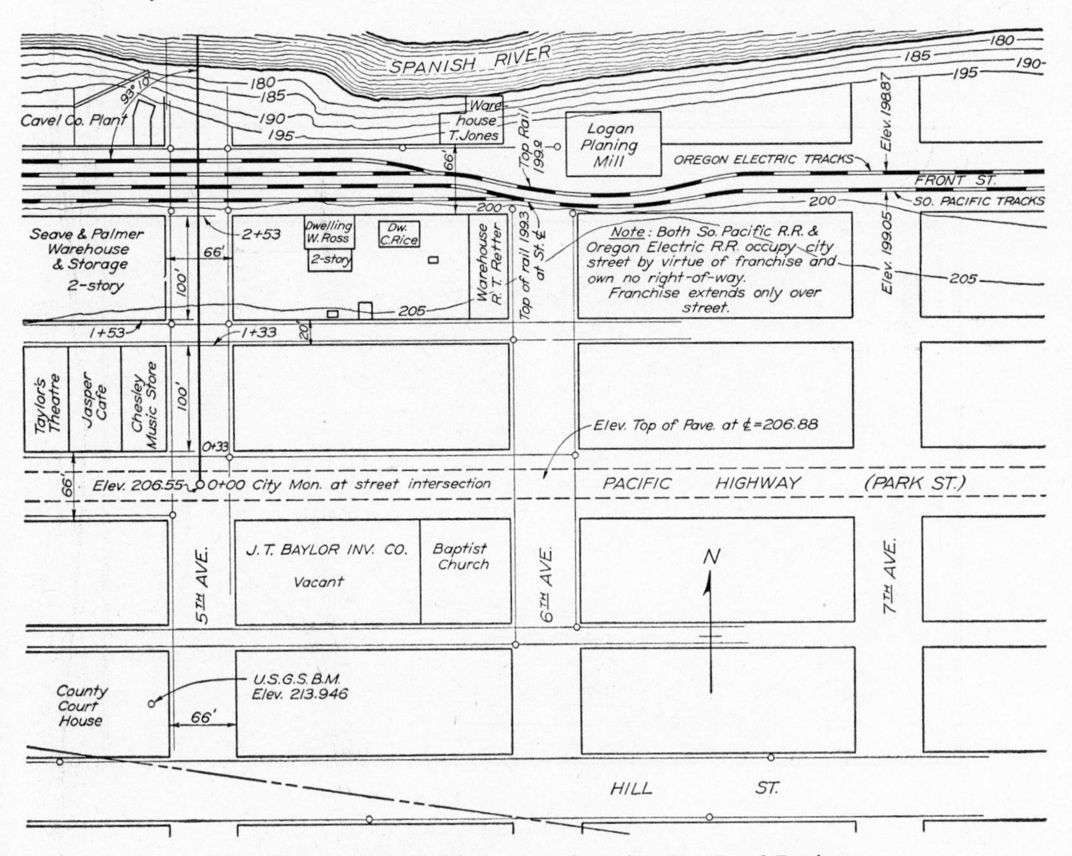

Fig. 1094 A City Plan for Location of a New Road Project.

cial symbols. A reference list of the most commonly-used map symbols is given in Appendix 31, and the student should refer to this list for identification of the symbols used in the figures of this chapter.

653. City Maps. The special use of a map determines what features are to be shown. Maps of city areas may be put to many uses, some of which will be described. Fig. 1094, a city plan for location of new road construction, shows only those features

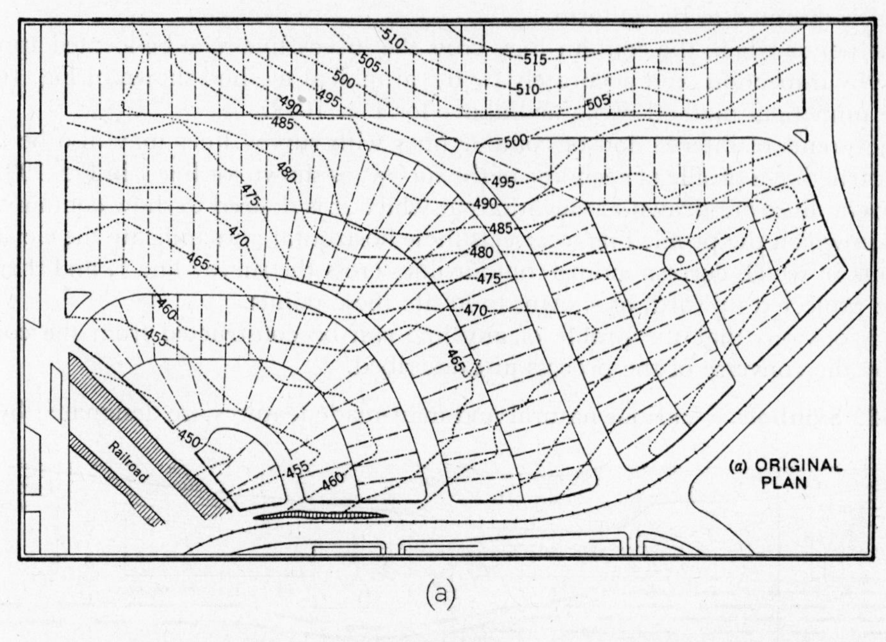

(a)

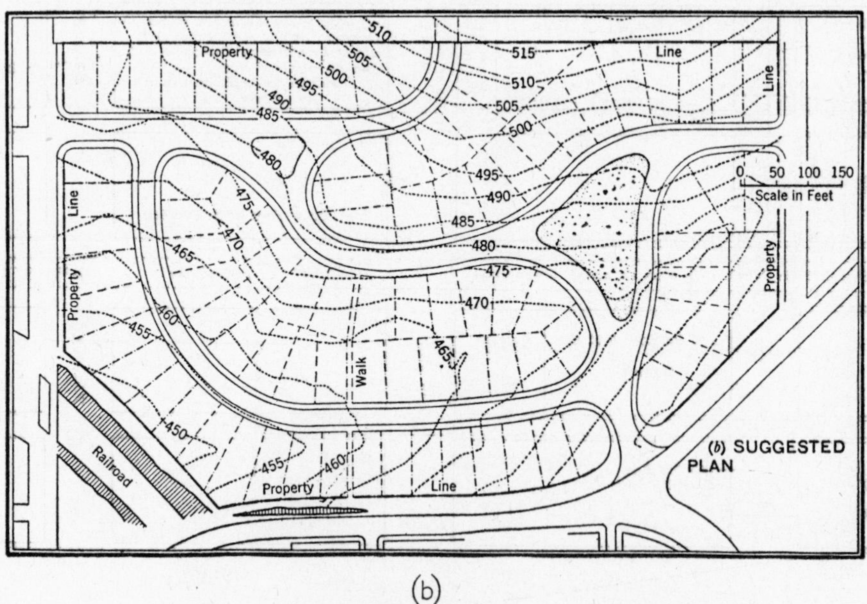

(b)

From Land Subdivision, *ASCE Manual No. 16 of Engineering Practice, 1939, p. 15.*

Fig. 1095 Adjustment of Streets to Topography.

of importance to the location and construction of the road. The transit line starts at the centerline intersection of Park St. and 5th Ave., and is marked as station 0 + 00. From here it extends north over the railroad yard to cross the river. Features near the transit line, such as buildings, are shown and identified by name. Street widths

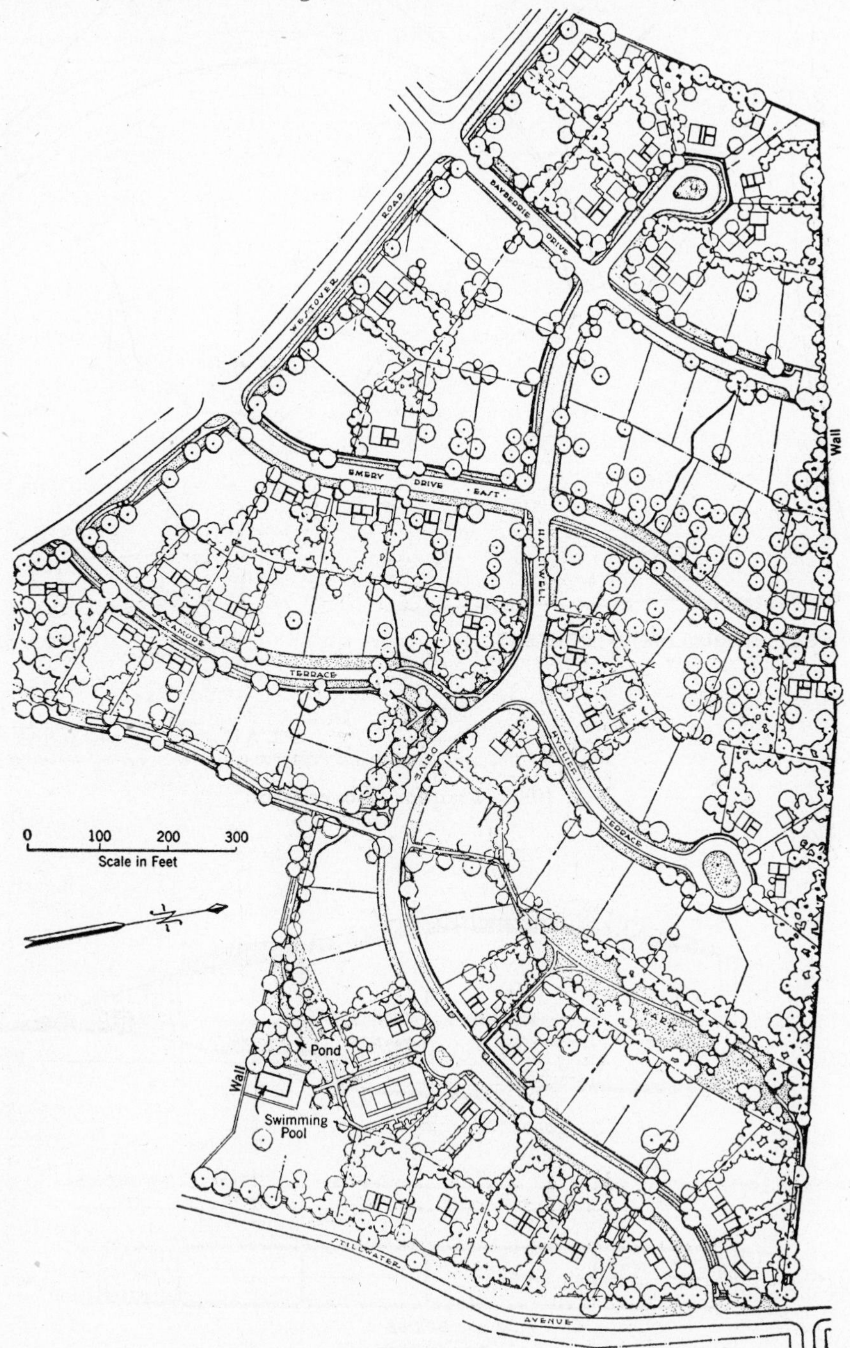

From Land Subdivision, *ASCE Manual No. 16 of Engineering Practice, 1939, p. 19.*

Fig. 1096 Land Subdivision Showing Use of *Culs-de-Sac*, (U-shaped streets), Stamford, Connecticut.

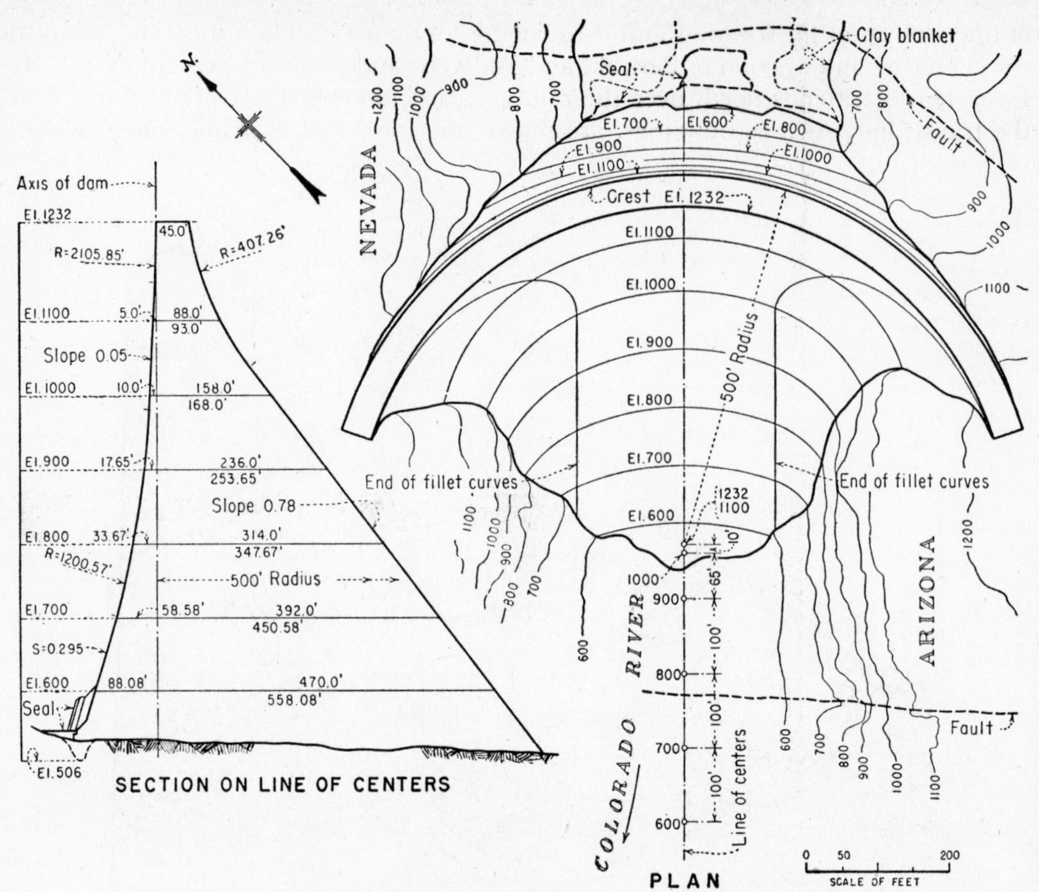

From Treatise on Dams. *Courtesy U.S. Dept. of the Interior, Bureau of Reclamation.*

Fig. 1097 Plan for Hoover Dam.

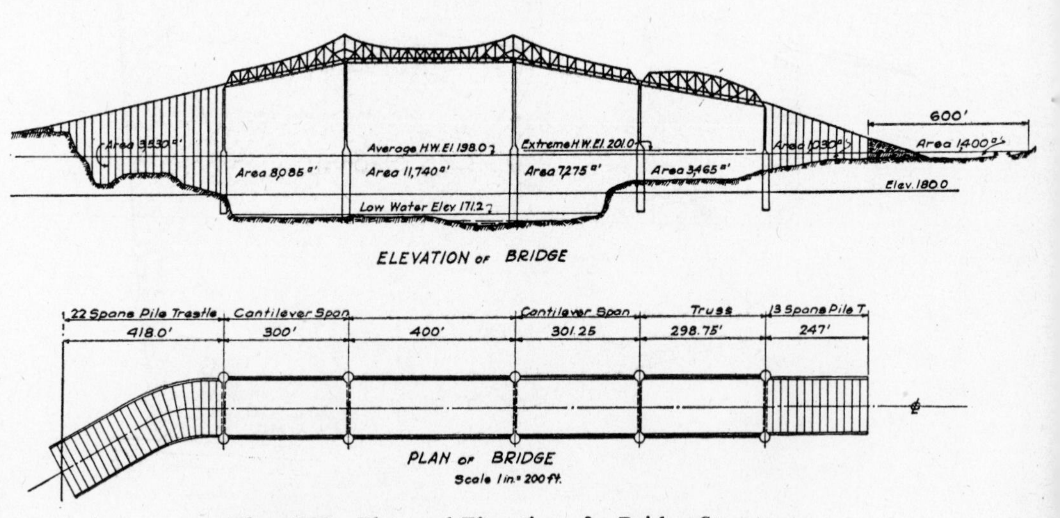

Fig. 1098 Plan and Elevation of a Bridge Structure.

are important and are shown. Contour lines between the railroad yard and the river indicate the steeply sloping terrain.

Maps perform an important function for those who plan the layout of lots and streets. For example, Fig. 1095 (a) shows an original layout of these features for a new residential area. An examination of the contours will show that this layout is not satisfactory, since the directions of the streets do not fit the natural ground slopes. Streets should be arranged so that the subdivision can be entered from a low point and so that a maximum number of lots will be above street grade. The layout at (b) is a decided improvement in that the streets curve to fit the topography and the entrance is located at a low point.

Maps have a definite use in landscape planning. Fig. 1096 is a landscape map showing a proposed layout of lots and trees for beautification of a new subdivision.

654. Structure Location Plans. When a construction project is contemplated in a particular area, the construction should be located most advantageously to fit

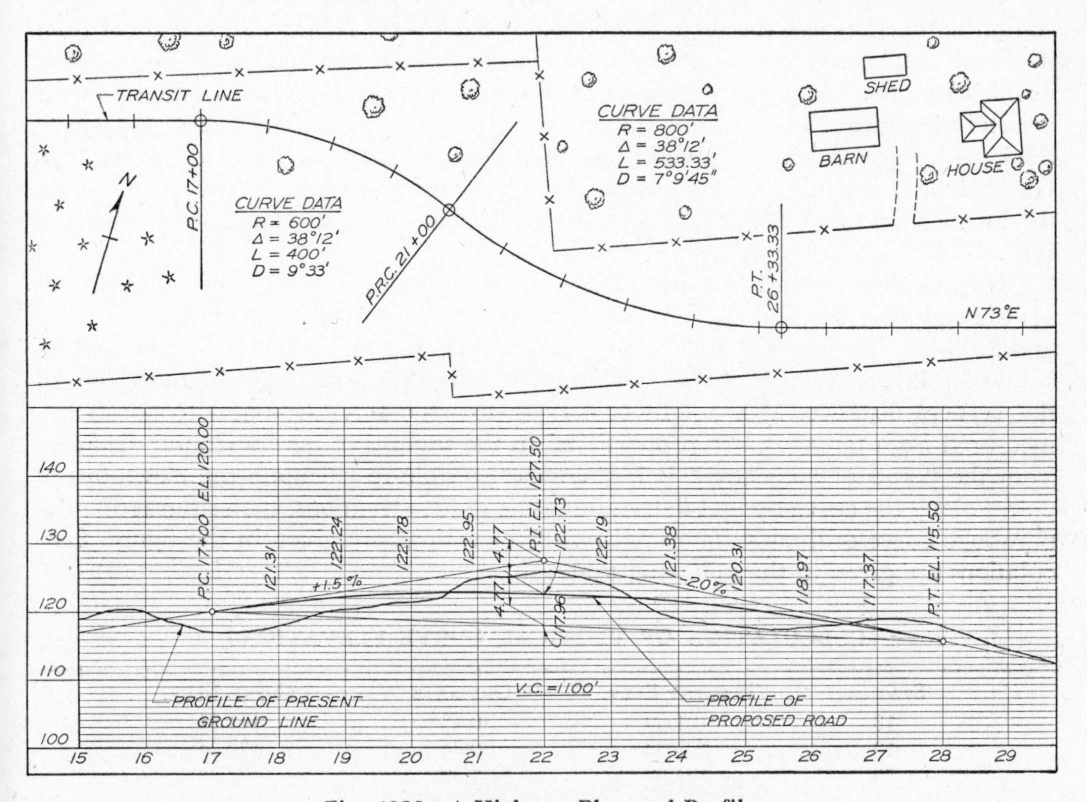

Fig. 1099 A Highway Plan and Profile.

the topography of the area. Thus a location plan on paper is of great advantage. A project location plan for a dam is shown in Fig. 1097. This map shows the important natural features, contours, a plan view of the structure, and a cross section through Hoover Dam.

Although many drawings, perhaps hundreds, comprise the complete detail drawings for a large bridge, one of the most important early drawings is a general arrangement plan and elevation in the form of a line diagram. As an example, Fig. 1098 shows a plan and elevation of a large bridge structure.

655. Highway Plans. Before highway construction starts, it is necessary first to plan the location and to arrange both the horizontal and vertical alignment most advantageously. Commonly, both plan and profile are drawn on the same sheet, as in Fig. 1099. In the plan at the top of the figure, the topography, consisting of such features as trees, fences, and farmhouses along the right of way, is shown. The transit line, locating the center line of the new road, is drawn with stations spotted at 100-foot distances apart. Data for laying in the horizontal curves in the field have been calculated and are listed. The point of curve (P.C.) at station 17 + 00 is the point at which the line begins to curve with a 600′ radius, for a curve length of 400′. The central angle is 38°12′, and the degree of curve (angle subtended by a 100′ chord) is shown as 9°33′. The reverse curve of 800′ radius begins at station 21 + 00. Note also the north point and the bearing N73°E of the transit line.

The vertical alignment is shown in profile below the plan. Note the listing of station numbers below the profile. The scale of this view is larger in a vertical than in a horizontal direction in order to show the elevations clearly. The existing ground profile along the centerline of the road is shown, as well as the profile of the proposed vertical alignment. The symbol P.I. denotes the point of intersection of the grade lines, and grade slopes are given in percentages. A one per cent grade would rise vertically one foot in each 100′ of horizontal distance. Station 22 + 00 is therefore the intersection point of an up-grade of 1.5 per cent and a down-grade of −2.0 per cent.

To provide a smooth transition between these grades, a vertical curve (V.C.) of 1100′ length is used. This curve is parabolic and tangent to grade at stations 17 + 00 and 28 + 00. The straight line joining these points has an elevation 117.96 directly below the P.I. At this point the parabolic curve must pass through the mid-point of the vertical distance, at a height of 4.77′ below the P.I. Ordinates to parabolas, measured from tangents, are proportional to the squares of the horizontal distances from the points of tangency.* It is therefore possible to calculate the elevations of points along the curve by first determining the grade elevations, then subtracting the parabolic curve ordinates. The final profile elevations are given in the figure. Calculations are given in the following table:

CALCULATION OF VERTICAL CURVE ELEVATIONS

Station	Tangent Elev.	Ordinate	Curve Elev.
18	121.50	0.19*	121.31
19	123.00	0.76	122.24
20	124.50	1.72	122.78
21	126.00	3.05	122.95
22	127.50	4.77	122.73
23	125.50	3.31	122.19
24	123.50	2.12	121.38
25	121.50	1.19	120.31
26	119.50	0.53	118.97
27	117.50	0.13	117.37

$*\dfrac{(100)^2\,4.77'}{(500)^2} = 0.19'$

656. United States Maps. United States maps, prepared by the U.S. Coast and Geodetic Survey and the U.S. Geological Survey are excellent examples of topographic mapping. A small section of a typical U.S. Geological Survey map is shown in Fig. 1100. Such maps cover large areas, the largest scale used being 1:62,500, or

Courtesy U.S. Geological Survey

Fig. 1100 A Typical Map of the U.S. Geological Survey.

very nearly one mile to the inch. The contour interval in this example is 20′. To so small a scale it would be impossible to show clearly vegetation, fences, etc. Therefore such maps can show only the main features of the terrain, such as contours, roads, railroads, rivers, lakes, and streams. These maps are very reliable, as they are based upon precise surveying.

657. Topographic Drawing Problems. The following problems are given to afford practice in topographic drawing. The drawings are designed for a 11″ × 17″

or Size B sheet. The position and arrangement of the titles should conform approximately to that of Fig. 1090.

Prob. 1. Draw symbols of six of the common natural surface features (streams, lakes, etc.), and six of the common development features (roads, buildings, etc.) shown in Appendix 31.

Prob. 2. Draw, to assigned horizontal and vertical scales, profiles of any three of the six lines shown in Fig. 1092.

Prob. 3. Assuming the slope of the ground to be uniform, and assuming a horizontal scale of 1″ = 200 ft. and a contour interval, §651, of 5 ft., plot, by interpolation, the contours of Fig. 1092.

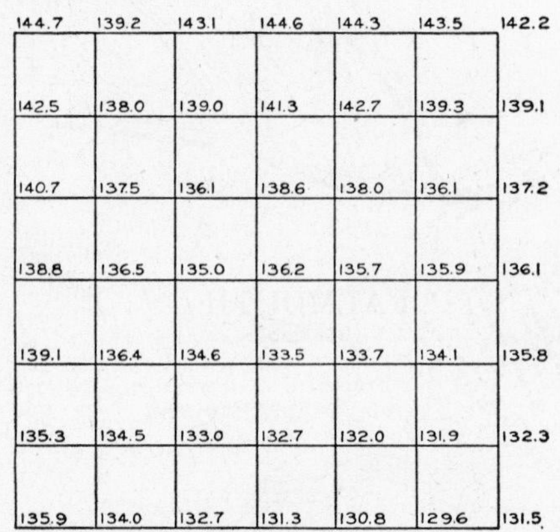

Fig. 1101 (Problem 5). To Draw Contours.

Prob. 4. Using the elevations shown in Fig. 1093 (a) and a contour interval of 1 ft., plot the contours to any convenient horizontal and vertical scales, and draw profiles of lines 3 and 5 and of any two lines perpendicular to them. Check, graphically, the points in which the contours cross these lines.

Prob. 5. Using a contour interval of 1 ft. and a horizontal scale of 1″ = 100 ft., plot the contours from the elevations given, Fig. 1101, at 100-ft. stations; check, graphically, the points in which the contours cross one of the horizontal lines and one of the vertical lines, using a vertical scale of 1″ = 10 ft.; sketch, approximately, the drainage channels.

Prob. 6. Draw a plat of the survey shown in Fig. 1090 to as large a scale as practicable. Use an engineers scale to set off distances, and a protractor to set off bearings; if the drawing is accurate, the plat will close.

Prob. 7. Draw a topographic map of a country estate, similar to that shown in Fig. 1091.

Prob. 8. Calculate profile elevations for a vertical curve 800′ long, to join grades of + 3.00% and − 3.00%. Assume grade elevations at points of tangency to be 100.00.

CHAPTER 24

PIPING DRAFTING

by
*D. G. Reid and L. A. Anderson**

658. Introduction. Pipe is used for transporting liquids and gases, and for structural elements such as columns, handrails, etc. The choice of the type of pipe is determined by the purpose for which it is to be used.

659. Kinds of Pipe. Pipe is made of aluminum, brass, clay, concrete (concrete made with ordinary aggregates and in combination with asbestos, etc.), copper, glass, iron, lead, plastics, rubber, wood, and other materials or combinations of them. Cast iron, steel, wrought iron, brass, copper, and lead pipes are most commonly used for transporting water, steam, oil, or gases.

660. Steel and Wrought-Iron Pipe. Steel and wrought-iron pipe is in common use for water, steam, oil, and gas. Up to the early 1930's it was available in three weights known as "standard," "extra strong," and "double extra strong" only. At that time increasing pressures and temperatures, particularly for steam service, made the availability of more diversified wall thicknesses desirable. American Standards Association Sectional Committee B36 has developed dimensions for ten different Schedules of pipe. See Appendix table 36. It will be observed, in this table, that dimensions are established for nominal sizes from $\frac{1}{8}''$ to 24″, and that dimensions have not been established for all Schedules. In the different Schedules, the outside diameters (O.D.) are maintained for each nominal size to facilitate threading and the uniform use of fittings and valves.

Certain of the Schedule dimensions correspond to the dimensions of "standard" and "extra strong" pipe. These are shown in **bold face** type in the appendix. The Schedule dimensions so shown for Schedules 30 and 40 correspond to standard pipe,

*Mechanical engineers, Sargent & Lundy.

and those for Schedule 80 correspond to extra strong pipe. There are no Schedule dimensions corresponding to "double extra strong" pipe. Pipe corresponding to all of the established Schedule dimensions is not always commercially available, and should be investigated before specifying pipe on drawings. Generally Schedules 40, 80, and 160 are readily available; others may or may not be.

Note that the actual outside diameter of pipe in nominal sizes $\frac{1}{8}''$ to 12″ inclusive is larger than the nominal size, whereas the outside diameter of pipe in nominal sizes 14″ and larger corresponds to the nominal size. This pipe in nominal sizes 14″ and larger is commonly referred to as O.D. pipe.

Pipe is available as welded or as seamless pipe. Welded pipe is available in Schedules 40 and 80 in the smaller sizes. Lap-welded pipe is made in sizes up to and including 2″. Butt-welded pipe is available as furnace-welded material, where a formed length is heated in a furnace and then welded, in sizes up to and including 3″. Butt-welded pipe is also available as continuous welded pipe, where the finished pipe is continuously heated, formed, and welded from a roll of strip steel, in sizes up to and including 4″. Seamless pipe is made in both small and large sizes.

Many of today's applications require the use of alloys to withstand the pressure-temperature conditions without having to be excessively thick. Numerous alloys in both ferritic and austenitic material are available. Reference should be made to the various specifications of the American Society for Testing Materials (ASTM)* for these alloys and for dimensional tolerances.

Steel pipe can be obtained as *black pipe* or as *galvanized pipe*. Galvanized pipe is not always available from the original producer, due to lack of facilities at the pipe mill.

Steel and wrought-iron pipe is available in lengths up to about 40 feet in the small sizes, the length decreasing with increasing size and wall thickness.

661. Cast-Iron Pipe. Cast-iron pipe is generally used for water or gas service and as soil pipe. For water and gas pipe, it is available in sizes from 3″ to 60″ inclusive and in standard lengths of 12 feet. Various wall thicknesses, depending on the internal pressure, are available. The dimensions and pressure ratings of the various Classes are shown in Appendix table 37.

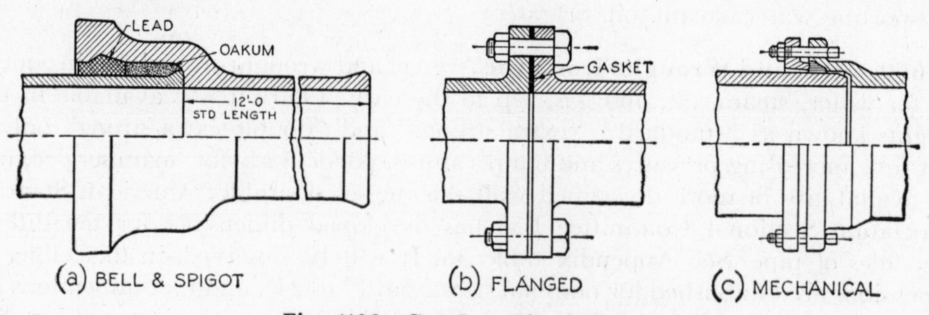

Fig. 1102 Cast-Iron Pipe Joints.

Generally speaking, water and gas pipes are connected with *bell and spigot* joints, Fig. 1102 (a), or *flanged* joints, (b), although other types of joints, (c), are also used.

As soil pipe, cast-iron pipe is available in sizes 2″ to 15″ inclusive, in standard lengths of 5 feet, and in *service* and *extra heavy* weights. Soil pipe is generally connected

*1916 Race St., Philadelphia 3, Pa.

with bell and spigot joints, but soil pipe with threaded ends is available in sizes up to 12″.

In using cast-iron pipe, the designer must consider both the internal pressure and the external loading due to fill and other loadings, such as roads, tracks, etc. Cast-iron pipe is brittle, and settlement can cause fracture unless sufficient flexibility is provided in the joints. For this reason, flanged joints are not usually employed for buried pipes unless adequately supported.

662. Seamless Brass and Copper Pipe. Pipe made of brass and copper, and having approximately the same dimensions as "standard" and "extra strong" steel pipe, is available. Such pipe is suitable for plumbing, including supply, soil, waste drain, and vent lines. It is also particularly suitable for process work where scale and oxidation of steel pipe is objectionable. Brass and copper pipe is available in straight lengths up to 12 feet.

Brass pipe is generally known as *red brass pipe*, which is an alloy of approximately 85 per cent copper and 15 per cent zinc. Copper pipe is practically pure copper with less than 0.1 per cent of alloying elements.

Brass and copper pipe should be joined with fittings of copper base alloy in order to avoid galvanic action resulting in corrosion. Where screwed joints are used, fittings similar to cast or malleable iron fittings, Fig. 1106, are available. Flanged fittings of brass and copper are of different dimensions than those made of ferrous material. Reference should be made to dimensional standards published by the American Standards Association (ASA B16.24) for dimensions of such fittings.

663. Copper Tubing. Where non-ferrous construction in sizes below 2″ is used, copper tubing is frequently employed. Such tubing is suitable for process work as mentioned in §662, for plumbing (particularly supply lines for hot and cold water), and for heating systems (particularly radiant heating).

Copper tubing is made as *hard temper* and as *soft tubing*. Hard temper tubing is much stiffer than soft tubing and should be used where rigidity is desired. Soft tubing can be easily bent, and therefore is generally used where bending during assembly is required. Neither hard nor soft temper tubing has the rigidity of iron or steel pipe, and consequently must be supported at much more frequent intervals than the latter. Where multiple runs of parallel tubes are employed for long distances (20 feet or more), it is common practice to use soft tubing and to lay the parallel runs in a trough construction, thus obtaining continuous support.

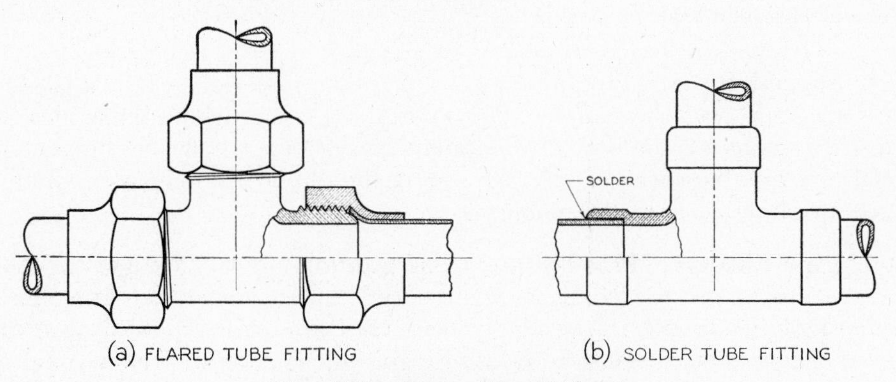

Fig. 1103 Copper Pipe Fittings.

Copper tubing joints are usually made with *flared* joints, Fig. 1103 (a), or with *solder* joints, (b). There are several types of flared joints, but the basic design of making a metal-to-metal joint is common to all. Fittings, such as tees, elbows, and couplings, are available for flared joints.

Solder joints are also known as *capillary* joints because the annular space between the tube and the fitting is so small that the molten solder is drawn into the space by capillary action. The solder may be introduced through a hole in the fitting, Fig. 1104, through the outer end of the annular space, Fig. 1103 (b), or fittings having a factory assembled ring of solder in the fitting can be purchased. Solder joints may be made with soft solder (usually 50-50 or 60-40 tin and lead) or with silver solder. This latter material has a higher melting point than soft solder, makes a stronger joint, and is suitable for higher operating temperatures.

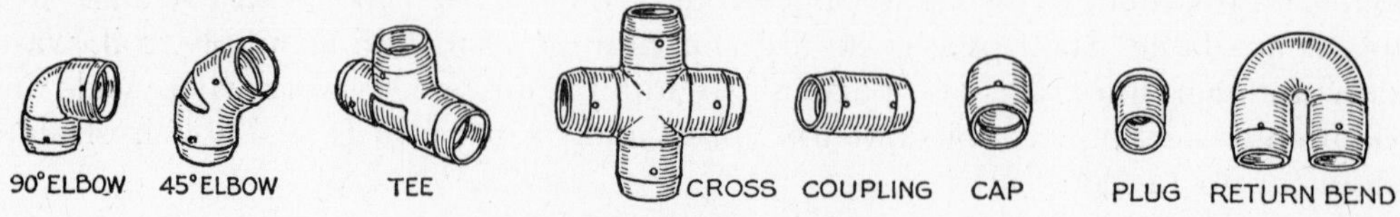

Fig. 1104 Solder Fittings.

Copper pipe or tubing has an upper operating temperature limit of 406°F. If solder fittings are used, the upper temperature limit will be dependent on the softening point of the solder rather than the limit of temperature of the base material.

Copper tubing can be connected to threaded pipe or fittings by means of *adapters*. These adapters are available with either male or female pipe threads, and with either flared or solder connections for the tubing. Two types of such adapters are shown in Fig. 1105.

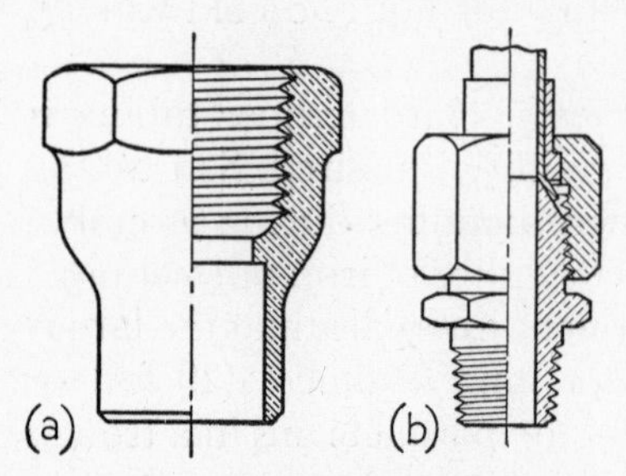

Fig. 1105 Adapters-Copper Tube to Threaded Pipe.

Copper tubing is available in straight lengths up to 20 feet or in coils of 60 feet for soft temper material. Hard temper material is available in straight lengths only, since it cannot successfully be coiled. Installation costs of coiled material are lower than for straight material, because of the fewer number of joints to be made. Copper tubing is available in both O.D. and nominal sizes. American Society for Testing Materials Specifications B88 and B251 give details of dimensions of such tubing.

664. Special Pipes. Pipe and tubing of other materials, such as aluminum, stainless steel, etc., are also available. Pipe and tubing of plastics are being increasingly used as these materials undergo development. Service for which these miscellaneous materials are suitable varies widely, because of the physical properties and temperature-pressure limitations of the materials.

665. Pipe Fittings. Pipe fittings are used to join adjacent lengths of pipe and frequently also to provide changes of direction, to provide branch connections at different angles, or to effect a change in size. They are made of cast iron, malleable iron, cast or forged steel, non-ferrous alloys, and other materials for special applications. They can be obtained in various weights that should be matched to the pipe with which they are to be used. Ferrous fittings are made for threaded, welded, or flanged joints. Non-ferrous fittings are made for threaded, solder, flared, or flanged joints. The common types of fittings for threaded joints are shown in Fig. 1106, for welded joints in Fig. 1107, for flanged joints in Fig. 1108, and for solder joints in Fig. 1104.

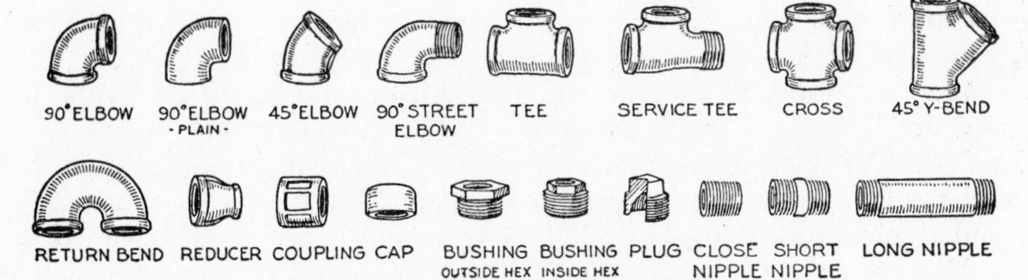

Fig. 1106 Screwed Fittings.

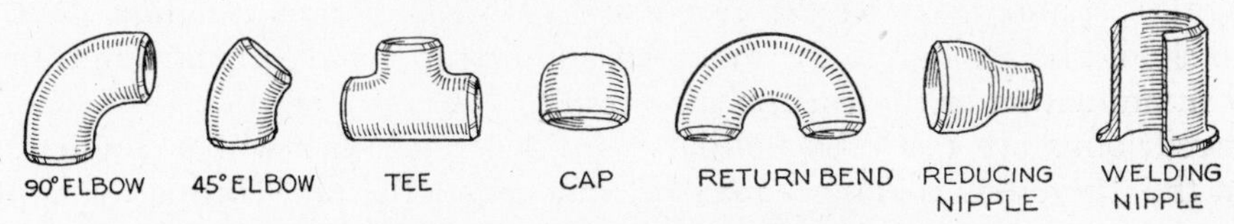

Fig. 1107 Butt-Welded Fittings.

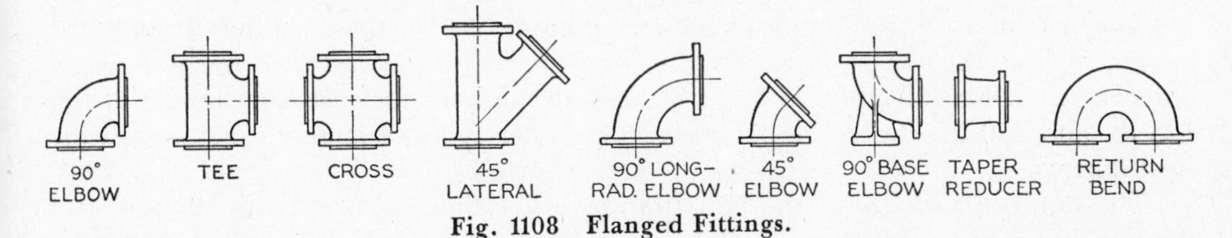

Fig. 1108 Flanged Fittings.

Where both or all ends of a fitting are of the same nominal size, the fitting is designated by the nominal size and the description; e.g., a *2″ screwed tee*. Where two or more ends of a fitting are not the same nominal size, the fitting is designated as a *reducing fitting*, and the dimensions of the run precede those of the branches, and the dimension of the larger opening precedes that of the smaller opening; e.g., a *2″ × $1\frac{1}{2}$″ × 1″ screwed reducing tee*. See Fig. 1109 for typical designations.

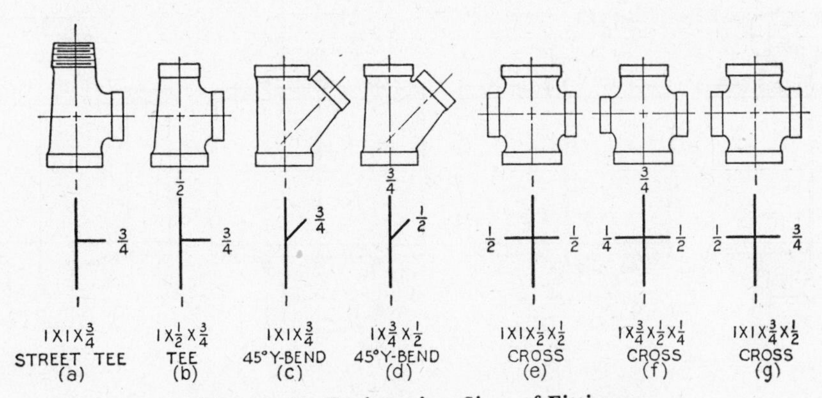

Fig. 1109 Designating Sizes of Fittings.

The threads of screwed fittings conform to the pipe thread with which they are to be used, either male or female as the case may be. See §666.

Dimensions of 125 lb. cast-iron screwed fittings, 250 lb. cast-iron screwed fittings, 125 lb. cast-iron flanged fittings, and 250 lb. cast-iron flanged fittings are shown in Appendix tables 38, 39, 40, and 43. See §666 for reference to steel fittings.

666. Pipe Joints. The joints between pipes, fittings, and valves may be *screwed, flanged, welded,* or for non-ferrous materials may be soldered.

The American Standard pipe threads are illustrated in Figs. 676-678, and tabular dimensions are shown in Appendix table 36. The threads of the American Petroleum Institute (A.P.I.) differ somewhat from the American Standard pipe threads. Refer to the A.P.I. Standards for these differences.

Threaded joints can be made up lightly by simply screwing the cleaned threads together. However, it is common practice to use *pipe compound* in making such joints, as this provides lubrication for the threads and enables them to be screwed together more tightly. It also serves to seal irregularities, thus providing a tighter joint. Such material should be applied to the male thread only, to avoid forcing it into the pipe where contamination or obstruction may result.

Flanged joints are made by bolting two flanges together with a resilient *gasket* between the flange faces. Flanges may be attached to the pipe, fitting, or appliance by means of a screwed joint, by welding, by lapping the pipe, or by being cast integrally with the pipe, fitting, or appliance.

The faces of the flanges between which the gasket is placed have different standard *facings,* such as *flat face,* $\frac{1}{16}''$ *raised face,* $\frac{1}{4}''$ *raised face, male and female, tongue and groove,* and *ring joints.* Flat face and $\frac{1}{16}''$ raised face are standard for cast-iron flanges in the 125 lb. and 250 lb. classes, respectively. The other types of facing are standard for steel flanges.

The number and size of the bolts joining these flanges varies with the size and the working pressure of the joint. Bolting for Class 125 cast-iron and Class 250 cast-iron flanges is shown in Appendix tables 41 and 44, respectively.

For dimensions of the various flange facings and for flange and bolting dimensions of the various sizes and pressure standards of steel flanges, refer to the American Standard for Steel Pipe Flanges and Flanged Fittings (A.S.A. B16.5), which is too voluminous to be included here. Some typical types of flanged joints are shown in Fig. 1110.

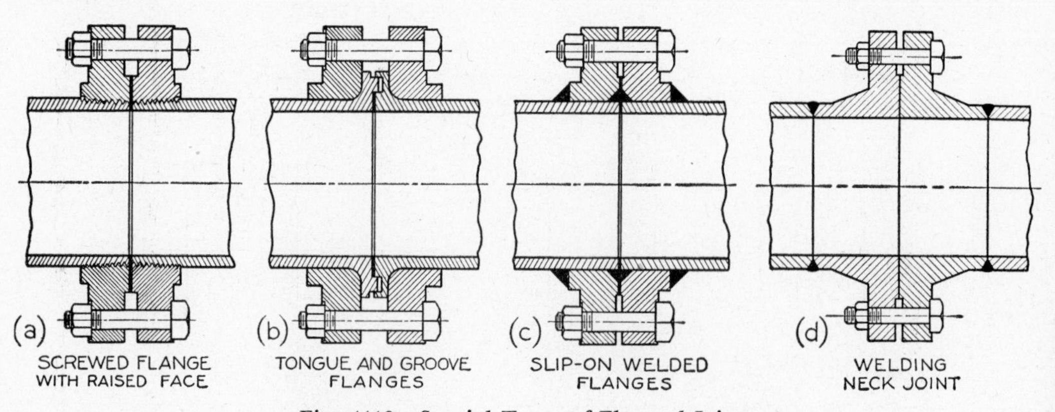

Fig. 1110 Special Types of Flanged Joints.

Piping construction employing welded joints is in almost universal use today, particularly for higher pressure and temperature conditions. Such joints may either be *socket welded* or *butt welded*, Fig. 1111. Socket-welded joints are limited to use in small sizes. The contours of the butt-welded joints shown at (b) and (c) are those shown in American Standard B16.25–1955.

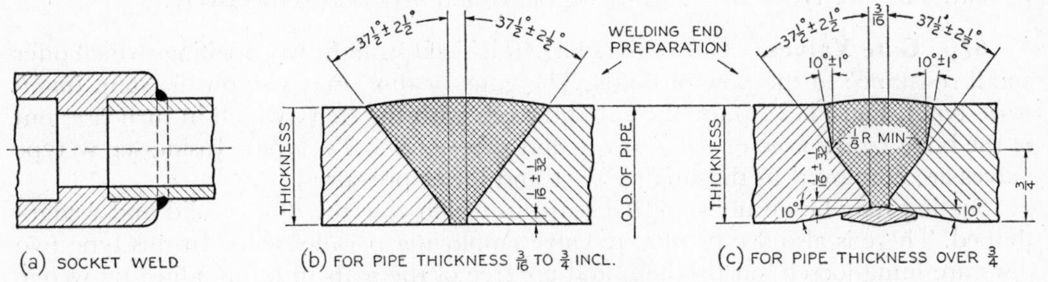

Fig. 1111 Welded Joints.

667. Valves. Valves are used to stop or to regulate the flow of fluids in a pipe line. The more common types are *gate valves*, *globe valves*, and *check valves*. Other types of valves, such as *pressure reducing valves* and *safety valves*, are special devices used to maintain automatically a desired lower pressure on the downstream side of the valve or to prevent automatically undesirable overpressure, respectively.

668. Globe Valves. Globe valves have approximately spherical bodies with the seating surface at either a right or an acute angle to the center line of the pipe, Fig. 1112 (a). In such a valve the flowing fluid must make abrupt turns in the body, thus resulting in considerably higher pressure loss than for a gate valve.

Globe valves are commonly used where close regulation of flow is desired, because they lend themselves better to this type of regulation and are less subject to cutting action in throttling service than gate valves.

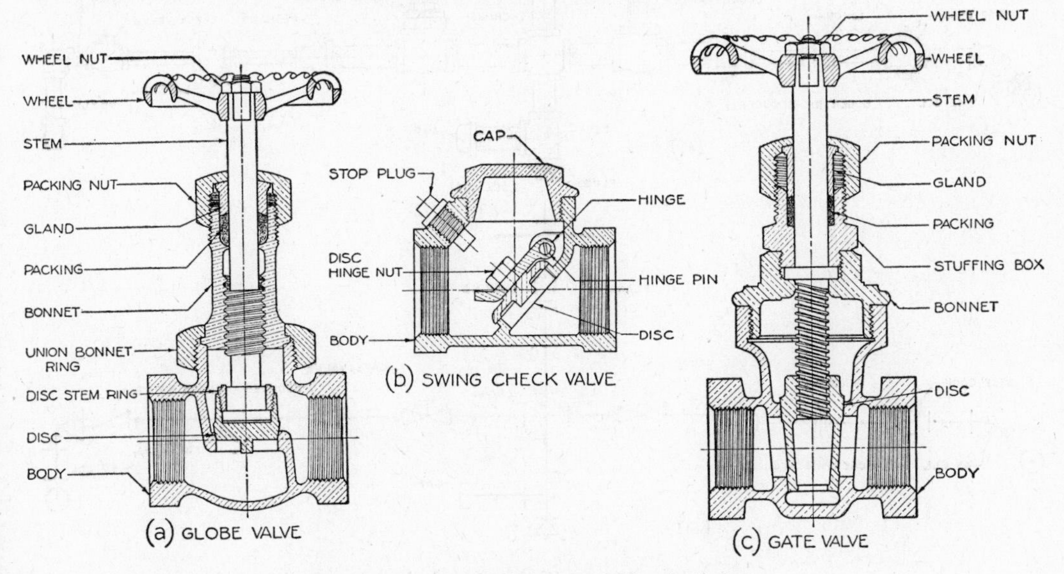

Fig. 1112 Gate, Globe, and Check Valves.

Valves of the *inside screw* and *O S & Y* types are available, §670. Angle valves and needle valves are special designs of the general class of globe valves.

669. Check Valves. Check valves are used to limit the flow of fluids to one direction only. The disc may be hinged so as to swing partially out of the stream, Fig. 1112 (b), or it may be guided in such a manner that it can rise vertically from its seat. The two types are called *swing checks* and *lift checks* respectively.

670. Gate Valves. Gate valves have full-sized straightway openings which offer small resistance to the flow of fluids. The gate, or disc, may rise on the stem (*inside screw* type), Fig. 1112 (c), or the gate may rise with the stem, which in turn rises out of the body (*rising stem*, or *outside screw* and *yoke*—*O S & Y*—type). Inside-screw type valves are employed in the smaller sizes and lower pressures.

Seating may be on non-parallel seats, in which case the disc is solid and wedge-shaped. There is also a type of gate valve employing parallel seats. In this type two discs are hung loosely on the stem and are free of the seats until an adjusting wedge reaches a lug at the closed position of the valve, when further movement of the stem causes the wedge to spread the discs and form a tight joint on the parallel seats. Such valves are used only on low pressure and temperature services.

671. American Standard Code for Pressure Piping. The American Standards Association has adopted an American Standard Code for Pressure Piping (ASA B31.1). This is a compilation of recommended practices and minimum safety standards covering various types of piping, such as Power Piping, Industrial Gas and Air

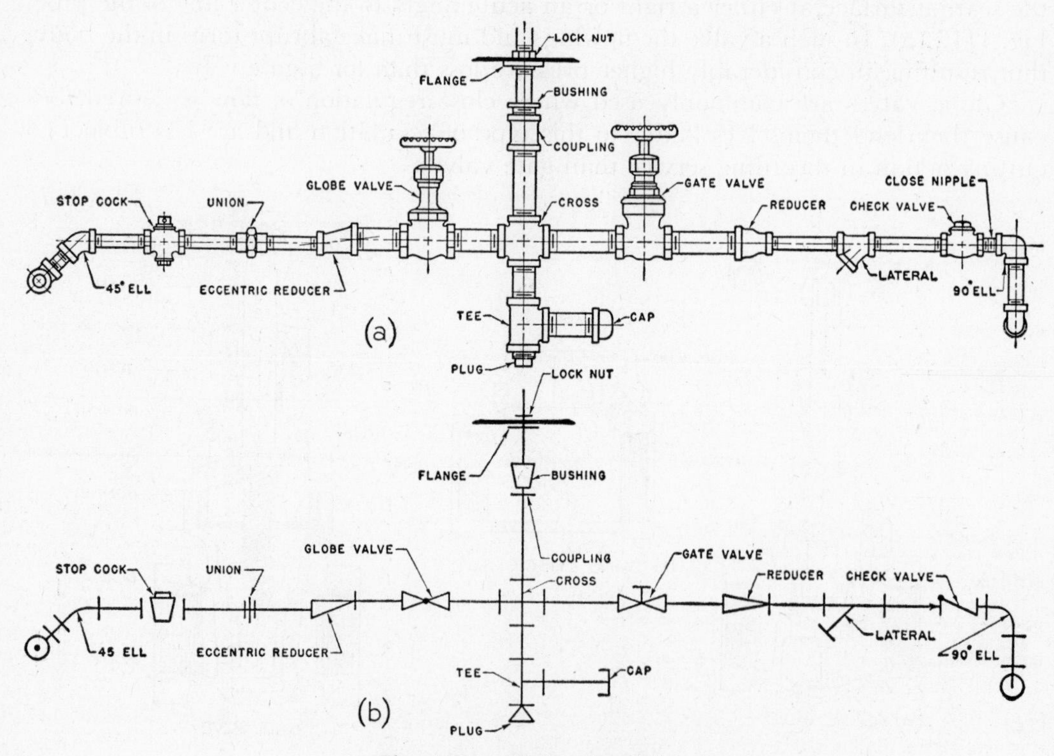

Fig. 1113 Piping Symbols.

Piping, Oil Refinery Piping, Oil Transportation Piping, Refrigerating Piping, Chemical Industry Process Piping, and Gas Transmission and Distribution Piping.

672. Piping Drawings. To simplify the preparation of working drawings of piping systems, the set of symbols shown in Appendix 32 has been developed to represent the various pipe, fittings, and valves in common use. An application of these symbols in a piping drawing is shown in Fig. 1113 (b).

Drawings of piping systems may be made as *single-line* drawings, Figs. 1113 (b) to 1116, or as *double-line* drawings, Figs. 1113 (a), 1118, and 1119. Either type of drawing may be made as multiview projections, Fig. 1114 (b), 1118, and 1120; as axonometric projections, Figs. 1114 (a) and 1121; or as oblique projections, Figs. 1116 and

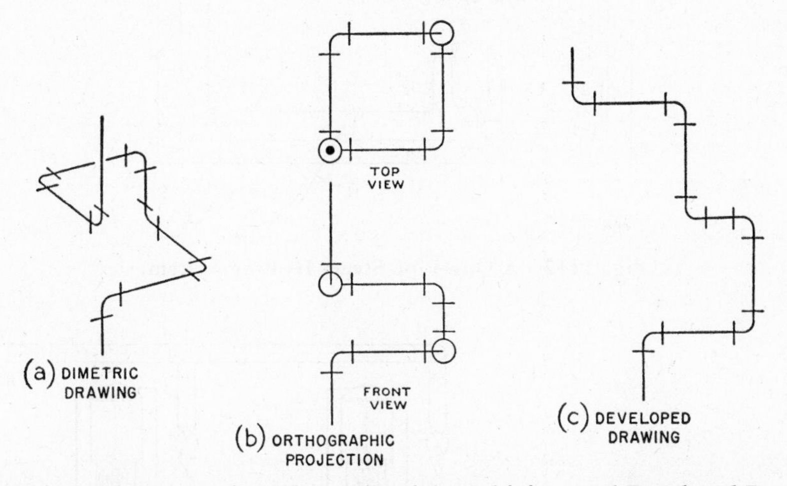

Fig. 1114 A Pipe Expansion Joint—Pictorial, Multiview, and Developed Drawing.

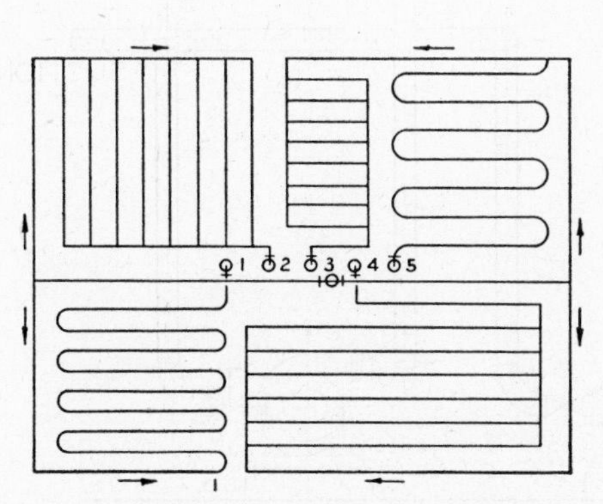

Fig. 1115 Pipe Grids and Serpentine Coils for a Panel Radiant Heating System.

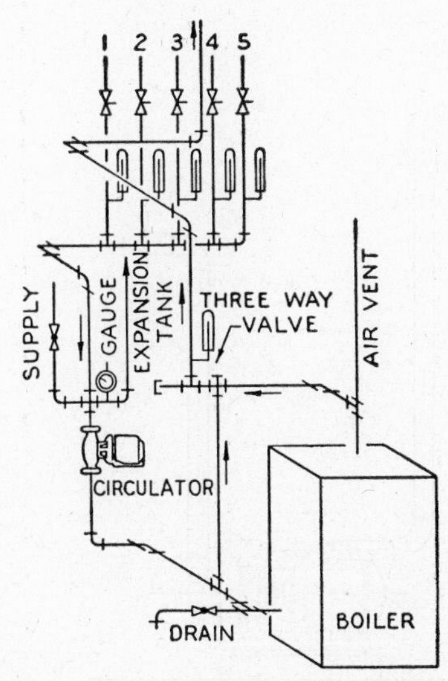

Fig. 1116 Schematic Drawing of Piping Connecting Boiler to Heating Coils.

1117. The oblique projection in Fig. 1117 is a modified form of oblique projection generally used in representing the piping arrangement for heating systems. In these cases, the pipe mains are shown in plan and the risers in oblique projection in various directions so as to make the representation as clear as possible.

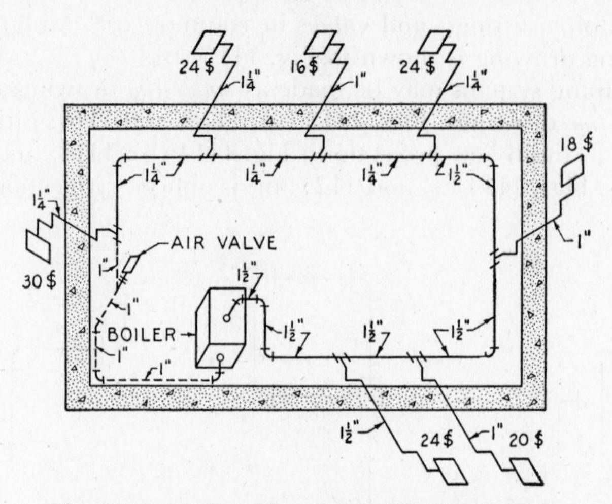

Fig. 1117 A One-Pipe Steam Heating System.

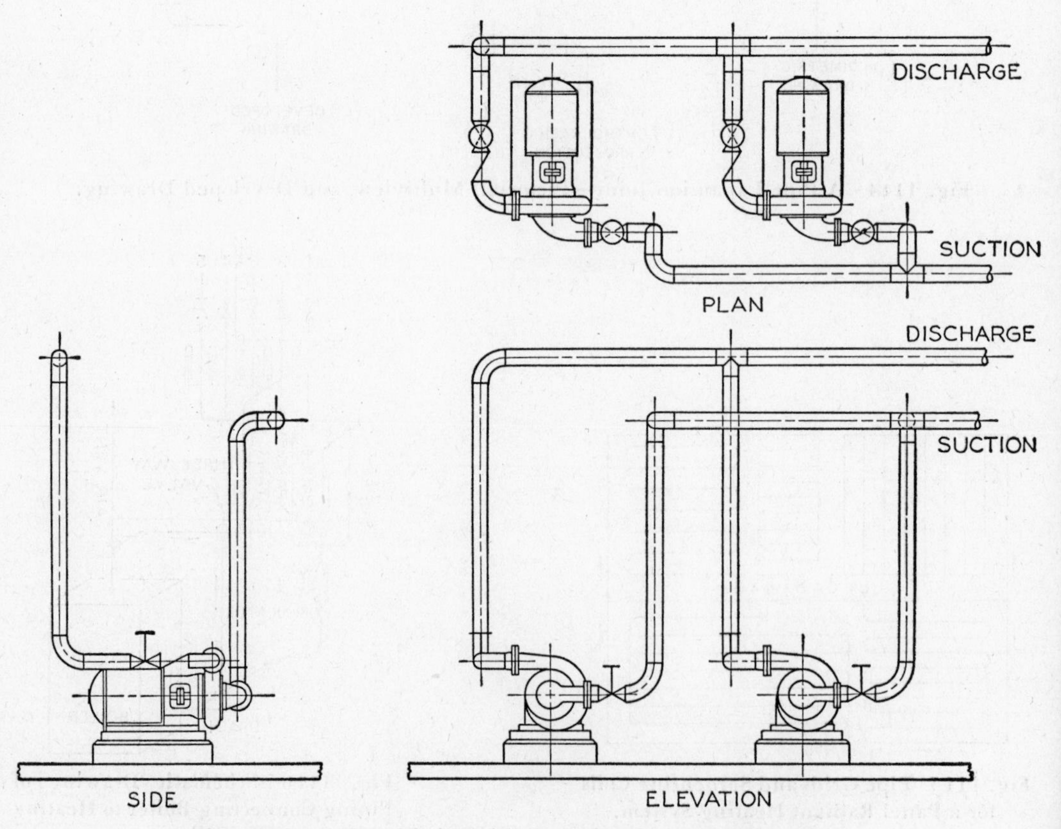

Fig. 1118 A Two-Line Piping Drawing for a Pumping Plant.

In most installations, some pipes are vertical and some are horizontal. If the vertical pipes are assumed to be revolved into the horizontal plane or the horizontal pipes revolved into the vertical plane by turning some of the fittings, Fig. 1114 (c), the entire installation can be shown in one plane. Such a drawing is a *developed piping drawing*.

In complicated systems where a large amount of piping of various sizes is run in close proximity, and where clearances are important, the use of double-line multi-view drawings, made accurately to scale, is desirable. The use of such drawings, which show the relative positions of component parts in all views, greatly reduces the probability of interferences when the piping is erected, and is almost a necessity where piping components are prefabricated in a shop and sent to the job in finished dimensions. Such method of fabrication is universal in large systems using large piping, most piping $2\frac{1}{2}''$ and larger being shop fabricated. See Figs. 1118 and 1119.

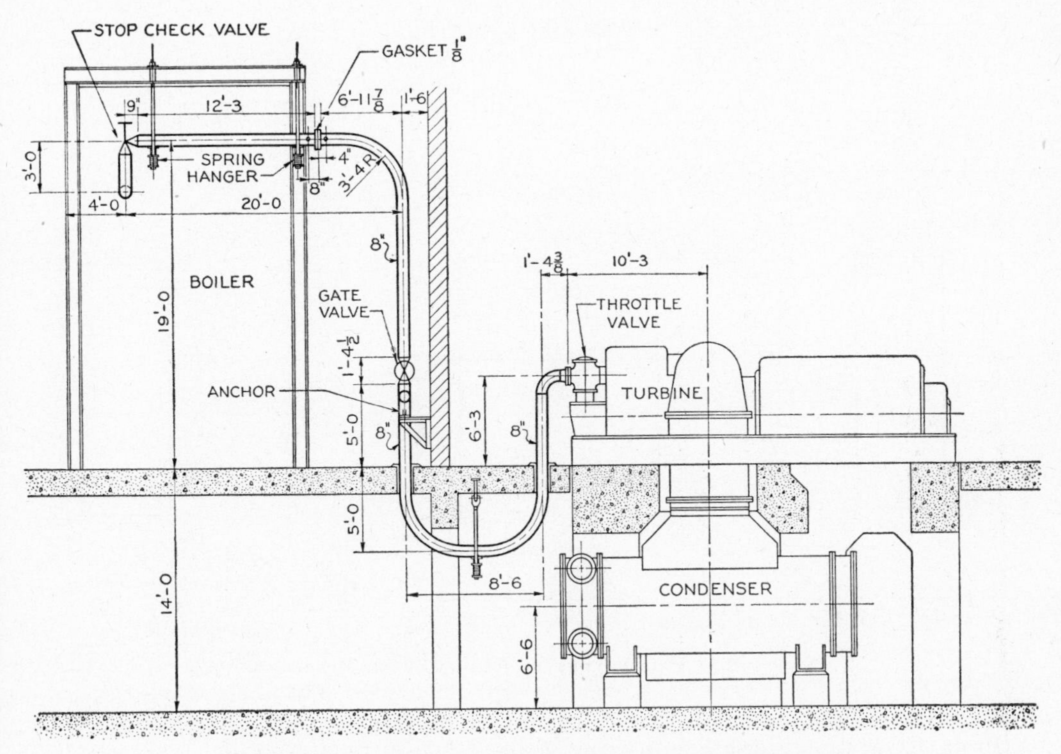

Fig. 1119 A Dimensioned Piping Drawing—Side View.

673. Dimensioning. In dimensioning a piping drawing, distances from center-to-center (c to c), center-to-end (c to e), or end-to-end (e to e) of fittings or valves and the lengths of all straight runs of pipe should be given, Fig. 1119. Fully-dimensioned single-line drawings need not be drawn to scale. Allowances in pipe lengths for make-up in fittings and valves must be made in preparing a bill of materials. All double-line drawings should have center lines if they are to be dimensioned. The size of the pipe for each run is shown by a numeral or by a note at the side of the pipe, with a leader when necessary.

674. Piping Drawing Problems. The drawings for the first six problems are to be three times as large as the corresponding illustrations in the book, unless otherwise specified.

Prob. 1. Make a double-line drawing, similar to Fig. 1113 (a) showing the following fittings: a union, a 45° Y-bend, an eccentric reducer, a globe valve, a tee, a stopcock, and a 45° ell. Use $\frac{1}{2}''$ and 1″ wrought-steel pipe and 125 lb. C.I. screwed fittings.

Prob. 2. Make a single-line drawing, similar to Fig. 1113 (b), showing the following fittings: a 45° ell, a union, a 45° Y-bend, an eccentric reducer, a tee, a reducer, a gate valve, a plug, a cap, and a cross.

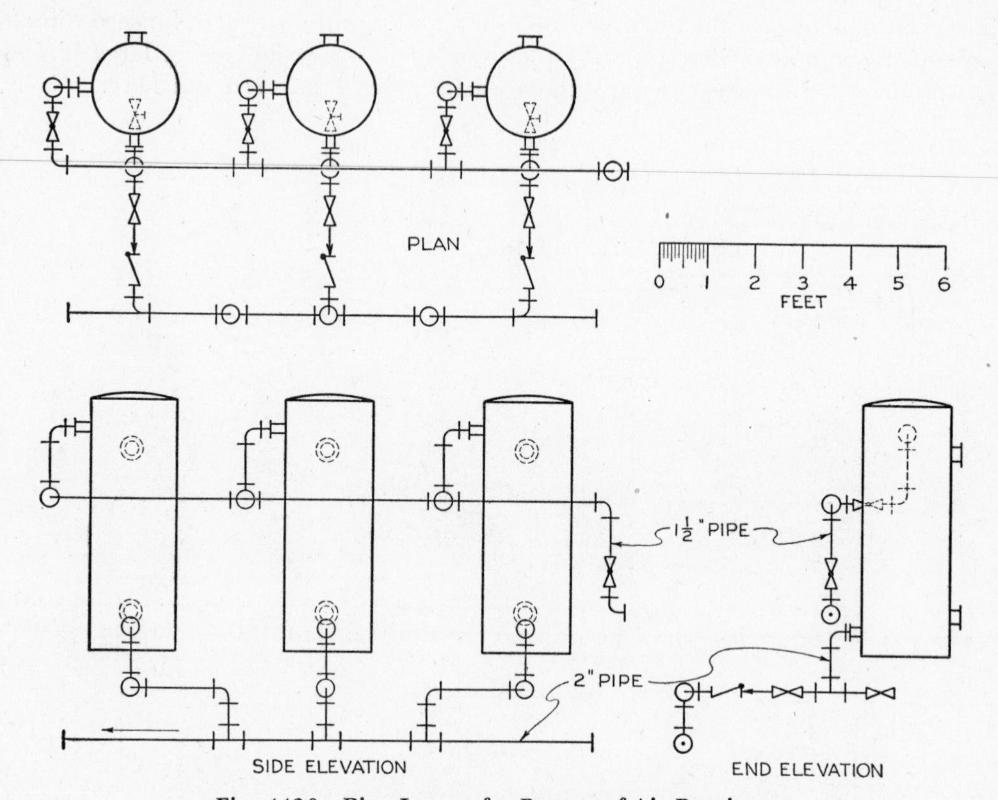

Fig. 1120 Pipe Layout for Battery of Air Receivers.

Prob. 3. Make a single-line drawing of the system of pipe coils and grids shown in Fig. 1115; show, by their respective standard symbols, the elbows and tees that must be used to connect pipes meeting at right angles if welding is not used to make the joints.

Prob. 4. Make an oblique projection, similar to that shown in Fig. 1116, of the one-pipe steam heating system shown in Fig. 1117. Show the pipes by single lines, the fittings by their standard symbols, and the boiler and radiators as parallelepipeds.

Prob. 5. Make a single-line isometric drawing of the piping layout shown in Fig. 1120. Use a scale of $\frac{3}{4}''=1'$-0, and a 17″x22″ sheet.

Prob. 6. Make a single-line multiview drawing of the piping layout shown in Fig. 1121. Use a scale of 1″=1′-0 and a 17″x22″ sheet.

Prob. 7. Make a double-line multiview drawing of the piping layout shown in Fig. 1120, to a scale selected by the student.

Prob. 8. Make a double-line multiview drawing of the piping layout shown in Fig. 1121, to a scale selected by the student. Use Schedule 80 wrought-steel pipe throughout, with Class 250 C.I. flanged fittings where pipe is larger than 2″, and Class 250 C.I. screwed fittings where pipe is 2″ and smaller.

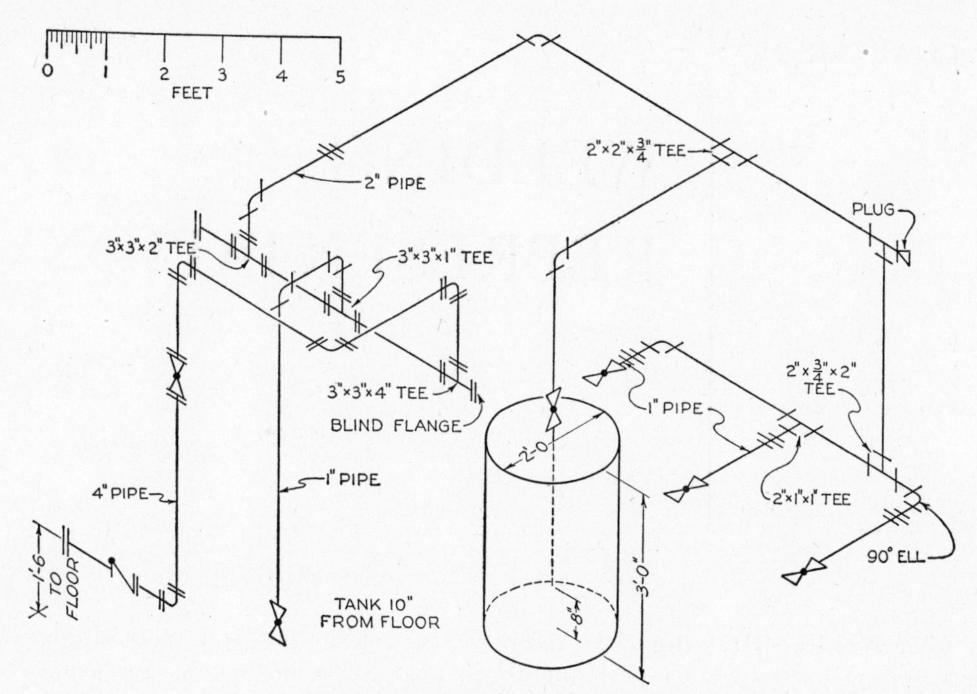

Fig. 1121 Isometric Pipe Layout.

CHAPTER 25

WELDING REPRESENTATION*

675. Welding Drawings. In recent years, welding has been increasingly used for fastening parts together permanently in place of bolts, screws, rivets, or other fasteners. Welding is also being used extensively in fabricating machine parts or other structures that formerly would have been formed by casting or forging, and is used to a considerable extent in the erection of structural steel frames for buildings, ships, and other structures.

Since welding is used so extensively, and for so large a variety of purposes, it is essential to have an accurate method of showing on the working drawings of machines or structures, the exact types, sizes, and locations of welds desired by the designer. The old practice of simply lettering a note on the drawing, "To be welded throughout," or "To be completely welded," which actually shifted responsibility for welding control to the welding shop, is not only dangerous, but may be unnecessarily expensive because shops will usually "play safe" by welding more than necessary.

To provide a means for placing complete welding information on the drawing in a simple manner, a system of welding symbols was developed by the American Welding Society, and published in 1947 under the title "Standard Welding Symbols."† Shortly afterwards, these symbols were adopted by the American Standards Association, and published as ASA Z32.2.1-1949.

A typical welding drawing is shown in Fig. 1122. It is an assembly drawing in the sense that it is composed of a number of separate pieces fastened together as a unit. The welds themselves are not drawn, but are clearly and completely indicated by the welding symbols. The joints are all shown as they would appear before welding. Dimensions are given to show the sizes of the individual pieces to be cut from

*The text matter and most of the illustrations in this chapter are based on ASA Z32.2.1-1949.

†American Welding Society, 33 West 39th St., New York 18, N.Y.

stock. Each component piece is identified by encircled numbers, and by specifications in the parts list, as shown.

676. Welding Processes. Three of the principal methods of welding are the oxyacetylene method, generally known as *gas welding*, the electric-arc method, generally known as *arc welding*, and electric-resistance welding, generally called *resistance welding*.

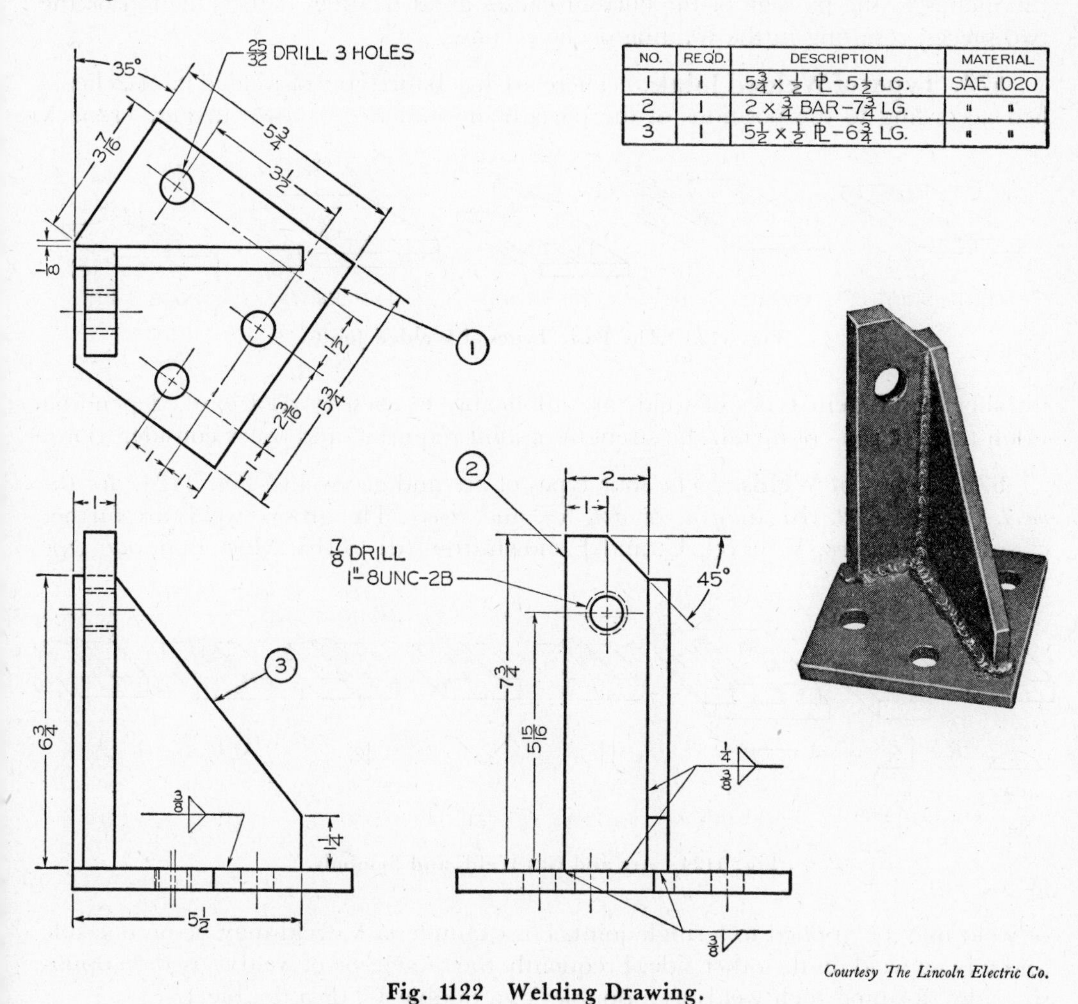

NO.	REQ'D.	DESCRIPTION	MATERIAL
1	1	$5\frac{3}{4} \times \frac{1}{2}$ PL-$5\frac{1}{2}$ LG.	SAE 1020
2	1	$2 \times \frac{3}{4}$ BAR-$7\frac{3}{4}$ LG.	" "
3	1	$5\frac{1}{2} \times \frac{1}{2}$ PL-$6\frac{3}{4}$ LG.	" "

Courtesy The Lincoln Electric Co.

Fig. 1122 Welding Drawing.

Gas welding was originated in 1895, when the French chemist, Le Chatelier, discovered that the combustion of acetylene gas with oxygen produced a flame of very high temperature—high enough to melt metals. This discovery was soon followed by the development of practical methods of producing and transporting oxygen and acetylene, and by the construction of suitable torches and welding rods.

In arc welding, the heat of an electric arc is used to fuse the metals that are to be welded or cut. The first arc welding was done in 1881 by De Meritens in France, and for a number of years the development of arc welding was very slow. During World War I, the U.S. Navy used welding to a limited extent for repairing machinery.

Since that time, as the result of extensive research, basic improvements have been made in the manufacture of electrodes and in the mechanical equipment used for welding, so that now arc and gas welding are among the most important construction processes in industry.

In resistance welding, two pieces of metal are held together under some pressure, and a large amount of electric current is passed through the parts. The resistance of the metals to the passage of the current causes great heating at the junction of the two pieces, resulting in the welding of the metals.

677. Types of Welded Joints. There are five basic types of welded joints, classified according to the positions of the parts being joined, as shown in Fig. 1123. A

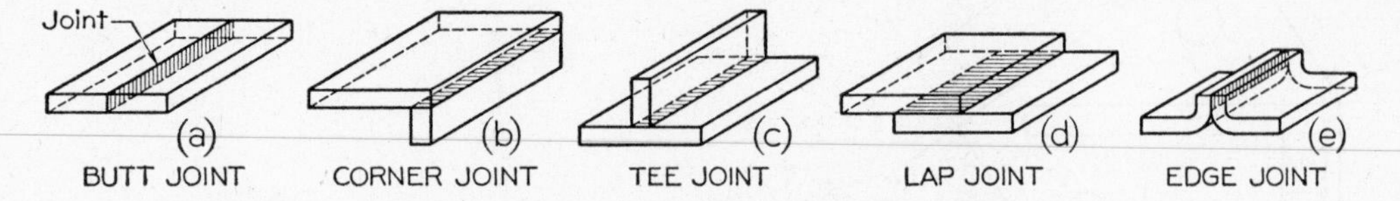

Fig. 1123 The Basic Types of Welded Joints.

number of different types of welds are applicable to each type of joint, depending upon the thickness of metal, the strength of joint required, and other considerations.

678. Types of Welds. The four types of arc and gas welds, Fig. 1124, are the *bead*, (a), the *fillet*, (b), the *plug* or *slot*, (c), and *groove*. The groove welds are further classified as square, V, bevel, U, and J, shown from (d) to (h). More than one type

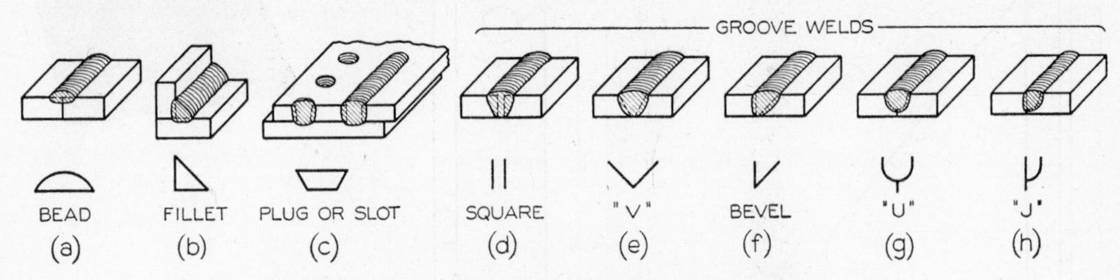

Fig. 1124 Arc and Gas Welds and Symbols.

of weld may be applied to a single joint. For example, a V weld may be on one side and a bead weld on the other side. Frequently the same type of weld is used on opposite sides, forming such welds as a double-V, a double-U, or a double-J.

The four basic resistance welds are the *spot weld*, *projection weld*, *seam weld*, and *flash* or *upset weld*. The corresponding weld symbols are shown in Fig. 1125. Supplementary symbols are shown in Fig. 1126.

TYPE OF WELD			
SPOT	PROJECTION	SEAM	FLASH OR UPSET
✳	X	XXX	I

Fig. 1125 Resistance Weld Symbols.

WELD ALL AROUND	FIELD WELD	FLUSH CONTOUR	CONVEX CONTOUR
○	●	—	⌒

Fig. 1126 Supplementary Symbols.

679. Welding Symbols. The basic element of the symbol is the "bent" arrow, Fig. 1127 (a). The arrow points to the joint where the weld is to be made, (b), and attached to the reference line, or shank, of the arrow is the weld symbol for the desired weld. The symbol would be one of those shown in Figs. 1124 or 1125. In this case a fillet weld symbol is shown.

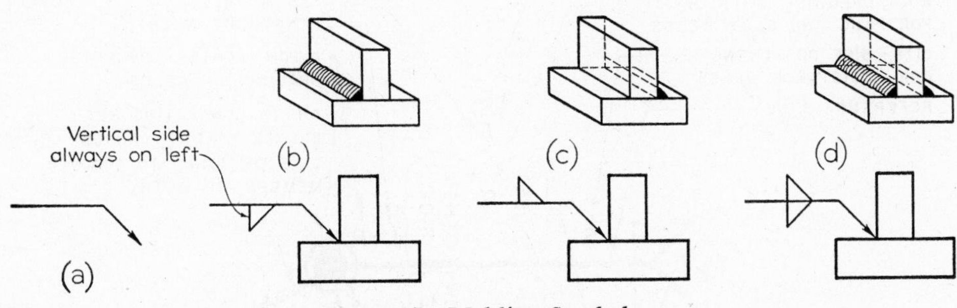

Fig. 1127 Welding Symbols.

The weld symbol is placed below the reference line if the weld is to be on the *arrow side* of the joint as at (b), or above the reference line if the weld is to be on the *other side* of the joint as at (c). If the weld is to be on both the *arrow side* and the *other side* of the joint, weld symbols are placed on both sides of the reference line, (d). This rule for placement of the weld symbol is followed for all the arc or gas weld symbols.

When a joint is shown by a single line on a drawing, as in the top and side views of Fig. 1128 (b), the *arrow side* of the joint is regarded as the "near" side to the reader of the drawing, according to the usual conventions of drafting.

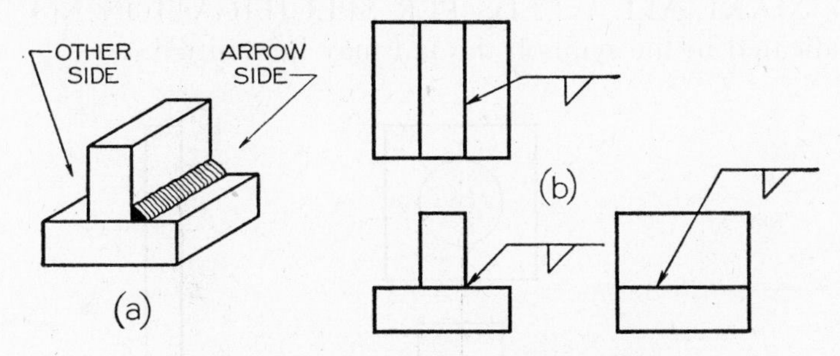

Fig. 1128 Arrow-Side and Other-Side.

In plug, slot, seam, and projection welding symbols, the arrow points to the outer surface of one of the members at the center line of the weld. In such cases, the arrow-side of the joint is the one touched by the arrow, or the "near" side to the reader. See Figs. 1140 and 1144.

Note that for all fillet or groove symbols, the *perpendicular side of the symbol is always drawn on the left*, Fig. 1127 (b).

For best results, the welding symbols should be drawn mechanically, but in certain cases where necessary, the symbols may be drawn freehand.

The complete welding symbol, enlarged, is shown in Fig. 1129. In practice, some companies will need to use only a simple symbol composed of the minimum elements, the arrow and the weld symbol; others will make use of additional components.

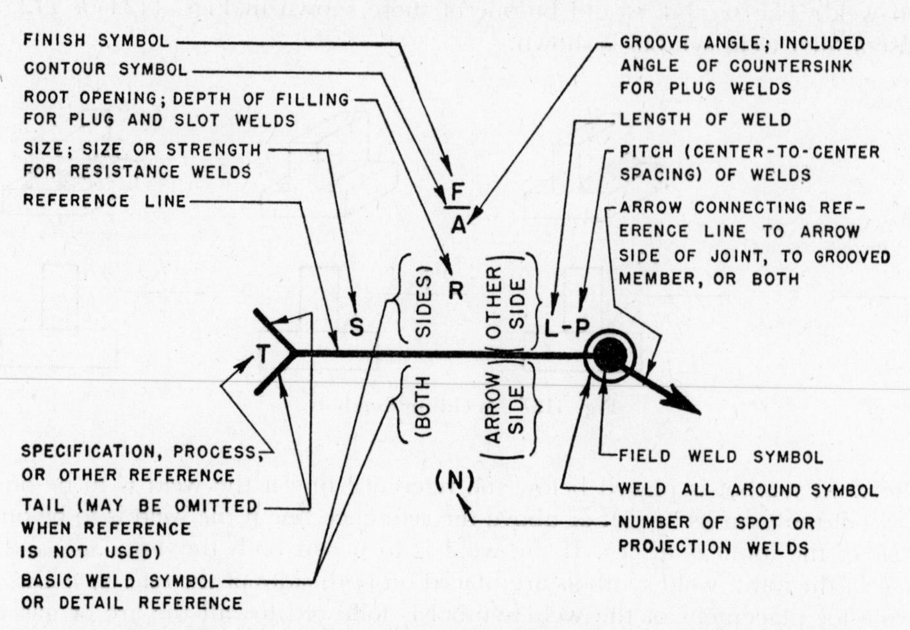

Fig. 1129 The Standard Locations of the Elements of a Welding Symbol.

Reference to a specification, process, or other supplementary information is indicated by any desired symbol in the tail of the arrow, Fig. 1130 (a). Otherwise, a general note may be placed on the drawing, such as UNLESS OTHERWISE INDICATED, MAKE ALL WELDS PER SPECIFICATION NO. XXX. If no reference is indicated in the symbol, the tail may be omitted.

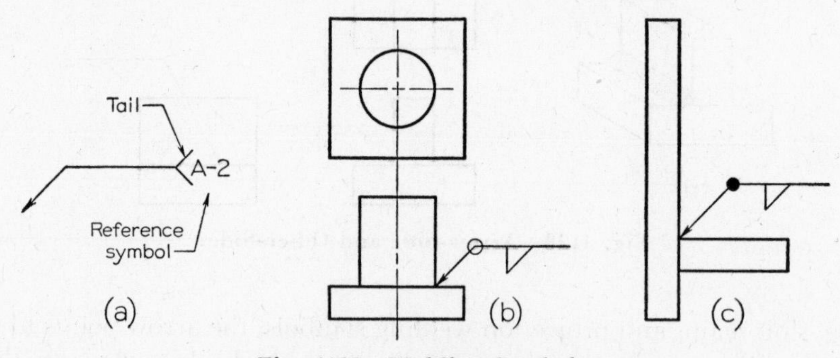

Fig. 1130 Welding Symbols.

To avoid repeating the same information on many welding symbols on a drawing, general notes may be used, such as FILLET WELDS $\frac{5}{16}''$ UNLESS OTHERWISE INDICATED, or ROOT OPENINGS FOR ALL GROOVE WELDS $\frac{3}{16}''$ UNLESS OTHERWISE INDICATED.

Welds extending completely around a joint are indicated by an open circle around the elbow of the arrow, as shown at (b). *When the weld-all-around symbol is not used, the welding symbol is understood to apply between abrupt changes in direction of the weld, unless otherwise shown.* A solid round dot at the elbow of the arrow indicates a weld to be made "in the field" (on the site) rather than in the fabrication shop, as shown at (c).

Spot, seam, flash, or upset symbols do not have *arrow-side* or *other-side* significance, and are simply centered on the reference line of the arrow, as shown in Fig. 1131 (a) to (c). Spot and seam weld symbols may be shown directly on the drawing at the desired locations, as shown at (d) and (e).

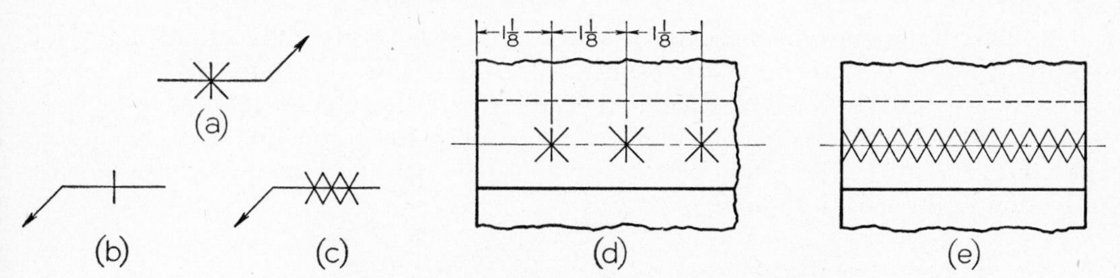

Fig. 1131 Spot, Seam, and Flash Weld Symbols.

For bevel- or J-groove welds, the arrow should point with a definite change of direction, or break, *toward the member* that is to be beveled or grooved, as shown in Fig. 1132 (a) and (b). In this case, the upper member is grooved. The break is omitted if the location of the bevel or groove is obvious.

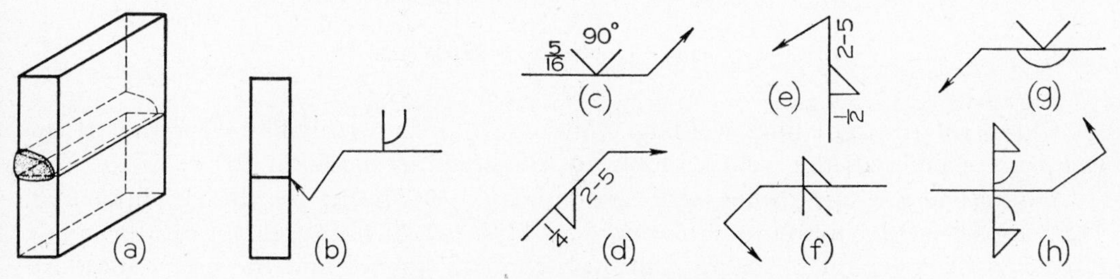

Fig. 1132 Welding Symbols.

Lettering in symbols should be placed to read from the bottom or from the right side of the drawing in accordance with the aligned system, as shown at (c) to (e).

When a joint has more than one weld, the combined symbols are shown, as at (f) to (h).

680. Fillet Welds. The usual fillet weld has equal sides, Fig. 1133 (a). The size of the weld is the length of one side, and is indicated, (b), by a dimension figure at

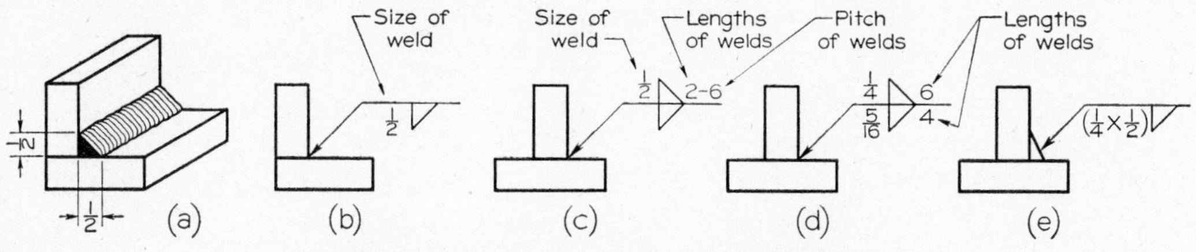

Fig. 1133 Dimensioning of Fillet Welds.

the left of the weld symbol. For fillet welds on both sides of a joint, the dimensions may be indicated on one side or both sides of the reference line, (c). The lengths of the welds, and the pitch (c to c spacing of welds) are indicated as shown. When the welds on opposite sides are different in size, the sizes are given as shown at (d). If a fillet weld has unequal sides, the weld orientation is shown on the drawing, if necessary, and the lengths of the sides given in parenthesis to the left of the weld symbol, as at (e). If a general note is given on the drawing, such as ALL FILLET WELDS $\frac{5}{16}''$ UNLESS OTHERWISE NOTED, the size dimensions are omitted from the symbols.

No length dimension is needed for a weld that extends the full distance between abrupt changes of direction. For each abrupt change in direction, an additional arrow is added to the symbol, except when the weld-all-around symbol is used.

Lengths of fillet welds may be indicated by symbols in conjunction with dimension lines, Fig. 1134 (a). The extent of fillet welding may be shown graphically by means of section lining, (b), if desired.

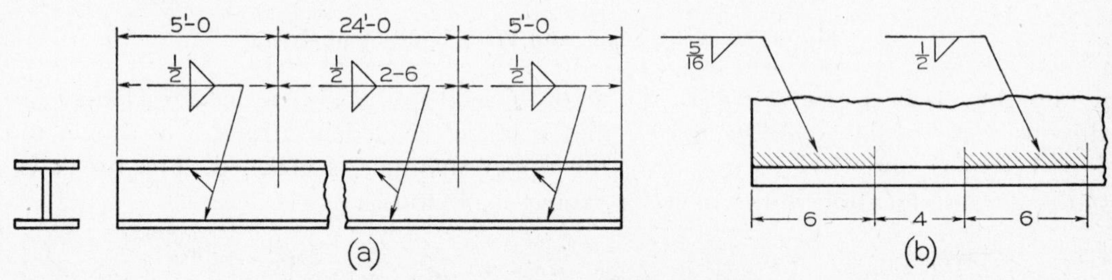

Fig. 1134 Lengths of Fillet Welds.

Chain intermittent fillet welding is indicated as shown in Fig. 1135 (a). If the welds are staggered, the weld symbols are staggered as shown at (b).

Unfinished flat-faced fillet welds are indicated by adding the flush symbol, Fig. 1126, to the weld symbol, as shown in Fig. 1136 (a). If fillet welds are to be made flat-faced by mechanical means, the flush-contour symbol and the user's standard

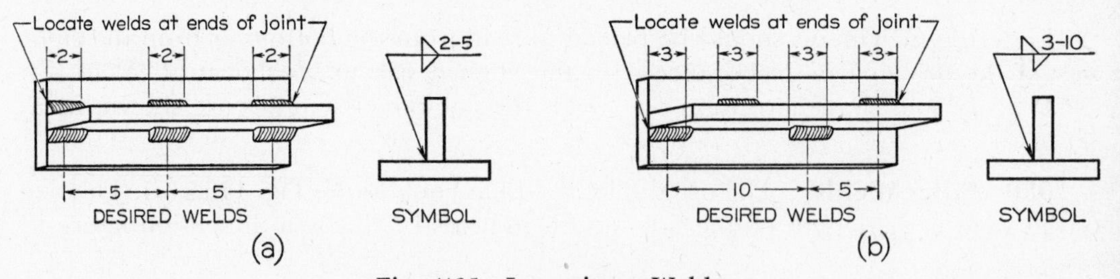

Fig. 1135 Intermittent Welds.

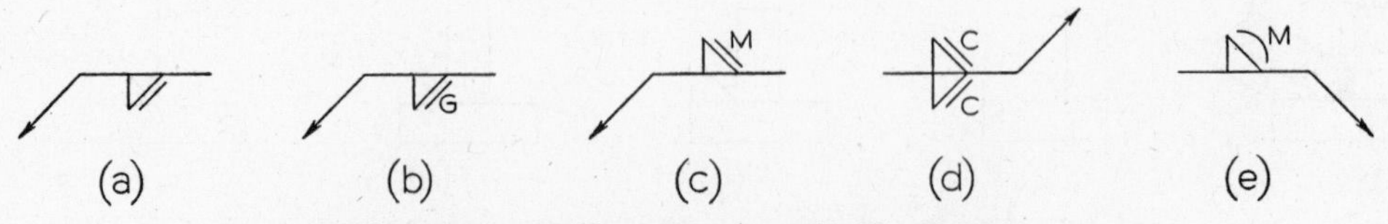

Fig. 1136 Surface Contour of Fillet Welds.

finish symbol are added to the weld symbol, as shown at (b) to (d). These finish symbols indicate the method of finishing (C = chipping, G = grinding, M = machining), and not the degree of finish. If fillet welds are to be finished to a convex contour, the convex-contour symbol is added, together with the finish symbol, as shown at (e).

681. Groove Welds. In Fig. 1137, various groove welds are shown above, and the corresponding symbolic representations below. The sizes of the groove welds (depth of the V, bevel, U, or J) are indicated at the left of the weld symbol. For example, at (a) the size of the V-weld is $\frac{1}{2}''$, at (b) the sizes are $\frac{1}{4}''$ and $\frac{7}{8}''$, at (c) the

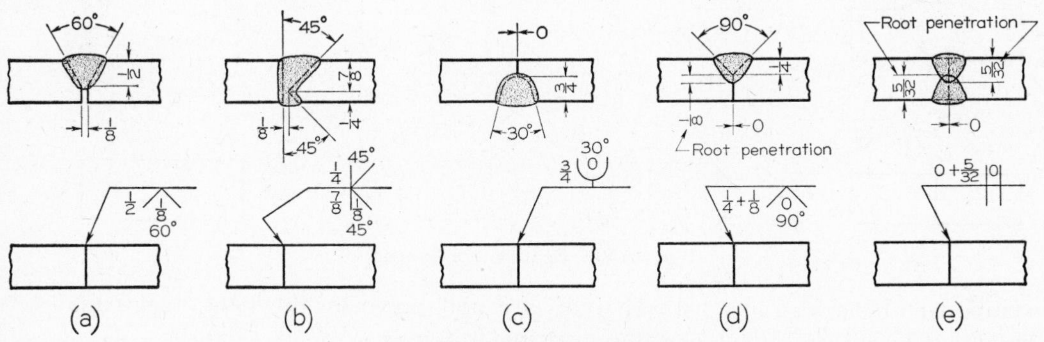

Fig. 1137 Groove Welds.

size is $\frac{3}{4}''$, and at (d) the size is $\frac{1}{4}''$. In this symbol at (d), the size is followed by $\frac{1}{8}''$, which is the additional "root penetration" of the weld. At (e), the root penetration is $\frac{5}{32}''$ from zero, or from the outside of the members. Note the overlap of the root penetration in this case.

The root opening or space between members, when not covered by a company standard, is shown inside the weld symbol. At (a) and (b), the root openings are $\frac{1}{8}''$. At (c) to (e) the openings are zero.

The groove angles, when not covered by a company standard, are shown just outside the openings of the weld symbols, as shown at (a) to (d).

A general note may be used on the drawing to avoid repetition on the symbols, such as ALL V-GROOVE WELDS TO HAVE 60° GROOVE ANGLE UNLESS OTHERWISE SHOWN. However, when the dimensions of one or both of two opposite welds differ from the general note, both welds should be completely dimensioned.

When single-groove or symmetrical double-groove welds extend completely through, the size need not be given in the welding symbol. For example, in Fig. 1137 (a), if the V-groove extended entirely through the joint, the depth or size would be simply the thickness of the stock, and would not need to be given in the welding symbol.

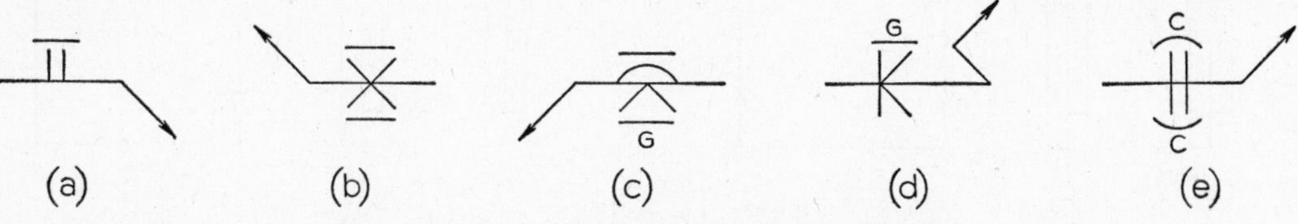

Fig. 1138 Surface Contour of Groove Welds.

When groove welds are to be approximately flush without finishing, the flush-contour symbol, Fig. 1126, is added to the weld symbols as shown in Fig. 1138 (a) and (b). If the welds are to be machined, the flush-contour symbol and the user's standard finish symbol are added to the weld symbol as shown at (c) and (d). These finish symbols indicate the method of finishing (C = chipping, G = grinding, M = machining), and not the degree of finish. If a groove weld is to be finished with a convex contour, the convex-contour and finish symbols are added, as at (e).

682. Bead Welds. Bead welds used as back or backing welds on single-groove welds are indicated, Fig. 1139 (a), by a single bead symbol opposite the groove weld

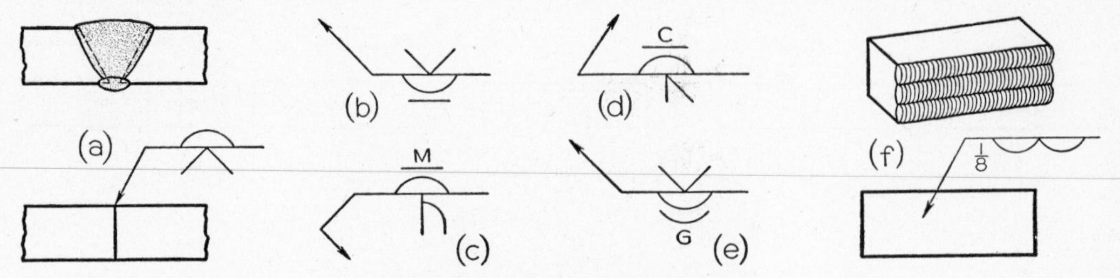

Fig. 1139 Bead Weld Symbols.

symbol. Dimensions of such bead welds are not shown on the symbol, but may be shown, if necessary, directly on the drawing.

When back or backing welds are to be approximately flush without machining, the flush-contour symbol is added to the weld symbols, as shown at (b). If they are to be machined, the user's finish symbol is added, (c) and (d). If the welds are to be finished with a convex contour, the convex-contour symbol and the finish symbol are added to the weld symbol, as at (e).

The dual-bead symbol is used to indicate a surface to be built up by welding, whether by single- or multiple-pass bead welds, as shown at (f). Since this symbol does not indicate a welded joint, there is no arrow-side or other-side significance; hence the symbol is always drawn below the reference line. The minimum height of the weld deposit is indicated at the left of the weld symbol, as shown at (f), except where no specific height is required. When a specific area of a surface is to be built up, the dimensions of the area are given directly on the drawing.

683. Plug and Slot Welds. As shown in Fig. 1124 (c), the same symbol is used for either plug welds or slot welds. A hole or a slot is made in one member to receive

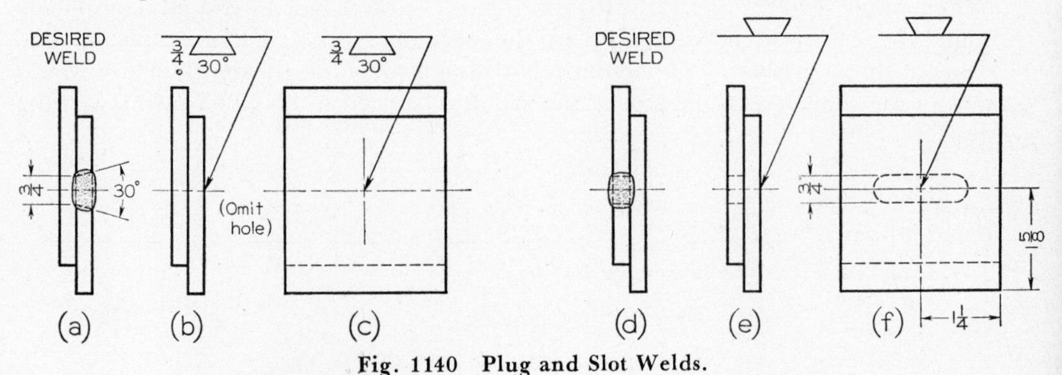

Fig. 1140 Plug and Slot Welds.

the weld, as shown in Fig. 1140 (a) and (d). If the hole or slot is in the arrow-side member, the weld symbol is placed below the reference line, (b) and (c); if in the other-side member, the weld symbol is placed above the line, (e) and (f).

The size of a plug weld, which is the smallest diameter of the hole if countersunk, is placed at the left of the weld symbol, as shown at (b). If the included angle of countersink of plug welds is in accordance with the user's standard, it is omitted from the welding symbol; otherwise it is indicated adjacent to the weld symbol, as shown at (b).

A plug weld is understood to fill the depth of the hole unless the depth of plug is indicated by a number, in inches, inside the weld symbol, Fig. 1141 (a). The pitch

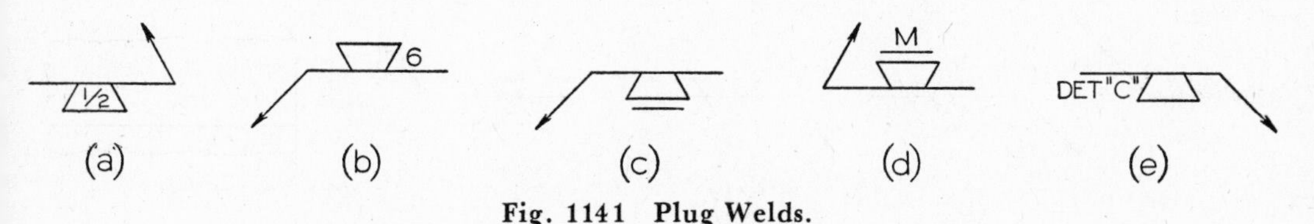

Fig. 1141 Plug Welds.

(center-to-center spacing) of plug welds is shown in inches at the right of the weld symbol, as shown at (b). If the weld is to be approximately flush without finishing, the flush-contour symbol is added, (c). If the weld is to be made flush by mechanical means, a finish symbol is added, as at (d). The flush-contour and finish symbols are used in the same manner for slot welds and for plug welds.

The depth of filling of slot welds is indicated in the same manner as for plug welds, (a). The size and location dimensions of slot welds cannot be shown on the welding symbol, and must be shown directly on the drawing, Fig. 1040 (f), or by a detail with a reference to it on the welding symbol, as shown in Fig. 1041 (e).

684. Spot Welds. Spot weld symbols may be shown directly in place on the drawing, Fig. 1131 (d), or by means of the welding symbol, Fig. 1142. The weld sym-

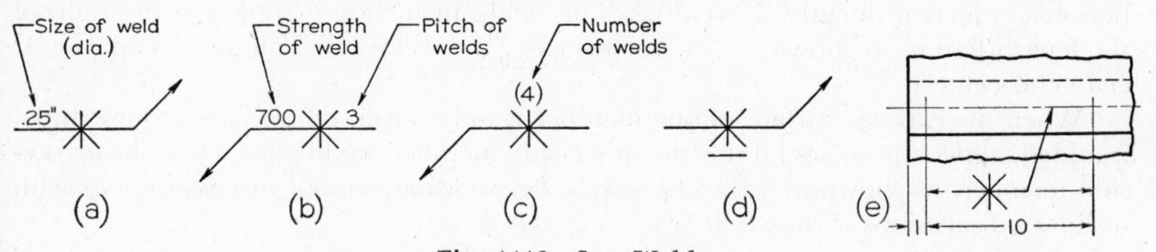

Fig. 1142 Spot Welds.

bol is centered on the reference line, since there is no arrow-side or other-side significance.

The size of a spot weld is its diameter. This value, expressed in hundredths of an inch, may be shown, with inch marks, at the left of the weld symbol on either side of the reference line, as shown at (a). If it is desired to indicate the minimum acceptable shear strength in pounds per spot, instead of the size of the weld, this value is placed at the left of the weld symbol, as shown at (b). The pitch (center-to-center spacing) is indicated, in inches, at the right of the weld symbol, (b). In this case the spot welds are 3″ apart. If a joint requires a certain number of spot welds, the number is given in parenthesis above or below the symbol, as at (c). If the exposed surface of one member is to be flush, the flush-contour symbol is added, above the symbol if it is the other-side member, and below if it is the arrow-side member, as shown at (d). The use of the welding symbol in conjunction with ordinary dimensions is shown at (e). When spot weld symbols are shown directly on the drawing, the spacing is shown by dimensions, as in Fig. 1131 (d).

685. Seam Welds. Seam weld symbols may be shown directly in place on the drawing, Fig. 1131 (e), or by means of the welding symbol, Fig. 1143. The weld sym-

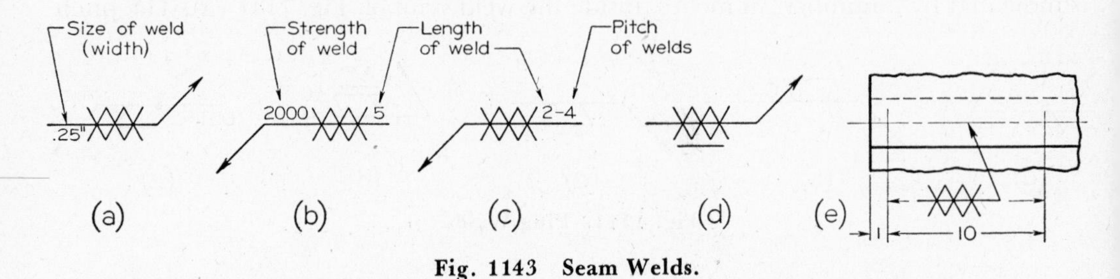

Fig. 1143 Seam Welds.

bol is centered on the reference line, since there is no arrow-side or other-side significance.

The size of a seam weld is its width. This value, expressed in hundredths of an inch, may be shown, with inch marks, at the left of the weld symbol, on either side of the reference line, as shown at (a). If it is desired to indicate the minimum acceptable shear strength in pounds per linear inch, instead of the size of the weld, this value is placed at the left of the weld symbol, as shown at (b). The length of a seam weld may be shown, in inches, at the right of the weld symbol, (b). In this case, the seam weld is 5″ long. If the weld extends the full distance between abrupt changes of direction, no length dimension in the symbol is given.

The pitch (center-to-center spacing) of intermittent seam welding is the distance between centers of lengths of welding. The pitch, in inches, is shown at the right of the length figure, as shown at (c). In this case, the welds are 2″ long and spaced 4″ center-to-center.

When the exposed surface of one member is to be flush, the flush-contour symbol is added, above the symbol if it is the other-side member, and below if it is the arrow-side member, as shown at (d). The use of the welding symbol in conjunction with ordinary dimensions is shown at (e).

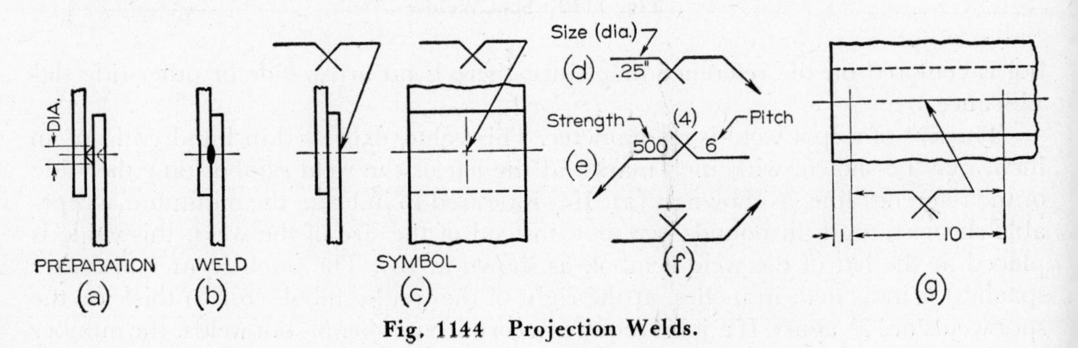

Fig. 1144 Projection Welds.

686. Projection Welds. In projection welding, one member is embossed in preparation for the weld, as shown in Fig. 1144 (a). When welded, the joint appears in section as at (b). The weld symbols, in this case, are placed below the reference lines, as shown at (c), to indicate that the arrow-side member is the one that is embossed. The weld symbols would be placed above the lines if the other member were embossed.

Projection welds are dimensioned by either size or strength. The size is the diameter of the weld in hundredths of an inch. This value is shown, with inch marks, to the left of the weld symbol, as shown at (d). If it is desired to indicate the minimum acceptable shear strength in pounds per weld, the value is placed at the left of the weld symbol, as at (e). The pitch (center-to-center spacing) is indicated, in inches, at the right of the weld symbol, as at (e). In this case, the welds are spaced 6″ apart. If the joint requires a definite number of welds, the number is given in parenthesis, as shown at (e). If the exposed surface of one member is to be flush, the flush-contour symbol is added, as shown at (f). The use of the welding symbol in conjunction with ordinary dimensions is shown at (g).

687. Flash and Upset Welds. Flash and upset weld symbols have no arrow-side or other-side significance, but the supplementary symbols do. A flash-welded joint is shown in Fig. 1145 (a), and an upset welded joint at (b). The joint after

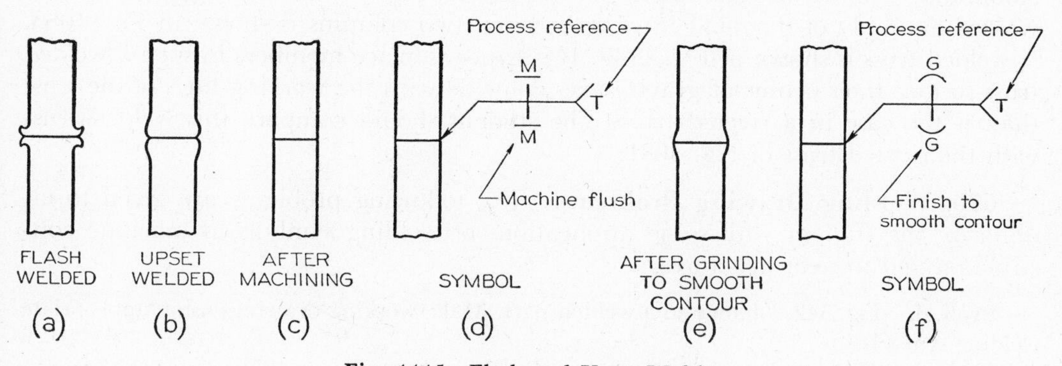

Fig. 1145 Flash and Upset Welds.

machining flush is shown at (c). The complete symbol at (d) includes the weld symbol together with the flush-contour and machining symbols.

If the joint is ground to smooth contours, (e), the resulting welding drawing and symbol would be constructed as shown at (f), which includes convex-contour and grind symbols. At either (d) or (f), the joint may be finished on only one side, if desired, by indicating the contour and machining symbols on the appropriate side of the reference line. The dimensions of flash and upset welds are not shown on the welding symbol. Note that the process reference must be placed in the tail of the symbol.

688. Welding Templates. To simplify the drawing of welding symbols, and to speed up the drafting, welding templates are available. The template shown in Fig. 1146 has all the forms needed for drawing the arrow, the weld symbols, and the supplementary symbols, as well as an illustration of the complete composite welding symbol for quick reference.

The symbols may be drawn in pencil or in ink. For the latter, the Leroy Pen is recommended, Fig. 207 (d).

689. Welding Applications. A typical example of welding fabrication for machine parts is shown in Fig. 1122. In many cases, especially when only one or

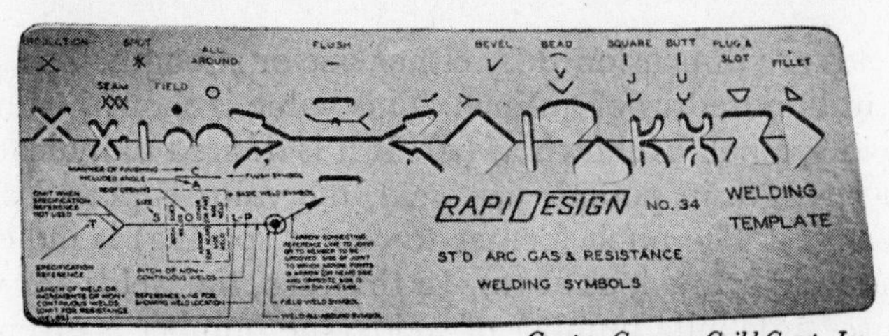

Courtesy Gramercy Guild Group, Inc.

Fig. 1146 Welding Template.

only a few identical parts are required, it is cheaper to produce by welding than to make patterns, sand castings, and do the necessary machining. Thus, welding is particularly adaptable to custom-built constructions.

Welding is also suitable for large structures that are difficult or impossible to fabricate entirely in the shop, and is coming into greater use for steel structures, such as building frames, bridges, ships, etc. A welded beam is shown in Fig. 1082, and a welded assembly of diagonal bracing between two columns is shown in Fig. 1083. A welded truss is shown in Fig. 1147. It is easier to place members in such a welded truss so that their center-of-gravity axes coincide with the working lines of the truss than is the case in a riveted truss. The student should compare this welded truss with the riveted truss of Fig. 1081.

690. Welding Drawing Problems. The following problems are given to familiarize the student with some applications of welding symbols to machine construction and to steel structures.

Prob. 1. Fig. 342. Change to a welded part. Make working drawing, using appropriate welding symbols.

Prob. 2. Fig. 345. Same instructions as for Prob. 1.

Prob. 3. Fig. 350. Same instructions as for Prob. 1.

Prob. 4. Fig. 361. Same instructions as for Prob. 1.

Prob. 5. Fig. 368. Same instructions as for Prob. 1.

Prob. 6. Fig. 380. Same instructions as for Prob. 1.

Prob. 7. Fig. 389. Same instructions as for Prob. 1.

Prob. 8. Fig. 393. Same instructions as for Prob. 1.

Prob. 9. Fig. 398. Same instructions as for Prob. 1.

Prob. 10. Make a half-size drawing of the joint at the center of the lower chord of the truss in Fig. 1147 where the chord is supported by two vertical angles. The chord is a structural tee, cut from an $8 \times 5\frac{1}{4} - 17$ lb. wide-flange shape. Draw the front and side views, and show the working lines, the two angles, the structural tee, and all welding symbols.

Prob. 11. Make a half-size front view, showing the welding symbols, of any joint of the truss in Fig. 1147 in which three or four members meet.

Prob. 12. Draw half-size front, top, and left-side views of the end joint of the truss in Fig. 1147, showing the welding symbols.

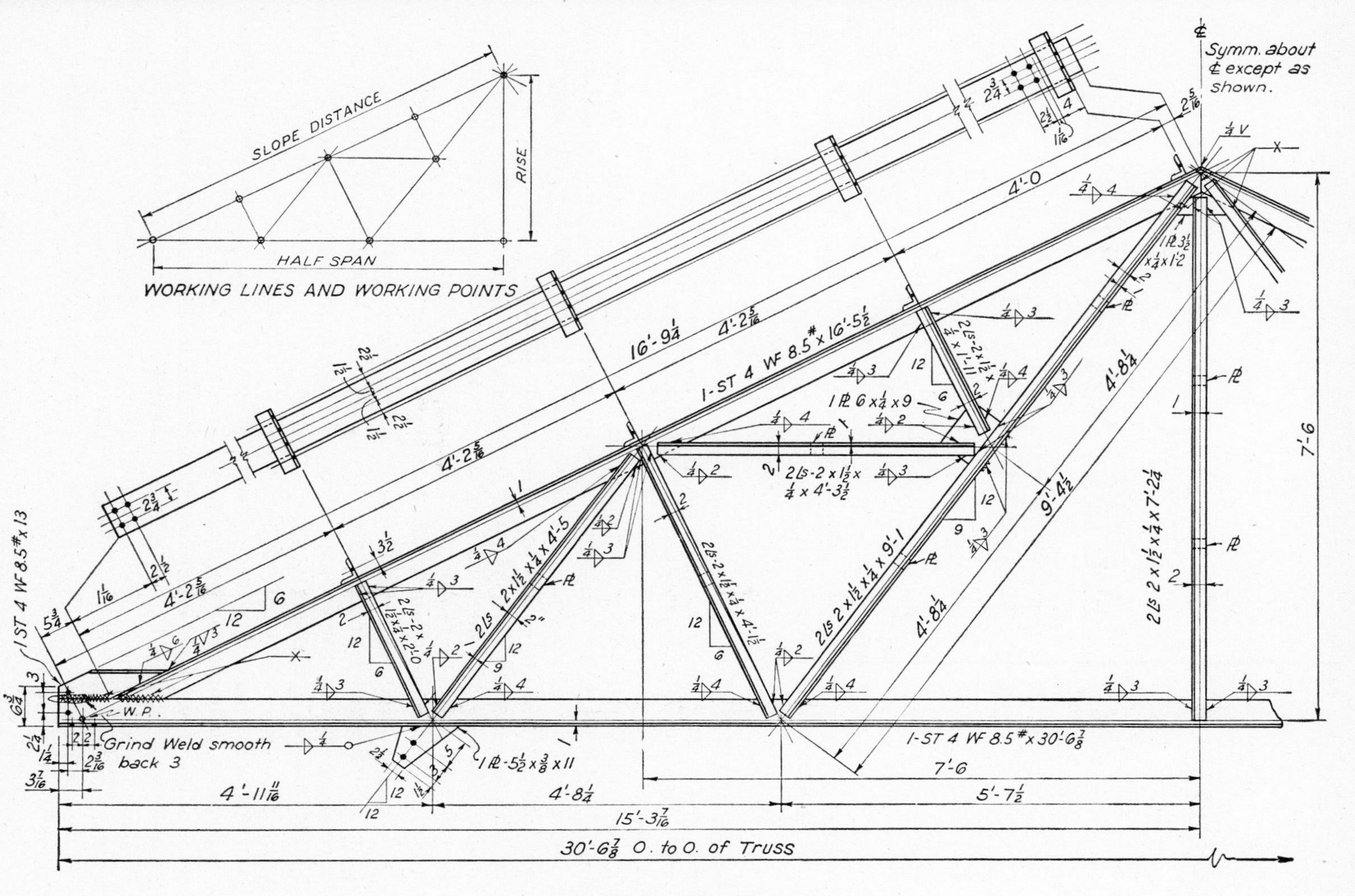

Fig. 1147 A Welded Truss.

CHAPTER 26

AERONAUTICAL DRAFTING

by
*William N. Wright**

691. Airplane Development. The aircraft industry is composed of a large number of professions and skills, each of which is integrated as nearly intact as possible into a single major operation which has become one of the most strategic of modern developments. Of course, the aircraft industry as a whole includes many supporting industries from which a vast amount of raw materials and equipment is drawn. Aircraft drafting is therefore a composite of mechanical, electrical, sheet-metal, and structural drafting in which deviations are made from each in order to establish uniformity in a single basic system. Drafting systems of individual aircraft firms, while distinct in themselves, must also serve the supporting industries, and are thus built upon the basic principles of drafting as set forth in earlier chapters.

An example of the magnitude of the aircraft industry is afforded in the development of a single heavy bomber during World War II. The combined efforts of over 2000 engineers and draftsmen, expending 3,000,000 man-hours in the preparation of 17,000 drawings containing designs for 90,000 different parts, were required to put this one airplane into production on a scale sufficient for the needs of our armed forces. This effort required not only accurate and workable designs but also a systematic accounting for the use of each part.

Present-day heavy bombers are much larger and far more complicated. A Boeing Stratofortress is shown in Fig. 1148. A distinctive feature of aeronautical drafting is the planning of logical assembly and installation sequences, as well as designing for maximum strength and minimum weight. The "exploded assembly" in Fig. 1149 shows the production breakdown for the Boeing Stratofreighter, and illustrates the need for careful planning.

*Aeronautical Engineer, Boeing Airplane Co.

Courtesy Boeing Airplane Co.

Fig. 1148 Stratofortress on Flight Test.

Now, with the frontiers of knowledge being pushed back by discoveries and developments in the field of physical science, a vast new area lies ahead, and challenges our ingenuity to exploit it for man's use. The aircraft industry is adapting itself to this challenge. It is to be expected that the transition from a subsonic to a supersonic environment will involve considerable changes in many aspects of present activities, but the technical drawing will still be the base of operation. It must remain sufficiently flexible in application to serve an expanding and diversified industry. This chapter contains some examples of present aeronautical drafting practice which permit coordination of effort, not only between engineering and manufacturing departments, but among different companies engaged in building the same model of airplane under a comprehensive production program. The resulting need for interchangeability of parts and assemblies has led to the development of many special types of gages, tools, jigs, and fixtures, and requires the engineer to be increasingly vigilant in planning for production as well as for performance. He must be careful to allow for the tooling phase of his design in the determination of dimensional tolerances, since only a portion of the total tolerance is available for the tools, the remainder being for the shops. Designers in some companies are responsible for both airplane and tool design because of their similarity, but in other companies the two types of design are kept separate because of their differences.

692. Requisites for an Aeronautical Engineer. The most important requisites for aeronautical drafting are accuracy and attention to detail. Lack of care in either will result in expensive revisions, waste of time and material in the shop and, in many cases, serious consequences during actual operation of the aircraft.

A good engineer is one who can design and complete the drawing of a part to meet shop production conditions, a drawing which is well thought out and properly part-numbered. The task of keeping part numbers and dimensions in order on related drawings, and seeing that all parts will properly match is one requiring patience and ability to handle details.

Less than ten per cent of the total engineering work in airplane design involves the use of higher mathematics, performance calculations, model testing, etc.; the remainder is careful attention to basic drafting practice in detail design. *A good aircraft*

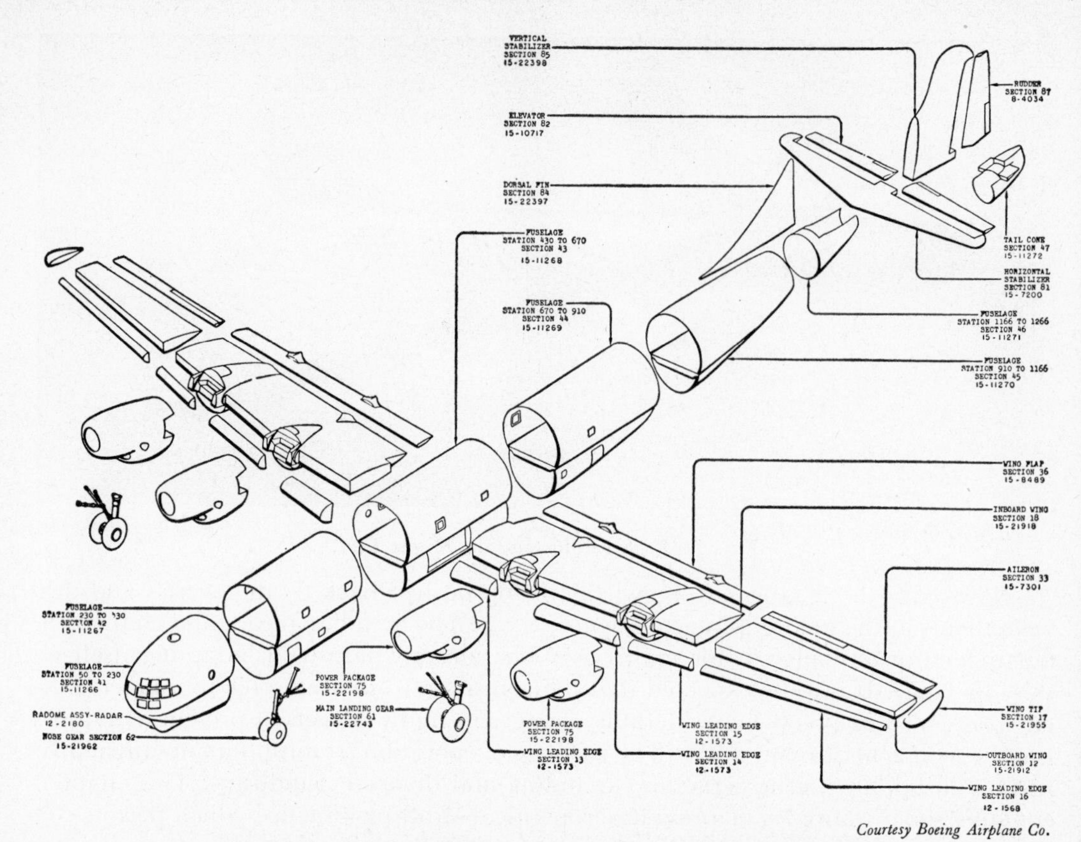

Courtesy Boeing Airplane Co.

Fig. 1149 Structural Breakdown of Boeing Stratofreighter.

engineer must know projections thoroughly and will find that a thorough knowledge of descriptive geometry is even more useful than the calculus. Neatness in drafting and lettering, plus the ability to make pencil drawings whose lines are dark enough to print sharply, are much in demand.

693. Drafting Equipment. Practically all aircraft drafting is done in pencil. A draftsman needs one small and one large sturdy bow compass, §43, an assortment of triangles, decimal and fractional scales, §35, French curves, and ship curves, §65. Ink work is required at plants where U.S. Navy contracts are on hand, since inked cloth tracings are specified for preparation of Van Dyke prints. However, the use of pencil cloth is growing, and may supplant the tedious effort of ink work. Drafting machines, §68, have been adopted by many large plants.

694. Dimensioning. In general, the principles of dimensioning outlined in Chapters 11 and 12 apply to aircraft drawings. With a few exceptions, dimensions are given in inches instead of in feet and inches. Practice differs in various factories as to the use of the fractional or decimal system of dimensioning, but the trend is toward adoption of the complete decimal system. The inch symbol is always omitted.

Many aircraft standards include the use of the unidirectional system of dimensioning, as shown in Fig. 589 (b), because many drawings are large and are difficult to read from the side.

695. Lettering. Lettering is usually of the vertical Gothic style shown in Fig. 122, but some firms use upper-case and lower-case inclined letters, as shown in Figs.

123 and 131. Upper-case lettering is usually preferred on drawings, and in most factories is not underlined. For dimensions and notes, lettering ranges from $\frac{3}{32}''$ to $\frac{1}{8}''$ in height; titles of special views are $\frac{3}{16}''$ or $\frac{1}{4}''$ high. The minimum height of letters on master layouts, Fig. 1158, is $\frac{3}{16}''$ to allow for one-half size reduction in the photographic process.

696. Scale of Drawings. Wherever possible, it is desirable to make all layout and shop drawings full size. However, when this is not practicable, the drawings may range fractionally to half, quarter, or eighth sizes. General arrangement drawings, Fig. 1150, and proposal drawings, §701, are made to scales of $\frac{1}{10}$, $\frac{1}{20}$, $\frac{1}{40}$, etc.

697. Placing of Views. Wherever practicable, the principal view of any part or assembly on a drawing is taken from the left side of the airplane with the nose pointing toward the left border. Sections are taken as projected views when possible; but, if removed, care must be exercised to maintain the direction of sight.

698. Design Practice. The design of an airplane is based upon a carefully prepared set of *specifications* which represents the requirements of a prospective customer. These specifications outline the required performance, power plants, type of construction, equipment, and related items which form the basis of the detail design. In the course of the preliminary development, the original conception of the design may be changed in some respects to assure more efficient performance as the result of wind tunnel or towing basin tests. Scale models of the proposed airplane are built and tested under conditions that simulate actual flight, and from these tests the designers determine the accuracy of their calculations. These models are very expensive, but they represent a necessary effort to detect and correct basic errors before actual construction is begun. When this information has been compiled, it is distributed to the engineers who are concerned with the detail development of the design. Usually a project engineer assumes this responsibility, and the engineers assigned to assist him are divided into groups, each group being guided by a group engineer. The airplane is thus developed in sections, but all sections are developed simultaneously. Since the airplane would not yet be in existence, no measurements of actual parts could be made, so the drawings produced by one group must be available to all other groups to avoid duplication or interference. In addition to cross-checking among the groups, a careful check is also kept on the weight and strength of each part to prevent the design development from getting out of control. A weight "budget" is assigned to each group, and definite strength requirements, as determined by the structures section of the engineering department, must be met and kept within the weight budget. Thus, the weight section knows in advance the trend of the design development, and can predict an increase or decrease in the weight of the finished airplane. Since many contracts require the payment of penalties for overweight airplanes, an airplane manufacturer could not continue long in business without a system by which he could predict and control weights.

It is evident that the designer's problem is quite complex in that he must not delay the design through his inability to make basic decisions quickly and to carry them out efficiently. He cannot sacrifice strength to save weight or add excessive weight to obtain adequate strength. An airplane design must always be a compromise in which the maximum possible strength must be obtained with the minimum possible weight.

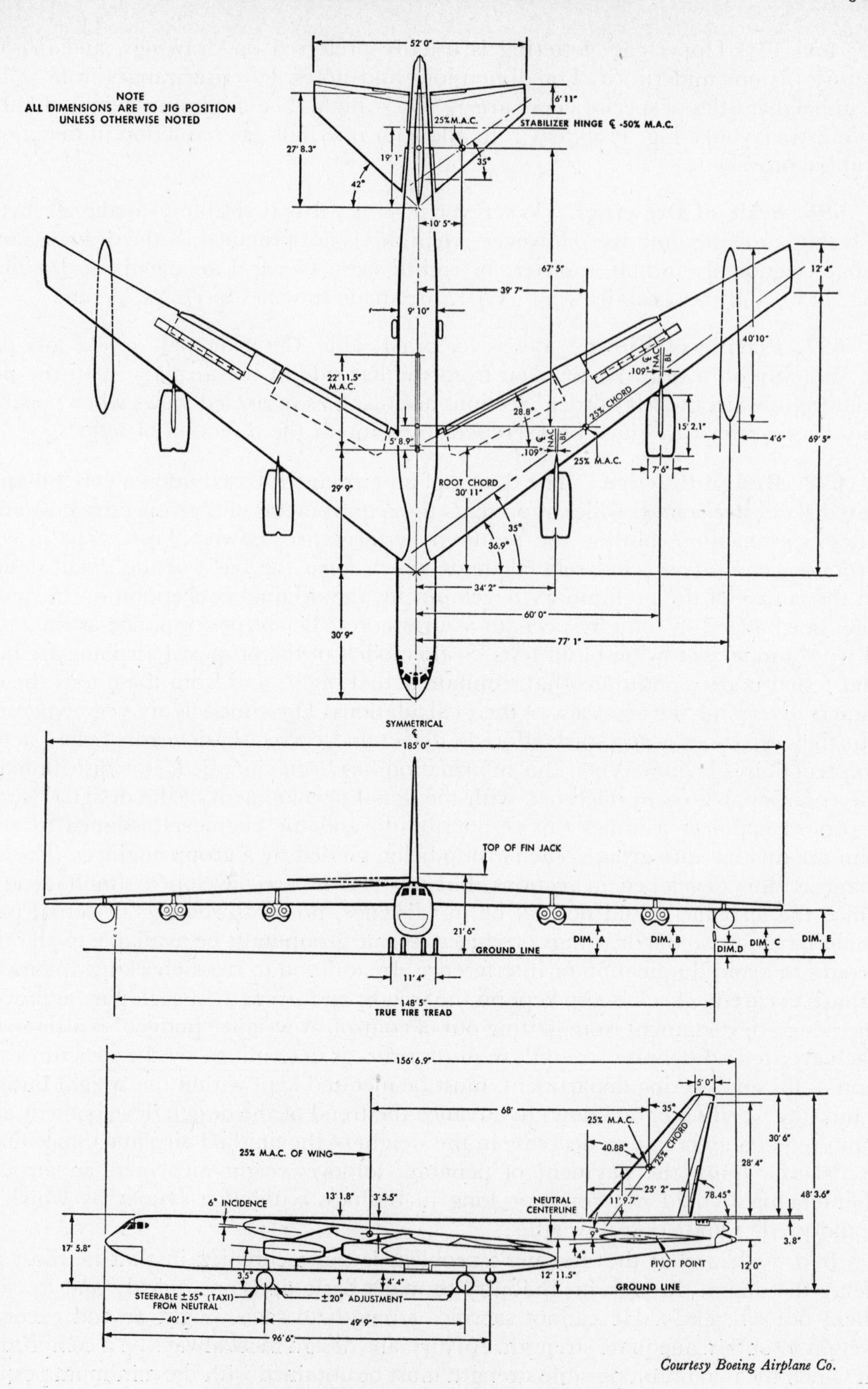

Courtesy Boeing Airplane Co.

Fig. 1150 General Arrangement Drawing.

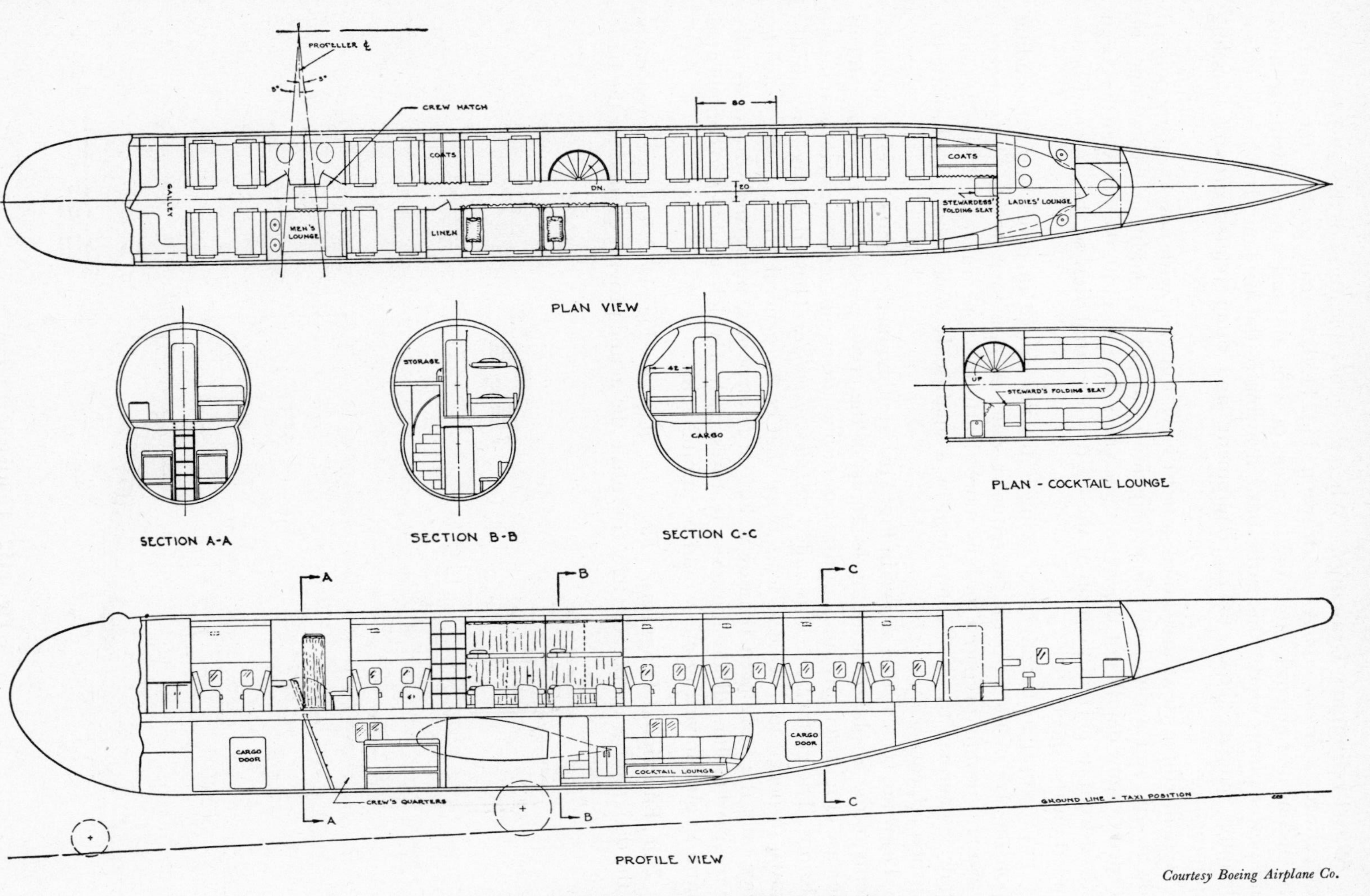

Courtesy Boeing Airplane Co.

Fig. 1151 Inboard Profile and Other Sections.

699. Preliminary Drawings. When the general features of the new design are determined, a *general arrangement* drawing, Fig. 1150, is made to show the principal dimensions and general data, such as areas of wing and tail surfaces, wing sections, and power plants used. This is the basic drawing of the airplane. It is also used for planning hangars, jacks, towing equipment, and other related ground handling equipment.

Usually three views are shown, a front view, top view with nose of plane pointed downward, and a right-side view (left side of airplane). Additional views may be needed to show details not shown on these views. In Fig. 1150, the side view has been moved from its normal position at the right of the front view so that it could be included on the sheet.

The general arrangement drawing is usually drawn in pencil and, depending upon the size of the airplane, to reduced scales of $\frac{1}{10}$th to $\frac{1}{40}$th size. A perspective drawing is usually prepared to show more clearly the appearance of the finished airplane. Sometimes this is accompanied by a small-scale wood model to illustrate the finished appearance.

Drawings such as that shown in Fig. 1151 are then made to show interior arrangements of the airplane, including locations of the crew, passengers, baggage, and in the case of military airplanes, guns, bombs, and the equipment required for military missions. The cutaway view is called an *inboard profile*. These drawings are usually drawn in pencil to a scale at least twice that of the general arrangement drawing, since only the body need be shown. For large airplanes, many such drawings may be required to show different interior features, each of which would be so complicated that a single inboard profile would be unintelligible.

700. Functional Diagrams. *Functional diagrams*, Fig. 1152, are prepared early in the design program because they constitute the actual proving ground for the basic

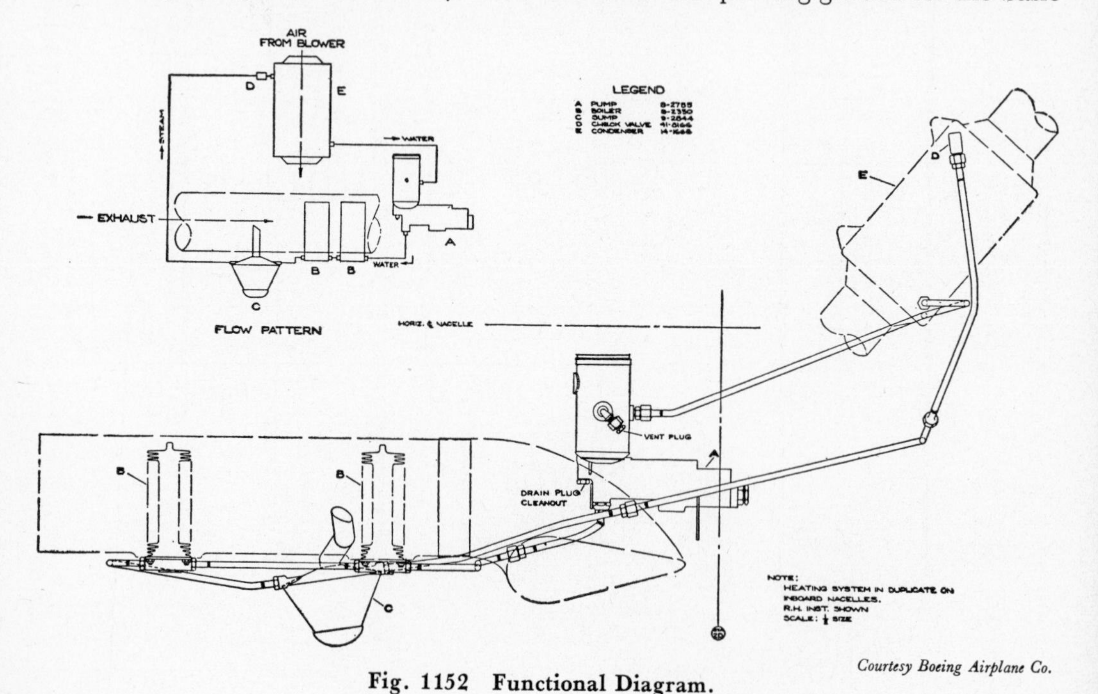

Courtesy Boeing Airplane Co.

Fig. 1152 Functional Diagram.

design. They are made first in rough pencil form, and are revised and refined as the design progresses.

In this diagram, the method of steam generation is shown. The system consists of a sump C which is provided with a means of introducing exhaust flames into its center in order to cause the water within it to rise as steam into the condenser, where its heat is transferred to the incoming cabin air. The condensed steam or water is then forced into the boilers B by a hydraulically-operated feed water pump A. The boilers mounted within the exhaust stack generate additional steam for a repetition of the cycle.

701. Proposal Drawings. A *proposal drawing* describes the airplane in sufficient detail to give the prospective customer a clear idea of the airplane proposed for his consideration. It is not necessary that it be complete in every detail, since it is used for developing the general scheme of structure, controls, equipment, air conditioning, fuel storage, armament, and other related features of the airplane. When these general features have been approved, the work of redesign in more detail and with greater care is begun.

702. Model Drawings. Wind tunnel testing is not confined to the preliminary design phase of airplane development, but continues throughout the detail design phase. In many instances it is necessary to revise the design as a result of flight testing when it becomes advisable either to correct or improve performances which were not anticipated in the earlier stages of wind tunnel testing.

Draftsmen who are engaged in preparing model drawings are usually assigned to the testing department. They design model structures and test equipment of many kinds. With the many models now being built with hollow metal structures, and filled with a maze of remotely operated controls, the work often becomes quite involved. Tank models are used for flying boats and seaplanes where testing for water characteristics is required.

703. Mock-Up Drawings. In the course of designing an airplane, it often becomes necessary for the shops to build a full-size wooden dummy of it to assist in the determination of space relationships which could not be accurately determined by orthographic layouts in the engineering department. This is especially true of the inhabited areas where crew and passenger comfort dictates design features to a great degree. *Mock-up drawings* are often made for the complete airplane, including the structure, so that detail designs in progress may be checked for interference with other parts. The mock-up itself is revised as the design progresses, and is usually inspected and approved by the customer before the final design is determined.

704. Design Layouts. When the general design of the airplane has been completed, its various components are developed in final detail form for fabrication in the shops. These final drawings are preceded by the *design layout*, which is a graphical study to determine the practical application of the general design. These layouts are not usually intended for shop use, but are made in sufficient detail for use in developing or revising the mock-up. Since the engineer who makes the layout very seldom makes the final shop drawings, he must be careful to include all the data necessary for another engineer or draftsman to make final drawings that faithfully describe the design. Major layouts are required for such components as body bulkheads, control cabin floors, wing spars, and flight control systems. From these, minor layouts are made of pulley brackets, fittings, and related parts. The minor layouts are usually in the category of *detail drafting*; so the draftsman must be well acquainted with the principles of descriptive geometry and orthographic projection, as explained in earlier chapters.

When the layout has been checked by the group engineer, and sometimes the project engineer, it is turned over to the detailer with instructions for detailing *production drawings* of the parts that may be desired on individual drawings. The layouts are usually made on high-quality vellum because of the ease with which it may be drawn upon and because of the small amount of shrinking and stretching of the vellum that will result from changes in humidity or from handling. Pencils of 4H to 6H hardness are used, and accuracy must be to at least $\pm$ 0.015″. The layouts are made full scale wherever practicable, and contain no dimensions, since they are meant to be scaled directly. Only necessary notes describing materials, rivets, bolts, etc., are included.

It is the draftsman's responsibility to determine which dimensions will be shown, but whatever they are, they must accomplish the intent of the layout.

705. Working Drawings. *Working drawings* are those which are officially released to the shops for manufacture of the airplane. They are broadly classified as *detail* drawings, *assembly* drawings, and *installation* drawings, according to the operations involved. The practice of some companies is to combine these basic types into composite drawings to reduce the total number of drawings required for a large project, while other companies keep them separated to adapt the drawings more readily to the shops involved. The statistics quoted on page 724 are based on the composite type. The 17,000 drawings would have been 100,000 if every part, assembly, and installation had been drawn separately. This type of drawing has the advantage of relating the parts to each other both pictorially and dimensionally so that emphasis may be placed on the dimensions and tolerances common to two or more parts. However, the draftsman must be very careful when dimensioning his drawings to see that the fabrication, assembly, and installation dimensions will be clearly distinguishable from each other. His choice of dimensions must be based upon shop operations rather than upon how he made the drawing. Reference lines in space present no problem to the draftsman, but they often require intricate jigs and fixtures in the shop, and complicate production operations. It would be well for the draftsman to remember that the shops depend upon him for information about his design and will try to do what he tells them to do. His responsibility to them requires him to do more than make attractive drawings; they must be also intelligently made.

706. Detail Drawings. *Detail drawings* are those which give complete information and instructions for making individual parts. Usually only one part is shown on a detail drawing, but in some cases it is convenient or desirable to show more than one. In such cases, the parts are not related in the sense that they are attached to each other, but only in the sense that they are similar or are properly identified by the title of the drawing. Castings and forgings are classified as details, although they are not actually finished parts. The drawing of the final machined forging or casting is likewise classified as a detail drawing. The Flap Swing Forging shown in Fig. 1153 is an example of a detail drawing which describes the part before it is machined into a finished part.

707. Assembly Drawings. *Assembly drawings* are those which give complete

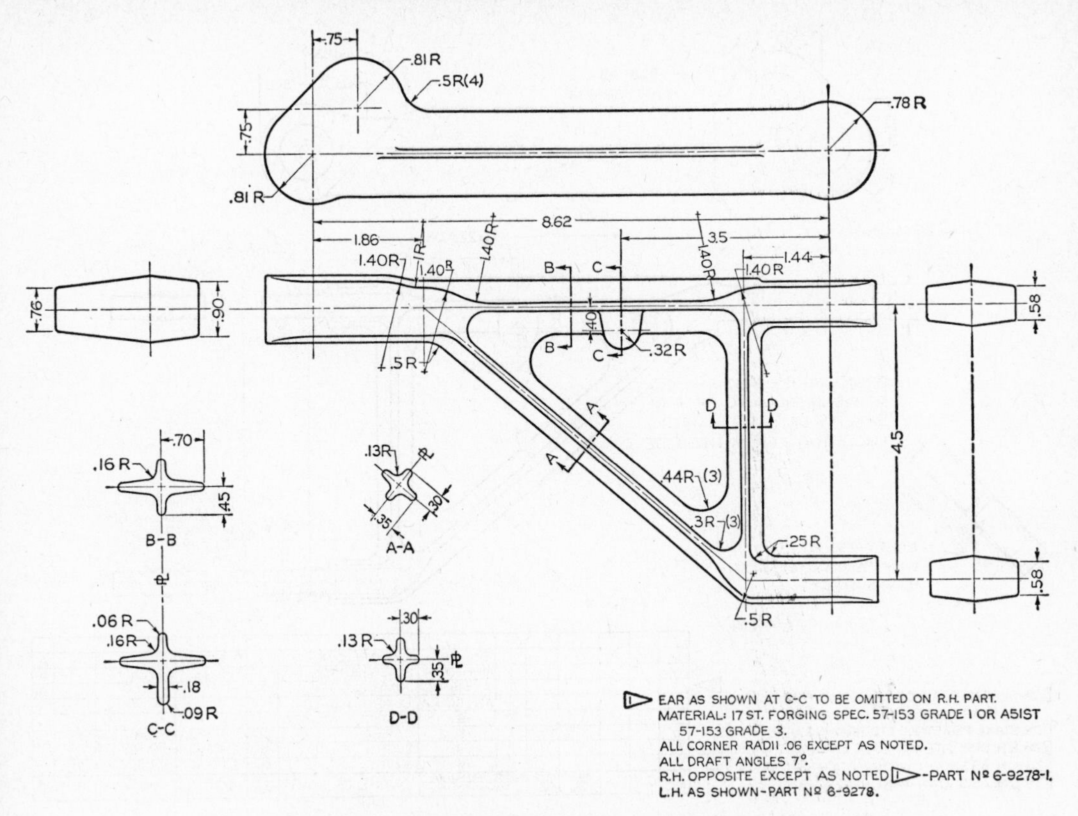

EAR AS SHOWN AT C-C TO BE OMITTED ON R.H. PART.
MATERIAL: 17 ST. FORGING SPEC. 57-153 GRADE 1 OR A51ST 57-153 GRADE 3.
ALL CORNER RADII .06 EXCEPT AS NOTED.
ALL DRAFT ANGLES 7°.
R.H. OPPOSITE EXCEPT AS NOTED -PART Nº 6-9278-1.
L.H. AS SHOWN-PART Nº 6-9278.
B-B
A-A
C-C
D-D

Fig. 1153 Forging-Flap Swing (Detail Drawing).

information and instructions for joining two or more detail parts to form a single compound part. As in detail drawings, only one assembly is usually shown on an assembly drawing, but in some cases it is convenient or desirable to show more than one. Fig. 1154 shows a typical assembly drawing which combines both detail and assembly information. The similarity between this drawing and the forging detail in Fig. 1153 is immediately apparent, since the forging is the part from which the Flap Support is made. The complete assembly, of course, requires the bearings and the electrical bonding jumper. In comparing the forging with the finished machined part, it will be noted that the assembly drawing does not show any of the forging dimensions, but only such dimensions as are necessary for the machining operation. Notice also that the designer has provided ample material in the forging to allow for the machine cuts shown on the assembly drawing, and he has dimensioned both parts from a common base line to reduce the chance of error in centering the bearings.

708. Installation Drawings. *Installation drawings*, Fig. 1155, are those which give complete information and instructions for placing detail parts or assemblies in their final positions in the airplane. They are easily distinguished from assembly and detail drawings by the presence of phantom lines which represent the airplane itself, and thus relate the installed parts to the airplane. It will be noted in the list of parts on this drawing that it covers both detail and assembly parts as well as instructions for their installation in the airplane. The parts that can be made from this drawing are identified in the parts list by one or two digit numbers in the "part number"

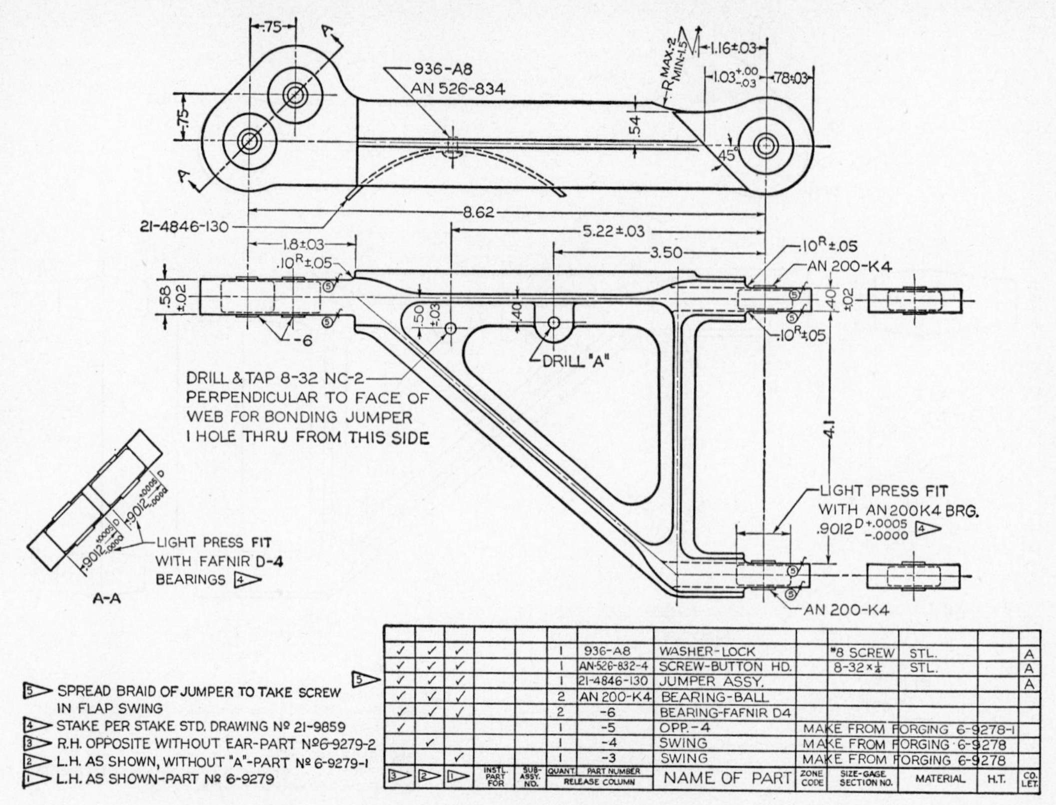

3	2	1	INSTL. PART FOR	SUB-ASSY. NO.	QUANT.	PART NUMBER	NAME OF PART	ZONE CODE	SIZE-GAGE SECTION NO.	MATERIAL	H.T.	CO. LET.
✓	✓	✓			1	936-A8	WASHER-LOCK		#8 SCREW	STL.		A
✓	✓	✓			1	AN-526-832-4	SCREW-BUTTON HD.		8-32×¼	STL.		A
✓	✓	✓			1	21-4846-130	JUMPER ASSY.					A
✓	✓	✓			2	AN 200-K4	BEARING-BALL					
✓	✓	✓			2	-6	BEARING-FAFNIR D4					
✓					1	-5	OPP.-4	MAKE FROM FORGING 6-9278-1				
	✓				1	-4	SWING	MAKE FROM FORGING 6-9278				
		✓			1	-3	SWING	MAKE FROM FORGING 6-9278				

Fig. 1154 Swing Assembly-Flap Support (Assembly Drawing).

column. These are known as *dash numbers*, since they are not complete part numbers. The complete part number is formed by adding the dash number to the drawing number. Thus the Bracket Assembly (2) is part number 6-8047-2, the -2 being the dash number. It is common practice in the aircraft industry to show only the dash number of a part on the drawing showing its design. As may be seen in the notes above the parts list, this drawing covers a left-hand and a right-hand installation, but only the left-hand installation is shown. The right-hand installation may be made by reversing the dimensions about a vertical axis in the same manner that left and right hands are reversed about a vertical axis in the body. This fact will account for the presence of some parts in the parts list which do not appear in the picture.

709. Undimensioned Drawings. When a dimensioned blueprint of a formed sheet-metal or *contoured* part is received in the shop, it is necessary for one or more flat templates to be made from the information provided. These templates are used for cutting out the flat material from which the part is to be made. In the case of contoured parts, the templates are used for making plaster patterns for the foundry where metal dies and punches are made for the drop hammers and stretch presses. Fig. 1164 shows how plaster patterns for contoured surfaces are built up from headers cut from flat sheets. In the aircraft industry, the making of flat pattern layouts is being transferred from the shops to the engineering department to eliminate the need for dimensioned drawings for these parts. See §712. An aeronautical draftsman must therefore be skilled in the technique of drawing flat pattern layouts of parts which otherwise

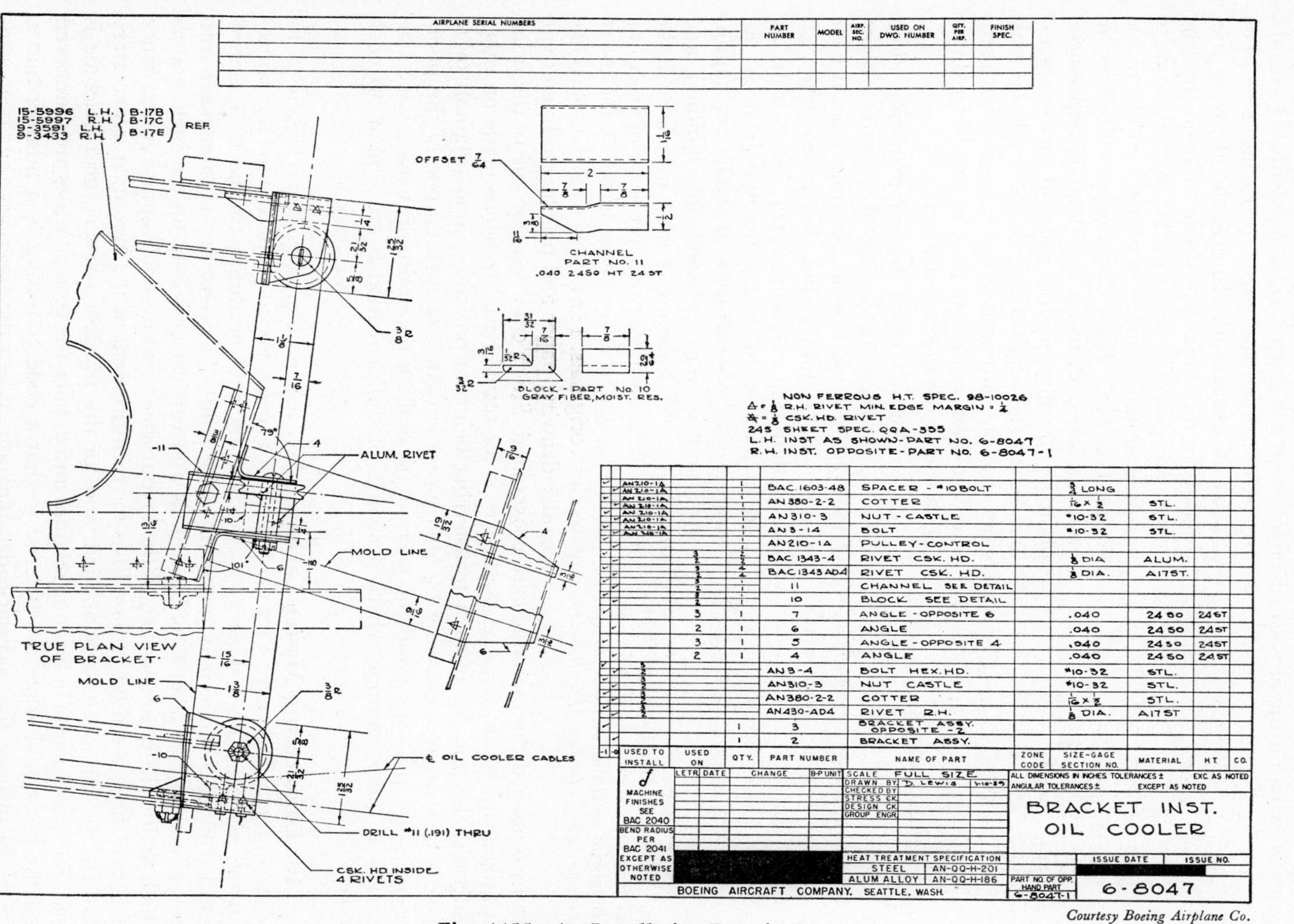

Courtesy Boeing Airplane Co.

Fig. 1155 An Installation Drawing.

would be dimensioned in the conventional manner, Fig. 1156. This technique includes drawing on metal with silver solder pencils, or on *glass cloth* with 9H pencils, instead of drawing on paper or tracing cloth. The absence of dimensions requires the drawing material to be dimensionally stable to avoid drafting errors, and likewise requires the reproductions to be dimensionally stable to avoid manufacturing errors. Fig. 1157 shows diagrammatically how an undimensioned drawing on metal is reproduced photographically for shop use and for blueprinting. When the original drawing is made on glass cloth, the photo reproduction process is not used, since the shop templates may be made by direct contact with the original drawing.

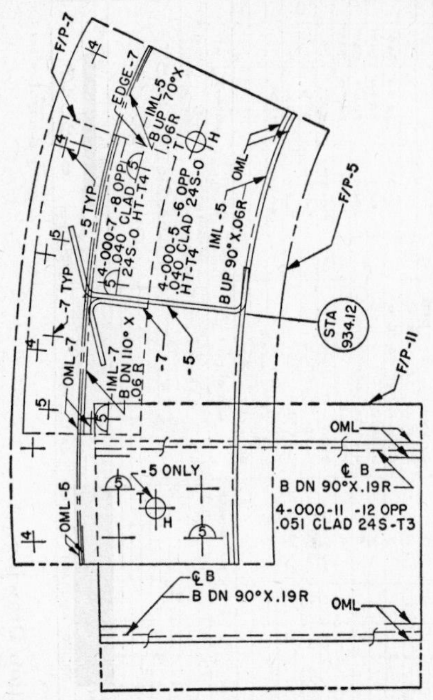

Fig. 1156 Undimensioned Drawing. *By permission from* A Simple Guide to Blueprint Reading, *by W. N. Wright. Copyright 1956, McGraw-Hill Book Co.*

Glass cloth is made by impregnating cloth woven from continuous glass fibers with a blend of Paraplex polyester resins. Not only does it have dimensional stability, but its transparency is greater than that of tracing paper. Many firms use it in preference to metal because it does not require costly photographic reproduction equipment, and is very economically handled and stored. The extensive use of huge templates in the aircraft industry presents a problem of handling and storing which is greatly reduced by the use of glass cloth originals. The material can be rolled like regular drawing paper and tracing cloth, and will not be damaged if reasonable care is exercised. Silver emulsions, or any of the usual sensitizing materials, may be applied to it for transferring drawings made on other materials, the purpose of such transferring being to provide second originals or to make a new original drawing by altering some details of the first one of the photo contact copy and thus avoid the necessity of making the entire drawing again. This practice is quite common where drawings of different models are similar to each other.

710. Engineering Master Layouts. When a *master layout*, Fig. 1158, is to be made on metal rather than on glass cloth, it is necessary to prepare it for the photocopying process. This is done by applying to the steel or aluminum sheet a specified number of coats of specially prepared white paint. The next step is to engrave into the painted surface a series of horizontal and vertical grid lines on ten-inch centers which divide the entire surface of the layout sheet into ten-inch square zones. Since these grid lines are used as base lines for measurement of the layout, it is necessary that they be very accurately applied, and for this purpose a special grid machine is used. For convenience in locating sections and details, the vertical zones are numbered and the horizontal zones are lettered so that a code consisting of a number and a letter will provide the necessary identification. When the sheet has thus been prepared, the draftsman may then proceed with his layout drawing, Fig. 1159. All lines and lettering are applied with silver solder or aluminum pencils in order to provide

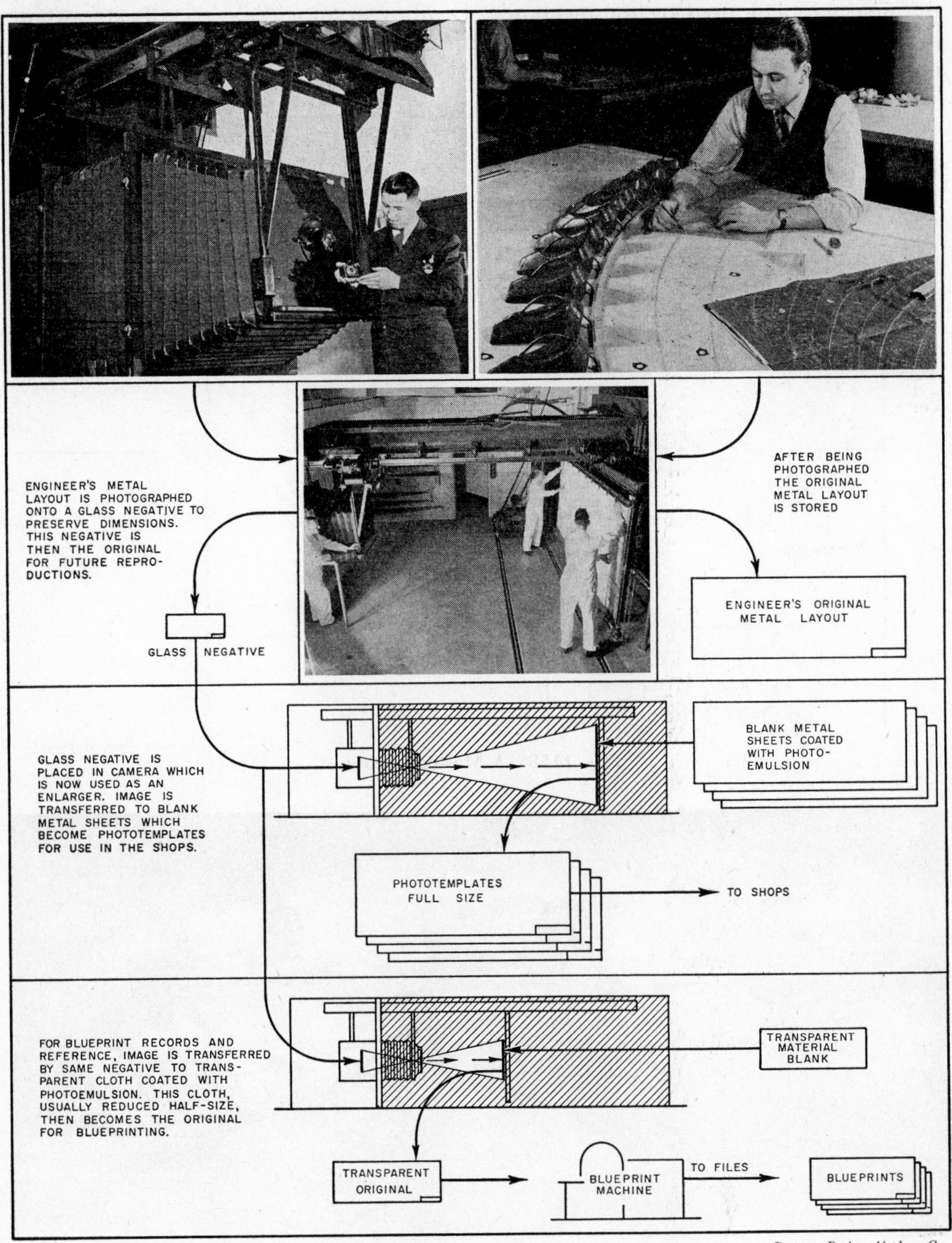

ENGINEER'S METAL LAYOUT IS PHOTOGRAPHED ONTO A GLASS NEGATIVE TO PRESERVE DIMENSIONS. THIS NEGATIVE IS THEN THE ORIGINAL FOR FUTURE REPRODUCTIONS.
AFTER BEING PHOTOGRAPHED THE ORIGINAL METAL LAYOUT IS STORED
GLASS NEGATIVE
ENGINEER'S ORIGINAL METAL LAYOUT
GLASS NEGATIVE IS PLACED IN CAMERA WHICH IS NOW USED AS AN ENLARGER. IMAGE IS TRANSFERRED TO BLANK METAL SHEETS WHICH BECOME PHOTOTEMPLATES FOR USE IN THE SHOPS.
BLANK METAL SHEETS COATED WITH PHOTO-EMULSION
PHOTOTEMPLATES FULL SIZE
TO SHOPS
FOR BLUEPRINT RECORDS AND REFERENCE, IMAGE IS TRANSFERRED BY SAME NEGATIVE TO TRANSPARENT CLOTH COATED WITH PHOTOEMULSION. THIS CLOTH, USUALLY REDUCED HALF-SIZE, THEN BECOMES THE ORIGINAL FOR BLUEPRINTING.
TRANSPARENT MATERIAL BLANK
TRANSPARENT ORIGINAL
BLUEPRINT MACHINE
TO FILES
BLUEPRINTS

Courtesy Boeing Airplane Co.

Fig. 1157 Reproduction of Undimensioned Drawings.

dense black lines which will produce sharp and clear lines on the negative. It is imperative that the draftsman exercise extreme care in keeping the lines sharp and exactly to scale, since two photographic steps lie between his original layout and the final template used by the mechanic. Drafting accuracy is especially important in view of the fact that the layout and template must be measured directly without the benefit of dimensions. It is the practice in some companies to have master layouts

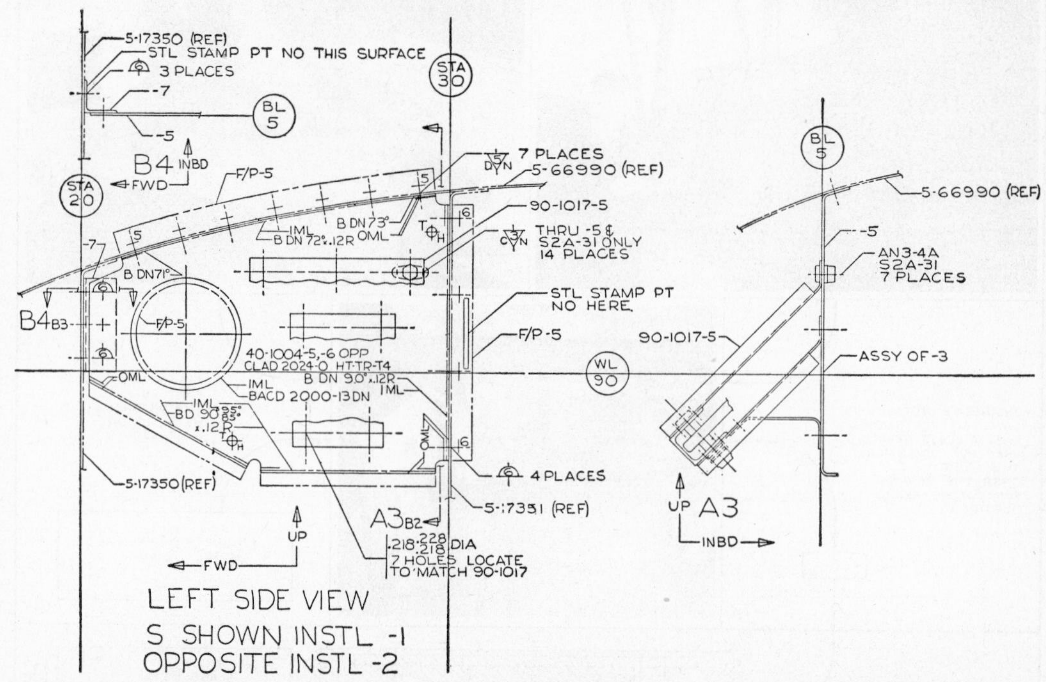

Courtesy Boeing Airplane Co.

Fig. 1158 A Master Layout.

Courtesy Boeing Airplane Co.

Fig. 1159 Starting a Master Layout.

made in a separate section of the engineering department to assure uniform quality control of the workmanship. In such cases the design engineers make dimensioned details of finished parts, and send them to the master layout draftsman, who then develops the flat pattern outlines on the metal sheet. This requires careful coordination and a thorough knowledge of the principles of descriptive geometry and orthographic projection as outlined in the earlier chapters. The usual allowable drafting tolerance on details is ± .010 inch, except on important structural details such as mating holes and parts to be jig-located, which must be accurate within ± .005 inch. When close tolerances are required, the dimension is added to the layout to assure proper control in the shop.

Under this arrangement, many of the vellum detail drawings are eliminated. All forgings and machined parts are detailed, but brackets and fittings that are fitted on a contour are laid down only on the master layout of the assembly affected. *Parts that can be conveniently shown on master layout assemblies are not detailed on separate sheets.*

711. Photo Templates. Upon completion and after being checked, the above-described master layouts go to the *photo template unit.* Here they are accurately photographed by a special large camera, Fig. 1157, fitted with sensitized glass plates. Glass negatives are required to prevent shrinking or stretching.

From this negative, by an enlargement process, other sensitized dural sheets are made to duplicate the original master layouts in the original full size. Metal sheets are also printed to be used as patterns.

If *mock-up patterns* are needed, sensitized plywood is printed to duplicate the master layout, and the shop then saws the wood to shape. Plywood reproductions are also placed on the shop bench, and the assembly is built up on them from the individual parts. Reproductions are also made of steel, aluminum, and masonite for various uses in the shop. For example, steel reproductions are used for drill and router templates, jig bases, and form templates; aluminum for reference and assembly table templates; and masonite for form blocks.

The photo template process is further employed to provide blueprints for reference in shop assembly work. Sensitized cloth is exposed from the negative of the master template. This results in a photo tracing which is one-half full size. This tracing is then sent to the blueprint department for reproduction of all needed prints for shop use and for the engineering file.

Since the enlargement made from the glass negative can be made to any desired fractional size, the photo process is often employed to make model drawings and patterns which may be made to a decimal-fraction scale.

Photo templates provide an additional advantage in the case of symmetrical and opposite-hand parts. It is necessary to draw only one side as an original master layout, since the opposite side can be produced by merely reversing the negative in the printing step. In a similar manner, intermediate wing ribs need not be drawn separately if the thickness ratio remains constant. All such ribs can be produced photographically from a drawing of only one rib by making the size of each successive copy in the proper proportion to its location in the plan form.

The prime advantage of the master layout photo template system is that it eliminates time usually lost in getting engineering information to the shop in the proper form. With this setup, the production department need not spend valuable months working up templates and data for shop use. The material is available just as fast as the photo unit can turn it out. This is important in getting new airplane designs into the air in the shortest possible time.

712. Flat Pattern Layouts. A complete description of the methods used in laying out flat patterns is beyond the scope of this chapter. However, the procedure for calculating bends is typical. If the two inner plane surfaces of an angle are extended, their line of intersection is called the IML or *inside mold line*, Fig. 1160 (a) to (c). Similarly, if the two outer plane surfaces are extended, they produce the OML or *outside mold line*. The *center line of bend* (℄ B) refers primarily to the machine on

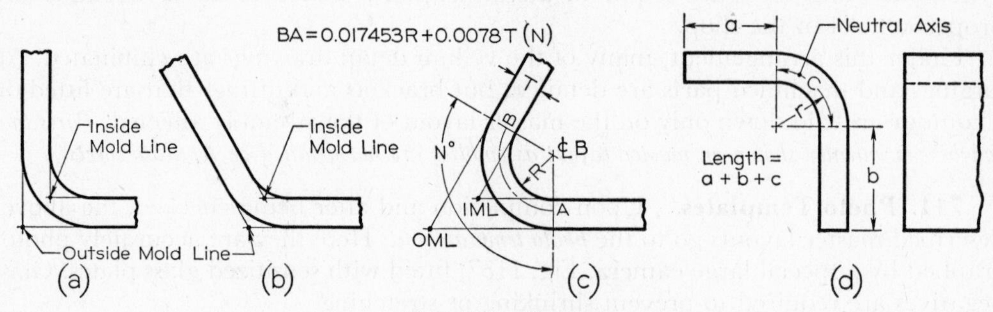

Fig. 1160 Bends.

which the bend is made, and is at the center of the bend radius. The bend radius for annealed dural is taken as double the gage, and equal to the gage for other metals. These three lines, as they appear on an undimensioned drawing, are shown in Fig. 1156. Note that the outside mold line is shown in two overlapping positions resulting from straightening out the bend.

The length, or *stretchout*, of the pattern equals the sum of the flat sides of the angle plus the distance around the bend measured along the *neutral axis*. The distance around the bend is called the *bend allowance*. When metal bends, it "crowds" on the inside and stretches on the outside. At a certain zone in between, the metal is neither

Courtesy Boeing Airplane Co.

Fig. 1161 Lofting.

compressed nor stretched. This is called the neutral axis. See Fig. 1160 (d). The neutral axis is usually assumed to be .44 of the thickness from the inside surface of the metal.

The developed length of material, or bend allowance (BA), to make the bend is computed from the empirical formula:

$$BA = 0.017453R + 0.0078T(N)$$

where R = radius of bend, T = metal thickness, and N = number of degrees of bend. See Fig. 1160 (c).

713. Lofting. *Lofting* is the drafting procedure used to develop a curved surface such that any cutting plane intersecting it will produce a smooth curved line. Lofting also is used to establish intersections of curved surfaces with each other. In view of the numerous curved surfaces in an airplane, it is evident that lofting is an important function of the engineering department. Ordinarily, lofting is done full size for the sake of accuracy, but sometimes it is necessary to reduce the scale for large airplanes. As shown in Fig. 1161, the work is done on painted metal sheets as described in §709, to maintain dimensional stability. The draftsmen cover their shoes with protective cloth slippers while working on the large surfaces. The same kinds of drawing tools are used in lofting as are used in master layouts, except that more extensive use is made of *ship curves*, *splines*, and *ducks* for laying out *faired lines*. Splines are flexible wood or plastic strips of varying widths which are held by the ducks to form curved guides for the draftsman's pencil. The ducks are simply lead weights fitted with metal prongs which rest on the splines and hold them firmly in place.

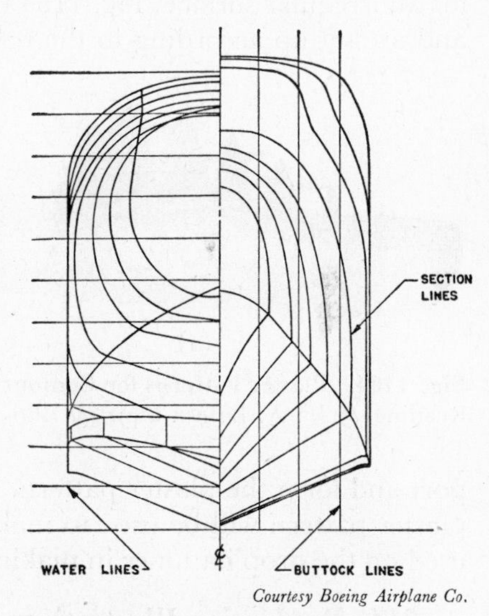

Courtesy Boeing Airplane Co.

Fig. 1162 A Set of Faired Lines.

To loft a set of faired lines of a flying boat hull, as for instance in Fig. 1162, the plan and elevation views are laid out in their proper relationship. *Station lines* are then drawn vertically in the elevation view to form vertical transverse reference planes. *Buttock lines* are then drawn parallel to the airplane centerline in the plan view to form vertical longitudinal reference planes, and finally, *water lines* are drawn perpendicular to the station lines in the elevation view to form horizontal reference planes. These are the basic airplane reference planes and are often used in place of dimensions for locating points. Since three dimensions are required to locate a point in space, the reference lines may be identified by numbers that represent distances from an established base line, and their intersection gives the location of the point in space. Thus a point could be located by naming the planes which intersect at that point, as follows:

STATION 157.63
WATERLINE 62.9
LEFT BUTTOCK LINE 15.1

Cross sections at certain stations where the shape is definitely known are laid out. For example, a cross section might be laid out for the hull at the control compartment, another at the trailing edge of the wing, and another at the tail of the airplane where space would be required for a gunner. Intermediate stations which are not known would be determined by establishment of smooth contours between the stations originally laid out. Fig. 1162 shows a typical set of faired lines in which the station contours are superimposed, those from the maximum section forward being on the right side and those from the maximum section aft on the left. Thus, all the station contours for the entire hull are contained in one view.

When the hull is properly lofted, its intersection with any plane will be a smooth curved line. When a few points are known on a curve, a spline is often used to develop the entire curve in the manner shown in Fig. 1161. The flexible spline is held down with ducks at the points definitely located. If it lies on all the points without being forced, the line is considered to be *faired*. A true test is to lift any duck on the spline except the end ones, and if the spline does not spring into another position, it is following a faired curve. Conversely, if it must be held in place by force, either the spline must be replaced by a more flexible one, or the points will have to be adjusted to bring them into a faired line.

Contours and sections developed by the loftsmen are used by the master layout draftsmen, and photo templates of the contours are made where frequent reference to them will be necessary. Fig. 1163 shows a photo template of a set of faired lines for an irregular surface. Fig. 1164 (a) shows how headers are made from the loft lines and are set up according to the reference planes to form a skeleton structure to sup-

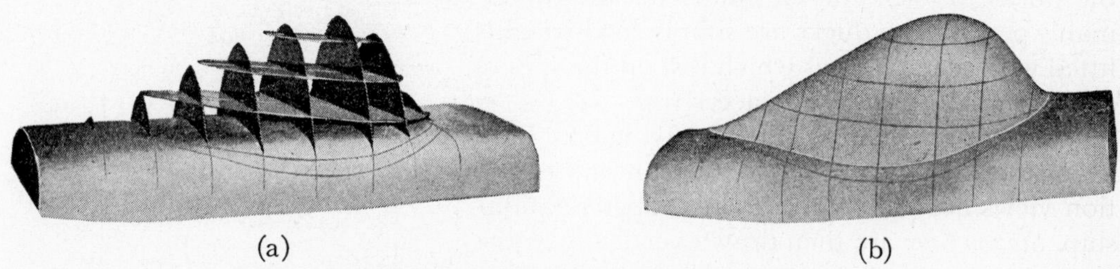

Fig. 1164 Plaster Patterns for Contoured Surfaces. *By permission, from* A Simple Guide to Blueprint Reading, *by W. N. Wright. Copyright 1956, McGraw-Hill Book Co.*

port and form the plaster pattern. The final contoured pattern is shown at (b). This plaster pattern will be used to make a mold for casting a metal punch and die to be used on the drop hammer in making the part whose contours are shown in Fig. 1163.

714. Production Illustration. In August, 1939, the Douglas Aircraft Company first used perspective drawings to provide a complete picturization of the manufacturing breakdown of a proposed airplane for a military bid. So useful were these drawings that the process was not only continued in the Douglas plant, but has been adopted throughout the country. Briefly, production illustrations are pictorial drawings (axonometric, oblique, or perspective), either in pencil or ink, of complete airplanes, mechanical systems of the airplane (such as the wiring or fueling systems), subassemblies (such as "Nose Wheel Assembly"), or of details (such as a "Hinge Bracket"). These drawings are employed in every way possible to clarify and amplify the working drawings. Their chief value is in assisting the large number of workers

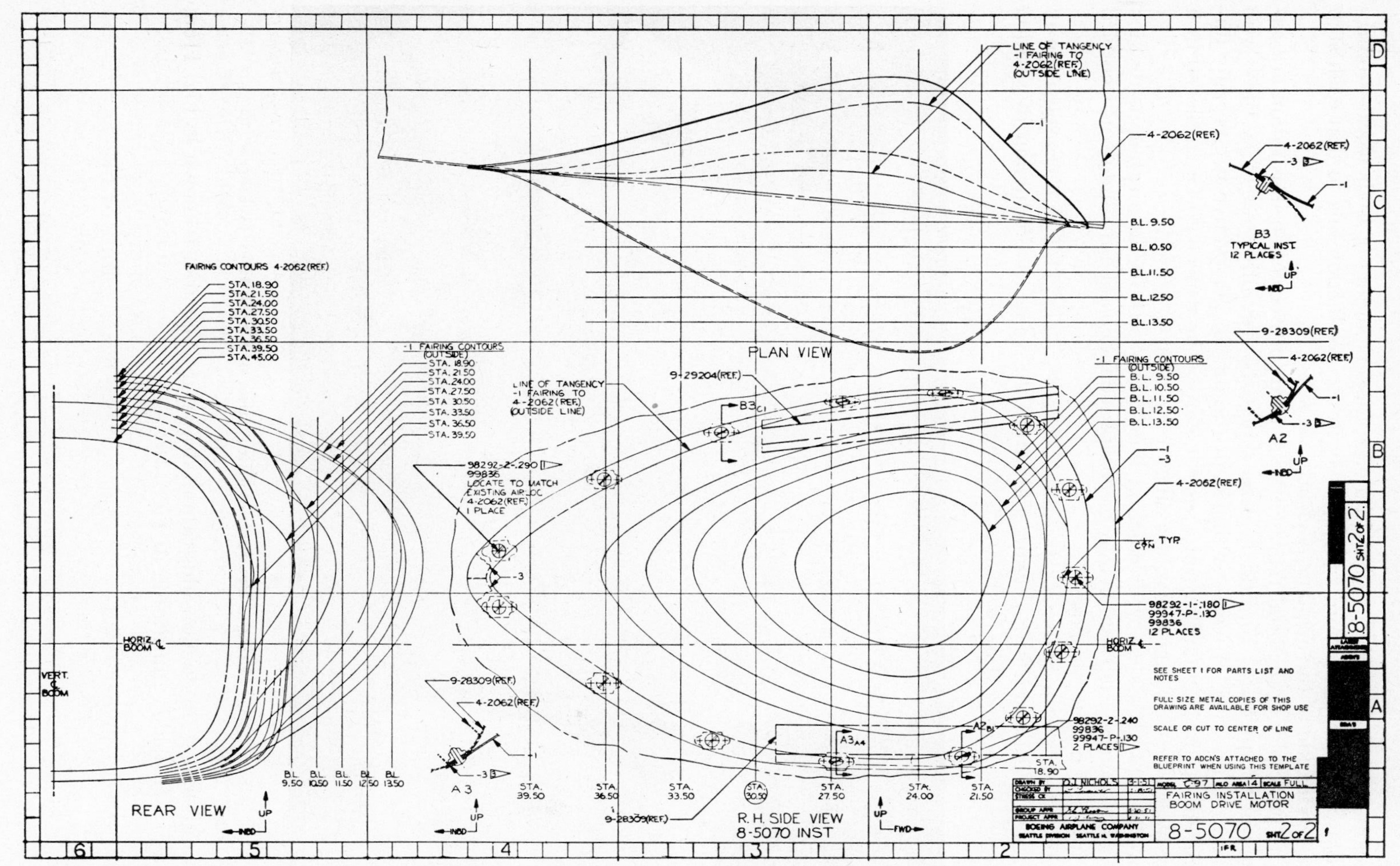

Fig. 1163 Loft Lines for an Irregular Surface. *By permission, from* A Simple Guide to Blueprint Reading, *by W. N. Wright. Copyright 1956, McGraw-Hill Book Co.*

who cannot read blueprints to obtain a clear understanding of their work. Production illustrations are therefore keyed closely to the production line so that each step in the fabrication of the airplane is clearly shown.

The principles of pictorial representation used in production illustration are explained in Chapters 15, 16, and 17, and should be carefully studied by those who are entering this type of work. A thorough training in perspective is especially important to the production illustrator, particularly if he expects to be employed in a company where true perspective drawings are required.

In Fig. 1149 is shown a production illustration of how an airplane is divided into sections so that each section may be completed independently and then assembled later into a complete airplane.

The drawing of a Stabilizer Jackscrew shown in Fig. 1165 is a combination of multiview and isometric projection, indicating the value of production illustration for design work as well as for manufacturing.

Courtesy Boeing Airplane Co.

Fig. 1165 Using a Production Illustration of a Stabilizer Jackscrew.

An unshaded production illustration, drawn in isometric, is shown in Fig. 1166, and a pencil-shaded illustration in Fig. 1167.

715. Aeronautical Drafting Problems. The following problems will afford the student some practice in problems that are typical of the aircraft industry.

Prob. 1. Draw to a full-size scale the *Forging-Flap Swing* shown in Fig. 1153. Rearrange as necessary to fit a Size B sheet.

Prob. 2. Draw to a full-size scale the *Swing Assembly-Flap Support* shown in Fig. 1154. Rearrange as necessary to fit a Size B sheet. To obtain dimensions not given, refer to the corresponding forging drawing in Fig. 1153.

Prob. 3. Draw to a full-size scale the *Terminal Assembly-Upper Drag Strut* shown in Fig. 718. Use Size B sheet, and prepare title strip similar to that shown.

Prob. 4. Draw to a full-size scale, the necessary views of the *Angle*, Part 4, shown in Fig. 1155.

Prob. 5. Draw to a full-size scale, the necessary views of the *Angle*, Part 5, which is the opposite to Part 4 in Fig. 1155.

Prob. 6. Draw to a full-size scale, a flat pattern of Part 5 (see Prob. 5), assuming a bend radius of twice the metal thickness. See the parts list in Fig. 1155 for the thickness.

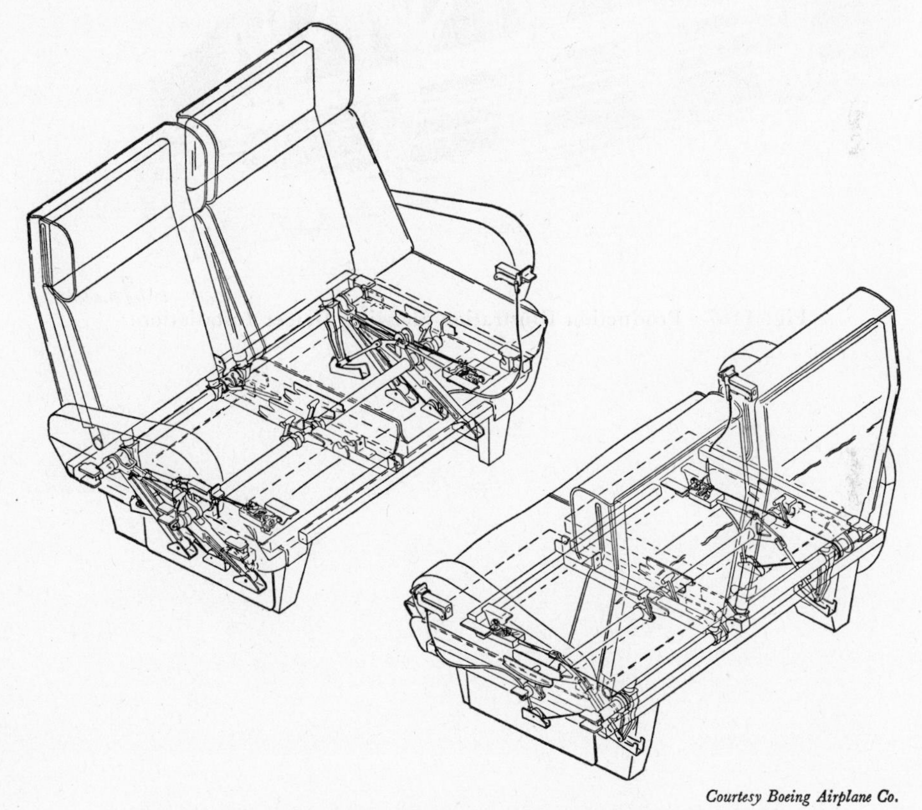

Courtesy Boeing Airplane Co.

Fig. 1166 Production Illustration Showing Seat Mechanism.

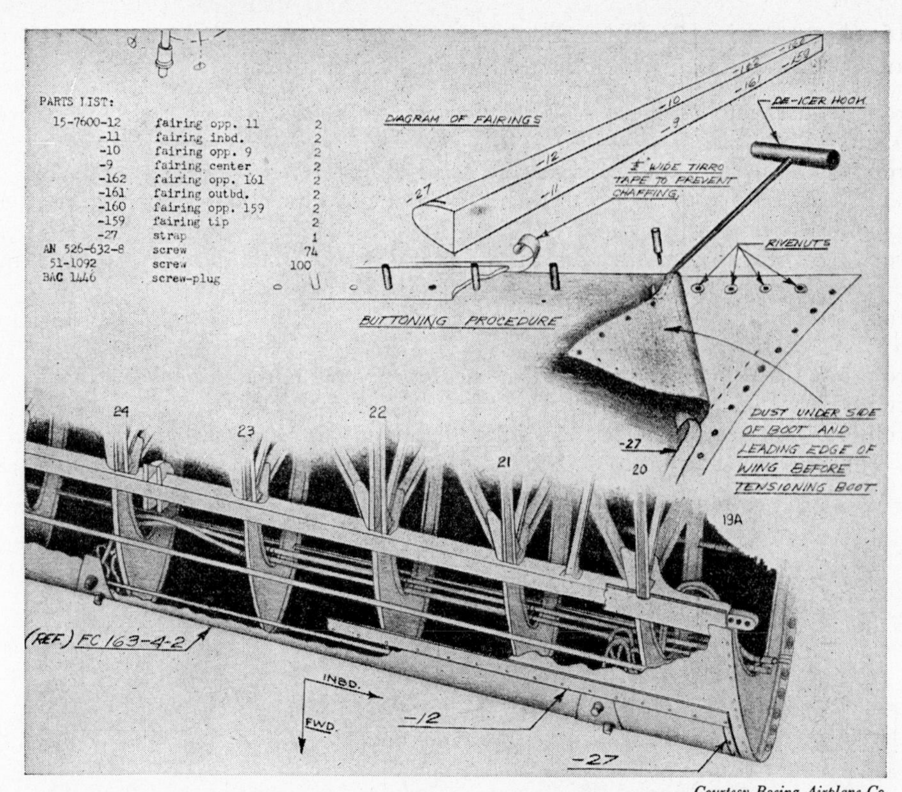

Courtesy Boeing Airplane Co.

Fig. 1167 Production Illustration Showing Deicer Installation.

CHAPTER 27

SHADING

716. Introduction. The effect of light can be utilized advantageously in describing the shapes of objects, and in the finish and embellishment of such drawings as display drawings, patent drawings (Chapter 28), etc. Ordinary working drawings are not usually shaded.

Figure 1168 illustrates the difference in appearance and clearness produced by the use of shade lines. The front view of the block at (a) conveys no idea as to whether

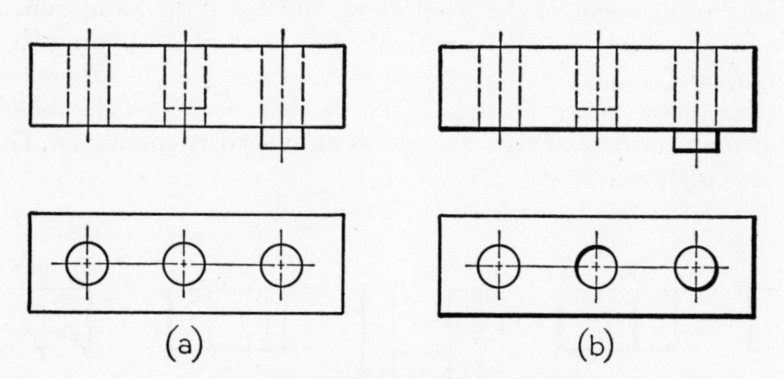

Fig. 1168 Effect of Shade Lines.

the pins in the holes are flush with the front face of the block, are countersunk, or project and form a boss on the front face. The front view at (b) shows, by the shade lines, that one pin is flush with the front face of the block, one is countersunk, and the other projects in front of the face. The sharpness, relief, and realistic effect produced by the shading are obvious.

The conventional source of light for shading drawings is behind, above, and to the left of the draftsman. The rays of light are assumed to be parallel to a diagonal of a cube whose faces are parallel to the planes of projection, Fig. 288 (a); therefore the conventional light rays make angles with the coordinate planes whose tangents are $1/\sqrt{2}$, and the projections of the rays make angles of 45° with the coordinate axes.

There are two systems in use for shading drawings: (a) *shade lines* and (b) *surface shading*.

717. Shade Lines. *Shade lines* are the lines separating illuminated faces or surfaces from those which are not illuminated. Usually the upper, front, and left-hand faces of an object are assumed to receive light from the conventional source, and are illuminated, while the lower, back, and right-hand faces are not illuminated; therefore the right-hand and lower edges of an object are ordinarily the shade lines. By this system of shading, the shade lines are made two or three times as wide as the other object lines of the drawing.

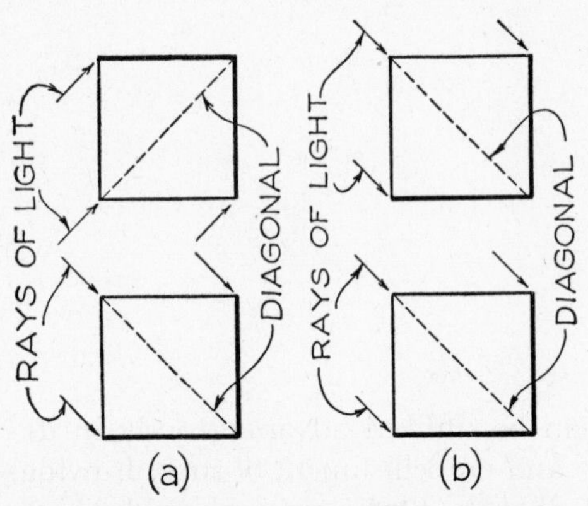

Fig. 1169 Conventional Rays of Light.

Figure 1169 (a) shows the top and front views of a cube, with conventional rays of light parallel to the dashed diagonal of the cube. It is obvious that these rays of light illuminate only the top, front, and left-hand faces of the cube, and therefore three vertical and three horizontal edges separate illuminated from non-illuminated faces. Two of these edges in the front and two others in the top view are *shade lines*, and are therefore made wider than the other object lines of the cube.

It is apparent from Fig. 1168 (a), according to the rule stated above, that the right-hand and lower edges of the front view, and the right-hand and upper edges of the top view are shade lines. To simplify the method of shading, it is common practice to *shade all views as if they were front views, to shade lines which separate illuminated surfaces from those which are not illuminated, and to shade the right-hand and lower edges or contour lines,* if the clearness of the drawing is enhanced by doing so. The cube will then be shaded as shown at (b).

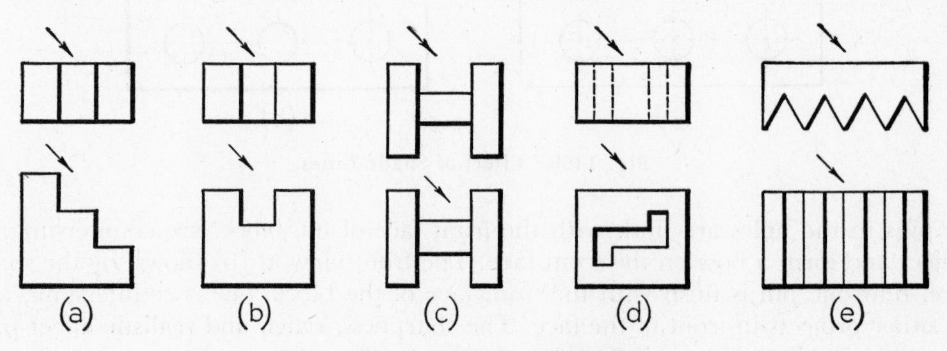

Fig. 1170 Application of Shading.

Figure 1170 illustrates the application of the preceding rule to the shading of various objects. Note that invisible edges at (d) are not shaded.

Surfaces parallel to conventional rays of light, Fig. 1171, are regarded as illuminated surfaces.

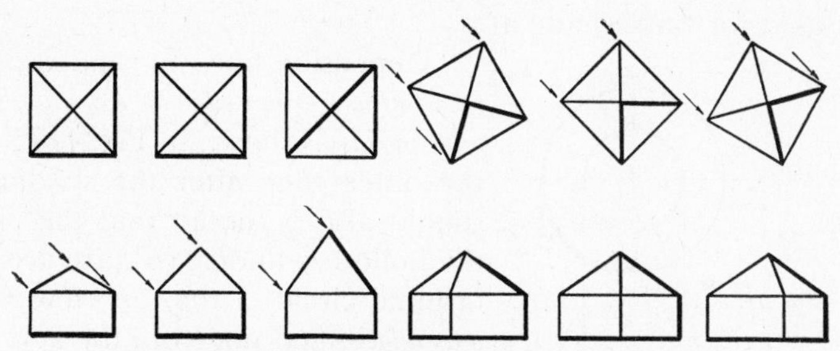

Fig. 1171 Surfaces Parallel to Rays of Light.

When two surfaces, inclined to the plane of projection, are located so that one surface is illuminated and the other is not, the edge of the dihedral angle formed by those surfaces is a shade line. If there is a series of such surfaces, all edges of the resulting angles are shade lines. The drawing will be clearer, however, if only alternate edges are shaded, as in Fig. 1170 (e).

Some draftsmen shade the edges of re-entrant dihedral angles, as illustrated in Fig. 1170 (e). Other draftsmen shade only the edges of projecting dihedral angles.

Shade lines generally do not add anything to the clearness of a pictorial view of an object, but they do improve the appearance of the drawing by adding variety and relief. The method of determining the shade lines on a pictorial view is the same as that explained above.

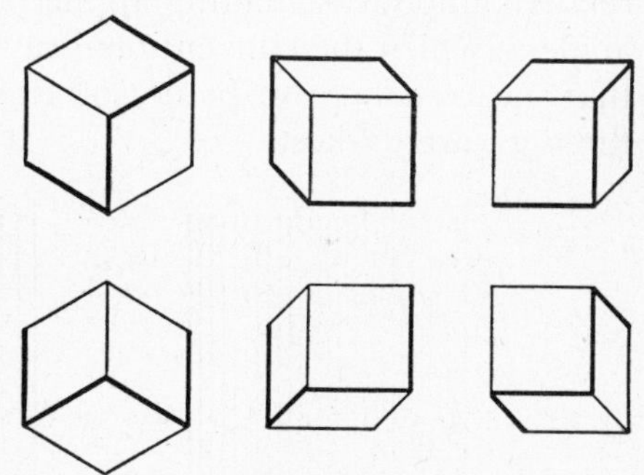

Fig. 1172 Shade Lines in Pictorial Drawings.

Figure 1172 shows the shade lines on a cube in isometric and in oblique projection according to this method; however, some draftsmen violate this convention for isometric, and shade the edges that intersect at the nearest corner, Fig. 1173.

Figure 1174 shows the shade lines on cylindrical solids. The circles in the top views and in the end view are shade lines only half-way around, the circle of the hole being shaded on the opposite side from that of the outer surface of the cylinder.

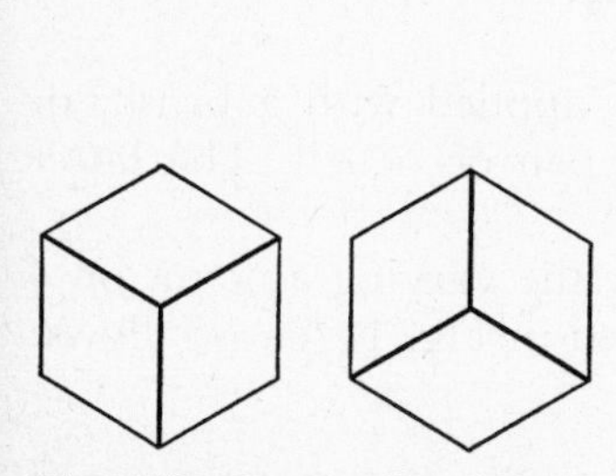

Fig. 1173 Conventional Shading of Isometric Drawings.

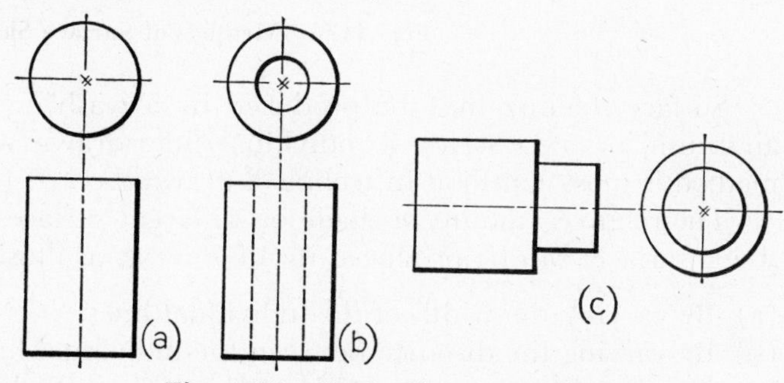

Fig. 1174 Shade Lines on Cylinders.

The shading of circles is done by shifting the center of the compass along a 45° line, and drawing a second semicircle with the same radius, as shown.

The additional width necessary to produce a shade line should be applied on the outside of the surface shaded; therefore, in cases like those shown in Fig. 1174 (c), the same center is used for both shade arcs.

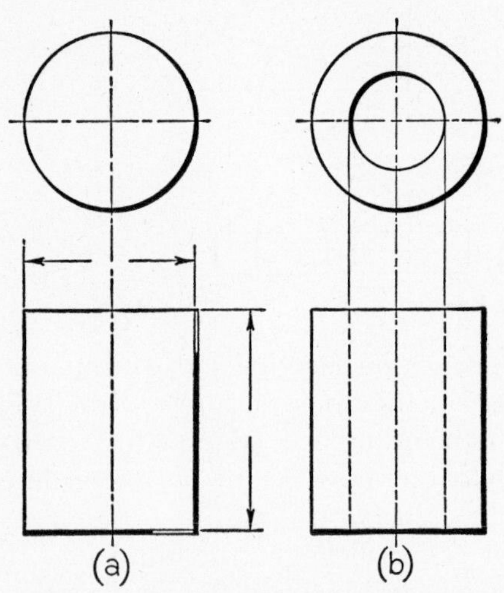

Fig. 1175 Dimensions on Shade Lines.

Dimension lines on shaded drawings should *read to the original edge of a surface before the shading was added*, as shown in Fig. 1175 (a), and not to the outer edge after the shading is added. It should also be noted that the invisible lines of the hollow cylinder, (b), are projected from the original circle in the top view *before the shading was added*, and not from the added shade circle.

718. Surface Shading. *Surface shading* is a means of utilizing further the effect of light in describing the shapes of objects. The conventional source of light and the direction of the rays are the same as those described above for shade lines.

A surface of an object appears lighter when the rays of light strike it at a large angle than when they strike at a small angle; therefore, the relative amount of shading applied to the surfaces of objects should depend upon the angle at which the conventional rays of light strike those surfaces. It is a general rule that, nearest the eye, plane surfaces in the light are lightest, and plane surfaces in the dark are darkest.

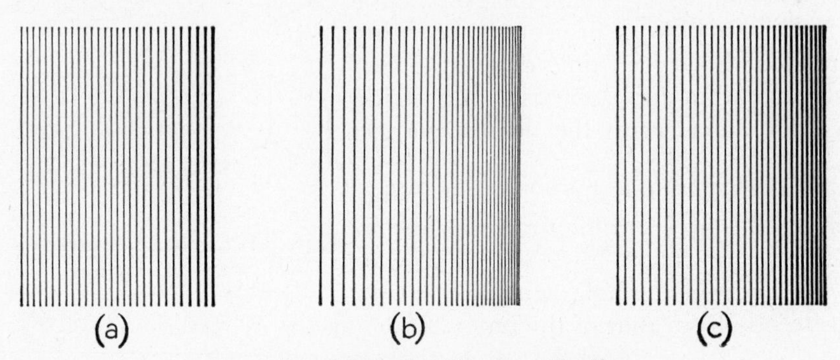

Fig. 1176 Methods of Surface Shading.

Surface shading may be produced by a wash or tint applied with a brush, or air-brush, or by a series of individual lines drawn with pen or pencil. The latter method is most common in technical drawing.

The relative amount of shade on different surfaces, or the varying amount on a single surface, can be produced in three ways, as illustrated in Fig. 1176, as follows:

(a) By varying the width of the individual lines.
(b) By varying the distance between the individual lines.
(c) By a combination of the first and second methods.

Figure 1177 shows different methods of applying surface shading on some common geometric solids. The gradual blending of the shade on these surfaces requires skill and practice.

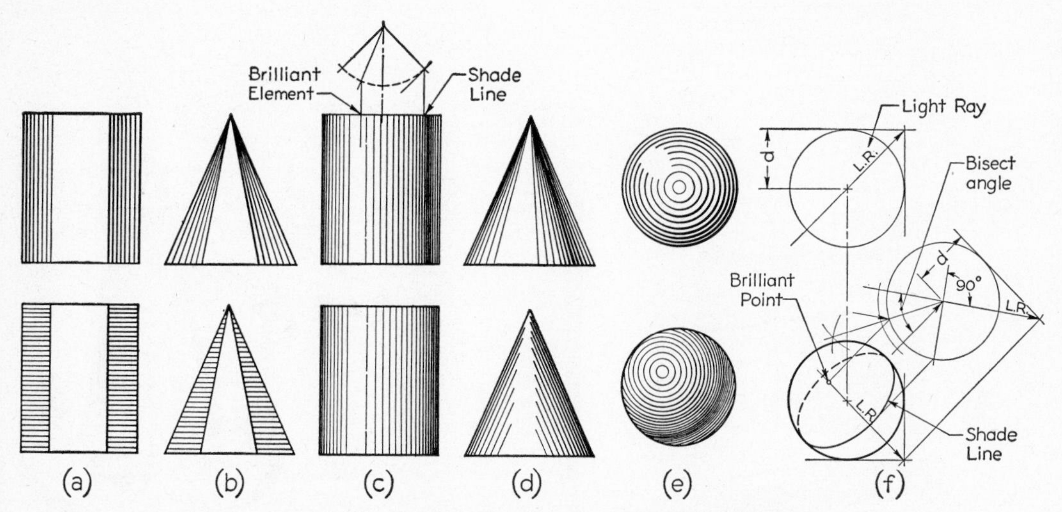

Fig. 1177 Surface Shading on Geometric Solids.

When surface shading is used on a drawing, the shade lines, §717, should be omitted unless the clearness of the drawing is enhanced by their use. The surface shading on a plane surface, inclined to the plane of projection, should grade from dark to light in the receding direction if the surface is in the shade, and from light to dark if the surface is in the light, as shown at the top in Fig. 1177 (a) and (b).

The surface shading on a cylinder and on a cone should be darkest at the element where the rays of light are tangent to the surface, and lightest at the *brilliant element*, as shown at (c) and (d). The brilliant element of a cylinder is the one from which the rays of light are reflected directly to the observer; it passes through the point at which the bisector of the angle between a ray of light and a visual ray through the center of the cylinder pierces the surface of the cylinder.

The surface shading of a sphere, (e), should be darkest where the rays of light are tangent to the sphere (on the shade line) and lightest at the *brilliant point*. The brilliant point on the surface of a sphere is the point in which the bisector of the angle between a ray of light and a visual ray through the center of the sphere pierces the surface of the sphere. The constructions for the determination of the brilliant element and the element of shade on the cylinder, and the brilliant point and the shade line on the surface of the sphere are shown in Fig. 1177 (c) and (f), respectively.

Line shading is used extensively in technical illustrations and in patent drafting with very satisfactory results, as shown in Figs. 1178, 1179, and 1185. See also Figs. 1180 through 1184.

719. Shading Exercises. The following problems are given to afford practice in determining the shade lines of objects, and in blending gradually the surface shading on curved surfaces. Use Size A sheets for Problems 1, 3, and 4, and Size B for Problem 2.

Prob. 1. Draw the two views of Fig. 873, Prob. 9. Shade the shade lines in both views, making them about twice as heavy as the other object lines. Omit the dimensions.

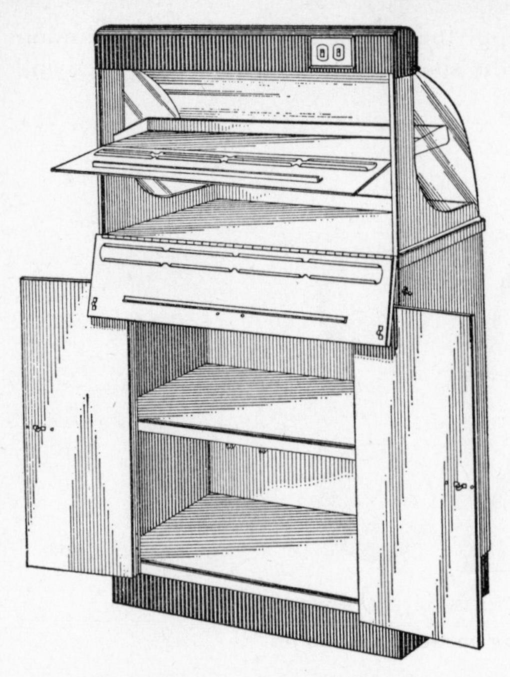

Fig. 1178 Surface Shading Applied to Pictorial Drawing of Display Case.

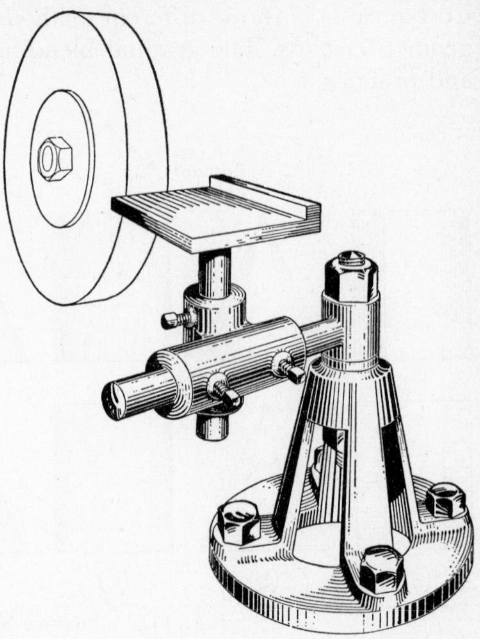

Courtesy A. M. Byers Co.

Fig. 1179 A Line-Shaded Drawing of an Adjustable Support for Grinding.

Prob. 2. Draw the two views of Fig. 577, and shade as in Prob. 1.

Prob. 3. Draw three rectangles about 2″ × 3″ each, the long sides being horizontal. Shade the surfaces of the rectangles similar to those of Fig. 1176, blending the shade of (a) gradually from dark at the top to light at the bottom, of (b) from dark at the bottom to light at the top, and of (c) from dark at the center to light at the top and bottom.

Prob. 4. Draw a right hexagonal prism, a right hexagonal pyramid, a right circular cylinder, and a right circular cone with the axes horizontal, similar to and about three times as large as those of Fig. 1177, and shade the surfaces.

Prob. 5. Select a part or an assembly from Figs. 772, 775, 777, 778, 794, and 873; draw it in oblique or axonometric projection, and shade it in a manner similar to that shown in Figs. 1178 and 1179.

CHAPTER 28

PATENT DRAFTING

720. Introduction. In view of the importance of properly executed drawings, it is essential that all drawings be mechanically correct and constitute complete illustrations of every feature of the invention claimed. The strict requirements of the Patent Office in this respect serve to facilitate the examination of applications and the interpretation of patents issued thereon.

To aid draftsmen in the preparation of drawings for submission in patent applications, the *Guide for Patent Draftsmen* was prepared and can be obtained from the Superintendent of Documents, U.S. Government Printing Office, Washington 25, D.C., at a cost of 15 cents per copy. The following instructions and illustrations are reproduced from that publication.

721. Drawings Required. The applicant for patent is required by statute to furnish a drawing of his invention whenever the nature of the case admits of it; this drawing must be filed with the application. Illustrations facilitating an understanding of the invention (for example, flow sheets in cases of processes, and diagrammatic views) may also be furnished in the same manner as drawings, and may be required by the Office when considered necessary or desirable.

722. Signature to drawing. Signatures are not required on the drawing if it accompanies and is referred to in the other papers of the application, otherwise the drawing must be signed. The drawing may be signed by the applicant in person or have the name of the applicant placed thereon followed by the signature of the attorney or agent as such.

723. Content of drawing. The drawing must show every feature of the invention specified in the claims. When the invention consists of an improvement on an old machine the drawing must when possible exhibit, in one or more views, the improved portion itself, disconnected from the old structure, and also in another view, so much only of the old structure as will suffice to show the connection of the invention therewith.

724. Standard for drawings. The complete drawing is printed and published when the patent issues, and a copy is attached to the patent. This work is done by the photolithographic process, the sheets of drawings being reduced about one-third in size. In addition, a reduction of a selected portion of the drawings of each application is published in the Official Gazette. It is therefore necessary for these and other reasons that the character of each drawing be brought as nearly as possible to a uniform standard of execution and excellence, suited to the requirements of the reproduction process and of the use of the drawings, to give the best results in the interests of inventors, of the Office, and of the public. The following regulations with respect to drawings are accordingly prescribed:

(*a*) *Paper and ink.* Drawings must be made upon pure white paper of a thickness corresponding to two-ply or three-ply Bristol board. The surface of the paper must be calendered and smooth and of a quality which will permit erasure and correction. India ink alone must be used for pen drawings to secure perfectly black solid lines. The use of white pigment to cover lines is not acceptable.

(*b*) *Size of sheet and margins.* The size of a sheet on which a drawing is made must be exactly 10 by 15 inches. One inch from its edges a single marginal line is to be drawn, leaving the "sight" precisely 8 by 13 inches. Within this margin all work must be included. One of the shorter sides of the sheet is regarded as its top, and, measuring down from the marginal line, a space of not less than $1\frac{1}{4}$ inches is to be left blank for the heading of title, name, number, and date, which will be applied subsequently by the Office in a uniform style.

(*c*) *Character of lines.* All drawings must be made with drafting instruments or by photolithographic process which will give them satisfactory reproduction characteristics. Every line and letter (signatures included) must be absolutely black. This direction applies to all lines however fine, to shading, and to lines representing cut surfaces in sectional views. All lines must be clean, sharp, and solid, and fine or crowded lines should be avoided. Solid black should not be used for sectional or surface shading. Freehand work should be avoided wherever it is possible to do so.

(*d*) *Hatching and shading.* Hatching should be made by oblique parallel lines, which may be not less than about one-twentieth inch apart.

Heavy lines on the shade side of objects should be used except where they tend to thicken the work and obscure reference characters. The light should come from the upper left-hand corner at an angle of 45°. Surface delineations should be shown by proper shading, which should be open.

(*e*) *Scale.* The scale to which a drawing is made ought to be large enough to show the mechanism without crowding when the drawing is reduced in reproduction, and views of portions of the mechanism on a larger scale should be used when necessary to show details clearly; two or more sheets should be used if one does not give sufficient room to accomplish this end, but the number of sheets should not be more than is necessary.

(*f*) *Reference characters.* The different views should be consecutively numbered figures. Reference numerals (and letters, but numerals are preferred) must be plain, legible and carefully formed, and not be encircled. They should, if possible, measure at least one-eighth of an inch in height so that they may bear reduction to one twenty-fourth of an inch; and they may be slightly larger when there is sufficient room. They must not be so placed in the close and complex parts of the drawing as to interfere with a thorough comprehension of the same, and therefore should rarely cross or mingle with the lines. When necessarily grouped around a certain part, they should be placed at a little distance, at the closest point where there is available space, and connected by lines with the parts to which they refer. They should not be placed upon hatched or shaded surfaces but when necessary, a blank space may be left in the hatching or shading where the character occurs so that it shall appear perfectly distinct and separate from the work. The same part of an invention appearing in more than one view of the drawing must always be designated by the same character, and the same character must never be used to designate different parts.

(*g*) *Symbols, legends.* Graphical drawing symbols for conventional elements may be used when appropriate, subject to approval by the Office. The elements for which such symbols are used must be adequately identified in the specification. While descriptive matter on drawings is not permitted, suitable legends may be used, or may be required in proper cases, as in diagrammatic views and flow sheets. The lettering should be as large as, or larger than, the reference characters.

(*h*) *Location of signature and names.* The signature of the applicant, or the name of the applicant and signature of the attorney or agent, may be placed in the lower right-hand corner of each sheet within the marginal line, or may be placed below the lower marginal line.

(*i*) *Views.* The drawing must contain as many figures as may be necessary to show the invention; the figures should be consecutively numbered if possible, in the order in which they appear. The figures may be plan, elevation, section, or perspective views, and detail views of portions or elements, on a larger scale if necessary may also be used. Exploded views, with the separated parts of the same figure embraced by a bracket, to show the relationship or order of assembly of various parts are permissible. When necessary a view of a large machine or device in its entirety may be broken and extended over several sheets if there is no loss in facility of understanding the view (the different parts should be identified by the same figure number but followed by the letters *a*, *b*, *c*, etc., for each part). The plane upon which a sectional view is taken should be indicated on the general view by a broken line, the ends of which should be designated by numerals corresponding to the figure number of the sectional view and have arrows applied to indicate the direction in which the view is taken. A moved position may be shown by a broken line superimposed upon a suitable figure if this can be done without crowding, otherwise a separate figure must be used for this purpose. Modified forms of construction can only be shown in separate figures. Views should not be connected by projection lines nor should center lines be used.

(*j*) *Arrangement of views.* All views on the same sheet must stand in the same direction and should if possible, stand so that they can be read with the sheet held in an upright position. If views longer than the width of the sheet are necessary for the clearest illustration of the invention, the sheet may be turned on its side. The space for a heading must then be reserved at the right and the signatures placed at the left, occupying the same space and position on the sheet as in the upright views and being horizontal when the sheet is held in an upright position. One figure must not be placed upon another or within the outline of another.

(*k*) *Figure for Official Gazette.* The drawing should, as far as possible, be so planned that one of the views will be suitable for publication in the Official Gazette as the illustration of the invention.

(*l*) *Extraneous matter.* An agent's or attorney's stamp, or address, or other extraneous matter, will not be permitted upon the face of a drawing, within or without the marginal line, except that the title of the invention and identifying indicia, to distinguish from other drawings filed at the same time, may be placed below the lower margin.

(*m*) *Transmission of drawings.* Drawings transmitted to the Office should be sent flat, protected by a sheet of heavy binder's board, or may be rolled for transmission in a suitable mailing tube; but must never be folded. If received creased or mutilated, new drawings will be required.

725. Informal drawings. The requirements of rule 84 relating to drawings will be strictly enforced. A drawing not executed in conformity thereto may be admitted for purpose of examination, but in such case the drawing must be corrected or a new one furnished, as required. The necessary corrections will be made by the Office upon applicant's request and at his expense.

726. Draftsman to make drawings. Applicants are advised to employ competent draftsmen to make their drawings.

The Office may furnish the drawings at the applicant's expense as promptly as its draftsmen can make them, for applicants who cannot otherwise procure them.

727. Return of drawings. The drawings of an accepted application will not be returned to the applicant except for signature.

A photographic print is made of the drawing of an accepted application.

728. Use of old drawings. If the drawings of a new application are to be identical with the drawings of a previous application of the applicant on file in the Office, or with part of such drawings, the old drawings or any sheets thereof may be used if the prior application is, or is about to be, abandoned, or if the sheets to be used are cancelled in the prior application. The new application must be accompanied by a letter requesting the transfer of the drawings, which should be completely identified.

729. Design Patents. The design must be represented by a drawing made in conformity with the rules laid down for drawings of mechanical inventions and must contain a sufficient number of views to constitute a complete disclosure of the appearance of the article. Appropriate surface shading must be used to show the character or contour of the surfaces represented.

Note: For mechanical and electrical symbols on patent drawings, see *Guide for Patent Draftsmen.*

A typical patent drawing is shown in Fig. 1185.

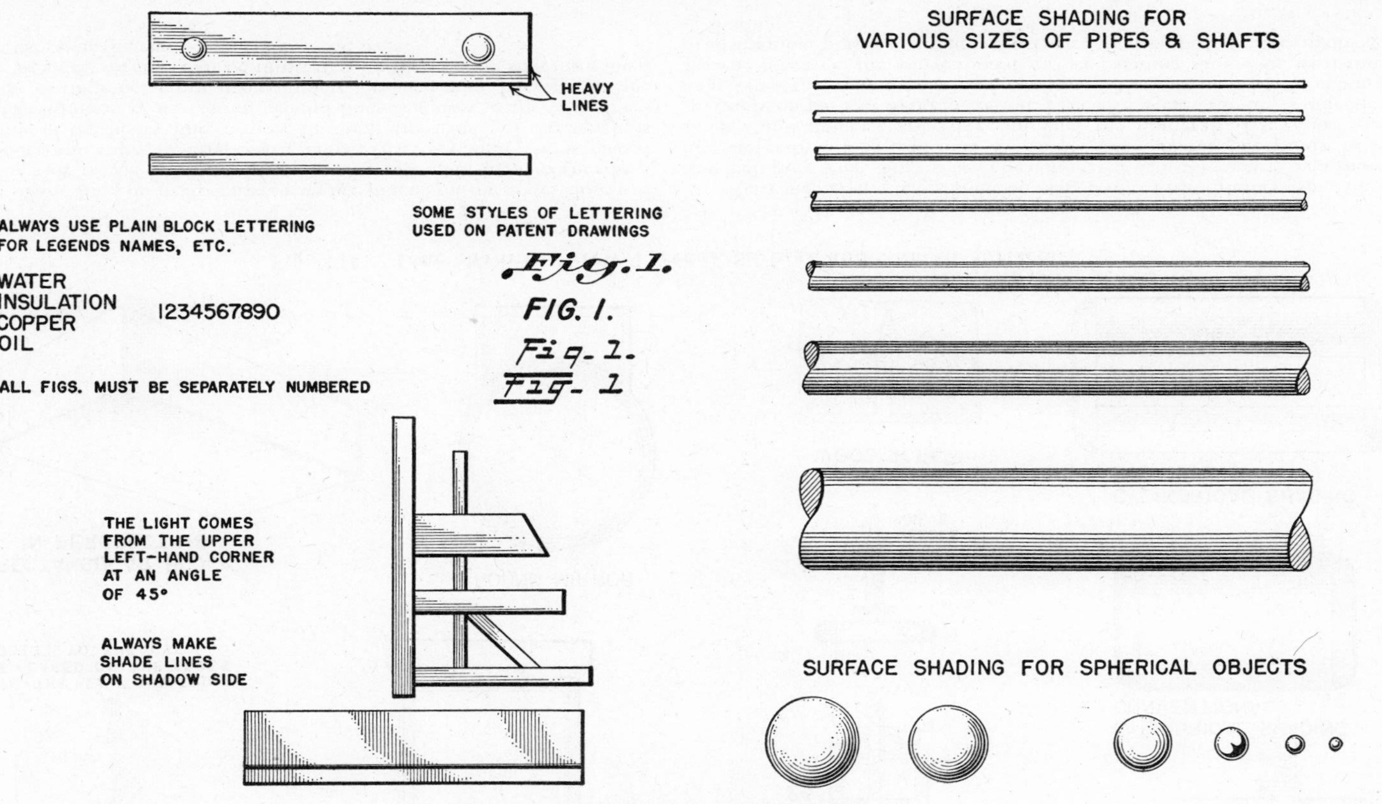

Fig. 1180 Line Shading and Lettering.

Letters and figures of reference must be carefully formed. Several types of lettering and figure marks are shown; however, the draftsman may use any style of lettering that he may choose. Place heavy lines on the shade side of objects, assuming that the light is coming from the upper left-hand corner at an angle of 45°. Make these heavy lines the same weight throughout the various views on the drawing. Descriptive matter is not permitted on patent drawings. Legends may be applied when necessary but only plain black lettering should be used. The different views should be consecutively numbered.

Surface delineations should be shown by proper shading. The figures show various types of surface shading. The amount of shading necessary depends on the size of the diameter of the shaft, etc. Note that a single heavy line on the shadow side is sufficient shading for small pipes, rods and shafts. When more than one shade line is used on cylindrical surfaces, the shading is blended from the second line. Note that the outer line is a light line. This rule on shading applies to spherical as well as cylindrical surfaces. Make all lines clear and sharp so that they will reproduce properly. India ink alone must be used for pen drawings to secure perfectly black solid lines.

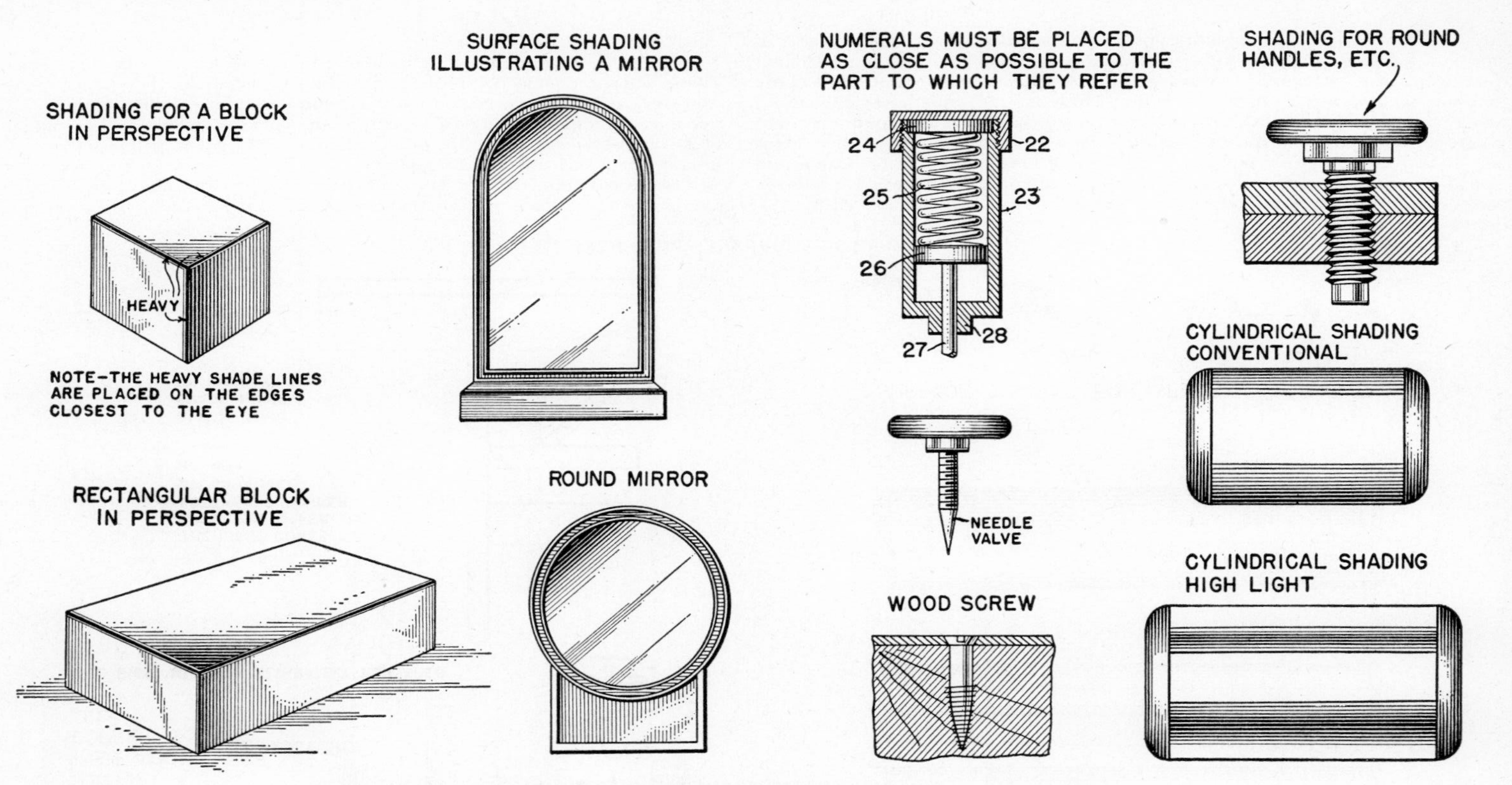

Fig. 1181 Line Shading, Screw Threads, Springs, and Curved Surfaces.

Heavy shade lines on perspective views are placed on the edges closest to the eye. The rule of the light coming from the upper left-hand corner at a 45° angle does not apply to perspective views. If a very light line is placed on either side of the heavy line, a more finished appearance of the article is obtained. The addition of horizontal ground lines as shown in the lower perspective view of the block emphasizes support for the same. The appearance of a mirror or shiny surfaces can be illustrated by the oblique shading shown on the two views on the right-hand side of the page.

Reference characters should be placed at a little distance from the parts to which they refer. They should be connected with these parts by a short lead line, never by a long lead line. When necessary, blank spaces must be left on shaded and hatched areas for applying the numerals. Use wood graining sparingly on parts of wood in section. Excessive wood graining is objectionable as it blurs the view and is very confusing. Various methods of shading are shown; however, the conventional surface shading should be used until the draftsman has obtained enough experience to attempt the more involved types of shading.

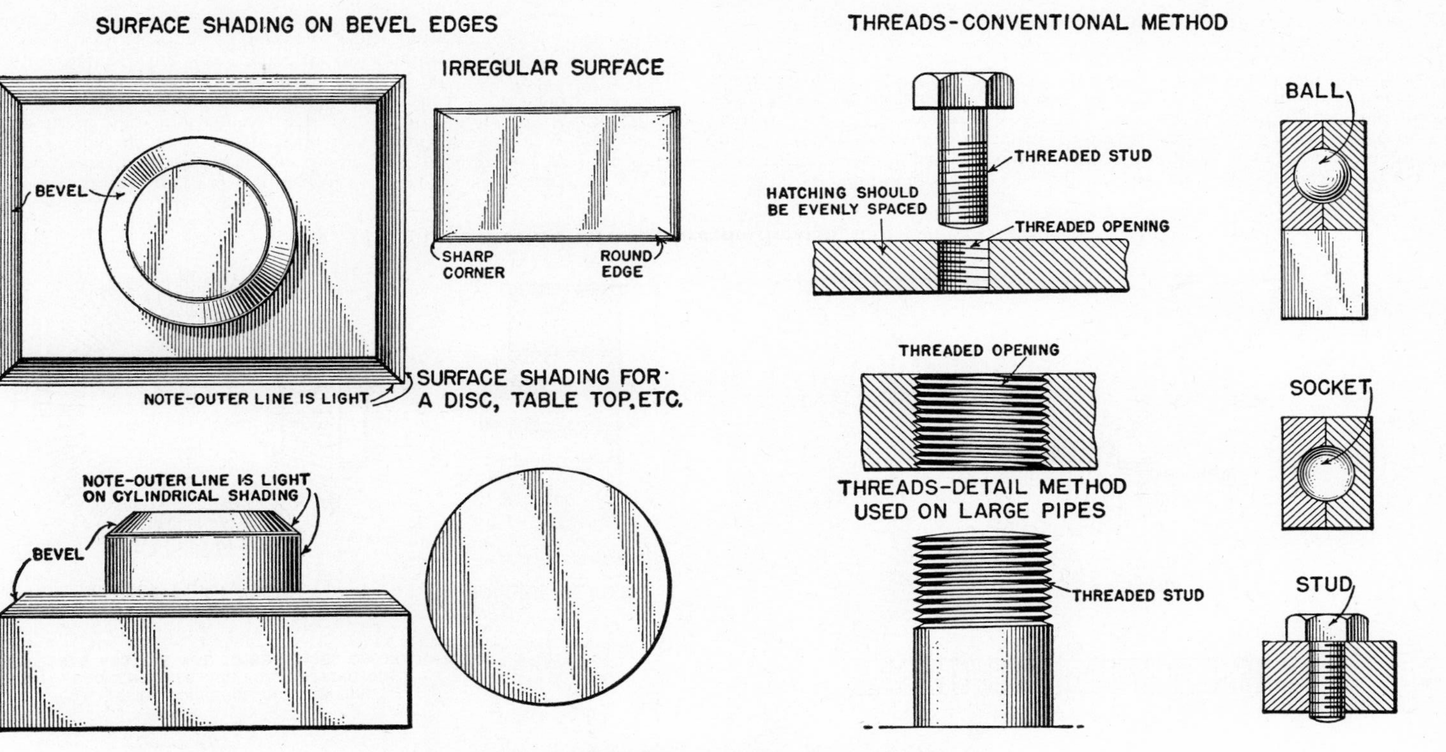

Fig. 1182 Inclined Surfaces, Threads, and Balls.

Inclined surfaces are distinguished from flat surfaces by using the shading shown on the illustration in the upper left-hand corner of the page. You will note that the outer line is always a light line. This gives a slanting effect to the surface as the heavy line is placed on the edge of the upper plane. The surface shading is blended from this heavy line giving the desired appearance. The other figures on the page show various methods of shading. Flat, shiny surfaces may be shown as illustrated in the circular figure. The scale to which a drawing is made ought to be large enough to show the mechanism without crowding.

Several methods of illustrating threads are shown in the figures above. The conventional thread may be shown on small bolts and openings. The detail method should be used on large pipes and threaded portions. Solid black shading as shown is very effective in illustrating the threads, but care should be used in applying same. Convex and concave surfaces are defined by the shading shown in the illustrations of the ball and socket. Plan the views properly so that one figure is not placed upon another or within the outline of another.

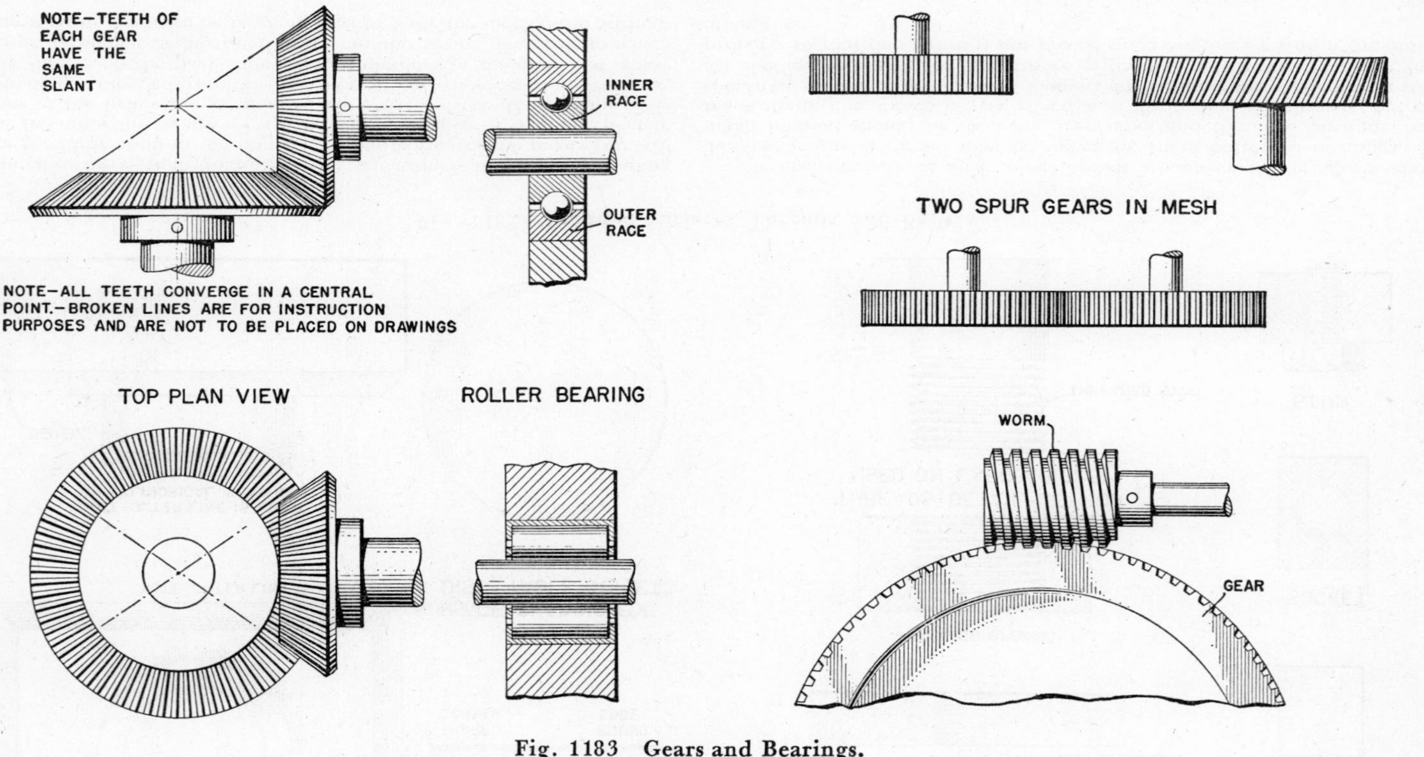
NOTE—TEETH OF EACH GEAR HAVE THE SAME SLANT
INNER RACE
OUTER RACE
TWO SPUR GEARS IN MESH
NOTE—ALL TEETH CONVERGE IN A CENTRAL POINT.—BROKEN LINES ARE FOR INSTRUCTION PURPOSES AND ARE NOT TO BE PLACED ON DRAWINGS
TOP PLAN VIEW
ROLLER BEARING
WORM
GEAR

Fig. 1183 Gears and Bearings.

The conventional method of illustrating bevel gears is clearly shown on the two figures on the left-hand side of the page. Particular care must be given to the correct spacing between the gear teeth and also to the weight of the shade lines used. Both must be correctly shown to obtain the desired effect. Two types of bearings are also shown. The roller bearing is clearly disclosed by the use of the conventional cylindrical shading. The fanciful black shading shown on the ball bearing is very effective in bringing out the idea of an object being shiny as well as round. The use of white pigment to cover lines is not acceptable.

The conventional method of showing a spur gear and a helical gear is shown on the two illustrations at the top of the page. The proper spacing between the gear teeth is essential in illustrating gears in mesh as shown on the central figure. A worm and worm gear in mesh is clearly illustrated in the lower figure on the page. Do not add the legends on the drawing. Every line and letter must be absolutely black. This direction applies to all lines however fine, to shading, and to lines representing cut surfaces in sectional views.

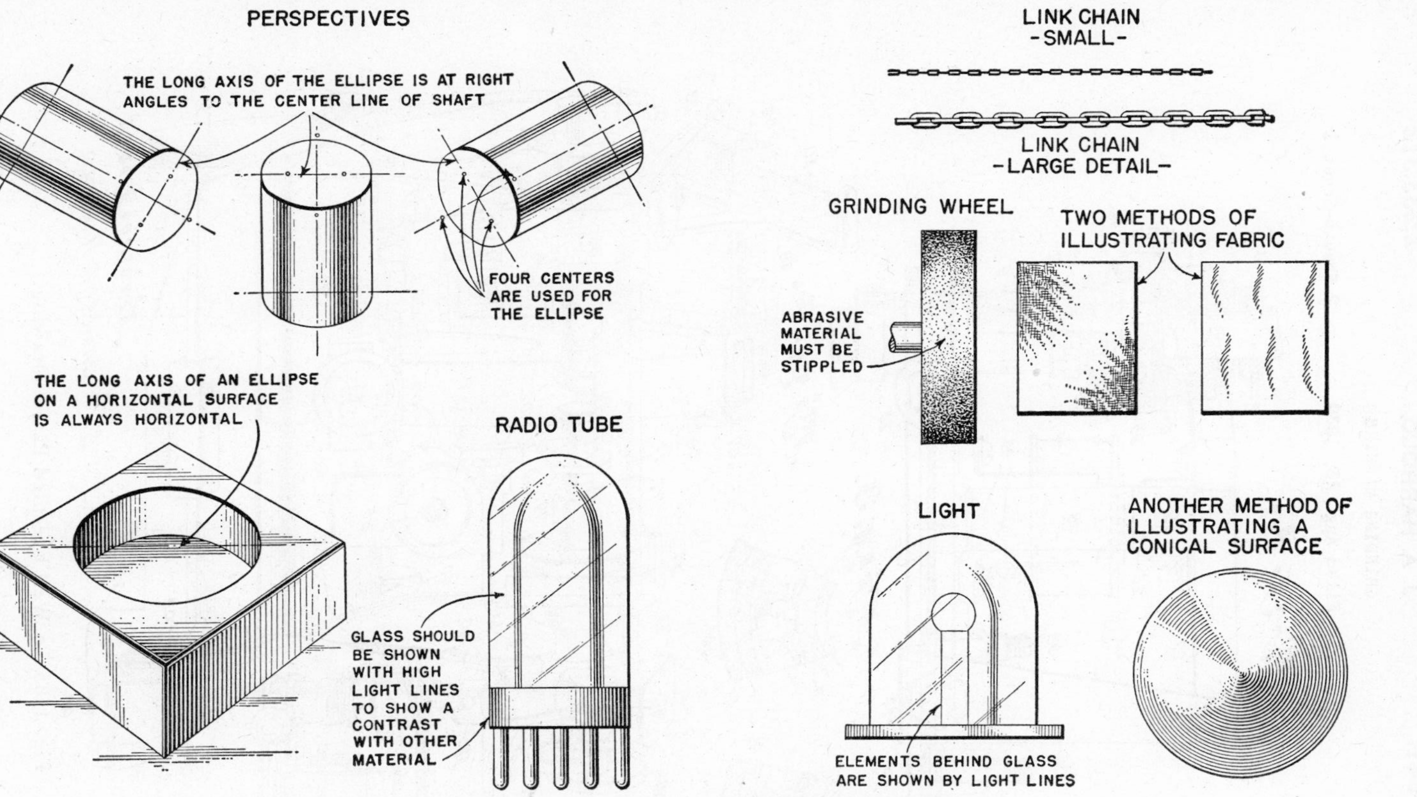
PERSPECTIVES
THE LONG AXIS OF THE ELLIPSE IS AT RIGHT ANGLES TO THE CENTER LINE OF SHAFT
FOUR CENTERS ARE USED FOR THE ELLIPSE
THE LONG AXIS OF AN ELLIPSE ON A HORIZONTAL SURFACE IS ALWAYS HORIZONTAL
RADIO TUBE
GLASS SHOULD BE SHOWN WITH HIGH LIGHT LINES TO SHOW A CONTRAST WITH OTHER MATERIAL
LINK CHAIN -SMALL-
LINK CHAIN -LARGE DETAIL-
GRINDING WHEEL
ABRASIVE MATERIAL MUST BE STIPPLED
TWO METHODS OF ILLUSTRATING FABRIC
LIGHT
ELEMENTS BEHIND GLASS ARE SHOWN BY LIGHT LINES
ANOTHER METHOD OF ILLUSTRATING A CONICAL SURFACE

Fig. 1184 Shading Perspective and Other Surfaces.

The four pictorial figures clearly explain the fundamental rules for determining the position of the long axis of the ellipse. Do not add the center lines as they are not permitted on patent drawings. These have been shown for instructive purposes only. Different types of shading are used when it is desired to show a contrast between materials as shown in the illustration of the radio tube. All drawings must be made with drafting instruments and every line and letter must be absolutely black. Free-hand work should be avoided wherever it is possible to do so.

Two illustrations of link chains are shown at the top of the page, the size of the view being the guiding factor in determining the correct showing. Abrasive material must be stippled as shown in the illustration of the grinding wheel. Irregular surfaces and objects that are impossible to properly show up with line shading must be stippled to bring out the desired effect. Free-hand shading should be used to designate fabric material. All elements behind glass should be shown in light full lines. The light oblique shade lines across the glass give the desired effect.

Feb. 25, 1941. J. A. HARROLD 2,233,312

GRINDING MACHINE

Filed Aug. 23, 1939 2 Sheets-Sheet 1

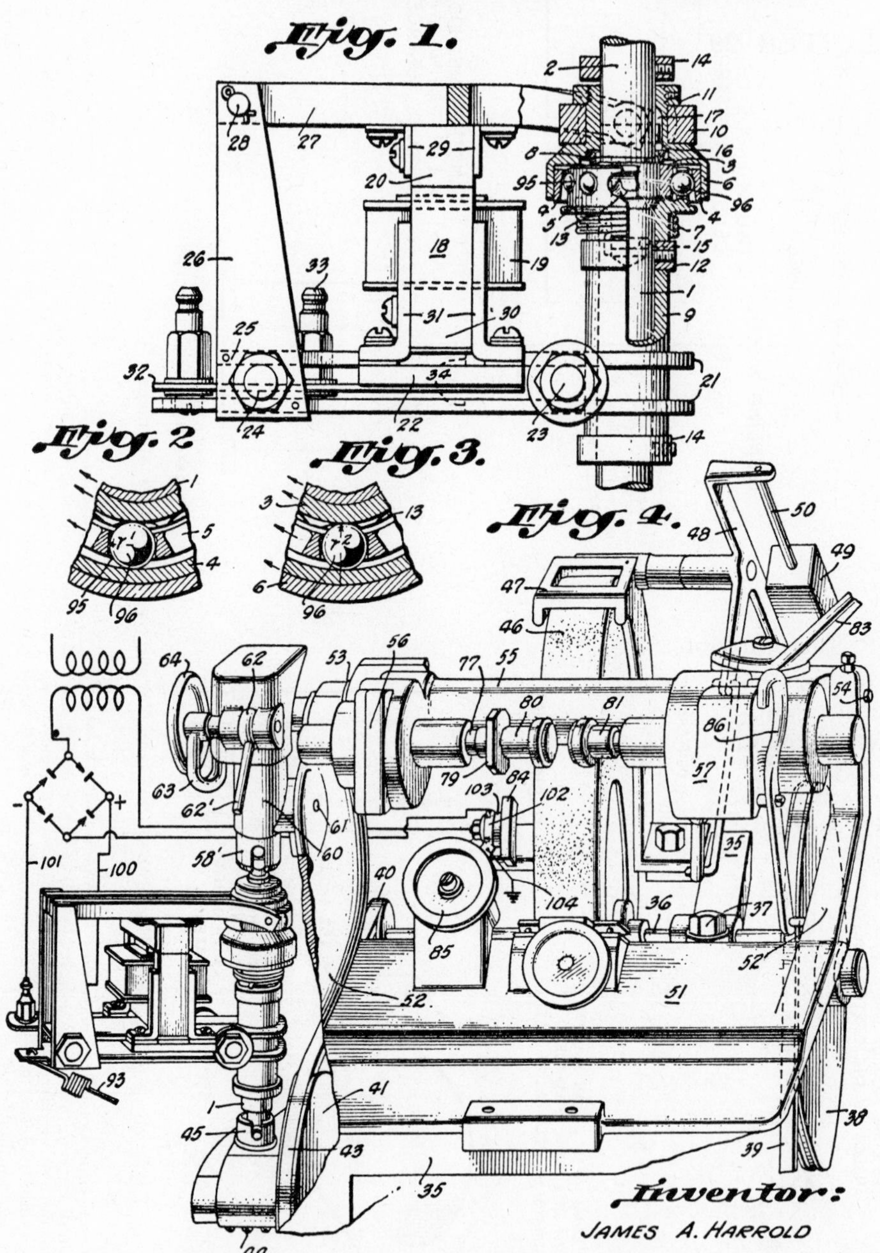

Fig. 1185 A Well Executed Patent Drawing.

CHAPTER 29

REPRODUCTION OF DRAWINGS

730. Introduction. After the drawings of a machine or other structure have been completed, it is usually necessary to preserve one or more copies for future reference and to supply copies to many different individuals and firms; it is therefore obvious that some means of exact, rapid, and economical reproduction must be employed.

731. Blueprint Process. Of the several processes in use for reproduction, the *blueprint* process is still probably the most generally used. Sir John Herschel discovered the process in 1837, and it was introduced into the United States in 1876 at the Philadelphia Centennial Exposition. It is essentially a photographic process in which the original tracing is the negative.

In all methods of reproducing drawings, the paper or cloth upon which the drawing is to be printed is coated or sensitized with a chemical preparation that is affected by the action of light. When such paper or cloth is exposed to light in a printing frame with the tracing so that the light must pass through the tracing to reach the sensitized paper, a chemical reaction is produced in all parts of the print except those which are protected by the opaque lines of the drawing.

After the paper has been exposed a sufficient length of time, it is removed from the frame, or the blueprint machine, and subjected to a developing bath and a fixing bath, or to a fixing bath only, according to the method employed.

When this process of printing was first employed, a *sun frame* was used, as shown in Fig. 1186.

In sunlight printing, the printing surface should be at right angles to the rays of light in order that the surface may receive the greatest amount of light and in order that the light may be diffused least under the lines of the drawing.

When only a small number of prints are required and better facilities are not available, prints can be made by exposing the tracing and print paper to sunlight while they are held against a window pane or under a piece of glass. The tracing

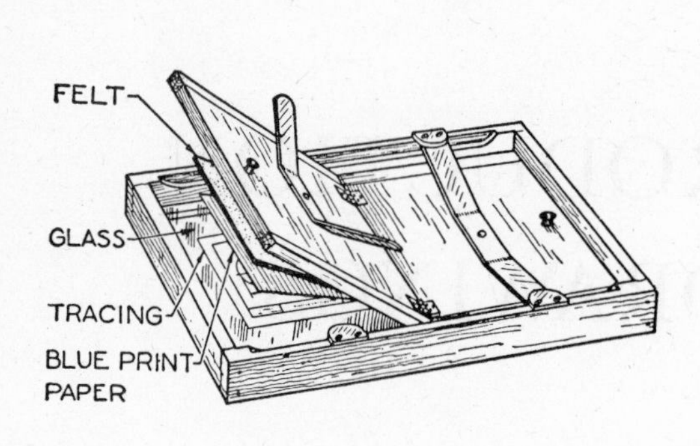

Fig. 1186 Sun Frame.

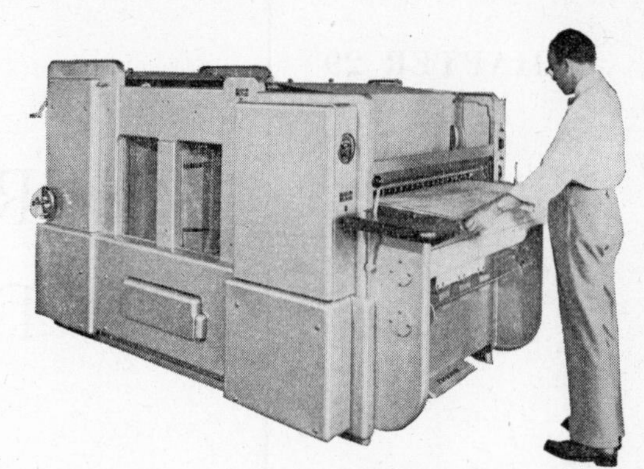

Courtesy Paragon-Revolute Corp.

Fig. 1187 Automatic Blueprinting Machine.

must be against the glass with its face toward the light, and the sensitized surface of the print paper against the back of the tracing.

Modern blueprint machines are available in *non-continuous* types in which cut sheets are fed through the blueprint machine for exposure only, and then washed in a separate washer. The *continuous* blueprint machine, Fig. 1187, combines exposure, washing, and drying in one continuous operation.

Blueprint papers are made by applying a coating of a solution of potassium ferricyanide and ferric ammonium citrate. In the old days, draftsmen often made their own blueprint paper, applying the solutions with a brush. Now blueprint papers are purchased in any desired quantity from manufacturers or dealers. They are available in various speeds and in rolls of various widths, or may be supplied in sheets of specified size. The coated side of fresh paper is a light greenish-yellow color. It will gradually turn to a grey-bluish color if not kept carefully away from light, and may eventually be rendered useless.

The length of exposure necessary depends not only upon the kind of paper used and the intensity of the light, but also upon the age of the paper. "The older the paper the quicker it prints and the longer to wash; the fresher the paper the slower it prints and the quicker to wash."

A print apparently ruined by overexposure may be saved by being washed in a solution of potassium dichromate. Hydrogen peroxide may be used for the same purpose.

Notations and alterations can be made on blueprints with any alkaline solution of sufficient strength to destroy the blue compound; for instance, with a 1.5 per cent solution of caustic soda.

Blueprints can be made from a typewritten sheet if carbon paper has been used with the carbon side turned over, so as to produce black imprints on both sides of the sheet.

Although best results are obtained when the original tracing is drawn in ink on cloth or vellum, excellent prints may be made from penciled drawings or tracings if the tracing paper or pencil tracing cloth is of good quality and if the draftsman has made all required lines and lettering jet black.

732. Vandyke Prints and Blue-Line Blueprints. A *negative Vandyke print* is composed of white lines on a dark brown background made by printing, in the same manner as for blueprinting, upon a special thin Vandyke paper from an original pencil or ink tracing. This negative Vandyke is then used as an original to make positive blueprints or "blue-line blueprints." In this way the Vandyke print replaces the original tracing as the negative, and the original is not subjected to wear each time a run of prints is made. The blue-line blueprints have blue lines on white backgrounds, and are often preferred because they can be easily marked upon with an ordinary pencil or pen. They have the disadvantage of soiling easily in the shop.

733. Diazzo-Moist Prints. A black-and-white print, composed of nearly black lines on a white background, may be made from ordinary pencil or ink tracings by exposure in the same manner as for blueprints, directly upon special black-print paper, cloth, or film.

Exposure may be made in a blueprint machine or any machine using light in a similar way. However, the prints are not washed as in blueprinting, but must be fed through a special developer which dampens the coated side of the paper with a developing solution.

A popular combination printer (exposer) and developer, the Bruning *Copyflex*, is shown in Fig. 1188. Two operations are involved: the tracing and BW paper are fed into the printer slot; and when they emerge, the BW paper is then fed through the developer slot. Within a minute or two after developing, the prints are practically dry and are ready for use.

BW colored-line prints in red, brown, or blue lines on white backgrounds may be made on the same machine simply by using the appropriate paper in each case.

Courtesy Charles Bruning Co., Inc.

Fig. 1188 Bruning Copyflex.

Courtesy General Aniline & Film Corp.

Fig. 1189 Ozalid Streamliner.

These prints, together with *diazzo-dry* prints, §734, are coming into greater use, and are largely replacing the more cumbersome blueprint process.

734. The Diazzo-Dry Process. The *diazzo-dry* process is based on the sensitivity to light of certain dyestuff intermediates which have the characteristic of decomposing into colorless substances if exposed to actinic light, and of reacting with coupling components to form an azo dyestuff upon exposure to ammonia vapors. It is a contact method of reproduction, and depends upon the transmission of light through the original for the reproduction of positive prints. The subject matter may be pen or pencil lines, typewritten or printed matter, or any opaque image. There is no negative step involved; positives are used to obtain positive prints. Sensitized materials can be handled under normal indoor illumination.

The diazzo whiteprint method of reproduction consists of two simple steps—exposure and dry development by means of ammonia vapors. Exposure is made in a printer equipped with a source of ultra-violet light, a mercury vapor lamp, or carbon arc, or even by sunlight. The light emitted by these light sources brings about a photochemical decomposition of the light-sensitive yellow coating of the paper except in those places where the surface is protected by the opaque lines of the original. The exposed print is developed dry in a few seconds in a dry-developing machine by the alkaline medium produced by ammonia vapors.

A popular exposer and developer combined in one machine is the *Ozalid Streamliner* shown in Fig. 1189. Two operations are involved: (1) the tracing and the sensitized paper are fed into the printer slot for exposure to light as shown in Fig. 1190 (a); and (2) the paper is then fed through the developer slot for exposure to ammonia vapors, as shown at (b). If it is desired to remove the ammonia odor completely, the

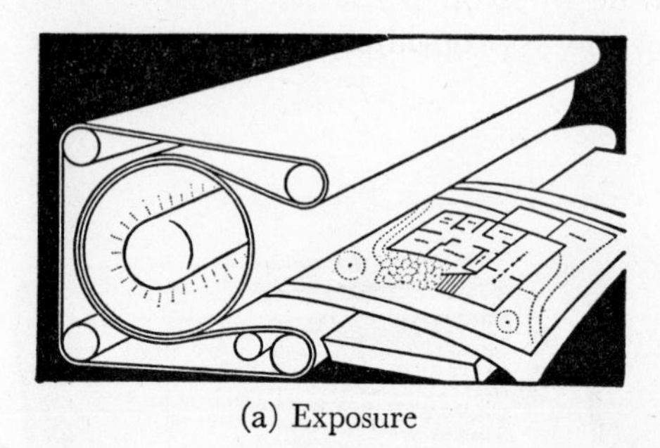

(a) Exposure

(b) Ammonia Development

Courtesy General Aniline & Film Corp.

Fig. 1190 Exposure and Development.

print is then fed through the printer with the back of the sheet next to the warm glass surrounding the light.

Ozalid prints may have black, blue, or red lines on white backgrounds, according to which paper is used. All have the advantage of being easily marked upon with pencil, pen, or crayon.

Ozalid "intermediates" are made in the same manner as regular prints, but upon special translucent paper, cloth, or foil (transparent cellulose acetates). These are used in place of the original to produce regular prints. They may be used to save wear on the original or to permit changes to be made. Changes may be made by painting out parts with correction solution, Fig. 1191 (a) and (b), and then drawing the new lines or lettering directly on the intermediate in pencil or ink, (c). Special masking and cut-out techniques can also be used.

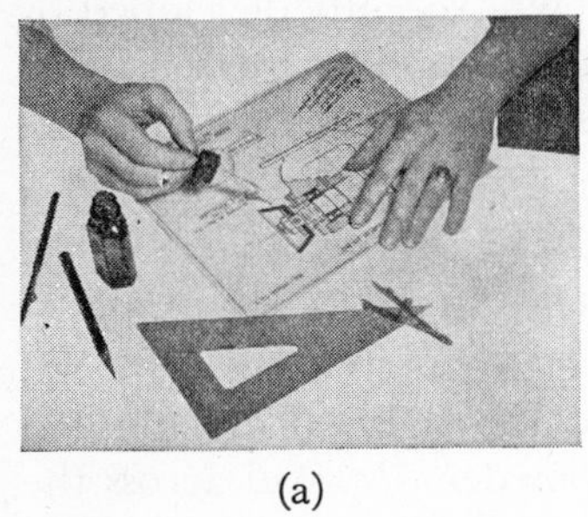
(a)

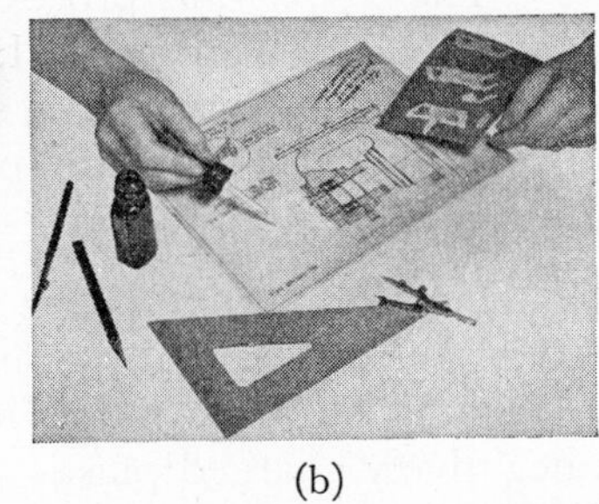
(b)

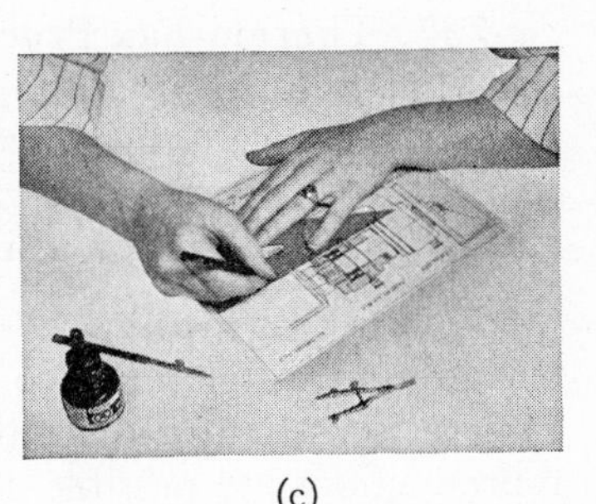
(c)

Courtesy General Aniline & Film Corp.

Fig. 1191 Changing a Drawing.

Several types of Ozalid foil intermediates are available, including matte surfaces to facilitate pen and pencil additions. By means of foils, many new procedures are possible, such as a composite print in which a wiring system is superimposed over a drawing.

Most manufacturers of reproduction equipment now carry ammonia-developing machines of this type.

735. Photographic Contact Prints. Either transparent or opaque drawings may be reproduced the same size as the original by means of *contact printing*. The original is pressed tightly against a sheet of special photographic paper, either by mechanical spring pressure or by means of suction as in the "vacuum printer." The paper is exposed by the action of the transmitted or reflected light, and the print is developed in the manner of a photograph in a dark or semi-dark room. To give a negative, a positive print is made from the negative by again contact printing on the equipment.

The *Portograph*, manufactured by Remington Rand Inc., is a popular machine for making contact prints of this type. The Portograph is available in various sizes with copying areas from 10″ × 15″ up to 40″ × 60″.

Excellent duplicate tracings can be made on paper, on film, and on either opaque map cloth or transparent tracing cloth through this process. Poor pencil drawings or tracings can be duplicated and improved by intensifying the lines so as to be much better than the original. Pencil drawings can be transformed into "ink-like" tracings. Also, by this process, a reproduction can be made directly from a blueprint.

The Eastman Kodak Company has developed a new "Kodagraph Autopositive" paper by means of which an excellent positive print can be made directly *from either a transparent or opaque original on an ordinary blueprint machine, ammonia-developing machine, Copyflex machine, or any similar machine, without use of a darkroom or costly photographic equipment.*

736. Duplicate Tracings on Cloth. Specially-prepared tracing cloth is available upon which a drawing may be reproduced from a negative Vandyke print of the original. Exposure of the duplicate tracing cloth may be made in a regular blueprint machine or in any machine employing light in the same way.

Excellent duplicate tracings can also be produced on the Ozalid machine with special Ozalid materials, on the Bruning machine with special Bruning materials, or on any of the several types of photographic print machines, as described in §735.

737. Thermo-Fax Process. The *Thermo-Fax* process was recently developed by the Minnesota Mining and Manufacturing Company. In this process the sensitized paper is sensitive to heat instead of light, the heat being supplied by a 2500 W tungsten lamp. The prints are easily and rapidly made in special Thermo-Fax printers, and have black lines on a light amber background.

738. Xerography. *Xerox* prints are positive prints with black lines on a white background. A selenium-coated and electrostatically charged plate is used. A special camera is used to project the original onto the plate; hence, reduced or enlarged reproductions are possible. A negatively charged plastic powder is spread across the plate and adheres to the positively charged areas of the image. The powder is then transferred to paper by means of a positive electric charge and then baked onto the surface to produce the final print. A large number of inexpensive prints can be made, and plates can be re-used.

739. Microfilm. *Microfilm* is coming into considerable use, especially where large numbers of drawings are involved, to furnish complete duplication of all drawings in a small filing space. The films may be 16 mm., 35 mm., or 70 mm., and may be produced in rolls or individually mounted on cards. For quick reference, the film can be easily projected in special readers or projectors. In many cases, the principal reason for microfilming is to provide duplicate copies for security reasons. Enlargement copies can be made from the film through photographic, electrostatic, or photocopy methods.

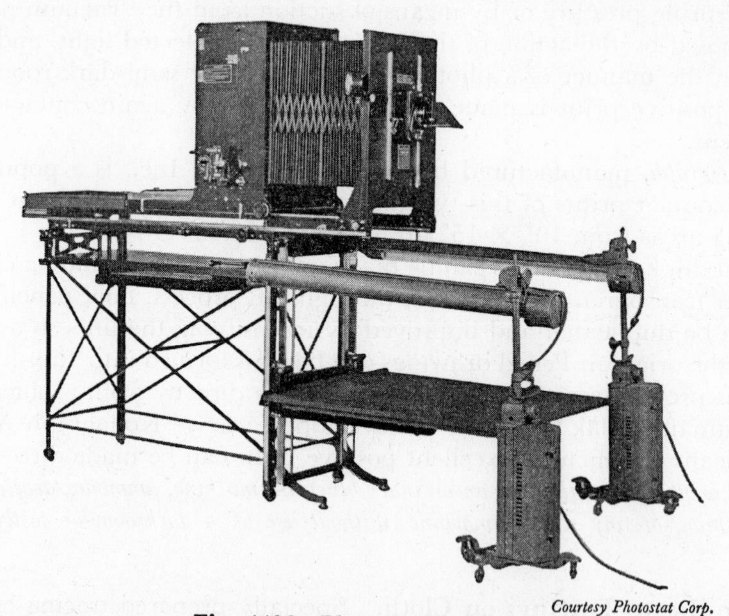

Courtesy Photostat Corp.

Fig. 1192 Photostat Machine.

740. Photostats. The *Photostat* machine, Fig. 1192, is essentially a specialized camera. A photostat print may be the same size, or larger or smaller than the original, while photographic contact prints, §735, must be the same size.

The original may be transparent or opaque. It is simply fastened in place, the camera is adjusted to obtain the desired size of print, and the print is made, developed, and dried in the machine (no darkroom is required). The result is a negative print with white lines on a near-black background. A positive print having near-black lines on a white background is made by photostating the negative print.

741. Line Etching. *Line etching* is a photographic method of reproduction. The drawing, in black lines on white paper or on tracing cloth, is placed in a frame behind a glass and photographed. This photographic negative is then mounted on a pane of glass and is printed upon a sheet of planished zinc or copper. After the print has been specially treated to render the lines acid-resisting, the plate is washed in a nitric acid solution, which eats away the metal between the lines, leaving them standing above the surface of the plate like type. The plate is then mounted upon a hardwood base, which can be used in any printing press as are blocks of type. All the line drawings in this book were reproduced by this process.

742. Mimeographing and Hectographing. While *mimeographing* is especially adaptable for reproducing typed material, it can also be very satisfactory in reproducing small and relatively simple drawings. The excellence of the reproduction of such drawings will depend upon the skill of the draftsman in drawing upon the stencil. The A. B. Dick Company, of Chicago, manufacturers of the Mimeograph, have developed a photochemical process by means of which a complicated drawing may be reduced and incorporated into the stencil, which is then used to produce very satisfactory prints.

In the *hectographing* process, an original is produced by typing on plain paper through a special carbon paper, or drawing with a special pencil or ink. This sheet is then brought into contact with a gelatin pad which absorbs the coloring from the lines of the original. The original is then removed, and prints are produced by bringing sheets of blank paper in contact with the gelatin. A number of different machines using this basic principle are available.

743. Offset Printing. The *Photolith*, *Multilith*, and *Planograph* methods are generally known as *offset printing*. A camera is used to reproduce the original, enlarged or reduced if necessary, upon an aluminum or zinc sheet. This master plate is then mounted on a rotary drum which revolves in contact with a rubber roller which picks up the ink from the image and transfers it to the paper. The prints are excellent positive reproductions.

APPENDIX

CONTENTS OF APPENDIX

1. BIBLIOGRAPHY OF TECHNICAL DRAWING AND RELATED SUBJECTS

Aeronautical Drafting

Anderson, Newton. *Aircraft Layout and Detail Design.* McGraw-Hill
Katz, Hyman H. *Aircraft Drafting.* Macmillan
Le Master, C. A. *Aircraft Sheet Metal Work.* American Technical Society
Liming, Roy. *Practical Analytic Geometry with Applications of Aircraft.* Macmillan
Meadowcroft, Norman. *Aircraft Detail Drafting.* McGraw-Hill
SAE Aeronautical Drafting Manual. Society of Automotive Engineers, Inc., 485 Lexington Ave., New York 17, N. Y.
Svensen, C. L. *A Manual of Aircraft Drafting.* D. Van Nostrand

American Standards

American Standards Association, 70 East 45th St., New York 17, N. Y. For complete listing of standards, see ASA *Price List and Index.*

Abbreviations
Abbreviations for Use on Drawings, Z32.13-1950
Abbreviations for Scientific and Engineering Terms, Z10.1-1941

Bolts and Screws
Hexagon Head Cap Screws, Slotted Head Cap Screws, Square Head Set Screws, and Slotted Headless Set Screws, B18.6.2-1956
High-Strength High-Temperature Internal Wrenching Bolts, B18.8-1950
Plow Bolts, B18.9-1950
Round Head Bolts, B18.5-1952
Slotted and Recessed Head Screws, Machine and Tapping Types, B18.6-1947
Slotted and Recessed Head Wood Screws, B18.6.1-1956
Socket Head Cap Screws and Socket Set Screws, B18.3-1954
Square and Hexagon Bolts and Nuts, B18.2-1955
Track Bolts and Nuts, B18.10-1952

Charts and Graphs
A Guide for Preparing Technical Illustrations for Publication and Projection, Y15
Time-Series Charts, Manual of Design and Construction, Z32.2-1937 (R 1947)

Dimensioning and Surface Finish
Preferred Limits and Fits for Cylindrical Parts, B4.1-1955
Rules for Rounding Off Numerical Values, Z25.1-1940 (R 1947)
Scales to Use with Decimal-Inch Dimensioning, Z75.1-1955
Surface Roughness, Waviness and Lay, B46.1-1955

Drafting Standards Manual
Sec. 1 Size and Format, Y14.1-1957
Sec. 2 Line Conventions, Sectioning and Lettering, Y14.2-1957
Sec. 3 Projections, Y14.3-1957
Sec. 4 Pictorial Drawing, Y14.4-1957
Sec. 5 Dimensioning and Notes, Y14.5-1957
Sec. 6 Screw Threads, Y14.6-1957

The following sections are in preparation (as of June, 1958):
- Sec. 7 Gears, Splines and Serrations, Y14.7
- Sec. 8 Castings, Y14.8
- Sec. 9 Forging, Y14.9
- Sec. 10 Metal Stamping, Y14.10
- Sec. 11 Plastics, Y14.11
- Sec. 12 Die Castings, Y14.12
- Sec. 13 Springs, Helical and Flat, Y14.13
- Sec. 14 Structural Drafting, Y14.14
- Sec. 15 Electrical Diagrams, Y14.15
- Sec. 16 Tools, Dies and Gages, Y14.16
- Sec. 17 Hydraulic and Pneumatic Diagrams, Y14.17

Gears
- Fine-Pitch Straight Bevel Gears, B6.8-1950
- Gear Nomenclature, Terms, Definitions, and Illustrations, B6.10-1955
- Letter Symbols for Gear Engineering, B6.5-1954
- Spur Gear Tooth Form, B6.1-1932
- System for Straight Bevel Gears, B6.13-1955
- 20-Degree Involute Fine-Pitch System, B6.7-1956

Graphical Symbols
- Graphical Symbols for Architectural Plans, Y32.9-1943
- Graphical Symbols for Electrical Diagrams, Y32.2-1954
- Graphical Symbols for Heat-Power Apparatus, Z32.2.6-1950 (R 1956)
- Graphical Symbols for Heating, Ventilating, and Air Conditioning, Z32.2.4-1949 (R 1953)
- Graphical Symbols for Plumbing, Y32.4-1955
- Graphical Symbols for Pipe Fittings, Valves, and Piping, Z32.2.3-1949 (R 1953)
- Graphical Symbols for Use on Maps and Profiles, Y32.7-1957
- Graphical Symbols for Welding, Z32.2.1-1949 (R 1953)
- Mathematical Symbols, Z10f-1928

Keys and Pins
- Machine Pins, B5.20-1954
- Woodruff Keys, Keyslots, and Cutters, B17f-1930 (R 1954)

Piping
- Cast-Iron Pipe Centrifugally Cast in Sand-Lined Molds, A21.8-1953
- Cast-Iron Pipe Flanges and Flanged Fittings (WSP of 25 lb), B16b2-1931 (R 1952)
- Cast-Iron Pipe Flanges and Flanged Fittings, Class 125, B16.1-1948 (R 1953)
- Cast-Iron Pipe Flanges and Flanged Fittings, Class 250, B16b-1944 (R 1953)
- Cast-Iron Screwed Fittings, 125 and 250 lb, B16.4-1949 (R 1953)
- Ferrous Plugs, Bushings, and Locknuts with Pipe Threads, B16.14-1949 (R 1953)
- Malleable-Iron Screwed Fittings, 150 lb, B16.3-1951
- Malleable-Iron Screwed Fittings, 300 lb, B16.19-1951
- Steel Butt-Welding Fittings, B16.9-1951
- Steel Pipe Flanges and Flanged Fittings, B16.5-1953
- Wrought-Steel and Wrought-Iron Pipe, B36.10-1950
- (For additional piping standards, see ASA *Price List and Index*)

Rivets
- Large Rivets ($\frac{1}{2}$ Inch Nominal Dia. and Larger), B18.4-1950 (R 1957)
- Small Solid Rivets, B18.1-1955

Small Tools and Machine Tool Elements

Jig Bushings, B5.6-1941 (R 1949)
Machine Tapers, Self-Holding and Steep Taper Series, B5.10-1953
Milling Cutters, Nomenclature, Principal Dimensions, etc., B5.3-1950
Reamers, B5.14-1949 (R 1955)
T-Slots—Their Bolts, Nuts, Tongues, and Cutters, B5.1-1949
Taps, Cut and Ground Threads, B5.4-1948
Twist Drills, Straight Shank and Taper Shank, B5.12-1950

Threads

Acme Screw Threads, B1.5-1952
Buttress Screw Threads, B1.9-1953
Nomenclature, Definitions, and Letter Symbols for Screw Threads, B1.7-1949 (R 1953)
Pipe Threads, B2.1-1945
Stub Acme Screw Threads, B1.8-1952
Unified and American Screw Threads for Screws, Bolts, Nuts, and Other Threaded Parts, B1.1-1949

Washers

Lock Washers, B27.1-1950
Plain Washers, B27.2-1953

Miscellaneous

Knurling, B5.30-1953
Preferred Thicknesses for Uncoated Thin Flat Metals, B32.1-1952

Architectural Drawing

Field, W. B. *Architectural Drawing*. McGraw-Hill
Kenny, J. E. and McGrail, J. P. *Architectural Drawing for the Building Trades*. McGraw-Hill
Martin, C. L. *Architectural Graphics*. Macmillan
Morgan, S. W. *Architectural Drawing*. McGraw-Hill
Ramsey, C. G. and Sleeper, H. R. *Architectural Graphic Standards*. John Wiley
Sleeper, H. R. *Architectural Specifications*. John Wiley

Blueprint Reading

DeVette, W. A. and Kellogg, D. E. *Blueprint Reading for the Metal Trades*. Bruce Pub. Co., Milwaukee
Heine, G. M. and Dunlap, C. H. *How to Read Electrical Blueprints*. American Technical Society, Chicago
Ihne, R. W. and Streeter, W. E. *Machine Trades Blueprint Reading*. American Technical Society, Chicago
Kenny, J. E. *Blueprint Reading for the Building Trades*. McGraw-Hill
Lincoln Electric Co. *Simple Blueprint Reading* (Welding). Cleveland, O.
Owens, A. A. and Slingluff, B. F. *How to Read Blueprints*. John C. Winston Co., Philadelphia
Spencer, H. C. and Grant, H. E. *The Blueprint Language*. Macmillan
Svensen, C. L. and Street, W. E. *A Manual of Blueprint Reading*. D. Van Nostrand
Wright, William N. *A Simple Guide to Blueprint Reading* (Aircraft). McGraw-Hill

Cams

Furman, F. Der. *Cams, Elementary and Advanced*. John Wiley
Rothbert, H. A. *Cams*. John Wiley

Charts and Graphs

Haskell, A. C. *How to Make and Use Graphic Charts.* Codex Book Co. New York, N. Y.
Karsten, K. G. *Charts and Graphs.* Prentice-Hall
Leicey, N. W. *Graphic Charts,* Lefax order No. 11-248; Lefax, Philadelphia
Lutz, R. R. *Graphic Presentation Simplified.* Funk and Wagnalls
Schmid, C. *Handbook of Graphic Presentation.* Ronald Press

Descriptive Geometry

Grant, H. E. *Practical Descriptive Geometry.* McGraw-Hill
Hood, G. J. and Palmerlee, A. S. *Geometry of Engineering Drawing.* McGraw-Hill
Howe, H. B. *Descriptive Geometry.* Ronald Press
Johnson, L. O. and Wladaver, I. *Elements of Descriptive Geometry.* Prentice-Hall
Levens, A. S. and Eggers, H. *Descriptive Geometry.* Harper
Pare, E. G., Hill I. L. and Loving, R. O., *Descriptive Geometry.* Macmillan
Rowe, C. E. and McFarland, J. D. *Engineering Descriptive Geometry.* D. Van Nostrand
Street, W. E. *Technical Descriptive Geometry.* D. Van Nostrand
Warner, F. M. *Applied Descriptive Geometry.* McGraw-Hill
Wellman, B. L. *Technical Descriptive Geometry.* McGraw-Hill

Descriptive Geometry Workbooks

Grant, H. E. *Practical Descriptive Geometry Problems,* H. E. Grant, Washington Univ.
Howe, H. B. *Problems for Descriptive Geometry.* Ronald Press
Johnson, L. O. and Wladaver, I. *Elements of Descriptive Geometry* (Problems). Prentice-Hall
Pare, E. G., Hill, I. L. and Loving, R. O., *Descriptive Geometry Worksheets.* Macmillan
Rowe, C. E. and McFarland, J. D. *Engineering Descriptive Geometry Problems.* D. Van Nostrand
Street, W. E., Perryman, C. C. and McGuire, J. G. *Technical Descriptive Geometry Problems.* D. Van Nostrand
Warner, F. M. *Applied Descriptive Geometry Problem Book.* McGraw-Hill
Wellman, B. L. *Problem Layouts for Technical Descriptive Geometry.* McGraw-Hill

Drawing Instruments and Supplies

Alteneder, Theo. & Sons, Philadelphia, Pa.
Charles Bruning Co., Chicago, Ill.
Eugene Dietzgen Co., Chicago, Ill.
Frederick Post Co., Chicago, Ill.
Gramercy Guild Group, Inc., New York 4, N. Y.
Keuffel & Esser Co., Hoboken, N. J.
V & E Manufacturing Co., Pasadena, Calif.

Electrical Drawing

Bishop, C. C. *Electrical Drafting and Design.* McGraw-Hill
Carini, L. F. D. *Drafting for Electronics.* McGraw-Hill
Kocher, S. E. *Electrical Drafting.* Int. Textbook Co.
Van Gieson, D. W. *Electrical Drafting.* McGraw-Hill

Engineering as a Vocation

Carlisle, N. D. *Your Career in Engineering.* E. P. Dutton & Co.
Grinter, L. E., Spencer, H. C., et al. *Engineering Preview.* Macmillan

McGuire, J. G. and Barlow, H. W. *An Introduction to the Engineering Profession*. Addison-Wesley Press, Inc.
Smith, R. J. *Engineering as a Career*. McGraw-Hill
Williams, C. C. *Building an Engineering Career*. McGraw-Hill

Engineering Drawing

French, T. E. and Svensen, C. L. *Mechanical Drawing*. McGraw-Hill
French, T. E. and Vierck, C. J. *Engineering Drawing*. McGraw-Hill
Giesecke, F. E., Mitchell, A. and Spencer, H. C. *Technical Drawing*, Macmillan
Katz, H. H. *Handbook of Layout and Dimensioning for Production*. Macmillan
Lent, Deane. *Machine Drawing*. Prentice-Hall
Luzadder, W. J. *Fundamentals of Engineering Drawing*. Prentice-Hall
Orth, H. D., Worsencroft, R. R. and Doke, H. B. *Theory and Practice of Engineering Drawing*, Wm. C. Brown Co., Dubuque, Iowa
Spencer, H. C. *Basic Technical Drawing*. Macmillan
Zozzora, F. *Engineering Drawing*. McGraw-Hill

Engineering Drawing Workbooks

Giesecke, F. E., Mitchell, A. and Spencer, H. C. *Technical Drawing Problems*, Macmillan
Hoelscher, R. P. and Springer, C. H. *Engineering Drawing and Geometry*. John Wiley
Johnson, L. O. and Wladaver, I. *Engineering Drawing Problems*. Prentice-Hall
Levens, A. S. and Edstrom, A. E. *Problems in Engineering Drawing*. McGraw-Hill
Luzadder, W. J., et al. *Problems in Engineering Drawing*. Prentice-Hall
McNeary, M., Weidhaas, E. R. and Kelso, E. A. *Creative Problems for Basic Engineering Drawing*. McGraw-Hill
Orth, H. D. Worsencroft, R. R. and Doke, H. B. *Problems in Engineering Drawing*. Wm. C. Brown Co. Dubuque, Iowa
Pare, E. G. and Hrachovsky, F. *Graphic Representation*. Macmillan
Pare, E. G. and Tozer, E. F. *Engineering Drawing Problems*. D. Van Nostrand
Spencer, H. C. and Grant, H. E. *Technical Drawing Problems, Series 2*. Macmillan
Turner, W. W., Buck, C. P. and Ackert, H. P. *Basic Engineering Drawing*. Ronald Press
Vierck, C. J., Cooper, C. D. and Machovina, P. E. *Engineering Drawing Problems*. McGraw-Hill
Zozzora, F. *Engineering Drawing Problems*. McGraw-Hill

Engineering Drawing and Graphics Texts

(*With engineering drawing or descriptive geometry*, *or both. In most cases*, *workbooks are also available.*)

Arnold, J. N. *Introductory Graphics*. McGraw-Hill
French, T. E. and Vierck, C. J. *Graphic Science*. McGraw-Hill
Levens, A. S. *Graphics in Engineering and Science*. John Wiley
Luzadder, W. J. *Graphics for Engineers*. Prentice-Hall
Pare, E. G. *Engineering Drawing*. Dryden Press
Rising, J. S. and Almfeldt, M. W. *Engineering Graphics*. Wm. C. Brown Co., Dubuque, Iowa
Rule, J. T. and Watts, E. F. *Engineering Graphics*. McGraw-Hill
Shupe, H. W. and Machovina, P. E. *Engineering Geometry and Graphics*. McGraw-Hill

Graphical Computation

Adams, D. P. *An Index of Nomograms*. John Wiley
Allcock, H. J., Jones, J. P. *The Nomogram*. Pitman
Davis, D. S. *Chemical Engineering Nomographs*. McGraw-Hill

Davis, D. S. *Empirical Equations and Nomography*. McGraw-Hill
Douglass, R. D., Adams, D. P. *Elements of Nomography*. McGraw-Hill
Heacock, F. A. *Graphic Methods for Solving Problems*. Edwards Bros., Inc.
Hoelscher, R. P., Arnold, J. N., and Pierce, S. H. *Graphic Aids in Engineering Computation*. McGraw-Hill
Johnson, L. H. *Nomography and Empirical Equations*. John Wiley
Kulmann, C. A. *Nomographic Charts*. McGraw-Hill
Levens, A. S. *Nomography*. John Wiley
Lipka, J. *Graphical and Mechanical Computation*. John Wiley
Mackey, C. O. *Graphical Solutions*. John Wiley
Mavis, F. T. *The Construction of Nomographic Charts*. International Textbook Co.
Runge, C. *Graphical Methods*. Columbia Univ. Press
Running, T. R. *Graphical Mathematics*. John Wiley

Handbooks

American Soc. of Heating and Ventilating Engineers. *A.S.H.V.E. Guide*. 51 Madison Ave., New York 10, N. Y.
ASME Handbook (4 vols.). McGraw-Hill
Colvin, F. H. and Stanley, F. A. *American Machinists Handbook*. McGraw-Hill
Eshbach, O. W. *Handbook of Engineering Fundamentals*. John Wiley
Huntington, W. C. *Building Construction*. John Wiley
Kent, W. *Mechanical Engineers Handbook*. John Wiley
Kidder, F. E. and Parker, H. *Architects and Builders Handbook*. John Wiley
Knowlton, A. E. *Standard Handbook of Electrical Engineers*. McGraw-Hill
Marks, L. S. *Mechanical Engineers Handbook*. McGraw-Hill
Oberg, E. and Jones, F. D. *Machinery's Handbook*. Industrial Press
O'Rourke, C. E. *General Engineering Handbook*. McGraw-Hill
Perry, J. H. *Chemical Engineers Handbook*. McGraw-Hill
SAE Handbook. Society of Automotive Engineers, 29 West 39th St., New York 18, N. Y.
SAE Automotive Drafting Standards. Society of Automotive Engineers, 29 West 39th St., New York 18, N. Y.
Smoley's New Combined Tables. C. K. Smoley & Sons, Scranton, Pa.
Tweney, C. F. and Hughes, L. E. C. *Chambers Technical Dictionary*. Macmillan
Urquhart, L. C. *Civil Engineers Handbook*. McGraw-Hill

Lettering

French, T. E. and Meiklejohn, R. *Essentials of Lettering*. McGraw-Hill
French, T. E. and Turnbull, W. D. *Lessons in Lettering, Books 1 and 2*. McGraw-Hill
George, R. F. *Modern Lettering for Pen and Brush Poster Design*. Hunt Pen Co., Camden, N. J.
Giesecke, F. E., Mitchell, A. and Spencer, H. C. *Lettering Exercises*. Macmillan
Hornung, C. P. *Lettering from A to Z*. Ziff-Davis Pub. Co., N. Y.
Ogg, Oscar. *An Alphabet Source Book*. Harper
Spencer, H. C. and Grant, H. E. *Technical Lettering Practice*. Macmillan
Svensen, C. L. *The Art of Lettering*. D. Van Nostrand

Machine Design

Albert, C. D. *Machine Design and Drawing Room Problems*. John Wiley
Berard, S. J., Watters, E. O. and Phelps, C. W. *Principles of Machine Design*. Ronald Press
Faires, V. M. *Design of Machine Elements*. Macmillan
Jefferson, T. B. and Brooking, W. J. *Introduction to Mechanical Design*. Ronald Press

Norman, C. A., Ault, S. and Zabrosky, I. *Fundamentals of Machine Design*. Macmillan
Spotts, M. F. *Elements of Machine Design*. Prentice-Hall
Vallance, A. and Doughtie, V. L. *Design of Machine Members*. McGraw-Hill

Mechanism

Ham, C. W., Crane, E. J. and Rogers, W. L. *Mechanics of Machinery*. McGraw-Hill
Keon, R. M. and Faires, V. M. *Mechanism*. McGraw-Hill
Schwamb, P., Merrill, A. L. and James, W. H. *Elements of Mechanism*. John Wiley

Map Drawing

Deetz, C. H. *Elements of Map Projection*. U.S. Government Printing Office
Hinks, A. R. *Maps and Surveys*. Macmillan
Manual of Surveying Instructions for the Survey of the Public Lands of the United States. U.S. Government Printing Office
Sloane, R. C. and Montz, J. M. *Elements of Topographic Drawing*. McGraw-Hill
Whitmore, G. D. *Advanced Surveying and Mapping*. International Texbtook Co.

Patent Drawings

Guide for Patent Draftsmen. U.S. Government Printing Office
Radzinsky, H. *Making Patent Drawings*. Macmillan

Perspective

Freese, E. I. *Perspective Projection*. Reinhold
Lawson, P. J. *Practical Perspective Drawing*. McGraw-Hill
Lubchez, B. *Perspective*. D. Van Nostrand
Morehead, J. C. and Morehead, J. C. Jr. *A Handbook of Perspective Drawing*. D. Van Nostrand
Norling, E. *Perspective Made Easy*. Macmillan
Turner, W. W. *Simplified Perspective*. Ronald Press

Piping Drawing

Crane & Co., Chicago (Catalog)
Crocker, S. *Piping Handbook*. McGraw-Hill
Day, L. J. *Standard Plumbing Details*. John Wiley

Production Illustration

Farmer, J. H., Hoecker, A. J. and Vavrin, F. F. *Illustrating for Tomorrow's production*. Macmillan
Hoelscher, R. P., Springer, C. H. and Pohle, R. F. *Industrial Production Illustration*. McGraw-Hill
Treacy, J. *Production Illustration*. John Wiley

Sheet Metal Drafting

Dougherty, J. S. *Sheet-Metal Pattern Drafting and Shop Problems*. Manual Arts Press
Giachino, J. W. *Basic Sheet Metal Practice*. International Textbook Co.
Jenkins, Rolland. *Sheet Metal Pattern Layout*. Prentice-Hall
Kidder, F. S. *Triangulation Applied to Sheet Metal Pattern Cutting*. Sheet Metal Pub. Co., N. Y.
Neubecker, William. *Sheet Metal Work*. Amer. Technical Society, Chicago

Neubecker, William. *The Universal Sheet Metal Pattern Cutter, Vols. I and II.* Metal Pub. Co., N.Y.
O'Rourke, F. J. *Sheet Metal Pattern Drafting.* McGraw-Hill
Paull, J. H. *Industrial Sheet Metal Drawing.* D. Van Nostrand

Shop Processes and Materials

Arc Welding in Design, Manufacturing and Construction. Lincoln Arc Welding Foundation, Cleveland, O.
Begeman, M. L. *Manufacturing Processes.* John Wiley
Boston, O. W. *Metal Processing.* John Wiley
Campbell, H. L. *Metal Castings.* John Wiley
Clapp, W. H. and Clark, D. S. *Engineering Materials and Processes.* International Textbook Co.
Colvin, F. H. and Haas, L. L. *Jigs and Fixtures.* McGraw-Hill
Dubois, J. H. *Plastics.* Amer. Technical Society, Chicago
Hesse, H. C. *Engineering Tools and Processes.* D. Van Nostrand
Hinman, C. W. *Die Engineering Layouts and Formulas.* McGraw-Hill
Johnson, C. G. *Forging Practice.* Amer. Technical Society, Chicago
Wendt, R. E. *Foundry Work.* McGraw-Hill
Young, J. F. *Materials and Processes.* John Wiley

Sketching

Guptill, A. L. *Drawing with Pen and Ink.* Reinhold Pub. Corp.
Guptill, A. L. *Sketching and Rendering in Pencil.* Reinhold Pub. Corp.
Jones, F. D. *How to Sketch Mechanisms.* The Industrial Press
Katz, H. H. *Technical Sketching.* Macmillan
Kautsky, T. *Pencil Broadsides.* Reinhold Pub. Corp.
Turner, W. W. *Freehand Sketching for Engineers.* Ronald Press
Zipprich, A. E. *Freehand Drafting.* D. Van Nostrand

Structural Drafting and Design

Bishop, C. T. *Structural Drafting.* John Wiley
Breed, C. B. and Hosmer, G. L. *Elementary Surveying* (Vol. I). John Wiley
Manual of Standard Practice for Detailing Reinforced Concrete Highway Structures. American Concrete Institute, Detroit 19, Mich.
Manual of Standard Practice for Detailing Reinforced Concrete Structures. American Concrete Institute, Detroit 19, Mich.
Parker, H. *Simplified Design of Structural Steel.* John Wiley
Parker, H. *Simplified Design of Structural Timber.* John Wiley
Steel Construction Manual. American Institute of Steel Construction, 101 Park Ave., New York 17, N. Y.
Structural Shop Drafting (Vols. 1-3). American Institute of Steel Construction, 101 Park Ave., New York 17, N. Y.

Tool Design

Bloom, R. R. *Principles of Tool Engineering.* McGraw-Hill
Cole, C. B. Tool Design. *American Technical Society*

Welding

Procedure Handbook of Arc Welding Design and Practice. Lincoln Electric Co., Cleveland, O.

2. TECHNICAL TERMS

"The beginning of wisdom is to call things by their right names."

—CHINESE PROVERB

n means a *noun;*
v means a *verb.*

Acme (*n*)—Screw thread form, pp. 356, 362.

Addendum (*n*)—Radial distance from pitch circle to top of gear tooth, pp. 581, 585.

Allen Screw (*n*)—Special set screw or cap screw with hexagon socket in head, p. 382.

Allowance (*n*)—Minimum clearance between mating parts, p. 336.

Alloy (*n*)—Two or more metals in combination, usually a fine metal with a baser metal.

Aluminum (*n*)—A lightweight but relatively strong metal. Often alloyed with copper to increase hardness and strength.

Angle Iron (*n*)—A structural shape whose section is a right angle, p. 670.

Anneal (*v*)—To heat and cool gradually, to reduce brittleness and increase ductility, p. 293.

Arc-weld (*v*)—To weld by electric arc. The work is usually the positive terminal.

Babbitt (*n*)—A soft alloy for bearings, mostly of tin with small amounts of copper and antimony.

Bearing (*n*)—A supporting member for a rotating shaft.

Bevel (*n*)—An inclined edge, not at right angle to joining surface.

Bolt Circle (*n*)—A circular center line on a drawing, containing the centers of holes about a common center, p. 314.

Bore (*v*)—To enlarge a hole with a boring bar or tool in a lathe, drill press, or boring mill, Figs. 564 (b) and 568 (b).

BOSS

Boss (*n*)—A cylindrical projection on a casting or a forging.

Brass (*n*)—An alloy of copper and zinc.

Braze (*v*)—To join with hard solder of brass or zinc.

Brinell (*n*)—A method of testing hardness of metal.

Broach (*n*)—A long cutting tool with a series of teeth that gradually increase in size which is forced through a hole or over a surface to produce a desired shape, p. 281.

Bronze (*n*)—An alloy of eight or nine parts of copper and one part of tin.

Buff (*v*)—To finish or polish on a buffing wheel composed of fabric with abrasive powders.

Burnish (*v*)—To finish or polish by pressure upon a smooth rolling or sliding tool.

Burr (*n*)—A jagged edge on metal resulting from punching or cutting.

Bushing (*n*)—A replaceable lining or sleeve for a bearing.

Calipers (*n*)—Instrument (of several types) for measuring diameters, p. 285.

Cam (*n*)—A rotating member for changing circular motion to reciprocating motion, p. 589.

Carburize (*v*)—To heat a low-carbon steel to approximately 2000° F. in contact with material which adds carbon to the surface of the steel, and to cool slowly in preparation for heat treatment, p. 293.

Caseharden (*v*)—To harden the outer surface of a carburized steel by heating and then quenching, p. 293.

Castellate (*v*)—To form like a castle, as a castellated shaft or nut.

Casting (*n*)—A metal object produced by pouring molten metal into a mold, p. 271.

Cast Iron (*n*)—Iron melted and poured into molds, p. 271.

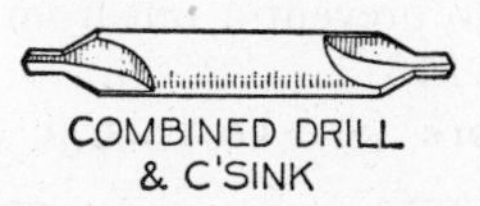
COMBINED DRILL & C'SINK

Center Drill (*n*)—A special drill to produce bearing holes in the ends of a workpiece to be mounted between centers. Also called a "combined drill and countersink," §360.

Chamfer (*n*)—A narrow inclined surface along the intersection of two surfaces.

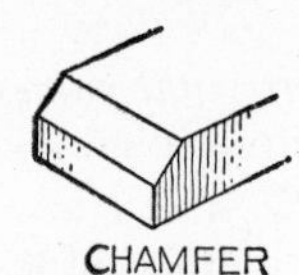

Chase (*v*)—To cut threads with an external cutting tool, Fig 568 (d).

Cheek (*n*)—The middle portion of a three-piece flask used in molding, p. 272.

Chill (*v*)—To harden the outer surface of cast iron by quick cooling, as in a metal mold.

Chip (*v*)—To cut away metal with a cold chisel.

Chuck (*n*)—A mechanism for holding a rotating tool or workpiece.

Coin (*v*)—To form a part in one stamping operation.

Cold Rolled Steel (CRS) (*n*)—Open hearth or Bessemer steel containing 0.12% to 0.20% carbon which has been rolled while cold to produce a smooth, quite accurate stock.

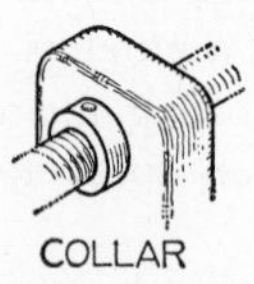

Collar (*n*)—A round flange or ring fitted on a shaft to prevent sliding.

Colorharden (*v*)—Same as *caseharden*, except that it is done to a shallower depth, usually for appearance only.

Cope (*n*)—The upper portion of a flask used in molding, p. 272.

Core (*v*)—To form a hollow portion in a casting by using a dry-sand core or a green-sand core in a mold, p. 272.

Coreprint (*n*)—A projection on a pattern which forms an opening in the sand to hold the end of a core, p. 272.

Cotter Pin (*n*)—A split pin used as a fastener, usually to prevent a nut from unscrewing, pp. 379, 813.

Counterbore (*v*)—To enlarge an end of a hole cylindrically with a *counterbore*, p. 284.

Countersink (*v*)—To enlarge an end of a hole conically, usually with a *countersink*, p. 284.

Crown (*n*)—A raised contour, as on the surface of a pulley.

Cyanide (*v*)—To surface-harden steel by heating in contact with a cyanide salt, followed by quenching.

Dedendum (*n*)—Distance from pitch circle to bottom of tooth space, pp. 581, 585.

Development (*n*)—Drawing of the surface of an object unfolded or rolled out on a plane, p. 553.

Diametral Pitch (*n*)—Number of gear teeth per inch of pitch diameter, p. 581.

Die (*n*)—1. Hardened metal piece shaped to cut or form a required shape in a sheet of metal by pressing it against a mating die. 2. Also used for cutting small threads. In a sense is opposite to a tap.

Die Casting (*n*)—Process of forcing molten metal under pressure into metal dies or molds, producing a very accurate and smooth casting, p. 275.

Die Stamping (*n*)—Process of cutting or forming a piece of sheet metal with a die.

Dog (*n*)—A small auxiliary clamp for preventing work from rotating in relation to the face plate of a lathe.

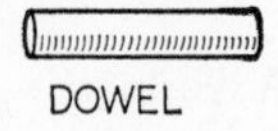

Dowel (*n*)—A cylindrical pin, commonly used to prevent sliding between two contacting flat surfaces.

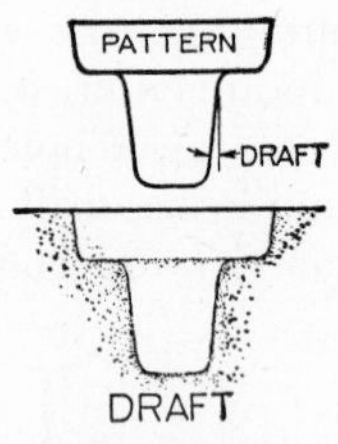

Draft (*n*)—The tapered shape of the parts of a pattern to permit it to be easily withdrawn from the sand or, on a forging, to permit it to be easily withdrawn from the dies, pp. 272, 292.

Drag (*n*)—Lower portion of a flask used in molding, p. 272.

Draw (*v*)—To stretch or otherwise to deform metal. Also to temper steel.

Drill (*v*)—To cut a cylindrical hole with a drill. A *blind hole* does not go through the piece, p. 282.

Drill Press (*n*)—A machine for drilling and other hole-forming operations, p. 278.

Drop Forge (*v*)—To form a piece while hot between dies in a drop hammer or with great pressure, p. 291.

Face (*v*)—To finish a surface at right angles, or nearly so, to the center line of rotation on a lathe, Fig. 568 (a).

FAO—Finish all over, p. 308.

Feather Key (*n*)—A flat key, which is partly sunk in a shaft and partly in a hub, permitting the hub to slide lengthwise of the shaft, p. 384.

File (*v*)—To finish or smooth with a file.

Fillet (*n*)—An interior rounded intersection between two surfaces, §§239, 302.

Fin (*n*)—A thin extrusion of metal at the intersection of dies or sand molds.

Fit (*n*)—Degree of tightness or looseness between two mating parts, as a *loose fit*, a *snug fit*, or a *tight fit*, pp. 303, 334-338.

Fixture (*n*)—A special device for holding the work in a machine tool, *but not for guiding the cutting tool*, §322.

Flange (*n*)—A relatively thin rim around a piece.

Flash (*n*)—Same as *fin*.

Flask (*n*)—A box made of two or more parts for holding the sand in sand molding, p. 272.

Flute (*n*)—Groove, as on twist drills, reamers, and taps.

Forge (*v*)—To force metal while it is hot to take on a desired shape by hammering or pressing, p. 291.

Galvanize (*v*)—To cover a surface with a thin layer of molten alloy, composed mainly of zinc, to prevent rusting.

Gasket (*n*)—A thin piece of rubber, metal, or some other material, placed between surfaces to make a tight joint.

Gate (*n*)—The opening in a sand mold at the bottom of the *sprue* through which the molten metal passes to enter the cavity or mold, p. 272.

Graduate (*v*)—To set off accurate divisions on a scale or dial.

Grind (*v*)—To remove metal by means of an abrasive wheel, often made of carborundum. Used chiefly where accuracy is required, p. 279.

Harden (*v*)—To heat steel above a critical temperature and then quench in water or oil, p. 293.

Heat-treat (*v*)—To change the properties of metals by heating and then cooling, p. 293.

Interchangeable (*adj.*)—Refers to a part made to limit dimensions so that it will fit any mating part similarly manufactured, p. 334.

Jig (*n*)—A device *for guiding a tool* in cutting a piece. Usually it holds the work in position, p. 290.

Journal (*n*)—Portion of a rotating shaft supported by a bearing.

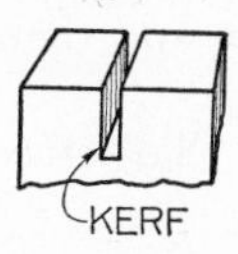

Kerf (*n*)—Groove or cut made by a saw.

Key (*n*)—A small piece of metal sunk partly into both shaft and hub to prevent rotation, p. 384.

Keyseat (*n*)—A slot or recess in a shaft to hold a key, p. 384.

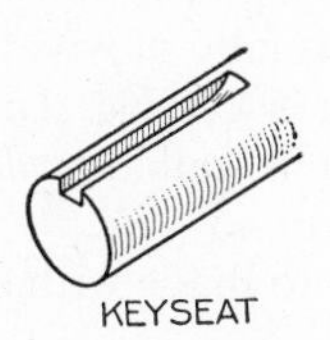

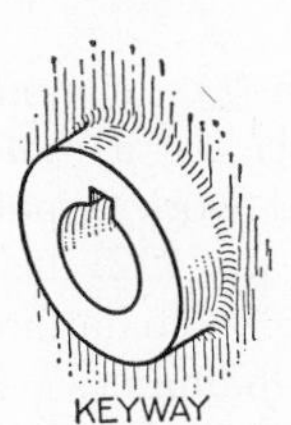

Keyway (*n*)—A slot in a hub or portion surrounding a shaft to receive a key, p. 384.

Knurl (*v*)—To impress a pattern of dents in a turned surface with a knurling tool to produce a better hand grip, p. 324.

Lap (*v*)—To produce a very accurate finish by sliding contact with a *lap*, or piece of wood, leather, or soft metal impregnated with abrasive powder.

Lathe (*n*)—A machine used to shape metal or other materials by rotating against a tool, p. 277.

Lug (*n*)—An irregular projection of metal, but not round as in the case of a *boss*, usually with a hole in it for a bolt or screw.

Malleable Casting (*n*)—A casting which has been made less brittle and tougher by annealing.

Mill (*v*)—To remove material by means of a rotating cutter on a milling machine, p, 278.

Mold (*n*)—The mass of sand or other material which forms the cavity into which molten metal is poured, p. 271.

MS (*n*)—Machinery steel, sometimes called *mild steel* with a small percentage of carbon. Cannot be hardened.

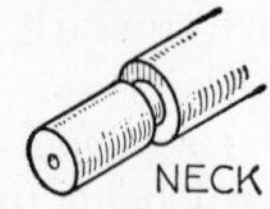

Neck (*v*)—To cut a groove called a *neck* around a cylindrical piece.

Normalize (*v*)—To heat steel above its critical temperature and then to cool it in air, p. 293.

Pack-harden (*v*)—To *carburize*, then to *case-harden*, p. 293.

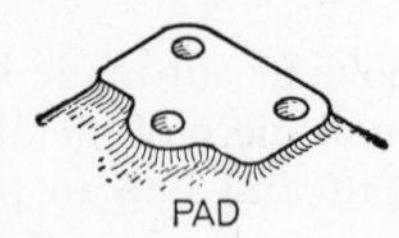

Pad (*n*)—A slight projection, usually to provide a bearing surface around one or more holes.

Pattern (*n*)—A model, usually of wood, used in forming a mold for a casting. In sheet metal work a pattern is called a *development*, pp. 272, 553.

Peen (*v*)—To hammer into shape with a ball-peen hammer.

Pickle (*v*)—To clean forgings or castings in dilute sulphuric acid.

Pinion (*n*)—The smaller of two mating gears, p. 580.

Pitch Circle (*n*)—An imaginary circle corresponding to the circumference of the friction gear from which the spur gear was derived, pp. 580, 581.

Plane (*v*)—To remove material by means of the *planer*, p. 279.

Planish (*v*)—To impart a planished surface to sheet metal by hammering with a smooth-surfaced hammer.

Plate (*v*)—To coat a metal piece with another metal, such as chrome or nickel, by electrochemical methods.

Polish (*v*)—To produce a highly finished or polished surface by friction, using a very fine abrasive.

Profile (*v*)—To cut any desired outline by moving a small rotating cutter, usually with a master template as a guide.

Punch (*v*)—To cut an opening of a desired shape with a rigid tool having the same shape, by pressing the tool through the work.

Quench (*v*)—To immerse a heated piece of metal in water or oil in order to harden it.

Rack (*n*)—A flat bar with gear teeth in a straight line to engage with teeth in a gear, p. 583.

Ream (*v*)—To enlarge a finished hole slightly to give it greater accuracy, with a *reamer*, p. 283.

Relief (*n*)—An offset of surfaces to provide clearance for machining.

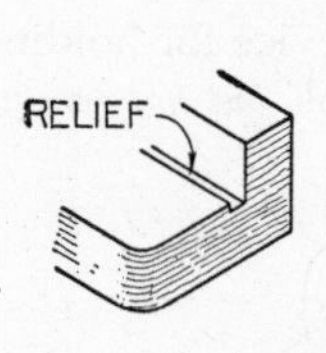

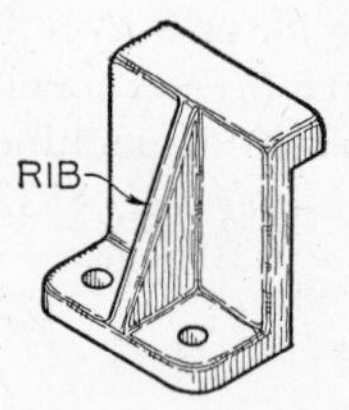

Rib (*n*)—A relatively thin flat member acting as a brace or support.

Rivet (*v*)—To connect with rivets, or to clench over the end of a pin by hammering, p. 386.

Round (*n*)—An exterior rounded intersection of two surfaces, §§239 and 302.

SAE—Society of Automotive Engineers.

Sandblast (*v*)—To blow sand at high velocity with compressed air against castings or forgings to clean them.

Scleroscope (*n*)—An instrument for measuring hardness of metals.

Scrape (*v*)—To remove metal by scraping with a hand scraper, usually to fit a bearing.

Shape (*v*)—To remove metal from a piece with a *shaper*, p. 279.

Shear (*v*)—To cut metal by means of shearing with two blades in sliding contact.

Sherardize (*v*)—To galvanize a piece with a coating of zinc by heating it in a drum with zinc powder, to a temperature of 575° to 850° F.

Shim (*n*)—A thin piece of metal or other material used as a spacer in adjusting two parts.

Solder (*v*)—To join with solder, usually composed of lead and tin.

Spin (*v*)—To form a rotating piece of sheet metal into a desired shape by pressing it with a smooth tool against a rotating form.

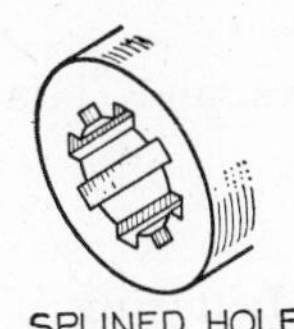

SPLINED HOLE

Spline (*n*)—A keyway, usually one of a series cut around a shaft or hole.

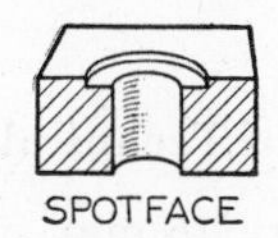

SPOTFACE

Spotface (*v*)—To produce a round *spot* or bearing surface around a hole, usually with a *spotfacer*. The spotface may be on top of a boss or it may be sunk into the surface, pp. 184, 284.

Sprue (*n*)—A hole in the sand leading to the *gate* which leads to the mold, through which the metal enters, p. 272.

Steel Casting (*n*)—Like cast-iron casting except that in the furnace scrap steel has been added to the casting.

Swage (*v*)—To hammer metal into shape while it is held over a *swage*, or die, which fits in a hole in the *swage block*, or anvil.

Sweat (*v*)—To fasten metal together by the use of solder between the pieces and by the application of heat and pressure.

Tap (*v*)—To cut relatively small internal threads with a *tap*, p. 284.

Taper (*n*)—Conical form given to a shaft or a hole. Also refers to the slope of a plane surface, p. 322.

TAPER PIN

Taper pin (*n*)—A small tapered pin for fastening, usually to prevent a collar or hub from rotating on a shaft.

Taper Reamer (*n*)—A tapered reamer for producing accurate tapered holes, as for a taper pin, pp. 322, 385.

Temper (*v*)—To reheat hardened steel to bring it to a desired degree of hardness, p. 293.

Template or **Templet** (*n*)—A guide or pattern used to mark out the work, guide the tool in cutting it, or check the finished product.

Tin (*n*)—A silvery metal used in alloys and for coating other metals, such as tin plate.

Tolerance (*n*)—Total amount of variation permitted in limit dimension of a part, p. 334.

Trepan (*v*)—To cut a circular groove in the flat surface at one end of a hole.

Tumble (*v*)—To clean rough castings or forgings in a revolving drum filled with scrap metal.

Turn (*v*)—To produce, on a lathe, a cylindrical surface parallel to the center line, p. 277.

Twist Drill (*n*)—A drill foı use in a drill press, p. 282.

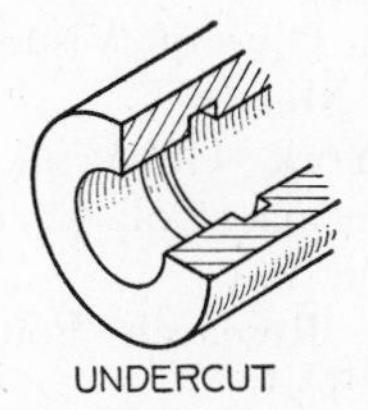

UNDERCUT

Undercut (*n*)—A recessed cut or a cut with inwardly sloping sides.

Upset (*v*)—To form a head or enlarged end on a bar or rod by pressure or by hammering between dies.

Web (*n*)—A thin flat part joining larger parts. Also known as a *rib*.

Weld (*v*)—Uniting metal pieces by pressure or fusion welding processes, pp. 290, 710.

WOODRUFF KEYS

Woodruff Key (*n*)—A semicircular flat key, p. 385.

Wrought Iron (*n*)—Iron of low carbon content useful because of its toughness, ductility, and malleability.

3. VISUAL AIDS FOR TECHNICAL DRAWING

The following visual aids are recommended. However, the instructor should preview each visual aid before using it in class, in order to determine its suitability for his purposes. All motion pictures listed are on 16 mm. film. An asterisk indicates that a follow-up film-strip is also available. The sources for the aids listed below are as follows:

BH —Bell & Howell Co., 7100 McCormick Ave., Chicago, Ill.
C —Chicago Board of Education, 228 No. LaSalle St., Chicago, Ill.
CF —Castle Films, Inc., 30 Rockefeller Plaza, New York 20, N. Y.
FP —Film Production Co., 3650 Fremont Ave. N., Minneapolis, Minn.
JH —Jam Handy Organization, 2900 E. Grand Blvd., Detroit 11, Mich.
McG—McGraw-Hill Book Co., 330 West 42nd St., New York 36, N. Y.
MIT—Massachusetts Institute of Technology, Div. of Visual Ed., Cambridge, Mass.
NY —New York University, College of Engineering, College Heights, New York, N. Y.
P —Purdue Research Foundation, Lafayette, Ind.
PSC —Pennsylvania State College, Film Library, State College, Pa.
U —United World Films, Inc., 1445 Park Ave., New York 29, N. Y.
UC —University of California, Educ. Films Dept., Univ. Extension, Los Angeles 24, Cal.

The Graphic Language

The Draftsman–11 min. sound movie–C.
According to Plan*–9 min. sound movie–McG.

Freehand Sketching

Freehand Drafting–13 min. silent movie–P.
Pictorial Sketching*–11 min. sound movie–McG.

Mechanical Drawing

Testing of T-Square and Triangles–13 min. silent movie–P.
T-Squares and Triangles, Parts 1 and 2–filmstrips–JH.
Use of T-Squares and Triangles–20 min. silent movie–P.

Lettering

Capital Letters–20 min. sound movie–P.
Lower Case Letters–13 min. sound movie–P.
Technical Lettering–filmstrips (a series of five)–JH.

Geometry of Technical Drawing

Applied Geometry–16 min. silent movie–P.
Geometric Construction, Parts 1 and 2–filmstrips–JH.

Views of Objects

Behind the Shop Drawing–20 min. sound movie–JH.
Orthographic Projection–30 min. silent movie –P.
Shape Description–25 min. sound movie–P.
Orthographic Projection*–18 min. sound movie–McG.
Visualizing an Object–10 min. sound movie–U.

Techniques and Applications

Drafting Tips–28 min. sound movie–PS.

Inking

Ink Work and Tracing–30 min. silent movie–P.

Dimensioning

Drawing an Anchor Plate–filmstrip–JH.
Selection of Dimensions*–18 min. sound movie–McG.

Shop Processes

Basic Machines: The Drill Press–10 min. sound movie–CF.
Basic Machines: The Lathe–15 min. sound movie–CF.
Basic Machines: The Shaper–15 min. sound movie–CF.
Cutting a Keyway on a Finished Shaft–13 min. sound movie–BH.
Fixed Gages–17 min. sound movie–BH.
Graphic Representation of Machine Operations–2 reels–MIT.
Laying Out Small Castings–16 min. sound movie–BH.
Machining a Tool Steel V-Block–21 min. sound movie–BH.
The Milling Machine–18 min. sound movie–CF.
The Drawings and the Shop*–15 min. sound movie–McG.
Shop Terms and Methods–2 reels–NY.
Shop Work–25 min. silent movie–P.

Sectional Views

Sections and Conventions*–15 min. sound movie–McG.
Sectional Views–20 min. silent movie–P.

Auxiliary Views

Auxiliary Views–18 min. silent movie–P.
Auxiliary Views: Single Auxiliaries*–23 min. sound movie–McG.
Auxiliary Views: Double Auxiliaries–13 min. sound movie–McG.

Threads and Fasteners

Cutting an External Acme Thread–16 min. sound movie–BH.
Cutting an External National Fine Thread–13 min. sound movie–BH.
Cutting Threads with Taps and Dies–20 min. sound movie–BH.
Screw Threads–23 min. silent movie–P.

Pictorial Drawings

Perspective Drawing–8 min. sound movie–UC.
Pictorial Drawing (Isometric)–21 min. silent movie–P.
Pictorial Sketching*–11 min. sound movie–McG.

Developments and Intersections

Descriptive Geometry–Line of Intersection of Two Solids–22 min. sound movie–U.
Development of Surfaces–15 min. silent movie –P.
Intersection of Surfaces–15 min. silent movie –P.
Oblique Cones and Transition Developments*–11 min. sound movie–McG.
Simple Developments*–11 min. sound movie –McG.

Structural Drawing

Structural Drawing–20 min. silent movie–P.

Blueprint Reading

Shop Drawing–15 min. sound movie–FP.

4. AMERICAN STANDARD ABBREVIATIONS FOR USE ON DRAWINGS

(From ASA Z32.13—1950)

A

Absolute . . . . ABS
Accelerate. . . . . ACCEL
Accessory . . . . ACCESS.
Account. . . . . ACCT
Accumulate. . . . . . . ACCUM
Actual . . . . ACT.
Adapter . . . . ADPT
Addendum . . . . ADD.
Addition . . . . ADD.
Adjust. . . . . ADJ
Advance. . . . . ADV
After. . . . . AFT.
Aggregate. . . . . AGGR
Air Condition. . AIR COND
Airplane . . . . APL
Allowance. . . . . ALLOW
Alloy . . . . ALY
Alteration. . . . . ALT
Alternate. . . . . ALT
Alternating Current. . . . . AC
Altitude . . . . ALT
Aluminum. . . . . AL
American Standard. . . . AMER STD
American Wire Gage. . AWG
Amount. . . . . AMT
Ampere . . . . AMP
Amplifier. . . . . AMPL
Anneal . . . . ANL
Antenna . . . . ANT.
Apartment. . . . . APT.
Apparatus. . . . . APP
Appendix. . . . . APPX
Approved. . . . . APPD
Approximate. . . . . APPROX
Arc Weld. . . . . ARC/W
Area . . . . A
Armature. . . . . ARM.
Armor Plate. . . . . . ARM-PL
Army Navy. . . . . AN
Arrange. . . . . ARR.
Artificial. . . . . ART.
Asbestos. . . . . ASB
Asphalt. . . . . ASPH
Assemble. . . . . ASSEM
Assembly. . . . . ASSY
Assistant . . . . ASST
Associate . . . . ASSOC
Association . . . . ASSN
Atomic. . . . . AT
Audible . . . . AUD
Audio Frequency . . . . AF
Authorized . . . . AUTH
Automatic. . . . . AUTO
Auto-Transformer . . AUTO TR
Auxiliary . . . . AUX
Avenue . . . . AVE
Average . . . . AVG
Aviation. . . . . AVI
Azimuth. . . . . AZ

B

Babbitt . . . . BAB
Back Feed . . . . BF
Back Pressure. . . . . BP
Back to Back. . . . . B to B
Backface. . . . . BF
Balance . . . . BAL
Ball Bearing. . . . . BB
Barometer. . . . . BAR
Base Line. . . . . BL
Base Plate. . . . . BP
Bearing. . . . . BRG
Bench Mark . . . . BM
Bending Moment. . . . . M
Bent. . . . . BT
Bessemer . . . . BESS
Between . . . . BET.
Between Centers. . . . . BC
Between Perpendiculars. . BP
Bevel . . . . BEV
Bill of Material . . . . B/M
Birmingham Wire Gage. . . . . BWG
Blank. . . . . BLK
Block. . . . . BLK
Blueprint . . . . BP
Board. . . . . BD
Boiler. . . . . BLR
Boiler Feed. . . . . BF
Boiler Horsepower. . . . . BHP
Boiling Point . . . . BP
Bolt Circle. . . . . BC
Both Faces. . . . . BF
Both Sides . . . . BS
Both Ways . . . . BW
Bottom . . . . BOT
Bottom Chord. . . . . BC
Bottom Face . . . . BF
Bracket . . . . BRKT
Brake . . . . BK
Brake Horsepower. . . . . BHP
Brass. . . . . BRS
Brazing. . . . . BRZG
Break . . . . BRK
Brinell Hardness. . . . . BH
British Standard . . . BR STD
British Thermal Units. . BTU
Broach . . . . BRO
Bronze. . . . . BRZ
Brown & Sharpe. . . . . . B&S (Wire Gage, same as AWG)
Building. . . . . BLDG
Bulkhead . . . . BHD
Burnish . . . . BNH
Bushing. . . . . BUSH.
Button . . . . BUT.

C

Cabinet. . . . . CAB.
Calculate. . . . . CALC
Calibrate. . . . . CAL
Cap Screw. . . . . . . CAP SCR
Capacity. . . . . CAP
Carburetor. . . . . CARB
Carburize. . . . . CARB
Carriage. . . . . CRG
Case Harden . . . . CH
Cast Iron CI
Cast Steel. . . . . CS
Casting. . . . . CSTG
Castle Nut . . . . . . CAS NUT
Catalogue . . . . CAT.
Cement . . . . CEM
Center. . . . . CTR
Center Line. CL
Center of Gravity. . . . . . . CG
Center of Pressure . . . . . . CP
Center to Center . . . . C to C

Centering. . . . . CTR
Chamfer. . . . . CHAM
Change. . . . . CHG
Channel. . . . . CHAN
Check. . . . . CHK
Check Valve. CV
Chord. . . . . CHD
Circle. . . . . CIR
Circular. . . . . CIR
Circular Pitch. . . . . CP
Circumference. . . . . . . CIRC
Clear. . . . . CLR
Clearance. CL
Clockwise. . . . . CW
Coated. . . . . CTD
Cold Drawn. . . . . CD
Cold Drawn Steel. . . . . CDS
Cold Finish. . . . . CF
Cold Punched. . . . . CP
Cold Rolled. . . . . CR
Cold Rolled Steel. . . . . CRS
Combination. . . . . . . COMB.
Combustion. . . . . COMB
Commercial. . . . . COML
Company. . . . . CO
Complete. . . . . COMPL
Concentric. . . . . CONC
Concrete. . . . . CONC
Condition. . . . . COND
Connect. . . . . CONN
Constant. . . . . CONST
Construction. . . . . . CONST
Contact. . . . . CONT
Continue. . . . . CONT
Corner. . . . . COR
Corporation. . . . . CORP
Correct. . . . . CORR
Corrugate. . . . . CORR
Cotter. . . . . COT
Counter. . . . . CTR
Counter Clockwise. . . . CCW
Counterbore. . . . . . . CBORE
Counterdrill. . . . . . CDRILL
Counterpunch. . . CPUNCH
Countersink. . . . . CSK
Coupling. . . . . CPLG
Cover. . . . . COV
Cross Section. . . . . . XSECT
Cubic. . . . . CU
Cubic Foot. . . . . CU FT
Cubic Inch. . . . . CU IN.
Current. . . . . CUR
Customer. . . . . CUST
Cyanide. . . . . CYN

D

Decimal. . . . . DEC
Dedendum. . . . . DED
Deflect. . . . . DEFL
Degree. . . . . (°) DEG
Density. . . . . D
Department. . . . . DEPT
Design. . . . . DSGN
Detail. . . . . DET
Develop. . . . . DEV
Diagonal. . . . . DIAG
Diagram. . . . . DIAG
Diameter. . . . . DIA
Diametral Pitch. . . . . DP
Dimension. . . . . DIM.
Discharge. . . . . DISCH
Distance. . . . . DIST
Division. . . . . DIV
Double. . . . . DBL
Dovetail. . . . . DVTL
Dowel. . . . . DWL
Down. . . . . DN
Dozen. . . . . DOZ
Drafting. . . . . DFTG
Draftsman. . . . . . . DFTSMN
Drawing. . . . . DWG
Drill. . . . . DR
Drive. . . . . DR
Drive Fit. . . . . DF
Drop. . . . . D
Drop Forge. . . . . DF
Duplicate. . . . . DUP

E

Each. . . . . EA
East. . . . . E
Eccentric. . . . . ECC
Effective. . . . . EFF
Elbow. . . . . ELL
Electric. . . . . ELEC
Elementary. . . . . ELEM
Elevate. . . . . ELEV
Elevation. . . . . EL
Engine. . . . . ENG
Engineer. . . . . ENGR
Engineering. . . . . . . ENGRG
Entrance. . . . . ENT
Equal. . . . . EQ
Equation. . . . . EQ
Equipment. . . . . EQUIP
Equivalent. . . . . EQUIV
Estimate. . . . . EST
Exchange. . . . . EXCH
Exhaust. . . . . EXH
Existing. . . . . EXIST.
Exterior. . . . . EXT
Extra Heavy. . . . . . . X HVY
Extra Strong. . . . . . . X STR
Extrude. . . . . EXTR

F

Fabricate. . . . . FAB
Face to Face. . . . . F to F
Fahrenheit. . . . . F
Far Side. . . . . FS
Federal. . . . . FED.
Feed. . . . . FD
Feet. . . . . (′) FT
Figure. . . . . FIG.
Fillet. . . . . FIL
Fillister. . . . . FIL
Finish. . . . . FIN.
Finish All Over. . . . . . FAO
Flange. . . . . FLG
Flat. . . . . F
Flat Head. . . . . FH
Floor. . . . . FL
Fluid. . . . . FL
Focus. . . . . FOC
Foot. . . . . (′) FT
Force. . . . . F
Forged Steel. . . . . FST
Forging. . . . . FORG
Forward. . . . . FWD
Foundry. . . . . FDRY
Frequency. . . . . FREQ
Front. . . . . FR
Furnish. . . . . FURN

G

Gage or Gauge. . . . . GA
Gallon. . . . . GAL
Galvanize. . . . . GALV
Galvanized Iron. . . . . GI
Galvanized Steel. . . . . . . GS
Gasket. . . . . GSKT
General. . . . . GEN
Glass. . . . . GL
Government. . . . . GOVT
Governor. . . . . GOV

Grade . . . . GR
Graduation . . . . GRAD
Graphite . . . . GPH
Grind . . . . GRD
Groove . . . . GRV
Ground . . . . GRD

H

Half-Round . . . . ½ RD
Handle . . . . HDL
Hanger . . . . HGR
Hard . . . . H
Harden . . . . HDN
Hardware . . . . HDW
Head . . . . HD
Headless . . . . HDLS
Heat . . . . HT
Heat Treat . . . . HT TR
Heavy . . . . HVY
Hexagon . . . . HEX
High-Pressure . . . . HP
High-Speed . . . . HS
Horizontal . . . . HOR
Horsepower . . . . HP
Hot Rolled . . . . HR
Hot Rolled Steel . . . . . . HRS
Hour . . . . HR
Housing . . . . HSG
Hydraulic . . . . HYD

I

Illustrate . . . . ILLUS
Inboard . . . . INBD
Inch . . . . (″) IN.
Inches per Second . . . . . . IPS
Inclosure . . . . INCL
Include . . . . INCL
Inside Diameter . . . . ID
Instrument . . . . INST
Interior . . . . INT
Internal . . . . INT
Intersect . . . . INT
Iron I
Irregular . . . . IRREG

J

Joint . . . . JT
Joint Army-Navy . . . . . . JAN
Journal . . . . JNL
Junction . . . . JCT

K

Key . . . . K
Keyseat . . . . KST
Keyway . . . . KWY

L

Laboratory . . . . LAB
Laminate . . . . LAM
Lateral . . . . LAT
Left L
Left Hand . . . . LH
Length . . . . LG
Length Over All . . . . . . LOA
Letter . . . . LTR
Light . . . . LT
Line L
Locate . . . . LOC
Logarithm . . . . LOG.
Long . . . . LG
Lubricate . . . . LUB
Lumber . . . . LBR

M

Machine . . . . MACH
Machine Steel . . . . MS
Maintenance . . . . MAINT
Malleable . . . . MALL
Malleable Iron . . . . MI
Manual . . . . MAN.
Manufacture . . . . MFR
Manufactured . . . . MFD
Manufacturing . . . . MFG
Material . . . . MATL
Maximum . . . . MAX
Mechanical . . . . MECH
Mechanism . . . . MECH
Median . . . . MED
Metal . . . . MET.
Meter . . . . M
Miles . . . . MI
Miles per Hour . . . . . . MPH
Millimeter . . . . MM
Minimum . . . . MIN
Minute . . . . (′) MIN
Miscellaneous . . . . MISC
Month . . . . MO
Morse Taper . . . . . . MOR T
Motor . . . . MOT
Mounted . . . . MTD
Mounting . . . . MTG
Multiple . . . . MULT
Music Wire Gage . . . . MWG

N

National . . . . NATL
Natural . . . . NAT
Near Face . . . . NF
Near Side . . . . NS
Negative . . . . NEG
Neutral . . . . NEUT
Nominal . . . . NOM
Normal . . . . NOR
North . . . . N
Not to Scale . . . . NTS
Number . . . . NO.

O

Obsolete . . . . OBS
Octagon . . . . OCT
Office . . . . OFF.
On Center . . . . OC
Opposite . . . . OPP
Optical . . . . OPT
Original . . . . ORIG
Outlet . . . . OUT.
Outside Diameter . . . . . . OD
Outside Face . . . . OF
Outside Radius . . . . OR
Overall . . . . OA

P

Pack . . . . PK
Packing . . . . PKG
Page . . . . P
Paragraph . . . . PAR.
Part . . . . PT
Patent . . . . PAT.
Pattern . . . . PATT
Permanent . . . . PERM
Perpendicular . . . . PERP
Piece . . . . PC
Piece Mark . . . . PC MK
Pint . . . . PT
Pitch . . . . P
Pitch Circle . . . . PC
Pitch Diameter . . . . PD
Plate . . . . PL
Plumbing . . . . PLMB
Point . . . . PT
Point of Curve . . . . PC
Point of Intersection . . . . . PI
Point of Tangent . . . . PT
Polish . . . . POL
Position . . . . POS
Potential . . . . POT.

Pound..................LB
Pounds per Square Inch.PSI
Power..................PWR
Prefabricated.....PREFAB
Preferred..............PFD
Prepare..............PREP
Pressure...........PRESS.
Process..............PROC
Production..........PROD
Profile..................PF
Propeller............PROP
Publication...........PUB
Push Button............PB

Q

Quadrant..........QUAD
Quality............QUAL
Quarter..............QTR

R

Radial..............RAD
Radius...................R
Railroad..............RR
Ream..................RM
Received...........RECD
Record..............REC
Rectangle..........RECT
Reduce..............RED.
Reference Line.....REF L
Reinforce..........REINF
Release..............REL
Relief................REL
Remove..............REM
Require..............REQ
Required...........REQD
Return..............RET.
Reverse..............REV
Revolution...........REV
Revolutions per
Minute............RPM
Right....................R
Right Hand...........RH
Rivet.................RIV
Rockwell Hardness.....RH
Roller Bearing.........RB
Room..................RM
Root Diameter.........RD
Root Mean Square....RMS
Rough..............RGH
Round................RD

S

Schedule.............SCH
Schematic.........SCHEM
Scleroscope Hardness....SH
Screw.................SCR
Second...............SEC
Section..............SECT
Semi-Steel.............SS
Separate.............SEP
Set Screw..............SS
Shaft..................SFT
Sheet..................SH
Shoulder...........SHLD
Side......................S
Single....................S
Sketch..................SK
Sleeve.................SLV
Slide....................SL
Slotted.............SLOT.
Small...................SM
Socket................SOC
Space..................SP
Special................SPL
Specific................SP
Spot Faced.............SF
Spring.................SPG
Square.................SQ
Standard..............STD
Station................STA
Stationary............STA
Steel...................STL
Stock..................STK
Straight...............STR
Stree....................ST
Structural............STR
Substitute............SUB
Summary...........SUM.
Support..............SUP.
Surface...............SUR
Symbol..............SYM
System................SYS

T

Tangent.............TAN.
Taper.................TPR
Technical...........TECH
Template..........TEMP
Tension............TENS.
Terminal..........TERM.
Thick.................THK
Thousand.................M
Thread...............THD
Threads per Inch......TPI
Through............THRU
Time......................T
Tolerance.............TOL
Tongue & Groove....T&G
Tool Steel..............TS
Tooth....................T
Total..................TOT
Transfer..........TRANS
Typical................TYP

U

Ultimate..............ULT
Unit.....................U
Universal...........UNIV

V

Vacuum..............VAC
Valve....................V
Variable..............VAR
Versus..................VS
Vertical............VERT
Volt......................V
Volume..............VOL

W

Wall.....................W
Washer............WASH.
Watt.....................W
Week..................WK
Weight................WT
West.....................W
Width....................W
Wood...................WD
Woodruff............WDF
Working Point.........WP
Working Pressure......WP
Wrought............WRT
Wrought Iron..........WI

X, Y, Z

Yard..................YD
Year..................YR

5. RUNNING AND SLIDING FITS[1]

Limits are in thousandths of an inch.

Limits for hole and shaft are applied algebraically to the basic size to obtain the limits of size for the parts.

Symbols H5, g5, etc., are Hole and Shaft designations used in ABC System.

Nominal Size Range Inches	CLASS RC 1			CLASS RC 2			CLASS RC 3			CLASS RC 4		
	Limits of Clearance	Standard Limits		Limits of Clearance	Standard Limits		Limits of Clearance	Standard Limits		Limits of Clearance	Standard Limits	
Over To		Hole H5	Shaft g4		Hole H6	Shaft g5		Hole H6	Shaft f6		Hole H7	Shaft f7
0.04– 0.12	0.1 0.45	+0.2 0	−0.1 −0.25	0.1 0.55	+0.25 0	−0.1 +0.3	0.3 0.8	+0.25 0	−0.3 −0.55	0.3 1.1	+0.4 0	−0.3 −0.7
0.12– 0.24	0.15 0.5	+0.2 0	−0.15 −0.3	0.15 0.65	+0.3 0	−0.15 −0.35	0.4 1.0	+0.3 0	−0.4 −0.7	0.4 1.4	+0.5 0	−0.4 −0.9
0.24– 0.40	0.2 0.6	+0.25 0	−0.2 −0.35	0.2 0.85	+0.4 0	−0.2 −0.45	0.5 1.3	+0.4 0	−0.5 −0.9	0.5 1.7	+0.6 0	−0.5 −1.1
0.40– 0.71	0.25 0.75	+0.3 0	−0.25 −0.45	0.25 0.95	+0.4 0	−0.25 −0.55	0.6 1.4	+0.4 0	−0.6 −1.0	0.6 2.0	+0.7 0	−0.6 −1.3
0.71– 1.19	0.3 0.95	+0.4 0	−0.3 −0.55	0.3 1.2	+0.5 0	−0.3 −0.7	0.8 1.8	+0.5 0	−0.8 −1.3	0.8 2.4	+0.8 0	−0.8 −1.6
1.19– 1.97	0.4 1.1	+0.4 0	−0.4 −0.7	0.4 1.4	+0.6 0	−0.4 −0.8	1.0 2.2	+0.6 0	−1.0 −1.6	1.0 3.0	+1.0 0	−1.0 −2.0
1.97– 3.15	0.4 1.2	+0.5 0	−0.4 −0.7	0.4 1.6	+0.7 0	−0.4 −0.9	1.2 2.6	+0.7 0	−1.2 −1.9	1.2 3.6	+1.2 0	−1.2 −2.4
3.15– 4.73	0.5 1.5	+0.6 0	−0.5 −0.9	0.5 2.0	+0.9 0	−0.5 −1.1	1.4 3.2	+0.9 0	−1.4 −2.3	1.4 4.2	+1.4 0	−1.4 −2.8
4.73– 7.09	0.6 1.8	+0.7 0	−0.6 −1.1	0.6 2.3	+1.0 0	−0.6 −1.3	1.6 3.6	+1.0 0	−1.6 −2.6	1.6 4.8	+1.6 0	−1.6 −3.2
7.09– 9.85	0.6 2.0	+0.8 0	−0.6 −1.2	0.6 2.6	+1.2 0	−0.6 −1.4	2.0 4.4	+1.2 0	−2.0 −3.2	2.0 5.6	+1.8 0	−2.0 −3.8
9.85–12.41	0.8 2.3	+0.9 0	−0.8 −1.4	0.8 2.9	+1.2 0	−0.8 −1.7	2.5 4.9	+1.2 0	−2.5 −3.7	2.5 6.5	+2.0 0	−2.5 −4.5
12.41–15.75	1.0 2.7	+1.0 0	−1.0 −1.7	1.0 3.4	+1.4 0	−1.0 −2.0	3.0 5.8	+1.4 0	−3.0 −4.4	3.0 7.4	+2.2 0	−3.0 −5.2

[1]From ASA B4.1–1955. For larger diameters, see the standard.

6. RUNNING AND SLIDING FITS (continued)[1]

Nominal Size Range Inches	Class RC 5			Class RC 6			Class RC 7			Class RC 8			Class RC 9		
	Limits of Clearance	Standard Limits		Limits of Clearance	Standard Limits		Limits of Clearance	Standard Limits		Limits of Clearance	Standard Limits		Limits of Clearance	Standard Limits	
Over To		Hole H7	Shaft e7		Hole H8	Shaft e8		Hole H9	Shaft d8		Hole H10	Shaft c9		Hole H11	Shaft
0.04– 0.12	0.6	+0.4	−0.6	0.6	+0.6	−0.6	1.0	+1.0	− 1.0	2.5	+1.6	− 2.5	4.0	+ 2.5	− 4.0
	1.4	0	−1.0	1.8	0	−1.2	2.6	0	− 1.6	5.1	0	− 3.5	8.1	0	− 5.6
0.12– 0.24	0.8	+0.5	−0.8	0.8	+0.7	−0.8	1.2	+1.2	− 1.2	2.8	+1.8	− 2.8	4.5	+ 3.0	− 4.5
	1.8	0	−1.3	2.2	0	−1.5	3.1	0	− 1.9	5.8	0	− 4.0	9.0	0	− 6.0
0.24– 0.40	1.0	+0.6	−1.0	1.0	+0.9	−1.0	1.6	+1.4	− 1.6	3.0	+2.2	− 3.0	5.0	+ 3.5	− 5.0
	2.2	0	−1.6	2.8	0	−1.9	3.9	0	− 2.5	6.6	0	− 4.4	10.7	0	− 7.2
0.40– 0.71	1.2	+0.7	−1.2	1.2	+1.0	−1.2	2.0	+1.6	− 2.0	3.5	+2.8	− 3.5	6.0	+ 4.0	− 6.0
	2.6	0	−1.9	3.2	0	−2.2	4.6	0	− 3.0	7.9	0	− 5.1	12.8	0	− 8.8
0.71– 1.19	1.6	+0.8	−1.6	1.6	+1.2	−1.6	2.5	+2.0	− 2.5	4.5	+3.5	− 4.5	7.0	+ 5.0	− 7.0
	3.2	0	−2.4	4.0	0	−2.8	5.7	0	− 3.7	10.0	0	− 6.5	15.5	0	−10.5
1.19– 1.97	2.0	+1.0	−2.0	2.0	+1.6	−2.0	3.0	+2.5	− 3.0	5.0	+4.0	− 5.0	8.0	+ 6.0	− 8.0
	4.0	0	−3.0	5.2	0	−3.6	7.1	0	− 4.6	11.5	0	− 7.5	18.0	0	−12.0
1.97– 3.15	2.5	+1.2	−2.5	2.5	+1.8	−2.5	4.0	+3.0	− 4.0	6.0	+4.5	− 6.0	9.0	+ 7.0	− 9.0
	4.9	0	−3.7	6.1	0	−4.3	8.8	0	− 5.8	13.5	0	− 9.0	20.5	0	−13.5
3.15– 4.73	3.0	+1.4	−3.0	3.0	+2.2	−3.0	5.0	+3.5	− 5.0	7.0	+5.0	− 7.0	10.0	+ 9.0	−10.0
	5.8	0	−4.4	7.4	0	−5.2	10.7	0	− 7.2	15.5	0	−10.5	24.0	0	−15.0
4.73– 7.09	3.5	+1.6	−3.5	3.5	+2.5	−3.5	6.0	+4.0	− 6.0	8.0	+6.0	− 8.0	12.0	+10.0	−12.0
	6.7	0	−5.1	8.5	0	−6.0	12.5	0	− 8.5	18.0	0	−12.0	28.0	0	−18.0
7.09– 9.85	4.0	+1.8	−4.0	4.0	+2.8	−4.0	7.0	+4.5	− 7.0	10.0	+7.0	−10.0	15.0	+12.0	−15.0
	7.6	0	−5.8	9.6	0	−6.8	14.3	0	− 9.8	21.5	0	−14.5	34.0	0	−22.0
9.85–12.41	5.0	+2.0	−5.0	5.0	+3.0	−5.0	8.0	+5.0	− 8.0	12.0	+8.0	−12.0	18.0	+12.0	−18.0
	9.0	0	−7.0	11.0	0	−8.0	16.0	0	−11.0	25.0	0	−17.0	38.0	0	−26.0
12.41–15.75	6.0	+2.2	−6.0	6.0	+3.5	−6.0	10.0	+6.0	−10.0	14.0	+9.0	−14.0	22.0	+14.0	−22.0
	10.4	0	−8.2	13.0	0	−9.5	19.5	0	−13.5	29.0	0	−20.0	45.0	0	−31.0

[1]From ASA B4.1–1955. For larger diameters, see the standard.

7. CLEARANCE LOCATIONAL FITS[1]

Limits are in thousandths of an inch.

Limits for hole and shaft are applied algebraically to the basic size to obtain the limits of size for the parts.

Symbols H6, h5, etc., are Hole and Shaft designations used in ABC System.

Nominal Size Range Inches	CLASS LC 1			CLASS LC 2			CLASS LC 3			CLASS LC 4			CLASS LC 5		
	Limits of Clearance	Standard Limits		Limits of Clearance	Standard Limits		Limits of Clearance	Standard Limits		Limits of Clearance	Standard Limits		Limits of Clearance	Standard Limits	
Over To		Hole H6	Shaft h5		Hole H7	Shaft h6		Hole H8	Shaft h7		Hole H9	Shaft h9		Hole H7	Shaft g6
0.04– 0.12	0 0.45	+0.25 – 0	+ 0 –0.2	0 0.65	+0.4 – 0	+ 0 –0.25	0 1	+0.6 – 0	+ 0 –0.4	0 2.0	+1.0 – 0	+ 0 –1.0	0.1 0.75	+0.4 – 0	–0.1 –0.35
0.12– 0.24	0 0.5	+0.3 – 0	+ 0 –0.2	0 0.8	+0.5 – 0	+ 0 –0.3	0 1.2	+0.7 – 0	+ 0 –0.5	0 2.4	+1.2 – 0	+ 0 +1.2	0.15 0.95	+0.5 – 0	–0.15 –0.45
0.24– 0.40	0 0.65	+0.4 – 0	+ 0 –0.25	0 1.0	+0.6 – 0	+ 0 –0.4	0 1.5	+0.9 – 0	+ 0 –0.6	0 2.8	+1.4 – 0	+ 0 –1.4	0.2 1.2	+0.6 – 0	–0.2 –0.6
0.40– 0.71	0 0.7	+0.4 – 0	+ 0 –0.3	0 1.1	+0.7 – 0	+ 0 –0.4	0 1.7	+1.0 – 0	+ 0 –0.7	0 3.2	+1.6 – 0	+ 0 –1.6	0.25 1.35	+0.7 – 0	–0.25 –0.65
0.71– 1.19	0 0.9	+0.5 – 0	+ 0 –0.4	0 1.3	+0.8 – 0	+ 0 –0.5	0 2	+1.2 – 0	+ 0 –0.8	0 4	+2.0 – 0	+ 0 –2.0	0.3 1.6	+0.8 – 0	–0.3 –0.8
1.19– 1.97	0 1.0	+0.6 – 0	+ 0 –0.4	0 1.6	+1.0 – 0	+ 0 –0.6	0 2.6	+1.6 – 0	+ 0 –1	0 5	+2.5 – 0	+ 0 –2.5	0.4 2.0	+1.0 – 0	–0.4 –1.0
1.97– 3.15	0 1.2	+0.7 – 0	+ 0 –0.5	0 1.9	+1.2 – 0	+ 0 –0.7	0 3	+1.8 – 0	+ 0 –1.2	0 6	+ 3 – 0	+ 0 –3	0.4 2.3	+1.2 – 0	–0.4 –1.1
3.15– 4.73	0 1.5	+0.9 – 0	+ 0 –0.6	0 2.3	+1.4 – 0	+ 0 –0.9	0 3.6	+2.2 – 0	+ 0 –1.4	0 7	+3.5 – 0	+ 0 –3.5	0.5 2.8	+1.4 – 0	–0.5 –1.4
4.73– 7.09	0 1.7	+1.0 – 0	+ 0 –0.7	0 2.6	+1.6 – 0	+ 0 –1.0	0 4.1	+2.5 – 0	+ 0 –1.6	0 8	+4 – 0	+ 0 –4	0.6 3.2	+1.6 – 0	–0.6 –1.6
7.09– 9.85	0 2.0	+1.2 – 0	+ 0 –0.8	0 3.0	+1.8 – 0	+ 0 –1.2	0 4.6	+2.8 – 0	+ 0 –1.8	0 9	+4.5 – 0	+ 0 –4.5	0.6 3.6	+1.8 – 0	–0.6 –1.8
9.85–12.41	0 2.1	+1.2 – 0	+ 0 –0.9	0 3.2	+2.0 – 0	+ 0 –1.2	0 5	+3.0 – 0	+ 0 –2.0	0 10	+5 – 0	+ 0 –5	0.7 3.9	+2.0 – 0	–0.7 –1.9
12.41–15.75	0 2.4	+1.4 – 0	+ 0 –1.0	0 3.6	+2.2 – 0	+ 0 –1.4	0 5.7	+3.5 – 0	+ 0 –2.2	0 12	+6 – 0	+ 0 –6	0.7 4.3	+2.2 – 0	–0.7 –2.1

[1]From ASA B4.1–1955. For larger diameters, see the standard.

8. CLEARANCE LOCATIONAL FITS (continued)[1]

Nominal Size Range Inches	Class LC 6			Class LC 7			Class LC 8			Class LC 9			Class LC 10			Class LC 11		
	Limits of Clearance	Standard Limits		Limits of Clearance	Standard Limits		Limits of Clearance	Standard Limits		Limits of Clearance	Standard Limits		Limits of Clearance	Standard Limits		Limits of Clearance	Standard Limits	
Over To		Hole H8	Shaft f8		Hole H9	Shaft e9		Hole H10	Shaft d9		Hole H11	Shaft c11		Hole H 12	Shaft		Hole H 13	Shaft
0.04– 0.12	0.3	+0.6	−0.3	0.6	+1.0	−0.6	1.0	+1.6	− 1.0	2.5	+2.5	−2.5	4	+ 4	− 4	5	+ 6	− 5
	1.5	− 0	−0.9	2.6	− 0	−1.6	3.6	− 0	− 2.0	7.5	− 0	−5.0	12	− 0	− 8	17	− 0	− 11
0.12– 0.24	0.4	+0.7	−0.4	0.8	+1.2	−0.8	1.2	+1.8	− 1.2	2.8	+3.0	−2.8	4.5	+ 5	−4.5	6	+ 7	− 6
	1.8	− 0	−1.1	3.2	− 0	−2.0	4.2	− 0	− 2.4	8.8	− 0	−5.8	14.5	− 0	−9.5	20	− 0	− 13
0.24– 0.40	0.5	+0.9	−0.5	1.0	+1.4	−1.0	1.6	+2.2	− 1.6	3.0	+3.5	−3.0	5	+ 6	− 5	7	+ 9	− 7
	2.3	− 0	−1.4	3.8	− 0	−2.4	5.2	− 0	− 3.0	10.0	− 0	−6.5	17	− 0	− 11	25	− 0	− 16
0.40– 0.71	0.6	+1.0	−0.6	1.2	+1.6	−1.2	2.0	+2.8	− 2.0	3.5	+4.0	−3.5	6	+ 7	− 6	8	+10	− 8
	2.6	− 0	−1.6	4.4	− 0	−2.8	6.4	− 0	− 3.6	11.5	− 0	−7.5	20	− 0	− 13	28	− 0	− 18
0.71– 1.19	0.8	+1.2	−0.8	1.6	+2.0	−1.6	2.5	+3.5	− 2.5	4.5	+5.0	−4.5	7	+ 8	− 7	10	+12	− 10
	3.2	− 0	−2.0	5.6	− 0	−3.6	8.0	− 0	− 4.5	14.5	− 0	−9.5	23	− 0	− 15	34	− 0	− 22
1.19– 1.97	1.0	+1.6	−1.0	2.0	+2.5	−2.0	3.0	+4.0	− 3.0	5	+6	− 5	8	+10	− 8	12	+16	− 12
	4.2	− 0	−2.6	7.0	− 0	−4.5	9.5	− 0	− 5.5	17	− 0	− 11	28	− 0	− 18	44	− 0	− 28
1.97– 3.15	1.2	+1.8	−1.2	2.5	+3.0	−2.5	4.0	+4.5	− 4.0	6	+7	− 6	10	+12	− 10	14	+18	− 14
	4.8	− 0	−3.0	8.5	− 0	−5.5	11.5	− 0	− 7.0	20	− 0	− 13	34	− 0	− 22	50	− 0	− 32
3.15– 4.73	1.4	+2.2	−1.4	3.0	+3.5	−3.0	5.0	+5.0	− 5.0	7	+9	− 7	11	+14	− 11	16	+22	− 16
	5.8	− 0	−3.6	10.0	− 0	−6.5	13.5	− 0	− 8.5	25	− 0	− 16	39	− 0	− 25	60	− 0	− 38
4.73– 7.09	1.6	+2.5	−1.6	3.5	+4.0	−3.5	6	+6	− 6	8	+10	− 8	12	+16	− 12	18	+25	− 18
	6.6	− 0	−4.1	11.5	− 0	−7.5	16	− 0	−10	28	− 0	− 18	44	− 0	− 28	68	− 0	− 43
7.09– 9.85	2.0	+2.8	−2.0	4.0	+4.5	−4.0	7	+7	− 7	10	+12	− 10	16	+18	− 16	22	+28	− 22
	7.6	− 0	−4.8	13.0	− 0	−8.5	18.5	− 0	−11.5	34	− 0	− 22	52	− 0	− 34	78	− 0	− 50
9.85–12.41	2.2	+3.0	−2.2	4.5	+5.0	−4.5	7	+8	− 7	12	+12	− 12	20	+20	− 20	28	+30	− 28
	8.2	− 0	−5.2	14.5	− 0	−9.5	20	− 0	−12	36	− 0	− 24	60	− 0	− 40	88	− 0	− 58
12.41–15.75	2.5	+3.5	−2.5	5	+6	− 5	8	+9	− 8	14	+14	− 14	22	+22	− 22	30	+35	− 30
	9.5	− 0	−6.0	17	− 0	−11	23	− 0	−14	42	− 0	− 28	66	− 0	− 44	100	− 0	− 65

[1]From ASA B4.1–1955. For larger diameters, see the standard.

9. TRANSITION LOCATIONAL FITS[1]

Limits are in thousandths of an inch.

Limits for hole and shaft are applied algebraically to the basic size to obtain the limits of size for the mating parts.

"Fit" represents the maximum interference (minus values) and the maximum clearance (plus values).

Symbols H8, j6, etc., are Hole and Shaft designations used in ABC System.

Nominal Size Range Inches	CLASS LT 1			CLASS LT 2			CLASS LT 3			CLASS LT 4			CLASS LT 6			CLASS LT 7		
	Fit	Standard Limits		Fit	Standard Limits		Fit	Standard Limits		Fit	Standard Limits		Fit	Standard Limits		Fit	Standard Limits	
Over To		Hole H7	Shaft j6		Hole H8	Shaft j7		Hole H7	Shaft k6		Hole H8	Shaft k7		Hole H8	Shaft m7		Hole H7	Shaft n6
0.04– 0.12	−0.15 +0.5	+0.4 − 0	+0.15 −0.1	−0.3 +0.7	+0.6 − 0	+0.3 −0.1							−0.55 +0.45	+0.6 − 0	+0.55 +0.15	−0.5 +0.15	+0.4 − 0	+0.5 +0.25
0.12– 0.24	−0.2 +0.6	+0.5 − 0	+0.2 −0.1	−0.4 +0.8	+0.7 − 0	+0.4 −0.1							+0.7 +0.5	+0.7 − 0	+0.7 +0.2	−0.6 +0.2	+0.5 − 0	+0.6 +0.3
0.24– 0.40	−0.3 +0.7	+0.6 − 0	+0.3 −0.1	−0.4 +1.1	+0.9 − 0	+0.4 −0.2	−0.5 +0.5	+0.6 − 0	+0.5 +0.1	−0.7 +0.8	+0.9 − 0	+0.7 +0.1	−0.8 +0.7	+0.9 − 0	+0.8 +0.2	−0.8 +0.2	+0.6 − 0	+0.8 +0.4
0.40– 0.71	−0.3 +0.8	+0.7 − 0	+0.3 −0.1	−0.5 +1.2	+1.0 − 0	+0.5 −0.2	−0.5 +0.6	+0.7 − 0	+0.5 +0.1	−0.8 +0.9	+1.0 − 0	+0.8 +0.1	−1.0 +0.7	+1.0 − 0	+1.0 +0.3	−0.9 +0.2	+0.7 − 0	+0.9 +0.5
0.71– 1.19	−0.3 +1.0	+0.8 − 0	+0.3 −0.2	−0.5 +1.5	+1.2 − 0	+0.5 −0.3	−0.6 +0.7	+0.8 − 0	+0.6 +0.1	−0.9 +1.1	+1.2 − 0	+0.9 +0.1	−1.1 +0.9	+1.2 − 0	+1.1 +0.3	−1.1 +0.2	+0.8 − 0	+1.1 +0.6
1.19– 1.97	−0.4 +1.2	+1.0 − 0	+0.4 −0.2	−0.6 +2.0	+1.6 − 0	+0.6 −0.4	−0.7 +0.9	+1.0 − 0	+0.7 +0.1	−1.1 +1.5	+1.6 − 0	+1.1 +0.1	−1.4 +1.2	+1.6 − 0	+1.4 +0.4	−1.3 +0.3	+1.0 − 0	+1.3 +0.7
1.97– 3.15	−0.4 +1.5	+1.2 − 0	+0.4 −0.3	−0.7 +2.3	+1.8 − 0	+0.7 −0.5	−0.8 +1.1	+1.2 − 0	+0.8 +0.1	−1.3 +1.7	+1.8 − 0	+1.3 +0.1	−1.7 +1.3	+1.8 − 0	+1.7 +0.5	−1.5 +0.4	+1.2 − 0	+1.5 +0.8
3.15– 4.73	−0.5 +1.8	+1.4 − 0	+0.5 −0.4	−0.8 +2.8	+2.2 − 0	+0.8 −0.6	−1.0 +1.3	+1.4 − 0	+1.0 +0.1	−1.5 +2.1	+2.2 − 0	+1.5 +0.1	−1.9 +1.7	+2.2 − 0	+1.9 +0.5	−1.9 +0.4	+1.4 − 0	+1.9 +1.0
4.73– 7.09	−0.6 +2.0	+1.6 − 0	+0.6 −0.4	−0.9 +3.2	+2.5 − 0	+0.9 −0.7	−1.1 +1.5	+1.6 − 0	+1.1 +0.1	−1.7 +2.4	+2.5 − 0	+1.7 +0.1	−2.2 +1.9	+2.5 − 0	+2.2 +0.6	−2.2 +0.4	+1.6 − 0	+2.2 +1.2
7.09– 9.85	−0.7 +2.3	+1.8 − 0	+0.7 −0.5	−1.0 +3.6	+2.8 − 0	+1.0 −0.8	−1.4 +1.6	+1.8 − 0	+1.4 +0.2	−2.0 +2.6	+2.8 − 0	+2.0 +0.2	−2.4 +2.2	+2.8 − 0	+2.4 +0.6	−2.6 +0.4	+1.8 − 0	+2.6 +1.4
9.85–12.41	−0.7 +2.6	+2.0 − 0	+0.7 −0.6	−1.0 +4.0	+3.0 − 0	+1.0 −1.0	−1.4 +1.8	+2.0 − 0	+1.4 +0.2	−2.2 +2.8	+3.0 − 0	+2.2 +0.2	−2.8 +2.2	+3.0 − 0	+2.8 +0.8	−2.6 +0.6	+2.0 − 0	+2.6 +1.4
12.41–15.75	−0.7 +2.9	+2.2 − 0	+0.7 −0.7	−1.2 +4.5	+3.5 + 0	+1.2 −1.0	−1.6 +2.0	+2.2 − 0	+1.6 +0.2	−2.4 +3.3	+3.5 − 0	+2.4 +0.2	−3.0 +2.7	+3.5 − 0	+3.0 +0.8	−3.0 +0.6	+2.2 − 0	+3.0 +1.6

[1]From ASA B4.1–1955. For larger diameters, see the standard.

10. INTERFERENCE LOCATIONAL FITS[1]

Limits are in thousandths of an inch.

Limits for hole and shaft are applied algebraically to the basic size to obtain the limits of size for the parts.

Symbols H7, p6, etc., are Hole and Shaft designations used in ABC System.

Nominal Size Range Inches	Class LN 2			Class LN 3		
	Limits of Interference	Standard Limits		Limits of Interference	Standard Limits	
Over To		Hole H7	Shaft p6		Hole H7	Shaft r6
0.04–0.12	0	+0.4	+0.65	0.1	+0.4	+0.75
	0.65	− 0	+0.4	0.75	− 0	+0.5
0.12–0.24	0	+0.5	+0.8	0.1	+0.5	+0.9
	0.8	− 0	+0.5	0.9	− 0	+0.6
0.24–0.40	0	+0.6	+1.0	0.2	+0.6	+1.2
	1.0	− 0	+0.6	1.2	− 0	+0.8
0.40–0.71	0	+0.7	+1.1	0.3	+0.7	+1.4
	1.1	− 0	+0.7	1.4	− 0	+1.0
0.71–1.19	0	+0.8	+1.3	0.4	+0.8	+1.7
	1.3	− 0	+0.8	1.7	− 0	+1.2
1.19–1.97	0	+1.0	+1.6	0.4	+1.0	+2.0
	1.6	− 0	+1.0	2.0	− 0	+1.4

Nominal Size Range Inches	Class LN 2			Class LN 3		
	Limits of Interference	Standard Limits		Limits of Interference	Standard Limits	
Over To		Hole H7	Shaft p6		Hole H7	Shaft r6
1.97– 3.15	0.2	+1.2	+2.1	0.4	+1.2	+2.3
	2.1	− 0	+1.4	2.3	− 0	+1.6
3.15– 4.73	0.2	+1.4	+2.5	0.6	+1.4	+2.9
	2.5	− 0	+1.6	2.9	− 0	+2.0
4.73– 7.09	0.2	+1.6	+2.8	0.9	+1.6	+3.5
	2.8	− 0	+1.8	3.5	− 0	+2.5
7.09– 9.85	0.2	+1.8	+3.2	1.2	+1.8	+4.2
	3.2	− 0	+2.0	4.2	− 0	+3.0
9.85–12.41	0.2	+2.0	+3.4	1.5	+2.0	+4.7
	3.4	− 0	+2.2	4.7	− 0	+3.5
12.41–15.75	0.3	+2.2	+3.9	2.3	+2.2	+5.9
	3.9	− 0	+2.5	5.9	− 0	+4.5

[1]From ASA B4.1–1955. For larger diameters, see the standard.

11. FORCE AND SHRINK FITS[1]

Limits are in thousandths of an inch.
Limits for hole and shaft are applied algebraically to the basic size to obtain the limits of size for the parts.
Symbols H7, s6, etc., are Hole and Shaft designations used in ABC System (Appendix I).

Nominal Size Range Inches	Class FN 1			Class FN 2			Class FN 3			Class FN 4			Class FN 5		
	Limits of Interference	Standard Limits		Limits of Interference	Standard Limits		Limits of Interference	Standard Limits		Limits of Interference	Standard Limits		Limits of Interference	Standard Limits	
Over To		Hole H6	Shaft		Hole H7	Shaft s6		Hole H7	Shaft t6		Hole H7	Shaft u6		Hole H7	Shaft x7
0.04– 0.12	0.05 0.5	+0.25 – 0	+0.5 +0.3	0.2 0.85	+0.4 – 0	+0.85 +0.6				0.3 0.95	+0.4 – 0	+ 0.95 + 0.7	0.5 1.3	+0.4 – 0	+ 1.3 + 0.9
0.12– 0.24	0.1 0.6	+0.3 – 0	+0.6 +0.4	0.2 1.0	+0.5 – 0	+1.0 +0.7				0.4 1.2	+0.5 – 0	+ 1.2 + 0.9	0.7 1.7	+0.5 – 0	+ 1.7 + 1.2
0.24– 0.40	0.1 0.75	+0.4 – 0	+0.75 +0.5	0.4 1.4	+0.6 – 0	+1.4 +1.0				0.6 1.6	+0.6 – 0	+ 1.6 + 1.2	0.8 2.0	+0.6 – 0	+ 2.0 + 1.4
0.40– 0.56	0.1 0.8	+0.4 – 0	+0.8 +0.5	0.5 1.6	+0.7 – 0	+1.6 +1.2				0.7 1.8	+0.7 – 0	+ 1.8 + 1.4	0.9 2.3	+0.7 – 0	+ 2.3 + 1.6
0.56– 0.71	0.2 0.9	+0.4 – 0	+0.9 +0.6	0.5 1.6	+0.7 – 0	+1.6 +1.2				0.7 1.8	+0.7 – 0	+ 1.8 + 1.4	1.1 2.5	+0.7 – 0	+ 2.5 + 1.8
0.71– 0.95	0.2 1.1	+0.5 – 0	+1.1 +0.7	0.6 1.9	+0.8 – 0	+1.9 +1.4				0.8 2.1	+0.8 – 0	+ 2.1 + 1.6	1.4 3.0	+0.8 – 0	+ 3.0 + 2.2
0.95– 1.19	0.3 1.2	+0.5 – 0	+1.2 +0.8	0.6 1.9	+0.8 – 0	+1.9 +1.4	0.8 2.1	+0.8 – 0	+ 2.1 + 1.6	1.0 2.3	+0.8 – 0	+ 2.3 + 1.8	1.7 3.3	+0.8 – 0	+ 3.3 + 2.5
1.19– 1.58	0.3 1.3	+0.6 – 0	+1.3 +0.9	0.8 2.4	+1.0 – 0	+2.4 +1.8	1.0 2.6	+1.0 – 0	+ 2.6 + 2.0	1.5 3.1	+1.0 – 0	+ 3.1 + 2.5	2.0 4.0	+1.0 – 0	+ 4.0 + 3.0
1.58– 1.97	0.4 1.4	+0.6 – 0	+1.4 +1.0	0.8 2.4	+1.0 – 0	+2.4 +1.8	1.2 2.8	+1.0 – 0	+ 2.8 + 2.2	1.8 3.4	+1.0 – 0	+ 3.4 + 2.8	3.0 5.0	+1.0 – 0	+ 5.0 + 4.0
1.97– 2.56	0.6 1.8	+0.7 – 0	+1.8 +1.3	0.8 2.7	+1.2 – 0	+2.7 +2.0	1.3 3.2	+1.2 – 0	+ 3.2 + 2.5	2.3 4.2	+1.2 – 0	+ 4.2 + 3.5	3.8 6.2	+1.2 – 0	+ 6.2 + 5.0
2.56– 3.15	0.7 1.9	+0.7 – 0	+1.9 +1.4	1.0 2.9	+1.2 – 0	+2.9 +2.2	1.8 3.7	+1.2 – 0	+ 3.7 + 3.0	2.8 4.7	+1.2 – 0	+ 4.7 + 4.0	4.8 7.2	+1.2 – 0	+ 7.2 + 6.0
3.15– 3.94	0.9 2.4	+0.9 – 0	+2.4 +1.8	1.4 3.7	+1.4 – 0	+3.7 +2.8	2.1 4.4	+1.4 – 0	+ 4.4 + 3.5	3.6 5.9	+1.4 – 0	+ 5.9 + 5.0	5.6 8.4	+1.4 – 0	+ 8.4 + 7.0
3.94– 4.73	1.1 2.6	+0.9 – 0	+2.6 +2.0	1.6 3.9	+1.4 – 0	+3.9 +3.0	2.6 4.9	+1.4 – 0	+ 4.9 + 4.0	4.6 6.9	+1.4 – 0	+ 6.9 + 6.0	6.6 9.4	+1.4 – 0	+ 9.4 + 8.0
4.73– 5.52	1.2 2.9	+1.0 – 0	+2.9 +2.2	1.9 4.5	+1.6 – 0	+4.5 +3.5	3.4 6.0	+1.6 – 0	+ 6.0 + 5.0	5.4 8.0	+1.6 – 0	+ 8.0 + 7.0	8.4 11.6	+1.6 – 0	+11.6 +10.0
5.52– 6.30	1.5 3.2	+1.0 – 0	+3.2 +2.5	2.4 5.0	+1.6 – 0	+5.0 +4.0	3.4 6.0	+1.6 – 0	+ 6.0 + 5.0	5.4 8.0	+1.6 – 0	+ 8.0 + 7.0	10.4 13.6	+1.6 – 0	+13.6 +12.0
6.30– 7.09	1.8 3.5	+1.0 – 0	+3.5 +2.8	2.9 5.5	+1.6 – 0	+5.5 +4.5	4.4 7.0	+1.6 – 0	+ 7.0 + 6.0	6.4 9.0	+1.6 – 0	+ 9.0 + 8.0	10.4 13.6	+1.6 – 0	+13.6 +12.0
7.09– 7.88	1.8 3.8	+1.2 – 0	+3.8 +3.0	3.2 6.2	+1.8 – 0	+6.2 +5.0	5.2 8.2	+1.8 – 0	+ 8.2 + 7.0	7.2 10.2	+1.8 – 0	+10.2 + 9.0	12.2 15.8	+1.8 – 0	+15.8 +14.0
7.88– 8.86	2.3 4.3	+1.2 – 0	+4.3 +3.5	3.2 6.2	+1.8 – 0	+6.2 +5.0	5.2 8.2	+1.8 – 0	+ 8.2 + 7.0	8.2 11.2	+1.8 – 0	+11.2 +10.0	14.2 17.8	+1.8 – 0	+17.8 +16.0
8.86– 9.85	2.3 4.3	+1.2 – 0	+4.3 +3.5	4.2 7.2	+1.8 – 0	+7.2 +6.0	6.2 9.2	+1.8 – 0	+ 9.2 + 8.0	10.2 13.2	+1.8 – 0	+13.2 +12.0	14.2 17.8	+1.8 – 0	+17.8 +16.0
9.85–11.03	2.8 4.9	+1.2 – 0	+4.9 +4.0	4.0 7.2	+2.0 – 0	+7.2 +6.0	7.0 10.2	+2.0 – 0	+10.2 + 9.0	10.2 13.2	+2.0 – 0	+13.2 +12.0	16.0 20.0	+2.0 – 0	+20.0 +18.0
11.03–12.41	2.8 4.9	+1.2 – 0	+4.9 +4.0	5.0 8.2	+2.0 – 0	+8.2 +7.0	7.0 10.2	+2.0 – 0	+10.2 + 9.0	12.0 15.2	+2.0 – 0	+15.2 +14.0	18.0 22.0	+2.0 – 0	+22.0 +20.0
12.41–13.98	3.1 5.5	+1.4 – 0	+5.5 +4.5	5.8 9.4	+2.2 – 0	+9.4 +8.0	7.8 11.4	+2.2 – 0	+11.4 +10.0	13.8 17.4	+2.2 – 0	+17.4 +16.0	19.8 24.2	+2.2 – 0	+24.2 +22.0

[1]From ASA B4.1–1955. For larger diameters, see the standard.

12. AMERICAN STANDARD UNIFIED AND AMERICAN THREADS[1]

Nominal Diameter	Coarse[2] NC UNC		Fine[2] NF UNF		Extra Fine[3] NEF UNEF	
	Thds. per in.	Tap Drill[4]	Thds. per in.	Tap Drill[4]	Thds. per in.	Tap Drill[4]
0 (.060)			80	3/64		
1 (.073)	64	No. 53	72	No. 53		
2 (.086)	56	No. 50	64	No. 50		
3 (.099)	48	No. 47	56	No. 45		
4 (.112)	40	No. 43	48	No. 42		
5 (.125)	40	No. 38	44	No. 37		
6 (.138)	32	No. 36	40	No. 33		
8 (.164)	32	No. 29	36	No. 29		
10 (.190)	24	No. 25	32	No. 21		
12 (.216)	24	No. 16	28	No. 14	32	No. 13
1/4	**20**	No. 7	**28**	No. 3	32	7/32
5/16	**18**	F	**24**	I	32	9/32
3/8	**16**	5/16	**24**	Q	32	11/32
7/16	**14**	U	**20**	25/64	**28**	13/32
1/2	13*	27/64	**20**	29/64	**28**	15/32
1/2	**12**	27/64				
9/16	**12**	31/64	**18**	33/64	24	33/64
5/8	**11**	17/32	**18**	37/64	24	37/64
11/16					24	41/64
3/4	**10**	21/32	**16**	11/16	**20**	45/64
13/16					**20**	49/64
7/8	**9**	49/64	**14**	13/16	**20**	53/64
15/16					**20**	57/64

Nominal Diameter	Coarse[2] NC UNC		Fine[2] NF UNF		Extra Fine[3] NEF UNEF	
	Thds. per in.	Tap Drill[4]	Thds. per in.	Tap Drill[4]	Thds. per in.	Tap Drill[4]
1	**8**	7/8	**12**	59/64	**20**	61/64
1 1/16					18	1
1 1/8	**7**	63/64	**12**	1 3/64	18	1 5/64
1 3/16					18	1 9/64
1 1/4	**7**	1 7/64	**12**	1 11/64	18	1 3/16
1 5/16					18	1 17/64
1 3/8	**6**	1 7/32	**12**	1 19/64	18	1 5/16
1 7/16					18	1 3/8
1 1/2	**6**	1 11/32	**12**	1 27/64	18	1 7/16
1 9/16					18	1 1/2
1 5/8					18	1 9/16
1 11/16					18	1 5/8
1 3/4	**5**	1 9/16			**16**	1 11/16
2	**4 1/2**	1 25/32			**16**	1 15/16
2 1/4	**4 1/2**	2 1/32				
2 1/2	**4**	2 1/4				
2 3/4	**4**	2 1/2				
3	**4**	2 3/4				
3 1/4	**4**					
3 1/2	**4**					
3 3/4	**4**					
4	**4**					

[1]ASA B1.1–1949. Bold type indicates Unified threads. For 8-, 12-, and 16-pitch thread series, see Table 13.
[2]Classes 1A, 2A, 3A, 1B, 2B, 3B, 2, and 3. [3]Classes 2A, 2B, 2, and 3.
[4]For approximate 75% full depth of thread. For decimal sizes of numbered and lettered drills, see Table 14.
*The 1/2″ Coarse thread with 13 threads per inch is optional American Standard, but is not Unified.

13. AMERICAN STANDARD UNIFIED AND AMERICAN THREADS[1]

Nominal Diameter	8-Pitch Series 8N		12-Pitch Series 12N and 12UN		16-Pitch Series 16 N and 16UN		Nominal Diameter	8-Pitch Series		12-Pitch Series		16-Pitch Series	
	Thds. per in.	Tap Drill[2]	Thds. per in.	Tap Drill[2]	Thds. per in.	Tap Drill[2]		Thds. per in.	Tap Drill[2]	Thds. per in.	Tap Drill[2]	Thds. per in.	Tap Drill[2]
1/2			**12**	27/64			2 1/16					16	2
9/16			**12**	31/64			2 1/8	8	2	12	2 3/64	16	2 1/16
5/8			12	35/64			2 3/16					16	2 1/8
11/16			12	39/64			2 1/4	8	2 1/8	**12**	2 11/64	**16**	2 3/16
3/4			12	43/64	**16**	11/16	2 5/16					16	2 1/4
13/16			12	47/64	**16**	3/4	2 3/8			12	2 19/64	16	2 5/16
7/8			12	51/64	**16**	13/16	2 7/16					16	2 3/8
15/16			**12**	55/64	**16**	7/8	2 1/2	8	2 3/8	**12**	2 27/64	**16**	2 7/16
1	8	7/8	**12**	59/64	**16**	15/16	2 5/8			12	2 35/64	16	2 9/16
1 1/16			**12**	63/64	**16**	1	2 3/4	8	2 5/8	**12**	2 43/64	**16**	2 11/16
1 1/8	8	1	**12**	1 3/64	**16**	1 1/16	2 7/8			12		16	
1 3/16			**12**	1 7/64	**16**	1 1/8	3	8	2 7/8	**12**		**16**	
1 1/4	8	1 1/8	**12**	1 11/64	**16**	1 3/16	3 1/8			12		16	
1 5/16			**12**	1 15/64	**16**	1 1/4	3 1/4	8		**12**		**16**	
1 3/8	8	1 1/4	**12**	1 19/64	**16**	1 5/16	3 3/8			12		16	
1 7/16			**12**	1 23/64	**16**	1 3/8	3 1/2	8		**12**		**16**	
1 1/2	8	1 3/8	**12**	1 27/64	**16**	1 7/16	3 5/8			12		16	
1 9/16					16	1 1/2	3 3/4	8		**12**		**16**	
1 5/8	8	1 1/2	12	1 35/64	16	1 9/16	3 7/8			12		16	
1 11/16					16	1 5/8	4	8		**12**		**16**	
1 3/4	8	1 5/8	**12**	1 43/64	**16**	1 11/16	4 1/4	8		**12**		**16**	
1 13/16					16	1 3/4	4 1/2	8		**12**		**16**	
1 7/8	8	1 3/4	12	1 51/64	16	1 13/16	4 3/4	8		**12**		**16**	
1 15/16					16	1 7/8	5	8		**12**		**16**	
2	8	1 7/8	**12**	1 59/64	**16**	1 15/16	5 1/4	8		**12**		**16**	

[1] ASA B1.1–1949. Boldface type indicates Unified threads. Classes 2A, 2B, 2, and 3.
[2] For approximate 75% full depth of thread.

14. TWIST DRILL SIZES

Size	Drill Diameter	Size	Drill Diameter	Size	Drill Diameter	Size	Drill Diameter	Size	Drill Diameter
1	.2280	17	.1730	33	.1130	49	.0730	65	.0350
2	.2210	18	.1695	34	.1110	50	.0700	66	.0330
3	.2130	19	.1660	35	.1100	51	.0670	67	.0320
4	.2090	20	.1610	36	.1065	52	.0635	68	.0310
5	.2055	21	.1590	37	.1040	53	.0595	69	.0292
6	.2040	22	.1570	38	.1015	54	.0550	70	.0280
7	.2010	23	.1540	39	.0995	55	.0520	71	.0260
8	.1990	24	.1520	40	.0980	56	.0465	72	.0250
9	.1960	25	.1495	41	.0960	57	.0430	73	.0240
10	.1935	26	.1470	42	.0935	58	.0420	74	.0225
11	.1910	27	.1440	43	.0890	59	.0410	75	.0210
12	.1890	28	.1405	44	.0860	60	.0400	76	.0200
13	.1850	29	.1360	45	.0820	61	.0390	77	.0180
14	.1820	30	.1285	46	.0810	62	.0380	78	.0160
15	.1800	31	.1200	47	.0785	63	.0370	79	.0145
16	.1770	32	.1160	48	.0760	64	.0360	80	.0135

LETTER SIZES

Size	Drill Diameter	Size	Drill Diameter	Size	Drill Diameter	Size	Drill Diameter	Size	Drill Diameter
A	.234	G	.261	L	.290	Q	.332	V	.377
B	.238	H	.266	M	.295	R	.339	W	.386
C	.242	I	.272	N	.302	S	.348	X	.397
D	.246	J	.277	O	.316	T	.358	Y	.404
E	.250	K	.281	P	.323	U	.368	Z	.413
F	.257								

All dimensions are in inches.

Drills designated in common fractions are available in diameters 1/64″ to 1¾″ in 1/64″ increments, 1¾″ to 2¼″ in 1/32″ increments, and 2¼″ to 3½″ in 1/16″ increments. Drills larger than 3½″ are seldom used, and are regarded as special drills.

15. AMERICAN STANDARD GENERAL PURPOSE ACME THREADS[1]

Size	Threads per Inch	Size	Threads per Inch	Size	Threads per Inch	Size	Threads per Inch
¼	16	¾	6	1½	4	3	2
5/16	14	⅞	6	1¾	4	3½	2
⅜	12	1	5	2	4	4	2
7/16	12	1⅛	5	2¼	3	4½	2
½	10	1¼	5	2½	3	5	2
⅝	8	1⅜	4	2¾	3	...	..

[1]ASA B1.5–1952.

16. AMERICAN STANDARD SQUARE AND HEXAGON BOLTS AND NUTS AND HEXAGON HEAD CAP SCREWS[1]

Nominal Size D Body Dia. of Bolt	Regular Bolts					Fin. Hex. Bolts & Hex. Cap Screws[2]		Heavy Bolts		
	Width across Flats W		Height H			Width across Flats W	Height H	Width across Flats W	Height H	
	Sq.	Hex.	Unfin. Sq.	Unfin. Hex.	Semifin. Hex.				Unfin. Hex.	Semifin. & Fin. Hex.
1/4	3/8	7/16	11/64	11/64	5/32	**7/16**	**5/32**	.....	.....	.....
5/16	1/2	1/2	13/64	7/32	13/64	**1/2**	**13/64**	.....	.....	.....
3/8	9/16	9/16	1/4	1/4	15/64	**9/16**	**15/64**	.....	.....	.....
7/16	5/8	5/8	19/64	19/64	9/32	**5/8**	**9/32**	.....	.....	.....
1/2	3/4	3/4	21/64	11/32	5/16	**3/4**	**5/16**	**7/8**	**7/16**	**13/32**
9/16	.....	.....	.....	.....	.....	**13/16**	**23/64**	.....	.....	.....
5/8	15/16	15/16	27/64	27/64	25/64	**15/16**	**25/64**	**1 1/16**	**17/32**	**1/2**
3/4	1 1/8	1 1/8	1/2	1/2	15/32	**1 1/8**	**15/32**	**1 1/4**	**5/8**	**19/32**
7/8	1 5/16	1 5/16	19/32	37/64	35/64	**1 5/16**	**35/64**	**1 7/16**	**23/32**	**11/16**
1	1 1/2	1 1/2	21/32	43/64	39/64	**1 1/2**	**39/64**	**1 5/8**	**13/16**	**3/4**
1 1/8	1 11/16	1 11/16	3/4	3/4	11/16	1 11/16	11/16	**1 13/16**	**29/32**	**27/32**
1 1/4	1 7/8	1 7/8	27/32	27/32	25/32	1 7/8	25/32	**2**	**1**	**15/16**
1 3/8	2 1/16	2 1/16	29/32	29/32	27/32	2 1/16	27/32	**2 3/16**	**1 3/32**	**1 1/32**
1 1/2	2 1/4	2 1/4	1	1	15/16	2 1/4	15/16	**2 3/8**	**1 3/16**	**1 1/8**
1 5/8	2 7/16	.....	1 3/32	.....	.....	.....	.....	2 9/16†	1 9/32†	.....
1 3/4	.....	2 5/8	.....	1 5/32	1 3/32	2 5/8	1 3/32	**2 3/4**	**1 3/8**	**1 5/16**
1 7/8	.....	.....	.....	.....	.....	.....	.....	2 15/16†	1 15/32†	.....
2	.....	3	.....	1 11/32	1 7/32	3	1 7/32	**3 1/8**	**1 9/16**	**1 7/16**
2 1/4	.....	3 3/8	.....	1 1/2	1 3/8	3 3/8	1 3/8	3 1/2	1 3/4	1 5/8
2 1/2	.....	3 3/4	.....	1 21/32	1 17/32	3 3/4	1 17/32	3 7/8	1 15/16	1 13/16
2 3/4	.....	4 1/8	.....	1 13/16	1 11/16	4 1/8	1 11/16	4 1/4	2 1/8	2
3	.....	4 1/2	.....	2	1 7/8	4 1/2	1 7/8	4 5/8	2 5/16	2 3/16
3 1/4	.....	4 7/8	.....	2 3/16	2	.....	.....	.....	.....	.....
3 1/2	.....	5 1/4	.....	2 5/16	2 1/8	.....	.....	.....	.....	.....
3 3/4	.....	5 5/8	.....	2 1/2	2 5/16	.....	.....	.....	.....	.....
4	.....	6	.....	2 11/16	2 1/2	.....	.....	.....	.....	.....

Bold face type indicates unified product.

[1]ASA B18.2–1955. All dimensions are given in inches. For thread series, minimum thread lengths, and bolt lengths, see page 375.

[2]ASA B18.6.2–1956. Hexagon head cap screw sizes 1/4 to 1 1/2 only.

†Heavy hexagon only.

AMERICAN STANDARD SQUARE AND HEXAGON HEAD BOLTS AND NUTS AND HEXAGON HEAD CAP SCREWS (Continued)

REGULAR NUTS					FIN. HEX. NUTS		HEAVY NUTS		
Width across Flats W		Thickness T			Width across Flats W	Thickness T	Width across Flats W	Thickness T	
Sq.	Hex.	Unfin. Sq.	Unfin. Hex.	Semifin. Hex.				Unfin. Sq. & Hex.	Semifin. Hex.
7/16	7/16	7/32	7/32	13/64	**7/16**	**7/32**	1/2	1/4	15/64
9/16	9/16	17/64	17/64	1/4	**1/2**	**17/64**	9/16	5/16	19/64
5/8	5/8	21/64	21/64	5/16	**9/16**	**21/64**	11/16	3/8	23/64
3/4	3/4	3/8	3/8	23/64	**11/16**	**3/8**	3/4	7/16	27/64
13/16	13/16	7/16	7/16	27/64	**3/4**	**7/16**	**7/8**	**1/2**	**31/64**
....	7/8	....	1/2	31/64	**7/8**	**31/64**	15/16*	....	35/64
1	1	35/64	35/64	17/32	**15/16**	**35/64**	**1 1/16**	**5/8**	**39/64**
1 1/8	1 1/8	21/32	21/32	41/64	**1 1/8**	**41/64**	**1 1/4**	**3/4**	**47/64**
1 5/16	1 5/16	49/64	49/64	3/4	**1 5/16**	**3/4**	**1 7/16**	**7/8**	**55/64**
1 1/2	1 1/2	7/8	7/8	55/64	**1 1/2**	**55/64**	**1 5/8**	**1**	**63/64**
1 11/16	1 11/16	1	1	31/32	1 11/16	31/32	**1 13/16**	**1 1/8**	**1 7/64**
1 7/8	1 7/8	1 3/32	1 3/32	1 1/16	1 7/8	1 1/16	**2**	**1 1/4**	**1 7/32**
2 1/16	2 1/16	1 13/64	1 13/64	1 11/64	2 1/16	1 11/64	**2 3/16**	**1 3/8**	**1 11/32**
2 1/4	2 1/4	1 5/16	1 5/16	1 9/32	2 1/4	1 9/32	**2 3/8**	**1 1/2**	**1 15/32**
....	2 7/16*	....	....	1 25/64	.....	.....	2 9/16*	1 5/8*	1 19/32
....	2 5/8*	....	....	1 1/2	2 5/8	1 1/2	**2 3/4**	**1 3/4**	**1 23/32**
....	2 13/16*	....	....	1 39/64	.....	.....	2 15/16*	1 7/8*	1 27/32
....	3 *	....	....	1 23/32	3	1 23/32	**3 1/8**	**2**	**1 31/32**
....	3 3/8*	....	....	1 59/64	3 3/8	1 59/64	3 1/2	2 1/4	2 13/64
....	3 3/4*	....	....	2 9/64	3 3/4	2 9/64	3 7/8	2 1/2	2 29/64
....	4 1/8*	....	....	2 23/64	4 1/8	2 23/64	4 1/4	2 3/4	2 45/64
....	4 1/2*	....	....	2 37/64	4 1/2	2 37/64	4 5/8	3	2 61/64
....	.....	....	....	....	.....	.....	5	3 1/4	3 3/16
....	.....	....	....	....	.....	.....	5 3/8	3 1/2	3 7/16
....	.....	....	....	....	.....	.....	5 3/4	3 3/4	3 11/16
....	.....	....	....	....	.....	.....	6 1/8	4	3 15/16

*Semifinished hexagon only. See ASA B18.2–1955 for jam nuts, slotted nuts, thick nuts, thick slotted nuts, castle nuts, and stove bolt nuts.

For methods of drawing bolts and nuts and hexagon-head cap screws, see Figs. 685, 686 and 689 (a).

17. AMERICAN STANDARD SLOTTED[1] AND SOCKET HEAD[2] CAP SCREWS[3]

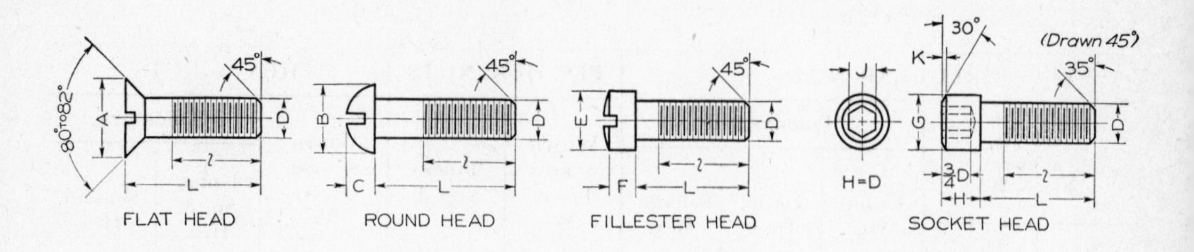

Nominal Size	Flat Head[1]	Round Head[1]		Fillister Head[1]		Socket Head[2]		
D	A	B	C	E	F	G	J	K
0	....	....	....	....	....	.096	.05	.055
1	....	....	....	....	....	.118	.05	.067
2 (.086)	....	....	....	....	....	.140	1/16	.079
3 (.099)	....	....	....	....	....	.161	5/64	.091
4 (.112)	....	....	....	....	....	.183	5/64	.103
5 (.125)	....	....	....	....	....	.205	3/32	.115
6 (.138)	....	....	....	....	....	.226	3/32	.127
8 (.164)	....	....	....	....	....	.270	1/8	.150
10 (.190)	....	....	....	....	....	5/16	5/32	.174
12 (.216)	....	....	....	....	....	11/32	5/32	.198
1/4	1/2	7/16	.191	3/8	11/64	3/8	3/16	.229
5/16	5/8	9/16	.245	7/16	13/64	7/16	7/32	.286
3/8	3/4	5/8	.273	9/16	1/4	9/16	5/16	.344
7/16	13/16	3/4	21/64	5/8	19/64	5/8	5/16	.401
1/2	7/8	13/16	.355	3/4	21/64	3/4	3/8	.458
9/16	1	15/16	.409	13/16	3/8	13/16	3/8	.516
5/8	1 1/8	1	7/16	7/8	27/64	7/8	1/2	.573
3/4	1 3/8	1 1/4	35/64	1	1/2	1	9/16	.688
7/8	1 5/8	....	....	1 1/8	19/32	1 1/8	9/16	.802
1	1 7/8	....	....	1 5/16	21/32	1 5/16	5/8	.917
1 1/8	2 1/16	....	....	....	....	1 1/2	3/4	1.031
1 1/4	2 5/16	....	....	....	....	1 3/4	3/4	1.146
1 3/8	2 9/16	....	....	....	....	1 7/8	3/4	1.260
1 1/2	2 13/16	....	....	....	....	2	1	1.375

[1]ASA B18.6.2-1956. [2]ASA B18.3-1954. [3]See Hexagon Head Screws on pages 380 and 802. For methods of drawing cap screws, screw lengths, and thread data, see Fig. 869.

18. AMERICAN STANDARD MACHINE SCREWS[1]

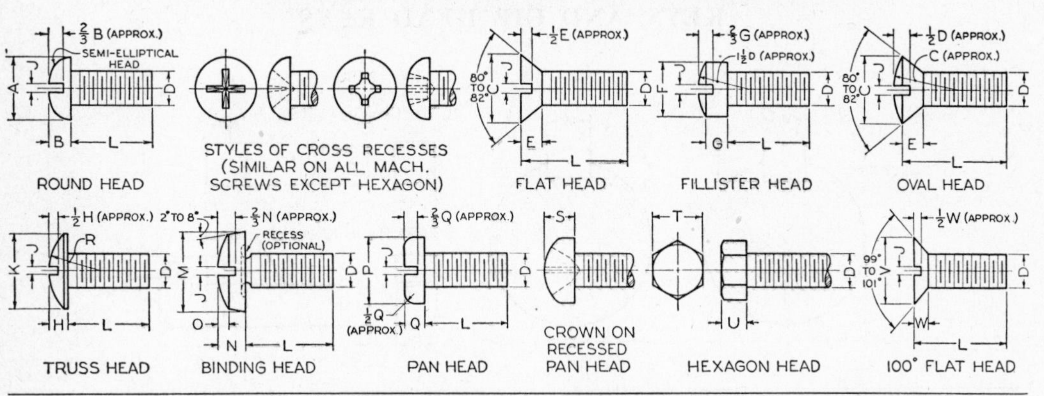

Nom. Size	Max. Diam. D	Round Head		Flat Heads & Oval Head		Fillister Head		Truss Head			Width of Slots
		A	B	C	E	F	G	K	H	R	
0	0.060	0.113	0.053	0.119	0.035	0.096	0.045	....	....	....	0.023
1	0.073	0.138	0.061	0.146	0.043	0.118	0.053	....	....	....	0.026
2	0.086	0.162	0.069	0.172	0.051	0.140	0.062	0.194	0.053	0.129	0.031
3	0.099	0.187	0.078	0.199	0.059	0.161	0.070	0.226	0.061	0.151	0.035
4	0.112	0.211	0.086	0.225	0.067	0.183	0.079	0.257	0.069	0.169	0.039
5	0.125	0.236	0.095	0.252	0.075	0.205	0.088	0.289	0.078	0.191	0.043
6	0.138	0.260	0.103	0.279	0.083	0.226	0.096	0.321	0.086	0.211	0.048
8	0.164	0.309	0.120	0.332	0.100	0.270	0.113	0.384	0.102	0.254	0.054
10	0.190	0.359	0.137	0.385	0.116	0.313	0.130	0.448	0.118	0.283	0.060
12	0.216	0.408	0.153	0.438	0.132	0.357	0.148	0.511	0.134	0.336	0.067
1/4	0.250	0.472	0.175	0.507	0.153	0.414	0.170	0.573	0.150	0.375	0.075
5/16	0.3125	0.590	0.216	0.635	0.191	0.518	0.211	0.698	0.183	0.457	0.084
3/8	0.375	0.708	0.256	0.762	0.230	0.622	0.253	0.823	0.215	0.538	0.094
7/16	0.4375	0.750	0.328	0.812	0.223	0.625	0.265	0.948	0.248	0.619	0.094
1/2	0.500	0.813	0.355	0.875	0.223	0.750	0.297	1.073	0.280	0.701	0.106
9/16	0.5625	0.938	0.410	1.000	0.260	0.812	0.336	1.198	0.312	0.783	0.118
5/8	0.625	1.000	0.438	1.125	0.298	0.875	0.375	1.323	0.345	0.863	0.133
3/4	0.750	1.250	0.547	1.375	0.372	1.000	0.441	1.573	0.410	1.024	0.149

Nom. Size	Max. Diam. D	Binding Head			Pan Head			Hexagon Head		100° Flat Head		Width of Slots
		M	N	O	P	Q	S	T	U	V	W	
2	0.086	0.181	0.046	0.018	0.167	0.053	0.062	0.125	0.050	....	....	0.031
3	0.099	0.208	0.054	0.022	0.193	0.060	0.071	0.187	0.055	....	....	0.035
4	0.112	0.235	0.063	0.025	0.219	0.068	0.080	0.187	0.060	0.225	0.048	0.039
5	0.125	0.263	0.071	0.029	0.245	0.075	0.089	0.187	0.070	....	....	0.043
6	0.138	0.290	0.080	0.032	0.270	0.082	0.097	0.250	0.080	0.279	0.060	0.048
8	0.164	0.344	0.097	0.039	0.322	0.096	0.115	0.250	0.110	0.332	0.072	0.054
10	0.190	0.399	0.114	0.045	0.373	0.110	0.133	0.312	0.120	0.385	0.083	0.060
12	0.216	0.454	0.130	0.052	0.425	0.125	0.151	0.312	0.155	....	....	0.067
1/4	0.250	0.513	0.153	0.061	0.492	0.144	0.175	0.375	0.190	0.507	0.110	0.075
5/16	0.3125	0.641	0.193	0.077	0.615	0.178	0.218	0.500	0.230	0.635	0.138	0.084
3/8	0.375	0.769	0.234	0.094	0.740	0.212	0.261	0.562	0.295	0.762	0.165	0.094

[1]ASA B18.6–1947.

Length of Thread: On screws 2″ long and shorter, the threads extend to within two threads of the head and closer if practicable; longer screws have minimum thread length of 1¾″.

Points: Machine screws are regularly made with plain sheared ends, not chamfered.

Threads: Either Coarse or Fine Thread Series, Class 2 fit.

Recessed Heads: Two styles of cross recesses are available on all screws except hexagon head.

19. SQUARE AND FLAT KEYS, PLAIN TAPER KEYS, AND GIB HEAD KEYS

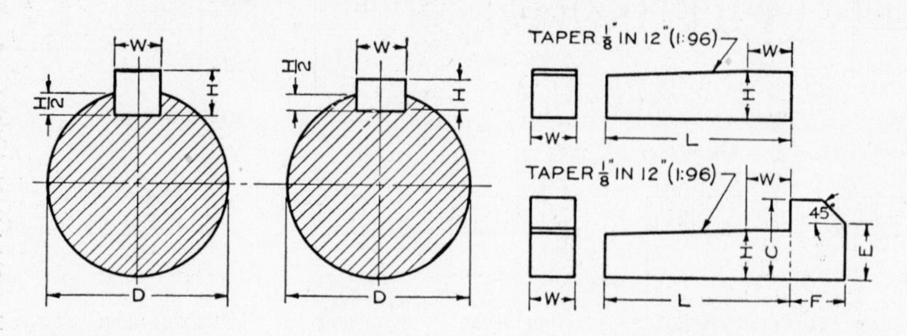

Shaft Diameters	Square Stock Key	Flat Stock Key	Gib Head Taper Stock Key					
			Square			Flat		
			Height	Length	Height to Chamfer	Height	Length	Height to Chamfer
D	W = H	W × H	C	F	E	C	F	E
1/2 to 9/16	1/8	1/8 × 3/32	1/4	7/32	5/32	3/16	1/8	1/8
5/8 to 7/8	3/16	3/16 × 1/8	5/16	9/32	7/32	1/4	3/16	5/32
15/16 to 1 1/4	1/4	1/4 × 3/16	7/16	11/32	11/32	5/16	1/4	3/16
1 5/16 to 1 3/8	5/16	5/16 × 1/4	9/16	13/32	13/32	3/8	5/16	1/4
1 7/16 to 1 3/4	3/8	3/8 × 1/4	11/16	15/32	15/32	7/16	3/8	5/16
1 13/16 to 2 1/4	1/2	1/2 × 3/8	7/8	19/32	5/8	5/8	1/2	7/16
2 5/16 to 2 3/4	5/8	5/8 × 7/16	1 1/16	23/32	3/4	3/4	5/8	1/2
2 7/8 to 3 1/4	3/4	3/4 × 1/2	1 1/4	7/8	7/8	7/8	3/4	5/8
3 3/8 to 3 3/4	7/8	7/8 × 5/8	1 1/2	1	1	1 1/16	7/8	3/4
3 7/8 to 4 1/2	1	1 × 3/4	1 3/4	1 3/16	1 3/16	1 1/4	1	13/16
4 3/4 to 5 1/2	1 1/4	1 1/4 × 7/8	2	1 7/16	1 7/16	1 1/2	1 1/4	1
5 3/4 to 6	1 1/2	1 1/2 × 1	2 1/2	1 3/4	1 3/4	1 3/4	1 1/2	1 1/4

Plain taper square and flat keys have the same dimensions as the plain parallel stock keys, with the addition of the taper on top. Gib head taper square and flat keys have the same dimensions as the plain taper keys, with the addition of the gib head.

Stock lengths for plain taper and gib head taper keys: The minimum stock length equals 4W, and the maximum equals 16W. The increments of increase of length equal 2W.

20. SQUARE AND ACME THREADS

Size	Threads per Inch	Size	Threads per Inch	Size	Threads per Inch	Size	Threads per Inch
3/8	12	7/8	5	2	2 1/2	3 1/2	1 1/3
7/16	10	1	5	2 1/4	2	3 3/4	1 1/3
1/2	10	1 1/8	4	2 1/2	2	4	1 1/3
9/16	8	1 1/4	4	2 3/4	2	4 1/4	1 1/3
5/8	8	1 1/2	3	3	1 1/2	4 1/2	1
3/4	6	1 3/4	2 1/2	3 1/4	1 1/2	over 4 1/2	1

21. AMERICAN STANDARD WOODRUFF KEYS[1]

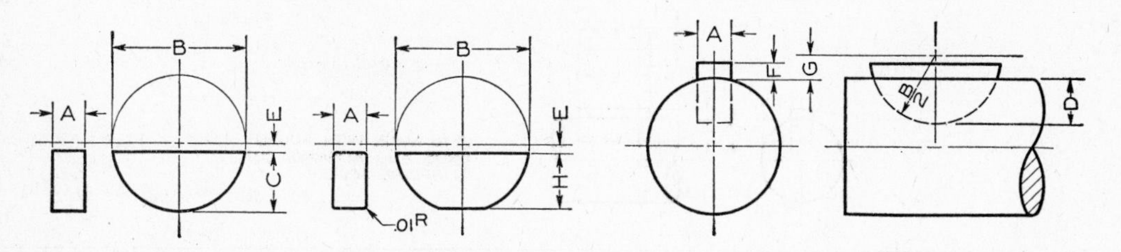

Key No.[2]	Nominal Sizes				Maximum Sizes			Key No.[2]	Nominal Sizes				Maximum Sizes		
	A × B	E	F	G	H	D	C		A × B	E	F	G	H	D	C
204	1/16 × 1/2	3/64	1/32	5/64	.194	.1718	.203	808	1/4 × 1	1/16	1/8	3/16	.428	.3130	.438
304	3/32 × 1/2	3/64	3/64	3/32	.194	.1561	.203	809	1/4 × 1 1/8	5/64	1/8	13/64	.475	.3590	.484
305	3/32 × 5/8	1/16	3/64	7/64	.240	.2031	.250	810	1/4 × 1 1/4	5/64	1/8	13/64	.537	.4220	.547
404	1/8 × 1/2	3/64	1/16	7/64	.194	.1405	.203	811	1/4 × 1 3/8	3/32	1/8	7/32	.584	.4690	.594
405	1/8 × 5/8	1/16	1/16	1/8	.240	.1875	.250	812	1/4 × 1 1/2	7/64	1/8	15/64	.631	.5160	.641
406	1/8 × 3/4	1/16	1/16	1/8	.303	.2505	.313	1008	5/16 × 1	1/16	5/32	7/32	.428	.2818	.438
505	5/32 × 5/8	1/16	5/64	9/64	.240	.1719	.250	1009	5/16 × 1 1/8	5/64	5/32	15/64	.475	.3278	.484
506	5/32 × 3/4	1/16	5/64	9/64	.303	.2349	.313	1010	5/16 × 1 1/4	5/64	5/32	15/64	.537	.3908	.547
507	5/32 × 7/8	1/16	5/64	9/64	.365	.2969	.375	1011	5/16 × 1 3/8	3/32	5/32	8/32	.584	.4378	.594
606	3/16 × 3/4	1/16	3/32	5/32	.303	.2193	.313	1012	5/16 × 1 1/2	7/64	5/32	17/64	.631	.4848	.641
607	3/16 × 7/8	1/16	3/32	5/32	.365	.2813	.375	1210	3/8 × 1 1/4	5/64	3/16	17/64	.537	.3595	.547
608	3/16 × 1	1/16	3/32	5/32	.428	.3443	.438	1211	3/8 × 1 3/8	3/32	3/16	9/32	.584	.4065	.594
609	3/16 × 1 1/8	5/64	3/32	11/64	.475	.3903	.484	1212	3/8 × 1 1/2	7/64	3/16	19/64	.631	.4535	.641
807	1/4 × 7/8	1/16	1/8	3/16	.365	.2500	.375	....	....	..	..	..	....	....	....

[1]ASA B17f–1930 (R 1955).

[2]Key numbers indicate nominal key dimensions. The last two digits give the nominal diameter B in eighths of an inch, and the digits before the last two give the nominal width A in thirty-seconds of an inch.

22. WOODRUFF KEY SIZES FOR DIFFERENT SHAFT DIAMETERS[1]

Shaft Diameter	5/16 to 3/8	7/16 to 1/2	9/16 to 3/4	13/16 to 15/16	1 to 1 3/16	1 1/4 to 1 7/16	1 1/2 to 1 3/4	1 13/16 to 2 1/8	2 3/16 to 2 1/2
Key Numbers	204	304 305	404 405 406	505 506 507	606 607 608 609	807 808 809	810 811 812	1011 1012	1211 1212

[1]Suggested sizes; not standard.

23. PRATT AND WHITNEY ROUND-END KEYS

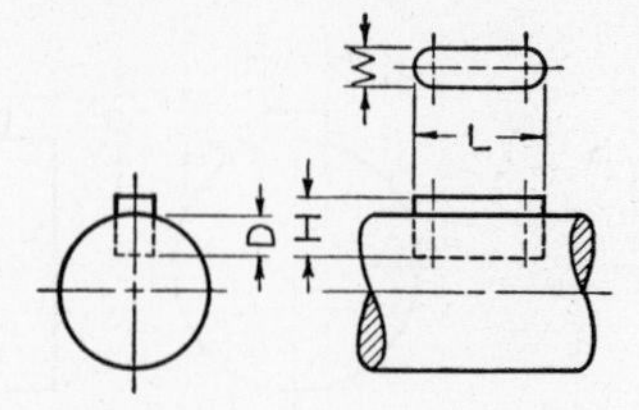

KEYS MADE WITH ROUND ENDS AND KEYWAYS CUT IN SPLINE MILLER

KEY NO.	L	W OR D	H	KEY NO.	L	W OR D	H
1	1/2	1/16	3/32	22	1 3/8	1/4	3/8
2	1/2	3/32	9/64	23	1 3/8	5/16	15/32
3	1/2	1/8	3/16	F	1 3/8	3/8	9/16
4	5/8	3/32	9/64	24	1 1/2	1/4	3/8
5	5/8	1/8	3/16	25	1 1/2	5/16	15/32
6	5/8	5/32	15/64	G	1 1/2	3/8	9/16
7	3/4	1/8	3/16	51	1 3/4	1/4	3/8
8	3/4	5/32	15/64	52	1 3/4	5/16	15/32
9	3/4	3/16	9/32	53	1 3/4	3/8	9/16
10	7/8	5/32	15/64	26	2	3/16	9/32
11	7/8	3/16	9/32	27	2	1/4	3/8
12	7/8	7/32	21/64	28	2	5/16	15/32
A	7/8	1/4	3/8	29	2	3/8	9/16
13	1	3/16	9/32	54	2 1/4	1/4	3/8
14	1	7/32	21/64	55	2 1/4	5/16	15/32
15	1	1/4	3/8	56	2 1/4	3/8	9/16
B	1	5/16	15/32	57	2 1/4	7/16	21/32
16	1 1/8	3/16	9/32	58	2 1/2	5/16	15/32
17	1 1/8	7/32	21/64	59	2 1/2	3/8	9/16
18	1 1/8	1/4	3/8	60	2 1/2	7/16	21/32
C	1 1/8	5/16	15/32	61	2 1/2	1/2	3/4
19	1 1/4	3/16	9/32	30	3	3/8	9/16
20	1 1/4	7/32	21/64	31	3	7/16	21/32
21	1 1/4	1/4	3/8	32	3	1/2	3/4
D	1 1/4	5/16	15/32	33	3	9/16	27/32
E	1 1/4	3/8	9/16	34	3	5/8	15/16

The length L may vary from the table, but equals at least 2W.
Max. length of slot is 4″ + W. Note that key is sunk two-thirds into shaft in all cases.

24. AMERICAN STANDARD PLAIN WASHERS[1]

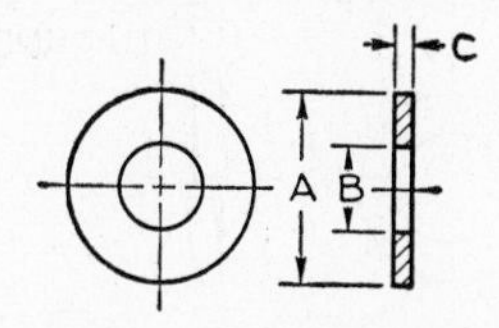

Inside Diameter A	Outside Diameter B	Nominal Thickness C	Inside Diameter A	Outside Diameter B	Nominal Thickness C	Inside Diameter A	Outside Diameter B	Nominal Thickness C
5/64	3/16	0.020	13/32	13/16	0.065	15/16	2	0.165
3/32	7/32	0.020	7/16	7/8	0.083	15/16	2 1/4	0.165
3/32	1/4	0.020	7/16	1	0.083	15/16	3 3/8	0.180
1/8	1/4	0.022	7/16	1 3/8	0.083	1 1/16	2	0.134
1/8	5/16	0.032	15/32	59/64	0.065	1 1/16	2 1/4	0.165
5/32	5/16	0.035	1/2	1 1/8	0.083	1 1/16	2 1/2	0.165
5/32	3/8	0.049	1/2	1 1/4	0.083	1 1/16	3 7/8	0.238
11/64	13/32	0.049	1/2	1 5/8	0.083	1 3/16	2 1/2	0.165
3/16	3/8	0.049	17/32	1 1/16	0.095	1 1/4	2 3/4	0.165
3/16	7/16	0.049	9/16	1 1/4	0.109	1 5/16	2 3/4	0.165
13/64	15/32	0.049	9/16	1 3/8	0.109	1 3/8	3	0.165
7/32	7/16	0.049	9/16	1 7/8	0.109	1 7/16	3	0.180
7/32	1/2	0.049	19/32	1 3/16	0.095	1 1/2	3 1/4	0.180
15/64	17/32	0.049	5/8	1 3/8	0.109	1 9/16	3 1/4	0.180
1/4	1/2	0.049	5/8	1 1/2	0.109	1 5/8	3 1/2	0.180
1/4	9/16	0.049	5/8	2 1/8	0.134	1 11/16	3 1/2	0.180
1/4	9/16	0.065	21/32	1 5/16	0.095	1 3/4	3 3/4	0.180
17/64	5/8	0.049	11/16	1 1/2	0.134	1 13/16	3 3/4	0.180
9/32	5/8	0.065	11/16	1 3/4	0.134	1 7/8	4	0.180
5/16	3/4	0.065	11/16	2 3/8	0.165	1 15/16	4	0.180
5/16	7/8	0.065	13/16	1 1/2	0.134	2	4 1/4	0.180
11/32	11/16	0.065	13/16	1 3/4	0.148	2 1/16	4 1/4	0.180
3/8	3/4	0.065	13/16	2	0.148	2 1/8	4 1/2	0.180
3/8	7/8	0.083	13/16	2 7/8	0.165	2 3/8	4 3/4	0.220
3/8	1 1/8	0.065	15/16	1 3/4	0.134	2 5/8	5	0.238
...	...	....	...	...	....	2 7/8	5 1/4	0.259
...	...	....	...	...	....	3 1/8	5 1/2	0.284

[1]From ASA B27.2–1953.

For parts lists, etc., give inside diameter, outside diameter, and the thickness; for example: 5/16 × 3/4 × 0.065 PLAIN WASHER.

25. AMERICAN STANDARD LOCK WASHERS[1]

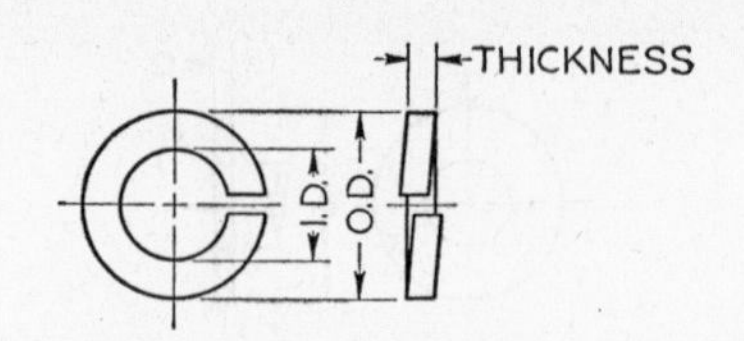

Nominal Size (Inside Dia.)	Inside Dia. Min.	Light		Medium		Heavy		Extra Heavy	
		Outside Dia. Max.	Thick-ness Min.	Outside Dia. Max.	Thick-ness Min.	Outside Dia. Max.	Thick-ness Min.	Outside Dia. Max.	Thick-ness Min.
0.086 (No. 2)	0.088	0.165	0.015	0.175	0.020	0.185	0.025	0.211	0.027
0.099 (No. 3)	0.102	0.188	0.020	0.198	0.025	0.212	0.031	0.242	0.034
0.112 (No. 4)	0.115	0.202	0.020	0.212	0.025	0.226	0.031	0.256	0.034
0.125 (No. 5)	0.128	0.225	0.025	0.239	0.031	0.225	0.040	0.303	0.045
0.138 (No. 6)	0.141	0.239	0.025	0.253	0.031	0.269	0.040	0.317	0.045
0.164 (No. 8)	0.168	0.280	0.031	0.296	0.040	0.310	0.047	0.378	0.057
0.190 (No. 10)	0.194	0.323	0.040	0.337	0.047	0.353	0.056	0.437	0.068
0.216 (No. 12)	0.221	0.364	0.047	0.380	0.056	0.394	0.063	0.500	0.080
1/4	0.255	0.489	0.047	0.493	0.062	0.495	0.077	0.539	0.084
5/16	0.319	0.575	0.056	0.591	0.078	0.601	0.097	0.627	0.108
3/8	0.382	0.678	0.070	0.688	0.094	0.696	0.115	0.746	0.123
7/16	0.446	0.780	0.085	0.784	0.109	0.792	0.133	0.844	0.143
1/2	0.509	0.877	0.099	0.879	0.125	0.889	0.151	0.945	0.162
9/16	0.573	0.975	0.113	0.979	0.141	0.989	0.170	1.049	0.182
5/8	0.636	1.082	0.126	1.086	0.156	1.100	0.189	1.164	0.202
11/16	0.700	1.178	0.138	1.184	0.172	1.200	0.207	1.266	0.221
3/4	0.763	1.277	0.153	1.279	0.188	1.299	0.226	1.369	0.241
13/16	0.827	1.375	0.168	1.377	0.203	1.401	0.246	1.473	0.261
7/8	0.890	1.470	0.179	1.474	0.219	1.504	0.266	1.586	0.285
15/16	0.954	1.562	0.191	1.570	0.234	1.604	0.284	1.698	0.308
1	1.017	1.656	0.202	1.672	0.250	1.716	0.306	1.810	0.330
1 1/16	1.081	1.746	0.213	1.768	0.256	1.820	0.326	1.922	0.352
1 1/8	1.144	1.837	0.224	1.865	0.281	1.921	0.345	2.031	0.375
1 3/16	1.208	1.923	0.234	1.963	0.297	2.021	0.364	2.137	0.396
1 1/4	1.271	2.012	0.244	2.058	0.312	2.126	0.384	2.244	0.417
1 5/16	1.335	2.098	0.254	2.156	0.328	2.226	0.483	2.350	0.438
1 3/8	1.398	2.183	0.264	2.253	0.344	2.325	0.422	2.453	0.458
1 7/16	1.462	2.269	0.273	2.349	0.359	2.421	0.440	2.555	0.478
1 1/2	1.525	2.352	0.282	2.446	0.375	2.518	0.458	2.654	0.496

[1]ASA B27.1–1950.

26. STANDARDS FOR WIRE GAGES[1]

Dimensions of Sizes in Decimal Parts of an Inch[2]

No. of Wire	American or Brown & Sharpe for Non-Ferrous Metals	Birmingham, or Stubs' Iron Wire	American S. & W. Co.'s (Washburn & Moen) Std. Steel Wire	American S. & W. Co.'s Music Wire	Imperial Wire	Stubs' Steel Wire	Steel Manufacturers' Sheet Gage[2]	No. of Wire
7–0's	.651354	....	.4900	....	.500	....	.....	7–0's
6–0's	.580049	....	.4615	.004	.464	....	.....	6–0's
5–0's	.516549	.500	.4305	.005	.432	....	.....	5–0's
4–0's	.460	.454	.3938	.006	.400	....	.....	4–0's
000	.40964	.425	.3625	.007	.372	....	.....	000
00	.3648	.380	.3310	.008	.348	....	.....	00
0	.32486	.340	.3065	.009	.324	....	.....	0
1	.2893	.300	.2830	.010	.300	.227	.....	1
2	.25763	.284	.2625	.011	.276	.219	.....	2
3	.22942	.259	.2437	.012	.252	.212	.2391	3
4	.20431	.238	.2253	.013	.232	.207	.2242	4
5	.18194	.220	.2070	.014	.212	.204	.2092	5
6	.16202	.203	.1920	.016	.192	.201	.1943	6
7	.14428	.180	.1770	.018	.176	.199	.1793	7
8	.12849	.165	.1620	.020	.160	.197	.1644	8
9	.11443	.148	.1483	.022	.144	.194	.1495	9
10	.10189	.134	.1350	.024	.128	.191	.1345	10
11	.090742	.120	.1205	.026	.116	.188	.1196	11
12	.080808	.109	.1055	.029	.104	.185	.1046	12
13	.071961	.095	.0915	.031	.092	.182	.0897	13
14	.064084	.083	.0800	.033	.080	.180	.0747	14
15	.057068	.072	.0720	.035	.072	.178	.0763	15
16	.05082	.065	.0625	.037	.064	.175	.0598	16
17	.045257	.058	.0540	.039	.056	.172	.0538	17
18	.040303	.049	.0475	.041	.048	.168	.0478	18
19	.03589	.042	.0410	.043	.040	.164	.0418	19
20	.031961	.035	.0348	.045	.036	.161	.0359	20
21	.028462	.032	.0317	.047	.032	.157	.0329	21
22	.025347	.028	.0286	.049	.028	.155	.0299	22
23	.022571	.025	.0258	.051	.024	.153	.0269	23
24	.0201	.022	.0230	.055	.022	.151	.0239	24
25	.0179	.020	.0204	.059	.020	.148	.0209	25
26	.01594	.018	.0181	.063	.018	.146	.0179	26
27	.014195	.016	.0173	.067	.0164	.143	.0164	27
28	.012641	.014	.0162	.071	.0149	.139	.0149	28
29	.011257	.013	.0150	.075	.0136	.134	.0135	29
30	.010025	.012	.0140	.080	.0124	.127	.0120	30
31	.008928	.010	.0132	.085	.0116	.120	.0105	31
32	.00795	.009	.0128	.090	.0108	.115	.0097	32
33	.00708	.008	.0118	.095	.0100	.112	.0090	33
34	.006304	.007	.0104	....	.0092	.110	.0082	34
35	.005614	.005	.0095	....	.0084	.108	.0075	35
36	.005	.004	.0090	....	.0076	.106	.0067	36
37	.004453	....	.0085	....	.0068	.103	.0064	37
38	.003965	....	.0080	....	.0060	.101	.0060	38
39	.003531	....	.0075	....	.0052	.099	.....	39
40	.003144	....	.0070	....	.0048	.097	.....	40

[1]Courtesy Brown & Sharpe Mfg. Co. [2]Now used by steel manufacturers in place of old U.S. Std. Gage.

The difference between the Stubs' Iron Wire Gage and the Stubs' Steel Wire Gage should be noted, the first being commonly known as the English Standard Wire, or Birmingham Gage, which designates the Stubs' soft wire sizes, and the second being used in measuring drawn steel wire or drill rods of Stubs' make.

27. AMERICAN STANDARD TAPER PINS[1]

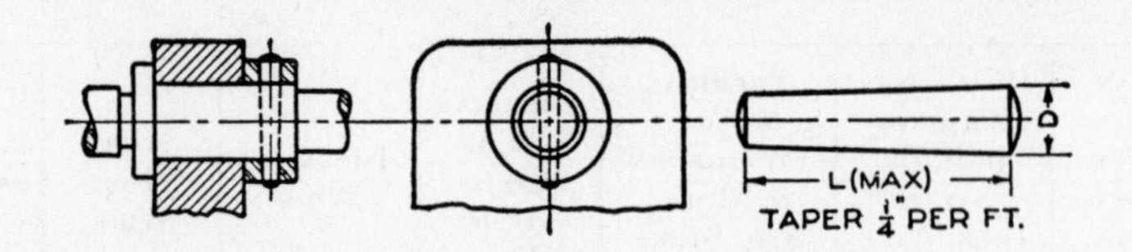

Number	7/0	6/0	5/0	4/0	3/0	2/0	0	1	2	3	4	5	6	7	8
Size (Large End)	0.0625	0.0780	0.0940	0.1090	0.1250	0.1410	0.1560	0.1720	0.1930	0.2190	0.2500	0.2890	0.3410	0.4090	0.4920
Shaft Dia. (Approx.)[2]		7/32	1/4	5/16	3/8	7/16	1/2	9/16	5/8	3/4	13/16	7/8	1	1 1/4	1 1/2
Drill Size (Before Reamer)[2]		.0595	.0785	.0935	.104	.120	.1405	.1495	.166	.189	.213	1/4	9/32	11/32	13/32
Length, L															
0.375	X	X													
0.500	X	X	X	X	X	X	X								
0.625	X	X	X	X	X	X	X								
0.750		X	X	X	X	X	X	X	X	X					
0.875					X	X	X	X	X	X					
1.000			X	X	X	X	X	X	X	X	X	X			
1.250						X	X	X	X	X	X	X	X		
1.500							X	X	X	X	X	X	X		
1.750								X	X	X	X	X	X		
2.000								X	X	X	X	X	X	X	X
2.250									X	X	X	X	X	X	X
2.500									X	X	X	X	X	X	X
2.750										X	X	X	X	X	X
3.000										X	X	X	X	X	X
3.250													X	X	X
3.500													X	X	X
3.750													X	X	X
4.000													X	X	X
4.250															X
4.500															X

[1]ASA B5.20–1954. [2]Suggested sizes—not American Standard.

All dimensions are given in inches. For sizes 9 and 10, see the Standard. Standard reamers are available for pins given above the line.

Pins Nos. 11 (size 0.8600), 12 (size 1.032), 13 (size 1.241), and 14 (1.523) are special sizes—hence their lengths are special.

To find small diameter of pin, multiply the length by 0.02083 and subtract the result from the large diameter.

28. AMERICAN STANDARD COTTER PINS[1]

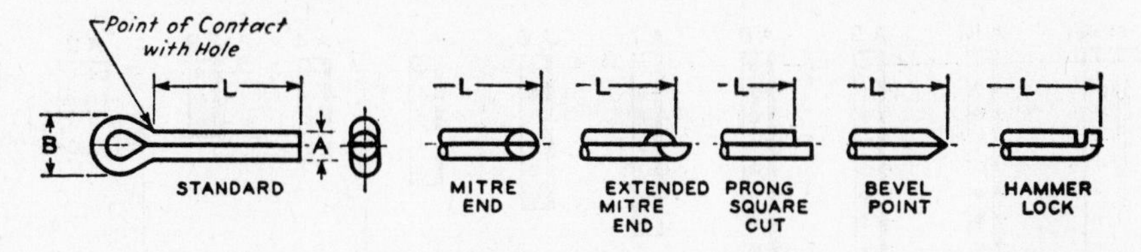

Diameter Nominal	Diameter A		Outside Eye Diameter B	Hole Sizes Recommended	Diameter Nominal	Diameter A		Outside Eye Diameter B	Hole Sizes Recommended
	Max.	Min.	Min.			Max.	Min.	Min.	
0.031	0.032	0.028	1/16	3/64	0.188	0.176	0.172	3/8	13/64
0.047	0.048	0.044	3/32	1/16	0.219	0.207	0.202	7/16	15/64
0.062	0.060	0.056	1/8	5/64	0.250	0.225	0.220	1/2	17/64
0.078	0.076	0.072	5/32	3/32	0.312	0.280	0.275	5/8	5/16
0.094	0.090	0.086	3/16	7/64	0.375	0.335	0.329	3/4	3/8
0.109	0.104	0.100	7/32	1/8	0.438	0.406	0.400	7/8	7/16
0.125	0.120	0.116	1/4	9/64	0.500	0.473	0.467	1	1/2
0.141	0.134	0.130	9/32	5/32	0.625	0.598	0.590	1 1/4	5/8
0.156	0.150	0.146	5/16	11/64	0.750	0.723	0.715	1 1/2	3/4

[1]ASA B5.20–1954. All dimensions are given in inches.

29. STANDARD BEAM CONNECTIONS WITH ⅞″ RIVETS*

STANDARD TWO-ANGLE CONNECTIONS "A" SERIES

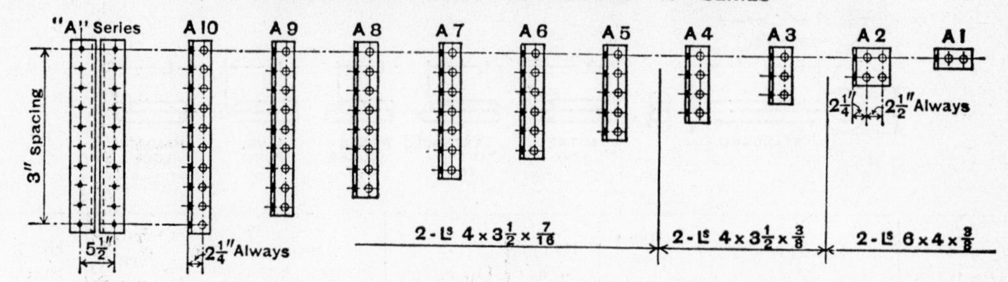

STANDARD TWO-ANGLE CONNECTIONS "H" AND "HH" SERIES

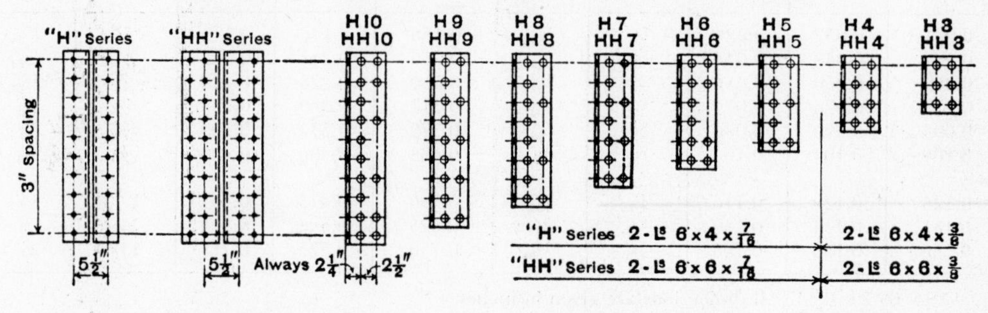

30. STANDARD BEAM CONNECTIONS WITH ¾″ RIVETS*

STANDARD TWO-ANGLE CONNECTIONS "B" SERIES

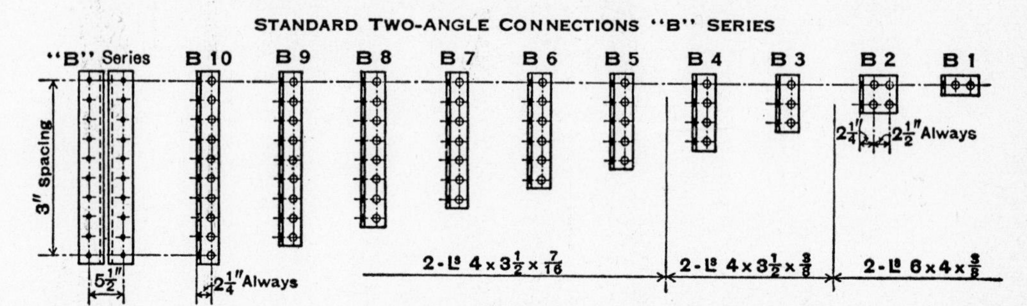

STANDARD TWO-ANGLE CONNECTIONS "K" AND "KK" SERIES

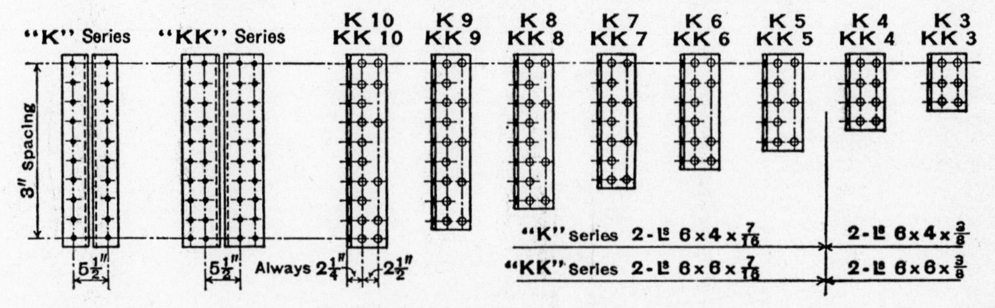

*From *Steel Construction Manual*, courtesy of American Institute of Steel Construction, 101 Park Ave., New York 17, N. Y.

31. TOPOGRAPHIC SYMBOLS

Board of Surveys and Maps

Highway
Railroad
Highway Bridge
Railroad Bridge
Drawbridges
Suspension Bridge
Dam
Telegraph or Telephone Line
Power-Trans. Line
Buildings in General
Capital
County Seat
Other Towns
Barbed Wire Fence
Smooth Wire Fence
Hedge
Oil or Gas Wells
Windmill
Tanks
Canal or Ditch
Canal Lock
Canal Lock (POINT UPSTREAM)
Aqueduct or Water Pipe

National or State Line
County Line
Township or District Line
City or Village Line
Triangulation Station
BM 1232 Bench Mark and Elevation
Any Location Station (WITH EXPLANATORY NOTE)
Streams in General
Lake or Pond
Falls and Rapids
Contours
Hachures
Sand and Sand Dunes
Marsh
Woodland of Any Kind
Orchard
Grassland in General
Cultivated Fields
Commercial or Municipal Field
Airplane Landing Field Marked or Emergency
Mooring Mast
Airway Light Beacon (ARROWS INDICATE COURSE LIGHTS)
Auxiliary Airway Light Beacon, Flashing

32. AMERICAN STANDARD PIPING SYMBOLS

(From ASA Z32.2.3–1949)

	FLANGED	SCREWED	BELL & SPIGOT	WELDED	SOLDERED
1. Joint					
2. Elbow—90°					
3. Elbow—45°					
4. Elbow—Turned Up					
5. Elbow—Turned Down					
6. Elbow—Long Radius					
7. Reducing Elbow					
8. Tee					
9. Tee—Outlet Up					
10. Tee—Outlet Down					
11. Side Outlet Tee—Outlet Up					
12. Cross					
13. Reducer, Concentric					
14. Reducer, Eccentric					
15. Lateral					
16. Gate Valve, Elev.					
17. Globe Valve, Elev.					
18. Check Valve					
19. Stop Cock					
20. Safety Valve					
21. Expansion Joint					
22. Union					
23. Sleeve					
24. Bushing					

33. HEATING, VENTILATING, AND DUCTWORK SYMBOLS

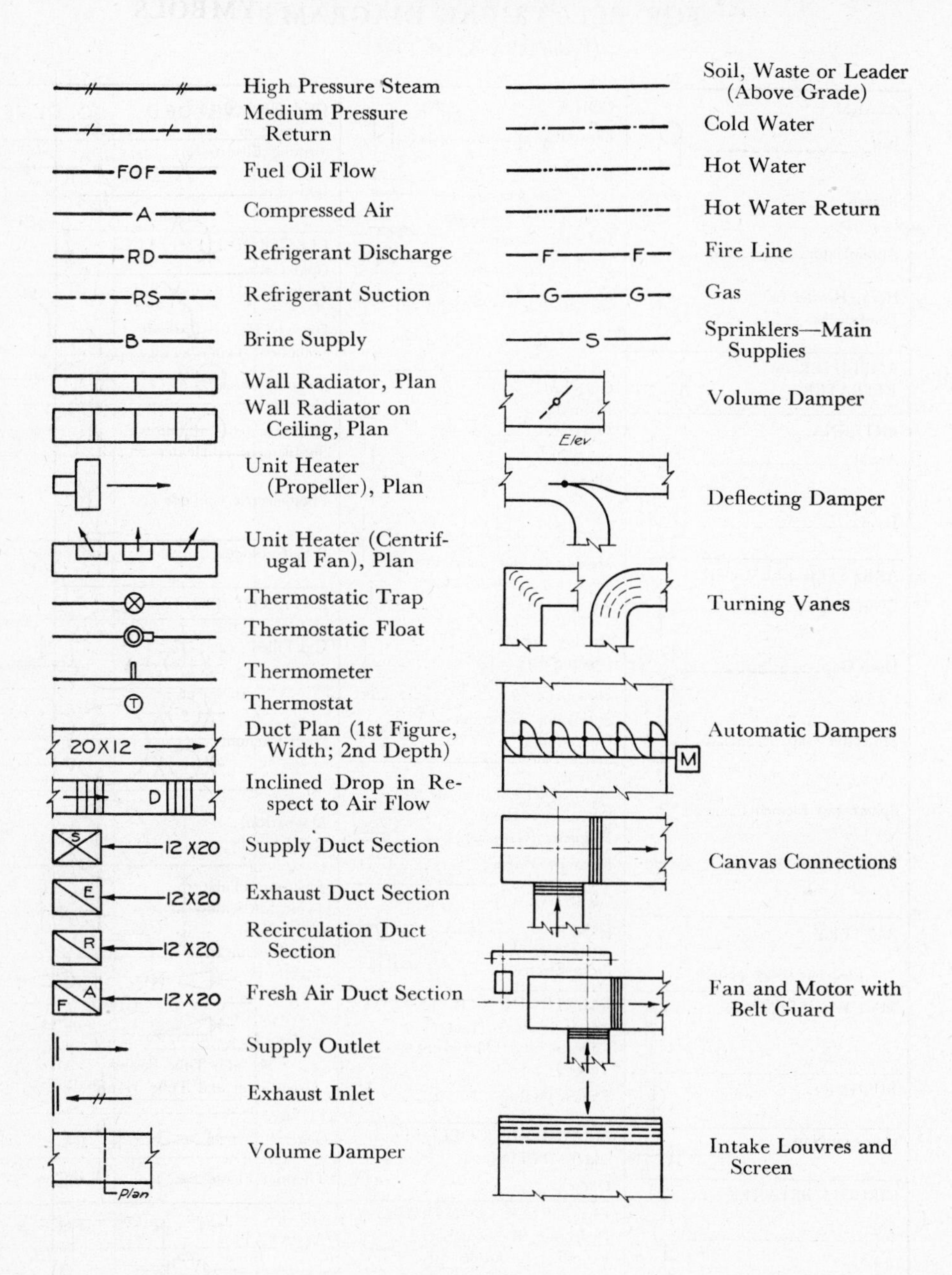
High Pressure Steam
Medium Pressure Return
FOF
Fuel Oil Flow
A
Compressed Air
RD
Refrigerant Discharge
RS
Refrigerant Suction
B
Brine Supply
Wall Radiator, Plan
Wall Radiator on Ceiling, Plan
Unit Heater (Propeller), Plan
Unit Heater (Centrifugal Fan), Plan
Thermostatic Trap
Thermostatic Float
Thermometer
T
Thermostat
20X12
Duct Plan (1st Figure, Width; 2nd Depth)
D
Inclined Drop in Respect to Air Flow
S
12 X20
Supply Duct Section
E
12X20
Exhaust Duct Section
R
12 X20
Recirculation Duct Section
F
A
12X20
Fresh Air Duct Section
Supply Outlet
Exhaust Inlet
Plan
Volume Damper
Soil, Waste or Leader (Above Grade)
Cold Water
Hot Water
Hot Water Return
F
F
Fire Line
G
G
Gas
S
Sprinklers—Main Supplies
Elev
Volume Damper
Deflecting Damper
Turning Vanes
M
Automatic Dampers
Canvas Connections
Fan and Motor with Belt Guard
Intake Louvres and Screen

34. AMERICAN STANDARD GRAPHICAL SYMBOLS FOR ELECTRICAL DIAGRAMS

(From ASA Z32.2–1954)

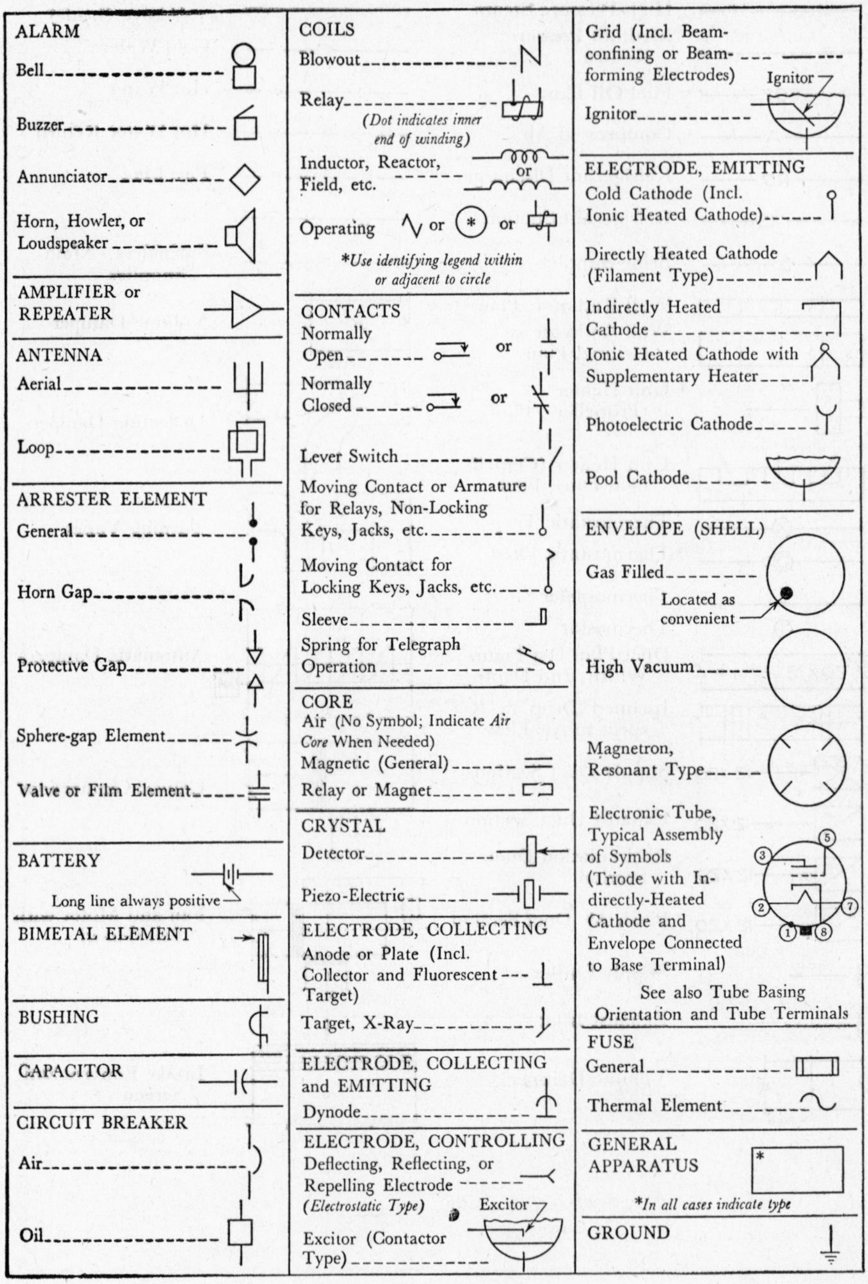

ALARM
- Bell
- Buzzer
- Annunciator
- Horn, Howler, or Loudspeaker

AMPLIFIER or REPEATER

ANTENNA
- Aerial
- Loop

ARRESTER ELEMENT
- General
- Horn Gap
- Protective Gap
- Sphere-gap Element
- Valve or Film Element

BATTERY
- Long line always positive

BIMETAL ELEMENT

BUSHING

CAPACITOR

CIRCUIT BREAKER
- Air
- Oil

COILS
- Blowout
- Relay (*Dot indicates inner end of winding*)
- Inductor, Reactor, Field, etc. (or)
- Operating (or, or)
  - **Use identifying legend within or adjacent to circle*

CONTACTS
- Normally Open (or)
- Normally Closed (or)
- Lever Switch
- Moving Contact or Armature for Relays, Non-Locking Keys, Jacks, etc.
- Moving Contact for Locking Keys, Jacks, etc.
- Sleeve
- Spring for Telegraph Operation

CORE
- Air (No Symbol; Indicate *Air Core* When Needed)
- Magnetic (General)
- Relay or Magnet

CRYSTAL
- Detector
- Piezo-Electric

ELECTRODE, COLLECTING
- Anode or Plate (Incl. Collector and Fluorescent Target)
- Target, X-Ray

ELECTRODE, COLLECTING and EMITTING
- Dynode

ELECTRODE, CONTROLLING
- Deflecting, Reflecting, or Repelling Electrode (*Electrostatic Type*)
- Excitor (Contactor Type) — Excitor
- Grid (Incl. Beam-confining or Beam-forming Electrodes)
- Ignitor — Ignitor

ELECTRODE, EMITTING
- Cold Cathode (Incl. Ionic Heated Cathode)
- Directly Heated Cathode (Filament Type)
- Indirectly Heated Cathode
- Ionic Heated Cathode with Supplementary Heater
- Photoelectric Cathode
- Pool Cathode

ENVELOPE (SHELL)
- Gas Filled — Located as convenient
- High Vacuum
- Magnetron, Resonant Type
- Electronic Tube, Typical Assembly of Symbols (Triode with Indirectly-Heated Cathode and Envelope Connected to Base Terminal) — 1, 2, 3, 5, 7, 8

See also Tube Basing Orientation and Tube Terminals

FUSE
- General
- Thermal Element

GENERAL APPARATUS
- **In all cases indicate type*

GROUND

35. AMERICAN STANDARD GRAPHICAL SYMBOLS FOR ELECTRICAL DIAGRAMS (Continued)

(From ASA Z32.2–1954)

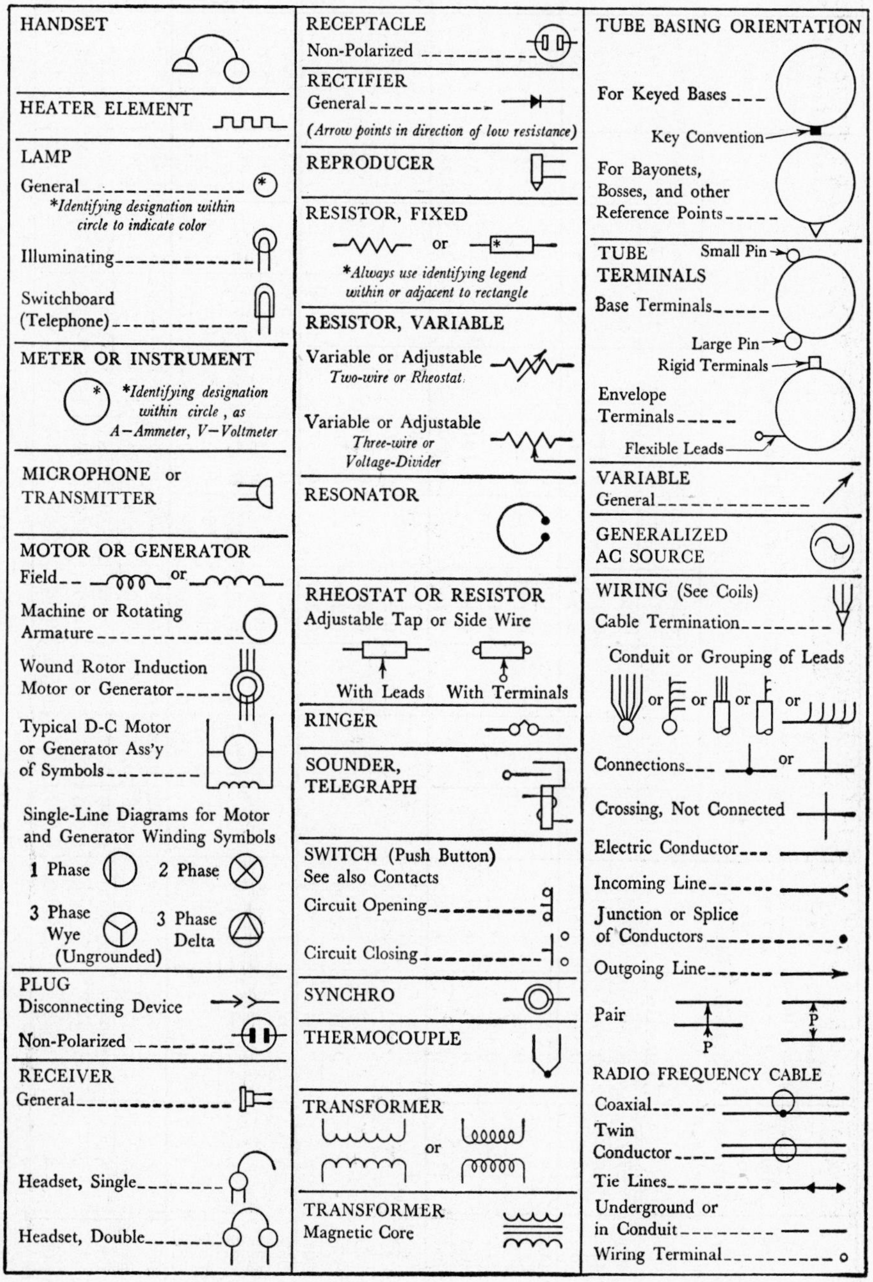

HANDSET

HEATER ELEMENT

LAMP
General
*Identifying designation within circle to indicate color
Illuminating
Switchboard (Telephone)

METER OR INSTRUMENT
*Identifying designation within circle, as A–Ammeter, V–Voltmeter

MICROPHONE or TRANSMITTER

MOTOR OR GENERATOR
Field — or
Machine or Rotating Armature
Wound Rotor Induction Motor or Generator
Typical D-C Motor or Generator Ass'y of Symbols
Single-Line Diagrams for Motor and Generator Winding Symbols
1 Phase 2 Phase
3 Phase Wye (Ungrounded) 3 Phase Delta

PLUG
Disconnecting Device
Non-Polarized

RECEIVER
General
Headset, Single
Headset, Double

RECEPTACLE
Non-Polarized

RECTIFIER
General
(Arrow points in direction of low resistance)

REPRODUCER

RESISTOR, FIXED
or
Always use identifying legend within or adjacent to rectangle

RESISTOR, VARIABLE
Variable or Adjustable *Two-wire or Rheostat*
Variable or Adjustable *Three-wire or Voltage-Divider*

RESONATOR

RHEOSTAT OR RESISTOR
Adjustable Tap or Side Wire
With Leads With Terminals

RINGER

SOUNDER, TELEGRAPH

SWITCH (Push Button)
See also Contacts
Circuit Opening
Circuit Closing

SYNCHRO

THERMOCOUPLE

TRANSFORMER
or

TRANSFORMER
Magnetic Core

TUBE BASING ORIENTATION
For Keyed Bases
Key Convention
For Bayonets, Bosses, and other Reference Points

TUBE TERMINALS
Small Pin
Base Terminals
Large Pin
Rigid Terminals
Envelope Terminals
Flexible Leads

VARIABLE
General

GENERALIZED AC SOURCE

WIRING (See Coils)
Cable Termination
Conduit or Grouping of Leads
or or or or
Connections — or
Crossing, Not Connected
Electric Conductor
Incoming Line
Junction or Splice of Conductors
Outgoing Line
Pair
P P

RADIO FREQUENCY CABLE
Coaxial
Twin Conductor
Tie Lines
Underground or in Conduit
Wiring Terminal

36. AMERICAN STANDARD WROUGHT STEEL PIPE[1] AND TAPER PIPE THREADS[2]

Nominal Pipe Size	D Outside Diameter of Pipe	Number of Threads per Inch	L_1[3] Normal Engagement by Hand between External and Internal Threads	L_2[3] Length of Effective Thread	Nominal Wall Thickness										Length of Pipe, Feet, per Square Foot External Surface[6]	Length of Standard-weight Pipe, Feet, Containing 1 Cu. Ft.[6]
					Sched. 10	Sched. 20	Sched. 30[4]	Sched. 40[4]	Sched. 60[5]	Sched. 80[5]	Sched. 100	Sched. 120	Sched. 140	Sched. 160		
⅛	0.405	27	0.180	0.2639	....	....	....	**0.068**	....	**0.095**	....	....	....	....	9.431	2,533.8
¼	0.540	18	0.200	0.4018	....	....	....	**0.088**	....	**0.119**	....	....	....	....	7.073	1,383.8
⅜	0.675	18	0.240	0.4078	....	....	....	**0.091**	....	**0.126**	....	....	....	....	5.658	754.36
½	0.840	14	0.320	0.5337	....	....	....	**0.109**	....	**0.147**	....	....	....	0.187	4.547	473.91
¾	1.050	14	0.339	0.5457	....	....	....	**0.113**	....	**0.154**	....	....	....	0.218	3.637	270.03
1	1.315	11½	0.400	0.6828	....	....	....	**0.133**	....	**0.179**	....	....	....	0.250	2.904	166.62
1¼	1.660	11½	0.420	0.7068	....	....	....	**0.140**	....	**0.191**	....	....	....	0.250	2.301	96.275
1½	1.900	11½	0.420	0.7235	....	....	....	**0.145**	....	**0.200**	....	....	....	0.281	2.010	70.733
2	2.375	11½	0.436	0.7565	....	....	....	**0.154**	....	**0.218**	....	....	....	0.343	1.608	42.913
2½	2.875	8	0.682	1.1375	....	....	....	**0.203**	....	**0.276**	....	....	....	0.375	1.328	30.077
3	3.500	8	0.766	1.2000	....	....	....	**0.216**	....	**0.300**	....	....	....	0.438	1.091	19.479
3½	4.000	8	0.821	1.2500	....	....	....	**0.226**	....	**0.318**	....	....	....	....	0.954	14.565
4	4.500	8	0.844	1.3000	....	....	....	**0.237**	....	**0.337**	....	0.438	....	0.531	0.848	11.312
5	5.563	8	0.937	1.4063	....	....	....	**0.258**	....	**0.375**	....	0.500	....	0.625	0.686	7.199
6	6.625	8	0.958	1.5125	....	....	....	**0.280**	....	**0.432**	....	0.562	....	0.718	0.576	4.984
8	8.625	8	1.063	1.7125	....	0.250	**0.277**	**0.322**	0.406	**0.500**	0.593	0.718	0.812	0.906	0.443	2.878
10	10.750	8	1.210	1.9250	....	0.250	**0.307**	**0.365**	**0.500**	0.593	0.718	0.843	1.000	1.125	0.355	1.826
12	12.750	8	1.360	2.1250	....	0.250	**0.330**	0.406	0.562	0.687	0.843	1.000	1.125	1.312	0.299	1.273
14 O.D.	14.000	8	1.562	2.2500	0.250	0.312	0.375	0.438	0.593	0.750	0.937	1.093	1.250	1.406	0.273	1.065
16 O.D.	16.000	8	1.812	2.4500	0.250	0.312	0.375	0.500	0.656	0.843	1.031	1.218	1.438	1.593	0.239	0.815
18 O.D.	18.000	8	2.000	2.6500	0.250	0.312	0.438	0.562	0.750	0.937	1.156	1.375	1.562	1.781	0.212	0.644
20 O.D.	20.000	8	2.125	2.8500	0.250	0.375	0.500	0.593	0.812	1.031	1.281	1.500	1.750	1.968	0.191	0.518
24 O.D.	24.000	8	2.375	3.2500	0.250	0.375	0.562	0.687	0.968	1.218	1.531	1.812	2.062	2.343	0.159	0.358

All dimensions are in inches except those in last two columns.
[1]From ASA B36.10–1950. [2]From ASA B2.1–1945. [3]Refer to §411 and Fig. 676. [4]Bold-face figures correspond to "standard" pipe.
[5]Bold-face figures correspond to "extra strong" pipe. [6]Calculated values for Schedule 40 pipe.

37. THICKNESSES AND WEIGHTS OF CAST IRON PIPE[1]

Size in.	Thickness in.	Outside Diam. in.	16-ft. Laying Length — Weight (lb) Based on Avg. per Foot[2]	16-ft. Laying Length — Weight (lb) Based on Per Length
Class 50—50-psi Pressure—115-ft. Head				
3	0.32	3.96	12.4	195
4	0.35	4.80	16.5	265
6	0.38	6.90	25.9	415
8	0.41	9.05	37.0	590
10	0.44	11.10	49.1	785
12	0.48	13.20	63.7	1,020
14	0.48	15.30	74.6	1,195
16	0.54	17.40	95.2	1,525
18	0.54	19.50	107.6	1,720
20	0.57	21.60	125.9	2,015
24	0.63	25.80	166.0	2,655
30	0.79	32.00	257.6	4,120
36	0.87	38.30	340.9	5,455
42	0.97	44.50	442.0	7,070
48	1.06	50.80	551.6	8,825
Class 100—100-psi Pressure—231-ft. Head				
3	0.32	3.96	12.4	195
4	0.35	4.80	16.5	265
6	0.38	6.90	25.9	415
8	0.41	9.05	37.0	590
10	0.44	11.10	49.1	785
12	0.48	13.20	63.7	1,020
14	0.51	15.30	78.8	1,260
16	0.54	17.40	95.2	1,525
18	0.58	19.50	114.8	1,835
20	0.62	21.60	135.9	2,175
24	0.68	25.80	178.1	2,850
30	0.79	32.00	257.6	4,120
36	0.87	38.30	340.9	5,455
42	0.97	44.50	442.0	7,070
48	1.06	50.80	551.6	8,825
Class 150—150-psi Pressure—346-ft. Head				
3	0.32	3.96	12.4	195
4	0.35	4.80	16.5	265
6	0.38	6.90	25.9	415
8	0.41	9.05	37.0	590
10	0.44	11.10	49.1	785
12	0.48	13.20	63.7	1,020
14	0.51	15.64	80.7	1,290
16	0.54	17.80	97.5	1,560
18	0.58	19.92	117.2	1,875
20	0.62	22.06	138.9	2,220
24	0.73	26.32	194.0	3,105
30	0.85	32.00	275.4	4,405
36	0.94	38.30	365.9	5,855
42	1.05	44.50	475.3	7,605
48	1.14	50.80	589.6	9,435
Class 200—200-psi Pressure—462-ft. Head				
3	0.32	3.96	12.4	195
4	0.35	4.80	16.5	265
6	0.38	6.90	25.9	415
Class 200—200-psi Pressure—462-ft. Head (*Cont'd*)				
8	0.41	9.05	37.0	590
10	0.44	11.10	49.1	785
12	0.48	13.20	63.7	1,020
14	0.55	15.65	86.4	1,380
16	0.58	17.80	104.0	1,665
18	0.63	19.92	126.5	2,025
20	0.67	22.06	149.1	2,385
24	0.79	26.32	209.8	3,355
30	0.92	32.00	297.8	4,765
36	1.02	38.30	397.1	6,355
42	1.13	44.50	512.3	8,195
48	1.23	50.80	637.2	10,195
Class 250—250-psi Pressure—577-ft. Head				
3	0.32	3.96	12.4	195
4	0.35	4.80	16.5	265
6	0.38	6.90	25.9	415
8	0.41	9.05	37.0	590
10	0.44	11.10	49.1	785
12	0.52	13.20	68.5	1,095
14	0.59	15.65	92.7	1,485
16	0.63	17.80	112.9	1,805
18	0.68	19.92	136.3	2,180
20	0.72	22.06	160.1	2,560
24	0.79	26.32	209.8	3,355
30	0.99	32.00	318.4	5,095
36	1.10	38.30	425.5	6,810
42	1.22	44.50	549.5	8,790
48	1.33	50.80	684.5	10,950
Class 300—300-psi Pressure—693-ft. Head				
3	0.32	3.96	12.4	195
4	0.35	4.80	16.5	265
6	0.38	6.90	25.9	415
8	0.41	9.05	37.0	590
10	0.48	11.10	53.1	850
12	0.52	13.20	68.5	1,095
14	0.59	15.65	92.7	1,485
16	0.68	17.80	121.0	1,935
18	0.73	19.92	145.4	2,325
20	0.78	22.06	172.2	2,755
24	0.85	26.32	224.3	3,590
Class 350—350-psi Pressure—808-ft. Head				
3	0.32	3.96	12.4	195
4	0.35	4.80	16.5	265
6	0.38	6.90	25.9	415
8	0.41	9.05	37.0	590
10	0.52	11.10	57.4	920
12	0.56	13.20	73.8	1,180
14	0.64	15.65	99.8	1,595
16	0.68	17.80	121.0	1,935
18	0.79	19.92	156.2	2,500
20	0.84	22.06	184.2	2,945
24	0.92	26.32	241.1	3,860

[1]From ASA A21.8-1953.
[2]Average weight per foot based on calculated weight of pipe before rounding.

38. AMERICAN STANDARD 125-LB. CAST-IRON SCREWED FITTINGS[1]

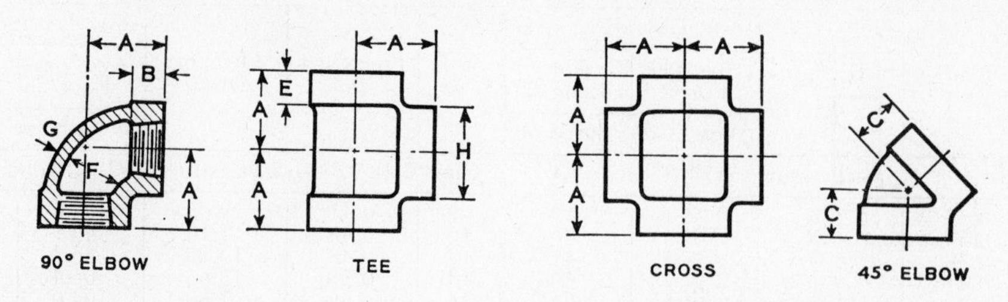

Dimensions of 90- and 45-Deg. Elbows, Tees, and Crosses (Straight Sizes)

Nominal Pipe Size	Center to End, Elbows, Tees, and Crosses A	Center to End, 45-Deg. Elbows C	Length of Thread, Min. B	Width of Band, Min. E	Inside Diameter of Fitting F		Metal Thick-ness G	Diameter of Band, Min. H
					Max.	Min.		
¼	0.81	0.73	0.32	0.38	0.584	0.540	0.110	0.93
⅜	0.95	0.80	0.36	0.44	0.719	0.675	0.120	1.12
½	1.12	0.88	0.43	0.50	0.897	0.840	0.130	1.34
¾	1.31	0.98	0.50	0.56	1.107	1.050	0.155	1.63
1	1.50	1.12	0.58	0.62	1.385	1.315	0.170	1.95
1¼	1.75	1.29	0.67	0.69	1.730	1.660	0.185	2.39
1½	1.94	1.43	0.70	0.75	1.970	1.900	0.200	2.68
2	2.25	1.68	0.75	0.84	2.445	2.375	0.220	3.28
2½	2.70	1.95	0.92	0.94	2.975	2.875	0.240	3.86
3	3.08	2.17	0.98	1.00	3.600	3.500	0.260	4.62
3½	3.42	2.39	1.03	1.06	4.100	4.000	0.280	5.20
4	3.79	2.61	1.08	1.12	4.600	4.500	0.310	5.79
5	4.50	3.05	1.18	1.18	5.663	5.563	0.380	7.05
6	5.13	3.46	1.28	1.28	6.725	6.625	0.430	8.28
8	6.56	4.28	1.47	1.47	8.725	8.625	0.550	10.63
10	8.08*	5.16	1.68	1.68	10.850	10.750	0.690	13.12
12	9.50*	5.97	1.88	1.88	12.850	12.750	0.800	15.47

[1]From ASA B16.4–1949; reaffirmed 1953.

All dimensions given in inches.

Fittings having right- and left-hand threads shall have four or more ribs or the letter "L" cast on the band at end with left-hand thread.

*This applies to elbows and tees only.

39. AMERICAN STANDARD 250-LB. CAST-IRON SCREWED FITTINGS[1]

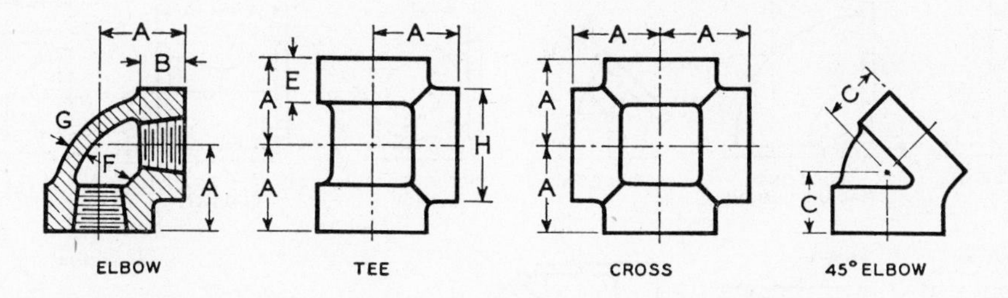

Dimensions of 90- and 45-Deg. Elbows, Tees, and Crosses (Straight Sizes)

Nominal Pipe Size	Center to End, Elbows, Tees, and Crosses A	Center to End, 45-Deg. Elbows C	Length of Thread, Min. B	Width of Band, Min. E	Inside Diameter of Fitting F Max.	Inside Diameter of Fitting F Min.	Metal Thickness[1] G	Outside Diameter of Band, Min. H
1/4	0.94	0.81	0.43	0.49	0.584	0.540	0.18	1.17
3/8	1.06	0.88	0.47	0.55	0.719	0.675	0.18	1.36
1/2	1.25	1.00	0.57	0.60	0.897	0.840	0.20	1.59
3/4	1.44	1.13	0.64	0.68	1.107	1.050	0.23	1.88
1	1.63	1.31	0.75	0.76	1.385	1.315	0.28	2.24
1 1/4	1.94	1.50	0.84	0.88	1.730	1.660	0.33	2.73
1 1/2	2.13	1.69	0.87	0.97	1.970	1.900	0.35	3.07
2	2.50	2.00	1.00	1.12	2.445	2.375	0.39	3.74
2 1/2	2.94	2.25	1.17	1.30	2.975	2.875	0.43	4.60
3	3.38	2.50	1.23	1.40	3.600	3.500	0.48	5.36
3 1/2	3.75	2.63	1.28	1.49	4.100	4.000	0.52	5.98
4	4.13	2.81	1.33	1.57	4.600	4.500	0.56	6.61
5	4.88	3.19	1.43	1.74	5.663	5.563	0.66	7.92
6	5.63	3.50	1.53	1.91	6.725	6.625	0.74	9.24
8	7.00	4.31	1.72	2.24	8.725	8.625	0.90	11.73
10	8.63	5.19	1.93	2.58	10.850	10.750	1.08	14.37
12	10.00	6.00	2.13	2.91	12.850	12.750	1.24	16.84

[1]From ASA B16.4–1949; reaffirmed 1953.

All dimensions given in inches.

The 250-lb. standard for screwed fittings covers only the straight sizes of 90- and 45-deg. elbows, tees, and crosses.

40. AMERICAN STANDARD CLASS 125 CAST-IRON FLANGES AND FITTINGS[1]

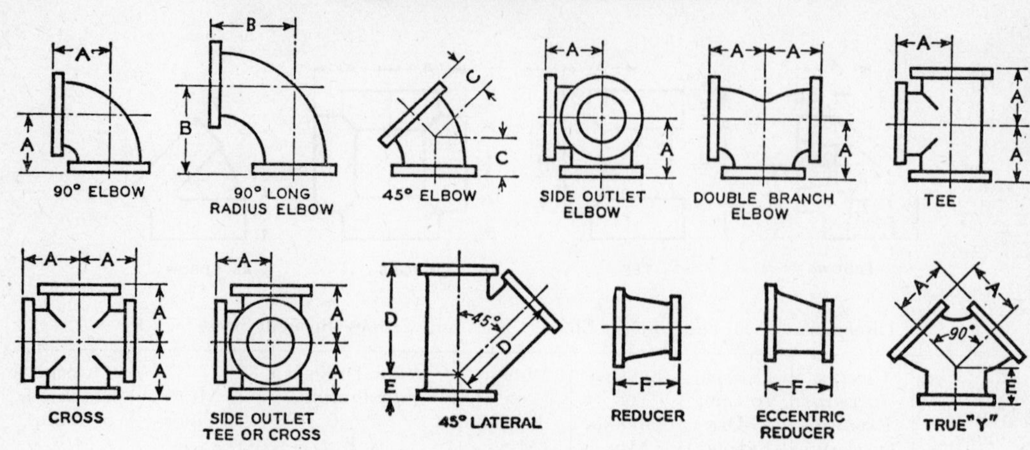

Dimensions of Elbows, Double Branch Elbows, Tees, Crosses, Laterals, True Y's (Straight Sizes), and Reducers

Nominal Pipe Size	Inside Diam. of Fittings	Center to Face 90 Deg. Elbow, Tees, Crosses True "Y" and Double Branch Elbow A	Center to Face 90 Deg. Long Radius Elbow B	Center to Face 45 Deg. Elbow C	Center to Face Lateral D	Short Center to Face True "Y" and Lateral E	Face to Face Reducer F	Diam. of Flange	Thickness of Flange (Min)	Wall Thickness
1	1	3½	5	1¾	5¾	1¾	....	4¼	7/16	5/16
1¼	1¼	3¾	5½	2	6¼	1¾	....	4⅝	½	5/16
1½	1½	4	6	2¼	7	2	....	5	9/16	5/16
2	2	4½	6½	2½	8	2½	5	6	⅝	5/16
2½	2½	5	7	3	9½	2½	5½	7	11/16	5/16
3	3	5½	7¾	3	10	3	6	7½	¾	⅜
3½	3½	6	8½	3½	11½	3	6½	8½	13/16	7/16
4	4	6½	9	4	12	3	7	9	15/16	½
5	5	7½	10¼	4½	13½	3½	8	10	15/16	½
6	6	8	11½	5	14½	3½	9	11	1	9/16
8	8	9	14	5½	17½	4½	11	13½	1⅛	⅝
10	10	11	16½	6½	20½	5	12	16	1 3/16	¾
12	12	12	19	7½	24½	5½	14	19	1¼	13/16
14 OD	14	14	21½	7½	27	6	16	21	1⅜	⅞
16 OD	16	15	24	8	30	6½	18	23½	1 7/16	1
18 OD	18	16½	26½	8½	32	7	19	25	1 9/16	1 1/16
20 OD	20	18	29	9½	35	8	20	27½	1 11/16	1⅛
24 OD	24	22	34	11	40½	9	24	32	1⅞	1¼
30 OD	30	25	41½	15	49	10	30	38¾	2⅛	1 7/16
36 OD	36	28	49	18	....	....	36	46	2⅜	1⅝
42 OD	42	31	56½	21	....	....	42	53	2⅝	1 13/16
48 OD	48	34	64	24	....	....	48	59½	2¾	2

[1]ASA B16.1-1948; reaffirmed 1953. All dimensions in inches.

41. AMERICAN STANDARD CLASS 125 CAST-IRON FLANGES, DRILLING FOR BOLTS AND THEIR LENGTHS[1]

Nominal Pipe Size	Diam. of Flange	Thickness of Flange (Min.)	Diam. of Bolt Circle	Number of Bolts	Diam. of Bolts	Diam. of Bolt Holes	Length of Bolts
1	4¼	7/16	3⅛	4	½	⅝	1¾
1¼	4⅝	½	3½	4	½	⅝	2
1½	5	9/16	3⅞	4	½	⅝	2
2	6	⅝	4¾	4	⅝	¾	2¼
2½	7	11/16	5½	4	⅝	¾	2½
3	7½	¾	6	4	⅝	¾	2½
3½	8½	13/16	7	8	⅝	¾	2¾
4	9	15/16	7½	8	⅝	¾	3
5	10	15/16	8½	8	¾	⅞	3
6	11	1	9½	8	¾	⅞	3¼
8	13½	1⅛	11¾	8	¾	⅞	3½
10	16	1 3/16	14¼	12	⅞	1	3¾
12	19	1¼	17	12	⅞	1	3¾
14 OD	21	1⅜	18¾	12	1	1⅛	4¼
16 OD	23½	1 7/16	21¼	16	1	1⅛	4½
18 OD	25	1 9/16	22¾	16	1⅛	1¼	4¾
20 OD	27½	1 11/16	25	20	1⅛	1¼	5
24 OD	32	1⅞	29½	20	1¼	1⅜	5½
30 OD	38¾	2⅛	36	28	1¼	1⅜	6¼
36 OD	46	2⅜	42¾	32	1½	1⅝	7
42 OD	53	2⅝	49½	36	1½	1⅝	7½
48 OD	59½	2¾	56	41	1½	1⅝	7¾

[1]ASA B16.1–1948; reaffirmed 1953.

42. SHAFT CENTER SIZES

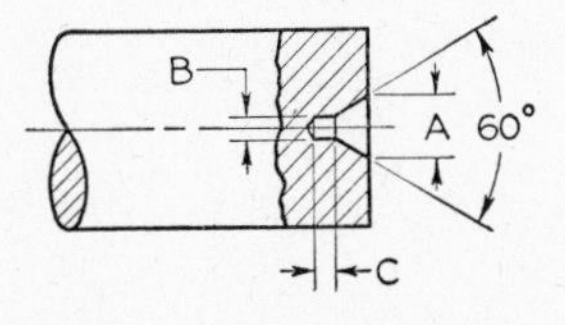

Shaft Diameter D	A	B	C	Shaft Diameter D	A	B	C
3/16 to 7/32	5/64	3/64	1/16	1⅛ to 1 15/32	5/16	5/32	5/32
¼ to 11/32	3/32	3/64	1/16	1½ to 1 31/32	⅜	3/32	5/32
⅜ to 17/32	⅛	1/16	5/64	2 to 2 31/32	7/16	7/32	3/16
9/16 to 25/32	3/16	5/64	3/32	3 to 3 31/32	½	7/32	7/32
13/16 to 1 3/32	¼	3/32	3/32	4 and over	9/16	7/32	7/32

43. AMERICAN STANDARD CLASS 250 CAST-IRON FLANGES AND FITTINGS[1]

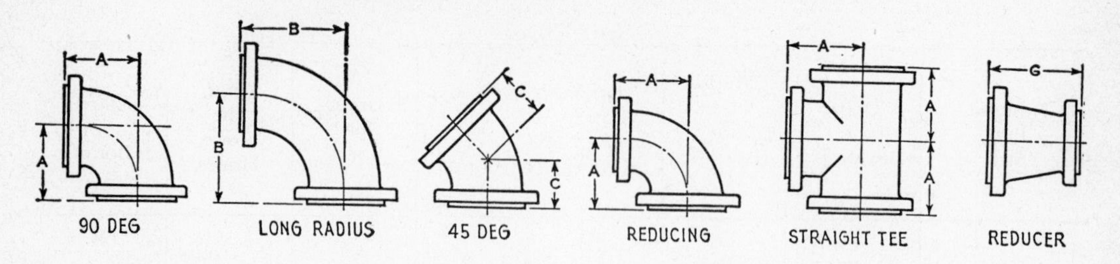

Dimensions of Elbows, Tees, and Reducers

Nominal Pipe Size	Inside Diam. of Fitting (Min.)	Wall Thickness of Body	Diam. of Flange	Thickness of Flange (Min.)	Diam. of Raised Face	Center-to-Face Elbow and Tee A	Center-to-Face Long Radius Elbow B	Center-to-Face 45 Deg. Elbow C	Face-to-Face Reducer G
2	2	7/16	6 1/2	7/8	4 3/16	5	6 1/2	3	5
2 1/2	2 1/2	1/2	7 1/2	1	4 15/16	5 1/2	7	3 1/2	5 1/2
3	3	9/16	8 1/4	1 1/8	5 11/16	6	7 3/4	3 1/2	6
3 1/2	3 1/2	9/16	9	1 3/16	6 5/16	6 1/2	8 1/2	4	6 1/2
4	4	5/8	10	1 1/4	6 15/16	7	9	4 1/2	7
5	5	11/16	11	1 3/8	8 5/16	8	10 1/4	5	8
6	6	3/4	12 1/2	1 7/16	9 11/16	8 1/2	11 1/2	5 1/2	9
8	8	13/16	15	1 5/8	11 15/16	10	14	6	11
10	10	15/16	17 1/2	1 7/8	14 1/16	11 1/2	16 1/2	7	12
2	12	1	20 1/2	2	16 7/16	13	19	8	14
14 OD	13 1/4	1 1/8	23	2 1/8	18 15/16	15	21 1/2	8 1/2	16
16 OD	15 1/4	1 1/4	25 1/2	2 1/4	21 1/16	16 1/2	24	9 1/2	18
18 OD	17	1 3/8	28	2 3/8	23 5/16	18	26 1/2	10	19
20 OD	19	1 1/2	30 1/2	2 1/2	25 9/16	19 1/2	29	10 1/2	20
24 OD	23	1 5/8	36	2 3/4	30 1/4	22 1/2	34	12	24

All dimensions given in inches.
[1]ASA B16b–1944; reaffirmed 1953.

44. AMERICAN STANDARD CLASS 250 CAST-IRON FLANGES, DRILLING FOR BOLTS AND THEIR LENGTHS[1]

Nominal Pipe Size	Diam. of Flange	Thickness of Flange (Min.)	Diam. of Raised Face	Diam. of Bolt Circle	Diam. of Bolt Holes	Number of Bolts	Size of Bolts	Length of Bolts	Length of Bolt Studs With Two Nuts
1	4⅞	11/16	2 11/16	3½	¾	4	⅝	2½	.....
1¼	5¼	¾	3 1/16	3⅞	¾	4	⅝	2½	.....
1½	6⅛	13/16	3 9/16	4½	⅞	4	¾	2¾	.....
2	6½	⅞	4 3/16	5	¾	8	⅝	2¾	.....
2½	7½	1	4 15/16	5⅞	⅞	8	¾	3¼	.....
3	8¼	1⅛	5 11/16	6⅝	⅞	8	¾	3½	.....
3½	9	1 3/16	6 5/16	7¼	⅞	8	¾	3½	.....
4	10	1¼	6 15/16	7⅞	⅞	8	¾	3¾	.....
5	11	1⅜	8 5/16	9¼	⅞	8	¾	4	.....
6	12½	1 7/16	9 11/16	10⅝	⅞	12	¾	4	.....
8	15	1⅝	11 15/16	13	1	12	⅞	4½	.....
10	17½	1⅞	14 1/16	15¼	1⅛	16	1	5¼	.....
12	20½	2	16 7/16	17¾	1¼	16	1⅛	5½	.....
14 OD	23	2⅛	18 15/16	20¼	1¼	20	1⅛	6	.....
16 OD	25½	2¼	21 1/16	22½	1⅜	20	1¼	6¼	.....
18 OD	28	2⅜	23 5/16	24¾	1⅜	24	1¼	6½	.....
20 OD	30½	2½	25 9/16	27	1⅜	24	1¼	6¾	.....
24 OD	36	2¾	30¼	32	1 11/16	24	1½	7¾	9½

[1]ASA B16b–1944; reaffirmed 1953.

INDEX

(NUMBERS LISTED ARE PAGE NUMBERS)

SHEET LAYOUTS

A convenient code to identify sheet sizes and forms, for use of instructors in making assignments, is shown here.

Three *sizes* of sheets are adopted: *Size A*, Fig. 1193, *Size B*, Fig. 1197, and *Size C*, Fig. 1198.

Eight *forms* of lettering arrangements are adopted, known as *Forms 1, 2, 3, 4, 5, 6, 7,* and *8*, as shown below.

The term *layout* designates a sheet of certain size plus a certain arrangement of lettering. Thus *Layout A-1* is a combination of *Size A*, Fig. 1193, and *Form 1*, Fig. 1194. *Layout C-678* is a combination of *Size C*, Fig. 1198, and *Forms 6, 7,* and *8*, Figs. 1201 to 1203 inclusive. Any other combinations may be employed as assigned by the instructor.

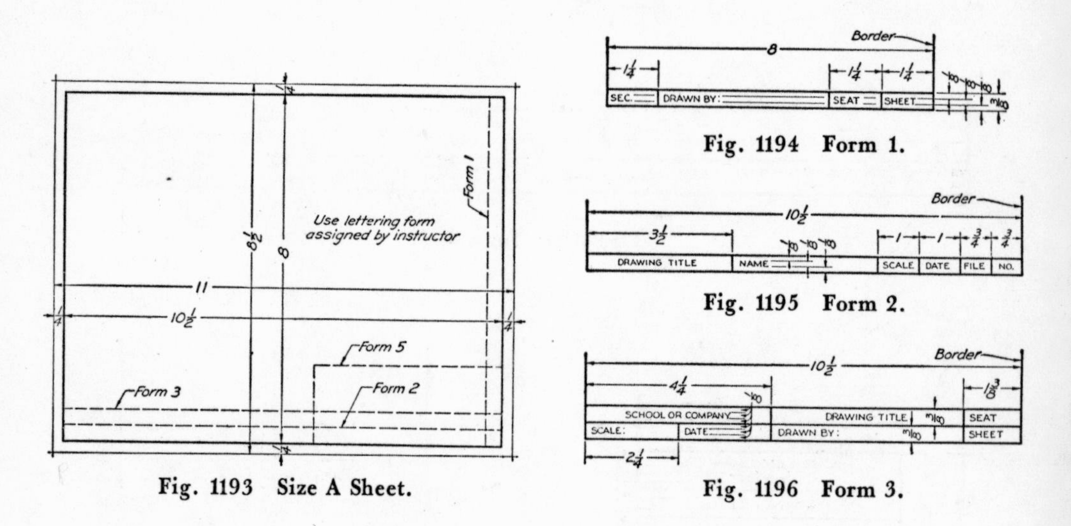

Fig. 1193 Size A Sheet.

Fig. 1194 Form 1.

Fig. 1195 Form 2.

Fig. 1196 Form 3.

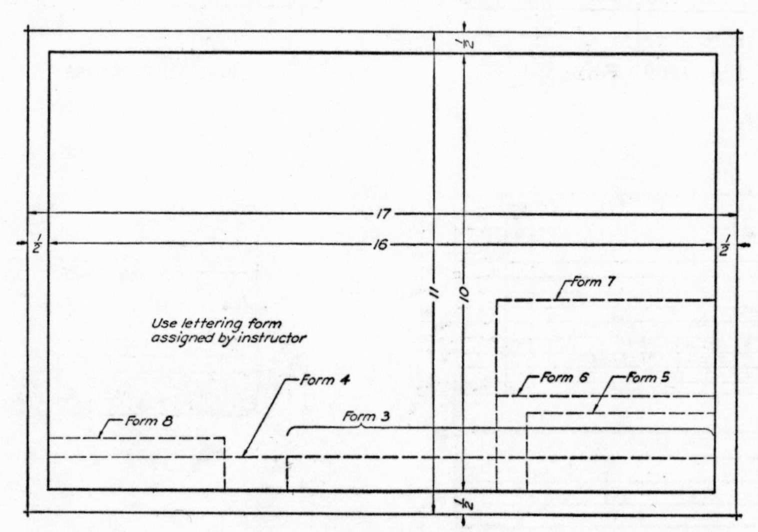

Fig. 1197 Size B Sheet.